HIGHWAY ENGINEERING

DR. L. R. KADIYALI
B.E. (Hons.), (Bombay),
P.G. Diploma in Highway and
Traffic Engineering (Newcastle-Upon-Tyne),
Ph.D. (Kakatiya University)
Formerly Chief Engineer, (Roads Wig),
Ministry of Road Transport & Highways, New Delhi
&
Study Director, Road User Cost Study,
Central Road Research Institute, New Delhi

KHANNA BOOK PUBLISHING CO. (P) LTD.
Publisher of Engineering and Computer Books
4C/4344, Ansari Road, Darya Ganj, New Delhi-110002
Phone : 011-23244447-48 **Mobile:** +91-9910909320
E-mail : contact@khannabooks.com
Website : www.khannabooks.com

Dear Readers,

To prevent the piracy, this book is secured with HIGH SECURITY HOLOGRAM on the front title cover. In case you don't find the hologram on the front cover title, please write us to at contact@khannabooks.com or whatsapp us at +91-99109 09320 and avail special gift voucher for yourself.

Specimen of Hologram on front Cover title:

Moreover, there is a SPECIAL DISCOUNT COUPON for you with EVERY HOLOGRAM.

How to avail this SPECIAL DISCOUNT:

Step 1: Scratch the hologram
Step 2: Under the scratch area, your "coupon code" is available
Step 3: Logon to www.khannabooks.com
Step 4: Use your "coupon code" in the shopping cart and get your copy at a special discount
Step 5: Enjoy your reading!

Price: ₹ 525.00

ISBN: 978-93-86173-13-3

Highway Engineering
by L.R. Kadiyali

First Edition: 2018
Reprint Edition: 2021, **2025**

To view complete list of books,
Please scan the QR Code:

Published by:
Khanna Book Publishing Co. (P) Ltd.
CIN: U22110DL1998PTC095547

Visit us at: **www.khannabooks.com**
Write us at: **contact@khannabooks.com**

Printed in India.

CONTENTS

Chapter

DEVELOPMENT OF HIGHWAYS

1.1. DEFINITIONS

Some terms like 'highways', 'roads', 'streets' and 'expressways' have precise meanings, though they are often used loosely in practice, A 'highway' is an arterial road facility designed for high speed and high volume traffic in the non-urban areas. For example, the national road network of a country is called the National Highway Network.

A 'road' is a lower order facility, designed for relatively lower speed and lower volume of traffic in the non-urban areas. For example, they can be district roads or village roads. A 'street' is an urban road facility.

An Expressway or Express Highway is a superior type of highway facility with full or partial control of access, generally a divided carriageway and catering to very high speeds.

1.2. EARLY ROADS

The origin of roads dates back to the period before the advent of recorded history. With the desire to hunt animals for food, the ancient man began to form pathways and tracks to facilitate his movements. As civilization advanced, the growth of agriculture took place and human settlements began to be formed. From one settlement to another, tracks were formed. These tracks might have been the skeletal framework of the modern highways.

Man soon learnt the use of animals as beasts of burden. Oxen, horses, donkeys, dogs, camels and elephants were the beasts of burden. These moved on the tracks on which men used to travel.

The next major event to revolutionarise transport was the invention of the wheel (approx. 3500-5000 BC) in Mesopotamia. Man soon saw the advantages of an axle joining two wheels and began to build two-wheeled and four-wheeled carts and chariots. The art of road-building soon began with the need to provide a hard durable surface to withstand the abrading effect of wheels.

Many civilizations have been known for their excellence and attainments in road building. The abundance of slave labour facilitated the construction of roads in some civilizations. The streets of the city of Babylon are known to have been paved perhaps one or two thousand years before

Christ. The royal procession route of that city was just about one kilometer in length and 10-20 m in width. It was paved with stone slabs which rested on a few layers of bricks jointed with bitumen. The construction of the Great Pyramid in Egypt, about 3000 BC, was facilitated because of a good road for transporting huge stone blocks. In Europe, there is evidence of a hard surfaced road in the island of Create. The surface consisted of stone slabs set in clay-gypsum mortar.

In the Bible, there is a reference to the construction of a road between Egypt and Babylon about 500 years before the Christian Era.

The Roman Civilization is well-known for the good road system it built. The Roman Empire extended over vast regions. The 100,000-km road network served military and administrative purposes. Some of the modern highways of Europe are aligned generally along the routes of the Roman Era. Rome was the focal point from where 29 major roads radiated in all directions. This is the basis of the saying: "All roads lead to Rome". The main features of Roman roads were direct alignment, good drainage, good road level and good pavement structure. The top layers of the pavement consisted of flat stones. Lime mortar was used to cement the stones. With the decline of the Roman Empire, road building received a set-back in Europe.

The Persians, under King Darius 1, who ruled in the fifth century B.C, built the Royal Persian Road, traversing from Turkey to the Persian Gulf. This road served both trade and military purposes.

The Chinese civilization is also known to have practiced the art of road building on a large scale.

As far as India is concerned, the Indus Valley Civilization (dating back to 5,000 years B.C.) flourished with well-planned towns having an elaborate street and drainage system. The streets were laid out in regular order, in straight lines (modern grid-iron pattern). The biggest street in Mohenjo Daro was about 800 m long and about 9 m wide. It is likely that wheeled carts were in existence then.

In India, the Mauryan Kings (4th-2nd century BC) built very good roads. Chandragupta, one of the well-known rulers, constructed a 2400 km long across the subcontinent, connecting Patliputra (modern Patna) to Takshila (now in Pakistan). It touched cities like Varanasi, Kausambhi, Mathura, Indraprastha (modern Delhi) and Kurukshetra. This road has now been developed into a National Highway on more or less the same alignment and is also known as the Grand Trunk Road. Ujjaini (modern Ujjain) was connected by a spur route to the south. Records show that Rajapathas (main highways) and Banjjapathas (trade routes) were constructed. Kautilya, a great administrator of the times, and the author of Arthashastra (finance), laid down, standard widths for various classes of roads. Magesthenes, a Greek ambassador in India during this period, mentions that the Mauryan rulers were great road builders. Special care was taken to repair the roads. Trees were planted all along the road to give shade. Wells were dug and rest houses provided for the comfort of the travelers. At large river crossings, ferry services were provided by the State.

The Gupta period (3rd-5fh century AD), is famous for its patronage of arts, industry, science and literature. Under the benevolent rule of the Guptas, roads received a great impetus. Fehien, a Chinese pilgrim who visited India in 405 A.D, travelled widely on India's roads and has left impressive accounts of the road system. He was particularly impressed by the security and comforts provided to the travellers. Another Chinese traveller, Hiuen-Tsang, visited India in the first part of the seventh century A.D. during the reign of Harshavardhna. He has referred to the good maintenance that the roads received. The King personally travelled on the roads on a wide scale. And this was responsible for the care, the roads received.

When the Muslim rule set in, the road system got further consolidated. Roads were needed for extending the sovereignty and administering the conquered regions. The travel accounts of Al Beruni and Ibn Battuta in the eleventh and the fourteenth centuries respectively, have left flattering accounts of the roads during those periods. Ibn Battuta narrates that the important roads were provided with tree plantation on both sides, rest houses, postal stations and deep wells for water supply.

The credit to revive the declining road system in India must go to Sher Shah, the Afghan monarch who ruled in the 16th century. The Grand Trunk Road, running from East to West in the Gangetic plains, was developed by him. It was named "Sarak-e-Azam", the great road. He planted shady trees, built "sarais" (rest houses) en route and erected the equivalent of the modern milestones. The latter were called "Kos Minars", one "Kos" measuring 2690 m approximately. Some of these exist even to this day.

The Mughals, who ruled India for 300 years, recognised the need for good roads as their empire extended almost over the entire Indian sub-continent. The empire was served by a good system of military roads. The Mughal Roads were paved with "Kankar", limestone aggregates.

1.3. EARLY DEVELOPMENTS IN HIGHWAY ENGINEERING IN EUROPE

The Industrial Revolution in Europe created accelerated demand for transport in that continent. Wheeled coaches began to make their appearance on the roads in the sixteenth century. The disruption of the road bed caused by the movement of animal drawn passenger coaches and goods wagons gave a spurt to scientific design of roads.

Pierre Tresaguet, the Inspector General of Roads in France, was the first to recognise the importance of drainage of roads and its methodical maintenance. He appreciated the role of moisture in soils and pavements and how moisture affects the performance of road beds. Camber began to be introduced in roads during his time. Thus, he can be rightly called the father of modern highway engineering.

The name of John Metcalf is associated in Britain with the art of building good and stable roads in the latter part of the eighteenth century. He used boulders to achieve strong foundations for roads and spread gravel as a surface layer. He pioneered the construction of good roads on soft ground, using a sub-base of bundles of heather (Various shrubs).

Thomas Telford is yet another illustrious name in highway engineering, immortalized by naming the hand-packed boulder foundation of roads as Telford base. The construction technique held the sway for nearly 150 years since Telford introduced it in the early part of the nineteenth century. In fact, it used to be practiced in India till recently.

A run of names of eminent highway engineers is incomplete without John Mc Adam's. He was a Scottish road builder who has influenced road construction so profoundly that the term "macadam" is frequently used in pavement specifications even to this day. His two important principles of good road construction were:

(*i*) It is the native soil that supports the traffic load ultimately, and when the soil is maintained in a dry state it can carry heavy loads without settlement.

(*ii*) Stones which are broken to small angular pieces and compacted can interlock with each other and form a hard surface.

Thus, Mc Adam's specifications were at variance with Telford's in that smaller pieces of stones with angular faces were favoured than larger hand-packed boulders. He is reported to have given a practical hint to engineers is selecting the size of stones ; the size is good if the stone can be put into the mouth. How valid his advice is even this day! Apart from the innovative specifications he introduced, Mc Adam is remembered for his foresight in urging the creation of a central highway authority to advise and monitor all matters relating to roads in Britain. His recommendation is valid even now in India.

Fig. 1.1 gives the cross-sections of some of the early roads.

A significant development which revolutionized road construction during the nineteenth century was the steam road roller introduced by Eveling and Barford. The development of Portland cement in the first few decades of the nineteenth century by Aspdin and Johnson facilitated modern bridge construction and use of concrete as a pavement material. Tars and asphalts began to be used in road construction in the 1830s, though it was the pneumatic tyred vehicle which gave real push to the extensive use of bituminous specifications.

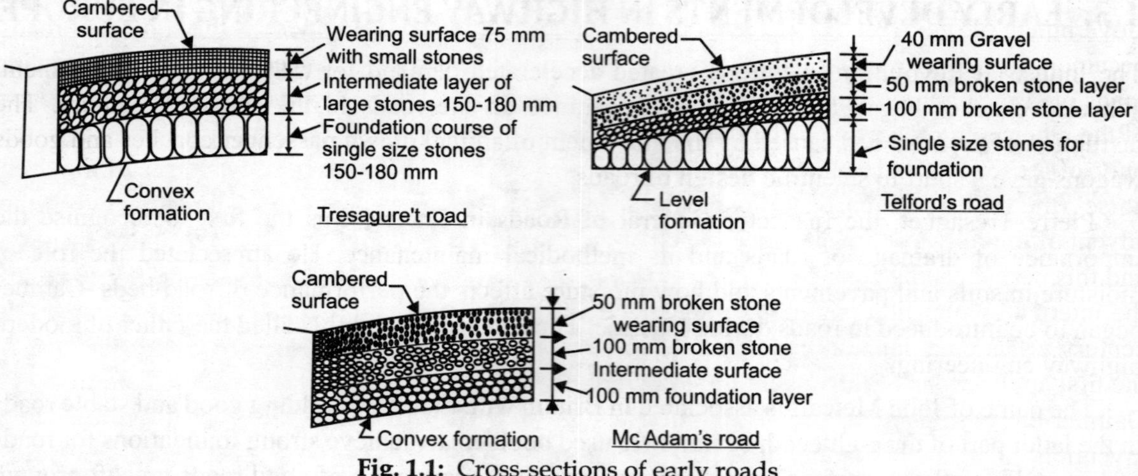

Fig. 1.1: Cross-sections of early roads

The automobile had its slow development in the nineteenth century, but the First World War, 1914-18, gave an impetus to its growth. Thus, the road was given a new lease of life.

1.4. DEVELOPMENT OF ROADS IN INDIA DURING THE BRITISH PERIOD

The period covering the decline of the Mughals and the beginning of the British rule was a period of neglect of the road system in India. The early British interest was only in maintaining roads of military importance and dire administrative requirements. William Bentinck took steps to build the modern Grand Trunk Road from Calcutta to Delhi, generally along Sher Shah's alignment, but with permanent bridges and good stone or 'Kankar' bed. The road was later on extended to Peshawar.

Lord Dalhousie gave further impetus to road building by organizing the provincial PWDs in place of the Military Board in 1885. Engineering Colleges were established to train civil engineers. The important roads were provided with metalled beds and were bridged. Specifications were framed for the construction of roads. By the end of the nineteenth century, these efforts had resulted in the establishment of a good system of trunk roads in the country.

A set-back to the rapid pace of road development was the introduction of the railways. The first railway line was opened in 1853 in Bombay and in the immediate decades to follow the railway network was rapidly extended to penetrate the entire country. The railways were then owned and operated by British companies. The railways served to carry the raw materials from the interior of the country to the major ports for shipment to the newly established industries in Britain for processing and manufacture. They also served to distribute the processed and manufactured goods from Britain. The needs of the administration and military were adequately served by the railways. The improvement of the roads was confined only to the feeder roads that led to and supported the railways. The railway lines were laid along well defined lines governed by the geographical and physical features of the land, and thus coincided with the alignment of the trunk routes already in existence. The improvement of these parallel roads served no purpose to the British interests then and was thus sadly neglected. With the transfer of most of the responsibilities for road work to the local governments in the middle of the nineteenth century, the construction of the local and district roads received some impetus but the roads languished for want of government attention. The Government of India Act of 1919 put a further damper on road development as the subject of roads became purely a provincial charge and the Central Government remained concerned only with roads of strategic importance. Another reason for the total neglect of roads during the later stages of the nineteenth century was that the road transport technology was itself primitive and consisted only, of horse carts and bullock carts. The authorities were quick to see the advantages of the 'iron horse' (the railway engine) and the quick means of transport it represented. The lag between the advent of the railways and the appearance of the motor vehicle in India was nearly half a century, and this period can be considered to be a dark period in road history of India. The development of the internal combustion engine by French and German scientists in the middle of the nineteenth century was a great landmark in the industrial history of the world. The motor vehicle was one of the first applications of this invention. Benz fitted the first petrol engine to a road vehicle in 1885. Daimler began to manufacture cars in the late 1880s, but it was the Americans who exploited the invention fully. The beginning of the twentieth century saw the first motor vehicles on the Indian roads. Road transport has never looked back since then.

The pneumatic tyred vehicles caused rapid deterioration of the Indian roads, which were already in a sad state of neglect due to lack of Government interest and inadequate finance at the command of the local bodies.

1.5. JAYAKAR COMMITTEE

The first World War period and that immediately following it registered rapid growth in motor transport. The clamour for better roads found expression in the Council of States. In November 1927, following a debate on the subject in the Council, the Government of India appointed a committee called the Road Development Committee consisting of members from both the Houses of the Central Legislature and Mr. M.R. Jayakar as Chairman to enquire into the matter. The Committee, Known popularly as the Jayakar Committee, was required to:

(*i*) examine the desirability of developing the road system of India and the means by which such development could be achieved.

(*ii*) Examine the possibility, having regard to the distribution of functions between the Central and Provincial Governments, of coordinating the activities of the different governing authorities in the country by the formation of a Central Road Board or otherwise.

The Committee's Report is a major land-mark in the planned development of roads in India.

The Committee was emphatic regarding the inadequacy of the Indian road system and urged that further development of the system was desirable for the general welfare of the country as a whole, and in particular:

(*a*) for the better marketing of agricultural produce.

(*b*) for the social and political progress of the rural population which will be advanced by the increased use of motor transport, and

(*c*) as a complement to railway development.

The Committee was of the opinion that road development in India was beyond the financial capacity of the Local Governments and that it was a matter of national interest. As such, it was a proper charge on Central revenues.

Since additional demands and requirements were created by the growth of motor transport, the Committee suggested imposition of additional taxation on motor transport. According to the Committee, any well-balanced scheme for such additional taxation should include:

(*a*) a duty on motor spirit,

(*b*) vehicle taxation, and

(*c*) licence fees for vehicles plying for hire.

The additional funds from motor spirit duty were to go to the Central revenue as Road Development Fund, intended for road development.

The Committee did not consider it necessary to create a Central Road Board, but recommended the appointment of a Road Engineer attached to the department dealing with roads.

The Government took steps to implement the various recommendations. The Central Road Fund was created with the additional duty levied on petrol and was intended for road development. Ten per cent of the same was detained by the Central Government and the remainder was distributed among the Provinces in the ratio which the consumption of petrol in each Province bore to the total consumption of India.

1.6. INDIAN ROADS CONGRESS

One of the recommendations of Jayakar Committee was the holding of periodic Road Conferences to discuss inter alia questions relating to road construction and development. The first of such Conferences was held in 1930. These Conferences created great interest in roads and paved the way for the establishment of the Indian Roads Congress (IRC) in 1934. The IRC is a body of professional highway engineers in India, having the following objectives:

(*a*) To promote and encourage the science and practice of road building and maintenance.

(*b*) To provide a forum for expression of collective opinion of its members on matters affecting roads.

(*c*) To promote the use of standard specifications and practices.

(*d*) To advise regarding education, experiment and research connected with roads.

(*e*) To hold periodic meetings to discuss technical questions regarding roads.

(*f*) To suggest legislation for the development, improvement and protection of roads.

(*g*) To suggest improved methods of administration, planning, design, construction, operation, use and maintenance of roads.

(*h*) To establish, furnish and maintain libraries and museums for furthering the science of road making.

During the last 80 years or so of its existence, the Indian Roads Congress has lived upto the expectations of its founding fathers admirably well. It has published a number of Standards and Codes of Practice dealing with various aspects of roads, bridges and traffic engineering.

Its technical publications have served to disseminate technical knowledge and stimulate interest in the science of road engineering. Its views on road development, planning, administration and financing are heard with respect by the policy making bodies of the Government.

Recognising the need fostering research on highway matters, the Indian Roads Congress has established a Highway Research Board. The Board identifies the various research needs and monitors the progress on schemes which are taken up for execution. Dissemination of the results of research is given due importance.

1.7. NAGPUR PLAN

The Second World War saw a rapid growth in road traffic which caused serious deterioration in the condition of roads. This situation called for immediate attention. Recognising this, the Government convened a Conference of Chief Engineers of Provinces at Nagpur in 1943. The objective was to discuss the needs of postwar road development and evolve a road plan for the next twenty years. The results of the Conference is famous as the Nagpur Plan. This Plan was a major attempt in planning for road development in a comprehensive and scientific manner.

Some of the major recommendations of the Nagpur Plan are:

1. Road development plan for the post-war period should consist of short-term measures to make good the damage caused by war conditions and a long-term programme to give adequate road communication throughout India.
2. Though 'Roads' was a Provincial subject, transport was vital to the national economy. Hence, it was time for the Centre to take active part in road development. Central lead was necessary to achieve balanced and co-ordinated development.
3. The roads should be divided into four classes:

   (*i*) National Highways, which would traverse Provinces or States, and be of national importance for strategic, administrative and other purposes.

(*ii*) Provincial and State Highways, which would be the other main roads of a Province or State.

(*iii*) District Roads, which would take traffic from the main roads to the interior of the districts. According to their importance, some of these should be considered as Major District Roads and the remaining as Other District Roads.

(*iv*) Village Roads, which would link the villages to the road system.

4. All classes of roads should be regarded as of importance to national welfare, and one class should not progress at the expense of the other, There should be a balanced development of all classes.
5. National Highways should be the framework within which the road system of the country should be developed. The Centre should bear full responsibility for their construction and maintenance, and should have an effective say in the use and control of these roads.
6. Adequate rural communication system, consisting of both village and district roads, was important. The roads should be designed and constructed by the highway department of Province or State. The system of construction or maintenance by the district boards or other local or village authorities was wasteful and inefficient.
7. The National and Provincial Highways and Major District Roads should mostly be provided with a durable hard pavement crust.
8. The Other District Roads, Village Roads and National Trails should be provided with a properly engineered earth surface, but improvements such as gravelling, soil stabilization, macadam or trackways could be considered where necessary.
9. All National Highways, Provincial Highways and Major District Roads should be adequately bridged. The permissible likely interruptions on submersible bridges or causeways should not generally be more than twelve hours at a time, or more than six times a year.

The Nagpur Plan proposed a formula for calculating the road length of different categories of roads, taking into consideration the geographical, agricultural and population conditions. The following formula has been suggested to give the length of National and Provincial Highways and Major District Roads:

Mileage of National and Provincial Highways and Major District Roads

$$= \frac{A}{5} + \frac{B}{20} + N + 5T + D - R$$

Where A = Agricultural area of the Province or State concerned in sq. miles

B = Non-agricultural area in sq. miles

N = Number of towns and villages having a population of 2000-5000

T = Number of towns and villages having a population of over 5000

D = An allowance for agricultural and industrial development during the next 20 years (about 15 per cent.)

R = Railway Mileage in the area under consideration.

The above formula ensures that in highly developed agricultural area, no village will be more than 5 miles from a main road and the average distance will be less than 3 miles in many cases. Similarly, in the case of non-agricultural area, no village will be more than 20 miles from a main road and the average distance will be 6 or 7 miles.

The formula for the mileage of Other District Roads or Village Roads is as under: Mileage of Other District Roads and Village Roads

$$= \frac{V}{5} + \frac{Q}{2} + R + 2S + D$$

Where V = Number of villages with population of 500 or less

Q = Number of villages with population of 501 -1000

R = Number of villages with population of 1001-2000

S = Number of villages with population of 2001-5000

D = Allowance for agricultural and industrial development during the next 20 years.

According to the formula given above, the mileage of roads (for post-partition India) targetted to be achieved in a 20-year period works out as given in Table 1.1.

Table 1.1: Targets laid down for road length in the Nagpur plan

	Category	Targetted road length (km)
1.	(*a*) National Highways	26,715
	(*b*) National Trails	6,680
2.	State Highways	86,825
3.	Major District Roads	80,145
4.	Other District Roads	1,33,580
5.	Village Roads	1,98,755
	Total	5,32,700

The implementation of the road development was taken up during the post-years and the five year plans.

1.8. DEVELOPMENT OF ROADS IN INDIA AFTER INDEPENDENCE

The Nagpur Plan provided a rational basis for road development in India to upgrade the war-ravaged road system.

The first major achievement was the taking over in 1947 by the Central Government of the complete financial responsibility for the development and maintenance of the National Highways. Under the new Constitution of India, the National Highways became a central subject and the Central Government statutorily took over the roads classified by them as National Highways. This was enabled by the National Highways Act, 1956 The salient features of this Act are:

- The National Highways vest in the Union (*i.e.*, The Central Government).

- It is the responsibility of the Central Government to develop and maintain, in-proper repair, all National Highways.
- The Central Government may, by notification, declare any highway to be a National Highway.
- The Central Government may give directions to the Government of any State for carrying out the work of development and maintenance of National Highways in that State.
- The Central Government may make rules for the levy of fees for services rendered in relation to use of ferries, temporary bridges or tunnels.

This Act was amended in 1995 to enable the Central Government to enter into an agreement with any person in relation to the development and maintenance of any part of a National Highway. Such a person is also entitled to collect and retain fees for services or benefits rendered by him. Thus, this amendment enables privately financed Toll Roads to be built and operated.

After independence, India launched a series of development plans, known as the Five Year Plans. Eleven such plans have already been implemented, and the 12th Five Year Plan (2012-2017) is coming to a close.

During the post-independence period, shortage of resources was a major constraint in road development. The long-term plans for road development (Nagpur Plan, Bombay Plan and Lucknow Plan) provided genera) guide-lines, but the implementation of these Plans was hampered due to acute shortage of funds. The Highway Engineering profession had, therefore, taken recourse to various innovative and cost-saving solutions. Some of these are discussed below:

1.8.1. Use of stage construction strategies

Since resources were limited, the engineers planned and constructed roads in stages, adopting the minimum design standards and specifications to begin with when traffic was low, and upgrading the standards and specifications gradually as traffic increased.

1.8.2. Adoption of intermediate design standards

Instead of adopting full design standards, a recourse was taken to adopt intermediate standards as an intervening stage between the minimum and desirable standards. For example, single lane pavements were not widened straight-away to two-lanes, but an intermediate lane option of 5.5 m width was considered.

1.8.3. Adoption of low-cost specifications

In order to bring about maximum economy, low cost specifications were adopted. Examples are : water-bound macadam instead of wet-mix macadam ; stone aggregate base courses in lieu of bitumen-bound bases ; surface dressing and thin premix carpet instead of dense bituminous concrete.

1.8.4. Use of local materials

Wherever local materials were available, the engineers preferred them to costlier and more durable materials available at long leads. Examples are : brick pavements; over-burnt brick ballast; murram, etc.

1.8.5. Preference for labour oriented technology

Since labour was available in plenty, the engineers preferred to carry outworks using labour force to the maximum possible extent and reserving use of equipment only to those specifications where quality control is of paramount importance.

1.8.6. Use of short design life

The road design was generally for a short design life (5 years). This was done to effect economy.

1.8.7. Low initial cost-the sole criterion

Low initial cost was the sole criterion, disregarding long-term effects of vehicle operating costs and maintenance costs.

1.8.8. Ad-hoc and tentative norms

Since indigenous research had not progressed, the engineers borrowed heavily from design criteria developed abroad and selected certain norms based on past experience.

Examples : Use of TRRL design curves for flexible pavements ; tentative capacity norms ; ad-hoc norms for maintenance of roads.

1.9. TWENTY YEAR ROAD DEVELOPMENT PLAN (1961-81)

1.9.1. Need

The length of roads envisaged under the Nagpur Plan was found to have been achieved by the end of the Second Plan, but the road system was deficient in many respects. The changed economic, industrial and agricultural conditions in the country warranted a review of the Nagpur Plan targets in the light of the future needs of the country's rapidly growing economy. Accordingly, the Estimates Committee of the Parliament recommended that the Government prepare a perspective plan for road development to meet the country's requirements for the next 25 to 30 years. Accordingly, a 20-year plan was drafted by the Roads Wing of the Government of India, and was adopted by the Chief Engineers in 1959 It is also popularly known as the *Bombay Plan.*

1.9.2. Objectives

The broad objectives of the Bombay plan are as under:

1. Provision of good communications in the rural areas is essential to check increasing urbanisation.

2. The future road system should, besides serving the highly developed and agricultural areas, also take into account the needs of the semi-development areas, administrative headquarters, places of pilgrimage and cultural institutions, important industrial and commercial centres, big railway junctions and ports. Strategic needs of the country should also receive due consideration.
3. The road length should be increased so as to give a road density of 32 km per 100 sq. km of area.
4. Every village.
   (*i*) in a developed area should be brought within 6.5 km (4 miles) of a metalled road and 2.4 km. (1.5 mile) of any road.
   (*ii*) in a semi-developed area should be brought within 13 km (8 miles) of a metalled road and 5 km (3 miles) of any road.
   (*iii*) In an undeveloped and uncultivable area should be brought within 20 km (12 miles) of any road.

1.9.3. Road length targets

Road length of different categories of road (in miles) should be determined by the following formula:

1. National Highways

$$\text{Mileage} = \frac{A}{40} + \frac{B}{50} + \frac{C}{60} + \left(20\,K + \frac{10\,M}{2}\right) + D$$

2. National Highways and State Highways and Major District Roads

$$\text{Mileage} = \frac{A}{12.5} + \frac{B}{15} + \frac{C}{20} + (30\,K + 15M + 7N + P) + D$$

3. National Highways, State Highways and Major District Roads

$$\text{Mileage} = \frac{1}{5}\left(A + \frac{B}{2} + \frac{C}{3}\right) + (30K + 15M + 7N + 6P) + 4Q + 1.5R) + D$$

4. National Highways, State Highways, Major District Roads and Other District Roads

$$\text{Mileage} = \frac{3}{10}\left(A + \frac{B}{2} + \frac{C}{3}\right) + (30K + 15M + 7N + 6P) + 8Q + 2.5\,R + \frac{S}{2} + \frac{T}{5}) + D$$

5. National Highways, State Highways, Major District Roads, Other District Roads and Village Roads

$$\text{Mileage} = \frac{2}{3}\left(A + \frac{B}{2} + \frac{C}{3}\right) + (30K + 15M + 7N + 6P) + 8Q + 3.75R + S + \frac{T}{2.5} + \frac{V}{8}) + D$$

Where A = Developed and agricultural area in sq. miles

B = Semi-developed area in sq. miles

C = Undeveloped and uncultivable area in sq. miles

K = Number of towns with population over 1,00,000

M = Number of towns with population between 1,00,000 and 50,000

N = Number of towns with population between 50,000 and 20,000

P = Number of towns with population between 20,000 and 10,000

Q = Number of towns with population between 10,000 and 5,000

R = Number of towns with population between 5.000 and 2,000

S = Number of villages with population between 2,000 and 1,000

T = Number of villages with population between 1,000 and 500

V = Number of villages with population below 500

D = Allowance (assumed 5 per cent) for future development and other unforeseen factors.

The targets calculated on the above basis are given in Table 1.2 below:

Table 1.2: Road length targetted in 1961-81 (Bombay Plan)

S. No.	Category	Road length		Cost Rs.
		Miles	Km	Crores
1.	National Highways	32,000	51,500	980
2.	State Highways	70,000	1,12,650	1,580
3.	Major District Roads	1,50,000	2,41,400	1,360
4.	Other District Roads	1,80,000	2,89,680	650
5.	Village Roads (classified)	2,25,000	3,62,100	630
Total		6,57,000	10,57,330	5,200

1.9.4. Cost

The cost of the Plan was estimated to be Rs. 5,200 crores. as indicated in the last column of Table 1.2, to be spread over 20 years.

1.9.5. Other recommendations

Some of the other recommendations of the plan are given below:

1. Specifications to be adopted should permit 'stage construction' in view of limited funds.
2. Funds for road construction and maintenance should come not only from direct beneficiaries (motor vehicles taxation), but also indirect beneficiaries. Some of the sources which may be tapped for raising additional resources are betterment levy, road cess on land revenue, tax on vehicles other than motor vehicles, and on diesel oil used for motor vehicles, and tolls on major bridges and high class roads.
3. In view of the dearth of trained technical staff with autonomous local bodies *e.g.* Panchayat

and Community Development Administration, provincialisation of the engineering staff of the local bodies should improve matters considerably.

4. Classified Village Roads may be under the Panchayats, to whom the State PWDs should give necessary technical advice.
5. The Highway Department should have a cell to deal with traffic engineering, road standards, etc.
6. A two-year professional course should be given to the fresh entrants to service.
7. Research should be given adequate attention.
8. Arterial roads should be fully bridged and two-lane wide to withstand trucktrailerisation.
9. Maintenance of roads should be accorded high priority. The annual maintenance expenditure is expected to increase from Rs. 30 crores in 1961 to Rs. 130 crores in 1981.

The 1961–81 Plan has been a guiding frame-work for highway planners during the Third, Fourth and Fifth Five Year Plans.

1.10. ROAD DEVELOPMENT PLAN 1981-2001 (LUCKNOW PLAN)

The Chief Engineers of States and the Indian Roads Congress have formulated a long-term Road Development Plan for India for the period 1981-2001). The Plan is also known as the Lucknow Plan. The goals and policies contained in the Plan are :

1. Road network should be developed to provide accessibility to all villages with a population of above 500 by the turn of the century.
2. Roads should be built in less industrialized areas so as to attract the growth of industries.
3. Roads should be a major choice of construction programme to generate employment.
4. Long-term master plans for determining the road network must be prepared at the sub-divisional, district, state and national level.
5. Transport requirements of small and medium towns and cities must be studied and these should be provided with good roads.
6. The National Highway length should be increased so as to form a square grid of 100 km side.
7. Expressways should be constructed on major traffic corridors to provide speedy travel.
8. Energy conservation through road improvements must be given a high priority.
9. The environmental quality of roads and the area through which they pass must be maintained and improved.
10. Road safety measures must be undertaken to contain and bring down road accident rates.
11. State Highways should be extended to serve district head-quarters, sub-divisional head-quarters, major industrial centres, places of commercial interest, places of tourist attraction, major agricultural market centres and ports.
12. Major District Roads should serve and connect all towns and villages with a population of 1500 and above.
13. Other District Roads should serve and connect villages with a population of 1000-1500.'

14. Maintenance of roads already constructed should receive attention.

1. National Highways (NH)

$$\text{Length (in km)} = \frac{Area\,(in\;sq.\,km)}{50}$$

2. State Highways (*SH*)

$$\text{Length (in km)} = \frac{Area\,(in\;sq.\,km)}{25}$$

Or

$$\text{Length (in km)} = 62.5 \times \text{Number of towns} - \frac{Area\,(in\;sq.\,km)}{50}$$

3. Major District Roads (MDR)

$$\text{Length (in km)} = \frac{Area\,(in\;sq.\,Km)}{12.5}$$

Or

Length (in km) = 90 × Number of towns

4. Total Road Length

Length (in km) = 4.74 × Number of villages and towns.

5. Rural Road length (Other District Roads and Village Roads)

The length may be obtained by subtracting the lengths of N.H., S.H. and M.D.R. from the total road length.

The basis of the formula for determining the NH length is that the. Plan envisages NH system to have a square grid of 100 km side. This gives a length of 66,000 km.

The basis of the first formula for determining the SH length is that it should be twice the NH length. The second formula is derived from the criterion that the NH and SH must pass through all towns (population above 5000). Since there are 3364 towns, the size of the square grid to achieve this is:

$$\sqrt{\frac{\text{Area of the country}}{\text{Number of towns}}} = \sqrt{\frac{2,87,782}{3,364}} = \sqrt{977.3} = 31.25 \text{ km}$$

$$\begin{aligned}
\therefore \text{ Length of NH and SH} &= 2 \times 31.25 \times \text{Number of towns}\\
&= 62.5 \times \text{Number of towns}\\
&= 62.5 \times 3364 = 2,10,250\\
\therefore \text{ Length of SH} &= 2,10,250 - 66,000\\
&= 1,44,250, \text{ say } 1,45,000 \text{ km.}
\end{aligned}$$

The basis of the first formula for determining the length of MDR is that it should be twice the length of SH. The second formula for MDR is empirical.

The basis of the formula to determine the total road length is that every village and town must be connected by a road. Since there are 5,83,563 villages and towns, the average area per human settlement is:

$$= \frac{\text{Area of the country}}{\text{Number of villages and towns}} = \frac{32,87,782}{5,83,563}$$

$\therefore$ Square grid size $= 5.634$ sq. km.

Side of square $= \sqrt{53634} = 2.37$ km

.'. Length of road $= 2 \times 2.37 \times$ Number of villages and towns

$= 4.74 \times$ Number of villages and towns

The above criteria result in the country having a road density of 0.82 km per sq. km, as against a density of 0.46 in 1981.

Example 1.1. The area of Maharashtra is 3,08,000 sq. km. The number of towns as per 1981 census is 276. The number of villages is 41,833. Calculate the length of various categories of roads.

1. N.H.

   Length $= \frac{\text{Area}}{50} = \frac{3,08,000}{50} = 6,160$ k m.

2. S.H.

   (*a*) Length of SH $= \frac{\text{Area}}{25} = 12.230$ km.

   (*b*) Length of SH and NH $= 62.5 \times 276 = 17.250$ km.

   $\therefore$ Length of SH $= 17,250 - 6,160 = 11,090$ km.

3. MDR

   (*a*) Length of MDR $= \frac{\text{Area}}{12.5} = 14,640$ km.

   (*b*) Length of MDR $= 90 \times$ Number of towns

   $= 90 \times 276 = 24,480$ km.

4. Length of ODR and VR

   Number of villages and towns $= 276 + 41,833 = 42,109$

   Total length of all categories of roads

   $= 4.74 \times 42,109 = 1,99,597$ km

   Length of ODR and VR $= 1,99,597 - (6,160 + 12,320 + 24. 640)$

   $= 1,99,597 - 43,120 = 1,56, 477$ km.

The lengths recommended for the country are indicated in Table 1.3

Table 1.3. Length of roads recommended in the road development plan 1981-2001

S. No.	Category	Road length recommended (km)
1.	N H	66,000
2.	SH	1,45,000
3.	MDR	3,00,000
4.	ODR and V R	21,89,000
	Total	27.00,000

The outlay recommended for various categories of roads are given in Table 1.4.

Table 1.4. Outlay for roads in 1981-2001 road development plan

S. No.	Category	Outlay (Rs. crores)
1.	N H	15,950
2.	S H	10,664
3.	MDR	12,759
4.	Rural Roads (ODR and VR)	24,879
	Total	64.252 Say Rs. 64,250 crores

Some other recommendations made in the Plan are briefly given below:

1. Expressways should be constructed on major traffic corridors to promote speedy travel.
2. Widening of roads should be taken up in stages as traffic develops.
3. It is not possible to make all Rural Roads to all-weather standards. To begin with, they may be made as earthen roads or gravel roads.
4. Selection of specifications should be based on (*i*) their amenability to stage construction (*ii*) the need to adopt appropriate technology (*iii*) the use of local materials (*iv*) The use of soil-stabilization techniques (*v*) the use of alternative binders (*vi*) the use of-cement concrete pavements wherever economically feasible and (*vii*) the need to conserve bitumen.
5. Ribbon development should be checked by suitable legislative and regulatory measures.
6. Research and Development activities should receive attention.
7. Data based on traffic flows and connected fields should be strengthened,
8. Additional sources of highway financing such as toll roads and private sector financing should be explored.
9. Road-making machinery technology should be upgraded.
10. It is important to impart adequate training to fresh entrants and in-service engineers.
11. Highway contracting industry should be strengthened.
12. A good Management Information System (MIS) should be set up to handle the large amount of data in roads departments.
13. Highway departments should modernize themselves by adopting computer-aided designs.
14. The Highway Departments should not feel shy of referring special problems to private consulting firms.

1.11. ROAD DEVELOPMENT PLAN VISION 2021

The Indian Roads Congress have prepared a Vision-2021 document for road development in India. The salient features of the Plan are:

1. The road network shall be expanded as under :

	Achievement till 2000 (Km)	Targets for 2021 (Km)
1. Expressways	–	15,766
2. National Highways	57,700	80,000
3. State Highways	124,300	160,000
4. Major District Roads		320,000
5. Other District roads and Village Roads	29,94,000	No target suggested

2. Half the National length should have four/six lenes, and the remaining half should have two lanes carriageway with hard shoulders.
3. 10,000 Km of State Highway should have four-lanes and the balance should have two lanes.
4. Forty per cent of the major District Roads should have two lane carriageways
5. The targets for basic access to village shall be as under:
   Villages with population above 1000 Year 2003
   Villages with population 500-1000 Year 2007
   Villages with population below Year 2010
6. Maintenance of existing assets should receive adequate attention.
7. Research and Development activities in the roads sector should receive good attention.
8. Alternative sources of funding such as toll financing, creation of a dedicated Road Fund through additional levies on fuel.
9. Upgradation of construction technology through adoption of innovative procedures and specifications should be favoured.
10. Road safety should be enhanced through engineering measures.
11. Environmental concerns caused by road and road traffic should be addressed.
12. Training of engineers should receive attention.
13. Greater recourse to Public-Private-Partnership in road projects should be taken.

1.12. RURAL ROADS, VISION: 2025

The Indian Roads Congress have prepared a Rural Road Development Plan, Vision 2025 features The salient features of the Plan are :

1. Master Plans should be prepared for Rural Roads showing the Core Network which gives accessibility to each village. All future programmes should strictly conform to this network.
2. All habitations with a population of above 100 will be connected by all weather roads.
3. It is estimated that a length of 2,90,000 Km of new roads will be needed to achieve full connectivity. Out of this 40,000 Km will be black-topped and the remaining 250,000 Km

will be gravel. The outlay required is Rs. 26,000 crores, besides Rs. 66,000 crores already taken up.

4. Upgradation of existing Rural Roads (about 1,237,000 Km length) shall be taken up, at a cost of Rs. 1,64,000 crores.
5. The maintenance of the Rural Road network will require Rs. 7,500 crores every year.
6. Stategies to cut the cost of Rural Roads shall be worked out, which include maximum use of locally available materials, soil-stabilisation, adoption of gravel road and use of low cost water crossings, and adoption of stage construction concepts.
7. Greater emphasis shall be given to the adoption of tractor-bound technology for construction and maintenance.

1.13. ROADS IN THE WORLD TODAY

Roads are the principal arteries of traffic, in the present-day world. The right indicator of a country's prosperity is its road length and vehicle ownership Table 1.5 gives an international comparison of road length in some selected countries. The following inferences can be drawn:

- America has the largest network of roads (63 lakh Km)
- India, with its 33 lakh Km of network comes second.
- The density of roads (Km/sq km) is very high in countries like Germany and Japan which are small in area.
- In countries which are large in area the density is low. India has a density of 1 km/sq km, USA 0.67 km/ sq km and China 0.12 km/sq km.
- The percentage of roads paved is very high in countries like UK. (100 per cent), Germany (99 per cent) and USA (91 per cent). Nearly the entire length is paved.
- In India, the percentage of paved roads is 50:

In modem times, Europe saw the beginnings of the Expressway system of World War II. Italy, under Mussolini, started the "Autostrade". The famous German "Autobahns" were planned in the late 1920s, and Hitler accelerated their completion. The Autobahns became a key part of the war-time infrastructure for the movement of tanks and other military vehicles.

UK started its Motorway construction rather late, in the 1950s.These form the arterial road grid of the country linking London to major cities like Manchester, Liverpool, Hull, Bristal, Edinburgh and Newcastle.

Table 1.5. International comparison of road statistics

	Country	Road length (km)	Road density (Km/sq km)	Per cent paved
1.	USA	6,300,000	0.67	91
2.	INDIA	2,009,600	0.63	50
3.	BRAZIL	1,030,000	0.23	9

4.	CHINA	1,157.000	0.12	90
5.	JAPAN	1,136,346	3.77	73
6.	GERMANY	650,700	5.97	99
7.	INDONESIA	372,414	0.19	47
8.	U.K.	366,999	1.50	100
9.	MALAYSIA	93,975	0.29	75
10.	THAILAND	62,000	0.12	97
11.	NIGERIA	32,810	0.04	83

* All values are for 1998

**India's road length now is 3.3 million km and the road density is I km/sq km.

Perhaps the largest arterial system, the US Interstate, was started after World War II as a national defence system. The construction of the 41,000 mile system was approved in 1956. It was funded by the Federal Government to an extent of 90 per cent, the balance being State's matching share. It linked all the major cities of the nation. It is toll-free. USA also pioneered the modern superhighway, a limited access, high-speed facility. The Bronx River Parkway constructed in 1925 was the forerunner to many such to come later.

1.13.1. The Inter-State system of U.S.A.

The world's best road system is perhaps in the U.S.A. The Interstate system was taken up after the Second World War as a defence system. It is now functional. The USA now has a length of 88,400 km of Expressways, of which 5,000 km was tolled. The remaining length is toll-free.

1.13.2. Autobahns of Germany

Germany began constructing its Expressways, which were known as Autobahns in the late 1920s. Before the start of the Second World War, Germany had about 4,000 km of Expressways. The country has now 11,238 km of Expressways most of which are non-toll

1.13.3. Expressways in France

France started the construction of its Expressways in the 1950s. The work was carried out through semi-government public companies. Private companies were involved in the work in the 1970s. The network was developed as atoll system. By 1996, the country had a network of 8,768 km of Expressways, 72 per cent of which are tolled.

QUESTIONS

1. (*a*) Define the terms
   (*i*) Highway
   (*ii*) Roads
   (*iii*) Streets
   (*iv*) Expressways

   (*b*) Trace the history of early roads with special reference to Indian roads
2. Discuss the development of roads in India during British period
3. Briefly state the objective of the following:
   (*a*) Jayakan Committee
   (*b*) Indian Roads congress
4. Give the major recommendation of the Nagpur Plan. State the formula used to determine the length of
   (*i*) National, Provincial and Major District Road
   (*ii*) Village Road or Other District Road
5. List the achievements of the Five Year Plans in road development. What innovative and cost saving solutions were taken up to overcome the acute shortage of funds.
6. What were the objectives of the Twenty Year Road Development Plan (1961 -81) and its targets. Briefly state other recommendations.
7. Enumerate the goals and the policies set for the Lucknow Plan (1981-201)
8. State the salient features of Road Development Plan Vision 2021 and Rural Roads Vision 2025.
9. Where India stands in comparison to road lengths in USA, UK, Germany, and China ?

Chapter

Highway Administration and Finance

2.1. ROADS AS INFRASTRUCTURE FOR PUBLIC USE

Roads, along with railways power, telecommunication, water supply 'sanitation' sewerage and airports, are an essential infrastructure for the economic well-being of a country. For a developing country, in particular, roads facilitate quick economic development.

Since roads are an essential infrastructure for the economy of a country, they are for public use. Hence it is incumbent upon the government to invest on roads. Thus, traditionally roads have been built and maintained by the government over the past many years.

2.2. FUNCTIONAL CLASSIFICATION OF ROADS

It is necessary to classify functionally the road system of a country for the following reasons;

1. To facilitate administration of the roads, assigning responsibility to administer roads of different organizations.
2. To assign appropriate design standards to different classes.
3. To facilitate the planner to develop an integrated and harmonious network for the country.

Universally, roads are classified into two broad categories:

1. Urban
2. Non-urban (rural)

Urban roads are those which are located within cities, towns and other inhabited areas and come under the jurisdiction of municipalities, cantonment boards and port trust.

Rural roads are those that are located in the non-urban areas and connect cities and towns across the country.

Both the urban and non-urban roads are further classified functionally into the following broad hierarchy :

1. Arterial roads
2. Sub-arterial roads

3. Collector roads
4. Access or local roads

Each country has developed its own classification system on the above lines to suit its individual requirements.

2.3. CLASSIFICATION OF ROADS IN INDIA

In India, urban roads are classified into the following classes :

1. Expressways
2. Arterial streets
3. Sub-arterial streets
4. Collector streets
5. Local streets

A future description of each class is given in chapter 18. Non-urban roads are classified into five categories in India, from the administrative and functional considerations. The classification was originally suggested in the Nagpur plan and is in force even now. Table 2.1 give the classification. In India, a term "Rural Roads" is sometimes loosely used to describe roads serving villages. This term includes the other district and village roads of Table 2.1.

Table 2.1: Classificatoin of non-urban roads in India

S. No.	Class	Description
1.	National Highway (N.H.)	These road traverse the length and breadth of national importance for strategic, administrative and other purposes. They connect capital of states, ports, foreign highways, large towns and industrial centres. They can be deemed to be the primary road network of the country.
2.	State Highways (S.H)	These roads are the main arteries of traffic within a state, serving important cities, and connected to the National Highways system and important highways of adjacent states.
3.	District Roads (i) Major District Roads (M.D.R.) (ii) Other District Roads (O.D.R).	These roads take off from the main roads in a state to the interior of the districts. According to their importance, some of these are considered as Major District Roads (M.D.R.) and the remaining as Other District Roads (O.D.R.)
4.	Village Roads (V.R.)	These roads connect villages and groups of villages with one another and to market places and district roads.

In, the Roads development plan (1981-2001) (Lucknow plan) a slightly modified road classification system has been suggested. It is as under;

(*A*) Primary system

(1) Expressways (2) National Highways

(*B*) Secondary system

(1) State Highways (2) Major District Roads

(*C*) Tertiary system

(1) Other District Roads (2) Village Roads.

2.4. ADMINISTRATION OF NATIONAL HIGHWAYS

The National Highways fall directly within the sphere of responsibility of Central Government. After consultation with Provincial Governments and discussions at meetings of the Transport Advisory council, the Government of India accepted with effect from 1 st April 1947, complete financial liability for the development and maintenance of certain roads provisionally approved by them as suitable for inclusion in a system of National Highways. Later, in, 1956, National Highways Act was passed declaring the National Highways and empowering the Central Government to declare any other highways to be a National Highways or omit any highways from the list of highways declared as National Highways. The National Highways initially included in the Schedule to the act were all extra municipal. That is, they do not include such parts as were situated in municipal areas, a municipal area being defined as one with a population of 20,000 or more and under the control and management of a municipal committee, town area committee, or any other authority. The Central Government is, however, enabled by the Act to enter into an agreement with the authority in control of the municipal area in relation to the development or maintenance of any such part of a highway situated within the municipal area on the basis of expenditure thereof.

The Act vests in the Central Government not only the National Highways lengths but also demarcated land appurtenant thereto, all structures constructed on, or across the National Highways such as bridges, culverts, tunnels, causeways, carriageways etc. and all trees, posts, fences and boundary and kilometer stones.

Under the Act, the Central Government may levy fees for services or benefits rendered in relation to the use of barriers, temporary bridges and tunnels on any National Highway. The Act has recently been amended to permit fees on roads also.

While control and management of the National Highway lie with the Central Government, the latter may delegate any of its function in relation to the development and maintenance of the National Highways to the Government of the State in which the National Highways is situated or to any authority subordinate to the Central or State Government.

At Present, actual works on construction and maintenance of National Highways are executed by the State Public Works departments on an agency basis. For this, the State PWDs are paid agency charges which are 9 per cent currently. A few stretches of the National Highways are also being constructed and maintained by the Central Public Works Department and the Border Roads Organisation.

In consultation with the State PWDs, the Central Roads Wing draws up plans for the development of the National Highways. The projects so drawn up are put forward to the Niti Udyog before finalisation of the Five Year plan. The selection of the schemes is done by the Niti Udyog within the overall framework of the Transport Sector plan and in the end, the plan for all sectors. An annual review of the plan is conducted by the Niti Udyog where adjustments in the approved outlay are made according to the financial situation and the progress of works achieved. Detailed plans and estimates for the scheme included in the approved programmes are prepared by the State PWDs. These schemes are scrutinized in the Central Roads Wing and are sanctioned from time to time. The actual funds are allotted on a yearly basis, within the Central budget.

As regard maintenance of the National Highways system, the state PWDs assess the needs based on traffic, terrain and environment conditions. The yearly programme of surface renewals, ordinary maintenance and special repairs is formulated by the state PWDs and approved by the Roads Wing.

The National Highways system is now about 66.900 km in length and is the main artery of traffic. Though constituting only about 2 per cent of the roads in the country, it carries nearly 58 per cent of freight moved by road and 36 per cent of passenger moved by road.

2.5. ADMINISTRATION OF ROADS UNDER OTHER CENTRAL MINISTRIES

In addition to the National Highways, there are other roads under the administrative control of central ministries.

Roads in the military areas are constructed and maintained by the Military Engineering Service (M.E.S.). These roads are financed by the ministry of Defence.

Roads in the cantonment areas, serving civil needs, are constructed and maintained by cantonment Board. These roads are also financed by the ministry of Defence.

Roads in areas around railway lines are constructed, maintained and financed by the Ministry of Railways.

In order to accelerate the economic development of the Northern, North Eastern and Western Border, the Border Roads Development Board has been constituted With the Prime Minister as chairman and the Defence Minister as Deputy chairman. The Border Roads Organization is the technical organisation set up in 1960 in charge of actual construction and maintenance of these roads. This organization has expanded considerably and is perhaps the largest single departmental undertaking for the construction of roads any where and is self-sufficient in manpower, machinery and equipment ant is know for its efficiency and speed of execution. In all 12,878 km of roads are being maintained by the Border Roads Organisation presently. in addition to road building work, the organization has the resources and expertise to undertake civil engineering works pertaining to hydel and irrigation projects, air-fields, buildings etc. It also provides technical expertise and personnel to certain public sector undertakings for such works abroad.

The Ministry of Rural Development acts as the nodal organization looking after the rural roads (village roads) programme in the country.

The ministry of Urban Development looks after urban transport, and hence urban roads. But the urban roads themselves are under the jurisdiction of municipalities and local bodies.

2.6. ROADS OF INTER-STATE OR ECONOMIC IMPORTANCE

In order to facilitate Inter-state road communications, and to help the State in their economic development through construction of roads and bridges a central scheme is being implemented by the Roads Wings of the Ministry of Road Transport and Highways since 1954.

An amount of Rs. 85 crores has been disbursed under this programme during the past 25 years. Till 1979. central financial assistance was being provided by way of 100 per cent loan for both roads of Inter-State importance and economic importance. However following the recommendations of the National Development council, this pattern of assistance has been changed since 1970s so as to provide 100 per cent loan assistance only for project of Inter-state importance. In respect of roads of economic importance, the central loan assistance is now restricted to 50 per cent of the cost, the remaining 50 per cent cost being met by the State Government.

The construction and maintenance of these roads is done by the State PW.Ds.

2.7. MINISTRY OF ROAD TRANSPORT AND HIGHWAYS

Road matters of the Central Government are handled by Ministry of Transport and Highways. It is headed by a Director General (Road Development), who is also ex-officeo Special Secretary to the Government of India. He is assisted by Chief Engineers, Superintending Engineers Executive engineers and Assistant Executive Engineers. Apart from the combined secretariat and technical organization at the Centre, the Roads Wing has a Superintending Engineer for the North Eastern Region at Guwahati and liaison-inspectorate organization (Regional Office)consisting of Superintending Engineer and Executive Engineer in various States. The function of the Ministry Road Transport and Highways are as under:

(*i*) to administer funds approved by Centre for the development of National Highways.

(*ii*) to formulate plans for development and maintenance of National Highways in consultation with the executing agencies (State PWDs).

(*iii*) to examine technically the projects pertaining to roads and bridges, prepared by the executing agencies (State PWDs) and recommend the same for financial sanction the Government.

(*iv*) to oversee technically the quality of works executed by the agencies.

(*v*) to review problems affecting the development of road project through periodic meeting with the State Chief Engineers either on the occasion of General Session of the Indian Roads Congress held annually or at the time of the meeting of their Council which takes place during the mid-year.

(*vi*) to administer the Central Road Fund.

(*vii*) to administer the Central Road Programme other than National Highways in the Union Territories.

(*viii*) to administer road programme such as Roads of Inter-State and Economic Importance, Strategic Roads and roads in the sensitive border areas.

(*ix*) to coordinate matters pertaining to road research and administer Central funds allocated for this purpose.

(*x*) to administer the Indian Roads Construction Corporation, a public sector undertaking.

(*xi*) to serve as a repository of technical information and know ledge on roads and bridges.

(*xii*) to evolve standards and specifications for road and bridge works.

2.8. NATIONAL HIGHWAYS AUTHORITY OF INDIA (NHAI)

The NHAI was established under the National Highway Authority of India Act. 1988, but was operationalised only in February 1995. The Authority is an autonomous body with executive responsibility for the development, maintenance and operation of National Highways and associated facilities and is under the administrative control of the Ministry of Road Transport and Highways.

The responsibilities of the Authority would, within the strategic policy framework, be eventually extended to cover the exiting NH network, as well as development of new NHs entrusted to it, The stated mission of the NHAI is as follows:

To meet the Nation's need for the provision of a National Highway network development to world standards within the strategic policy framework set by the Government of India and thus promote economic well-being and the quality of life of the people.

Towards these objectives the Authority plans to :

(*a*) Take responsibility of development and maintenance of all NHs in the country within a period of 5 years.

(*b*) To manage the NH network in a manner that provides safety and comfort to those who use it.

(*c*) Improve and extend the NH network in an efficient and environmentally sensitive manner.

(*d*) Seek visible ways to involve the private sector in the provision, maintenance and operation of roads.

(*e*) Encourage research and develop effective liaison with institution involved in research on road and road-related issues.

(*f*) Improve road safety including road geometries.

(*g*) Provide on-route facilities for road users.

(*h*) To provide a congenial environment including landscaping and pleasant roadsides to ensure safe, comfortable and pleasant driving.

(*i*) To promote the scheme of tree plantation along the roads as well as to beautify all major intersections and junctions.

2.9. CENTRAL ROAD FUND

One of the recommendations of the Indian Road Development (Jayakar) Committee, 1927 was that the duty on patrol should be increased from 25 to 40 paise per gallon, and proceeds from the additional duty be kept apart as the Road Development fund. The Government accordingly created the Central Road Fund (C.R.F.) in 1929. The rate of additional duty was raised from 6 paise to 10 paise per gallon in 1931. The rate was 3.5 paise per litre of petrol till 1988. Out of the actual accruals to C.R.F., 80 per cent is allocated to the State Government on the basis of actual consumption of petrol within their territories and is known as C.R.F. (Allocations) The balance 20 per cent is set apart as C.R.F. (Ordinary) Reserve. The cost of administering the account is met out of C.R.F. (Ordinary) Reserve and the full C.R.F. (Allocations) are utilised on schemes approved by the Central Government relating to :

(*i*) Road research
(*ii*) Traffic and economic surveys
(*iii*) Training of young engineers
(*iv*) Road and bridge schemes of all-India importance
(*v*) Expenditure on construction of parallel service roads on bypasses
(*vi*) Construction of rest places for drivers and conductors and way-side amenities
(*vii*) Purchase of machinery

The CRF scheme has been recently reviewed (1988). It has now been decided to expand its scope. The features of the revised scheme are:

1. Both petrol and diesel are brought under its purview.
2. The accruals will henceforth be 5 per cent of the basic price out of the duty of customs and of excise levied on petrol and diesel.
3. The accruals will be solely earmarked for road development
4. $\frac{1}{2}$ per cent of the accruals will be used for defraying the cost of administering the fund.
5. 35 $\frac{1}{2}$ per cent shall be utilized by the Central Government for National Highways.
6. 64 per cent shall be utilized by the States for State roads to be distributed on the basis of petrol and diesel consumption in each State.

In 1998, the Government of India announced a cess of Re. 1 per litre on the sale of petrol, thus generating an amount of Rs. 7900 million (Rs. 790 crores) for the development of National Highways.

In 1999, the Government of India, in its budget, imposed a further cess of Rs. 0.50 per litre of diesel sold. It is estimated that an additional annual revenue of 5,000 crores (Rs. 50 billion) would be available. However, the Government has announced that only 50 per cent (Rs. 2,500 crores, or Rs. 25 billion) would be available for the development of highways. This amount will be further distributed as under :

60 per cent (Rs 1500 crores or 15 billion) will be set apart for the maintenance and development of National Highways

30 per cent (Rs. 750 crores) for the maintenance and development of the State Highways and Major District Roads.

Balance 10 per cent (Rs. 250 crores) for the Railways for the construction of road-Over-Rail-Bridges and safety works at unmanned rail crossings.

Thus, it may be noted that from 1999, the additional accrual to the highway sector would be annually Rs. 3.300 crores. An amount of Rs. 2,500 crore has also been set apart for the development of low volume roads in rural areas (ODRs and VRs). It has been targetted to provide all-weather road access to all villages with a population of 1000 and above by 2003 and to all villages with a population of 500 and above by 2007.

It has been estimated that the Government expenditure (Centre and State) over the years has only been in the range of 30-40 per cent of the proceeds.

2.10. ADMINISTRATION OF STATE ROADS

The State Government have the responsibility of development and maintenance of State Highways. District Roads and Village Roads. The State Public Works Department (P.W.Ds.) generally look after those roads. The State P.W.Ds. are also the executing agencies of the Central Government works pertaining to the National Highways. C.R.F. roads, some Strategic roads, sensitive border area roads and roads of Inter-State and Economic Importance.

Some States have a separate department to look after highways. This has proved to be very beneficial. Some States have even a separate organization to look after the National Highways Works. The Central Government has now set up a separate organization to deal with National Highway works. This organization is known as the National Highway Authority of India (NHAI)

In the matter of the administration of rural roads (Other District Roads and Village Roads) a wide diversity is noticed. In some States, rural roads are handled by the public Works Departments. In some other States, they are looked after by the Zilla Parishads, Panchayat Samitis and Panchayats. Other organizations like Community Development Blocks, Agricultural Department also look after some of these roads in many States. There is a need for a single unified agency for the planning and execution of rural roads programme to have efficient management and effective monitoring.

Since the traffic on roads in the country has grown very fast, the problems of traffic engineering, traffic management, traffic control and traffic safety have also increased. In order to attend to these matters, the Central Government and Indian Roads Congress have been requesting the State P.W.Ds. to set up Traffic Engineering Cells. Some of the States have established such cells. The Central Government provides some financial assistance to the States in setting up such cells.

2.11. ROAD'S RESEARCH

Road research has played an important role in India's roads development. The problems of planning, design, construction and maintenance of roads in India are so unique and challenging that the ready-made solution from research in other countries are not found feasible and economical. Indigenous solutions have to be devised to suit our local needs and resources. This has been aided considerably by road research, which has carried out technical capabilities to the frontiers of knowledge in this field.

2.11.1. Central Road Research Institute (CRRI)

Pioneering work on road research in the country has been carried out by the Central Road Research Institute, Delhi, set up in the chain of national laboratories under Council of Scientific Research. The Institute has the following objectives:

(*i*) to carry out basic and applied research for investigation, design, construction and maintenance of different types of roads and runways.

(*ii*) to carry research on the utilization of locally available materials for construction and maintenance of roads and runways economically.

(*iii*) to carry out research on road traffic and transportation, including traffic safety and transport economics.

(*iv*) to develop labour intensive methods and manual aids for low-cost-all-weather roads.

(*v*) to develop appropriate machinery, tools, equipment and instruments for adopting technologies as related to highway engineering and as relevant to the country to achieve indigenous self-sufficiency.

(*vi*) to render technical advice and consultancy services to various organizations.

(*vii*) to arrange for utilization of results of research by extension unit, display centre etc.

(*viii*) to disseminate results of research by publication of scientific material.

(*ix*) to conduct refresher and training courses for staff of other Research Institutions, Universities and Highway Departments.

(*x*) to provide library and documentation service.

The Institute is headed by a Director and has the following specialist divisions:

1. General project
2. Geotechnical engineering
3. Extension
4. Flexible pavements
5. Rigid pavement
6. Traffic and transportation
7. Environment and safety
8. Workshop
9. Electronic instrumentation
10. Documentation
11. Highway planning and management.

2.11.2. State Highway Research Laboratories

Some of the State Governments have Highway Research laboratories for carrying out research and testing and quality control during highway and bridge construction. Notable among them are :

1. Highway Research Station, Chennai.
2. Maharashtra Engineering Research Institute, Nasik.
3. Gujarat Engineering Research Institute, Vadodra.

2.11.3. Highway Research Board (HRB)

In order to coordinate all matters pertaining to highway research, a Highways Research Board has been formed under the Indian Roads Congress in 1973. The board is a recommendatory body, without any finance of its own for sponsoring research. The actual funds are earmarked by the Central and State Government. This is a serious drawback in the existing functioning of the Board. The following are the objectives of the Board:

(*i*) To identify highway research needs.

(*ii*) To recommend priorities and schedule of research

(*iii*) To recommend specific research problems to selected organizations or institutions.

(*iv*) To evaluate, monitor, interpret and disseminate research findings.

(*v*) To act as an advisory organization to the Central and State Government and research Organization and assist them in the selection of research projects.

(*vi*) To stimulate highways research.

(*vii*) To prepare bibliographies and state-of-art report.

2.12. HIGHWAY STAFF TRAINING INSTITUTE

Highway engineering is a subject which is fast developing. In view of the serious shortcomings in the present road system in the country, vast improvements and developments are going to take place in the years to come. This is a challenging job to the highway engineering profession. The success depends upon the technical expertise which the engineers can develop. In this task, the role of adequate training at entry to the profession and periodic in service training assumes considerable importance. Every year a large number of fresh engineers enter the Central and State Highway Departments but there are no facilities at present to impart training to them. In order to overcome this deficiency the National Institute for Training of Highway Engineers (NITHE) has been set up and is functioning from NOIDA (U.P.)

2.13. ROAD MAKING MACHINERY

Road building is increasingly becoming a highly specialised subject, which demands very strict control over quality of work. The quantum of work to be done within a specified time is also very high. All these factors necessitate the use of various kinds of road making machinery. These include earthmoving equipment of all kinds (excavators, shovels, bulldozers, tractors etc.), compacting equipment (road rollers, sheepsfoot rollers, vibratory rollers etc.) compressors, bituminous equipment (bitumen pressure distributors, heaters, hotmix plants cold- mix plant, paver finisher etc.) and bridge construction equipment (cranes, derrick, etc.) It is difficult for the State PWDs. to purchase the equipment from their own finance. The Central Government has been accordingly procuring heavy, sophisticated and allied items of machinery directly for use on National Highways and other Central works. These items of machinery are owned by the Central Government, but are managed by the State P.W.Ds. as executing agents for Central works. In addition, the Ministry also sanctions loan assistance to the States for purchasing other general items of road construction machinery, whenever the State cannot arrange the same from other State funds or through contractors.

2.14. PROPOSAL FOR A ROAD BOARD

The road transport sector has earned for the Government, over the years, a fair amount of revenue by way of taxation on vehicle, spare parts, fuel, bitumen and toll charges and octroi levies. It has been the feeling in the road transport industry that the share of road and road transport in the Government outlay on development has been much less than the revenue it yields. With a view to ensure all round development of roads and road transport on a just and equitable basis, a view exist in the country that a separate Roads Board should be constituted, having over ail control of development of roads. There is another view that a National Transport Commission should be set up dealing with all modes of transport. Such a set up is expected to secure better coordination of all modes of transport.

2.15. NATIONAL RURAL ROAD DEVELOPMENT AGENCY

The Prime Minister announced a nation-wide Rural Roads programme, which is known by the name Pradhan Mantri Gram Sadak Yojna (PMGSY). Under the PMGSY, it is proposed to connect every habitation with a population of over 1000 persons through good all-weather roads by the year 2003 and those with a population of more them 500 persons by the year 2007. The programme is expected to cost nearly Rs. 60,000 crores and is expected to provide connectivity to about 1,00,000 habitations. The programme also aims to upgrade, to specifications, about 5 lakh kilometers of existing rural roads. For a successful implementation of this ambitious programme, a National Rural Road development Agency (NRRDA) has been set up under the Ministry of Rural Development, Govt, of India.

2.16. HIGHWAY FINANCING AND TAXATION IN INDIA

The construction of highways and their maintenance is carried out by the Govt, in almost every country in the world. The funds necessary for the purpose are realized from taxes on road users, and in special cases, through levy of tolls.

It has often been said that the level of taxation on roads and road transport in India is very high, especially when compared to the level of outlays on roads. A number of committees have gone into the question, and the general finding of those committees has been that the road taxation system in India is complicated and needs to be simplified. The origin of the high levels of taxation can be traced to the Government's anxiety as owners of Railways, to regulate road transport so that it does not become a competitor to Railways.

Road transport is a favourite source of tax collection by the government for mobilising its general resources. This is evident from the fact that the Government collected Rs. 26,000 crores as taxes in 2000-01, but spent only Rs. 14,500 crores on construction, maintenance, administration, policing and research on roads. There is no economic justification or rationality behind this, but the Government has to mop up revenues from all possible sources, and road transport happens to be a source which has historically been taxed disproportionately to its legitimate charge. This policy can have crippling effect not only on this mode of transport, but also on the economy as a whole since road transport affects almost every sector of the economy.

Table 2.2 gives the total tax revenue collected by the Government and the expenditure incurred by it on roads (both construction and maintenance), for the past few years. It is seen that in the fifties, the expenditure on roads was above 65 percent of the tax collection. This figure dropped down to 35-45 per cent in the seventies and eighties and is currently 42 per cent. This is an indication that the expenditure on roads is progressively falling down in relation to the earning from taxes.

Table 2.2. Tax collection from road transport and expenditure on roads (Construction and maintenance)

Year	Tax/User charge Revenue (crores)	Expenditure on roads (Rs. Crores)	Expenditure on roads a s percentage of tax collection
1950-51	47.37	34.47	72.7
1960-61	166.94	109.76	65.7
1970-71	683.12	257.60	37.7
1984-85	4400.00	1824.00	41.5
2001–02	5000.00	2100.00	42.0

2.16.1. Constitutional position

Under Entry 35 of the concurrent list in the Seventh Schedule of the Constitution, parliament has the power to legislate in respect of mechanically propelled vehicles, including the principles on which taxes on such vehicles are to be levied. But such legislation cannot determine the quantum of these taxes. Taxes on vehicles fall exclusively under the competence of State legislature.

Entry 56 of the State list in the Seventh Schedule provides for taxes on goods and passengers carried by road or on inland waterways. Entry 57 is for taxes on vehicles and Entry 59 is for tolls. Thus, each State is free to determine the type of taxes on motor transport and their level. This has given rise to a taxation system of wide variety in the States and hardly an effort has gone in to evolve a uniform system of taxation, conducive to the development of the road transport industry.

The Central Government has constitutional authority to levy custom and excise duty on vehicles, spare parts; tyres and tubes and oil products. The Central Government has also the authority to levy tolls on a bridge, road, ferry or tunnel forming part of the National Highway system.

As regards inter-State movement, although parliament can legislate under Entry 42 of the Union list for inter-state trade and commerce, practically no attempt has been made to propose principles of taxation applicable to such transport.

2.16.2. Multiplicity of taxes

One of the unhappy features of the taxation system on road transport in India is the multiplicity of types of taxes, the authorities imposing them and the places of collection. Almost all the committees appointed by the Central Government have pointed out this irksome position, but little has been done to simplify the system.

2.16.3. Types of taxes and levies

The taxes and levies on roads transport can be grouped broadly under Central and State collection. The various taxes and levies are enumerated below:

(A) Central

1. Import duty on motor vehicles and spares
2. Import duty on petroleum products
3. Excise duty on motor vehicles, spares and tyres and tubes
4. Excise duty on petroleum products
5. Toll at selected bridges, tunnels and roads on National Highways.

(B) State

1. Sales Tax on motor vehicles, spares and tyres and tubes
2. Sales Tax on petroleum products
3. Motor Vehicles Tax
4. Taxes on passengers and goods
5. permit fees
6. Drivers and conductors' licence fee.
7. Octroi
8. Toll levied by State
9. Wheel Tax

The total tax collection from the road transport sector in the country for 1984-85 was Rs. 4400 crores, composed of the break-up indicated in table 2.3.

Table 2.3. Taxes on road transport (1984-85)

A. Central		Rs. Crores
1.	Taxes on vehicles	11
2.	Taxes on goods and passengers	23
3.	Excise duty on	
	(*a*) Petrol	678
	(*b*) Diesel used in road transport	298
	(*c*) Motor vehicles	382
	(*d*) Tyres	410
4.	Import duty on	
	(*a*) Vehicles	133
	(*b*) Petroleum products	120
	Total 'A'	2055

	B. States	
5.	Sales Tax on fuel	273
6.	Taxes on vehicles	705
7.	Taxes on goods and passengers	494
8.	Sales Tax on spares and tyres	140
	Total 'B'	1612
	C. Invisible Taxes (Estimated)	
9.	Raw materials used in vehicles and spares	430
10.	Raw materials in tyres	300
	Total 'C	730
	Total A + B + C	4397
	Say Rs.	4400 crores

The taxes on petroleum products by the Centre and the sales tax on these products by the States yield nearly one-third the total taxes on road transport in the country. Petrol is very heavily taxed when compared to diesel. This is clearly intended to show disfavour on the users of private cars and other personalized mode'. The taxes on petroleum products also indicate the concern of the Government on the growing oil import bill and the energy shortages. The Central import duty and excise on motor vehicles and spare parts constitute about one-fourth of the total taxes. The States, in addition, levy sales tax. It has been roughly assessed that nearly 40 percent of the 'on-road: price of new vehicle are taxes. The same situation holds good for spare parts and accessories. The tyre industry bears a great burden of taxes. It contributes 10 per cent of the total road transport taxes to the Centre. In addition, sales tax is also charged by States at various rates.

The Motor Vehicles Tax is also known as token tax or road tax and is levied by the State Governments on all kinds of motor vehicles, including cars, taxis, buses, trucks, three-wheelers and two-wheelers. The tax rates vary from State to State. There is obviously a need to have a uniform tax structure in all the States.

Passenger taxes are levied in the States at rates varying from 10 to 35 per cent of the fares charged. In some States this tax is compounded with the Motor Vehicles Tax. The tax is in the nature of a surcharge. Since the calculations involved in the determination of the tax are rather tedious, there is every appeal in compounding this tax with the Motor Vehicles Tax.

Goods taxes are levied in States other than Tamil Nadu, West Bengal and Sikkim. The amount of goods tax collected is generally a lump sum amount or calculated as a percentage of freight charged. There is again a wide variation in the tax collected from State to State, giving rise to the practice of operators registering their vehicles in States where the taxes are more favourable. Obviously there is a need to have uniformity.

The permit fees charged for operating vehicles vary with inter-State, Zonal or National permits. The fees also vary from State to State. Octroi is levied by local municipal authorities in some States. Some progressive States have already abolished this levy. Some of the objections to Octroi are :

(*i*) It violates the principle of free flow of trade across local boundaries.

(*ii*) It encourages concentration of industries in metropolitan areas, thus acting against dispersal of industries.

(*iii*) It breeds corruption.

(*iv*) It causes unnecessary delays and hold-ups causing loss of valuable time of trucks and the crew.

(*v*) It causes extra fuel consumption due to idling of vehicles.

In view of the vexatious nature of the tax, almost every Committee appointed by the Government has recommended its abolition. The implementation has been slow, because it involves the reimbursement of foregone revenue to the local bodies by the State Government.

The Jha Committee has estimated octroi revenue for 1967-77 at Rs. 180 crores and the-estimate for 1977-78 at Rs. 270 crores. Thus the amount involved is substantial. The Pande Committee has estimated that 30-46 per cent of effective travel time of vehicles on trunk routes is lost at check-points and the socioeconomic cost of Octroi is much more than the revenue it yields.

2.16.4. Tolls

Tolls are levied by Central Government on selected bridges and roads on National Highways as part of a newly introduced scheme to generate resources for taking up road schemes. According to this scheme, a toll of Rs. 10 per loaded truck, Rs. 5 per unloaded truck and Rs. 2 per passenger car is levied on selected projects till such time as the initial cost of the scheme is recovered. The amount so realised is available for further road schemes on National Highways in that State where toll was collected. The scheme has proved quite successful. Revenue to the tune of Rs. 0.5-1.0 crore can be realised at a single point on the trunk route system. It should, however, be ensured that the location of toll collection points is not too frequent and only worthwhile schemes which appeal to the ordinary road user as a real benefit should be selected.

State Governments also charge toll on State roads at selected points on the above principle. An example is the Thana Creek Bridge in Maharashtra. There is a grand scheme now being considered for bridge or tunnel to link Mumbai with the mainland, and a system of toll is being thought of.

Many countries have modernized their road system by constructing Expressways on the basis of toll financing. Notable among them are France, Japan, Korea and Mexico. The acute shortage of funds for roads has recently led to a thinking by the Government of India to explore possibilities of toll-financed road projects. The intention is to involve both the private and public sector in this venture. Chapter 20 gives various forms of Toll taxes being adopted. The advantages of toll roads are:

1. Toll roads get completed quickly since the companies raise funds at high rates of interest.
2. Toll financing creates a new revenue structure, independent of uncertainties of future availability of funds.
3. Tolling is an efficient method of charging the actual road user and is an equitable form of user taxation.
4. Toll roads are very well maintained.
5. Toll companies enjoy complete autonomy, and thus they provide excellent project control and good wayside facilities.

6. Toll roads can be used as a traffic restraint measure to decongest roads.

The disadvantages of toll roads are :

1. Since toll roads are constructed with borrowed capital, their costs are high.
2. Toll collection is an additional cost component.
3. Delays are caused at toll collection plazas.
4. When traffic shies away from a toll road, it uses the substandard toll-free road, often causing higher costs to society.
5. For the same reason as above, the full potential of toll roads is not utilised.
6. Toll companies raise capital at a higher rate of interest than what the Government pays when borrowing directly. This is because of the risks involved.

QUESTIONS

1. Describe the highway classification system in India, both for urban streets and rural highways.

2. Describe the National Highway system in India (Hint: Give definition, length, legal status role of Ministry of Road Transport and Highways, role of State P.W.Ds).

3. Write short notes on :

(a) Central Road Fund (*b*) Highway Research Board

(c) Central Road Research Institute

(*d*) National Institute for Training of Highway Engineers

(*e*) Octroi. (*f*) National Highways Authority of India.

4. Comment upon the road transport taxation system in India.

5. What are Toll Roads? What are their advantages and disadvantages?

Chapter

ROAD USER AND THE VEHICLE

3.1. INTRODUCTION

A. highway engineer is required to design road facilities which will be used by pedestrians, cyclists, animal-drawn vehicles and a variety of motor-powered vehicles. Thus there is a great variation in the characteristics of vehicles and behaviour of the users and an understanding of the same is an essential pre-requisite for an adequate design of the facility. Human factors that govern the behaviour of the driver, schist and the pedestrian affect significantly the design elements. A knowledge of how this behaviour is influenced by various external conditions will be useful in designing the facility. The characteristics of the different types of vehicles will influence the geometric design elements of the highway and will determine the safety of traffic using the road.

3.2. HUMAN FACTORS GOVERNING ROAD USER BEHAVIOUR

3.2.1. Human body as a complex system

The human body has a complex mechanism exhibiting varied reactions to external stimuli. When dealing with highway engineering design, human behavior can be studied under the following groups:

1. Physiological
   (*i*) Vision
   (*ii*) Hearing

2. Psychological
   (*i*) Perception ,
   (*ii*) Intellection
   (*iii*) Emotion
   (*iv*) Volition

3.2.2. Vision

Pedestrians, cyclists and drivers are able to use the road safely because of the help received by the eyes in seeing the road traffic thereon and in evaluating the size, shape, colour, distances and speed of approach of various objects on the road. Safety of traffic depends upon the ability of the road users in seeing traffic lights, traffic signs, vehicles on the road, safe gaps and safe crossing places. The drivers are able to cross, overtake, stop, accelerate and decelerate their vehicles on seeing the road conditions, the traffic conditions and the environmental conditions affecting safe traffic movement.

Vision is achieved through light rays from the object getting refracted through the eye lens and coming into focus on the retina. An image forms on the retina, which its transmitted to the brain through the optic nerve, enabling the vision of the object.

The sharper vision is concentrated in the zone formed by a cone whose angle is 3 degrees about the centre of the retina. This zone is known as the zone of acute vision or visual acuity. For distant vision, the objects should be within this narrow cone if satisfactory perception has to be achieved. However the vision is still satisfactory when the angle of the cone is upto 10 or 12 degrees. For most persons, visual acuity is reasonably sharp within a central angle upto about 10 degrees. Beyond 10 or 12 degree angle the vision is less well defined. Visual acuity declines as the age of the person increases. Considerations of visual acuity are important in locating traffic signs, sign posts and information message at the road-side edge. According to AASHTO for every decade after age 25 years, drivers need twice the brightness at night to receive visual information. Thus, by the age 65, some drivers may need 16 times the brightness they did at age 25 years.

Peripheral vision deals with the total visual field for the two eyes. The angle of peripheral vision is about 160 degrees in the horizontal field and 115 degrees in the vertical direction. Within this zone, the eyes of a normal person are able to see objects, but without clear detail and colour, For example, the driver can be alerted by a movement or a bright light in the zone of peripheral vision. He then turns in the direction of the object or light and with the help of acute vision can see the object clearly. The angle or peripheral vision falls with speed. It has values of 110 degrees at 30 K.P.H. and 40 degrees at 100 K.P.H. This emphasises the need for travel at safe speeds.

Colour vision is important to enable a driver or a pedestrian to discern traffic lights and colour schemes in traffic signs. Colour-blindness, by itself is no serious disability, since the human can develop other means for recognizing signs and colour.

Glare from head-lights and street lights cause discomfort when driving. Different persons have different capacity to adapt to glare. Age of the driver is an important factor in glare problems. The glare recovery time generally lies in the range of 3-6 seconds.

The road user has to judge the depth and distance of an object stereoscopically and its speed. The vision of the road user plays an important part in such matters.

3.2.3. Hearing

For safe driving, cycling and walking, sound is an invaluable aid. Horns can alert the road user.

Similarly the sound of a nearing vehicle or the sound of skidding may alert the road user and avert an accident. Efforts are being made to take measures for the control of noise pollution on roads.

3.2.4. Perception, Intellection, Emotion and Volition

The psychological traits of a road user are extremely important to a highway engineer. The characteristics which are important are perception, intellection, emotion and volition, abbreviated as HEV. The time taken for these processes is known as PIEV time.

Perception is the process of perceiving the sensations received through the sensory organs, such as the eyes, ears, skin, nose and the brain. It is recognition of the sensations and becoming aware of the information transmitted.

The next stage is intellection, which means the identification of the stanuli by the development of new thoughts and ideas. It is slightly different from simple recognition by past experiences, which is part of the normal perception process. When a person receives certain stimuli, new thoughts and is as may form leading to better understanding of the stimuli.

Emotion is the individual trait of the person governing his decision making process.

Volition is the will to react to a situation.

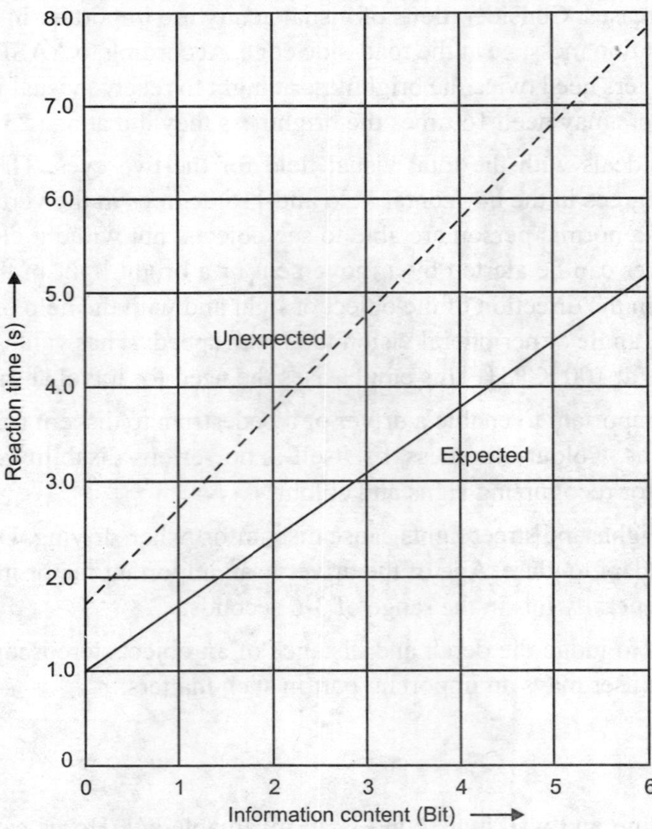

Fig. 3.1. The 85th Percentile driver reaction time, expected and unexpected information

In highway design practice, the time that elapses between the perception of a danger by a road user and his decision to take action is an important consideration, In the design if sight distances, the perception and brake-reaction time is used. The perception time is that time required for a driver to come to the realization that brakes must be applied. The brake reaction time is that time lag between the perception of danger and the effective application of brakes. The AASHO practice is to use a reaction time i.e., combined perception and brake-reaction time of 2.5 sec. The IRC practice is to use combined perception and brake-reaction time of 2.5 sec for roads other than in the hilly region and to use a value of 3 sec for hill roads. It may be recognized that reaction times can vary significantly with expected and unexpected information ; drivers can take longer to respond when decisions are complex or features are unexpected. Fig. 3.1 shows the 85th percentile driver reaction time to expected and unexpected information. In Fig. 3.1, 'bit' is a term used to quantify the amount of information needed to make a decision.

3.3. PEDESTRIAN CHARACTERISTICS

3.3.1. Speed

Speed of walk of pedestrians is needed for the design of traffic signals and other pedestrian facilities. The average walking speeds range from 0.75 m to 1.8 m per sec.

The rate assumed in the Manual on Uniform Traffic Control Devices for timing pedestrian signals is 1 .2 m per sec.

The designers may keep in mind that many pedestrians consider themselves as not being governed by any laws; also any regulations pertaining to the movement of pedestrians are not being enforced. For instance, many pedestrians have a tendency to take the shortest distance between two points ratler than use underpass or overpass, as may be provided. Age is another important factor; while the elderly have physical limitation, the very young tend to be careless and rash. Provision of adequate sidewalks is a great help in reducing accidents involving pedestrians and whenever site conditions permit, the movement of pedestrians may be confined by a pedestrian guard rail, having openings only at designated pedestrian crossings.

3.3.2. Space occupied by pedestrians

For design of sub-way. foot bridges and other facilities, the space required by a pedestrian is generally taken as an ellipse with a major axis of 0.6 m and minor axis of 0.45 m. The : pacing between pedestrians while walking is generally assumed to be 2.5 m. which roughly corresponds to a time headway of 2 sec.

3.4. VEHICLE CHARACTERISTICS

3.4.1. Size

Table 3.1 gives the sizes of vehicles commonly used in India.

Table 3.1 : Sizes of vehicles commonly found in India

S. No.	Vehicle Type	Breadth (*m*)	Length (*m*)	Height (*m*)	Wheelbase (*m*)
1.	Maruti 800 Car	1.34	3.30	1.41	2.18
2.	Large Car	1.68	4.31	1.60	2.50
3.	Jeep	1.75	3.73	1.70	2.31
4.	Single Unit Truck	2.4-2.5	6.97	3.10.	4.23
5.	Bus	2.5	7.0-9.5	3.22	4.0-5.2
6.	Three-wheeler Autorickshaw	1.30	2.49	1.30	2.0

3.4.2. Power performance of vehicles

The power developed by the engine of a vehicle moving at a speed ha' to overcome all resistance to motion such as:

(*a*) Rolling resistance
(*b*) Air resistance
(*c*) Grade resistance
(*d*) Inertia forces during acceleration and deceleration
(*e*) Transmission losses

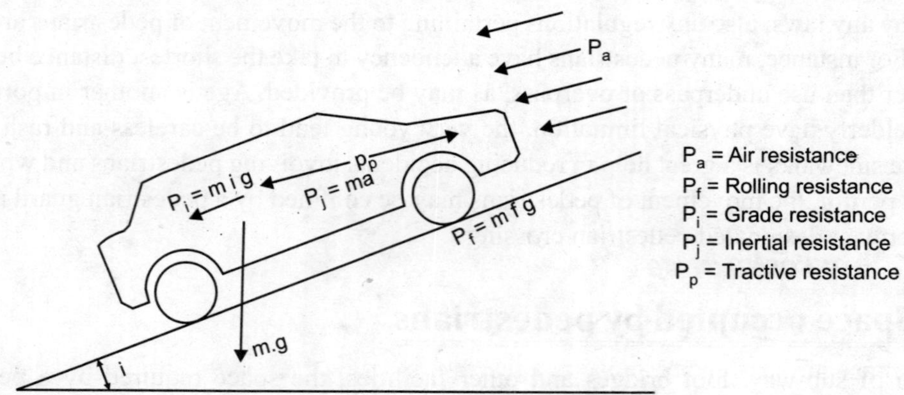

Fig. 3.2: Forces acting on a vehicle

3.4.2.1. Rolling resistance

When vehicle wheels roll over the road surface, the tyres get deformed due to the irregularities and she roughness of the surface. The road itself may deflect, if the wheel load is high. When in speed, shocks and impact are caused due to motion on an irregular surface. All these factors cause hindrance to motion. The smoother the surface the less is the resistance caused. A study conducted in India has yielded the results given in Table 3.2.

Table 3.2 : Values of rolling resistance coefficient from Indian studies

S. No.	Type of surface	Coefficient of rolling resistance
1.	Asphaltic concrete	0.01
2.	Premixed carpet in good condition	0.016
3.	Premixed carpet in bad condition	0.022
4.	Water-bound macadam in good condition	0.025
5.	Water-bound macadam in bad condition	0.037
6.	Gravel/ Earth	0.046

The rolling resistance is given by

$$P_f = m.f.g. \tag{3.1}$$

where m = mass of the vehicle in kg

f = coefficient of rolling resistance

g = acceleration due to gravity in m/sec^2

P_f = rolling resistance in N

The coefficient of rolling resistance also varies with speed, increasing in value at higher speeds. The following approximate equation accounts for this increase .

$$f_u = f_o [1 + 0.01 (V - 50)] \tag{3.2}$$

where f_u = coefficient of rolling resistance at speed V (K.P.H.)

V = Speed (K.P.H.)

f_o = coefficient of rolling resistance assumed constant upto a speed of 50 K.P.H., and having values as given in Table 3:2.

3.4.2.2. Air resistance

The motion of a vehicle at speed is resisted in the following ways by air:

(*i*) Since air has density, it exerts a reaction pressure against the front of the vehicle when it moves at speed.

(*ii*) The friction of air against the sides of the vehicle body causes resistance.

(*iii*) The eddying of the air stream behind the vehicle, under the body and around the wheels causes power loss.

(*iv*) The flow of air through the vehicle for ventilating the cooling causes resistance to motion.

The following formula can be used to determine the air resistance:

$$P_a = C_a A v^2 \tag{3.3}$$

where P_a = air resistance in N

A = projected front area of the vehicle in sq. *m.* on a plane at right angles to the direction of motion for Indian vehicles, the value can be taken from Table 3.3.

v = speed of the vehicle relative to air in m/sec

C_a = coefficient of air resistance, having value set forth in Table 3.3 Kg/ m^3

Table 3.3 : Values of Frontal area and Coefficient of air resistance

S. No.	Type of vehicle	Frontal area (m^2)	Mass (Kg)	Coefficient of air resistance C_a (Kg/m^3)
1.	Maruti 800 Car	1.54	880	0.40
2.	Premier Padmini Car	1.63	1065	0.42
3.	Ambassador Car	2.15	1365	0.39
4.	Jeep	2.38	1200	0.37
5.	Tata Truck	5.37	6120 (Empty)	0.48
6.	Ashok Leyland Beaver Haulage Truck	5.37	8125 (Empty)	0.48

3.4.2.3. Grade resistance

When a vehicle which was moving on level stretch at a particular speed has to move up an incline, additional work has to be done in keeping the vehicle at the same speed as on the level stretch. The additional work is equal to the work that will be needed to lift the vehicle through a height represented by the inclination. If the horizontal distance is unity (1 metre) and if the inclination is i per cent, the rise will be $\frac{i}{100}$ metres. If the mass of the metres. If the mass of the vehicle is m kg, the additional force to move the vehicle up the incline, P_i is given by:

$$P_i = \frac{m.i.g}{100} \tag{3.4}$$

3 4.2.4. Inertia forces during acceleration

When a vehicle moves on a road, it has to frequently change its speed by accelerating or decelerating, which involves additional force. Similarly, if a vehicle has to start from a sopped position, additional force is needed to accelerate. The additional force P_j is given by:

$$P_j = \text{Mass } x \text{ Acceleration} \tag{3.5}$$

Hence $(\pm) P_j$ = m.a.

where P_j = Force to overcome inertia while accelerating or decelerating, N

m = mass, *kg*

a = acceleration or deceleration is m/see^2

3.4.2.5. Transmission losses

Power losses occur in a vehicle in a number of ways. Some loss occurs in the transmission system, in the gear system itself. At the start of the vehicle, high power is needed. Similarly, a high engine power is needed when climbing uphill, which is accomplished by driving at a low speed, For movement along a good road, where the resistance to motion will be small, a high gear is generally used. The highest forward gear will be usually 1:1, representing direct drive. A further gear reduction is made at the rear axle. The total effect of the above is to consume about 10-15 per cent of the engine horse-power which may be as high as 25 per cent in case of trucks in their lower gear.

3.5. POWER REQUIREMENTS OF THE VEHICLE

The mechanical power developed by the engine is transmitted to the driving wheels by the transmission system. The torque developed at the flywheel is converted to a torque at the rear axle. The following equation holds good:

Rear-axle torque. $T_a = K.\,T_e.G_tG_a$ (3.6)

where T_a = rear axle torque

K = efficiency of the transmission system, representing the losses between the engine and the driving wheels, and can have a value of 0.80 to 0.90

T_e = engine torque at the flywheel

G_t = transmission gear ratio

G_a = rear-axle gear ratio

The rear axle torque imparts a tractive force P_p at the contact of the tyre and the road (Fig. 3.3). This tractive force, known as the tyre-rim pull, is given by the following equation:

$$P_p = \frac{\text{Rear axle torque}}{\text{Radius of the rolling drive}} = \frac{K.T_e G_t G_a}{r_w} \quad (3.7)$$

r_w is related to the radius of the tyre r_o by the following formula:

$$r_w = r_o\lambda$$

where λ = tyre deformation factor (will have a value 0.945–0.950 for the high pressure air tyres and 0.930-0.935 for low pressure tyres, on hard surfaces

The power output corresponding to the tractive effort P_p when the vehicle moves at a speed of v m/sec is:

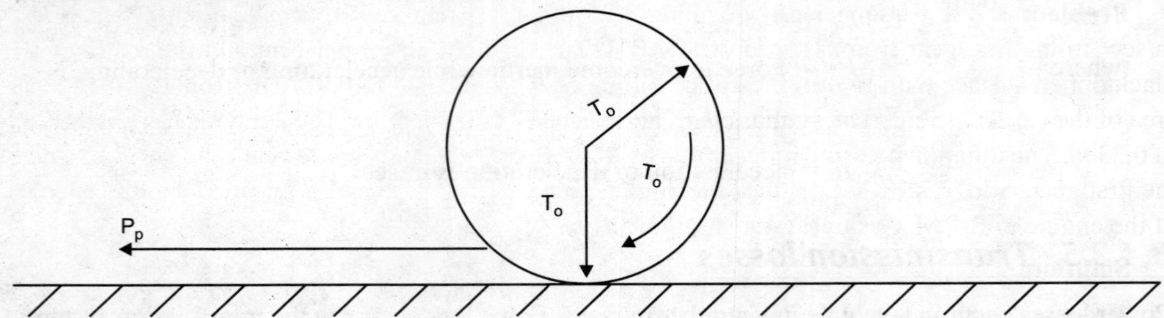

Fig. 3.3: Tractive effort at the driving wheel

$$\text{Power output} = P_p.v \tag{3.9}$$

$$v = \frac{V \times 1000}{3600} = \frac{V}{3.6} \tag{3.10}$$

where V = speed in K.P.H.

$$\therefore \quad \text{Power output} = P_p \frac{V}{3.6} \tag{3.11}$$

Also
$$v = \frac{2.\pi.r_w n}{60.G_t.G_a} \tag{3.12}$$

where n = Engine speed in R.P.M.

$$v = \frac{V}{3.6} = \frac{2.\pi.r_w.n}{60.\,G_t.G_a}$$

$$\therefore \quad V = \frac{0.377\, r_w.n}{G_t.G_a} \tag{3.13}$$

Substituting (3.13) in (3.11)

$$\text{Power output} = \frac{P_p \times 0.377 \times r_w \times n}{G_t.G_a}$$

$$\text{Engine power} = \frac{P_p \times 0.377 \times r_w \times n}{G_t.G_a.K} \tag{3.14}$$

where K = transmission efficiency

Equation (3.14) gives engine power in watts. To convert into Horse-Power,

$$\text{Engine Horse Power (Metric)} = \frac{\text{Engine Power in watts}}{735} \tag{3.15}$$

It may be noted that power developed at the wheels should be able to overcome the various resistances.

Thus: P_p = Rolling resistance + Air resistance + Grade resistance + inertia forces due to acceleration

$$= P_f + P_a \pm P_i \pm P_j \tag{3.16}$$

Problem 3.1. A passenger car weighing 2.25 tonnes is required to accelerate at a rate of 2 m/sec^2 in the first gear from a speed of 8 K.P.H. The gradient is + 3 per cent and the road has a blacktopped surface with the coefficient of rolling resistance equal to 0.02. The frontal projection area of the car is 2.15 m^2. The coefficient of air resistance is 0.39 kg/m^3 The car tyres have a radius of 0.33 *m*. The inflation pressure reduces this by a factor 0 935. The rear axle gear ratio is 3.82:1 and the first gear ratio is 2.78:1. Calculate die engine power (in Watts and H.P.) needed and the speed of the engine in R.P.M. Assume a transmission efficiency of 0.85.

Solution:

Tractive force needed, $P_p = P_f + P_a \pm P_i \pm P_j$

From Equation 3.1

$$P_f = m.g.f.{=}2250 \times 9.81 \times 0.02 = 441.45 \text{ N}$$

Average speed $= 8$ K.P.H.

$$P_a = 0.39 \times 2.15 \times \frac{8}{3.6} \times \frac{8}{3.6} = 4.14 \text{ N}$$

$$P_i = \text{m.g.i.} = 2250 \times 9.81 \times \frac{3}{100} = 662.18 \text{ N}$$

$$P_j = \text{m.a} = 2250 \times 2 = 4500 \text{ N}$$

$$P_p = 441.45 + 4.14 + 662.18 + 4500 = 5607.77 \text{ tN}$$

Power output $= 5607.77 \times \frac{8}{3.6} = 12461.71$ W

$$= \frac{12461.71}{735} \text{ P.H.} = 16.95 \text{ (Metric)}$$

Since transmission efficiency is 0.85,

Engine horse power $= \frac{16.95}{0.85} = 19.94$

$$V = \frac{0.377 \times r_w.n}{G_t.G_a} \quad (3.17)$$

$$r_w = r_o \lambda \quad (3.18)$$

$$= 0.33 \times 0.935 = 0.308$$

$$\therefore \quad n = \frac{8 \times 3.82 \times 2.78}{0.308 \times 0.377} = 732 \text{ R.P.M.}$$

QUESTIONS

1. Develop an equation for the tractive force needed to keep a vehicle in motion up an incline, overcoming air and rolling resistance and accelerating.
2. A truck weighing 15 tonnes is required to accelerate at a rate of 0.5 m/sec^2 from rest to an initial speed of 10 K.P.H. It has to negotiate an upward gradient of 10 per cent. The road has a rolling resistance of 0.015. The frontal area is 5.37 m^2 and the coefficient of

air resistance is 0.48 kg/m^3 The transmission efficiency is 0.90 Calculate the horse-power needed. If the tyres have a radius of 0.5 m and inflation pressure reduces this by a factor 0.94, the rear axle ratio is 4.2 : 1 and the first gear ratio is 3.1 : 1, calculate the engine H.P. and R.P.M. **[Ans.** 51.3 HP (metric); 735 R.P.M.]

3. Write a short note on : PIEV.
4. Develop an equation for determining the braking distance of a vehicle in terms of initial speed and coefficient of friction.
5. A truck travelling at 50 K.P.H. on a wet bituminous surface (coefficient 0.3) is suddenly brought to rest by braking. Caculate the distance travelled in coming to a halt.

[Ans. 32.8m]

Chapter

GEOMETRIC DESIGN

4.1. INTRODUCTION

Geometric design of a highway deals with design of features such as the horizontal alignment, vertical profile, cross-sectional elements, access control, junctions, grade separated interchanges and median openings. The design should result in an efficient, safe and economic operation of transport. When considering cost, the initial cost plus the maintenance and user costs should be given due attention. In India, the standards are laid down by the Indian Roads Congress and these should govern.

4.2. HIGHWAY CLASSIFICATION

The geometric design is related to the highway type. In India, the highways in the rural (non-urban) areas are classified as under:

1. Expressways, which are superior facilities along arterial routes where only cars, buses, trucks and two-wheelers with engine capacity greater than 50 c.c. are allowed, all at-grade crossings are not allowed, separation of opposing traffic by a median is mandatory and rest-areas at convenient intervals are provided.
2. National Highways are the main highways traversing the length and breadth of the country, connecting state capitals, ports, large industrial and tourist centres and foreign highways and serve strategic and defence needs.
3. State Highways are the main trunk routes of a State, connecting up with National Highways or highways of adjacent States, ana linking the district headquarters and important cities within the State.
4. District Roads are the roads traversing each district, serving areas of production and markets and .connecting these with each other or with National Highways, State Highways, Railways or important navigational routes. They should be capable of taking road traffic into the heart of rural areas throughout the year with only minor interruptions. District Roads are divided into two classes on the basis of traffic as under:

    (*i*) Major District Roads (MDR) for the higher order of traffic (*ii*) Other District Roads (ODR) for the lower order of roads

5. Village roads linking villages to the nearest main road.

Village roads and ODRs are often grouped together as "Rural Roads" in India.

As regards the urban streets, the following classification is followed:

1. Expressways which are divided arterial highways for motorized traffic with full or partial control of access and generally provided with grade separation at intersections. Parking, loading and unloading of goods and pedestrian traffic are not permitted on them.
2. Arterial Streets are thee which form the principal network for through traffic, and serve intra-urban travel between the Central Business District (CBD) and outlying suburban centres. They are generally spaced at less than 1.5 Km in highly developed areas and the CBD at 8 Km or more in the suburban fringe areas. They are generally divided carriageways facilities with full or partial control of access. Parking and loading and unloading of goods are usually restricted and regulated. Pedestrians are allowed to cross only at designated locations.
3. Sub-arterial Streets are of somewhat lower level of travel mobility than arterial streets. Spacing varies from about 0.5 Km in the CBD to 3 to 5 Km in the suburban fringe. Parking and loading and unloading of goods are restricted and regulated. Pedestrians are allowed to cross only at designated locations.
4. Collector Streets are intended for collecting and distributing traffic to and from local streets and providing access to arterial streets. Full access is allowed on these streets from abutting properties. There are few parking restrictions.
5. Local streets are those intended for access to residences, business premises or abutting property. Parking is unrestricted. Pedestrian movements are not controlled.

4.3. TERRAIN

The topography and the physical features of the area through which the road passes governs the design elements. The terrain is classified as under:

Terrain Classification	Percent of cross-slope of the Country
Level	0–10
Rolling	10–25
Mountainous	25–60
Steep	Greater than 60

4.4. DESIGN SPEED

The speed for which a road is designed influences the geometric features of the road. The design speed should be appropriate for the class of the road and the terrain. Usually, the 95th or the 98th percentile speed (from the speed distribution curve) is selected as the design speed. The design speeds recommended by the Indian Roads Congress tor Rural Highways and Urban Streets are given in Tables 4.1 and 4.2 respectively.

Table 4 1: Suggested design speeds in India for rural highways

Classification	Plain Terrain		Rolling Terrain		Mountainous Terrain		Steep Terrain	
	Ruling	Minimum	Ruling	Minimum	Ruling	Minimum	Ruling	Minimum
National Highways and State Highways	100	80	80	65	50	40	40	30
Major District Roads	80	65	65	50	40	30	30	20
Other District Roads	65	50	50	40	30	25	25	20
Village Roads	50	40	40	35	25	20	25	20

Table 4.2 : Suggested design speeds in India for urban streets

Classification of Roads	Speeds in K.P.H.
Arterial	80
Sub-arterial	60
Collector streets	50
Local streets	30

A design speed of 120 K.P.H. is being adopted for expressways in India.

4.5. DESIGN VEHICLE

Different types of vehicles use the road. The geometric design has to cater to the critical vehicle which governs the radii, width of pavements, clearances and parking geometries. In India, the Indian Roads Congress standards are followed. For important roads catering to high volume of traffic, the largest vehicle-such as a single unit truck, semi-trailer and semi-trailer/ full trailer combination will govern. For the design of junctions in residential areas, passenger car will govern. The turning radii of design vehicles suggested by AASHTO are given in Table 4.3.

Table 4.3 : Turning radii (m) AASHTO design vehicle

	Passenger car	Single unit truck	Semi-trailer	Semi-trailer/ Full-trailer
Minimum turning radius (m)	7.32	12.8	13.72	13.75
Minimum inside radius (m)	4.65	8.61	6.00	6.85

These are illustrated in Fig 4. 1

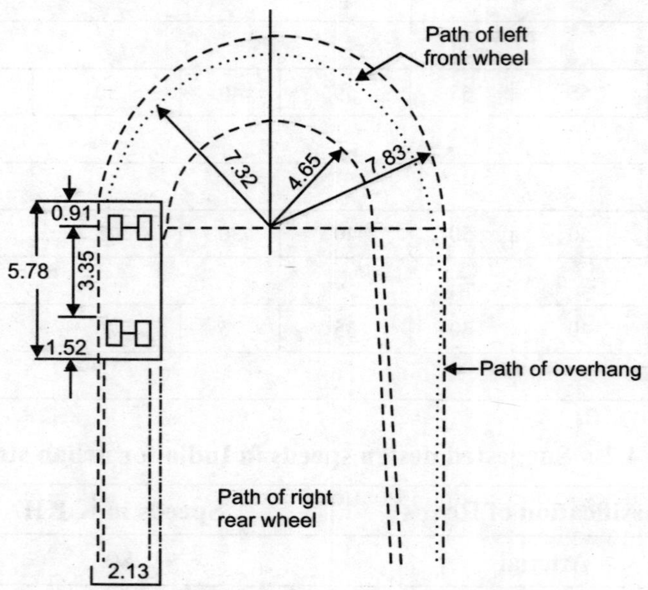

Passenger vehicle

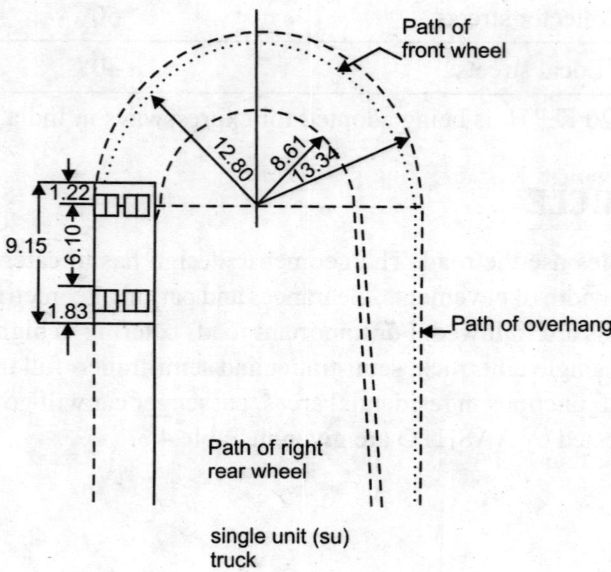

single unit (su)
truck

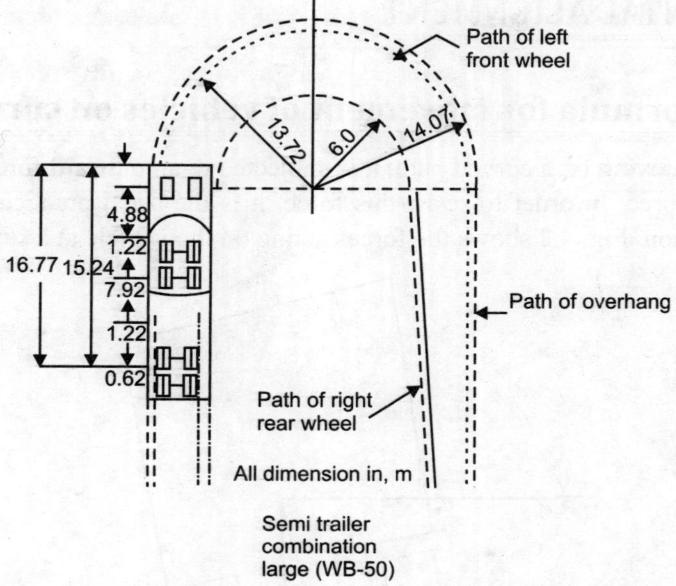

Fig. 4.1: Turning radii of some AASHTO-Design Vehicles

The relevant details from the IRC Standard are given below:

1. **Maximum Width** : 2.5 m
2. **Height**
   Buses and Trucks for normal application : 3.8 m
   Trucks carrying containers : 4.2 m
   Double decker bus length : 4.75 m
3. **Length**
   Single unit truck (2 or more axles) : 11 m (exclusive of front and rear bumpers)
   Single unit bus (2 or more axles) : 12 m (exclusive of front and rear bumpers)
   Truck-tractor semi-trailer combination : 16 m (exclusive of front and rear bumpers)
   Truck-trailer combination : 18 m (exclusive of front and rear bumpers)
4. **Weight**
   Single axle with dual wheels : 10.2 Tonnes
   Tandem axle (spacing between axles not less than 1.2 m and more than 2.5 m) : 18 Tonnes

4.6. HORIZONTAL ALIGNMENT

4.6.1. Basic formula for movement of vehicles on curves

When a vehicle is moving on a curved path, it is subjected to an outward force, commonly known as the centrifugal force. In order to resist this force, it is the usual practice to super-elevate the roadway cross-section. Fig. 4.2 shows the forces acting on the vehicle at a super-elevated section.

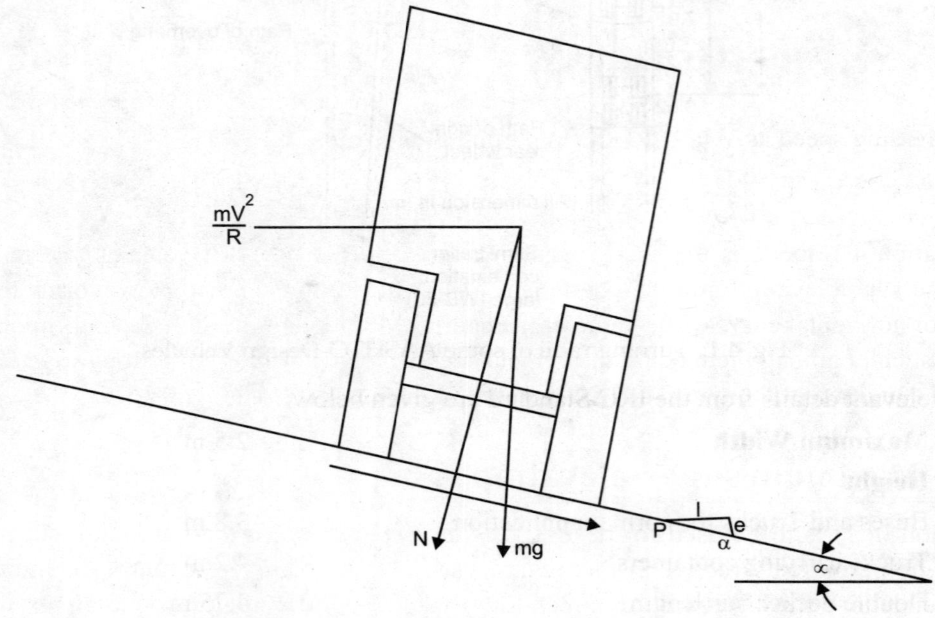

Fig. 4.2: Forces acting on a vehicle at a super elevated section

Let m = mass of the vehicle,

v = speed of the vehicle, m/sec

R = radius of the curve, m

g = acceleration due to gravity = 9.81 m/sec^2

p = side friction force resisting the centrifugal force = Nμ

N = normal force

μ = coefficient of lateral friction

α = angle of super-elevation

e = super elevation, tan α

The centrifugal force $= \dfrac{mv^2}{R}$

For equilibrium $\dfrac{mv^2}{R} \cos \alpha = m\, g \sin \alpha + P = m\, g \sin \alpha + N\mu$

$$= mg \sin \alpha + \left(mg \cos \alpha + \frac{mgv^2}{R} \sin \alpha \right)$$

Therefore, $$\frac{v^2}{gR} = \tan \alpha + \mu + \frac{\propto v^2}{gR} \tan \alpha$$

The term $\frac{\propto v^2}{gR}$ tan α is small and can be neglected. Thus,

$$\frac{v^2}{gR} = e + \mu$$

Expressing speed as V in K.P.H., and putting g = 9.81,

$$\frac{V^2}{127\,R} = \text{e} + \mu \qquad (4.1)$$

Equation 4.1 above is the basic equation relating to the speed of vehicles, the radius of the curve, the superelevation and the coefficient of lateral friction. The equation forms the basis of design for horizontal curves. The value of the coefficient of lateral friction depends upon a number of factors, chief among them being the vehicle speed, type and condition of roadway surfaces, and type and condition of the tyre. The IRC recommends a constant value of 0.15.

4.6.2. Maximum super-elevation value

If equation (4.1) is to be used for design, it is desirable to know the maximum super-elevation that can be permitted. Practice in this regard varies from country to country. In India, the IRC recommendations for hill roads are a maximum value of 0.07 for snowbound areas and 0.10 for area not affected by snow. In other cases, a value of 0.07 is considered maximum.

4.6.3. Maximum radii of curves

Equation (4.1) can be rewritten as:

$$R = \frac{V^2}{127\,(e + \propto)} \qquad (4.2)$$

Adopting the value of $e = 0.07$ or 0.10 and $\mu = 0.15$ as per current Indian practice the formula becomes:

$$R = 0.0357\ V^2 \text{ (for plain and rolling terrain)} \qquad (4.3)$$

$$R = 0.0315\ V^2 \text{ (for hill roads which are not snow bound)} \qquad (4.4)$$

$$R = 0.0357\ V^2 \text{ (for hill roads which are snow bound)} \qquad (4.5)$$

Since the current Indian Standards give two values for Design Speed (*i.e.* Ruling and Minimum), the above formulae yield two values for radius, the ruling radius and the absolute minimum radius. These values are indicated in Table 4.4.

Table 4.4 : Minimum radii of horizontal curves for different terrain condition

	Classification of road	Plain terrain		Rolling terrain		Mountainous terrain				Steep terrain			
						Areas not affected by snow		Snow based areas		Areas not effected by snow		Snow based areas	
		Ruling Minimum	Absolute Minimum	Ruling Minimum	Absolute Minimum	Ruling Minimum	Absolute minimum	Ruling Minimum	Absolute Minimum	Ruling Minimum	Absolute Minimum	Ruling Minimum	Absolute Minimum
1.	National Highways and State Highways	360	230	230	155	80	50	90	60	50	30	60	33
2.	Major District Roads	230	155	155	90	50	30	60	33	30	14	33	15
3.	Other District Roads	155	90	90	60	30	20	33	23	20	14	23	15
4.	Village Roads	90	60	60	45	20	14	23	15	20	14	23	15

4.6.4. Super-elevation rates

Super-elevating on curves is intended to counteract a part of the centrifugal force, the remaining part being resisted by the lateral friction. Also, super-elevating results in economy in maintenance. This is because skidding and unequal pressure on the wheels of vehicles, which result from high value of sideway force between the tyres and the roadway surface, give rise to pot-holes and necessitate frequent attention to the surface. Super-elevation can be provided either fully to counteract the centrifugal force or to counteract a fixed proportion of the centrifugal force.

As per Indian practice, super-elevation is calculated on the assumption that it should counteract the centrifugal force developed at three-fourth the design speed. Thus:

$$e = \frac{(0.75\,V^2)}{127\,R} = \frac{V^2}{225\,R} \tag{4.6}$$

The super-elevation calculated from Equation (4.6) is restricted to 0.07 and 0.10 as discussed in article 4.6.2.

4.6.5. Radii of curves beyond which no super-elevation is required

The normal cambered section of a highway can itself be continued on a curve where the super-elevation calculated is less than the camber. From formula (4.6).

$$R = \frac{V^2}{225\,e} \tag{4.7}$$

Substituting the value of the camber for e in Equation (4.7) the minimum radius beyond which no super-elevation is required is obtained. Table 4.5 gives the values for various design speeds and cambers.

Table 4.5 : Radii beyond which super-elevation is not essential for different cambers and speeds (Indian practice)

Design speed (K.P.H.)	Radius in metres beyond which no Super-elevation is needed for values of camber of				
	0.04	0.03	0.025	0.02	0.017
20	50	60	70	90	100
25	70	90	110	140	150
30	100	130	160	200	240
35	140	180	220	270	320
40	180	240	280	350	420
50	280	370	450	550	650
65	470	620	750	950	1100
80	700	950	1100	1400	1700
100	1100	1500	1800	2200	2600

In such cases where the radii is greater than those given in Table 4.5, it is however, desirable to remove the adverse crown in the outer half of the carriageway and super-elevate-at the normal crown slope.

4.6.6. Method of attainment of super-elevation

The normal cambered surface on a straight reach of road is changed into a super-elevated surface in two stages. In the first stage, outer half of the camber is gradually raised until it is level Fig. 4.3(*b*). It is desirable to accomplish the raising of the outer-half till it is level before the starting point of the transition curve. The raising of the outer edge should be done in a slope not exceeding 1 in 150 for plain and rolling terrain and 1 in 60 for hilly terrain.

In the second stage, three methods may be adopted to attain the full super-elevation:

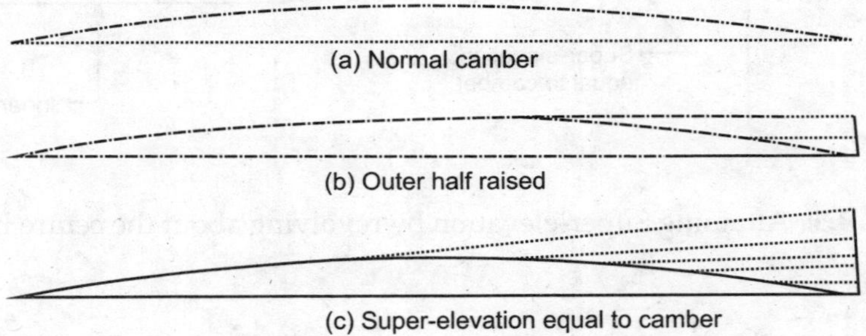

Fig. 4.3: First stage in attaining super elevation by bringing super-elevation equal to camber

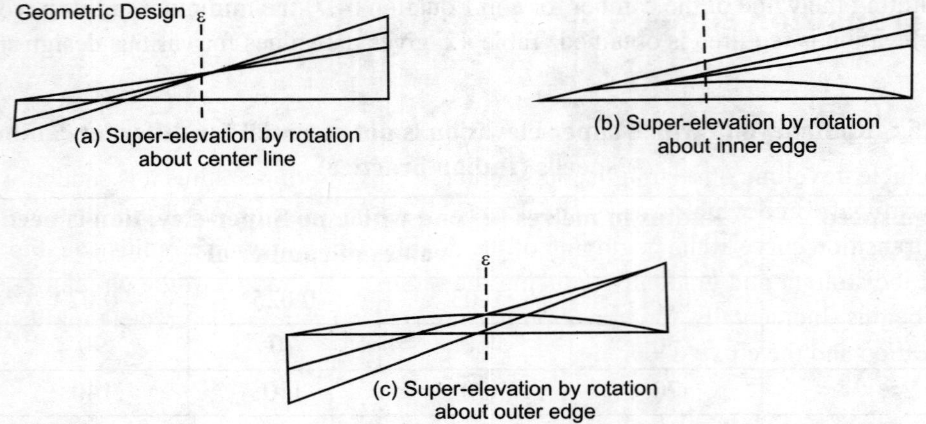

Fig. 4.4: Second stage in attaining super-elevation

(*i*) The surface of the road is rotated about the centre-line of the carriageway, gradually lowering the inner edge and raising the upper edge, keeping the level of the centre line constant [Fig. 4.4(*a*)].

(*ii*) The surface of the road is rotated about the inner edge, raising the centre and the outer edge [Fig. 4.4(*b*)].

(*iii*) The surface of the road is rotated about the outer edge, depressing the centre and the inner edge [Fig. 4.4(*c*)]

Method (*i*) is most generally used as it results in the least distortion of the pavement. Fig. 4.5 shows the method of attaining super-elevation as per method (*i*).

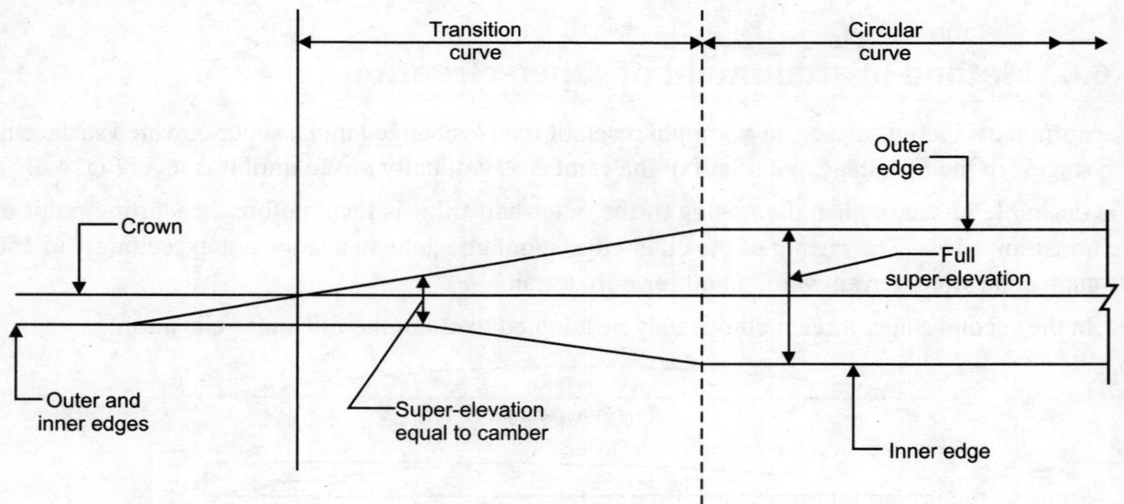

Fig. 4.5: Attaining Super-elevation by revolving about the centre line

4.6.7. Transition curves

4.6.7.1. Need for transitions

When a vehicle travelling on a straight course enters a curve of finite radius it is suddenly subjected to the centrifugal force which causes shock and sway. In order to avoid this it is customary to provide a transition curve at the beginning of the circular curve, having a radius equal to infinity at the end of the straight and gradually reducing the radius to the radius of the circular curve where the curve begins. Incidentally, the transition portion is also used for the gradual application of the super-elevation and the curve widening.

4.6.7.2. The Spiral

Various forms of curves are suitable for highway transitions, but the one most popular and recommended for use in this country is the Spiral. It is easy to set out in the field and the rate of change of acceleration is uniform throughout the length of the transition.

Fig. 4.6 indicates the main elements of a circular curve provided with spiral transitions at its two ends.

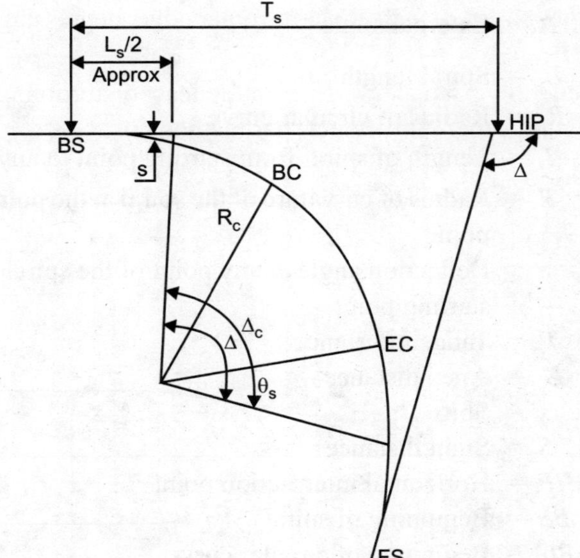

Fig. 4.6: Main elements of a circular curve provided with transitions

Some of the important properties of the spirals are given below:

(*i*) $$L_s R_c = LR = \text{constant} \quad (4.8)$$

(*ii*) $$L = M\sqrt{e} \quad (4.9)$$

Where M is a constant

$$M = \sqrt{2\,RL} \quad (4.10)$$

Also $L = 2Re$ (4.11)

(*iii*) $\theta = \left(\frac{L}{L_s}\right)^2 \theta_s$ (4.12)

(*iv*) $\theta_s = \frac{L}{2R_c}$ radians

$= \frac{28.65\, L_s}{R_c}$ degrees (4.13)

(*v*) $T_S = \frac{L_S}{2} + (R_c + s)\tan\frac{\Delta}{2}$ (4.14)

(*vi*) $S = \frac{L^2 s}{24\, R_c}$ (4.15)

(*vii*) $E_s = (R_c + s)\sec\frac{\Delta}{2} - R_e$ (4.16)

The following nomenclature may be noted:

θ_s = Spiral angle

Δ_c = Angle of the circular curve

Δ = External angle

L_s = Spiral length

R_c = Radius of circular curve

L = Length of spiral from starting point to any point

R = Radius of curvature of the spiral at the point distant L from starting point

e = Deflection angle at any point of the spiral distant L from the starting point

T_s = Tangent distance

E_s = Apex distance

s = Shift

S = Sight distance

HIP = Horizontal intersection point

BS = Beginning of spiral

BC = Beginning of circular curve

EC = End of circular curve

ES = End of spiral

4.6.7.3. Length of transition

The length of the transition should be determined from the following two considerations:

(*i*) The rate of change of centrifugal acceleration adopted in the design should not cause discomfort to the drivers.

If C is the rate of change of acceleration

$$C = \frac{V^2}{R_c} \div t$$

$$= \frac{V^2}{R_c} \div \frac{L_s}{V} = \frac{V^3}{R_c L_s} \qquad (4.17)$$

where R_c and L_s have the same meaning as in 4.6.7.2,

t = time in second to travel a distance of L_s

v = speed in m/sec

Thus,

$$L_s = \frac{V^3}{(3.6)^3 \, CR_c} = \frac{0.0215\,V^3}{R_c} \qquad (4.18)$$

where V = Speed in K.P.H.

the value of C is give by the following equation:

$$C = \frac{80}{75+V} \qquad (4.19)$$

Subject to a maximum of 0.8 and minimum of 0.5.

(*ii*) The rate of change of super elevation should be such as not to cause higher gradients and unsightly appearances. This could be kept 1 in 150 for roads in plain and rolling terrain and 1 in 60 in hilly terrain.

The higher of the values given in considerations (*i*) or (*ii*) above may be used. Suggested values are given in Table 4.6.

Table 4.6 : Minimum transition lengths for different speeds and curve radii

Plain and Rolling Terrain							**Mountainous and Steep Terrain**					
Curve radius, R (metres)	**Design speed (KPH)**						**Curve radius (metres)**	**Design speed (KPH)**				
	100	**80**	**65**	**50**	**40**	**35**		**50**	**40**	**30**	**25**	**20**
	Transition length-metres							**Transition length-metres**				
45					NA	70	14				NA	30
60				NA	75	55	20				35	20
90				75	50	40	25			NA	25	20
100			NA	70	45	35	30			30	25	15
150			80	45	30	25	40		NA	25	20	15
170			70	40	25	20	50		40	20	15	15
200		NA	60	35	25	20	55		40	20	15	15
240		90	50	30	20	NR	70	NR	30	15	15	15

300	NA	75	40	25	NR		80	55	25	15	15	NR
360	130	60	35	20			90	45	25	15	15	
400	115	55	30	20			100	45	20	15	15	
500	95	45	25	NR			125	35	15	15	NR	
600	80	35	20				150	30	15	15		
700	70	35	20				170	25	15	NR		
800	60	30	NR				200	20	15			
900	55	30					250	15	15			
1000	50	30					300	15	NR			
1200	40	NR					400	15				
1500	35						500	NR				
1800	30											
2000	NR											

NA-Not applicable

NR-Transition not required

4.6.8. Widening on curves

Widening of pavements is needed on curves for the following reasons:

(*i*) On curves the vehicles occupy a greater width because the rear wheels track inside the front wheels (Fig. 4.7).

On curves drivers have difficulty in steering their vehicles to keep to the centre line of the line.

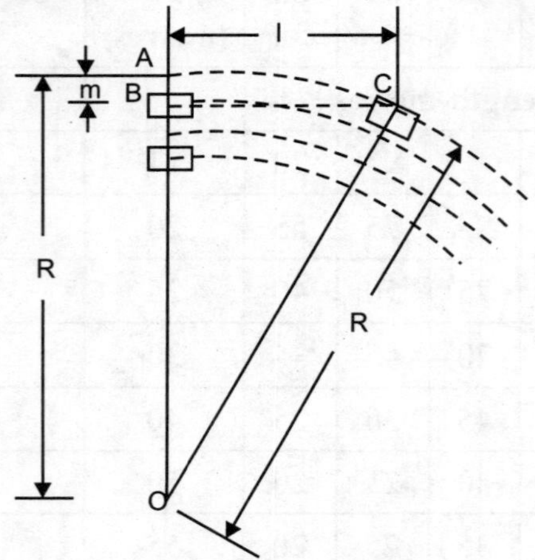

Fig. 4.7: Widening on curves

From Fig. 4.7, considering the right angled triangle OBC,

$$l^2 + (R-m)^2 = R^2$$

or $$l^2 = 2\,Rm - m^2$$

Neglecting m^2, $$l^2 = 2\,Rm$$

$$\therefore \quad m = \frac{l^2}{2R} \tag{4.20}$$

Assuming a wheel base of 6 m for a vehicle corresponding to AASHTO Single Unit (SU).

Widening in metres

$$\text{m} = \frac{6^2}{2R} = \frac{18}{R} \tag{4.21}$$

where R = radius in metres

The widening due to psychological reasons is a function of speed and can be assumed to be given by the empirical formula:

$$\frac{0.1V}{\sqrt{R}} \tag{4.22}$$

where V = speed in K.P.H

The total widening for pavements of those width which are in excess of two lanes is given by the formula:

$$W_e(\text{in m}) = n\,\frac{1}{R_c} + 0.1\,\frac{V}{\sqrt{R_c}} \tag{4.23}$$

where n = number of lanes

R_c = radius of the curve in metres

Table 4.7 gives the widening necessary for single lane and 2-lane pavements.

Table 4.7 : Extra width of pavement at horizontal curve

Radius of curve (m)	**upto 20**	**21 to 40**	**41 to 60**	**61 to 100**	**101 to 300**	**above 300**
Extra width (m)						
Two-lane	1.5	1.5	1.2	0.9	0.6	Nil
Single-lane	0.9	0.6	0.6	Nil	Nil	Nil

4.6.9. General controls for horizontal alignment

The following general controls for horizontal alignment should be kept in view in a sound design practice.

(*i*) Alignment should be as directional as possible.

(*ii*) The alignment should be consistent with the topography and should conform generally to the natural contours.

A line cutting across the contours involves high fills and large cuts, mars the landscape and is difficult for maintenance.

(*iii*) The number of curves should, in general, be kept to a minimum.

(*iv*) The alignment should be fluent and avoid abrupt turns. Widening alignment consisting of short curves should be avoided, since it is the cause of erratic operation.

(*v*) A sharp curve at the end of a long tangent is extremely hazardous and should be avoided. If sharp curvature is unavoidable over a portion of the route selected, it is preferable that this portion of the road be preceded by successfully sharper curves.

(*vi*) Short curves giving the appearance of kinks should be avoided, especially for small deflection angles. The curves should be sufficiently long to provide pleasing appearance and smooth driving on important highways. They should be at least 150 metres long for a deflection angle of 5 degrees, and the minimum length should be increased by 30 metres for each 1 degree decrease in the deflection angle.

(*vii*) For a particular design speed, as large a radius as possible should be adopted. The minimum radii should be reserved only for the critical locations.

(*viii*) The use of sharp curves should be avoided on high fills. In the absence of cut slopes, shrubs, trees, etc., above the roadway, the drivers may have difficulty in estimating the extent of curvature and fail to adjust to the conditions.

(*ix*) While abrupt reversal in curvature is to be avoided, the use of reverse curves becomes unavoidable in hilly terrain. When they are provided, adequately long transitional curves should be inserted for super-elevation run-off.

(*x*) Curves in the same direction separated by short tangents, say 300-500 metres long are called broken-back curves.

They should be avoided as they are not pleasing inappearance and are hazardous.

(*xi*) Compound curves may be used in difficult topography in preference to a broken-back arrangement, but they should be used only if it is impossible to fit in a single circular curve. To ensure safe and smooth transition from curve to curve, the radius of the flatter curve should not be disproportional to the radius of the sharper curve. A ratio of 2:1 or preferably 1.5:1 should be adopted.

(*xii*) The horizontal alignment should blend with the vertical harmoniously. General controls for the combination of horizontal and vertical alignments are indicated in section 4.8.

4.7. VERTICAL ALIGNMENT

4.7.1. Gradients

4.7.1.1. Maximum grades

One of the important considerations in designing a highway is the gradient. The cost of operation of vehicles and the capacity of a highway are profoundly affected by the grades provided. On mountainous and steep terrain, the grades are not only influenced by the ability of vehicles to negotiate them, but also by the altitude of a road above the sea level since rarefaction of air causes loss of engine power. A peculiar consideration for Indian conditions is the presence of animal drawn traffic.

The grades should be selected so as to encourage uniform operation as far as possible for a given terrain and the vertical profile should be as smooth as economically possible. Roads once constructed to a given vertical profile are very difficult to be upgraded to flatter gradients at a subsequent stage without great loss to the initial investment.

Table 4.8 gives the Indian practice as regards to maximum gradients. In Table 4.8 gradients up to the ruling may be used as a matter of course in design. The limiting gradients may be used where the topography of the area compels this or where the adoption of gentler gradients would enormously add to the cost. Exceptional gradients are meant to be used only in very exceptional situations and for short lengths prescribed in. Table 4.8.

Table 4.8: Gradients for different terrain types-Indian practice

S. No.	Terrain type	Ruling gradient (%)	Limiting gradient %	Exceptional gradient
1.	Plain	3.3 (1 in 30)	5 (1 in 20)	6.7 (1 in 15) (for short distances not exceeding 100 m at a stretch)
2.	Rolling	3.3 (1 in 30)	5 (lin 20)	6.7 (1 in 15) (for short distances not exceeding 100 mat a stretch)
3.	Mountainous	5 (1 in 20)	5 (1 in 16.7)	7 (1 in 14.3) (for a distance not more than 400 m
4.	Steep			
(*i*)	Upto 3000 m height above mean sea level	5 (1 in 20)	6 (1 in 16.7)	7 (1 in 14.3) (for short distances not exceeding 100 m)
(*ii*)	Above 3000 m height above mean sea level	6 (1 in 16.7)	7 (1 in 14.3)	8 (1 in 12.5) (for short distances not exceeding 100 m)

4.7.1.2. *Minimum gradients for drainage*

W here the drainage of a road is to be effected by means of longitudinal ditches, as in cut sections or hill roads, or along the curbed edge of pavements, a certain minimum gradient is necessary for efficient drainage. The values given in Table 4.9 are commonly adopted.

Table 4.9: Minimum gradients for drainage

S. No.	Site condition	Minimum gradients (%) (Desirable)
1.	Longitudinal drainage along curbed edge of pavements, (lined)	0.5
2.	Longitudinal drainage along road side ditches (unlined)	1.0

4.7.2. Vertical curves

4.7.2.1. *General considerations*

Vertical curves should be provided at points of change of grade so as to achieve gradual transition. If the point of vertical intersection (termed PVI) is above the road surface, the curve is a summit or crest curve and if it is below the road surface, the curve is a sag or valley curve. If the total change in gradient from one tangent to another does not exceed 0.5 per cent, vertical curves can be dispensed with. The parabola is a vertical curve most frequently used because of the ease with which it can be laid out.

4.7.2.2. *Length of summit curves*

For summit curves, the most important consideration in determining the length of the curve is the sight distance requirement. The following basic formulae can be used to determine the length:

Case I. When the sight distance (S) is less than the length of the curve (L):

$$L = \frac{AS^2}{(\sqrt{2h_1} + \sqrt{2h_2})^2} \tag{4.24}$$

Case II. When the sight distance (S) is greater than the length of the curve (L):

$$L = 2S - \frac{2(\sqrt{h_1} + \sqrt{h_2})^2}{A} \tag{4.25}$$

In the above formulae,

S = sight distance in metres

L = length of vertical curve, in metres

h_1 = height of eye above the roadway surface

h_2 = height of object above the roadway surface

A = algebraic difference in slopes expressed in absolute values in a decimal form

Adopting h = 1.2 m ;

h_2 = 0.15 m

as per current Indian practice Eqs. (4.24) and (4.25) transform as under for safe stopping sight distance:

When S is less than L,

When S is greater than L,

$$L = \frac{AS^2}{4.4} \tag{4.26}$$

When S is greater than L,

$$L = 2S - \frac{4.4}{A} \tag{4.27}$$

If the gradient is expressed in per cent, the above formulae become:

$$L = \frac{NS^2}{440}$$

$$L = 2S - \frac{440}{N}$$

For intermediate or overtaking sight distance, $h_1 = h_2 = 1.2$ m.

This gives the following relationships:

When L is greater than S,

$$L = \frac{AS^2}{9.6}$$

When L is less than S.

$$L = 2S - \frac{9.6}{A}$$

4.7.2.3. Sag or valley curves

For sag curves, one of the criteria for determining the length is rider comfort. The other criteria are vehicle headlight, sight distance, drainage control and general appearance. As per rider comfort as the basis of design of sag curves, α, the vertical radial acceleration, is given by,

$$\alpha = \frac{v^2}{R}$$

where α is in m/sec²

v = speed in m/sec

$$= \frac{V^2}{13R} \tag{4.28}$$

where V = speed in K.P.H.

Since the radial acceleration changes from zero to a in a length of L/2, where L = length of the sag curve, the rate of change of radial acceleration C is given by:

$$C = \frac{V^2}{13R} \div \frac{L \times 3.6}{2V}$$

$$= \frac{V^3}{23.4\,LR} \text{ m/sec}^3 \tag{4.30}$$

The recommended value of C for design is 0.6 m/sec. Substituting this in (4.29), we get

$$L = \frac{V^3}{14R} \tag{4.31}$$

$$\therefore \quad LR = \frac{V^3}{14} \tag{4.32}$$

For a transition curve of the form of spiral or a cubic parabola, it can be shown that

$$L = (2RLA)^{1/2} \tag{4.33}$$

where A = algebraic difference in grades of the straights, expressed as a decimal

Substituting Eq. (4.32), in (4.33)

$$L = \left(2\frac{V^3}{14}A\right)^{1/2} = \frac{V^{3/2}A^{1/2}}{2.65} \tag{4.34}$$

Another criterion for design of valley curves is that for night travel, the design must ensure that the roadway ahead is illuminated by vehicle headlights for a sufficient length, which enables the vehicle to brake to a stop, if necessary. This is known as the headlight sight distance and is equal to the safe stopping distance (Table 4.10). For designing valley curves for the above needs, the following points should be kept in view:

(*i*) height of headlight above the road surface is 0.75 m,

(*ii*) the useful beam of headlight is one degree upwards from the grade of the road; and

(*iii*) the height of the object is nil.

4.7.3. General controls for vertical alignment

The following general controls for vertical alignment should be kept in view in designing the vertical profile of a highway:

(*i*) The grade line selected should be smooth with gradual changes, consistent with the class of highway and terrain. Numerous brakes and short lengths of grades should be avoided.

(*ii*) The roller-coaster or hidden type of profile should be avoided as it is hazardous and aesthetically unpleasant.

(*iii*) Undulating grade lines, involving substantial lengths of momentum grades, should be appraised for their effect upon traffic operation. Such profiles permit heavy trucks to operate at higher overall speeds than when an upgrade is not preceded by a down-grade but may encourage excessive speeds of trucks with consequent hazard to traffic, (*iv*) A broken-back grade line (two vertical curves in the same direction separated by short section of tangent grade) should generally be avoided.

(*v*) On long continuous grades, it may be preferable to place the steepest grades at the bottom and flatter grades near the top. Alternatively, the long grades may be broken by short intervals of flatter grades.

(*vi*) Intersections on grades should be avoided as far as possible. Where unavoidable, the approach gradients and the gradient through the intersections should be flattened to the maximum possible extent.

4.8. COMBINATION OF HORIZONTAL AND VERTICAL ALIGNMENT

The following general controls should be kept in view in designing a highway for combination of horizontal and vertical alignment:

(*i*) The overall appearance of highway is enhanced with a judicious selection of horizontal and vertical elements. Best results are obtained when there is a proper balance between curvature and grades. Straight alignment or flat horizontal curves imposed on a road with steep or long grades or excessive curvature imposed on a road with a fiat grade are both poor designs. A compromise between the two offers the best in safety, capacity, ease and uniformity of operation.

(*ii*) Vertical curvature superimposed upon horizontal curvature and vice-versa, generally results in a pleasing effect.

(*iii*) From the safety point of view, sharp horizontal curvature should not be introduced at or near the top of a pronounced summit vertical curve. In such situations, the horizontal curve is made longer than the vertical curve.

(*iv*) Sharp horizontal curves should be avoided at or near the low point of a pronounced sag curve.

(*v*) On two-lane highways, the need for safe passing sections at frequent intervals and for an appreciable percentage of the length of the highway often supersedes the general desirability for combination of horizontal and vertical alignment. In these situations, it is desirable to have long straight sections to secure, sufficient passing sight distance in design.

(*vi*) Horizontal curvature and vertical profile should be made as flat as feasible at highway intersections when sight distance is important and vehicles may have to slow down or stop.

(*vii*) On divided highways, variation in the width of the median and the use of separate profiles and horizontal alignments should be considered to derive design and operational advantages of one-way roadways.

4.9. SIGHT DISTANCE

4.9.1. General

The design of a highway with adequate sight ahead of a travelling vehicle results in safe operation. The distance along the road surface at which a driver has visibility of objects, stationary or moving, at a specified height above the carriageway is known as the sight distance. A knowledge of the sight distance requirement is needed in designing vertical curves. It also governs the set-back of buildings, slopes fences and other obstructions adjacent to a carriageway on a horizontal curve. Sight distance requirement can be considered under two heads:

(*i*) Stopping sight distance and

(*ii*) overtaking (passing) sight distance.

4.9.2. Stopping sight distance

Stopping sight distance is the distance required by a driver of a vehicle travelling at a given speed to bring his vehicle to a stop after an object on the roadway becomes visible. Because of its importance to safety, all highways must be designed for the minimum stopping sight distance. It is made up of two components:

(*i*) the distance travelled during perception and brake reaction time; and

(*ii*) the distance travelled during the time the brakes are under application till the vehicle comes to a stop.

When sensations received through the eyes, ears or body are strong enough to be recognised and interpreted, they become perceptions. In the case of a motorist, it is the time which elapses between the instant the driver perceives the object on the carriageway and the instant that he realizes that braking of the vehicle is needed. The time lag or the brief interval between the perception of danger and the effective application of the brakes is called the *brake reaction time*. The perception time and the brake reaction time depend upon a variety of factors, viz., age. sex. alertness and visual acuity of the driver, visibility, .vehicle design, the size and type of the object etc.

As per current Indian Practice, a value of 2.5 second is considered reasonable for the perception time and brake reaction time taken together. The distance travelled during this interval is given by

$$d_1 = vt$$

or

$$d_1 = 0.278\,Vt = 0.278\,\text{V} \times (2.5)$$

$$= 0.695\,\text{V} \quad (4.35)$$

where d_1 = distance travelled in metres

v = speed in metres per second

V = speed in K.P.H.

t = perception and brake reaction time

= 2.5 sec

The braking distance is the distance in which a moving vehicle comes to a stop after the application of brakes. On a level roads, the distance is given by:

$$d_2 = \frac{V^2}{254\,f} \quad (4.36)$$

where d_2 = braking distance in metres

V = Speed in K.P.H.

f = coefficient of longitudinal friction between the tyres and the pavement.

The coefficient of friction is assumed to vary from 0.40 at 20 K.P.H. to 0.35 at 100 K.P.H. The above considerations yield the value given in Table 4.10, recommended by IRC.

Table 4.10: Stopping sight distance as per Indian practice

Design speed K.P.H.	Safe stopping sight distance (Metres)
20	20
25	25
30	30
40	45
50	60
60	80
65	90
80	120
100	180

Some slight adjustments are needed in the values of the braking sight distance to take into account the effect of grades. The following amended formula may be used to calculate d_2:

$$d_2 = \frac{V^2}{254(f \pm 0.01G)} \tag{4.37}$$

where G = longitudinal grade in per cent

The positive sign inside the bracket may be used if the gradient is upward and the negative sign is used if the gradient is downward. Correction for grade should not be applied on undivided roads with two-way traffic, but must invariably be considered for divided highway which have independently designed profiles.

The safe stopping sight distance is measured between two points, one 1.2 m above the carriageway representing the driver's eye and the other at 0.15 m height representing an object.

4.9.3. Overtaking sight distance

Overtaking becomes necessary because some vehicles travel fast while others travel slow. The highway should be designed so that the vehicles travelling at the design speed are afforded chance to overtake the slower vehicles.

Overtaking manoeuvre is possible only if the drivers have sufficient sight distance available to them so that the whole operation can be accomplished without danger. On undivided roads, overtaking involves the occupation of the road space normally used by opposing traffic. The overtaking sight distance is that distance which should be available to enable the driver to overtake another vehicles safely and comfortably without interfering with the speed of an oncoming vehicle travelling at the design speed should it come into view after the overtaking manoeuvre is started.

For computing the minimum overtaking sight distance on two-lane two-way highways, the following assumptions are made:

(*i*) The overtaken vehicle travels at a uniform speed which is 16 K.P.H. less than the design speed of the road.

(*ii*) The overtaking vehicle has reduced its speed and trails the overtaken vehicle as it enters the overtaking section.

(*iii*) The overtaking vehicle follows the vehicle ahead for a short while to perceive the clear road ahead before beginning the overtaking movement.

(*iv*) The overtaking is done by acceleration to the design speed and hurriedly returning to its own side of the road.

(*v*) When the overtaking vehicle returns to its lane, it arrives alongside an oncoming vehicle in the opposite lane.

[Some authorities such as AASHTO lay down that a certain clear distance should be available between the two]. Fig. 4.8 below indicates the elements that go to make up the overtaking sight distance.

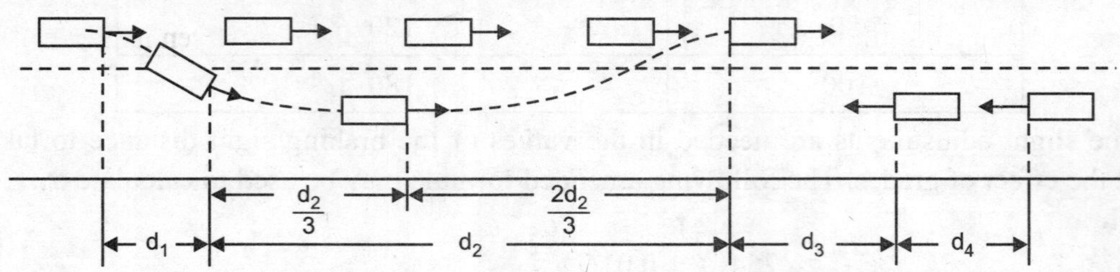

Fig. 4.8: Elements of overtaking sight distance

The dimension d_1 represents the distance travelled during the perception and reaction time and during the initial acceleration to the point of encroachment on the right lane. AASHTO practice gives a value of 3.6 to 4.5 seconds for this initial manoeuvre. The dimension d_2 represents the actual distance covered by the overtaking vehicle during the overtaking manoeuvre. The overtaking vehicles are assumed in AASHTO practice to take 9.3 to 10.4 seconds for this purpose.

The dimension d_3 is the distance between the overtaking vehicle at the end of its manoeuvre and the opposing vehicle and is known as the clearance length. It is assumed to vary from 34 to 90 m as per AASHTO practice. The distance d_4 represents the distance travelled by an opposing vehicle at the design speed while the overtaking manoeuvre is taking place. Strictly speaking it should be $d_1 + d_2$, but in practice it is taken as $\frac{2}{3}\ d_2$ on the reasoning that the overtaking vehicle can always return to its lane if the driver sees an opposing vehicle.

As per latest Indian practice, the calculation have been simplified by assuming that it takes about 9 to 14 seconds for the overtaking manoeuvre. About 2/3rd of this value is accounted for the distance travelled by the opposing vehicle. The final values as per Indian practice are given in Table 4.11.

Table 4.11: Overtaking sigh distance

Speed K.P.H.	Safe overtaking sight distance (metres)
40	165
50	236
60	300
65	340
80	470
100	640

The overtaking distance is measured between two points both 1.2 m above the carriageway, representing the position of the eye of the driver and the object.

As per Indian practice, for sections of roads where the overtaking sight distance cannot be provided, as a second preference an intermediate value between the safe stopping sight distance and the safe overtaking sight distance is recommended. An intermediate sight distance is considered to afford reasonable opportunities for overtaking. Recommended values are given in Table 4.12.

Table 4.12 : Intermediate sight distance for two-lane highways

Speed K.P.H.	Intermediate sight distance (metres)
20	40
25	50
30	60
40	90
50	120
60	160
65	180
80	240
100	360

4.9.4. Sight distance for multi-lane roads

Divided highways with 4 or more lanes need only be designed for a safe stopping sight distance. Undivided highways with 4 lanes have enough opportunities for overtaking within one-half of the carriageway. Such roads, therefore, need only be designed for safe stopping sight distance.

4.9.5. Set-back distance at obstructions of horizontal curves

On horizontal curves with obstructions on the inside, an important consideration is the lateral clearance so as to obtain the sight distance. Fig. 4.9 indicates the elements of design.

The following formulae give the value of the set-back distances:

$$M = \frac{S^2}{8R} \tag{4.38}$$

(for cases where the required sight distance is wholly within the curved road)

$$M = \frac{E(2L - S)}{8R} \tag{4.39}$$

(for cases where the required sight distance is greater than the available length of curve)

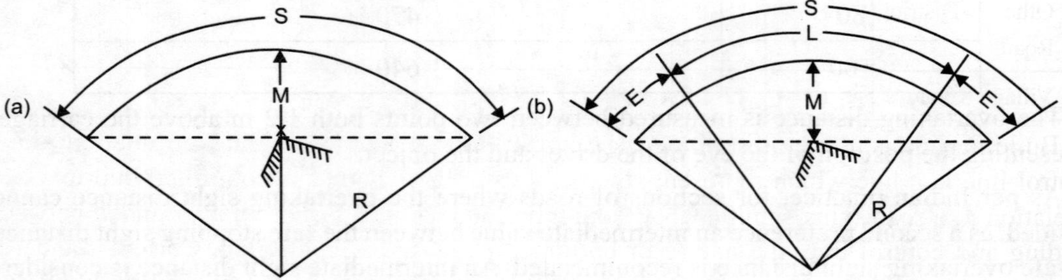

Fig. 4.9: Elements of set-back distance at obstructions of horizontal curves

4.10. CROSS-SECTIONAL ELEMENTS

4.10.1. General

The cross-sectional elements in the highway design pertain to those features which deal with its width. They cover aspects such as right-of-way width, roadway width, pavement width, central reservations (median), shoulder, camber, side slopes, horizontal and vertical clearance etc.

4.10.2. Right-of-way

The road land width or the right-of-way width is the width of land secured and preserved to the public for road purposes. The right-of-way should be adequate to accommodate all the elements that make up the cross-section of the highway and may reasonably provide for future development. Indian practice is summarised in Table 4.13. In order to prevent ribbon development along highways, it is sometimes necessary to establish building lines and control lines with the following definitions:

Table 4.13: Right-of-Way width of different classes of roads in India

Class of road	Right-of-way width in metres							
	Plain and rolling terrain				Mountainous and steep terrain			
	Rural areas		Urban areas		Rural areas		Urban areas	
	Normal	Range	Normal	Range	Normal	Exceptional	Normal	Exceptional
1. National and State Highways	45	30-60	30	30-60	24	18	20	18.
2. Major District Roads	25	25-30	20	15-25	18	15	15	12
3. Other District Roads	15	15-25	15	15-20	15	12	12	9
4. Village Roads	12	12-18	10	15-20	9	9	9	9

Building line is a line prescribing the nearest limits of future buildings in relation to a road. Control line is a line which represents the nearest limits of future uncontrolled building activity in relation to a road. This signifies that though building activity is not totally banned between the building and control line, the nature of buildings permitted is controlled. Current standards for building and control lines in India are given in Table 4.14. The terms right-of-way, building line and control lines are indicated in Fig. 4.10.

Table 4.14: Standards for building lines and control lines in India

Class of road	Plain and rolling terrain			Mountainous and Steep Terrain			
	Rural area		Urban and industrial area	Distance between building line and boundary (metres)			
				Rural areas		Urban areas	
	Width between building lines (overall width) (metres)	Width between control lines (overall width) (metres)	Distance between building line and road boundary (metres)	Normal	Exceptional	Normal	Exceptional
1. National and State Highways	80	150	3-6	5	3	5	3
2. Major District Roads	50	100	3	5	3	5	3
3. Other District Roads	25/30*	35	–	5	3	5	3
4. Village Roads	25	30	–	5	3	5	3

*If the land width is equal to the width between building lines indicated in this column, the building lines shall be set 2.5 m from the road boundary lines.

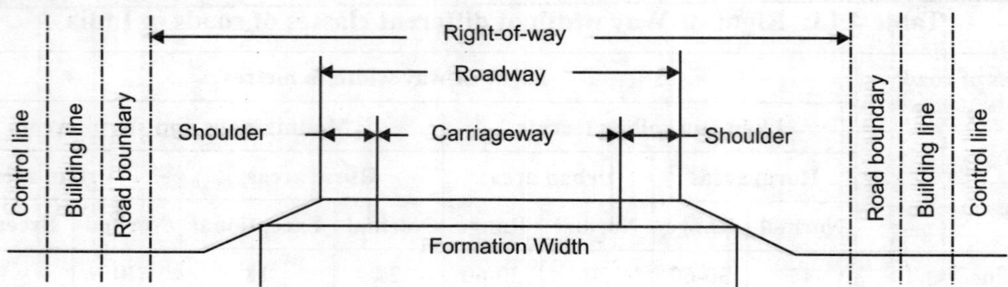

Fig. 4.10: Right of Way, Building line and control Line

4.10.3. Space standards for urban roads

In urban situations the space available is generally restricted and costly. Table 4.15 below represents current Indian practice.

Table 4.15: Space standards for urban roads

S. No.	Category of road	Minimum space in metres
1.	Expressway (6 lane divided)	60
2.	Arterial Street (4 lane divided with cycle tracks)	50-60
3.	Sub-arterial street (4 lane divided)	30-40
4.	Collector street (4 lane)	20-30
5.	Local street such as main residential street, shopping street in residential area, minor residential street and cul-de-sac	10-20

4.10.4. Roadway width

Current Indian practice as regards roadway width is summarised in Table 4.16.

Table 4.16: Standards for roadway width–Indian practice

Highway classification		Width of roadway (in metres)	
		Plain and rolling terrain	Mountainous and steep terrain
1.	**National and state highways**		
	(*a*) Single lane		6.25
	(*b*) Two lane	12.0	8.8
2.	**Major district roads**	9.0	

3.	**Other district roads**		
	(*i*) two lanes	9.0	4.75
	(*ii*) single lane	7.5	4.75
4.	**Village roads**	7.5	4.0

For roads with more than two lanes, the roadway width has to be worked out to suit the number of lanes, medians (if any) and shoulders.

4.10.5. Carriageway width

The width of a traffic lane governs the safety and convenience of traffic and has a profound influence on the capacity of a road. Generally, a width of 3–4 metres is adopted for each traffic lane. In India, single lane pavements are generally 3.75 metres wide, whereas two lane pavements (without raised curbs) are 7.0 metres wide. When raised curbs are provided, two lane pavements are made 7.5 metres in width. An intermediate width of 5.5 metres is commonly used in India for less important roads with two lanes. For multilane pavements Indian practice is to provide 3.5 metre lane. Single lane pavements in India are often provided with paved shoulders 0.9 metre wide on either side to facilitate crossing and overtaking manoeuvres. For village roads, a carriageway width of 3.0 metres is generally deemed sufficient.

4.10.6. Central reservation (Median)

A central reservation or a median is the longitudinal space separating dual carriageways. The functions of the central reservation are:

(*i*) To separate the opposing streams of traffic

(*ii*) To minimise head light glare

(*iii*) To include space for safe operation of crossing and turning vehicles at intersections at grade, *e.g.*, right turning pockets.

(*iv*) To provide a stopping area in case of emergencies.

(*v*) To provide space for future widening of the carriageways.

Indian practice is to provide a minimum width of 5.0 metres for rural areas, which is relaxed to 3.0 metres in restricted conditions. The current Indian practice is to provide raised median, with curbs, but experience has shown that raised median results in poor safety. For urban situations, a width of 5.0 metres is considered desirable, but a minimum width of 1.2 metres is accepted. American practice is to have a minimum value of 1.2 metres and a greater desirable dimension of 18.3 metres or more for rural highways. For urban freeways, a minimum of 7.3 metres is recommended, which may be further lowered to 5 metres in extremely restricted locations. On severely restricted arterial streets, where a narrow separator 0.6-1.2 metres is alone feasible, it may be desirable to have few, if any, openings in the median except at intersections.

In India, it is recommended that the minimum width of medians on rural highways should be 5 metres, but this could be reduced to 3 metres where land is restricted. On long bridges and viaducts, the width of median may be reduced to 1.5 metres, but in any case it should not be less than 1.2 m. As far as possible, the median should be of uniform width in a particular section of the highway. However, where changes are unavoidable, a transition of 1 in 15 to 1 in 20 must be provided.

4.10.7. Shoulders

The shoulder is that portion of the roadway contiguous with the travelled way and is intended for accommodation of stopped vehicles, emergency use and lateral support of the pavement courses. The width of the shoulder should be adequate for giving working space around a stopped vehicle. American practice recommends a metre width for high type facilities and a width 1.2 to 2.4 metres for low type facilities. U.K. practice for rural roads recommends widths ranging from 1.2 to 3.65 metres depending upon the type of road and the nature of the kerb side treatment. Current Indian practice for 2-lane roads suggests a shoulder width of 2.5 metres for roads in rural areas.

4.10.8. Curbs (Kerbs)

A curb is a vertical or sloping member along the edge of a pavement or shoulder forming part of a gutter, strengthening or protecting the edge, and clearly defining the edge to vehicle users. Its functions are:

(*i*) To facilitate and control drainage
(*ii*) To strengthen and protect the pavement edge
(*iii*) To delineate the pavement edge
(*iv*) To present a more finished appearance

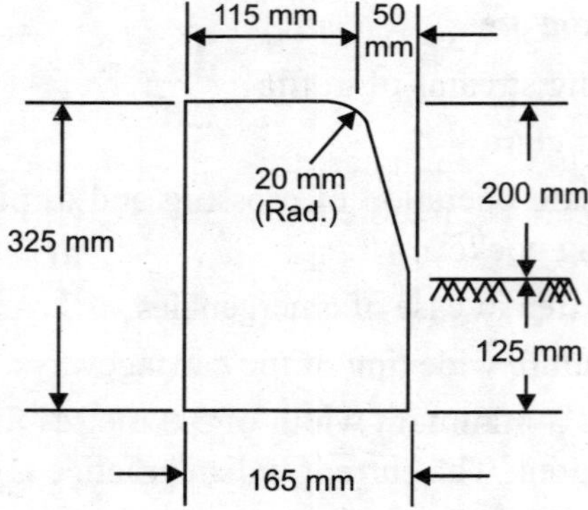

Fig. 4.11: Nearly vertical barrier curb

Curbs are generally needed on urban roads where raised footpaths are a common feature. Curbs are classified as barrier .and mountable. Barrier curbs are designed to discourage vehicles from leaving the pavement..

The face may be vertical or sloping and the height may vary from 150 to 350 mm. A typical design for a nearly vertical barrier curb is given in Fig. 4.11.

In Fig. 4.11 a smaller height (say 150 mm) may be adopted for pedestrian refuge islands. A larger height (say 325 mm) is adopted for bridges.

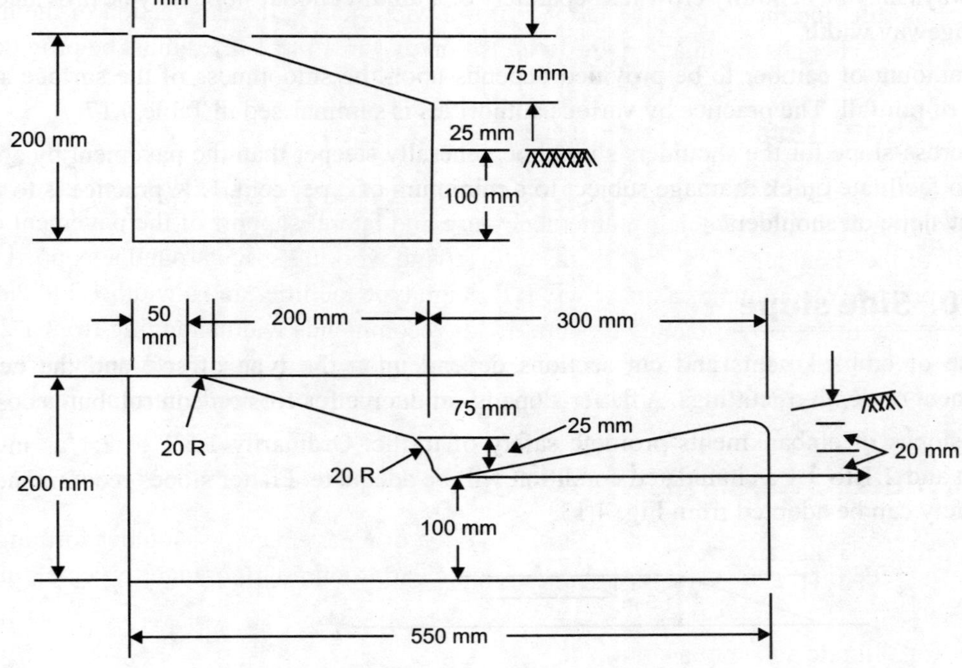

Fig. 4.12: Typical designs of mountable curbs

Mountable curbs are those which can be easily mounted by vehicles, if required. They are used at medians and channelizing islands. Two typical designs are shown in Fig. 4.12.

4.10.9. Camber

Camber, also known as cross-slope, facilitates drainage of the pavement laterally. The pavement can have a crown or a high point in the middle with slopes downwards toward both the edges. This

Table 4.17 : Camber prescribed by various authorities

No.	Authority	Type of surface	Camber (%)
1.	AASHTO	(*i*) High	1–2
		(*ii*) Intermediate	1.5–2
		(*iii*) Low	2–4
2.	U.K.		2.5
3.	I.R.C.	(*i*) Cement Concrete or high type bituminous	1.7–2.0
		(*ii*) Thin bituminous surfacing	2.0–2.5
		(*iii*) Water-bound macadam	2.5–3.0
		(*iv*) Earth	3.0–4.0

type is favoured on two-lane roads and wider undivided roads. On divided roads, the individual carriageways may be centrally crowned separately or a unidirectional slope may be provided across the carriageway width.

The amount of camber to be provided depends upon the smoothness of the surface and the-intensity of rainfall. The practice by various authorities is summarised in Table 4.17.

The cross-slope for the shoulders should be generally steeper than the pavement by about 0.5 percent to facilitate quick drainage subject to a minimum of 3 per cent. U.K practice is to provide 5 per cent slope on shoulders.

7.10.10. Side slope

The slope of embankments and cut sections depend upon the type of soil and the height of embankment or depth of cuttings. A flatter slope is conducive for erosion control, but is costly.

Flat slopes of embankments promote safety of traffic. Ordinarily 1.5:1 to 2:1 as inundated condition and 2:1 to 3:1 as inundated condition will be adequate. Flatter slopes recommended for traffic safety can be adopted from Fig. 4.13.

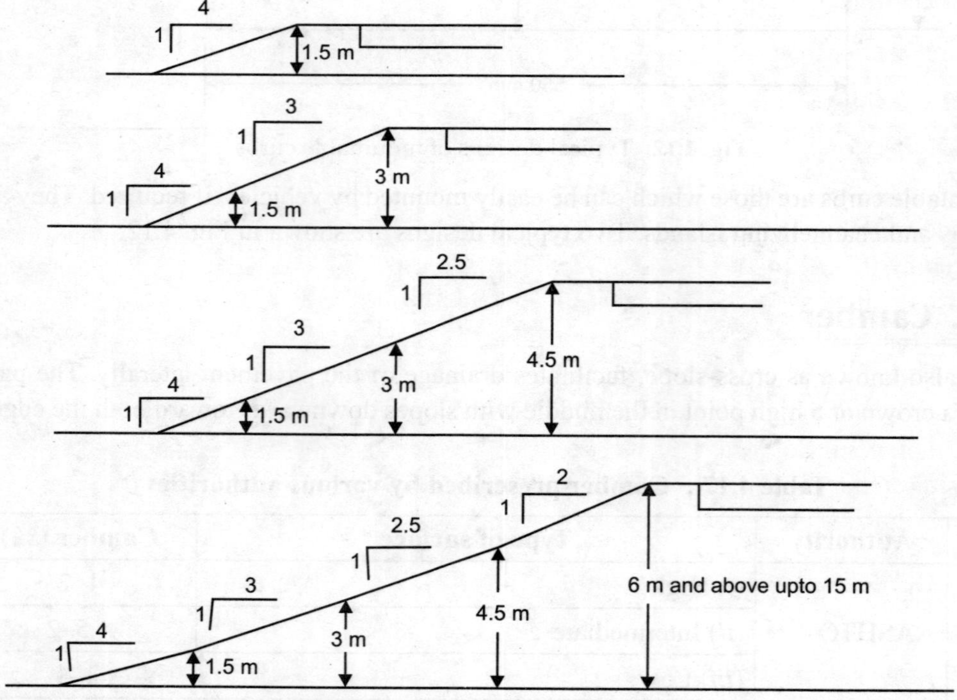

Fig. 4.13: Side slope for road embankments

4.10.11. Lateral and vertical clearances

Lateral clearance is the distance between the extreme edge of the carriageway to the face of a structure. Vertical clearance is the height above the highest point of the travelled way to the lowest point of the overhead structure.

Current Indian standards for lateral clearauce are summarised below:

Type of roads	Minimum lateral clearance
A. Rural Roads	
1. Single carriageway	
(*i*) National and State Highways	: Normal 2.5 metres, exceptional 2.0 metres
(*ii*) Major District Roads and Other District Roads	: Normal 2.0 metres, exceptional 1.5 metres
(*iii*) Village Roads	: Normal 1.5 metres, exceptional 1.0 metre

If footpaths are provided, the lateral clearance should be the width of the footpaths plus one metre.

2. Divided carriageway	
(*i*) Left hand clearance as in 1 above :	2.0 metres desirable
(*ii*) Right hand side clearance	1.5 metres minimum

B. Urban roads

1. **Single carriageways :** For kerbed roads, the carriageway width is widened on both sides under structures (underpass etc.) by 0.25 metre in the case of lower category urban roads and by 0.5 metre in the case of higher category urban roads to account for kerb shyness. If a footpath does not form the cross-section of the urban road, the minimum lateral clearance in addition to the above values shall be 0.5 metre for lower category roads and 1 metre for higher category roads. Where a raised footpath is provided, it is not necessary to have additional clearance beyond the width of the footpath.
2. **Divided carriageways :** The extra width to account for kerb shyness on kerbed roads is the same as in (1) above. The lateral clearance on the left side shall be same as in (1) above. Right hand clearance shall be, in addition to the factor to account for kerb shyness, 0.5 metre in the case of lower category roads and 1 metre in the case of higher category roads.
3. **Vertical clearance :** The minimum vertical clearance shall be at least 5 metres. However, in urban areas, this should be increased to 5.5 metres so that double decker buses could be accommodated.

4.10.12. Typical cross-sections

Typical cross-sections of different classes of highways are shown in Figs. 4.14 to 4.22. These illustrate the principles of designing cross-sectional elements discussed earlier. These are only illustrative cross-sections and a large variety of combinations is possible.

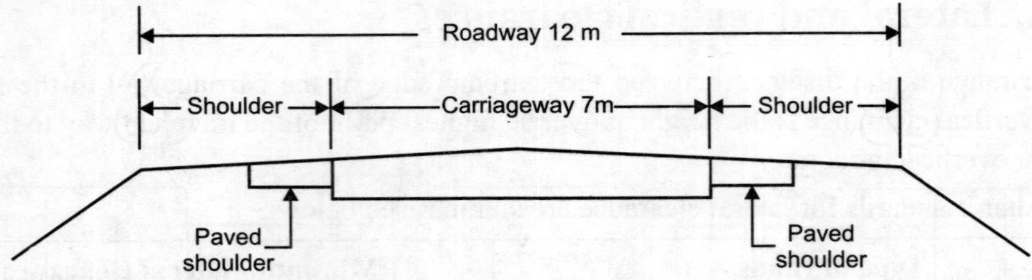

Fig. 4.14: Typical 2-lane National or State Highway (Rural Section)

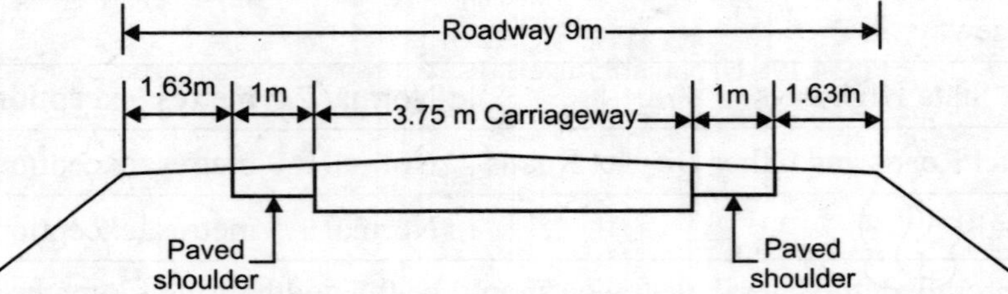

Fig. 4.15: Typical single lane road with paved shoulders (Major District Road or Other District Road)

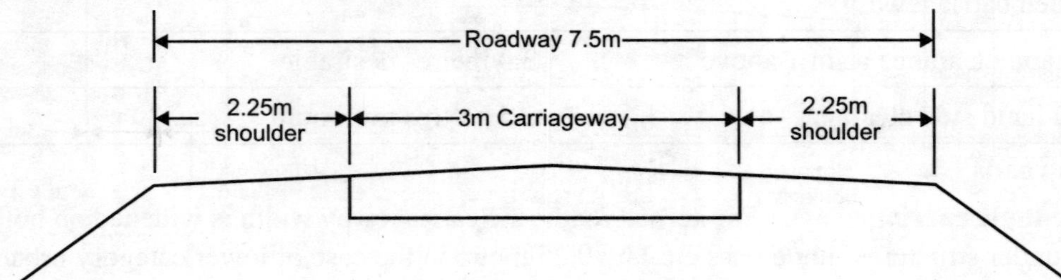

Fig. 4.16: Typical village road

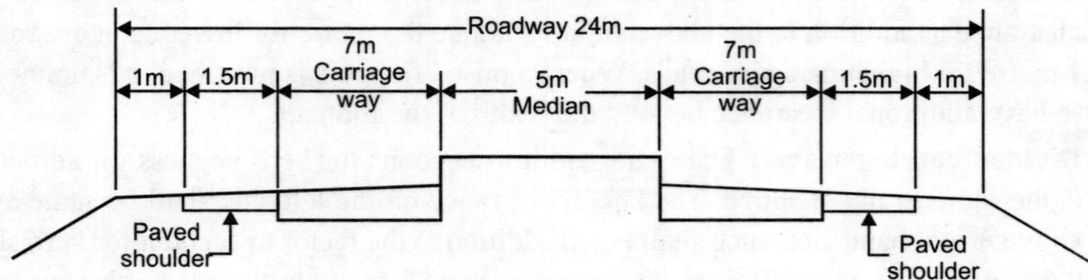

Fig. 4.17: Typical dual carriageway (2-Lane each)

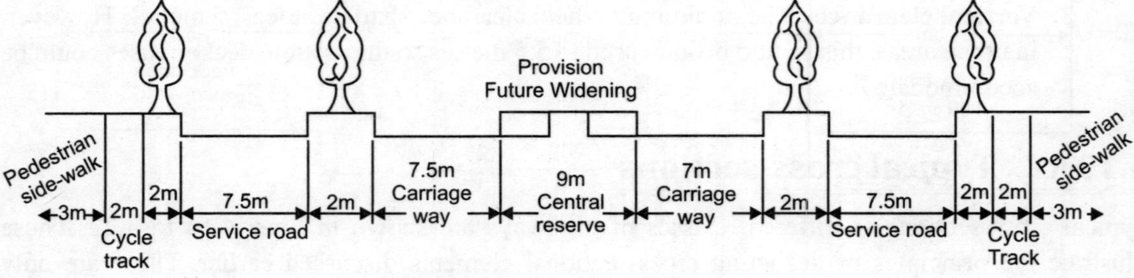

Fig. 4.18: Urban arterial street

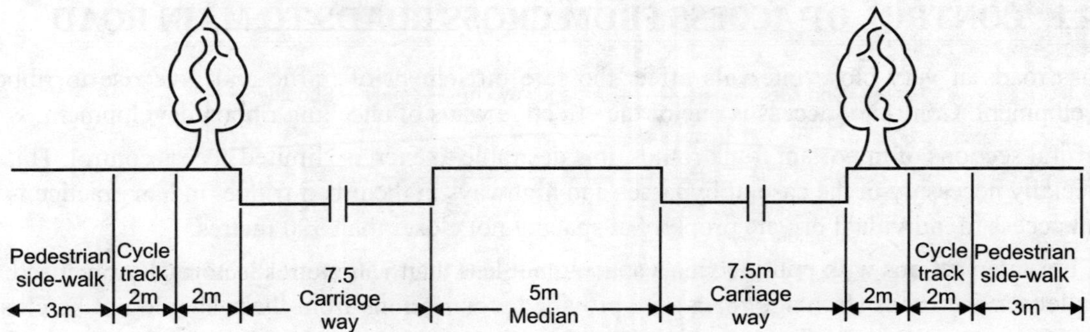

Fig. 4.19: Urban sub-arterial street through open area

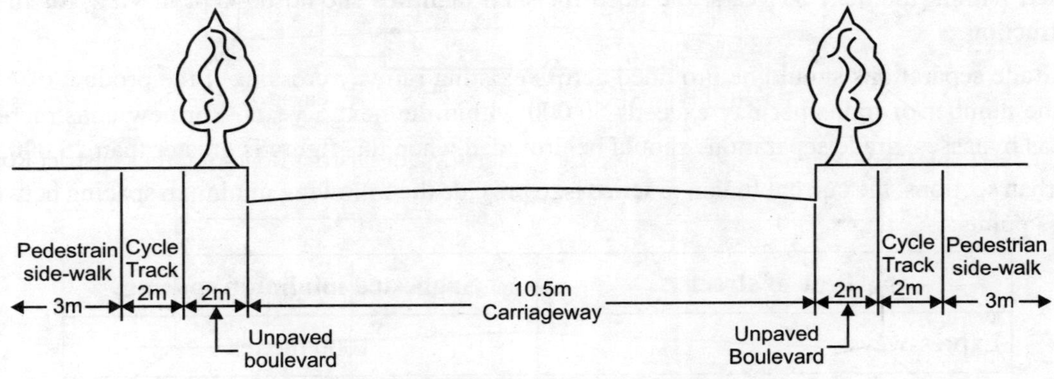

Fig. 14.20: Urban collector street through residential area

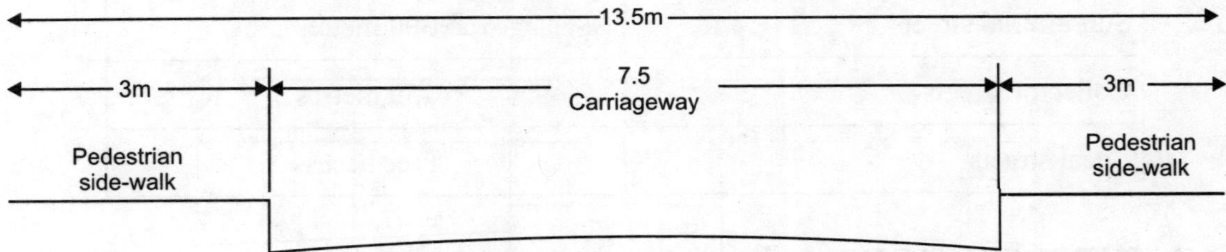

Fig. 4.21: Urban residential street

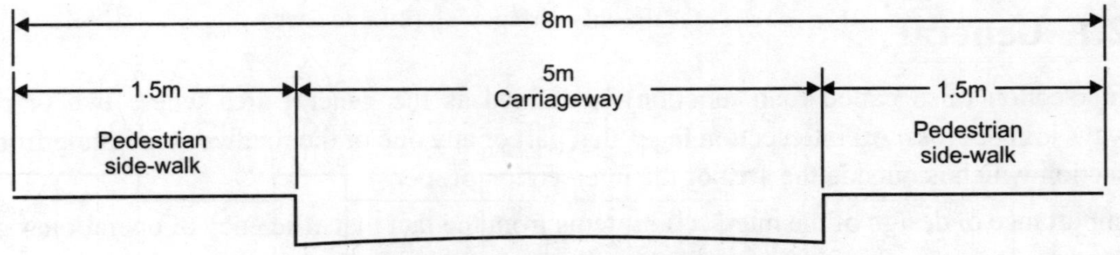

Fig. 4.22: Urban cul-de-sac

4.11. CONTROL OF ACCESS FROM CROSS ROADS TO MAIN ROAD

Gross roads at very close intervals affect the safe movement of traffic and give rise to ribbon development. Control of access is one of the effective ways of checking ribbon development.

On rural sections of important trunk routes, it is desirable to exercise limited access control. This is especially necessary in the case of bypasses and highways in the urban fringe. Indian practice is to limit access to individual private property at spacing not closer than 300 metres.

For intersections with public roads a spacing not less than 750 metres is normally practised. If parallel service roads (frontage roads) are provided, connections from these should not be closer than 750 metres. If the ADT (fast vehicles only) on the cross-road within the next 5 years exceeds 5000, grade separations should be provided at the intersections. Where this traffic figure will be reached within' the next 20 years, the need for such facilities should be kept in view for future construction.

Grade separations should be provided across existing railway crossing if the product of ADT and the number of trains per day exceeds 50,000 within the next 5 years. For new constructions such as bypasses, grade separations should be provided when this figure is greater than 25,000.

On urban sections, the current Indian practice is to provide the following minimum spacing between access points:

Type of street	Suggested minimum spacing
Expressways	1000 metres
Arterial streets	500 metres
Sub-arterial streets	300 metres
Collector streets	150 metres
Local streets	Free access

4.12. INTERSECTIONS

4.12.1. General

An intersection (also called road junction) is defined as the general area where two or more highways join or cross. An intersection leg is that part of any one of the roadways radiating from an intersection which is outside the area of the intersection proper.

The importance of design of the intersection stems from the fact that efficiency of operation, safety, speed, cost of operation and capacity are directly governed by the design.

Since an intersection involves conflicts between traffic in different directions, its scientific design can control accidents and delay and can lead to orderly movement of traffic. Intersections represent potentially dangerous locations from point of view of traffic safety.

It is believed that well over half the fatal and serious road accidents in built-up areas occur at junctions. The following principles should be considered in a good design:

(*i*) The number of intersections should be kept at minimum. If necessary, some minor roads may be connected with each other before joining a major road.

(*ii*) The geometric layout should be so selected that hazardous movement by drivers are eliminated. This can be achieved by various techniques such as channelizing and staggering.

(*iii*) The design should permit the driver to discern quickly, either from the layout or from traffic signs, the path he should follow and the actions of merging and diverging. This can be achieved by a good layout, traffic islands, signs and carriageway markings. Good visibility improves safety.

(*iv*) The layout should follow the natural vehicle paths. Smoothness, in contrast to abrupt and sharp corners, should guide minor streams into stopping or slowing down positions.

(*v*) The number of conflict points should be minimised by separating some of the many cutting, merging or diverging movements.

(*vi*) Vehicles that are forced to wait in order to cross a traffic stream should be provided with adequate space at the junctions.

4.12.2. At-grade and grade separated junctions

An intersection where all roadways join or cross at the same level is known as an at-grade intersection. An intersection layout which permits crossing manoeuvres at different levels is known as grade-separated intersection. The choice between an at-grade and grade-separated junctions at a particular site depends upon various factors such as traffic, economy, safety, aesthetics, delay etc. Grade-separated junctions generally are more expensive initially and are justified in certain situations. These are:

(*i*) On high type facilities such as expressways

(*ii*) Certain at-grade intersections which have reached the maximum capacity and where it is not possible to improve the capacity further by retaining the at-grade crossing.

(*iii*) At certain locations which have a proven record of bad accident history when functioning as at-grade junctions.

(*iv*) At junctions where the traffic volume is heavy and the delays and loss caused justify economically the provision of grade-separation.

(*v*) At certain specific topographical situations where it is logical to provide a grade-separated structure rather than an at-grade intersection, which may involve considerable earthwork or acquisition of land.

4.12.3. Basic forms of at-grade intersections

Fig. 4.23 gives the basic forms of at-grade intersections.

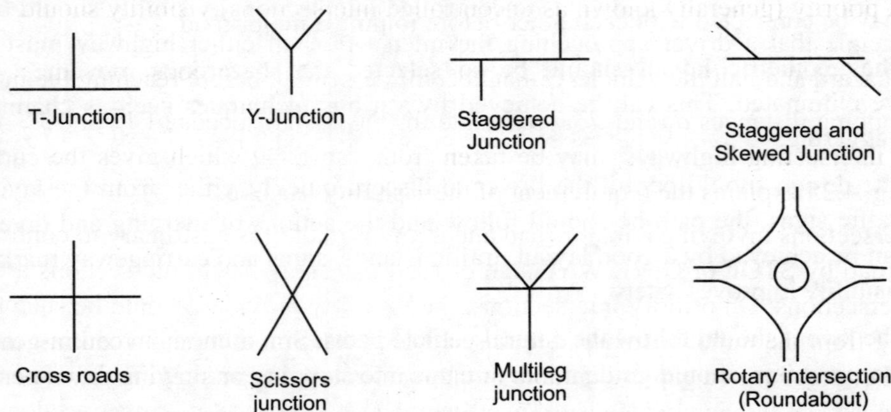

Fig. 4.23: Basic forms of At-grade intersections

4.12.4. Alignment and vertical profile

In view of the conflict and hazard involved at intersections, the alignment and vertical profile of the intersecting roads should permit the drivers to discern the danger and make the manoeuvres necessary for safety. This requires that the alignment be as straight and gradients as low as practicable. It is desirable that the intersection roads meet at or nearly at right angles. For this purpose, it may be necessary to modify the alignment of the roads as shown in Fig. 4.24.

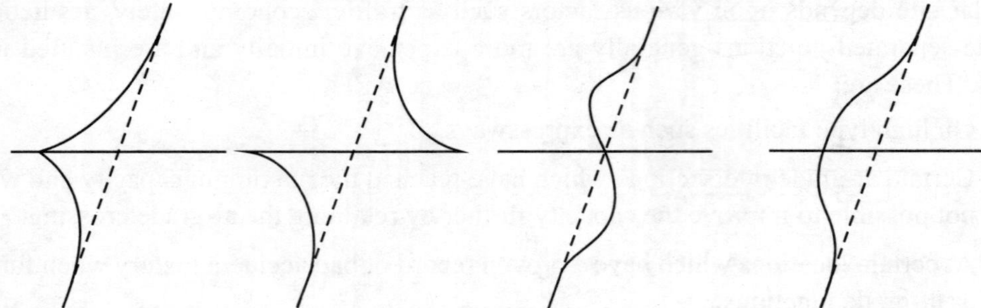

Fig. 4.24: Modification in alignment of intersecting roads to give a nearly right-angle crossing

4.12.5. Visibility at intersections

The safety of traffic can be ensured only if the visibility is full and unimpeded along both roads. Any obstruction should be clear of the minimum visibility triangle for a height of 1.2 metres above the roadway.

4.12.5.1. Rural road intersections

At intersection where the intersecting roads are of more or less equal importance and there is no established priority (generally known as uncontrolled intersections) visibility should be provided on the principle that a driver approaching the intersection on either highway must be able to perceive a hazard and halt the vehicle, if that becomes essential, before reaching the intersection.

The stopping distances d_1 and d_2 appropriate for the design speeds of V_1 and V_2 respectively in the two intersecting highways, may be taken from Table 4.9 which gives the current Indian practice. Fig. 4.25 explains the requirement of the visibility triangle.

On intersections involving a major road and a minor road, it is customary to control traffic on the minor road by STOP or GIVE WAY sign or road markings. Such intersections are known as priority intersections. On priority intersections, the visibility provided should be such that drivers approaching from the minor road are able to see the vehicles on the major road in good time and judge whether the required gap is available in the main road traffic stream for a safe crossing. For this purpose current Indian practice is to specify a visibility distance of 15 metres along the minor road (Fig. 4.26).

V_1, V_2 Design speed of road
d_1, d_2 Safe stopping sight distance

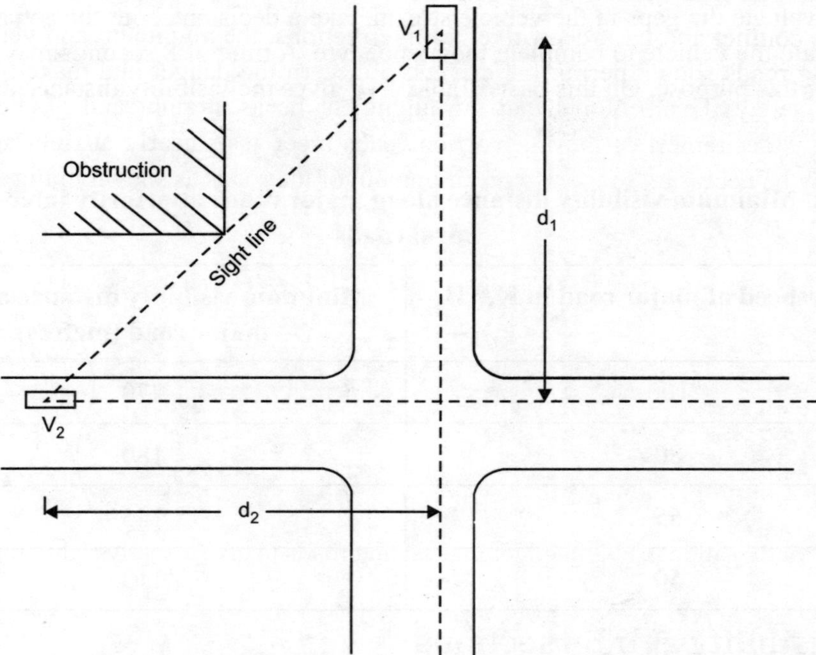

Fig. 4.25: Visibility triangle at uncontrolled intersections

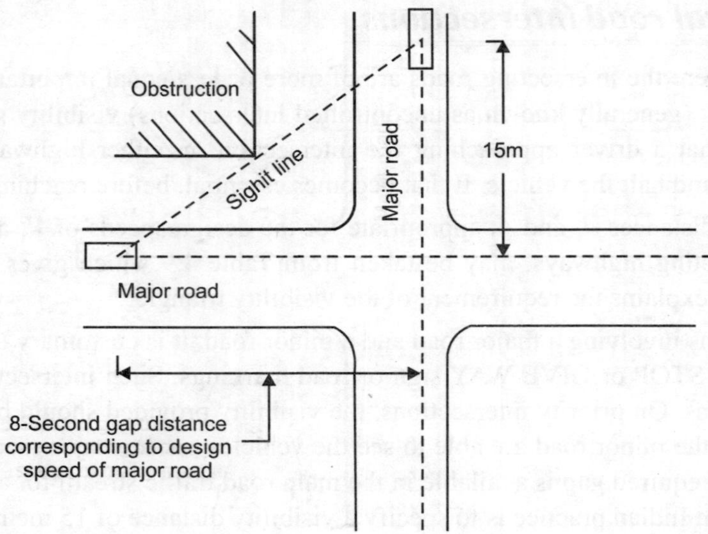

Fig. 4.26: Visibility triangle at priority intersections

On the major road of priority junctions, the visibility needed along the major road depends upon the time required by the driver on the minor road to perceive the traffic conditions on the intersection, evaluate the gaps in the vehicle stream, take a decision about the actual crossing and finally accelerate the vehicle to complete the manoeuvre. A time of 8 seconds may be considered appropriate for this purpose. On this basis. Table 4.18 gives the visibility distance along the major road.

Table 4.18: Minimum visibility distance along major roads at priority inter-sections on rural roads

Design speed of major road in K.P.H.	Minimum visibility distance along a major road (metres)
100	270
80	180
65	145
50	110

4.12.5.2. Urban Road Intersections

On urban intersections, the visibility distance on a liberal scale as on rural intersections is not possible and is perhaps not necessary due to reduced speed of operation. Current U.K. practice is to limit the visibility distance along the minor road in a priority intersection to about 4.5–9.0 metres. On the major road, the visibility distance is provided as shown in Table 4.19.

Table 4.19 : Visibility distance along major roads on priority junction

Type of road	Visibility distance along a major road (Metres)
1. All purpose primary distributor	120–150
2. District or local distributor	90
3. Access road	60

4.12.6. Radii of curves at intersections

The design of intersection curves has profound influence on speed. Curve radii are therefore, selected to control speed and ensure safety of operation. Generally, the approach to the design of curves falls under one of the following categories:

(*i*) Design of curves for a minimum speed and sharpest turn
(*ii*) Design of curves to permit different operating speeds

4.12.6.1. Sharpest turn curves for minimum speed

When it is necessary to provide for turning vehicles within minimum space, the minimum turning paths of design vehicles apply.

Table 4.20 : Minimum edge of pavement designs for turns at intersections

AASHTO Design vehicle	Angle of turn (degrees)	Radius of simple curve (m)	3-central compound curve symmetric		3-centred compound curve symmetric	
			Radii, (m)	Offset, (m)	Radii, (m)	Offset, (m)
P	30	18	–	–	–	–
SU		30	–	–	–	–
WB-40		45	–	–	–	–
WB-40		60	–	–	–	
P	45	15	–	–	–	–
SU		23	–	–	–	–
WB-40		36	–	–	–	–
WB-50		52	60-30-60	0.9	–	–

The turning radii for AASHTO design vehicles have been given earlier in Table 4.3. Considering these, the Minimum edge of pavement designs for sharpest turns for an angle of turns of 30° and 45° for AASHTO vehicles are given in Table 4.20.

Compound curves are generally three-centred, as shown in Fig. 4.27.

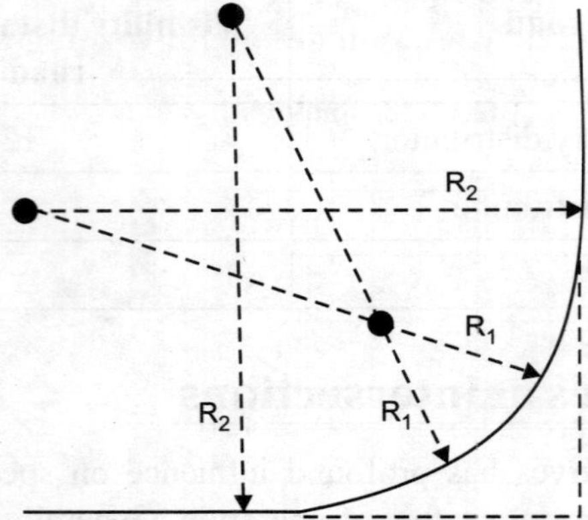

Fig. 4.27: Compound curve for corner design

The sharpest turns are usually provided for unchannelized intersections and for curbs of urban streets at intersection.

In residential streets, a curb radius of 6 metres is regarded as the minimum, though radii of 4.5–7.5 metres are satisfactory. For junctions using commercial vehicles, a radius of 9-15 metres will be necessary.

4.12.6.2. Minimum design for turning roadways

Where the inner edges of pavement for left turns at intersections are designed to accommodate semi-trailer combinations, or where the design permits passenger vehicles to turn at speeds 25 K.P.H. or more, the pavement area at the intersections may become excessively large for proper control of traffic. To avoid this a corner island is provided to form a separate turning roadway. A turning roadway is a connecting roadway for traffic turning between two intersection legs, Fig. 4.28.

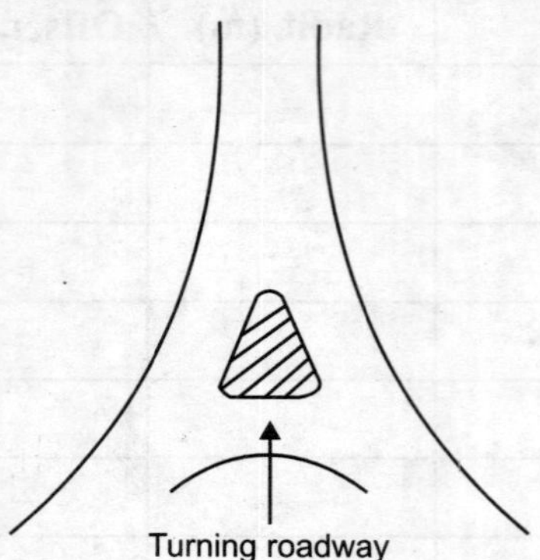

Fig. 4.28: Turning roadway

AASHTO standards for turning roadways are summarised in Table 4.21.

Table 4.21: Minimum design for turning roadways

Angle design of turn (Degrees)	Design designation	3-Centred compound curve		Width of lane (m)	Approximate island size (Sq. m)
		Radii (m)	Offset (m)		
75	A	45-22.5-45	1.5	4.25	5.5
	B	45-22.5-45	1.5	5.5	4.5
	C	55-27.5-55	1.1	6.0	4.5
90	A	45-15-45	0.9	4.25	4.5
	B	45-15-45	1.5.	5.5	7.5
	C	55-20-55	1.8	6.0	11.5
105	A	36-12-36	0.6	4.5	6.5
	B	30-11-30	1.5	6.5	4.5
	C	55-14-45	2.5	9.0	5.5
120	A	30-9-30	0.75	5.0	11.0
	B	30-9-30	1.5	7.5	8.5
	C	55-12-55	2.6	10.5	20.0
135	A	30-9-30	0.75	5.0	43.0
	B	30-9-30	1.5	8.0	34.0
	C	50-11-50	2.75	11.0	60.0
150	A	30-9-30	0.75	5.0	130.0
	B	30-9-30	1.8	9.0	110.0
	C	50-11-50	2.2.	11.5	160.0

Note: *A* – Primarily passenger vehicles; permits occasional design single unit trucks to turn with restricted clearances

B – Provides adequately for SU; permits occasional WB-50 to turn with slight encroachment on adjacent traffic lanes

C – Provides fully for WB-50

4.12.6.3. Curves designed for different speeds

Vehicles turning at intersections designed for the sharpest turn by design vehicles, as discussed in para 4.12.6.2 above, have to operate at low speed, perhaps less than 15 K.P.H. It is often desirable and feasible to design for vehicles operating at higher speeds, as in the case of rural intersections. The speeds for which intersection curves should be designed depend largely upon the vehicle speeds on the approach legs, type of intersection, and volumes of through'and turning traffic.

Generally, a desirable turning speed for design is the average running speed, which may be roughly taken as 0.7 of the design speed.

Design at such speeds offers little hindrance to smooth flow of traffic and may be justified on some intersections for certain movements involving little or no conflict with pedestrians or other vehicular traffic.

Table 4.22 gives the AASHTO practice for the design of such curves.

Table 4.22: Minimum radii for intersection curves for various turning speeds

Design turning speed V (K.P.H.)	25	30	40	50	60	65
Side friction factor, f	0.32	0.27	0.23	0.20	0.18	0.16
Assumed minimum superelevation, e	0.00	0.02	0.04	0.06	0.08	0.09
Total e+f	0.32	0.29	0.27	0.26	0.26	0.25
Calculated min. R (m) $R = \frac{V^2}{127(e+f)}$	14	28	47	70	96	129
Suggested minimum R (m)	15	27	45	70	95	130

4.12.7. Widths of carriageways in junctions

The carriageway width needs to be increased in curves as already explained in 4.6.8.

The following values of carriageway widths are as per IRC practice.

Table 4.23: Widths of carriageways at intersections

Inner radius (m)	Design speed (K.P.H.)	Single lane width (m)	Two lane width for one or two way traffie (m)
10.5	18	5.5	11.5
15	23	5.5	10.5
20	27	5.0	10.0
30	32	4.5	9.0
40	37	4.5	9.0
50	41	4.5	8.0

75	50	4.5	8.0
100	57	4.5	8.0
125	62	4.5	8.0
150	64	4.5	8.0
Straight		4.5	7.0

4.12.8. Speed change lanes

Drivers entering into or leaving an intersection have to necessarily change their speed. Drivers entering an intersection will reduce the speed to a safe value at which the intersection can be negotiated. While leaving the intersection, they will increase the speed until the desired open highway speed is reached. When such deceleration or acceleration is accomplished on the highway travelled way, traffic is disrupted and hazard may be caused. To avoid this, speed change lanes are normally provided on superior facilities such as expressways and are also frequently provided on other main highways. IRC practice is to provide acceleration lanes where the future traffic on the acceleration lane is expected to be more than 1000 PCUs per day. These lanes also increase the capacity of the junction.

Speed change lanes are either acceleration lanes or deceleration lanes. An acceleration lane is an auxiliary traffic-lane to enable a vehicle entering a road to increase its speed so that it can merge more safely with through traffic. A deceleration lane is an auxiliary' traffic lane to enable a vehicle leaving the through traffic stream to reduce speed without interfering with other traffic.

Acceleration and deceleration lanes are usually provided with taper, vide Fig. 4.29.

Table 4.24 gives the lengths of deceleration and acceleration lanes as per IRC practice.

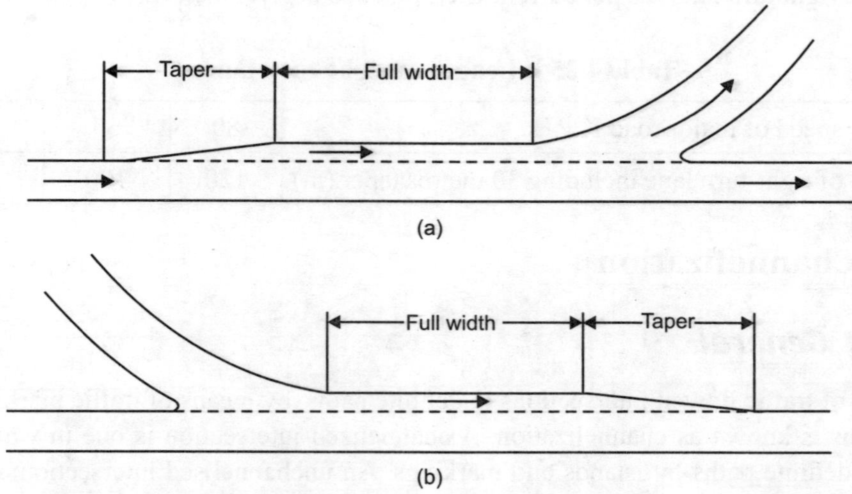

Fig. 4.29: Acceleration and deceleration lanes

Table 4.24 : Design length of deceleration and acceleration lanes (Grade 2 Per cent or less)

Design speed of turning roadway curve, K.P.H.		Stop condition	25	30	40	50	60	65	75	80
Minimum curve radius (m)		–	15	27	45	70	95	130	170	210
Design speed of highway (K.P.H.)	Length of taper (m)	Total length of deceleration lane, including taper (m)								
65	60	100	90	85	75	60	–	–	–	–
80	70	130	120	115	110	100	90	–	–	–
100	80	150	150	145	140	130	120	100	90	–
110	90	165	165	160	150	145	140	115	100	–
		Total length of acceleration lane, including taper								
65	60	–	100	75	70	–	–	–	–	–
80	70	–	210	190	180	150	120	–	–	–
100	80	–	340	320	300	270	240	180	120	–
110	90	–	465	450	420	400	370	300	250	175

The width of a speed change lane should not be less than a normal traffic lane, viz., 3.75 metres.

4.12.9. Right turn lanes

Storage lanes for right-turning traffic enable the vehicles to slow down to the necessary turning speed without impeding the flow of through traffic and for waiting when signals are provided. The length of such right turn lanes as per current U.K. practice is given below.

Table 4.25 : Length of right-turn lanes

Design speed of major road K.P.H.	80	65	50
Length of right-turn lane including 30 metres taper (m)	120	100	80

4.12.10. Channelization

4.12.10.1. General

The direction of traffic flow at intersections to definite paths, by means of traffic markings, islands or other means is known as channelization. A channelized intersection is one in which traffic is directed into definite paths by islands and markings. An unchannelised intersection, on the other hand, is one without islands for directing traffic into definite paths. An unchannelised intersection is the most simple type but is the most dangerous and inefficient. All important junctions should, therefore, provide for channelizations.

4.12.10.2 Purposes served

Channelization serves the following purposes:

(*i*) **Separation of conflicts:** To diminish the number of possible vehicle conflicts, to reduce the possible area of conflicts in the carriageway and to present drivers with only one decision at a time.

(*ii*) **Control of angle of conflict:** Small angles of crossing cause severe accidents if they occur. Severity is reduced if the angle of conflict is controlled.

(*iii*) **Control speed:** To reduce the speed of traffic entering the intersection and increase the speed of traffic leaving the intersection, bending or funnelling by suitable channelization techniques is resorted to, vide Figs. 4.30 and 4.31.

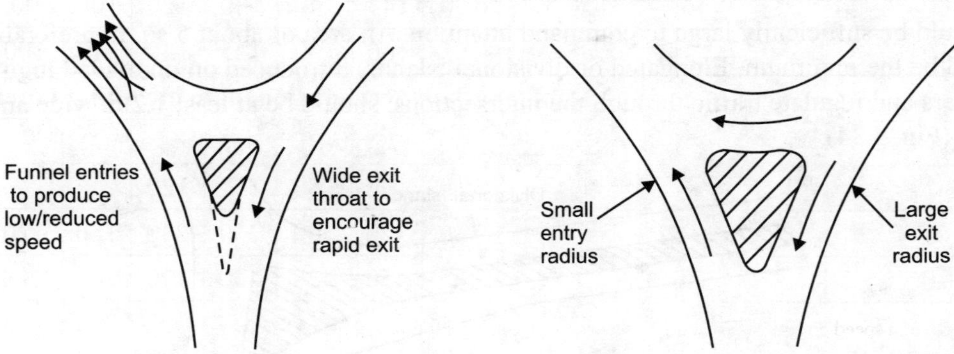

Fig. 4.30: Channelizing island **Fig. 4.31:** Control of radius of entry and exit for control of speed

(*iv*) **Protection of traffic for vehicles leaving or crossing the main traffic stream:** Separate storage pockets Tor right turning traffic at an intersection and the adjacent island, vide Fig. 4.32 can increase safety.

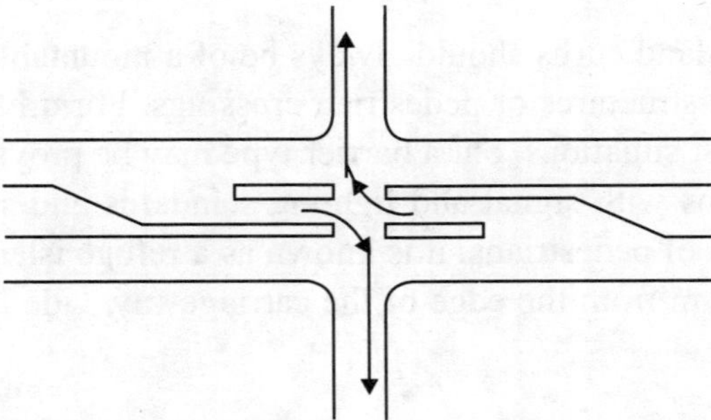

Fig. 4.32: Channelising to protect right turning traffic

(*v*) **Protection of pedestrians:** To provide a refuge for pedestrians between traffic flows. a channelizing such as in Fig. 4.32 serves as a refuge and makes the crossing much safer.

(*vi*) **Elimination of excessive intersectional areas:** Intersections with large corner radii and those at oblique angles have large paved areas, which permit and encourage hazardous uncontrolled vehicle movements. If these unused paved areas are converted into channelizing islands, orderly movement results and hazards are reduced vide Fig. 4.33.

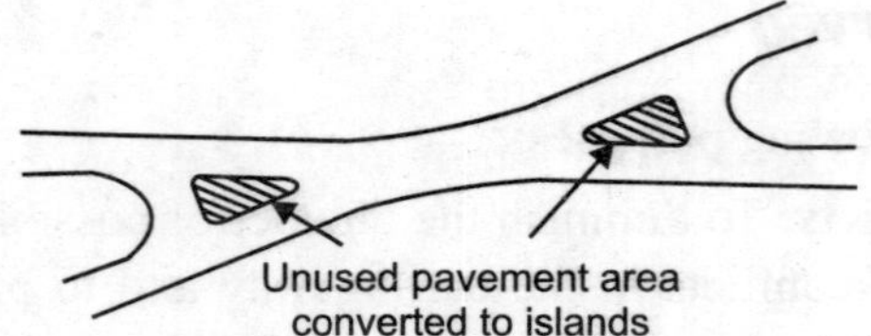

Fig. 4.33: Unused pavement areas converted to channelizing islands

(*vii*) **Location of Traffic Control Devices:** To provide space for traffic control devices such as direction indicators, reflectors, signs, etc.

4.12.10.3. Features of channelising islands

They should be sufficiently large to command attention. An area of about 5 sq. m preferably 7 sq. m, should be the minimum. Elongated or divisional islands, introduced on undivided highways to alert drivers and regulate traffic through the intersections, should be at least 1.2 m wide and 3.5 to 6 m long (Fig. 4.34).

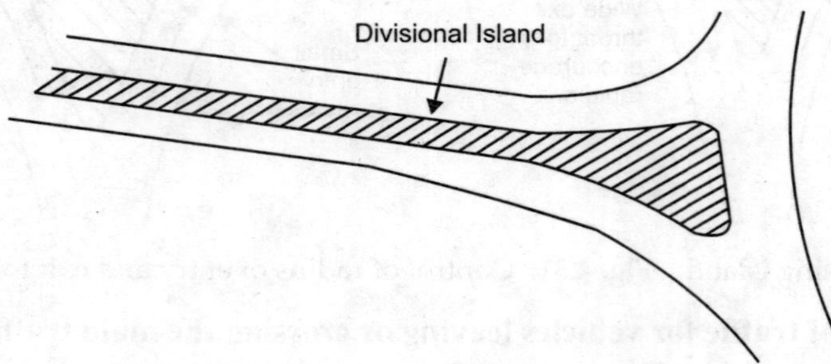

Fig. 4.34: Divisional island combined with a triangular channelizing island at entrance to a rotary junction

In rural intersections, island curbs should always be of a mountable type, except where there is a need for a barrier, as at structures or pedestrian crossings. For urban intersections mountable curbs are preferable for most situations, but a barrier type may be provided, where speeds are low and protection of pedestrians with signal and lighting standards and signs is needed. Where the island is used for protection of pedestrians, it is known as a refuge island. The edges of the island are usually offset 0.3 to 0.6 m from the edge of the carriageway, vide Fig. 4.35 to reduce the risk of overriding the islands.

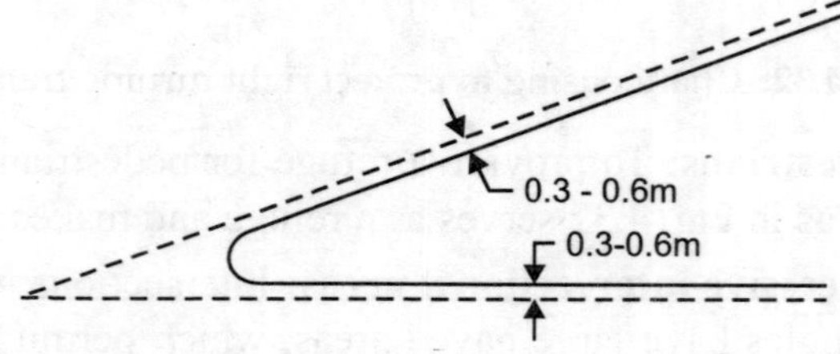

Fig. 4.35: Edge offset at channelizing island

The approach noses of islands should be rounded off to a minimum radius of 0.6–0.9 m and the merging end nose should be rounded off to a radius of at least 0.3 m. The approach to the nose should be marked for additional guidance of traffic by means of diagonal markings and chevrons (Figs. 4.36 and 4.37).

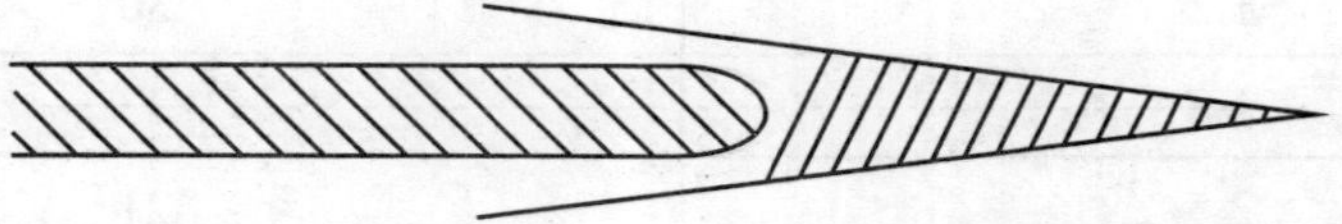

Fig. 4.36: Diagonal markings in advance of divisional island

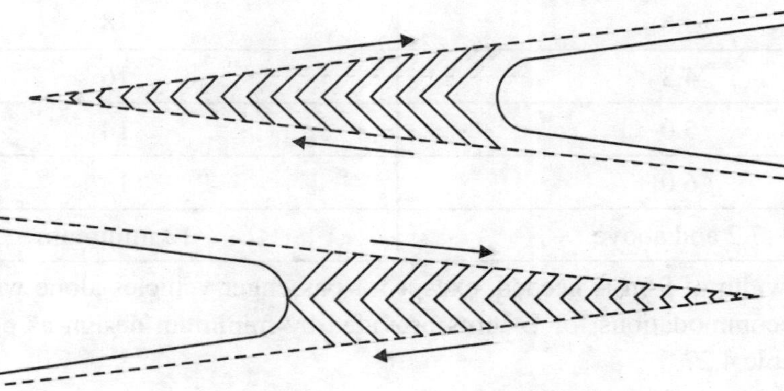

Fig. 4.37: Chevron markings

4.12.11. Median openings

4.12.11.1. Minimum length of median openings

The length of the opening depends upon the width of the median (Fig. 4.38). Table 4.26, adapted from AASHTO, gives the dimensions of the opening for SU design vehicles.

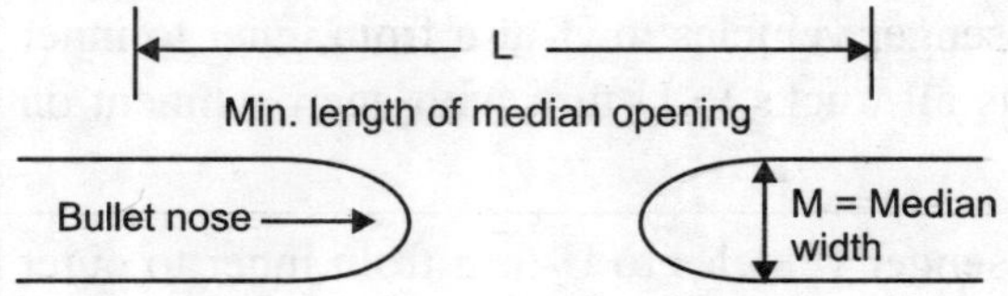

Fig. 4.38: Median opening with bullet nosed ends

Median opening for making U-turns are needed in the following locations:

(*i*) Near to all terminal junctions

(*ii*) At regular intervals, say 1.5 Km, between junctions to facilitate maintenance operation, policing, repair of stalled vehicles etc.

(*iii*) Beyond important intersections to enable drivers unfamiliar with the locality, to correct any mistake in manoeuvring.

(*iv*) For accommodating minor turning movements.

Table 4.26: Minimum length of median opening

Median width, M (in metres)	L= Minimum length of median opening (in m) (bullet nose ends)
1.2	30
1.8	23
2.5	21
3.0	19
3.6	18
4.3	16
5.0	14
6.0	13
7.2 and above	12 minimum

A minimum width of 9 m is needed, except for passenger vehicles alone where only 6 m is sufficient. The accommodations for U-turns provided by minimum design as per AASHTO are summarised in Table 4.27.

Table 4.27: Median opening for U-turns design

Median Width (m)	Type of manoeuvre accommodated on 4-Lane divided highway	AASHTO design vehicle provided for
18	Permits nearly all vehicles to turn from inner to inner lane	All vehicles
12	Permits all passenger vehicles to U-turn from inner to inner lane; permits some trucks to turn from outer to outer lane, with large trucks partially encroaching on shoulder.	P and SU
10	Permits passenger vehicles to U-turn from inner to inner lane, permits all trucks to U-turn with encroachment on shoulders	P and SU
6	Permits passenger vehicles to U-turn from inner to outer lane, U-turn impossible for large trucks without backing.	P only

4.12.11.2. Gap in median at junctions

To ensure that large vehicles can turn right without difficulty to or from a major road, the gap in the medians should normally extend 3 m beyond the continuation of both kerb lines of the minor road to the edge of the major road (Fig. 4.39) and should also be determined by 12–15 m radius control circles tangential both to the centre line of the minor road (or to the sides of any refuge or island) and the side of the central reserve away from the minor road.

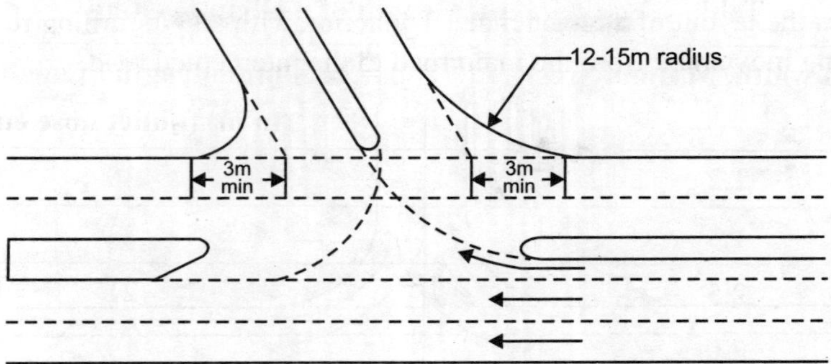

Fig. 4.39: Gap in median at junction

4.12.12. T-Intersection

Fig. 4.40 shows an unchannelised T-junction for very light traffic and minor and local roads. Obviously it is hazardous to use this in locations where speeds are high and turning movements are large.

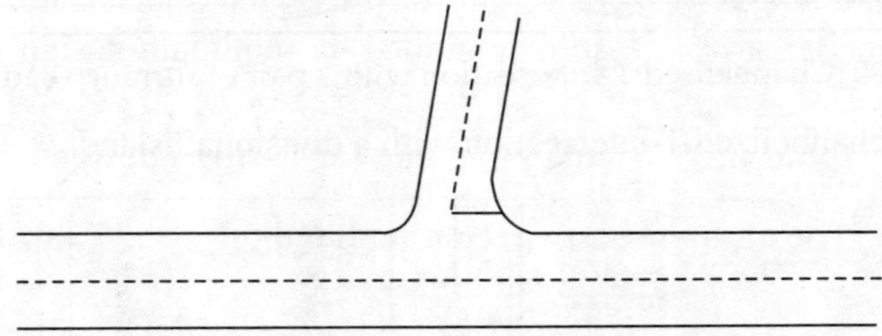

Fig. 4.40: Unchannelised T-junction

Fig. 4.40 is an improvement since the main road is flared to provide an additional lane to facilitate left turns.

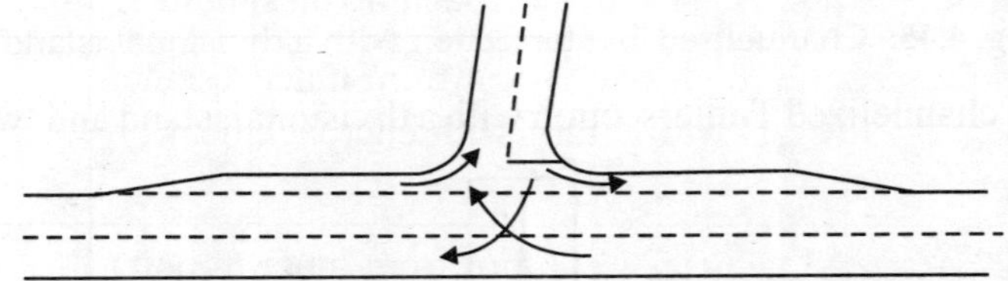

Fig. 4.41: Flared unchannelised T-intersection to facilitate left turning movements

If there is heavy right turning movement, the flaring can be done opposite to the intercepted road, as in Fig. 4.42.

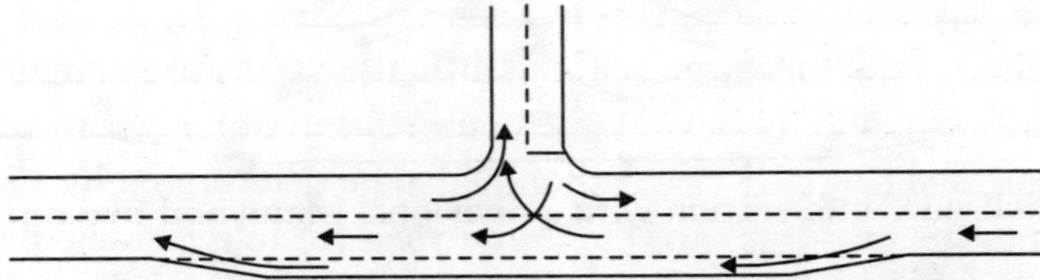

Fig. 4.42: Flared unchannelised intersection to facilitate right turning movements

Fig. 4.43 gives the layout of a channelized T-junction with single turning roadway, providing for large left turning movement from the main road to the intercepted road.

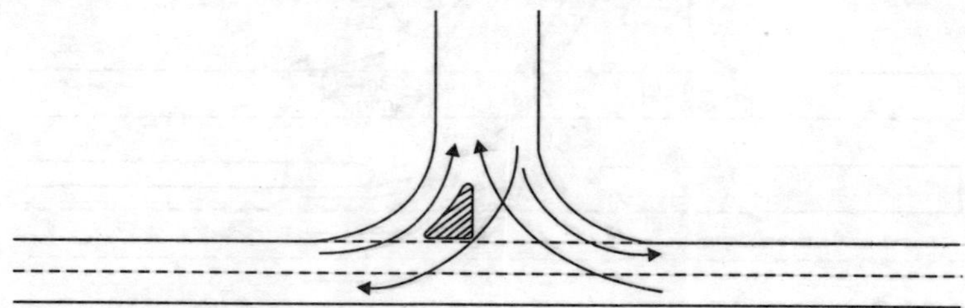

Fig. 4.43: Channelised T-inlersection with single turning roadways

Fig. 4.44 gives the layout of a channelized T-intersection with a pair of turning roadways.

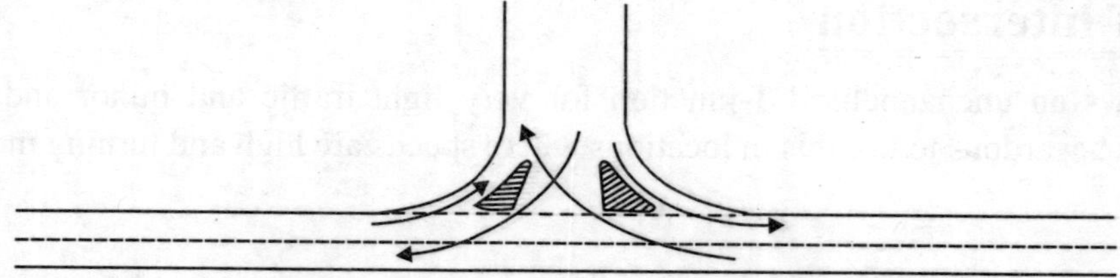

Fig. 4.44: Channelised T-intersection with a pair of turning roadways

Fig. 4.45 shows a channelized T-intersection with a divisional island.

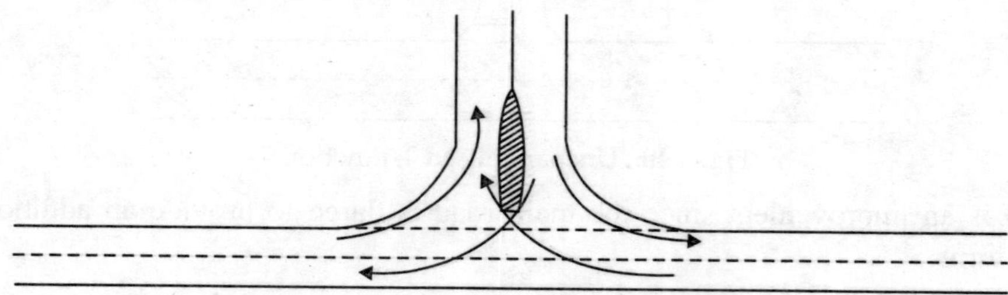

Fig. 4.45: Channelised T-intersection with a divisional island

Fig. 4.46 shows a channelized T-intersection with a divisional island and two turning roadways.

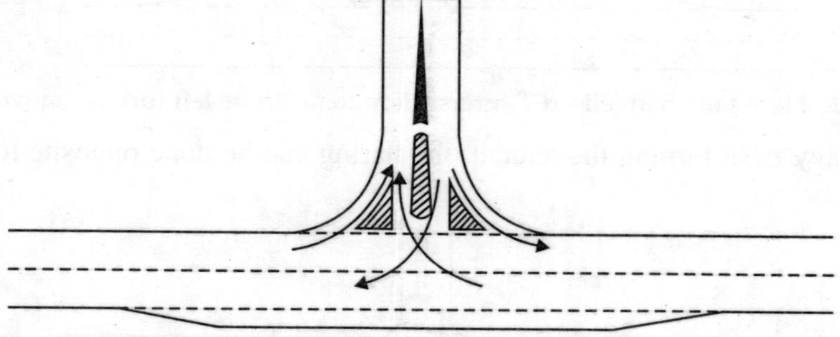

Fig. 4.46: Channelised T-intersection with a divisional island and two turning roadways

Fig. 4.47 shows a channelized T-junction with 2-way turning roadways. It is usually provided with signal control

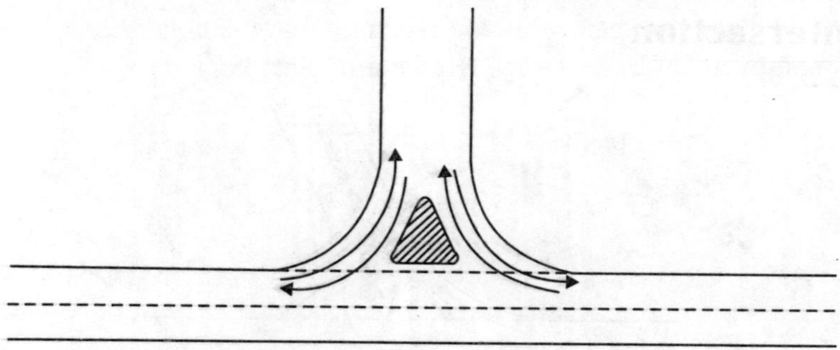

Fig. 4.47: Channelised T-intersection with 2-way turning roadways

Fig. 4.48 shows a channelized T-intersection on a divided highway facility providing for separate right turning lane.

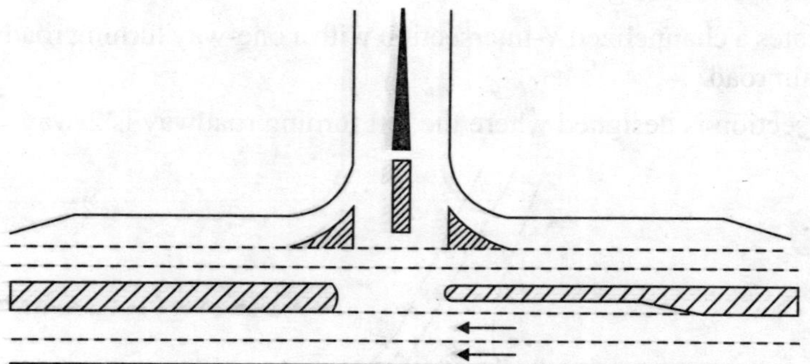

Fig. 4.48: Channelized T-Intersection on a divided highway

Fig. 4.49 shows a bulb-type channelized T-intersection suitable for high speeds. The median width on the through highway should be at least 12 m. The inner radius of the bulb should be at least 15 m, preferably 25 m.

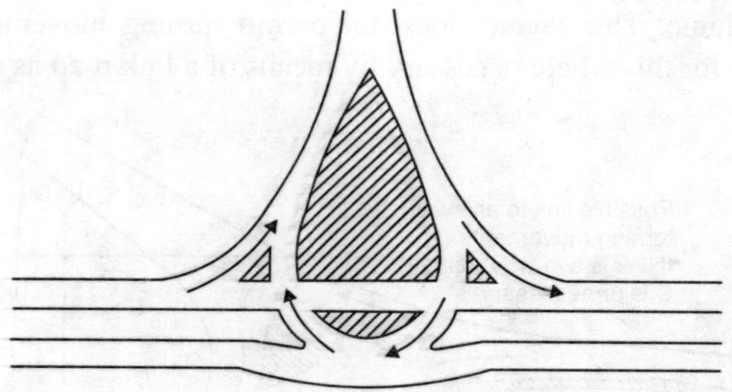

Fig. 4.49: Bulb-Type channelized T-intersection

4.12.13 Y-Intersection

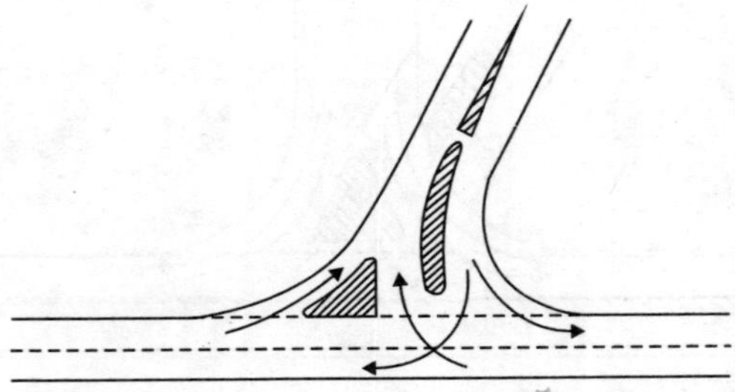

Fig. 4.50: Channelised Y-inter Section with one-way left turning roadway from the main road

For most Y-intersections, the design can suitably follow the same features as for T-intersections, except where the intersection angle is very acute.

Fig. 4.50 indicates a channelized Y-intersection with a one-way turning roadway for left turning traffic from the main road.

Fig. 4.51 intersection is designed where the left turning roadway is 2-way.

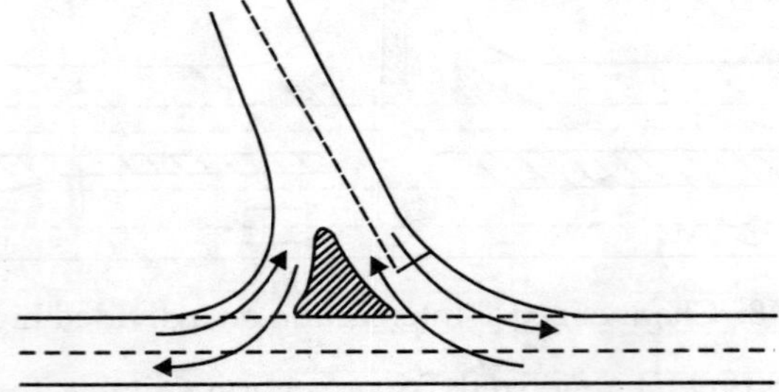

Fig. 4.51: Channelised Y-intersection with 2-way left turning roadway from the main road

Fig. 4.52 shows a high type channelized Y-intersection, where the branches of Y-intersection carry equal traffic volume. This layout does not permit turning movements from A to B, but provision can be made for this where necessary by means of a link road as shown.

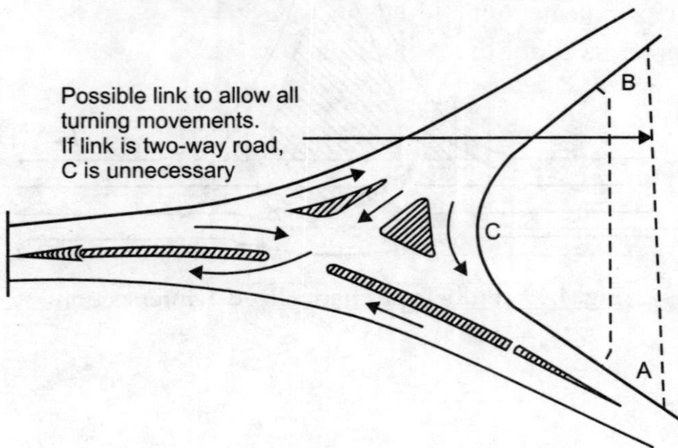

Fig. 4.52: High type channelizedi Y-intersection

Another high type channelized Y-intersection is shown in Fig. 7.55.

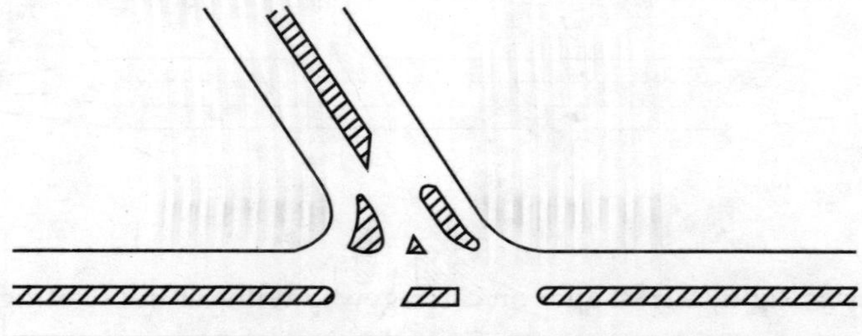

Fig. 4.53: High type channelized intersection of the Y-type

4.12.14. Staggered intersection

Many cross-roads can be conveniently converted into a properly designed staggered intersection. This will increase the safety. A simple right/ left stagger suitable for an unimportant junction is shown in Fig. 4.54. A minimum distance of 35 m is necessary between the staggered roads. A right/ left stagger is preferable to left/ right stagger where there are no separate right-turning lanes in the main road.

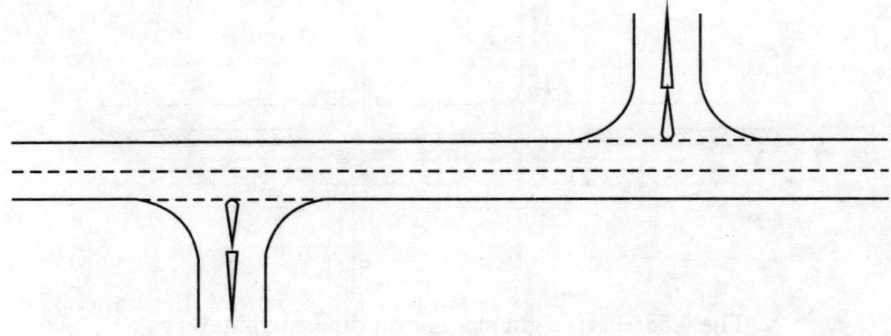

Fig. 4.54: Simple right/left stagger

Figs. 4.55 and 4.56 indicate suitable layout for staggered junctions on single carriageways, where a central lane is reserved for the use of right turning traffic.

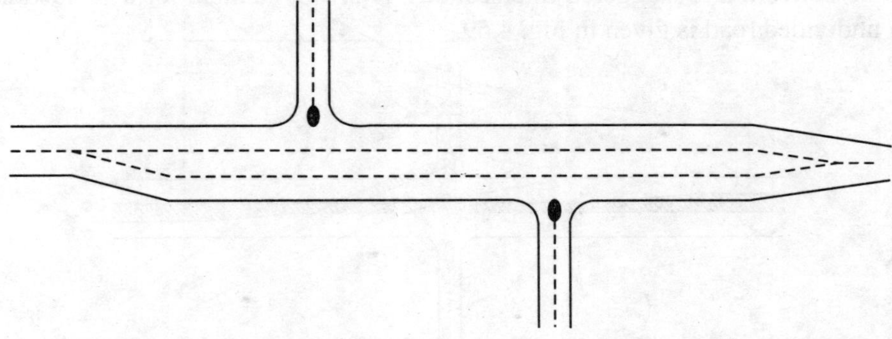

Fig. 4.55: Left/right stagger on single carriageway with central waiting areas

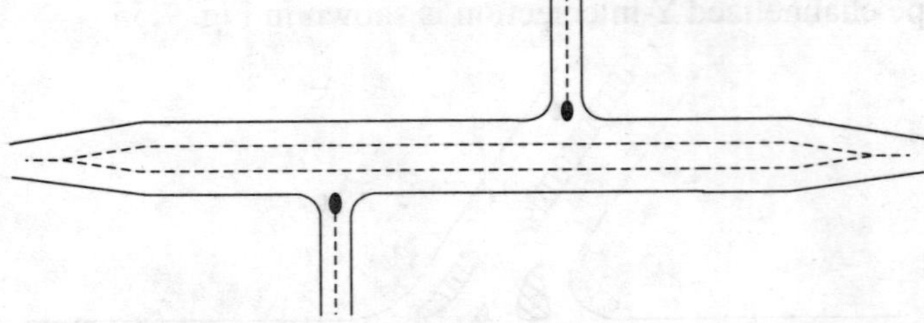

Fig. 4.56: Right/left stagger on carriageway with central waiting areas

For dual carriageways, the arrangements in Figs. 4.57 and 4.58 can be adopted.

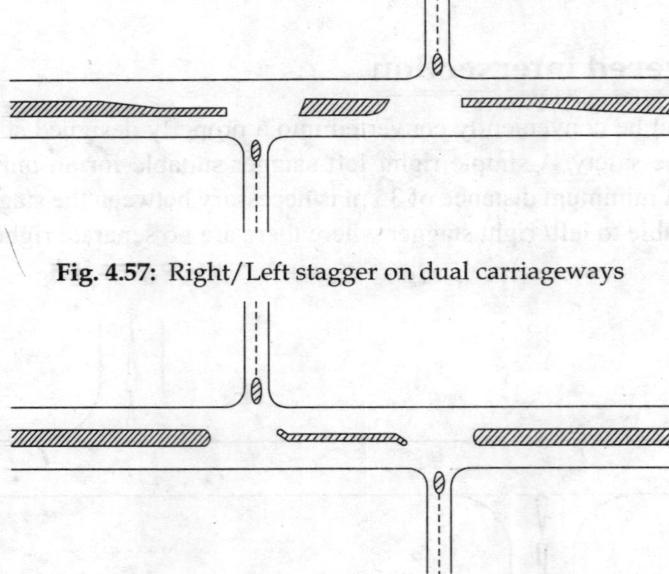

Fig. 4.57: Right/Left stagger on dual carriageways

Fig. 4.58: Left/right stagger on dual carriageways

4.12.15. Cross-roads

In general, uncontrolled cross-roads are unsatisfactory due to the high accident rate. If possible, they should be converted to staggered intersections. A layout suitable for a divided carriageway crossing on undivided road is given in Fig. 4.59.

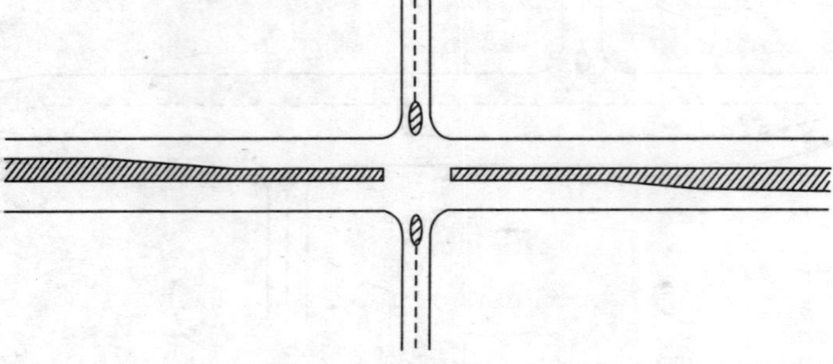

Fig. 4.59: Cross-road

4.12.16. Rotary intersection (Roundabout)

4.12.16.1. General

A rotary intersection is a specialised form of at grade intersection laid out for movement of traffic in one direction round a central island. The vehicles from the converging areas are forced to move around the central island in a clockwise direction (in countries like India where the left-side rule of traffic prevails) in an orderly manner and weave out of the rotary movement into their desired directions.

4.12.16.2. Definition of terms

A sketch of the rotary along with some of its elements is given in Fig. 4.60.

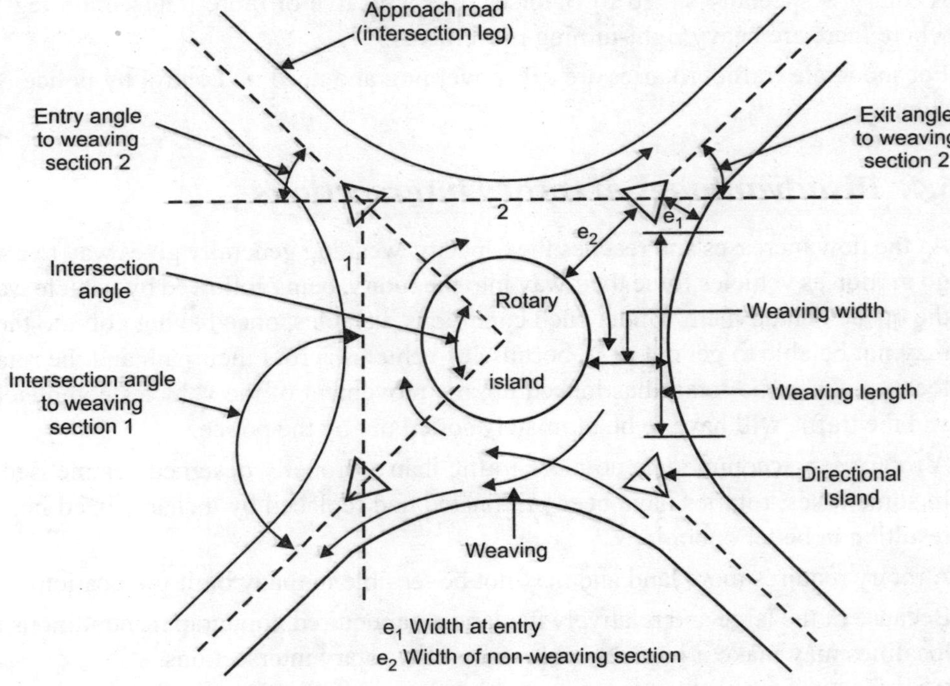

Fig. 4.60: Rotary elements

The definitions of some of the common terms in rotary design are given below:

Diverging	:	The dividing of a single stream of traffic into separate streams
Merging	:	The converging of separate streams of traffic into a single stream
Weaving	:	The combined movement of merging and diverging of traffic streams moving in' the same general direction
Weaving length	:	The length of a section of a rotary in which weaving occurs

2.16.3. Advantages of rotary intersections

advantages of rotary intersections are:

(*i*) An orderly and regimented traffic flow is provided by rotary one-way movement. Individual traffic . movements are subordinated in favour of traffic as a whole.

(*ii*) Normally, all traffic proceeds simultaneously and continuously at fairly uniform, though low speed. Frequent stopping and starting are avoided.

(*iii*) Weaving movements replace the usual angular crossing of typical at-grade intersections. Direct conflict is eliminated, all traffic streams merging or leaving at small angles. Accidents that occur from such movement are of minor nature, usually involving property damage only.

(*iv*) All turns can be made with ease, although little extra travel distance is required for all movements except left turns.

(*v*) A rotary is specially suited for intersections with five or more intersection legs, and/or where there are heavy right-turning movements.

(*vi*) For moderate traffic, rotaries are self-governing and need no control by police or traffic signals.

4.12.16.4. Disadvantages of rotary intersections

(*i*) As the flow increases and reaches the capacity, weaving generally gives way to a stop and go motion as vehicles force their way into the rotary, being followed by vehicle waiting in the queue behind them. Under such conditions, vehicles, once having got into the rotary, may not be able to get out of it, because of vehicles across their path and the rotary may lock up. Once the rotary has locked up, the movement of the vehicles completely stops and the traffic will have to be ultimately sorted out by the police.

(*ii*) A rotary can accommodate no more traffic than a properly designed channelised layout. In some cases, rotaries have been eliminated and replaced by a channelized intersection resulting in better operations.

(*iii*) A rotary requires more land and may not be feasible in many built-up locations.

(*iv*) Because of the large and relatively flat land area required, topographic conditions in some localities may make it impracticable to develop rotary intersections.

(*v*) Where pedestrian traffic is large, a rotary by itself is not sufficient to control traffic and has to be supplemented by traffic police.

(*vi*) When used on high speed roads, rotaries require extremely large size.

(*vii*) Where the angle of intersection between two roads is too acute, it becomes difficult to provide adequate weaving length.

(*viii*) When provided at close intervals they make travel troublesome.

(*ix*) Traffic turning right has to travel a little extra distance.

(*x*) A rotary requires many warning and directional signs for safety. The central island and entrances and exits must be well lighted at night. These tend to make it costly.

(*xi*) The rotary is not readily adaptable to stage development. Attempts at stage development generally result in some over-design when viewed from immediate traffic needs.

4.12.16.5. Guidelines for selecting a rotary type of intersection

Considering some of the above advantages and disadvantages, the following general guidelines may be kept in view when selecting a rotary type of intersection:

(*i*) Rotaries are most apt where volumes entering from different intersection legs are approximately equal.

(*ii*) A total volume of 3000 vehicles per hour entering from all the intersection legs appears to be the maximum practical capacity of high type rotaries.

(*iii*) A rotary design is most appropriate when the proportion of turning traffic is very high. It is considered that at a four-leg intersection, a rotary may need less land than traffic signals if right-turning traffic exceeds about 30 per cent of all approaching traffic.

(*iv*) A rotary is a good choice when there are more than four approaches to the junction.

(*v*) When the approach widths are restricted and it is impossible to provide separate lanes for through and turning traffic, a rotary may be desirable.

(*vi*) Rotaries are not generally warranted for intersections carrying very light traffic. Normally, the lowest traffic volume for which a rotary design should be considered is about 500 vehicles per hour.

(*vii*) Rotaries should be a good choice for moderately busy intersections in urban and suburban areas, where otherwise the alternative may be to go in for a complicated channelized layout or traffic signals.

4.12.16.6. Shape of rotary island

The shape and layout of the rotary island depends upon the layout of the intersecting roads and the traffic flow pattern. The following brief remarks about the different types in common use are intended to be general guidelines.

Type	Remarks
1. Circular	Suited where roads of equal importance carrying approximately equal volumes intersect at nearly equal angles.
2. Squarish with rounded edges	Suited for predominantly straight ahead flows.
3. Elliptical, elongated, oval or rectangular	Intended to favour through traffic. They may be provided to suit the geometry of the intersecting legs or to provide long weaving lengths. Elongated central island will be approximate for a scissors junction.
4. Irregular	Shape is dictated by the existence of a large number of approaches.

Figure. 4.61 indicates the different shapes.

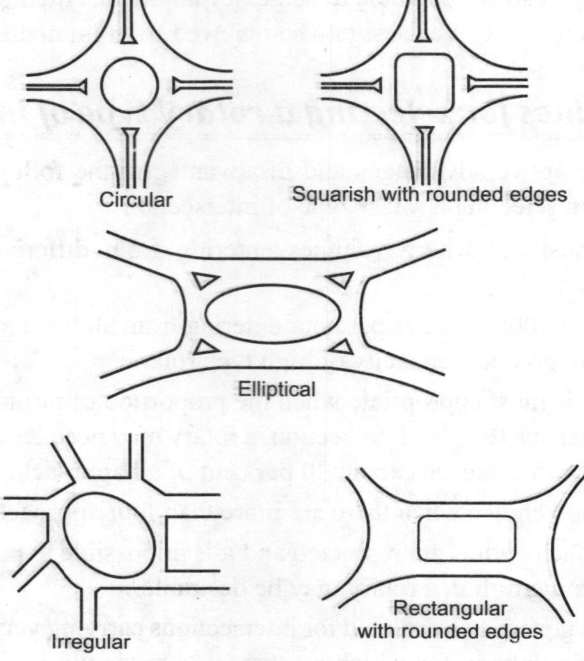

Fig. 4.61: Shapes of rotary island

4.12.16.7. Rotary design elements

4.12.16.7.1. Design speed

The design speed of a rotary governs the various elements such as radii and weaving lengths. To keep the dimensions within practical limits the design speed has to be necessarily low, as compared to the design speed of the intersecting highways. Current Indian practice is to design rotaries in rural areas for a speed of 40 K.P.H. and those in urban areas to a speed of 30 K.P.H.

4.12.16.7.2. Radius at entry

The radius at entry is determined by the design speed, super-elevation and coefficient of friction. The radius should result in the drivers lowering their speeds to a safe value, which is the design speed of the rotary. A range of 2.0-35 m is found to be suitable for rural designs arid a range of 15-25 m is suitable for urban design.

4.12.16.7.3. Radius at exit

The exit radius should be higher than the radius of the rotary island so that it favours a higher speed by drivers. This will enable the rotary to be cleared rapidly. The general practice is to keep the radius of exit curves 1.5 to 2 times the radius of the entry curves. If, however, there is a large pedestrian traffic across the exit road, radii similar to those at entrances may be provided to keep the exit speeds reasonably low.

4.12.16.7.4. Radius of the central island

The radius of the central island is governed by the rotary design speed and theoretically it should be equal to the radius at entry. In practice, however, the radius of the central island may be kept slightly larger than that of the curve at entry, this being an attempt to give a slight preference to the traffic already on the rotary and to slow the approaching traffic. A value of 1.33 times the radius of entry curve is probably adequate for this purpose.

4.12.16.7.5. Weaving lengths

The weaving length determines the ease with which the traffic can merge and diverge. The weaving length is decided on the basis of factors such as the width of the weaving section, the average width of entry, total traffic and the proportion of weaving traffic in it.

As a general rule, it is desirable to prevent direct traffic cuts and this can be achieved by making the ratio of the weaving length to the weaving width large enough. A ratio of 4:1 is regarded as a minimum. Current Indian practice as regards weaving length is summarised in Table 4.28.

Table 4.28: Minimum length of weaving section

Design speed (K.P.H.)	Minimum weaving length (m)
40	45
30	30

It is also necessary to put a limit to the maximum weaving length to discourage speeding. Twice the values given above can serve as the upper limits.

4.12.16.7.6. Width of carriageway at entry and exit

The carriageway width of the intersection legs is governed by the design year traffic entering and leaving the intersection.

A minimum width of carriageway of 5 m, with provision for extra widening due to curvature, may be provided for the entrance and exit. Table 7.32 summarises current Indian practice.

Table 4.29: Width of carriageway at entrance and exit

Carriageway width of approach road	Radius at entry (m)	Width of carriageway at entry and exit (m)
7 m (2 lanes)	25–35	6.5
10.5 m (3 lanes)		7.0
14 m (4 lanes)		8.0
21 m (6 lanes)		13.0
7 m (2 lanes)	15–25	7.0
10.5 m (3 lanes)		7.5
14 m (4 lanes)		10.0
21 m (6 lanes)		15.0

4.12.16.7.7. *Width of rotary carriageway*

Fig. 4.62 indicates the design elements.

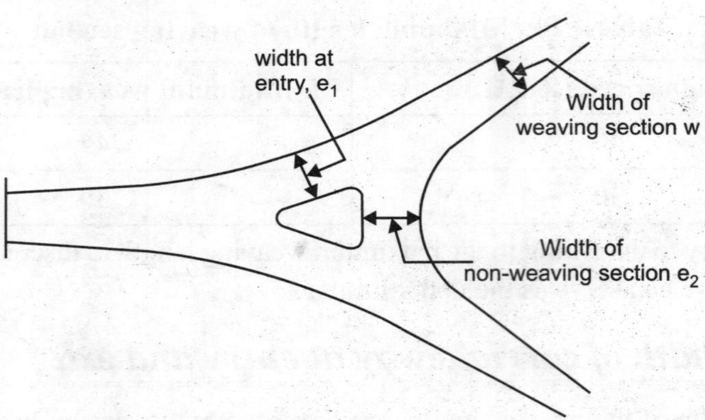

Fig. 4.62: Width of rotary carriageway

The width of the non-weaving section should be equal to be widest single entry into the rotary, and should generally be less than the width of the weaving section. The width of the weaving section (*w*) should be one traffic lane (3.5 *m*) wider than the mean entry width thereto.

Thus $w = \frac{e_1 + e_2}{2} + 3.5$ with reference to Fig. 4.62.

4.12.16.7.8. *Entry and exit angles*

Entry angles should be larger than exit angles and it is desirable that the entry angles should be about 60° if possible; the exit angles should be small, even tangential. A design with an entry angle of 60° and exit angle of 30° is shown in Fig. 4.63.

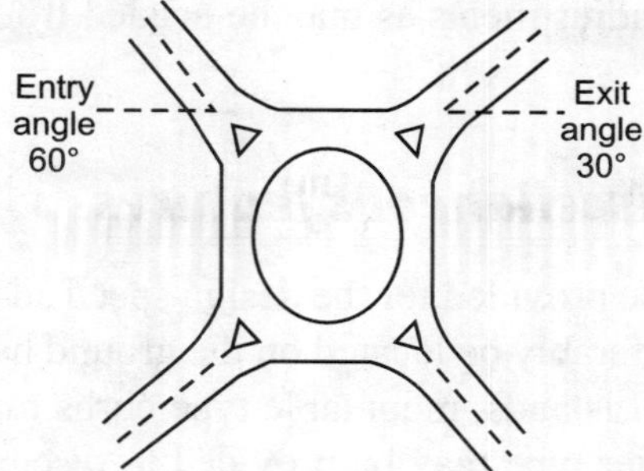

Fig. 4.63: An ideal layout with 60° entry angle and 30° exit angle

4.12.16.7.9. External curb line

The external curb line of the weaving sections should not be normally re-entrant, but consist of a straight or large radius curve, same as the entry and exit curves. Such an arrangement eliminates waste of area, which is unlikely to be used by traffic, Sec Fig. 4.64.

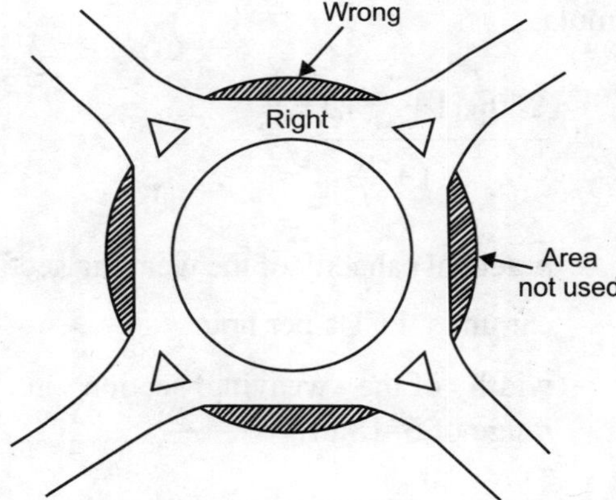

Fig. 4.64: External curb line of weaving sections

4.12.16.7.10 Super-elevation and camber

The rotary curvature being opposite to that of entry and exit, the super-elevation in the two portions will be opposite to each other. This causes difficulty to vehicles, especially double-decker buses and top heavy trucks, in changing from one cross-slope to another. To cause the least discomfort, the algebraic difference in the cross-slopes should be limited to about 0.07. The individual superelevations should be limited to the least amount consistent with speed. The crown-line, which is the line of meeting of the opposite cross-slopes should, as far as possible, be located such that the vehicles cross it while travelling along the common tangent to the reverse curve.

Channelizing islands should be situated on the peak with the road surfaces sloping away to all the sides. The cross-slope at entrance should be carried around on the outer edge of the rotary to the adjacent exit making slight adjustments as may be needed from the point of view of curvature and speed.

4.12.16.7.11. *Other miscellaneous features*

Stopping sight distance should be provided for the design speed adopted. Values may be taken from Table 4.9. The rotary should preferably be located on the ground having a slope flatter than 1 in 50. For the channelizing and central islands, mountable type curbs may be provided, whereas for the outer edges of the rotary a barrier type may be provided to discourage pedestrians from crossing over. Landscaping and drainage requirements should be considered.

4.12.16.8. *Capacity*

The capacity of a rotary is directly determined by the capacity of each weaving section. The capacity of a weaving section is determined by the geometric layout, including entrances and exits, and the percentage of weaving traffic. The Transport and Road Research Laboratory (U.K.) which has pioneered research on this aspect, recommends the following formula, which is a modification of the well-known Wardrop formula:

$$Q_p = \frac{280w\left(1+\frac{e}{w}\right)\left(1+\frac{p}{3}\right)}{1+\frac{w}{l}} \tag{4.40}$$

where Q_p = practical capacity of the weaving section of the rotary in passenger car units. PCUs per hour

w = width of the weaving section in metres (within the range of 6–18 m)

e = average entry width of the rotary in metres

$= \frac{e_1 + e_2}{2}$ (Ref. Fig. 4.62)

l = length of weaving section between the ends of the channelizing islands in metres

p = proportion of weaving traffic, *i.e.*, ratio of sum of crossing streams to the total traffic on the weaving section.

$= \frac{b+c}{a+b+c+d}$ as in Fig. 4.65

In the above formula, the equivalency factors recommended by TRRL and IRC are given in Table 4.30.

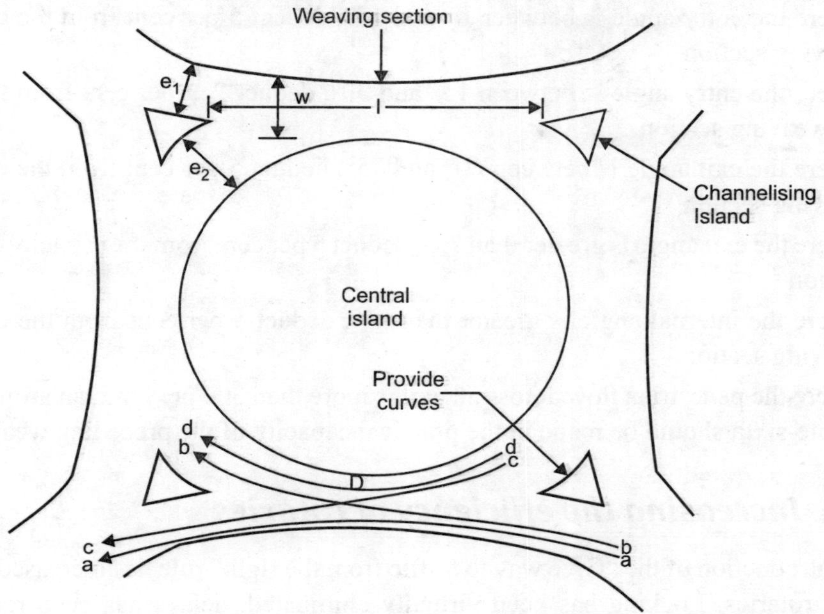

Fig. 4.65: Relevent dimensions of weaving section and proportion of weaving traffic for use in capacity formula for rotaries

Table 4.30: Equivalency factors for use in rotary design

	Vehicle Type	PCU
1.	Cars and light commercial vehicles, including 3-wheelers	1.0
2.	Buses and medium heavy commercial vehicles	2.8
3.	Motorcycles and scooters	0.75
4.	Pedal cycles	0.5
5.	Animal drawn vehicles	4 to 6

The above formula is valid under the following conditions:

(*i*) There are no standing vehicles on the approaches to the rotary

(*ii*) The site of the level and approach gradients do not exceed 1 in 25

(*iii*) $\frac{e}{w}$ should be between 0.4 and 1.0

(*iv*) $\frac{w}{l}$ should be between 0.12 and 0.40

(*v*) p should be between 0.4 and 1.0.

(*vi*) l should be between 18 and 90 m

In the above formula, certain adjustments need to be made to take into account the geometric layout. These are:

(*i*) Where the entry angle is between 0° and 15°, deduct 5 per cent from the capacity of the weaving section

(*ii*) Where the entry angle is between 15° and 30°, deduct $2\frac{1}{2}$ per cent from the capacity of the weaving section

(*iii*) Where the exit angle is between 60° and 75°, deduct 5 per cent from the capacity of the weaving section

(*iv*) Where the exit angle is greater than 75°, deduct 5 per cent from the capacity of the weaving section

(*v*) Where the internal angle is greater than 95°, deduct 5 per cent from the capacity of the weaving section

(*vi*) Where the pedestrian flow across an exit is more than 300 per hour, an arbitrary reduction of one-sixth should be made in the practical capacity of the preceding weaving section.

4.12.16.9. *Increasing the efficiency of rotaries*

In U.K., the introduction of the "Give way to traffic from the right" rule has increased the efficiency and safety of rotaries. Locking has been virtually eliminated, delays has been reduced and the accident rate has been reduced by more than 40 per cent. Signals on the approaches can also increase the capacity and improve operation.

4.12.16.10. *An example of the design of a rotary intersection*

Problem: Traffic flow in an urban section at the intersection of two highways in the design year are given below:

Approach	**Left turning**			**Straight ahead**			**Right turning**		
	Cars	**Commercial**	**Scooters**	**Cars**	**Commercial**	**Scooters**	**Cars**	**Commercial**	**Scooters**
N	200	50	100	250	100	150	150	50	80
E	180	60	80	220	50	120	200	40	120
S	250	80	100	150	50	90	160	70	90
W	220	50	120	180	60	100	250	60	100

The highways at present intersect at right angles and have a carriageway width of 15 m. Design a rotary intersection making suitable assumptions.

Solution. The rotary is located in an urban section and hence a design speed of 30 K.P.H. is appropriate. Since the intersection legs carry almost equal traffic, a round shaped central island will be adopted. The entrance and exit angles will be 45° each. A radius of 20 m at entry, 40 m at exit and 80 m for the central island will be adopted. These values conform to current Indian standards discussed earlier. The traffic in terms of PCUs from each leg is depicted in Fig. 4.66.

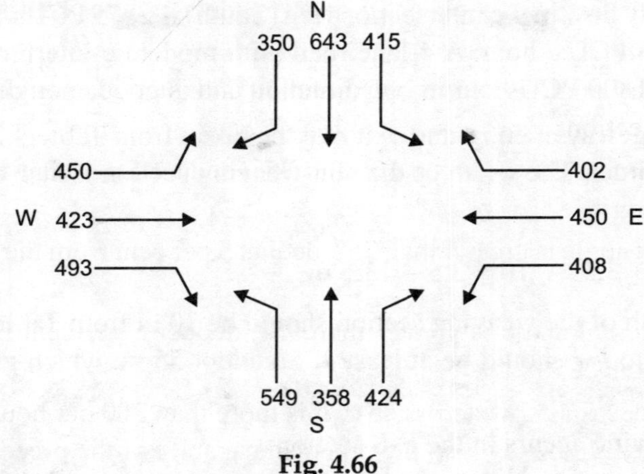

Fig. 4.66

The traffic in terms of PCUs on each leg is depicted in Fig. 4.67

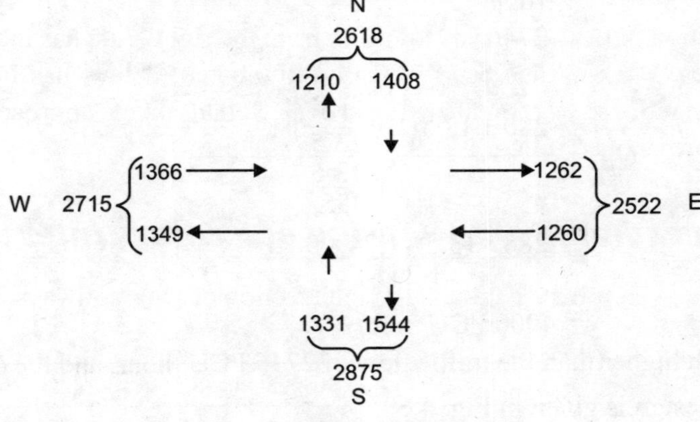

Fig. 4.67

The traffic is assigned to the network in Fig. 4.68

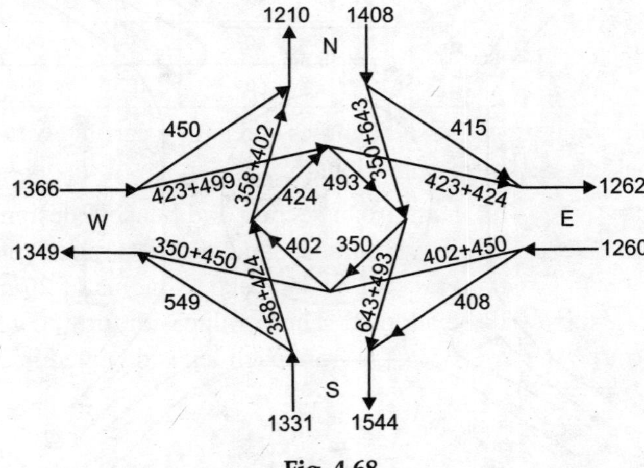

Fig. 4.68

The maximum 2-way flow in the intersections leg (South) is 2875 PCUs/hour, and the maximum in one direction is 1544 PCUs/ hour. A 4-lane road with moderate interference from cross-traffic has a capacity of 1400-1800 PCU/hour in one direction and is good enough.

The width of carriageway at entry and exit may be taken from Table 4.29. For a 4-lane road, a width of 10.0 m is . in order. The width of the non-weaving section is also kept 10.0 m. The width of the weaving section is:

$$\frac{e_1 + e_2}{2} + 3.5 = 10 + 3.5 = 13.5 \text{ m}$$

The minimum length of the weaving section should be 30 m from Table 4.28, but this fails the requirement that the ratio l/w should be at least 4. So adopt 55 m, which gives a l/w ratio slightly greater than 4.

The maximum weaving accurs in the E-S section

$$p = \frac{1136 + 852}{350 + 1136 + 852 + 408}$$

$$= \frac{1988}{2746} = 0.72$$

From Eq. (4.40)

$$Q_p = \frac{280 + 13.5\left(1 + \frac{10}{13.5}\right)\left(1 - \frac{0.72}{3}\right)}{1 + \frac{13.5}{55}}$$

$$= \frac{288 \times 13.5 \times 1.74 \times 0.76}{1.245}$$

$$= 4000 \text{ PCU/ hour}$$

This is very much higher than the traffic flow of 2746 PCU/ hour, and the design is acceptable. A sketch of the design is given in Fig. 4.69.

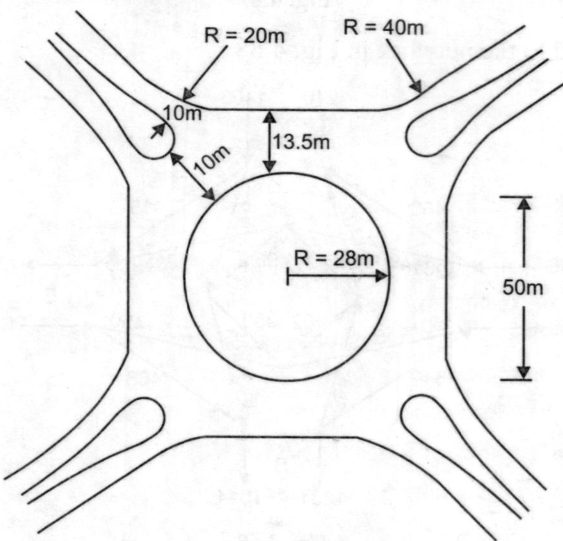

Fig. 4.69. Designed rotary intersection

4.12.17. Mini roundabouts

4.12.17.1. Introduction

A serious drawback of the conventional (large diameter island) roundabout is that very often the traffic locks itself, A locked roundabout causes large traffic delays and unlocking can only be achieved by means of police help. The conventional round-about is thus not truly self-regulatory at all times.

In the United Kingdom, which is one of the countries where roundabouts still retain their popularity, a major change in the approach to roundabout design was the off-shoot of the introduction of the priority from-the-right rule in 1966. This rule was intended to get over the drawback of locked roundabouts, but other advantages soon began to offer themselves. The absence of weaving under the new rule made possible a drastic reduction in the size of the central island and an availability of greater carriageway space with the existing layout. The width of approach at the entry to the junction could be increased considerably thus rendering multi-vehicle flow from Give-Way line. The mini roundabout has made its appearance and was soon to become very popular in many cities in the U.K. Experiments showed that the use of small diameter islands could bring in higher capacities than large diameter roundabouts. This has formed the basis of current U.K. practice.

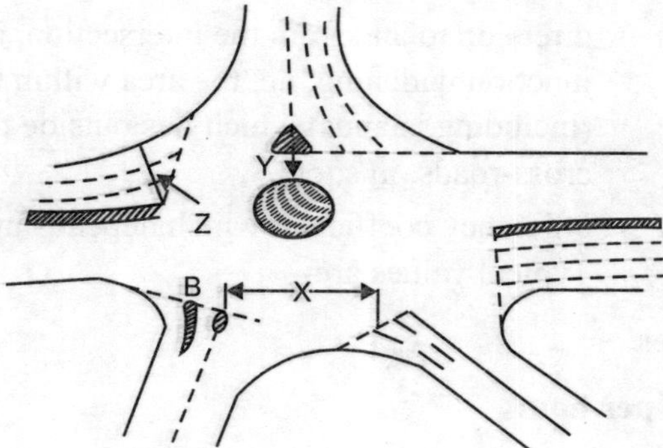

Fig. 4.70: Typical mini-roundabout as per U.K. practice

4.12.17.2. Principles of design

The basic principles of design of a mini-roundabout illustrated in Fig. 4.70 are:

(*i*) Provision of small diameter island of a diameter of about one-third than that of a hypothetical circle inscribed . within the outer carriageway boundaries, but normally not less than 8 m. For new layouts, .space for a larger diameter (15 to 25 m) may be reserved if such a provision becomes necessary in future.

(*ii*) An increase in the number of lanes at the Give-Way line.

(*iii*) A minimum stopping distance of 26 m between the Give-Way line and the point of conflict with a vehicle from the left. This is the dimension X in Fig. 4.70.

(*iv*) A width between traffic islands and the roundabout (shown as dimension Y in Fig. 4.70) which is not less than the total lane width at the entry preceding it, *i.e.*, dimension shown as Z in Fig. 4.70.

(*v*) An entry taper (about 1 in 6) that is about twice sharp as the exit taper (about 1 in 12).

(*vi*) A deflection island (shown as B in Fig. 4.70) is intended to ensure that straight-through movements do not

(*vii*) To maintain junction capacity, a single lane approach should be tapered to give at least three lanes at the junction and a 2-lane approach tapered to give at least four lanes at the junction. At the exits, the merge should normally be from four lanes to two lanes and two lanes to one lane.

4.12.17.3. Blackmore formula

Blackmore gives the capacity of a mini-roundabout as:

$$q = K\left(\sum w + a^{1/2}\right) \tag{4.41}$$

where q = total entry volume in PCU/hour

$\sum w$ = sum of the basic road width (not half widths) used by traffic in both directions to and from the intersection, in metres.

a = direction to and from the intersection, in metres, area of the junction widening, i.e. the area within the intersection outline (including islands) which lies outside the area of the basic cross-roads, in sq. m.

K = efficiency coefficient which depends upon site conditions. Typical values are:

	K Values (PCUs) per hour
3 way junction	80
3 way junction	70
≥ 5 way junction	65

The practical capacity is about 80 per cent of that given by the above equation.

4.12.17.4. Advantages of mini-roundabouts

(*i*) **Increase in capacity:** Wherever mini-roundabouts have replaced large diameter roundabouts, a general increase in the capacity of the junction has been noted. As compared to signalised junctions too, mini-roundabouts can bring about higher capacities.

(*ii*) **Accident rates:** Typical accident rates before and after conversion of junctions into mini-roundabouts are yet to be fully analysed and evaluated, but a trend in decrease in the number of fatal and serious injury accidents has been noticed.

4.12.17.4.5. *Limitations of mini-roundabouts*

As indicated earlier, mini-roundabouts are the result of the introduction of the priority-from-the-right rule. They can be introduced only in situations where such rules are in force and are strictly followed. Non-observance of this rule can only lead to chaos and serious accidents.

The mini-roundabout requires careful considerations on the design speed and ways of reducing speeds at junctions to this value. A very high standard of pavement markings and road signing, a great care in provision of guide islands, alignment of the kerbs and location of the central island are necessary for the satisfactory functioning of this type of junctions.

Adequate visibility to the approaching drivers is of prime importance. This alone will enable him to give way to the traffic on his right and to check on the traffic on his left for giving way to him.

4.13 GRADE SEPARATED INTERSECTIONS

4.13.1. Criteria for provision

The general site conditions favouring a grade-separated intersection have been enumerated earlier in Section 4.12.2. Current Indian practice requires grade separation across streets and highways to be provided as per the following criteria:

- **Urban streets.** Grade separation to be provided if the estimated traffic volumes within the next 5 years are in excess of the capacity of the intersection; when traffic projections show that volumes within the next 20 years will exceed the capacity of an at-grade intersection, the need for a grade separated facility should be kept in view for future construction.
- **Rural highways.** Grade separation should be provided at intcr-scctions of divided rural highways if the ADT (fast vehicles only) on the cross-road within the next 5 years exceeds 5000. Where this figure will be reached within the next 20 years, the need for such facilities should be kept in view.
- **Grade-separation is of great importance across railway lines.** Current Indian practice for rural highways requires grade separation to be provided across existing railway lines if the product of ADT (fast vehicles only) and the number of trains per day exceeds 50,000 within the next 5 years. For new constructions such as bypasses, grade separations should be provided when this figure is greater than 25,000.

4.13.2. Types of grade-separated intersections

Basically two types are met with:

(*i*) Grade-separated intersections without interchange

(*ii*) Grade-separated intersections with interchange

Interchange is a system whereby facility is provided for movement of traffic between two or more roadways at different levels in the grade separated junction. A structure without interchange is an over-bridge or underpass or flyover, whereby the traffic at different levels moves separately without a provision for an interchange between them.

The different forms of a grade-separated junction can be considered under the number of legs the intersection serves. Thus the interchanges can be classified as three-leg, four-leg and mult-eg, and these, in turn, can be subdivided into various types as below:

A. Three-leg interchange
1. *T* interchange
2. *Y* interchange
3. *A* partial rotary interchange

B. Four-leg interchange
1. Diamond interchange
2. Half clover leaf interchange
3. Clover leaf interchange
4. Rotary interchange
5. Directional interchange

C. Multi-leg interchange
1. Rotary interchange

These types are discussed below, in detail.

4.13.3. Three-leg Intetchange

If one of the intersection legs meets a highway at approximately a right angle but does not cross it, and if the junction is provided with interchanging facility, the junction is called a *T* interchange. If the intersection angle is acute or obtuse, the junction is called a *Y* interchange. A widely used form of a *T* or interchange is a trumpet, shown in Fig. 4.71(*a*) and (*b*). The usual form is (*a*) but (*b*) can also be adopted under particular site conditions, but this has the drawback that vehicles leaving the major road have to turn through a small radius.

These designs can be adopted for junctions of major streets with expressway.

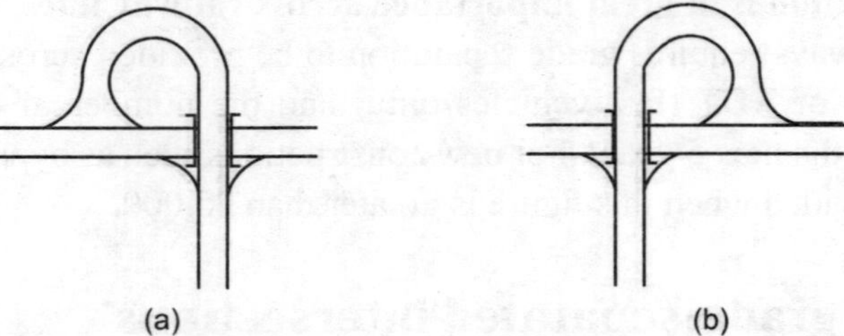

Fig. 4.71: Trumpet interchange

Designs shown in Fig. 4.72 (*a*), (*b*) and (*c*) are Y-shaped and these permit limited movements.

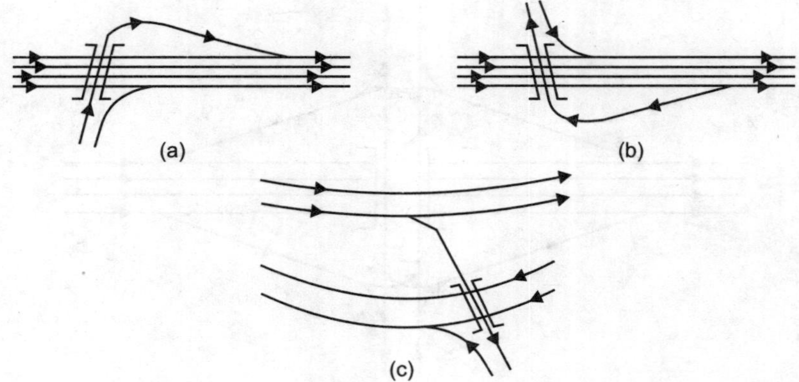

Fig. 4. 72: Y-shaped interchanges

A design which can be converted to full cloverleaf at a later date when the junction is likely to become four-leg, is indicated in Fig. 4.73. As an alternative to the arrangement suggested in Fig. 4.73, a partial bridged rotary intersection can be adopted, Fig. 4.74.

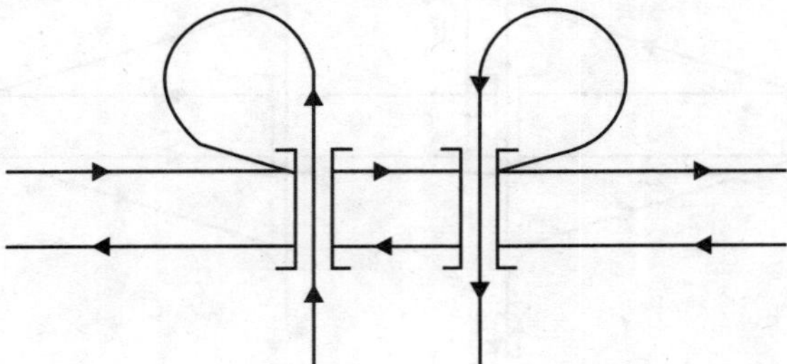

Fig. 4.73: Three-leg interchange which can be converted to a cloverleaf

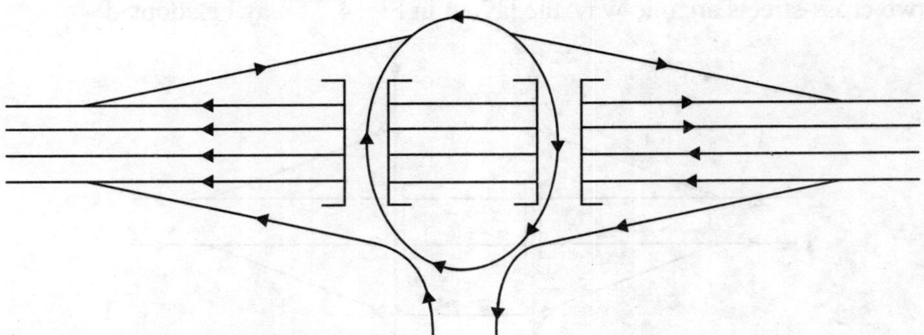

Fig. 4.74: Partial bridged rotary

4.13.4. Diamond interchange

A popular form of interchange in urban locations involving a major-minor crossing is a diamond interchange. It can be designed for a relatively narrow right-of-way of the major road. A simple form of diamond interchange is shown in Fig. 4.75. A split-diamond layout Fig. 4.76, can be adopted where two parallel cross-streets are available as connections to the main street.

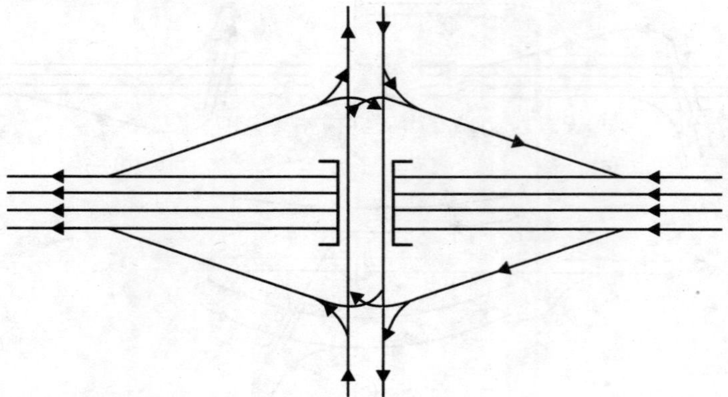

Fig. 4.75: Diamond interchange

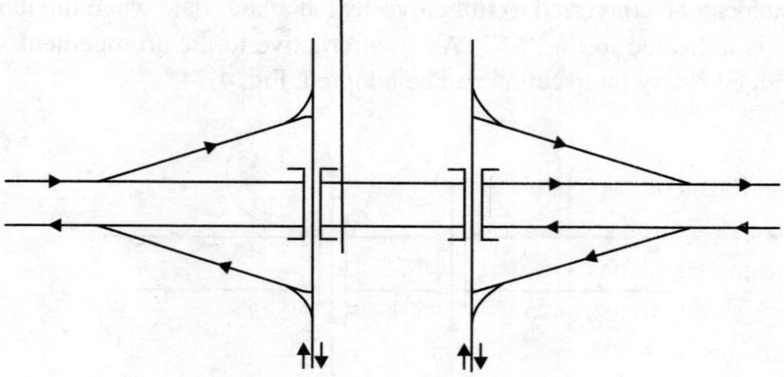

Fig. 4. 76: Split diamond interchange

If the two cross-streets arc one-way, the layout in Fig. 4.77 may be adopted.

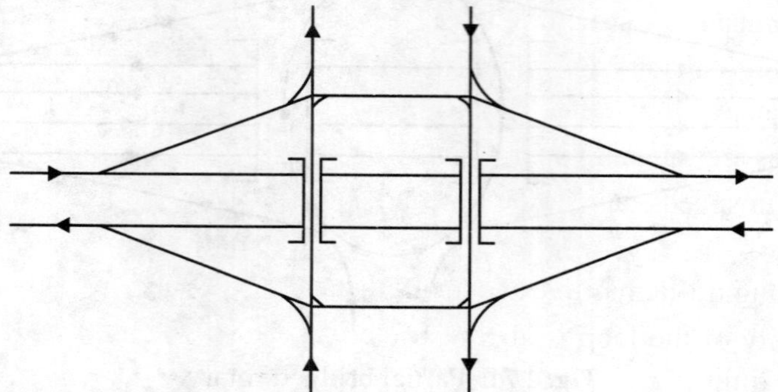

Fig. 4. 77: Split diamond interchange with one-way streets

4.13.5. Half cloverleaf interchange

When a major road crosses a minor road (not more than 3 lanes), it is possible to provide a partial cloverleaf. Two such designs are given in Fig. 4.78 (*a*) and (*b*).

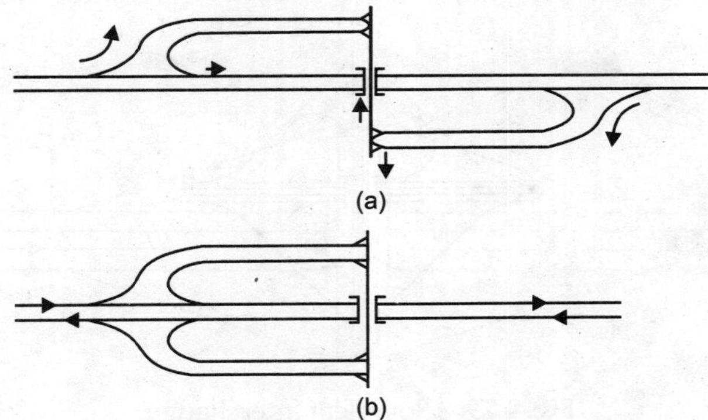

Fig. 4.78: Partial cloverleaf for major-minor Road crossing

4.13.6. Cloverleaf interchange

The cloverleaf interchange is a 4-leg interchange with a single structure and is very popular. It is used when two high volume and high speed facilities intersect each other. Some of its advantages are:

(*i*) Through traffic on both roads is unimpeded

(*ii*) Only one structure is required

(*iii*) Left-turning traffic has a direct path

(*iv*) It is very simple to use and does not confuse the drivers

The disadvantages of this type of interchange are:

(*i*) Relatively large area is needed

(*ii*) The carriageway area required is also higher than a rotary interchange

(*iii*) Weaving manoeuvres are involved, some of them on the roadway of the structure and some underneath the structure.

(*iv*) The weaving capacity is limited to about 1200 PCUs/hour, but may be increased by providing an ancillary lane.

(*v*) The U-turns are long and operationally difficult.

(*vi*) Loop design speeds have to be low and speeds above 50 KPH are likely to increase the cost.

(*vii*) Right-turning traffic has to travel extra distance.

(*viii*) The capacity of the loop is also restricted. A capacity of 800-1200 vehicles per hour is almost the limit.

Fig. 4.79 gives the layout of cloverleaf interchange.

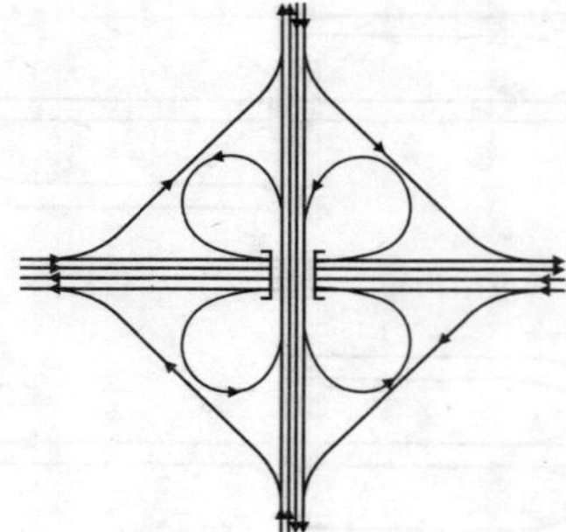

Fig. 4.79: Cloverleaf interchange

4.13.7. Rotary interchange

A rotary interchange is another form of 4-leg interchange popular in some countries, especially U.K. It can also be used for multi-leg interchange. Fig. 4.80 gives the layout.

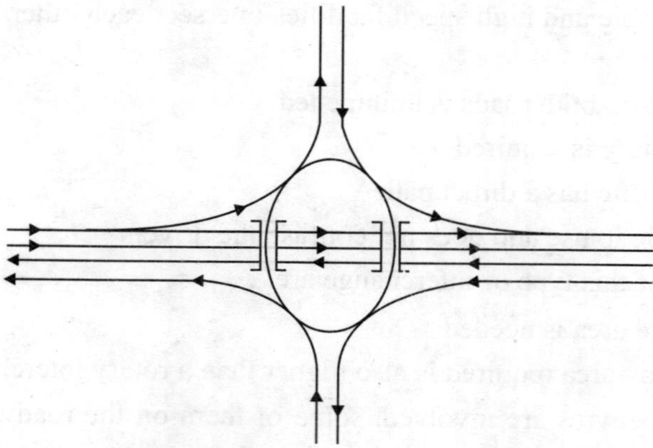

Fig. 4.80: Grade-separated rotary interchange

The advantages claimed for this type of interchange are:

(*i*) It occupies relatively less land area

(*ii*) The carriageway area is also less than many other types

(*iii*) U-turns are easy

(*iv*) Suitable for most sites

The disadvantages are:

(*i*) The capacity is limited by the capacity of the roundabout itself.

(*ii*) The straight through traffic on one road is required to weave with turning traffic from the other road.

While the design in Fig. 4.80 above is for a 2-level interchange, a further improvement can be made by having a 3-level rotary interchange (Fig. 4.81).

In this arrangement through traffic on each road is unimpeded and the weaving takes place in the rotary only for the turning traffic. The number of structures required is five.

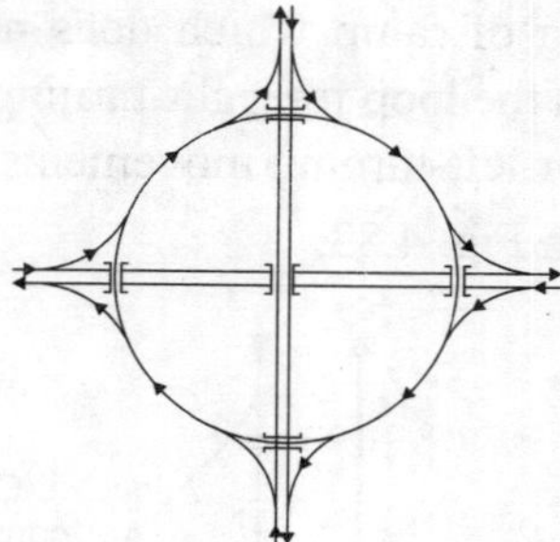

Fig. 4.81: Three-level rotary interchange

4.13.8. Directional interchange

An interchange which gives direct or semi-direct connections for the major right-turning movements is called a directional inter-change. These interchanges are rather complex and involve complicated multi-level structures.

Fig. 4.82 gives a typical design.

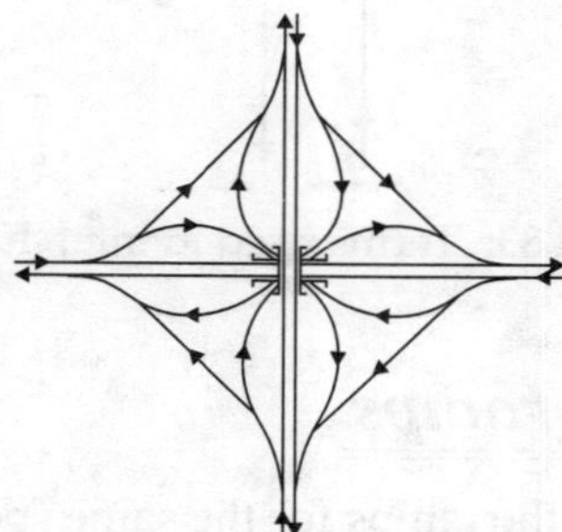

Fig. 4.82: Directional interchange

4.13.9. Design considerations of interchange

4.13.9.1. Definition of some common terms

In dealing with interchange design, some common terms are. met with, which are defined below:

1. ***Interchange Ramp*** **(Access Ramp or Slip road).** An interconnecting roadway of a traffic interchange or any connection between highway facilities at different levels, on which vehicles may enter or leave a designated roadway. It may be in the form of a loop, outer connection or direct connection.
2. ***Loop.*** A one-way turning roadway that curves about 270° to the left to accommodate a right-turning movement.
3. ***Outer Connection.*** A ramp used by traffic destined for left-turn movement from one of the through roadways separated by a grade separation structure to the second through roadway.

4. ***Direct Connection.*** A form of ramp which does not deviate much from the intended direction of travel. It avoids the loop for right-turning movements. An outer connection is a direct connection ramp for left-turning movements.

The above terms are illustrated in Fig. 4.83.

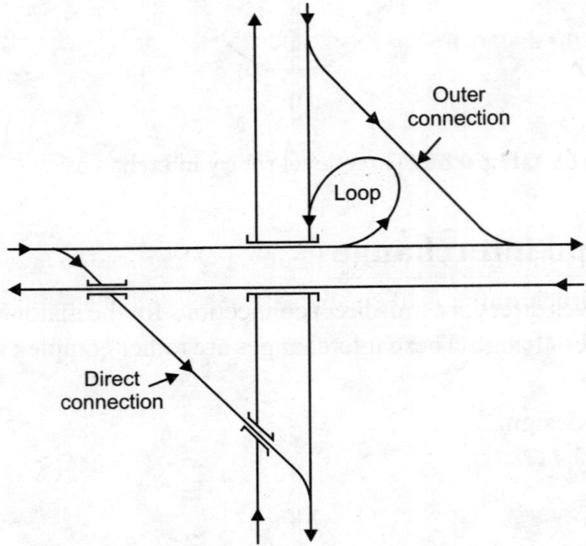

Fig. 4.83: Terms used in interchange

4.13.9.2. Design speed for ramps.

It is rarely, if ever, feasible to design the ramps for the same speed as the through roads. The design speeds recommended for ramps as per AASHTO are given in Table 4.31.

Table 4.31: Guide values for ramp design speeds

Highway design speed KP.H.	50	65	80	100	105	115	120	135
Ramp design speed KP.H.								
Desirable	40	55	70	80	90	100	100	105
Minimum	25	30	40	50	50	50	55	65
Corresponding minimum radius, m								
Desirable	45	90	170	210	260	320	320	380
Minimum	15	27	45	70	70	70	90	130

For design of loops, the speed should be about 40–50 K.P.H. Higher values are impractical in many locations. For direct connections, a design speed of 60–65 K.P.H. is appropriate.

4.13.9.3. Radü of curves on ramps

Table 4.31 indicates the radii corresponding to the design speed selected.

4.13.9.4. Sight distance on ramps

A minimum stopping sight distance appropriate for the design, speed, selected for the ramp design should he provided.

4.13.9.5. Gradients on ramps

AASHTO guidelines are:

(*i*) Ramp gradients should desirably be limited to 6 per cent in areas subject to snow and ice

(*ii*) Where heavy truck traffic is likely, the upward gradient should be limited to 4 per cent.

(*iii*) In exceptional cases, grades may be as steep as 10 per cent, on minor ramps and low volume ramps.

4.13.9.6. Width of ramps

Ramps should preferably carry one-way traffic only. The width of pavements should be selected as appropriate for different radii.

4.13.9.7. Capacity

The practical capacity of through lanes can be taken as 1500 PCUs per hour per lane for interchange design. For ramps, a capacity of 1200 PCUs per hour per lane is appropriate. The capacity of weaving sections may be calculated from the principles discussed in the Highway Capacity Manual, 1965.

4.14. SUMMARY OF DESIGN STANDARDS FOR RURAL EXPRESSWAYS

Expressways in rural sections are just making their entry into the Indian highway network. The following summarises the geometric design elements being adopted for these facilities:

1. Right-of-way : 90 m
2. Design Speed : 120 K.P.H.
3. Carriageway width : Dual carriageway 7.5 m, with provision for widening to 11 m.
4. Shoulders : 3.5 m wide, out of which 2.5 m width, is treated and 1.0 m is untreated.
5. Median : 6 m
6. Cross-fall: Carriageway ; 2.5 per cent Treated shoulder : 3.0 per cent Untreated shoulder: 4.0 per cent.
7. Safe stopping sight distance (*i*) desirable : 500 m
   (*ii*) minimum : 250 m

8. Horizontal curves : (*i*) Minimum Radius : 700 m
(*ii*) Desirable Radius : 2600 m
(*iii*) Minimum curve length : 240 m
9. Gradient: (*i*) Ruling : 1 in 50
(*ii*) Absolute maximum : 1 in 40
(*iii*) Absolute minimum : 1 in 800
10. Interchange design speed : (*i*) Desirable : 80 K.P.H.
(*ii*) Minimum : 60 K.P.H.
11. Interchange ramp gradient: (*i*) Desirable : 4 per cent
(*ii*) Absolute minimum : 6 per cent
12. Interchange carriageway : (*ii*) Desirable 7 m
(*ii*) Minimum 5.5 m

A. typical cross-section of an Expressway is given in Fig. 4.84.

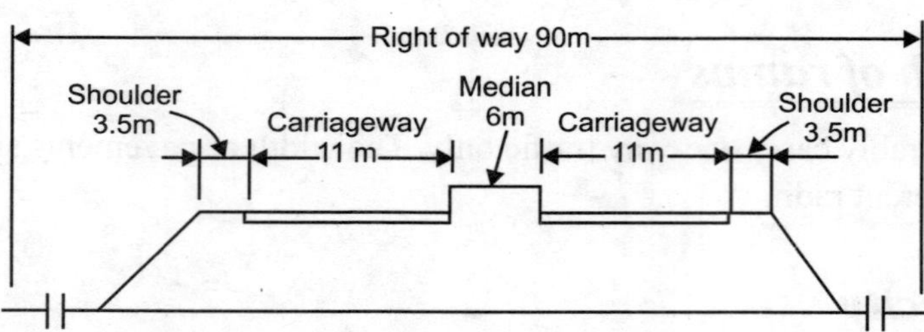

Fig. 4.84. Typical cross-section of an expressway.

QUESTIONS

1. In drawing up geometric design standards for the country, what are the considerations to be kept in view?
2. Describe the classification system of urban streets in India. Define the various classes of urban streets.
3. Describe the classification of rural highways adopted in India. Define the various classes.
4. Describe the basis for terrain classification in geometric design.
5. What is the Design Hour Volume?
6. What are the major dimensions and axle weights of the IRC Design Vehicle?
7. Define Design Speed in Geometric Design. What percentile speed is usually taken as Design Speed? What are the suggested Design Speeds in India for rural and urban conditions? What is the distinction between ruling and minimum design speeds and where are they adopted?
8. Determine the capacity of single lane (uni-directional) pavement on a rural highway in India for a design speed of 80 K.P.H. from theoretical considerations. The average length

of mixed vehicles of all types in 7 metres. The coefficient of friction is 0.5. Assume a value of perception-brake reaction time recommended by IRC.

[Ans. 1005 veh/hr/lane]

9. What is the Level of Service in capacity studies? What is the Level of Service recommended for Indian conditions for design of rural highways?

10. Draw a typical Speed-Flow curve and indicate the Levels of Service.

11. What are the recommended values of capacity as per IRC standards for rural highway of various pavement widths?

12. Derive the following equations used in the design of horizontal curves in India :

$R = 0.0357\ V^2$ (for plain and rolling terrains)

$R = 0.0315\ V^2$ (for hill roads, snowbound)

$R = 0.0357\ V^2$ (for hill roads, not snowbound)

13. Derive from fundamentals the equation used for calculating the super-elevation in India :

$$e = \frac{V^2}{225\,R}$$

What are the maximum values of super-elevation recommended in India for

(*i*) plain and rolling terrains

(*ii*) hilly areas which are snowbound

(*iii*) hilly areas which are not snowbound.

14. Design a horizontal curve for a National Highway in rolling terrain. Calculate the ruling minimum and absolute minimum radii. Make suitable assumptions.

[Ans. Assumptions:

Design speed	=	80 K.P.H. (Ruling); 65 K.P.H. (Minimum)
e	=	0.07, $\mu = 0.15$
Ruling radius	=	230 m
Absolute minimum radius	=	155 m

15. Calculate the super-elevation to be provided for a horizontal curve of radius 50 m for a design speed of 40 K.P.H. in snow-bound hilly terrain. What is the maximum super-elevation that can be provided and what will be the coefficient of friction then? Is the design safe? If it is not safe, what remedy do you suggest?

[Ans. e = 0.14, restricted to 0.07]

μ = 0.18, which is greater than 0.01, hence not safe. Remedy is to increase the radius or post a road sign restricting the speed]

16. Calculate the safe driving speed on a curve of radius 300 m and having a super-elevation of 0.07. Assume suitable values of friction. Is the curve meeting the NH standard in plain terrain? If the pavement is 7 m wide how much should the outer edge be raised if super-elevation is provided by rotating about the inner edge?

[Ans. Assume μ = 0.15, V = 91.6 K.P.H.;

Since minimum speed on NH is 80 K.P.H., curve is safe.

Raise outer edge by 0.49 m]

17. A National Highway located in rolling terrain has a radius of 250 m. Determine the length of the transition curve making suitable assumptions. (Assumptions, $V = 80$, $e = 0.07$ max, rate of attainment of $e = 1$ in 150, rotation of superelevation about centre.)

[Ans. 84.7 m and 80.3 m, adopt 84.7 m]

18. Calculate the extra width of a two lane pavement for a National Highway in hilly terrain (steep) for a design speed of 40 K.P.H. and radius of curve of 50 m. The vehicles using the road have a wheelbase of 7 m.

[Ans. 1.55 m]

19. What are the general controls to be kept in view in designing the horizontal alignment of a road?

20. Derive the formulae for determining the length of a summit curve (*i*) when the sight distance is less than the length of the curve and (*ii*) when the sight distance is greater than the length of the curve Modify the formulae when overtaking sight distances are involved.

21. Derive a formula for determining the length of a sag curve for fulfilling rider comfort criterion.

22. A summit curve is to be provided at the intersection of two gradients, + 1.5 per cent and –2.0 per cent. What is the length required (*i*) for stopping sight distance of 180 m and (*ii*) for overtaking sight distance of 640 m? What is the vertical distance between the Point of Vertical Intersection (PVT) and curve in either case?

[Ans.	Length	(*i*)	257.7 m, say 260 m
		(*ii*)	1493.3 m, say 1500 m
	Vertical Distance	(*i*)	1.14 m
		(*ii*)	6.56 m]

23. A sag curve- is to be designed where two gradients meet. The gradients are – 2.0 per cent and + 2.5 per cent. The design speed is 100 K.P.H. Find the length of the curve (*i*) for rider comfort and (*ii*) for headlight sight distance. The stopping sight distance is 180 m.

[Ans. (*i*) 80 m (*ii*) 147 m]

24. What are the general controls to be kept in mind in designing the vertical profile of a road?

25. What are the general controls to be kept in mind in designing a combination of vertical and horizontal alignment?

26. What are the standards for road gradient in India? Explain where (*i*) ruling, (*ii*) limiting and (*iii*) exceptional gradients are used?

27. Write short note on : Grade compensation on curves.

28. Derive the formula for stopping sight distance in terms of speed, perception and brake reaction time, coefficient of friction and gradient of road.

Calculate the safe stopping distance on a National Highway in plain terrain on a downward gradient of 3 per cent. Make your own assumptions.

Hint

Assumptions (*i*) V = 100 K.P.H., (*ii*) t = 2.5 sec, (*iii*) f = 0.35

Stopping sight distance = 173.1 m.

29. Derive formulae for determining the minimum set-back distance from centre of road to an obstruction, the road being a four lane carriageway, the length of curve and (*ii*) when sight distance is greater than length of curve.

Ans. (*i*) $$m = R - \left(R - \frac{3}{2}p\right)\cos\frac{\alpha}{2}$$

(*ii*) $$m = R - \left(R - \frac{3}{2}p\right)\cos\frac{\alpha}{2} + \frac{S-L}{2}\sin\frac{\alpha}{2}$$

Where $$\frac{\alpha}{2} = \frac{180\,S}{2\left(R - \frac{3}{2}p\right)} \text{ degrees}$$

30. Calculate the set-back distance from the inner edge of the curve in a four-lane divided carriageway, the length of the curve being 1500 m, the stopping sight distance 250 m, the lane width being 3.5 m and the radius of the curve being 400 m.

[Ans. 17.87 m]

31. Give the recommended right-of-way widths for various classes of roads in India. Explain the terms "building line" and "control line".

32. What are the current space standards for urban roads in India?

33. Explain the term "roadway width". What are the current standards for roadway width for various types of roads in India?

34. What are the functions of a median? What are the current Indian standards in this regard?

35. What are the functions of a curb? Describe' the "barrier" and "mountable" curbs.

36. What is the function of camber in a road surface? What are the recommended values of camber for various types of surfaces?

37. What considerations govern the side slopes of embankments and cutting of roads? What are the present standards in this regard?

38. Give sketches of various types of roads in (*i*) rural areas and (*ii*) urban areas.

39. What are Expressways? What are the geometric design standards for such facilities?

40. What are the general principles to be observed in designing intersections?

41. What locations justify grade-separated intersections?

42. What are the basic forms of at-grade intersections? Give sketches showing the details of each type.

43. What are the objectives of Channelisation? What are the features of channelising islands?

44. (*i*) What are the advantages and disadvantages of rotary intersections?

(*ii*) What are the guidelines for selecting a rotary type of intersection?

45. (*i*) What are the design considerations for a rotary intersection?

(*ii*) How is the capacity of a rotary determined?

46. Describe the features of mini-roundabouts, giving their advantages and limitations.

47. Describe with sketches the various types of grade-separated junctions and conditions under which they are provided.

Chapter

SURVEYS, SOILS AND AGGREGATES, INVESTIGATIONS AND TESTS

5.1. INTRODUCTION

Accurate surveys and investigations are needed for preparing a highway project which incorporates economy, safety, environmental sustainability and user comfort. Surveys and investigations of the following types are needed.

1. Traffic surveys
2. Soil surveys and testing
3. Material (Aggregates) surveys
4. Topographical surveys
5. Drainage surveys

5.2. TRAFFIC SURVEYS

5.2.1. Traffic counts

Information about traffic forms the basis for the design of the pavement, fixing the number of traffic lanes and economic appraisal of the project. Classified volume counts for a seven-day period should be conducted. Applying a seasonal correction factor, the Average Annual Daily Traffic (AADT) can be computed. Since traffic on Indian roads is of a mixed nature, it becomes necessary to convert the flow into equivalent Passenger Car Units (PCUs) by using certain equivalency factors. The PCU equivalents as per Indian practice are given.in Table 5.1.

Table 5.1: PCU Equivalency factors

S.No.	Vehicle type	PCU
1.	Passenger car, pick-up van or autorickshaw	1.00
2.	Motor cycle. scooter or cycle	0.50

3.	Agricultural tractor, light commercial vehicle	1.50
4.	Truck or bus	3.00
5.	Tractor-trailer, agricultural tractor-trailor	4.50
6.	Cycle rickshaw	2.00
7.	Hand cart	3.00
8.	Horse-drawn vehicle	4.00
9.	Bullock cart*	6.00–8.00

* For smaller bullock-carts, a value of 6 will be applicable

For geometric design, the commonly used unit for traffic is the 30th HV, defined as the hourly traffic that is exceeded by 29 hourly volumes in a year. For Indian conditions, 30th HV may be assumed to be 8-10 per cent of the AADT. For the design of carriageway width, the traffic in both directions is taken for a 2-lane road. For roads with more than 2 lanes, the design is done on the approximate assumption that the traffic in the predominant direction can be taken to be 67 per cent of the total traffic. The field data sheet prescribed for classified volume count in India is given in Table 5.2.

Table 5.2: Field data sheet for traffic census

Date and day of the week.....................
Direction of traffic
From.......To...........

Road Classification....................
Kilometrage/mileage
Route number (if any).......
District..................State...........

Type of Vehicles/ Hour of count	Cars, Jeeps, Vans, Three-wheelers	Buses	Trucks	Motor-cycles and scooters	Animal drawn vehicle**	Cycles	Others (specify)	Remarks, Including weather conditions
(1)	(2)	(3)	(4)	(5)	(6)	(7)	(8)	(9)
From–Hours* To–Hours								
Hourly total								
From–Hours* To–Hours								
Hourly total								

From–Hours* To–Hours								
Hourly total								
From–Hours* To–Hours								
Hourly total								

Notes:

Name and signature of enumerator with date

....

1. Record traffic-volume in columns 2 to 8 by making tallies in the form of verticalstrokes for first four vehicles and drawing an oblique stroke for every 5th as shown in within brackets (1111).
2. Some roads carry appreciable volume of other traffic like cycle rickshaws. Record the volume of such vehicles in column 8 after specifying the vehicle type.

* The hour of counts should be entered before the start of the enumeration. P.M. hours should be recorded after adding 12 to the actual hour, for example, 2 RM. should be recorded as 14.00 hours.

Name and signature of supervisor with date

....

** If felt necessary by Highway Authority, this column could be subdivided into two for recording the volume of 'pneumatic–tyred' vehicle separately.

5.2.2. Prediction of traffic for the design year

Highway facility is designed for a design period of 15-25 years, and a design period of 20 years is widely used as the basis. Future traffic estimates should consider:

1. Normal traffic growth
   - Diverted traffic
   - Induced traffic
2. Generated traffic
3. Development traffic

Rate of normal growth of traffic can be estimated by analysing the historical trend and fitting a best fit line. Thus:

$$T_n = T_0 (1 + r)^n \tag{5.1}$$

where T_n = Traffic in the nth year

n = number of years

r = rate of growth

Taking logs,

$$\text{Log}_e T_n = \text{Log}_e T_0 + n \log (1 + r) \quad (5.2)$$

If the past data is available on traffic for a number of years and the corresponding data on some economic indicators such as GNP, population etc. is also available, an econometric model of the following type can be fitted.

$$\text{Log}_e T = A_0 + A_1 \log_e \text{GNP} \quad (5.3)$$

where T = Traffic volume

GNP = Gross National Product

A_0 = Regression constant

A_1 = Regression coefficient

The value of A_1 is known as the Elasticity Coefficient, and is the factor by which the GNP growth rate is to be multiplied to arrive at the growth rate of traffic. The following values of Elasticity Coefficient are appropriate for Indian conditions.

Vehicle type	Elasticity coefficient
Car	1.8
Two-wheeler	2.1
Bus	1.5
Truck	1.3
Slow-moving vehicle	1.5

Thus if the country's GNP is expected to grow at 8 per cent per annum, the car population will grow at 8 × 1.8 = 14.4 per cent per annum.

The estimation of diverted, induced and development traffic requires a detailed study.

5.2.3. Speed and delay surveys

To understand how the road is performing under traffic conditions, it is necessary to conduct a speed and delay survey. The most popular method is to adopt the "Moving Observer" method, under which a car is driven along the road, noting its travel time, number of vehicles met with in the opposite direction and the number of vehicles overtaken. The survey yields both the journey speed and volume.

5.2.4. Axle load survey

Trucks in India are generally overloaded and cause severe damage to the pavement. In order to assess the damaging effect of such trucks, it is necessary to stop the trucks and move them on to a portable weigh bridge. Sampling has to be done since all trucks cannot be managed. The Vehicle Damage Factor is calculated by adopting the Fourth Power Rule as under:

$$F = \left(\frac{L}{L_s}\right)^4 \tag{5.4}$$

where F = Equivalent factor of load L

L = Load in kN

L_s = Standard axle load (80 kN)

Thus, an axle load of 120 kN causes a damage equivalent to:

$$F = \left(\frac{120}{80}\right)^4 = 5.06 \text{ times the standard axle load of 80 kN}$$

Axle load surveys can also be done by using a modern technology, Weigh-in-Motion (WIM) where the trucks do not have to be stopped.

5.2.5. Origin-Destination surveys

For assessing the bypassable traffic at towns or for planning new facilities like Expressways, bypasses, new ring roads etc origin destination surveys are conducted by methods like the registration number matching, road side interviews and post card questionnaire.

5.3. SOIL SURVEYS AND TESTING

5.3.1. Need for soil surveys

Road pavements are built on embankments which are constructed using soil borrowed from nearby locations. The properties of soil determine their suitability for use in the embankment or for use in soil-stabilisation. In hill roads the soil strata influences the stability of the cuts and their vulnerability of slips and landslides. In high altitude areas, frost action which can cause failure of pavements depends on the soil properties. Soil surveys thus fulfil the following requirements:

(*i*) determination of the physical properties of soil for use in embankment and location of suitable borrow areas

(*ii*) design of slopes of embankments and cuts

(*iii*) design of high embankments to safeguard against slope failure and settlement

(*iv*) determination of suitability for stabilisation with other additives

(*v*) determination of strength of the subgrade which controls the design of pavement thickness and composition

(*vi*) determination of surface and subsurface drainage requirements

(*vii*) classification of earthwork into ordinary soil, hard soil, soft rock, hard rock etc. to facilitate cost estimation

5.3.2. Soil sampling for low embankment sections

Soil samples at intervals of 200 m for new roads and 500 m for road widening projects shall be collected by digging trial pits 0.5 to 1.0 m square and of a depth 1.25–1.5 m more than the depth of borrow pits proposed.

5.3.3. Soil sampling for high embankments

Embankments of height greater than 6 m require soil sampling at 50–100 m intervals. The depth should be 15 m or 1½ times the height of the embankment. Undisturbed soil samples are collected by drilling holes.

5.3.4. Soil tests

The following tests are carried out on soil obtained from borrow areas:

1. Classifications and Identification of Soils for General Engineering Purposes (IS: 1498-1970)
2. Gradation Test (IS:2720-Part V)
3. Liquid Limit and Plastic Limit (IS:2720-Part V)
4. Density and Optimum Moisture Content (IS:2720-Part VII and VIII, Part VII being Standard Proctor Test, and Part VIII being Modified AASHTO Compaction)
5. Deleterious Constituents (only in salt infested areas, IS:2720-Part XXVII)
6. Shrinkage and Swelling Properties (1S:2720, Part XXXX)

5.3.5. Soil classification system in India

Indian soil classification system considers two types of soils:

(*a*) Coarse grained
- Gravels (*G*), with more than half coarser than 4.75 mm
- Sand (*S*), with size 0.075–4.75 mm

(*b*) Fine grained (with size less than 0.075 mm)
- Silts (M)
- Clays (C)
- Organic (O)

The soils are identified by following a system of two alphabets. The first alphabet G, S, M, C and O refer to the classes indicated above. The second alphabet denotes the basic engineering property of that soil. The following second alphabets describe the property:

W :Well graded

P : Poorly graded

L :Low plasticity

I :Intermediate plasticity

H :High plasticity

Thus, GW stands for well-graded gravel, ML stands for silt with low plasticity and CH stands for clay with high plasticity.

The following definitions of some of the properties (known as Atterberg Limits) are needed to understand the classification of soils:

Liquid Limit (LL) is the minimum water content at which the soil can flow under its own weight.

Plastic Limit (PL) is the minimum water content at which the soil can be rolled into a thread 3 mm in dia without breaking.

Shrinkage Limit (SL) is the water content at which further loss of moisture does not cause a decrease in the volume of the soil.

Plasticity Index (PI) is the range in water content at which the soil exhibits plastic behaviour.

Thus $$PI = LL - PL$$

Fig. 5.1 is a Plasticity chart which helps in classifying the soil.

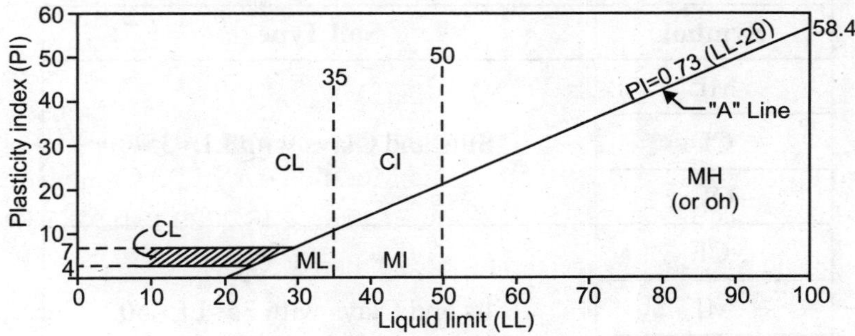

Fig. 5.1: Plasticity chart

Tables 5.3, 5.4 and 5.5 give the classification of gravels, sands and fine grained soils.

Table 5.3 : Classification of gravels

Symbol	Percent fines passing 0.075 sieve	Grading	Plasticity
GW	0–5	$U > 4$ $I < C_z < 3$	–
GP	0–5	Above conditions not satisfied	–
GM	> 12		Below A line (Fig. 5.1) or PI less than 4
GC	> 12		Above A line (Fig. 5.1) and PI>7

In the above Table, U is the Uniformity Coefficient defined as D_{60}/D_{10} where D_{60} is the diameter at which 60% of soil (by weight) is finer and D_{10} is the diameter at which 10% of soil is finer.

$C_z = D_{30}^2, / D_{60} \times D_{10}$, where D_{30} is the diameter at which 30% soil by weight is finer and D_{60} and D_{10} have already been described above

Table 5.4: Classification of sands

Symbol	Percent fines passing 0.075 sieve	Grading	Plasticity
SW	0 – 5	U > 60 I < Cz < 3	–
SP	0 – 5	Above conditions not satisfied	–
SM	> 12	–	Below "A" line in Fig. 5.1 or PI<4
SC	> 12	–	Above "A" line and PI>7

Table 5.5: Grading of fine-grained soils

Symbol	Soil Type
ML	Silts and Clays with LL<35
CL	
OL	
CI	Silts and Clays with 35<LL<50
MI	
OI	
CH	Silts and Clays with LL>50
MH	
OH	
Peat (Pt)	Highly Organic Soils

5.3.6. Weight-volume relationships of soil

A mass of soil has three phases:

(*a*) Solids (mineral particles)

(*b*) Air (gas)

(*c*) Liquid (water)

Fig. 5.2 shows these phases.

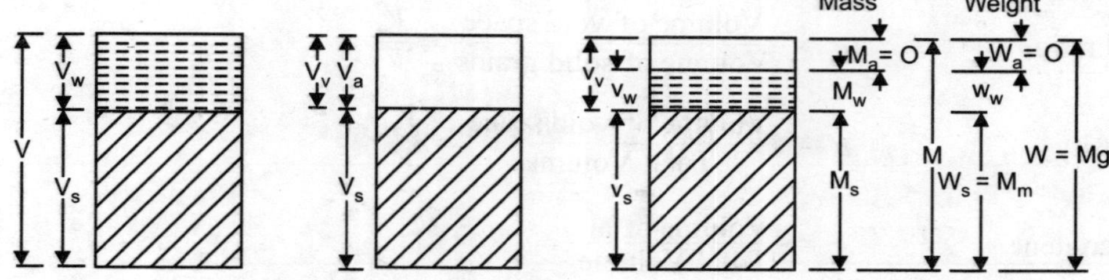

Fig. 5.2: Phases in soil mass

M	=	Total Mass	M_a	=	Mass of air (= 0)
M_w	=	Mass of water	M_s	=	Mass of solids
G_s	=	Specific Gravity of solid grains	W	=	Total Weight
W_a	=	Weight of air (= 0)	W_w	=	Weight of water
W_s	=	Weight of solids	V	=	Total volume
V_a	=	Volume of air	V_v	=	Volume of void space
V_w	=	Volume of water	V_s	=	Volume of solids
g	=	Gravitational acceleration = 9.81 m/sec^2			

The following definitions need to be noted:

Bulk density $$\rho = \frac{\text{Total Mass}}{\text{Total Volume}} = \frac{M}{V} \quad (5.6)$$

Bulk unit weight $$\gamma = \frac{\text{Total weight}}{\text{Total Volume}} = \frac{Mg}{V} \quad (5.7)$$

Thus $$\gamma = \rho g \quad (5.8)$$

Density of solids $$\rho_s = \frac{\text{Mass of solids}}{\text{Volume of solids}} = \frac{M_s}{V_s} \quad (5.9)$$

Unit weight of solid soil grains $$= \frac{W_s}{V_s} = \frac{M_s g}{V_s} = \rho_s g \quad (5.10)$$

Water content in soil mass $$= \frac{\text{Weight of water}}{\text{Weight of solid grains}}$$

Dry density $$\rho_s = \frac{\text{Mass of solids}}{\text{Total Volume}} = \frac{M_s}{V} \quad (5.11)$$

Dry unit weight $$= \rho_s g \quad (5.12)$$

Degree of saturation $$S_r = \frac{\text{Volume of water}}{\text{Volume of void space}} = \frac{V_w}{V_v} \quad (5.13)$$

Void ratio $e = \frac{\text{Volume of void space}}{\text{Volume of solid grains}} = \frac{V_v}{V_s}$ (5.14)

Porosity $n = \frac{\text{Volume of void space}}{\text{Total Volume}} = \frac{V_v}{V}$ (5.15)

Air content $A = \frac{\text{Volume of air}}{\text{Total Volume}} = \frac{V_a}{V}$ (5.16)

The following problem illustrates how the above relationships can be used.

Problem 5.1

A cylindrical specimen of soil sample has a volume of 85 cm^3 and weighs 162 gms in its natural condition. After drying completely in an oven, the specimen weighs 128 gms. The specific gravity of the solid particles is 2.70. Calculate the water content, void ratio and the degree of saturation.

$W = 162$ gms

$W_s = 128$ gms

Weight of water in the specimen = 162–128 = 34 gms

Volume of water = 34 cc

Water content $= \frac{34}{128} = 0.265$ or 26.5 per cent

Volume of solid $= \frac{\text{Weight of solids}}{\text{Specific gravity of solids}} = \frac{128}{2.70} = 47.4\ \text{cm}^3$

Volume of voids $= 85 - 47.4 = 37.6\ \text{cm}^3$

Void ratio $= \frac{\text{Volume of voids}}{\text{Volume of solids}} = \frac{37.6}{47.4} = 0.79$

Degree of saturation $= \frac{\text{Volume of water}}{\text{Volume of void space}} = \frac{34}{37.6} = 0.90$ or 90 percent

5.3.7. Compaction of Soil

Soil obtained from the borrow pits is in a loose state and needs to be compacted. The density to which a soil can be compacted depends on the soil type, moisture content and the compaction effort. Sandy, and gravelly soils can be compacted to a high dry density (in the range of 1.7 to 2.3 gms/ cc), whereas clays can only be compacted to a low density (in the range of 1.4 to 1.8 gms/ cc) and silty soils can be compacted to a density between that of sand and gravel and clayey soil.

The moisture content at which the soil is compacted determines the density that can be achieved. For a given type of soil, and a compactive effort, there is an optimum moisture content (known as OMC) at which a maximum value of density (MDD) can be achieved. The general shape of the curve is given in Fig. 5.3.

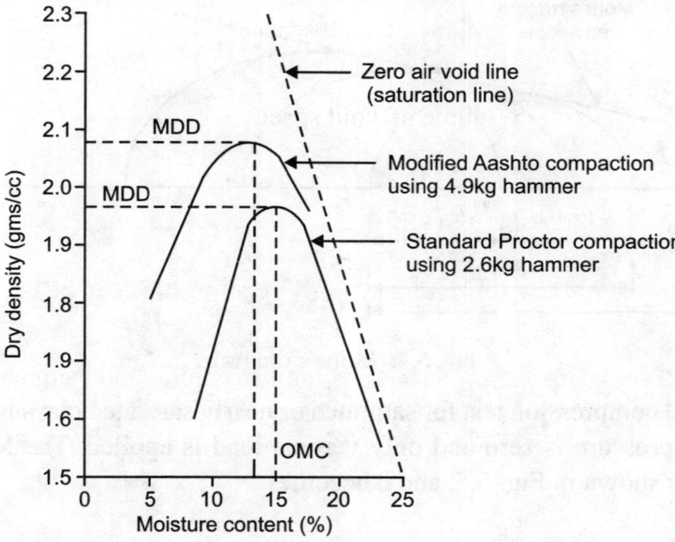

Fig. 5.3: Dry density moisture content relationship

Fig. 8.3 shows two curves, one for the Standard Proctor Compaction using 2.6 Kg hammer (IS:2720-Part VII) falling through a height of 310 mm and the other for Modified AASHTO compaction using 4.9 Kg hammer falling through a height of 450 mm (IS:2720-Part VIII). It is seen that as the compactive effort increases, the MDD increases and the OMC comes down. The Standard Proctor compaction (light compaction) is used for low volume roads, whereas for high volume roads Modified AASHTO compaction (heavy compaction) is used.

5.3.8. Determination of shear strength properties

The shear resistance of soils is determined by the Coulomb's law, which is expressed by the following relationship:

$$s = c + \sigma \tan \phi \tag{5.17}$$

where
s = shearing resistance
c = cohesion
σ = normal stress.
ϕ = angle of internal friction
$\tan \phi$ = coefficient of internal friction

The shearing resistance is found by:

(*i*) Direct Shear Test, using a shear box, where a normal load is applied and a horizontal force is applied to the upper box, and finding the horizontal force that causes a shear failure.

(*ii*) Triaxial Test, in which a cylindrical specimen is subjected to all-round pressure and an axial load is applied till the specimen fails in shear. The analysis is done by plotting the Mohr's diagram, Fig. 5.4.

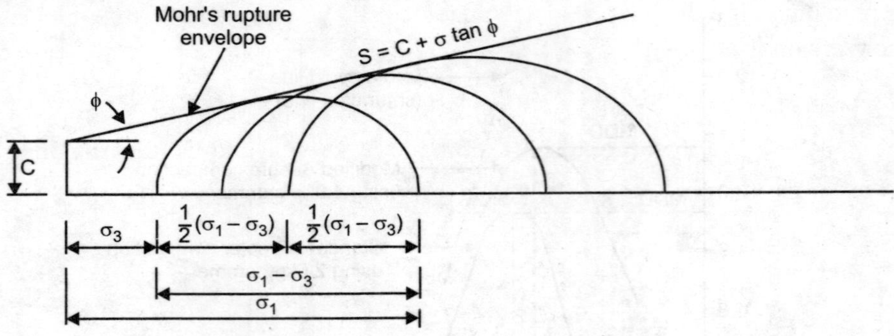

Fig. 5.4: Mohr's diagram

(*iii*) Unconfined compression test for saturated or nearly saturated clays is conducted in which the lateral pressure is zero and only vertical load is applied. The Mohr's diagram then becomes as shown in Fig. 5.5, and *c* becomes

$$C = \frac{q_u}{2} \tag{5.18}$$

where q_u is the unconfined compressive strength

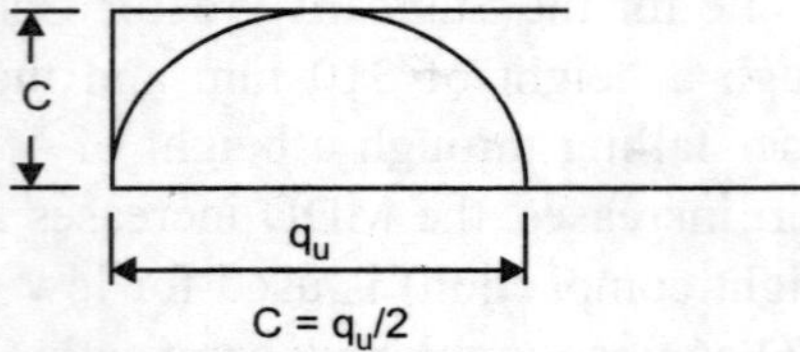

Fig, 5.5: Mohr's circle for $\phi = 0$ in an unconfined compression test

This test is the standard test for measuring the strength of soil-cement and soil-lime stabilised materials.

5.3.9. Plate bearing test

For designing rigid pavements, the Modulus of subgrade reaction is needed. This is obtained by a plate load test, using a 750 mm dia plate. The loading arrangement is shown in Fig. 5.6.

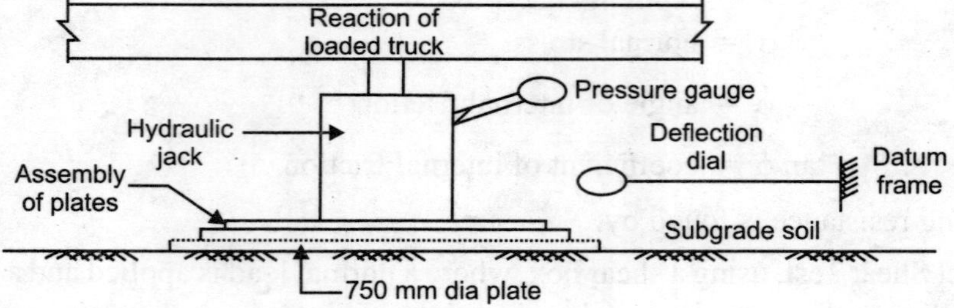

Fig. 5.6: Plate bearing test

The load settlement curve is generally a curve but a straight line passing through the origin and a settlement of 1.25 mm is drawn as shown in Fig. 5.7. The slope of this line gives the modulus of subgrade reaction *k*. If a dia plate is not available, a smaller plate of 300 mm dia is used and the following conversion is adopted.

$$K_{750} = K_{\phi}\,(1.21\,\phi + 0.078)$$

where K_{750} = Modulus of subgrade reaction using 750 mm dia plate

K_{ϕ} = Modulus of subgrade reaction using plate of dia ϕ.

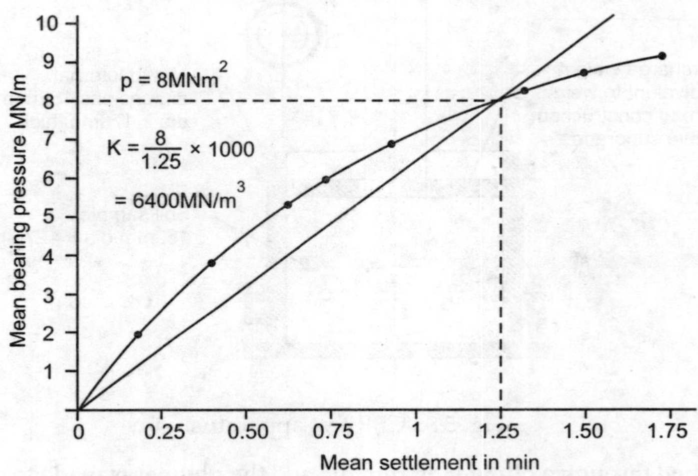

Fig. 5.7: Load settlement curve for plate bearing test

5.3.10. CBR test

The California Bearing Ratio (CBR) test is the most popular test for determining the strength of the subgrade on which the flexible pavement rests and is covered by IS:2720-Part XVI) It consists of a mould of 150 mm dia and 175 mm height in which the soil specimen is prepared to have the desired density *d* (gm/cc). The weight of the wet soil is calculated as under:

$$\frac{100+m}{100} \times \text{Volume of mould} \times d$$

The volume of the standard mould is 2244 c.c, and thus, the weight of the wet soil is:

$$\frac{100+m}{100} \times 2244 \times d$$

The density required for roads is given in the specification. Some specifications require that the specimen be soaked for 4 days and then the test done. The test set up is shown in Fig. 5.8.

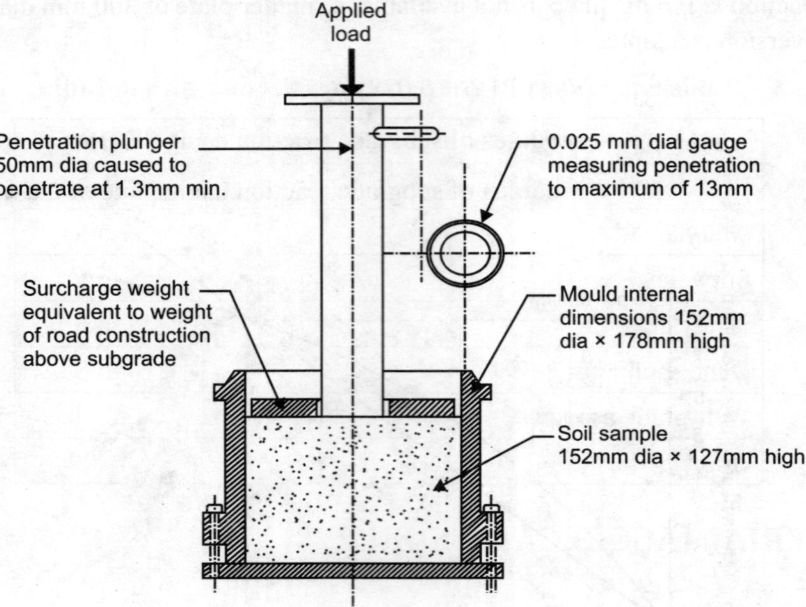

Fig. 5.8: CBR test apparatus

The load is applied through a plunger 50 mm dia and the plunger is made to penetrate at a rate of 1.3 mm/min. The loads at 2.5 mm and 5 mm penetration are recorded. The standard crushed stone penetration curve which gives a penetration of 2.5 mm at 6.895 MN/m^2 load and 5.0 mm at 10.343 MN/m2 load is shown in Fig. 5.9. If actual loading curve is upwards, a correction is required. The CBR value is expressed as a percentage of the actual load causing penetration of 2.5 mm and 5 mm to the standard crushed stone load values (6.895 MN/m^2 and 10.343 MN/m^2 as mentioned earlier). If the value at 2.5 mm penetration is greater than at 5 mm penetration, the former is adopted. If not, the test is repeated and if the new value at 5 mm is still greater, that value is adopted.

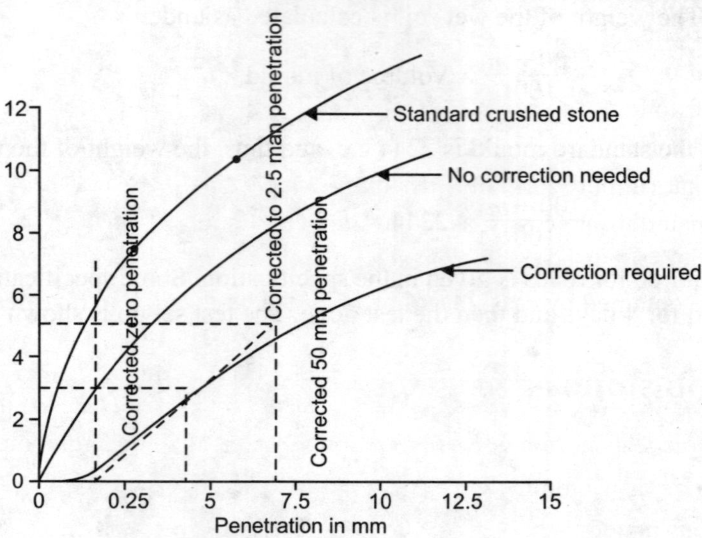

Fig. 5.9: Load-penetration curve

The typical value of CBR of Indian soils are given in Table 5.6.

Table 5.6: Typical CBR values of soils met with in India

Soil type	CBR
Black cotton soil and heavy clays	1 to 2
Alluvial silt	2 to 5
Silty clay	3 to 5
Sandy clay	4 to 6
Dune sand	6 to 8
Well-graded sand	8 to 20
Gravel (Murram)	8 to 20

5.4. MATERIAL (AGGREGATE) SURVEYS

5.4.1. Stone aggregates

Apart from soil, the other material widely used in road construction is stone aggregate. Stone aggregates are used in drainage layers, sub-bases, non-bituminous bases, bituminous layers and cement concrete. Road-making aggregates are obtained by crushing rock. The types of rock met with in India are:

Igneous Rocks

Granite

Basalt

Metamorphic Rock

Quartzite

Sedimentary Rocks

Limestone

Sandstone

Kankar (impure form of limestone)

Where stone materials are costly, as in the Gangetic Plains of North India, broken brick ballast from over burnt bricks are used. Slag from steel plants are also used as aggregates.

Aggregates are obtained by crushing stone obtained by blasting in quarries.

5.4.2. Tests on stone aggregates

The following are the tests normally carried out on samples of stone aggregates:

1. Particle size distribution
2. Deleterious materials content
3. Water absorption and bulk specific gravity

4. Flakiness and elongation
5. Aggregate impact test
6. Los Angeles abrasion test
7. Aggregate crushing test
8. Aggregate polishing test
9. Soundness test
10. Stripping test

5.4.3. Particle size distribution

To determine the gradation of stone aggregates, sieving is done. The standard set of sieves is prescribed in IS:460-1985 (Part 3). The sieving is done with dry aggregates if they do not contain dust or clay. If they do contain dust or clay, a wet sieve analysis is carried out.

5.4.4. Deleterious materials content

Deleterious materials in aggregates include clay lumps, organic matter and soft particles which are undesirable for road works. Normally, most specifications require that not more than 5 per cent of the sample should be finer than 0.075 mm. A test usually done for determining the proportion of silt/ clay/ organic matter is the Sand Equivalent Test, IS:2720 (Part XXXVII).

5.4.5. Water absorption and bulk specific gravity tests

Specific gravity of aggregates is a good indicator of its strength. The higher the specific gravity, the denser is the rock and hence stronger is the aggregate. Pores in the aggregates absorb water and when water enters the strength is adversely affected. The test is conducted by immersing the sample (2-3 Kg) in distilled water enclosing the sample in a wire mesh container. The weight of the container with the aggregates immersed in water is noted (W_1) The material is then surface dried and weighed in air, giving saturated weight (W_2) The material is then oven dried at a temperature 100–110°C and its dry weight noted (W_3).

$$\text{The percentage of water absorption} = 100 \times \frac{(W_2 - W_3)}{W_3}$$

$$\text{Bulk Specific Gravity of aggregates} = \frac{\text{Dry Weight of Aggregates}}{\text{Volume of Aggregates}}$$

$$= \frac{W_3}{\text{Weight of Water Displaced} \div \text{Specific Gravity of Water}}$$

$$= \frac{W_3}{W_2 - W_1}, \text{ since the specific gravity of water} = 1 \qquad (5.19)$$

Water absorption above 1.0 per cent is unsatisfactory for use in wearing courses, and values above 2.0 per cent are unsatisfactory for use in base courses.

5.4.6. Flakiness and elongation

Aggregates which are flaky and elongated are not conducive for interlocking. Flaky and elongated aggregates are difficult to compact and affect the workability and stability of mixes. Flakiness and elongation percentages are determined by testing the individual particles through a standard Flakiness gauge and Elongation gauge. The test is covered in IS:2386 (Part I). The latest MORTH specifications prescribe a combined Flakiness and Elongation Index. In this test procedure, the Flakiness Index is first found out and the flaky particles are removed and the balance quantity of aggregates are subjected to Elongation Test. The Elongation Index is then calculated. The sum of the two values (Flakiness index and Elongation Index) is the Combined Flakiness and Elongation Index. Normally, a combined Flakiness and Elongation Index value of 30 is specified for important works.

5.4.7. Aggregate impact test

In road pavements, stone aggregates in the various layers are subjected to impact. Hence the aggregates should have adequate resistance to impact and should not get crushed. The test has been standardised, and the equipment consists of an arrangement of making a weight of 13.5-14.0 Kg fall through a height of 380 mm over a cylindrical cup of dia 102 mm containing aggregates passing through 12.5 mm sieve and retained on 10 mm sieve. 15 blows through the hammer are given and the crushed aggregates are sieved through 2.35 mm sieve. The weight of material passing through this sieve, expressed as a percentage of the total weight of the sample gives the Aggregate Impact Value (AIV). For base courses, the AIV should not exceed 40, and for surface courses it should not exceed 30. The testing equipment is shown in Fig. 5.10.

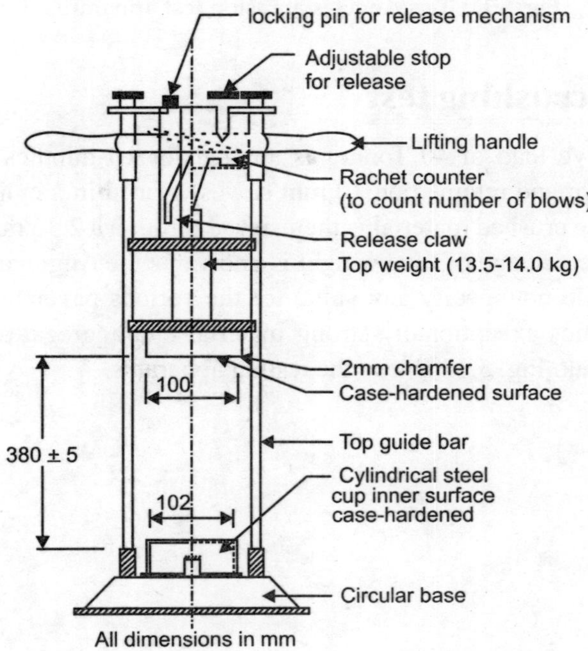

Fig. 5.10: Aggregate impact test equipment

5.4.8. Los Angeles abrasion test

This test measures the abrasion resistance of aggregates, and its important for the top layers of a pavement. The test equipment consists of a circular drum of internal diameter 700 mm and length of 500 mm mounted on a horizontal axis. About 5-10 Kg of aggregates are placed in the cylinder along with abrasive iron balls 48 mm dia and weight 390-445 gm. The cylinder is rotated at a speed of 30-33 revolutions per minute for 500-1000 revolutions, depending upon the material. After rotating for the specified number of revolutions, the material is sieved through 1.7 mm sieve. The weight of fines passing through the 1.7 mm sieve expressed as a percentage of the total weight of the sample is known as the Los Angeles Abrasion Value. The value specified for Wet Mix Macadam in India is 40 per cent (maximum), for Dense Bituminous Macadam the value is 35, for Bituminous Concrete the value is 30, and for Cement Concrete the value is 35 per cent. The equipment is shown in Fig. 5.11.

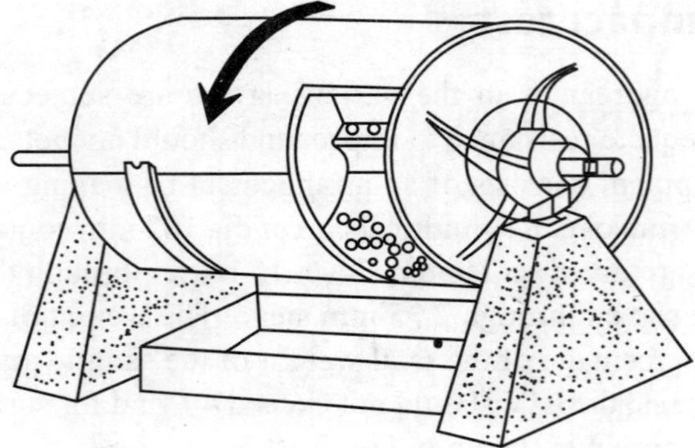

Fig. 5.11: Los Angeles abrasion test apparatus

5.4.9. Aggregate crushing test

In this test, a compressive load of 40 Tonnes is applied for 10 minutes on a coarse aggregate (passing through 12.5 mm and retained on 10 mm sieves) placed in a cylindrical mould 115 mm dia and 180 mm high. The crushed material is then sieved through a 2.36 mm sieve, and the weight of fines expressed as a percentage of total weight is known as the Aggregate Crushing Value. The MORTH Specifications do not specify any value for the various pavement sources. Generally, a value less than 10 signifies exceptionally strong material and aggregates with values above 35 would be regarded as weak. Fig. 5.12 shows the testing apparatus.

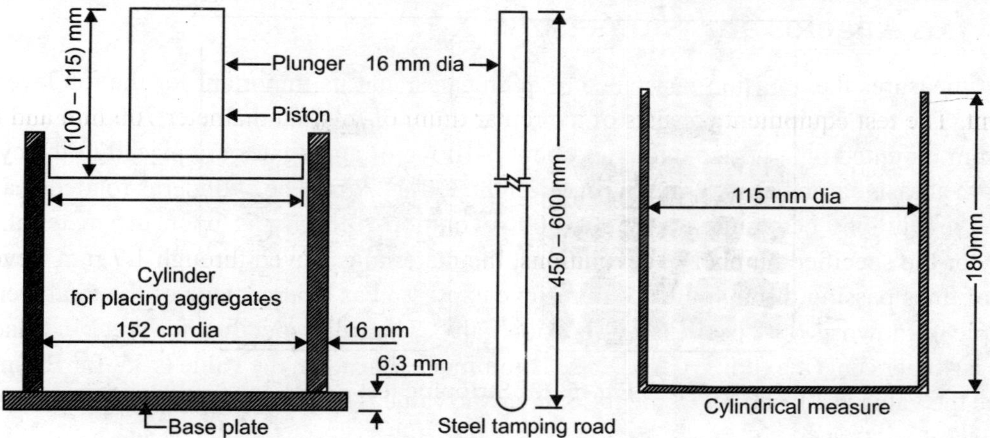

Fig. 5.12: Aggregate crushing strength apparatus

5.4.10. Aggregate polishing test

Aggregates in the surfacing courses should not get polished when tyres move on them. Polished aggregates cause skidding when the surface is wet. The test is standard in BS:812-114. In this method, stone aggregates are embedded in a curved mould in cement mortar and the mould is subjected to polishing by a rotating pneumatic wheel. The British Portable Tester is a quick method for testing the Polishing Stone Value (PSV). The machine consists of a rubber sliding shoe which is mounted at the end of a pendulum. When released, the rubber sliding brushes past the specimen and comes to a halt. A graduated disc measures the PSV directly. A minimum value of 55 is specified for surface courses.

5.4.11. Soundness test

Porous aggregates which absorb water to an extent of more than 2 per cent do not have good durability. They are then subjected to soundness test. Under this test, aggregates of a specified size are subjected to alternate cycles of wetting and drying, in a saturated solution of either sodium sulphate or magnesium sulphate. After 5 cycles, loss in weight due to fragmentation is determined by sieving all undersize particles. Generally the loss shall not be more than 12 per cent if sodium sulphate solution is used or 18 per cent if magnesium sulphate solution is used.

5.4.12. Stripping test

Bitumen coats the aggregates and imparts binding property. In the presence of water, the bond between the bitumen film and the stone aggregates gets affected, depending upon the type of aggregate. This phenomenon is known as stripping. The test for determining stripping prescribed by the Indian Standard is to immerse the freshly coated specimen in water at a controlled temperature for 24 hours. By visual inspection, the percentage of area uncoated is assessed. The maximum retained coating specified by IRC for Dense Bituminous Macadam and Bituminous Concrete is 95 per cent. Fig. 5.13 shows the testing set-up.

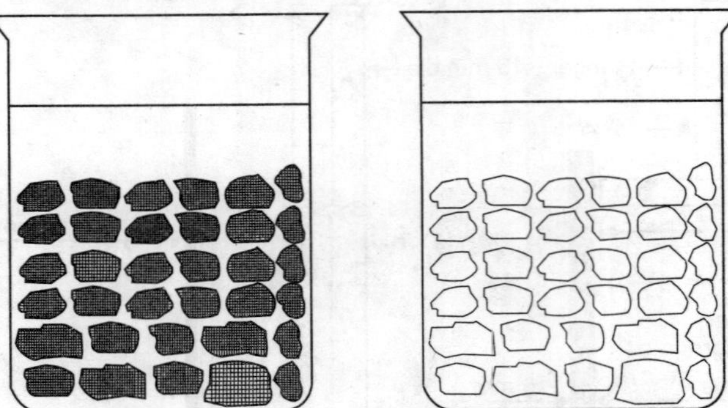

Fig. 5. 13: Stripping test

5.5. TOPOGRAPHIC SURVEY

5.5.1. Guidelines for selecting the alignment

The first stage in topographic survey is the selection of the alignment and locating the route. The following guidelines should be followed in selecting the alignment:

1. A direct alignment results in the least length of road, and hence the lowest initial cost, maintenance cost and road user cost.
2. The alignment should serve the maximum number of villages and towns, though it may involve construction of a few bypasses.
3. The alignment should keep the curvature and gradient to a low value consistent with the terrain.
4. The use of agricultural land should be minimal. If a road already exists and adequate right-of-way is available, the improvements suggested should be as far as possible within the available land.
5. The alignment should have minimal impact on the environment including fauna, flora and drainage system.
6. The alignment should be located so that suitable borrow areas are available close by and sources of gravel and stone materials are also within short leads.
7. Frequent crossing and recrossing railway lines should be avoided.
8. Religious places of worship, burial grounds and cemeteries should not be disturbed by the alignment.
9. Places of archaeological and historical importance should not be interfered with.
10. Proximity to educational institutions, hospitals, playgrounds should not be interfered with.
11. Where major rivers have to be crossed, a right-angled crossing results in the most economical bridge.
12. Areas liable to flooding and land-slides and which are low-lying and marshy should be avoided.

13. The alignment should not interfere with utility services like electric high tension lines, main water pipes and optical fibre cables which already exist.
14. In aligning hill roads, obligatory points such as passes, saddles and valleys dictate the alignment. Areas liable to landslides, snow drift, avalanches should be avoided. Hair-pin bends should be used only where other alternatives are costly. The alignment should be on the side of the hill exposed to sun during winter. Alignment along the course of a river will have easy gradients. Tunnels are costly to build, and if they are unavoidable, the tunnel length should be minimal.
15. In desert areas, the alignment should avoid shifting sand dunes. Where longitudinal dunes are present, the alignment should preferably be on the top of the dune or in the interdunal space. Alignment passing through areas which are having vegetation should be preferred.

5.5.2. Use of data from remote sensing and google earth images

Technological advances in recent times have greatly facilitated the selection of alignment. The Survey of India have remote sensing maps available and should be consulted. The Google earth images are a great help in developing contours of hills and facilitate the selection of proper alignment of hill roads. They are also of great use in selecting alignment in areas other than hilly terrain since they give information on location of settlements, existing tracks, drainage channels, wooded areas and other environmentally sensitive areas.

5.5.3. Use of topo sheets

In India, the Survey of India publish Topo sheets to a scale of 1:50,000, These should be studied to select possible alignments keeping in view the guidelines given in para 5.5.1.

5.5.4. Ground reconnaissance

The map study described in paras 5.5.2 and 5.5.3 above results in various possible alignments and route location. They need to be examined in the field by ground reconnaissance. The engineer should walk along the possible alignments collecting the following information:

1. Topography (plain, rolling or hilly)
2. Length along various alternatives
3. Bridging requirement
4. Railway line crossings
5. Right-of-way available, if there is already a track or road, and constraints on account of built-up area, monuments, educational institutions, places of religious importance etc.
6. Geology of the area, giving soil type, nature of hill slopes, landslides, rocky faces, places susceptible to flooding, snow-drift, avalanches, sand dunes and their disposition.
7. Temperature variation, annual rainfall, wind direction, exposure to sun, water-table
8. Availability of borrow area and stone aggregates
9. Availability of water

10. Value of land
11. Important towns and villages, and need for bypasses
12. Location of existing utilities
13. Availability of labour
14. Availability of suitable contractors
15. Traffic counts from existing records

The reconnaissance survey shall enable the highway authority to select the best route to be followed.

5.5.5. Preliminary survey

Once the best route to be adopted is known, the next stage is the Preliminary Survey. It consists of running a traverse line along the selected route. In hilly areas, it is necessary to cut a trace 1.0 to 1.2 m wide so that the instruments for the survey can be conveniently placed. The survey should be done by modern instruments such as theodolite or total station. Levels are taken at 50 m intervals along the chosen traverse line. The levels should be connected to GTS datum. Distances can be measured by metallic tape or chain, or with modern distance measuring instruments. An accuracy of 1 in 2000 should be aimed at. Cross-sections are taken at intervals of 100-250 m in plain terrain. Bench Marks should be established at every 250-500 m, connected to the GTS datum.

5.5.6. Preparation of drawings

Based on the survey details collected, plans and longitudinal sections are prepared. Contours at 1-3 m intervals can also be shown in the drawings. The scales to be adopted are:

(*i*) Built-up areas and stretches in hilly terrain

1:1000 for horizontal scale   1:100 for vertical scale

(*ii*) Plain and Rolling Terrain

1:2500 for horizontal scale   1:250 for vertical scale

5.5.7. Determination of final centreline in the office

Firstly, the maps prepared as part of Reconnaissance Survey are studied and a few alternatives are prepared. The best is then selected which satisfies the design standards, environmental impact and economy.

For the selected alignment, the horizontal curves are designed including transitions. The vertical profile is then determined. A sample drawing is given in Fig. 5.14.

The estimates based on the above are then prepared.

5.5.8. Final location survey

The final centreline is then laid out on the ground by a theodolite or Total Station. The Horizontal Intersection Points (HIPs) are fixed on the ground and suitably referenced. The salient points on curves such as Beginning of Spiral (BS), Beginning of Circular Curve (BC), End of Circular Curve (EC) and End of spiral (ES) are fixed on the ground and suitably referred. The final centre line is staked at 50 m intervals in plain terrain and 20 m in hilly terrain. These will facilitate the construction.

5.6. DRAINAGE SURVEYS

The drainage surveys include:

(*i*) determination of High Flood Level (HFL)

(*ii*) determination of water table

(*iii*) catchment area of rivers and streams where culverts and bridges are to be provided

(*iv*) cross-sections of the river/ streams at the proposed crossing and upstream and downstream

(*v*) rainfall details (maximum hourly rainfall)

QUESTIONS

1. What are the various types of surveys and investigations required in connection with a highway project?

2. What are the traffic studies required for a highway project?

3. What requirements are fulfilled by soil surveys? Give the soil classification system as followed in India.

4. Define the following and give relations to determine these:

(*i*) Bulk density
(*ii*) Bulk unit weight
(*iii*) Density of solid
(*iv*) Unit weight of solid soil grains
(*v*) Water control in soil mass
(*vi*) Dry density
(*vii*) Degree of saturation
(*viii*) Void ratio
(*ix*) Porosity
(*x*) Air content

5. Write short notes on

(*i*) Compaction of soil
(*ii*) Shear strength of soil
(*iii*) Plate bearing test.

6. Define CBR and explain the test used to determine it.

7. List the tests which are carried out on stone aggregates. Describe the following tests:

(*i*) Aggregate impact test
(*ii*) Los Angeles abrasion test
(*iii*) Aggregate crushing test

8. What guidelines should be followed to select the alignment of a road?

9. List the information which should be collected while carrying out ground reconnaissance.

10. What are the stages in highway route location and survey using conventional survey techniques? Describe them.

Chapter

DESIGN, DRAWINGS, ESTIMATES AND PROJECT REPORT

6.1 DESIGN

6.1.1. Need for assessment of various alternatives

Highway design involves the following aspects:

1. Geometric design.
2. Embankment and cut design.
3. Pavement design.
4. Drainage design.

The detailed procedure involved in the design of each of the above elements has been discussed in separate Chapters. In this Chapter, a brief account is given of the broader issues involved in highway design and how these are being solved by the present-day Highway Engineer.

The survey and investigation work carried out will yield very extensive data on all aspects affecting the final design of the highway. Using these, it is possible to develop a number of alternative strategies in design. The Highway Engineer of a few years ago had to select the final design on the basis of his intuition or previous experience. He did not have the tools for quickly assessing the relative advantages and disadvantages of a number of alternative solutions in a short time. Quite often; some of the most attractive solutions or some of the solutions which are economical in the long-run escaped his notice. The fast advent of computers and recent developments in evolving a Highway Design Model have revolutionised the scene and brought to the aid of the Highway Engineer the most comprehensive computer-aided design methodologies.

6.1.2. Computer aided highway design and its advantages

Computers are able to generate the most optimum highway design, consisting of:

1. The horizontal alignment.

2. Vertical alignment.
3. Road cross-section.
4. Super-elevation.
5. Computation of earthwork.
6. Total transportation cost.

The combination of serial photography, satellite imageries and computers has rendered it possible to improve economy in design. The engineer can now consider various alternatives and evaluate them, so as to produce the most economical option. Great savings in time are possible and the cost of engineering can thus be reduced very significantly.

Better designs are possible of achievement. Instead of the conventional circular, curve with two transitional ends it is now possible to design other alternatives such as (*i*) circular curve with unequal transitions and (*ii*) a cubic spiral, which is a curve wholly transitional, with radius continuously changing and passing through any desired obligatory points. Cubic spline is very versatile both for vertical and horizontal curves. The computer can give as output, the full setting out data for curves.

An example of a computer-aided highway design package popular in U.K. is BIPS (British Integrated Programme Suite). The MOSS now renamed as MX Roads is now popular in India. The Digital Terrain Model (DTM) already described in an earlier Chapter enables evaluation of a number of alternative vertical and horizontal alignments and select the best and give the quantities of earthwork. Computers can develop perspective views of the alignment, so that the one which is the most aesthetically pleasing and gives optimal comfort and safety can be selected. The computer can work out the ground cross-section and road cross-section and give them as. print-outs.

Highway design models can consider various design options, unit costs of construction, interest rates, service life, geology, rainfall, maintenance inputs and vehicle operating costs and determine the most optimal solutions. The two popular Highway Design Models currently in use are the World Bank's HDM and the TRRL's RTIM.

6.1.3. Steps Involved in Design

The Steps involved in design of a highway project, after the surveys and investigations, are as under:

1. Establishment of the class of the road and the terrain through which it passes.
2. Selection of the design speed appropriate to the class of the road and the terrain.
3. Selecting the design period.
4. Selection of the cross-sectional elements keeping in view the future traffic requirements.
5. Selection of geometric design standards appropriate to the class of the road, terrain and design speed.
6. Designing the alignment and profile.
7. Designing problematic high embankments and deep cuts.
8. Designing the pavement.
9. Designing surface and subsurface drainage, erosion control and cross-drainage structures.
10. Designing road junctions.

11. Estimating quantities and cost.
12. Estimating future maintenance cost.
13. Estimating road user cost.
14. Carrying out economic analysis for various alternatives and selecting the optimal solution.
15. Preparation of final drawings, estimates and project report for the selected alternative.

Some of the steps are iterative in nature, and the whole process lends itself admirably well to computer-aided design.

6.1.4. Stage construction

In India, due to paucity of funds, a system of stage construction is generally followed in road projects. If it is so, it is highly desirable to visualise the final specifications of the road and detail out the various stages to which it will be built according to the selected time frame. In such cases, the following points should be kept in mind:

1. The alignment and profile should be selected to suit the final design and these should be provided for in the initial stage itself. It is costly to rectify them later.
2. It is also preferable to build the roadway to the full cross-section at the beginning. It is difficult to add thin strips of earthwork at a later date.
3. The pavement width may be built in stages. To begin with a single lane (3.75 m) may be provided, which may then be widened to an intermediate width (5.5 in) or two lane width (7.0 m) as and when traffic grows.
4. The pavement structure can also be built up in stages, adding to it as and when traffic demands it.

6.2. DRAWINGS

Drawing are necessary for enabling the construction engineer to execute the project. It facilitates the preparation of detailed estimates and helps in inviting tenders. The bidders get an accurate idea of the work involved from the drawings. They prepare their tenders on the basis of the drawings. While executing the project, the drawings help to line out the works in the field and to carry out the work in the manner conceived by the designer. Accurate and comprehensive drawings thus form an important part of highway project preparation.

6.2.1. Types of drawings

A highway project should contain the following drawings

1. Locality map-cum-site plan.
2. Land acquisition plans.
3. Plan and longitudinal section.
4. Typical cross-section sheet.
5. Detailed cross-sections.

6. Drawings for cross-drainage works.
7. Road junction drawings.
8. Drawings for retaining walls and other structures.

All drawings are generally prepared on a size of 594 mm × 420 mm.

6.2.2. Locality map-cum-site plan

This is a combination of the key map to a scale of 1 : 250,000 and the index map to a scale of 1 : 50,000. It should show the location of the project in the State or Province and indicate the names of towns and cities. The index plan of the project should show details of other roads, railway line and rivers in the vicinity. The sheet contains the Legend to varioussymbols used in the detailed sheets contained in the project drawings folder. Fig. 6.1 gives a Locality Map.

6.2.3. Land acquisition plans

These are prepared on the village maps available with the Revenue authorities showing the boundaries of various fields and their survey number. Scales of 1 : 2000 to 1 : 8000 are suitable. The plans should show wells, structures, trees, land boundaries, nature of land (irrigated, diy, barren, forest) and the nature of crops. The road land boundary should be super-imposed on the land maps.

6.2.4. Plan and longitudinal section

The normal practice is to show both the plan and profile together in one sheet, one below the other. A length of 1 km fits well with the drawing size indicated earlier. This enables the following scale to be adopted:

Horizontal–1 : 2500

Vertical– 1 : 250

For hill roads, the scales may have to be brought down to:

Horizontal–1 : 1000

Vertical– 1 : 100

The Plan should show the final centre-line, right-of-way boundaries, existing roadway, existing and proposed structures, cross-road, railway lines, electric and telephone lines, cables, utilities, bearings of tangent lines, radii and other design data of horizontal curves, positions of points of transit (POTs), horizontal intersection points (HIPs), location of cross-sections, continuous chainages, contours, north point etc.

The longitudinal section should show the final road profile, existing road profile, details of grades, vertical curve data, super-elevation, cross-fall, CBR values of the pavement, HFL, location of cross-drainage structures intersecting roads etc.

A typical plan and profile is shown in Fig. 6.2

6.2.5. Typical cross-section sheet

A typical cross-section sheet for embankment and cutting should be attached, showing the cross-sectional elements such as road-way width, carriageway width, side slopes, shoulder width and specification, cross-fall etc. A typical sheet is shown in Fig. 6.3.

6.2.6. Detailed cross-sections

Cross-sections should be presented at every 50–100 m in plain terrain, 50–75 m in rolling terrain and 20 m in hilly terrain. Besides, they are needed at important points on curves. A convenient scale is 1 : 100. The cross-sections should show the existing ground profile, proposed road levels, areas of cut and fill, and thickness and composition of various pavement courses. A typical sheet is shown in Fig. 6.4.

6.2.7. Drawings for cross-drainage structures

A scale of 1 : 50 is suitable. Standard designs can be used, if available.

6.2.8. Road junction drawing

These are drawn to a scale of 1 : 500 or 1 : 600. They should show the existing road, proposed improvements, road sings, markings, islands etc.

6.2.9. Drawing for retaining walls and other structures

In hill roads, retaining walls, breast walls and check walls are needed. The drawings should show the details, along with the foundation soil on which they rest.

6.3. ESTIMATES

Highway cost estimation is an important part of a highway engineer's work. The estimate should be:

(*i*) detailed, comprehensive and should not miss any item.

(*ii*) accurate, both as regards quantities and rates.

(*iii*) containing adequate provision for quality control, supervision and overheads.

The Structure of the estimate consists of the following:

(*i*) Estimate of quantities

(*ii*) Rates

(*iii*) Abstract of cost.

The quantities for each item of work are estimated accurately. The units of measurement are generally standardised. The units in general usage in India for some of the principal items of work are listed in Table 6.1.

Table 6.1: Units of measurement for certain common highway work items

S. No	Item	Unit of Measurement
1	Clearing and grubbing	Hectares
2	Cutting trees of specified sizes	Number
3	Excavation and rock blasting	cu. m
4	Earthwork in embankment	cu. m
5.	Turfing	sq. m
6.	Sub-bases	cu. m
7.	Granular bases	cu. m
8.	Bituminous levelling course	cu. m
9.	Tack coat	sq. m
10.	Bituminous macadam	cu. m
11.	Bituminous penetration macadam	sq. m
12.	Built-up-spray-grout	sq. m
13.	Surface dressing	sq. m
14.	Premix carpet	sq. m
15.	Mix-seal surfacing	sq. m
16.	Semi-dense carpet	sq. m
17.	Asphaitic concrete	sq. m
18.	Seal coat	sq. m
19.	Cement concrete for structures and pavements	cu. m

Rates for various items of road work vary from State to State. Each State Public Works Department Circle has its own Schedule of Rates, which are updated from time to time. For items which are not included in the Schedule, detailed analysis should be done for arriving at realistic rates.

For National Highways, the Ministry of Shipping, Road Transport and Highways have brought out a Standard Data Book containing the analysis of rates for all the road work items. Both hard copy and soft copy are available. For Rural Roads, the Ministry of Rural Development have brought out a Standard Data Book containing the analysis of rates for all items of work involved in rural roads.

For National Highway Projects in India, provisions are made for contingencies, work-charged establishment, and agency charges as indicated in Table 6.2.

Table 6.2: Provision for contingencies, work-charged establishment and overheads on National Highway works :

Contingencies : 3 per cent

Agency charges : 9 per cent

Work-charged establishment charges:

S. No.	Particulars	Road projects	Bridge projects
1.	For works upto Rs. 25 lakhs	2%	2%
2.	For works Rs. 25–50 lakhs	1.75%	1.50%
3.	For works above Rs. 50 lakhs	1.50%	1.0%

6.4. EARTHWORK QUANTITIES

6.4.1. General

The most common item of work encountered in highway projects is earthwork. The estimation of earthwork quantities thus deserves detailed treatment. According to the MOST Specifications, the rate of earthwork includes all operations involved in it, including excavation in borrow areas, transporting to the site of embankment including all lifts and leads not greater than 50 m and compaction.

6.4.2. Computation of earthwork quantities

The quantities of earthwork in embankments and cuts are computed by the cross-sectional area method.

The area of earthwork in each cross-section is computed by graph-paper method or by planimeter. The volume of earth-work from the successive cross-section can be computed by two formulae:

1. Average end area formula (Trapezoidal formula)

$$V = \frac{(A_1 + A_2)}{2} \times L$$

where V = Volume in cu m

$A_1\ A_2$ are areas of the successive cross-sections, sq. m

L = Distance between the cross-sections, in m.

2. Prismoidal formula

$$V = \frac{L}{6}(A_1 + 4A_m + A_2)$$

where V and L have the same meaning as above

A_1, A_2 = End areas of cross-sections, sq. m.

A_m = Area of cross-section mid-way between the end areas.

The Average end area formula is simple and is generally preferred by Highway Engineers. MOST Specifications recommend this method. But, this formula is likely to be erroneous when the sections are not right prisms. This error is small when the sections do not change rapidly. The volume computed by this formula is likely to be higher than the true values in the case of sections changing rapidly. The prismoidal formula is more accurate.

If a highway design software such as MX Roads is used, the earthwork quantities are obtained as outputs.

6.4.3. Mass diagram

The earthwork for a highway project involves cuts and fills. The most balanced design is one which utilises the cut volume in the adjacent fills. This is achieved by a mass diagram which is a graphical representation of the amount of cut and the amount of fill and the manner in which the earth is to be hauled from cut to fill.

The principles of drawing a mass diagram are illustrated in Fig 6.5 and are as under:

1. The mass diagram is plotted directly below the longitudinal profile.
2. the horizontal distances are the chainages along the centre-line.
3. the ordinate at any station along the mass diagram indicates the earthwork quantity accumulated upto that point, and is the summation of the differences between cut and fill.

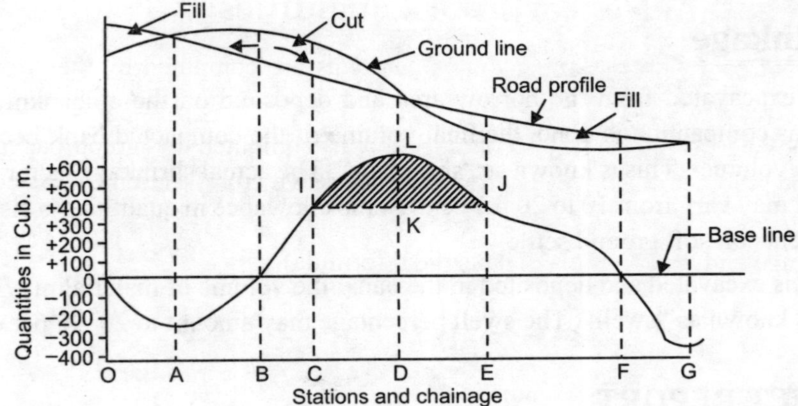

Fig. 6.5: Mass diagram

4. The maximum (+) ordinate indicates a change from cut to fill (D) and the minimum (–) ordinate indicates a change from fill to cut (A and G).
5. A rising curve at any point indicates an excess of excavation over fill at the point. A falling curve indicates the reverse.
6. If the curve has steep slopes it indicates heavy cuts or high fills. Flat slopes indicate small earthwork quantities.
7. A convex loop of the mass diagram indicates that the haul from cut to fill is from left to right. A concave loop indicates that the haul from cut to fill is from right to left. In Fig. 8.5, the arrows indicate the direction of haul.
8. A balance point is defined as a point where the volume in excavation balances the volume in embankment.
9. Any line drawn parallel to the base line and intersecting two points within the same curve indicates a balance of cut and fill between these two points (C and E). Such a line (HJ) is called a balancing line.
10. The area between a balancing line and the mass diagram is a measure of the haul between the balance . points. This area (hatched area in Fig. 6.5) divided by the maximum ordinate between the balance line and the curve (KL) gives the average distance of haulage of the cut material to make the fill.
11. When the earth excavation and embankment quantities balance at the end of the section, the mass diagram curve would end at the base line at the zero point.

A good amount of trial is needed to balance the cut and fill.

6.4.4. Trial haulage

A certain amount of trial haulage (free haulage) is included in the earthwork rate itself and the contractor is expected to carry earth over this haulage within his quoted rate. The MOST specifications give this initial haulage as 50 m. Any lead over this (overhaul) is paid for at extra rates.

6.4.5. Shrinkage

When earth is exacavated from the borrow area and deposited on the embankment, its volume increases. But as compaction is done, the final volume of the compacted bank becomes less than the borrow area volume-. This is known as "shrinkage". The actual shrinkage factor depends on the soil deposit and may vary from 10 to 20 per cent. Some allowance in quantities is also needed when high embankments on soft ground settle.

When rock is excavated and deposited in the bank, the volume of material may occupy alarger volume. This is known as "swell". The swell percentage may amount to 20–40 per cent.

6.5. PROJECT REPORT

The project documents contain the following:

1. The Project report
2. Estimate
3. Drawings.

The Project Report is an important component of the project documents and it should give a comprehensive idea of the nature of the work involved and an appreciation of the proposal. The Project Report should contain information on all aspects of the scheme and should be grouped as indicated below:

1. Preliminary
    - 1.1. Name of work and its scope
    - 1.2. Authority, plan provision
    - 1.3. History, geography, terrain, climate
    - 1.4. Necessity
2. Road features
    - 2.1. Route selection
    - 2.2. Alignment
    - 2.3. Environmental factors
    - 2.4. Traffic
    - 2.5. Cross-sectional elements.
3. Road design and specification
    - 3.1. Geometric design standards
    - 3.2. Road design
    - 3.3. Pavement design
    - 3.4. Masonry works
    - 3.5. Specifications.
4. Drainage facilities
    - 4.1. General conditions, HFL, water-table, seepage. flows
    - 4.2. Surface drainage, catch-water drains, longitudinal drains
    - 4.3. Sub-surface drainage, blanket courses, sub-drains
    - 4.4. Cross-drainage structures.
5. Material, labour and plant
    - 5.1. Sources of construction material, transport
    - 5.2. Labour availability, amenities
    - 5.3. Equipment.
6. Rates
    - 6.1. Schedule of rates
    - 6.2. Justifications for rates
7. Construction programming
    - 7.1. Working season

7.2. Schedule of completing the work.

8. Miscellaneous
    8.1. Rest houses, temporary quarters
    8.2. Diversions and haul roads
    8.3. Site amenities
    8.4. Traffic control device etc.
    8.5. Road-side plantation, turfing, landscaping
9. Economic appraisal
    9.1. Costs and benefits
    9.2. Time horizon
    9.3. Interest rate
    9.4. Results of appraisal.

6.6. STAGES IN PROJECT PREPARATION

Highway Project Preparation is carried out in the following stages:

1. Project identification/Pre-feasibility
2. Feasibility
3. Detailed Engineering
4. Preparation of tender documents

The first stage, known as Project Identification or Pre-Feasibility, involves preliminary reconnaissance, marking the alignment on topo sheets and estimating the cost on a very rough basis. This stage results in the identification of the project as a possible project to be taken up during the Five Year Plan.

The second stage, known as the Feasibility stage, involves a study of various alternative alignments, choice of pavement type and different design strategies, estimation of traffic, conducting investigations sufficiently in detail to arrive at a reasonably accurate cost and carrying out an economic analysis of the various alternatives. This facilitates the selection of the optimum design option. The Feasibility Report enables the administration to financially sanction the Project judging from its economic viability.

The third stage is known as detailed engineering. The final alignment is transferred to the ground and detailed alignment plans, longitudinal sections and cross-sections are prepared. Soil investigations-are carried out in detail. The design of the pavement is finalised. The design of the bridges is done. The detailed estimate and the Bill of Quantities are prepared.

In the next stage, the contract documents are prepared. These include:

1. Notice Inviting Tenders
2. General conditions of contract
3. Specifications
4. Bill of Quantities
5. Pre-Qualification document (for Contractor pre-qualification)

QUESTIONS

1. What are the various steps involved in the design of a highway project after the results of survey and investigation are available?
2. What is "stage construction" as practised by highway engineers in India?
3. Describe the various types of highway drawings, their sizes and scales adopted.
4. What are the widely used formulae for estimating earthwork quantities?
5. Describe the term "mass diagram". Explain the principles of preparing a mass diagram.
6. What is the format for a Highway Project Report?

Chapter

Economic Evaluation of Highway Projects

7.1. ROLE OF ECONOMIC EVALUATION

A developing country like India has serious shortages of resources needed for economic development. The outlay for various sectors of economic activity is decided by planning at the national level, keeping in view the national goals and policies. Within the allocation earmarked for the highway sector, a number of schemes can be taken up. Each enjoying its own urgency and attractiveness. It thus becomes necessary to screen and evaluate the various alternatives so that a wise decision can be reached on the most appropriate choice. This is achieved by modern techniques of economic evaluation of projects.

Economic evaluation is a rational approach at quantifying the future benefits and costs of proposed highway improvements with a view to determine the extent to which the projects will contribute to the goal of raisin" the living standard of the people and their general welfare. It provides for a systematic and unbiased procedure for selection of schemes for implementation under the Five Year Plans. It ensures that the most worthwhile projects are given the highest priority.

Economic evaluation of highway projects can also be carried out to weigh other alternative transport projects, such as railways, pipe-lines or inland water transport in order to select the most beneficial scheme.

The following are some of the specific objectives in carrying out an economic evaluation :

1. To decide whether the scheme under consideration is worth investment at all.
2. To rank schemes competing for scarce resources in order of priority.
3. To compare various alternative schemes and select the one most economical.
4. To assist in phasing the programme (stage construction) depending upon the availability of resources.

7.2. BASIC PRINCIPLES

Economic evaluation involves a number of basic principles listed below :

1. Economic evaluation makes it possible to choose the best of the various alternatives.

The question before the analyst is to suggest the most attractive of them. Often the choice is between 'do-nothing', and othert improvement schemes.

2. In economic evaluation, all past actions are irrelevant. What is of prime importance is the future flow of costs and benefits.
3. In highway projects, the appraisal is carried out from the view-point of the nation as a whole, and is not restricited to any sub-set like the highway agency, truckers, private motorists and bus operators.
4. Economic analysis should not be misunderstood with financial analysis.
5. Economic evaluation should take place within a set of established criteria such as minimum attractive rate return, interest rate etc.
6. Opportunity cost of capital and resources should be considered wherever they are important.
7. The period of analysis need not be too long in view of the uncertainties associated with the future traffic and benefits. In any case, the discounted cash flows of a distant future period are insignificant. For highway projects, it is enough if the analysis covers a period 15-25 years after opening to traffic.

7.3. TIME VALUE OF MONEY

The fundamental premise on which all methods of economic evaluation rests is that money earns income over a period of time. For example, Rs. 100 today will be worth Rs. 672. 75 at the end of 20 years if invested at 10 per cent compound rate of interest. So also, a sum of Rs. 672.75 which might become due to an individual after 20 years from today is worth only Rs. 100 at the present, assuming the same rate of interest. This points to the need for devaluing the future benefits and costs to the present time to determine their present worth. The process of calculating the present worth of a future payment is known as "discounting" and the interest rate used is called the "discount" rate.

The following formulae are very useful in dealing with the problems in economic evaluation :

1. The amount S to which Re. 1 will increase in n years with a compound interest rate of r,

$$S = (1 + r)^n \tag{7.1}$$

2. The present value A of Re. 1, n years hence, when discounted at interest rate r,

$$A = \frac{1}{(1+r)^n} = (1 + r)^{-n} \tag{7.2}$$

3. The present value A of an annuity of Re. 1 per year for n years when discounted at interest rate r,

$$A = \frac{(1+r)^n - 1}{r} \tag{7.3}$$

4. The amount P per annum for n years at interest rate r needed to accumulate Re. 1 at the end of n years (Sinking fund factor),

$$P = \frac{r}{(1+r)^n - 1} \tag{7.4}$$

5. The amount P per annum for n years at interest rate r needed for recovery of capital of Re. 1 including interest (Capital recovery facror),

$$P = \frac{r(1+r)^n}{(1+r)^n - 1} \tag{7.5}$$

6. Present worth A for uniform payments of Re. 1 per year for n years at interest rater (present worth factor),

$$A = \frac{(1+r)^n - 1}{r(1+r)^n} \tag{7.6}$$

Ready made Tables giving values of the above factors are available and are handy to use.

7.4. COSTS AND BENEFITS

In economic evaluation, the main objective is to compare the costs and benefits of various alternative schemes and select the one, most advantageous. The first step is, therefore, to determine the costs and benefits. There is a great deal of confusion in the designation of what constitutes "costs" and what constitutes "benefits". The most simple description is that the negative effects of a scheme constitute the costs. They indicate the outflows. On the other hand, the positive effects are called benefits and they represent in-flows. As long as sufficient care is taken to see that the signs are assigned properly , it is immaterial whether the economic consequence is labeled as 'costs' or benefits'.

Costs and benefits can be traced to the provider of facility (Highway department), the highway users and the non-users. In economic analysis, since all consequences are to be considered, the costs and benefits to all parties are to be reckoned.

Some consequences can be quantified into monetary terms whereas some cannot. The aim of the analyst should be to quantify as many elements as can be monetarily quantified. Those which cannot be ultimately quantified into monetary terms are kept separately apart and a judgement value can be accorded to them before a final decision is taken.

The economic evaluation of highway projects is generally done by computing the total transport cost which consists of the following components:

1. Cost of construction of the facility
2. Cost of maintenance of the facility
3. Road user cost
4. Cost to the society.

The Government, which is often the agency providing the facility, incurs expenditure on constructing a road. This includes land acquisition, earthwork, road pavement and structures. The Government also invests money on maintenance and upkeep annually.

The Road User Cost, which is borne by the actual user of the Highway facility (passenger, crew of vehicles, operator, consigner of goods, pedestrian, cyclist etc.), is composed of

A. Vehicle operating costs :

1. Fuel
2. Lubricants
3. Tyre
4. Spare parts
5. Maintenance labour
6. Depreciation

7. Fixed costs such as:
   - (*i*) Interest on capital
   - (*ii*) Insurance
   - (*iii*) Taxes
   - (*iv*) Registration fee
   - (*v*) Garaging charges
   - (*vi*) Fines, tolls, etc.
   - (*vii*) Permit charges
   - (*viii*) Loading and unloading charges
   - (*ix*) Commission on booking
   - (*x*) Overhead charges such as rent, salary, electricity, postal, telephone, stationery etc.
8. Crew costs.

B. Travel time costs

1. Time value of vehicle occupants
2. Time value of goods in transit
3. Time value of vehicles in transit.

C. Accident costs

1. Cost of fatality
2. Cost of injuries
3. Cost of damages to property.

Costs of society include the following:

1. Impact on the environment (noise pollution, air pollution, vibration).
2. Loss of aesthetics
3. Changes in land values
4. Land severance
5. Discomfort and inconvenience.

Benefits from highway projects in effect represent the difference in costs with the new facility and the old facility. Benefits can be grouped under the following heads:

(*i*) Benefits to the existing traffic, by way of reduced road user costs.
(*ii*) Benefits to the generated traffic.
(*iii*) Benefits to traffic diverted from other routes and modes.
(*iv*) Benefits to traffic operating on other routes and modes where reduction in traffic has been caused by the opening of the new facility.

Vehicle operating costs are affected by a number of factors such as:

A. Vehicle factors

1. Age
2. Make
3. Horse-power, engine capacity
4. Load carried
5. Condition of vehicle
6. Level of maintenance input
7. Type of fuel used
8. Type of tyres (rayon, nylon, radial ply, cross ply etc.).

B. Roadway factors

1. Roughness of the surface
2. Type of surface
3. Horizontal curvature
4. Vertical profile
5. Pavement width
6. Type and condition of shoulder
7. Urban or rural location
8. Number of junctions per km.

C. Traffic factors

1. Speed of travel
2. Traffic volume and composition,

D. Environmental factors

1. Altitude
2. Rainfall
3. Temperature.

The Road User Cost Study and its updation have established that the vehicle operating cost components are closely governed by (*i*) roadway factors such as roughness, pavement width, rise and fall and horizontal curvature, (*ii*) vehicle factors such as age and load carried and (*iii*) traffic factors such as speed and volume of traffic. It follows, therefore, that good roads result in lower vehicle operating costs.

Highway improvements result in speedier travel. Savings in travel time are enjoyed by occupants of vehicles, goods in transit and the vehicles in transit.

Road accident rates are governed to a certain extent by the condition of the road as described in Chapter 27 Highway improvements can thus bring about a reduction in road accidents. The cost of road accidents, which have been eliminated by highway improvements, represents a benefit.

When carrying out economic analysis, costs and benefits are considered exclusive of taxes. Taxes do not represent an economic cost and represent only a transfer within the community. Insurance premia are also excluded from economic analysis since the savings in accidents already account for this element.

In a developing country, there are certain resources which are scarcer than the others. The prevailing market prices, therefore, do not reflect the true economic value of the resources. In order to correct such distortions and imperfections, "shadow pricing" is done. A case in example is the cost of imported fuel in India. Since foreign exchange reserves are very precious, such imported items are shadow priced at a higher value than the market price when carrying out the economic evaluation. Similarly, unskilled labour is surplus in India and the prevailing wage rate (which is statutorily fixed) may not truly reflect this situation. A shadow-pricing of such labour at a slightly lower level would then be appropriate.

Inflation is disregarded in economic analysis, as it is generally assumed that all prices increase in the same proportion, but relative prices remain constant. But if differential inflation is expected to occur among commodities, necessary adjustments need to be made.

7.5. EVALUATION TECHNIQUES

The methods commonly adopted for economic evaluation are:

1. Net Present Value Method (NPV)
2. Benefit/Cost Ratio Method (B/C Ratio)
3. Interest Rate of Return (IRR) Method.

7.5.1. Net present value (NPV) method

The NPV method is based on the discounted cash flow (DCF) technique. In this method, the stream of costs and benefits associated with the project over its time horizon is calculated and is discounted at a selected discount rate to give the present value. Benefits are treated as positive and costs are treated as negative. Any project with a positive NPV is treated as acceptable. In comparing more than one project, a project with the highest NPV is selected.

The NPV is algebraically expressed as:

$$\text{NPV}_0 = (B_0 - C_0) + \frac{B_1 - C_1}{(1+i)} + \frac{B_2 - C_2}{(1+i)^2} \ldots + \frac{B_n - C_n}{(1+i)^n} \tag{7.7}$$

Where NPV_0 = Net present value in the year 0

B_t = Value of benefits which occur in the year t

C_t = Value of costs which occur in the year t

i = discount rate per annum

n = number of years considered for analysis.

The method is illustrated in the following problem.

Problem 7.1. The cost of improving an existing road, 25 km long, is Rs. 4.00 lakhs per km. The (*i*) road user costs, with and without the improvements, (*ii*) accident costs, with and without improvement, and (*iii*) maintenance costs, with and without the improvements are tabulated in Table 1 for a 10-year period after the completion of the improvements. Assuming a discount rate of 10 per cent, find out whether the project is economically justifiable. Use the NPV method.

Table 7.1. Computation of NPV (All amounts in Lakhs of Rupees)

Year (t) (1)	Rood User Costs		Accident Costs		Maintenance Costs		Benefits (Bt) Cols. [(3) + (5) + (7)] – coll [(2) + (4) + (6)] (8)	$B_t - C_t$ (9)	$\frac{B_t - C_t}{(1+0.1)^t}$ (10)
	With impr. (2)	Without impr. (3)	With impr. (4)	Without impr. (5)	With impr. (6)	Without impr. (7)			
0	–	–	–	–	–	–	– 100.0	–100.0	–100.0
1	105.5	126.5	1.1	3.1	3.5	2.5	22.0	22.0	+ 20.0
2	110.3	132.2	1.1	3.1	3.5	2.5	23.1	23.1	+ 19.1
3	115.8	138.9	1.2	3.5	3.5	2.5	24:4	24.4	+ 18.4
4	121.6	145.8	1.2	3.7	3.5	2.5	25.7	25.5	+ 17.6
5	127.6	153.0	1.3	3.8	3.5	2.5	26.9	26.9	+ 16.7
6	134.0	161.0	1.3	4.0	3.5	2.5	28.7	28.7	+ 16.2
7	140.7	168.9	1.4	4.0	3.5	2.5	30.0	30.0	+ 15.4

8	147.8	177.0	1.5	4.4	3.5	2.5	31.1	31.1	+ 14.5
9	155.1	186.2	1.6	4.7	3.5	2.5	33.2	33.2	+ 14.1
10	162,9	195.2	1.6	4.9	3.5	2.5	34.6	34.6	+ 13.4
									+ 165.4
									– 100.0
								NPV	= + 65.4

Solution: The calculations are tabulated in Table 7.1.

Cost of improvements = Rs. 25 × 4 = Rs. 100 lakhs

NPV = (+ Rs. 165.4 – Rs. 100) lakhs = + Rs. 65.4 lakhs

Since the NPV is positive, the project is economically justified

7.5.2. Benefit-Cost (B/C) ratio method

There are a number of variations of this method, but a simple procedure is to discount all costs and benefits to their present worth and calculate the ratio of the benefits to costs. Negative flows are considered costs, and positive flow as benefits. Thus the savings in the transport costs are considered as benefits. If the B/C ratio is more than one, the project is worth undertaking.

Problem 7.2. An existing single lane road, 30 km long, is to be widened to two lanes.

The cost of widening is Rs. 10 lakhs per km. The vehicle operating costs, accident costs and maintenance costs, with and without widening, for a 10 year period are tabulated in Table 7.2. The discount rate is 12 per cent. Is the project worthwhile?

In the AASHTO practice of road-user analysis, the B/C ratio expresses the ratio of the net annual benefits to the net annual costs The benefits are determined for a simple reference year, which for convenience can be the first year of operation after construction or the median year of the analysis period.

Table 7.2: Compulation of B/C ratio (All figures in Lakhs of Rupees)

Year (t)	Rood user costs		Accident costs		Maintenance costs		Benefits (B_t) cols. [(3) + (5) + (7)] – cols. [(2) + (4) + (6)]	Discounted benefits $\frac{B_t}{(1+0.12)^t}$
	With impr.	Without impr.	With impr.	Without impr.	With impr.	Without impr.		
(1)	(2)	(3)	(4)	(5)	(6)	(7)	(8)	(9)
1.	101.5	160.7	2.5	3.6	10.0	7.5	57.8	51.6
2.	105.6	168.2	2.6	3.7	10.0	7.5	61.2	48.8
3.	110.2	176.3	2.7	3.8	10.0	7.5	64.7	48.1

4.	116.2	185.2	2.8	3.9	10.0	7.5	64.6	43.0
5.	122.3	190.0	2.9	4.0	10.0	7.5	66.2	37.6
6.	128.4	199.0	2.9	4.0	10.0	7.5	69.7	35.3
7.	135.6	210.0	3.0	4.1	10.0	7.5	73.0	33.0
8.	143.2	219.5	3.1	4.2	10.0	7.5	74.9	30.3
9.	149.1	228.2	3.2	4.3	10.0	7.5	77.7	28.0
10.	154.6	240.1	3.2	4.3	10.0	7.5	84.1	27.1
							Total	382.8

Solution. Cost of project = Rs. 30 × 10 lakhs = Rs. 300 lakhs

Benefit/cost ratio $= \dfrac{382.8}{300}$ 1.27, greater than 1.

Hence the project is economically justified.

7.5.3. Interest rate of return (IRR) method

The interest rate of return is the discount rate which makes the discounted future benefits equal to the initial outlay. In other words, it is the discount rate which makes the stream of cash flows to zero. Equation (7.7) can be arranged as below, assuming $B_0 = 0$

$$C_0 = \frac{B_1 - C_1}{(1+i)} + \frac{B_2 - C_2}{(1+i)^2} + \ldots\ldots \frac{B_n - C_n}{(1+i)^n} \tag{7.8}$$

The solution to the above equation can be done by trial and error, a rather tedious process. With a computer programme, the work is rendered very simple.

If the interest rate of return calculated from the above formula is greater than the rate of interest obtainable by investing the capital in the open market, the scheme is considered acceptable.

7.5.4. Comparison of the various methods of economic evaluation

The three methods of economic evaluation described above have their own advantages and shortcomings.

The B/C ratio method is very widely used by the highway engineers. It however, suffers from the following drawbacks :

1. It requires an assumption of a discount rate, which should bear relation to the opportunity cost of capital. It is, however, rather difficult to know the opportunity cost of capital accurately.
2. The significance of the B/C ratio is ambiguous, and its relative value is difficult to understand and interpret. For instance, if there are two proposals, one with a B/C ratio of 1.05 and the other with a ratio of 1.10 the difference is very difficult to appreciate.

3. It is somewhat confusing and difficult to decide which items should be turned as costs and placed in the numerator.

The IRR method is popular with international lending agencies like the World Bank. It lends itself admirably well for use in a computer-aided design model. It avoids the need for selecting a discount rate initially. The rate derived from computations can be easily compared with the market rate of interest, with which economists, financial experts and banks are familiar. It disadvantage is that the computations are tedious and a solution can be obtained only by trial and error.

The NPV method suffers from the same disadvantage as in case of B/C ratio method in that a rate of discount has to be assumed.

7.5.5. Selection of discount rate

As seen from the discussion above, the selection of an appropriate discount rate (or interest rate) is crucial in the B/C ratio and NPV methods.

The choice of the discount rate is governed by a number of complex factors, and is dependent on the future availability of finance and the various opportunities for its use. The attitude of the society towards present consumption as against saving for future is an important factor. Will the present generation prefer to consume the resources now or conserve it for future use by the current or future generations ? The answer to this question will give the "social time preference rate of interest" . Another approach is to find out the social yield that the resources employed by a marginal public project would have otherwise generated. This determines the "social opportunity cost rate of interest". In a truly competitive economy, the two rates of interest would be equal and investments and consumption would then be ideally allocated. But such a situation is difficult to find, and more so in a developing country where capital is very scarce. In such situations, some general guidelines can be given for selecting an appropriate discount rate. Such a rate should not be less than the rate of borrowing or lending by the Govt, or the market rate of interest. A rate of 12 per cent is generally being adopted.

7.6. ECONOMIC EVALUATION OF RURAL ROAD PROJECTS

"Rural Roads" means the low-volume roads connecting villages. When a new road is built to give access to a remote village, the traditional method of quantifying benefits on account of vehicle operating cost savings, travel time saving and reduction in accident costs fails because there was no vehicular traffic before the construction of the road. In such cases, benefits can still be quantified by using the agricultural producer surplus approach and assigning monetary values to social benefits.

7.6.1. Producer surplus approach

This approach, pioneered by the World Bank nearly three decades ago, is as under :

(*a*) In the absence of any road, the isolated village community may be producing just the quantity needed for its consumption (subsistence farming). The transport costs are so high that the farmer does not find it remunerative enough to take it to the market and sell it at the prevailing market price.

(*b*) If a road is built, it may result in the following consequences:

- the transport costs come down and the farm-gate price (the price received by the farmer for his product at the farm gate) is high enough to induce him to grow more of the commodity and sell it, as seen from the shift in production from Q_1 to Q_2 as the farm-gate price increases from P_1 to P_2 in Marginal cost curve MC_1 (Fig 7.1)

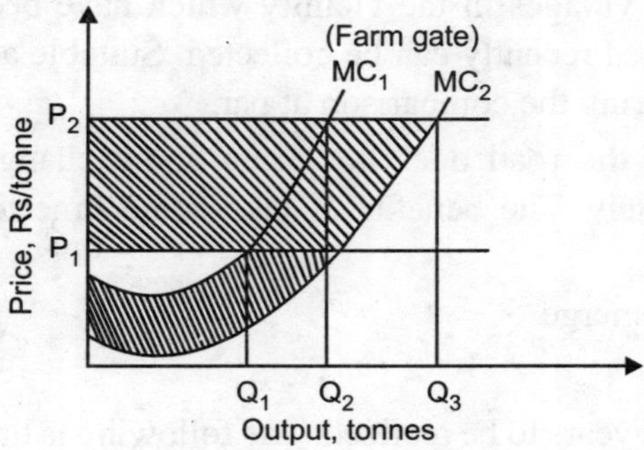

Fig. 7.1: Marginal cost curves

- as the transport costs come down, the cost of agricultural inputs also come down thus making the production costs come down. The marginal cost curve MC_1 thus shifts downwards to MC_2 increasing the production from Q_2 further to Q_3.
- the shaded area gives the farm-level benefits.
- the farner may diversify his agricultural operations, shifting from the traditional food-crops under subsistence farming to more remunerative crops like fruits, vegetables, flowers, mushrooms etc.

(*c*) Thus, the benefits can be evaluated for each crop. The steps involved are :

- determine the production of each crop without the project in the base year (Q_1)
- determine the local consumption of each crop without the project in the base year (Q_{1c})
- determine the farm-gate price without the project (market price minus transport cost) (P_1)
- determine the value of exports of each commodity without the project $[(Q_1 - O_{1c})P_1]$
- determine similarly the value of exports of each commodity with the project $[(O_2 - Q_{2c})\, P_2]$
- sum up for the *n* crops and find the difference between "with" and "without" which is the benefit (B_t)

$$B_1 = [\Sigma(Q_2 - Q_{2c})P_2] - [\Sigma(Q_1 - Q_{1C})P_1]$$

There are several other issues that need to be looked into. These are :

1. In the determination of the production of crops, what area of land should be considered ? The answer is: The land area which comes under the zone of influence of the road. This can be demarcated on a map showing the rural road network, the distances. Between farms and market centres, the terrain and the means of transport.

2. The production of crops can be determined from a base-line survey before the road is constructed. How can the cropping pattern and production be determined after the road is constructed at the time of project formulation ?

   If the post-project analysis is to be done, the data can be collected after the project is completed. But if pre-project analysis is to be done, data on the cropping pattern and production from the villages in the vicinity which have been already benefited from a road project completed recently can be collected. Suitable adjustments need to be made for the land area to bring the comparison at par.

3. The construction of the road does not bring about changes in cropping pattern and production immediately. The benefits B_t take some time to fructify. How can this be accounted for?

Three possible scenarios emerge

Scenario I (Optimistic)

The full benefits take three years to be realized. The following is the distribution of the benefits.

Year 1	50%
Year 2	80%
Year 3	100%

Scenario II

The full benefits take four years to be realized as under :

Year 1	25%
Year 2	50%
Year 3	75%
Year 4	100%

Scenario III

The full benefits are realized in a period of 7 years as under:

Year 1	2.5%
Year 2	5%
Year 3	10%
Year 4	25%
Year 5	40%
Year 6	65%
Year 7	100%

The above three Scenarios are shown in Fig. 7.2.

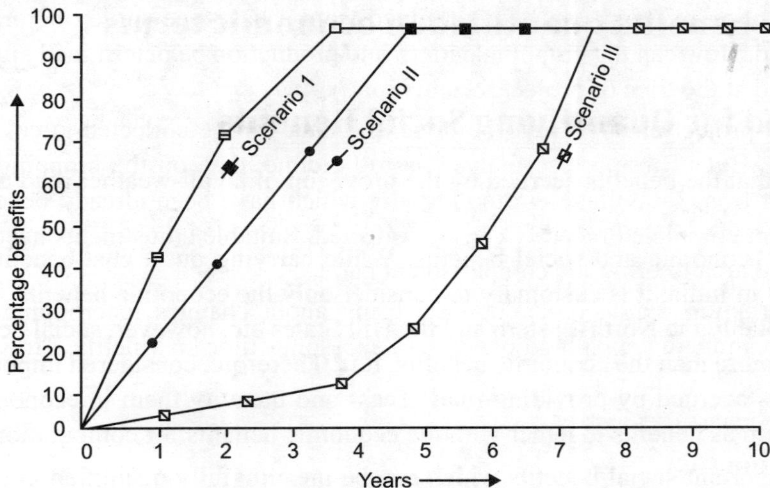

Fig 7.2: Percentage of benefits accrued due to the road with time

From among the three scenarios (I,II and III) considered above, Scenario 1 (wherein the full benefits accruing from the road are realized in 3 year's time) is found to be close to the ground reality.

7.6.2. Saving in vehicle operating cost

When good paved all-weather roads are provided to link villages, one tangible benefit is the reduction in vehicle operating cost (VOC). Traffic which was using earthen tracks before and hence incurring higher operating costs (fuel, lubricants, tyres, spare parts, depreciation etc) will perceive monetary benefits. The subject of VOC, as influenced by road condition (geometry, riding quality etc), has been well researched in India. In fact, standard tables giving the values of VOC components have been published by the Indian. Roads Congress and can be used.

It should, however, be noted that the benefit due to reduction in VOC of vehicles carrying agricultural produce to the market has already been accounted for through the increased quantity of goods produced (in the Producer Surplus approach) and should not be counted again. The benefit due to reduction in VOC of vehicles other than those carrying agricultural goods needs to be considered.

7.6.3. Savings in travel time of passengers

As roads get built, the travel time of villagers, particularly to the market centre, gets reduced. A value may be assigned to these savings approximately by the simple wage rate approach. Either the full wage rate or a fraction of it can be considered.

7.6.5. Social benefits quantified in economic terms

7.6.5.1. Need for Quantifying Social Benefits

It is well known that the benefits accrued by the provision of an all-weather road access to a rural habitation

Include both economic and social benefits. While carrying out a cost-benefit analysis for a rural road project in India, it is customary to consider only the economic benefits. In several parts of the country, notably the North-Eastern and the Hill States etc, however, social benefits can be as important, if not more than the economic benefits. It is. Therefore, considered important to identify the social benefits accrued by providing road access and quantity them in economic terms so as to incorporate them as benefits together with the economic benefits for computation of the EIRR.

The more important social benefits which can be meaningfully quantified in economic terms pertain to the increase in average annual income of households ; increase in the adult literacy level, thereby increasing the earning potential of workers and also bringing down the growth rate of population ; improved health standards which in turn lead to higher output etc.

7.6.5.2. Increase in Household Income

As an access road is provided to a village, the opportunities for better education and health, employment and marketing the non-agricultural output like handicrafts increase. Thus, if a comparison is made between the household income of an average family in a connected village and an unconnected village, the former is always higher. The different is the benefit. From this benefit, the agricultural producer surplus (para 7.5.1) can be deducted to arrive at the benefits due to other reasons.

QUESTIONS

1. Why is economic evaluation of highway schemes done ?
2. What are the basic principles of highway economic analysis ?
3. What are the common methods of economic analysis and their advantages and disadvantages ?
4. The cost of improving an existing road, 30 km long, is Rs. 2,000,000 per km spread equally over three years. The improvement consists of strengthening the existing pavement, providing a smooth bituminous concrete surface and improvement of geometrics. The present traffic consists of :

   Commercial Vehicles per day : 2000

   Cars per day : 1500

   Two-wheelers per day : 1200

   Total Vehicles per day : 4700

The cost of operation on the existing road and improved road is as under:

Cost of Operation (Rs. 1 Km)			
		Existing Road	Improved Road
(*a*)	Commercial Vehicle	5.00	4.00
(*b*)	Car	3:00	2.50
(*c*)	Two-wheeler	0.60	0.50

The accident rate on the existing road and improved road is as under:

Existing Road :1.5 per mil veh km

Improved Road :1.0 per mil veh km

The cost of one accident is Rs. 100, 000

The rate of growth of traffic is 10 per cent annum.

The annual maintenance costs per km Rs. 30,000 and Rs. 40,000 per km on the old and new surface respectively. The analysis period is 10 years after opening the road to traffic. The discount rate is 12 per cent. Carry out the economic analysis using B/C Ration method. **[Ans. B/C Ratio = 5.5]**

5. The cost of minor improvements to an existing roac/25 km long, is Rs. 4.0 million per Km. The cost is incurred in the first year itself. The (*i*) road user cost, with and without improvements; (*ii*) accident costs, with and without improvements and (*iii*) maintenance costs', with and without improvements are tabulated in Table 7.3 for a 10-year period after completion of the improvements. Assuming a discount rate of 10 per cent, find out whether the project is economically viable. Use the NPV Method.

Table 7.3: All Costs in Million Rupees

Year	Rood user costs		Accident costs		Maintenance costs	
	With impr.	Without impr.	With impr.	Without impr.	With impr.	Without impr.
0	100.9	100.9	1.1	1.1	2.5	2.5
1	105.5	126.9	1.1	3.1	3.5	2.5
2	110.3	132.2	1.1	3.1	3.5	2.5
3	115.8	138.9	1.2	3.5	3.5	2.5
4	121.6	145.9	1.2	3.7	3.5	2.5
5	127.6	153.0	1.3	3.8	3.5	2.5
6	134.0	161.0	1.3	4.0	3.5	2.5
7	140.7	168.9	1.4	4.0	3.5	2.5
8	147.8	177.0	1.5	4.4	3.5	2.5
9	155.1	186.2	1.6	4.7	3.5	2.5
10	162.9	195.2	1.6	4.9	3.5	2.5

[Ans. NPV = + 65.4 million rupees, project is viable]

Chapter

EMBANKMENT DESIGN AND CONSTRUCTION

8.1. IMPORTANCE OF PROPER DESIGN AND CONSTRUCTION OF EMBANKMENT

The road pavement rests on an embankment which is suitably kept above the natural ground level. In stretches where cutting of the natural ground is involved, the ground after cutting to the desired level also needs attention because it may not have the desired degree of compaction. The type, thickness and composition of the pavement depend upon the nature of the embankment soil. High embankments need special attention to ensure the stability of slopes and failures due to settlement.

8.2. TOP WIDTH OF EMBANKMENT

The top width of the embankment includes the carriageway width, median, if provided, and shoulders. This aspect has been discussed in Chapter 4.

8.3. HEIGHT OF EMBANKMENT

The height of the embankment is determined by the longitudinal profile selected. The general rule followed is that the bottom of the pavement is at least 0.6 m and preferably 1.0 m abode e the high flood level or water table.

8.4. SIDE SLOPES

Normally 2:1 (2 horizontal and 1 vertical) slope is adequate for low embankments in uninundated conditions and 3:1 for inundated conditions. For high embankments, the slopes given are discussed in Chapter 7.

8.5. STABILITY OF HIGH EMBANKMENT SLOPES

Side slopes of high embankments can fail due to rotational failure. The most popular method to evaluate slope failure is by the Swedish Circle method also referred to as the Fellenius method Two types of analysis are done:

(*i*) $\phi = 0$ analysis (Total stress analysis)

(*ii*) c, ϕ analysis (Effective Stress analysis)

(*i*) The total stress analysis is carried out when the embankment is built on saturated clay and the internal angle of friction $\phi = 0$. Fig. 8.1 gives the forces under $\phi = 0$ analysis.

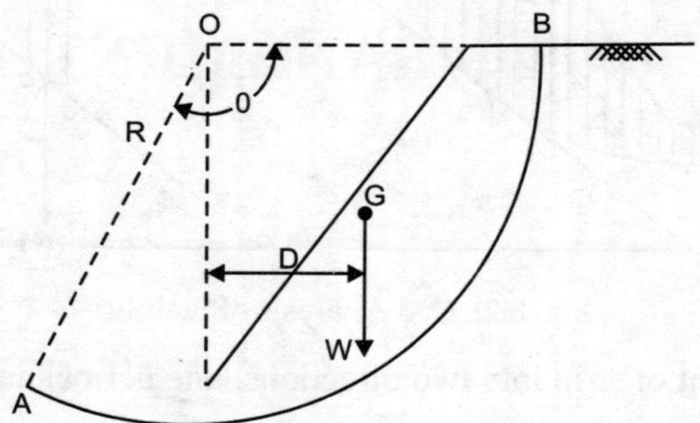

Fig. 8.1: $\phi = 0$ analysis of stability

R = radius of arc of circle

W = weight of soil in the failure zone acting at centroid G

D = Horizontal Distance between centre of circle and centroid G

s = shear strength per unit area

θ = angle subtended by the arc AB

Disturbing moment = W.D.

Restoring moments = s × length of arc AB × R

= s R × θ R

$$= sR^2\theta \tag{8.1}$$

When failure is about to take place disturbing moment is equal to the restoring moment.

$$\therefore WD = sR^2\theta \tag{8.2}$$

Factor of safety against rotational failure

$$= \frac{\text{Restoring moment}}{\text{Disturbing moment}} = \frac{sR^2\theta}{WD} \tag{8.3}$$

(*ii*) The c-ϕ analysis is done by the method of slices as illustrated in Fig. 8.2.

Let w = weight of strip abcd

θ = angle of inclination of tangent to horizontal

ϕ = angle of shearing resistance

c = apparent cohesion of soil

l = length of arc of the strip

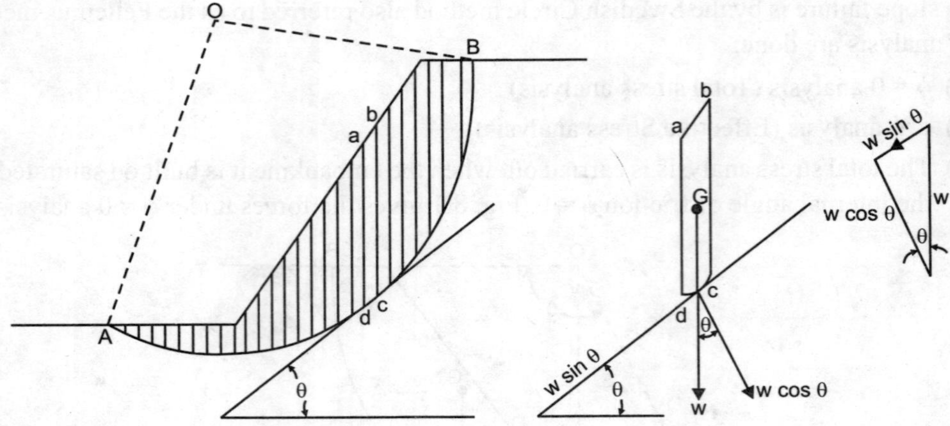

Fig. 8.2: C, ϕ Analysis of Stability

Resolving w the weight of strip into two directions, one normal and the other tangential to the arc cd.

$$\text{Disturbing force} = w \sin \theta \tag{8.4}$$

$$\text{Restoring force} = cf + w \cos \theta \tan \theta \tag{8.5}$$

$$\text{Factor of safety} = \frac{cf + w \cos \theta \tan \theta}{w \sin \theta} \tag{8.6}$$

The critical circle is the one which has the lowest factor of safety. It is determined by the Fellenius method shown in Fig. 8.3.

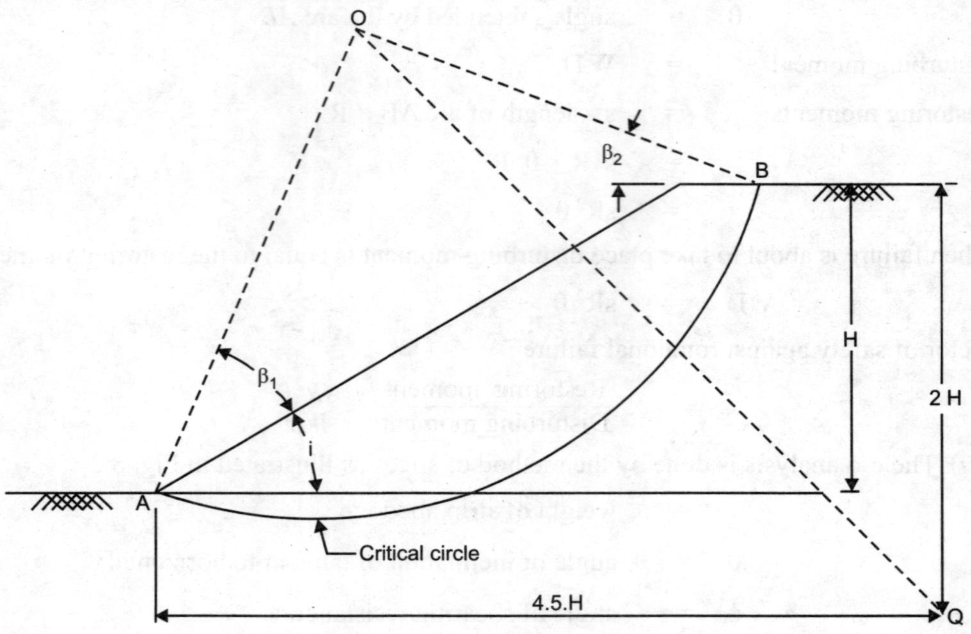

Fig. 8.3: Fellenius construction

In this method, the centre of the critical circle lies on the extension of the line through *O*, (when $\phi = 0$) from a point *Q* located at a depth of 2H from the top of the road embankment and 4.5 H horizontally from the toe of the embankment, *H* being the height of the embankment. The location 0 is determined by the angles β_1 and β_2 which are given in Table 8.1 for different embankment slopes.

Table 8.1 : Angles β_1 and β_2 for different embankment slopes

Slope	i°	β_1	β_2
1 : 0.58	60	29	40
1 : 1	45	28	47
1 : 1.5	33.8	26	35
1 : 2	26.6	25	35
1 : 3	18.4	25	35
1 : 5	11.3	25	37

8.6. SETTLEMENT OF EMBANKMENT ON CLAYEY STRATA

Clay is a highly compressible soil and hence embankments constructed on clayey strata undergo settlement. Settlement causes undulations in the pavement even though the body of the embankment is well compacted. The amount of settlement depends upon the decrease in the void ratio. The following analysis is used for determining the settlement of clayey strata under the embankment:

Let *H* be the thickness of the clay strata subjected to consolidation. Let e_0 be the initial void ratio which reduces to e as the pressure on clay increases from p_0 to $p_0 + \Delta p$ as a result of the load imposed by the embankment.

$$e_0 - e = a_v \Delta p$$

where a_v = coefficient of compressibility (8.7)

$$\therefore \quad a_v = \frac{e_0 - e}{\Delta p} \tag{8.8}$$

Decrease in porosity Δ_n is given by:

$$\Delta_n = \frac{\Delta e}{1+e_0} \tag{8.9}$$

$$= \frac{a_v}{1+e_0} \Delta p = m_v \Delta p \tag{8.10}$$

where m_v = coefficient of volume compressibility

$$= \frac{a_v}{1+e_0} \tag{8.11}$$

$$S = \text{settlement of clay stratum}$$

$$= H.\,\Delta p\, m_v \tag{8.12}$$

Assuming that the field consolidation for clay is a straight line on the semi-log scale (Fig. 8.4),

$$e = e_0 - C_c \log_{10} \frac{p_0 + \Delta p}{p_0} \tag{8.13}$$

where C_c = Compression Index

Combining 8.13 with 8.8 and 8.11, one gets

$$A_v = \frac{C_c}{\Delta p} \log_{10} \frac{p_0 + \Delta p}{p_0} \tag{8.14}$$

and

$$m_v = \frac{C_c}{\Delta p(1+e_0)} \log_{10} \frac{p_0 + \Delta p}{p_0} \tag{8.15}$$

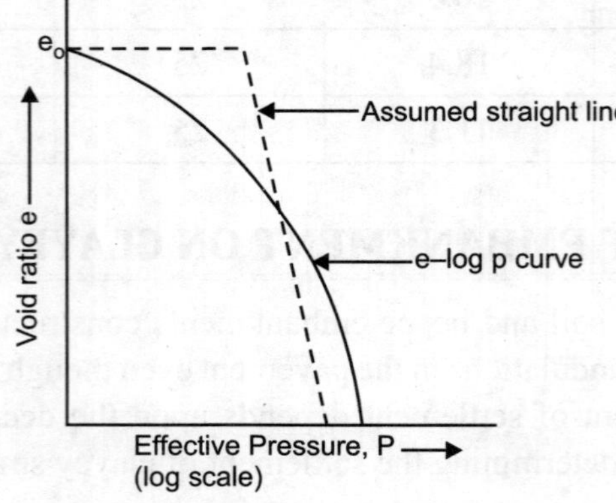

Fig. 8.4: e-log p curve

Substituting the value of m_v in 8.12,

$$S = H \frac{C_c}{1+e_0} \log_{10} \frac{p_0 + \Delta p}{p_0} \tag{8.16}$$

Time required for consolidation (t) can be found as under:

$$t = \frac{TH^2}{C_v} \tag{8.17}$$

where T = Time factor corresponding to the degree of consolidation

C_v = Coefficient of consolidation

H = Length of effective drainage path

$= D$ for one-way drainage

$= \frac{D}{2}$ for two-way drainage

D = Depth of stratum

8.7. METHODS OF EMBANKMENT CONSTRUCTION IN PROBLEMATIC LOCATIONS

Construction of embankment in highly slushy or soft soil can be done using one of the following methods:

(*i*) Use of light weight material like flyash, cinder etc.

(*ii*) Removal of problematic soil by excavation or bog blasting

(*iii*) Construction of the embankment in stages so that settlements can take place before the pavement is constructed.

(*iv*) Placing surcharge fill above the finally desired level of the embankment

(*v*) Use of sand drains (Fig. 9.5)

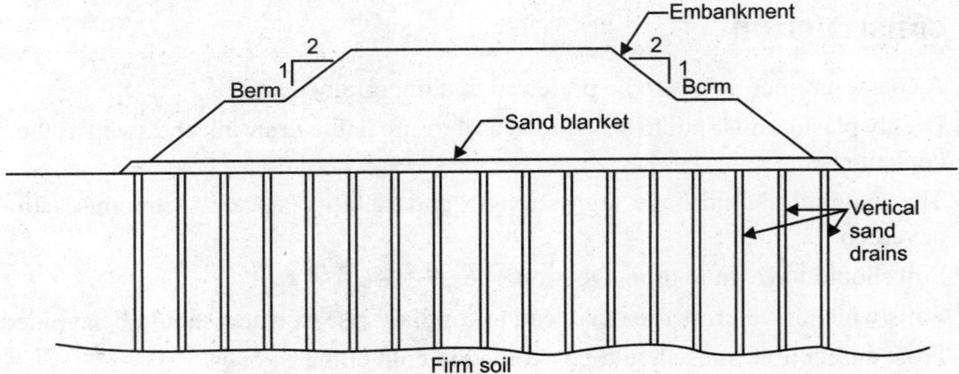

Fig. 9.5: Sand drain arrangement

8.8. SELECTION OF SOIL FOR EMBANKMENT CONSTRUCTION

8.8.1. Suitable fill materials

The following materials can be used in forming the embankment:

(*i*) Soil

(*ii*) Moorum

(*iii*) Gravel

(*iv*) Reclaimed material from pavement

(*v*) Fly ash and pond ash

8.8.2. Unsuitable fill materials

The following materials are unsuitable as fill material:

(*i*) Materials from swamps, marshes and bogs

(*ii*) Peat, log, stump and perishable material; any soil classified as OL, OI, OH or Pt in accordance with IS:1498

(*iii*) Materials susceptible to spontaneous combustion

(*iv*) Materials in a frozen condition

(*v*) Clay having liquid limit exceeding 50 and plasticity index exceeding 25

(*vi*) Materials with salts resulting in leaching in the embankment

(*vii*) Materials with total sulphate content (expressed as SO_2) exceeding 0.5 per cent by mass, if deposited within 500 m, or other distance prescribed in the Contract, or metallic items forming part of permanent works.

8.8.3. Guidelines for selection of materials for embankment construction

(*i*) A coarse-grained soil is to be preferred to a tine-grained soil.

(*ii*) Highly plastic soils should be avoided as they are liable to shrink and swell as the moisture content changes.

(*iii*) The material should have high strength and stability since it economises the cost of pavement.

(*iv*) Soil should have good drainage properties.

(*v*) Soils which have properties that lead to capillary rise of water should be avoided.

(*vi*) Frost-susceptible soils should be avoided in cold climate zones.

(*vii*) Soils containing highly organic matter should be avoided.

(*viii*) Soils containing sodium sulphate in excess of 0.2 per cent should be avoided.

(*ix*) Soils having a maximum dry density of less than 1.44 gm/cc are generally unsuitable.

(*x*) Soils having a maximum dry density of less than 1.52 gm/cc are to be avoided for use in embankments of height exceeding 3 m or in embankments of any height subject to long periods of inundation.

(*xi*) The top 500 mm of the embankment (subgrade) shall be constructed with soils having a maximum dry density greater than 1.65 gm/cc.

8.9. EMBANKMENT CONSTRUCTION PRACTICE

Before starting the embankment construction, the original ground shall be cleared and grubbed, removing all grass, shrubs and tree stumps. The top soil shall, thereafter, be stripped and stored for use on embankment slopes thus facilitating the growth of turf. The original ground, if soft and loose, and has a density of less than 90 per cent of Proctor's density, shall be loosened, watered and compacted in layers of 250 mm to obtain a density of 100 per cent of Proctor's density. This treatment shall cover a thickness of 500 mm.

Soil shall be obtained from borrow pits observing the following practices

(*i*) Borrow-pits shall be rectangular in shape

(*ii*) They shall be near to the boundary of the Right-of-Way

(*iii*) No borrow-pit shall be located with 5 m from the toe of the final embankment after accounting for any future widening.

(*iv*) The depth of borrow-pits should not exceed 1.5 m.

(*v*) The depth of borrow-pit and its location should be so regulated that it should not cut an imaginary line starting from the edge of the embankment and having a slope of 1:4 (Fig. 8.6).

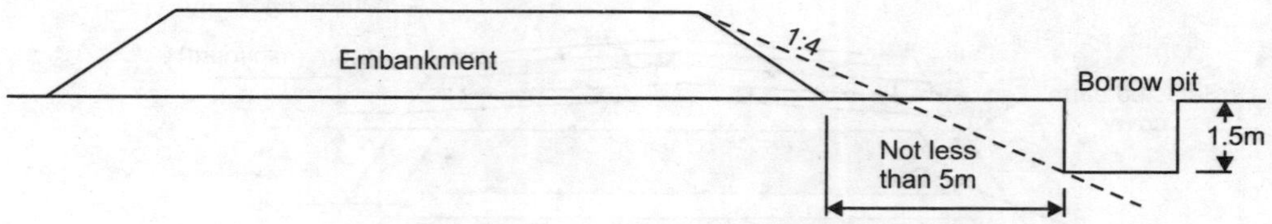

Fig. 8.6: Location of borrow pit

(*vi*) Borrow-pits should not be dug continuously. Ridges not less than 8 m wide should be left at intervals not exceeding 300 m.

(*vii*) If borrow-pits are dug in temporarily acquired land, their depth should not be greater than 450 mm.

(*viii*) Borrow-pits shall not be dug within 0.8 Km of habitations since stagnant water in them can breed mosquitoes.

The borrowed soil is deposited in layers and compacted by rollers. The depth of each compacted layer shall not exceed 250 mm. The compaction shall be done at Optimum Moisture Content (OMC) to achieve the maximum density as under.

Table 8.2: Compaction requirement of embankment

Type of work/material	Relative compaction as per cent of max. lab. dry density as per IS:272() (Part 8)
1. Sub-grade and shoulders	Not less than 97%
2. Embankment	Not less than 95%
3. Expansive clays	
Subgrade and 500 mm below subgrade	Not allowed
Remaining portion of embankment	90-95%

If the embankment is to be widened, and its slopes are steeper than 1 in 4, continuous horizontal benches at least 300 mm wide shall be cut to ensure adequate bond with the new material. If the slope is flatter than 1 in 4, the slope surface may be scarified before placing the new material.

8.10. CONSTRUCTION OF EMBANKMENT USING FLY-ASH

Fly-ash is an industrial waste and is available at no cost near thermal plants. Its disposal is an environmental problem. Fly-ash can be used in the construction of road embankment, and Government of India have made it mandatory that it should be used in road construction within 100 Km of a thermal plant. It is a light-weight material and can be used advantageously in high embankments on soft soil. Fly-ash embankments shall be covered on the sides and top by soil to prevent erosion. Typical cross-sections are given in Fig. 8.7 and 8.8. The side cover is laid along with each layer of fly-ash. The layer thickness can be in the range of 200-400 mm depending upon the compaction equipment used.

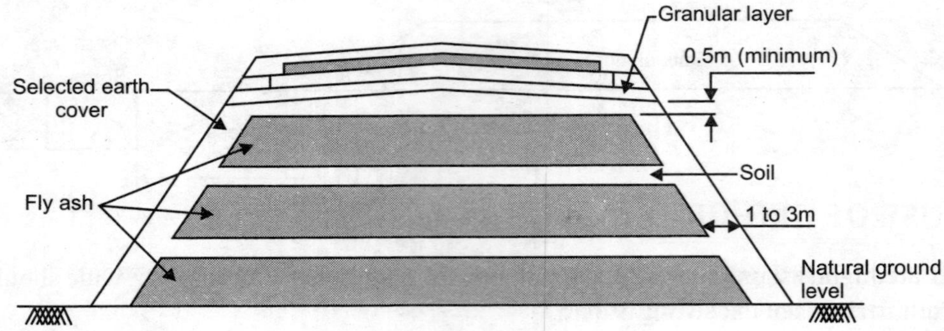

Fig. 8.7: Typical cross-section of embankment with alternate layer of fly-ash and soil

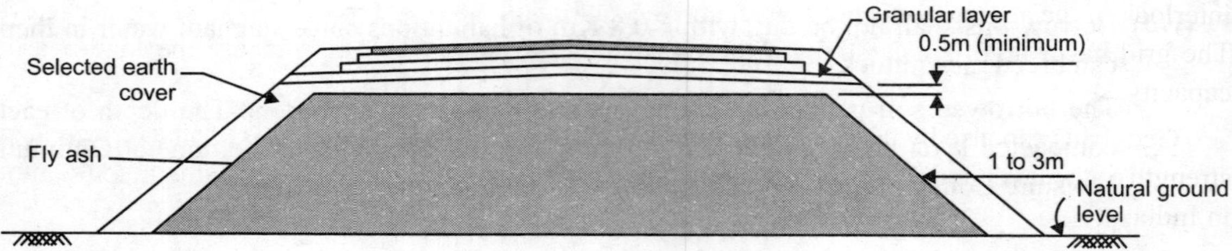

Fig. 8.8: Typical cross-section of em`bankment with core of fly-ash

8.11. USE OF GEOTEXTILES

One of the new developments in soil engineering is the use of Geotextiles. These are synthetic fabrics, made out of nylon, polyester, polypropylene or polyethylene. Being non-decomposable they find ready application in highway engineering problems. They are now manufactured in India. The main characteristics which are of use to a highway-engineer are their strength, permeability and non-decomposability. They are used for the following purposes :

(*i*) Separation of two widely different natural soils to prevent their mixing under the influence of water. For example, if a base course of stone aggregate is laid on a sub-grade layer of expansive soil, the latter can work up into the interstices of the former, thus imparing its strength. If a geotextile layer is interposed in between physical separation is ensured.

(*ii*) As a filtering medium, the fabric of a proper aperture size can prevent the loss of particles of size larger than the aperture. For this purpose, the equivalent opening size of the fabric should be less than D_{85} of the soil, D_{85} of the soil is the 85 per cent size, i.e 85 per cent being less than this size).

(*iii*) As a reinforcing layer, the fabric increases the tensile strength of the soil-fabric system. This permits greater loads on weak soils, thus resulting in economy in pavement construction.

Fig 8.9 illustrates the typical uses.

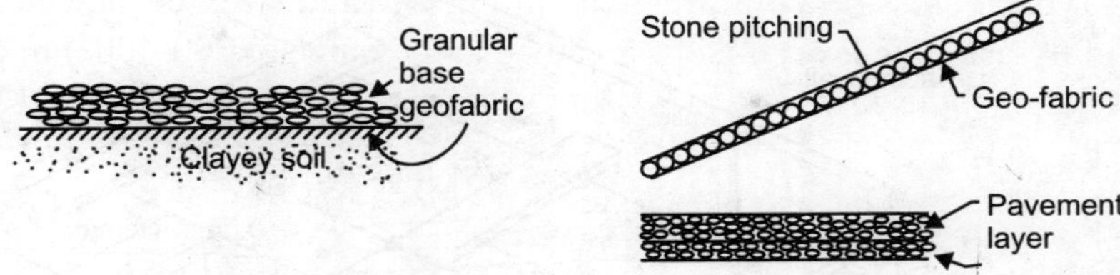

Fig. 8.9: Various uses of Geo-fabrics

8.12. USE OF GEO-GRID

Geo-grids are high strength plastic grids which are interposed either at the bottom of a layer or in a layer itself to enhance the strength. The material used is a polyethylene or polypropylene. When used at the bottom of a granular layer and above a sub-grade, the grid increases the load-bearing capacity. The grid supports the overlying granular layers at the points of contact. The mechanical interlock of the aggregates is held intact and the sinking of the aggregates into the soil is prevented. The grid has a high tensile strength through which the grid is able to increase the load-bearing capacity.

Geo-grids can also by used as reinforcing layers in bituminous layers. This results in increasing strength of the layer and thus bringing about about economy. Geo-grids are now being manufactured in India.

8.13. REINFORCED EARTH

8.13.1. General

Reinforced earth, often used in lieu of a conventional retaining wall, comprises of a compacted soil mass (generally coarse-grained in nature) within which tensile reinforcing elements are embedded. The principle of introducing tensile reinforcement in soil is not new. In the old days bamboo rods, straw etc. have been used to reinforcement in soil is not new-in the old days bamboo rods, straw etc. have been used to reinforce mud walls and bricks. Basically there are 3 components in a reinforced earth (Fig. 8.10 retaining wall):

(*a*) compacted earth fill, usually coarse-grained, with less than 15% passing 75 micron sieve,

(*b*) reinforcement in the form of metallic strips or rods with cross-section as per design requirements or geosynthetics-strips, sheets, geogrids etc., fastened to the facing unit and extending into the backfill to a length as per design requirement and (*c*) facing wall, generally consisting of precast reinforced concrete units or *U* -shaped steel section etc., to avoid soil erosion between reinforcements and to improve outside appearance. Within the reinforced earth wall, the soil mass derives its strength from the forces of friction developed between the soil reinforcing elements, the stresses within soil mass being transferred to reinforcing elements with adequate design tensile strength.

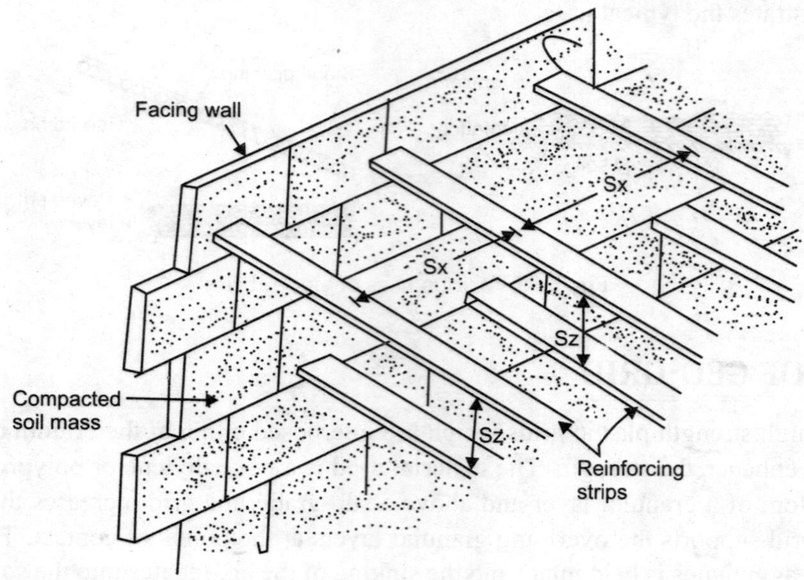

Fig 8.10: Arrangement of reinforcing strips in reinforced earth.

8.13.2. Design principles

The main considerations in reinforced earth design are (*i*) the tensile stresses which are to be carried by the reinforcing elements and (*ii*) the frictional resistance between the reinforcement and the soil.

In a reinforced earth retaining wall of height H (Fig. 8.11) the vertical stress in backfill material at any point at depth z, due to the soil column above it is given by :

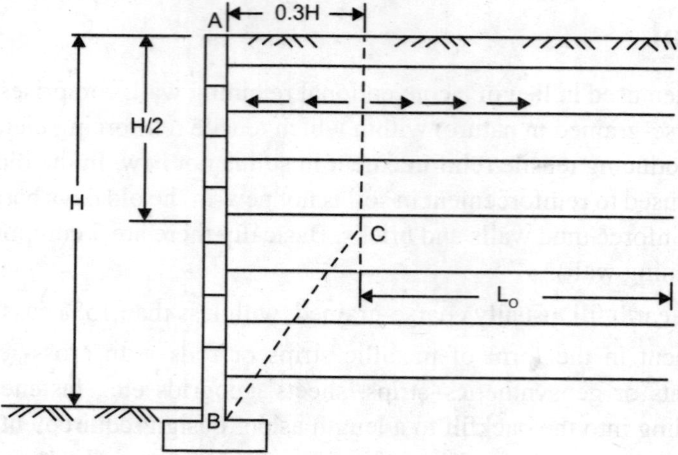

Fig. 8.11 Reinforced earth retaining wall

$$\sigma_z = \gamma z \tag{8.18}$$

Where γ is the unit weight of backfill material. However, the actual stress will be somewhat higher than this value, because of the total -active thrust of backfill material beyond the reinforcing elements.

The actual vertical stress has been given as:

$$\sigma_z = \gamma z \left[1 + K_A \frac{z^2}{L^2}\right] \tag{8.18}$$

Where K_A is the coefficient of active pressure and L, the length of reinforcing elements. The tensile force 'T' in the reinforcing element due to transfer of lateral strees from the soil to the element is given by :

$$T = (K\sigma_z) \times (S_x \times S_z) \tag{8.19}$$

Where S_x is the horizontal spacing in-between reinforcing elements, S_z is the spacing in the vertical direction and K is the coefficient of earth pressure at depth z, *i.e.*, it is the ratio of lateral earth pressure to the vertical earth pressure at depth z. Generally, S_x is of the order of 0.8 to 1.0 m and S_z in the range 0.2 m to 1.0 m.

Having compute the tensile force T as above and given the permissible tensile strength of the material, the required cross-section of the element can be determined.

Like in an unreinforced earth backfill behind a retaining wall, there is an active earth zone which for design purposes is looked upon as, in limiting equilibrium, a wedge of soil just behind the wall with a failure plane developing at an angle 45° + ϕ/2 to the horizontal. A simplified failure surface may be considered to be vertical upto a depth of H/2 from the top at a distance 0.3H from the back of the wall, while the lower half of the surface may be approximated to an inclined surface at an angle of 45° + ϕ/2 from the bottom of the wall.

The portion of reinforcing element falling within the wedge ADC is ignored and the remaining length (L_0) termed the 'effective length' is only taken into consideration as functional in carrying the tensile stresses. Based on experimental work, the maximum tensile stress in a reinforcing element occurs at a point within the reinforced soil, the position of this point varying with depth.

The frictional resistance (R) on the surface of an element considering only top and bottom surfaces is given by:

$$R = 2b\, L_0\, \gamma_z \tan\delta \tag{8.20}$$

Where b is the width of the element, L_0 the effective length of the element and δ is the angle of friction between soil and the element.

The factor of safety against bond failure (F_b) is given by:

$$F_b = R/T \tag{8.21}$$

The value of F_b should normally be not less than 2 for design purposes.

The shearing resistance between soil and reinforcement is determined by direct shear or full-scale pull-out tests. The factor of safety against tensile failure F_T is given by

$$\text{“}F_T = S/T\text{”} \tag{8.22}$$

Where S is the tensile strength of the material of reinforcing element and T is the tensile stress in a reinforcing element as given earlier to be equal to K σ_z S_x S_z.

8.14. GABION STRUCTURES

8.14.1. Gabions

Gabions are large steel wire mesh boxes filled with stone; galvanized/PVC coated wire, not less than 3 mm in diameter or a net of geosynthetic material These mesh boxes filled with stone form building blocks for structures requiring flexibility or form units which resist high water velocities. (Fig. 8.12)

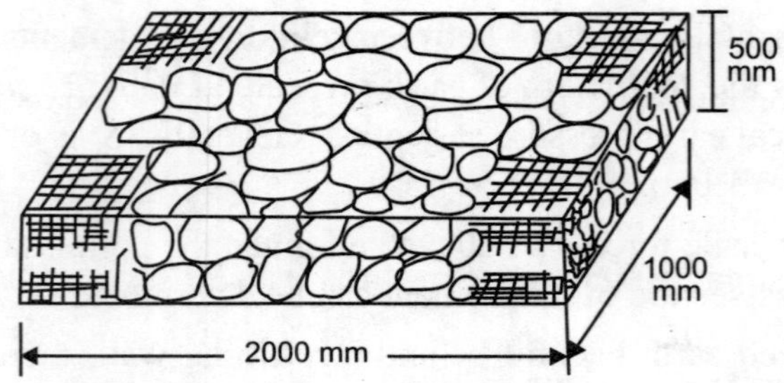

Fig. 8.12. Typical dimensions of a gabion

A retaining wall can be built with these blocks much in the same manner as a brick masonry retaining wall is built; the joints may be staggered in alternate courses.

A typical application of providing a retaining structure built with gabions to prevent any damage from a possible slip surface is shown in Fig. 8.13. Filter material preferably filter fabrics are placed behind the wall to drain water.

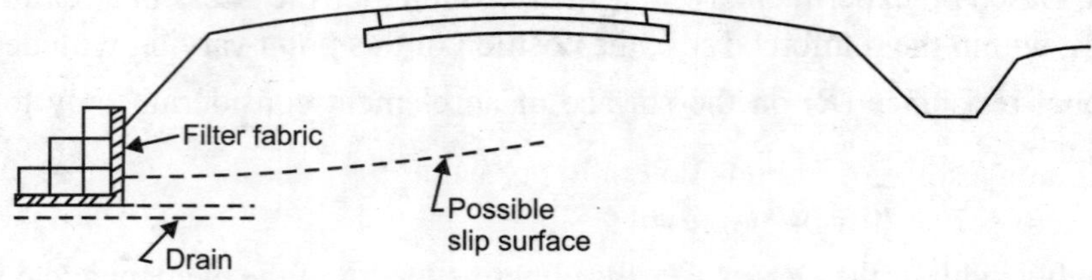

Fig. 8.13: Use of Gabions as a retaining structure

8.14.2. Design of gabion retaining wall

The various considerations in the design of a retaining wall using wire mesh boxes filled with stones as building blocks are given below:

(*i*) The resultant force R of the thrust P of backfill and weight W of the wall, should pass through the middle-third of base width B of the wall (Fig. 8.14)

(*ii*) The base width B of the retaining wall should be approximately two-thirds the height (H).

(*iii*) The top width (*t*) should not be less than 1 metre when there is no surcharge ; with surcharge it should not be less than 2 metres.

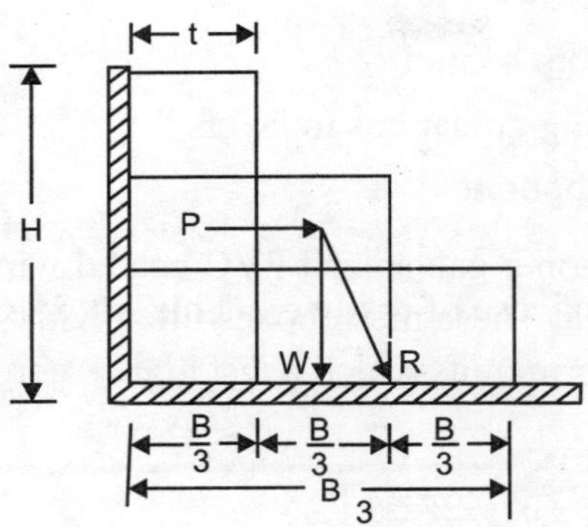

Fig. 8.14: A Gobion retaining wall

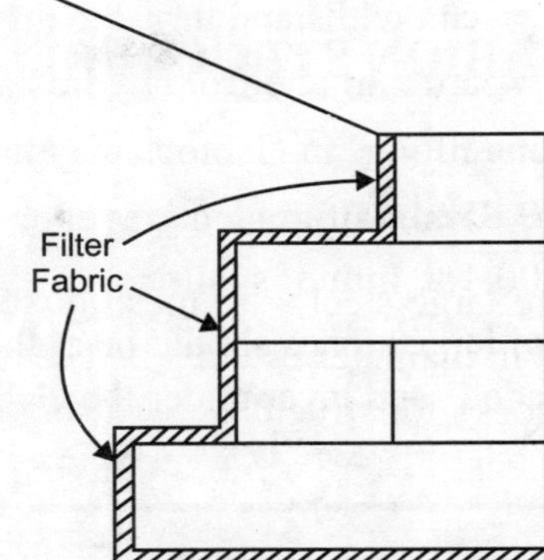

Fig. 8.15: A Gabion wall built in steps

(*iv*) If relatively higher size aggregates are used in filling, the wire-mesh boxes, the base width *B* should be suitably increased.

(*v*) For purpose of strengthening, counter, forts may have to be provided and taken into the body of the main wall.

(*vi*) For increased resistance against sliding, the base of the wall should be taken into the ground to provide sufficient key.

(*vii*) In the event of a slip/potential slip, the wall should be taken below the slip/potential slip plane on firm foundation.

(*viii*) Provide adequate drainage to carry water away from the base ; to prevent seepage into ground below the wall, put a thin concrete slab with collector drains, immediately below the base.

(*ix*) A Gabion wall may also be built in steps as shown in Fig. 8.15

8.14.3. Use of gabions as revetment

In order to maintain a soil slope in position or to protect it from erosion, gabions can be laid flat on the sloping face of soil as a facing of stone to form a monolithic matress. The function of such a revetment is to protect the soil and not to withstand its thrust. The bottom of the mattress should be supported by a gabion wall, as shown in Fig. 8.16

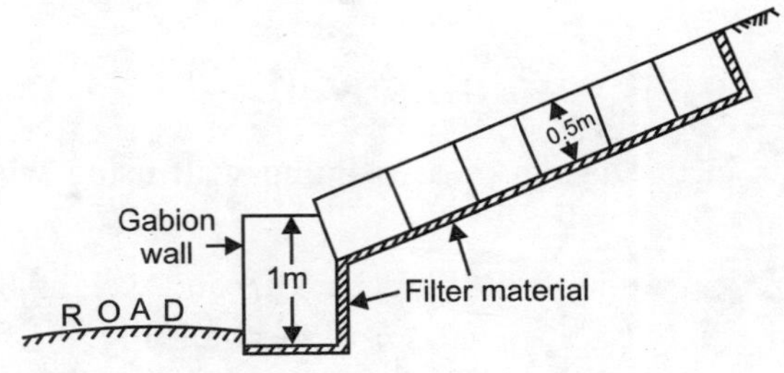

Fig. 8.16: Gabions as revetment

The main advantages in using Gabions as revetment are given below :

- Gabions are self draining
- They posses the needed flexibility
- can withstand thrust to an appreciable extent
- are immune from hydrostatic pressure from behind

The stone fillings in Gabions may have the following gradations in layers :

- 50-75 mm of graded gravel or crusher run at bottom
- 100-150 mm of spalls

Layer of large stones should be at the top. In case of ease of easily erodable silt, it is advisable to place graded sand layer under the Gabion.

QUESTIONS

1. Describe the Swedish circle method of analysis of high embankments by (*a*) total stress analysis and (*b*) effective stress analysis.

2. An embankment is constructed with clay having a cohesion 0.02 MN/m^2. The radius of the slip circle is found to be 12 m and the affected sector subtends an angle of 110 degrees. The density of the soil is 1.5 gm/c.c. The arc of the segment affected by slip is 15 m^2. The horizontal distance between the centre of the slip circle and the centre of gravity of the segment is 2.2 m. Using total stress analysis, find the factor of safety against overturning.

[Ans. 11.4]

3. What are the methods available for the design of high embankments over weak soils?

4. Write short notes on :

(*i*) blasting technique for embankment construction in marshy areas

(*ii*) vertical sand drains.

5. (*a*) What are the desirable properties of a soil for use as embankment material ?

(*b*) Discuss the various types of soils commonly available in India as regards their suitability for embankment construction.

(*c*) What are the IRC guidelines for selecting soils for embankment construction ?

6. Write short notes on:

(*i*) Geo-textiles

(*ii*) Rainforced earth retaining walls

(*iii*) Gabions

Chapter

TYPES OF PAVEMENT AND FACTORS GOVERNING PAVEMENT DESIGN

9.1. TYPES OF PAVEMENT

Pavement can be of the following types:

(*i*) Flexible Pavement which is a layered system consisting of several courses, and in which the wheel load is transmitted to the subgrade by the lateral distribution with increasing depth. The thickness is so selected that the stresses on the subgrade are kept within the safe bearing capacity.

(*ii*) Rigid Pavement, having high flexural strength, and transmitting the wheel load by beam action. Cement concrete pavement is a rigid pavement.

(*iii*) Semi-rigid pavement, which is an intermediate type between flexural and rigid, having a much lower flexural strength than a rigid and also deriving support by lateral distribution of load through the depth like a flexible pavement. Typical examples are soil-cement, lime-pozzolona concrete and lean concrete.

(*iv*) Some authorities identify a fourth type, composite pavement, which consists of layers of different composition. But strictly speaking all pavements are composite pavements since all layers have different composition.

9.2. REQUIREMENTS OF A GOOD PAVEMENT

A pavement should have the following requirements:

(*i*) It should be structurally adequate to withstand the loads imposed on it.

(*ii*) It should be thick enough to distribute the loads to a safe value on the subgrade.

(*iii*) It should provide a hard wearing surface and not get abraded by the wheels of vehicles.

(*iv*) It should be skid-resistant

(*v*) It should have a good riding quality so that the vehicles can travel at a good speed without discomfort and at a low vehicle operating cost.

(*vi*) It should be dust-proof

(*vii*) It should not create high levels of noise

(*viii*) Its surface should be impervious so that water does not percolate through the voids and soften the subgrade.

9.3. PAVEMENT COURSES

A pavement consists of various courses:

(*i*) At the top is the surfacing which provides a smooth, abrasion resistant, dust-free, noise-free surface, also resisting the ingress of water. In a concrete pavement, the Pavement Quality Concrete (PQC) slab is the top course.

(*ii*) The surfacing is supported by a base, which helps in distributing the loads in a flexible pavement.

(*iii*) Below the base is the sub-base, which further helps in distributing the loads. In a concrete pavement, the pavement rests on a sub-base, which is constructed as a Dry Lean Concrete (DLC) or soil-cement.

(*iv*) A drainage layer is needed to drain away the water percolating through the voids, cracks and joints in a pavement. In a flexible pavement, a part of the sub-base is made as the drainage layer.

9.4. COMPARISON BETWEEN FLEXIBLE AND RIGID PAVEMENTS

The following comparison gives the advantages and disadvantages of the two types of pavements:

Table 9.1: Comparison of flexible and rigis pavements

	Characteristics	Flexible pavement	Rigid (Concrete) pavement
1.	Life	Has a low life of 10-15 years	Has a long life of 30-40 years
2.	Maintenance	Requires heavy maintenance such as routine maintenance (filling up potholes, patch repairs etc,), renewal of surface once in 3-5 years and overlay once in 10-15 years.	Practically maintenance free. In a jointed plain concrete pavement, renewal of sealant in joints may be needed once in 5-10 years. A Continuously Reinforced Concrete Pavement does not require sealant renewal.
3.	Riding quality	Though a good riding quality can be achieved after initial construction, the riding quality deteriorates with the passage of traffic, thus entailing higher vehicle operating costs.	A good riding quality is maintained throughout its life.

4.	Skid-resistance	Whenever the surface becomes smooth, a thin bituminous layer can be laid to restore anti-skid property.	If the surface texture wears out, regrooving can be done to restore anti-skid property.
5.	Stage construction	Amenable to stage construction	Not amenable
6.	Binder Availability	Bitumen is derived from petroleum crude whose world-wide reserves are fast dwindling.	Cement is available in plenty in India and will continue to be available for many years in the future.
7.	Penetration of water	Water percolates through voids, causing stripping of binder from aggregates and softening the subgrade.	Very little chances of water penetrating and hence it is ideal for high rainfall areas.
8.	Consumption of stone aggregates	High when the pavement is built initially and aggregates are needed for maintenance also.	Lower consumption than flexible pavement in initial construction, and no consumption at all for maintenance.
9.	Fuel consumption of trucks	14 per cent higher consumption of diesel than concrete pavement.	Lower consumption of diesel.
10.	Consumption of electricity for street lighting	10 per cent higher consumption than concrete roads because of dark colour.	Being grey or whitish, 10 per cent less consumption of electricity for street lighting.
11.	Initial cost	May be slightly cheaper than concrete pavement in the initial cost.	May be marginally costlier (5-15 per cent) than flexible pavement.
12.	Life-cycle cost	Life cycle cost is 2 to 3 times that of concrete pavement.	Life cycle cost is one-third to one-half that of flexible pavement.
13.	Management of traffic during construction	Can be thrown open to traffic immediately after construction.	The concrete has to be cured for about 14 days to gain strength and hence traffic cannot use a freshly laid concrete road.

14.	Utilities	If any repairs to utilities lying beneath the concrete pavement is to be done, it is easy to cut the pavement and reinstate it after repairs.	Concrete pavement cannot be cut and reinstated for repair of utilities. Special arrangements for relocating utilities outside the concrete pavement have to be done.
15.	Environmental pollution	Since bitumen and aggregates have to be heated, emissions from the hot-mix plant cause environmental pollution.	Since heating is not involved, environment pollution is less.

9.5. FACTORS GOVERNING PAVEMENT DESIGN

The factors which govern pavement design are discussed below:

9.5.1. Wheel load

The design axle load in many countries, including India is 80 kN (18,000 lbs). Since vehicles are usually overloaded in India, it becomes necessary to express the damaging effect of higher axle loads in terms of the standard axle load. The fourth power rule is generally followed, which is given below:

$$F = \left(\frac{L}{L_s}\right)^4 \qquad (9.1)$$

where F = Equivalence Factor of a load L in terms of its damaging effect when compared to the standard axle load of 80 kN

Axle load spectrum studies are conducted and based on the fourth power rule, a multiplying factor known as VDF (Vehicle Damage Factor) is determined. If an axle load spectrum study is not conducted, the following values of VDF can be adopted.

Initial traffic volume in terms of commercial vehicles per day (CVPD)	**VDF Values for roads in**	
	Rolling/ plain terrain	**Roads in hilly terrain**
0–150	1.5	0.5
150–1500	3.5	1.5
More than 1500	4.5	2.5

9.5.2. Distribution of commercial traffic over carriageway

The number of lanes in a carriageway determines how much of the total two-way traffic should be considered as governing the traffic on the design lane. The following distribution is recommended in India:

(*i*) Single Lane Roads : Total number of CVPD

(*ii*) Two-lane Roads with bi-directional traffic : 50 per cent of the total two-way traffic (If the VDF in one direction is higher, the traffic in the direction of higher V.D.F is taken for design).

(*iii*) Four-lane Roads (undivided): 40 per cent of the total CVPD in both directions.

(*iv*) Dual Carriageway Roads: 75 per cent of the CVPD in each direction for dual two lane carriageway in each direction (*i.e.*, total four lanes). For a total six lane divided carriageway, the distribution factor is 60 per cent of the CVPD in each direction. For a total eight lane divided carriageway, the distribution factor is 45 per cent of CVPD in each direction.

9.5.3. Axle load repetitions for design

The number of axle load repetitions during the design period governs the design. Since traffic grows every year, say at an annual growth rate of r (expressed in decimals), the number of axle load repetitions is given by the formula:

$$N = \frac{365 \times |(1+r)^n - 1|}{r} \times \mathrm{A} \times \mathrm{D} \times \mathrm{F} \tag{9.2}$$

where

N = Number of repetitions of standard axle load of 80 kN

r = annual rate of growth of commercial vehicles in decimals (*i.e.*, for 5 per cent annual growth rate, $r = 0.05$)

A = Initial commercial vehicle traffic per day in the year of completion of construction

D = Lane distribution (Refer para 9.5.2)

F = VDF (Refer para 9.5.1)

9.5.4. Subgrade

As per current Indian practice, the 500 mm thick subgrade should have a 4-days soaked CBR of minimum 8 for roads having traffic of 450 CVPD. For low volume rural roads, the thickness recommended covers a wide range of CBR (ranging from 2 to 15).

9.5.5. Position of wheel load

For the design of jointed concrete pavements, the position of wheel loads in a slab panel is critical. These are discussed in Chapter 12.

9.5.6. Climatic factors

Rainfall and frost affect pavement performance. Rainfall is taken into account of providing pavement drainage.

Good pavement drainage also minimises the damages caused by frost action. Variation of temperature causes warping stresses in concrete pavements. These are discussed in Chapter 8

9.5.7. Properties of materials

The sub-base, base and surfacing layers have strengths which are the basis for determining the stresses and strains in the layers. These are discussed in Chapter 5 The strength of concrete which is dependent on the properties of materials forming the concrete is discussed in Chapter 5

QUESTIONS

1. List and define different types of pavements. What are the essential requirements of a good pavement?
2. Give in a tabular form the comparison between flexible and rigid pavements.
3. Describe the various factors which govern pavement design.

Chapter

PAVEMENT MATERIALS

10.1. MATERIALS USED FOR VARIOUS SPECIFICATIONS

The following are the pavement materials in common use:

(*A*) Flexible Pavements

1. Soil stabilised materials
2. Granular bases and sub-bases
3. Brick, stone and concrete blocks
4. Bituminous materials

(*B*) Rigid (Concrete) Pavements

1. Cement
2. Coarse aggregates
3. Sand
4. Admixtures and additives
5. Steel
6. Sealants

10.2. SOIL STABILISED MATERIALS

10.2.1. Need for soil stabilisation

As stone aggregates, which are the main materials used in pavement construction, are getting fastly depleted, alternatives have to be considered. One of the promising techniques is soil-stabilisation, where the local soil or marginal material is blended with a stabilising agent to improve its strength, eliminate undesirable properties like excessive swelling and shrinkage, reduce compressibility and settlements and bring about economy in road construction.

10.2.2. Types of soil-stabilisation techniques

The following are the broad soil-stabilisation techniques being adopted:

1. Mechanical stabilisation
2. Stabilisation by additives

These are discussed in detail in Chapter 13.

10.3. GRANULAR BASES AND SUB-BASES

10.3.1. Types of specifications

The various specifications used are:

1. Telford stone soling
2. Gravel bases and gravel sub-bases (GSB)
3. Water-bound-macadam (WBM)
4. Wet mix macadam (WMM)

10.3.2. Telford stone soling

This consists of laying single size stone boulders in thickness of 150-225 mm, and packing with spalls into the voids and then rolling with a smooth wheel roller weighing not less than 8T. This specification suffers from the disadvantage that it is purely labour-oriented and has not interlocking properties. It is thus not in use in India.

10.3.3. Gravel base and gravel sub-base (GSB)

Gravel (popularly known as murram in India) is a naturally occurring material. The maximum size of particles should be one-half to one-third the thickness of the compacted layer. The maximum thickness of the compacted layer is 150 mm.

The grading requirements of granular sub-base are given in Table 10.1.

Table 10.1. Grading for granular sub-base

IS sieve designation	Percent by weight passing IS sieve					
	Grading I	Grading II	Grading III	Grading IV	Grading V	Grading VI
75.0 mm	100	–	–	–	100	–
53.0 mm	80–100	100	100	100	80–100	100
26.5 mm	55–90	70–100	55–75	50–80	55–90	75–100

9.50 mm	35–65	50–80	–	–	35–65	55–75
4.75 mm	25–55	40–65	10–30	15–35	25–50	30–55
2.36 mm	20–40	30–50	–	–	10–20	10–25
0.85 mm	–	–	–	–	2–10	–
0:425 mm	10–15	10–15	–	–	0–5	0–8
0.075 mm	<5	<5	<5	<5	–	0–3

Gradings III and IV are preferably used in lower sub-base. Gradings V and VI are used as sub-base-cum-drainage layer. The thickness-of each layer shall not be less than 150 mm.

The physical requirements of materials for granular sub-base are given in Table 10.2.

Table 10.2: Physical requirements of materials for granular Sub-base

Aggregate impact value	40 maximum
Liquid limit	25 maximum
Plasticity index	6 maximum
CBR at 98% dry density as per IS:2720-Part VIII	Minimum 30

10.4. BRICK, STONE AND CONCRETE BLOCKS

10.4.1. Brick pavement

In the alluvial plains of North India, stone aggregates are to be obtained from long leads. Hence, bricks are used for:

(*i*) sub-base for important roads
(*ii*) base layer for middle category of roads
(*iii*) surfacing layer of unimportant roads like village roads
(*iv*) shoulders for single lane roads
(*v*) service lanes, footpaths and cycle tracks in towns and cities

Bricks, which should be over-burnt, should have a compressive strength of at least 27.5 MN/m^2. The water absorption should be in the range of 5-8 per cent.

Bricks are laid on a prepared sub-grade over which a cushion of 25-40 mm sand is introduced. The bricks can be laid as brick-on-edge or flat. The pattern of laying can be (*i*) header and stretcher bond (*ii*) herring bone bond or (*iii*) transverse bond.

10.4.2. Stone block pavement

Stones have a high hardness and stone blocks are used where heavy traffic is expected such as warehouses, godowns, harbours, railway yards etc. The blocks should have a minimum thickness of 150 mm and are laid over a layer of sand 50-63 mm thick. They can also be laid over a cement-sand mortar layer. The joints are filled with cement-sand mortar or sand-bitumen.

10.4.3. Interlocking concrete block pavement

Interlocking concrete block pavements are used for:

1. Footpaths and side-walks
2. Cycle tracks
3. Residential streets
4. Car parks
5. Fuel Stations
6. Rural Roads through villages
7. Roadside rest areas
8. Bus depots
9. Approaches to railway level crossings
10. City streets
11. Truck parking areas
12. Industrial floors
13. Warehouses and godowns
14. Container depots
15. Port wharf and roads
16. Roads in high altitudes

They have a long life, are practically maintenance-free, unaffected by oil spillage, can be laid quickly and do not require curing at site and can be opened to traffic quickly. Since they are factory made, their strength and quality are assured.

The compressive strength should be a minimum of 30 MPa. The suggested thickness of the blocks is as under:

1. Light traffic, (footpaths, cycle tracks, residential streets) 60 mm
2. For Medium traffic 80 mm
3. For heavy traffic in container yards, ports, warehouses 100-120 mm

The blocks are laid over a cushion of sand 20 to 40 mm thick. The joints are filled with fine sand.

The base and sub-base thickness depends upon the CBR value of the sub grade. Some guide lines are given in Table 10.3.

Table 10.3: Suggested thickness in mm of sub-base and base for interlocking concrete block pavements

	Type of Road	Layer	CBR%	
			Above 10	5–10
1.	Cycle tracks, Pedestrian footpaths	Base	200	200
2.	Residential streets and axle load repetitions less than 10 msa	Sub-base	200	250
3.	Collector streets, industrial streets, truck parking areas, axle load repetitions 10–20 msa	Base	250	250
		(WBM/WMM)	200	250
4.	Arterial streets, axle load repetitions 20–50 msa, Container yards, Ports	Sub-base	250	250
		Base		
		(WBM/WMM)	250	250
		or		
		WBM/WMM	150	150
		and		
		DLC over it	75	75

The blocks should have adequate edge restraint made with cement concrete strips or kerb or kerb and channel.

The water entering through the joints should be drained away. A typical arrangement is given in Fig. 10.1.

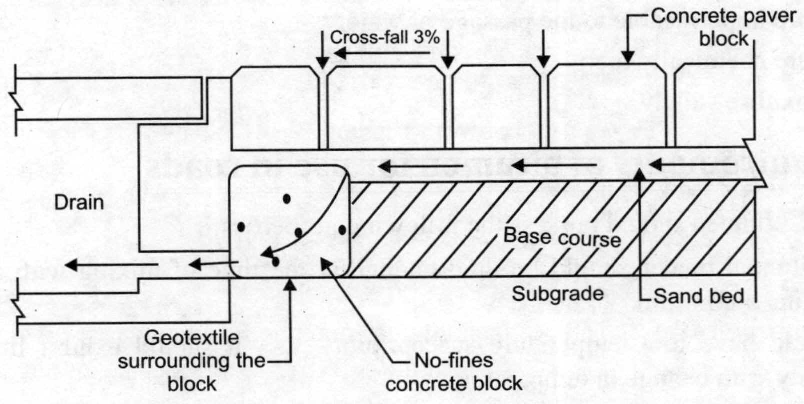

Fig. 10.1: Surface drainage in a block pavement

10.5. BITUMINOUS MATERIALS

10.5.1. Terms used

Bitumen is the residue obtained after the fractional distillation of petroleum crude. The term "bitumen" is used in the British and Indian Specifications, whereas the term "asphalt" is used in the American Specifications.

A "straight-run-bitumen" is a bitumen whose viscosity composition has not been adjusted by blending or softening by fluxing, cutting back oil or other treatment.

A "penetration grade bitumen" is a bitumen whose degree of hardness can be measured by the standard penetration test.

A "viscosity grade (VG)" bitumen whose properties are defined by its viscosity rather than by standard penetration test.

A"cut-back" is a bitumen is one whose viscosity has been reduced by a volatile diluent like kerosene, so that it can be used without heating.

An "emulsion" is a bituminous binder which is in a liquid form rendered so by the addition of an emulsifier.

A "cationic emulsion" is one having positively charged particles.

An "anionic emulsion" is one having negatively charged particles.

10.5.2. Properties of bitumen

The general properties of bitumen are given below:

1. They are predominantly hydrocarbons
2. They are soluble in carbon disulphide (CS_2)
3. They soften on heating
4. They are insoluble in water
5. They are impermeable to the passage of water
6. They are chemically inert
7. They oxidise slowly

10.5.3. Requirements of bitumen for use in roads

For use in roads, bitumen should possess the following properties:

1. The bitumen binder should be fluid enough at the time of mixing with aggregates by providing a thin film.
2. It should have low temperature susceptibility, *i.e.*, it should exhibit little change in viscosity with change in temperature.
3. It should have enough amounts of volatiles in it and it should not lose them excessively when subjected to higher temperature.
4. It should be ductile and not brittle.

5. It should have good affinity to the aggregates and should not cause stripping in the presence of water.
6. It should be capable of being heated to the design temperature without causing fire hazard.

10.5.4. Tests on bitumen

10.5.4.1. Viscosity

Viscosity is the resistance offered by the fluid to its flow. It is thus opposite to fluidity. It is measured by viscometers. As per the current IS, paving bitumen are available in four grades, VG 10, VG 20, VG 30 and VG 40. The VG 10 is the softest, whereas VG 40 is the hardest. VG 10 corresponding to penetration grade 80-100. VG 20 corresponds to penetration grade 60-80, VG 30 corresponds to penetration grade 50-70'and VG 40 corresponds to penetration grade 40-60.

10.5.4.2. Penetration

An indirect measure of viscosity is the amount of penetration of a standard needle under the standard prescribed load (100 gms), temperature (25°C) and time in which the penetration is recorded (5 sees) (Fig. 10.2).

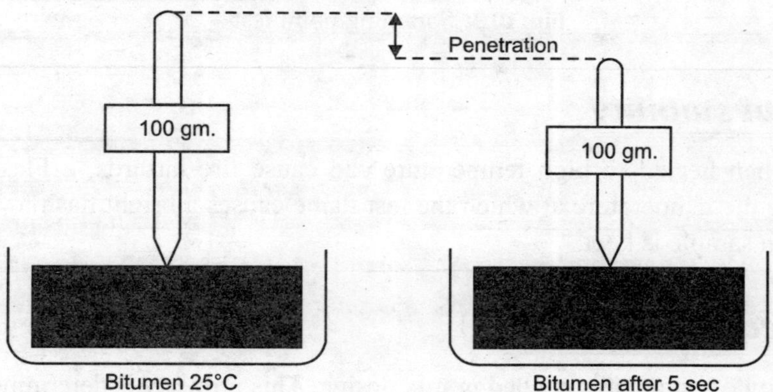

Fig. 10.2: Penetration test

10.5.4.3. Softening point

Bitumen loses its hardness on heating, but there is no defined softening point. The standard test to determine the softening point is the "Ring and Ball Test" under which a brass ring containing the bitumen is suspended in water or glycerine at a given temperature. A steel ball is placed on the disc of bitumen (Fig. 10.3). The liquid medium is then heated at 5°C per minute. The temperature at which the softened bituminous material touches the bottom plate at a specified distance below the ring is recorded as the softening point. Bitumen VG 40 has a softening point of 50–65'C and the softening point of VG 10 bitumen is in the range 30–45'C.

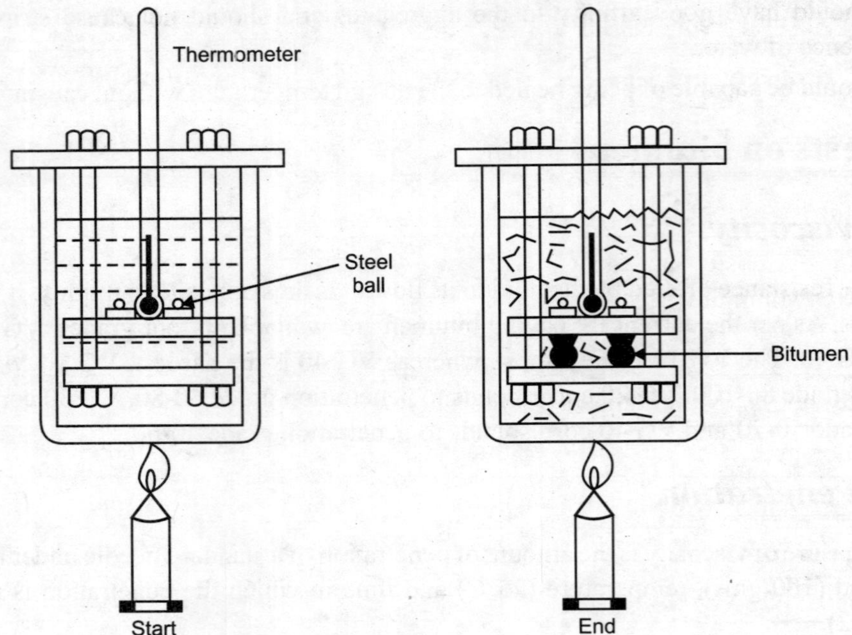

Fig. 10.3: Softening point test

10.5.4.4. Heat stability

Since bitumen when heated to high temperature can cause fire hazards, a Flash Point Test is conducted, noting the temperature at which the test flame causes a bright flash to appear in a cup where the bitumen sample is kept.

10.5.4.5. Specific gravity

The specific gravity of bitumen is needed in mix design. This property is determined by preparing a cube of the same. The value is usually in the range of 1.02–1.04.

10.5.4.6. Solubility

Bitumen is soluble in CS_2. Insoluble bitumen indicates the presence of mineral matter which is undesirable. Indian specifications require 99 per cent solubility.

10.5.4.7. Thin film oven test

When bitumen is mixed with aggregates in a hot-mixed plant it would tend to age harden. This is determined by the Thin Film Oven Test and Rolling Thin Film Oven Test (RTFOT). The requirements of Paving Bitumen in India are given in Table 10.4.

Table 10.4: Requirements of paving bitumen and its tests

	Property	Viscosity grade of paving bitumen				IS Test
		VG 10	VG 20	VG 30	VG 40	
1.	Absolute viscosity at 60°C, poise (Minimum)	800	1600	2400	3200	IS 1206 (Part 2)
2.	Kinematic viscosity at 135°C, centistoke (Minimum)	250	300	350	400	IS 1206 (Part 3)
3.	Flash point	220	220	220	220	IS 1209
4.	Solubility in CS_2 (Minimum %)	99.0	99.0	99.0	99.0	IS 1216
5.	Penetration at 25°C	80–100	60–80	50–70	40–60	IS 1203
6.	Softening point °C (Minimum)	40	45	47	50	IS 1205
7.	Tests on residue from thin film oven test (RTFOT)					
	(*i*) Viscosity ratio at 60°C (Max)	4.0	4.0	4.0	4.0	IS 1206 (Part 2)
	(*ii*) Ductility at 25°C (Min)	75	50	40	25	IS 1208

10.5.5. Cut-backs

In order to avoid heating bitumen to bring down its viscosity, a volatile diluant is added to make it suitable to coat the aggregates with a thin film. There are three types of cut-backs:

Rapid Curing (RC) .... Bitumen fluxed with naptha type of distillate

Medium Curing (MC) .... Bitumen fluxed with kerosene

Slow Curing (SC) .... A liquid residue produced in the refining process, containing little or no volatile constituents

The use of cut-backs is not favoured because the diluent can find its way to water bodies and contaminate it.

10.5.6. Emulsions

Emulsions are widely used for cold-mix specifications which do not require a hot-mix plant and find application in repair works and for cold-weather work. But they are costly and contain 50-65 per cent bitumen, the rest being water. On application to the road, the emulsion breaks down, changing its brown colour to black when the water evaporates leaving the bitumen to provide a film over the aggregates. The emulsions can be rapid setting (RS), medium setting (MS) or slow setting (SS). As mentioned earlier, emulsions can be anionic or cationic. Polymer modified emulsions are also available which enhance the properties of bitumen. The prefix *C* is added when the emulsions is cationic. Thus CRS indicates a rapid setting cationic emulsion.

10.5.7. Guide to the selection of bituminous binders

The following general guidelines give the correct bituminous binder for various types of work.

Table 10.5: Guidelines for selection of bituminous binder

	Type of binder		Application
1.	VG 40 (30/40 Penetration)		Hot mix bituminous macadam, bituminous concrete and penetration macadam in normal summer temperatures and in plains.
2.	VG 30 (60/70 Penetration)	1.	Hot mix bituminous macadam, bituminous concrete in normal summer temperatures and in plains.
		2.	Penetration macadam in winter and at high altitudes
3.	VG 10 (80/100 Penetration)	1.	Surface dressing, seal coat, premised chipping carpet in normal summer temperatures and in plains.
		2.	Penetration macadam in summer temperature and in plains.
		3.	Hot mix bituminous macadam and bituminous concrete in summer temperatures and in plains.
		4.	Hot mix bituminous macadam and bituminous concrete in winter temperatures and at high altitudes.
4.	Cut-back RC–3	1.	Surface dressing and seal coats at normal summer temperatures and in plains.
		2.	Surface dressing at winter temperatures and at high altitudes.
		3.	Cold mix macadam and bituminous carpets at summer and winter temperatures, both in plains and at high altitudes.
5.	Cut-back MC and SC		Priming
6.	Emulsions	1.	Prime coat and tack coat
		2.	Cold weather work
		3.	Wet condition of aggregates
		4.	Maintenance and repair works
		5.	Sealing of fine cracks
		6.	Work in remote areas

10.5.8. Modified bitumen

10.5.8.1. Advantages of modified bitumen

The properties of bitumen can be vastly enhanced by adding small quantities of modifiers. The advantages are:

1. Lower susceptibility to temperature variations
2. Higher resistance to cracking and reflective cracking
3. Higher resistance to deformation at higher temperature
4. Better age resistance properties
5. Better adhesion between aggregates and binder
6. Higher fatigue life
7. Overall improved performance

10.5.8.2. Types of bitumen modifiers

The bitumen modifiers can be classified as given in Table 10.6.

Table 10.6: Classification of rubber and polymer-based bitumen modifiers

Type of modifier	Examples	Indicative dose (Per cent by weight of bitumen)
Plastomeric	Polyethylene (PE), Ethylene vinyl acetate (EVA), ethylene butyl acrylate (EBA), Ethylene methyl acrylate copolymers (EMA)	3–6
Elastomeric	Styrene isoprene styrene (SIS), styrene-butadiene-styrene (SBS), styrene-butadiene rubber, ethlylene ter polymer (ETP)	3–5
Synthetic rubber latex	Styrene butadiene rubber latex and any other synthetic rubber	3–5
Natural rubber	Latex or rubber powder	2–4
Crumb rubber or Treated crumb rubber	Crumb rubber, treated crumb rubber	10–12

10.5.8.3. *Use of waste plastic*

Waster Plastic in the form of polythene bags and other non-bio-degradable waste like bottles etc. can be advantageously used to modify the properties of bitumen and when used can enhance the properties of bitumen as described in para 10.5.7.1. The waste plastic is shredded and added to the mix in the hot-mix plant.

10.6. BITUMINOUS COURSES IN HIGHWAY PAVEMENTS

10.6.1. Types of bituminous courses

Bituminous Courses is a flexible pavement are of the following types:

1. Prime coats and Tack coats
2. Surface dressing
3. Seal coats
4. Thin open-graded premix carpets
5. Base course
6. Dense surfacings

10.6.2. Prime coat

A prime coat is an initial application of low viscosity bituminous material to an absorbent surface like granular base, Wet Mix Macadam, Water Bound Macadam etc. with the objective of plugging the voids and to coat the loose mineral particles. The binder chosen is a slow setting emulsion. Medium setting (MC) or slow setting (SC) cutbacks can also be used with sub-zero temperature. The quantity of primer is 0.6–0.9 Kg per sq. m for surfaces with medium porosity and 1.2–1.5 Kg per sq. m for surfaces with high porosity.

10.6.3. Tack coat

A tack coat is provided for bonding the new bituminous layer to an old one. There is a strong feeling that a primed surface does not need a tack coat. The quantity of binder is 0.2–0.25 Kg per sq. m for a normal bituminous surface and 0.25–0.30 Kg per sq. m for a diy and hungry bituminous surface. The binder used is a slow setting emulsion. Cut–backs should be used at sites at sub-zero temperatures.

10.6.4. Surface dressing

Surface dressing is a process wherein a thin film of bituminous binder is sprayed on the road surface, covered with a coat of stone aggregates and rolled. It can be applied as one coat or two coats. The aggregates can be precoated with bitumen to reduce the whip-off due to traffic. The binder used is a low viscosity bitumen, VG-10. A rapid curing cut-back RC-3 and emulsion can also be used. The nominal sizes of aggregates normally specified are.

First coat of surface dressing	:	12 mm
Second coat of surface dressing	:	10 mm

The quantities of aggregates recommended by IRC are.

First coat : 0.14 – 0.15 m^3 per 10 sq. m

Second coat : 0.09 – 0.11 m^3 per 10 sq. m

The quantity of binder used in Kg per 10 sq. m is :

	Bitumen	Cut-back
First coat	17.2–19.5	19.5–22
Second coat	9.8–12.6	12.2–14.6

10.6.5 Open-graded premix carpet (PC)

This specification is used as a thin surfacing layer 20 mm thick and is a popular specification in India for roads carrying low and medium traffic. The carpet is provided with a seal coat to protect water entering through the voids. The quantities of aggregates and binder are given below:

Table 10.7: Materials for 20 mm carpet (Per 10 Sq m)

	Material	Quantity
1.	Aggregates for premixing	
	12 mm size	0.18 cu m
	10 mm size	0.09 cu m
2.	Aggregates for seal coat	
	Low rainfall area	0.06 cu m
	High rainfall area	0.09 cu m
3.	Bitumen for premixing	14.6 Kg
4.	Bitumen for premixed seal coat in low rainfall area	6.5 Kg
5.	Bitumen for liquid seal coat in high rainfall area	9.8 Kg

10.6.6. Built-up spray grout (BUSG)

This specification was popular in India till recently, but is not recommended any more now. The work consists of the following operations:

1. Apply a tack coat
2. Spread coarse aggregates 50 mm down at the rate of 0.5 cu m per 10 sq m and roll it.
3. Apply binder at 12.5 to 15.0 Kg per 10 sq m
4. Apply second layer of coarse aggregates at 0.5 cu m per 10 m^2 and compact it.
5. Apply binder at 12.5 to 15 Kg per 10 sq. m.
6. Spread key aggregates (20 mm down) at 0.13 cu m per 10 sq. m and roll.
7. Provide a wearing course of 20 mm premix carpet.

10.6.7. Bituminous macadam

Bituminous Macadam (BM) is an open-graded mix where aggregates are mixed with bitumen and is used as base and binder course.

The following two gradings are used.

Table 10.8: Grading of aggregates and bitumen content for bituminous macadam

Grading	1	2
Nominal maximum aggregate size (mm)	40	10
Layer thickness (mm)	80–100	50–75
I.S. sieve (mm)	Cumulative % by weight of total aggregates passing	
45	100	
37.5	90–100	
26.5	75–100	100
19	–	90–100
13.2	35–61	56 88
4.75	13–22	16–36
2.36	4–19	4–19
0.3	2–10	2–10
0.075	0–8	0–8
Bitumen content (per cent by mass of total mix)	3.3	3.4

The mixing is done in a hot-mix plant. The mixing, laying and rolling temperatures are given in Table 10.9.

Table 10.9: Mixing, laying and rolling temperatures of bituminous mixes (Degree Celsius)

Bitumen Grade	Bitumen temperature	Aggregate temperature	Mixed material temperature	Laying temperature	Rolling temperature
VG 40	160–170	160–175	160–170	150 Min	100 Min
VG 30	150–165	150–170	150–165	140 Min	90 Min
VG 20	145–165	145–170	145–165	135 Min	85 Min
VG 10	140–160	140–165	140–160	130 Min	80 Min

The rolling is done in the following sequence:

(*i*) Initial or break-down rolling with 8–10 tonne static weight smooth wheel roller.

(*ii*) Intermediate rolling with 8–10 tonne static weight or vibratory roller or pneumatic tyre roller of 12 to 15 tonnes wight with a tyre pressure of at least 0.56 MPa.

(*iii*) Finish rolling with 6–8 tonnes and tandem roller.

10.6.8. Dense bituminous macadam (DBM)

Dense Bituminous Macadam is used for base courses.

The following two gradings are commonly adopted for DBM.

Table 10.10. Composition of DBM

Grading	1	2
Nominal size of aggregate (mm)	37.5	26.5
Layer thickness (mm)	75–100	50–75
I.S. Sieve (mm)	Comulative % by weight of total aggregate passing	
45	100	
37.5	95–100	100
26.5	63–93	90–100
19	–	71–95
13.2	55–75	56–80
9. 5	–	–
4.75	38–54	38–54
2.36	28–42	28–42
1.18	–	–
0.6	–	–
0.3	7–21	7–21
0.15	–	–
0.075	2–8	2–8
Bitumen Content % by mass of total mix	Min 4.0	Min 4.5

The Bitumen Content is determined from the Marshall Mix Design procedure. The requirements of the mix are given in Table 10.11.

Table 10.11: Requirements of dense bituminous macadam

Properties	Viscosity grade bitumen	Modified bitumen	Cold climate
Compaction	75 Blows on each face of specimen		
Min. stability (kN at 60°C)	9.0	12.0	10.0
Marshall flow (mm)	2–4	2.5–6	3.5–5
Marshall Quotient $\left(\frac{\text{Stability}}{\text{Flow}}\right)$	2–5	2.5–5	2.5–5
% Air voids	3–5	3–5	3–5
% Voids filled with bitumen	65–75	65–75	65–75
Coating of aggregates	95% Min	95% Min	95% Min
Tensile strength ratio	80% Min	80% Min	80% Min

The mixing and laying procedures are the same as given in Para 10.6.7.

10.6.9. Bituminous concrete (BC)

Bituminous Concrete is used as the surface layer. The composition of two commonly adopted gradings are given in Table 10.12.

Table 10.12: Composition of bituminous concrete layers

Grading	1	2
Nominal aggregate size (mm)	19	13.2
Layer thickness (mm)	50	30–40
I.S. Sieve (mm)	Cumulative % by weight of total aggregate passing	
26.50	100	
19.00	90–100	100
13.20	59–79	90–100
9.50	52–72	70–88
4.75	35–55	53–71
2.36	28–44	42–58

1.18	20–34	34–48
0.60	15–27	26–38
0.30	10–20	18–28
0.15	5–13	12–20
0.075	2–8	4–10
Bitumen content % by mass of total mix	Min 5.2	Min 5.4

The mix is designed as per the Marshall method. The requirements are the same as given in Table 10.10 for DBM. The mixing and laying procedures are the same as given in para .10.6.7.

10.6.10. Penetration macadam

This specification consists of a compacted layer of stone aggregates into which a bituminous binder is allowed to penetrate and fill the voids. After the bitumen is applied, a layer of key aggregates is spread on the surface and rolled. A seal coat is provided to make the surface impervious to water. The specification is used where hot-mix plant is not available. The quantity of bitumen is 50 Kg per 10 sq. m for 75 mm layer. The quantity of aggregates is 0.6 cu m per 10 sq. m of layer of thickness 50 mm, and 0.9 cu m per 10 sq m of layer of thickness 50 mm, and 0.9 cu m per 10 sq m of layer thickness 75 mm. The quantity of key aggregates is 0.15 cu m and 0.18 cu m (per 10 sq m) for a layer of thickness 50 mm and 75 mm respectively.

10.6.11. Semi-dense bituminous concrete (SDBC)

Semi-dense bituminous concrete is an intermediate type of surfacing between the open textured premix carpet (PC) (para 10.6.5) and the Bituminous concrete (BC) (para 10.6.9). The voids in the mixture is in the range of 5–10 per cent as against 3–5 per cent in the BC. The specification is not favoured for high traffic roads, and is adopted for low traffic roads. SDBC is laid in thickness of 20 mm or 25 mm. The gradings commonly adopted conform to Type III (*a*) and III (*b*) mixes of the Asphalt Institute and are given in Table 10.13.

Table 10.13: Gradation of aggregates for semi-dense bituminous concrete

Sieve designation	Percent by weight passing	
	20 mm thickness	25 mm thickness
20 mm	–	100
12.5 mm	100	75–100
10 mm	75–100	60–85
4.75 mm	35–55	35–55
2.36 mm	20–35	20–35

600 micron	10–22	10–22
300 micron	6–16	6–16
150 micron	4–12	4–12
75 micron	2–8	2–8

The binder content is in the range of 4.5 to 6.0. per cent.

10.6.12. Mastic asphalt

Mastic asphalt is a mixture of bitumen, miner filler, fine stone aggregates in such a proportion that a voidless compact mass is produced. The mixture is fluid at about 200°C, but solidifies on cooling. Because of its toughness, it is used in city streets carrying high volume of traffic and in bridge decks. The bitumen is an industrial grade bitumen, and its quantity is in the range of 14–17 per cent by weight of total mix. Fine aggregates shall pass 2.36 mm sieve. The filler is usually limestone powder. The coarse aggregates shall conform to the following grading for thickness of 25–50 mm and its per cent in the total mix by weight shall be 30–50.

Table 10.14: Grading of coarse aggregates of mastic asphalt

I.S. Sieve	Per cent passing IS Sieve
19 mm	100
13.2 mm	88–96
2.36 mm	0–5

The mastic asphalt is prepared in a cooker at a temperature of 170–180°C, and laid, in a width of 1 m confined between angle irons. To impart skid resistance, precoated chips of size 9.5 mm–13.2 mm are spread at a rate of 0.05 per 10 sq in and pressed into the surface when the mastic has a temperature of 80–100°C.

10.6.13. Stone matrix asphalt (SMA)

Stone matrix asphalt, also known as Stone mastic asphalt, is a gap graded mixture with about 70–80 per cent stone content, 6–7 percent binder, 8–12 per cent of cellulose filler and 0.3 per cent of fibres. It is laid to a thickness of 30–75 mm. It is rut resistant and is used for heavily trafficked sections. The mixing is done in a hot-mix plant.

10.6.14. Slurry seal

Slurry Seal is a treatment used for maintenance of bituminous surface. It is a mixture of slow setting cationic emulsion, stone aggregates, filler (cement) and water. It is prepared by mixing the ingredients in a truck mounted equipment and laid by the spreader box which is a part of the truck mounted machine. Rolling is done by pneumatic tyred roller.

10.6.15. Microsurfacing

Microsurfacing is a mixture of modified bitumen emulsion, stone aggregates, water and additives and is used for preventive maintenance as well as periodic renewals. The mixing and laying is done in a truck-mounted equipment.

10.6.16. Bituminous cold mix

Bituminous Cold Mixes are used in locations where hot-mix plants are not available, in remote areas and for maintenance operations. They can be used for preparing Bituminous Macadam (for base courses and pavement strengthening), Semi Dense Bituminous Concrete and storable mixes for maintenance operations (pot-hole filling and patch repairs).

10.6.17. Mix-seal surfacing

This specification is an open-graded surfacing about 20 mm thick. The gradation and bitumen content are as under:

Table 15.15: Mix seal specifications

Gradation of aggregates		
Sieve size (mm)	**Cumulative per cent by weight passing sieve**	
	Type A	**Type B**
13.2	–	100
11.2	100	88–100
5.6	52–88	31–52
2.8	14–38	5–25
90 micron	0–5	0–5
Quantity of Aggregates (cu m per 10 sq m)	0.27	0.27
Quantity of Binder (Kg per 10 sq m)	22.0	19.0

10.7. MATERIALS FOR RIGID (CONCRETE) PAVEMENTS

10.7.1. Materials used for concrete pavements

The following materials are used for concrete pavements Cement Aggregates

1. Cement
2. Aggregates
   - Coarse aggregates
   - Fine aggregates

3. Admixtures and additives
4. Steel
5. Sealants

10.7.2. Cement

The following varieties of cement are used:

1. Ordinary Portland Cement (OPC) 53 Grade, IS: 12269
2. Ordinary Portland Cement (OPC) 43 Grade, IS:8112
3. Portland Pozzolana Cement (PPC), IS: 1489 (Part 1) with flyash content not more than 20 per cent by weight of PPC)
4. Portland Slag Cement (PSC), IS:455 (with Granulated Blast Furnace Slag Content not more than 50 per cent by weight of PSC)

10.7.3. Coarse aggregates

The coarse aggregates shall be obtained by crushing rock. The water absorption shall not be more than 2 per cent. The maximum size shall be 26.5 mm for Dry Lean Concrete and 31.5 mm for pavement concrete. The Los Angeles Abrasion Value shall not exceed 35 per cent.

10.7.4. Fine aggregates

Fine aggregates shall be natural sand or crushed stone sand, and shall conform to IS:383.

10.7.5. Combined gradation of coarse and fine aggregates

The coarse and fine aggregates shall be mixed to give a combined gradation as given in Table 10.16.

Table 10.16: Combined gradation of coarse and fine aggregates for concrete pavements

I.S. Sieve	Percentage by weight passing the Sieve	Per cent passing IS Sieve
	For dry lean concrete	For Pavement concrete
31.5 mm	–	100
26.5 mm	100	85–95
19.0 mm	75–95	68–88
9.5 mm	50–70	45–65
4.75 mm	30–55	30–55
2.36 mm	17–42	–
600 micron	8–22	8–30

300 micron	7–17	–
150 micron	2–12	5–15
75 micron	0–10	0–5

10.7.6. Admixtures and additives

10.7.6.1. Admixtures

Plasticisers and Superplasticisers are used for improving the workability of concrete. The maximum quantity of these shall be 2 per cent by weight of cementitious materials. Silica Fume upto 10 per cent by weight of cementitious materials is added to increase the abrasion resistance of the concrete.

10.7.6.2. Additives

The most common additive used is flyash. The quantity of flyash shall be upto a maximum of 20 per cent by weight of the cementitious material. Site mixing may be allowed if proper facility is available. Granulated Blast Furnace Slag (GBFS) may be used upto 50 per cent by weight or Portland slag cement provided it is mixed in the factory.

10.7.7. Steel

Steel used for dowel bars shall be MS round bars with yield strength of 240 MPa. Steel used for tie bars shall be deformed/ plain bars and of Grade Fe 500. Steel for CRCP can be Grade Fe 415 or TMT Bars of Fe-500.

10.7.8. Sealant

Sealants can be hot-poured, cold-poured or pre-formed. Hot-poured sealant can be rubberised bitumen. Cold-poured sealants commonly used are Polysulphide and Silicone. Preformed sealant strips are also used by inserting the strip into the groove.

QUESTIONS

1. List the various pavement materials in common use.
2. What are the two broad soil stabilization techniques?
3. List the various types of granular bases and sub-bases and explain them
4. State the uses of interlocking concrete block pavements Draw a sketch showing surface drainage in a block pavement.
5. (*a*) Define the following terms:
   (*i*) Straight-run bitumen
   (*ii*) Penetration grade bitumen

(*iii*) Viscosity grade bitumen

(*iv*) Cut-back

(*v*) Emulsion

(*vi*) Cationic emulsion

(*b*) Emumerate the general properties of bitumen and its requirements for use in roads.

6. List the various tests specified for bitumen and explain any three of them.

7. State the guidelines for selection of bituminous builder for various types of road work

8. Write short notes on :

(*i*) Prime coats and tack coats

(*ii*) Surface dressing

(*iii*) Seal coats

9. Give the specifications for the following :

(*i*) Bituminous macadam

(*ii*) Dense bituminous macadam

(*iii*) Bituminous concrete.

10. Write short notes on the following :

(*i*) Penetration macadam

(*ii*) Mastic aphalt

(*iii*) Mix seal surfacing

(*iv*) Bituminous cold mix

11. List the various types of materials used for concrete (Rigid) pavements and write their specifications.

Chapter

DESIGN OF FLEXIBLE PAVEMENTS

11.1. CLASSIFICATION OF METHODS

Design of pavements has been mainly a matter of experience in the distant past, but during the past 30-40 years, the design methodology has undergone rapid changes. The trend now is to design the pavements based on an understanding of the stresses and strains in the layers. In this Chapter, the design methods considered are :

1. Analytical methods
2. Empirical methods
3. Methods based on pavement performance

A mention is made of the Mechanistic-Emperical method now being proposed by AASHTO.

11.2. ANALYTICAL METHODS

11.2.1. General

A pavement structure is not so easily amenable to accurate structural analysis because the materials forming the flexible pavement layers and the soils supporting the pavement are not materials exhibiting uniform properties. While the performance of bituminous mixtures is now becoming more and more amenable to scientific analysis, the properties of granular bases, stabilised courses and soil subgrades are still areas of darkness. Thus, a completely satisfactory method of design, wholly based on theoretical considerations has not yet emerged. But modern computers can analyse complicated stress theories quickly and hence theoretical analysis is likely to be used more commonly in future than hitherto.

The theoretical approach is also known as the "analytical" or "rational" or "structural" or "mechanistic" design approach.

11.2.2. Stresses in homogeneous mass: Boussinesq's Theory

If the subgrade, sub-base, and surfacing layers in a pavement are assumed to form a homogeneous mass, the analysis of the stresses can be greatly simplified. In 1885, Boussinesq analysed the distribution of stresses in an ideal elastic, homogeneous and isotropic solid obeying Hooke's law and presented equations for horizontal and vertical stresses in such a material under a load. The vertical stress under a load at any horizontal section decreases from the maximum at the point located directly beneath the load to zero at a very large distance from this point. The pressure distribution is bulb-shaped as shown in Fig. 11.1

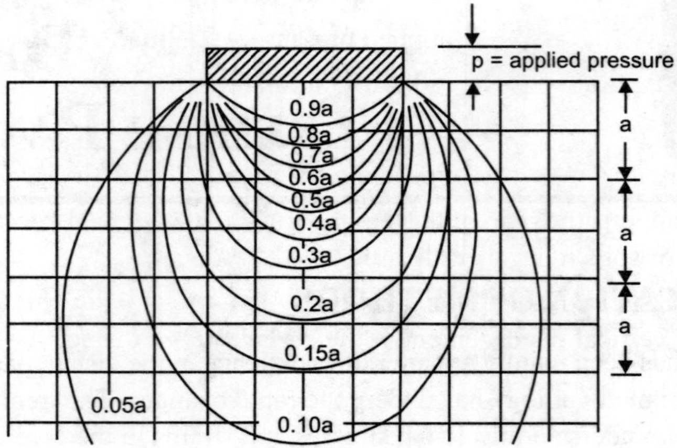

Fig. 11.1: Pressure distribution bulb

The vertical stress (σ_z), at any point below the surface due to a uniformly distributed load on a circular area, the point being located on the vertical axis passing through the centre of the circle, is given by (Fig. 11.2) :

$$\sigma_z = p\left[1 - \frac{z^3}{(a^2 + z^2)^{3/2}}\right] \tag{11.1}$$

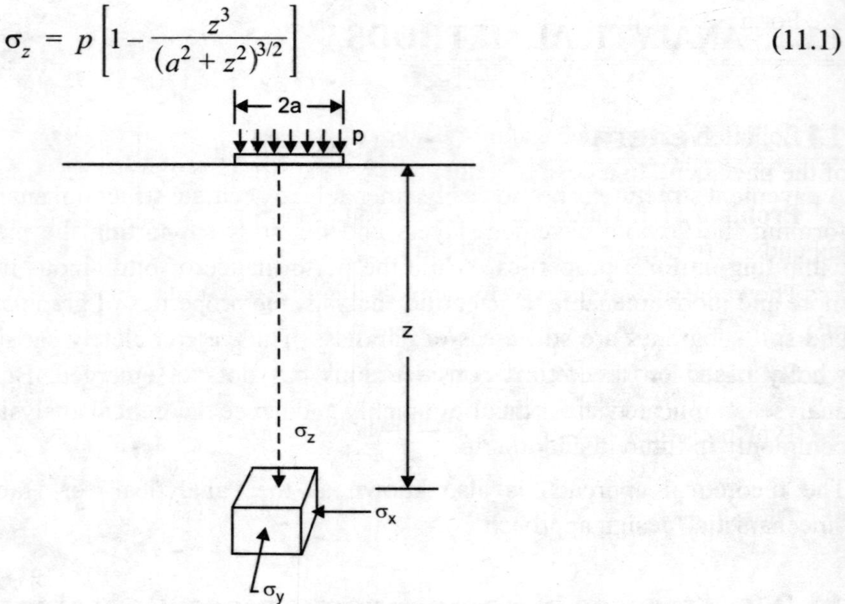

Fig. 11.2: Stresses below a circular loaded plate

The radial (horizontal) stress is given by

$$\sigma_x = \sigma_y = \frac{p}{2}\left[(1+2\propto) - \frac{2(1+\propto)z}{(a^2+z^2)^{1/2}} + \frac{z^3}{(a^2+z^2)^{3/2}}\right] \quad (11.2)$$

In the above formulae :

σ_z = vertical stress on a point on the z axis

σ_x = σ_y = radial or horizontal stress

P = applied pressure per unit area

a = radius of circular loaded plate

z = depth

μ = Poisson's ratio.

The Poisson's ration, μ is the ratio of the strain normal to the applied stress to the strain parallel to the applied stress. For soils it is generally around 0.5.

The Modulus of Elasticity E, of soil is the ratio of unit stress to the unit strain in the region of elastic behaviour. The vertical displacement at the surface ($z = 0$) under the centre of the applied load is given by :

$$\Delta = \frac{2pa}{E}(1-\mu^2) \quad (11.3)$$

Putting $\mu = 0.5$,

$$\Delta = 1.5\frac{pa}{E} \quad (11.4)$$

This equation is applicable for a flexible plate,

For a rigid plate,

$$\Delta = 1.8\frac{pa}{E} \quad (11.5)$$

Equation 11.4 can be used for design of a pavement by limiting the value of Δ, the deformation of the pavement, to a desired value.

Problem 11.1. Calculate the deflection at the surface of a pavement due to a wheel load of 40 kN and a tyre pressure off 0.5 MN/m^2.

The value of E of the pavement and subgrade may be assumed to be uniformly equal to 20 MN/m^2.

Solution: Radius of contact area a, is given by :

$$\text{Type pressure} = \frac{\text{Wheel load}}{\square\, a^2}$$

Therefore,

$$a = \sqrt{\frac{\text{Wheel load}}{\square \times \text{Tyre pressure}}} = \sqrt{\frac{40 \times 1000}{\square \times 0.5 \times 10^6}} = 15.95 \text{ m}$$

$$p = \frac{40 \times 1000}{\square \times 0.1595^2}$$

$$\therefore \text{Deflection} = 1.5\,\frac{p.a}{E} = 1.5 \times \frac{40 \times 1000}{\square \times 0.1595^2} \times \frac{0.1595}{20 \times 10^6}$$

$$= 0.005986 \text{ m} = 5.986 \text{ mm}$$

Say $= 6$ mm

The drawbacks in the simple Boussinesq's approach are:

(*i*) The assumption that soil is perfectly elastic and homogeneous is not true. Soil may be elastic only upto a certain limit.

(*ii*) The pavement consists of a number of layers, each with its own modulus of elasticity. Hence the assumption of one constant property for the entire mass is not justified.

(*iii*) The assumption that the load is uniformly distributed may not be correct.

Inspite of the drawbacks, application of Boussinesq's theory was the first attempt in analytical solution to pavement design.

11.2.3. Stresses in layered system: Burmister's Theory

Further refinements in analysing the stresses in pavements took the form of conceiving the pavement as a layered system.

Burmister proposed an analysis of a two-layer system, the top layer of finite thickness and the bottom layer of semi-infinite mass (Fig. 11.3). The top layer represents the surfacing, base and sub-base whereas the bottom layer represents the subgrade soil.

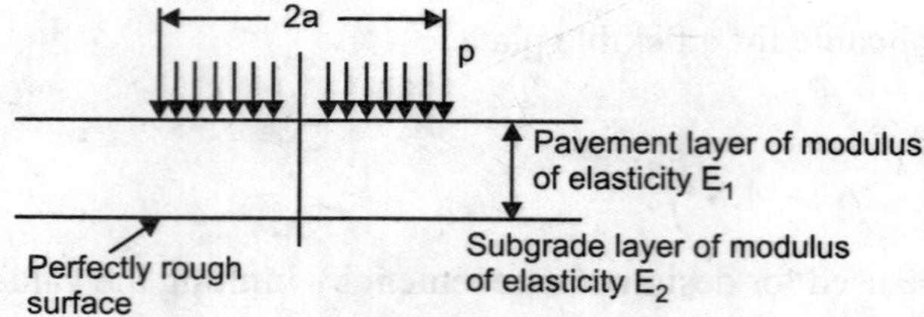

Fig. 11.3: Burmisler's two-layered system

The following equations give the deflection at the surface :

$$\Delta = F_w \frac{1.5pa}{E_2} \text{ (for a flexible plate)} \tag{11.6}$$

$$\Delta = F_w \frac{1.18\,pa}{E_2} \text{ (for a rigid plate)} \tag{11.7}$$

In the above equations,

Δ = deflection at the surface

P = load intensity on the circular plate

a = radius of plate

E_2 = Modulus of easticity of the lower layer

F_w = Displacement factor.

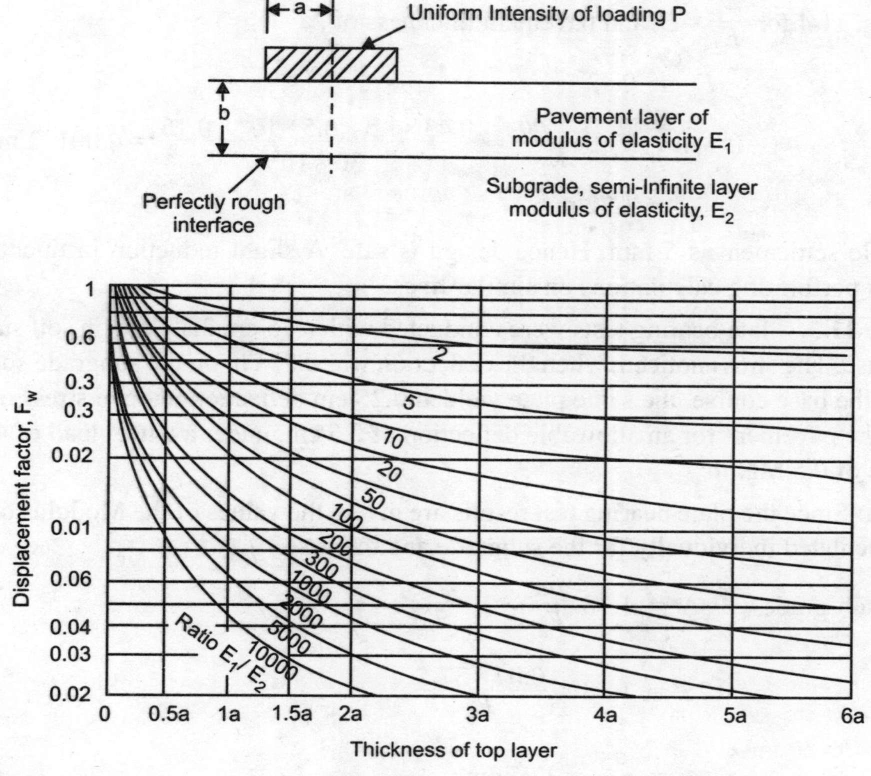

Fig. 11.4: Displacement factor F_w

The displacement factor, F_w can be obtained from graphs presented by Burmister in Fig. 11.4

Burmister suggested that the displacement under the wheel load can be limited to 5.0 mm for flexible pavements.

Problem 11.2. Design the thickness of a flexible pavement by Burmister's two layered analysis, for a wheel load of 40 kN and a tyre pressure of 0.5 MN/m² The modulus of elasticity of the top layer may be taken as 150 MN/m² and that of the subgrade as 30 MN/m². The value of F_w the displacement factor can be read from Fig. 11.4.

Solution: Radius of circular area, a, is given by

$$\text{Tyre pressure} = \frac{\textit{Wheel load}}{\pi a^2}$$

$$\text{Therefore,} \quad 0.5 \times 10^6 = \frac{40 \times 1000}{\pi a^2}$$

$$\text{Therefore,} \quad a = 15.95 \text{ cm, say } 160 \text{ mm}$$

Select a thickness of pavement of 2a, *i.e.* 320 mm

$$\frac{E_1}{E_2} = \frac{150}{30} = 5.$$

From Fig. 11.4 for $\frac{E_1}{E_2} = 5$, and pavement thickness of 2a

$$F_w = 0.43$$

$$\therefore \quad \Delta = F_w \frac{1.5\,pa}{E_2} = \frac{0.43 \times 1.5 \times 0.5 \times 10^6 \times 0.16}{30 \times 10^6} = 0.00172 \text{ m}$$

$$= 1.72 \text{ mm}$$

Allowable settlement is 5 mm. Hence design is safe. A slight reduction in thickness can be attempted by performing calculations on similar lines.

Problem 11.3. Plate bearing tests were conducted with a 75 cm dia plate on soil subgrade and a granular base. The stress noticed, when the deflection was 0.25 cm on the subgrade soil, was 0.07 MN/m^2. On the base course, the same plate yielded 0.25 cm deflection under a stress of 0.14 MN/m^2. Design the pavement for an allowable deflection of 0.5 cm, under a wheel load of 40 kN and a tyre pressure of 0.5 MN/m^2.

Solution: Since the plate bearing test results are given, the values of the Modulus of Elasticity, *E*, can be calculated individually for the subgrade and the base.

For the sub-grade, $\Delta = 1.18 \frac{pa}{E_2}$

$$0.25 = 1.18 \times \frac{0.07 \times 37.5}{E_2}$$

$$a = \frac{75}{2} = 37.5 \text{ cm} = \frac{37.5}{100} \text{ m}$$

$$\Delta = 0.25 \text{ cm} = \frac{0.25}{100} \text{ m}$$

$$E_2 = \frac{1.18 \times 0.07 \times 37.5 \times 100}{0.25 \times 100} \; 12.4 \text{ MN/m}^2$$

For the base, $\Delta = 1.18 \frac{p.a}{E_2} F_w$

$$\frac{0.25}{100} = \frac{1.18 \times 0.14 \times 37.5}{100 \times E_2}$$

$$F_w = \frac{0.25 \times 100 \times 12.4}{100 \times 1.18 \times 0.14 \times 37.5} = 0.50$$

Ratio of base course to radius $= \frac{15}{37.5} = 0.4$, if thickness of base course is 15 cm.

In Fig. 11.4, for a ratio of base course to radius of 0.4 and F_w of 0.5,

$$\frac{E_2}{E_1} = \frac{1}{100}$$

$$E_1 = E_2 \times 100 = 1240 \text{ MN/m}^2$$

$$a = \sqrt{\frac{\text{Wheel load}}{\square \times \text{Tyre pressure}}} = \sqrt{\frac{40 \times 1000}{\square \times 0.5 \times 10^6}} = 0.16 \text{ m}$$

Deflection under wheel load

$$= \frac{0.5}{100} = F_w \frac{1.5\, p.a}{E_2}$$

$$\frac{0.5}{100} = \frac{F_w \times 1.5 \times 0.5 \times 10^6 \times 0.16}{12.4 \times 10^6}$$

$$F_w = \frac{0.5 \times 12.4 \times 10^6}{100 \times 15 \times 0.5 \times 10^6 \times 0.16} = 0.52$$

From Fig. 11.4, for $F_w = 0.52, \dfrac{E_2}{E_1} = \dfrac{1}{100}$

$$\frac{\text{Pavement thickness}}{\text{Radius}} = 0.4$$

Pavement thickness $= 0.4 \times$ Radius $= 0.4 \times 16 = 6.4$ cm. Say 65 mm

11.2.4. Three-layered analysis

With quick computational facilities now available, the analysis of three or more layers is no more a difficult task. The three-layer system can be conceived as follows :

1. Top layer, representing all the bituminous layers taken together.
2. Second layer, representing the unbound road-base and sub-base.
3. Third layer, representing the subgrade.

The system is shown in Fig. 11.5. Jones and Peattie have analysed the stresses and strains in the system and have presented them in the form of Tables. Table 11.1 will facilitate design for some values which are commonly met with.

The materials in the three layers are assumed to be elastic and their mechanical property is characterised by the Modulus of Elasticity, *E*. In simpler treatments, the loading is assumed to be uniformly distributed over a circular contact area. The more commonly evaluated quantities are :

(*i*) The vertical compressive strains and stresses reaching the top of the layers representing the subgrade and unbound layers.

(*ii*) The horizontal and vertical stresses at the bottom of the unbound granular layer.

p = contact stress

$2a$ = diameter of circular loaded area

σ_{z_1} = vertical stress at first interface

σ_{z_2} = vertical stress at second interface

σ_{r_1} = horizontal stress in first layer, first interface

σ_{r_2} = horizontal stress in second layer, first interface

σ'_{r_2} = horizontal stress in second layer, second interface

σ_{r_3} = horizontal stress in third layer, second interface

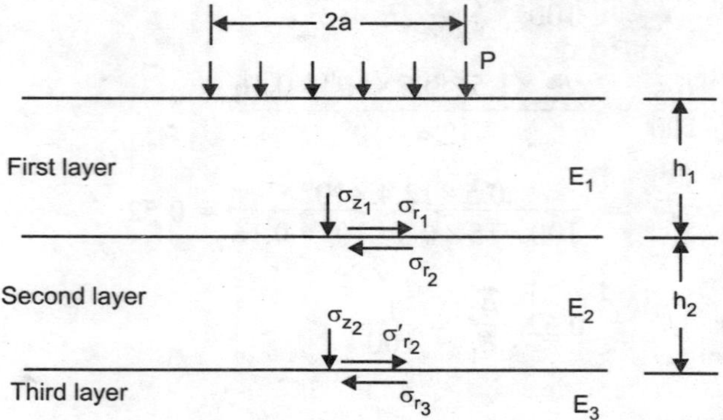

Fig. 11.5: Three layered system

(*iii*) The horizontal tensile strain at the bottom of the bituminous bound layer.

(*iv*) Surface deflection.

The following non-dimensional parameters are used in the analysis of the system :

$$a_1 = \frac{a}{h_2} \qquad H = \frac{h_1}{h_2} \qquad \text{(See Fig. 11.5)}$$

$$K_1 = \frac{E_1}{E_2} \qquad K_2 = \frac{E_2}{E_3}$$

For design of pavements, values of E are needed. They can be determined from laboratory tests. A rough formula used for determining E is:

$$E = 10 \text{ CBR } (\text{MN/m}^2) \tag{11.8}$$

The pavement analysis procedure consists of the following steps :

1. Determine the wheel load and contact pressure.
2. Select a pavement with a top bituminous layer of thickness h_1 and a bottom layer of granular base of thickness h_2.
3. Evaluate the Modulus of Elasticity of all the three layers.
4. Calculate the vertical and horizontal stresses and strains at the two interfaces from standard tables.
5. Compare the stresses and strains with allowable values for the materials selected.
6. Make adjustments if necessary.

An example of the use of the design procedure is given in Problem 11.4.

Problem 11.4. A three-layer pavement structure consists of 300 mm of bituminous layer on 150 mm of crushed stone road-base. The subgrade has a CBR of 5. A single wheel load with a tyre pressure of 0.5 MN/m^2 is applied over a circular contact area of 120 mm radius. Calculate the vertical and horizontal stresses and strains on both sides of each interface.

$$E_1 = 2000 \text{ MN/m}^2$$
$$E_2 = 100 \text{ MN/m}^2$$
$$E_3 = 50 \text{ MN/m}^2$$

Poisson's ratio = 0.5 for all layers.

Solution:

$$a = 120 \qquad h_1 = 300 \text{ mm} \qquad h_2 = 150 \text{ mm}$$

$$a_1 = \frac{120}{150} = 0.8 \qquad H = \frac{300}{150} = 2$$

$$K_1 = \frac{2000}{150} = 20 \qquad K_2 = \frac{100}{50} = 2$$

$$P = \text{Contact stress} = 0.5 \text{ MN/m}^2$$

From Table 11.1, the values of the stress and strain factors can be read out and when multiplied by the actual stress (0.5 MN/m^2), these yield the stresses. The strains can be obtained by multiplying the stress factors by $\frac{P}{E}$ for vertical strain and $\frac{P}{2E}$ for horizontal strain.

Table 11.1: Stress and strain factors in three-layered system (H = 2.0, K_1 = 20)

α_1	σz_1	$\sigma z_1 - \sigma r_1$	$\sigma z_1 - \sigma r_2$	σz_2	$\sigma z_2 - \sigma r_2$	$\sigma z_2 - \sigma r_3$
0.1	0.00134	0.00968	0.00048	0.00108	0.00014	0.00068
0.2	0.00533	0.03839	0.00192	0.00429	0.00055	0.00273
0.4	0.02100	0.14845	0.00741	0.01702	0.00216	0.01078
0.8	0.07950	0.52414	0.02621	0.06576	0.00820	0.04101
1.6	0.26613	1.41720	0.07085	0.23188	0.02740	0.13698
3.2	0.67382	2.38258	0.11913	0.63006	0.06384	0.31919
K_2 = 2.0						
0.1	0.00059	0.01219	0.00061	0.00033	0.00051	0.00025
0.2	0.00235	0.04843	0.00242	0.00130	0.00203	0.00101
0.4	0.00922	0.18857	0.00943	0.00518	0.00803	0.00401
0.8	0.03412	0.68382	0.03419	0.02023	0.03093	0.01547
1.6	0.10918	2.04134	0.10207	0.07444	0.10864	0.05432
3.2	0.29183	4.60426	0.23021	0.23852	0.30709	0.15354
K_1 = 20.0						

0.1	0.00027	0.01987	0.00095	0.00002	0.00131	0.00001
0.2	0.00106	0.07675	0.00384	0.00007	0.00524	0.00001
0.4	0.00406	0.30182	0.01509	0.00082	0.02085	0.00010
0.8	0.01396	1.13555	0.05678	0.00110	0.08180	0.00041
1.6	0.03538	3.83254	0.19163	0.00431	0.30676	0.00153
3.2	0.06182	11.55403	0.57770	0.01644	1.04794	0.00524

Stresses

A. Vertical Stresses

First layer, $\sigma_{z_1} = 0.03412 \times 0.5 = 0.01706$ MN /m² (Compressive)

Second layer $\sigma_{z_2} = 0.02023 \times 0.5 = 0.010115$ MN / m² (Compressive)

B. Horizontal Stresses

Horizontal stress at top of first interface

$$\sigma_{z_1} - \sigma_{T_1} = 0.68382 \times 0.5$$

$$\sigma_{z_1} - \sigma_{T_1} = -0.68382 \times 0.5$$

$$0.5\,(0.03412 - 0.68382) = -0.64970 \times 0.5$$

$$= -0.32485 \text{ MN/m}^2 \text{ (Tensile)}$$

Horizontal stress at bottom of first interface

$$\sigma_{z_1} - \sigma_{r_2} = (0.03419) \times 0.5$$

$$\therefore \quad \sigma_{r_2} = \sigma_{z_2} - 0.03419 \times 0.5$$

$$0.03412 \times 0.5 - 0.03419 \times 0.5 = -0.00007 \times 0.5$$

$$= -0.000035 \text{ MN/m}^2 \text{ (Tensile)}$$

Horizontal stress at top of second interface

$$\sigma_{z_2} - \sigma_{r_2} = (0.03093) \times 0.5$$

$$\therefore \quad \sigma_{z_1} - \sigma_{r_2} = -(0.03093 - 0.02023) \times 0.5 = -0.01070 \times 0.5$$

$$= -0.00535 \text{ MN/m}^2 \text{ (Tensile)}$$

Horizontal stress at bottom of second interface

$$\sigma_{z_2} - \sigma'_{r_2} = 0.01547 \times 0.5$$

$$\therefore \quad \sigma_{z_2} - \sigma_{r_3} = -(0.01547 - 0.02023) \times 0.5 = +0.00476 \times 0.5$$

$$= -0.00238 \text{ MN/m}^2 \text{ (Compressive)}$$

C. Strains

First Interface

Vertical strain in first layer $= \frac{P}{E_1}(\sigma_{z_1} - \sigma_{r_1}) = \frac{1}{2000} \times 0.68382 \times 0.5$

$= 0.000170955$ (Compressive)

Vertical strain in second layer $= \frac{P}{E_2}(\sigma_{z_1} - \sigma_{r_1}) = \frac{1}{100} \times 0.3419 \times 0.5$

$= 0.000171$ (Compressive)

Radial strain in first layer $= -\frac{P}{2E_1}(\sigma_{z_1} - \sigma_{r_1}) \times 0.5 = -\frac{1}{2 \times 2000}(0.68382) \times 0.5$

$= -0.0000855$ (Tensile)

Radial strain in second layer $= -\frac{P}{PE_2}(\sigma_{z_1} - \sigma_{r_2}) = -\frac{1}{2 \times 100} \times 0.03419 \times 0.5$

$= -0.0000855$ (Tensile).

D. Strains at Second Interface

Vertical strain in second layer $= \frac{P}{E_2}(\sigma_{z_2} - \sigma_{r_2}) = \frac{0.5}{100} \times 0.03093$

$= -0.000155$ (Compressive)

Vertical strain in third layer $= \frac{0.5}{E_3}(\sigma_{z_2} - \sigma_{r_3}) = \frac{0.5}{50} \times 0.01547$

$= -0.000155$ (Compressive)

Horizontal strain second layer $= -\frac{0.5}{2 \times E_2} \times (\sigma_{z_2} - \sigma_{r_2}) = -\frac{0.5}{2 \times 100} \times 0.03093$

$= -0.000077$ (Tensile)

Horizontal strain in third layer $= -\frac{0.5}{2 \times E_3}(\sigma_{z_2} - \sigma_{r_3})$

$= -\frac{0.5}{2 \times 50} \times 0.01547$

$= -0.000077$ (Tensile).

11.3. EMPIRICAL METHODS

11.3.1. CBR Method

Soil strength can be evaluated by a variety of ways. Plate bearing tests, penetration tests and triaxial tests are commonly used for evaluating soil strength. The plate bearing test gives the modulus of sub-grade reaction, which is commonly used in the design of rigid pavements. The penetration test most commonly employed is the CBR test, which is widely used in the design of flexible pavements. The triaxial test gives the shear strength of soils, but this method is not now commonly used for design purposes, though it is commonly used for evaluating embankment stability and for designing pavement mixtures. The CBR method will be described fully.

The CBR test is a laboratory test to determine the resistance of a material against penetration of a standard plunger. The test has been described in Chapter 5. The design curves based on this test were originally developed by the California State Highway Department for a 7000 lb wheel load. The curve for a 12000 lb was developed by interpolation. The curves were based on the Department's experience over a number of years on pavements which had and had not failed. The curves are based on the principle that a meterial of a certain CBR value requires a certain minimum thickness of construction over it.

The original curves have been used by the TRRL, U.K., for developing their own design procedure, and are contained in the original edition of Road Note No. 29. These curves have also been adopted by the Indian Roads Congress. In these curves, the wheel loads, corresponding to the California Highway Department's design procedure are as below (See Fig. 11.6).

Curve	Wheel load (lb)
A	4000
B	5000
C	7000
D	9000
E	12000
F	14000
G	16000

The advantage with the CBR design curves is that it is possible to design the total thickness of pavement as well as the thickness of the individual courses, provided the CBR values of the materials forming the individual courses are known in addition to the CBR value of the subgrade.

The criticism levelled against the CBR method is on the following considerations:

1. The CBR test is an ad-hoc and arbitrary test and the result does not give a direct indication of the strength of a soil.
2. The conditions of moisture to be used for the test, especially whether the specimen is to be soaked for four days or not, are also arbitrary and great caution is needed in using the test procedure for pavement design. In arid climates, such a condition may lead to over-design.
3. The curves used in the design procedure were developed by each individual organisations to suit its own conditions. The application of such curves by other organisations has to be carried out with great caution.
4. The CBR method of design does not represent a reliable method of design in the case of certain semi-rigid materials, especially when the CBR values are in excess of 60, and in that case, the compressive strength of the material would be a more reliable guide.

11.3.2. IRC Guidelines

The IRC guidelines for flexible pavements, published in 1970 were the basis for design in India till recently. The soil strength is measured by the CBR tests. The selection of moisture content is very important, and the procedure laid down has been described earlier in Chapter 5.

The thickness is given by 7 curves (Fig. 11.6). The traffic is in terms of commercial vehicles per day, the laden weight of vehicles exceeding 3 tonnes. The traffic is the total number of vehicles in both directions (irrespective of whether the design is for a two-lane or a dual carriageway). The future traffic is estimated on the basis of the existing traffic and probable growth rate. A growth rate of 7.5 per cent can be assumed in the absence of accurate data. For single lane roads, because of the concentration of traffic on one lane only, the traffic intensity is taken to be twice that for two-lane roads.

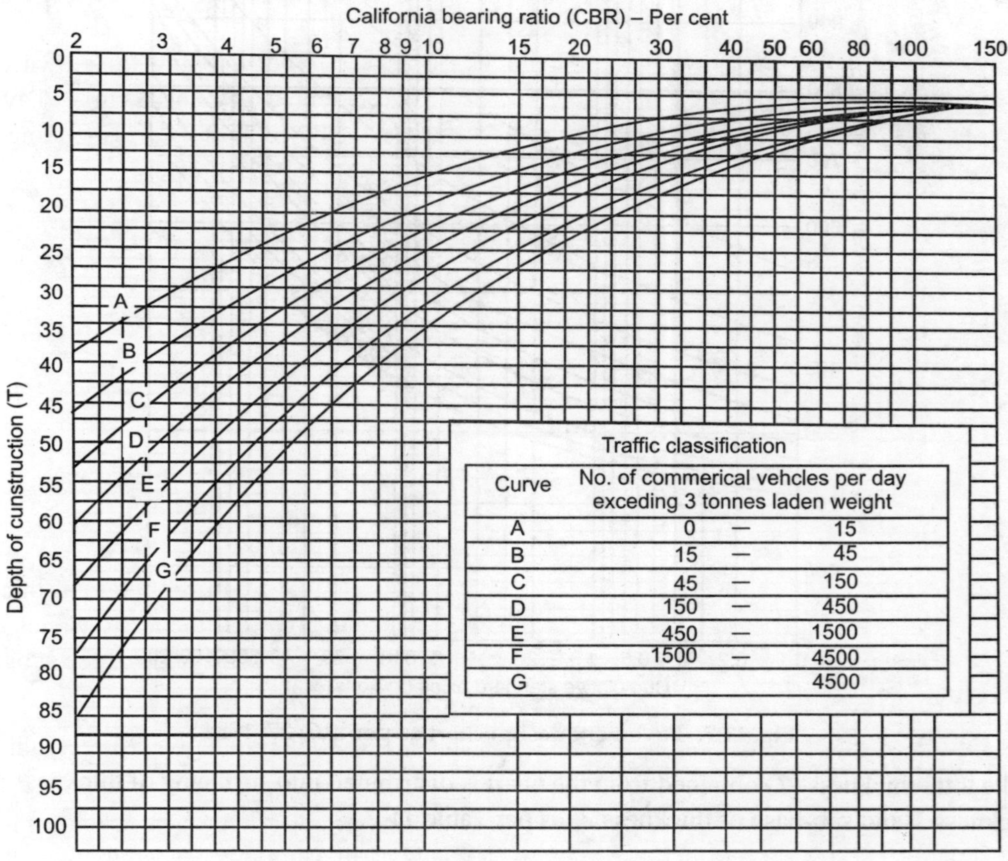

Curve	No. of commerical vehcles per day exceding 3 tonnes laden weight		
A	0	—	15
B	15	—	45
C	45	—	150
D	150	—	450
E	450	—	1500
F	1500	—	4500
G			4500

Fig. 11.6: CBR design curves as per IRC–37 (1970)

The IRC guide-lines were revised in 1984. The features of the new practice are :

(*i*) A new set of curves for determining pavement thickness relating the cumulative standard axles and the CBR value to the thickness.

(*ii*) Revised curves relating the commercial vehicle traffic and the CBR value to the thickness.

(*iii*) Recommendations on types of pavement materials suitable for various layers.

The method of computing the cumulative standard axles has been described in Chapter 9. The number of standard axles per commercial vehicle is designated as Vehicle Damage Factor (VDF). The pavement design chart extrapolated upto 200 million standard axles is given in Fig. 11.7.

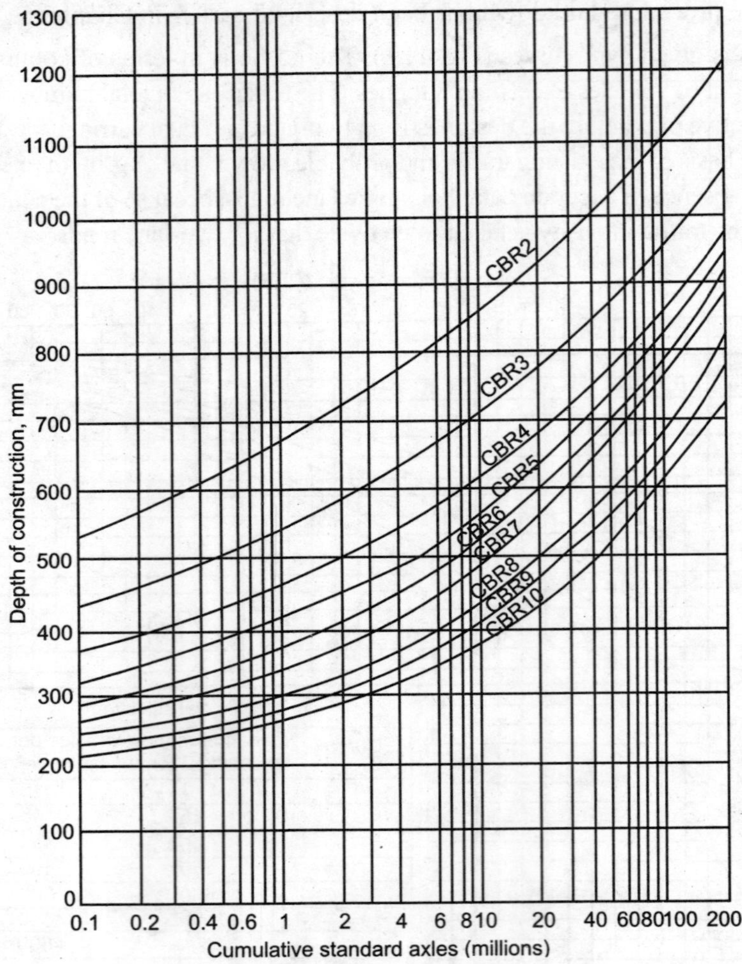

Fig. 11.7: Pavement design chart as per IRC: 37–1984

The total thickness (T) obtained from the chart is distributed into surfacing of thickness X, base of thickness Y and sub-base of thickness Z, as per Table 11.2.

Table 11.2: Composition of pavement as per IRC (1984) guidelines.

Cumulative standard axles (million)	Minimum thickness of layers (mm)		
	Surfacing (X)	Base (Y)	Sub-base (Z)
0.5	20 mm PC or	150	T— 50

	2 coats SD		Minimum thickness 100 mm on subgrades of CBR less 2 than 20 per cent.
0.5–2	20 mm PC or MS	225	T–225 Minimum thickness 150 mm on subgrades of CBR less than 20 per cent.
2–5	20 mm PC/MS/ SDC over 50 mm/75 mm BM	250	T–300/325 Minimum thickness 750 mm on subgrades of CBR less than 30 per cent.
5–10	20 mm SDC/BC over 60–80 mm DBM	250	T–355 to 355
10–15	40 mm BC over 65–80 mm BM	250	T–355 to 370
15–20	40 mm BC over 80-100 mm BM	250	T—370 to 390
20–30	40 mm BC over 100–115 mm BM	250	T–390 to 405

SD–Surface dressing, PC–Premix carpet
MS–Mix seal SDC–Semi-dense carpet
BC–Bituminous concrete BM–Bituminous macadam
DBM—Dense bituminous macadam.

11.3.3. Revised IRC guidelines (IRC: 37:2001)

The Indian Roads Congress have brought out a revised version, IRC : 37-2001. The features of this method are :

(*i*) The design catalogue is based on the results of the Ministry of Surface Transport (MOST) Study : "Analytical Design Approach for Flexible Pavements".

(*ii*) The flexible pavement has been designed as a four-layer structure.

(*iii*) The catalogues cover soils having CBR values in the range of 2–10.

(*iv*) Traffic upto 150 million standard axles (msa) has been considered.

(*v*) The damaging effect of axle loads can be taken from Table 11.3.

(*vi*) For loads upto 10 msa, the thickness can be taken from Fig. 11.8. Fig. 11.9. gives the pavement design thickness for loads of 10–150 msa.

(*vii*) The design catalogues are given seprately for each CBR.

Table 11.3: Equivalence factors and damaging power of different axle loads

Gross axle weight Kg	Load equivalency factors	
	Single axle	Tandem axle
(1)	(2)	(3)
900	0.0002	0.0000
11810	0.002	0.0002
2720	0.009	0.001
3630	0.031	0.003
4540	0.08	0.006
5440	0.176	0.013
6350	0.35	0.024
7260	0.61	0.043
8160	1.00	0.070
9070	1.55	0.110
9980	2.30	0.166
10890	3.27	0.242
11790	4.48	0.348
12700	5.98	0.470
13610	7.8	0.633
14520	10.0	0.834
15420	12.5	1.08
16320	15.5	1.38
17230	19.0	1.73
18140	83.0	2.14
19051	27.7	2.61
19958	33.0	3.16
20865	39.3	3.79
21772	46.5	4.49
22680	55.0	5.28
23587	–	6.17

24494	–	7.15
25401	–	8.20
26308	–	9.4
27216	–	10.7
28123	–	12.1
29030	–	13.7
29937	–	15.4
30844	–	17.2
31752	–	19.2
32660	–	21.3
33566	–	23.6
34473	–	26.1
35380	–	28.8
36288	–	31.7

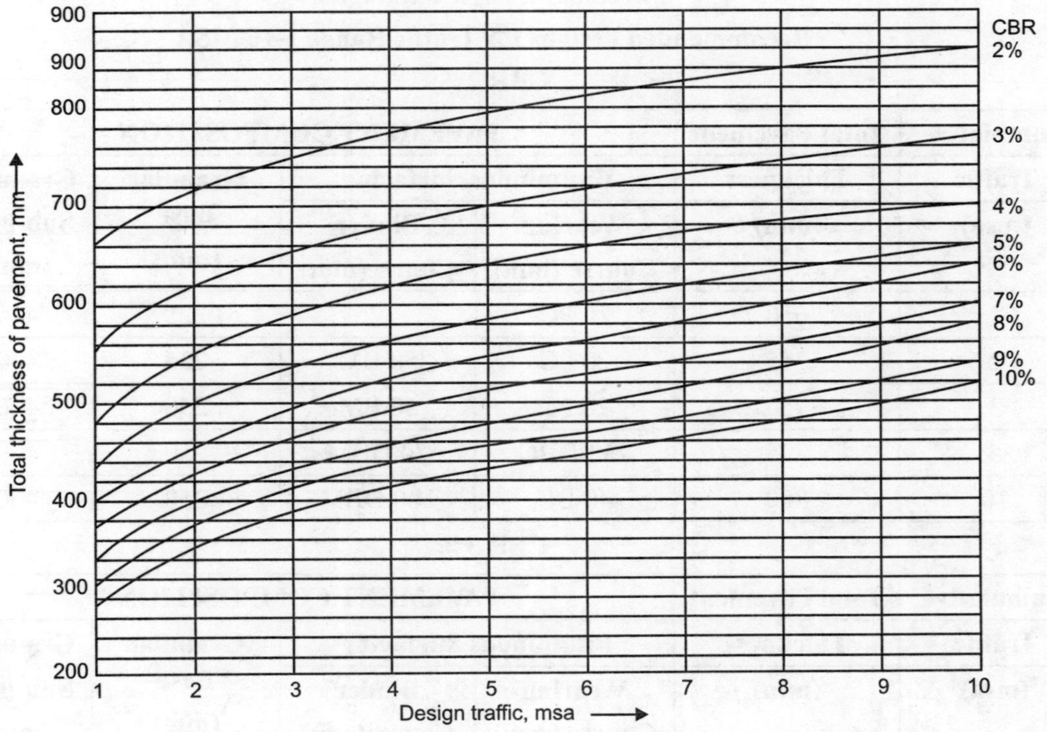

Fig. 11.8: Pavement thickness design chart for traffic 1–10 MSA. (IRC: 37: 2001)

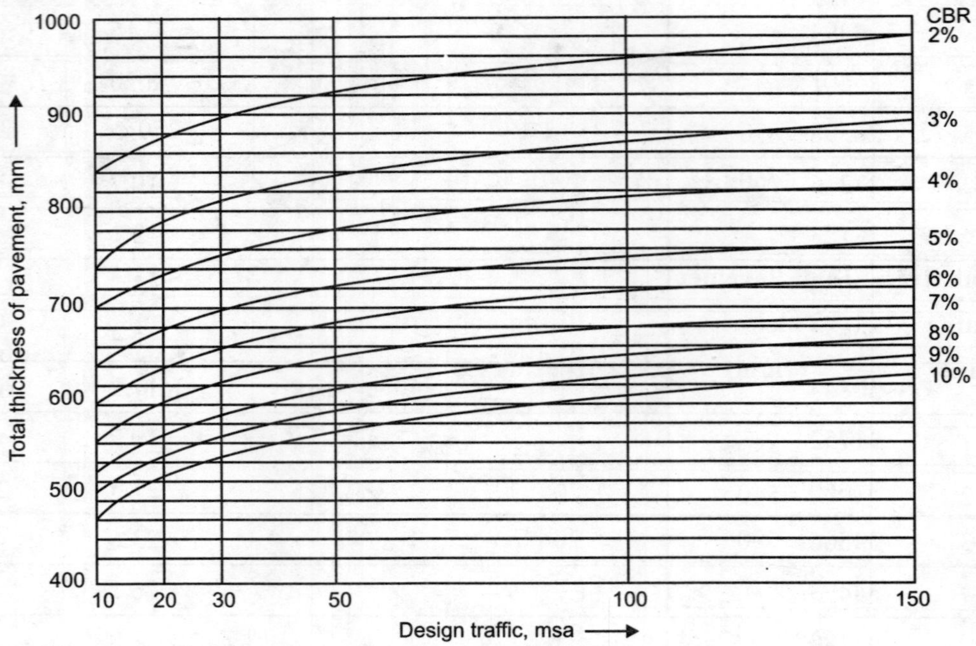

Fig. 11.9: Pavement thickness design chart for traffic 10–150 MSA. (Source: IRC: 37: 2001)

Pavement Design Catalogue

Recommended designs for Traffic Range 1-10 MSA

CBR 2%

Cumulative Traffic (msa)	Total Pavement Thickness (mm)	PAVEMENT COMPOSITION			
		Bituminous Surfacing		Granular Base (mm)	Granular Sub-base (mm)
		Wearing Course (mm)	Binder Course(mm)		
1	650	20 PC		225	435
2	715	20 PC	50 BM	225	440
3	750	20 PC	60 BM	250	440
5	795	25 SDBC	70 DBM	250	450
10	850	40 BC	100 DBM	250	460

CBR 3%

Cumulative Traffic (msa)	Total Pavement Thickness (mm)	PAVEMENT COMPOSITION			
		Bituminous Surfacing		Granular Base (mm)	Granular Sub-base (mm)
		Wearing Course (mm)	Binder Course(mm)		
1	550	20 PC		225	435

2	610	20 PC	50 MB	225	335
3	645	20 PC	60 BM	250	335
5	690	25 SDBC	60 DBM	250	335
10	760	40 BC	90 DBM	250	380

CBR 4%

Cumulative Traffic (msa)	Total Pavement Thickness (mm)	PAVEMENT COMPOSITION			
		Bituminous Surfacing		Granular Base (mm)	Granular Sub-base (mm)
		Wearing Course (mm)	Binder Course(mm)		
1	480	20 PC		225	255
2	540	20 PC	50 BM	225	265
3	580	20 PC	50 BM	250	280
5	620	25 SDBC	60 DBM	250	285
10	700	40 BC	80 DBM	250	330

CBR 5%

Cumulative Traffic (msa)	Total Pavement Thickness (mm)	PAVEMENT COMPOSITION			
		Bituminous Surfacing		Granular Base (mm)	Granular Sub-base (mm)
		Wearing Course (mm)	Binder Course(mm)		
1	430	20 PC		225	205
2	490	20 PC	50 BM	225	215
3	530	20 PC	50 BM	250	230
5	580	25 SDBC	55 DBM	250	250
10	660	40 BC	70 DBM	250	300

CBR 6%

Cumulative Traffic (msa)	Total Pavement Thickness (mm)	PAVEMENT COMPOSITION			
		Bituminous Surfacing		Granular Base (mm)	Granular Sub-base (mm)
		Wearing Course (mm)	Binder Course(mm)		
1	390	20 PC		225	165
2	450	20 PC	50 BM	225	175
3	490	20 PC	50 BM	250	190
5	535	25 SDBC	50 DBM	250	210
10	615	40 BC	65 DBM	250	260

CBR 7%

Cumulative Traffic (msa)	Total Pavement Thickness (mm)	PAVEMENT COMPOSITION			
		Bituminous Surfacing		Granular Base (mm)	Granular Sub-base (mm)
		Wearing Course (mm)	Binder Course(mm)		
1	375	20 PC		225	150
2	425	20 PC	50 BM	225	150
3	460	20 PC	50 BM	250	160
5	505	25 SDBC	50 DBM	250	180
10	580	40 BC	60 DBM	250	230

CBR 8%

Cumulative Traffic (msa)	Total Pavement Thickness (mm)	PAVEMENT COMPOSITION			
		Bituminous Surfacing		Granular Base (mm)	Granular Sub-base (mm)
		Wearing Course (mm)	Binder Course(mm)		
1	375	20 PC		225	150
2	425	20 PC	50 BM	225	150
3	450	20 PC	50 BM	250	150
5	475	25 SDBC	50 DBM	250	150
10	550	40 BC	60 DBM	250	200

CBR 9% AND 10%

Cumulative Traffic (msa)	Total Pavement Thickness (mm)	PAVEMENT COMPOSITION			
		Bituminous Surfacing		Granular Base (mm)	Granular Sub-base (mm)
		Wearing Course (mm)	Binder Course(mm)		
1	375	20 PC		225	150
2	425	20 PC	50 BM	225	150
3	450	20 PC	50 BM	250	150
5	475	25 SDBC	50 DBM	250	150
10	550	40 BC	60 DBM	250	200

Pavement Design Catalogue II

Recommended Designs For Traffic Range 10-150 MSA

CBR 2%

Cumulative Traffic (msa)	Total Pavement Thickness (mm)	PAVEMENT COMPOSITION		
		Bituminous Surfacing		Granular Base (mm) and Sub-base (mm)
		BC (mm)	DBM (mm)	
10	850	40	100	
20	880	40	130	
30	990	40	150	Base = 250
50	925	40	175	
100	955	50	195	Sub-base = 460
150	975	50	215	

CBR 3%

Cumulative Traffic (msa)	Total Pavement Thickness (mm)	PAVEMENT COMPOSITION		
		Bituminous Surfacing		Granular Base (mm) and Sub-base (mm)
		BC (mm)	DBM (mm)	
10	760	40	90	
20	790	40	120	
30	810	40	140	Base = 250
50	830	40	160	
100	860	50	180	Sub-base = 460
150	890	50	210	

CBR 4%

Cumulative Traffic (msa)	Total Pavement Thickness (mm)	PAVEMENT COMPOSITION		
		Bituminous Surfacing		Granular Base (mm) and Sub-base (mm)
		BC (mm)	DBM (mm)	
10	700	40	80	
20	730	40	110	
30	750	40	130	Base = 250
50	780	40	160	
100	800	50	170	Sub-base = 460
150	820	50	190	

CBR 5%

Cumulative Traffic (msa)	Total Pavement Thickness (mm)	PAVEMENT COMPOSITION		
		Bituminous Surfacing		Granular Base (mm) and Sub-base (mm)
		BC (mm)	DBM (mm)	
10	660	40	70	
20	690	40	100	
30	710	40	120	Base = 250
50	730	40	140	
100	750	50	150	Sub-base = 460
150	770	50	170	

CBR 6%

Cumulative Traffic (msa)	Total Pavement Thickness (mm)	PAVEMENT COMPOSITION		
		Bituminous Surfacing		Granular Base (mm) and Sub-base (mm)
		BC (mm)	DBM (mm)	
10	615	40	65	
20	640	40	90	
30	655	40	105	Base = 250
50	675	40	125	
100	700	50	140	Sub-base = 460
150	720	50	160	

CBR 7%

Cumulative Traffic (msa)	Total Pavement Thickness (mm)	PAVEMENT COMPOSITION		
		Bituminous Surfacing		Granular Base (mm) and Sub-base (mm)
		BC (mm)	DBM (mm)	
10	580	40	60	
20	610	40	90	
30	630	40	110	Base = 250
50	650	40	130	
100	675	50	145	Sub-base = 460
150	695	50	165	

CBR 8%

Cumulative Traffic (msa)	Total Pavement Thickness (mm)	PAVEMENT COMPOSITION		
		Bituminous Surfacing		Granular Base (mm) and Sub-base (mm)
		BC (mm)	DBM (mm)	
10	550	40	60	
20	575	40	85	
30	590	40	100	Base = 250
50	610	40	120	
100	640	50	140	Sub-base = 460
150	660	50	160	

CBR 9%

Cumulative Traffic (msa)	Total Pavement Thickness (mm)	PAVEMENT COMPOSITION		
		Bituminous Surfacing		Granular Base (mm) and Sub-base (mm)
		BC (mm)	DBM (mm)	
10	540	40	50	
20	570	40	80	
30	585	40	95	Base = 250
50	605	40	115	
100	635	50	135	Sub-base = 460
150	655	50	155	

CBR 10%

Cumulative Traffic (msa)	Total Pavement Thickness (mm)	PAVEMENT COMPOSITION		
		Bituminous Surfacing		Granular Base (mm) and Sub-base (mm)
		BC (mm)	DBM (mm)	
10	540	40	50	
20	565	40	80	
30	580	40	90	Base = 250
50	600	40	110	
100	630	50	130	Sub-base = 460
150	650	50	150	

11.3.4. Empirical methods using no soil strength criteria

Empirical methods of flexible pavement design are based on accumulated experience on the performance of pavements on different types of subgrades. For evaluating the subgrades, some simple classification tests can be conducted. Elaborate tests for determining soil strength are not carried out in some of the simpler methods of pavement design.

An example of this method of design is the U.S. Highway Public Road Administration (PRA) system known as Group Index Method. In the Group Index method a soil is classified into seven groups A–1 to A–7. The formula for Group Index has been given below:

$$GI = 0.2\,a + 0.005\,ac + 0.01\,bd$$

where a = portion of percentage passing 75 micron sieve greater than 35 and not exceeding 75 per cent expressed as a positive whole number from 1 to 40.

b = portion of percentage passing 75 micron sieve greater than 15 and not exceeding 55, expressed as a positive whole number from 1 to 40.

c = proportion of LL greater than 40 and not exceeding 60, expressed as a positive whole number from 1 to 20.

d = portion of the numerical plasticity index greater than 10 and not exceeding 30, expressed as a positive whole number between 1 and 20.

The Group Index is expressed nearest to the whole number and is written in brackets after the sub-group or Group Number. The classification of soils and soil – aggregate mixtures is given in Table 11.4

Table 11.4. Classification of soils and soil-aggregate mixtures (PRA)

General classification	Granular Material (35% or less passing 0.075 mm)							Silt- clay materials more than 35% passing 0.075 mm			
Group classification	A-1		A-3	A-2				A-4	A-5	A-6	A-7
	A-1-a	A-1-b		A-2-4	A-2-5	A-2-6	A-2-7				A-7-5 A-7-6
Sieve analysis, Per cent passing:											
2.00 mm (No. 10)	50 max.	–	–	–	–	–	–	–	–	–	–
0.425 mm (No. 40)	30 max.	50 max.	50 min.	–	–	–	–	–	–	–	–
0.075 mm (No. 200)	15 max	25 max.	10 max.	35 max.	35 max.	35 max.	35 max.	36 min.	36 min.	36 min.	36 min.
Characteristics of fraction											

Passing 0.425 mm (No. 40)											
Liquid limit	–	–	–	40 max.	41 min.	40 max.	41 min.	40 max.	41 min.	40 max.	41 min.
Plasticity Index	6 max.		NP	10 max.	10 max.	11 min.	11 min.	10 max.	10 max.	11 min.	11 min.*
Usual types of significant constituent materials.	Stone fragm- ents. gravel and sand.		Fine sand silty or clayey gravel and Sand					Silty soils		Clayey soils	
Group Index	0	0	0	0	0	4 max.	4 max.	8	12 max.	16 max.	20 max.
General rating as subgrade.	Excellent to good							Fair to poor			

*Plasticity index of A–7–5 subgroup is equal to or less than LL minus 30. Plasticity index of A–7–6 subgroup is equal to or greater than LL minus 30.

NP signifies non-plastic.

It should be noted that the only tests required to calculate *GI* are the soil particle-size determination, liquid limit and plasticity index. The higher the *GI*, the poorer is the soil.

The design curves which were presented by Steele in 1945 are given in Fig. 11.10. The curves give the thickness of (*i*) sub-base (Curve A) and (*ii*) total thickness of sub-base, base and surfacing for light traffic (Curve B), medium traffic (Curve C) and heavy traffic (Curve D). Another curve, Curve E, gives the additional thickness of base if sub-base is to be omitted.

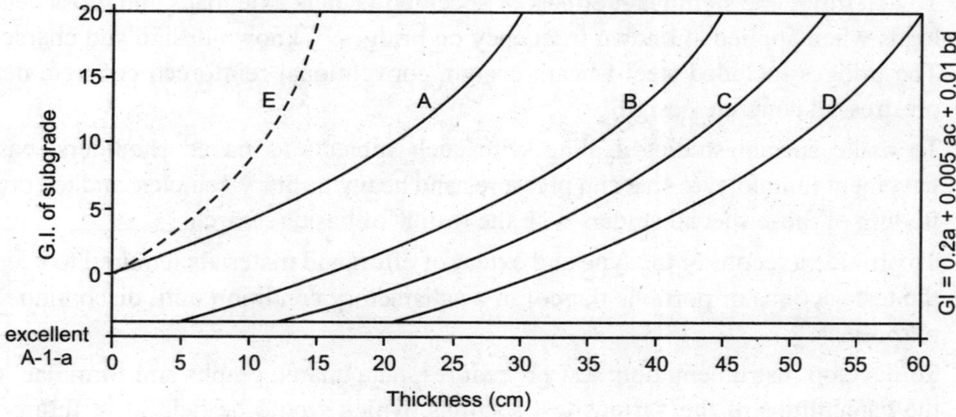

Fig. 11. 10: Pavement thickness design based on group index

The G.I. Method is only of historical importance now. It has rarely been used on a large scale.

CURVE A Total thickness of sub = base

CURVE B Total thickness of sub-base, base and surface, light traffic (less than 50 com. veh./ day)

CURVE C Total thickness of sub-base, base and surface, medium traffic (50–300 com. veh./day)

CURVE D Total thickness of sub-base, base and surface, heavy traffic (above 300 com. veh./day)

CURVE E Thickness of additional base which may be substituted for sub-base of curve A

11.4. AASHO ROAD TEST

AASHO Road Test was the largest and most comprehensive highway research project undertaken so far. It was the most ambitious attempt ever made to solve the perplexing problems that face designers of modern highway pavements. Estimated to cost about 27 million dollars (about 130 crores of rupees), the test was started in November 1958 and ended in November 1960 and during this intervening period of full two years, the various sections of the Test .track were subjected to test traffic. It took nearly 10 years to plan, design and actually carry out this ambitious programme. The findings of the Test have made a significant contribution to the engineers' concepts of design and performance of highway pavements.

11.4.1. Objectives of the test

The specific objectives of the test were:

1. To determine the significant relationships between the number of repetitions of specified axle loads of different magnitude and arrangement and the performance of different thicknesses of uniformly designed and constructed asphaltic concrete, plain portland cement concrete and reinforced portland cement concrete surfaces on different thicknesses of bases and sub-bases when laid on a basement soil of known characteristics.
2. To determine the significant effects of specified vehicle axle loads and gross vehicle axle loads when applied at known frequency on bridges of known design and characteristics. The bridges included steel I-beam design, conventional reinforced concrete design and prestressed concrete design.
3. To make special studies dealing with such subjects as paved shoulders, base types, pavement fatigue, tyre size and pressures and heavy military vehicles, and to correlate the finding of these special studies with the results of basic research.
4. To provide a record of the type and extent of effort and materials required to keep each of the test sections or portions thereof in a satisfactory condition until discontinued for test purposes.
5. To develop instrumentation, test procedures, data charts, graphs and formulae, to reflect the capabilities of the various test sections, which would be helpful in future highway design, in the evaluation of the load carrying capabilities of existing highways and in determining the most promising areas for further highway research.

11.5. METHODS BASED ON PAVEMENT PERFORMANCE

11.5.1. AASHTO method

1. The AASHTO guide for design of pavement structures is an updated version (1993) of the guidelines issued in the past. The recommended procedure is based on the extensive AASHO Road Tests. Further research carried out has facilitated in formulating the present design guidelines.
2. ***Time horizon for design:*** The guide recommends that the highway pavements be designed for a sufficiently long analysis .period. Suggestions are given in Table 11. 5.

Table 11.5: Analysis period suggested by AASHTO

Highway conditions	Analysis period (Years)
High volume, urban	30-50
High volume, rural	20-50
Low volume, paved	15-25
Low volume, aggregate surface	10-20

During the above analysis period, consideration may be given for rehabilitation. The period that elapses between initial construction or reconstruction or rehabilitation and the time when the pavement reaches its terminal serviceability is known as the Performance Period.

3. ***Traffic:*** The design guide is based on the total number of equivalent standard axle loads (ESAL), the standard axle load being 80 kN. The ESAL is determined from the axle load studies, and converting the actual axle loads to 80 kN axle loads, using equivalency factors.

   The equivalency factors are dependent upon :

   - Whether the pavement is flexible or rigid.
   - Structural Number (SN) of the flexible pavement and thickness of the rigid pavement.
   - Terminal pavement serviceability
   - Index, whether the axles are single, tandem or tridem.

   The 1993 A ASHTO guide gives separate Tables to account for the above variables.

   The distribution of traffic in the design traffic lane is as given in Table 11.6.

Table 11.6: Lane distribution factor

Number of lanes in each direction	Percentage of ESAL in design lane
1	100
2	80-100
3	60-80
4	50-75

4. ***Reliability:*** In the past, design methods followed the "deterministic" approach, wherein fixed values were assigned to design factors based on past experience. The present practice is to follow a "probabilistic" approach, whereby each design factor is assigned an appropriate factor of safety, with a mean and variance. The AASHTO guide uses this approach.

   The following definition of reliability may be noted.

   - "The reliability of a pavement design performance is the probability that a pavement section designed using the process will perform satisfactorily over the traffic and environmental conditions for the design period."

   The reliability levels suggested by AASHTO for various types of roads are given in Table 11.7.

Table 11.7: Reliability levels suggested by AASHTO

Highway class	Recommended reliability	
	Urban	Rural
Interstate and other freeways	85–99.9	80–99.9
Principal arterials	80–99	75–95
Collectors	80–95	75–95
Local	50–80	50–80

When stage construction strategy is adopted, the reliability at each stage (R State) can be calculated from :

$$R_{\text{(stage)}} = [R_{\text{(overall)}}]^{1/n} \tag{11.8}$$

Where n is the number of stages.

Thus if the design is done in 3 stages, $n = 3$ and if the overall reliability desired is 90%, the reliability required at each of the two stages is :

$$(0.90)^{1/3} = 0.965$$

The overall standard deviation for use in the design is another important input. A standard deviation of 0.45 is appropriate for a flexible pavement and 0.35 for rigid pavements.

5. ***Serviceability:*** Newly constructed pavements have a PSI value of more than 4. The expected number of repetitions dictates the terminal PSI. A value of 2.5 is commonly adopted for major highways whereas a smaller value of 2.0 is tolerated for highways with lower traffic.
6. ***AASHTO Design Equations:*** The AASHO Road test yielded certain equations. Using equations, the AASHTO guide has derived the following equation for flexible pavements (in FPS units):

$$\text{Log}_{10}(W_{18}) = Z_R \times S_0 + 9.36 \times \log_{10}(SN+1) - 0.20 + \frac{\log_{10}\left\{\frac{\Delta PSI}{4.2-1.5}\right\}}{0.40 + \frac{1094}{(SN+1)^{5.19}}} + 2.32 \times \log_{10}(\text{M}_\text{R}) - 8.07 \quad (11.9)$$

where W_{18} = Predicted number of 18,000 lb ESALs

Z_R = Standard normal deviate

S_0 = combined standard error of the traffic prediction and performance prediction

Δ PSI = difference between the initial design serviceability index, p_o, and the design terminal serviceability index, p_t.

M_R = Resilient modulus (psi)

$$SN = a_1 D_1 + a_2 D_2 + a_3 D_3 \quad (11.10)$$

where a_1, a_2, a_3 are layer coefficients for the surface, base and sub-base respectively, and D_1, D_2 and D_3 are the thickness (in inches) of the surface, base and sub-base layers respectively. For taking into account the local drainage characteristics, SN can be modified as :

$$SN = a_1 D_1 + a_2 D_2 m_2 + a_3 D_2 m_3 \quad (11.11)$$

where m_2 = drainage coefficient for base course

m_3 = drainage coefficient for sub-base course.

The layer coefficient a_1 for asphalt concrete surface course used in the AASHO Road Test was 0.44. The layer coefficient depends upon the resilient modulus of the material. The values of a_1 given in Table 11.8, can be used:

Table 11.8. Values of a_1 for different Resilient Modulus

Resilient modulus (psi)	a_1
450,000	0.44
400,000	0.42
300,000	0.36
200,000	0.30
100,000	0.20

For granular base courses,

$$a_2 = 0.249 (\log E_2) - 0.977.$$

The approximate values of a_2 for granular base courses can also be used from Table 11.9

Table 11.9. Values for a_2

CBR	a_2
100	0.14
55	0.12
45	0.11
30	0.09
20	0.07

For granular sub-base courses, the layer coefficient a_3 can be calculated from:

$$a_3 = 0.227 (\log E_3) - 0.839,$$

where E_3 = Resilient modulus

The following Table 11.10, for granular sub-base courses, can also be used to determine a_3.

Table 11.10. Values for a_3

CBR	a_3
100	0.14
40	0.12
30	0.11
25	0.10
15	0.09
10	0.08

The values of m_2 and m_3, the modifiers for layer coefficient for drainage conditions, can be taken from Table 11.11

Table 11.11. Modifiers, m_2 and m_3, for layer coefficients for drainage conditions,

Quality of drainage	Percentage of time, the pavement is exposed to moisture levels approaching saturation			
	Less than 1%	1–5%	5–25%	Greater than 25%
Excellent	1.40–1.35	1.35–1.30	1.30–1.20	1.20
Good	1.35–1.25	1.25–1.15	1.15–1.00	1.00
Fair	1.25–1.15	1.15–1.05	1.00–0.80	0.80
Poor	1.15–1.05	1.05–0.80	0.80–0.60	0.60
Very poor	1.05–0.95	0.95–0.75	0.75–0.40	0.40

The value of Z_R, the standard normal deviate, can be taken from statistical tables. Values for different levels of probability are given in Table 11.12.

Table 11.12. Standard Normal-Deviate for different values of reliability

Reliability (%)	Standard Normal Deviate (Z_R)
50	0.000
60	–0.253
70	–0.524
85	–0.841
90	–1.282
95	–1.645
99	–2.327
99.9	–3.090

The solution to the above equation can be by trial and error. Computer programmes are available for getting results quickly.

A Nomograph given in Fig. 11.11 offers a quick solution of equation 11.9.

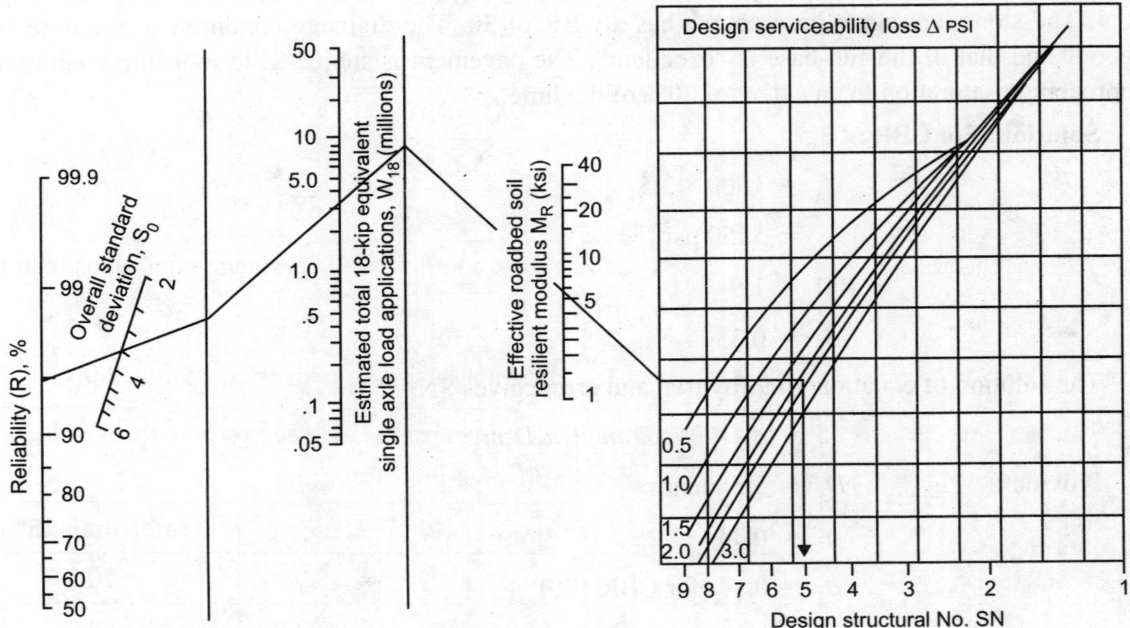

Fig. 11.11: Design chart for flexible pavements based on using mean values for each input

The minimum thickness values of asphalt surface and aggregate base courses can be taken from Table 11.13.

Table 11.13: Minimum thickness of asphalt surface and aggregate base courses

Traffic ESAL	Thickness (inches)	
	Asphalt concrete	Aggregate base
Less than 50,000	1.0	4
50,001–150,000	2.0	4
150,001–500,000	2.5	4
500,001–2,000,000	3.0	6
2,000,001–7,000,000	3.5	6
Greater than 7,000,001	4.0	6

The following problem illustrates the use of the procedure.

Problem. 11.5 Design a flexible pavement for a subgrade CBR valve of 3.5. The expected number of repetitions of the standard axle load is 5 million. The reliability needed is 95%. The standard deviation is 0.35.

The loss in serviceability expected is 1.9. The layer coefficient far asphalt surface course is 0.14. The stone aggregate base course has a CBR of 30. The drainage condition of the base is "good" and that of the sub-base is "excellent". The pavement is subjected to moisture condition approaching saturation to an extent of 20% of the lime.

Solution: For CBR of 3.5

$$M_R = 1500 \times 3.5$$

$$= 5250 \text{ psi}$$

$$\Delta PSI = 1.9$$

$$S_0 = 0.35$$

The solution of equation 16.9 (by trial and error) gives SN = 5.0

$$5 = a_1 D_1 + a_2 D_2 m_2 + a_3 D_3 m_3$$

Provide $D_1 = 3.5$ inches

$$a_1 = 0.44$$

$$a_2 = 0.14 \text{ (for CBR 100)}$$

$$a_3 = 0.11 \text{ (for CBR 30)}$$

$$m_2 = 1.00 \text{ to } 1.15$$

Take $m_2 = 1.15$

$m_3 = 1.20$ to 1.30

Take $m_3 = 1.3$

Try $D_2 = 6$ inches

$\therefore$ $5 = 0.44 \times 3.5 + 0.14 \times 6\ 1.15 + 0.11 \times D_3 \times 1.3$

$\therefore$ $D_3 = \dfrac{5 - 1.54 - 0.966}{0.11 \times 1.3} = 17.4$ inches, Say 17.5 inches.

11.5.2. Asphalt Institute method

The Asphalt Institute (USA) have published various Manuals and Handbooks dealing with different aspects of bituminous pavements. One well-known publication in this series is the Thickness Design Asphalt Pavements for Highways and Streets, Manual Series No. 1 (MS-1), issued in 1991.

The method is based on the results of various pavement performance tests.

The principle behind the method is given below :

1. The asphalt pavement is characterised as a multi-layered system.
2. The design procedure is based on established theory, experience, test data and a computer programme.
3. The final design is presented in the form of design charts.
4. For developing the design charts, two specific stress-strain conditions are considered :
   (*i*) The horizontal tensile strain, E_t on the underside of the lowest asphalt-bound layer should be within the permissible value. If the strain is excessive, fatigue of the layer will result.
   (*ii*) The vertical compressive strains E_c at the surface of the subgrade layer should be such that excessive permanent deformation (rutting) should not take place due to overloading.
5. The equation for fatigue criterion is :

$$N_f = 0.0796\,(E_t)^{-3.291}\,(E)^{-0.854}$$

where N_f = allowable number of repetitions of standard axle load to control fatigue cracking.

E_t = horizontal tensile strain at the bottom of the asphalt layer.

E = dynamic modules of the asphalt mixture.

6. The equation for limiting the pavement deformation is:

$$N_d = 1.365 \times 10^{-9}\,(E_c)^{-4.477}$$

where N_d = allowable number of repetitions to limit rutting to 12.7 mm for the design traffic.

E_c = vertical compressive strain on the surface of the subgrade.

7. The traffic during the design period is expressed in terms of equivalent standard Axles (FSAL) on the design lane. The conversion factors are AASHTO factors.
8. The distribution of trucks in the design lane is as under:

Number of traffic lanes (Two directions)	Percentage of trucks in design lane
2	50
4	45
6 or more	40

9. The soil strength is expressed in terms of CBR Resistance Value (*R*). From these, the subgrade resilient modules (M_r) is calculated as under:

M_r (MPa) = 10.3 CBR

M_r (PSI) = 1500 CBR

M_r (MPa) = 8.0 + 3.8 (*R* value)

M_r (PSI) = 1155 + 555 (*R* value)

The design M_r is then selected based on the traffic, as under:

Traffic ESA, design subgrade value (per cent)

Traffic ESA	Design subgrade value (per cent)
104 or less	60
104 – 106	75
More than 106	40

10. The charts are given for three Mean Annual Air Temperature (MAAT):

$\leq 7°C$ 15.5°C $\geq 24°C$

Charts give thickness of pavement in terms of full depth asphalt or a combination of asphalt layer and granular layer.

The maximum traffic catered to is 20 million standard axles, and thus this method is of limited application in India when the load repetitions are large (30–200 million standard axles).

11.6. PROPOSED MECHANISTIC–EMPIRICAL DESIGN METHOD BY AASHTO

AASHTO considered the revision of the 1993 Design Guide and proposed a new Mechanical Empirical (ME) method which is now available for trial as NCHRP-2003. This method is supposed to be an improvement over the 1993 guide, as it includes the computation of stresses and strains, and the assessment of the damage caused to the pavement. The three stage design process is given in Fig. 11.12.

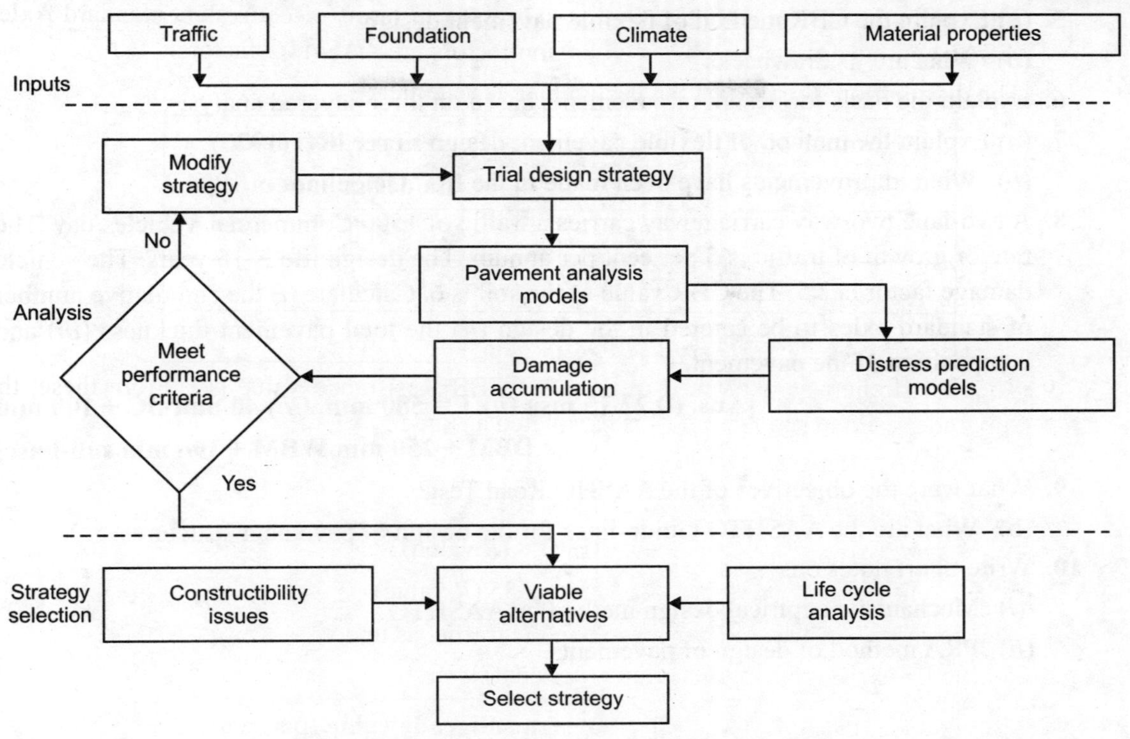

Fig. 11.12: NCHRP 2003 design process for flexible pavements

A software and user manual are available for use.

QUESTIONS

1. (*a*) Explain the Boussinesq's theory and how it can be used for design of highway pavements.

   (*b*) What are its limitations ?

2. Calculate the deflection under a wheel load of 40 kN. The tyre pressure is 0.6 MN/m^2 and E of pavement and subgrade is 25 MN/m^2. **[Ans. 0.524 cm]**

3. Explain Burmister's two-layer theory for design of pavements.

4. Determine the thickness of a flexible pavement by Burmister's two layer theory for a wheel load of 40 kN and a tyre pressure of 0.5 MN/m^2. The Modulus of Elasticity of the pavement material is 120 MN/m^2 and that of the subgrade is 12 MN/m^2. The value of F_w for E_1/E_2 of 10 can be taken as under:

Thickness of top layer	F_w
0.5 a	0.8
1.0 a	0.5
2.0 a	0.3

The allowable deflection is 0.5 cm. **[Ans. 16 cm]**

5. (*a*) Explain the CBR method of flexible pavement design.

(*b*) What are its drawbacks ?

6. Explain the basis for design of pavements based on three-layered analysis.

7. (*a*) Explain the method of flexible pavement design as per IRC (1970).

(*b*) What improvements have been made in the IRC Guidelines in 1984 ?

8. A two-lane two-way carriageway carries a traffic of 1500 Commercial Vehicles/day. The rate of growth of traffic is 5 per cent per annum. The design life is 15 years. The vehicle damage factor is 2.5. The CBR value of the soil is 6. Calculate (*i*) the cumulative number of standard axles to be catered in the design (*ii*) the total pavement thickness (*iii*) and composition of the pavement.

[Ans. (*i*) 22.15 msa (*ii*) T = 580 mm. (*iii*) 40 mm BC + 100 mm DBM + 250 mm WBM + 190 mm sub-base]

9. What were the objectives of the AASHO Road Test?

(b) What are the AASHTO Guide-lines for the design of pavement structures.

10. Write Short notes on:

(*i*) Mechanistic enpirical design method by AASHTO.

(*ii*) PRA method of design of pavements.

Chapter

Rigid Pavements

12.1. FACTORS GOVERNING DESIGN

The following factors govern the design of rigid pavements:

A. Loading

1. Wheel load and its repetitions
2. Area of contact of wheel load
3. Location of load with respect to panel of a slab

B. Properties of the supporting medium

1. Strength and properties of soil subgrade
2. Provision or non-provision of sub-base

C. Properties of concrete

1. Strength
2. Modulus of elasticity
3. Poisson's ratio
4. Shrinkage properties
5. Fatigue behaviour

D. External Conditions

1. Temperature variation
2. Friction between slab and sub-base

E. Joints

1. Whether joints are provided or not
2. Arrangement of joints
3. Dowel bars and Tie bars

F. Reinforcement

1. Quantity of reinforcement in jointed pavement
2. Quantity of reinforcement in continuously reinforced concrete pavements

12.2. WHEEL LOAD, AREA OF CONTACT AND ITS REPETITIONS

Wheel load is the most important factor that affects design. In India, though the legal axle load is 10.2 tonnes for a single axle, 19.0 tonnes for a tandem axle and 24.0 tonnes for a tridem axle, overloading is common. In order to get over this phenomenon, an axle load spectrum study is done and the 98th percentile axle load is considered for design.

The tyre inflation pressure determines the area of contact. A tyre pressure of 0.8 MPa is adopted, for design. The contact area is assumed to be circular. The distribution of commercial traffic varies across the carriageway width and has been discussed in Chapter 9.

The number of repetitions of the load affects the consumption of fatigue of the concrete and depends upon the design life. A design life of 30 years is commonly adopted. The rate of growth of traffic is determined as explained in Chapter 9. A rate of 7.5 per cent is considered in the Guidelines.

In jointed concrete pavements the location of the wheel load determines the stresses induced. The three critical locations are given in Fig. 12.1. Interior loading produces tensile stresses in the slab. Edge loading produces tensile stresses at the bottom of the slab and a small tensile stress at the top. Comer loading produces tensile stresses at the top of the slab due to cantilever action.

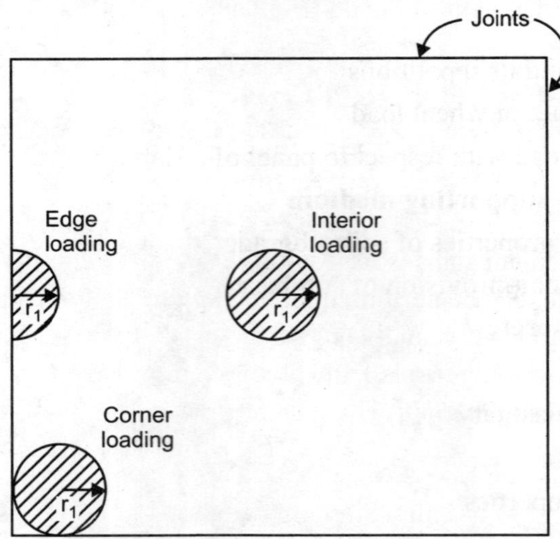

Fig. 12.1: critical loading positions in a concrete slab panel

12.3. STRENGTH AND PROPERTIES OF SOIL SUBGRADE

The strength of the soil subgrade determines how well it can support the slab. The strength is determined by the modulus of subgrade reaction, described in Chapter 9. The drainability and susceptibility to frost action affect the design.

12.4. PROVISION OR NON-PROVISION OF SUB-BASE

It is customary to provide a sub-base below the concrete slab and above the subgrade. The sub-base performs the following functions:

1. It provides a uniform and firm support to the pavement.
2. It prevents mud-pumping, which can occur due to water seeping through the joints, reaching the subgrade, moistening it and when a load passes over the slab causing the water and mud to come up through the joints. This causes loss of support, which is the most important requirement for the design of a concrete slab.
3. It reduces frost action.
4. It provides a firm support to the pavers which lay the concrete slab.
5. It acts as a capillary cut-off in subgrades liable to capillary rise of water.

The materials used for sub-base are:

(*i*) Dry Lean Concrete having a 7-day average compressing strength of 10 MPa and a thickness of 150 mm.

(*ii*) Cement-treated sub-bases having a minimum 28 day compressive strength of 7 MPa and thickness in the range of 100-200 mm.

(*iii*) Untreated granular sub-bases of thickness in the range of 150-300 mm.

It is customary to provide a debonding interlayer of polythene sheet of minimum thickness of 125 mm to allow smooth movement of the concrete slab caused by temperature changes.

12.5. PROPERTIES OF CONCRETE

12.5.1. Strength

The design of concrete pavement slabs is based on the 28 day strength. A 90 day strength, which is 1.1 times the 28 day strength, can be permitted in view of the fact that during the initial 90 days, the number of repetitions of load is very small and has negligible effect on cumulative fatigue damage of concrete. For design of the concrete slab, the flexural strength is required, which can be obtained after conducting the beam test. Alternatively, it can be derived as per the following relationship.

$$F_{cr} = 0.7\sqrt{f_{ck}} \tag{12.1}$$

where F_{cr} = Flexural strength (modulus of rupture), MPa

F_{ck} = cube strength, MPa

In no case the 28 day flexural strength of pavement quality concrete shall be less than 4.5 MPa. The target mean flexural strength to be achieved while designing the mix shall be such that there is 95 per cent probability that the characteristic strength would be achieved while the mix is produced in the field. Thus,

$$f'_{cr} = f_{cr} + 1.96\,\sigma \tag{12.2}$$

where f'_{cr} = target mean flexural strength at 28 days, MPa

f_{cr} = characteristic flexural strength at 28 days, MPa

σ = standard deviation of field test samples, MPa

1.96 = factor accounting to the desired confidence level for 5 per cent tolerance level

12.5.2. Modulus of elasticity and Poisson's ratio

The value of Modulus of elasticity, *E*, is generally taken as 30,000 MPa. The Poisson's ratio is taken as 0.15.

12.5.3. Coefficient of thermal expansion

Since a concrete slab expands and contracts due to changes in temperature, stresses are induced which depend upon the coefficient of thermal expansion, which is taken as 10×10^{-6} per °C.

12.5.4. Shrinkage

Concrete expands slightly during setting due to hydration of cement, but on drying it shrinks. The stresses induced are small and are generally opposed to those of temperature and are normally not critical for design.

12.5.5. Fatigue behaviour

Due to repetitive application of loads, progressive deterioration of the internal structure of concrete takes place.

The ratio of the flexural stress and the flexural strength, known as the stress ratio (*SR*), is an indication of the ability of concrete to withstand the repetitive loads. The IRC guidelines recommend that if *SR* is less than 0.45, the concrete is expected to sustain infinite number of repetitions. As the stress ratio increases, the number of load repetitions required to cause cracking decreases. The following relationships are given.

$$N = \text{Fatigue life} = \text{Unlimited for SR} < 0.45 \tag{12.3}$$

$$N = \left[\frac{4.2577}{SR - 0.4325}\right]^{3.268} \text{ for } 0.45 \le \text{SR} \le 0.55 \tag{12.4}$$

$$N = \frac{0.9718 - SR}{0.0828} \text{ for SR} > 0.55 \tag{12.5}$$

12.6. EXTERNAL CONDITIONS (TEMPERATURE VARIATION)

Charges in temperature cause stresses in the slab when they curl upwards or downwards. The curling stresses at the edge of the slab are to be added to the wheel load stress. The warping stress at the edge has been calculated by Bradbury as per the following formula:

$$f_{te} = \frac{E\,\alpha\,\Delta t}{2}\,C \tag{12.6}$$

where f_{te} = temperature stress at the edge

E = Modulus of elasticity

α = Coefficient of thermal expansion of concrete

Δt = change in temperature,

C = Bradbury's Coefficient depending upon $\frac{W}{\lambda}$ or $\frac{L}{\lambda}$ where

W = width of panel of slab

L = Length of panel

l = radius of relative stiffness (Refer equation 12.15)

The values of Bradbury's Coefficient C are given in Table 12.1

Table 12.1: Bradburys coefficient, C.

$\frac{L}{l}$ or $\frac{W}{l}$	C
1	0.000
2	0.040
3	0.175
4	0.440
5	0.725
6	0.920
7	1.030
8	1.077
9	1.080
10	1.075
11	1.050
12	1.000

The temperature variation for various regions in India are given in Table 12.2

Table 12.2: Recommended temperature differentials for concrete slabs

Zone	State/ Regions	Max. temperature differential °C in slab thickness			
		150 mm	200 mm	250 mm	300 mm to 400 mm
(*i*)	Hilly region Uttaranchal, West Bengal, Jammu & Kashmir, Himachal Pradesh and Arunachal Pradesh	12.5	13.1	14.3	15.8

(*ii*)	Punjab, UP, Uttaranchal, Gujarat, Rajasthan, Haryana and North M.P., excluding hilly regions	12.5	13.1	14.3	15.8
(*iii*)	Bihar, Jharkhand, West Bengal, Assam and Eastern Odisha excluding hilly regions and coastal areas.	15.6	16.4	16.6	16.8
(*iv*)	Maharashtra, Karnataka, South M.P., Chhattisgarh, Andhra Pradesh, West-ern Odisha and North Tamil Nadu, excluding hilly regions and coastal areas.	17.3	19.0	20.3	21.0
(*v*)	Kerala and South Tamil Nadu, excluding hilly regions and coastal areas.	15.0	16.4	17.6	18.1
(*vi*)	Coastal areas bounded by hills.	14.6	15.8	16.2	17.0
(*vii*)	Coastal areas unbounded by hills.	15.5	17.0	19.0	19.2

12.7. FRICTION BETWEEN SLAB AND SUB-BASE

The friction between slab and the sub-base determines the restraint on expansion and contraction of the slab panel due to temperature changes. It is used in the design of longitudinal and transverse steel according to the following formula:

$$A = \frac{LfW}{2S} \tag{12.7}$$

where
A = Area of steel in cm^2 per m width or length of slab
L = distance in m between transverse joints (for longitudinal steel) or free longitudinal joints (for transverse steel)
W = Weight of slab in Kg/m^2
f = coefficient of friction
S = Allowable working stress in steel in Kg/cm^2 (usually taken as 50 to 60 per cent of the maximum yield stress of steel)

12.8. JOINTS

The following are the types of joints which are provided in a concrete pavement :

1. Expansion Joint
2. Contraction Joint
3. Longitudinal Joint
4. Construction Joint

Details of various types of joints are given in Fig. 12.2.

Expansion Joints are provided to allow for the expansion of slabs. They are full depth joints provided transversely only where the slab abuts with a bridge or culvert.

Contraction joints are provided in jointed concrete pavement for relieving the tensile stresses in the concrete due to drying, shrinkage, for preventing the formation of irregular cracks due to restraint in free contraction of concrete and for relieving stresses due to warping.

They are sawn to a depth of one-fourth to one-third the slab thickness, facilitating the formation of a natural crack extending to the full depth of the slab. Dowel bars are provided for load transfer.

Longitudinal joints are provided in multi-lane pavements and when the pavement is more than 5 m wide.

They are formed by sawing to a depth of one-third to one-half of the slab depth. Tie bars are provided to keep the adjoining slabs together.

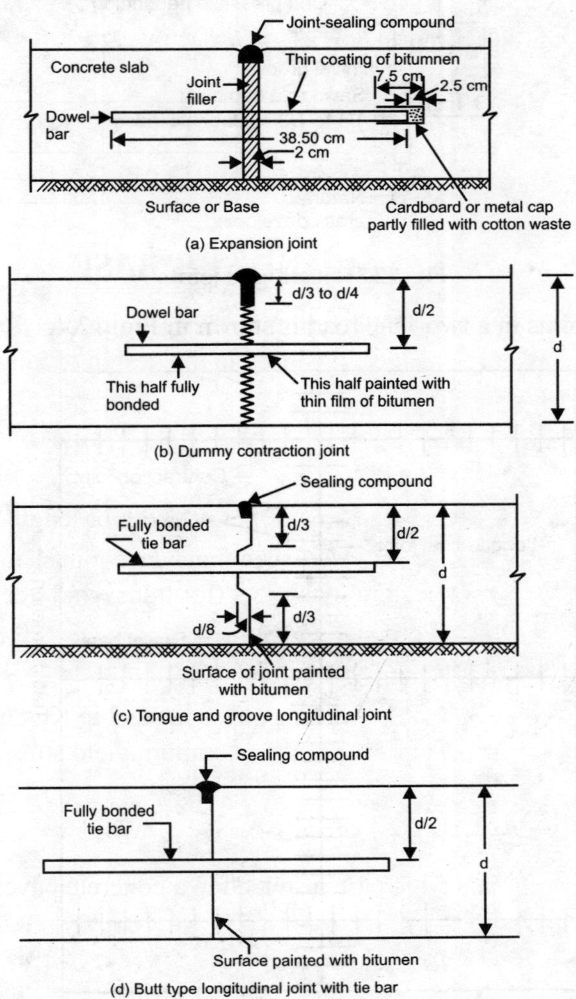

Fig. 12.2: Typical cross-sections of joints

Construction Joints are needed when the day's work is stopped at a location other than a contraction or expansion joint.

The joints are sealed with a sealant. Sealant can be:

(*i*) hot-poured (rubberised bitumen) (*ii*) cold-poured (polysulphide or silicon)
(*iii*) preformed

The details of sealing of joints are shown in Fig. 12.3.

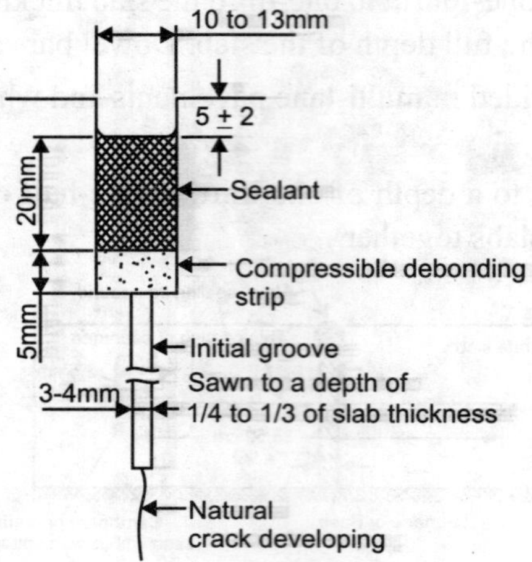

Fig. 12.3: Sealing of joints

The arrangement of joints in a two-lane road is shown in Fig. 12.4.

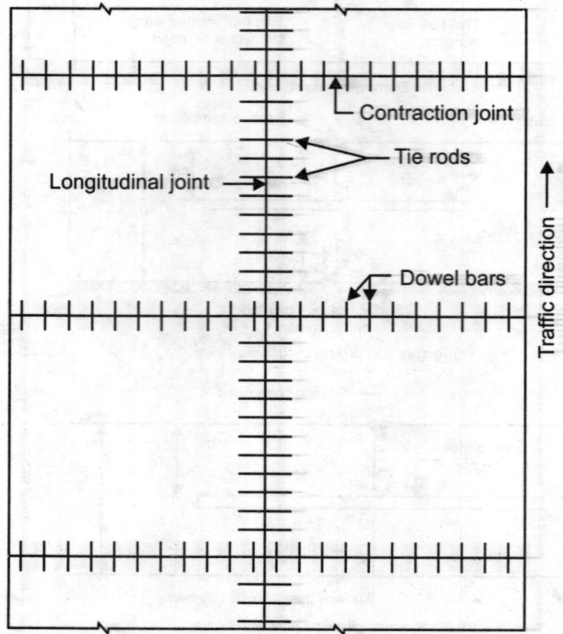

Fig. 12.4: Joints configuration of two lane road

12.9. DOWEL BARS

Dowel Bars are provided in transverse joints for transfer of load. From the experience gained all over the world, it is found that it is only the bearing stress in concrete that is responsible for the performance of dowel bars. The maximum bearing stress (F_b max) between the concrete and the dowel bar is given by the equation:

$$F_{b\,\max} = \frac{K_{mds}\, P_t\,(2+\beta Z)}{4\beta^3 E} \tag{12.8}$$

where β = relative stiffness of the bar embedded in concrete, mm

$$= \sqrt{\frac{K_{mds}\, b_d}{4\, EI}} \tag{12.9}$$

K_{mds} = modulus of dowel support, MPa/m, whose value may be taken as 4,15,000

b_d = diameter of the dowel, mm

Z = joint width (5 mm for contraction joint and 20 mm for expansion joint) in mm

E = Modulus of elasticity of dowel bar, MPa

I = Moment of inertia of the dowel, mm^4

P_t = Load transferred by the dowel bar, kN

The allowable bearing stress on concrete is calculated from the formula:

$$M_w = \frac{(101.6 - b_d) f_{ck}}{95.25} \tag{12.10}$$

12.10. TIE BARS

The area of steel required per metre length of joint may be calculated from:

$$A_s = \frac{bfw}{S_{st}} \tag{12.11}$$

where A_s = Area of steel in mm^2 per metre length of joint

b = lane width in metres

f = coefficient of friction between slab and sub-base, which may be taken as 1.5

w = weight of slab in kN/m^2

s_{st} = Allowable working stress of steel in MPa (125 MPa for plain bars and 200 MPa for deformed bars)

The length of tie bar is calculated from the formula:

$$L = \frac{2S\ A}{BP} \tag{12.12}$$

where L = Length of tie bar in mm

S = Allowable working stress in steel (values as given in Formula 12.11)

A = cross-sectional area of one tie bar, mm^2

B = Permissible bond stress MPa (1.75 MPa for plain bars and 2.46 MPa for deformed bars)

P = Perimeter of tie bar (mm)

12.11. IRC RECOMMENDATIONS FOR DOWEL BARS AND TIE BARS

The diameter and spacing of dowel bars and tie bars recommended by IRC are given in Tables 12.3 and 12.4 respectively.

Table 12.3: Dowel bar dimensions

Slab thickness (mm)	Dowel bar dimensions		
	Diameter (mm)	**Length (mm)**	**Spacing (mm)**
200	25	360	300
230	30	400	300
250	32	450	300
280	36	450	300
300	38	500	300
350	38	500	300

Table 12.4: Dimensions of Tie bars

Slab thickness (mm)	Dimensions (mm)				
	Diameter	**Max. spacing**		**Min. length**	
		Plain bars	**Deformed bars**	**Plain bars**	**deformed bars**
150	8	330	530	440	480
	10	520	830	510	560
200	10	390	620	510	560
	12	560	900	580	640

250	12	450	720	580	640
300	12 16	370 660	600 1060	580 720	640 800
350	12 16	320 570	510 910	580 720	640 800

Problem 12.1: *Design a tie-bar system for a cement concrete pavement, given:*

1. Slab thickness, $h = 20$ mm.
2. Slab width, $b = 3.35$ m.
3. Number of lanes to be tied = 2.
4. Coefficient of friction between slab and subgrade = 1.5.
5. Weight of slab = 480 kg/m^2
6. Allowable working stress in steel = 1400 kg/cm^2.
7. Maximum permissible bond stress
   (*i*) Plain bars: 17.5 kg/cm^2
   (*ii*) Deformed bars : 24 kg/cm^2

Solution: $A_s = \dfrac{bfw}{S_{st}} = \dfrac{3.35 \times 15 \times 480}{1400} = 1.72$ cm^2/m.

Assuming dia of bar to be 10 mm,

A, area of cross-section of one-bar = 0.785 cm^2.

P = Perimeter of one bar = 3.142 cm

No. of tie bars required per m,

$$N = \frac{A_s}{A} = \frac{1.72}{0.785}$$

$$\text{Spacing} = \frac{100}{N} = \frac{100 \times 0.785}{1.72} = 45.6 \text{ cm, say 45 cm.}$$

$$\text{Length of tie bar} = \frac{2SA}{BP} = \frac{2 \times 1400 \times 0.785}{B \times 3.142} = \frac{700}{B}$$

$$= \frac{700}{17.5} = 40 \text{ cm for plain bars}$$

$$= \frac{700}{24} = 29.2 \text{ cm for deformed bars.}$$

The length can be made 45 cm for plain bars and 35 cm for deformed bars, keeping an additional 5 cm margin for inaccuracy in placing.

12.12. REINFORCEMENT

12.12.1. Function of reinforcement in jointed concrete pavement

The reinforcement, if at all provided in jointed concrete pavement, is not intended to increase the flexural strength. Its role is to control cracking.

12.12.2. Quantity of reinforcement in jointed concrete pavement

The quantity of reinforcement (longitudinal and traverse) is calculated from the following formula:

$$A = \frac{L_d f w}{2S} \qquad (12.13)$$

where

A = Area of steel in mm^2 required per m width or length of slab

L_d = Distance in *m* between free transverse joints (for longitudinal steel) and free longitudinal joints (for transverse steel)

f = coefficient of friction between pavement and sub-base, taken a 1.5

w = Weight of slabs in kN/m^2

S = Allowable working stress in steel in MPa (usually taken as 50 to 60 per cent of the minimum yield stress of steel)

The steel is placed 50-60 mm below the surface.

12.12.3. Quantity of reinforcement in continuously reinforced concrete pavements (CRCP)

As per Indian practice, the quantity of longitudinal steel for normal work is 0.7 per cent of the cross-sectional area of the pavement. The longitudinal steel is placed 100-115 mm below the surface for slab thickness of 250-330 mm. Transverse steel reinforcement is calculated from the following formula:

$$\text{\% of transverse steel, } P_t = \frac{\gamma_c W_s f}{2 f_s} \qquad (12.14)$$

where

P_t = transverse steel, %

γ_c = unit weight of concrete (kN/m^3)

W_s = total pavement width (m)

f = coefficient of friction taken as 1.5.

f_s = allowable working stress in steel (taken as 75% of yield strength)

12.13. ANALYSIS OF STRESSES IN A CONCRETE PAVEMENT

The pioneering work in the design of concrete pavement was done by Westergaard who presented his results in 1925. The analysis is based on the following assumptions:

1. The concrete slab is homogeneous and isotropic and has uniform elastic properties.
2. The reaction of the subgrade is only vertical and is proportional to the deflection of the slab. In other words, the support provided by the subgrade is similar to that given by a dense fluid and the subgrade has no shear strength.
3. The reaction of the subgrade at a point is equal to k times deflection at that point, k being the Modulus of Subgrade Reaction (see Fig. 12.5).

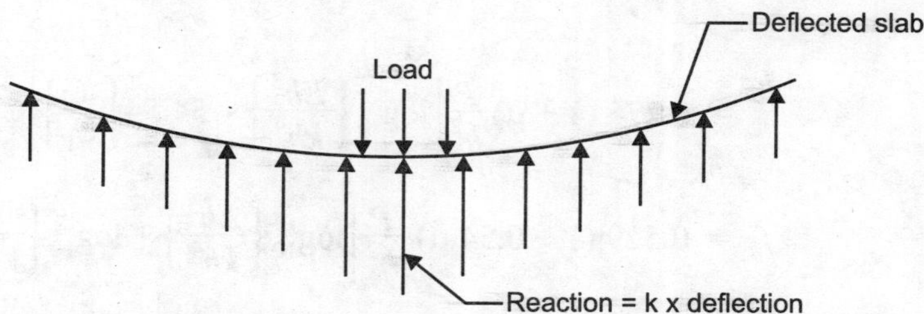

Fig. 12.5: Reaction of subgrade under loaded concrete slab

4. The slab is of uniform thickness
5. The loading positions are as given in Fig. 12.1.

The radius of relative stiffness (ℓ) which Westergaard used in his analysis is given by the formula:

$$\ell = \sqrt[4]{\frac{Eh^3}{12(1-\mu^2)k}} \tag{12.15}$$

where ℓ = radius of relative stiffness (in)

E = Modulus of Elasticity (psi)

h = pavement thickness (in)

μ = Poisson's ratio, 0.15

k = Modulus of subgrade (lb/in^3)

The three original equations proposed by Westergaard in 1925 are:

$$f_i = 0.31625\,\frac{P}{h^2}\left[4\log_{10}\left(\frac{\ell}{b}\right) + 1.0693\right] \tag{12.16}$$

$$f_e = 0.57185\,\frac{P}{h^2}\left[4\log_{10}\left(\frac{\ell}{b}\right) + 0.3593\right] \tag{12.17}$$

$$f_c = \frac{3P}{h^2}\left[1 - \left\{\frac{a\sqrt{2}}{\ell}\right\}^{0.6}\right] \tag{12.18}$$

where f_i = tensile stress at the bottom due to interior loading (psi)

f_e = tensile stress at the bottom due to edge loading (psi)

f_c = tensile stress at the top of the slab due to corner loading (psi)

P = Wheel load

ℓ = radius of relative stiffness (Eq^n 12.15)

h = slab thickness in inches

a = radius of contact area of wheel in inches

$b = (1.6\, a^2 + h^2)^{1/2} - 0.675\, h$

The above fonnulae were subsequently modified by Westergaard himself and by Teller and Sutherland and are as under:

$$f_i = 0.275\,(1 + \mu)\frac{P}{h^2}\left[\log_{10}\left\{\frac{Eh^3}{kb^4}\right\} - 54.54\left\{\frac{\ell}{C_1}\right\}\right]C_2 \quad (12.19)$$

$$f_e = 0.529\,(1 + 0.54\,\mu)\,\frac{P}{h^2}\left[\log_{10}\left\{\frac{Eh^3}{kb^4}\right\} + \log_{10}\left\{\frac{b}{1-\alpha^2}\right\} - 1.0792\right] \quad (12.20)$$

$$f_c = \frac{3P}{h^2}\left[1 - \left\{\frac{12(1-\alpha^2)k}{Eh^3}\right\}^{0.3}(a\sqrt{2})^{1.2}\right] \quad (12.21)$$

In the above formula, P, μ, h, E, a and ℓ are as described earlier and

$$b = (1.6\, a^2 + h^2)^{1/2} - 0.675\, h \text{ where } a < 1.724\, h \quad (12.22)$$

and $$b = a \text{ where } a \geq 1.724\, h \quad (12.23)$$

C_1 and C_2 and correction factors to allow for redistribution of subgrade reactions and can be taken as

$$C_1 = 5\,\ell \text{ and } C_2 = 0.2$$

In the above equation, the radius of contact area is found by dividing the wheel load by the tyre pressure and treating the area as a circle.

Thus $$a = \sqrt{\frac{P}{Q\square}}$$

where Q = Tyre pressure in MPa

P = Load in N

a = radius of contact area in mm

For dual tyre, the total area of the two tyres plus the rectangular area between them is considered and a is obtained as above.

The above analysis was the procedure adopted in India till 2011, when the procedure was revised based on the Finite Element Method and a software [IT SLAB-II developed by the I.I. T. Kharagpur. The analysis has been done for (*i*) bottom-up cracking involving combination of load and positive non-linear temperature differential and (*ii*) top-down cracking involving combination of load and negative linear temperature differential. Single axles, tandem and tridem axles have been considered. Several regression equations have been presented making the designer's task easy. Charts have also been presented giving the flexural stress developed for various combination of axle loads, temperature differential and concrete slab thickness.

12.14. CONTINUOUSLY REINFORCED CEMENT CONCRETE PAVEMENT (CRCP)

Recently, concrete slabs are being built with continuous reinforcement, thus eliminating the transverse joints. This is achieved by providing adequate reinforcement.

The CRCP slabs do develop numerous fine transverse cracks, but they are generally at the surface and do not affect the structural performance. Such cracks are also found at the end sections (60-150 mm) and are not present in the central portion.

The reinforcement of CRCP slabs is generally very high, and the mesh is not lighter than 5.5 kg/cm^2 and longitudinal bars are not less than 650 mm^2 per metre width of road. The longitudinal reinforcement is generally not less 0.60 per cent of the cross-sectional area and consists of deformed bars placed at 1/3 rd to $\frac{1}{2}$ the slab depth. The transverse steel also consisting of deformed bars, is placed over the entire width, and the steel in tied to the longitudinal bars. The area of steel is calculated from Eq. 12.14 to counteract temperature stresses.

The CRCP is used for very heavy traffic conditions and is far costlier than ordinary reinforced cement concrete slabs with joints. British practice is to provide a bituminous surfacing of thickness not less than 90 mm over CRCP,

12.15. FIBRE REINFORCED CONCRETE

Fibre-reinforced concrete (FRC) is a recent addition to the family of reinforced concretes. It differs fundamentally from the earlier members of the group, viz, reinforced concrete and prestressed concrete, in that the amount and location of the reinforcement are pot governed by loading conditions and the location and magnitude of tensile stresses. The reinforcement aims at uniformly improving the structural quality of the material as a whole. This is achieved by dispersing randomly oriented short length fibres in the concrete. The fibres are of steel, polypropylene, polyethylene or nylon. FRC provides a crack-arresting mechanism and improves the tensile strength and fatigue resistance of concrete. In road pavements, these qualities have a high premium.

12.16. PRESTRESSED CONCRETE PAVEMENTS

The use of prestressing enhances the available tensile strength of concrete by imparting compressive forces to it initially. The increase in flexural strength enables the load carrying capacity of the pavement to be greatly increased. It also enables joints to be placed at far greater intervals than in plain concrete pavements. The high load carrying capacity enables a reduction in the pavement thickness and reduced joints lead to better riding quality.

QUESTIONS

1. What are the factors affecting the design of rigid pavements?
2. (*a*) What are the functions of a sub-base in rigid pavements?

(*b*) What are the specifications adopted for sub-bases for rigid pavements?

3. Explain Westergaard's analysis of stresses for rigid pavements.

4. Give Bradbury's formula for temperature stresses in a concrete slab.

5. What are the functions of reinforcement in a concrete slab?

6. Describe IRC guidelines for Rigid Pavement Design.

7. Design the reinforcement of a cement concrete slab 225 mm thick for the following conditions:

(*i*) Density of concrete : 2.35 gm/cc.

(*ii*) Coefficient of friction : 1.5.

(*iii*) Transverse joint spacing : 12.5 m.

(*iv*) Pavement width : 3.75 m.5.

(*v*) Working stress in steel : 130 MN/m^2.

[Ans. 3.67 cm, 2. 10 mm bars at 25 cm centres].

9. Explain briefly the fatigue behaviour of concrete and how it is utilised in the PCA design method.

10. (*a*) Explain the need for joints in cement concrete pavements.

(*b*) What are the various types of joints?

(*c*) Sketch a typical expansion joint.

(*d*) Sketch a typical contraction joint.

(*e*) Sketch a tongue and grooved warping joint.

(*f*) Sketch a longitudinal joint.

(*g*) What are the IRC guide-lines for joint spacing?

11. (*a*) What is the formula used for design of a dowel bar ?

(*b*) What is the formula used for design of tie-bars?

12. Write note on :

(*a*) CRCP

(*b*) Fibre-reinforced concrete for pavements

(*c*) Prestressed concrete pavements.

Chapter

SOIL STABILISATION

13.1. DEFINITION

Soil stabilization is a process of treating a soil in such a manner so as to maintain, alter or improve the performance of the soil as a construction material. The changes in the soil properties are brought about either by the incorporation of additives or by mechanical blending of different soil types.

13.2. OBJECTIVES OF SOIL STABILIZATION

Soil-stabilization is practised in road construction with one or more of the following objectives :

(*i*) To improve the strength of sub-bases, bases and, in the case of low-cost roads, surface courses.

(*ii*) To bring about economy in the cost of a road.

(*iii*) To make use of locally available soils and other materials which are otherwise inferior.

(*iv*) To eliminate or improve certain undesirable properties of soils, such as excessive swelling or shrinkage, high plasticity, difficulty in compacting etc.

(*v*) To control dust.

(*vi*) To facilitate compaction and increase load-bearing capacity.

(*vii*) To reduce frost susceptibility.

(*viii*) To reduce compressibility and thereby settlements.

(*ix*) To improve permeability characteristics.

13.3. TYPES OF STABILISATION TECHNIQUES

Broadly, soil-stabilization takes the following forms :

(*i*) Mechanical stabilization, whereby the stability of the soil is increased by blending the available soil with imported soil or aggregate so as to obtain a desired particle-size distribution, and by compacting the mixture to a desired density. Compacting a soil at an appropriate moisture content is itself a form of mechanical stabilisation, which has been dealt with in chapter 10.

(*ii*) Stabilisation by additives such as lime, cement, sodium silicate, calcium chloride, bituminous materials and resinous materials. Chemical stabilisation is the general term implying the use of chemicals for bringing about stabilisation.

13.4. MECHANICAL STABILISATION

13.4.1. Principles

Mechanical stabilisation is achieved by intelligently blending locally occurring materials so as to obtain a desired grading. Compaction of soil is also a form of mechanical stabilisation. It is well-known that a dense, well-graded mass offers high resistance to lateral displacement under a load. If the well-graded material is compacted, densification of the mass takes place. The mechanical strength of the mass is due to the internal friction and the cohesion. Internal friction is supplied by the coarser particles (gravels, sands and silts) whereas cohesion is due to the clay fraction.

13.4.2. Applications

The application of the principle of mechanical stabilisation is evident in the following specifications:

(*i*) Soil-aggregate mixtures
(*ii*) Sand-clay mixtures
(*iii*) Sand-gravel mixtures
(*iv*) Stabilisation of soil with soft aggregates.

13.4.3. Soil-aggregate mixtures

As the name itself implies, a soil-aggregate mixture is a material in which soil and aggregate particles are mixed in suitable proportions such that the resulting mixture conforms to a dense and stable mix when properly compacted. Fig. 13.1 shows the soil-aggregate matrix. This technique is used in countries abroad for the construction of base courses, and in the case of low-traffic roads for the surface course as well. Its use in India is not popular, mainly because of the extensive use of water-bound macadam, though the Ministry of Surface Transport Specifications include the process.

Fig. 13.1. Soil aggregate matrix.

The particle-size distribution of the mixture is a major factor in determining the stability. The aggregates should be so graded that a grain-to-grain contact exists, producing internal friction. For a high value of density to result, the grain-size distribution should follow the Fuller's curve given by the equation :

$$p = 100\left(\frac{d}{D}\right)^n \qquad (13.1)$$

where p = percentage passing any sieve

d = aperture of the sieve in question

D = the maximum size of the aggregate

n = exponential, whose value can be taken as 0.5.

Typical Indian specifications for sub-base/base courses are given in Table 13.1.

Table 13.1: Specifications for soil-aggregate sub-base courses

Sieve designation	Per cent by weight passing the sieve for a nominal maximum size of:		
	80 mm	40 mm	20 mm
80 mm	100	–	–
40 mm	80–100	100	–
20 mm	60–80	80–100	100
10 mm	45–65	55–80	80–100
4.75 mm	30–50	40–60	50–75
2.36 mm	–	30–50	35–60
600 micron	10–30	15–30	15–35
75 micron	0–10	0–10	0–10

Notes:

1. Not less than 10 per cent should be retained between each pair of successive sieves specified for use excepting the largest pair.
2. The material passing 425 micron sieve when tested according to IS: 2720 (Part V) shall have liquid limit and plasticity index of not more than 25 per cent and 6 per cent respectively.
3. Some authorities specify that the percentage passing 75 micron sieve should be 5–15, so as to supply cohesion.

It may be noted from the above Table that the plasticity of the binder is also an important factor contributing to the satisfactory performance of the specification. A maximum value of 25 for L.L and 6 for P.I is usually specified for sub-base mixtures. For surfacing mixtures, a slight relaxation is allowed so as to provide greater cohesion and help offset the moisture lost by evaporation. Thus L.L should not exceed 35 and the P.I should lie between 6 and 9 in surfacing mixtures.

The construction is accomplished in lift thickness of 100–200 mm and compaction to 100 per cent of the laboratory maximum density is sought after. The moisture at the time of compaction generally is 1 per cent above to 1 per cent below the O.M.C.

When used as a surfacing course, which is to be maintained for some time without bituminous surface treatment, it is necessary to specify that a minimum of 8 per cent pass the 75 micron sieve.

13.4.4. Sand-clay mixtures

A sand-clay road is composed of a favourable mixture of clay, silt and sand. If some coarser materials such as gravel are also present, the mixture will perform still better. In order to get over the undesirable characteristics of clay, the blending of clay with a proportion of sand can alter the properties significantly. If sand is available at economical leads, the specification can be very cheap. Sand-clay mixtures are constructed to a thickness of about 200 mm and used as a surfacing course for low-traffic roads. The mixture can also serve as a good sub-base and base. When used as a subbase, the mix should have a minimum soaked CBR of 20, whereas when used as a base for heavily trafficked roads, the minimum soaked CBR of 20, whereas when used a base for heavily trafficked roads, the minimum soaked CBR should be normally 80. A somewhat smaller value is permissible for low-traffic roads.

The gradings given in Table 13.2 are recommended.

Table 13.2: Specifications for sand-clay mixtures

Sieve designation	Percent Passing by Weight		
25 mm	100	–	–
10 mm	–	100	–
4.75 mm	70–100	80–100	100
2.36 mm	–	50–80	80–100
2.00 mm	55–100	–	–
1.18 mm	–	40–65	50–80
600 micron	–	–	30–60
425 micron	30–70	–	–
300 micron	–	20–40	20–45
75 micron	8–25	10–25	10–45

The requirements of liquid limit and plasticity index given under soil-aggregate mixtures apply to sand-clay roads also. The mixture is compacted to 100 per cent of the maximum dry density at a moisture content of 1 per cent above to 2 per cent below the O.M.C.

Experiments on sand-clay bases in the Calcutta test-track have proved the efficacy of the specification. Since then, sand-clay mixtures have been used in some road projects in the country. More extensive use of the specification in future can be visualised if proper equipment is used and quality control is exercised.

13.4.5. Sand-gravel mixtures

In general, gravel denotes a material having predominantly coarse particles 2.0–60 mm in size and resulting from disintegration of rock. Gravel occurs as a natural deposit in a river bed if the disintegrated rock particles are transported by water. In this case, the particles are often rounded. The material also occurs in pits, when it is found mixed with sand. In tropical countries, the materials obtained from the disintegration of laterite are extensively found in a natural admixture of clay and coarse fractions and is known as lateritic gravel. In India, this material is locally known as moorum. Often, the natural moorum contains too much of plastic materials which lowers its value as a road pavement material. A suitable admixture of moorum with sand will not only result in a better gradation and increased strength, but also will reduce the plasticity. Sand-moorum mixtures find ready application in sub-bases and bases in India The grading and plasticity requirements given in para 13.4.3 can be adopted for sand-moorum mixtures.

13.4.6. Stabilisation of soil with soft aggregates

In India, there are many areas where an abundant supply of natural soft aggregates can be had. These aggregates are gravel, moorum and kankar (an impure form of limestone mixed with clay and earth). In the alluvial plains of the North, aggregates of over-burnt bricks (jhama bricks) are also used for road construction. A method of stabilisation of soil using these soft aggregates has been introduced successfully in India and is known as Mehra's method.

The principle behind this method is to embed in a soil mortar, coarse aggregates roughly one-third of the total volume. The aggregates are normally soft aggregates such as over-burnt brick ballast, kankar, moorum or laterite. Because of the larger proportion of the soil mortar, the resulting material has no grain-to-grian contact in the coarse aggregates, which merely float in the soil, vide Fig. 13.2. Each aggregate is thus enveloped all round in the compacted soil and is thus protected from the crushing effect of traffic. This enables the soft aggregate to retain its strength and angular character for an indefinite period About 10 per cent of the coarse aggregates which are collected are not mixed with the soil, but are saved and spread on the layer of the soil-aggregate mixture before rolling.

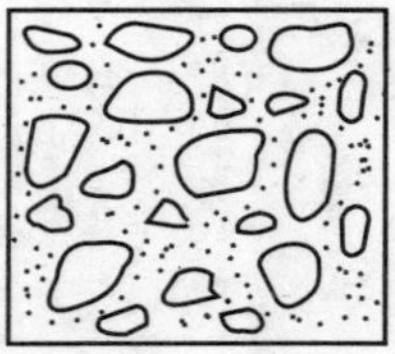

Fig. 13.2. Soil-aggregate matrix in Mehra's method of soil stabilisation.

The soil is required to have a P.I. value 8–11 and a minimum sand content of 33 per cent. The soft aggregates should have a maximum Wet Aggregate Impact value of 50 per cent when used as a sub-base, 40 per cent when used as a base-course with bituminous surfacing and 30 per cent when used as a surfacing course.

The stabilised soil layer is suitable as a surfacing course without any bituminous treatment for very light traffic (about 50 tonnes per day). For light traffic (about 200 tonnes per day), a light bituminous surfacing is needed. For medium traffic (about 500 tonnes per day), a thin stone grafting (about 25 mm thick) is given on the soil-soft aggregate mixture while compacting, and a thin bituminous surfacing is provided. For areas with a high rainfall, it is necessary to provide two coats of surface dressing.

The soil-aggregate mixture is compacted to maximum dry density at optimum moisture content.

13.4.7. Combining materials to obtain required gradation

(A) In dealing with mechanical stabilisation, it is often found necessary to combine different materials to obtain the finally desired gradation. Rothfuch's graphical method is a reasonably quick, accurate and simple method, and is used for design of cement concrete mixes, bituminous mixes and granular mixes. The method consists of the following stages :

(*i*) Using the desired aggregate gradation, a distribution curve is plotted with the percentage passing as linear ordinates and the sieve sizes on the horizontal scale. In order to mark the sieve sizes on the horizontal scale, an inclined line (OA in Fig. 13.3) is first of all drawn. By marking the known percentages passing each size sieve on this line and dropping vertically the intersection point to the horizontal axis, the location of the sieve size on the horizontal axis is determined.

(*ii*) The particle size distributions of the given materials to be blended are plotted on this scale. The distribution curves will not generally be straight lines (Lines OB, CDE and CF GA in Fig. 13.3).

(*iii*) With the aid of a transparent straight edge, straight lines are drawn representing the particle size distribution in the best possible manner (Lines *HJ* and *KA*). This means that the areas enclosed between the distribution curve and the straight line should be minimum and are balanced about the line.

(*iv*) The opposite ends of these lines are joined together (Lines *BH* and *JK*).

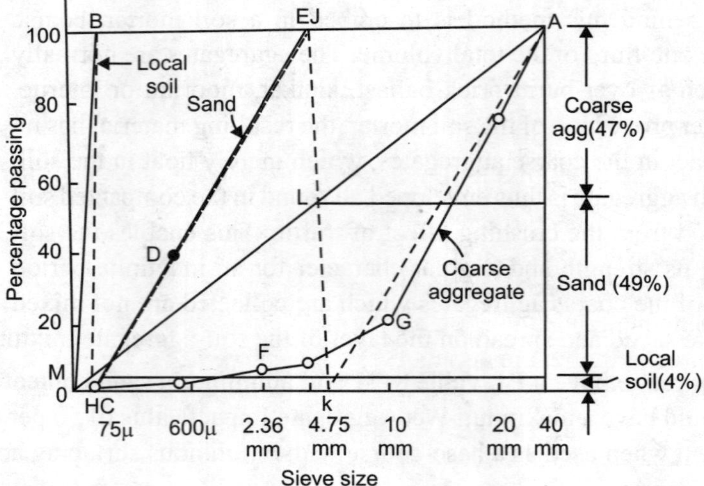

Fig. 13.3. Rothfuch's graphical method of combining aggregates.

(*v*) The proportions for blending can be read off from the points where the joining lines cross the straight line representing the mixture. These points are *L* and *M*.

The method is illustrated by the following example.

Cols. 1 and 2 in the Table 13.3 give the gradation limits for various sieve sizes of a stabilised soil mixture. Three materials are available, whose gradations are given in Cols. 4, 5 and 6. Work out the blending proportion.

Solution. Col. 3 in Table 13.3 gives the average percentage passing.

Table 13.3: Blending of materials

Sieve designation	Percentage passing					
	Required gradation limits	Average	Materials available			
			Coarse agrregates (A)	Sand (B)	Local soil (C)	47%A 49%B 4%C
(1)	(2)	(3)	(4)	(5)	(6)	(7)
40 mm	100	100	100		–	100
20 mm	80–100	90	75		–	88
10 mm	55–80	67.5	20		–	62
4.75 mm	40–60	50	8	100	–	53
2.36 mm	30–60	40	6	80	–	42
600 mm	15–30	22.2	2	40	–	21
75 micron	0–10	5	Nil	2	100	5

A reference to Fig. 13.3 will illustrate the stages involved in the graphical method. The percentages of various materials as scaled out are :

Material (*A*): 47%

Material (B): 49%

Material (C): 4%.

The gradation of the final mixture on the basis of the above blending is indicated in Col. 7 of the Table.

(B) Another graphical method in which 3 materials can be blended is by means of a triangular chart. The use of this method is illustrated by an example in which three materials, *A*, *B* and *C* are to be blended to obtain a gradation shown in Cols. 2 and 3 of Table 13.4. Cols. 4, 5 and 6 give the gradation of the three individual materials.

Table 13.4: Blending of materials

Sieve designation	Gradation limits	Average	Percentage passing		
			Material A	Material B	Material C
1	2	3	4	5	6
20 mm	100	100	100		–
10 mm	80–100	90	80		–

4.75 mm	50–75	63	25	100	–
2.36 mm	35–60	48	10	80	–
600 mm	15–35	25	5	40	–
75 micron	0–10	5	Nil	Nil	100

From the Table 13.4, the gravel, sand and silt-clay fractions of the three materials are as under :

Material	Percentage of		
	Gravel	Sand	Silt-clay
A	90	10	0
B	20	80	0
C	0	0	100

In the triangular chart, Fig. 13.4, each side of the equilateral triangle represents percentages (0–100) of the three materials, viz., gravel, sand and silt-clay.

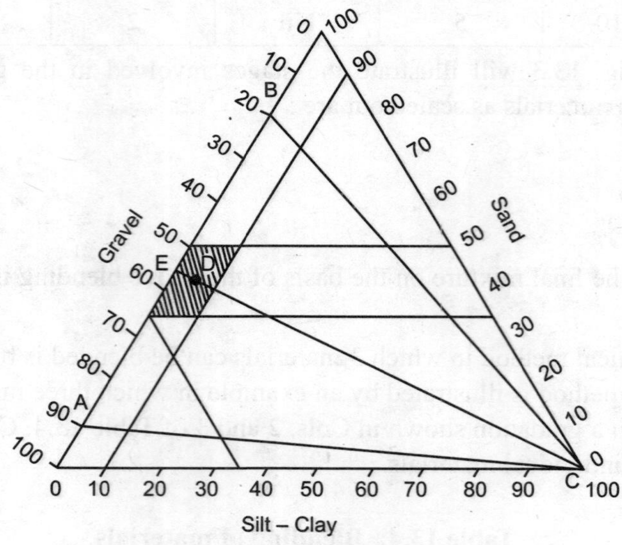

Fig. 13.4: Triangular blending chart.

The gradation limits of the desired mix is composed of the three fractions as follows :

Gravel	:	50 – 65
Sand	:	35 – 50
Silt-clay	:	0–10

Plotting these on the triangular chart, one obtains a hatched parallelogram which will contain all combinations of the three materials which will fulfil the gradation requirements.

A point "*D*" is selected inside this parallelogram about its centre.

Point "*A*" represents the material "*A*", point "*B*" represents material "*B*" and point "C" represents material "C". The triangle ABC represents all possible combinations of the three materials.

Join *C* to *D* and produce *CD* to meet *AS* at *E*.

The following ratio of lengths is then determined by scaling off:

$$\frac{AE}{AB} = 0.5$$

$$\frac{EB}{AB} = 0.5$$

$$\frac{ED}{EC} = 0.058, \text{ say } 0.06 \textit{ i.e. } 6 \text{ per cent.}$$

The ratio $\frac{ED}{EC}$ represents the proportion of *C* in the total mixture.

Proportion of A and B = 1 – 0.06 = 0.94

Proportion of A = 0.5 × 0.94 = 0.47, *i.e.* 47 per cent.

Proportion of B = 0.5 × 0.94 = 0.47, i.e. 47 per cent.

The gradation obtained as a result is tabulated in Table 13.5.

Table 13.5. Blending of materials

Sieve designation	Gradation limits	Percentage passing						
		Material A	Col. 3 × 0.47	Material (B)	Col. 5 × 0.47	Material (C)	Col. 7 × 0.06	Tatal Cols. 4 + 6 + 8
(1)	(2)	(3)	(4)	(5)	(6)	(7)	(8)	(9)
20 mm	100	100	47	100	47	100	6	100
10 mm	80–100	80	38	100	47	100	6	91
4.75 mm	50–75	25	12	100	47	100	6	65
2.36 mm	35–60	10	5	80	38	100	6	49
600 mm	15–35	5	3	40	19	100	6	28
75 micr	0–10	–	–	–	–	100	6	6

It is seen that the final grading as given in Col. 9, satisfies the limits given in Col. 2.

(C) In some cases, it is possible to arrive at the blending proportion by simple calculations. As an example, consider the gradation desired in Table 13.6. Col. 2 gives the desired gradation limits and Col. 3, the mean. The gradation of Materials A, B and C are given in Cols, 4, 5 and 6. By examination it is seen that material retained on 4.75 mm size sieve is to be fully supplied by Material A. The quantity of such material in Material A is 100–23 = 77 per cent, whereas the desired quantity in the final mix is 50 per cent.

Table 13.6

Sieve designation	Percentage passing				
	Gradation limits	Mean	Material (A)	Material (B)	Material (C)
(1)	(2)	(3)	(4)	(5)	(6)
40 mm	100	100	100		
20 mm	80–100	90	82		
10 mm	55–80	68	48		
4.75 mm	40–60	50	23	100	
2.36 mm	30–50	40	11	91	
1.18 mm	–	–	6	34	100
425 micron	15–30	23	3	10	84
150 micron	–	–	1	3	59
75 micron	5–15	10	0	2	36

Therefore, the required proportion of Material *A* in the final mix

$$= 100 \times \frac{50}{77} = 65 \text{ per cent.}$$

The combined proportion of *B* and C will be 100 – 65 = 35 per cent.

It is seen that as regards material passing 425 micron sieve, Material A forming 65 per cent of the total mix, can supply only :

$$\frac{65}{100} \times 3 = 2 \text{ per cent.}$$

But the requirement being 23 per cent, 21 per cent will have to be supplied by Materials *B* and *C*. The combined proportion of Materials *B* and *C* in the final mixture being 35 per cent, it is obvious that these two materials must be blended suitably. Let *b* and *c* denote the percentage of Materials *B* and *C* respectively.

$$b + c = 35$$

$$b \times \frac{10}{100} + c \times \frac{84}{100} = 21$$

$$0.10\,b + 0.84c = 21$$

But $b + c = 35$

Solving the simultaneous equations,

$$b = 11, c = 24$$

With the above blending proportion, the resulting gradation is tabulated in Table. 13.7.

Table. 13.7: Resulting gradation

Sieve designation	Gradation limits	Material (A)	65% of col. 3	Material (B)	11% of col. 5	Material (C)	24% of col. 7	Total cols. 4 + 6 + 8
(1)	(2)	(3)	(4)	(5)	(6)	(7)	(8)	(9)
40 mm	100	100	65	100	11	100	24	100
20 mm	80–100	82	53	100	11	100	24	88
10 mm	55–80	48	31	100	11	100	24	66
4.75 mm	40–60	23	15	100	11	100	24	50
2.36 mm	30–50	11	7	91	10	100	24	41
1.18 mm	–	6	4	34	4	100	24	32
425 micron	15–30	3	2	10	1	84	20	23
150 mircon	–	1	1	3	–	59	14	15
75 micron	5–15	0	0	2	–	36	9	9

It is seen that the gradation in Col. 9 satisfies the requirements in Col. 2.

(D) When only two materials are to be combined, a simple graphical method described below can be followed:

It is required to blend sand (A) and silt-clay (*B*) to obtain the following gradation:

Sieve designation	**Percentage passing**	**Material (A)**	**Material (B)**
4.75 mm	100	100	
2.36 mm	80–100	91	
1.18 mm	50–80	34	100
425 micron	30–60	10	84
300 micron	20–45	3	59
75 micron	10–25	2	36

(i) On a convenient size of rectangular graph, scale off top linearly from 100 to 0 as percentage of *A* in mix. Scale off the base correspondingly from 0 to 100 as percentage of *B* in max. Scale off the vertical ordinate from 100 at top to 0 at bottom on the left and 0 to top to 100 at bottom on the right, representing the percentages passing sieve (Fig. 13.5)

(*ii*) On the left ordinate, mark off percentages passing the given sieves for material A. On the right ordinate mark off percentages passing the given sieves for material B.

(*iii*) For each sieve size, connect by a straight line the points, representing the respective percentages passing each material. The intersection of each sieve line by any vertical line will define the combined grading of the two materials mixed in the proportions shown in the top and bottom horizontal scales.

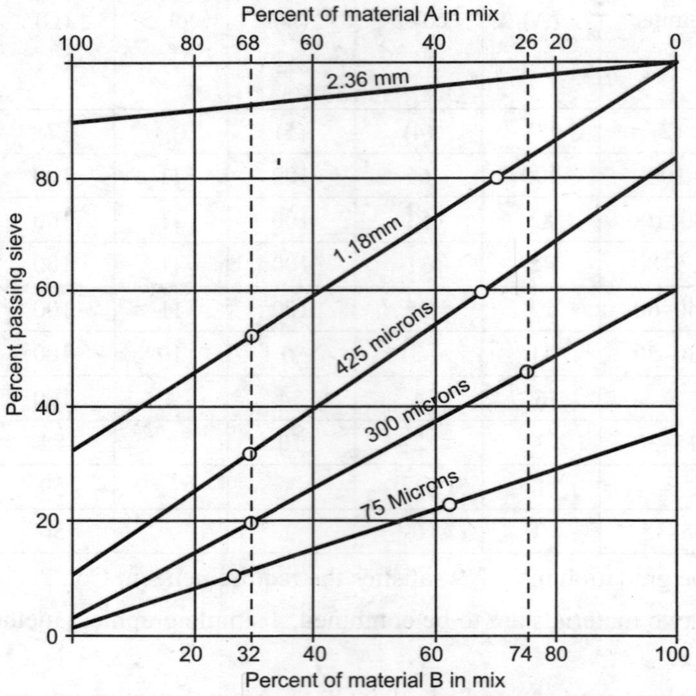

Fig. 13.5. Graphical method of combining two aggregates.

(*iv*) On each sieve line, the specified percentage passing limits are marked by small circles. The intercept lying between the circles represents, for any particular sieve line, the range of proportion that will comply the specification.

(*v*) If mixtures within the specification limits are possible, vertical lines can be erected such that their intersection points with all sieve lines lie on the acceptable intercepts. The hightest and the lowest percentages of either material at which this can be done represent the limiting mixtures which conform to the specifications. The mid-point between the limiting mixtures will usually provide the best mixture.

(*vi*) In Fig. 13.5. the limits indicated are (*i*) 68 per cent A, 32 per cent B and (*ii*) 26 per cent A, 74 per cent B. In practice, the acceptable properties would lie in the range of A : B = 2:1 to A :B = 1: 3. Probably a ratio of A : B = 1 : 1 would be the best.

13.4.8. Combining for plasticity

The specifications for mixtures usually give an upper limit for the *L.L* and *P.I* values. While the individual materials may have unacceptable *L.L* and *P.I* values, the correct blend of materials can be arrived at to result in acceptable *L.L* and *P.I* values. For this purpose, some formulae will be useful.

Consider materials A and B being combined to form a mixture C.

Let L_a, L_b, L_c = Liquid Limit of A, B, C respectively.

I_a, I_b, I_c = Plasticity Index of A, B, C respectively.

P_a, P_b, P_c = Percentage passing 425 micron sieve for A, B, C respectively.

X_a = Percentage of material A in mix.

Then, if (L_a, L_b, P_a P_b or (I_a, I_b, P_a, P_b) are known, the value of X_a to give a desired value of L_c or I_c can be determined from the following formulae :

$$X_a = \frac{100\, P_b (L_c - L_b)}{P_a (L_a - L_c) + P_b (L_c - L_b))} \tag{13.2}$$

$$X_a = \frac{100\, P_b (I_c - I_b)}{P_a (I_a - I_c) + P_b (I_c - I_b)} \tag{13.3}$$

Alternatively, if the *L.L.* and *P.I.* of the constituent materials are known, and the percentage of material A in the mixture has been selected previously to meet the particle size distribution criteria, the values of L_c and I_c can be determined from the following formulae :

$$L_c = \frac{X_a (P_a L_a - P_b L_b)}{X_a (P_a - P_b) + 100\, P_b} \tag{13.4}$$

$$I_c = \frac{X_a (P_a I_a - P_b I_b) + 100\, P_b I_b}{X_a (P_a - P_b) + 100\, P_a} \tag{13.5}$$

The following examples illustrate the use of these formulae :

Problem 13.1. It is proposed to construct a sand-silty clay road conforming to the following gradation specifications :

Table 13.8

Sieve designation (1)	Percentage gradation limits (2)	Passing mean (3)	Sand material (4)	Silt-clay material (5)
4.75 mm	100	100	100	
2.36 mm	80–10	90	91	
1.18 mm	50–80	65	34	100
425 micron	30–60	45	10	84
300 micron	20–45	33	3	59
75 micron	10–25	18	2	36

Sand (*A*) and silty clay (*B*) are available, whose gradation is indicated in Cols. 4 and 5 respectively in Table 13.8. Gradation requirements indicate that a 1 : 1 ratio of the two materials would be adequate. L.L. and P.I. of Materials A and B are as under :

	A	B
L.L.	25	38
P.I.	2	10

What will be the *L.L* and *P.I* of the mixture ? If the maximum *L.L* and *P.I* are to be respectively 35 and 9, what should be the proportion of A and B in the mix ?

$L_a = 25, I_a = 2, L_b = 38, I_b = 15, P_a = 10, P_b = 84, X_a = 50$

Substituting in formulae 13.4 and 13.5,

$$L_c = \frac{X_a(P_aL_a - P_bL_b) + 100\ P_bL_b}{X_a(P_a - P_b) + 100\ P_b}$$

$$= \frac{50(10 \times 25 - 84 \times 38) + 100 \times 84 \times 38}{50(10 - 84) + 100 \times 84}$$

$$= \frac{50 \times 10 \times 25 + 50 \times 84 \times 38}{500 + 50 \times 84} = \frac{12,500 + 1,59,600}{4700}$$

$$= \frac{1,72,100}{4,700} = 36.6$$

$$I_c = \frac{X_a(P_aI_a - P_bI_b) + 100\ P_bI_b}{X_a(P_a - P_b) + 100\ P_b}$$

$$= \frac{50(10 \times 2 - 84 \times 10) + 100 \times 84 \times 10}{50(10 - 84) + 100 \times 84}$$

$$= \frac{50 \times 10 \times 2 + 50 \times 84 \times 10}{500 \times 84 + 50 \times 10} = \frac{43,000}{4,700} = 9.2$$

These values do not satisfy the desired L.L and P.I values. The proportion can be worked out by the equation 13.2.

$$X_a = \frac{100 \times P_b(L_c - L_b)}{P_a(L_a - L_c) + P_b(L_c - L_b)}$$

$$= \frac{100 \times 84 \times (35 - 38)}{10 \times (25 - 38) + 84(35 - 38)} = \frac{100 \times 84 \times 3}{10 \times 13 + 84 \times 3}$$

$$= \frac{25200}{382} = 66\%.$$

Using formula 13.3, $X_a = \dfrac{100 \times P_b(I_c - I_b)}{P_a(I_a - I_c) + P_b(I_c - I_b)}$

$$= \frac{100 \times 84 \times (9 - 10)}{10 \times (2 - 9) + 84(9 - 10)} = \frac{100 \times 84}{70 + 84}$$

$$= \frac{100 \times 84}{154} = 55\%$$

Obviously, the values worked out by the earlier formula, *i.e.* 66% of A would be suitable for the *P.I.*, criterion also.

Hence a practical mix of $A : B = 2 : 1$ would be suitable. This mix, it may be noted from para 17.4.7, also satisfies the gradation requirements.

13.5. SOIL-LIME STABILISATION

During the last twenty-five years, the use of soil-lime stabilisation has gained popularity in the U.S.A., Africa, Australia and India. Soil-lime mixtures are used as sub base or base courses. Because of the favourable climatic conditions in India and the occurrence of clayey soils in large areas, this technique offers considerable scope.

13.5.1. Mechanism of lime-soil interaction

When lime (CaO) is added to a fine-grained soil, a number of reactions take place. Some of them occur immediately while others are slow to occur. One of the early reactions is base-exchange (ion exchange). Clay particles are usually negatively charged with exchangeable ions of sodium, magnesium, potassium or hydrogen adsorbed on the surface. The strong positively charged ions of calcium present in lime replace the weaker ions of sodium, magnesium, potassium or hydrogen, resulting in a preponderance of positively charged calcium ions on the surface of the clay particles. This in turn reduces the plasticity of the soil. The clay particles tend to agglomerate into large sized particles (flocculation), imparting friability to the mixture.

After the above first stage reactions are complete, any additional quantity of lime will react chemically with the clay minerals. The aluminous and siliceous materials in the clayey soils will react with lime in the presence of water to form cementitious gels, which increase the strength and durability of the mixture. These pozzolanic reactions are slow and extend over a long period of time, several years in some instances.

Another possible source of strength is the formation of calcium carbonate due to the absorption of carbon dioxide from air.

13.5.2. Soils amenable to treatment

Clayey soils are most amenable to lime treatment. The fraction passing 425 μ should be at least 15 per cent and the clay content should be at least 10 per cent. The P.I. of the soil should be at least 10. These conditions are satisfied by many soils in India. The aluvial silty soils of the northern plains, the clayey soils of the deltas, the black-cotton soils and the moorums found in many parts are eminently suitable for soil-lime stabilisation.

13.5.3. Quantity of lime

The strength of a soil lime-mixture is greatly influenced by the lime content. A concentration of lime less than 2 per cent is not generally amenable to proper mixing and is not recommended for use. A quantity of 3 to 10 per cent by weight of dry soil is normally required to stabilise most soils.

$Ca(OH_2)$ (*i.e.* hydrated lime) in a powder form is preferred to CaO (quick lime) because of the danger from burns that can be caused to unprotected workmen when handling quicklime.

13.5.4. Properties of soil-lime

Due to the aggregation of smaller particles into bigger ones, one of the early effects of adding lime is to make the grains coarser.

Lime brings about a substantial reduction in plasticity. The liquid limit generally decreases and the plastic limit increases, thus causing a reduction in the Plasticity Index of the soil, vide Fig. 13.6.

The soil swells and shrinkage potential is significantly reduced by the addition of lime. This is highly important in dealing with expansive soils, which swell in volume when water is added and shrink in volume when the moisture content is reduced.

The strength of soil increases substantially when treated with lime. The unconfined compressive strength of typical fine-grained soils range from 0.18 to 0.7 MN/m^2. Increases in strength of 28 days cured lime-soil specimens may be of the order of 0.7 to 1.7 MN/m^2.

The benefits of addition of lime depend to a large extent on factors such as purity of lime, the fineness of lime, the degree of pulverization of the soil, the compaction imparted, the time between final mixing and compaction and the curing conditions. For pulverisation, it is generally specified that at least 80 per cent should pass through 4.75 mm sieve and all the particles should pass through 25 mm sieve.

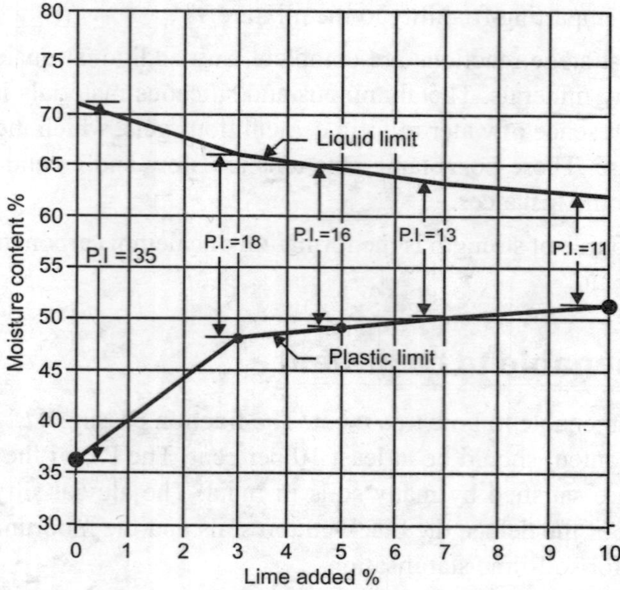

Fig. 13.6. Effect of lime on plasticity index.

Provided the soil-lime mixtures are designed properly, the durability characteristics of the mixture are not in doubt. Although some reduction in strength due to cyclic freeze-thaw or prolonged moisture effects is possible, the residual strength itself is sufficient to meet the design requirements adequately.

13.5.5. Design of pavement layers

The CBR method of design is often used to determine the thickness of the various layers in a flexible system. For use as a sub-base layer, a minimum CBR value of 20–30 is normally desired. For use in base courses, CBR of 80–100 is normally stipulated. To account for the uncertainties in mixing in the field, current Indian practice is to assume 45–60 per cent of the laboratory CBR value as the CBR under field conditions.

13.5.6. Relevance of lime stabilised soil in the Indian context

The main advantage of soil-lime is that lime is a cheap material which is locally available in several parts of this country. The manufacture of lime and the soil lime stabilisation techniques are amenable to labour-intensive technology and are ideally suited to our country. However, the experience in lime-soil work has not been so far very successful. This is mainly because lime of a high degree of purity is needed and a great deal of control over quality is necessary in execution.

13.5.7. Lime-cement-soil stabilisation

Both lime and cement produce cementitious products in the presence of clay minerals and water. A mixture of lime and cement is sometimes used for stabilisation. The combined lime and cement content can normally be around 10–15 per cent, and the ratio of lime to cement can be between 1: 3 and 3 : 2 depending upon the soil type and the strength desired. Working with highly expansive clays, lime is added initially upto about 3 per cent to render the soil more friable. Additional quantity of lime and cement are added subsequently for the development of required strength.

13.5.8. Lime-pozzolana stabilisation

Pozzolana is a siliceous material, which, while in itself possessing no cementitious properties, will in a finely divided form and in the presence of water, react with calcium hydroxide and form cementitious compounds. Pozzolana can be a naturally occurring volcanic ash or industrial waste such as fly-ash, or can be produced by calcining clay. When flyash is used, the mixture is known as lime-flyash stabilised mixture, or in an abbreviated form LFA stabilised mixture.

Not all soils possess enough quantity of clay minerals with which lime can react to form cementitious products. If, therefore, a pozzolanic material is added to such soils, the stabilisation can easily take place. Silts, sandy soils, gravels, crushed stone and slags are some of the material types where lime-pozzolana stabilisation can be successful. The alluvial silts of northern India fall into this category of soils which can be stabilised with lime-pozzolana.

The ratio of lime to pozzolana depends upon a number of factors and can vary widely from 1: 1 to 1: 9. The combined quantity of lime-pozzolana in a mixture can vary from 10 to 25 per cent.

Lime-pozzolana-aggregate mixtures can be used for the superior strength road bases. A layer of this material has great structural strength and behaves more like a semi-rigid pavement.

The use of lime-pozzolana in our country is still in an early stage, but in view of the problem of disposal of huge quantities of fly-ash from thermal plants, the future may see more and more use of this material.

13.6. SOIL-CEMENT STABILISATION

13.6.1 General

The addition of cement to soil to improve its strength is now in vogue for the past forty years or so but its use in India has not caught on mainly due to the shortage of cement. The principal advantages with soil-cement are that almost all soils are amenable to this technique. It is a scientifically designed engineering material and cement itself is a standard material whose quality is tested and assured. Because of its very high flexural strength, it has a vary high load spreading property. Thus soil-cement is able to spread the load over a wider area and bridge over locally weak spots of the underlying subgrade or sub-base. In view of its high flexural rigidity, it is often classed as a semi-rigid pavement, something which is intermediate between a flexible pavement and a rigid pavement. The durability of soil cement is of a high order and its strength is known to increase with age. The main disadvantages are the cost being higher than lime-soil and the need for a high degree of quality control. Because of volumetric changes that take place when cement hydrates, early shrinkage cracks are formed in soil-cement layers, affecting their overall performance.

13.6.2. Action involved in cement-soil stabilisation

When water is added to cement, major cementitious products like calcium silicate hydrates and calcium aluminium hydrates are produced. In stabilisation of granular materials with cement, these cementitious materials provide the bond between the mineral particles. In the case of fine-grained soils, the cementitious bond provided by the calcium silicate hydrates and the calcium aluminate hydrates is further strengthened by the secondary hydrous calcium silicates and aluminates formed by the reaction of free lime in the cement paste and the clay mineral particles. The reaction phenomenon between the free lime and the clay mineral particles is just the same as in the case of lime-soil stabilisation. Base-exchange and flocculation also take place, rendering the soil more friable and reducing its plasticity.

13.6.3. Factors affecting strength of soil-cement mixes

(*i*) Cement Content

The cement content necessary for effective stabilisation varies with the soil type. The strength of a soil-cement mix for a particular soil type varies with the cement content. As a rough guide, the cement content, expressed as a percentage by weight of the dry soil, varies between 4 and 14. For preliminary estimation purposes, a value of 10 per cent seems reasonable. The cement content is generally selected to obtain the desired compressive strength. The criterion in the past was a 7 day unconfined compressive strength of 1.7 MN/m^2 with moist-cured cylindrical specimens having a height to diameter ratio of 2 : 1. Recent practice is to adopt a strength of 2.8 MN/m^2 Table 13.9 gives the range of cement requirements as per American practice.

Table 13.9: Cement requirements for various soil types

Soil type (PRA classification)	Usual range of cement requirement by weight (percent)
A–1–a	3–5
A–1–6	5–8
A–2	5–9
A–3	7–11
A–4	7–12
A–5	8–13
A—6	9—15
A–7	10–16

From the above Table, it is seen that the quantity of cement needed to stabilise gravelly soil is much less than that required to stabilise silty and clayey soils.

It is observed that for the range of cement contents normally employed in stabilisation work, the strength of the mixture increases with increase in cement content.

Ordinary. Portland Cement is used for the majority of soil-stabilisation works. A rapid-hardening cement can be used if high strengths are desired initially.

(*ii*) Moisture content

Since hydration of cement takes place only in the presence of water, the importance of water is obvious. Water also improves the workability of the soil and facilitates compaction.

The exact amount of water to be added is governed by many considerations. One important factor is that the soil-cement mixtures exhibit the same type of moisture-density relationship as an ordinary soil. Thus, for a given compactive effort, there is an "optimum moisture content" at which the maximum density is obtained.

The best moisture content for maximum density may not necessarily be the optimum moisture content for maximum strength. It is generally seen that highest compressive strength can be obtained with specimens compacted slightly below the optimum for maximum density

Some of the water is taken up by the cement for hydration. The moisture necessary for maximum compaction is sufficient to provide for this.

(*iii*) Soil

Soil type has a profound influence on the success of stabilisation with cement. It is often claimed that almost any type of soil can be stabilised with cement. Though this is true in a large measure, certain soil types cannot be stabilised with cement at economical costs.

Soils with a low organic matter are generally preferred. A safe upper limit is 2 per cent, though soils with 3 to 4 per cent organic matter have also been successfully stabilised with cement.

It is well-known that the presence of sulphates has a harmful effect on the life of cement concrete. For the same reasons, the presence of sulphates in the soil has to be viewed with suspicion. For cohesive soils, a maximum sulphate content of 0.25 per cent is usually specified, though for non-cohesive materials an upper limit of 1.0 per cent may be all right.

The presence of a small amount of clay in the soil is beneficial to cement stabilisation, but large clay content brings in problems of mixing and pulverising. It is desirable if the clay content is restricted to 5 per cent. A thumb rule often employed is that the practical upper limit for stabilisation with machinery is when the P.I. multiplied by the percentage finer than 425 μ is less than 250.

As the plasticity of the soil increases, the amount of cement needed to effectively react increases. Highly plastic soils cannot, therefore, be economically stabilised with cement. An upper limit of 45 for L.L. (Liquid Limit) and 20 for P.I. (Plasticity Index) is generally observed. More plastic soils can be treated with cement after being pre-treated with lime. As regards the grading of the soils, it is recognised that a well-graded mixture requires less of cement and is preferred. British practice indicates the following grading as suitable, Table 13.10.

Table 13.10: Grading of materials for soil-cement

Sieve size (nearest) equivalent IS sieve	Percentage by weight passing
50 mm	100
40 mm	95
20 mm	450
10 mm	35
4.75 mm	25
600 micron	8
300 micron	5
75 micron	0

It is also specified that the uniformity coefficient (*i.e.* ratio of the particle size for which 60 per cent is finer to the particle size for which 10 per cent is finer) should not be less than 5.

Though British practice limits soil-cement work to well-graded materials below 50 mm size, experience elsewhere has shown that sandy and gravelly soils containing 10–35 per cent combined silt and clay, sandy soils deficient in fines, silty soils and clayey soils can also be effectively stabilised. Single sized sands with a low uniformly coefficient (less than about 3) present problems for cement stabilisation.

***(iv)* Degree of pulverisation in mixing**

The presence of lumps of soil inhibits effective stabilisation. Pulverization of soils, especially clays, must be carried out before mixing. The following requirements are often laid down, Table 13.11.

Table 13.11. Requirements for pulverisation for stabilisation with cement

Sieve designation	Percentage by weight of soil passing the sieve after pulverisation
25 mm	100
4.75 mm	80

(*v*) **Mixing**

For best results, cement should be uniformly distributed and mixed throughout the material. The addition of water helps the cement to adhere to the particles of the soil and prevents segregation.

(*vi*) **Compacting**

The hydration of cement starts as soon as water is added, and it therefore is desirable to compact the material as soon as mixing is completed. Any delay is likely to result in the loss of the cementing action of the additive and in the need for extra compactive effort to break down the cement bonds that have already formed. A serious loss in strength can follow. For this purpose, it is often stipulated that compaction should be completed within two hours of mixing

(*vii*) **Curing**

As in the case of cement concrete, soil cement requires the presence of sufficient moisture to meet the needs of chemical reactions. A seven days' moist curing is necessary.

13.6.4. Design criteria

The most popular design criterion for soil-cement is in terms of the unconfined compressive strength after 7 days' moist curing. A value of 1.7 MN/m^2 with cylindrical specimens (ratio of height to diameter of 2:1) is specified for Indian conditions. A minimum strength of 2.76 MN/m^2 is generally desirable for heavily trafficked roads and for higher layers of the pavement structure (*i.e.* base course), whereas for lightly trafficked roads and for bottom layers of the pavement structure (*i.e.* sub-base course) a minimum strength of 1.7 MN/m^2 is probably adequate.

The CBR method of design is also applied to soil-cement layers. For this purpose, the specimens are initially cured for 7 days and soaked in water for 4 days prior to testing. For use as sub-bases, a CBR value of 20–30 is desirable, whereas a value of 80–100 is desirable for use as a base course. Since imperfections in mixing in the field can yield lower strength values, it is desirable to design the laboratory mix to yield a higher CBR value. Normally, it can be assumed that the field mix can yield a strength which is 60 per cent of the laboratory strength. When determining the thickness of the layer, an equivalency factor of 1.5 can be assumed vis-a-vis unbound granular layers.

If a small quantity of cement, say in the range of 2–3 per cent, is added to a soil, the soil properties can be improved appreciably, though not to the same extent as a soil-cement mixture. Such a mixture can be used as a sub-base, and is known by the name of cement-modified soil. The soil should possess characteristics similar to those required for soil-cement. The design is generally based on CBR method. A CBR value to 20–30 in the field is generally suitable for sub-base.

13.7. CHLORIDE STABILISATION

Granular soils lack stability when they are too dry. If their moisture content can be stabilised by the addition of some chemicals, then these soils can be used successfully. Chlorides of calcium and sodium are two of the most popular salts used for this purpose.

13.7.1. Calcium Chloride

Calcium chloride has been used extensively as a dust-palliative for more than half a century. The material has also been used as a dust palliative in India, and on an experimental basis in the body of the subgrade and sub-bases.

Calcium chloride has deliquesent and hygroscopic properties. By the former is meant the ability of a material to absorb moisture from the air and thus to dissolve and become liquid. By the latter is meant the ability of a material to absorb and retain moisture without necessarily becoming liquid. These properties render the material ideally suitable as a dust-palliative on untreated low cost roads. In dry climate regions, the moisture evaporates from the road during the day, but if calcium chloride is present, the moisture can be regained in the night. Another property of calcium chloride is that it lowers the vapour pressure of water in which it is dissolved. This reduces the rate of evaporation. An increase in the surface tension of water is noticed when calcium chloride is present in a soil-water mixture. As the surface tension of the pore water rises, the rate of evaporation falls. When some evaporation takes place the pore water content is itself reduced, and this in turn causes surface tension to rise further. The water films then close on the soil particles and grip them together. If calcium chloride is added to a soil, it is observed that its unit weight increases for a given compactive effort. In other words, to obtain a desired density, less compactive effort is needed. This is mainly because of the lubricating effect of the chemical. It is well known that chlorides dissolved in water lower the freezing temperature of water. This property makes the chemical extremely useful in frost-susceptible locations.

Calcium chloride is obtained as a waste product in the manufacture of ammonia, ammonium carbonate, potassium chlorate and sodium carbonate. Its disposal was considered a problem, and its use as a soil stabiliser in such situations is to be welcomed.

The rate of application of the materials as a dust palliative for untreated roads is about 4.8–1.5 kg/sq. m. per year. The material is applied in a dry form and the road surface is bladed.

As an admixture to the soil to obtain greater density and strength, it is used at a small rate of about 0.5 per cent by dry weight of soil.

Work carried out in India has shown that calcium chloride can retain the moisture in the surface only when the relative humidity of the atmosphere is above 31 per cent. This precludes its use in extremely dry areas.

13.7.2. Sodium Chloride

Sodium chloride (common salt) is available in a natural state as rock salt and sea water. Its use as a stabiliser is derived from the many properties listed under calcium chloride. It is, however, less hygroscopic and inferior to calcium chloride as a lubricant for aiding compaction. An important beneficial effect is the crystallisation of the salt forming a compact and hard surface which improves the stability of the layer and prevents evaporation. The quantity of sodium chloride is roughly the same as calcium chloride, viz., about 0.5 per cent by weight.

13.8. OTHER CHEMICALS/MATERIALS STABILISATION

A number of other chemicals/materials have been used for chemical stabilisation of soils. Some of them are :

1. Sodium silicate 2. Lignin 3. Resins 4. Molasses.

Sodium silicate reacts in aqueous solution with soluble calcium salts, forming insoluble and gelatinous calcium silicates. Calcium needed for the reaction can either be present in the soil itself (as in lime-stone aggregates or chalky soils) or be added in aqueous solutions. The amount of chemical needed may vary from one to ten per cent. Experimental work carried out in India on stabilisation of sand with sodium silicate has yielded satisfactory results.

The natural binding material that holds together the fibres in wood is lignin. The material is a major byeproduct in paper manufacturing industry. Lignin is available from the paper manufacturing process in a water solution known as calcium lignosulphonic acid. Calcium lignosulphate (or lignin sulphate or simply lignin) is the constituent which is used as a road binder. About 0.5 to 1 per cent by weight of dry soil is used for stabilisation. Stabilisation is by the cementing bond that develops between the soil particles due to the presence of the binder. The material also closes the voids and thus reduces penetration of water through the layer. It retards the rate of evaporation of water and arrests loss of moisture.

Natural or processed resins can also be used for soil stabilisation. Vinsol resin and resin or derivatives of resin are commonly used. Resins are wood products. Resin-treated soils reduce water absorption, facilitate compaction and increase the stability of the treated mixtures. Their drawback is their susceptibility for micro-biological attack by bacteria and fungi, but this handicap can be identified and surmounted. A quantity of 1 to 3 per cent by weight of soils is normally sufficient.

Molasses is a waste-product from the process of manufacturing sugar from sugarcane. A thick syrupy liquid, it is hygroscopic and can be used as a dust-palliative and as a binder for incorporation during compaction. It is easily leached out by rain water. But some additives have been tried to make it insoluble. If water is prevented from entering the mixture, as by means of an impermeable bituminous surfacing, it can be expected to last long.

Various new chemical stabilizing agents have come to the market recently. The following are some of them, which have been used with good results in India and abroad :

1. RBI Grade 81
2. Renolith
3. Fujibeton

13.9. SOIL BITUMEN STABILISATION

13.9.1. Basic principles

The addition of a bituminous binder to a soil improves its properties considerably. Firstly, if the soil lacks cohesion, the bitumen coats the soil particles, binds them together and supplies cohesion. Example of such a stabilisation is a sand bitumen in which sand and bitumen are mixed and laid. Secondly, bitumen being a waterproofing material, the mixture becomes less prone to the adverse effects caused by ingress of water. Soil-aggregate mixtures or cohesive soils can be made to benefit from this action. A third procedure is to spray a bituminous binder on a dry surface of a low-cost road (earth/gravel), with a view to prevent dust and to stop the entry of moisture into the road.

The success of the above principles really lies in selecting the right quantity and type of the binder. An excess of the binder will result in too thick a binder film around the soil particles and destroy part of the internal friction. When aiming at waterproofing, it is seldom necessary to fill up the entire void space. The gradation of the soil particles also has an important bearing on the satisfactory performance of the stabilisation technique. These factors will be discussed in detail under the different types of soil-bitumen stabilisation processes.

13.9.2. Types of soil-bitumen stabilisation processes

The following are the variations in the bituminous stabilisation techniques :

(*i*) Sand-bitumen (*ii*) Soil-bitumen.

(*iii*) Soil-aggregate-bitumen

(*iv*) Spraying bitumen on earth/gravel roads (oiling).

13.9.3. Sand-Bitumen stabilisation

There are regions where sand is the only predominant road building material available within economical leads and stone aggregate or gravel has to be conveyed over a long distance. There are also areas where there is acute scarcity of water, and specifications such as water bound macadam or compacted gravel will have to be discarded. Under such conditions, a successful specification is sand-bitumen. The arid desert region of Rajasthan where dune sand is met with and the coastal plains of the South where beach sand is available, are examples of such areas where the technique has been tried successfully. Based on the limited experience in the country, the Indian Roads Congress have come out with specifications for use of sand-bitumen as a base-course.

13.9.3.1. Gradation of sand

A wide range of sands can be successfully stabilised with bitumen. Though a well-graded sand-bitumen mixture will have a higher stability, poorly graded sands and single size sands have been satisfactorily stabilised with bitumen. The latter include the wind-blown dune-sands. Indian practice limits the use of this specification to sands having less than 10 per cent of material passing 75 microns sieve A typical gradation of the Rajasthan desert sand is given in Table 13.12.

Table 31.12. Typical gradation of Rajasthan sand for sand-bitumen stabilisation

Sieve designation	Percentage passing
600 microns	100
300 microns	98–100
150 microns	60–65
75 microns	4–5

13.9.3.2. Type of binder

The types of binder used for sand-bitumen work are :

(*i*) Penetration grade bitumen
(*ii*) Road tar.
(*iii*) Cold application cut-back
(*iv*) Hot application cut-back
(*v*) Emulsions

Penetration grade bitumen 30/40 or 80/100 can be used provided the sand is pre-heated and dried to a suitable temperature (135–165°C). A small quantity of flux such as kerosene can be beneficial to make the bitumen workable. Road Tar of grade RT-3 has also been used.

The use of cold- application cut-backs has been very popular for sand-bitumen stabilisation. A medium-curing cut-back such as MC-1 and MC-2 are most common, though both rapid-curing and slow-curing (RC-1, RC-2 and SC-1 and SC-2) cutbacks have also been frequently used. In India, RC-3 has been tried successfully. A hot application cut-back is also suitable. Emulsions are ideal for dry conditions in desert regions in the tropics, since they provide the optimum fluids content for compaction. The use of emulsions for sand- bitumen under wet conditions is, however, difficult.

13.9.3.3. Quantity of binder

For sand-bitumen stabilisation, the quantity of binder is selected after carrying out stability tests with various binder contents. The optimum binder content at which the stability is maximum is determined. The range of binder contents found to be satisfactory is 4–10 per cent by weight of total mix.

13.9.3.4. Incorporation of hard aggregates

Blending of sand with some quantity of hard crushed aggregate will improve the stability of sand-bitumen mixtures, especially if the sand is single-size. The proportion of coarse aggregates can be about 30 per cent by weight.

13.9.3.5. Design criteria

For satisfactory performance as a base course, the Hubbard-Field stability values should be as below :

Table 13.13: Minimum values of Hubbard-Field stability of sand-bitumen for base course

	Type of mix	Specimens	Test Method	Stability (kg)
1.	Sand-bitumen with penetration grade bitumen	5 cm dia × 2.5 cm height	Testing at 60°C	360
2.	Sand-bitumen with cut-back bitumen	5 cm dia × 5 cm height	Testing at 25°C	540
3.	Sand-bitumen with in-corpora-tion of coarse-aggregates	15 cm dia × 7 cm	Testing at 60°C	900
4.	Sand bitumen with incorporation of coarse aggregate	15 cm dia × 7 cm height	Testing at 25°C	1100

13.9.3.6. Surfacing

Sand-bitumen base courses are generally laid to a compacted thickness of 100 mm. They require a surfacing course to provide a good running surface and to protect the base from abrasive action of traffic. A 20 mm premix chipping carpet has been found to be suitable for this purpose in India. Such roads have been found to be satisfactory for light to medium traffic in India.

13.9.4. Soil-bitumen stabilisation

The term soil-bitumen is generally used for stabilising cohesive fine-grained soils. The incorporation of bitumen in a cohesive soil results in water-proofing the layer. Such soils have good bearing capacity at low moisture contents, but they tend to lose the bearing capacity when the moisture content increases. The bituminous binder ensures that the moisture content never reaches beyond the safe limit. Soil-bitumen layers are used as sub-bases and bases.

13.9.4.1. Types of soils suitable for stabilisation

The following gradation limits are generally prescribed

Sieve designation	Percentage passing
4.75 mm	50
425 micron	35–100
75 micron	10–50

The liquid limit should be less than 40 and the plasticity index should be less than 18, Highly plastic soils are hard to stabilise, because of the difficulty in dispersing the binder.

13.9.4.2. Typical amount of binder

The binders which find ready application for soil-bitumen are the cut-backs, road tars and emulsions. A medium-curing cut-back is preferable, but slow-curing cut-backs can be used with relatively highly plastic soils. Rapid curing cut-backs can be used with sandy soil. Road Tars in the grades RT-3, RT-4, RT-5 and RT-6 can also be used. Emulsions can be used with relatively less plastic soils and in dry climates, where the natural moisture content of the soils is not high. Slow-setting emulsions are appropriate.

The amount of binder depends upon the moisture content of the soil, the type of soil and type of binder used. The combined volume of bitumen and water must not exceed the pore space in the soil at the desired density. The normal range is 4 to 8 per cent by weight of dry mix. It is usually found that if the clay mineral content is high, a greater quantity of binder is needed. On the other hand, if the iron and aluminium content is high, a smaller quantity of binder is needed. The binder content is selected after carrying out tests such as density, fluids content, strength, water absorption and swelling.

13.9.4.3. Design criteria

Soil-bitumen stabilisation is yet in a developmental stage and further research is needed to understand its behaviour. Design criteria can thus be very tentative. A common requirement is that the Hubbard-Field Stability Value on 5 cm dia × 5 cm height specimens should not be less than 180 kgs after standing the specimens in water for 7 days. The stability value of the specimen before saturation should be at least 450 kgs. A further requirement is that the swelling should not exceed 5 per cent and water absorption should not exceed 7 per cent.

13.9.4.4. Usage

Soil-bitumen is generally used as a sub-base. The thickness provided varies from 100 mm to 200 mm.

13.9.5. Soil-aggregate-bitumen stabilisation

13.9.5.1. Principles

Soil-aggregate-bitumen mixture implies the addition of a suitable quantity of bituminous binder to a granular mixture which is fairly well graded and has good internal friction but lacks cohesion. The binder not only supplies cohesion, but also waterproofs the layer. The addition of binder prevents softening and loss of strength in a mixture containing fines.

13.9.5.2. Gradation

The gradation of the granular materials should be generally similar to those indicated under mechanical stabilisation (Table 13.1) The maximum P.I. values can, however, be in the range of 10 to 15.

13.9.5.3. Binder

A rapid curing cut-back (RC-1, RC-2 or RC-3) or a slow-curing emulsion or Road Tar (RT-4, RT-5 or RT-6) is generally suitable. The quantity to be used will generally be in the range of 2-6 percent by weight of dry materials. The binder should be able to coat adequately the individual particles of the fine fraction.

13.9.5.4. Design criteria

For soil-aggregate-bituminous mixtures, the CBR tests can be conveniently used for design. The CBR values of the untreated and treated mixtures can both be determined and compared before selecting the final design.

13.9.6. Spraying bitumen on earth/gravel roads (oiling)

13.9.6.1. Principles

Earth/gravel surfaces are dusty in the dry season when the moisture which binds the soil particles is absent. In the wet season, water easily enters the: surface and softens the same. The conditions can be improved by spraying a binder which will coat the dusty particles and hold them together, at the same time providing a reasonably water-proof surface. The binder is of a low viscosity so that it is able to penetrate the compacted surface by gravity. The process is rather cheap and economical and is suitable for low-volume low-cost roads.

13.9.6.2. Binder

The binder is usually a medium curing or a slow-curing cut-back. MC-0, MC-1, SC-0 or SC-1 cut-backs are generally suitable. Slow-curing cut-backs have the advantage that the volatiles do not evaporate too rapidly, thus giving more time for the binder to seep into soil.

A rate of application of about 5 litres per sq.m. in two or three applications will usually result in a penetration of about 15 mm to 25 mm into soil and be found to be satisfactory. A light dressing of sand will render the surface non-slippery in wet weather.

13.10. CONSTRUCTION OF SOIL-STABILISED ROADS

13.10.1. General

The constructional practice in soil-stabilisation varies with the type of stabilisation, but there are certain steps and procedures which are common. It is, therefore, convenient to deal with the construction practice for all types of stabilisation together. The minor variations needed for each type of stabilisation will be indicated at the appropriate place.

The construction technique to be adopted for a given situation depends upon a number of factors, viz.

(*i*) Type of stabilisation
(*ii*) Type of soil
(*iii*) Type of binder, if any, to be added
(*iv*) Leads involved for the materials
(*v*) Magnitude of the project
(*vi*) Availability of equipment
(*vii*) Availability of labour.

Broadly, the following three construction techniques can be identified :

(*i*) Labour intensive methods
(*ii*) Machinery intensive methods

Labour intensive techniques are indicated for the following conditions :

(*i*) Availability of cheap labour, as in developing countries, making it more economical to use labour-intensive techniques than equipment intensive techniques.
(*ii*) Small magnitude of the work, which is also scattered. This condition is prevalent in developing countries where the construction of link roads to villages is given emphasis.
(*iii*) Equipment is not manufactured indigenously, and the level of skill needed for operation and maintenance of imported equipment has not been developed.

Equipment intensive techniques are indicated for the following conditions :

(*i*) The equipment is produced indigenously.
(*ii*) Labour is scarce, and it becomes more economical to use equipments.
(*iii*) The work is of a large magnitude, fairly concentrated, and the time schedule for compaction is tight.

Intermediate or appropriate technology is an intelligent blend of labour and machinery. Under this, it is recognised that small implements and tools and simple mechanical equipment can raise the productivity of labour and aid in obtaining good quality of work. India is a country which has already a good industrial base for manufacturing and servicing simple tools and equipment, and at the same time has surplus labour. Intermediate technology can be applied with good benefit in India under the prevailing conditions.

13.10.2. Labour intensive methods

The various operations involved can be discussed under the following heads :

(*i*) Collection of materials
(*ii*) Preparation of the subgrade
(*iii*) Pulverisation, where necessary
(*iv*) Mixing
(*v*) Spreading
(*vi*) Compaction.

The materials (soil, sand gravel etc.) are collected on the sides of the subgrade in requisite proportions and stacked in the form of windrows.

The subgrade is well-compacted to the required density and true to grades and the desired cross-profile.

If clay is one of the soil materials to be used, it is necessary to pulverise it. The clods are broken with the help of pick-axes or rammers. Application of a country plough driven by a tractor/bullock can also be tried. If a power roller is available, the same can be passed over the layer of clods a number of times, with frequent raking of the crushed material.

If the materials to be mixed are soil, sand and gravel they are mixed dry by labour using spades or shovels. The required quantity of water is added and the materials are wet mixed by manual labour. If an additive such as lime or cement is to be added, the soil is first spread to a uniform thickness and the bags of lime or cement are spotted at the desired spacings. The bags are then opened and the contents spread by manual means to cover the calculated area, which should be marked by strings. Water to the required quantity is added in stages and the soil and lime are mixed till the mixture has a uniform colour and the desired moisture content. If a bituminous binder is to be added, the mixing should preferably be done in a paddle type mixer, for a period of about 1 to 2 minutes.

Bitumen stabilised mixtures are spread to a uniform thickness in loose layers not exceeding 15 cm. Mechanically stabilised mixes, soil-lime mixes and soil-cement mixes are spread to a thickness which will give a compacted thickness of not more than 150 mm. The thickness of any stabilised layer should not, normally, be less than 100 mm.

The cement soil mix should be compacted within 2 hours of the mixing. When cut-back bitumen is used as a binder, rolling should start only after the mix has cured. The curing time depends upon the type of cut-back used and varies from 1 to 7 days. With penetration grade mixes, rolling can start as soon as the mix is laid and spread. When emulsions are used, the rolling can start after $\frac{1}{2}$ hour.

Rolling of stabilised mixtures should be by 8–10 tonne power rollers. When sand-bitumen and soil-bitumen stabilisation is used, it is preferable to carry out initial rolling by means a light pneumatic tyred roller. Rolling is carried out till 100 per cent of laboratory density is achieved.

Traffic is allowed on sand-bitumen layers only after 24 hours. Only light pneumatic vehicles are allowed initially. Normal traffic is allowed only after a month. Soil-lime and soil-cement layers are moist cured for a period of 7 days. Curing is achieved by providing or covering the surface with damp sand, straw or hessian.

13.10.3. Machinery intensive methods

Three basic construction methods are available when machinery is employed, viz.,

(*i*) Mix-in-place

(*ii*) Travelling plant

(*iii*) Stationary plant.

13.10.3.1. Mix-in-place method

In this method, a train of machines is run over the soil to be processed. For breaking and pulverising the soil, rippers, cultivators, rotary tillers, ploughs, scarifiers or disc harrows are used. Water is then added to the loose soil from a water tanker. If the stabiliser is liquid, it is distributed by a spraying tanker. Dry powder is either spread manually or from bulk spreaders. Mixing is carried out by means of disc harrows or pulvi-mixers. Dry mixing is initially done in two to three passes of the machines and is followed by wet mixing with the addition of water. A single-pass stabiliser is also used, and it performs the various operations such as cutting into the soil, pulverising and mixing, all in one operation. Compaction is done by rollers which follow the machines for laying the mix.

13.10.3.2. Travelling plant method

This method involves the use of a travelling plant which travels along the job site picking up the soil and stabiliser, mixing it in a mixer, discharging the mix on the ground. Compaction is done separately by rollers which follow the travelling plant.

13.10.3.3. Stationary plant method

This method is based on the process of mixing the ingredients in a centrally located plant, conveying the mix to the site, laying and compacting the same. The central mixing plant can be of the batch type or continuous type.

13.10.3.4. Advantages and disadvantages

The advantages and disadvantages of the three types are summarised below.

Table 13.14: Advantages and disadvantages of types of stabilisation techniques using equipment

Type		Advantages		Disadvantages	
1.	Mix-in-place	(*i*)	Plant is simple, cheap and easily transported.	(*i*)	It is difficult to obtain a uniform thickness of lift, because of the difficulty of setting the machines to a given depth.
		(*ii*)	The number of machines can be adjusted to suit the quantum of work. Flexibility is available.	(*ii*)	The mixing is not uniform as with travelling plant or stationary plant.
		(*iii*)	The whole processed section is ready for compaction at the same time.	(*iii*)	Heavy rain is likely to spoil the whole section.

		(iv)	A large out-put may be maintained.	(iv)	In a dry climate, water lost by evaporation is difficult to replace.
		(v)	If excess moisture is to be got rid of as, for example in a wet area, this is the only suitable method.		
2.	Travelling plant	(i)	Accurate proportioning of added water possible.	(i)	The cost of plant initially is high.
		(ii)	Uniform mixing obtained.	(ii)	Is suitable for concentrated and large quantum of work.
		(iii)	Short mixing time is involved	(iii)	Minor breakdowns can cause considerable dislocation.
		(iv)	Uniform surface can be obtained.		
		(v)	Depth of lift can be accurately controlled.		
		(vi)	It has the highest output for given expenditure on plant and labour.		
3.	Stationary Plant	(i)	Accurate proportioning of mixture and water is possible.	(i)	Expensive if in-situ soil is to be processed.
		(ii)	The depth can be easily and accurately controlled.	(ii)	Material must be compacted as delivered and not as a complete section.
		(iii)	Concrete mixers can be used.		
		(iv)	Losses of moisture during mixing and transporting are small.		
		(v)	Suitable for locations where formwork is needed, as in the case of sandy layers where vibrators are needed.		
		(vi)	No additional haulage in soil has to be taken from a borrow pit.		

13.10.4. Intermediate or appropriate technology for soil stabilisation

The use of simple tools, implements and equipment can be beneficial in soil-stabilisation work in many ways. Firstly, it can lend itself to a reasonable control over the quality of the work, which is so essential for the success of the specification. Secondly, it can be suitable for a large quantum of work which is to be completed in a tight schedule. Thirdly, it does not do away with labour totally, and hence is not inappropriate to labour-surplus economies.

The implements that are frequently used are the agricultural attachments such as disc harrows, disc ploughs, grader blades, rotillors etc. which can be conveniently towed by a small agricultural tractor. Water tankers for adding water can be pneumatic-wheeled and pulled by bullocks. The Central Road Research Institute has developed a simple equipment known as the Rotillor which is a versatile multi-purpose machine suitable for agriculture as well as for road making. For road making, the machine scarifies the top soil upto the required depth, pulverises the soil and mixes the soil and stabiliser. The equipment is towed by an agricultural tractor.

QUESTIONS

1. (*a*) Define the term soil stabilisation.
   (*b*) What is the purpose of soil stabilisation ?
   (*c*) What are the broad categories of soil stabilisation techniques available ?

2. (*a*) What is meant by mechanical stabilisation ?
   (*b*) What are the principles behind mechanical stabilisation ?
   (*c*) What are the specifications commonly used under mechanical stabilisation ?

3. (*a*) What is Fuller's curve ?
   (*b*) What are the Indian specifications for soil-aggregate mixtures ?

4. (*a*) What is a sand-clay road ?
   (*b*) What are the IRC specifications for such a road ?

5. Describe Mehra's method of stabilisation using soft aggregates.

6. Describe the Rothfuch's graphical method of combining different materials to obtain a desired gradation.

7. Describe the triangular chart method of blending materials for stabilisation.

8. (*a*) Describe the mechanism of soil-lime stabilisation.
   (*b*) What are the soils amenable to lime stabilisation ?
   (*c*) What is the usual quantity of lime needed ?
   (*d*) What are the improvements in soil properties brought about by soil-lime stabilisation?

9. Write short note on :
   Soil-Lime-Pozzolana stabilisation.

10. (*a*) Describe the actions involved in soil-cement stabilisation.
   (*b*) What are the advantages and disadvantages of soil-cement stabilisation?

(*c*) What are the factors affecting soil-cement mixes?

(*d*) What is the cement content needed for different types of soil?

(*e*) What design criteria are normally followed for soil- cement mixes?

11. (*a*) Describe the technique of soil stabilisation using Calcium Chloride and Sodium Chloride.

(*b*) What are the other chemicals that have been used for soil stabilisation?

12. (*a*) What are the basic principles behind soil-bitumen stabilisation?

(*b*) What are the different specifications possible under soil-bitumen stabilisation?

13. (*a*) Describe sand-bitumen stabilisation, giving gradation of sand, types of binder and design criteria.

(*b*) Why is a surfacing needed on a sand-bitumen layer?

14. Write short note on :

Oiling earth/gravel roads.

15. Describe (*a*) labour-intensive and (*b*) equipment-oriented soil-stabilisation techniques. What are the conditions under which each of the above two techniques are adopted?

Chapter

GRANULAR SUB-BASES, BASES AND SURFACE COURSES

14.1. GENERAL REMARKS

Granular materials such as large sized stone boulders, broken stones, brick ballast or broken kankar have been extensively used in the past as sub-bases and bases of road pavements and even as wearing courses to a limited extent. The entire thickness of a flexible pavement to withstand the superimposed load due to traffic need not consist of costly materials such as bituminous concrete or macadam, but as one goes lower down from the surface, cheaper and lower quality materials may be substituted. This is because the stresses go on diminishing at lower depths. Granular bases and sub-bases are extensively used taking advantage of their low cost, local availability and the diminishing stresses in the lower layers. In this country, for economic reasons, many an unimportant road is provided with only a granular wearing course such as water-bound macadam without any bituminous surfacing on top. The construction of granular courses involves the least of mechanical equipment and this is a great advantage to a developing country like ours where labour is plentiful and some machines still have to be imported. The use of granular materials in base, sub-base and wearing course is thus bound to hold the ground for many more years to come in this country, though in many of the Western countries some of these techniques and specification are fast disappearing and are only of historical interest.

14.1.1. Types of specifications

The term granular base, sub-base and wearing course is a very general one and it embraces many types of specifications, many of which do not apparently have much in common, except that use is made of granular materials such as broken stone, crushed aggregate, brick-ballast, slag etc. The following types of specifications will be discussed in this chapter :

S.No.	Name of specification	Principal materials that can be used
1.	Telford stone soling (Boulder stone soling or pitched base course)	Stone boulders
2.	Gravel courses	Gravel

3.	Water-bound macadam	Broken stone aggregates, brick ballast, slag, kankar
4.	Dry-bound macadam	Broken stone aggregates, gravel
5.	Wet-mix macadam	Crusher run, stone aggregates
6.	Lime-fly-ash concrete	Broken stone aggregates
7.	Lean cement concrete	Broken stone aggregates
8.	Cement-bound-granular material	Broken stone aggregates/gravel

14.2. TELFORD STONE SOLING

14.2.1. General

Thomas Telford introduced in England, early in the nineteenth century, a road with a hard crust consisting of a base of large sized stone boulders hand laid and closely packed and a wearing course of broken stones. This base course specification is named after him and is in limited use in this country as a base or sub-base course, though it has become obsolete in countries abroad, mainly because of the manual labour involved in laying it. Even in India, its use is disfavoured now.

14.2.2. Materials

Any stone which is hard and durable is suitable for Telford stone soling. Granite, basalt (trap), limestone, sand-stone, laterite, epidiorite, quartzite, etc. are usually used in its construction. In areas where superior types of stone such as granite or basalt are difficult to get, soling stones of inferior variety are used, conserving the costly types to the upper layers of the pavement. The stone need not be very hard ; normally, a little crushing of the stones under the roller during consolidation is helpful in getting a compact mass with a minimum of voids between the stones. But the stone shall not be so soft as to get powdered under the roller. The Los Angeles Abrasion value should not be more than 60.

14.2.3. Size of stones

The soling layer is usually constructed in thickness of 150–225 mm. The thickness of the stones should be equal to the thickness of course specified. Stones larger than 250 mm in the longest dimension are not permitted. The base area of the stones should not be less than 250 sq. cm nor more than 500 sq. cm.

14.2.4. Laying

The Telford soling layer is generally laid to a width 0.3 m more than that of the wearing course. When it is to be laid on clay, fine sand or such soils as may be forced up into the voids, a layer of filler (sand, moorum, stone screenings etc.) not less than 25 mm in thickness shall first be spread upon the prepared subgrade. This layer of filler shall not be considered as part of the depth of the base course.

Telford had specified that the soling course should be laid on a subgrade which is horizontal across the roadway and the required camber should be attained by making the thickness of the layer slightly more at the centre than at the edges. The practice of making the subgrade itself cambered and providing a uniform thickness of soling stone layer is equally good. The camber and superelevation, as may be needed, should be the same as that of the surfacing.

Before the stones are laid, firm and stable shoulders of earth of sufficient width must be formed, watered and rolled.

The soling stones are laid by manual labour with the largest face downwards, as nearly perpendicular to the subgrade as possible, in close contact with each other, but not leaning on each other, breaking joints as far as possible. The thickness of the course should be made up by single size stones.

14.2.5. Wedging and packing

After laying a short length of about 30 m of soling stones, spalls should be inserted and driven by hammer into the interstices so as to fill up the voids as far as possible. The spalls should consist of sledgings of the larger stones collected for the base course.

14.2.6. Rolling and filling

All sharp projecting points should be napped to produce an even surface. Rolling is then carried out thoroughly with a smooth wheel roller weighing not less than 8 tonne. The rolling should commence at the edges and gradually proceed towards the centre (or from the inner edge to the outer edge in case of super-elevated curves), the rolling proceeding in a longitudinal direction. Any depressions or surface irregularities resulting from rolling should be made good by stone spalls. The whole surface is then gradually blinded with an approved filler, which may consist of fines resulting from the sledging of larger stones, sand, gravel (murram) or screenings, the maximum size not exceeding 20 mm. Rolling is then continued while the filler is swept into the interstices by means of brooms and surface watered to facilitate the voids being filled. Rolling, filling and watering should be continued until the base is thoroughly compacted and all rocking and settlement has eased. When completed, the surface shall be smooth, even and true to cross- section and grade and when tested with 3 m straight edge or template, the departure from the true surface should not be more than 12 mm.

14.2.7. Quantity of materials required

The quantity of materials normally required for a 10 sq. m area of stone soling of 150 mm and 225 mm depth are given below :

Table 14.1. Quantity of Materials for 10 sq. m of soling area

S. No.	Materials	Quantity (cu. m)	
		150 mm layer	225 mm layer
1.	Soling stones	1.5	2.25
2.	Spalls	To be sledged from soling stones	To be sledged from soling stones
3.	Filler	0.4–0.5	0.6–0.7

From the foregoing remarks, it can be concluded that there cannot be two opinions about the distinct superiority of a graded stone macadam course over a Telfored stone soling base, but all the same due to economic factors this specification can be adopted in this country for low-cost and intermediate type of roads, where stones of the required quality are locally available.

14.3. GRAVEL COURSES

14.3.1. Definition of gravel

Gravel is a naturally occurring material consisting of small pebbles, stones, or fragments of stones intermixed with finer materials such as powdered rock, sand, loam, silt or clay. Sometimes by the term gravel is also meant rounded or water-borne stones or pebbles which have no fine material in them, and it is then known more popularly as shingle. In this chapter, only such material as has finer particles also in their natural state is covered by the term "gravel."

For use in pavement layers, naturally occurring gravel may have to be processed to meet the gradation, strength requirements and other properties mentioned in para 14.3.4.

14.3.2. Types of gravel, naturally occurring in different parts of the country

One of the gravel types most commonly occurring in this country is moorum. Moorums are (red) gravels and are the result of disintegration of laterite and other rocks containing iron. This kind of moorum is available almost throughout the country though in isolated patches in some of the States.

Soil kankar mixtures are available extensively in many parts of the country and these can also be termed as gravels.

14.3.3. Uses of gravel in road pavement

In a flexible pavement structure, the lower layers are subjected to much lesser stresses than the upper ones and hence inferior quality of materials, which are also less expensive, can be used in the lower layers. Naturally occurring gravel can ideally be used as improved subgrades, sub-bases and bases since they are cheap and have just the adequate strength to withstand the stresses coming in the lower layers of a pavement.

Gravel has been extensively used in this country as a surfacing material in low-cost roads. In future also this specification will continue to be adopted on less important roads.

Another important use of gravel has been in the construction of stable shoulders to pavements. Most of the road length in India is at present having only a single-lane carriageway. This presents great hardship for vehicles crossing each other or overtaking one another, especially in the rainy season when the embankment soil is very moist and slushy. Provision of shoulders, about 1.0–1.5 m wide on either side of the carriageway, with materials such as naturally occurring gravel, will not only protect the edges of the pavement but also facilitate the crossing and overtaking of vehicles on single-lane roads.

14.3.4. Desirable qualities of gravel

14.3.4.1. Grading

Since gravel is, in general, a mixture of coarse, medium and fine particles, it is available in its natural state in a variety of gradings. The material, which is well-graded, containing all sizes from top to bottom, so as to result in a dense layer when compacted, is the best. A good grading secures good interlock of the particles and this is very necessary to enable the layer to support the superimposed loads without detrimental deformation. The maximum size of particles that can be permitted will depend upon thickness of the layers in which the gravel will be compacted. A desirable practice is to limit the maximum size of particles to one-half to one-third of the thickness of the compacted layer. Since layers more than 150 mm compacted thickness are not desirable, the maximum size of particles should not exceed 80 mm. Tables 14.2, to 14.6 give the typical grading requirement of gravel with 80 mm, 50 mm, 40 mm, 25 mm and 20 mm maximum sizes. If the aggregates are of a crushable nature, the grading need not be strictly followed.

Table 14.2. Gradation requirements of gravel: maximum size 80 mm

Sieve designation	Per cent by weight passing		
	Coarse graded	Intermediate graded	Close graded
80 mm	100	100	100
40 mm			80–100
20 mm	55–75	55–75	60–80
10 mm			45–65
4.75 mm	10–30	15–45	30–50
2.36 mm			20–40
600 micron			10–30
75 micron	Less than 10	Less than 12	5–15

Table 14.3. Gradation requirements of gravel: maximum size 50 mm

Sieve designation	Per cent by weight passing		
	Coarse graded	Intermediate graded	Close graded
50 mm	100	100	100
40 mm			70—100
25 mm	50–80	50–80	55–85
20 mm			50–80
10 mm			40–70
4.75 mm	15–35	20–50	30–60
2.36 mm			20–50
600 micron			10–30
75 micron	Less than 10	Less than 12	5–15

Table 14.4. Gradation requirements of gravel: maximum size 40 mm

Sieve designation	Per cent by weight passing		
	Coarse graded	Intermediate graded	Close graded
40 mm	100	100	100
20 mm			80–100
10 mm			55–80
4.75 mm	20–40	25–55	40–60
2.36 mm		15–45	30–50
600 mm			15–30
75 micron	Less than 10	Less than 12	5–15

14.3.4.2. Strength of coarser fractions

In order that the coarser fractions do not get crushed or easily worn out under traffic, it is necessary that the Aggregate Impact Value should not be more than 50 per cent, 40 per cent and 30 per cent respectively for sub-base, base and surfacing courses . The 4 days soaked CBR value of gravel when used as a sub-base should not be less than 20 and when used as a base/surfacing course should not be less than 80. For unimportant roads, a lesser CBR than 80 has to be tolerated for base/ surfacing courses.

14.3.4.3. Properties of finer fractions

The material passing 425 micron sieve should have adequate binding properties to hold the larger particles together. The Liquid Limit of the material passing 425 micron size sieve, should not be more than 25 and the Plasticity Index should not be more than 6. Some little amount of plasticity is definitely desirable and an absolutely non-plastic binder is not good.

Table 14.5. Gradation requirements of gravel: maximum size 25 mm

Sieve designation	Per cent by weight passing		
	Coarse graded	Intermediate graded	Close graded
25 mm	100	100	100
20 mm			70–100
10 mm			50–80
4.75 mm	25–45	30–60	35–65
2.36 mm		20–50	25–50
600 micron			15–30
75 micron	Less than 10	Less than 12	5–15

Table 14.6. Gradation Requirements of gravel: maximum size 20 mm

Sieve designation	Per cent by weight passing		
	Coarse graded	Intermediate graded	Close graded
20 mm	100	100	100
10 mm			80–100
4.75 mm	40–65	45–70	50–75
2.36 mm		30–55	35–60
600 micron			15–30
75 micron	Less than 10	Less than 12	5–15

14.3.4.4. Other properties

Coarser particles which are smooth are inferior to those having sharp edges as far as load bearing capacity is concerned, This is because of the good interlocking and frictional resistance obtained by the sharp-edged particles.

The material should be free from vegetable matter and roots and lumps or balls of clay or such other deleterious matter.

14.3.5. Construction methods in traffic compacted gravel course

Where the gravel course is intended for an unimportant road and where road rollers for compaction are not available, the gravel course is generally laid in layers of small thickness and allowed to be compacted by the traffic itself. The gravel is laid in layers not exceeding 5 mm thickness and is uniformly distributed by means of a drag or other levelling device. The traffic is then allowed to use the surface. The traffic should predominantly be pneumatic-tyred as steel tyred traffic is likely to cause rutting. The surface is kept in shape periodically and is sprinkled with water at least twice a day. As each layer gets compacted, successive layers are added up.

14.3.6. Construction methods in rolled gravel course

There is no denying the fact that the best gravel road is one which is constructed in small layers, say maximum 150 mm depth, and rolled adequately. The optimum moisture : intent which gives the maximum density for the gravel should be determined and the necessary quantity of water should be added before rolling. The number of passes of the roller necessary to obtain the desired density should be determined previously by trial. The roller most frequently used is the 8–10 Tonne diesel powered smooth-wheeled roller. The materials are spread uniformly by manual labour or by means of a drag or a blade-grader. The rolling should begin at the edges and proceed towards the centre, except in superelevated curves where it should start from the inner edge and proceed towards the outer edge. On completion of the rolling, the surface, when tested by means of a 3 m straight edge along the centre-line of the road. should not show deviations more than 15 mm. The maximum variation in the cross-profile should be less than 12 mm.

The camber which is usually necessary for untreated gravel roads is about 4 per cent for areas with rainfall more than 100 cm annually and 3 per cent for areas with rainfall less than 100 cm annually. The cambers suggested will ensure that the rain-water is carried away to the side-drains as soon as precipitation occurs. However, when the gravel surface is subsequently upgraded to bituminous standards, the excessive camber should be rectified.

14.3.7. Characteristics of a gravel surface

14.3.7.1. Action of clay binder

The clay binder in a gravel surface acts as a moisture regulator and binds the coarser fractions together. When the atmospheric conditions are dry, the clay fraction retains a thin film of moisture which binds the particles together. During the first rains, the moisture causes the clay to expand causing the pores to close and prevents further water entering the material. Only the top surface becomes slushy, which, however, dries up subsequently, causing little harm to the structure of the layer. If the clay binder is absent, the particles are not held up together and the surface soon disintegrates. If the layer has been built up without adequate watering, the clay fraction is dry and has no binding properties and the surface disintegrates. With a clay which is highly plastic, the ingress of rain water causes swelling of the binder and consequent dislocation of the particles in the framework.

The gravel roads are very dusty, especially in the dry season. In the wet season they tend to become slushy, especially if a highly plastic clay binder is used. The iron-tyred traffic crushes the softer particles at the top surface and rutting is very easily caused. Pneumatic-tyred traffic sucks the dry and loose particles and thus loosens the coarser fractions, which soon start getting dislodged. Corrugations are also caused in gravel roads due to pneumatic traffic much more quickly than in a water-bound macadam road or in a blacktopped road.

Gravel surfaced roads can be used for light traffic, on Village Roads and Other District Roads.

14.3.8. Improvement of gravel surface

The main drawback of a gravel surface is that it is too dusty. This can be avoided only if the surface is blacktopped. Sometimes a cheap blacktopped specification such as two coats of surface dressing or a thin premix chipping carpet is laid directly over a gravel surface. The gravel surface would need to be primed before this is done. Though there are many instances of such roads in this country, and such roads are standing well under low traffic, it is desirable to provide a crushed stone macadam surface, 75 mm compacted thickness, before attempting to lay a bituminous treatment, especially if the traffic is likely to increase. In other countries, low cost dust palliatives have been successfully used on gravel surfacings to control dusty conditions in the dry season.

14.3.9. Maintenance of gravel surfaces

Gravel roads deteriorate because of a number of factors :

(*i*) The iron-tyred traffic crushes the particles and loosens them. This leads to rut formation.

(*ii*) The pneumatic tyred and iron-tyred traffic cause wear by abrasion.

(*iii*) The smaller particles at the surface are sucked away by the pneumatic wheels and are lost with fast traffic. This results in the coarser particles being gradually loosened. Rutting and pot-holing are then caused.

(*iv*) Wind also blows away the lighter particles from the surface.

(*v*) When the gravel surface becomes dry in summer months and it is not kept moistened, the moisture evaporates from the clay binder which then loses its binding property. The particles, as a result, get loosened and dislodged.

(*vi*) Corrugations are frequently caused in gravel roads due to pneumatic traffic.

Maintenance of gravel roads is similar to earth roads. Careful sectioning of the road should be done preferably after a rain when the surface is not dusty. Additional material, required to fill in potholes and ruts and to rectify the corrugations, should be spread and consolidated when moisture conditions are favourable. About 75 mm (loose) thickness of additional material may be needed annually, depending on traffic conditions.

14.4. WATER-BOUND MACADAM

14.4.1. Historical

Mc Adam was a Scottish engineer who introduced, in the early nineteenth century, the idea of constructing roads composed of small size stones held together by means of a binding material. This concept had revolutionised the road building science then, aided as it was by the invention of the stone crusher in 1858 by Blake, the steam road roller by Aveling in 1867 and by the use of bituminous materials early in the twentieth century. With the advent of bituminous materials, the term macadam was qualified by "water bound" macadam if the stone materials are held together by the addition of water and filler; by "dry-bound" macadam if the aggregates are held together by mechanical interlock; 'wet-mix' macadam if graded stones are mixed with water and compacted; by "penetration" macadam if a bituminous material is sprayed over the stones and allowed to penetrate into the course and by "premix" macadam if the bituminous material is mixed with the aggregates prior to laying.

14.4.2. General description of the specification

Water bound macadam may be defined as a dense and compact course of a road pavement composed of stone aggregates bound together by a thin film of cementing medium consisting of fine mineral filler (such as stone screenings or gravel) with cementitious properties and containing a minimum laden moisture to impart to the binder necessary cohesive and adhesive properties to enable it to bind the aggregates together. The strength of a water-bound macadam course is thus primarily due to :

(*i*) the thorough mechanical interlock in the aggregate particles. This interlock structure imparts interlocking resistance and frictional resistance and prevents displacement under load ; and

(*ii*) the cohesion between the aggregate particles due to the cementitious film of soil-moisture binder.

14.4.3. Materials generally used for water-bound macadam

The most common and durable material for use as aggregates in water-bound macadam is broken stone aggregates. The stones may consist of hard varieties such as granite, basalt, diorite, quartzite, etc. or softer varieties such as sandstones, limestones, kankar, epidiorite, laterite etc. The harder varieties are generally preferred for the upper layers whereas the softer ones can be used for the lower layers. Crushed slag is also used as an aggregate. Another material very commonly used in the Indo-Gangetic plains of India, where the stone materials are hard to get, is the over-burnt brick ballast, also known as jhama metal. The last named material is generally used as a lower course to be followed by an upper course using hard stones, if the road is an important one and has heavy traffic. For light and medium trafficked roads, the jhama metal can itself be used as a wearing course provided it is protected by a light bituminous premix carpet.

14.4.4. Size and grading requirements of coarse aggregates

14.4.4.1. Normal size of aggregates

The main source of strength of a water-bound macadam surface is due to the mechanical interlock in the aggregates and it is thus apparent that the aggregates should be well graded. Well graded aggregates can be obtained only by a crusher whereas hand breaking can yield single size aggregates. In India the practice so far has been to use single size stones in water-bound macadam mainly because of the necessity to accept hand-broken materials. Recent specifications however, specify some grading for the stone metal. Three types of gradings are generally indicated vide Table 14.7.

Table 14.7. Grading requirements of coarse aggregates

Grading No.	Size range	IS sieve designation	Per cent by weight passing
1.	90 mm to 45 mm	125 mm	100
		90 mm	90–100
		63 mm	25–60
		45 mm	0–15
		22.4 mm	0–5
2.	63 mm to 45 mm	90 mm	100
		63 mm	90–100
		53 mm	25–75
		45 mm	0–15
		22.4 mm	0–5
3.	53 mm to 22.4 mm	63 mm	100
		53 mm	95—100
		45 mm	65—90
		22.4 mm	0—10
		11.2 mm	0—5

For soft aggregates such as kankar, laterite or brick ballast which get crushed excessively under roller, the above suggested grading is not very important.

The I.R.C. specification also provides that Grading Nos. 1, 2 and 3 may be used for surface-treated water-bound macadam but for water-bound macadam which is not provided with any surface treatment Grading Nos. 2 and 3 above should be used. Grading No. 1 consisting of over-size metal is generally used as a substitute for the Telford stone base which is now getting rarer. The maximum size of the aggregate to be used would depend upon the type of aggregate available and the total consolidated thickness of the layer. With softer varieties of aggregates, a larger size is generally indicated and similarly with a higher consolidated thickness desired, a larger size of aggregates may be used. The following table may be used as a general guide in selecting the grading depending upon the thickness of course :

Table 14.8. Relation between IRC specification grading No. and thickness of course of W.B.M.

Grading No. of IRC Specification	Size range	Consolidated thickness of course for which suitable
1.	90 mm to 45 mm	100 mm
2.	63 mm to 45 mm	75 mm
3.	53 mm to 22.4 mm	75 mm

14.4.5. Strength and other requirements of crushed stone or slag aggregates

The crushed or broken stones should be hard, durable and free from excess of flat, elongated, soft and disintegrated particles, dirt and other objectionable matter. The combined flaky and elongated particles of crushed aggregates (as per I.S. 2386-Part I) should be below 30 per cent. The Los Angeles Abrasion Value (IS : 2386-Part IV) should not be more than 50 per cent for sub bases, not more than 40 per cent for surface treated base courses and not more than 30 per cent for water-bound macadam which is not provided with any surfacing. Alternatively, the Aggregate Impact Value (IS : 2386-Part IV) should not be more than 40 per cent for sub-base courses, not more than 30 per cent for surface treated base course and not more than 20 per cent for water-bound-macadam which is not provided with any surfacing. The crushed slag shall be made from air-cooled blast furnace slag and shall be angular in shape, reasonably uniform in quality, density and be reasonably free from thin, elongated and soft pieces, dirt and either objectionable matter. It shall not contain more than 20 per cent of glassy material and shall not weigh less than 1120 kg per m^3.

14.4.6. Requirements of brick aggregates

The brick aggregate should be from over-burnt bricks or brick-bats. The Los Angeles Abrasion Value should not be more than 50 per cent.

14.4.7. Requirements of screenings and binding material

The screenings, also known as "choke" materials, fill in the voids left in the coarse aggregates after they are consolidated and help to cement the stone aggregates together. To effectively perform these functions, the screenings should be properly graded and also should have some plastic material in them to impart cementitious properties. Excess of plasticity is harmful since, under the influence of moisture, the material may lose its stability. The screening material ideally suited for this purpose is naturally occurring moorum which has coarse fractions upto 12.5 mm size as well as enough fines and clay content. If the moorum is having too much of plastic clay, a mixture of stone screenings (grit or stone dust) and murram makes a good material. Hard kankar nodules make excellent material for screenings because of the high cementitious property of limestone in presence of water. In locations where none of the naturally occurring screening materials are available, a mixture of stone grit or dust (12.5 mm down) with a small amount of clayey or silty soil is equally effective. In no case should the filler material consist entirely of ordinary earth.

The grading requirements of the screenings, as given in the I.R.C. specifications, are reproduced in Table 14.9:

Table 14.9. Grading for screenings

Grading classification	Size of screenings	IS sieve designation	Per cent by weight passing the IS sieve
A	13.2 mm	13.2 mm	100
		11.2 mm	95–100
		5.6 mm	15–35
		180 micron	0–10
B	11.2 mm	11.2 mm	100
		5.6 mm	90–100
		180 micron	15–35

Screenings of type A are suggested for Grading Nos. 1 and 2 of coarse aggregates and of type *B* for Grading No. 3 of coarse aggregates.

The plasticity index of the material should not be more than 6 for surface treated W.B.M. and 4 to 9 for W.B.M., not surfaced.

Screening materials may be dispensed with in case of soft aggregates such as kankar, laterite, brick ballast etc.

14.4.8. Thickness of courses

The water-bound macadam is constructed by spreading loose metal which gives a consolidated thickness of 75 mm–100 mm. A compacted layer less than 75 mm thickness is not desirable and a compacted layer more than 100 mm is equally undesirable. If the thickness of the base is more than the above value, the construction is done in multiple layers.

14.4.9. Quantity of materials necessary

The loose thickness of aggregates will be about 120 to 140 per cent of the compacted thickness. The lower value may be taken for well graded coarse aggregate and the higher value for a poorly graded one. Thus for a 10 sq m of 75 mm consolidated thickness of W.B.M., the quantity of loose aggregate would vary between

$$10 \times \frac{75}{1000} \times 1.2 = 0.90 \text{ cu m and } 10 \times \frac{75}{1000} \times 1.4 = 1.05 \text{ cu m}$$

The quantity of screenings and binding material will depend upon the voids left in the coarse aggregates after rolling. The coarse aggregates have voids of 40–50 per cent in their loose state and after rolling this reduces to 30–40 per cent. Thus if a quantity of 0.90 cu m of loose metal with 40 per cent initial voidage is consolidated to have a voidage of 30 per cent, the volume of screening necessary to fill in the voids is :

$$0.90 \times \frac{60}{100} \times \frac{30}{70} = 0.232 \text{ cu m.}$$

If the coarse aggregate having a volume of 1.05 cu m arid a voidage of 50 per cent in the loose state is consolidated to have a voidage of 40 per cent, volume of screenings necessary to fill the voids is:

$$1.05 \times \frac{50}{100} \times \frac{40}{60} = 0.35 \text{ cu m.}$$

Thus the range of the requirement of screenings and binding materials would be from 0.232 cu m to 0.35 cu m. When screenings and binding materials are collected separately, screenings should form about 50–75 per cent of the total volume and the binding material should form the balance. The approximate quantities of coarse aggregates and screenings required for 100/75 mm compactedthickness of WBM as recommended by the IRC are given in Table 14.10.

Table 14.10. Approximate quantities of coarse aggregates and screenings required for 100/75 mm compacted thickness of water bound Macadam (WBM) sub-base/base course for 10 m² area

Classification	Size range	Compacted-thickness	Loose Qty.	Screenings			
				Stone Screening		Crushable type such as Moorum or gravel	
				Grading classification and size	For WBM sub-base/ base course (Loose quantity)	Grading classification and size	Loose qty.
Grading 1	90 mm to 45 mm	100 mm	1.21 to 1.43 m^3	Type A 13.2 mm	0.27 to 0.30 m^3	Not uniform	0.30 to 0.32 m^3
Grading 2	63 mm to 45 mm	75 mm	0.91 to 1.07 m^3	Type A 13.2 mm	0.12 to 0.15 m^3	–do–	0.22 to 0.24 m^3
—do—	—do—	—do—	—do—	Type B 11.2 mm	0.20 to 0.22 m^3	–do–	–do–
Grading 3	53 mm to 22.4 mm	75 mm	–do–	–do–	0.18 to 0.21 m^3	–do–	–do–

The quantity of binding material should be as follows :

Table 14.11. Quantity of binding material for 10 sq m area

Compacted thickness	Course	Quantity of binding material 10 sq m area (m^3)
75 mm	Sub-base/Base	0.06–0.09
	Surfacing	0.10–0.15
100 mm	Sub-base/Base	0.08–0.10
	Surfacing	0.12–0.16

14.4.10. Camber

The camber to be provided for water bound macadam course to be surfaced will be the same as that of the surfacing. But if the water-bound macadam course itself is to serve as a surfacing course, the camber may be provided as per Table 14.12.

Table 14.12. Camber for W.B.M surface course

Camber for water bound macadam surface course	Average annual rainfall	
	Less than 100 cm	Over 100 cm
	3 per cent	4 per cent

14.4.11. Preparation of subgrade, sub-base or base

If the water bound macadam is to be laid over a base of Telford stone soling or of water-bound macadam constructed by using over-size broken stones, a thin layer of murram (gravel) should be spread over the base as a cushion course. The purpose of this cushion course is to ensure that the stone aggregates of the water bound macadam caught between the roller wheel and the hard soling stones beneath do not get unduly crushed. If the WBM is to be laid down over an existing blacktopped surface, diagonal furrows (at 45° angle) 50 mm × 50 mm should be dug at 1 metre intervals. The general experience in the country is that when WBM is to be laid over an existing blacktopped surface, a minimum thickness of overlay of 150 mm in two layers should be attempted.

14.4.12. Shoulders

For providing lateral confinement to the water-bound macadam course during rolling, it is absolutely essential that firm and stable shoulders are formed before-hand. They should be adequately compacted by watering and rolling. Since it is sometimes difficult to roll these narrow strips of shoulders, the entire shoulder width is often constructed along with the water-bound macadam.

14.4.13. Spreading metal

14.4.13.1. Manual method

In this country, the spreading of metal for water-bound macadam is so far being done manually though in future mechanised spreading may make its appearance. The aggregates are spread by labourers using baskets and smartly turning the hands during emptying to scatter the stones uniformly. The aggregates should not be dumped in heaps directly over the base within the area over which it is to be spread. The material should be collected in stock-piles along the side of the pavement in the shoulders. The aggregates should be spread to proper profile by means of templates placed across the road about 6 metres apart.

14.4.13.2. Mechanical spreaders

Mechanical spreaders are to be preferred to hand spreading because hand-spreading can be costly and uneven. The spreaders should be driven at a slow speed without jerks so that the deposition of the aggregates is uniform. The gate openings should be properly adjusted.

After the materials are spread, the surface is checked with a straight-edge 3 metre long and by camber boards and the variation should not be more than 12 mm. Some hand packing may be necessary where the spreading has been defective.

The coarse aggregates should not normally be spread in length more than 3 day's average work in advance of the rolling.

14.4.14. Rolling of aggregates

14.4.14.1. Dry rolling

Rolling of the water-bound macadam is usually done in this country by means of 3-wheel power rollers of 8–10 Tonne capacity. Vibratory rollers are also used wherever available. The dry rolling of the coarse aggregates shall begin with the outside rear wheel covering equal parts of the shoulder and metal and the roller shall be run forward and backward until the shoulder and the metal are bound together firmly. The rolling shall then progress gradually from the edges to the centre (or to the outer edge in the case of superelevated curves), parallel to the centre-line of the road, and lapping uniformly each preceding rear wheel track by half its width, until the entire course has been rolled by the rear wheel. The rolling should continue until the road metal is thoroughly keyed and creeping of the stone ahead of the roller is no longer visible. This may take 3 or 4 traverses of the roller over the entire area. Slight sprinkling of water to facilitate the interlocking may be resorted to, if required. The surface should be checked by means of templates and camber boards and any surplus material removed and depressions made good by fresh materials and rolled again. Rolling should not be done in lengths less than 100 m. During working of the roller, it should be ensured that the speed is not more than 5 KMPH and that the roller is not started or stopped with jerks. The shifting of the transverse position of the roller on each successive trip should be done outside the length being rolled.

14.4.14.2. *Wet rolling*

The rolling is then continued accompanied by copious watering till the creeping or movement of metal ahead of the roller would have ceased.

14.4.15. Application of screenings, watering and rolling

In many locations it is possible to collect the screenings in two parts, viz., (*i*) the choke material or the granular screenings which fill in the voids in the coarse aggregates and key them in position and (*ii*) the binder or filler material which has a little plasticity and helps to cement the materials together aided by the presence of water film. If these materials are collected separately, the sequence of construction is as under :

(*i*) After rolling of the aggregates, the choke material is applied in thin layers manually by smartly turning the spreading baskets and brooming the material into the interstices. The roller is again worked during the above operation so that its jarring effect will cause the materials to settle into the voids of the coarse aggregates. Successive thin layers of the choke material are applied after each rolling till no more of the material can be forced into the voids.

(*ii*) The surface is then sprinkled copiously with water, swept and rolled. Hand brooms are used to sweep the wet material into the voids and distribute the same. The sprinkling, sweeping and rolling shall be continued and additional screenings applied, where necessary, until the coarse aggregates are well-keyed and bonded for the entire depth and until a slurry of screening and water forms into a wave in front of the roller wheels. Excessive quantity of water likely to damage the subgrade should be avoided.

(*iii*) The binding material is then applied at a uniform and slow rate in two or more successive layers. After each application, the surface is copiously sprinkled with water and swept with brooms to fill in the voids and the surface rolled. The roller wheels are kept moistened to prevent the binding material that may stick to them. The spreading of the binding material, sweeping and rolling shall continue till all voids are filled in and the slurry that results out of the filler and water forms a wave ahead of the roller wheels.

If the choke material and binding material are not collected separately, the material collected is applied, water sprinkled, the surface swept and rolled in the same manner as detailed above.

The number of passes of a roller required for wet rolling of the coarse aggregates and the dry and wet rolling of the screenings will be about 4.

A 8–10 Tonne roller can achieve an output of about 50–75 cu m of loose thickness of metal which corresponds to a compacted volume of about 35–50 cu m per day. About 100–200 length of road pavement 3.5 m wide and of 75–100 mm compacted thickness can be consolidated in a day.

14.4.16. Finishing

If the water-bound macadam surface is to be allowed to be opened to traffic, it is preferable if a layer of coarse sand is spread on the surface which has been allowed to dry for some hours, the surface slightly sprinkled and rolled to get a smooth finish. The purpose of this operation is to prevent the gravel, when wet, from sticking to the wheels of the vehicles and being pulled away, dislodging in turn some of the interlocked stones. The quantity of sand required is about 0.06 cu m per 10 sq m.

14.4.17. Curing, protection and opening to traffic

In dry weather, the surface should be slightly sprinkled with water and kept moist for about 15 days. The macadam sets in about 2 to 3 days and no traffic should be allowed till then. During this intervening period, the road should be protected by barricading.

14.4.18. Tolerances specified for surface finish

In this country, water-bound macadam courses are either left unsurfaced or are surfaced with a bituminous treatment. The bituminous treatment may consist of surface dressing or a thin procoated chipping carpet or a dense-graded asphaltic concrete 20–25 mm thick. Because of thin or no surfacing provided on top, the surface finish of the macadam itself should be of a high order if the irregularities are not to be reflected in the surface. The I.R.C. Specification lays down that the variation in finish of surface should not be more than 15 mm and 12 mm when tested with a 3 m long straight edge respectively for WBM with oversize metal (40–90 mm) and WBM with normal size metal (20–50 mm and 40–63 mm). The variation in the cross-profile should not exceed 12 mm.

14.4.19. Multiple layered construction

When the total depth of construction is to be built up in more than one layer, the same degree of care and refinement is necessary in all the layers.

14.4.20. Water-bound macadam using Jhama Brick aggregates, Kankar or Laterite

For consolidating water-bound macadam using jhama brick aggregates, the roller should be light and should not exceed 8 tonnes in weight. Screenings may not be necessary for consolidating the course as the aggregates will get crushed during rolling. Local earth wherever suitable may be used as a binder material. After consolidation, the surface should be blinded with earth or sand till the surface dries up and is surface-treated.

For consolidating kankar, only a hand rammer or a vibrator is used if the material is soft. For hard varieties, a lighter roller 6–8 Tonnes may be used. No screenings may be necessary during consolidation if the kankar is soft. A thin layer of earth spread over the surface when it has dried helps to improve that surface considerably.

Laterite also does not need screenings to be applied separately. The roller should preferably be light, say 6–8 Tonnes. Excessive rolling is to be avoided.

14.4.21. Defects and likely causes

Table 14.13 lists out the defects and likely causes for the same in a WBM layer.

Table 14.13. Defects and likely causes in water-bound macadam

Defects	1. Formation of ruts	2. Formation of pot holes	3. Corrugations	4. Ravelling	5. Damaged edge	6. Roughness of surface	7. Wear of surface	8. Knobbiness
Indequate crust thickness	×	×					×	
Predominance of iron-tyred traffic	×						×	
Excessive camber	×							
Use of sandy binder	×	×		×		×		
Use of clay binder		×		×				
Use of too little binder		×						
Use of excessive binder		×		×				
Use of soft aggregates							×	
Insufficient compaction				×		×	×	×
Poor subgrade soil					×	×		
Defective rolling			×					
Effect of vibrations up by pneumatic tyred vehicles			×					
Effect of braking			×					
Effect of changes in grade and curves with inadequate super-elevation			×					
Use of excessive blindage material			×					
Transverse picking of W.B.M			×					
Evaporation of moisture in binder material		×		×				
Improper shoulder maintenance					×			
Improper grading of metal		×		×		×		
Use of oversize metal		×		×		×		
Heavy rainful		×		×	×	×	×	
Use of hetergenous metal								×

14.4.22. Suitability of surface course for traffic

Water-bound macadam surface is suitable for about 1000 Tonnes of traffic per lane per day, 50 per cent of which may consist of iron-tyred traffic. Its life can be enhanced if it is provided with a bituminous surfacing.

14.4.23. Life

The life of unsurfaced water-bound macadam varies from 1 year to 8 years depending upon the aggregates used and the type and volume of traffic.

Water-bound macadam roads provided with surface dressing or thin premix chipping carpets not more than 20 mm thick have a life varying from 2 to 10 years, depending upon the type of aggregates used, type of surfacing provided and the traffic using the road.

14.4.24. Advantages and disadvantages

Compared to the Telford stone soling base course, the water-bound macadam is far superior in quality because the materials are carefully graded and the resulting compacted mass is almost voidless. The materials that fill the voids are also selected and there is no possibility of plastic materials of the subgrade entering into the voids. The interlock of the aggregate particles of a macadam course is of a high order imparting it adequate strength. From the point of view of cost, the water-bound macadam course using hard stones can be about double the cost of Telford stone soling course of same thickness.

When compared to bituminous base courses, the advantage of water- bound macadam lies in its lesser cost. A base course of bituminous penetration macadam can cost about 2–3 times the cost of water-bound macadam course of same thickness. A premix macadam base course can cost about 3–4 times that of the water-bound macadam. But it should be remembered that if a bitumen-bound base course is used in place of a water-bound macadam, the thickness of the course can be reduced to two-thirds or even half.

As a surface course, water-bound macadam is popular in this country now only for Village Roads, Other District Roads and some Major District Roads. The main disadvantage of this surface is its dustiness and its vulnerability to disruption under iron-tyred traffic which wears the stones into dust. The dust in turn is carried away by wind and by suction of the pneumatic tyred vehicles.

To conclude, water-bound macadam specification is very popular as a base course in this country and is likely to remain so in future also because of cheap labour available. Even if spreading is mechanised in future, this specification will remain popular. As a surface course, without any blacktopped treatment, it is suitable for only minor roads.

14.5. DRY BOUND MACADAM

As the name itself implies, dry bound macadam is a crushed- aggregate course which develops its strength only through the interlock between the particles. Unlike water-bound macadam, no water is added during compaction. This method of construction demands thorough compaction, which is generally achieved by vibratory compactors. After first compacting the coarse aggregates, screenings are added and vibrated or rolled into the voids till the aggregate is keyed. If the course is built directly on a subgrade, a layer of screenings is used as a blanket course and also to serve as an inverted choke.

Dry-bound macadams are not commonly used in this country. They have a good scope of being used in the desert region where there is a scarcity of water for the consolidation of the water-bound macadam course and also in sub-zero temperatures where water might freeze.

14.6. WET-MIX MACADAM

14.6.1. General

Wet-Mix macadam is a specification in which a well-graded aggregate is mixed with water in a mechanical mixer and the resultant mixture is laid by pavers and compacted. The aggregate is generally crusher-run, and includes fines also. Because of the close grading, the course will have good interlock with excellent density. The specification is in use in U.K. but has recently been introduced in India.

14.6.2. Grading

Common practices regarding grading are given in Table 14.14.

Table 14.14. Grading for Wet-mix Macadam

Sieve designation	Per cent by weight passing	
	(40 mm nominal size)	(25 mm nominal size)
53.00 mm	100	
45.00 mm	95–100	
26.50 mm	–	100
22.40 mm	60–80	50–100
11.20 mm	40–60	–
4.75 mm	25–40	35–55
2.36 mm	15–30	–
600 micron	8–22	10–30
75 micron	0–8	2–9

14.6.3. Moisture content

The optimum moisture content for mixing is determined by conducting suitable density tests. The moisture content during mixing is maintained at this optimum ± 0.5 per cent. The moisture content is usually in the range 2–5% by weight.

14.6.4. Construction procedure

The mixing can be done in a suitable mechanical mixer. Specially designed mixers can be fabricated for this specification. Otherwise, a bituminous macadam plant can be used. Ordinary concrete mixers can also be used. Laying is done by paver-finishers and compaction by 8–10 Tonne smooth wheel rollers.

14.6.5. Advantages and disadvantages

The main advantage of wet-mix macadam over water-bound macadam is that it is composed of a well-graded mixture. This ensures good interlock and high stability. Addition of water while mixing facilitates the handling of the mixture. The operation of laying is much simpler than that of water-bound macadam, where the screenings and binding material have to be added in stages and forced into voids. If a crusher-run material is used, there is no possibility of plastic fines entering into the mixture. The compaction is greatly facilitated by the moisture added which lubricates the individual particles.

One disadvantage of the wet-mix macadam is that it is slightly costlier than water-bound macadam. This is because the specification involves the use of mixing plant and paver. On the

other hand water-bound macadam has been traditionally a labour-briented specification in this The aggregates for wet mix macadam will have to be crusher-run, whereas the aggregates for water-bound macadam are generally hand-broken in this country.

14.7. DRAINAGE LAYER

The need for draining the water, that enters the pavement layers, away from the subgrade is now recognised as a standard practice of pavement design. A typical sketch showing a drainage layer is given in Fig. 14.1. The drainage layer extends over the full width of the embankment.

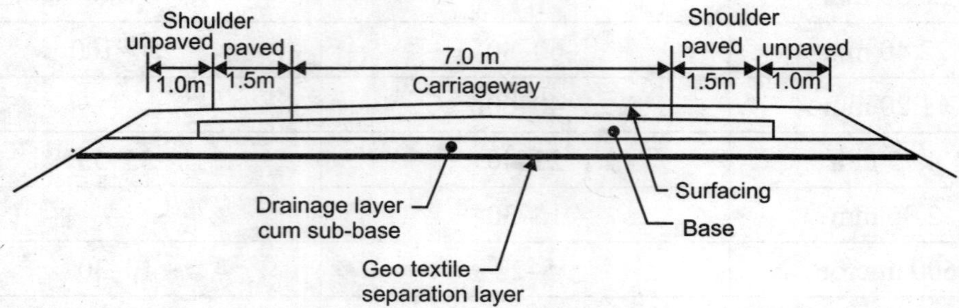

Fig. 14.1. Pavement drainage layer.

The drainage layer must permit the drainage of the water, and hence it must be sufficiently open-graded. Table 18.15 gives three gradings which have an acceptable coefficient of permeability in the range of 300 to 900 m per day

Table 14.15. Grading of drainage layer

Sieve Size	Percent passing		
	1	2	3
22.40	100	100	100
13.20	81.5	79.5	75
11.20	72.5	69.5	63
4.75	49	43.5	32
3.35	29.5	22	5.8
2.36	25	17	0
1.18	9.8	0	0
Coefficient of permeability (m per day)	300	600	900

14.8. USE OF GEO-TEXTILES AND GEO-GRIDS

The use of geo-textiles in bringing about separation between the soil subgrade and the granular layer and thereby preventing intrusion of the soil into the stone matrix is becoming popular. Geotextiles also reinforce the granular layer and help in reducing its thickness. Geo-grids, which are available in a range of thicknesses of the mesh and opening sizes, achieve the reinforcing action more effectively.

QUESTIONS

1. (*a*) Describe the Telford stone soling practice.
   (*b*) What are its advantages and disadvantages ?

2. (*a*) State the use of gravel as a pavement course.
   (*b*) What are the desirable properties of gravel ?
   (*c*) Give the gradation requirements of gravel for use as a pavement course.

3. (*a*) Describe briefly the construction of a gravel road.
   (*b*) How is a gravel road maintained?

4. (*a*) How does a water-bound-Macadam layer derive its strength ?
   (*b*) What are the materials generally used for water-bound-Macadam ?
   (*c*) Give the gradation requirements for water-bound-Macadam.

5. Give a detailed specification for water-bound-Macadam layer.

6. Describe the constructional practice in laying a water-bound-Macadam layer.

7. What are the defects commonly noticed in water-bound-Macadam and what are the likely causes?

8. What are the advantages and disadvantages of water-bound-Macadam ?

9. Write short notes on :
   1. Dry-bound Macadam   **2.** Wet -mix Macadam.

Chapter

Brick, Stone and Cement Concrete Block Pavements

15.1. BRICK PAVEMENT

15.1.1. General

Bricks have been used for paving streets and roads over a number of years, both in India and abroad. Their use in countries abroad has generally declined, but in India, brick pavements are popular in the northern plains. The scarcity of the stone aggregates and the suitability of the local soils for the manufacture of bricks, have made this specification popular there.

15.1.2. Usage

In India, brick paving is adopted for the following uses :

(*i*) As sub-base layer for important roads

(*ii*) As base layer for the middle category of roads

(*iii*) As a surfacing layer for very unimportant roads (village roads etc.)

(*iv*) As a specification for shoulders, especially of single-lane roads.

(*v*) As a surfacing layer for service lanes, foot-paths and cycle-tracks in towns and cities.

15.1.3. Advantages and disadvantages

The following are the advantages of brick-pavements for conditions in India :

(*i*) In the alluvial plains of the north, the scarcity of stone materials compels the engineer to use bricks in any component of the pavement structure.

(*ii*) The manufacture and laying of bricks as practised in India is a highly labour intensive process and should find favour with the engineers for many years to come. Practically no mechanical equipment is needed.

(*iii*) The salvage value of the bricks is high. They can be reused after dismantling for any reason.

(*iv*) If used as a surface course, the brick pavement is dust-free. This property can be used to advantage in paving city lanes, foot paths and cycle-tracks at a cheap cost.

(*v*) A tolerably firm shoulder of bricks is a great advantage for single lane roads. Such shoulders ensure firm lateral support to the pavement, increase the capacity of the single lane road and provide a stable surface for overtaking and crossing manoeuvres.

(*vi*) Traffic can be allowed as soon as the pavement is laid. There is no curing time or setting time.

(*vii*) The maintenance of the brick pavement is relatively easy and cheap. Some of the disadvantages are :

(*i*) Brick pavement lacks the interlocking stability obtained from graded aggregates. In fact, it is argued that the stability of a layer of broken brick aggregates, well compacted, is much higher than that of brick paving.

(*ii*) As a surfacing layer, the abrasion by iron-tyred traffic causes quick wear, leading to ruts. With pneumatic tyred traffic also, unprotected brick surfaces wear out quickly. This disadvantage is overcome by providing a thin bituminous surfacing such as a chipping carpet.

15.1.4. Quality of bricks

The success of the specification primarily hinges upon the quality of bricks. Not all soils yield good bricks. The bricks produced in the northern alluvial plains of India in the States of Punjab, Haryana, Rajasthan, Uttar Pradesh, Bihar and West Bengal are generally suitable. Good quality bricks of high strength can only be produced in well-designed kilns where the temperature of burning is high. Extra burning results in over-burnt bricks (vitrified bricks) which have not only higher strength but also lower water absorption. Results in Utter Pradesh have indicated that overburnt bricks of compressive strength greater than 27.5 MN/m^2 can be produced in Bull's kiln. The water absorption of such bricks is about 5–8 per cent. Where iron-tyred traffic is expected, vitrified bricks are used.

15.1.5. Construction practice

In India, bricks are laid on a prepared subgrade or sub-base. A cushion of 25–40 mm thickness of sand is introduced below the bricks to take up any inequalities in the supporting layer. Bricks laid on sand-cement mortar or sand-bitumen have been observed to perform satisfactorily but this has to be discarded in most cases from considerations of economy. The thickness of layer needed determines whether the bricks are to be laid flat in one layer, or flat in two layers, or on edge or flat in one layer followed by one layer on edge. The patterns of ' arrangement of the bricks are illustrated in Fig. 15.1. Brick paving can be constructed with the joints filled in with bitumen-sand mixture or bitumen alone. This improves the performance of the specification, but increases the cost. For the latter reason, it is not commonly followed in this country. The general practice in this country is to fill the joint space with fine sand. Rolling with a light tandem roller (3 Tonnes) beginning from the edges and working towards the centre is recommended.

15.1.6. Surfacing

The provision of a thin bituminous surfacing such as a single coat of surface dressing or a thin premix carpet protects the brick pavement when laid as a surface course and prolongs its life. Before laying such a thin specification, priming of the brick surface is generally desirable.

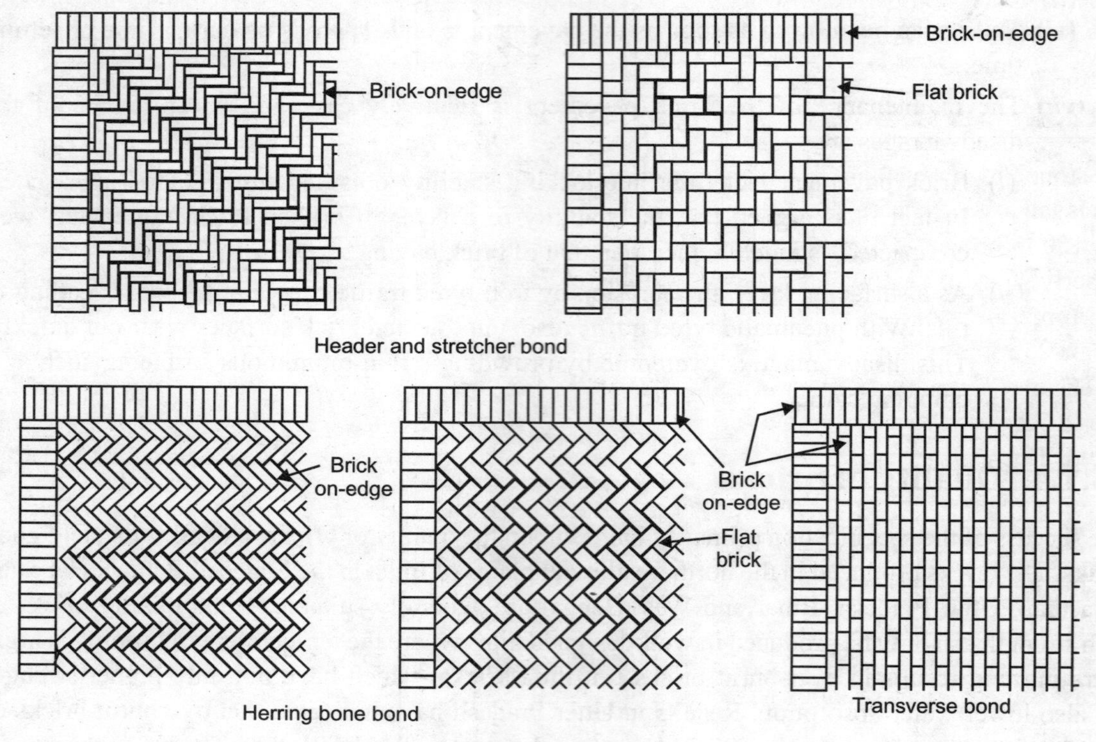

Fig. 15.1. Brick pavement patterns.

15.2. STONE BLOCK PAVEMENT

Stone block pavement consists of a series of rectangular or square stone blocks laid over a firm concrete base. Because of the hardness of the stones (granite, basalt etc.), the pavement is highly wear-resistant and can take a very heavy traffic. For this purpose, this specification is often used for paving the roads in the vicinity of harbours, railway yards, warehouses, transport godowns etc.

15.2.1. Advantages and disadvantages

The following are the advantages of the stone block pavements :

(*i*) In view of the hard stones at the road surface, the pavement can withstand very heavy traffic. Iron-tyred traffic can be catered to very effectively.

(*ii*) The pavement requires practically no maintenance.

(*iii*) Life of the pavement is exceptionally long.

(*iv*) Since manual method of laying is involved, it is suitable for labour surplus economy. The dressing of stones also involves manual labour.

Some of the disadvantages of the specification are :

(*i*) Some stones tend to become slippery due to wear.

(*ii*) Unless properly constructed and maintained, the stone block pavement can be unsanitary. Dirt and filth can enter the open joints and cause stench.

15.2.2. Construction practice

The stone block pavement is usually laid on a firm base, about 150 mm thick. A layer of sand 50–63 mm is laid over the base for acting as a cushion for the stone blocks. The blocks are generally of a size 150 mm deep × 125 mm wide × 200–300 mm long. The sides of the blocks must be dressed properly so as to have rectangular faces and right angular corners. Bulges or depressions greater than 6 mm are not generally permitted. The width of a joint should not be more than 12 mm at top, or 25 mm at bottom. The blocks are laid manually in straight parallel courses, with the long dimension perpendicular to the curbs. The joints in one row should break the joints in another by at least 75 mm. After laying, the surface is tamped by a rammer 30–35 kg until the blocks are well seated. The joints are filled by a suitable material like bitumen-sand or cement mortar.

15.3. INTERLOCKING CEMENT CONCRETE BLOCK PAVEMENT

15.3.1. Uses

Precast cement concrete block pavements are used in the following locations :

(*i*) Pedestrian footpaths.

(*ii*) Heavily trafficked locations such as ports and container terminals.

(*iii*) Roads subjected to wear due to studded tyres used in winter months.

(*iv*) Desert roads.

(*v*) Locations subject to settlement.

(*vi*) Locations where utilities are laid underground.

(*vii*) Airfield taxiways and aprons.

15.3.2. Advantages of concrete block paving

The use of concrete block paving technology has been very rapidly growing during the last two or three decades. This is due to the following advantages offered by the technique :

1. It is ideal for locations where frequent repair or remodelling of services and utilities becomes necessary. Thus it is ideal for city streets.
2. The blocks can be made to any shape, pattern and colour, rendering the pavement aesthetically attractive.
3. At locations where settlements are likely to occur, the blocks can be taken out and relaid.

4. The blocks have a high resistance to wear, thus making them suitable for heavy traffic.
5. Since concrete is unaffected by oil spillage, the specification is ideal for parking areas, bus stops, bus stands, bus depots and airfield taxiways and aprons.
6. Blocks are manufactured in the factory at controlled conditions, thus ensuring good quality and speed of construction.

15.3.3. Dimensions and shape

Rectangular, square, hexagonal and proprietary shapes have been used (Fig. 15.2). The dimensions should permit manual handling. Thicknesses in the range of 50–120 mm have been used, depending upon traffic. Chamfered edges are preferred to sharp edges. A compressive strength of about 50 N/mm^2 is generally preferred, though strengths in the range 25–35 N/mm^2 are also used.

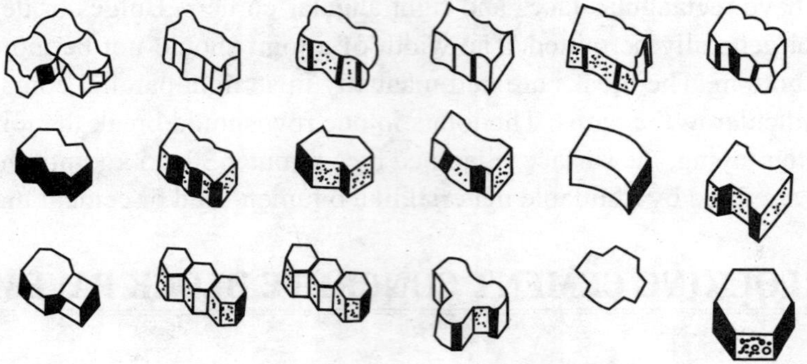

Fig. 15.2. Shapes of concrete blocks.

15.3.4. Construction practice

The blocks are laid over a cushion layer of sand 30–50 mm thick, which in turn is laid over a base layer of lean concrete, granular material, cement-bound material etc. (Fig. 15.3). The sand layer needs to be confined by means of edge supports. After the blocks are laid by hand, they are compacted into the sand by a plate compactor. The open spaces in the joints are filled by fresh sand. The width of the joint itself should not be more than 5 mm.

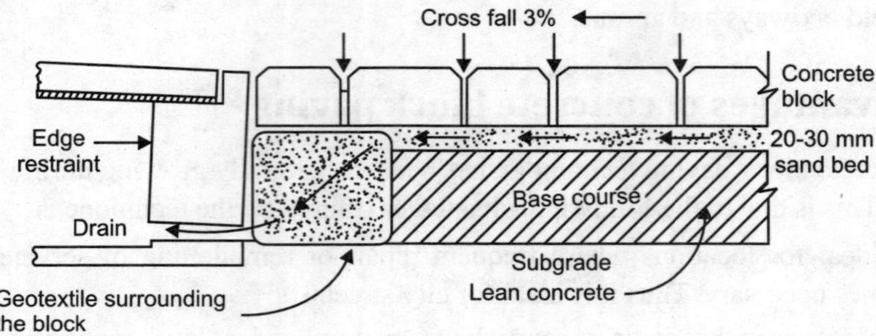

Fig. 15.3. Details of bedding of concrete blocks.

QUESTIONS

1. (*a*) What are the uses of bricks as a pavement layer?
   (*b*) What are the advantages and disadvantages of brick pavements?
   (*c*) What are the requirements for the quality of bricks?
   (*d*) Describe the construction practice for brick pavements.
2. (*a*) What are the advantages and disadvantages of stone block pavements?
   (*b*) Describe the construction practice in laying stone block pavements.
3. (*a*) Describe the use of interlocking cement concrete blocks in pavements.
   (*b*) Draw a sketch to show the details of bedding of concrete blocks.

Chapter 16 HILL ROADS

The general guide lines for the location of hill roads and the selection of their alignment have already been dealt with eanlier. In this chapter cerain distanct of hill road design and problem encountered will be discussed.

16.1. GEOMETRIC FEATURES OF HILLY TERRAIN

Hilly terrain in India is classified under two groups as under:

Percent cross-slope of country	
Mountainous	25–60
Steep	greater than 60

The Himalayan region of the north and the Western Ghats of the south come under this category.

16.1.1. Design speed

Table 16.1 gives the design speeds suggested for adoption in India

Table 16.1: Suggested design speeds (KPH) for hill roads in India

Road class		Mountainous terrain		Steep terrain	
		Minimum	Ruling	Minimum	Ruling
1.	National Highways and State Highways	50	40	40	30
2.	Major District Roads	40	30	30	20
3.	Other District Roads	30	25	25	20
4.	Village Roads	25	20	25	20

16.1.2. Minimum radii of horizontal curves

The basic equation for the design of horizontal curves in a road is:

$$\frac{v^2}{gR} = e + f \quad (20.1)$$

$$R = \frac{V^2}{127(e+f)} \quad (20.2)$$

where v = speed in metres per second

V = speed in Km/hr

e = super elevation in decimals (maximum 0.07 for snow bound areas and 0.10 for areas not affected by snow)

f = coefficient of side friction between tyre and road surface, taken as 0.15

R = Radius in metres

Table 16.2 gives the minimum radii for various classes of roads on the basis of the above formula.

Table 16.2: Minimum radii of horizontal curves for various classes of hill roads

Road class	Minimum Radius (m)							
	Mountainous terrain				Steep terrain			
	Areas not affected by snow		Snow-bound areas		Areas not affected by snow		Snow-bound areas	
	Ruling Min	Absolute Min	Ruling Min	Absolute Min	Ruling Min	Absolute Min	Ruling Min	Absolute Min
National Highways and	50	50	90	60	50	30	60	33
State Highways								
Major District Roads	50	30	60	33	30	14	33	15
Other District Roads	30	20	33	23	20	14	23	15
Village Roads	20	14	23	15	20	14	23	15

16.1.3. Gradients

The gradients recommended for Hill Roads are given in Table 16.3.

Table 16.3: Gradients for hill roads

Classification of gradient	Mountainous terrain and steep terrain more than 3000 m above mean sea level	Steep terrain upto 3000 m above mean sea level
Ruling	1 in 20 (5%)	1 in 16.7 (6%)
Limiting	1 in 16.7 (6%)	1 in 14.3 (7%)
Exceptional	1 in 14.3 (7%)	1 in 12.5 (8%)

Ruling gradients are used as a matter of course. Limiting gradients are used where the adoption of gentler gradients would add enormously to the cost. Exceptional gradients are to be adopted only in very difficult situations and in short lengths not exceeding 100 m.

The cumulative rise/ fall in elevation over 2 Km length shall not exceed 100 m in mountainous terrain and 120 m in steep terrain.

Grades shall be eased at horizontal curves by an amount known as "grade compensation" calculated by the formula:

$$\text{Grade compensation} = \frac{30 + R}{R} \quad \text{(per cent)} \tag{20.3}$$

(Subject to a maximum of $\frac{v^2}{gR}$)

where R = Radius in metres

16.1.4. Width of carriageway

The width of carriageway is dependent on the projected traffic volume in the design year. Table 16.4 gives the width and design service volume.

Table 16.4: Carriageway width and capacity of hill roads

Lane class	Carriageway width (m)	Design service volume PCU/day	
		For low curvature (0-200 degrees/Km)	For high curvature (above 200 degrees/Km)
Single lane	3.75	1,600	1,400
Intermediate lane	5.5	5,200	4,500
Two lane	7	7,000	5,000

The Roadway width including carriageway and shoulders are given in Table 16.5.

Table 16.5: Widths of carriageway and shoulders for hill roads

Road Class	Carriageway width (m)	Roadway width (m)	Shoulders (m)	Total (m)
1.	National Highways and State Highways			
	(*i*) Single lane	3.75	2 × 1.25	6.25
	(*ii*) Double lane	7.00	2 × 0.9	8.8
2.	Major District Roads and Other district roads	3.75	2 × 0.5	4.75
3.	Village Roads	3.00	2 × 0.5	4.00

16.2. HAIR PIN BENDS

In aligning a hill road, it becomes sometimes necessary to attain height at a particular location, without substantial covering of horizontal distance. In such cases, it is customary to introduce a series of hair-pin bends.

The design of such hair-pin bends is a special problem. The geometric standards set forth are not possible of adoption in view of the restricted space. The current Indian standards for the design of hair-pin bends are given below :

1. Minimum design speed — 20 km/hr.
2. Minimum width at apex
   (*i*) National Highways and State Highways
   Double lane — 11.5 m
   Single lane — 9.0 m
   (*ii*) Major District Roads
   Other District Roads — 7.5 m
   (*iii*) Village Roads — 6.5 m
3. Minimum radius of inner curve — 14.0 m
4. Minimum length of transition — 15.0 m
5. Gradient
   Maximum — 1 in 40
   Minimum — 1 in 200
6. Superelevation — 1 in 10

The distance between two consecutive hair-pin bends should be a minimum of 60 m.

The maximum gradient of 1 in 40 (2.5 per cent) stipulated above seems to be rather restrictive and it may not be always possible to adhere to it

A design for a hair-pin bend is shown in Fig. 16.1.

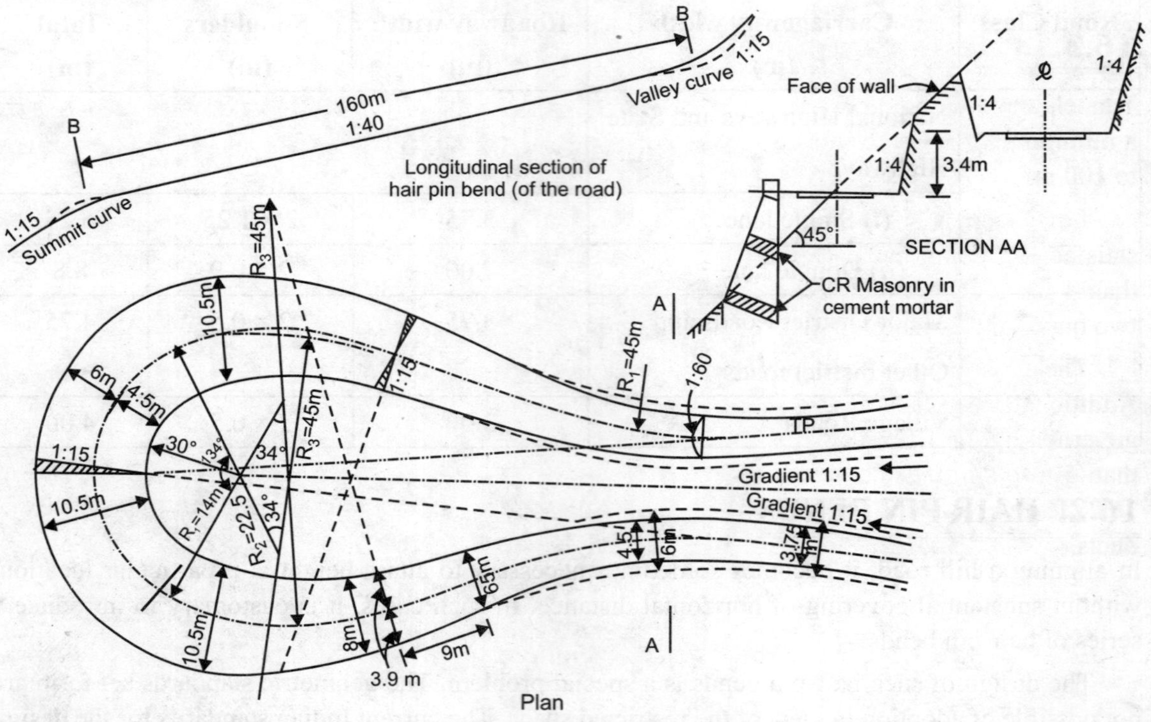

Fig. 16.1. Typical hair pin bend.

16.3. TUNNELS

Tunnels become necessary under the following circumstances :

1. In the development of an alignment along a hill face, a sudden vertical rocky cliff, is encountered.
2. The cost of an open cut through a high rocky outcrop is prohibitively high when compared to a tunnel.
3. When crossing mountain range, the normal development of alignment entails steep climb and descent over a long stretch of road, and it is considered more economical to cut across the mountain range with a tunnel (Example : Jawahar Tunnel on Jammu-Srinagar Road).
4. The normal route development covers an area of avalanches, rock falls and snow-cover and it is possible to avoid these with a tunnel located at a stable geological formation and at a lower altitude.
5. Tunnels are also constructed under channels and rivers.

The disadvantages of a tunnel alternative are :

1. High cost of construction.
2. Problem of maintenance of lighting and ventilation.
3. Problem of clearing seepage water if met with.

Thus, before a tunnel is selected, its economics should be thoroughly studied.

16.3.1. Design standards

Tunnels should preferably be aligned on straight sections. If horizontal curves become unavoidable, a minimum radius of 200 m should be adopted. Only in exceptional cases should this be reduced to 100 m.

Long tunnels should have a good longitudinal gradient for effective drainage; 2 per cent is satisfactory. Long tunnels, which are more than 300 m length, should not have a gradient steeper than 4 per cent. With tunnels less than 300 m long a straight fall gradient should be adopted, but two outward falling gradients are permitted for a tunnel over 300 m long.

The cross-section should permit a minimum carriageway of two lanes, which is 7.0 m. Additional side-walks 0.75 m on either side are usually necessary. The height of the tunnel is governed by the IRC standards for clearances which specify that a vertical clearance of not less than 5 m be provided. A cross-section incorporating the above requirements is indicated in Fig. 16.2. The semicircular portion at the top can be used for housing the inlet and exhaust ventilation ducts.

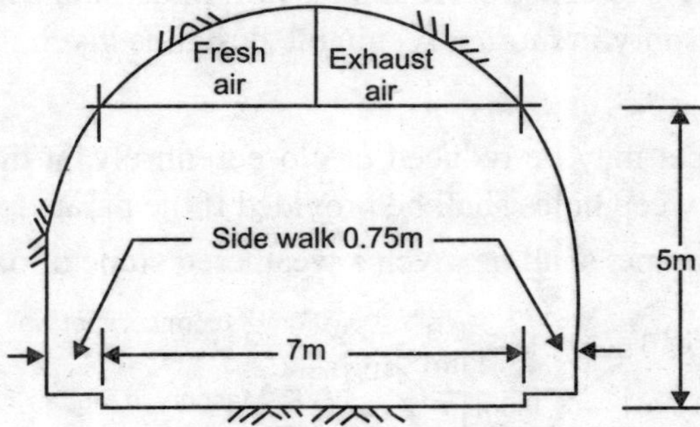

Fig. 16.2. Tunnel cross-section.

16.3.2. Ventilation

Road vehicles passing through a tunnel emit fumes which need to be cleared quickly. Artificial ventilation becomes necessary in tunnels above 400 m length. The ventilation system should be capable of producing a fresh air input of 0.5 m^3 per metre length of the tunnel. The speed of air is generally limited to 5 m per sec.

16.3.3. Illumination

The lighting of tunnels calls for special consideration. The driver entering a tunnel should be able to adjust himself to the artificial lighting conditions inside the tunnel. Similarly, the driver leaving a tunnel should be able to adjust himself to the natural light outside the tunnel. For achieving these requirements, the tunnel is divided into zones of different lighting intensities.

16.4. MISCELLANEOUS STRUCTURES IN HILL ROADS

16.4.1. Types of structures

A variety of structures is met with in hill road design. Of these, the following are common :

1. Retaining walls
2. Parapet walls
3. Breast walls
4. Check walls
5. Catch-pits
6. Snow-chutes.

16.4.2. Retaining walls

Retaining walls are needed to retain the fill portion of the highway cross-section. Due to the warps of the hill faces, it often becomes necessary to take the road partly in filling and partly in cutting, fully in cutting or fully in filling. In such circumstances, retaining walls become necessary.

Since retaining walls are costly to build, a great deal of attention is needed in its design. Obviously, a cement concrete or stone-masonry in mortar specification will work out to be too costly. The usual expedient is to construct retaining wall in dry stone masonry, with horizontal and vertical bands in stone masonry in mortar. A suitable design is given in Fig. 16.3.

Note:

1. The gap in parapet may be reduced or closed entirely, at the discretion of Engineer-in-charge. Suitable weep holes shall be provided if the parapet wall is built without gaps.
2. Top surface of parapet shall be given a weathered slope of one inch.

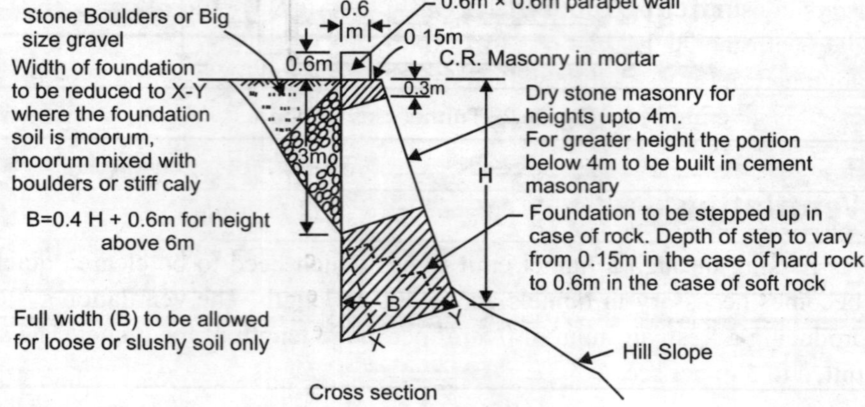

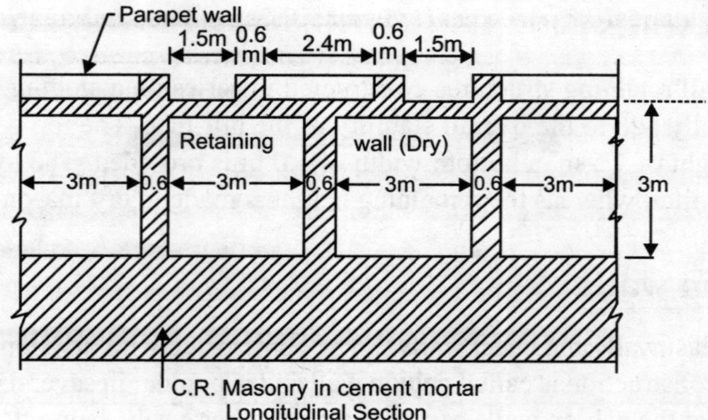

Fig. 16.3. Typical design for a retaining wall.

The backside of the wall should be filled with boulders so that drainage of water is facilitated.

16.4.3. Parapet walls

Parapet walls are needed to give protection, psychologically and physically, to the motorists while travelling on roads with steep valley slopes. They need not be made continuous. A design of a parapet wall popular in India is given in Fig. 16.3. It is seen that the parapet wall is 0.6 m × 0.6 m in section and is in dry stone masonry, except the 0.6 m end portions. The walls are 3.6 m long, with gaps of 1.5 m.

16.4.4. Breast walls

Breast walls are constructed to buttress the uphill slopes of the road cross-section. They should be stout enough to withstand the earth pressure of the soil behind, along with the surcharge, caused by the slope. Obviously, its cross-section will depend upon height and the surcharge angle of the slope. A simple design is given in Fig. 16.4.

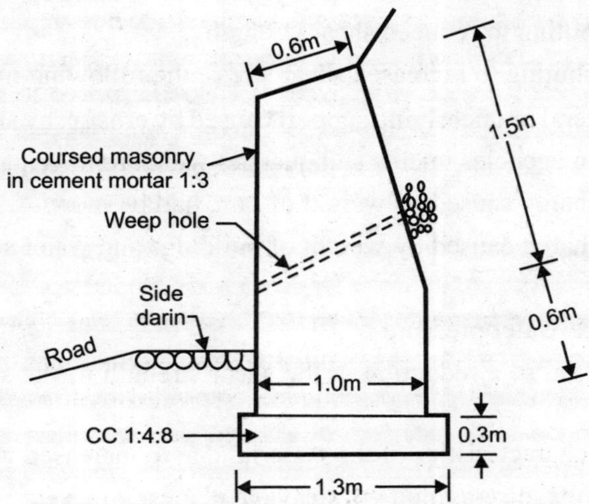

Fig. 16.4. Breast wall.

16.4.5. Check walls

Checkwalls are small retaining structures constructed in series on a sloping hill face to check the slides and to generally add to the overall stability of the hill-face. The top width is generally kept 0.6 m and for a height of 1.5 m, a bottom width of 1.0 m is provided. The top and bottom 0.15 m are set in cement mortar, whereas the remaining height is made in dry masonry.

16.4.6. Gabion walls

Retaining walls, breast walls and checkwalls can be constructed with dry stone masonry encased in wire mesh. Such a construction is called gabion wall and is popular in several developing countries. The advantage is that the gabion walls can adjust themselves easily, being flexible in nature, to the settlement or disturbances that normally take place and hence do not get damaged. Gabion walls are also used as toe-protection walls where the road runs parallel to a stream.

16.5. LANDSLIDES

16.5.1. General statement of the problem

Landslides are a common problem encountered in hill roads, and more so in the Himalayan hill ranges in India. This is because these hill ranges are very young geologically and are subjected to adjustments and settlements. The moment a road formation is cut in these hills, the natural conditions get disturbed and slips, subsidences and land-slides are caused.

16.5.2. Factors contributing to land-slides

The factors contributing to land-slides can be discussed under the following groupings :

1. Factors contributing to increased shear stress
2. Factors contributing to reduced shear strength.

Of the factors contributing to increased shear stress, the following may be noted :

1. Removal of lateral or underlying support caused by erosion by streams, rivers and glaciers.
2. Work of human agencies in cuts and quarries which remove partly the lateral support.
3. Increased surcharge caused by weight of rain, hail or snow.
4. Increased surcharge caused by weight of buildings and water stored in reservoirs.
5. Earthquakes.
6. Increase in pore-water pressure.

Of the factors that lead to a reduction in the shear strength of the soil, the following may be mentioned :

1. Change in the characteristics of the material due to increased water content.
2. Presence of faults, discontinuities, cleavage planes, joints etc.
3. Weakening due to progressive creep.

4. Action of tree roots and burrowing animals.
5. Increased percolation of water caused by clearing of vegetation by grazing animals and deforestation.

16.5.3. Correction of slides

Land-slides can be corrected in a number of ways. The efficacy of each treatment depends upon the particular conditions prevailing at site. Generally, trial and error approach has to be followed. A combination of some of the well-known treatments work in most situations. Some of the remedies are :

1. Improvement in the surface drainage by catchwater drains, side drains and cross-drainage structures .
2. Improvement in the sub-surface drainage by burried French drains.
3. Afforestation.
4. Prevention of grazing of animals by fencing.
5. Grouting, rock-bolting, dowelling.
6. Construction of check-walls, breast-walls and toe walls.
7. Asphalt-mulch treatment of the slopes and growth of vegetation.
8. Jute-netting and wire-netting.
9. Chemical treatment.

16.6. SNOW REMOVAL

16.6.1. Need for snow removal

Snow not only blocks the traffic, but also causes loss of speed. When snow melts, moisture enters the pavement and can cause softening of the subgrade. Icy surfaces, caused by compacted snow or freezing of melted snow, are extremely dangerous from the point of view of traffic safety. Hence, it is necessary to remove snow and ice as expeditiously as possible.

16.6.2. Control of snow drift through highway location

It is possible to control the drift of snow by proper design of the highway. Some of the points that deserve consideration in this regard are :

1. When crossing valleys, it is better to choose locations passing through tall forests so that the trees can shield the road from winds and trap the snow.
2. In rolling country, the leeward side of the natural slopes should be avoided and the road should be located on the windward side or at the crest, if possible.
3. The finished grade-line should be higher than the height of the vegetation and the top of the snow deposition on the adjoining ground.

4. The highway cross-section should have as little obstruction as possible. Even highway appurtenances like curbs, guard-rails, traffic signs etc. can cause accumulation of snow.
5. In special problematic areas, snow fences, located perpendicular to the direction of the wind on the upwind side, are erected. The distance of such fences from the highway should be 10–15 times the height of the fence plus 5 m. Rows of fences can be provided instead of a single row.

16.6.3. Snow clearance

Manual methods of snow clearance are slow and cannot cope up with the needs of high altitude snow removal. Machinery is the only answer. The machinery employed can be of the following types :

1. One-way blade plough
2. Two-way blade plough
3. Rotary ploughs.

Conventional earth-moving equipment such as bull-dozers and graders can also be used.

16.6.4. Salt application

Salt application is commonly used for treatment in icy conditions. The salt brings about melting of ice at temperatures below the normal freezing point *i.e.*, 0°C. Sodium chloride and calcium chloride are commonly used. Normally 90–180 kg of salt is needed per km length of the two-lane road.

QUESTIONS

1. Give the geometric standards of hill roads as per Indian standards.
2. What is a hair pin bend? State the current Indian standards for its design. Draw a typical sketch of a hair pin bend.
3. In what circumstances provision of tunnel becomes necessary? State the design standards.
4. How are land-slides caused in hill roads and how can they be prevented ?
5. Write notes on:
   (*a*) Retaining walls
   (*b*) Breast walls
   (*c*) Control of snow in hills roads.

Chapter

RURAL ROADS (LOW VOLUME ROADS)

17.1 INTRODUCTION

17.1.1. Common terms

Low Volume Roads are a special category of roads subjected to a relatively small number of heavy loads. In view of their relatively low usage and the relatively low level of risk involved, AASHTO recommends a low level of reliability (50%) for design. 'Reliability" may be defined as "the probability that any particular type of distress (or combination of distress manifestations) will remain below or within the permissible level during the design life". In terms of the number of commercial vehicles plying per day, low volume roads are generally considered to be receiving not more than 50 commercial vehicles per day, as in the case of rural roads with gravel surfacing in India. Somewhat lower geometric design standards for low volume roads are generally acceptable in comparison to medium and high traffic intensity highways.

17.1.1.1. Rural roads

The rural roads in India are essentially low volume roads covering the categories of Other District Roads (ODR) and Village Roads (VR). Under the Pradhan Mantri Gram Sadak Yojana (PMGSY), announced by the Govt, of India in the year 2000. It is proposed that all the villages in India be connected by a road by the year 2019. Ideally, all rural roads should be constructed with such surfacings and high-level cross-drainage structures that these are usable by traffic all through the year in all seasons without any interruption.

17.1.1.2. All-weather roads

At cross drainage structures, the duration of overflow or interruption at one stretch does not exceed 12 hours for ODRs and 24 hours for VRs in hilly terrain, and 3 days in the case of roads in plains. The total period of interruption during the year should not exceed. 10 days for ODRs and 15 days for VRs. The pavement should consist of metalling (WBM) or higher type where rainfall is more than 150 cm/year and should atleast be of material better than local soil, such as moorum, gravel, kankar, laterite etc. where the rainfall is less than 150 cm/year.

17.1.1.3. Fair-weather roads

Roads not satisfying the minimum requirement specified above for all-weather roads, these roads should be taken to be in a stage of development to be improved subsequently for conversion into all-weather type.

17.1.1.4. Low cost water crossings

The all-weather character of a low-volume rural road is largely determined by the types of cross-drainage structures provided *i.e.*, the structures for crossing of water barriers like streams along the road alignment. Broadly, there are two types of cross-drainage structures : the submersible types such as fords, paved dips, causeways etc. and the non-submersible or high-level structures like the culverts, bridges etc. Since the cost of cross-drainage structures constitutes a significant percentage of the total cost, such designs of these structures should be worked out which are low-cost but at the same time meet all the technical and functional requirements.

17.1.2. Rural road network planning

17.1.2.1. Basic approach

In order to avoid haphazard, unplanned growth of all-weather roads in the rural areas, it is a matter of necessity that a scientific approach to planning is adopted. Towards an optimal utilisation of available resources, it is necessary to effectively plan all road development activities in a comprehensive and coordinated manner taking the long range aspect of planning into consideration. To ensure that each village has road access to larger centres for marketing, health, education and various other commercial, social and welfare activities, a basic approach suggested that the road network is developed in the form of rooted spanning trees, each tree having its root either at the market centre or at an existing main road. Following steps are involved :

Step 1. Prepare a plan showing the location of various villages, market centres, main roads etc.

Step 2. The market centres, main roads and villages of various population levels are considered to possess the following magnitude of charges, for assessing the force with which traffic is attracted:

- Individual village : Population divided by 100, subject to maximum of 100 units
- Market centres : 100 units
- Main roads : 80 units.

Step 3. Connect the villages closest to the market centres/main roads and then proceed towards the interior depending on the Weight' of the road length. This 'weight' (W) is given by :

$$W = \frac{Q}{D} \tag{17.1}$$

where D = Length of the link option

Q = Total traffic attracting force situated at the market centre/main road/village already connected, as the case may be. Q = 100 or 80, if the village is to be directly connected to the market centre or main road respectively. Where a village under consideration is to be connected to a village already connected by rooted tree, such an 'already connected village' is termed as 'attracting village'. Q is calculated from the following equation :

$$Q = p + \frac{m}{\sqrt{d}} + \sum_{i=1}^{n} \frac{q_i}{r_i} \tag{17.2}$$

where p = self-charge of the attracting village

m = charge situated at the tree root (market centre or main road)

d = distance of the attracting village from tree root

q_i = charge (= population divided by 100) of the ith village situated at a distance r_i from the attracting village.

The procedure is illustrated in Fig. 17.1.

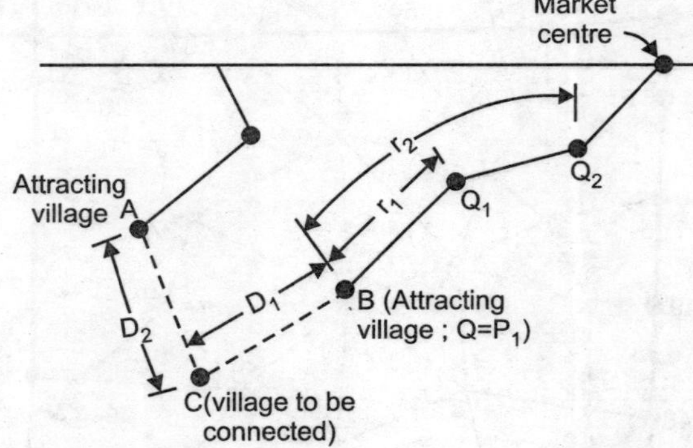

Fig. 17.1. Procedure for determining the road links.

The generation of various rooted minimal spanning trees will start from their roots. For identifying the various tree roots, start connecting the villages which are closest to the market centres/main road by selecting the links of maximum weight (W) calculated from equation (17.1) . In this way, all possible roots are identified. Side by side, the various minimal spanning trees are generated by adding links of maximum weight (W) and connecting village by village. The procedure is continued till all the villages are connected.

Step 4. The system thus developed will be an idealized one without any regard to the existing rural roads. For making use of the existing road network to the maximum extent possible, the generated network is superimposed on the existing one and modifications are made in the planned network by considering the existing roads to have maximum weight. The proposed links need to be examined in the light of actual field conditions.

17.1.2.2. Planning model

Adopting the basic approach of developing the network in form of rooted spanning trees as outlined in para 17.1.2.1, a planning model has been suggested for which a computer programme has been developed in Fortran 77 language which can be worked with a microprocessor/computer. In this model, the 'market centres' are considered to be the attractors of traffic generated from adjacent villages using rural roads for trips to the market centre either directly or via the existing main roads connecting the market centres (Fig. 17.2). A number of alternative networks can be generated, each connecting all villages but using different link options. For each network generated, there will be different levels of construction and travel costs. If construction costs are reduced by selecting links of minimum length, the distances to be travelled in order to reach root nodes may become very long with consequent increases in travel costs. On the other hand, if links providing the shortest distances to root nodes are selected then each node will obtain more or less direct connection to the root nodes, consequently increasing the construction costs. The optimum network is the one for which the total cost *i.e.*, construction plus travel costs are a minimum (Fig. 17.3.).

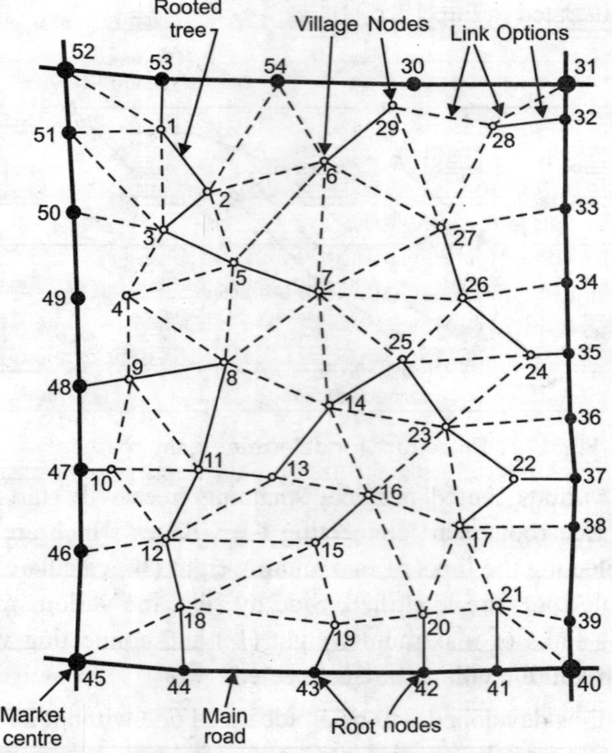

Fig. 17.2. A graph formed by villages, market centres, main roads and rural road link options.

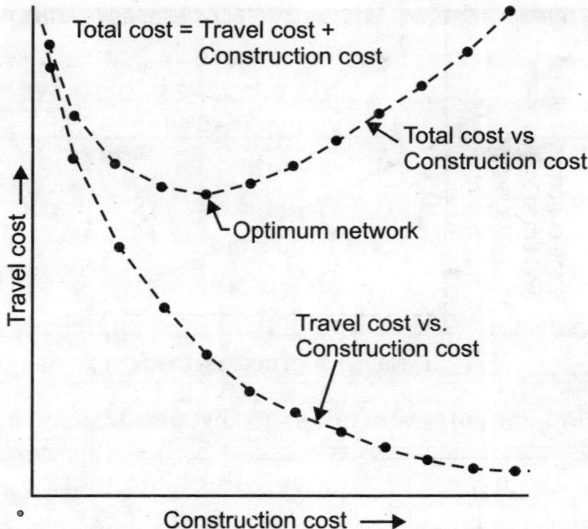

Fig. 17.3. Optimum network from total cost concept.

17.1.3. Socio-economic aspects

It is well known that an adequate network of roads in rural areas contributes very significantly to overall rural development. The agricultural production in general, increases, especially in respect of perishable produce like fruits, vegetables, milk etc. because of faster, more efficient and economical road transportation facilities to the market centre. The farmer gets a higher farm gate price for his crops because of reduced transportation costs to the market centre and benefits immensely by the availability of fertilizer, improved seeds and pesticides and exposure to better agricultural facilities through visits of agricultural extension workers. In the areas of education and health, there is tremendous improvement since the walking distance of children going to school and patients going to health centres reduces very significantly. The villagers are able to visit towns/big villages, market centres/growth centres, block or district headquarters more frequently because of road transportation facilities. What is more, a good road network in rural areas checks the exodus of villagers to towns and cities. Under the aegis of the Indian Roads Congress, a study was undertaken in 9 selected districts of the country with the objective of quantifying the socio-economic impact of roads in rural areas on the overall rural development. Some of the salient findings of this study are shown in Figs. 17.4. to 17.7.

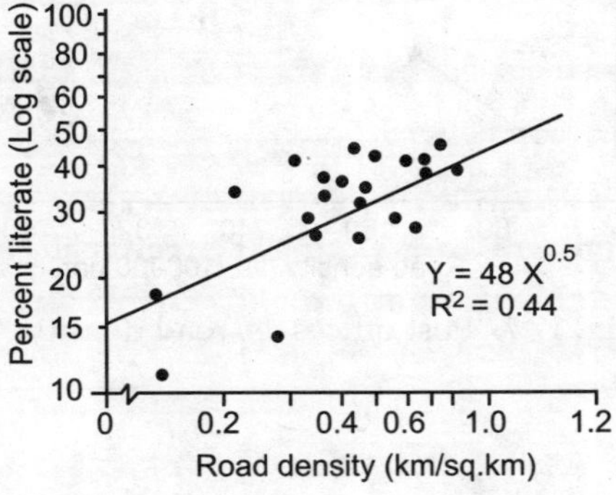

Fig. 17.4. Percent literate vs. road density.

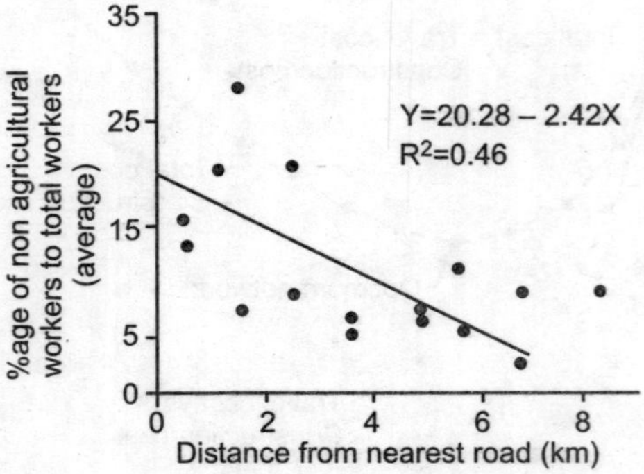

Fig. 17.5. Non-agricultural workers vs distance from nearest road.

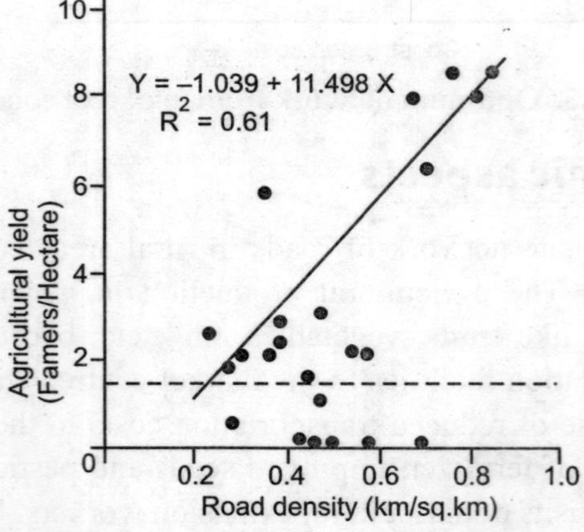

Fig. 17.6. Agricultural yield vs. road density.

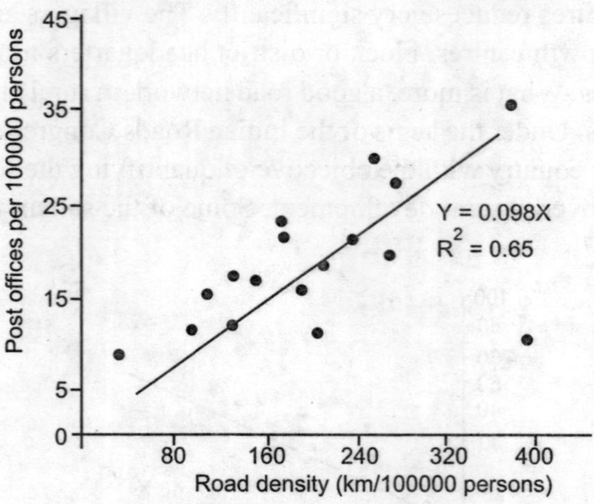

Fig. 17.7. Post offices vs. road density.

17.2. TYPES OF RURAL ROAD PAVEMENTS

17.2.1. Earth Roads

An earth road is the inferior most type of road facility and often marks the first stage in the development of a rural road. In fact, it is in the form of a track which may be a bull-dozed or graded track or may even be hand-made. Such a road is made out of the soil available at the site itself or from borrow areas close to the site. The broad requirements of soils for both the base and wearing courses are that 60 to 80% should be of sand fraction ; the silt content may be 10 to 30% for the base and 5 to 15% for the wearing course, while the clay content should be less than 5% for the base and 10 to 15% for the wearing course. The liquid limit value for the fraction passing 425 micron size should be less than 35% for both the base and wearing courses while the plasticity index should be less than 6% for the base and 5 to 10% for the wearing course. The central crown of the earth road surface should be raised so as to give a camber of 1 in 20 to 1 in 30, steep enough to help the road surface to be free of water and at the same time to avoid erosion by rain water. In stretches, where

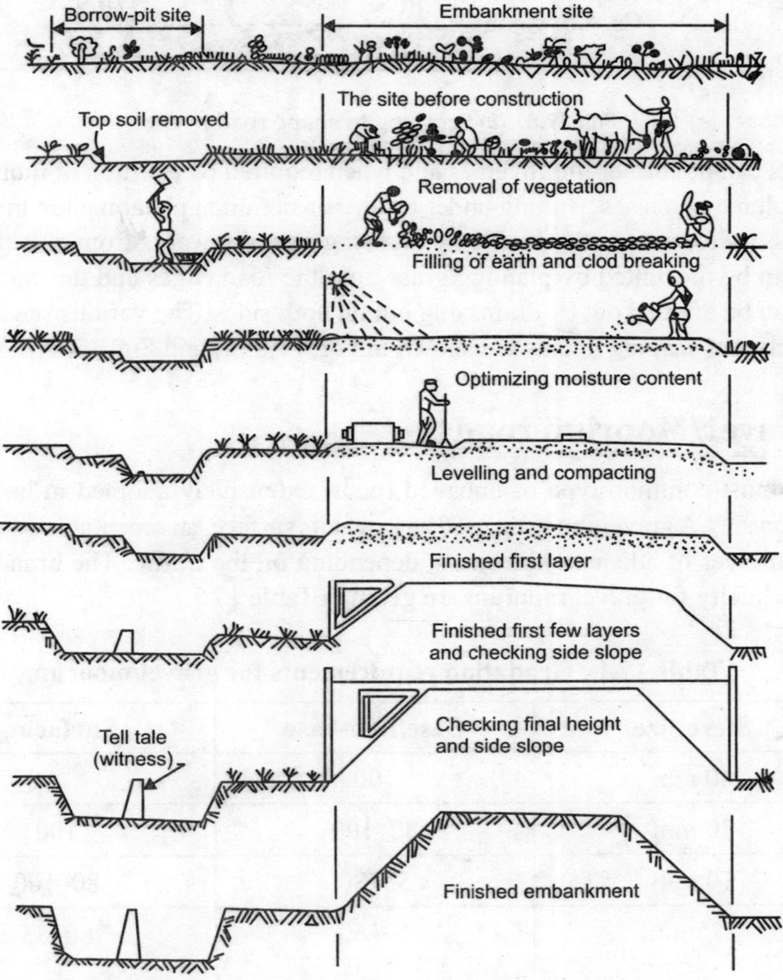

Fig. 17.8. (*a*) Stages of embankment construction.

necessary, local gravel/moorum may have to be provided. The carriageway and roadway widths should normally be not less than 3.5 m and 7 m wide respectively or meet the design requirements.

The three essential pre-requisites for the success of earth roads are: adequate drainage, adequate compaction (upto atleast 30 cm in depth) and periodic maintenance. The borrow pits should ordinarily be outside the land width and any grass/vegetation at the surface should be removed before excavation. The preparation of subgrade should be done carefully, comprising site clearance, excavating and construction of the fill to the desired grade and the shaping of subgrade. The borrowed soil (which may involve mixing of two local soils in the required proportions) should be roller compacted to atleast 95% of the maximum dry density (obtained from standard Proctor compaction test), at optimum moisture conditions in relatively thin layers of 15 cm loose thickness. After drying out for a few days, earth roads may be opened to traffic.

Fig. 17.8. (*b*) Scraping to shape road in cuts.

Earth roads can be further improved as and when required by addition of more moorum/gravel when traffic volume increases. Rutting under traffic is a common phenomenon in earth roads and it becomes necessary to reshape the profile by blending of soil inwards from outside the road edges. Soil erosion can be prevented by planting grass upto the road edges and the water collected from the roadway can be drained out by drains dug out on both sides. The various stages of construction in embankments and shaping in cuts are shown in Fig. 17.8 (*a*) and Fig. 17.8 (*b*) respectively.

17.2.2. Gravel/Moorum roads

These are the most common type of unpaved roads, extensively adopted in both developed and developing countries. A gravel/moorum road has, over its surface, an acceptable quality of compacted gravel/moorum layer of adequate thickness, depending on the traffic. The broad requirements of grading and plasticity for gravel/moorum are given in Table 17.1.

Table 17.1. Gradation requirements for gravel/moorum

Sieve size	Base/Sub-base	Surfacing
40 mm	100	
20 mm	80–100	100
10 mm	55–80	80–100
4.75 mm	40–60	60–85

2.36 mm	30–50	45–70
600 micron	15–30	–
300 micron	–	20–40
75 micron	5–15	10–25

The liquid limit should not be more than 27% for base/sub-base and not more than 35% for surfacing while the Plasticity Index should be less than 6% for base/sub-base and 6 to 9% for surfacing. The gravel/moorum should be hard, tough and angular, free from soft pieces (like shale), etc. and meet the specification on restriction placed on the proportion of flaky and elongated particles.

A gravel/moorum is, however, suitable only if an acceptable quality of gravel/moorum, as outlined above, is easily available at reasonable cost, the traffic is light and although an all-weather facility is required, the needed funds for providing a WBM layer are not available. Generally, when the traffic exceeds about 15 vehicles per day, an earth road has to be provided with a maximum 15 cm compacted thickness of gravel. A cross-slope of 1 in 20 should be provided for ready drainage of surface water.

For purposes of providing a gravel/moorum layer, a box cutting should be made in the compacted subgrade and the soil obtained in the process of making the box cutting used in the side shoulders. The select gravel/moorum is spread to give the required crossfall (1 in 20) inside the trench. Compaction of the gravel/moorum layer is carried out by a power roller commencing from the edges and progressing towards the centre. The compaction should be carried out to the specified dry density (generally not less than 95% of the maximum dry density obtained in a standard compaction test: IS 2720 Part 8). When the traffic is very light, thinner layers of 5 cm compacted thickness of gravel/moorum may be perioded which get compacted by traffic. These thin layers are renewed, whenever the surface irregularities and traffic warrant such renewals. Over a period of time, a hard base is thus built up in layers.

Depending on the traffic and other site conditions, a certain amount of loss of gravel/moorum occurs due to wear caused by traffic, requiring the surface coat to be renewed at relatively short intervals of time. For very heavy traffic, where gravel/moorum renewals become too expensive, upgrading to metalled standards are to be considered. If the natural soil is a heavy clay, introduction of an intermediate layer of a better soil in-between the heavy clay subgrade and the select gravel/moorum layer surfacing will be necessary. Where the road is to be built on soft ground which is swampy/marshy, a 20 cm compacted thickness of sand or stones over fascines/geotextile should be provided before placing the 15 cm thick select gravel/moorum layer.

Two of the main disadvantages of gravel/moorum roads are the penetration of water through the surfacing in wet weather and the tendency to become dusty in dry weather. To overcome these two shortcomings, it is necessary to pay due attention to the surface drainage by adopting a suitable camber of 1 in 20 and providing adequately designed side drains) and resorting to dust palliatives, as may be warranted by traffic and other site conditions.

17.2.3. Metalled roads

The low volume rural roads are generally built up in one or more stages, as the traffic increases or more funds become available. The road can be built first upto the sub-base level, deferring the metalling to the second stage or the road can be constructed upto the base course level, leaving the bituminous surfacing for the next stage, after the traffic has built up.

Metalling comprises the placing of a Water Bound Macadam (WBM) course, consisting of broken aggregates conforming to a specified grading ; compacted with a power roller, the grading and quantity requirements being so worked out as to develop interlock between the aggregate pieces and bring about bonding with stone dust and water.

Normally, WBM requires the use of hard stone aggregates when used in higher types of road facilities, but for low volume rural roads, if hard stone is not easily available, relatively inferior aggregates that are locally available may be used. However, typically, the following physical requirements are set forth.

(*a*) Sub-base Course

Los Angeles Abrasion Value : Max. 60%

Or

Aggregate Impact Value : Max. 50%

(*b*) Base Course

Los Angeles Abrasion Value : Max. 50%

Or

Aggregate Impact Value : Max. 40%

Flakiness Index : Max. 30%

If the sub-base, generally WBM (over size) with stone aggregate size 90-40 mm (Grading I) is used, maximum 100 mm compacted thickness in one layer can be laid. For the base course, generally WBM with stone aggregate size 63-40 mm size range is used and maximum 75 mm compacted thickness in one layer can be laid.

17.2.4. Bituminous surface treated roads

As the traffic increases on a gravel/moorum road or metalled read, there is an increased amount of wear due to abrasion by traffic causing the individual pieces of surface material to gradually get dislodged. Also, due to the loss of binding material (which gets 'sucked up' by pneumatic tyred traffic), the pavement surface deteriorates rapidly. Additionally, the entry of rain water through the surface and base into the subgrade, softens the subgrade leading to the weakening of the pavement foundation, consequently worsening the pavement condition. With such traffic and/or climatic conditions, an unpaved road surface deteriorates at such a fast rate that very frequent renewals become necessary to keep the surface in a satisfactory condition, It is believed that if the number of high speed pneumatic tyred vehicles exceed 50 per day and/or the average annual rainfall in the region is more than 150 cm, a bituminous surface treatment is warranted.

Although providing a bituminous surface treatment is expensive (in certain situations, can cost nearly as much as the rest of the pavement crust), a stage comes when excessive maintenance costs overtake the cost of bituminous treatment.

The most common types of bituminous surface treatment for low volume roads are : surface dressing (single or double-coat) or 20 mm thick open-graded pre-mixed bituminous carpet. The choice of type of treatment depends on the type and volume of traffic, availability of funds and needed plant and equipment for construction.

17.2.5. Roller compacted cement concrete paving

In lieu of a bituminous surfacing, the alternative of providing roller compacted cement concrete pavement can be considered for low volume rural roads, in view of ready availability of cement and also simple equipment requirements for its construction. Besides, in comparison to bituminous construction works, the adverse effects on the environment in respect of cement concrete construction works are much less.

The basic mix design procedure for roller compacted concrete is the same as for conventional concrete, except for the lower water-cement ratio and higher density in case of roller compacted concrete. The maximum size of the aggregate is generally limited to 20 mm, since larger sizes give rise to the segregation problem ; also, lower sizes increase costs. Flyash as partial replacement of cement is increasingly being considered, with consequent reduction in material costs.

During construction the main difference, in comparison to conventional cement concrete, is the tendency for segregation because of the low workability mix. Due to the low water content, speed of placement is an important consideration ; however, adoption of bituminous paving techniques in roller compacted cement concrete brings in good speed. Wherever needed, the joints can be made by sawing only.

17.2.6. Brick pavements

The option of adopting brick pavements becomes attractive in situations where road aggregates of acceptable quality are not available within reasonable leads and where quality bricks are available from kilns, close to the site of construction. The bricks to be used in the pavement should be hard and tough enough to resist abrasion and impact of the traffic. The crushing strength (on the flat) should not be less than 60 kg/cm^2 and water absorption after 48 hours of immersion in water should not be more than 1/500 of its dry weight.

While laying a brick pavement, the subgrade or bottom of excavation over which the bricks are to be laid, should be well compacted and prepared to the desired profiles. Especially where the subgrade is soft, a cushion of sand, generally 7.5 cm to 15 cm in thickness, should be provided before laying the bricks, to function both as a drainage layer and as a good foundation. The bricks should be laid flat, on edge or in two layers, depending on the thickness desired. Edging of the pavement is provided with bricks on end, parallel to the road. If a second layer is to be laid, a cushion of sand is provided between the two layers and a herringbone pattern adopted. The joints between the bricks should be blinded with sand.

17.3. USE OF LOW-GRADE AGGREGATES AND STABILIZED SOIL

17.3.1. General

The single most important aspect of developing low volume roads in rural areas is to aim at providing basic access at minimum cost covering as many villages as possible, within the allocated funds. To keep the developmental costs to a minimum, yet at appropriate levels of serviceability, choice of materials is an important consideration. If the same quality of materials is insisted upon, as specified for higher road facilities, the cost of materials and consequently the high initial cost of construction will defeat the very objective of adopting low cost techniques for low volume rural roads. Experience all over the world has shown that it is possible to incorporate locally available materials at low cost and yet provide a road facility at an appropriate, acceptable level of serviceability. The local materials may be incorporated as such, as they occur in nature, or may be suitably processed at low cost to render them suitable for use in road works. However, caution has to be exercised in a judicious selection of materials otherwise incorporating any local material without a proper evaluation can have disastrous consequences. The first essential requirement towards a scientific and judicious selection is to carry out a rigorous materials survey, obtain representative samples of various materials and obtain the necessary data for evaluating their suitability for use as such or after processing.

17.3.2. Survey of locally available materials

The locally available materials that can be incorporated in pavement design include various types of soils encountered; sands from streams/river beds and other sources ; moorums which may even be available at shallow depths below the ground level or from other quarries ; gravels, well spoils, soft aggregates like laterite, kankar, dhandla, river sand-gravel mixes etc. While locations of the PWD/ Zilla Parishad approved quarries of stone metal can be obtained from the district headquarters, the quarries of locally available materials like the ones mentioned above are generally not known, but local enquiries from villagers etc. can help a great deal in this regard. Samples of other locally available materials like lime that could possibly be used as soil stabilizers should also be collected.

It is to be emphasized here that special care has to be exercised in collection of samples of these locally available materials since they can be quite variable in their engineering properties. So much so, the locally available moorums and gravels from the same source can give widely different strength values. This naturally has significant implications in the construction and performance of roads. The site quality control of materials can be exercised meaningfully only if sampling and testing of these locally available materials has been carefully done and the design values arrived at judiciously based on the laboratory strength values. Besides the variable quality of materials, another important aspect is that of the quantities available from each source ; these should be carefully estimated and invariably taken note of, during the field surveys.

In India, a comprehensive study on the locally available materials has recently been undertaken and related maps of all locally available materials in some parts are now available. These maps can be made use of in locating locally available materials for any road project.

17.3.3. Low-grade aggregates

Although it is widely accepted that for use in low volume roads, the same quality of materials as specified for higher types of road materials need not be insisted upon, caution must be exercised in ensuring that the low-grade aggregates incorporated atleast meet the minimum requirements.

Low-grade aggregates may be defined as those which lose strength by more than 15% on wetting, as determined by the Aggregate Impact Value test on the aggregate in the dry and in wet states.

For the use of low-grade aggregates in the sub-base, base and surface courses, the following requirements of wet Aggregate Impact Value should be satisfied :

Pavement layer	**Wet aggregate impact value**
Sub-base course	Max. 50%
Base course	Max. 40%
Surface course	Max. 30%.

These low grade aggregates should also be subjected to soaked CBR value and a minimum volume of 20 should be ensured.

If the low-grade aggregates meet the above requirements and are available within economic leads in comparison to conventional hard aggregates, their use in low volume rural road pavements should be maximised in order to cut down costs. These may be incorporated in the following ways :

- as coarse aggregate in WBM construction
- as graded granular material if their particle size distribution lies within the broad limits given in Table 17.2.

Table 17.2. Gradation of low-grade aggregates

Sieve Size	Percent passing by weight		
	Grading 1	Grading 2	Grading 3
90 mm	100	100	100
63 mm	90–100	90–100	90–100
4.75 mm	35–70	40–90	50–100
75 micron	0–20	0–25	0–30

The plasticity index (PI) of the material should be normally less than 6 when used in the sub-base or base course and between 6 and 9 when used as surfacing. However, depending on the climate, the liquid limit and plasticity index requirements for use in the wearing course are as in Table 17.3.

Table 17.3. Liquid limit and plasticity index of low-grade aggregates

Climate	Liquid Limit not to exceed	Plasticity index (range)
(*i*) Moist temperate and wet tropical	35	6–9
(*ii*) Seasonal wet tropical	40	6–15
(*iii*) Arid	55	15–30

17.3.4. Soil-aggregate mixes

These may occur as such in their natural form (*e.g.* soil-gravel/moorum) or may be specially proportioned and blended with suitable aggregate fractions. Acceptance of these mixtures is based on the grading and plasticity requirements given in Table 17.4.

Table 17.4. Gradation requirements of soil-aggregate mixes

Sieve Size	Percent passing by weight				
	← Sub-base and base courses →			←Sub-base/Base/Surfacing→	
90 mm	100				
40 mm	80–100	100			
20 mm	60–80	80–100	100		
10 mm	45–65	55–80	80–100	100	
4.75 mm	30–50	40–60	50–75	80–100	100
2.36 mm	–	35–50	35–60	50–80	80–100
1.18 mm	–	–	–	40–65	50–80
600 micron	10–30	15–30	15–35	–	30–60
300 micron	–	–		20–40	20–45
75 micron	5–15	5–15	5–15	10–25	10–20
CBR	|←Not less than 30 →			|←Not less than 20→	
Plasticity Index	|←Less than 6 →			|←6–9→	

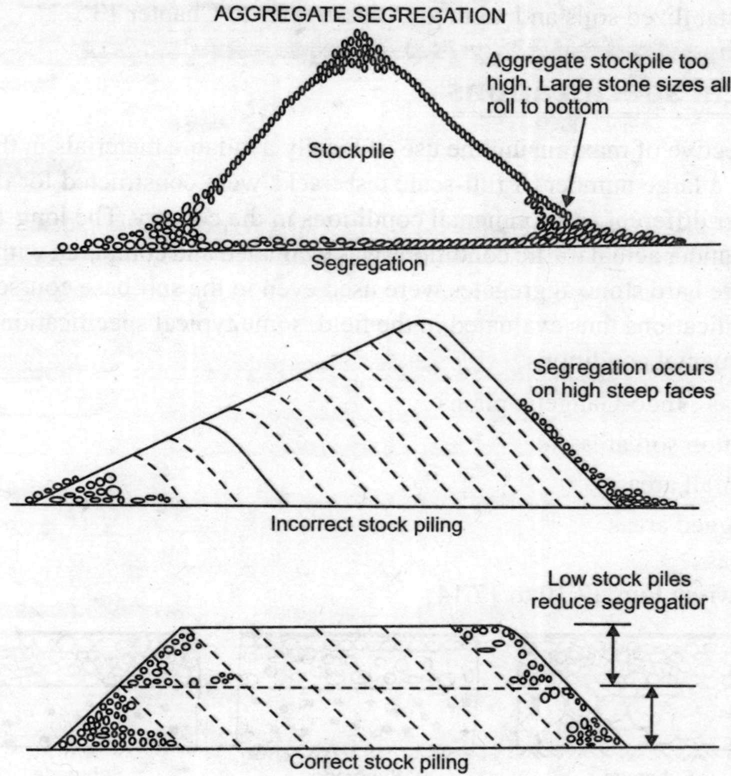

Fig. 17.9. Segregation in stockpiling of soil-aggregate mixes and its correction

For use in the wearing course, the requirements of liquid limit and plasticity index may be relaxed, depending on climatic conditions as given in para 17.3.3.

In order to avoid the problem of segregation in soil-aggregate mixes, correct stockpiling is necessary as shown in Fig. 17.9.

17.3.5. Stabilized soils

If the locally available soils, gravels and aggregates as they occur in their natural form, do not meet the various requirements of grading, plasticity, strength and durability, these can be suitably processed to render them suitable for use in various layers.

The simplest and most commonly adopted method of processing is the mechanical stabilization technique to meet the grading and plasticity requirements. This technique is widely used for sub-base courses, where the local sandy soils may be mixed with clay and clay soils with sand. For such mixtures, the sand content should normally be not less than 50% and soaked CBR value should be evaluated for pavement design.

Clayey soils and moorums with overly plastic fines can be effectively and economically modified/stabilized with small percentages of lime, significantly bringing down the plasticity index and making them more workable. Granular soils can be more effectively stabilized by small percentages of cement.

For details of stabilized soils and methods, please refer to Chapter 13.

17.3.6. Typical specifications

With the main objective of maximising the use of locally available materials in the construction of low volume roads, a large number of full-scale test tracks were constructed for different subgrade soil types and under different environmental conditions in the country. The long-term performance of these test tracks under actual traffic conditions was evaluated and compared with the conventional specifications where hard stone aggregates were used even in the sub-base courses. With a number of alternative specifications thus evaluated in the field, some typical specifications emerged for the following environmental conditions:

- Plain areas ; Indo-Gangetic plains
- Black cotton soil areas
- High rainfall areas
- Water-logged areas
- Hilly areas

These are shown in Figs 17.10 to 17.14

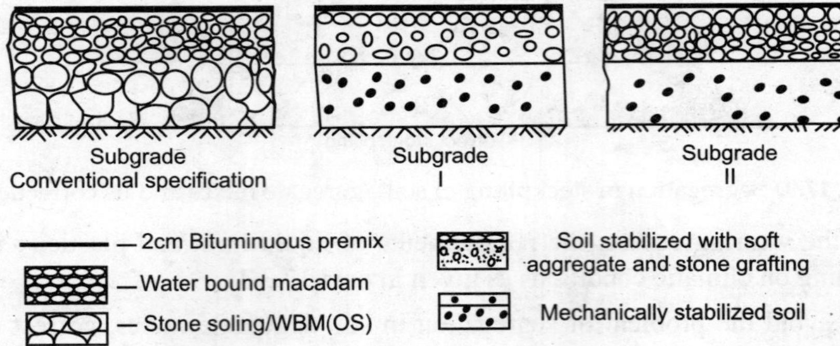

Fig. 17.10. Alternative specifications for plain areas : Indo-Gangetic plain

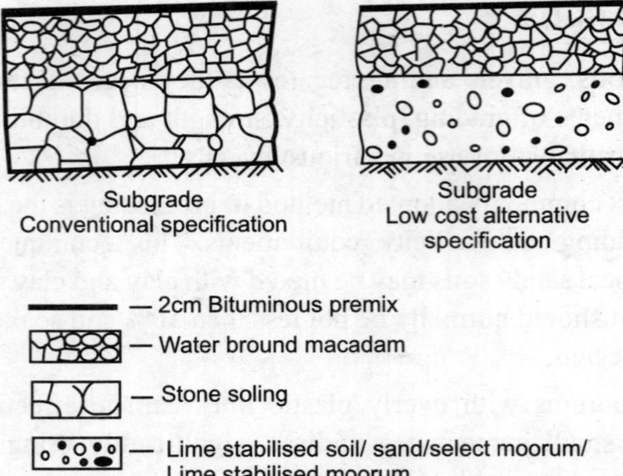

Fig. 17.11. Alternative specifications for black cotton soil areas.

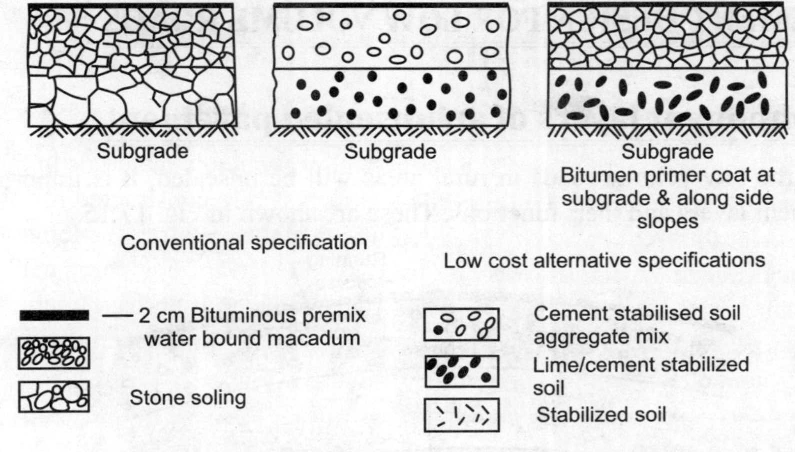

Fig. 17.12. Alternative specifications for high rainfall areas.

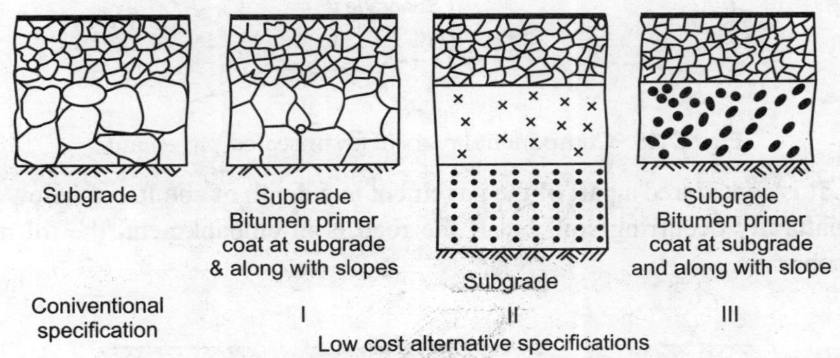

The thickness will vary with subgrade strength, traffic composition and volume, etc.

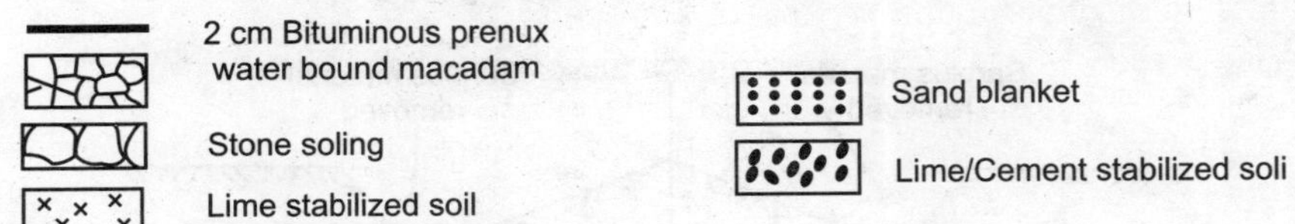

Fig 17.13. Alternative specifications for water logged areas.

Fig. 17.14. Alternative specifications for hilly areas.

17.4. PAVEMENT DESIGN FOR LOW VOLUME ROADS

17.4.1. Component layers of an unsealed pavement

Since many of the low-volume roads in rural areas will be unsealed, it is important to note the various component layers and their functions. These are shown in Fig. 17.15.

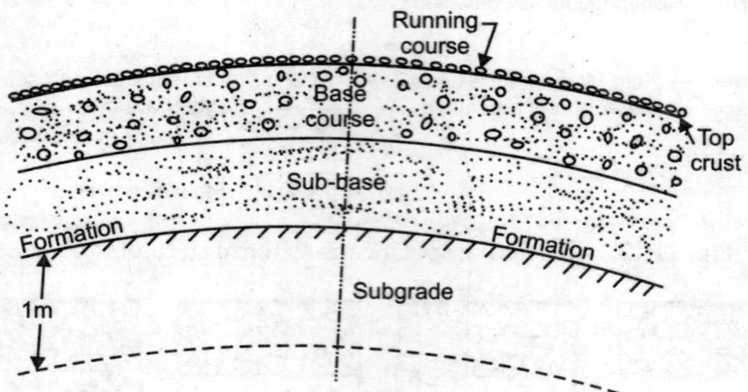

Fig. 17.15. Component layers of an unsealed pavement.

Subgrade. It is considered a part of the pavement to a depth of about 1 m below the formation. It consists of naturally occurring soil, but if the road is in embankment, the fill material would constitute the subgrade.

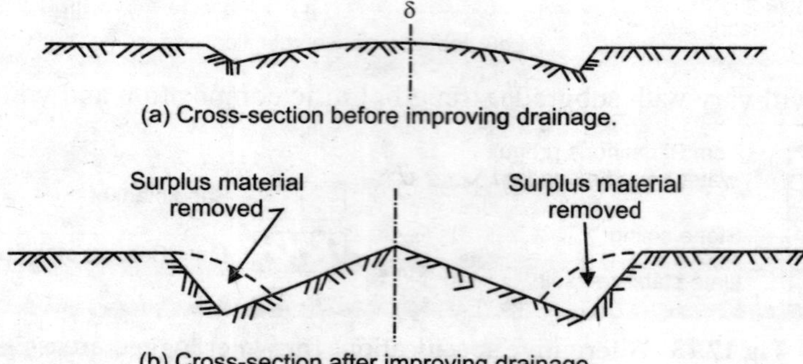

Fig. 17.16. Improved subgrade.

It may have to be improved by compaction ; lime or cement stabilization, excavating poor material and replacing with select material, all adequately compacted ; or by extra drainage as shown in Fig. 17.16.

Sub-base. It is an optional layer, but being inferior in material quality to base course material, incorporating a sub-base helps in cost reduction.

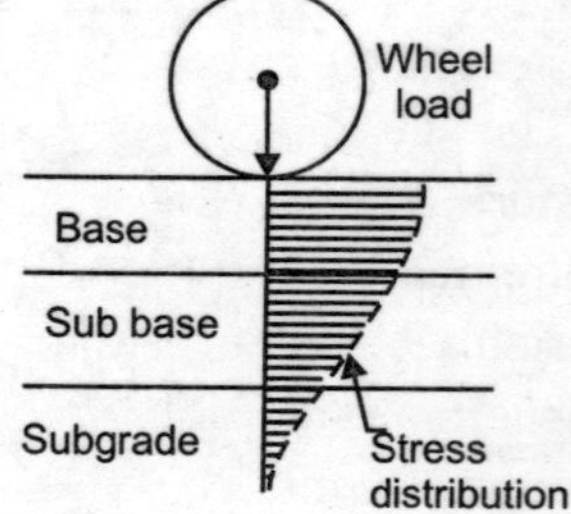

Fig. 17.17 Stress variations with depth.

Base course and top crust. Since max. stress intensity is at the base level, highest quality of material is incorporated. On the surface of base course is put a water-proof top crust, to drain away the rain water with the help of crossfall.

Running course is a thin layer (12 mm thick) of broken stone or gravel, which serves as a protection against erosion by climate or vehicle loadings. The stress distribution within the layered system under a wheel load is shown in Fig. 17.17.

17.4.2. CBR method of design

In the CBR method of flexible pavement design, traffic is considered in terms of commercial vehicles (pneumatic tyred) like trucks, buses etc. per day in the design year. However, in the rural areas of India, there could be a sizable number of solid-wheeled carts on the roads, especially in the harvesting season, which can cause considerable damage to the pavement. While the choice of CBR design curves A, B and C (Fig. 17.18) would cover the range of commercial vehicles on roads in the rural areas, this design method does not take into account the solid-wheeled cart traffic.

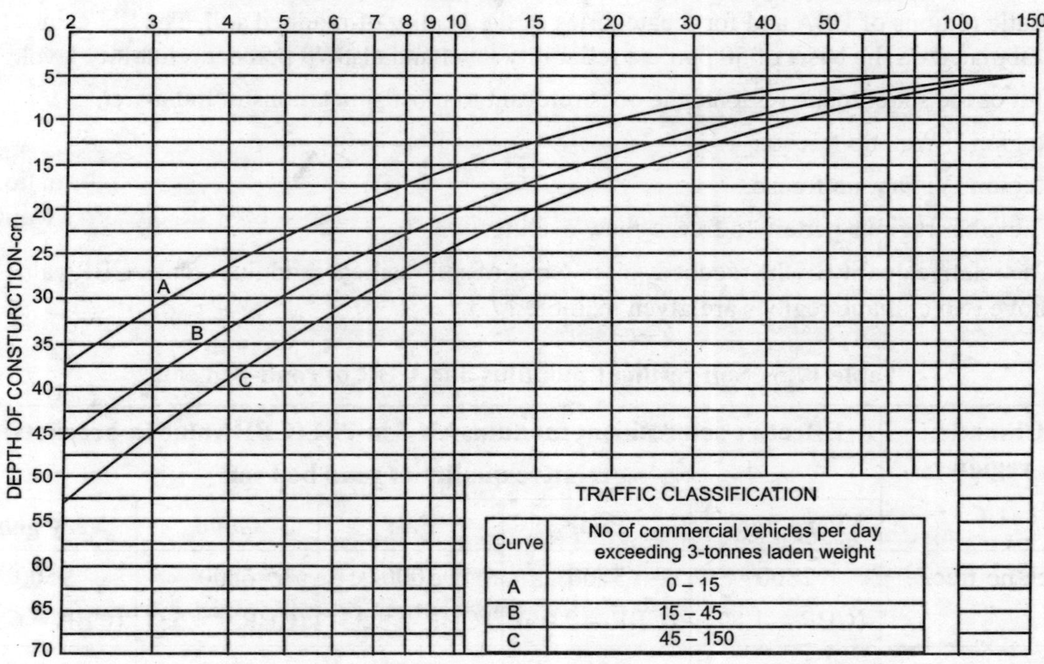

Fig. 17.18. CBR curves for flexible pavement design.

The subgrade strength is determined in terms of the CBR value. The compacted density and moisture content of specimens very significantly affect the CBR value. For rural roads, it is a requirement that upto a depth of atleast 300 mm below the formation, the subgrade soil should invariably be compacted to atleast 95% of the maximum dry density obtained in a standard Proctor compaction test. For carrying out the CBR tests for subgrade strength, therefore, the dry density of specimens should be 95% of max. Proctor density.

The moisture content of the specimens at which the CBR test value is to be determined depends on a number of local environmental factors such as depth of water table and its fluctuations, rainfall, the permeability of soil, drainage conditions and the extent of water proofing by the surfacing. In respect of rural roads, which usually have thin pavements comprising granular materials, generally unsealed (and even when sealed, only a thin bituminous treatment is provided which by no standards can be considered completely waterproof), with unsealed shoulders, box type construction etc., it is reasonable to assume that in most cases, the subgrade will be nearly saturated atleast during the rainy season, except for roads in arid regions. The recommended general practice is therefore to cater to the worst condition of saturation by soaking the soil specimens under water for a period of 4 days before testing for the CBR value.

Lastly, it is to be recognized that in the use of CBR design curves, there is an in-built resulting minimum level of serviceability which may be at a higher level than what may be considered as the acceptable minimum for the low volume roads in rural areas.

17.4.3. AASHTO design method for low volume roads

The AASHTO design catalógues for low-volume roads give the Structural Number (SN) values for 6 climatic regions of USA and for 5 categories of the quality of roadbed soil. The SN values have been tabulated on the basis of 50 percent reliability level and also 75 percent reliability level.

Out of the six climatic regions, the ones relevant to most situatations in India are :

Region I: Wet, no freeze.

Region IV: Dry, no freeze.

The other regions pertain to freeze-thaw cycling,

The relative quality of the roadbed soil in terms of soil resilient modulus values/CBR values for the above two climatic regions are given in Table 17.5.

Table 17.5. Soil resilient modulus and CBR of road-bed soil

Climatic region	Effective soil resilient modulus 'M_R' in PSI (CBR value in brackets)* relative quality of road bed soil				
	Very poor	Poor	Fair	Good	Very good
Wet, no freeze	2800 (CBR = 1.87)	3700 (CBR = 2.46)	5000 (CBR = 3.33)	6800 (CBR = 4.53)	9500 (CBR = 6.33)
Dry, no freeze	3200 (CBR = 2.13)	4100 (CBR = 2.73)	5600 (CBR =3.73)	7900 (CBR = 5.27)	11,700 (CBR = 7.80)

*Using the relationship M_R (PSI) = 1500 × CBR.

The range of SN values for each condition is based on a specified range of 80 kN equivalent single axle load applications at each of the following 3 traffic levels :

High	7,00,000 to 1,000,000
Medium	4,00,000 to 6,00,000
Low	50,000 to 3,00,000

17.4.4. Design method proposed by IRC : SP : 72-2007

The Indian Roads Congress have come out in the year 2007 with guidelines for the design of Low Volume Rural Roads The salient features of the guidelines are :

1. A design life of 10 years is adopted.
2. The number of repetitions of 80 kN standard axle loads over the design life is estimated. The damaging effect of indivisual axles is determined from the fourth power rule:

$$\text{Damaging effect} = \left(\frac{W}{W_S}\right)^4$$

where W = Single axle load in kN

W_s = Standard axle load of 80 kN

3. Based on the Cumulative Equivalent Standard Axle Load (ESAL) repetitions over the design life, traffic is divided into seven categories as under :

Traffic category	Cumulative ESAL applications
T_1	10,000 – 30,000
T_2	30,000 – 60,000
T_3	60,000 – 1,00,000
T_4	1,00,000 – 2,00,000
T_5	2,00,000 – 3,00,000
T_6	3,00,000 – 6,00,000
T_7	6,00,000 – 1,000,000

4. The subgrade strength is evaluated in terms of CBR value. Five subgrade strength classes are defined as under :

Quality of subgrade	Class	CBR
Very poor	S_1	2
Poor	S_2	3-4
Fair	S_3	5-6
Good	S_4	7-9
Very good	S_5	10-15

5. The Pavement design is obtained from the catalogue given in Fig. 17.19.

17.5. INTERMEDIATE TECHNOLOGIES IN RURAL ROAD CONSTRUCTION AND MAINTENANCE

17.5.1. Relevance of intermediate technologies

One of the biggest bottlenecks in the development of rural road network at appropriate quality level has been the lack of availability of the needed equipment. There has been a tendency to try to deploy the same high capacity equipment/plant as is being used for the higher road categories..

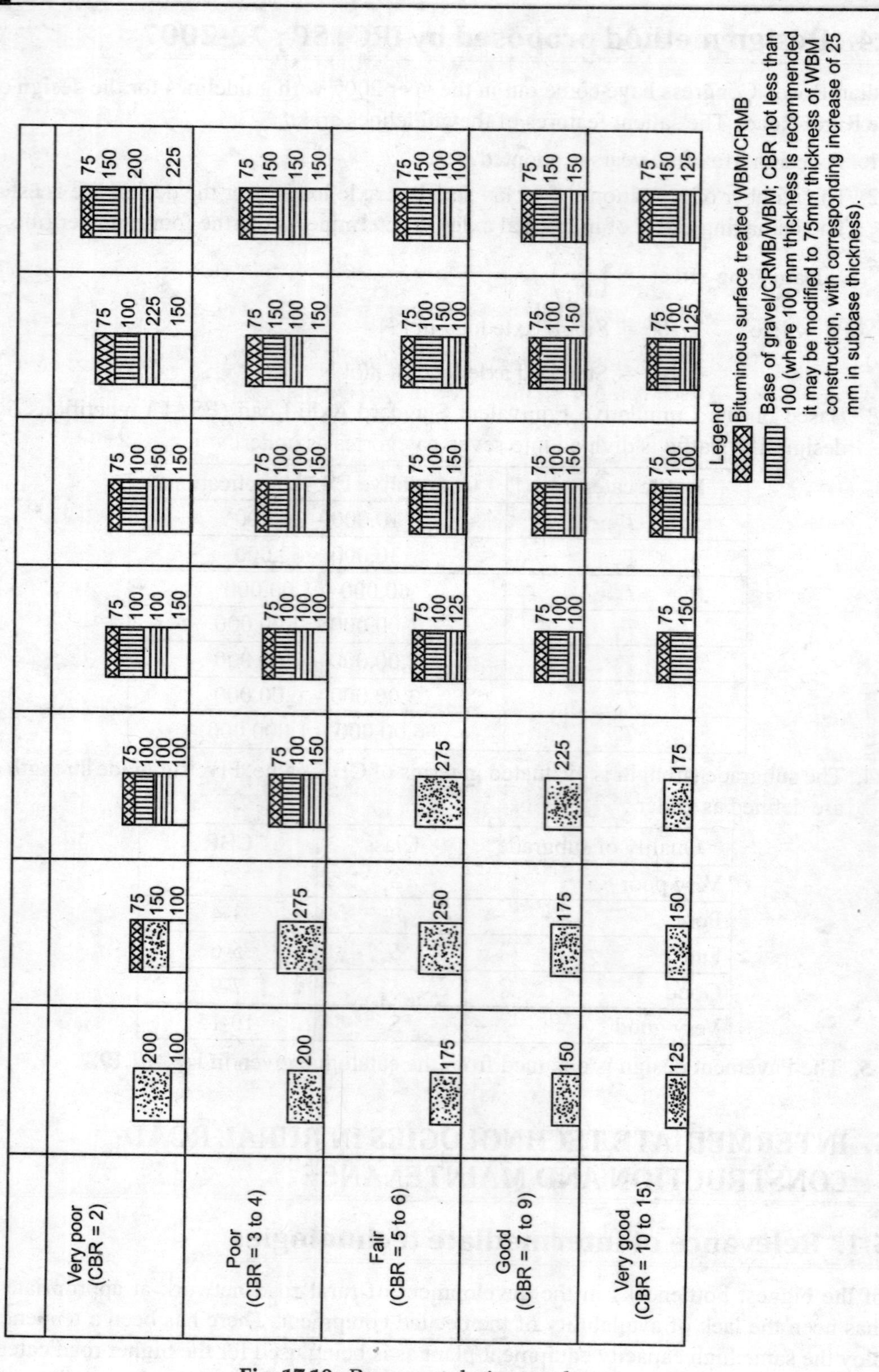

Very poor (CBR = 2)	200 / 100	75 / 150 / 100	75 / 100 / 100 / 100	75 / 100 / 100 / 150	75 / 100 / 150 / 150	75 / 100 / 225 / 150	75 / 150 / 200 / 225
Poor (CBR = 3 to 4)	200	275	75 / 100 / 150	75 / 100 / 100 / 100	75 / 100 / 100 / 150	75 / 150 / 100 / 150	75 / 150 / 150 / 150
Fair (CBR = 5 to 6)	175	250	275	75 / 100 / 125	75 / 100 / 150	75 / 100 / 150 / 100	75 / 150 / 100 / 100
Good (CBR = 7 to 9)	150	175	225	75 / 100 / 100	75 / 100 / 150	75 / 100 / 150	75 / 150 / 150
Very good (CBR = 10 to 15)	125	150	175	75 / 150	75 / 100 / 100	75 / 100 / 125	75 / 150 / 125

Legend

Bituminous surface treated WBM/CRMB

Base of gravel/CRMB/WBM; CBR not less than 100 (where 100 mm thickness is recommended it may be modified to 75mm thickness of WBM construction, with corresponding increase of 25 mm in subbase thickness).

Fig. 17.19. Pavement design catalogue

However, several of the sites may be inaccessible to the high capacity equipment and even where such equipment is available and the sites are accessible, their use turns out to be uneconomical ; proper maintenance of the equipment, when such equipment goes out of order presents an additional problem. All these factors eventually result in unsatisfactory situations where the pavement layers remain under compacted, mixing of materials with water and other admixtures remains unsatisfactory, processing of aggregates and other materials cannot be carried out satisfactorily etc.

It is a foregone conclusion that based on past experiences in India and several other developing countries, the technologies best suited to rural road development, work out to be somewhere in between the labour intensive and equipment intensive technologies. These are known as intermediate technologies ; their development can be brought about in the following two ways :

(*a*) Starting from the equipment-intensive end, the size and production capacity of the equipment (as used for higher road facilities) can be reduced in steps (thus reducing their capital cost) and for any given task, the labour content can be increased, till the most suitable blend is attained.

(*b*) Starting from the labour-intensive end, the labour content can be decreased in steps substituting capital equipment for labour so that productivity both in magnitude and quality improves, till a happy blend between the two is achieved.

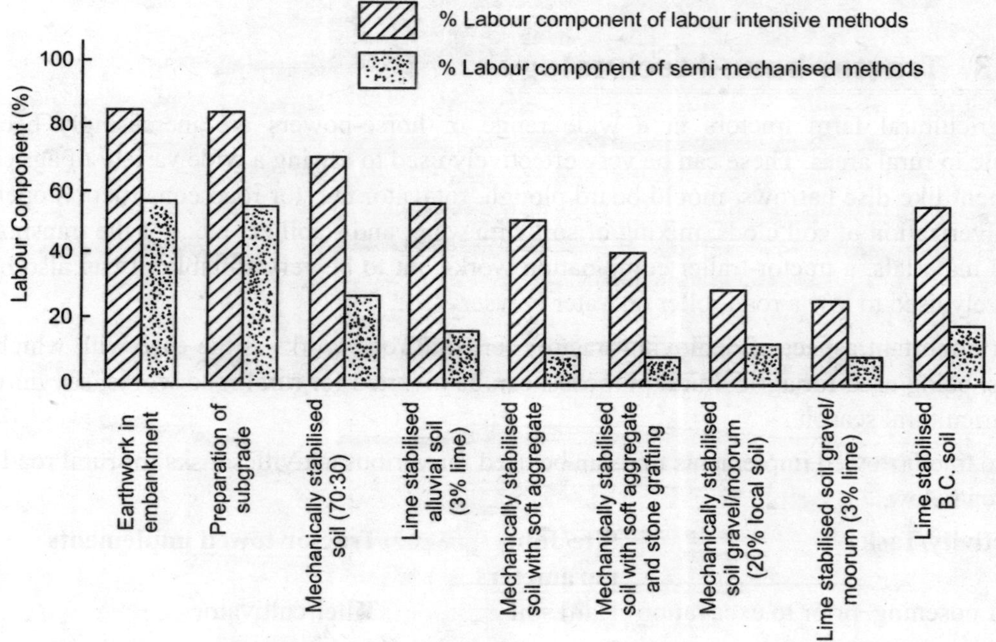

Fig. 17.20. Percentage labour component using labour intensive and tractor bound techniques.

The percentages of labour component using labour intensive and tractor bound techniques for typical items of rural road development are shown in Fig. 17.20.

17.5.2. Appropriate compaction equipment

Compaction of pavement materials to the desired density at controlled moisture content is an essential requirement for all rural road works. However, it is not always possible to have a conventional 8/10 tonne power roller or a vibratory roller and water bowser for the entire duration of the works. As a result, the under compacted pavement layers continue to give trouble for years to come under traffic and heavy maintenance costs have to be borne.

Adopting the rather simplistic but down-to-earth approach that some compaction is better than no compaction, some simple designs of tractor towed/animal drawn road roller and water bowser were worked out at CRM. For purposes of design of animal drawn equipment, the draft capacity of animal is an important design parameter. An animal's draft capacity varies from country to country and even from region to region ; in this respect, a pair of buffaloes, have a draft capacity of about one and a half times that of a bullock. In view of this design parameter, a road roller was designed with 0.5 to 1.2 tonne weight, to be towed by one or two animals depending upon the soil condition and the depth of loose soil to be compacted. For the design of such a road roller, the other factors like availability of local construction materials, machines, tools, skills, infrastructural facilities in rural areas were also considered to bring about economy in its fabrication cost and cost of operation and minimum dependence on higher levels of skills which are generally not available in rural areas.

17.5.3. Tractor bound technology

The agricultural farm tractors in a wide range of horse-powers are increasingly becoming available in rural areas. These can be very effectively used to towing a wide variety of agricultural equipment like disc harrows, mould board plough, rotavator etc. for road construction operations like pulverization of soil clods, mixing of soil with water and stabilizer etc. For the transportation of road materials, a tractor-trailer combination works out to be very suitable. It can also be very effectively used to tow a road roller or water bowser.

An important aspect of deploying tractors for rural road works is the ease with which these tractors, being used for agricultural purposes, can be diverted for rural road works, say during the lean agricultural season.

The tractor-towed implements that can be used for various activities/tasks for rural road works are given below.

Activity/Task	Site/Job parameters	Tractor-towed implements
• Loosening, prior to excavation excavation	Hard soil	Tiller/cultivator
• Excavation	Soft and hard soils	Disc plough and mould board plough
• Site clearance		Hydrodozer plate
• Loading, hauling and unloading unloading		Tractor-towed dozer, scraper and trailer
• Pulverization of soil clods	Soft and hard soil	Disc harrows/offset disc harrows: Rotavator

• Mixing	Blending different Soil types, water: lime/cement additives	Rotavator
• Watering		Tractor-towed water bowser
• Spreading		Tractor-towed grader or dozer
• Levelling		Tractor-towed grader or dozer

Experiences all over the developing world indicate the great potential of tractor bound technology in the development of low volume rural roads. Particularly impressive, in this context, is the utilization of agricultural tractors in towing a wide variety of implements in the construction and maintenance of low volume roads in rural areas, both unpaved and paved.

Some of the small capacity equipment developed in China which may find wide applications in developing countries can be adapted with certain modifications, for the construction and maintenance of low volume rural roads in India. Similarly, for the construction and maintenance of paved roads, a wide variety of tractor-towed equipment has been developed which appear to be pre-eminently suited to developing countries.

17.6. PAVEMENT COURSES FOR RURAL ROADS

17.6.1. Granular sub-base

Granular sub-base consists of materials such as natural sand, moorum, gravel, crushed stone, crushed slag, brick metal, Kankar or combination thereof. Three gradings as given in Table 17.6. are prescribed.

Table 17.6. Grading for sub-base courses

IS Sieve	Per cent by weight passing		
	Grading I	Grading II	Grading III
75 mm	100	–	–
53 mm	–	100	–
26.5 mm	55-75	50-80	100
9.5 mm	–	–	–
4.75 mm	10-30	15-35	25-45
2.36 mm	–	–	–
425 micron	–	–	–
75 micron	< 10	< 10	< 10

The material is spread and rolled at optimum moisture content so as to achieve 100 per cent of maximum dry density as per IS 2720 (Part 7).

17.6.2. Water bound macadam

Water bound Macadam can be used as a sub-base or base or surfacing course. Broken stone, crushed slag or over-burnt aggregates are used.

Three grading as given in Table 17.7 can be used. Grading I is restricted to sub-base only, and the compacted thckness is 100 mm. Gradings 2 and 3 can be used for base and sufacing courses, the compacted thickness being 75 mm. The gradings of screening have been given in Table 14.9.

17.6.3. Gravel/Soil-aggregate base and surface course

For Rural Roads, naturally occurring gravel or soil-aggregate mixture are used for base and surface courses. The gradation requirements for base courses and those for surface courses are given in Table 17.7.

Table 17.7. Grading requirements for different courses

IS Sieve	Per cent by weight passing			
	Base Courses			Surface course
	A	B	C	
53 mm	100	–	–	–
37.5 mm	97-100	100	–	–
26.5 mm	–	97-100	100	100
19 mm	67-81	–	97-100	97-100
9.5 mm	–	56-70	67-79	–
4.75 mm	33-47	39-53	47-59	41-71
425 micron	10-19	12-21	12-21	12-28
75 micron	4-8	4-8	4-8	9-16

17.6.4. Other local materials

A variety of local and low-grade material can be used for pavement courses for Rural Roads. They are:

1. Naturally occurring (limestone), laterite or Dhandla
2. Mechanically stabilized materials
3. Lime - soil stabilized materials
4. Cement - soil stabilized materials

5. Lime - flyash - soil stabilized material
6. Lime - cement - soil stabilized materials
7. Brick pavement
8. Stone set pavement

QUESTIONS

1. Define the following terms:
   (*a*) Low volume roads
   (*b*) Rural roads
   (*c*) All - weather roads
   (*d*) Fair - weather roads
   (*e*) Low cost water crossings

2. State the basic approach adopted in rural road network planning, giving the steps involved in planning.

3. Write start notes on:
   (*a*) Rural road network planning model
   (*b*) Socio economic aspects

4. (*a*) List the various types of rural road pavements.
   (b) State the broad requirement of soils for earth roads and the prerequisites for the success of earth roads.
   (*c*) Draw a sketch showing the various stages of embankment construction.

5. Give the salient features of the:
   (*i*) Gravel Moorum roads
   (*ii*) Metalled roads

6. Write short notes on:
   (*a*) Bituminous surface treated roads
   (*b*) Roller compacted cement concrete pavement
   (*c*) Brick pavements

7. Discuss the use of following materials in the construction of rural roads
   (*a*) Low grade aggregate
   (*b*) Soil aggregate mixes
   (*c*) Stabilized soils.

8. Give the typical specifications of rural roads for the following environmental conditions:
   (*a*) Plain areas – Indo - Gangetic plains
   (*b*) Black cotton soil areas
   (*c*) High rainfall areas
   (*d*) Water logged areas.
   (*e*) Hilly areas

9. (*a*) What are the component layers of unsealed pavements in rural areas?
   (*b*) Describe the CBR method of design.
10. State the salient features of design of rural roads recommended by IRC.
11. Discuss the relevance of Intermediate Technologies in the construction and maintenance of rural roads
12. Write short notes on the following pavement courses for rural roads
    (*a*) Granular sub-base
    (*b*) Water bound Macadan.
    (*c*) Gravel soil aggregate base and surface course
    (*d*) Other local materials

Chapter

URBAN ROADS

18.1. FUNCTIONAL CLASSIFICATION OF ROADS IN URBAN AREAS

18.1.1. Urban and rural areas

AASHTO defines urban areas as those areas 'within boundaries set by the responsible state and local officials having a population of 50,000 or more'. The urban areas are further subdivided into 'urbanised areas (population of 50,000 and over)' and 'small urban areas (population between 5,000 and 50,000)'. Rural areas are those areas which are outside the boundaries of urban areas. The population should be projected for the design year.

18.1.2. Functional classification

While the urban highway classification system in India has been listed in Chapter 4, the American classification system is briefly outlined below :

(*a*) The Principal Arterial System caters to the needs of the major centres of activity of urbanised areas constituting high traffic volume corridors, the trips entering and leaving the urban area, the traffic bypassing the central city, trips between central business districts (CBDs) and residential areas and provides continuity for rural arterials skirting round the urban area. The principal arterials are not designed to provide direct access to abutting land ; in fact, practically all fully and partially controlled access facilities are part of the principal arterial system. Its spacing may be less than 1 km in highly developed areas but may be as much as 5 km or more in areas with very low levels of urban development.

(*b*) The Minor Arterial System lays higher emphasis on land access than the principal arterial, caters to trips of moderate length in comparison to high trip lengths of the principal arterial, carries local bus routes, interconnects with the principal arterial system, provides intra-community continuity but does not enter into the residential or other localities. Its spacing may very from 0.2 to 1 km in CBD to 2 km or so in the suburbs.

(*c*) The Collector Street System collects traffic from local streets in residential localities, commercial and industrial areas and feeds that into the arterial system. In the CBD and other areas of high urban development, the collector system may include the whole street grid system.

(*d*) The Local Street System is designed for direct access to abutting land and provides connection to the Collector Street System and further on to the arterial system.

18.2. PEDESTRIAN FACILITIES ON URBAN ROADS

In urban areas, the roadway designs must incorporate facilities for the safety and convenience of pedestrians. Such pedestrian facilities include sidewalks/footways, crosswalks, separations for pedestrian crossings and some street furniture items.

(*a*) **Sidewalks.** For pedestrian safety, sidewalks should be provided on either side of an urban road with a capacity of 30 to 50 persons per minute per metre width after deducting 0.9m 'dead width' in shopping areas and 0.45 m elsewhere. A 'level of service' approach has also been suggested for the design of sidewalks as under :

Level-of-service A: It provides an average of 12 m^2 or more, allowing each person to choose a desired walking speed and to avoid conflicts with other persons.

Level-of-service B : In the range 4 to 12 m^2 per persons, they become aware of other pedestrians.

Level-of-service C : In the range 2 to 4 m^2 per person, adjustments are needed by a pedestrian to avoid conflicts.

Level-of-service D: In the range 1.5 to 2 m^2 per person, frequent changes in speed and position are required.

Level-of-service E : A range of 0.5 to 1.5 m^2 per person implies that the speed of practically all pedestrians gets reduced.

Level-of-service F: At less than 0.5 m^2 per person, a person is likely to be stationary and is able to walk only by shuffling.

Along any street or highway where no shoulders have been provided, sidewalks need to be constructed even if the pedestrian traffic is light. Wherever pedestrians move along a main or high speed road, a side-walk should be provided. In all urban and sub-urban situations, a border area should separate the roadway from buildings/residences/businesses with a minimum border width of 2.4 m. In residential areas, sidewalks may be 1.2 to 2.4 m in width and a planted strip, if provided between the side-walk and carriageway curb should be atleast 0.6 m wide for proper maintenance. If, however sidewalks are provided adjacent to the curb, the sidewalk should be 0.6 m wider than the sidewalk where a planted strip is provided between the curb and the sidewalk. The sidewalk should be paved preferably with precast blocks or tiles; in any case the surface should not only be all-weather but also inviting enough to induce pedestrians to use sidewalks. The widths of footways as per British practice are given in Table 18.1.

Table 18.1. Recommended footway widths

Type of road	Recommended minimum footway widths
Primary distributor Urban motorway	No footway
All purpose road	2.7 m*
District distributor	2.7 m in principal business and industrial districts* 2.4 m in residential districts.*
Local distributor	2.7 m in principal business and industrial districts.* 1.8 m in residential districts.*
Access road	Principal means of access : 2.7 m in principal business districts.* 1.8 m in industrial districts.* 1.8 m normally in residential districts.* 3.7–4.5 m adjoining shopping frontages. Secondary means of access : 0.9 m instead of footway on roads in principal business and industrial districts. 0.6 m verge instead of footway on roads in residential districts.

*If no footway is required, provide at least 0.9 m verge.

For the protection of pedestrians from moving vehicles on the carriageway, a barrier type curb is usually sufficient for low- speed streets. However, for speeds exceeding 70 km/hour, where a possibility may exist for a vehicle to mount the curb at a flat approach angle, a suitably designed pedestrian guard rail of adequate height (Fig. 18.1) should be provided to separate the sidewalk/walkway and the vehicle lanes. These longitudinal pedestrian guard rails should have openings only at designated pedestrian crossings (Fig. 18.2).

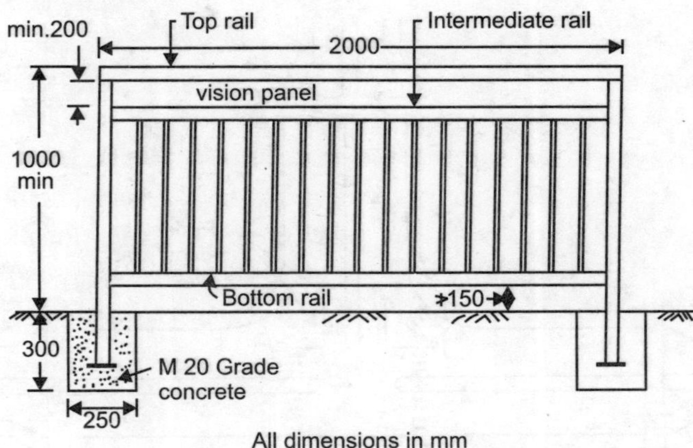

Fig. 18.1. Typical dimensions of a pedestrian guard rail (with vision panel).

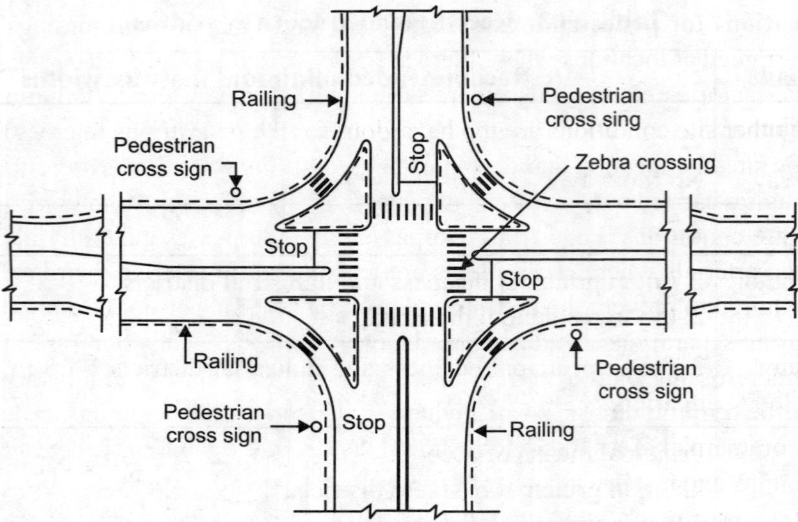

Fig. 18.2. Type design of four arm channelised intersection showing arrangement of zebra crossing and pedestrian barrier railings.

(*b*) **Crosswalks.** At an intersection or at other points within the roadway, crosswalks define the pedestrian walking areas, delineated by pavement markings which should be visible both during the day and night. The width of crosswalks should be atleast as much as that of the sidewalk. The markings should preferably be "zebra" type, consisting of equally spaced white stripes running parallel to the direction of movement of traffic, width of stripes being around 60 cm, the width in-between the stripes also being 60 cm. Depending on the site conditions at an intersection, pedestrian safety zones and islands including refuge islands, loading zones, safety zones should be provided for the use of pedestrians. Typical locations of refuge islands are shown in Fig. 18.3. Where the pedestrian volume at an intersection is very high, pedestrian signals need to be provided. These pedestrian signals are installed exclusively for the purpose of directing pedestrian traffic at signalised locations. The intervals of pedestrian signals can be combined with vehicular signal intervals.

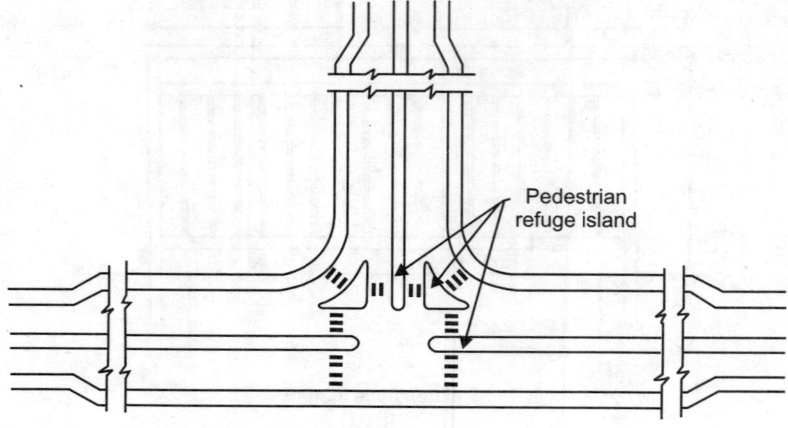

Fig. 18.3. Pedestrian refuge islands.

(*c*) **Separations for pedestrian crossings.** Undercrossings or overcrossings for pedestrians are warranted at locations where there are peak volumes of pedestrians *e.g.* from schools, offices, factories etc. and also there is medium to heavy traffic volume on the road or where other site conditions are too hazardous for the pedestrians to cross the road. At any specific site, the location and design of undercrossing or overcrossing should be based on field studies carried out on volume of pedestrians, traffic volume, intersection capacity and other site conditions. Pedestrian overpasses or underpasses generally become necessary at urban arterials where cross streets are terminated. Pedestrian guard rails are however, still required to prevent the pedestrians from crossing the arterial despite the provision of pedestrian separations. Studies abroad have shown that pedestrians are more reluctant to use undercrossing than overcrossings. The width of walkway at a pedestrian separation should be a minimum of 2.4 m ; where higher pedestrian volumes are expected like in busy commercial areas, the width should be suitably increased. It is desirable to provide both ramps and a stairway, whenever possible.

The flows should not normally exceed 20 persons per 0.3 m width per minute on level or upto 1 in 20 gradient and 14 persons per 0.3 m width per minute on stairs or ramps steeper than 1 m 20 The gradients of continuous ramps should not be steeper than 1 in 10. If a landing is provided at mid-height, a maximum slope of 1 in 7 can perhaps be allowed. A minimum height of 2.3 m and width 2.4 in should be provided for subways upto about 2.3 m length, and for longer subways, the dimensions need to be increased to 2.6 m height and 2.75 m width.

(*d*) **Street furniture.** On most of the urban roads and streets in the cities and towns of India, besides the increase in traffic volume, the number of pedestrians is also increasing rapidly. Therefore, not only well-desigried sidewalks, crosswalks and pedestrian separations are needed, attention must also be paid to a number of street furniture items for the convenience of pedestrians. These items include drinking water kiosks, roadside toilets, public telephone booths etc. Such items of street furniture should be located carefully, so that, they do not obstruct the smooth flow of pedestrians along a sidewalk and at the same time are under the vigil of the public eye to prevent vandalism. An appropriate location for such street furniture items is near the farther edge of the sidewalk away from the urban road or street. The designs of such items should be aesthetically appealing in the interest of a pleasant streetscape. Some typical designs of a roadside drinking water kiosk, a roadside toilet and a public telephone booth are shown in Fig. 18. 4.

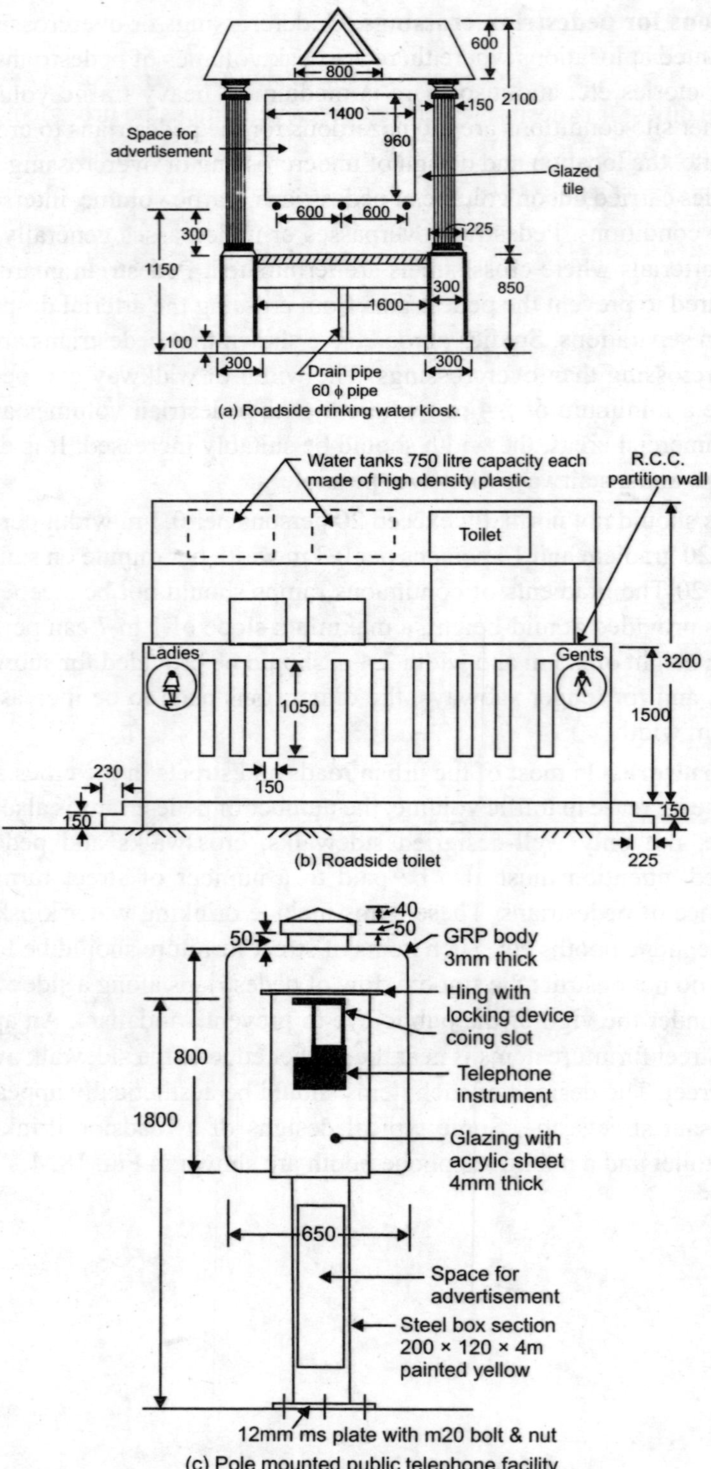

Fig. 18.4. Some street furniture items for the convenience of pedestrians.

18.3. SEPARATE CYCLE TRACKS

Segregation of cycle tracks promotes safety. In view of the large number of cyclists on the streets of Indian towns and cities, it is desirable that separate cycle tracks be provided, wherever feasible.

The design criteria for cycle tracks are laid down by the Indian Roads Congress. i. The design criteria commonly adopted are :

	Design element	Guidelines	
1.	Justification for separate cycle tracks.	Where the peak hour cycle traffic is 400 or more on routes with a vehicular traffic of 100–200 veh/hour. Where the motor vehicle traffic is more than 200 veh/hour, separate cycle tracks are justified even if the cycle traffic is only 100 per hour.	
2.	Capacity (per day)	One-way traffic	Two-way traffic
		2-lanes 2000–5000	500–2000
		3-lanes > 5000	2000–5000
		4-lanes >	5000
3.	Horizontal curves (Radius)	Minimum	
		(*i*) Minimum 10 m	
		(*ii*) Where the gradient is steeper than 1 in 40, minimum radius shall be 15 m.	
4.	Vertical curves	Maximum radius	
		(*i*) Summit curves : 200 m	
		(*ii*) Valley curves : 100 m	
5.	Gradients	Gradient	Maximum length (m)
		1 in 30	90
		1 in 35	125
		1 in 40	160
		1 in 45	200
		1 in 50	250
		1 in 55	300
		1 in 60	360
		1 in 65	425
		1 in 70	500
		(Gradients of 1 in 20 and 1 in 30 may be allowed for short lengths of 20 m and 50 m respectively.)	

6.	Sight distance	Clear view 25 m. In gradient of 1 in 40 or steeper, clear view shall not be less than 60 m.
7.	Lane width	1 metre per lane
8.	Width of pavement	Minimum : 2 m (for 2 lanes)
9.	Clearance	(*i*) Vertical: 2.25 m minimum ; 2.50 m desirable (*ii*) Horizontal: 0.25 m

Other general considerations for the design of separate cycle tracks are :

(*i*) Preferably cycle tracks should be provided on both sides of the road.

(*ii*) They should be separated from the main carriageway by a verge or a berm, which should be preferably 0.5–1.0 m wide.

(*iii*) Cycle tracks should be constructed and maintained with care and should have good riding qualities. Black-topped and concrete surfaces provide smooth riding qualities, whereas water-bound macadam and earthern tracks are not inviting for the cyclists to ride upon.

(*iv*) The tracks should be clear of obstructions such as hedges, ditches, tree roots, kerbs etc. by at least 0.5 m.

18.4. BUS FACILITIES

18.4.1. Bus stops

The following guidelines for the location of bus stops may be kept in view :

(*i*) Since bus traffic near intersections interfere with the traffic flow, if not properly designed for, it is desirable that the bus stops should be located at least 75 m away from the intersection.

(*ii*) It is preferable to locate bus stops on the exit side of the intersections.

(*iii*) Bus stops on opposite sides of two-way undivided carriageways should be staggered, preferably so that buses stop tail to tail and move off from each other. A suggested stagger is 60–90 m.

(*iv*) Bus stops should be as far as possible, so located as to disembark the passengers at safe places such as curbs or islands.

(*v*) For buses intending to turn right at an intersection, if it becomes necessary to locate the bus stop before the intersection, the stop should be sufficiently away from the intersection so that the bus can be manoeuvered from the curb to the extreme right lane, well before the intersection.

18.4.2. Bus bays

It is desirable to recess the curb sufficiently for accommodating the bus-bays. Typical layout of a single bus bay (with fencing) is shown in Fig. 18.5.

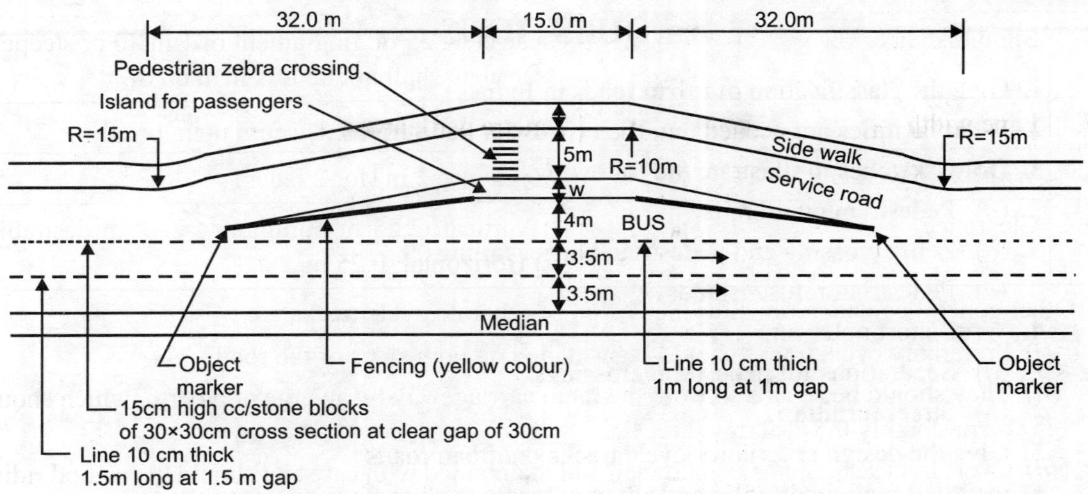

Fig. 18.5. Typical layout of a single bus bay (with fencing).

18.4.3. Driveways

Driveways are roads providing access to abutting property owners and utilities. An important example is the access road to a petrol-filling facility. The geometries of the driveways can have a profound influence on the traffic flow and safety of travel on the through road. The entrance and the exit should be wide with easy curves. The layout suggested by the I.R.C. for motor-fuel filling station is given in Fig. 18.6.

Minimum Distance to Intersection :

60 m for main roads and 45 m for others

X Not less than 13 m where no cycle tracks are required now or in future ; not less than 20 m where cycle tracks exist or may be required in future

Y Half the roadway width but not less than 6 m for National and State Highways and 5 m for others.

Z Not less than 9 m.

R Radius not less than 11 m.

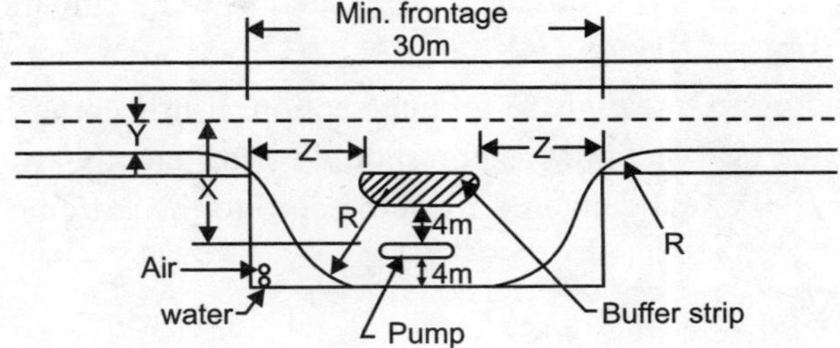

Fig. 18.6. Layout of a motor-fuel filling station.

QUESTIONS

1. Give the classification of urban roads in India.
2. What facilities are needed on urban roads for pedestrians. Discuss them briefly.
3. Draw sketches to illustrate the following:
   (*a*) Pedestrian guard rails
   (*b*) Zebra crossing and pedestrian barrier railings.
   (*c*) Pedestrian refuse islands
4. Write short notes on:
   (*a*) Separations for pedestrian crossings
   (*b*) Street furniture
5. Give the design criteria for cycle tracks on urban roads.
6. (a) What guidelines will you follow to locate the bus stops on urbon roads.
   (*b*) Draw a typical layout of a single bus bay (with fencing)
   (*c*) Sketch the layout for a motor-fuel filling station.

Chapter

EXPRESSWAYS

19.1. COMMON TERMS

Expressways, Motorways and Freeways are terms denoting superior type of highway facilities designed for high speeds, high traffic volumes and safety. The term Motorway is common in U.K. In Germany, it is known as Autobahn. In Italy, it is known as Autostrade. In France, it is called by the name Autoroute.

The word commonly used in USA is Freeway. In India, the term Expressway is commonly used.

Expressways can be Urban Expressways or Inter-city Expressways.

19.2. FEATURES OF EXPRESSWAYS

Expressways have some special features which differentiate them from ordinary roads. These are :

1. They cater to high speeds. A design speed of 120 km/hr is common.
2. They have dual carriageways separated by a median, so that each carriageway is exclusively reserved for one-direction traffic. However, as a first stage, there are some two-lane undivided Expressways too.
3. They have grade separation with the provision of overbridges or underpasses for roads crossing the Expressway.
4. They have access control, which permits traffic to enter or leave the facility only at designated locations.
5. Slow-moving traffic is not allowed. Thus cycles, slow speed agricultural tractors, low capacity two-wheelers are excluded. Pedestrians are similarly not allowed to enter.
6. The traffic is guided and serviced through adequate provision of signs, markings, safety barriers, telephone communications, rest areas and rescue services.
7. The pavement construction and maintenance are of superior standards that permit high speeds.

19.3. DESIGN STANDARDS

Design Standards for Expressways have not been evolved yet in India. The following are the typical values being used in the interim period on some of the projects being planned.

1. Right-of-way — 90–100 m

2. Design Speed — Plain terrain : 120 km/hr
Rolling terrain : 85–100 km/hr ;
Hilly terrain : 60–70 km/hr

3. Carriageway width
- 2 × 2 lane : 2 × 7.5 m
- 2 × 3 lane : 2 × 11.25 m

4. Paved shoulder
- Left side (including 0.5 m edge strip) — 2.5–3.0 m
- Right side (including 0.5 m edge strip) — 0.7–1.0 m

5. Unpaved shoulder
- Left side — 1.0–1.5 m

6. Median width
- 2 × 2 lane: 12–15 m (so as to accommodate future six-laning)
- 2 × 3 lane: 4–5 m

A typical cross-section of the Mumbai-Pune Expressway is given in Fig. 19.1.

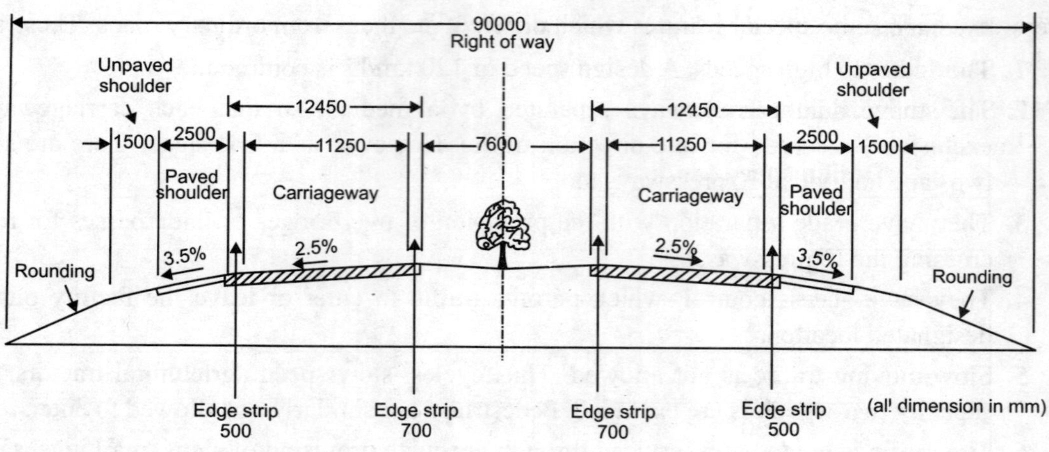

Fig. 19.1. Cross-section of 2 × 3 lane expressway.

7. Gradient

Plain terrain

- Ruling: 2.0 per cent

- Maximum: 2.5 per cent

Rolling terrain

- Ruling: 2.0 per cent
- Maximum: 3.0 per cent

Hilly terrain

- Ruling: 4.0 per cent
- Maximum : 5.0 per cent

8. Radius of horizontal curve

Plain terrain

- Desirable: 2600 m
- Minimum: 700 m

Rolling

- Desirable: 450 m
- Minimum : 300 m

Hilly

- Desirable : 750 m
- Minimum : 125 m

9. Superelevation

Plain and rolling terrain : 7 per cent

Hilly terrain: 8 per cent

10. Camber

Bituminous pavement: 2.5 per cent

Cement concrete

Pavement: 2.0 per cent

Shoulder: 1.0 per cent more than of pavement

11. Safe stopping sight distance

Design speed km/hr	Safe stopping sight distance (m)
60	80
70	100
80	120
90	150
100	180
110	220
120	250

12. Interchanges

Design Speed for interchange elements (km/hr)

Terrain	Left-turning ramps	Right-turning ramps/loops
Plain/Rolling	70–90	50
Hilly	40–50	30–40

Gradient

Ramps/Diagonal connections : 4 per cent

Loops : 5 per cent

Width of Ramps

1. Roadway : 12 m
2. Carriageway : 5.5 m for traffic upto 16,000 PCUs/day
   7.5 m for traffic above 16,000 PCUs/day and upto 32,000 PCUs/day
3. Paved shoulder: 1.5 m either side
4. Radius of curvature:

Loop/Ramp design speed Km/hr	Radius (m)
90	300
70	180
60	130
50	85
40	55

13. Speed change lanes

Table 19.1 gives the length of speed change lanes.

Table 19.1. Length of speed change lanes

Design Speed of turning roadway curve (km/hr)		Stopped condition	25	30	40	50	60	65 and above
Design speed of highway (km/hr)	Length of taper (m)	Total length of deceleration lane including taper (m)						
65	60	100	90	85	75	60	–	–
80	70	130	120	115	110	110	85	–
100	80	150	150	145	140	130	120	100
110	90	165	165	160	150	145	140	115
120	100	180	180	170	160	150	145	125

Design speed of highway (km/hr)	Length of taper (m)	Total length of acceleration lane including taper (m)						
65	60	–	100	75	70	–	–	–
80	70	–	210	190	180	150	120	–
100	80	–	340	320	300	270	240	180
110	90	–	465	450	420	370	300	250
120	100	–	550	500	450	400	350	300

14. Bridges and viaducts

The width should be the same as that of the approach pavement.

15. Culverts

Culverts should accommodate the full expressway including carriageway, median and shoulders.

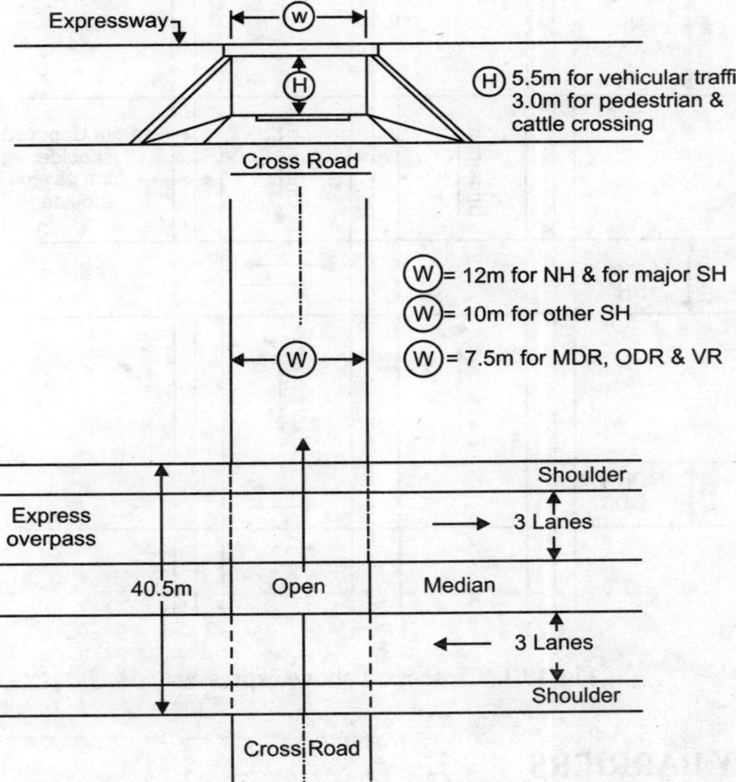

Fig. 19.2. Underpass for cross-roads in an expressway.

16. Clearance

    The minimum vertical clearance shall be 5.5 m.

17. Width of underpass for accommodating cross-roads ; Fig. 19.2 gives a sketch.

18. Width of structure with expressway at ground level.

    Fig. 19.3 gives a typical cross-section.

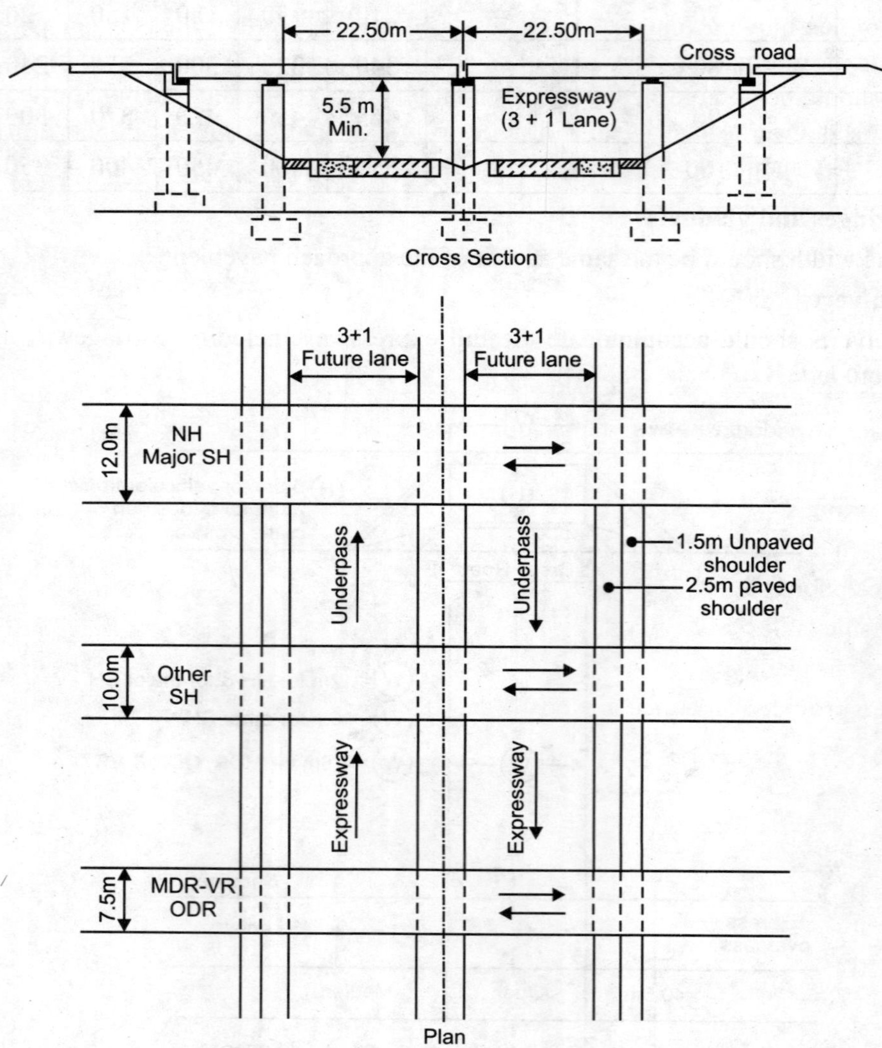

Fig. 19.3. Cross-road above expressway.

19.4. SAFETY BARRIERS

Since Expressways are high speed facilities, it is desirable to provide safety barriers Edge barriers are provided on road embankments governed by the height and slope of the embankments. The warrants can be taken from Fig. 19.4. Median barriers are provided depending upon the traffic and the median width. Fig. 19.5 gives the criteria.

The barriers can be of steel beam type, concrete or wire type. A steel beam barrier is shown in Fig. 19.6. The typical New Jersey type concrete barrier is shown in Fig. 19.7.

19.5 EXPRESSWAY SIGNS AND MARKINGS

Since expressways cater to high speeds, the signs and markings should be visible and easily read by the drivers of fast-moving vehicles. As expressways are access-controlled and are cut-off from the local sources of information, there should be adequate information signs to facilitate the traveller. The letters should be of large size, symbols should be large and legends should be short. The signs are reflectorised. Overhead signs, even if reflectorsed, do not get sufficient light from the beam of a car. Hence they should be illuminated.

It is convenient to provide overhead gantry signs for effective visibility.

All edge strips should be marked with reflectorised continuous line and lane markings shall be of broken reflectorised type.

19.6. REST AREAS

Rest areas are provided all along major highways, particularly Expressways. The spacing is about 160 km. The facilities are :

- Parking
- Open areas : Gardens/Parks
- Telephone
- Picnic tables
- Repair facility
- Toilets
- Restaurant
- First aid
- Fueling station

They are provided on either side of the carriageway. A typical layout of a rest area is given in Fig. 19.8.

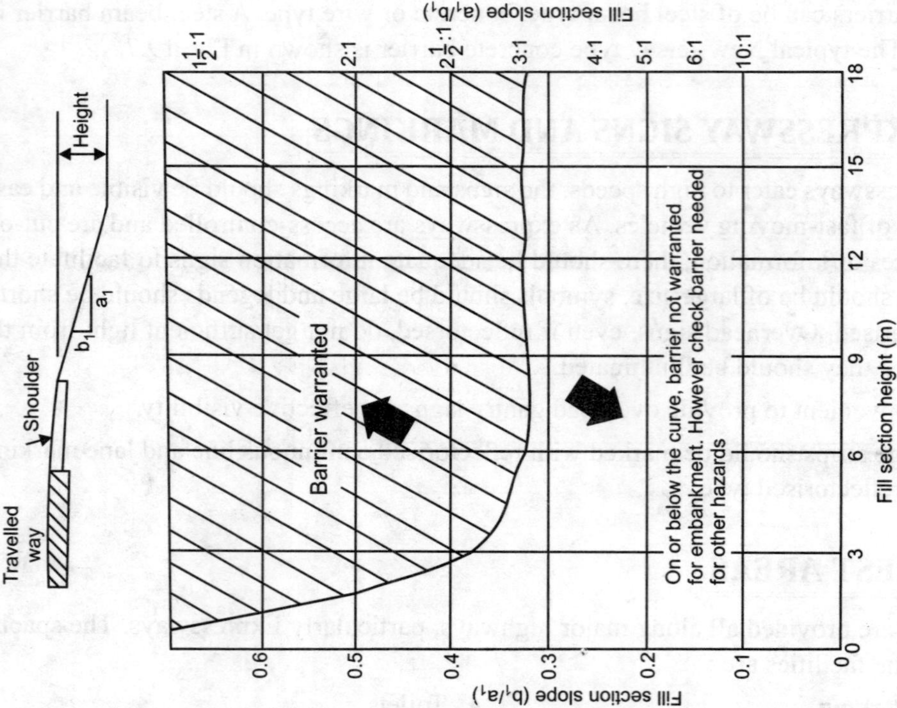

Fig. 19.4. Warrants for fill section embankments.

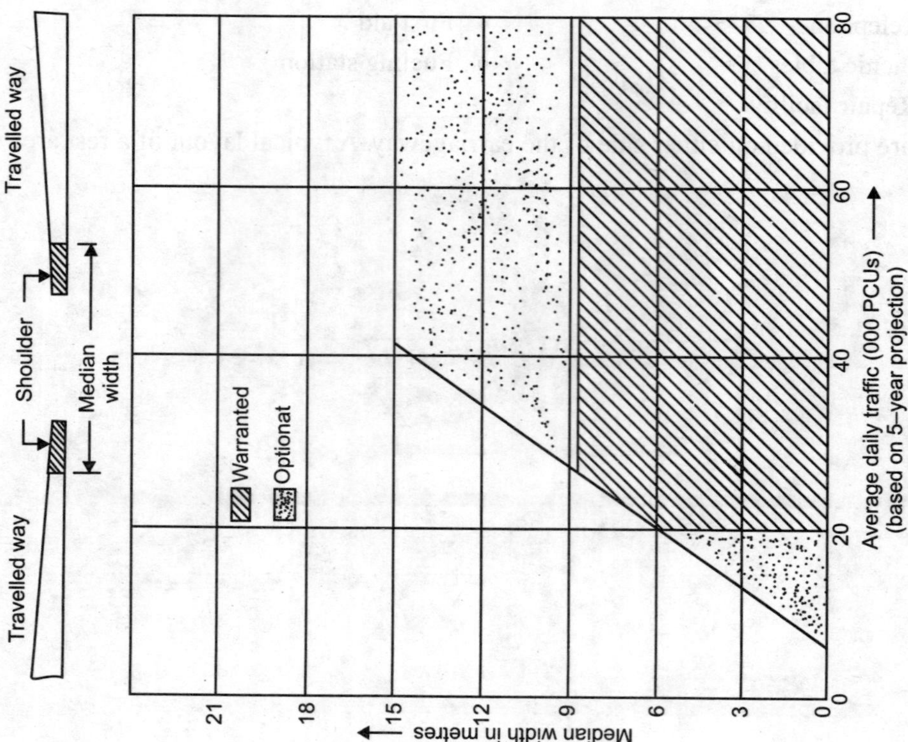

Fig. 19.5. Warrants for median barriers.

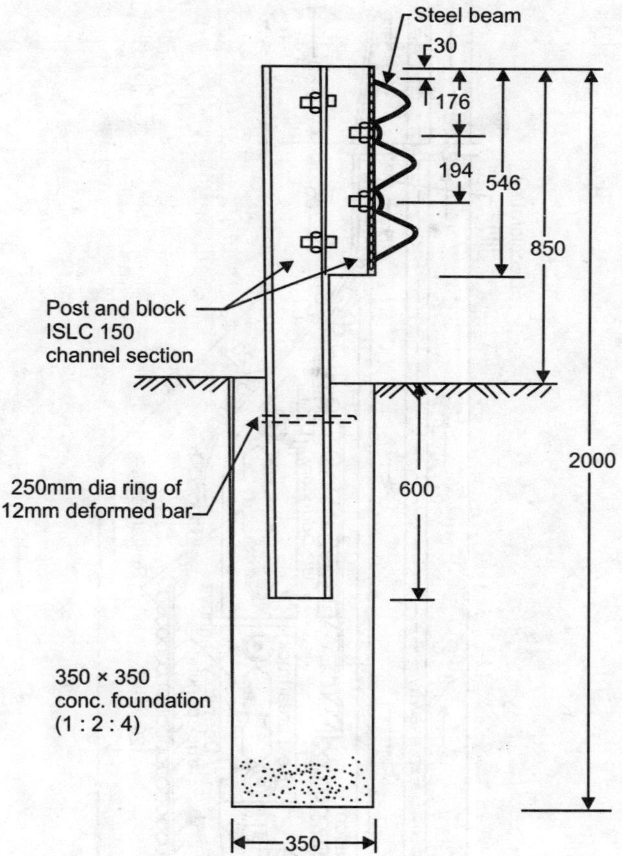

Fig. 19.6. Typical details of steel beam section.

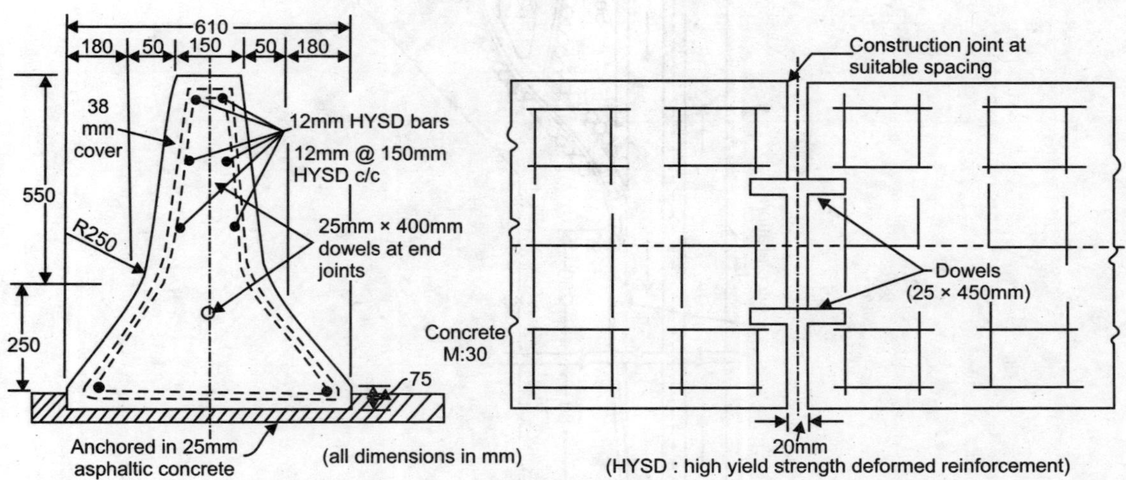

Fig. 19.7. Median barrier cast-in-design.

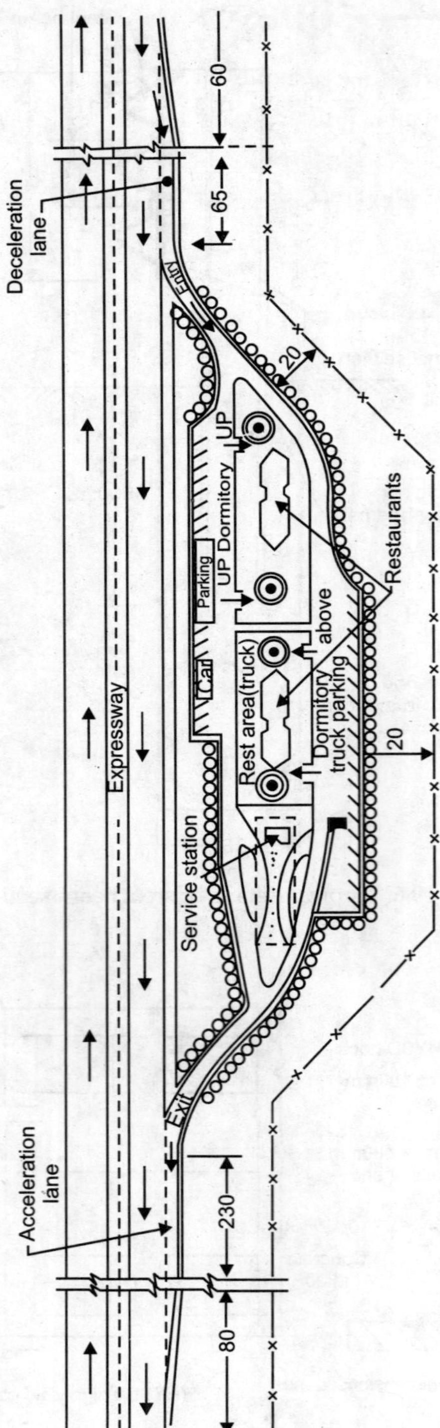

Fig. 19.8. Typical layout of rest area serving traffic from one direction.

QUESTIONS

1. Enumerate the salient features of expressways.
2. Draw a cross section of an expressway and give the typical values for the design standards for right-of-way, design speed, carriageway width, shoulders, median width.
3. Write down the design standards for the following.
   (*a*) Gradient
   (*b*) Radius of horizontal curves
   (*c*) Superelevation
   (*d*) Camber
   (*e*) safe stopping sight distances
   (*f*) Interchanges
4. Draw sketches for the following .
   (*a*) Underpass for cross-roads in an expressway
   (*b*) Cross-road above expressway
   (*c*) Typical steel beam section
5. What facilities are provided in a rest area on an expressway. Draw typical layout of rest area serving traffic from one direction.

Chapter

TOLL ROADS

20.1. BACKGROUND

Many countries have modernised their road system through the involvement of the private sector in funding, designing, constructing, operating and maintenance of highway facilities. The private entrepreneur or the government body levy a fee on the road user for using the facility. Roads where the costs are recovered through such user fees are known as Toll Roads.

Toll roads have been in existence in many countries since a long time. In U.S.A. early toll roads were known as Turnpikes.

20.2. ADVANTAGES AND DISADVANTAGES OF TOLL ROADS

Advantages

1. Since the financing of Toll Roads is done by private entrepreneurs, time is the essence of a project. Hence, such projects get completed in a short time.
2. Toll financing is an additional source of funding.
3. Tolling is an efficient method of charging the actual road user and is thus an equitable form of user taxation.
4. Toll roads are designed to high standards and are well maintained so as to attract clientele.
5. Toll road companies provide good way-side facilities.

Disadvantages

1. Toll roads are constructed with borrowed capital, and their costs tend to be high. If government borrows money, the interest rates tend to be lower.
2. The cost of toll collection is an additional cost component.
3. Delays are often caused at Toll plazas.
4. When a toll-free road exists as an alternative, the full potential of toll roads is not utilised.
5. Toll projects are financially viable over a long period. This period is known as the concession period, and is usually 20–30 years. A number of uncertainties and risks are

involved in such long-term projects. The most important risk is the accuracy of traffic projections. Others like political and official interferences are equally important risks.

20.3. FORMS OF TOLL PROJECTS

Private sector funded infrastructure projects can take many forms. Some of these are :

BOT. Build, Operate and Transfer. The entrepreneur builds the facility, operates it and collects toll, and transfers it at the end of the concession period back to the government. Most highway projects in India follow the BOT.

BOLT. Build, Operate, Lease and Transfer. The operator builds, and leases the facility from the Government, operates it for a certain period and transfers it at the end.

The Government gives him certain fixed lease charges for building and operating the services. Indian Railways have been using this system.

BOO. Build, Own, Operate. This is similar to BOT, but without transfer of ownership to the private entrepreneur.

DFBO. Design, Fund, Build and Operate. This system has been tried in U.K. and is being tried in India also.

LDO. Lease, Develop, Operate, in which the Government retains ownership of an existing facility, and receives payments as specified in the lease agreement with a private lessee, who in turn finances development and operates the facility.

ROT. Rehabilitation, Operate, Transfer. This is similar to BOT, but the work involves the rehabilitation of an existing facility rather than building a new one.

In addition to the above, there are some other forms of Public-Private-Partnership (PPP) in financing roads. These are :

1. **Annuity Payments**

   Under this scheme, the Concessionaire is required to meet the entire upfront cost. The client makes a fixed payment at regular intervals (say, six monthly or yearly) to the concessionaire for an agreed period, which enables the Concessionaire to recover his initial investment, interest thereon and operation and maintenance costs. The client may or may not collect toll, and if toll is collected, it goes to the client's revenue kitty.

2. **Viability Gap Funding (VGF)**

   If it is found that the project is not fully commercially attractive on the basis of collection of toll alone, the client agrees to give a grant (currently restricted to 40 per cent in India on National Highway Projects) to cover the viability gap.

In some cases, where the traffic is very high and toll collections are expected to be high, the Concessionaire does not seek a VGF, and, on the contrary, agrees to give a premium (also called negative grant) to the client.

Some other forms of raising revenues for highway projects have been recently tried. One of these, pioneered, by China, is "asset securitisation", by which the existing highway assets can be leveraged to raise fresh funds, particularly from international lending institutions like the World Bank and the Asian Development Bank.

Special Purpose Vehicle (SPV) is a newly created financial and construction company, having equity participation from the government, financial institutions and construction company. The Moradabad bypass in India is an example where the National Highways Authority of India and the ILFS have agreed to form an SPV.

In India, the recently amended National Highways Act permits the levy of toll on roads. Till then, tolls were permitted only on bridges. The Government of India have recently enunciated the policy on Private Sector Participation on National Highways. The salient features of the Policy are :

1. The policy of privatisation will be implemented by the NHAI and in exceptional cases by the State Public Works Departments.
2. The basic principles in identification of NH projects for private investments would be that the project is an approved project of the Ministry of Surface Transport and it is capable of yielding adequate financial returns.
3. The Government will provide support by carrying out all preparatory works such as
   - Feasibility Study
   - Land for right of way and roadside facilities
   - Relocation of utility services
   - Cutting of trees
   - Removal of encroachments and resettlemenst and rehabilitation of persons affected by the projects

   Depending upon financial viability, the cost for the above activities may be recouped from the project.
4. Suitable traffic support/guarantee will be provided on case to case basis.
5. Exemption will be given from import duty on identified high quality construction plant and equipment.
6. Foreign Direct Investments upto 74% would be permitted automatically. Beyond that, proposals would need clearances of the Foreign Investment Promotion Board.
7. A five year tax holiday is available followed by a further five year period of 30% exemption and this concession can be availed of in any 10 consecutive years during the first 20 years of operation. (Section 80 IA (12) (ca) of the IT act 1961).
8. Upto 40% may be excluded from Income Tax on the income derived by financial institutions from finance that they provide for infrastructure projects.
9. Exemption is available from Income Tax on the income from dividend and interest on long term capital gains derived from investments in the form of shares or long term finance to any enterprise set up to develop, maintain and operate an infrastructure facility.
10. Subscription to equity shares and debentures are eligible for deductions under section 99 of the Income Tax Act.
11. Real estate development can be made an integral part of BOT projects to enhance their financial viability.
12. NHAI can provide capital grants to the developers of a road project, on a case to case basis.

13. NHAI can participate upto 30% of total equity of a company floated to develop a road project.
14. The ownership of the land for the highway construction and roadside facilities would continue to vest in the Government. Mortgaging and subleasing of this land for raising finances is not allowed. However, land will be given on lease to entrepreneurs.
15. There will be provision of unified check barriers at the Inter-State borders for the use of all governmental authorities. Such barriers would be located outside the right of way with proper entry/exit layout.
16. Dispute resolution and arbitration would be under the Indian Arbitration and Conciliation Act, 1996.
17. Entrepreneurs would be protected against force majeure situations including political, non-political and legislative changes.
18. There will be a standard concession agreement, which addresses the concerns of various stakeholders in the project.

The following are some of the recent projects under implementation in India on a BOT basis :

1. Durg Bypass (16 km long)
2. Hubli-Dharwar Bypass (30 km long)
3. Bangalore-Mysore Expressway (200 km long)
4. Mumbai-Pune Expressway (120 km long).

More projects of bridges and four-laning of two-lane National Highways are now under implementation.

A Model Concession Agreement has been prepared and is being adopted for all National Highway Projects

QUESTIONS

1. What are the advantages and disadvantages of Toll roads?
2. State the various forms of toll projects.
3. What is PP in road financing and how it can be achieved?
4. List the salient features of private sector participation in National Highways.

Chapter

DESERT ROADS

21.1. DESERT REGIONS IN INDIA

India has some areas which are deserts. The western portion of Rajasthan is part of the That desert. Due the physiography and climate characteristics of the area, the design, construction and maintenance of roads in this region present a variety of problems. A part of the coastal area of Tamil Nadu has low rainfall and the soil is sandy. Some of the problems discussed here equally apply to such areas.

21.2. SAND DUNES, THEIR FORMATION AND TYPES

IN the desert region strong winds sweep through the area in the hot summer months, setting sand particles in motion. The sand particles are very fine, their diameter being less than 0.3 mm. A wind velocity of about 10–15 km/hour is sufficient to carry such small particles. As the wind comes across minor irregularities in the land relief, eddies and stagnation zones are caused In zones of eddies, the sand is carried in aerial suspension and in the stagnation zones the particles are deposited. This gives rise to the formation of sand dunes, which can be as high as 100 m above the ground.

When the prevailing winds have a constant direction throughout the year, sand-dunes are formed stretching out in the direction of the wind. Such sand-dunes are called longitudinal sand-dunes. Since the prevailing winds in Rajasthan are in the South-West to North-East direction, longitudinal dunes in the same directions are formed

Transverse sand-dunes are at right-angles to the wind direction, which is South-West in Rajasthan. These are formed when the prevailing winds change their main direction. They also occur at the leeward end of a chain of longitudinal sand dunes (Fig. 21.1)

Barchan sand-dunes consist of mounds of sand, crescent-shaped in plan (Fig. 21.1) Barchan sand-dunes shift quickly.

21.3. PRINCIPLES OF ROAD LOCATION IN DUNE AREAS

The following principles should be observed in locating a road in dune areas

1. Avoid sections where sand is loose and unstable
2. Align road where sand is covered by vegetation
3. Preference should be given to areas having coarse sands.
4. If a road has to be aligned parallel to a sand ridge, avoid locating the road on the slopes of the sand-ridge. Locate the road either at the top of the ridge itself or in the interdunal space (Fig. 21.2).
5. It is a good practice to locate the roads away from sand ridges and Barchans to a minimum distance of twice their height.

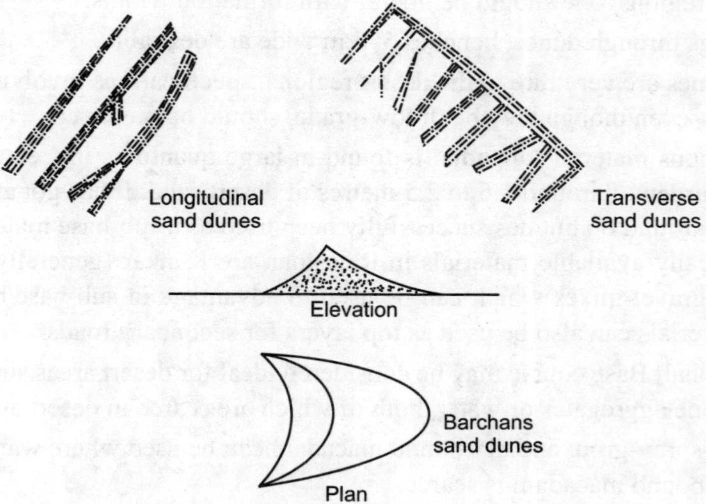

Fig. 21.1. Types of sand-dunes

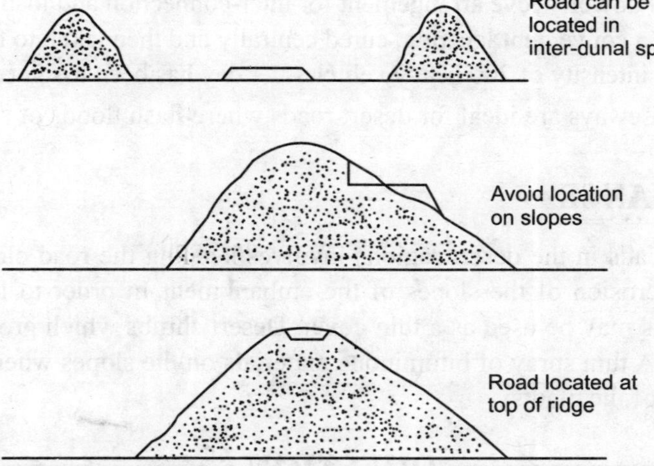

Fig. 21.2. Possible locations of road in sand ridge.

21.4. GUIDELINES FOR DESIGN

Some of the guidelines to be observed in designing roads in desert regions are indicated below :

1. Sections with partial cut and partial fill should be avoided since they are difficult to maintain.
2. Cuttings should be avoided as far as possible, though this ideal is very difficult of achievement.
3. In a level country, subjected to strong winds, it is inadvisable to have high road embankments.
4. In sandy regions, one should be liberal with formation widths.
5. In cuttings through dunes, benches 5–8 m wide are desirable.
6. Since stones are very rare in the desert regions, specifications involving locally available materials, even though they be of low-grade, should be preferred.

   A calcareous material 'dhandla' is found in large quantities in Central Rajasthan under an overburden of around 1.5 to 2.5 metres of desert sand. It has got an Aggregate Impact Value of around 60 but has successfully been used as a sub-base material in desert areas. Other locally available materials in Rajasthan are Kankar (generally hard and nodular) and soil-gravel mixes which can be used to advantage in sub-base/base courses. These local materials can also be used as top layers for secondary roads.
7. Sand Asphalt Base course may be considered ideal for desert areas since it does; not need either stone aggregates or water, both of which are scarce in desert areas.
8. Built-up-spray-grout and dry bound macadam can be used where water for consolidation of water bound macadam is scarce.
9. A relatively recent pavement system for desert areas comprises precast cement concrete blocks laid over a well compacted sandy subgrade. The blocks are ribbed and provided with a special dowel-sleeve arrangement for inter-connection and load transfer. The blocks are precast at a convenient location, cured centrally and then taken to the site for paving. A design traffic intensity of 200 heavy vehicles per day has been assigned for the system.
10. Dips and causeways are ideal for desert roads where flash floods of short duration occur.

21.5. MAINTENANCE

The maintenance of roads in the desert regions involves keeping the road clear of drifting sands and preventing wind erosion of the slopes of the embankment In order to foster the growth of vegetation, clayey soils may be used as a thin cover. Desert shrubs which grow locally should be planted on the slopes. A thin spray of bituminous materials on the slopes where plants are planted will foster the growth of the plants.

QUESTION

1. How are sand dunes formed and what are their types?

(*b*) List the principles of road location in dune areas.

2. Give the design considerations for the design of roads in desert areas.

Chapter

Roads in Swampy and Water-logged Areas and in Black Cotton Soils

22.1. SWAMPY GROUND

Swampy or marshy ground implies unstable ground composed of peat or soft clay, generally in a saturated condition. Peats and soft clays are highly compressible and have very low shear strengths. Fills as low as 0.5 to 1.0 m are known to settle on such soft ground ; some of the marsh deposits, may even settle under their own weight. Fills of greater height can rupture the marsh. A typical swampy area in India is the Rann of Kutch.

22.1.1. Road construction in swampy areas

(*a*) **By total excavation.** In swampy areas, where the depth of peat or soft clay is shallow, say less than 3 m or so, the most suitable and economical solution lies in completely removing the soft material with the help of excavators. The work of excavation is started at one edge of the swamp, the excavators working on firm ground beneath the peat or soft clay and the process of filling kept closely behind the process of excavation. The excavated peat or soft clay is dumped on the side slopes of the embankment.

(*b*) **By partial excavation.** In situations, where the peat or soft clay is more than 3 m, total excavation is no longer practicable and partial excavation, restricted to only about 2.5 to 3.0 m depth is resorted to. The remaining lower part is displaced by the weight of the fill material over it; very often, additional surcharge above the finished level of the fill is applied to expedite the process of displacement. After the desired displacement has taken place, any additional surcharge left over and above the finished level of the fill is removed.

(*c*) **Displacement by blasting.** The blasting methods are resorted to, only if the methods described at (*a*) and (*b*) above are not found suitable. In the blasting method, first the peat or soft clay is displaced under the weight of the fill and then the remainder by blasting. The bog blasting technique has been described in Chapter 8. The volume displaced by peat or soft clay is occupied by the embankment material. Sand or gravelly soil should be preferred for embankment construction in such situations.

(*d*) **Accelerating consolidation in compressible clay layer.** An embankment constructed over a highly compressible clay layer can be rendered suitable within a relatively short period of a few months for laying a road pavement without the risk of post-construction settlements and any/or long-term instability of the embankment, by the installation of suitably designed vertical sand drains. The vertical sand drains facilitate quick pre-construction settlement of the embankment and bring about a rapid increase in the shear strength of the saturated clay, so as to ensure no post-construction settlements or instability of the embankment.

22.2. WATER-LOGGED AREA, ITS FEATURES

A problem which has been noticed in India recently with large scale irrigation is waterlogging. Water-logged areas are those where the level of the sub-soil water table is within 1 m below the ground level, thus favouring capillary rise of water into the subgrade. The rise in water-table is generally accompanied by the surfacing of injurious salts such as sulphates, chlorides and carbonates of sodium, calcium and magnesium. These salts increase in volume when they crystallise under suitable hygroscopic and temperature conditions. The repeated volume changes of salts due to alternate hydration and dehydration leads, in due course, to breaking of the pavement. Salt concentration in soils above 0.2 per cent is considered to be harmful.

Similar conditions prevail where stagnant water accumulates as a result of flooding.

22.2.1. Treatment of roads in water-logged areas

One of the most effective measures is to depress the water-table. This can be achieved by a system of sub-surface drains leading to suitable natural drainage channels.

Where it is too expensive to provide such elaborate drainage facilities, the other alternative is to keep the road embankment sufficiently high. The aim should be to help the bottom of the pavement at least 1.5 m above the highest water-table

A capillary cut-off, placed 0.15 m above the ground level or the standing water level, can effectively arrest the rise of water. Capillary cut-offs can take the form of sand-blanket, bituminous impregnation using a primer, heavy duty tar felt, polythene envelope or bituminous stabilised soil. Chapter 12 gives further details.

The bituminous primer found suitable for this purpose consists of 50 per cent straight-run bitumen (80/100) with 50 per cent high speed diesel oil or its equivalent, in two applications of 9.75 kg per 10 sq. m. Sufficient time should be allowed after the first application for the primer to penetrate before the second application is given.

The pavement should also be designed adequately for the adverse conditions. The thickness should be determined for saturated conditions of the subgrade soil. The bottom 10 cm thickness of the sub-base should be made up of such material as stabilised soil which will be stable when wet.

A suitable type and grade of geotextile provided over the subgrade in water logged areas effectively serves as a cutoff for the capillary rise of ground water and also as a salt migration cutoff.

Where flooding is a possibility, it is desirable to keep the embankment sufficiently high such that the top of the subgrade is at least 30 cm above the highest recorded flood level.

22.3. ROADS IN BLACK COTTON SOILS

Black cotton soils are expansive clays. They are found in various parts of the country. It has been observed that nearly one-sixth of the total area of India consists of these soils. Since these soils exhibit considerable volume changes when the moisture content changes, the road surface in such soils has a tendency to become wavy.

The main reasons for an unsatisfactory performance of road pavements on block cotton soil subgrades are :

1. Intrusion of the granular sub-base/base into the soft subgrade or ingress of the soft subgrade material into the voids of the subbase/base, resulting in the admixing of the subgrade and subbase/base materials (Fig. 22.1).
2. Very poor subgrade drainage properties resulting in an increase of moisture in the pavement crust.

Fig. 22.1. Admixing of subgrade and subbase/base materials.

3. Very poor load bearing power of subgrade, which has to withstand stresses imposed by traffic loads.

The most effective way of dealing with such soils is to provide a suitable layer between the soil and the stone layer such that the soil does not work up. A minimum thickness of 10 cm is adequate for this purpose. The material for the layer can be gravel (murram) of low P.I., sand or local soil stabilised with lime. The need for such a layer may be evaluated by the criterion that detrimental intrusion of the soil occurs when the ratio D_{15}/D_{85} is greater than 5, where :

D_{15} = particle size wherein 15 per cent of the base/sub-base course particles are smaller than this size.

D_{85} = particle size wherein 85 per cent of the subgrade soil particles are smaller than this size.

It is also found that expansive soils may be improved by compaction at water contents somewhat over the optimum.

With the advent of the use of geotextiles in road construction the provision of a carefully selected geotextile laid over the soft expansive clay subgrade is a promising solution, with the following advantages :

1. The geotextile serves as a separator between the subgrade and the pavement crust, preventing the intrusion of granular subbase/base into the soft subgrade as also the ingress of soft subgrade into the granular subbase/base course.
2. The geotextile brings about rapid drainage of water within the pavement, thus preventing any flooding of the pavement.
3. The geotextile with its inherent tensile strength makes it possible for traffic loads to be spread over a wider area, which enhances the load bearing power of the subgrade.

Since the Black Cotton soils exhibit very significant volume changes with changes in moisture content, both surface and sub surface drainage of a road pavement are of paramount importance and should be given special attention.

QUESTIONS

1. What is a swampy ground? Discuss the various methods of road construction in such areas.
2. Discuss the features of water logged areas. Suggest ways of treatment of roads in water logged areas.
3. Explain black cotton soils. Give reasons for unsatifactory performance of road pavements in such areas. Suggest effective ways of dealing with such soils, particularly by the use of geo-textiles.

Chapter

ROAD CONSTRUCTION MACHINERY

23.1. ROLE OF LABOUR VS. MACHINERY IN ROAD CONSTRUCTION

Labour has been used to a large extent in the construction of roads in India hitherto. The reasons are :

1. Labour, especially unskilled labour, is plentiful and cheap in a labour-surplus economy like India. In fact, employment generation is one of the attractions of investments on roads.
2. Road construction schemes are a popular measure of affording succour and relief to flood and drought affected persons.
3. Several road construction specifications currently being followed in the country are amenable to labour-intensive methods. Examples are : Water-bound macadam, penetration macadam, built-up-spray-grout, surface dressing, thin chipping carpet etc.
4. The principle of stage construction being followed in the country on account of scarcity of resources fits in well with labour-intensive specifications.
5. The country has so far not taken up construction of new express highways which are facilities involving huge investments, high specifications and high design standards. The works far has generally been of rehabilitation, strengthening and widening of pavements. Labour-intensive techniques are ideally suited for such low-cost solutions.
6. Because of lack of demand for modern highway construction machinery, the indigenous industry in this area has not developed to levels obtaining in the western countries. Only a few years ago, earth-moving equipment, hot-mix plants and paver-finishers started being manufactured on a modest scale.

Inspite of the lack of demand for machinery-oriented construction methods in the highway industry so far in the country, the future prospects for the use of machinery are bright. Organisations like the Border Roads employ machines such as dozers, tractors and hot-mix plants. This has been found to be necessary in view of the scarcity of labour in the remote areas where they operate.

The advantages of machinery are :

1. Machinery is useful for working in remote and inaccessible areas where labour is scarce.
2. Machinery is useful where a large quantity of work is to be done in a concentrated area in a limited time. Examples are : urban streets, new road construction involving superior specifications, approaches to mighty river bridges etc.
3. Control over quality can be exercised better if machinery is used.
4. Certain specifications such as bituminous macadam and asphaltic concrete are possible only with machinery.

23.2. EARTHWORK MACHINERY

23.2.1. Uses of earthwork machinery

Earthwork involves a number of operations and a variety of equipment can be put to use. Table 23.1 below indicates the possible uses of machinery for earthwork.

Table 23.1. Use of machinery for earthwork

S.No.	Type of job	Machinery normally employed
1.	Clearing and grubbing	Tractor dozer, rooter or ripper
2.	Excavation of soil	Shovels, draglines, scrapers, dozers
3.	Transportation of soil	Scrapers, dumpers, trucks
4.	Spreading earth	Grader, dozer
5.	Watering	Water tankers, sprinklers, pumps and piping system
6.	Compaction	Rollers of different types.

In this country, machinery so far introduced are dozers, tractors, scrapers and graders. These will be described here. Transporting water and compaction are described subsequently.

23.2.2. Volume Changes

When dealing with earthwork by machinery, it is necessary to understand the property of swell and shrinkage of soils. The soil in the natural state in the borrow area has a certain volume. On excavation, the soil increases in volume. This increase may be termed as 'swell'. Table 23.2 gives some typical values of 'swell' in different soil types.

The soil, upon compaction to the desired density in an embankment, loses its volume. This reduction in volume may be termed as 'shrinkage'. It is generally found that the compacted volume is about 10 per cent less than the volume of the soil in the natural state. In order to get over these difficulties, the contractor is usually paid on the basis of compacted volume of the embankment.

Table 23.2. Swell of soils

S. No.	Soil	Swell (percent) range	Average value which can be used for calculations
1.	Clean sand or gravel	5–15	10
2.	Top soil	10–25	20
3.	Loamy soil	10–35	20
4.	Common earth	20–45	25
5.	Clay	30–60	35
6.	Solid rock when blasted	50–80	60
7.	Murram	25–35	30

23.2.3. Clearing and Grubbing

Clearing and grubbing is done by dozers, rippers and rooters. A tractor of 50 H.P. –130 H.P. tows the ripper and rooter. The ripper consists of a heavy T-shaped beam of welded plates mounted on two steel wheels. The beam carries tyres, generally three in number. Replaceable hardened ripper shoes fit on to the ends of the tyres and these shoes do the ripping and loosening. The penetration of the tyres is about 0.3–0.6 m. The output of a ripper is about 3000 cu m per hour in ordinary soil, 1500 cu m per hour in hard dry clay and 500 cu m per hcur in murram.

Grubbing the top soil, generally to an extent of 15 cm, is done by a dozer blade. A tractor-dozer of HP 90-120 is used. For heavy clearance, a bigger capacity of 160-200 HP is needed The tractors are preferably crawler type. About 0.2 hectares of land can be grubbed under average conditions per hour.

23.2.4. Tractors

A tractor is a versatile equipment having a variety of uses in road construction, such as :

1. To pull rippers and rooters
2. To pull towed scrapers
3. To pull sheepsfoot rollers
4. To push load scrapers
5. To pull towed pneumatic rollers
6. To perform miscellaneous jobs such as pulling down trees, acting as power unit for winches and hoists etc.

There are two main types of tractors :

1. Crawler type
2. Pneumatic wheel type.

Crawler tractors have a low maximum speed, around 10 kmph and are used primarily where high speeds are sacrificed in order to obtain good traction and high draw-bar pull. They are also preferred where the ground is not firm. The pressure under the tracks is as low as 0.4 to 0.5 kg/cm^2. Pneumatic wheeled tractors are used for moving at high speeds (upto 50 kmph) on firm ground. The sizes of tractors and their applications are indicated in Table 23.3.

Table 23.3. Tractor sizes and applications

S. No.	Size	Applications
1.	40–50 HP	Towing one double drum sheepsfoot roller and towing smaller rippers (1 Tonne)
2.	70–90 HP	Towing two double-drum sheepsfoot rollers, 7–9 cu m scrapers or 3 T-rippers ; push-loading scrapers.
3.	110–130 HP	Towing 10-13 cu m scrapers or 6 T-rippers and push loading scrapers.

23.2.5. Tractor Dozers

A tractor dozer, popularly called bull-dozer, is a tractor with a 3–3.5 m long, 0.9–1.2 m high blade mounted in front of it. If the blade of the equipment is set at an angle, it is called an angle dozer. The blade of some dozers can be tilted in the vertical plane to a tilt of about 1 in 10. The equipment is very versatile and can perform the following operations :

1. Clearing and grubbing land of vegetation and tree stumps.
2. Removing top soil from borrow areas.
3. Opening up pilot road formations in steep hill sides to such a width that other equipment can move in and complete the work.
4. Moving earth for short distances, say upto 100 m.
5. Box-cutting a formation for laying pavement layers.
6. Moving earth in situations where the turning space for other plant such as scrapers is not available.
7. Spreading earth in layers.
8. Clearing construction sites of debris.

Tractor-dozers can be crawler mounted or pneumatic wheel type. The relative advantages of each type have been indicated when dealing with tractors. Crawler-mounted dozers have a digging and travelling-with-load speed of about 2.5 kmph. This may be about 4 to 5 kmph for wheeled dozers.

The tractor-dozers are used for moving earth over short leads, say upto 100 m. They are never economical for hauls longer than this.

Dozers are also used for spreading earthwork.

The tractor-dozers are available in sizes ranging from 65 to 200 H.P. The output of dozers in moving earthwork can be seen from Table 23.4.

Table 23.4. Output of tractor-dozers in earthwork

Type / Size	Output in cu m per hour in borrow-pit measure				
	Lead 15 m	Lead 30 m	Lead 60 m	Lead 90 m	Lead 120 m
Crawler type					
65 HP	45	30	15	10	8
80 HP	75	50	25	18	15
130 HP	85	55	20	20	18
Wheeled type					
165 HP	90	60	35	25	20

23.2.6. Scrapers

Scrapers dig their own load, as they move forward. They combine the operations of digging, loading, hauling and discharging. A cutting blade, which can be raised or lowered upto 20 cm, is pulled through the earth causing it to travel up the face of the blade into the bowl of the scraper. Some of the earth falls forward into a carrying apron. When the bowl is full, the apron is lowered to prevent spillage and the cutting edge is raised. After haulage, the material is dumped by lowering the cutting edge to the desired height above the fill and opening the front apron. The blade then serves as a strike off beam.

Table 23.5. Output of scrapers in cu m per hour in terms of borrow-pit measure

Soil type	8 cum struck capacity towed scraper					12 cum struck capacity motorised scraper						
	One-way lead of					One-way lead of						
	150 m	175 m	250 m	300 m	350 m	300 m	450 m	600 m	750 m	900 m	1200 m	1500 m
1. Light loam or crumbly soil	45	40	35	30	25	60	55	45	40	35	30	25
2. Sand and gravel	45	40	35	30	25	60	55	45	40	35	30	25
3. Good common earth	45	40	35	30	25	58	50	45	40	35	30	25
4. Clay, hard and tough	40	38	33	28	23	55	50	40	38	35	28	23
5. Murram	40	38	33	28	23	55	50	40	38	35	28	23
6. Hard murram	40	38	33	28	23	55	50	40	38	35	28	23
7. Clay, wet and sticky	40	38	33	28	23	55	50	40	38	35	28	23

Two types of scrapers can be identified, viz. (*i*) towed scrapers and (*ii*) motorised scrapers. Towed scrapers are used for short hauls, say 150–500 m, at a maximum speed of 10 kmph For longer hauls, in the range of 500–1500 m, motorised scrapers are used. These attain speeds of 30 kmph or more. Since motorised scrapers cannot develop a high tractive effort, a pusher tractor-dozer is used in combination. Usually one dozer works in combination with 3–5 scrapers.

Towed scrapers are available in sizes of 7–12 cu m struck capacity. Motorised scrapers are available in sizes upto 25 cu m, though a size of 8–10 cu m struck capacity is common for highway works.

A towed scraper of 8 cu m struck capacity requires a 90 B.H.P. crawler tractor. A motorised scraper of 8–10 cu m struck capacity needs a dozer of 165 HP for push loading.

The output of a scraper with an operating efficiency of 65 per cent, normal swell factors and a 50 min working hour has been tabulated in Table 23.5.

23.2.7. Graders

Graders are used for a number of purposes :

1. For spreading heaped earth into layers.
2. For shaping the cross-section during construction.
3. For maintaining the cross-section of the embankment.
4. For maintaining gravel surfaces.

- Graders are either towed or self-propelled. The latter, also called autopatrol or motor grader, are more common. The grader has a blade of about 3.5 m, but its effective length during spreading becomes 2.75 m. The blade can be set at any angle. A 100–110 HP motor grader is a popular size. The normal grading speed is 3 kmph. The output of a grader for spreading the earth, which is achieved generally in 4 passes, is about 1300 sq m per hour, assuming 65% operating efficiency and a 50 min working hour. Assuming a normal compacted thickness of layer of 15 cm, the output in terms of compacted volume becomes about 200 cu m per hour.

23.3. ROCK EXCAVATION MACHINERY

Rock excavation is achieved by drilling and blasting. The equipment used for this work consists of air compressors and rock drills.

23.3.1. Air compressors

Compressed air is used for (*i*) drilling rock, (*ii*) operating pneumatic equipment such as pavement breakers, clay cutters etc. and (*iii*) well sinking. Portable compressors worked by electrical power or dissel engine are common.

The capacity of an air-compressor is expressed in terms of the volume of air delivered by it in cubic feet per min (CFM) in the F.P.S. system or in cubic metres per minute (CMM) in metric system. The common sizes used are 3, 4, 5 and 6 CMM. The compressor delivers air at gauge pressures in the range 5–8 kg/cm^2.

23.3.2. Drilling

Drilling into rock is accomplished by jack hammers or wagon drills (drifters). Jack hummer is a pneumatic percussion type drill, small enough to be handled by a man. The weight ranges from 4.5 kg to 35 kg. The hammer gives 2000–2500 blows per minute. The motion of the piston is transmitted to a tungsten carbide drilling bit through the drilling rod. They can drill holes upto 6 m depth.

23.3.3. Blasting

Blasting is done by explosives. The normal explosive used is dynamite of a concentration of about 40–60 per cent or gelatine which is a waterproof jelly-like dynamite. Detonators are used for charging.

23.4. PRODUCTION OF AGGREGATES

The production of aggregates involves collection of large-sized stones (by blasting or from natural deposits) and crushing them to smaller sizes.

23.4.1. Crushers

Rock crushers break large stones into smaller pieces. This is often accomplished in stages, the first stage crushing being called primary crushing, the second and third stages being called secondary and tertiary crushing. The ratio of reduction of the size of a stone in a crusher is a very important aspect that deserves careful consideration. A ratio larger than 4 : 1 is likely to result in flaky aggregates being produced.

Crushers are of different types such as :

1. Jaw crushers
2. Gyratory crushers
3. Cone crushers
4. Roll crushers
5. Impact or hammer mills
6. Rod and ball mills.

Jaw crusher is the oldest and the most familiar type. It consists of a fixed jaw and a swinging, reciprocating jaw. The distance between the two jaws is gradually reduced as the stone travels downward. Jaw crushers are called granulators when they are used for the production of aggregates of size 20 mm and down. Special jaw plates are needed for this purpose and they are usually set for 16 mm.

Cone crushers and roll crushers are used for secondary or tertiary crushing to produce small-sized stones in large quantities. Rod and ball mills are used to produce very fine aggregates such as sand from stones.

23.4.2. Screens

Screens can be of vibratory or revolving type. The former is used for large plants and the latter for small plants. Revolving screen consists of a long cylinder perforated with holes of largest size at one end and smallest size at the other, the intermediate ones being in the middle. Vibratory screens are of multiple deck, the sizes of the openings progressively decreasing at each level.

23.5. TRANSPORTING EQUIPMENT

Apart from scrapers, the other transporting equipment commonly used are trucks and dumpers.

Trucks manufactured in India have rated pay-load of $7\frac{1}{2}$ tonnes, though loading upto 12 tonnes is common. The weight of some of the common road materials can be taken from Table 23.6.

Table 23.6. Weight of common road materials

S. No.	Material	Weight kg/m^3
1.	Earth, loose	1300–1600
2.	Sand, loose	1500
3.	Sand, wet	1900
4.	Murram, loose	1500
5.	Stone boulders	
	Limestone	1300
	Granite	1500
	Basalt	1600
6.	Stone metal (25 mm–75 mm)	1500–1600
7.	Stone chips (below 20 mm)	1700

Conveying material by trucks involves loading and unloading. Table 23.7 gives the number of trips per day that can be achieved with a truck for various leads, assuming an average speed of 25 kmph and 65 per cent operational efficiency.

Table 23.7. Number of trips per day by trucks

One-way lead (km)	No. of trips per day
1.5	9
3	7
5	6
6.5	5
8	4

10	4
15	3
20	3
30	2

23.5.3. Dumpers are used for coveyance of earth and other materials on a large scale. They are loaded by shovels. Rear dumpers and bottom dumpers are two popular verieties. Capacities from 3 cu m to 20 cu m struck volume are available. They travel at a speed of 20–30 kmph. An economic lead is 2.5 km.

23.6. WATERING EQUIPMENT

23.6.1. Water tankers

Water-tankers of 4000–5000 litre capacity are widely used for watering earthwork and in water-bound macadam work. A sprinkler attachment with a bar provides for uniform distribution at any desired speed. A water sprinkler can cater to about 1000 cu m of compacted earthwork per day.

23.6.2. Water pumps

Small pumps of 2.5–5 H.P., diesel driven, are extremely useful for conveying water for roadway work. Alkathene pipes 50 mm–100 mm dia are commonly used for this purpose.

23.7. COMPACTION EQUIPMENT

Compaction is needed for various layers of the embankment and pavement. The different types of compacting equipment and their applications are indicated in Table 23.8.

Table 23.8. Compacting equipment and their applications

S. No.	Compacting Equipment	Applications	
1.	Road roller, 8–10 T three-wheeled	1.	Earthwork rolling
		2.	Soil stabilisation
		3.	Rolling granular bases (WBM, murram etc.)
		4.	Rolling surface dressing, premix carpet, mix seal, bituminous macadam, built-up-spray grout.
		5.	Break-down rolling of asphaltic concrete

2.	Road roller, 6 T or 8 T tandem	1.	Surface dressing
		2.	Finish rolling of asphaltic concrete
3.	Sheepsfoot rollers	1.	Rolling cohesive soils
4.	Pneumatic-tyred roller	1.	Rolling non-plastic silts and silty soils
		2.	Intermediate rolling of asphaltic concrete
5.	Vibratory roller	1.	Compacting sand and cohesionless soils
		2.	Compacting all types of soils for obtaining high densities
		3.	Compacting granular bases and sub-bases to obtain high densities
		4.	Compacting bituminous bases and surfaces to obtain high degree of compaction
6.	Frog rammer	1.	Compacting sand and cohesionless soils
		2.	Compacting in restricted space.

23.7.1 Three-wheeled road roller 8-10 T

This is the most common rolling equipment and is versatile in applications. It is diesel powered, the B.H.P. being around 35. The diameter of the front roll is around 105 cm, its width being around 100 cm. The diameter of the rear roll is around 145 cm, its width being around 50 cm. The rolling width is around 2 metres. The front roll gives a load of around 35–40 kg/cm width and the front roll gives a load of 70–80 kg/cm width The speed of rolling is in the range 1.5–6.0 kmph.

23.7.2. Sheepsfoot rollers

Sheepsfoot rollers consist of hollow circular drums of steel 1.2–1.5 m long and 0.9–1.2 m dia, with legs or tamping feet on the circumferential area at the rate of 12-18 per square metre of area. They are assembled in units of 1,2,3, or 4 drums. The tamping feet on the drums are staggered into rows. The length of the feet should be atleast sufficient to pass through three-fourths' of the thickness of the loose layer and is usually 15–20 cm. The rollers can be ballasted with water. The weight of a single drum is in the range 1200-1800 kg when empty and 2200–2800 kg when ballasted. A tractor of 45 HP can pull a single drum and one of 70–90 HP can pull double drum. A speed of 4 kph is common.

Sheepsfoot rollers are suitable for cohesive soils and the moisture content of the soils should be preferably near their plastic limit. The number of passes of sheepsfoot rollers depend upon the type of soil, moisture content and density desired. Generally 8–16 passes are needed.

23.7.3. Pneumatic tyred rollers

Pneumatic tyred rollers consist of a box mounted over two axles, the rear axle having one more wheel than the front and the wheels of the front axles so arranged that they are located in plan in between the rear wheels. Generally there are four wheels in the front and five in the rear. The assembly is loaded to get a pressure of around 300 kg per linear metre measured along the axle. Weights in the range 12–45 Tonnes are common.

When compacting soils, the layer thickness should be 10–20 cms. The roller is suitable for compacting non-plastic soils and silty soils. The optimum speed is 4 kmph. They are either self-propelled or towed by a tractor.

Pneumatic tyred rollers are also used for intermediate rolling of dense asphaltic concrete. Rollers of 12–18 T are used. The tyre pressure should be a minimum of 0.5 MN/m^2. The number of passes required is about 8–12. One pneumatic roller is generally required for an output of 25 cu m/hour.

23.7.4. Vibratory rollers

Vibratory rollers have become very popular in highway engineering applications in recent years, with the growing need to compact pavement layers and subgrade to a high degree of densification. It is well-known that considerable economy can be obtained in pavement construction if the subgrade and the pavement layers are compacted to high densities. Compacting to such high densities by static rollers is very difficult and costly. Since vibratory rollers induce oscillations, they are able to (*i*) achieve break-down of internal friction between particles of road construction materials and (*ii*) bring about better orientation of particles caused by cyclic deformation. Thus, these rollers achieve high compaction with a few passes. Both cohesive and non-cohesive materials can be compacted by vibratory rollers. The layer thickness can be significantly increased, resulting in economy. For example, as against a layer thickness of 250 mm for soils compacted by a static roller, thickness of over 1000 mm can be easily compacted by vibratory rollers. Similarly, as against a layer thickness of 50 mm, for bituminous layers compacted by a static roller, layer thickness of 250 mm can be compacted by a vibratory roller.

Vibratory rollers are now being manufactured in India. For compacting soils and granular layers, a frequency of 1500–2500 vibrations per minute (VPM) and an amplitude of 0.8–1.5 mm are recommended. For bituminous courses, a frequency in the range of 2000–3000 VPM and an amplitude in the range of 0.4–0.8 mm are recommended .

23.7.5. Combination rollers

Some modern types of road rollers are a combination of two different types of road rollers. For example, one vibrating drum in front and the rear axle having 3 or 4 pneumatic tyres, with a weight range 4 to 15 tonnes. These road rollers combine the advantages of both pneumatic and vibratory types of road rollers.

23.7.6. Hand operated vibratory rollers

For compaction in small and irregularly shaped areas, where ordinary rollers cannot find access, small capacity compactors can be used. A typical design comprises a double drum walk behind roller with two vibrating drums and weight range of only 0.4 to 1.0 tonne.

23.7.7. Vibrating plate compactors

Being light in weight, very mobile and easy to handle by hand, the vibrating plate compactor is a popular tool for compaction in small works, such as pothole patching. Commonly, the vibrating effect is achieved by installation of a rotating eccentric weight in the system. The plate compactor is normally provided with a long steel handle so that the operator can walk behind it and control the operation. For working in very tight and inaccessible areas, vibratory plate compactors can be equipped with a remote control arrangement.

23.7.8. Frog rammers

Frog rammers are used for compacting cohesionless soils and soils in inaccessible and restricted locations where ordinary rollers cannot work. They are worked by pneumatic system or by a diesel engine. The weight varies from 30 kg to one tonne.

23.8. BITUMINOUS EQUIPMENT

In bituminous construction, the following categories of equipment are used :

1. Bitumen storage equipment
2. Bitumen heating equipment
3. Bitumen spraying equipment
4. Bitumen mixers
5. Paver finishers
6. Compacting equipment

The compacting equipment has already been described earlier.

23.8.1. Bitumen storage equipment

Bitumen is stored in drums or in special bulk handling depots. The capacity of one drum is 155–162 kg of bitumen. Bulk handing has the following advantages over packed bitumen:

1. The steel sheets used for making drums are imported, entailing an expenditure of nearly Rs. 36 crores in foreign exchange annually.
2. The price difference between bulk bitumen and packed bitumen is nearly Rs. 500 per tonne, and thus bulk bitumen is considerably cheaper than packed bitumen.
3. There is less wastage of bitumen when handled in bulk than when handled in drums.
4. The cost of drums is avoided if bulk bitumen is used.

In the bulk system of handling, hot bitumen is filled in rail tank-wagons or tank lorries at the refinery. These have special arrangement for heating bitumen when in transit The bitumen rail-wagons or tank-lorries deliver hot bitumen directly to the works or to special storage tanks.

A bitumen tank-lorry has a capacity in the range of 6–12 T (Fig. 23.1). The tank is insulated with material such as glass wool, asbestos magnesia etc. There are heating flues running along the length of the tank. A pump with a capacity of about 200–300 litres per minute is provided.

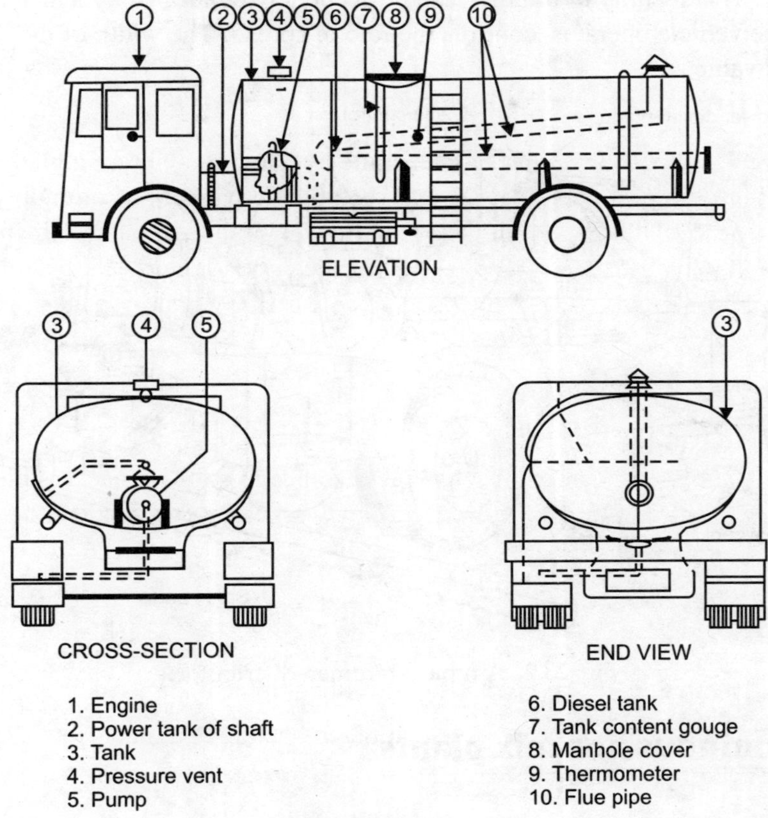

Fig. 23.1. Bitumen tanker.

A bitumen tank trailer, towed by a tractor or a truck, is also used for moving bulk bitumen to the work-site. The capacity is of the order of 3–6 T. The tank is adequately insulated and an arrangement for heating and pumping is provided.

Stationary bulk bitumen storage tanks are installed in bulk bitumen depots. A capacity of 5/15 T is economical and convenient. Arrangement for heating and pumping is provided.

23.8.2. Bitumen boilers

Bitumen boilers are needed for heating bitumen obtained in packed form. Boilers of a wide range of capacity are available. 100–10,000 litres capacity boilers are standardised by B.I.S. Small boilers are fired by coal or firewood, but large boilers are fired by oil-burners.

23.8.3. Bitumen pressure distributors

A bitumen pressure distributor (Fig. 23.2) is a tank of capacity 5000 to 20,000 litres mounted on a lorry, having oil-fired burners and a pump. A spray bar unit with nozzles is fitted at the rear of the tank just above the road surface. Bitumen can be applied at a pressure on to the road surface in connection with specifications like surface dressing, grouted macadam, built-up-spray grout, seal coat, tack coat etc. The quantity of bitumen can be accurately controlled by a metering device. The speed at which the vehicle operates, controls the rate of spread. The width of the spray bar can be set to any desired value.

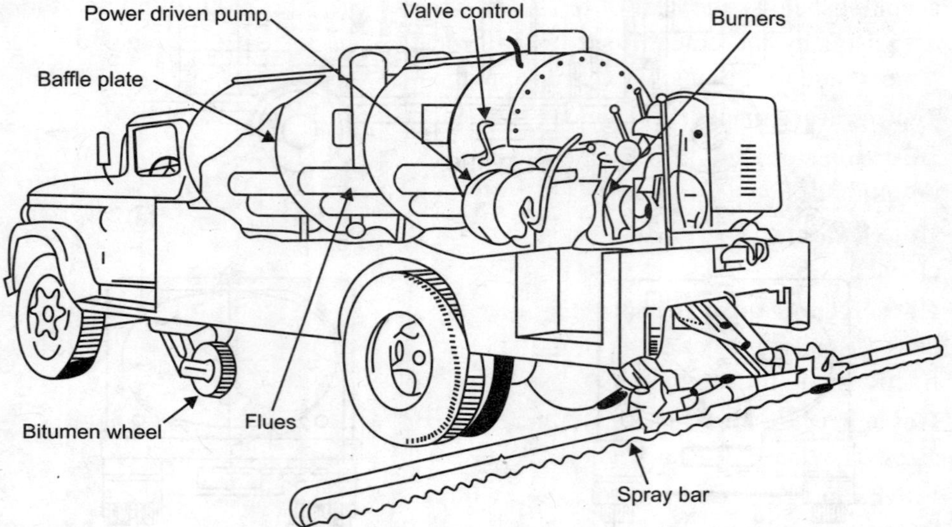

Fig. 23.2. A typical bitumen distributor.

23.8.4. Bituminous hot-mix plants

23.8.4.1. Plant types

For the production of a hot bituminous paving mixture, aggregates are blended, heated and dried and then mixed thoroughly with bitumen. The two types of mixing facilities which are widely used can broadly be classified as under :

A. Batch mixing plant

B. Drum mixing plant

Any of the above types can also be broadly classified as stationary *i.e.*, permanently situated and not required to be dismantled and moved or portable *i.e.*, capable of being dismantled and moved by road or rail.

(*A*) *Batch mixing plant*

In a batch mixing plant, various sizes of hot aggregates in storage bins are withdrawn in pre-determined quantities to make one 'batch' before mixing with a predetermined quantity of bitumen. It comprises the following :

(*a*) **Cold aggregate storage bins of different aggregate sizes:** These should be atleast four in number with discharge gates to control the flow of aggregates of different sizes in the desired amounts. Filler is not included in these bins, and is handled differently, since it is to be protected from moisture.

(*b*) Conveyor and cold elevator: The cold aggregate feeding system discharges onto a conveyor in pre-set quantities to give the required grading and then into the cold elevator which discharges into the Dryer.

(c) Dryer: The main functions of a Dryer are (*i*) removal of moisture from the aggregates, accomplished by the heat of the dryer which vaporises the moisture and the vapour is drawn off by the draught and (*ii*) to heat the aggregates so that they are of the right temperature for mixing with bitumen in the pug mill.

(*d*) Temperature measuring device : It is essential for the control of quality of the mix.

(*e*) Dust collector: Provided for the prevention of pollution as well as to reduce loss of fines, upsetting design grading.

(*f*) Hot elevators: These consist of a system of bucket elevators housed within a covered chamber.

(*g*) Hot screening unit: The hot materials carried by the elevators are discharged over a multi-deck (usually a 4-deck) vibrating table screen which separates the different aggregate fractions into the different hot bins.

(*h*) Hot aggregate bins : These are the temporary storage for the hot aggregates of different sizes. They have discharge gates opening at the bottom.

(*i*) Filler silo : Because of its fineness, the filler material is stored separately in a filler silo and, after weighing, the hopper discharges the filler material into the mixer, generally after bitumen has been admitted into the mixer.

(*j*) Bitumen supply arrangement : It consists of (*i*) bitumen storage (*ii*) bitumen heating and circulation and (*ii*) bitumen metering. For each batch, bitumen may either be pumped into a special bucket of known weight and weighed on a scale or it may be measured in volume by a meter.

(*k*) Measuring and mixing of aggregate and bitumen : The aggregate is drawn from hot bins in predetermined quantities and dropped into a pug mill mixer in batches ; the required amount of bitumen is added and mixed with aggregate. The hot aggregates and the hot bitumen are thoroughly mixed in the pug mill mixer.

(*j*) Discharge into truck or hot storage silo : After the mixing operation has been completed, the final mixture is discharged from the bottom of the pug mill mixer directly to the hauling truck below or taken to a hot storage silo for temporarily storing the mix, before haulage.

A schematic diagram of the working of a typical Batch Plant is shown in Fig. 23.3 (*a*).

(*B*) *Drum mixing plant*

When compared to the Batch Mixing Plant described above, the Drum Mixing Plant is without the hot aggregate bins and (the pug mill mixer. It only consists of a cold aggregate feed system ; a rotating drum mixer (the drum mixer not only dries the aggregate but also blends the aggregate and bitumen into the designed mix), the heating unit or flame being at the upper end of the drum with cold aggregate entering the drum at the flame end), a bitumen supply system, a dust collector system and a surge silo. In a continuous plant like the drum mixing plant, provision of surge or storage bins is an absolute necessity.

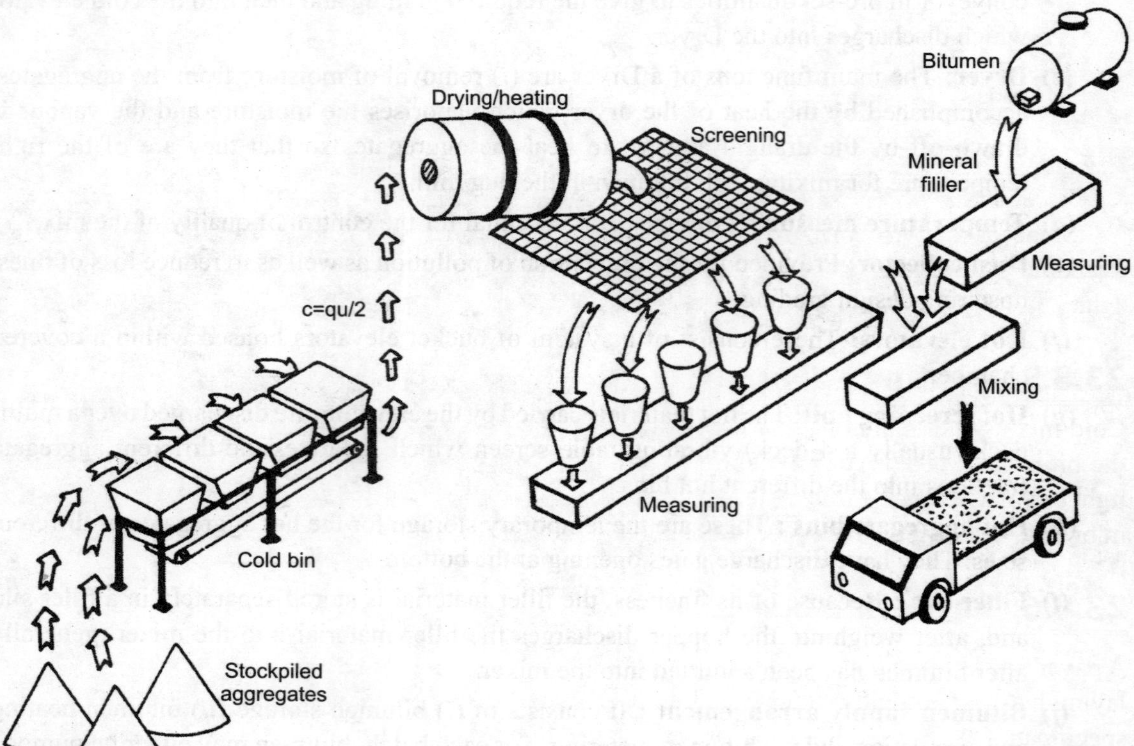

(*a*) Batch mix type

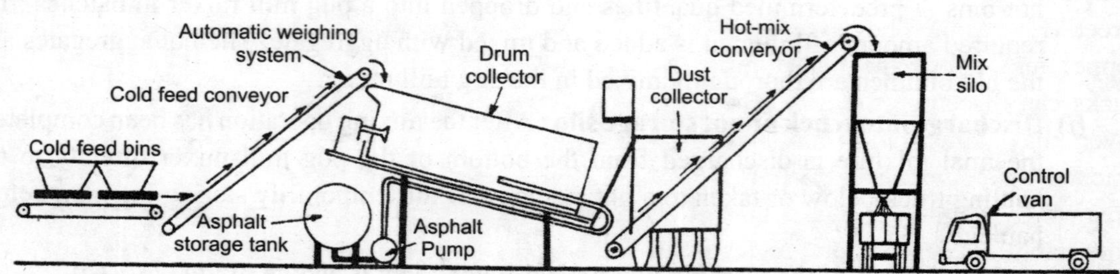

(*b*) Drum mix type.

Fig. 23.3. Schematic diagrams of the operation of hot mix plants

A schematic diagram of the working of a typical drum mix plant is shown in Fig. 23.3 (*b*). Recently, there have been a number of modifications in the conventional design of a Drum Mixer, making its use more popular in comparison to a batch mixing plant, the main advantages being savings in energy and reduced pollution.

Hot-mix plants of various capacities are covered by an Indian Standard. The capacities are :

20—30 T/hour

30—45 T/hour

40—60 T/hour

60—90 T/hour

80—120 T/hour

Small capacity mixing plants, where an arrangement is made for drying the aggregates, are available in the country. They are called mini-hot-mix plants. The capacity can be 3–5 T/hr or 6–10 T/hr.

23.8.5. Cold-mix plants

Cold mix asphalt plants are those where there is no arrangement for heating the aggregates. Only the bitumen is heated, when required and added to the mixer. Cold mixing is generally done for light chipping carpet, slurry seals etc. The plants available in India have a capacity of 0.15–0.20 cu m or 0.30–0.35 cu m. They are mounted on a steel chassis having pneumatic wheels.

23.8.6. Paver finisher

A paver finisher is indispensable for laying hot-mix hot-laid bituminous specifications. The equipment is self-propelled and is capable of laying the bituminous material to any desired thickness and partially compact it by means of a vibrating screed (Fig. 23.4). The paver has a hopper into which the rear-dump trucks can discharge the mix. The paver may be crawler mounted (tracked) or equipped with rubber tyres which permit a greater degree of freedom for movement. The crawler-mounted machines are more stable and can support greater width of the screed. The screed width can be adjusted, generally in the range 2 to 5 m. The bituminous mix discharged from a tipper lorry into the receiving hopper in the front portion of the paver is carried along the conveyor through flow control gates to the augers which distribute the material in front of the screed, to the full width of the screed. A strike off beam, which also tamps the mat, controls the layer thickness. The paver finisher operates at speeds 1.5–10 m/min. A speed of 3—5 m/min. will be found generally acceptable. The width of the mat can be adjusted in the range 2–5 m. The cross-profile can be controlled by adjusting the screws of the strike-off beam.

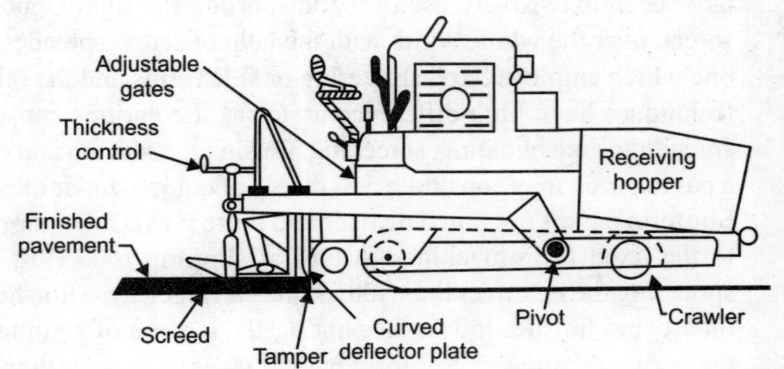

Fig. 23.4. Paver finisher

Paver finishers of a capacity 45–75 T/hr are generally used for roadworks.

23.9. CEMENT CONCRETE ROAD-MAKING EQUIPMENT

The traditional method of concrete road construction in India has been labour-intensive. But then concrete road construction itself has been done on modest scale in India so far. If larger scale work is to be done, a good deal of mechanisation would be needed. The equipment used can be considered under the following heads :

1. Batching and mixing plants
2. Spreading machines
3. Compacting and finishing machines.

23.9.1. Batching and mixing equipment

The ingredients for concrete are always proportioned by weight in a weigh batching equipment.

Mixing of concrete is done by concrete mixers which are available in a variety of sizes. Mixing is continued for a period of not less than one and half minutes after all the materials are introduced.

Tilting drum type mixers are available in small sizes, whereas not-tilting drum type mixers are available in large sizes.

23.9.2 Transportation, placing and compaction of concrete

Generally, 5 to 10 m^3 tippers are ideal for transporting and placing concrete. Transit mixers also have been extensively used for transporting the mix. Concrete discharged on the carriageway is spread over the whole width with the help of screw spreader. There are two types of pavers in use : one which employs fixed shuttering or side forms and the other using the slipform technique. Both techniques have little difference as for as the various construction operations are concerned *i.e.*, spreading, consolidating, screening, finishing, texturing and curing. Depending on the requirements, a paving train may constitute 5–6 pieces of equipment or these may be clubbed to give 2 to 3 pieces. Slipform pavers are much heavier and more powerful equipments which move on tracks compared to the fixed form machines. A typical slipform train (Fig. 23.5) consists of paver, intended for spreading the mix over the width of the carriageway, a finisher which vibrates, forms an appropriate mould and finishes the surface and lastly, a piece of equipment capable of texturing and spraying the curing compound. Slipform paving is more popular than fixed form paving.

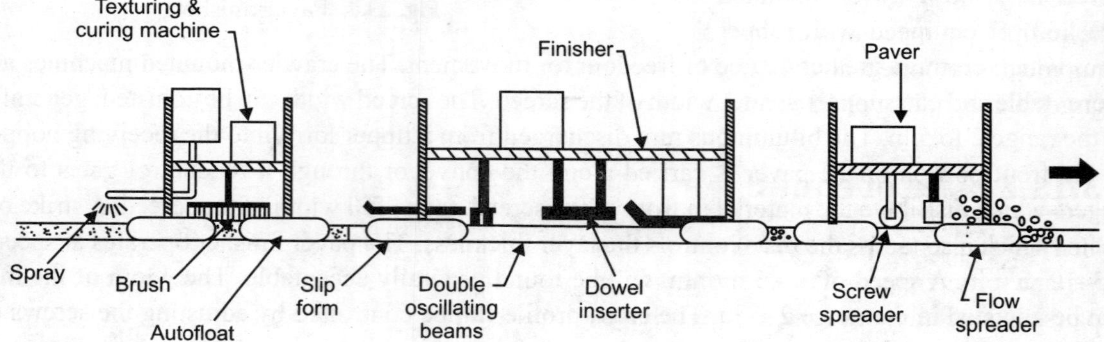

Fig. 23.5. Main components of slipform paving train for construction of a cement concrete pavement.

23.10. BRIDGE CONSTRUCTION AND INSPECTION EQUIPMENT

For the construction of major bridges, a wide range of equipment is needed.

This includes :

(*a*) Cranes (*b*) Grabs

(c) Sand pumps (*d*) Diving equipment

(*e*) Piling equipment (*f*) Sheet piling equipment

(*g*) Concrete batching and mixing plants

(*h*) Slip form equipment (*i*) Bridge builders

(*j*) Girder launching equipment.

Most of the equipment is locally manufactured in India. The technology of bridge building has been developed to a high degree of sophistication in the country.

Periodic inspection of bridges will be beneficial in identifying the defects in time and in taking up immediate remedial measures. Manual inspection of the underside and sides of girders and bearings is extremely tedious since scaffolding will have to be constructed. In countries abroad, a mobile bridge inspection unit is used for this purpose. The unit is mounted on a road vehicle. By means of turn tables, telescoping booms and hydraulic arrangements, the inspection cabin can be manoeuvered to any location

23.11. EQUIPMENT USAGE RATES

The costing of equipment usage rates is necessary for estimating the cost of works and for maintenance or proper accounts of machinery in use. Factors which govern the usage rates of equipment are :

1. Initial cost of equipment
2. Life of equipment
3. Salvage value
4. Method of depreciation
5. Interest charges
6. Maintenance and repairs cost
7. Cost of operation
8. Overhead charges.

23.11.1. Initial cost of equipment

The initial cost of equipment can be easily determined by enquiries in the local market. If any equipment is imported, all the expenses incurred in connection with freight, customs duty, clearance charges and handling charges are to be considered.

23.11.2. Life of equipment

When an equipment is used, it wears out in due course, demanding replacement to worn-out parts and maintenance. As the equipment becomes older and older, the cost of maintenance and repairs also increases. A time may eventually come when the hourly cost of owning and operating the machine will be greater than the average hourly cost upto that time. This indicates that the equipment has reached the end of its economic life and should be scrapped.

Tyres wear out more frequently than the main equipment. Life of tyres ranges from 10,000 to 50,000 km, depending upon the type of surface on which they operate. It is, therefore, customary to delete the cost of tyres from the cost of the new equipment and treat tyres separately.

23.11.3. Salvage value

Salvage value of an equipment is its resale value after it is used for its full working period equal to its economic life. A salvage value of 20 per cent is generally considered in the calculations.

23.11.4. Depreciation

During the process of using the equipment over its economic life, the equipment progressively loses its value from its initial cost to salvage value. Since the equipment was used on various works during its life-time, those works should bear the initial cost (minus the salvage value) of the equipment in proportion to the actual use. In other words, the owner should recover the cost of equipment from the various works on which it has been used.

There are various methods of calculating depreciation, but the one most widely used viz., the straight line method, is the most simple. According to this method, the useful life of the equipment is assumed in some units (hours or kilometres etc.) and the depreciation per unit (hour or kilometre) is obtained by dividing the total initial cost (minus salvage value) by its useful life.

Storage charges may be added at 1 per cent of depreciation.

23.11.5. Interest charges

The capital invested on the equipment would have earned some interest had it been invested elsewhere. Thus it is appropriate to include the interest on capital in determining the usage charge.

23.11.6. Repairs and maintenance charges

The equipment gets worn out due to use. Spare parts need to be replaced when they get worn out. Labour is involved in such repairs and in periodic maintenance. It is generally found that the repairs and maintenance charges represent 40 to 50 per cent of the investment charges (initial cost less the salvage value).

23.11.7. Cost of operation

The cost of operation consists of wages of crew and the cost of consumables. The cost of crew wages should include all benefits such as leave salary, provident fund, medical facilities, etc. The cost of consumables includes fuel and lubricants. A rough and ready method of calculating the fuel consumption is by the following formula :

Quantity of diesel: 0.15 litres per B.H.P. per hour (23.1)

Thus, a diesel grader, with a B.H.P. of 115, working at 70 per cent efficiency, on a 50 minutes hour, will consume :

$0.15 \times 115 \times 0.7 \times \frac{50}{60} = 10$ litres of diesel per hr.

The lubricating oil needed will depend upon the B.H.P., the capacity of crank-case and the number of hours between oil changes etc. The following formula gives the quantity of the lubricating oil:

$$q = \text{H.P.} \times 0.7 \times 0.003 + \frac{c}{t} \tag{23.2}$$

where q = quantity of lubricating oil in litres per hour

H.P. = Rated H.P.

0.7 = operating factor

c = crank case capacity in litres

t = time taken for changing of oil in crank case, which is generally about 100 hours.

Thus a 100 H.P. engine, with a crank-case capacity of 15 litres, requiring change every 100 hours, will consume lubricating oil at the rate of:

$$q = 100 \times 0.7 \times 0.003 + \frac{15}{100} = 0.21 + 0.15$$

$$= 0.36 \text{ litres per hour.}$$

In addition to lubricating oil for the main engine, other oil such as brake oil and gear oil, grease and cotton waste are also needed. These may cost as much as the engine oil.

The following problem indicates how the usage charges can be computed :

Problem 23.1. Calculate the hourly usage charges for a truck, given the following particulars :

1. Initial cost : Rs. 12,00,000
2. Salvage value (excluding tyres) : Rs. 152,000
3. Life : 40,000 hrs.
4. Cost of 6 tyres : Rs. 48,000
5. Life of tyres : 8,000 hours
6. Repairs and maintenance : 100 per cent of initial cost less salvage value
7. Diesel consumption : 10 litres per hour
8. Engine oil consumption : 0.4 litres per hour
9. Cost of diesel : Rs. 50 per litre
10. Cost of engine oil : Rs. 115 per litre
11. Interest : 10 per cent.

Solution.

I. Ownership charges

1. Depreciation

(*a*) Cost of truck less tyres : Rs. 11,52,000

(*b*) Salvage value : Rs. 152,000

(*c*) $a - b$: Rs. 10,00,000

(*d*) Economic life (hrs.) : 40,000

(*e*) Depreciation per hour : $\frac{c}{d} : \frac{Rs. 10,00,000}{40,000}$

= Rs. 25

(*f*) Depreciation per hour of tyres

$= \frac{48000}{8000} =$ Rs. 6

Total $e + f$ = Rs. 31

2. Storage

1% of e = Rs. 0.25

3. Interest at 10 per cent per annum (assuming 4000 hrs/year)

$12,00,000 \times \frac{10}{100} \times \frac{1}{4000} =$ Rs. 30.00

Total (1), (2), (3) = Rs. 31 + 0.25 + 30 = Rs. 61.25

II. Operation charges

4. Repairs and maintenance at
   100 per cent of (*e*) : Rs. 25
5. Fuel
   Diesel : 10 litres @ Rs. 50 : Rs. 500
6. Engine oil
   0.4 litres @ Rs. 115 : Rs. 45
7. Other lubricants and cotton
   waste same as (6) : Rs. 45

Total (4), (5), (6), (7) : Rs. 25 + 500 + 45 + 45

: Rs. 615

Total I and II : Rs. 615 + 61.25 = Rs. 676.25

Add 5% overhead charges Rs. 33.80

Total Rs. 710.05

Assuming truck operates at 50 k.m. per hour, charges per km = **Rs. 14.20**

QUESTIONS

1. Discuss the role of labour versus machinery in road construction.
2. What are the various earth-moving equipments and their usage?
3. What are the various types of equipment used for transporting road materials?
4. What are the various types of equipment used for compacting earthwork and pavement?
5. Describe the working of a hot-mix plant.
6. Describe the working of a paver-finisher.
7. How is the usage rate of a road-making equipment calculated?

Chapter

ROAD CONSTRUCTION PROGRAMMING AND MANAGEMENT

24.1. NEED FOR ROAD CONSTRUCTION PROGRAMMING

Highway projects involve huge investments of public funds. Where resources are scarce, these funds have competing demands. Their channelisation into the roads sector is at the cost of other sectors of the economy. This underlines the need to utilise the allotted funds in the shortest time possible so that long gestation periods are avoided. Delays in construction involve locking up of scarce resources and inconveniencing the public. Delays in construction lead to cost escalation and claims by the contractor. Such delays should be avoided at all costs. This can be achieved by careful programming and monitoring of the works.

Highway construction today is a complex process, involving a variety of subdisciplines and specialisation. A number of suppliers and contractors may have to be working on different sub-components of the main work. There is a need to effectively co-ordinate the activities such that no bottlenecks arise.

The timely availability of labour, materials and equipment is crucial to the successful implementation of a project. Labour availability in India is seasonal since the rural labour is gainfully employed in agricultural operations in the appropriate seasons. Materials like cement, bitumen and steel are scarce commodities in a developing country and a good deal of advance planning and procurement is needed. Equipment is equally scarce. Though indigenous capacity exists in the country to manufacture the road-making equipment, it is necessary to order them in advance and plan their utilisation. A good construction programme can take care of the constraints.

Due to seasonal characteristics, certain specifications and items of work cannot be carried on throughout the year. For example, bituminous specifications have to be suspended in the rainy season. On the other hand, water-bound-macadam can be ideally undertaken during rainy season. Well-sinking and foundation works in mid-stream have to be suspended in the rainy season. These seasonal constraints have to be identified and taken care of in a construction programme.

Many of the activities in a project are interdependent. For example, land acquisition is a pre-requisite for road construction. Many of the activities need to be synchronised. For example, the construction of bridge and its approaches should be completed at the same time. One without the other cannot be used.

24.2. REQUIREMENTS OF A PROGRAMME

A good construction programme should :

1. Identify all elements of activities
2. Estimate the time taken for each activity accurately
3. Recognise the sequence of operations and mutual interdependence of critical activities
4. Recognise seasonal and other constraints
5. Be flexible and have capabilities for updation
6. Be easy of achievement and neither too tight nor too lax.

24.3. BAR CHART

A bar chart is the most simple and satisfactory form of presentation of a programme. It is also known as a Gantt chart, after Gantt who evolved it and popularised it. It is highly amenable to programming and monitoring of road works.

The bar-chart, a sample of which is given in Fig. 24.1, is prepared on the following lines :

1. The work is broken up into major activities.
2. These activities are listed one below the other, roughly in the sequence of operation in which they are found necessary. In the chart, these activities are entered on the left side. The quantities of works are also listed, preferably.
3. Time frame is indicated on the horizontal axis. Most convenient unit of time is a month, though a week or a quarter can also be used.
4. The beginning of each activity is selected depending upon the resource availability, seasonal constraints and financial limitations. These are marked against each activity.
5. The length of the bar is decided based on the manpower, equipment and financial resources and the possible rate of progress.
6. The percentages of each activity are scaled off and marked on the bar.
7. The financial progress is indicated at the bottom.
8. As the activity progresses, it is covered by hatching the blank space within the bar.

When a work is completed, the hatched area is covered by a full patch or a cross hatch.

24.4. CPM

Some of the best known methods of modern project monitoring are the PERT, (Programme Evaluation and Review Technique) and the CPM (Critical Path Method). The latter is particularly suited to monitor highway and bridge projects.

The CPM is a diagrammatic representation of a project by means of a network. The network is constructed of various activities and depicts the sequence and interplay of these activities. The path in the network which is critical to the project is determined and is known as the Critical Path.

The project is broken up into smaller activities and the duration of each of these is determined. The earliest finishing time (EFT) and latest finish time (LFT) of each event are then determined. The lines connecting the events where the EFT and LFT are the same, form the CPM.

The advantages of the CPM to a highway project are :

1. It enables the planner to chalk out a logical programme with inter-dependence of the various activities and the restraints.
2. A thorough and detailed examination of the project is compulsorily done.
3. The resources can be scheduled to the best advantage.
4. The activities which are likely to cause delays and bottlenecks are identified in the beginning itself and the Engineer can keep a vigil over them.
5. The CPM can bring about considerable savings in cost and time.
6. The CPM can be used for planning, design, construction and maintenance, which are the normal activities conducted by a Highway Department.

24.5. CONSTRUCTION MANAGEMENT

Whether the road-work is executed departmentally or through contracts, the management of the construction is an important responsibility. The management of the work starts much before the actual commencement of the work. It includes activities such as invitation of tenders, selection of contractors, mobilisation and actual execution. The following aspects deserve careful consideration in construction management:

1. Management of materials
2. Management of labour
3. Management of equipment
4. Financial management.

Materials required for road and bridge works cover a wide variety. More important among them are cement, steel, bitumen, stone aggregates, sand and gravel. Efficient management of the materials includes activities such as assessment of requirements ; location of sources of supply, purchase, transport, storage and issue on works. The procurement should be so phased that works do not suffer at any stage due to lack of materials and at the same time, the stock of materials is not unnecessarily high.

Labour availability for large scale road construction is seasonal in some parts of India, depending upon the agriculture activities. Hence careful thought should be given to ensure an adequate supply of labour. Amenities to labour such as temporary housing, medical facilities and creches for children are provided by the contractor as part of the conditions of contract. Their wages are governed by the Minimum Wages Act.

Road construction equipment is costly to purchase and operate. Its efficient management is thus of prime importance. Important points to be kept in view are :

1. Selection of proper size, number and specifications of equipment to do the work in hand.
2. Preparation of an utilisation programme.

3. Care in operation, by experienced operators.
4. Adequate maintenance, which needs a back-up of spare parts and skilled mechanics.
5. Safety apsects.

Financial management covers budgeting, keeping proper accounts, ensuring adequate flow of funds and keeping watch over the financial progress.

QUESTIONS

1. What is the need for a road construction programming?

2. State the requirements of a good construction programme.

3. Write short notes on:

(*a*) Bar chart

(*b*) C.P.M

4. What aspects deserve careful consideration in road construction management?

Chapter

HIGHWAY DRAINAGE

25.1. IMPORTANCE OF HIGHWAY DRAINAGE

Water is important for the construction of embankments, pavement layers and structures in a highway. But once the road is constructed, water is detrimental to the good performance of a road and must be drained off quickly. It is well said "cheap drainage is preferable to costly pavement"

25.2. MECHANISM OF DAMAGE TO HIGHWAYS DUE TO FAULTY DRAINAGE

Lack of drainage affects the performance of highways in a number of ways. They are briefly discussed below :

1. If the water falling on the pavement surface is not quickly drained away, it finds its way into the pavement layers through surface cracks, voids and joints (in cement concrete pavements). The water gets entrapped in the spaces between successive lifts of bituminous layers. The pavement is then in the "bathtub" condition, which may cause uplift pressures and result in reduced supporting power. Water may cause stripping of bitumen from aggregates and lower the binding strength of the pavement. In cement concrete slabs, "piping" is caused, which causes loss of fines in the soil and void space. A classic example of this phenomenon of entrapped water frequently met with in India is the raising of pavements attempted by earthwork or granular bases over a black-topped surface, which is left intact. This interface acts as an impervious layer to entrap water and causes "bathtub" condition. In the bitumenised road pavements, the development of 'potholes' is a common phenomenon mostly caused by the ingress of surface water, as illustrated in Fig. 25.1.

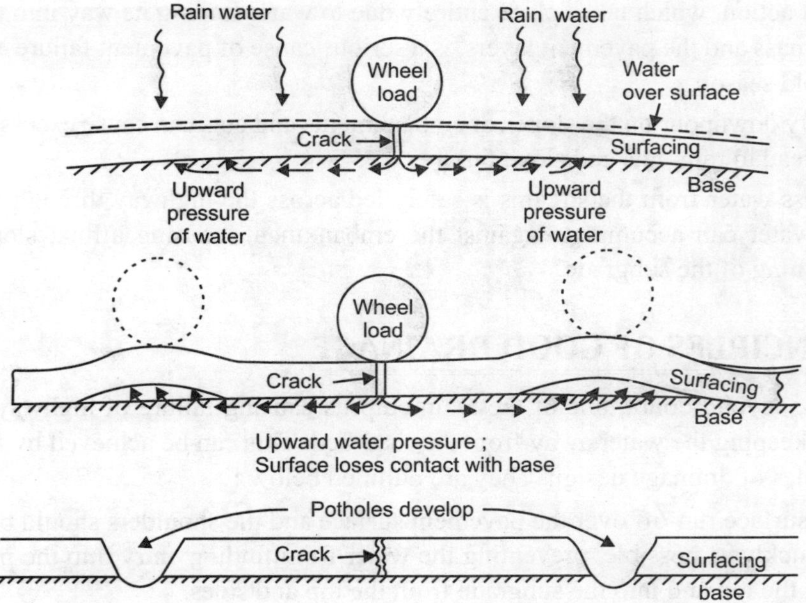

Fig. 25. 1. Development of potholes due to ingress of surface water.

2. The surface run-off or capillary water reaches the subgrade soil, softens it and reduces its supporting power. Clayey soils are particularly affected by the ingress of moisture.
3. Shoulders are rarely, if ever, surfaced. The bare earthen shoulders offer ready ingress of water. Lack of adequate outward slope presents longer opportunity for water to enter the subgrade through the sides than if the shoulders were adequately sloped.
4. In India, a commonly present road feature is the shoulders at a higher level than the road surface due to neglect of blading and dressing of the shoulders. During heavy downpour, water on the road surface does not get a free outlet and accumulates. Apart from finding its way through the cracks and voids in the pavement surface, the pavement edge at its junction with the earthen shoulder provides a possible entry point to the water. The water thus enters the subgrade from the sides of the pavement.
5. Clayey soils, which are susceptible to high volume change when changes in moisture content take place, swell and shrink alternatively with, wet and dry spells. Deep cracks are formed in the subgrade. Heaving up and deterioration of the pavement follow. Black-cotton soils, frequently found in India, suffer this distress condition.
6. In high embankments, stability of slopes is jeopardised when water enters the soil mass, causing an increase in its weight and at the same time reducing its shear strength properties.
7. In cut sections, seepage water can build up sufficient pressure to cause the failure of slopes.
8. Landslides and subsidences are caused by a general lack of drainage. The Himalayan hill region abounds in locations where serious landslides and subsidences are caused, totally disrupting road traffic. The National Highways to Srinagar and Gangtok are typical examples.

9. Frost action, which takes place entirely due to water finding its way into the pores of the soil mass and the pavement layers, is a serious cause of pavement failure at high altitudes in cold season.
10. Heavy downpour on the slopes of embankments and cuts can cause excessive soil erosion and lead to rain-cuts and slope failures.
11. Unless water from the streams is safely led across the highway through adequate vents, the water can accumulate against the embankment, causing afflux, slope erosion and softening of the subgrade.

25.3. PRINCIPLES OF GOOD DRAINAGE

Since water is, beyond doubt, one of the main culprits causing failure of highways, the designer should aim at keeping the water away from the road-bed. This can be achieved by following some simple principles of drainage design. They are outlined below :

1. The surface run-off over the pavement surface and the shoulders should be drained away as quickly as possible, preventing the water from finding entry into the pavement layers from the top and into the subgrade from the top and sides.
2. Precipitation over the open land adjoining the highway should be led away from the highway through natural drainage channels or artificial drains. Suitable cross-drainage channels should be provided to lead the water across the highway embankment which may be cutting cross to the natural drainage courses.
3. Consideration should be given to deal with the precipitation on the embankment and cut slopes such that erosion is not caused.
4. Seepage and sub-surface water is detrimental to the stability of cut slopes and bearing power of subgrades. Similarly, it can be of great importance in preventing frost action. An effective system of sub-surface drainage is a guarantee against such failures.
5. Landslide-prone zones deserve special investigations for improving drainage.
6. Poor embankment soils can perform satisfactorily if drainage is considered in the design.
7. Water-logged-and flood-prone zones demand detailed consideration for improving the overall drainage pattern of the area through which the highway is aligned.

25.4. SURFACE DRAINAGE

The rain water that comes over a pavement should be led away to side ditches as quickly as possible. If allowed to remain on the pavement, it can percolate through the voids, cracks and joints in the pavement layers, reach the subgrade, soften it and reduce its supporting power. A film of water on a pavement can cause skidding. The surface drainage of a pavement is affected by adequate camber and longitudinal slope discussed earlier.

The Indian Roads Congress recommends that a return period of 10 years should be adopted for the estimation of run-off in rural areas and 25 years in urban areas.

The storm duration is chosen equal to the 'time of concentration' which is the time taken for the water from the remote part of the catchment to reach the point of exit. The run-off from a catchment is given by the Rational Formula given below:

$$Q = 0.028\, P I_C A \tag{10.1}$$

where Q = Discharge in cu m per sec

P = Factor depending upon nature of surface (varying from 0.20 for sandy soils to 0.90 for rock, concrete or bituminous surface

I_C = Critical intensity of rainfall in cm per hour during the time of concentration

A = Catchment area in hectares

25.4.1. Design of cross-section of drainage channel (Open drain)

The discharge on the road is led to drainage channels on both sides of the road. The hydraulic design of the channel is carried out on the basis of the well-known Mannings formula:

$$V = \frac{R^{2/3} S^{1/2}}{n} \tag{10.2}$$

where V = Velocity in m/sec; which is 6 for lined channels and 1-2 in unlined earthen channels

S = Bed slope

R = Hydraulic mean depth = $\frac{\text{Area of flow}}{\text{Wetted perimeter}}$

n = Manning's Rugosity Coefficient, which varies from 0.012 for paved drain to 0.025 for soils

A trapezoidal side ditch is generally provided for roads in open areas.

25.4.2. Step-wise procedure for design of open drains

The design of the channel is done in the following step-by-step manner :

1. From the known soil type, arrive at the value of Manning's Rugosity coefficient, side slopes and the maximum permissible velocity.
2. Determine the slope from the topography.
3. For the given discharge, calculate the hydraulic mean depth from Manning's Formula.
4. Find out the cross-sectional area from the given discharge and the maximum permissible velocity.
5. From the results of steps 3 and 4, solve the simultaneous equations to obtain the bottom width and depth.
6. Calculate the critical depth and determine whether the flow is tranquil or turbulent. If it is tranquil, add a freeboard to the depth and finalise the cross-section. If the flow is turbulent, it may be necessary to line the channel or decrease the longitudinal slope.

25.4.3. Illustrative design problem

Problem 25.1. A longitudinal channel with a trapezoidal cross-section is to be constructed in a cut section. The longitudinal slope is 1 in 2500. The soil is clay, with Manning's Rugosity Coefficient of 0.024. The maximum allowable velocity is 0.6 m/sec. Design the channel for a discharge of 3 cu m/sec.

Solution. $$V = \frac{R^{2/3} S^{1/2}}{n}$$

$$0.6 = \frac{R^{2/3} \times (0.0004)^{1/2}}{0.024} = \frac{R^{2/3} \times 0.02}{0.024}$$

$$\therefore \quad R = \left(\frac{0.6 \times 0.024}{0.02}\right)^{3/2} = 0.61 \text{ m}$$

$$\text{A} = \frac{Q}{V} = \frac{3.0}{0.6} = 5 \text{ m}^2$$

Perimeter $$P = \frac{A}{R} = \frac{5}{0.61} = 8.20 \text{ m}$$

$$A = d(b + 2d) = 5$$

$$P = b + 2d(1 + 2^2)^{1/2} = \text{b} + 4.48\, d = 8.20.$$

Thus $$bd + 2d^2 = 5$$

$$b + 4.48\, d = 8.20$$

Multiplying the second equation by d and subtracting the first from the product,

$$2.48\, d^2 - 8.20\, d + 5 = 0$$

$$d = 0.81\ m \text{ and } b = 4.57 \text{ m}$$

Critical depth, $$\frac{V^2}{g} = \frac{0.6^2}{9.81} = 0.037 \text{ m}.$$

Since the actual depth of 0.81 m is greater than the critical depth (0.037 m), the flow is tranquil. No lining is necessary. Provide a freeboard of 0.3 m and adopt a depth of 1.11 m.

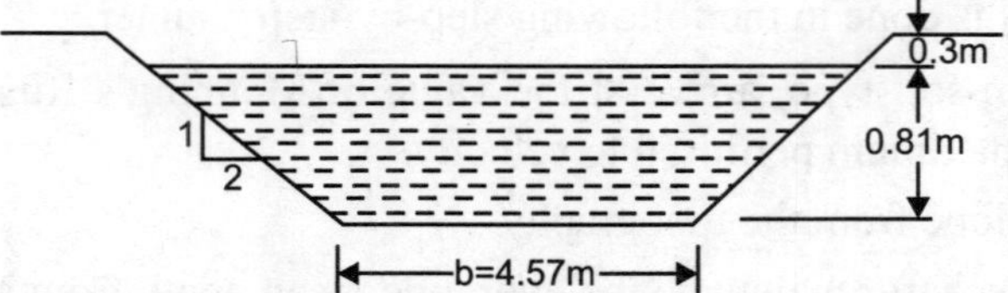

Fig. 25.2. Sketch of trapezoidal drain.

Some typical designs of side drain on hill roads are shown in Fig. 25.3.

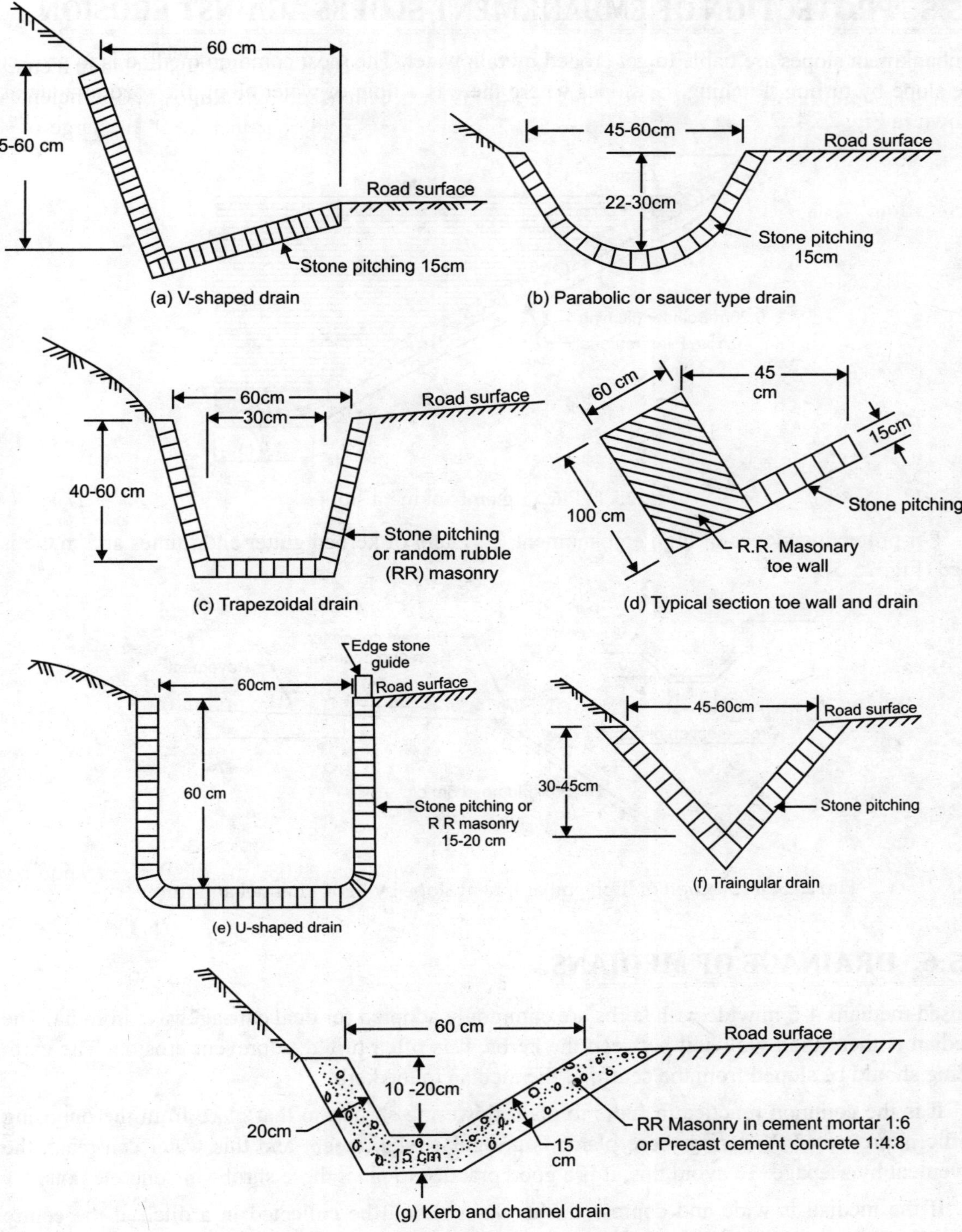

Fig. 25.3. Typical side drains on hill roads

25.5. PROTECTION OF EMBANKMENT SLOPES AGAINST EROSION

Embankment slopes are liable to get eroded by rain water. The most common method is to protect the slope by turfing. Pitching the slopes where there is a flow of water along the road is done as shown in Fig. 25.4.

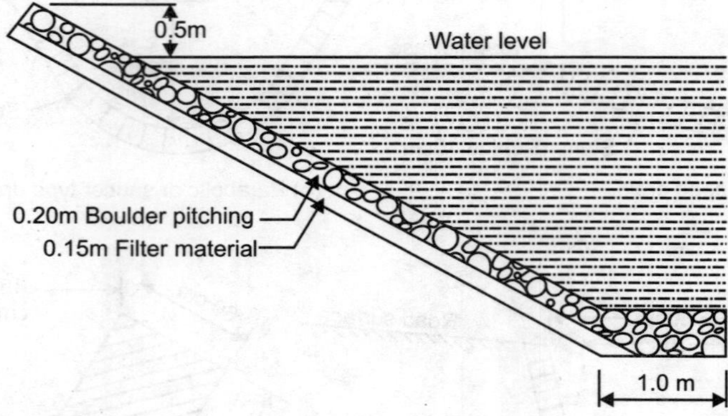

Fig. 25.4: Pitching embankment slopes

For protecting slopes of high embankment, a system of kerbed gutter and flumes at 5 m c/c is used (Fig. 25.5).

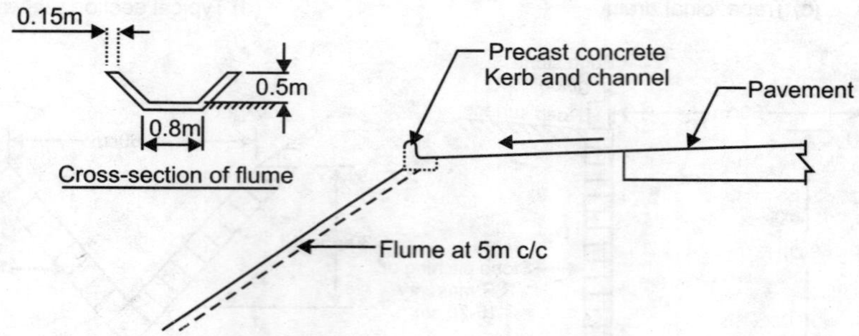

Fig. 25.5: Protection of high embankment slope by kerb, channel and flumes

25.6. DRAINAGE OF MEDIANS

Raised medians 4-5 m wide with kerbs are commonly adopted for dual carriageways in India. The median is then filled with soil between the kerbs. It is often turfed to prevent erosion. The earth filling should be sloped from the centre of the median to the kerb.

It is the common practice in India to plant flowering shrubs so that glare from the opposing traffic is prevented. Watering these plants is a regular requirement and this water can reach the pavement by seepage. To avoid this, it is a good practice to plant these shrubs in concrete pots.

If the median is wide and depressed, the water should be collected in a ditch at the centre and led away at suitable intervals by pipes laid below the carriageways and made to discharge eventually at the side ditches.

25.7. DRAINAGE OF INTERSECTIONS

In roundabouts a part of the carriageway slopes towards the central island. This water needs to be collected in a drain around the island, and led away to an outlet by a pipe. Fig. 25.6 shows the arrangement.

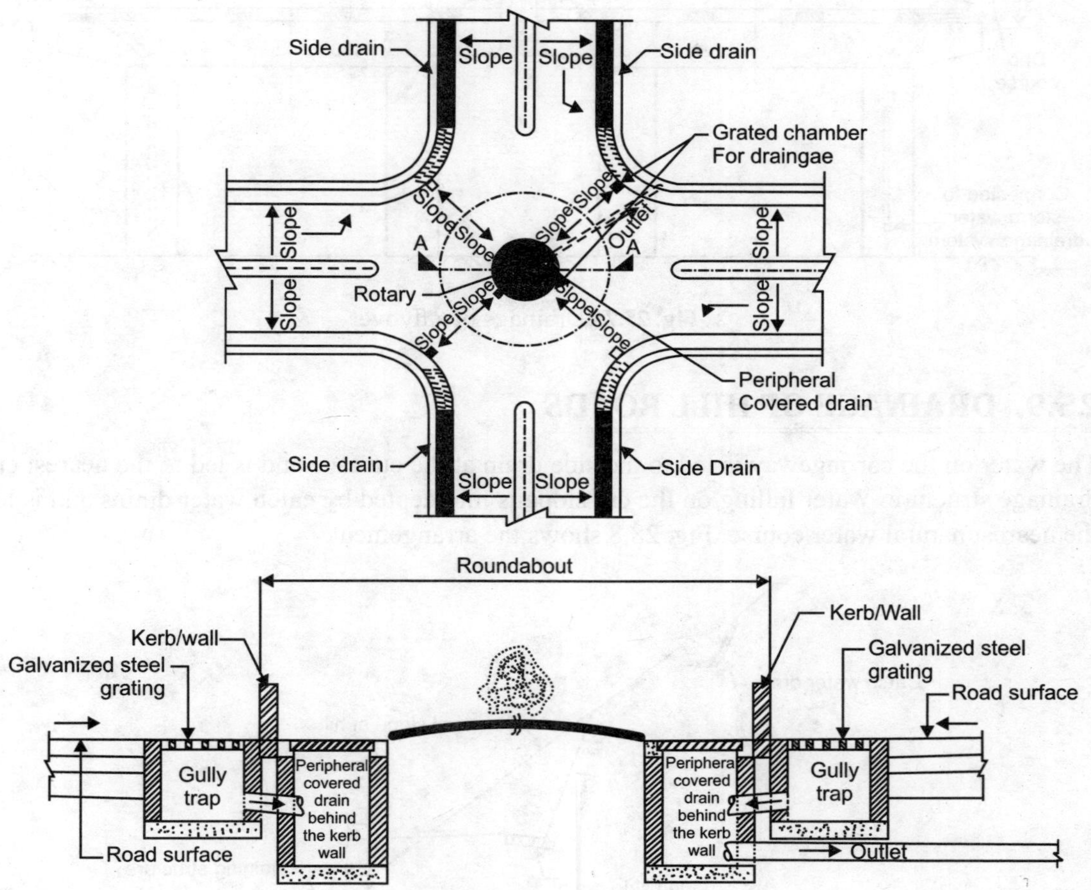

Fig. 25.6: Drainage arrangement at a roundabout

25.8. DRAINAGE OF FLYOVERS

Rain water which is received on the carriageway has to be led away. Providing outlets which discharge the water to the side carriageways below will fall on the pedestrians and other road users and should be avoided. The proper arrangement is to collect the water at the carriageway above at pier locations and provide pipes at the sides of the pier and discharge the water to the underground drainage system. (Fig. 25.7)

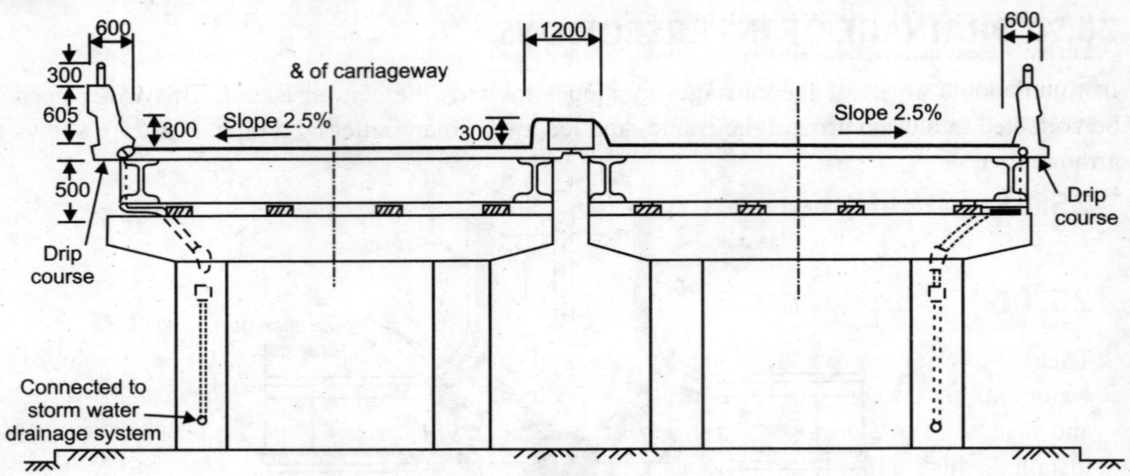

Fig. 25.7. Drainage of a flyover

25.9. DRAINAGE OF HILL ROADS

The water on the carriageway is led to the side drain at the cut face and is led to the nearest cross-drainage structure. Water falling on the cut slope is intercepted by catch water drains and is led to the nearest natural water course. Fig. 28.8 shows the arrangement.

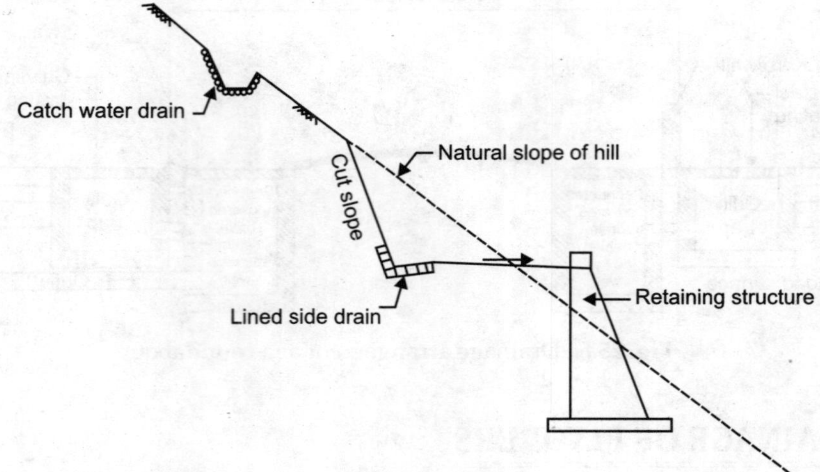

Fig. 25.8. Side drain and catch water drain in hill roads

25.10. DRAINAGE OF URBAN ROADS

Urban drainage is provided by the storm water drains below the carriageway. They are provided generally at the carriageway edges or beneath the medians. Inlets can be grated inlets which discharge the water directly into manholes at suitable intervals or grated inlets in the vertical face of kerbs. In the latter case, short pipes take the water from the inlets into manholes. The possibility of ground water harvesting should be investigated so that the recharging of the fast depleting ground water is facilitated.

25.11. SUBSURFACE DRAINS

25.11.1. The need

There are a number of situations where merely providing surface drainage is not enough *e.g.*, water moving through pervious subsurface layers of soil or high water table causing soft subgrades and instability of slopes etc. To combat such situations, it is necessary that subsurface drains be installed which will intercept and carry subsurface water to a stream or open drainage system.

25.11.2. Drains in cut slopes

In cut slopes which are rendered unstable due to seepage water finding its way to the slope face can be remedied by installing a subsurface drain at the back of the cut face.

A typical subsurface drain is shown in Fig. 25.9 installed to a depth below the pervious soil layer and backfilled with a filter material. Such a drain will arrest the water coming down through the pervious layer before it gets to the cut face and with proper design, the water will be led to a suitable outlet say at a stream. If, however, the depth of the pervious soil layer had been excessive so as to make it practically impossible to install a drain at that depth, installation of horizontal drains would have been a better solution to the problem. Horizontal drains are perforated metallic or PVC pipes driven into the slope at a suitable gradient and at design intervals to arrest the seepage water and drain it down at the roadside drain. A proper design for the installation of horizontal drains has to be worked out on the basis of detailed field and laboratory investigations.

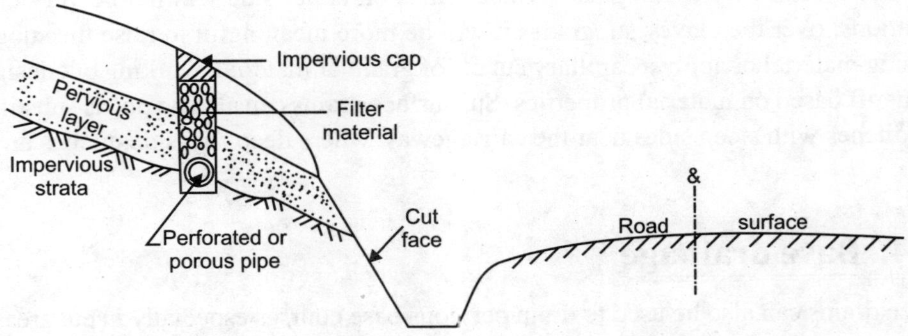

Fig. 25.9. Subsurface drain at cut slopes.

25.11.3. Subgrade drainage

If there is ground water in a layer close to the roadway, which can render the subgrade soft, reducing its strength and resulting in an unsatisfactory performance of the pavement, the installation of a subsurface drain can provide the needed subgrade drainage. As shown in Fig. 25.10 the drain in installed in the pervious layer beside the road, to intercept the ground water before it can reach the subgrade.

When the natural ground water level is near the ground surface, subsurface drains can be placed along each side of the road to lower the water level, as shown in Fig. 25.11.

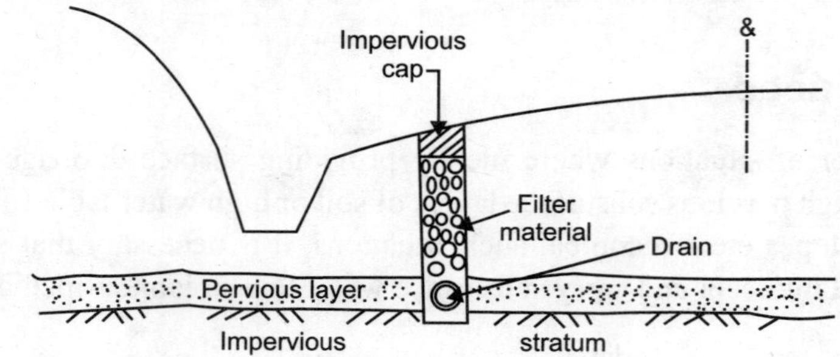

Fig. 25.10. Subdrain for subgrade drainage.

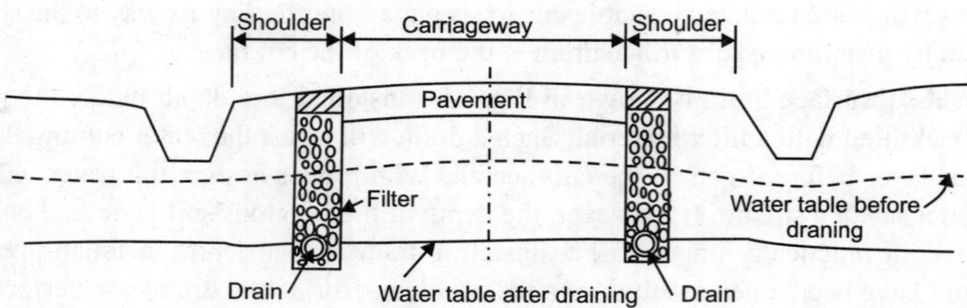

Fig. 25.11. Subdrains for lowering ground water table.

However, if the subgrade is very clayey in nature, the water will not drain readily from such soils and this system of providing subsurface drains on either side will not be very effective. For such situations, over the clayey subgrades it will be more meaningful to raise the subgrade with a free draining material or apply a capillary cut off of granular material, working out design thickness of such cut off based on material properties. Subsurface drains can also be put as substitutes to deep roadside ditches with steep sides near the carriageway, where deep roadside ditches are considered a hazard.

25.11.4. Base drainage

Subsurface drains can also be used to drain pervious base courses especially in cut areas and where it is not practical to carry the base through the shoulder.

These subsurface drains do not lower the water table but carry off water which permeates through the surface rather than running off into ditches (Fig. 25.12). Where subgrade drains are used, they also drain the base course. If the base course comprises of dense graded aggregates, subsurface drains are not of much use as dense graded aggregates do not drain freely.

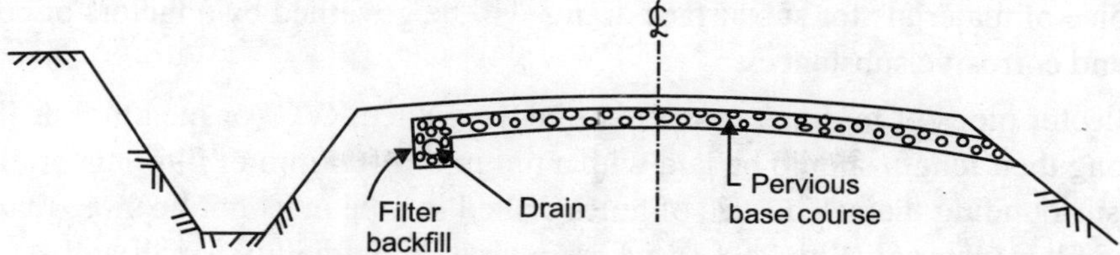

Fig. 25.12. Subdrains for base drainage

25.11.5. Filter media and drain pipe

Subsurface drain may comprise of perforated pipe, porous concrete pipe, solid pipe laid with open joints or simply a free draining material in a trench with no pipe. The most satisfactory type is the perforated pipe or porous pipe (no-fines concrete pipe) laid in a trench and backfilled with a granular, free draining material covered at the surface with a relatively impervious soil.

For a graded filter material, gradation requirements are given by the piping and permeability ratios as under :

$$\text{Piping ratio} = \frac{D_{15}\,(\text{filter})}{D_{85}\,(\text{soil to be drained})} < 5$$

$$\text{Permeability ratio} = \frac{D_{15}\,(\text{filter})}{D_{15}\,(\text{soil to be drained})} > 5$$

An example of a suitable filter material gradation (*B*) for the soil to be drained with gradation (*A*) is shown in Fig. 25.13. Starting from the soil to be drained and its desired filter grading,the gradings of successive layers in the graded filter can be worked out on the basis of "Piping" and "Permeability" ratios. Where river sand meets the grading requirements, the same may be used as graded filter material.

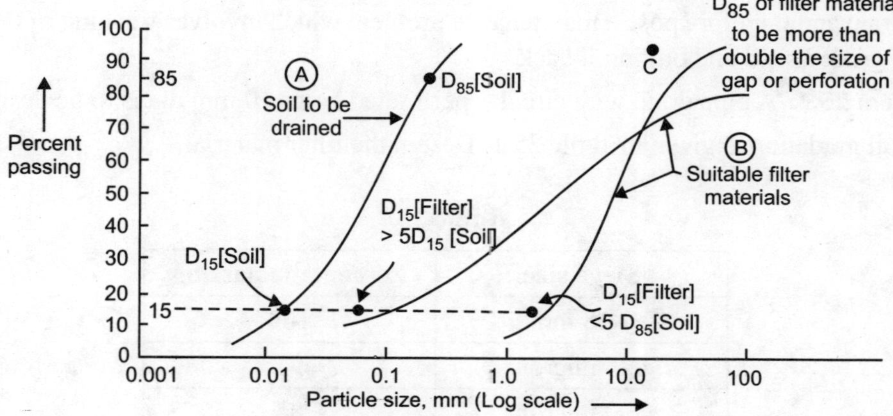

Fig. 25.13. Gradation for filter materials.

The perforated pipe is usually 100 mm to 150 mm diameter with holes in two or more lines on one side of the pipe, the perforated side of the pipe should be placed downwards to avoid clogging of the pipe by fine grained materials.

The choice of materials for subsurface drains will be governed by a factors of cost, foundation conditions and corrosive substances.

The collector pipes of no-fines concrete or heavy walled PVC (or metallic) drilled at frequent intervals along their length should be laid with a minimum 100 mm of filter material (or river sand if suitable) surrounding them. The size of holes drilled in pipe must not be more than half the 85% grain size of filter material (Fig. 25.13); where hydraulic gradient is small and seepage rate low, 5 mm diameter holes are used if restricted to lower 60° arc. Solid pipes with open joints may also be used but are likely to permit silt and fine sand to enter the pipe unless special precautions are taken. It is, therefore, necessary that open joints should be covered with burlap or tarpaper or surrounded with gravel or crushed stone before backfilling.

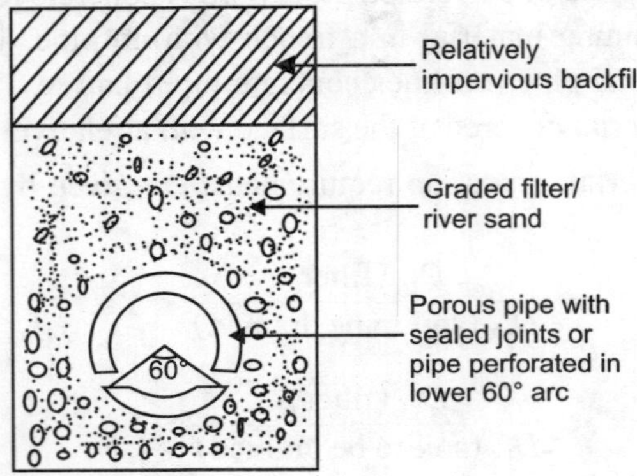

Fig. 25.14. Perforated pipe in subdrain.

Conventionally 'French Drains' are constructed with no open conduit in them and the flow of water takes place through the porous backfill, generally consisting of graded sand. This type of drain is quite satisfactory so long as the filter material is not clogged. However, experience all the world over shows that such drains are more likely to get plugged with fine material, when compared to a pipe drain and therefore pose a maintenance problem which involves washing of the "clogged" backfill of graded sand and putting it back.

Problem 25.2. A pipe drain with circular perforated holes 10 mm dia is to be designed. The soil gradation is given in Table 25.1. Design the filter material.

Table 25.1.

Sieve size	Percentage passing
1.18 mm	100
425 micron	93
300 micron	85
150 micron	60
75 micron	15
53 micron	7

D_{15} (protected soil) $= 75$ micron $= 0.075$ mm

$\therefore D_{15}$ (filter) $\geq 5 \times 0.075 \geq 0.375$ mm.

D_{85} (protected soil) $= 300$ micron $= 0.3$ mm

$\therefore D_{15}$ (filter) $\leq 5 \times 0.3 \leq 1.5$ mm

These two points have been plotted in Fig. 25.15. as A and B.

Size of perforated hole = 10 mm dia.

$\therefore D_{85}$ (filter) $> 2 \times 10$ mm > 20 mm.

This point is marked in the figure as *C*. *A* suggested grading is freely sketched to lie to the right of point *C* and passing in between points *A* and *B*.

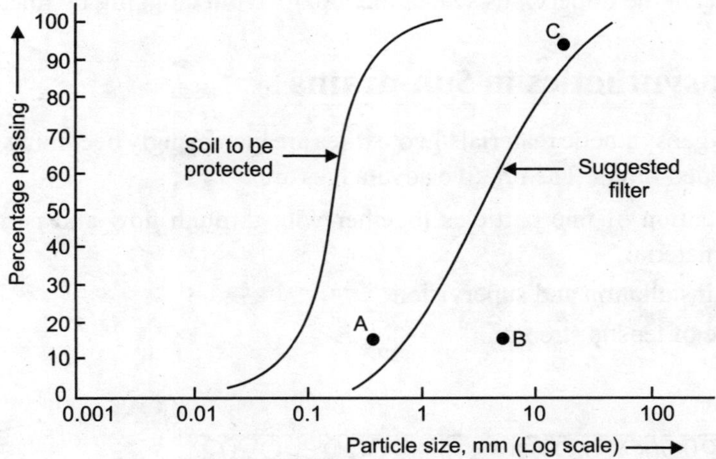

Fig. 25.15. Filter design.

25.11.6. Location of subsurface drain

Ordinarily, the location of a subsurface drain should be decided on the basis of field observations, except that the location of base drains can be decided beforehand being independent of the location of water in the subgrade.

If the subsurface drain is to perform the function of intercepting moving water in a pervious layer, it must be placed at the bottom of the layer (trench to be dug 100-150 mm below the pervious layer) through which water is flowing. The drain should extend the full length of the problematic layer of soils and lead to a suitable outlet to be carefully located.

For purposes of protecting slopes, the subsurface drains are placed well back of the cut face, the trench dug from a point above the top of the slope. For intercepting water under a road, such drains are generally installed under the shoulder but could also be at the roadside ditch line, depending on site conditions. For lowering the ground water table, the subdrains should be installed close to the carriageway edges ; if installed across the road, settlements over the drains are likely to cause a series of dips in the road pavement surface. It is only in coarse soils that subsurface drains can be effective in lowering the ground water table, and should be taken to a depth, not less than 1.25 m. The gradient towards the outlet should be at least 0.15%.

The outlet should not be submerged or plugged with debris but should be free like an open ditch, culvert or stream. The bottom of the pipe is kept above high water in the ditch and the end protected with a grate or screen.

For the installation of subsurface drains, firm continuous bedding is essential to prevent settlements and/or crushing of the pipe. The filter backfill should be brought up equally on both sides of the pipe and thoroughly compacted.

In deep subdrains, meant for intercepting water in a pervious layer, the filter backfill may be of the order of 0,6 m only above the pipe, the remainder of the trench being backfilled with the soil, removed from the trench or any suitable fill. It is, however, desirable to use filter material for the entire backfill, except the impervious seal at the top, to avoid clogging by fines.

25.11.7. Geosynthetics in Sub-drains

Suitable grades of geosynthetic materials/geotextiles are increasingly becoming popular as preferred alternatives to graded filters. The relative advantages are :

(*i*) high retention of fine particles together with through flow and permeability similar to graded material

(*ii*) Ease of installation and supervision

(*iii*) Presence of tensile strength.

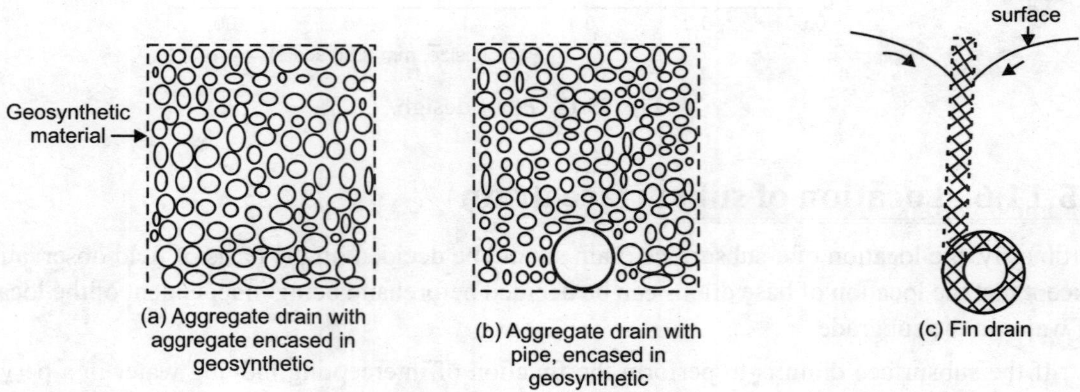

25.16. Geosynthetics in subdrains.

Several varieties of Geosynthetic filter fabrics are available and the selection of Geotextile/ Geocomposite fabric is based on the following :

— Survivability requirement

— Prevention of piping

— Permeability requirement

— Uniformity requirement

Fig. 25.16. shows the common types of subsurface drains where geosynthetics are used : (*a*) aggregate drains with the aggregate encased in geosynthetic (*b*) aggregate drain with pipe, encased in geosynthetic and (*c*) fin drain.

Geosynthetic of suitable grade performs two functions when incorporated in a subsurface drain : that of a filter and a separator. Fine soil particles are prevented from entering the drain while water passes through, thus lowering the water level in the surrounding soil. Simple drain construction with geosynthetic consists of : excavation of trench and lining with geosynthetic ; laying and levelling of the perforated drainage pipe ; infilling with clean stone/aggregate ; and overlapping of the geosynthetic and covering with selected backfill.

A fin drain refers to a drain of limited thickness ; it comprises a geonet sandwiched between two layers of geotextile. The fin drain is easily handled and installed in a narrow trench with or without a geopipe. The excavated material can be used as backfill. Fin drains have special relevance where drainage stone is expensive or if access is limited. A network of subsurface collection drains can prevent water infiltration into foundation areas while cutoff drains can arrest flow of water from a preferential direction.

25.12. CAPILLARY CUT-OFF TREATMENT

In special situations such as roads in water-logged areas, there is a possibility of water rising up from the water-table by capillary action and softening the subgrade. In such cases, a capillary cut-off treatment is normally resorted to. Such treatment is generally expensive.

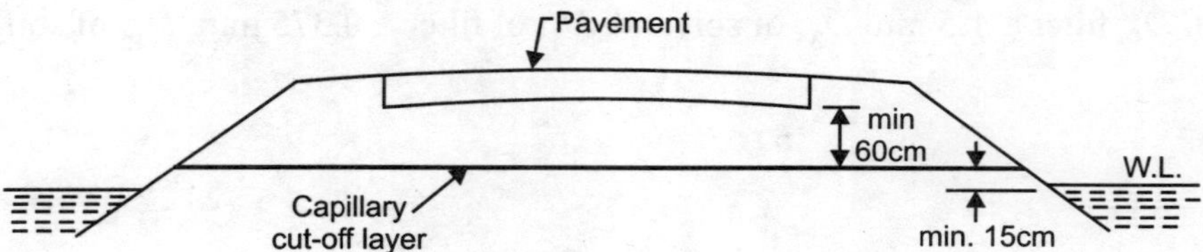

Fig. 25.17. Capillary cut-off.

Some of the suitable capillary cut off layers are :

(*i*) Sand blanket, extending over the full embankment width.

(*ii*) Bituminous impregnation using primer treatment.

(*iii*) Heavy duty tar felt

(*iv*) Polythene envelope

(*v*) Bituminous stabilised soil.

The capillary cut-off should be placed at least 15 cm above the ground level or standing water level, whichever is higher, and it should not be placed closer than 60 cm from the subgrade level (Fig. 25.17).

QUESTIONS

1. Discuss the importance of drainage in highway engineering.

2. What are the methods of achieving good highway drainage ?

3. A longitudinal drain is to be constructed along a highway. The discharge is 4 cu m/sec. The longitudinal slope is 1 in 2000. The soil is clayey, with Manning's Rugosity Coefficient of 0.030. The maximum allowable velocity is 0.6 m/sec. Design a trapezoidal section.

[Ans. $b = 4.85$, $d = 0.98$ m
Side slope 1 in 2,
Freeboard: 0.3 m]

4. What are the methods available for protection of slopes of embankments against erosion?

5. Write short notes on :

(*i*) Catch-water drains (*ii*) French drains.

6. An embankment is protected by pitching laid over a filter layer. The embankment soil gradation is as under:

Sieve Size	Percentage passing
1.18 mm	100
425 micron	95
300 micron	85
150 micron	60
75 micron	15
53 micron	9

Design the filter material.

[**Ans.** D_{15} filter ≤ 1.5 mm D_{85} of soil and, D_{15} of filter ≥ 0.375 mm, D_{15} of soil]

Chapter

CROSS–DRAINAGE STRUCTURES

26.1. IMPORTANCE OF CROSS-DRAINAGE

The adequate functioning of a road depends to a large extent on the effectiveness of cross-drainage. A road has to be so aligned as to cut across natural drainage channels and major rivers. At times, the road cuts across man-made channels, intended for irrigation. Some artificial channels are also intended to take care of natural drainage. The function of the cross-drainage structures is to ensure that the run-off water is discharged across the road from one side to the other, as quickly as possible, without causing undue ponding, parallel flow along the road embankment, overtopping of the road embankment or erosion of the portions of the road. Quick drainage prevents water from penetrating the soil in the embankment. A dry subgrade has greater bearing strength than a wet subgrade.

26.2. TYPES OF CROSS DRAINAGE STRUCTURES

Cross-drainage structures are divided into the following categories :

1. Culverts, which have a total waterway upto 6 m.
2. Minor bridges, which have a waterway in the range 6–30 m.
3. Medium size bridges, which have a waterway in the range 30–100 m.
4. Major bridges, which have a waterway greater than 100 m.
5. Causeways, which allow the water to flow over the roadway.

The terms minor bridges and medium sized bridges are not exactly defined, and they may be clubbed together as minor bridges.

Bridges may be high level or submersible. High level bridges are those designed for the maximum possible flood discharge and which have their deck level sufficiently above the flood level likely to occur under such a flood. All important bridges on arterial routes are designed. as high level bridges. Generally a return period of 50 years of floods is considered. A high level submersible bridge is one which has adequate clearance above the level of the highest flood likely to occur once in 10 years, and the bridge is likely to be submerged 2 or 3 times in the course of a 10-year period. A low level submersible bridge is provided on low-cost roads. The deck level of such a bridge is fixed so that it does not get submerged except under high floods. The period of interruption to traffic should not exceed three days at a time or a total period of 15 days in a year.

26.3. SELECTION OF SITE

The selection of site of cross-drainage is an art and requires cónsiderable investigations. A few points that deserve attention are listed below :

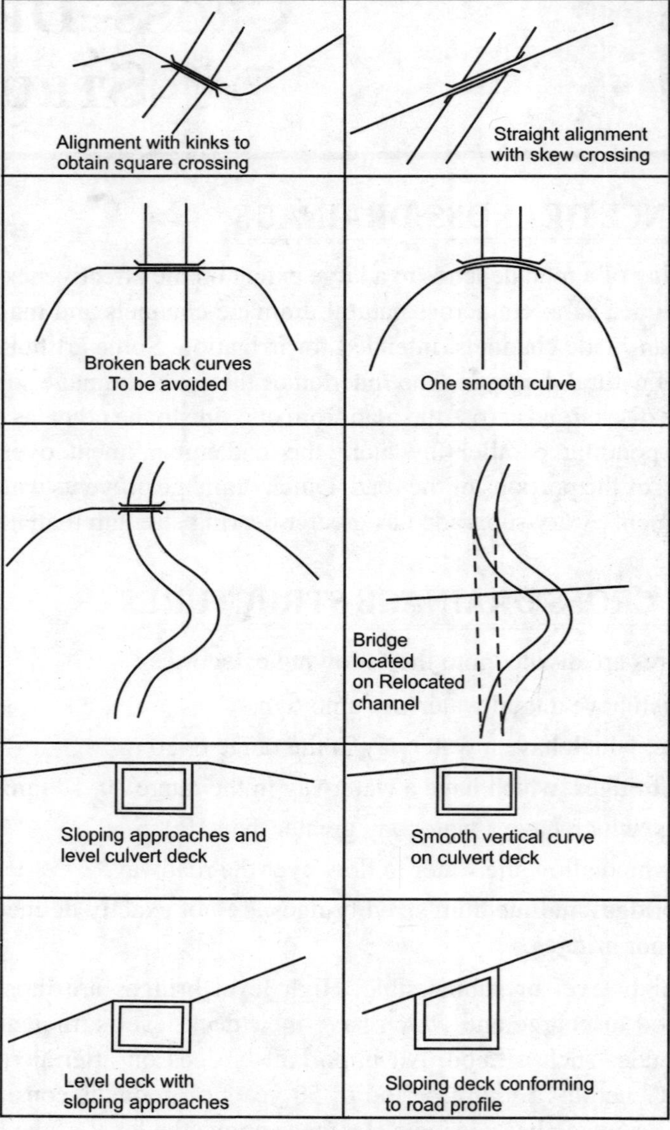

Fig. 26.1. Location of culverts.

1. A site which is on a straight reach of the stream should be selected. The site should be sufficiently below bends.

2. The site should be sufficiently away from the confluence of large tributaries. This will ensure that the eddying effects, scour problems, and cross-currents are sufficiently beyond the selected site.
3. A site which has well defined and stable banks results in accurate determination of the waterway and should be preferred to one with ill-defined banks.
4. A site which offers a square crossing should be selected as far as possible. A square crossing gives the minimum length of structure. While this requirement should be insisted upon for major bridges and the road alignment consequently subordinated to bridge siting, the geometries of the approaches should get precedence over siting the culverts and minor bridges. Some examples of faulty and good locations of culverts and minor bridges are given in Fig. 13.1.
5. A site which offers more advantageous foundation conditions should be preferred to one with unfavourable foundation conditions. Thus if bed-rock is exposed, the minimum expenditure will be involved in foundations.
6. A site which is sufficiently away from landslides and subsidences should be favoured.
7. The site should not interfere adversely with any other bridges, weirs, regulators, dams etc. in the vicinity.
8. A site which blends with the topography and landscape will be aesthetically pleasing.

26.4. HYDROLOGY OF CROSS-DRAINAGE STRUCTURES

26.4.1. Estimation of design discharge

The most important step in the fixation of waterway of a cross-drainage structure is to determine the design flood discharge. An accurate determination of the discharge will:

(*i*) govern the safety of the structure ;
(*ii*) influence the selection of design of foundations, Waterway and protective works and
(*iii*) affect the cost of the structure.

There are many ways of estimating the flood discharge. They may be grouped into four categories:

(*i*) Methods based on empirical formulae
(*ii*) Estimation based on Rational Formula
(*iii*) Method based on bed slope and area of cross-section of flow
(*iv*) Method based on flood marks on existing structures.

26.4.2. Empirical formulae for run-off

In India, there are three popular empirical formulae for determining the maximum run-off, all of them based on the catchment area. These formulae have been developed on the basis of past experience in different parts of the country. They are simple and reasonably accurate, provided sufficient care is exercised in determining the catchment area accurately and selecting the appropriate coefficient. The three formulae are given below:

Dicken's Formula

$$Q = CM^{3/4} \tag{26.1}$$

where Q = the peak run-off in m^3/sec.

M = catchment area in sq. km.

C = a constant having the following values :

= 11 – 14 where the annual rainfall is 60 – 120 cm.

= 14 – 19 in Madhya Pradesh

= 22 in Western ghats

Ryve's Formula, devised for Channai (Madras)

$$Q = CM^{2/3} \tag{26.2}$$

where C = 6.8 for areas within 25 km of the coast

= 8.5 for areas between 25 and 160 km of the coast

= 10.0 for limited areas near the hills.

Inglis' Formula, devised for Mumbai (Bombay)

$$Q = \frac{125M}{\sqrt{M+10}} \tag{26.3}$$

26.4.3. Rational Formula

The rational formula is based on a knowledge of the intensity of rainfall and the catchment area. In its simple form, the formula is :

$$Q = 0.028\, P.A.f.I_c \tag{26.4}$$

where Q = Maximum run-off in m^3/sec

A = Area of catchment in hectares

I_c = Critical intensity of rainfall in cm per hour

P = Percentage coefficient of run-off for the catchment characteristics.

f = Factor depending on catchment area.

The values of f may be taken from Fig. 26.2

The critical intensity of rainfall, I_c, is the maximum intensity which can occur in a time interval .equal to the concentration time t_c of the catchment during the severest storm in the region of a given frequency. The time of concentration t_c, is the time taken by the run-off from the farthest point of the catchment to reach the site. The following simple-relationship determines t_c:

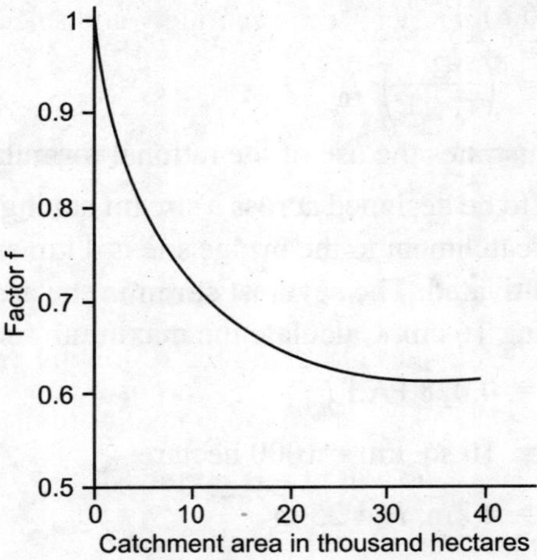

Fig. 26.2. Factor *f* in rational formula.

$$t_c = \left(0.87 \times \frac{L^3}{H}\right)^{0.385} \tag{26.5}$$

where t_c = time of concentration in hours

L = distance from the farthest point to the site in km.

H = the fall in level from the farthest point to the site, in m.

Knowing t_c I_c can be calculated from the maximum one-hour rainfall intensity I_0, in the region from past records for a selected return period. A return period of 50 years is very commonly adopted for culverts and bridges in India. It has been observed that the mean intensity (i) of a storm of shorter duration (t) can be higher than that (I) of a longer duration (T). The following formula expresses this relationship

$$\frac{i}{I} = \frac{T+1}{t+1} \tag{26.6}$$

If F is the total precipitation in T hours,

$$I = \text{average intensity} = \frac{F}{T} \tag{26.7}$$

Combining (13.6) and (13.7),

$$i = \frac{F}{T}\frac{(T+1)}{(t+1)}$$

If $t = 1$ hour, $i = I_0$ (one-hour rainfall)

$$\therefore \quad I_0 = \frac{F}{T}\left(\frac{T+1}{1+1}\right)$$

$$= \frac{F}{2}\left(\frac{T+1}{T}\right) \tag{26.8}$$

Since we are interested in the run-off in the time of concentration t_c,

$$I_c = \frac{F}{T}\left(\frac{T+1}{t_c+1}\right) \tag{26.9}$$

Combining (26.9) with (26.8),

$$I_c = \left(\frac{2}{t_c + 1}\right) I_0 \tag{26.10}$$

The following problem illustrates the use of the rational formula.

Problem 26.1. A bridge is to be designed across a stream having a catchment area of 10 sq. km. The farthest distance from the catchment to the bridge site is 4 km and the fall in level is 25 m. The soil is loamy and is largely cultivated. The severest storm in the area was of 2 hours duration, the precipitation in this period being 16 cm. Calculate the maximum run off. taking p = 0.3

Solution.

$$Q = 0.028 \text{ PA.f} I_c \qquad \text{Eq. (26.4)}$$

$$A = 10 \text{ sq. km} = 1000 \text{ hectares}$$

$$L = 4 \text{ km}, H = 25 \text{ m}$$

$$t_c = \left(0.87\frac{L^3}{H}\right)^{0.385} = \left(\frac{0.87 \times 64}{25}\right)^{0.385} 1.36 \text{ hr.}$$

From Fig. 26.2, $f = 0.97$ and $P = 0.3$

$$I_c = \left(\frac{2}{t_c + I}\right) I_0$$

$$I_0 = \frac{F}{T}\left(\frac{T+1}{1+1}\right) = \frac{16}{2}\left(\frac{2+1}{2}\right) = 12 \text{ cm/hr.}$$

$$I_c = \frac{2}{(1.36+1)} \times 12 = 10.2 \text{ cm/hr.}$$

$$Q = 0.028 \times 0.3 \times 1000 \times 0.97 \times 10.2 = 83.1 \text{ m}^3\text{/sec.}$$

26.4.4. Bed slope and cross-sectional area method

The most popular formula, Manning's formula has already been described in Chapter 25 when dealing with flow through channels.

26.4.5. Calculation of discharge from observed flood marks on nearby structures

A very accurate method of flood estimation is from observed flood marks on nearby structures.

26.4.6. Design discharge

The usual practice is to estimate flood discharge from two or three methods. The highest of the values obtained is used as the design discharge, provided it does not exceed the next highest by more than 50 per cent. If it does, the design discharge is restricted to that limit.

26.5. FIXATION OF WATERWAY

26.5.1. Alluvial streams

In alluvial streams, the practice in India is to determine the regime width by Lacey's equations:

$$W = 4.8\,Q^{1/2} \tag{26.11}$$

$$D = \frac{0.473\,Q^{1/3}}{f^{1/3}} \tag{26.12}$$

In the above equations,

W = Regime width, m

D = Scour depth, m

Q = Discharge, m^3/sec.

f = Silt factor (which depends on soil grain size) = $1.76\sqrt{m}$,

where m = mean dia of the particles in mm.

Values of silt factor can be taken from Table 26.1.

Table 26.1. Lacey's Silt Factor

Soil	Grain size (mm)	f
Very fine silt	0.081	0.50
Fine silt	0.120	0.60
Medium silt	0.233	0.85
Standard silt	0.323	1.00
Medium sand	0.505	1.25
Coarse sand	0.725	1.50

The general rule in alluvial streams is to provide a waterway equal to the Lacey's regime width. Contraction is normally avoided, as it will entail deeper foundations and costly river training works.

26.5.2. Streams not wholly alluvial

In streams which are not wholly alluvial, the linear waterway is made equal to the actual surface width of the stream, measured from edge to edge along the high flood level on the plotted cross-section.

26.5.3. Streams with rigid boundaries

The bridge should span from bank to bank, measured along the highest flood level.

26.5.4. Streams which overflow the banks

The fixation of waterway in the case of streams which overflow the banks and flood for long widths is a difficult problem. Lacey's regime width gives a good guide. River training works, such as protection of banks of the stream, guide bunds and pitching of the highway embankment slopes may be found to be necessary.

Problem 26.2. An alluvial river has a maximum flood discharge of 95 m³ / sec. The bed is of medium silt, with a particle size of 0.233 mm. Determine the linear waterway of the bridge.

Solution.

Silt factor, $f = 1.76\sqrt{m} = 1.76\sqrt{0.233} = 0.85$ (See also Table 26.1)

$Q = 95 \text{ m}^3/\text{sec.}$

$W = 4.8 \times Q^{1/2}$ Eq. (26.11)

$= 4.8 \times 95^{1/2} = 46.8 \text{ m}$

A waterway of 50 m may be adopted.

26.6. FOUNDATION DEPTH AND SPAN

The simple rules for fixing the depth of bridge foundations in India are :

Rule (1) Erodible beds : Take the foundation to a depth below the maximum high flood level, one-third greater than the calculated depth of maximum scour, subject to a minimum depth below the scour line of two metres for arch bridges and 1.2 m for other bridges.

Rule (2) Hard beds : When a substantial stratum of solid rock or other materials not erodible at the calculated maximum velocity is encountered at a level higher than given by Rule (1) above, the foundation should be securely anchored into rock (say about 0.3 m into rock and 0.6 m into other hard stratum).

Rule (3) The pressure on the foundation at the level chosen should be within the safe bearing capacity of the material.

Rule (4) Culverts and minor bridges may be provided with bed floors. In such cases, the foundations need not be taken to depths indicated in Rules (1) and (2).

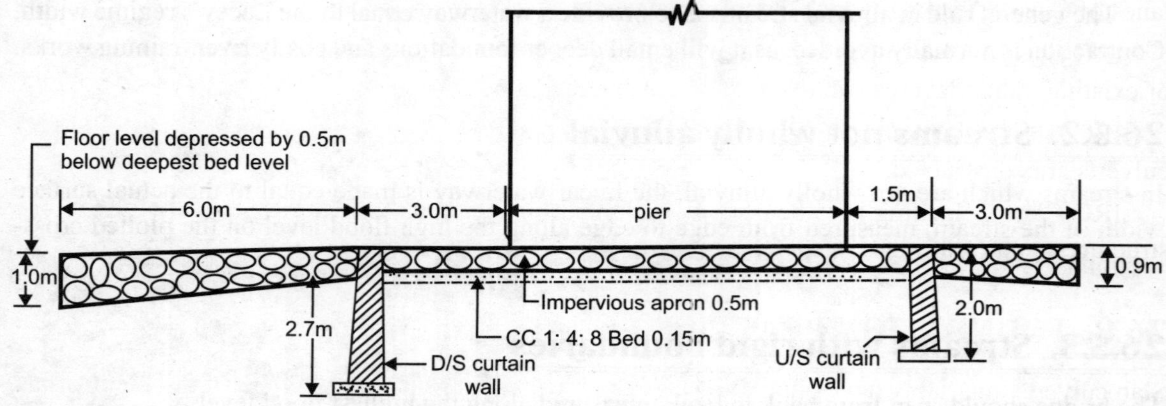

Fig. 26.3. Floor protection for a bridge.

The top of the bed floor should be about 0.3 m below the bed level. The foundations of the abutments should be taken to 1.25 m below the top of the floor. An upstream cut-off wall 1–2.5 m deep and a downstream cut-off wall 1.5–3.5 m deep should be provided. The thickness of the floor is kept about 0.3–0.6 m. A stone boulder talus is provided both on the upstream and downstream. A typical design adopted for minor bridges is given in Fig. 13.3

A well-known rule in bridge engineering in fixing the span of a structure is that the cost is the least when the cost of the superstructure equals the cost of the substructure. Keeping this criterion in view, the spans of R.C.C. slab bridges are kept 1.5 times the height of pier from the bottom of the foundation to its top.

26.7. BRIDGE LOADINGS

Loadings of bridges in India have been standardised by the Indian Roads Congress. The major features of the various loading classes are given below :

Class	Features
IRC Class 70 R	A tracked vehicle of 70 Tonnes load or a wheel load of 100 Tonnes. Individual axle load of 20 Tonnes and bogie axle load of 40 Tonnes.
IRC Class AA	A tracked vehicle of 70 Tonnes- or a wheel vehicle of 40 Tonnes. The maximum individual axle load of 20 Tonnes and bogie axle load of 40 Tonnes.
IRC Class A	A train of vehicles consisting of a main vehicle of 28.2 Tonnes load, followed by two trailers of 13.6 Tonnes load each. The maximum axle load is 11.4 Tonnes. The minimum nose-to-tail distance between two trains is 18.4 m.
IRC Class B	A train of vehicles consisting of a main vehicle of 16.8 Tonnes load, followed by two trailers of 8.2 Tonnes load each. The maximum axle load is 6.8 Tonnes. The minimum nose-to-tail distance between two trains is 18.4 m.

Bridges and culverts on National Highways are designed for one lane of Class 70 *R* or two lanes of Class A loading, whichever produces more severe stresses.

IRC Class AA loading is to be adopted within certain municipal limits, in certain contemplated or existing industrial areas, in other specified areas, or along specified highways.

IRC Class A loading is to be normally adopted on all roads on which permanent bridges and culverts are constructed.

IRC Class B loading is to be normally adopted for temporary structures (for example timber structures) and for bridges in specified areas.

26.8. FOUNDATION PRACTICE

Slab culverts invariably have open foundations with spread footings. The width of the foundations is adjusted to suit the safe bearing capacity of soils.

Small bridges, with RCC slab or beam-and-slab or prestressed concrete superstructure, are also generally provided with open foundations.

Large bridges on alluvial soil which is erodible, are provided with well foundations in India. The depth of such wells ranges from 5 m to 50 m. The well is generally circular in shape and is hollow. Other shapes such as doubled *D* and twin circles are also adopted. The well steining is generally constructed in concrete. A cutting edge, adequately reinforced, pierces through the soil stratum. The material inside the hollow is grabbed by special grab buckets and thrown out. The well sinks, because of its own weight as material is grabbed underneath the cutting edge. Additional load on the steining facilitates sinking. For deep wells, divers with compressed air helmets may be necessary. For still deeper wells, or where rock cutting is involved, pneumatic sinking may become necessary. A very good expertise has been developed in India in the design and sinking of well foundations.

26.9. BANK PROTECTION AND RIVER TRAINING WORKS

When the banks of a river are about to erode or fail, it is necessary to take immediate measures to protect the banks. The river bank may fail due to any one of the following reasons :

(*i*) erosion of soil by river currents or waves

(*ii*) excessive hydrostatic pressure in the materials in the bank

(*iii*) slip-circle failure

(*iv*) liquefaction and subsequent movement of soil mass.

Bank protection measures may be found necessary to deflect the course of rivers. The usual methods employed are :

(*i*) slope pitching along with adequate apron

(*ii*) lining with cement concrete blocks

(*iii*) construction of spurs of timber or wire sausages or boulders.

Guide bunds are generally provided to guide and confine the flow of large meandering rivers, when the flow has to be contracted through the openings in the bridges. The length, geometry, top width slopes, rip-rap pitching and boulder apron thickness should be selected after careful considerations. Hydraulic model studies also facilitate in the determination of these design elements.

26.10. TYPES OF CULVERTS

Culverts are constructed with various materials and to different designs. The popular forms are :

1. Temporary stone scuppers
2. Stone slab culverts with abutments and piers
3. R.C.C. slab culverts with abutments and piers
4. R.C.C. Box culverts
5. R.C.C. Pipe culverts
6. Armco Pipe (metal pipe) culverts
7. Masonry arch culverts.

26.10.1. Temporary Stone Scuppers

In hill roads, it is customary to cut the roads from the hill-face and to open up the road immediately thereafter to traffic. Further improvements such as construction of the pavement are taken up gradually. In such cases, it is the practice to construct temporary stone corbelled scuppers immediately after the road is cut (Fig. 26.4). These are eventually replaced by permanent structures.

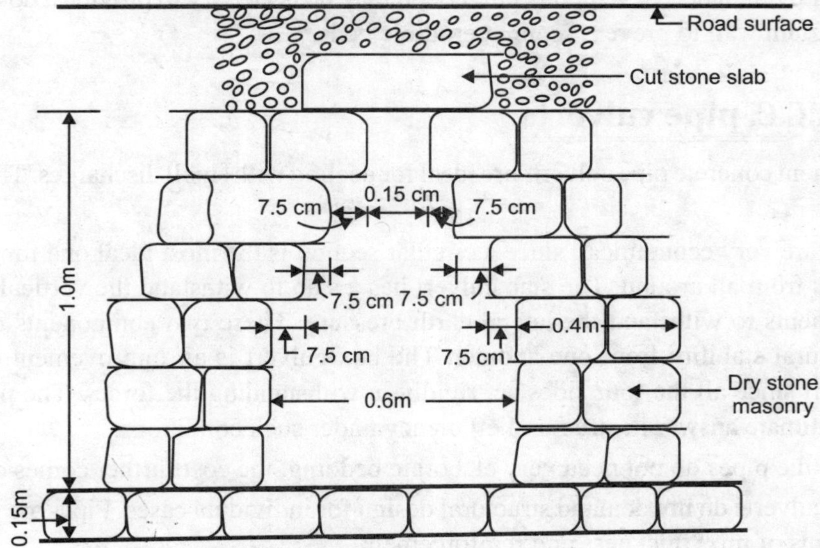

Fig. 26.4. Dry stone masonry scupper.

26.10.2. Stone slab culverts with abutments and piers

Where good quality stone slabs are available like granite slabs, they can be used for superstructure of culverts, with suitable masonry abutments. Spans of such slabs are necessarily small, say upto one metre. Such culverts can be constructed for rural roads (Other District Roads and Village Roads).

26.10.3. R.C.C. Slab culverts

R.C.C. slab culverts with masonry abutments and piers can be constructed for individual spans upto 6 m. The disadvantages of this type are :

1. Need for stout abutment sections which increase the cost of construction.
2. High cost of abutments when the difference between the road formation level and the stream bed is high.
3. Occurrence of humps in the longitudinal profile of roads at culvert locations.

The advantages of the R.C.C. slab culverts are :

1. They lend themselves easily for in-situ construction at distant locations.
2. Slabs and abutments are standardised and hence no specialised knowledge of design is needed by the field engineer.

26.10.4. R.C.C. box culverts

R.C.C. Box culverts are ideal for in-situ constructions for opening of size upto 5 m. They can be constructed in square boxes or rectangular boxes, depending upon site conditions. Single or multiple boxes can be constructed. The advantage lies in the elimination of the stout abutment sections which are unavoidable in slab construction. Standard designs can eliminate the process of designing for individual needs. With box culverts, it may be necessary to provide a downstream and an upstream curtain wall to prevent scour at erodible beds.

26.10.5. R.C.C. pipe culverts

Reinforced cement concrete pipe culverts are ideal for dealing with small discharges. The advantages are :

1. They are very economical, since a circular section is the most ideal one for withstanding forces from all around. The slab culvert has a slab to withstand the vertical load and the abutments to withstand the lateral earth pressure. These two components do not derive structural stability from one another. The box culvert is an improvement over the slab culvert since all the four sides act rigidly in withstanding the forces. The pipe culvert is the ultimate answer in structural efficiency under such conditions.
2. Since the pipes do not need very elaborate bedding, the cost further comes down.
3. Pipe culverts do not demand structural design for individual cases. Pipes are standardised, in terms of mix, thickness and reinforcement.
4. Pipes can be cast under strictly controlled conditions at a Central plant and thus their quality is assured.
5. They generally eliminate the humps in longitudinal profile.

The disadvantages with pipes are that they cannot be used for large openings. A minimum cover of at least half the diameter of the pipe is to be ensured over the pipes.

In India, R.C.C. pipes of the NP 3 class are suitable for highway loadings. For rural roads (Village Roads and Other District Roads), NP 2 pipes are suitable. Two conditions of bedding are followed :

1. Granular bedding, (Fig. 26.5), which consists of forming and grading the bed with sand or any granular material passing 4.75 mm sieve. The thickness of the bedding layer should be minimum 75 mm. Such a bedding is suitable for fills upto 4 m height.
2. Concrete cradle bedding (Fig. 26.6), consists of a concrete bed of M 15 grade concrete. The thickness of the bedding layer should be minimum 250 mm. Such a bedding is used for height of fills above 4 m.

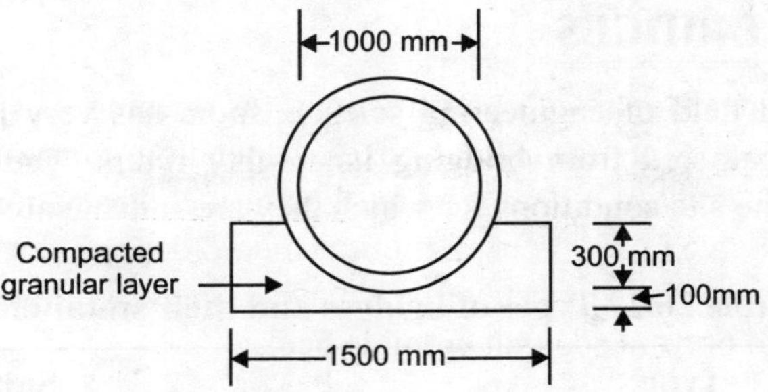

Fig. 26.5. Granular bedding for pipe culverts

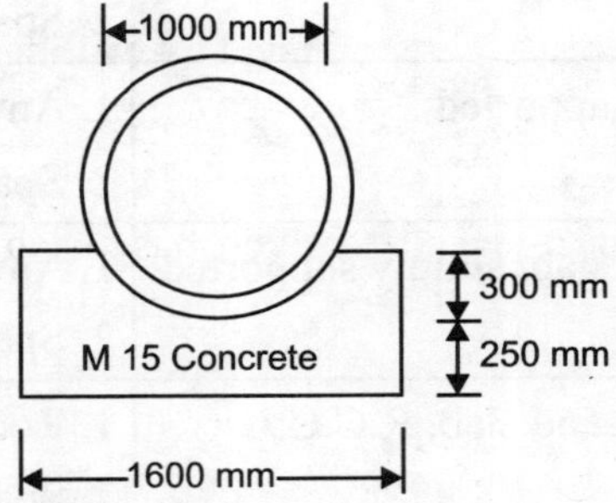

Fig. 26.6. Concrete cradle bedding for pipe culverts

Single pipe or a row of pipes can be used. The distance between pipes should be at least half the diameter of the pipe subject to a minimum of 450 mm. The pipes are kept sufficiently below the floor of the deepest bed. Pipe culverts are generally designed to flow full. Suitable head-walls are needed to protect the embankment fills. The slope of the floor is kept to a minimum of 1: 1000.

26.10.6. Corrugated metal pipes (Armco pipes)

Armco pipes are corrugated metal pipes. Culverts using such pipes are extensively used, especially for military roads and emergency operations since they are light to carry and easy to handle.

26.10.7. Masonry arch culverts

Brick or stone masonry arches were very popular forms for culverts till recently. The advent of R.C.C. construction has almost stopped this form of construction now. It is difficult to get skilled masons proficient in this art now. Culverts constructed in the 19th century and early decades of the twentieth century were mainly of arches. Inverted arches were used to form the floor protection. These structures, though constructed of bricks and lime mortar, are still in service on many roads. The disadvantage with this form is the unavoidable humps in the longitudinal profile that result. It can also give employment to skilled masons. It is a labour intensive form of construction.

26.11. TYPES OF BRIDGES

Bridging is a specialised field of engineering science. India has very high degree of skill and expertise in this field, gathered from bridging its mighty rivers. The types of superstructure commonly adopted and the site conditions for which they are suitable are indicated in Table 26.2.

Table 26.2. Types of bridges and their suitability

S.No.	Type	Suitability
1.	Masonry arch	1. Good foundations 2. Spans upto 15 m.
2.	RCC Slab, simply supported	1. Any type of foundations 2. Spans upto 10 m.
3.	R.C.C. T-beam and slab, simply supported	1. Any type of foundations 2. Spans in the range 10–25 m.
4.	Continuous T-beam and slab, R.C.C.	1. Rocky unyielding foundations 2. Spans in the range 10–20 m.
5.	R.C.C. Balanced Cantilever	1. Deep foundations 2. Spans in the range 25–60 m.
6.	Prestressed T-beam and slab, simply supported	1. Any type of foundations, generally deep. 2. Spans in the range 20–60 m.
7.	Prestressed segmental balanced cantilever	1. Deep foundations 2. Spans in the range 60–100 m.
8.	Cable-stayed girders	1. Deep foundations 2. Spans in the range 80–500 m.

A simple rule in determining the length of span of the super-structure of bridges is that the cost of the foundations and sub-structure should equal the cost of the superstructure. This is generally achieved by making the span length $1\frac{1}{2}$ to 2 times the height of foundations plus the sub-structure.

26.12. LOW COST CROSS DRAINAGE WORKS (FOR ODR'S AND VR'S ONLY)

Rural roads (comprising ODRs and VRs) are designed to carry only a very low volume of traffic, and providing high level cross-drainage facilities would prove to be too expensive for these low cost roads. Moreover, provision of such cross drainage structures as are required for all-year-round usage in all climatic conditions without any interruption whatsoever, is neither necessary nor feasible if all the yet unconnected villages are to be provided with a road access within the very limited funds available for the purpose, within the next few years. It should be possible, therefore, to tolerate interruptions to traffic for short periods of time which should not be considered to affect the all-weather characteristics of such low volume roads. For a quick selection of various alternatives of different types of cross-drainage structures for low volume roads in rural areas (ODRs and VRs), certain guidelines are given in Table 26.3.

Table 26.3. Guidelines for the choice of cross drainage structure for rural roads

SI. No.	Type of structure	Locations where the structure will be found suitable
1.	Ford (Fig. 26.7)	Shallow non-perennial stream where maximum depth of flow does not exceed 1.6 m and the period of interruption at one stretch does not exceed 12 hours for ODRs and 24 hours for VRs in hilly terrain, and 3 days in the case of roads in plains. Also, the total period of interruption should not exceed 10 days for ODRs and 15 days for VRs. suitable for roads in arid or isolated areas carrying very low traffic. Also for roads in a network where alternate route is available during the rainy season. Low in initial cost, and ideal for crossing of men on foot and cattle.
2.	Paved dip (or bed causeway) (Fig. 26.8)	Same as for SI. No. 1 except that the traffic on the road may be somewhat higher.
3.	Vented causeway (Fig. 26.9)	Shallow streams having permanent dry weather flow exceeding 0.6 m in depth. The same criterion indicated for SI. No. 1 as regards interruption equally apply to this case also.
4.	Submersible bridge	In cases where high level bridge is too expensive to construct.
5.	Culvert	This is a high level structure having total span up to 6 m. Suitable for narrow streams.
6.	High level bridge	Ordinarily high level bridges designed to cater for the maximum possible flood discharge should not be considered for rural roads. However, situations warranting such bridges are crossing of deep narrow streams or ravines in hilly terrain.
7.	Ferry and pontoon/ boat bridge	In case of large streams where it is not economical to provide a bridge, ferry service or pontoon/boat bridge may be appropriate.

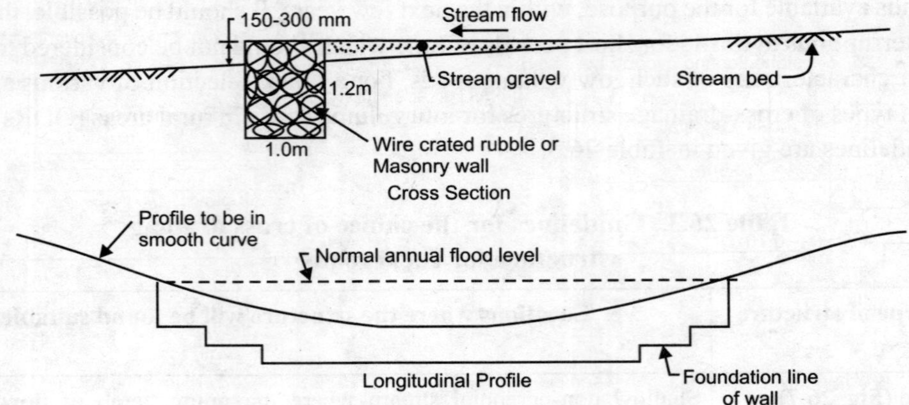

Fig. 26.7. Typical ford

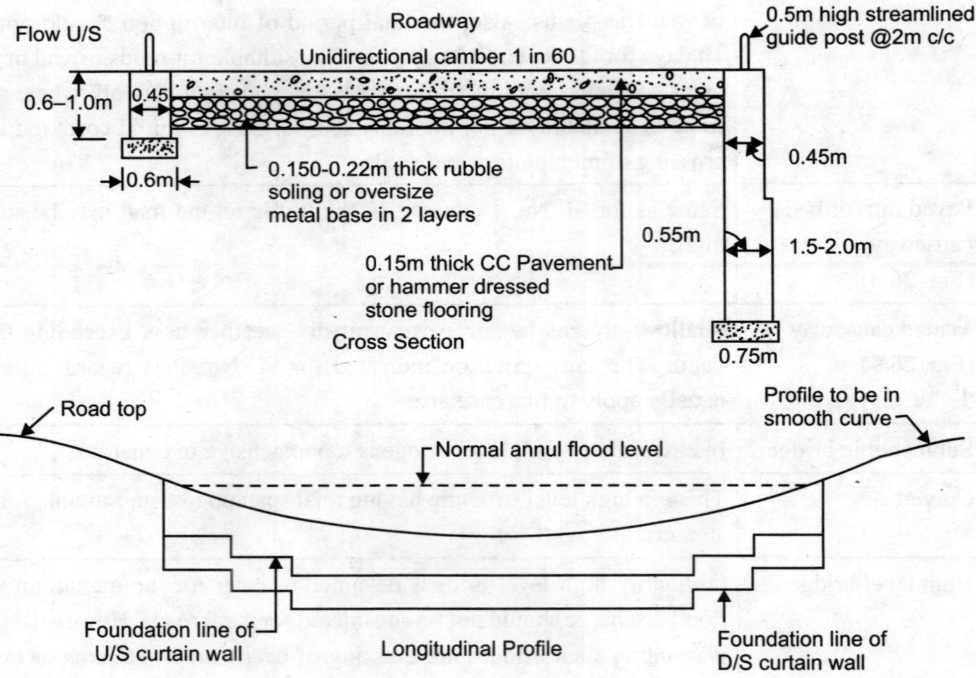

Fig. 26.8. Typical paved dip

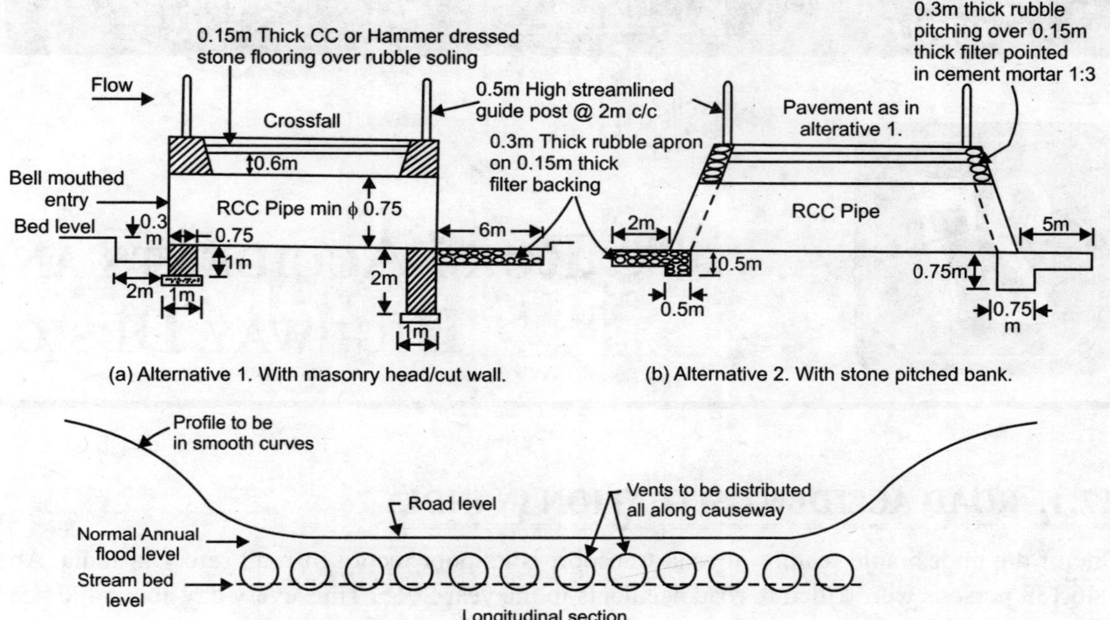

Fig. 26.9. Vented causeway with R.C.C. pipes for vents

QUESTIONS

1. What are the considerations to be kept in view in selecting the site for a cross-drainage structure?
2. What are the various types of highway culverts and the conditions for their use?
3. What are the various types of highway bridges and their suitability?

Chapter

Road Accidents and Highway Design

27.1. ROAD ACCIDENT SITUATION IN INDIA

One of the undesirable feature of road transport is its poor record of road safety in India. About 1,46,133 persons were killed in road accidents in the year 2015. Thus every day about 400 people are killed on Indian roads *i.e.* one person dies in road accident every 3.6 minutes. The road accidents rose by twenty percent in the year 2016.

It is seen that the vehicle population in the country has grown at around 14 per cent per annum, causing the number of persons killed to grow annually, at a rate of around 6.5 per cent per annum.

Road accidents cause heavy losses to the economy in the shape of loss of output, charges incurred in hospitalisation and treatment and damages to vehicles and property. In the year 2016 the economic loss due to various categories of road accidents works out to around Rs. 55000 crores or roughly 3 per cent of the national income of the country (GPD).

27.2. CAUSES OF ROAD ACCIDENTS

Road accidents, as the word 'accident' itself suggests, are randomly occurring events, caused by the interaction of a diverse set of factors. The causes can be broadly grouped under the following heads:

(*i*) The road (*ii*) The vehicle

(*iii*) The driver (*iv*) The road user other than the driver

(*v*) Environmental factors

In article 27.3, the causes attributed to the road and its design are considered in detail.

As regards the vehicle, the following causes are common:

1. Failure of brakes
2. Poor visibility due to faulty vehicle lighting system
3. Design of the body and features like:

- Shape and dimensions of the driver's seat
- Visibility of the driver from the driver's seat
- Presence of protruding objects
- Condition of tyres

Safety can be improved by the incorporation of features such as:

- Collapsible steering column
- Provision of seat belts and air bags
- Splinter-proof glass
- Anti-locking features during skidding

As regards the driver, various factors come into play. These are:

- Age
- Sex
- Training
- Drunken driving
- Driver judgement, skill and emotional make-up
- Fatigue caused by long hours of driving
- Whether wearing crash helmets or not (for two-wheeler drivers)
- Overspeeding and other wrong driving manoeuvres

The road users other than the motorists are:

- Pedestrians
- Cyclists
- Motor cyclist
- Drivers of animal-drawn vehicles

All the above categories are vulnerable road users and are commonly victims of road accidents.

Pedestrian safety can be enhanced by:

- Education, particularly of school children
- Provision of separate pedestrian facilities like footpaths, subways
- Fencing off footpaths, except at designated pedestrian crossings
- Provision of pedestrian signals at crossings
- Provision of "Zebra" crossings
- Provision of good street lighting
- Making cyclists obey rules such as prohibiting double-riding, using separate cycle paths where provided

It is commonly expressed that road safety can be enhanced by three "E"s. These are:

- Engineering
- Enforcement
- Education

The highway engineer is mostly concerned with Engineering. The next section discusses this aspect in detail.

27.3. HIGHWAY DESIGN AND ROAD SAFETY

27.3.1. Road safety in the various stages of highway system

Road safety can be enhanced by a highway engineer in various stages of the highway project. These stages are:

- Planning
- Designing
- Construction
- Maintenance
- Operation

27.3.2. Road safety incorporated at the planning stage

When planning the road network of a region or an urban area, the following measures will help improve safety:

- Careful planning of the land-use and zoning should be done, so that travel is minimised.
- Residential areas should be separated from industrial and commercial areas.
- Roads and streets should be classified into a systematic hierarchy.
- Direct access from individual plot holders to the main road should be minimised.
- Routes and networks should be planned such that areas are separated into self-contained zones (often called as neighbourhoods or environmental areas). All non-essential traffic within neighbourhoods should be avoided. The neighbourhoods should be self-sufficient in respect of school and shopping facilities.
- Towns and villages along the route should be bypassed or served through "spur" roads. The new bypasses should permit very limited access. If necessary, parallel service roads may be provided for local traffic.

27.3.3. Traffic calming

"Traffic calming" is a set of engineering measures to reduce speeds and volumes of motor vehicle traffic in local areas. The technique has been successfully used in developed countries and holds promise for the developing countries also. Some of the measures include:

- Speed breakers or humps
- Rumble strips
- Crash cushions
- Road narrowing
- Separate cycle tracks

Fig. 27.1. shows a speed control hump. The indiscriminate use of humps is to be deprecated. They are not favoured on main highways in rural sections, but are common on urban streets. They should be painted with white paint (preferably reflectorised) to make them visible at night. Suitable road signs should also be placed to forewarn the drivers.

Rumble strips contain a series of small humps which cause the vehicle to rattle, sending a clear message to the drivers to slow down. Fig. 27.2 shows a design of rumble strips.

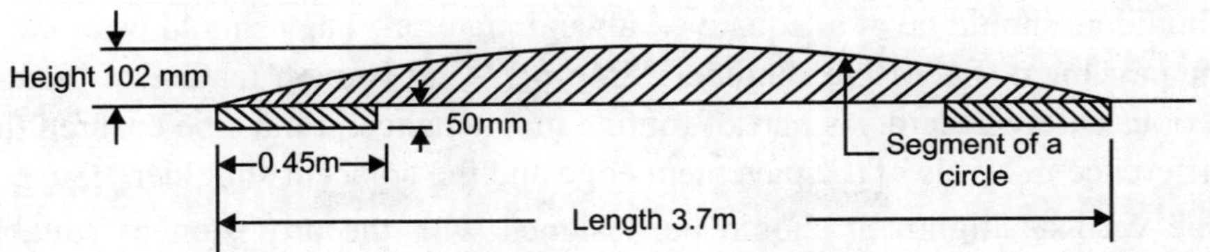

Fig. 27.1. A typical road hump or speed breaker

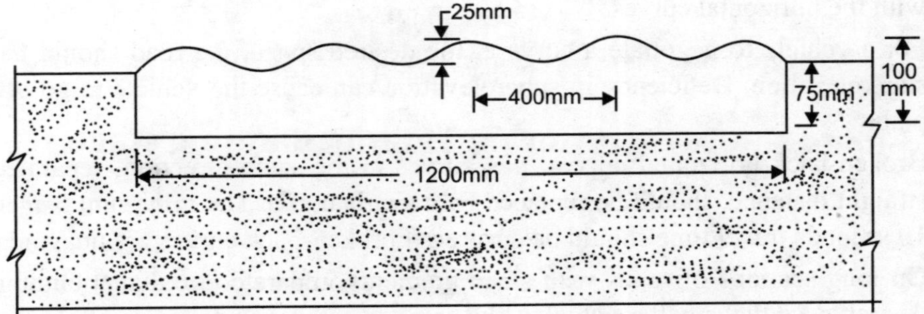

Fig. 27.2. Typical rumble strips

27.3.4. Highway design for road safety

Whenever new roads are constructed, or existing roads are taken up for improvement, the features of the highway must be designed with a clear objective of enhancing safety. Some of the aspects that need attention are listed below:

1. The design speed for the highway should be carefully selected, relating it to the terrain and road class. It is common to use the 85th percentile speed for design.
2. All roads must have the minimum stopping sight distance.
3. Overtaking sight distance is costly to provide. As large a percentage of the length as is consistent with economy should be designed for overtaking sight distance. Wherever overtaking is prohibited, suitable signs should be erected.
4. Horizontal curves are common accident spots. Sharp curves permit accident spots. Sharp curves permit low speeds and large radii curves cater to high speeds. It should be the endeavour of the designers to provide as large a radius as possible, taking recourse to minimum and ruling radii only when constrained by site conditions.
5. Transitional ends to circular curves enable the vehicle to enter the curve from the straight line gradually. They must be provided wherever they are needed.
6. For a vehicle to negotiate any curve at the desired speed, the road should have adequate superelevation. Deficiency in superelevation can cause serious accidents.
7. The minimum carriageway width for a two-way road is 7 m to cater for one lane in each direction. For low volume roads, single lane (3.75 m) and intermediate lane (5.5 m) pavements are in use in India. To enhance the safety to these narrow roads, good shoulders may be provided.

8. Shoulders should be of adequate width and material. They should be at the same level as the pavement. Excessive shoulder "drop-off 's, a common feature of Indian roads, are a serious safety hazard. As part of routine maintenance, it must be ensured that there is no difference in levels of the pavement edge and the adjacent shoulder.
9. The vertical alignment should be designed with the provision of suitable shock-free vertical curves. The vertical curves should be made to blend homogeneously with the horizontal curves. It is a good practice to design the vertical curves to be fully contained with the horizontal curves.
10. For a vehicle to negotiate a curve at the desired speed, the road should have adequate superelevation. Deficiency in superelevation can cause the vehicle to go off the desired path.
11. Broken-back horizontal curves, *i.e.*, curves in the same direction, separated by a short straight distance, should be replaced by a single curve. The minimum tangent length for 10 seconds travel time should be ensured if broken-back curves are unavoidable.
12. On long sections of roads with steep gradients, separate (auxiliary) climbing lanes are desirable, so that smaller vehicles like cars can safely overtake heavily loaded trucks.
13. Embankment side slopes should be as flat as possible so that errant vehicles accidentally leaving the road have a good chance of survival. For low embankments up to 1.5 m height, a flat slope of 4:1 is desirable. For embankments of large heights, a variable cross-section of side slopes starting form 2:1 at the top and ending at 4:1 at the ground level is suggested.
14. The texture of the pavement surface should be sufficiently rough to prevent skidding.
15. Road signs give advance warning to the driver of the road condition ahead. They should be made part of the road design for enhancing safety (see Chapter 28)
16. Road markings, reflectorised for night visibility, guide and assist the driver to safely negotiate the road (see Chapter 29).
17. Reflectorised road delineators are specially useful at night time.
18. Safety barriers are intended to prevent vehicles going off the road and deflect them back on their course. The barriers absorb the impact and deflect the vehicle. Barriers can be of a steel trough plate bolted to a steel post or of cement concrete. Barriers are provided at high embankments, on medians, and to protect vehicles from hitting electric posts, bridge abutments and piers.
19. Roads (particularly city streets) should be illuminated adequately. As per Indian practice, the average level of illumination on an important route carrying fast traffic should be 30 lux. For secondary routes with light traffic, it can be 4 lux.
20. Pedestrian sidewalks should preferably be raised above the road level by providing kerbs and pedestrian guard rails.
21. Junction designs should incorporate good features such as channelisation, acceleration and deceleration lanes, separate right turning pockets, signages and markings.
22. Traffic Impact Attenuators should be provided wherever warranted.

27.3.5. Road Safety Audit (RSA)

Road Safety Audit is a formal procedure for assessing accident potential of a specific design. The assessment is done objectively by an independent expert who is not a part of the original design team. The audit is carried out for various stages of the project such as:

- Feasibility
- Preliminary design
- Detailed design
- Pre-opening stage

The Auditor identifies any potential safety problems and provides helpful and constructive suggestions based on sound safety engineering practice, as to how any potential design difficulties can be resolved.

The RSA should cover various aspects of the highway facility including

1. Horizontal alignment
2. Vertical alignment
3. Visibility
4. Cross-sectional elements like medians, shoulders, kerbs
5. Intersection elements
6. Median openings
7. Roadside furniture
8. Plantation and landscaping
9. Embankment and earth profiles
10. Drainage
11. Road markings
12. Road signs
13. Road side facilities (bus-bays, truck lay-byes, rest areas, fuel stations etc.)

RSA is becoming an important activity on National Highways in India.

27.3.6. Safety during construction

The stripes are 10–15 cm in width and are alternately in reflectorised white and orange colours.

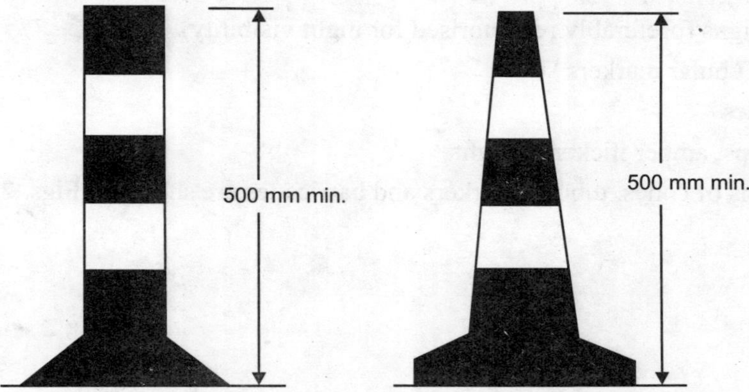

Fig. 27.3. Cones and tubular markers (with alternate white and orange stripes).

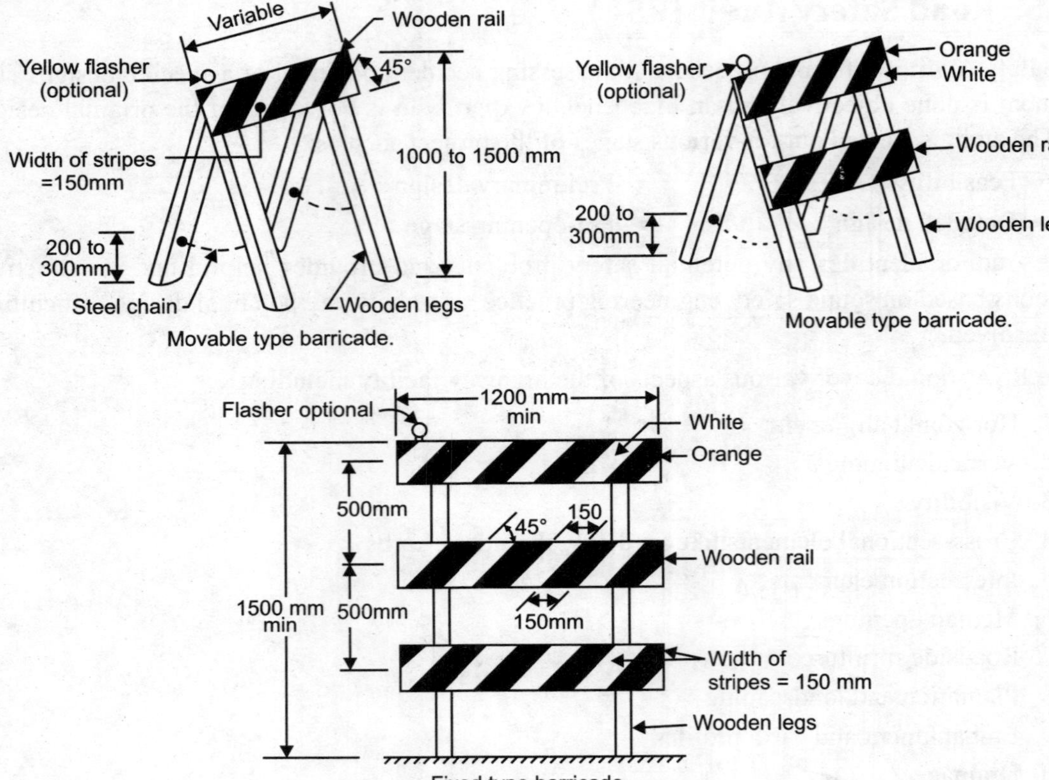

Fig. 27.4. Movable / fixed type barricades.

Construction or maintenance activities go on, even when traffic operates. This gives rise to serious hazards. Safety can be ensured through various measures such as:

- Installing proper traffic control devices
- Observing safety rules
- Making it legally mandatory for the contractor to follow good safety practices

Traffic control devices that are usually used for ensuring safety at construction sites are:

- Traffic signs (preferably reflectorised for night visibility)
- Cones, Tubular markers
- Barricades
- Red lamps, amber flickering light

Typical designs of cones, tubular markers and barricades are shown in Figs. 27.3 to 27.5.

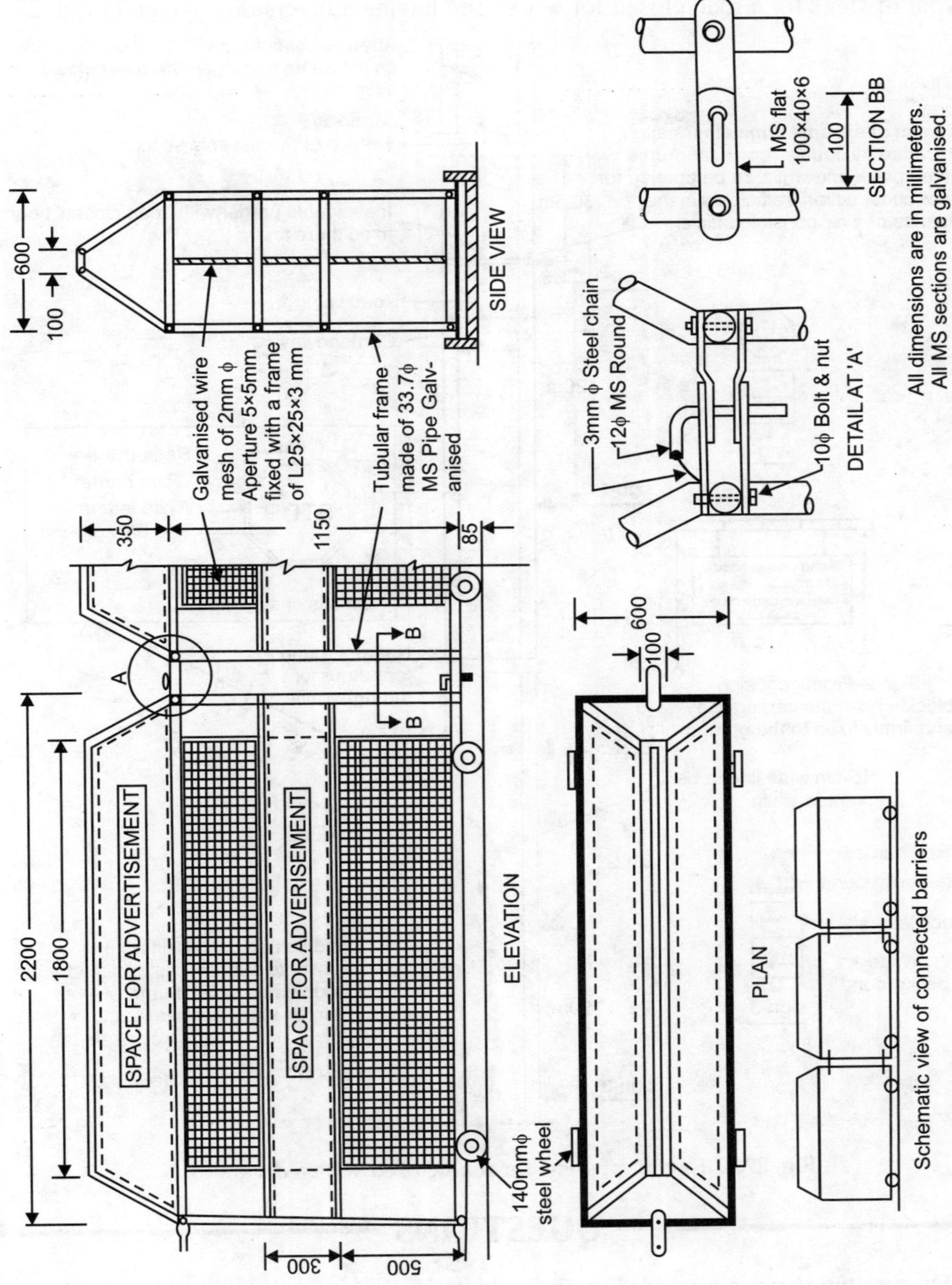

Fig. 27.5. Portable metallic barrier.

Good traffic management practices are a key to safety in construction zones. Various measures are possible, such as:

- Detouring (Diversion)
- Speed control
- Posting of flagmen
- Closure of pavement, partially or fully for works
- Proper signages

A layout of signs for a road closed for works and having a diversion is given in Fig. 27.6.

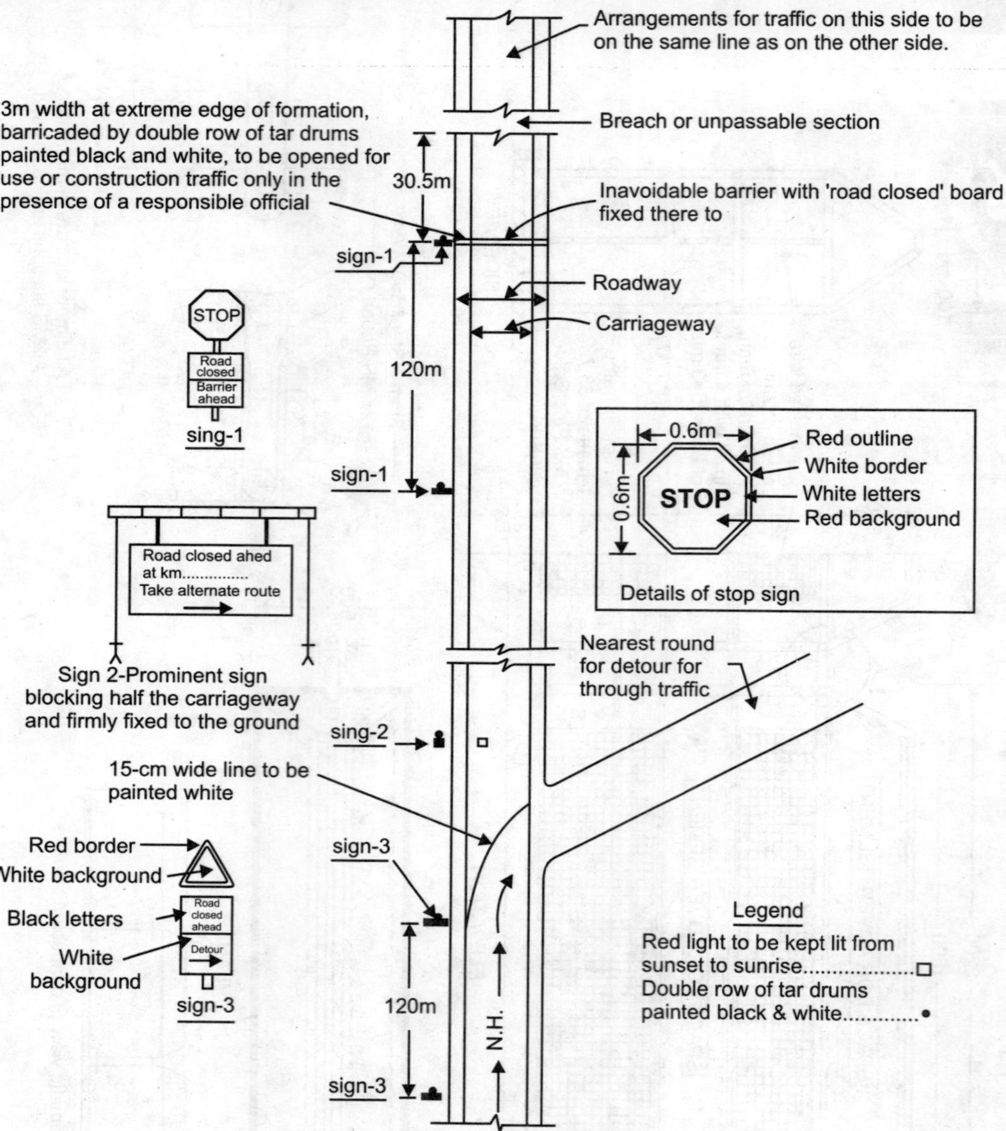

Fig. 27.6. Layout of signs for road closed with a diversion.

QUESTIONS

1. Discuss the various causes of road accidents.
2. Discuss how highway design can enhance road safety.
3. Write short notes on:
   1. Traffic Calming
   2. Road Humps
   3. Rumble Strips
   4. Safety Barriers
   5. Safety Audit
4. Discuss how safety can be enhanced during construction operations in a highway.

Chapter

ROAD SIGNS

28.1. ROLE OF ROAD SIGNS

Road signs are intended to serve the following purposes :

1. To give timely warning of hazardous situations when they are not self-evident.
2. To regulate traffic by imparting messages to the drivers when to stop, give way or limit their speeds, where parking is prohibited etc. etc.
3. To supply information on highway routes, directions and points of interest.

28.2. CLASSIFICATION

Road signs can broadly be classified into the following :

- **Cautionary or Warning signs** to warn the road users of certain hazardous situations on and adjacent to the roadway.
- **Mandatory/Regulatory signs** to inform the road users of certain laws, regulations, prohibitions or restrictions which have to be complied with ; any cases of non-compliance are liable to legal action.
- **Informatory signs** to provide guidance to the road users in respect of direction and place identification, facilities like roadside rest areas and petrol pumps, parking etc.

28.3. SITING OF ROAD SIGNS

- The road signs should normally be placed on the left side of the carriageway but may have to be placed or repeated above the carriageway as 'overhead signs' if the local conditions so warrant.
- On kerbed roads, no part of the sign should be less than 0.6 m away from the kerb edge. On unkerbed roads, it must be ensured that no part of the sign comes in the way of vehicular traffic ; no part of the sign should be less than 2 m from the edge of carriageway.
- The bottom edge of the lowest portion of the sign should not be less than 2 m above

the kerb on kerbed roads and not less than 1.5 m above the crown of the pavement on unkerbed roads.

- A sign may be placed on the median of a divided carriageway if placing it on left hand side is not considered effective enough for a given set of site conditions.
- The signs should be so sited as not to cause any obstruction to the movement of vehicular traffic on the carriageway and where placed on the shoulder/footpath/refuge island, the obstruction to pedestrians should be minimised.
- Signs may also be fixed to existing utility posts like electric poles etc. if conveniently located.

28.4. VISIBILITY

All roads signs, particularly Cautionary/Warning signs and Mandatory/Regulatory signs should have reflective material like luminous paints or reflective devices or should»be lighted for better visibility and legibility at night, except those in lighted streets of built-up areas.

28.5. CAUTIONARY/WARNING SIGNS

Fig. 28.1 shows the more common cautionary/warning signs, which are in the shape of an equilateral triangle, with the apex pointing upwards. These signs have red border and black symbols on white background. The side of the triangle is 900 mm for normal sized signs and 600 mm for small sized signs while the width of the border is 70 mm and 45 mm respectively. Details regarding sizes of symbols etc. are stipulated by IRC. In urban locations, these signs should be about 50 m away from the points of hazard, which may be adjusted to suit site conditions. In non-urban locations, these signs should be located at the following distances in advance of the hazard :

	Plain or Rolling Terrain	Hill Terrain
National Highways and State Highways	120 m	60 m
Major District Roads	90 m	50 m
Other District Roads	60 m	40 m
Village Roads	40 m	30 m

RIGHT HAND CURVE

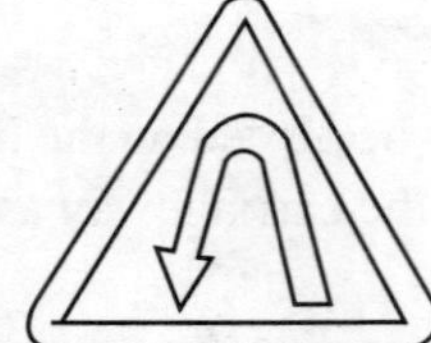

LEFT HAND HAIRPIN BEND

LEFT HAND CURVE

DANGEROUS DESCENT
NARROW BRIDGE
NARROW ROAD AHEAD
LOOSE GRAVEL
CYCLE CROSSING
PEDESTRIAN CROSSING
FALLING ROCKS
MAJOR ROAD AHEAD
MAJOR ROAD AHEAD
SIDE ROAD
IRREGULAR Y
INTERSECTIONS
STAGGERED INTERSECTION
ROUNDABOUT
RAILWAY CROSSING (UNGUARDED)
RIGHT HAND HAIRPIN BEND
ZIG-ZAG ROAD
DANGEROUS ASCENT

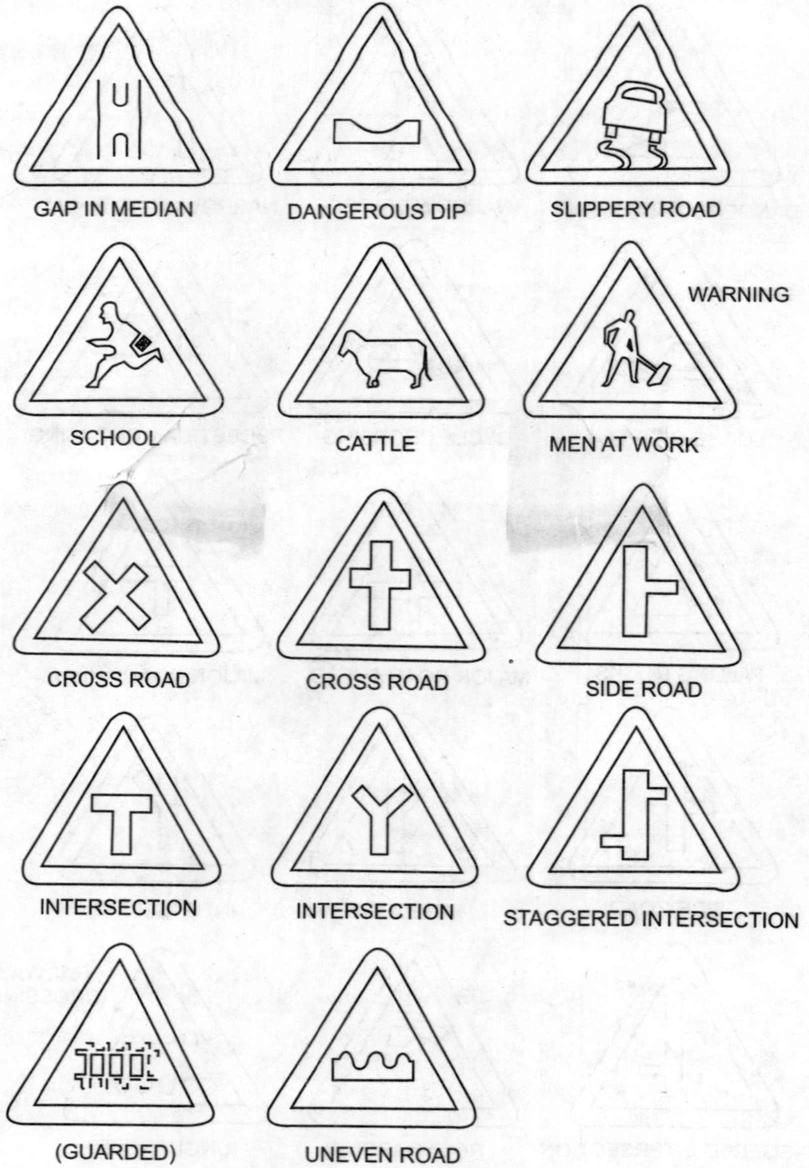

Fig. 28.1. Cautionary / Warning signs.

28.6. MANDATORY/REGULATORY SIGNS

Two of the most important of these signs are the "Stop" and "Give Way" signs. The "Stop" sign (900 mm × 900 mm) is octagonal in shape and is red in colour with a 30 mm white border. A small sized version is 600 mm × 600 mm with a 20 mm border. The "Give Way" sign is an equilateral triangle (having each side of 900 mm) in shape, white in colour with a 70 mm red border as shown in Fig. 28.2. A small sized version has a side of 600 mm with 45 mm border. Other mandatory/

regulatory signs shown in Fig. 28.3 are circular in shape with 600 mm diameter for normal size and 400 mm diameter for the small size and are white in colour with red border of 65 mm and 45 mm respectively ; the width of red oblique bar is 60 mm and 40 mm respectively. A smaller size of 300 mm diameter may be used in conjunction with traffic light signals or on bollards on traffic islands; widths of red border and red oblique bar in such cases is 35 mm and 30 mm respectively. The 'No Parking" and 'No Stopping* signs have a red border and red slanting bars but the background is blue instead of white, as in other signs of this type. Details on sizes of numbers and symbols in black are stipulated by IRC

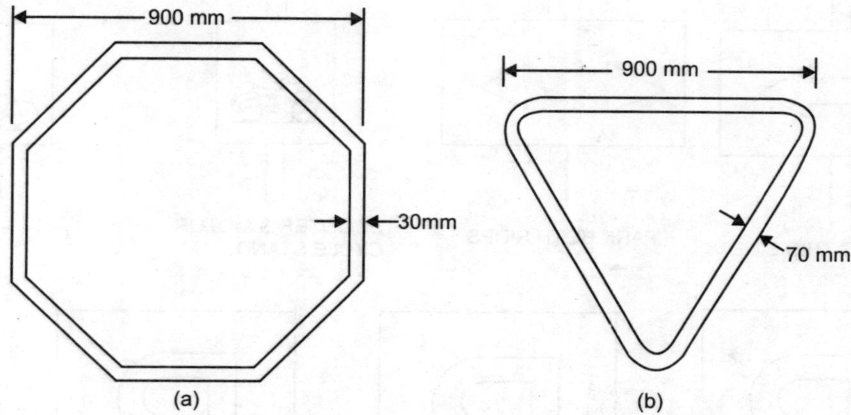

Fig. 28.2. STOP and GIVE WAY signs.

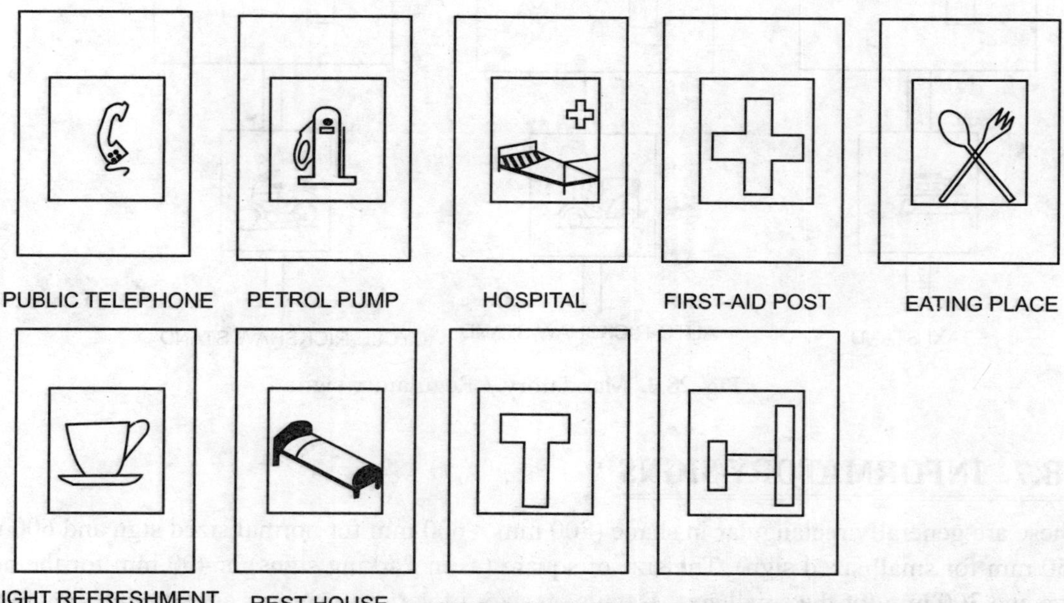

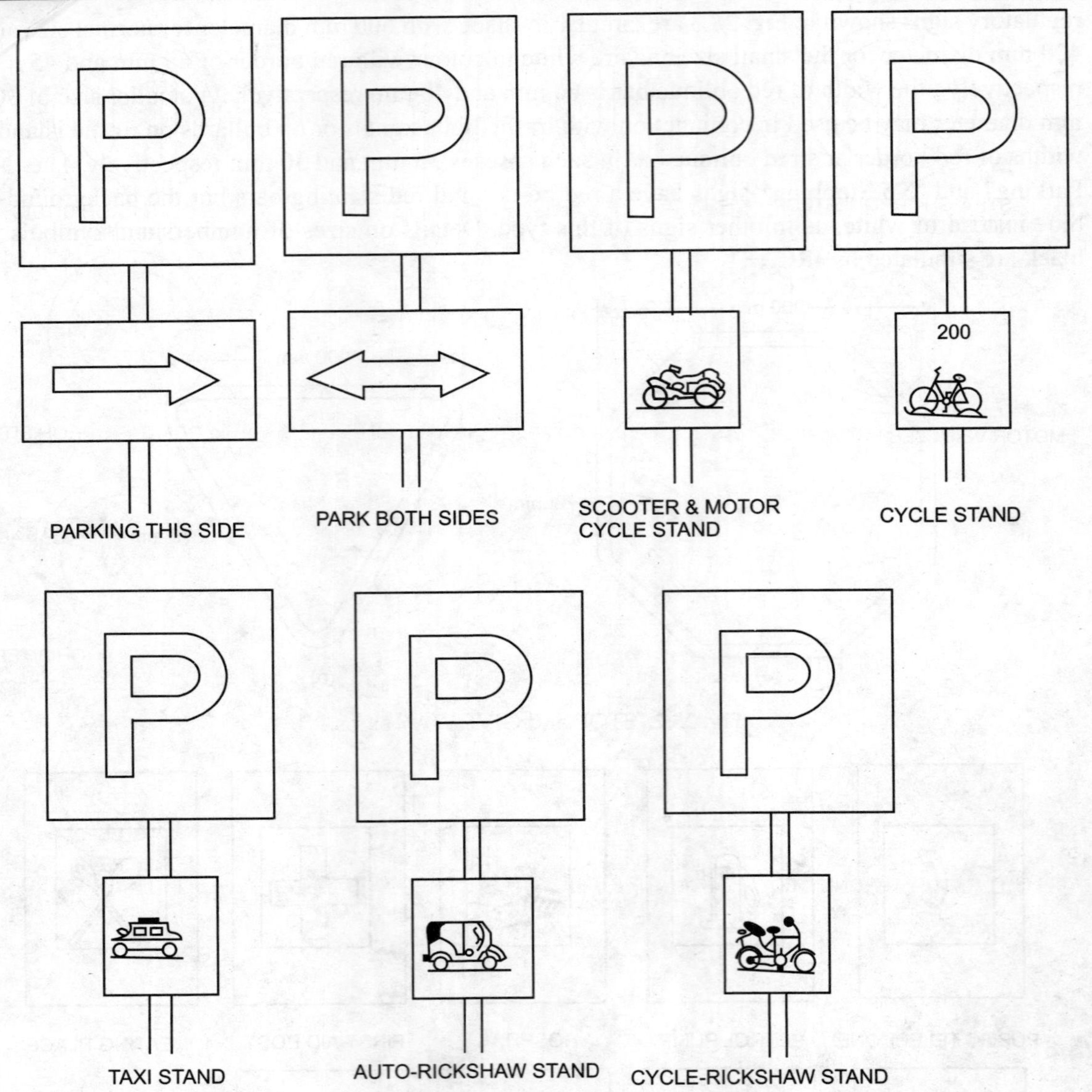

Fig. 28.3. Mandatory / Regulatory signs.

28.7. INFORMATORY SIGNS

These are generally rectangular in shape (800 mm × 600 mm for normal sized sign and 600 mm × 450 mm for small sized sign). The size of square (as in Parking signs) is 400 mm for the normal size and 300 mm for the small size. Details on sizes of various symbols are stipulated by the IRC. These signs have white background, black letters and arrows and black border. Some of the more common infor-matory signs are shown in Fig. 28.4. A typical direction and place identification sign is shown in Fig. 28.5 ; the size of these signs depends on the message and size of letters.

STRAIGHT PROHIBITED NO ENTRY

ONE WAY SIGNS

VEHICLES PROHIBITED BOTH DIRECTIONS

MOTOR VEHICLES PROHIBITED

TRUCKS PROHIBITED

BULLOCK CART & HAND CART PROHIBITED

BULLOCK CART PROHIBITED

TONGA PROHIBITED

HAND CART PROHIBITED

CYCLE PROHIBITED

RIGHT TURN PROHIBITED

LEFT TURN PROHIBITED

U-TURN PROHIBITED

OVERTAKING PROHIBITED

HORN PROHIBITED

NO PARKING

NO STOPPING

SPEED LIMIT

WIDTH LIMIT

HEIGHT LIMIT

LENGTH LIMIT

LOAD LIMIT

AXLE LOAD LIMIT

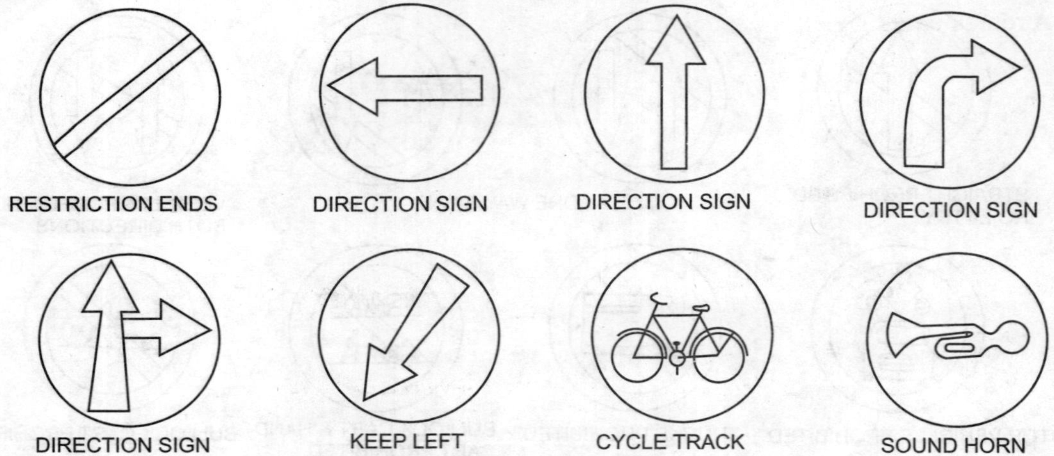

Fig. 28.4. Informatory signs.

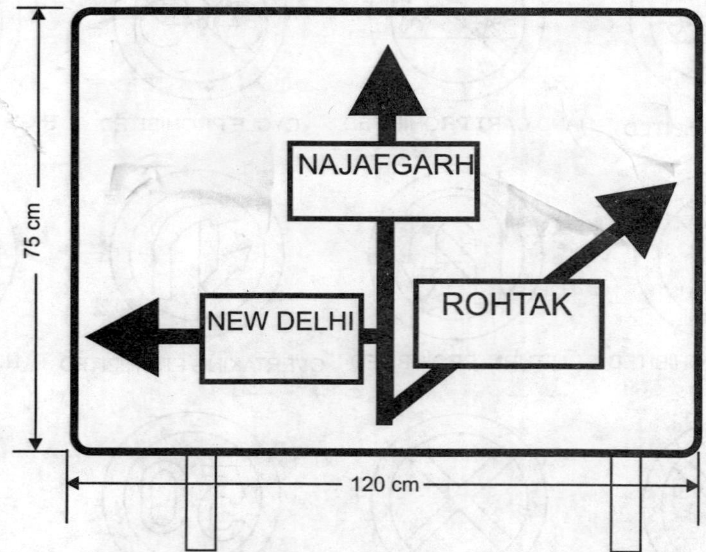

Fig. 28.5. A typical direction and place identification sign.

A route marker sign for National Highways is given in Fig. 28.6

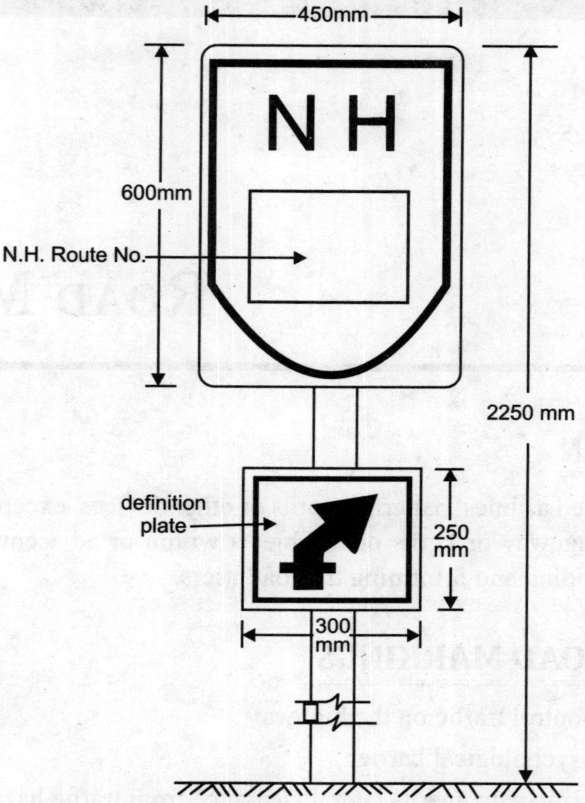

Fig. 28.6. Route marker sign for National Highways.

QUESTIONS

1. (*a*) What purposes are served by road signs?
   (*b*) Give the classifications of road signs.
   (*c*) Where can road signs be conveniently placed?
2. Describe the size, shape and colour of:
   1. Warning signs
   2. Prohibitory signs
   3. Mandatory signs
   4. Informatory signs.
3. Sketch two important road signs for the following types:
   (*a*) Mandatory signs
   (*b*) Prohibitory sings
   (*c*) Warning signs.

Chapter

Road Markings

29.1. DEFINITION

Road markings are defined as lines, patterns, words or other devices, except signs, set into, applied or attached to the carriageway or kerbs or to objects within or adjacent to the carriageway, for controlling, warning, guiding and informing the road users.

29.2. ROLE OF ROAD MARKINGS

(*a*) To guide and control traffic on the highway.

(*b*) To serve as a psychological barrier.

(*c*) To delineate traffic path and its lateral clearance from traffic hazards.

(*d*) To aid pedestrians and cyclists in channelising their movement into safe locations.

29.3. CLASSIFICATION

The road markings may be classified as:

A. Carriageway markings

(*i*) **Longitudinal markings.** These markings are placed along the direction of traffic for the guidance of drivers. These are white in colour except for the following, which are yellow in colour:

- Lines indicating parking restrictions
- Obstruction approach markings
- "No Overtaking" zone markings
- Continuous centre line may optionally be yellow.

Broken lines indicate that, traffic permitting, these may be crossed.

Solid lines, however, are restrictive and are not permitted to be crossed except for entry or exit from a premises or a side road or to avoid a stationary obstruction.

Double solid lines indicate maximum restrictions and are not to be crossed except in an emergency.

In a combination of broken and solid lines, a solid line may be crossed, with discretion, if the broken line of the combination is nearer to the direction of travel. Vehicles from opposite direction are not permitted to cross the solid line.

Figs. 29.1 (*a*) and (*b*) show some of the more important types of longitudinal markings.

The various longitudinal markings include the following:

- centre lines
- traffic lanes
- no passing zones
- warning signs
- border or edge line
- bus lane markings
- cycle lane markings

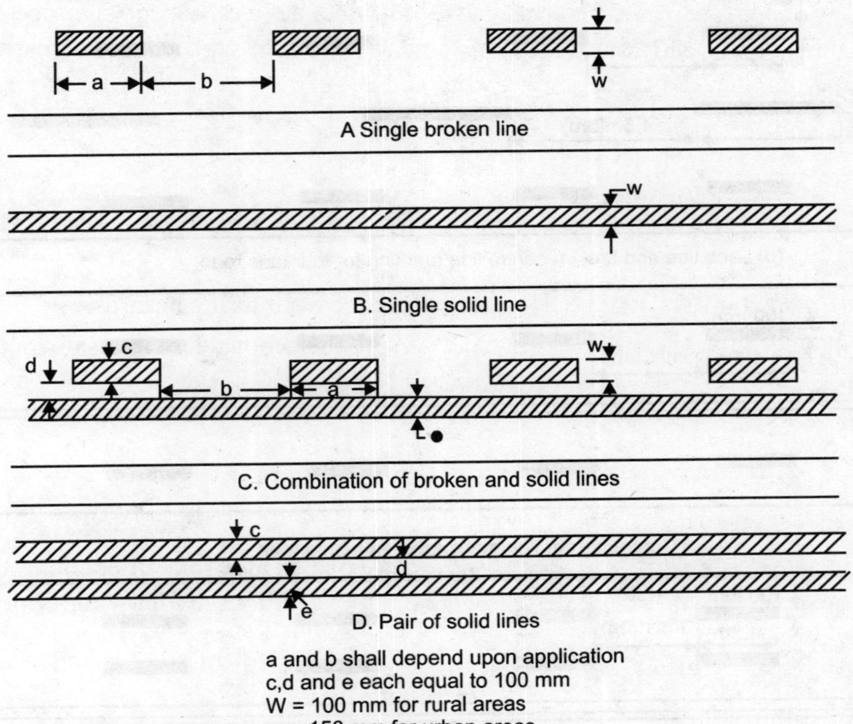

Fig. 29.1. (*a*) Recommended size of longitudinal markings.

(*ii*) **Markings at intersections.** The type of carriageway marking for a particular intersection is dependent on a number of variables like speed characteristics and availability of space.

The markings in this category may be further subdivided into markings on approaches to intersections and markings in the intersection area. The different types of markings under each of these two groups are as under:

(*a*) Markings on approaches to intersections:

- Stop lines
- Give way lines
- Pedestrian crossings
- Cyclist crossings
- Markings on speed change lane
- Direction arrows
- Protected right turn lanes

(*b*) Markings in intersection area

- Marking at rotaries
- Box markings
- Continuity lines

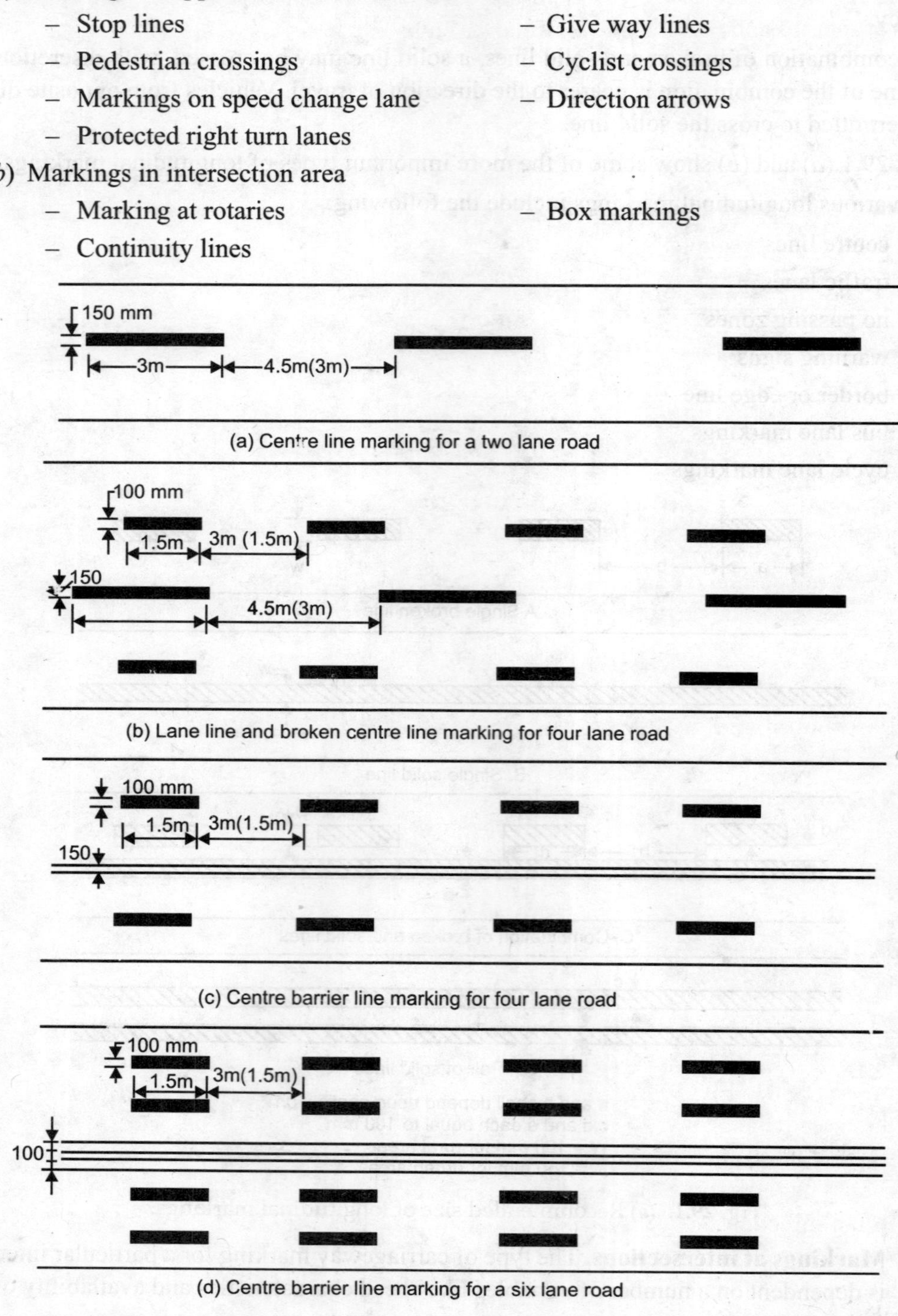

(a) Centre line marking for a two lane road

(b) Lane line and broken centre line marking for four lane road

(c) Centre barrier line marking for four lane road

(d) Centre barrier line marking for a six lane road

Note: Figure in brackets to be used on curves and approaches to intersections.

Fig. 29.1. (*b*) Centre line and lane line marking for urban areas.

Fig. 29.2. shows the usage of 'Give way' and 'Stop' lines while Fig. 29.3 shows typical markings at rotaries and Fig. 29.4. gives details of diagonal and chevron markings.

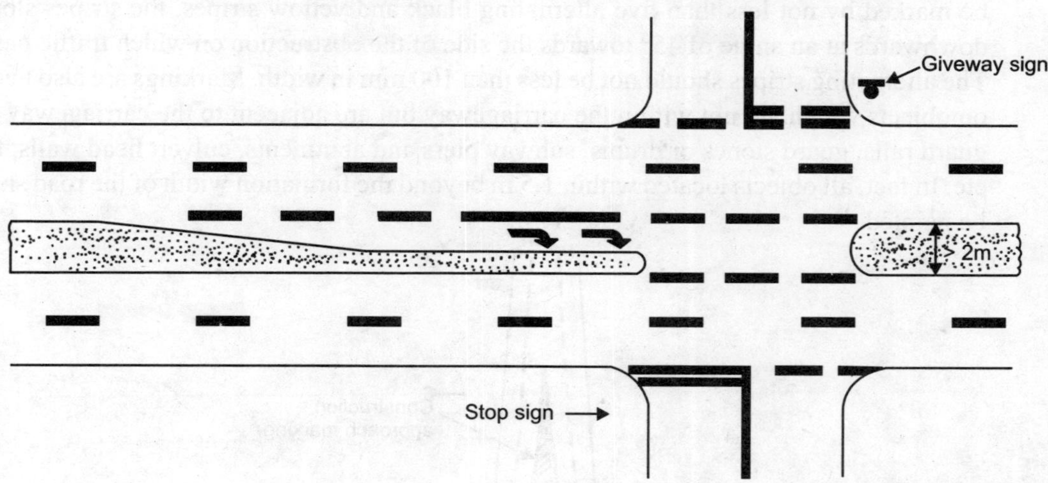

Fig. 29.2. Typical usage of giveway and stop lines.

(*iii*) **Markings on hazardous locations.** The more common road markings falling under this category include the following:

- Carriageway width transition markings
- Median width transition markings
- Obstruction approach markings
- Road-Rail level crossings

Fig. 29.5 shows road markings on transition from two lane undivided to four lane divided carriageway. Fig. 29.6 shows approach markings for obstructions in the roadway.

(*iv*) **Markings for parking.** These include the following:

- Parking space limits
- Parking restrictions
- Bus stops

Fig. 29.7 shows typical markings for parking spaces.

(*v*) **Word messages.** Word messages, painted on the pavement, like STOP ; SLOW ; BUS ; BUS LANE ; KEEP CLEAR ; SCHOOL ; RIGHT TURN ONLY; EXIT ONLY can also be used for providing information to drivers to guide, warn or regulate traffic. While the messages RIGHT TURN ONLY and EXIT ONLY should be used only to supplement the directional arrows, the legends SCHOOL, SLOW and STOP should be used only in support of standard signs. All these legends are painted on the pavement in white and should be reflectorised, except on well-illuminated roads. The letters should be elongated in the direction of traffic ; letter height in the direction of travel should be 1.25 m for speeds upto 50 km per hour and 2.5 m for speeds above 50 km per hour. There should be a minimum 0.3 m clear at either side of the lane/carriageway and in between letters.

Fig. 29.8 shows a 'Keep Clear' word message.

(*vi*) **Object markings.** Markings are also needed on physical obstructions within the carriageway, adjacent to the carriageway and on speed breakers as they may be serious hazards to the moving traffic *e.g.* piers and abutments, monuments, traffic islands, signs and signal supports, poles, trees etc. Obstructions within the carriageway should be marked by not less than five alternating black and yellow stripes, the stripes sloping downwards at an angle of 45° towards the side of the obstruction on which traffic passes. The alternating stripes should not be less than 100 mm in width. Markings are also needed on objects which are not within the carriageway but are adjacent to the carriageway e.g., guard rails, guard stones or drums, subway piers and abutments, culvert head walls, trees etc. In fact, all objects located within 1.5 m beyond the formation width of the road should be painted.

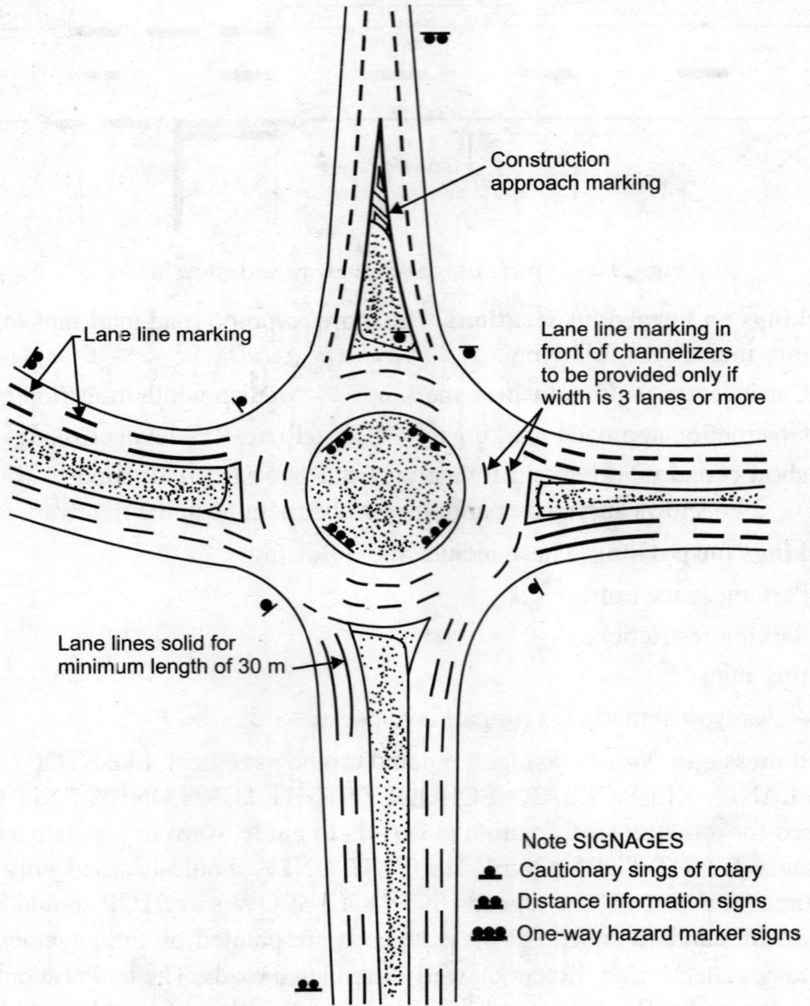

Fig. 29.3. Typical road markings at rotaries.

Poles close to the carriageway should be marked with alternate black and white horizontal stripes upto a height of 1.25 m above the road level, the stripes being uniform and not less than 100 mm in width. Objects which are not likely to be hit unless a vehicle runs off the carriageway should be painted solid white; on trees, marking should be upto a height of 1.25 m above the road level with a 300 mm band in black paint in the middle. Speed breakers where warranted should be painted in white to give visual warning to drivers in addition to posting suitable advance warning signs on the roadside located 40 m in advance of the speed breaker.

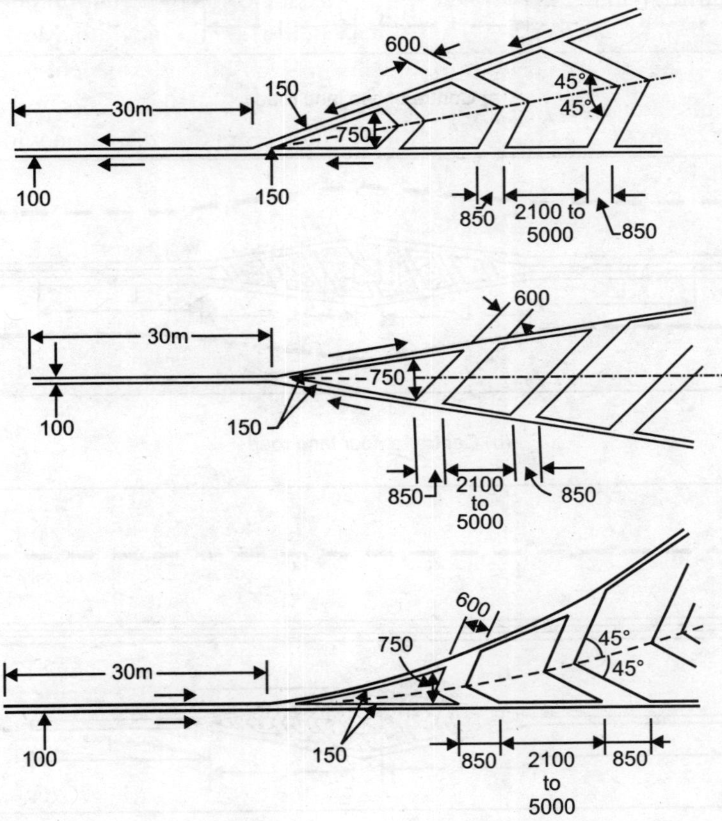

Fig. 29.4. Details of diagonal and chevron markings.

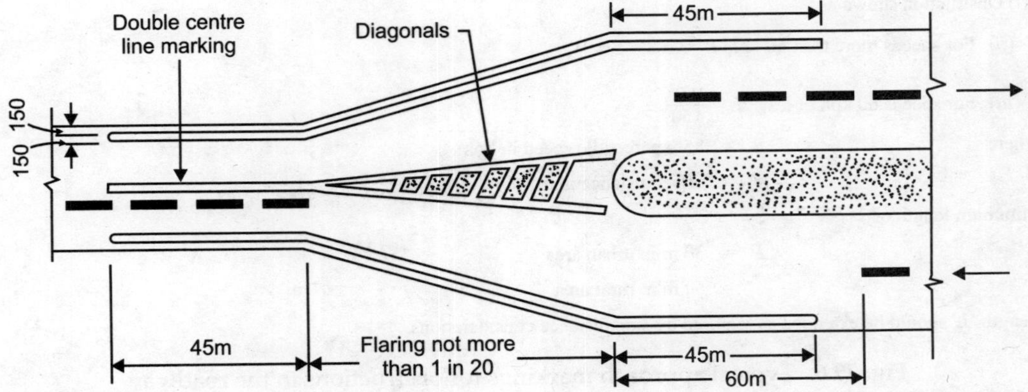

Fig. 29.5. Markings on transition from 2-lane undivided to 4-lane divided carriageway.

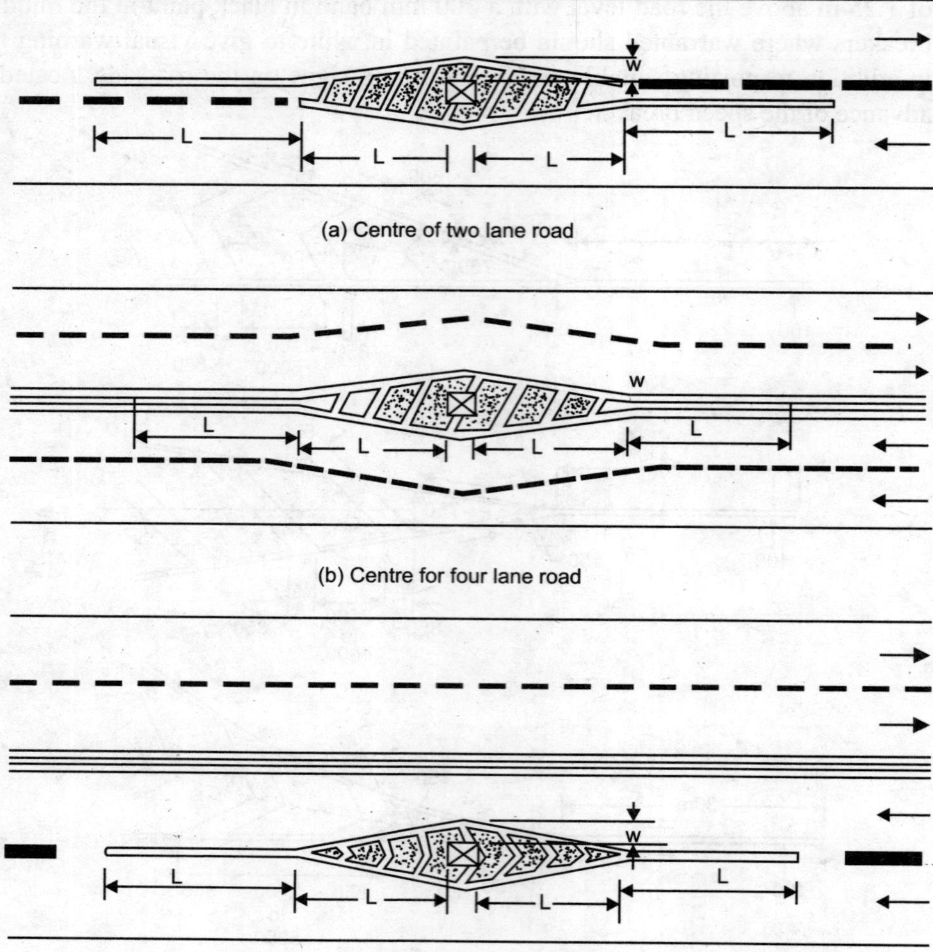

(a) Centre of two lane road

(b) Centre for four lane road

(c) Traffic passing on both sides of obstruction

Notes:(*i*) Obstruction shown as

(*ii*) For speeds more than 60 kph, $L = 0.63 \times S \times W$

(*iii*) For speeds 60 kph or less ; $L = \dfrac{WS^2}{150}$

where S = 85th percentile speed in kph

W = offset distance in m

Minimum length of

L = 30 m in urban area

= 60 m in rural area

Length '*L*' should be extended as required by sign distance considerations.

Fig. 29.6. Typical approach markings for obstructions in the roadway.

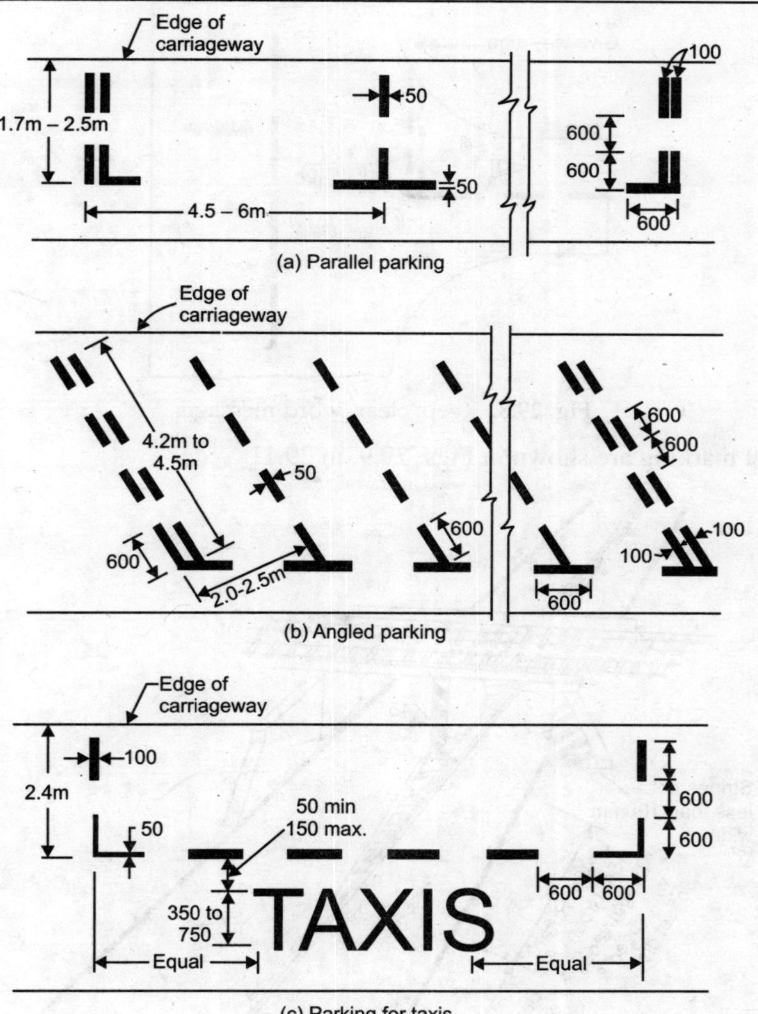

Note: Width of bay shall change as per site requirements and type of vehicles.

Fig. 29.7. Typical markings for parking spaces.

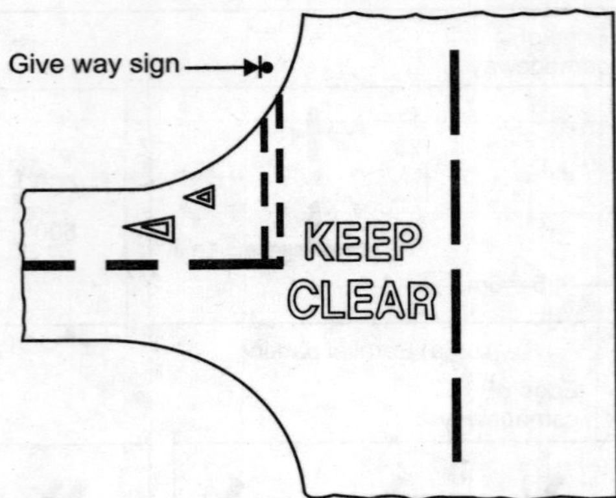

Fig. 29.8. Keep clear word message.

Typical object marking are shown in Figs. 29.9. to 29.11

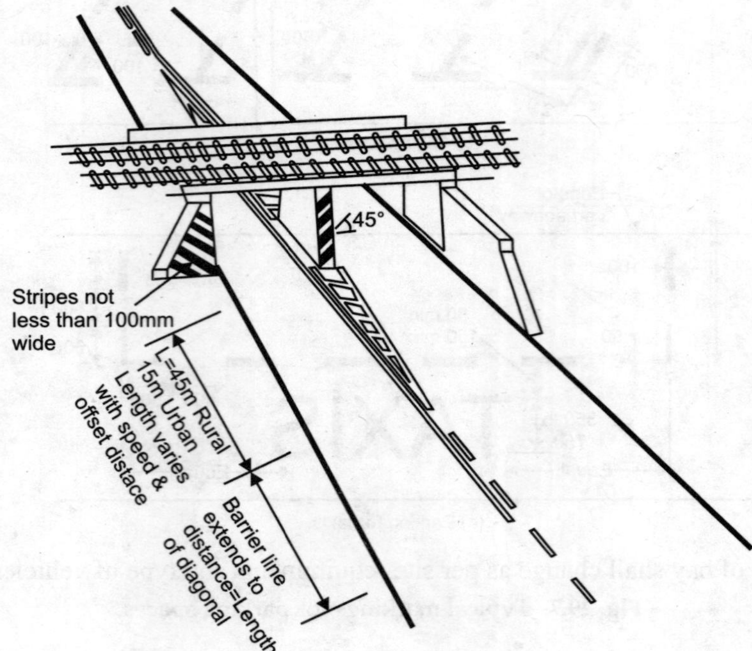

Fig. 29.9. Typical marking on objects in and adjacent to the roadway.

Fig. 29.10. Kerb markings.

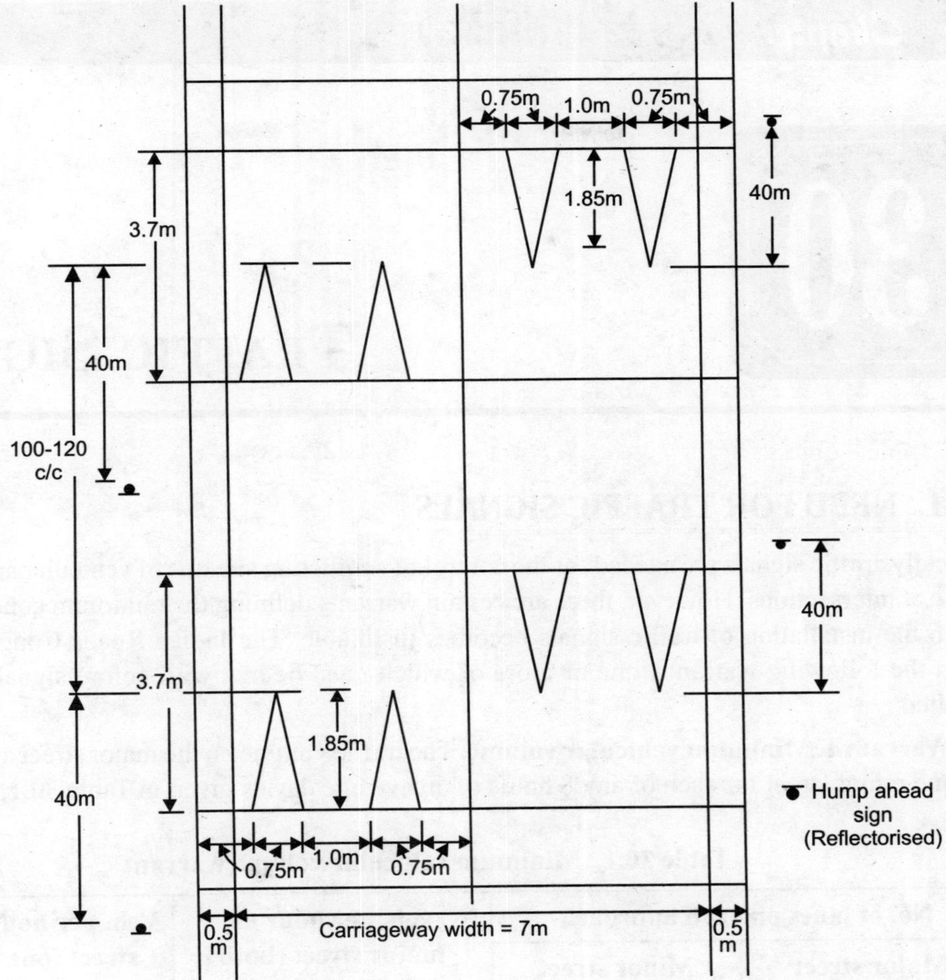

Fig. 29.11. Markings on speed breakers.

QUESTIONS

1. What is the role of road marking? Give its classification.
2. Draw sketches to show:
   (*i*) Longitudinal marking for a six lane divided carriageway
   (*ii*) Marking at rotaries
   (*iii*) Diagonal and chevron marking
   (*iv*) Angled parking
   (*v*) Kerb markings
   (*vi*) Markings on speed breakers

Chapter

TRAFFIC SIGNALS

30.1. NEED FOR TRAFFIC SIGNALS

Basically, traffic signals are needed for the control of conflicting streams of vehicular and pedestrian traffic at intersections. However, there are certain warrants defining the minimum conditions under which the installation of traffic signals becomes justifiable. The Indian Roads Congress has laid down the following warrants, one or more of which must be met with before signals need to be installed :

Warrant 1. Minimum vehicular volume. The traffic volume on the major street and the higher volume minor street for each of any 8 hours of an average day as given in Table 30.1, must exist.

Table 30.1. Minimum vehicular volume warrant

No. of lanes on each approach		Veh. per hour on major street (both approaches)	Veh. per hour on minor street (one direction only)
Major street	Minor street		
1	1	650	200
2 or more	1	800	200
2 cr more	2 or more	800	250
1	2 or more	650	250

Warrant 2. Interruption of continuous traffic. This warrant applies when traffic volume on a major street is so heavy that traffic on a minor intersecting street experiences excessive delay or hazard while entering the intersection. The traffic volumes on the major street and on the higher volume minor street for each of any 8 hours of an average day are given in Table 30.2.

Table 30.2. Interruption of continuous flow warrant

No. of lanes* on each approach		Veh. per hour on major street (both directions)	Veh. per hour on minor street (one direction only)
Major street	Minor street		
1	1	1000	100
2 or more	1	1200	100
2 or more	2 or more	1200	150
1	2 or more	1000	150

*These traffic lanes at intersection shall be minimum 2.8 m wide.

Warrant 3. Minimum pedestrian volume. For each of any hour of an average day, the following traffic volume must exist:

(*i*) On the major street, 600 or more vehicles per hour enter the intersection (both approaches); or where there is a raised median island 1.2 m or more in width, 1000 or more vehicles per hour (both directions) enter the intersection.

(*ii*) During the same 8 hours as in (*i*) above, there are 150 or more pedestrians per hour on the highest volume cross-walk crossing the major street.

Warrant 4. Accident experience. The accident experience warrant is satisfied when :

(*i*) Adequate trial of less restrictive remedies with satisfactory observance and enforcement have failed to reduce the accident frequency.

(*ii*) Five or more reported accidents, of types susceptible of correction by traffic signal control have occurred within a period of 12 months, each accident involving personal injury or property damage to an apparent extent of Rs. 2000 or more.

(*iii*) The signal installation will not seriously disrupt traffic flow.

30.2. ADVANTAGES AND DISADVANTAGES OF TRAFFIC SIGNALS

Traffic signals, when properly designed, located and operated, have one or more of the following advantages :

(*i*) They can provide for an orderly movement of traffic.

(*ii*) When proper geometric layouts and control measures are employed, they can increase the traffic-handling capacity of the intersection.

(*iii*) They can reduce the frequency of certain types of accidents, especially the right-angle type and pedestrian accidents.

(*iv*) Under favourable conditions, they can be co-ordinated to provide for continuous or nearly continuous movement of traffic at a definite speed along a given route.

(*v*) They can be used to interrupt heavy traffic at intervals to permit other traffic-vehicular or pedestrian—to cross.

(*vi*) Traffic signals dispense with police control and can thus be economical.

(*vii*) If properly designed and set, they can assign right-of-way impartially to traffic, unlike manual controls which can stop and interrupt traffic streams at the personal whim of the traffic controller.

As regards their disadvantages, the following may be mentioned, especially if the signals are installed improperly:

(*i*) Excessive delay to vehicles may be caused, particularly during off-peak hours.

(*ii*) Unwarranted signal installations tend to encourage the disobedience of the signal indications.

(*iii*) Drivers may be induced to use less adequate and less safe routes to avoid delays at signals.

(*iv*) Accident frequency, especially of the rear-end type, may increase.

(*v*) When the installations break down, due to any fault in the system, total and widespread confusion and difficulties can result.

Indian practice. The Indian practice is to have an amber period of 2 seconds as a transition interval between termination of related green movement and exhibition of a red indication or between termination of a red indication and commencement of related green movement.

Flashing amber. A flashing amber signal is a hazard identification beacon, normally used to warn of obstruction and intersections to supplement regulatory signs and to warn of midblock cross-walks.

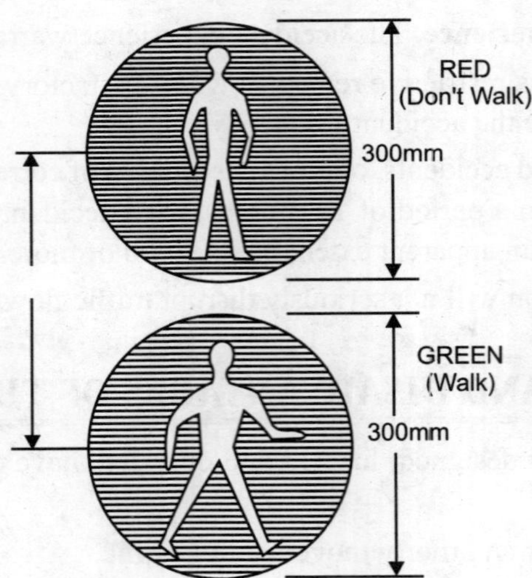

Fig. 30.1. Pedestrian signal Indications as per Indian practice

30.3. SIGNAL FACE

The minimum number of lenses in a signal face is three–red, amber and green, and the maximum number in American practice is five. The lenses in a signal face can be arranged in a vertical or horizontal straight line. The relative positions (top to bottom or left to right) are : red, amber and green. A simple signal face with three lenses in a vertical line is indicated in Fig. 30.2 A traffic signal has a signal face mounted on a vertical post (Fig. 30.3.) or provided overhead (Fig. 30.4.)

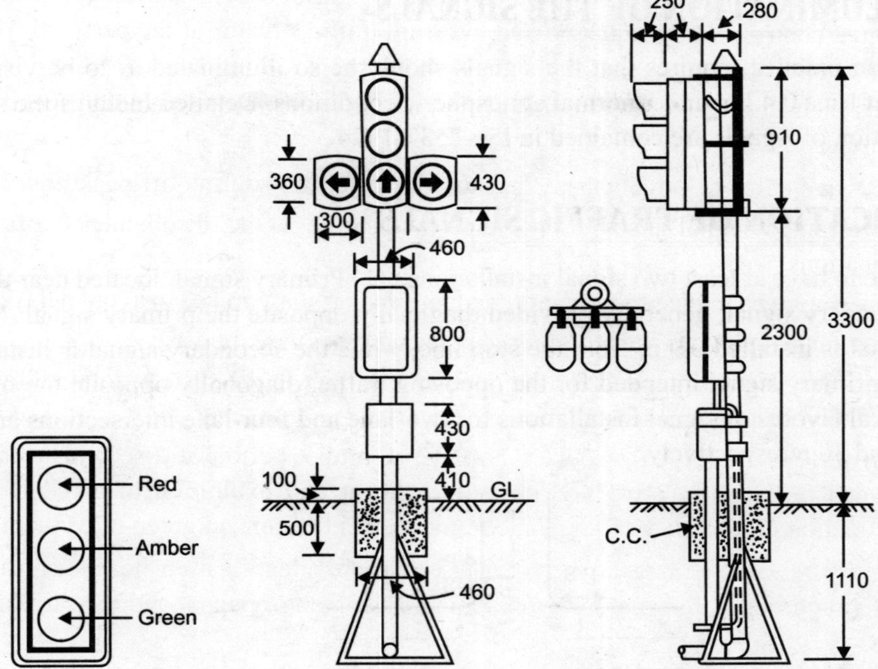

Fig. 30.2. Signal face.

Fig. 30.3. Signal post.

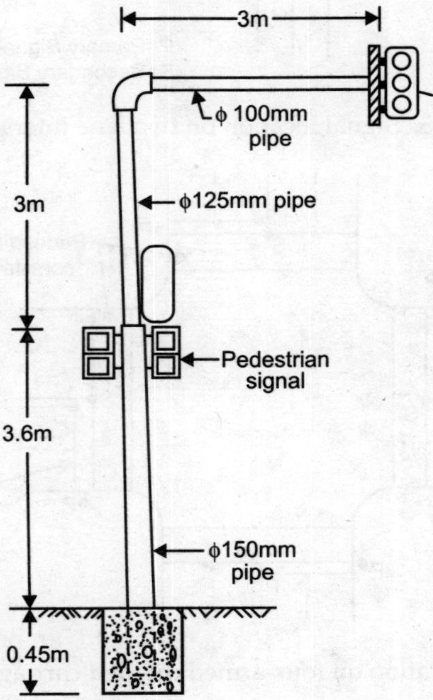

Fig. 30.4. Overhead signal post.

30.4. ILLUMINATION OF THE SIGNALS

The American practice requires that the signals should be so illuminated as to be visible from a distance of at least 0.4 km under normal atmospheric conditions. Detailed Indian specifications for the illumination of signals are contained in IS : 7537-1974

30.5. LOCATION OF TRAFFIC SIGNALS

It is desirable to have at least two signal installations : (1) Primary signal, located near the stop line and (2) Secondary signal, generally provided diagonally opposite the primary signal. Normally, a primary signal is installed 0.9 m from the stop line, while the secondary signal is installed on the back of the primary signal intended for the opposing traffic, diagonally opposite the first primary signal. Typical layouts of signal installations for two-lane and four-lane intersections are shown in Figs. 30.5 and 30.6 respectively.

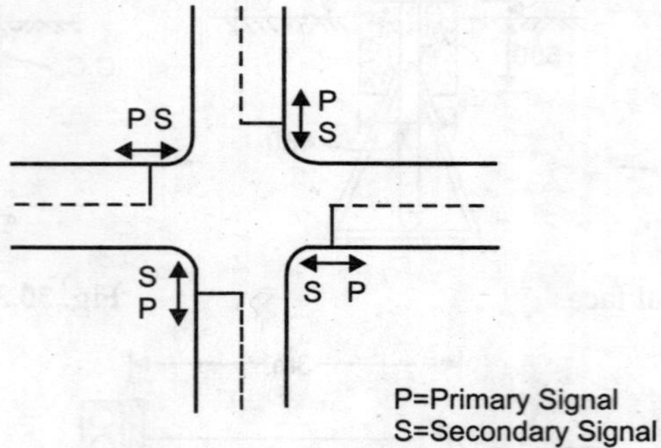

Fig. 30.5. Signal location on two lane intersection.

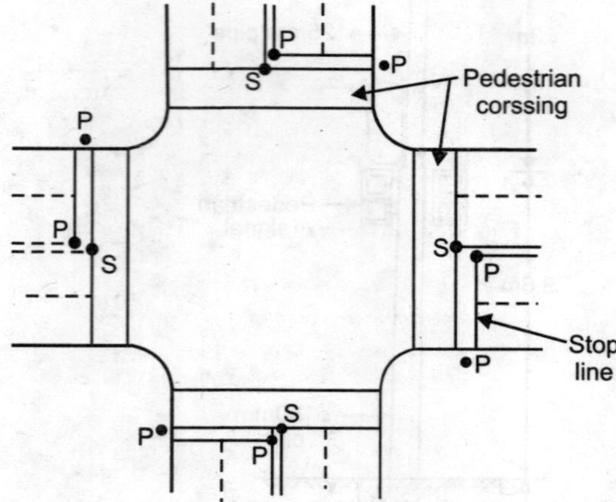

Fig. 30.6. Signal location on four-armed divided carriageway intersection.

30.6. DESIGN OF SIGNALS

In designing signals, the following definitions are relevant:

Saturation Flow. The maximum continuous flow of one vehicle behind the other beginning with the green aspect and ending with the amber/red aspect.

Cycle. The complete sequence of signal aspect at any one arm of an intersection beginning with the onset of the next red. The time taken is known as the cycle time.

Lost time. Time lost at the beginning and termination of green phases due to starting time and termination time.

Phase. Part of a time cycle allocated to any traffic movement receiving the right-of-way.

Intergreen Period. The time from the end of the green period of the phase losing right of way to the beginning of the green period of phase gaining right of way.

The following general considerations govern the design of signals :

The cycle time should be such that delay at all approaches is minimised. This is found from the formula:

$$C_0 = \frac{1.5\,L + 5}{1 - Y} \text{ seconds}$$

where C_0 = Optimum cycle time

L = Lost time per cycle in seconds.

$$Y = y_1 + y_2 ......+ y_m$$

$$y_1, y_2, y_3, ..., y_n = \frac{q_1}{s_1}, \frac{q_2}{s_2}, \frac{q_3}{s_3}, ..., \frac{q_n}{s_n}$$

where $q_1, q_2, ..., q_m$ = flow is phases 1, 2......., n

$s_1, s_2......., s_n$ = saturation flow in phases 1, 2,......, n

The maximum cycle time should be 120 seconds.

The apportionment of effective green time in a cycle can be made to the different phases such that:

$$g_1 : g_2 :...... g_n = y_1 : y_2 : ....: y_n$$

where, $g_1, g_2, g_2, ..., g_n$ = effective green time in phases 1,2......., n.

$$y_1, y_2, ..., y_n = \frac{q_1}{s_1}, \frac{q_2}{s_2}, ..., \frac{q_n}{s_n}$$

$q_1, q_2, q_3, ...q_n$ = flow in phase 1, 2, 3,......, n

$s_1, s_2, s_3,......, s_n$ = saturation flows in phases 1, 2,......, n

The amber time is two seconds.

The saturation flow is generally taken as :

$$S = 525\,w \text{ PCU/hour}$$

where, PCU = passenger car units

w = width of approach road in m.

Problem 30.1. A four-armed intersection has the following widths :

North arm : 7.5 m

South arm : 7.5 m

East arm : 5.5 m

West arm : 5.5 m

What is the saturation flow in each arm ?

Solution. Saturation flow = 525 w PCU/hr.

Saturation flow in the North/South arms = 525 × 7.5 = 3940 PCU/hr.

Saturation flow in East/West arms = 525 × 5.5 = 2890 PCU/hr.

Problem 30.2. A fixed time 2-phase signal is to be provided at an intersection having four arms. The design hour traffic and saturation flow are as under:

	North (N)	South (S)	East (E)	West (W)
Design hour flow	800	400	750	600
Saturation flow	2400	2000	3000	3000

The time lost per phase due to starting delays is 2 seconds. Calculate the optimum cycle time. Allocate the green times to the two phases.

Solution. $\frac{q_N}{S_N} = \frac{800}{2400} = 0.33; \frac{q_s}{S_s} = \frac{400}{2000} = 0.20$

∴ Maximum value of q/S in N-S direction = 0.33

$$\frac{q_E}{S_E} = \frac{750}{3000} = 0.25\ ; \frac{q_w}{S_w} = \frac{600}{3000} = 0.20$$

∴ Maximum value of q/S in E-W direction = 0.25

L = time lost due to delays = 2 + 2 + 2 + 2 = 8 sec.

$$\text{Optimum cycle time } C_0 = \frac{1.5L + 5}{1 - Y}$$

$$= \frac{1.5(8) + 5}{1 - (0.33 + 0.25)} = \frac{17}{0.42} = 40 \text{ seconds.}$$

Effective Green Time per cycle = $C_0 - L$ = 40 – 8 = 32 seconds

Effective Green Time per phase can be calculated as under:

$$g_{NS} = \frac{y_{NS}}{Y}(C_0 - L) = \frac{0.33}{0.58} \times 32 = 18 \text{ seconds}$$

$$g_{EW} = \frac{y_{EW}}{Y}(C_0 - L) = \frac{0.25}{0.58} \times 32 = 14 \text{ seconds}$$

The timing diagram is shown in Fig. 30.7.

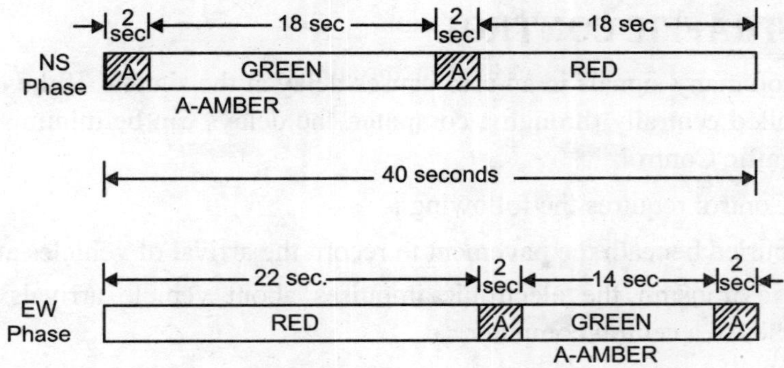

Fig. 30.7. Timing diagram.

30.7. FIXED-TIME SIGNALS AND VEHICLE-ACTUATED SIGNALS

Fixed time signals are those in which the green periods, and hence the cycle lengths are predetermined and of fixed duration. Vehicle actuated signals, on the other hand, are those in which the green periods vary and are related to the actual demands made by traffic. This is made possible by installing detectors on all the approaches. An intermediate type, semi-vehicle-actuated signals, is also available, in which the right of way normally rests with the main road and detectors are located only on the side roads.

The advantages and disadvantages of the three types are briefly given below :

Type		**Advantages**		**Disadvantages**	
1.	Fixed time	(*i*)	Simple in construction.	(*i*)	Inflexible and hence may cause avoidable delay.
		(*ii*)	Relatively inexpensive.	(*ii*)	Require careful setting.
		(*iii*)	Most successfully used in linked systems requiring a fixed cycle length for a given patten and speed of progression.		
2.	Vehicle-actuated	(*i*)	They are flexible and are able to adjust to changing traffic conditions automatically.	(*i*)	Require costly equipment such as detectors and sophisticated controllers.
		(*ii*)	Delay is held to a minimum and maximum capacity is achieved.	(*ii*)	Cannot provide signal co-ordination.
3.	Semi-vehicle actuated		Useful for junction of a side street having low traffic volume with a main street having heavy flow.		They are believed to cause high accident rates at times of light traffic.

Modern fixed time equipments are built for operation with different settings at certain periods of the day, to cover different conditions. This is achieved by providing time switches.

30.8. AREA TRAFFIC CONTROL

The presence of too many signals in an area causes delay at the signals. If the operations of these signals are controlled centrally through a computer, the delays can be minimised. This system is known as Area Traffic Control.

Area Traffic Control requires the following :

1. Sensor buried beneath the pavement to record the arrival of vehicles at the approaches.
2. Cables to transmit the electronic impulses about vehicle arrivals at various signal approaches to a central computer.
3. A central computer.
4. A pre-programmed signal setting plan, so as to minimise delays.

QUESTIONS

1. Why are traffic signals needed?
2. Draw sketches of a ground level and an overhead signal.
3. What is a 'saturation flow'?
4. Write short notes on:
   (*i*) Cycle in signal
   (*ii*) Phase in signal
   (*iii*) Optimum cycle of a signal
   (*iv*) Maximum cycle time.
5. What is Area Traffic Control?

Chapter

HIGHWAY SAFETY APPURTENANCES

31.1. BACKGROUND

With-the ever increasing speeds and higher volumes of traffic plying on highways, provision of safety appurtenances is becoming increasingly necessary. In the developed countries of the World, considerable amount of R & D work has been carried out in this direction. In India too, there is an increasing awareness for the urgent need of providing the essential safety appurtenances, if the high rate of accidents is to be brought down. The following safety appurtenances, which are of immediate relevance in the Indian context have been covered in this chapter:

- Delineators
- Traffic impact attenuators
- Roadside rest areas
- Traffic safety barriers
- Traffic aid posts

31.2. DELINEATORS

In general, delineators are light-retro-reflecting devices mounted by the side of the roadway to indicate the roadway alignment, especially for night driving. The three broad types of delineators are:

(a) Roadway indicators
(*b*) Hazard markers
(*c*) Object markers

31.2.1. Roadway indicators

These are installed to delineate the edges of the roadway (road formation) in order to guide drivers about any abrupt changes *e.g.*, sharp curves, abrupt reduction in pavement width (*e.g.*, from four-lane divided to two-lane) and where the alignment may be confusing for some similar reason. These roadway indicators could be in form of vertical posts or in form of ballasted drums. However, ballasted drums are not favoured mainly because of the damage that they can cause to an out-of-control vehicle. Some typical designs of the vertical post type roadway indicators are shown in Fig. 31.1. The roadway indicators should be placed not less than 0,6 m nor more than 1.8 m outside the outer edge of the shoulder. The decision to use roadway indicators is dictated by the importance of the road, volume of fast traffic, speed of traffic, accident experience and danger posed by any specific deficiency in road alignment. The spacing on straight stretches may be 50 m to 70 m in plain terrain; 35 m to 50 m in rolling terrain and 20 m to 35 m in mountainous terrain. Along curves, the spacing will be governed by the radius of the curve.

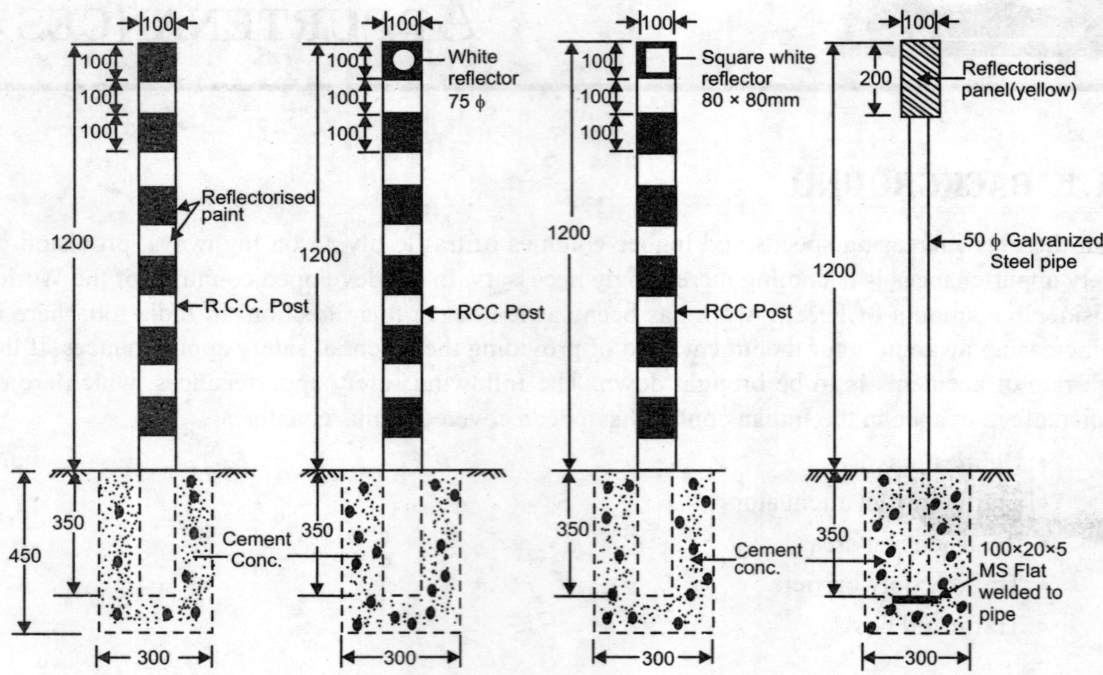

Fig. 31.1. Typical designs of vertical post type roadway indicator.

These are provided to define obstructions such as guard rails, abutments, piers, barrier rails, ends of diversions from main track etc. In its most common form, a hazard marker is a striped board painted alternately black-yellow conventionally; a colour combination of alternate white-orange (reflectorised) is, however, more suitable. A typical design is shown in Fig. 31.2. On rotary island and T-junctions, typical designs for one-way and two-way hazard markers are shown in Fig. 31.3.

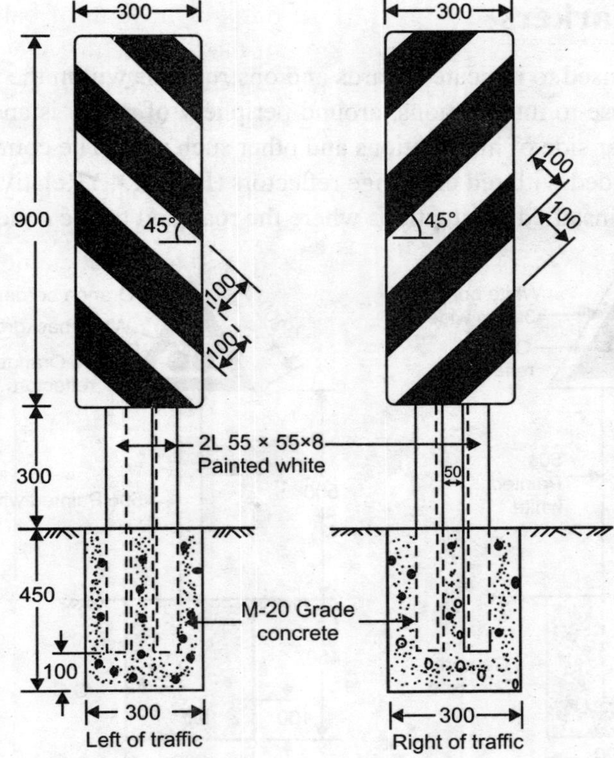

Note: The stripes slope down at 45° towards the side on which traffic is to pass.

Fig. 31.2. Typical design of hazard markers.

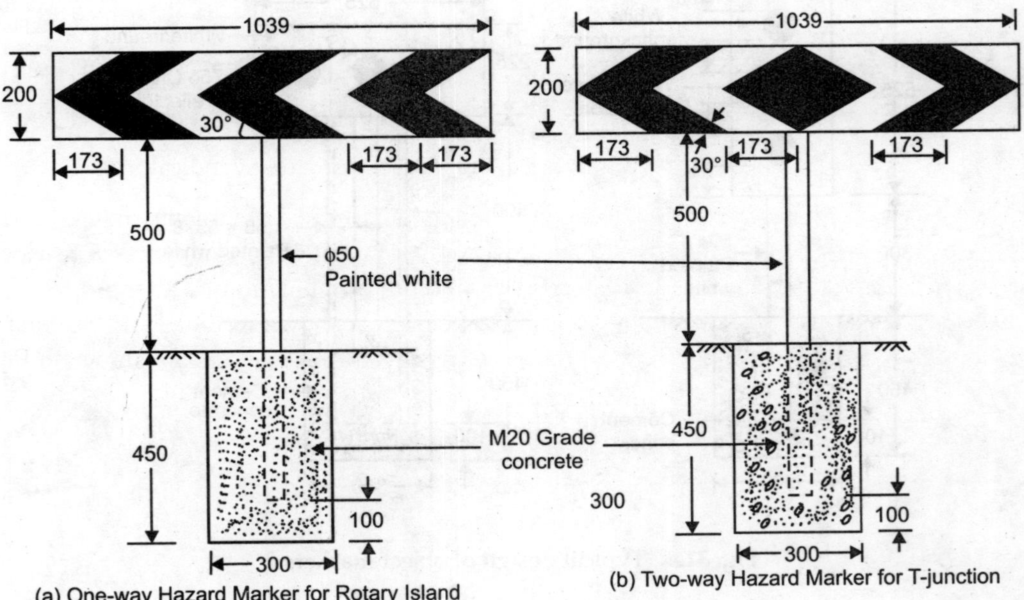

Fig. 31.3. Hazard marker design for rotary island and for T-junction.

31.2.3. Object Markers

The object markers are used to indicate hazards and obstructions within the vehicle flow path *e.g.*, channelising islands close to intersections; around periphery of rotary islands; median openings ; medians/islands at the far side of intersections and other such sites. The commonly adopted design is a triangular plate studded with red or orange reflectors (Fig. 31.4.) Relatively larger sized object markers (Fig. 31.5) are installed in situations where the road and traffic conditions so warrant.

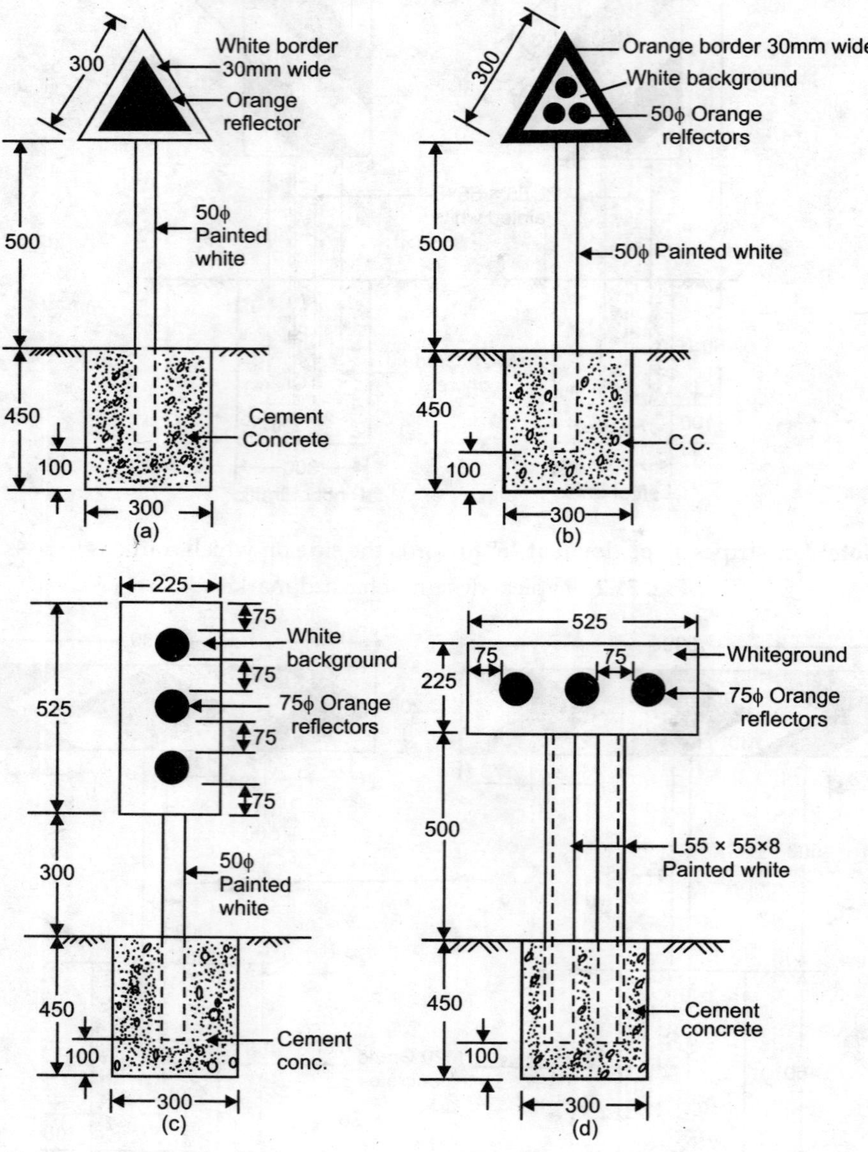

Fig. 31.4. Typical design of object markers.

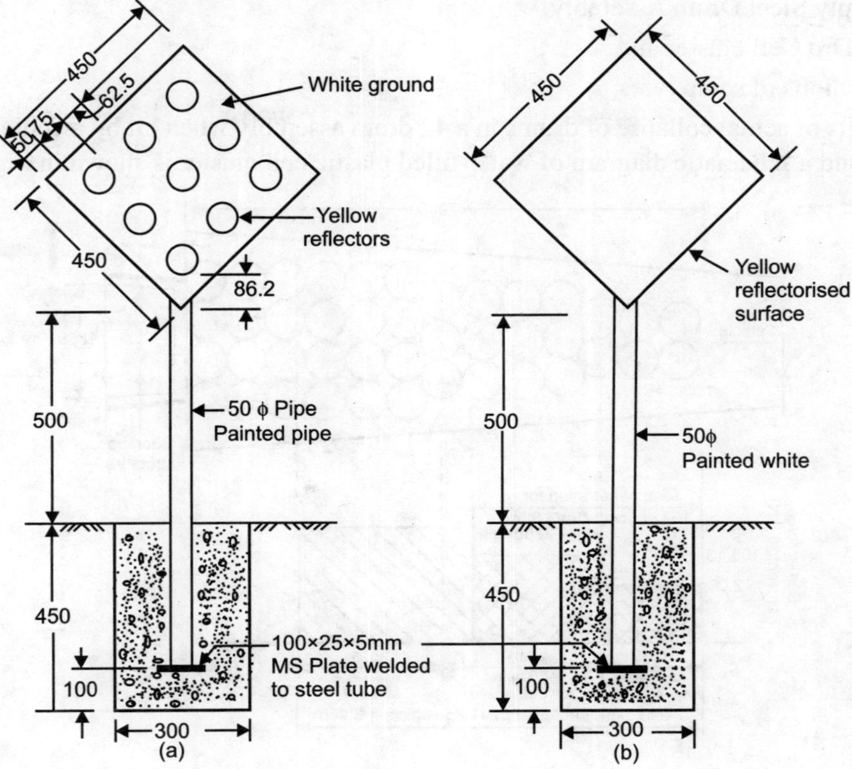

Fig. 31.5. Design of large reflectors.

31.3. TRAFFIC IMPACT ATTENUATORS

On modern highways like the Expressways and busy National Highways where vehicles travel at very high speeds, collision of speeding vehicles with fixed objects (like bridge abutments and structural columns etc.) is a common type of road accidents. The exit core area in interchanges (*i.e.*, the area downstream from the shoulder intersection point of the main carriageway and an exit road) is a common area where serious accidents are known to take place. Traffic Impact Attenuators or Crash Cushions are protective systems which prevent errant vehicles from impacting hazards by either smoothly decelerating the vehicle to a stop when hit head-on, or by redirecting it away from the hazard for glancing impacts. These barriers are used to shield rigid objects or hazardous conditions that cannot be removed, relocated or made breakway. The main design considerations for impact attenuators are.

- The attenuators should be designed to provide the additional stopping distance needed for crashes above 55 to 65 Km/hr., and
- The attenuators should provide a stopping force near, but slightly less than, the design frontal crush force on the vehicle.

There are many types of traffic impact attenuators in use in developed countries but the ones most relevant to Indian conditions are :

- Empty Steel Drum Assembly
- Hi-Dro Cell cluster and
- Cushions of scrap tyres.

The results of actual collapse of drums in a 42 drum assembly when hit by a vehicle are shown in Fig. 31.6 and a schematic diagram of water-filled plastic cell cluster is shown in Fig. 31.7.

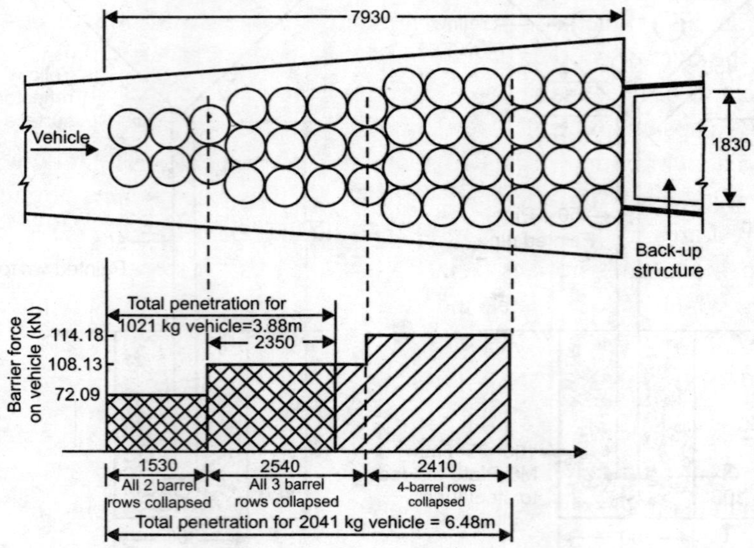

Fig. 31.6. Traffic impact attenuator using empty steel drums.

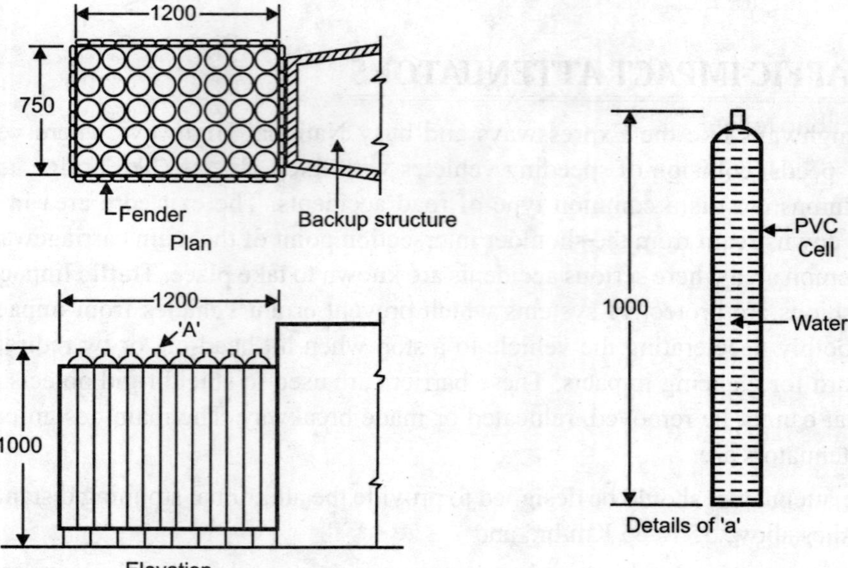

Fig. 31.7. Traffic impact attenuator using water-filled plastic cells.

31.4. ROADSIDE REST AREAS

31.4.1. The need

Specially designed studies abroad on high speed highways have shown that watchfulness or vigilance in driving declines after about 30 minutes of continuous driving. If the driver then receives some mental stimulation, his efficiency can be maintained for another 30 minutes. It is, therefore, very necessary from the overall consideration of highway safety that well planned and well designed rest areas adjacent to or near the highway be provided with facilities such as seats, shelter sheds, litter bins, toilets, maps, picnic tables, restaurant vending stands, adequate parking space etc. Such rest areas should also provide the needed soothing environment, preferably by the side of a lake or waterfall, in a garden with shady areas free from noise and other pollution. After spending some time in a rest area, drivers can resume their journey with the desired level of vigilance and watchfulness. In India, only limited facilities are available adjacent to the highway in form of roadsides *dhabas* which are often substandard and do not measure to the required standards.

31.4.2. Types of rest areas

Considering the very high volumes of traffic on Indian highways, the high percentage of trucks and also the different requirements of car/bus passengers and truck drivers, it is desirable, if not necessary, to separate out the rest area for car/bus passengers from those designed for truck drivers.

The various facilities to be provided in the rest area designed for the use of passenger cars, buses and two-wheelers, are as under :

- Parking for cars, buses and two-wheelers
- Open space and passageway
- Rest house
- Public toilet
- Public telephone
- Drinking water kiosks
- Garden
- Fuel station and repair shops
- Restaurant and vending stands
- Free rest shelters
- First aid room

A typical layout of the rest area with above facilities is shown in Fig. 31.8. On a high traffic density corridor on a National Highway, the spacing of such Rest Areas will be around 40 Km along the highway.

The various facilities to be provided in a Rest Area to cater to the needs of truck drivers should be essentially the same as for cars/buses and two-wheelers, except that there should be more of vending stands rather than a full fledged restaurant and more of free rest shelters rather than a full-fledged rest house. Also, the facilities of fuel station and repair shops may not be provided since these would be available at 40-Km intervals. The Rest Areas for truck drivers also may be provided at 40-Km intervals, to start with for high traffic density corridor on a National Highway ; however, as the demand grows, these may be provided at closer intervals.

The safety roadside rest areas which provide only limited facilities (like parking, garden/open space and passageway, public toilets, drinking water and limited eating facilities) for cars, buses and two-wheelers may have to be provided at closer intervals of 10 to 20 Km on a high traffic density corridor, as the demand grows.

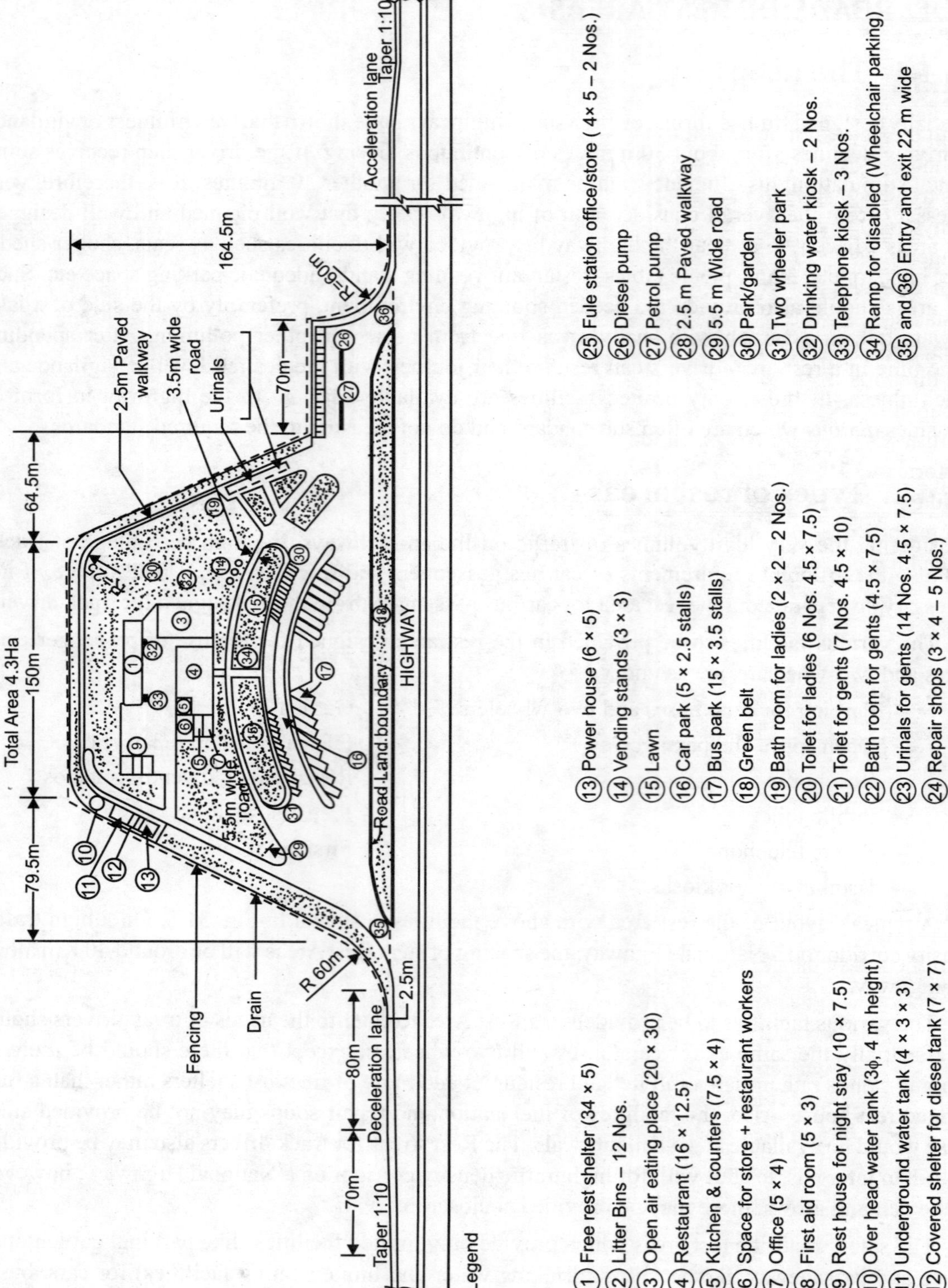

Fig. 31.8. Layout of a rest area for cars, buses and two-wheelers.

31.5. TRAFFIC SAFETY BARRIERS

31.5.1. Necessity of safety barriers

With the very significant improvements that are taking place in road technology as well as vehicle technology, the vehicle speeds on highways are also increasing rapidly. While the positive aspect of this trend is that there is a travel time cost saving, the negative aspect is that under such conditions, many accidents take place involving vehicles which lose control and run off the road. Especially dangerous are the road sections with sharp curves, approaches to restricted roadway bridges and high embankment sections. Vehicles which thus veer off are liable to collide with hazardous obstacles such as poles, trees and bridge structural elements. If, however, suitably designed and properly located safety barriers are provided, it is possible to redirect the vehicle nearly parallel to the direction of the barrier, contain within tolerable limits the, forces experienced by the vehicle occupants, minimise the severity of the accident and reduce the damage. On modern highway facilities, therefore, traffic safety barriers are being extensively used. Based on the mode of performance, the traffic safety barriers can be (*a*) flexible e.g., cable wire type barriers, (*b*) semi-rigid e.g. steel beam type barriers and (*c*) rigid *i.e.*, cement concrete barriers as shown in Fig. 31.9, the major difference among these being the amount of deflection that takes places in the barrier when hit by a vehicle.

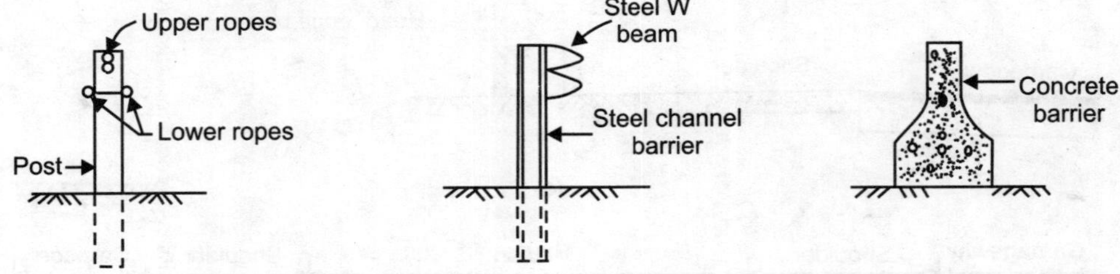

Fig. 31.9. Types of traffic safety barriers based on mode of performance

31.5.2. Requirements of a safety barrier

A traffic safety barrier must meet the following requirements:

- The space available behind the barriers must be adequate to permit dynamic deflection of the barrier.
- The barrier system should contain and redirect the vehicle at design conditions, not allowing it to penetrate or vault over the barrier.
- It should not cause sudden deceleration or spin of the vehicle.
- The vehicle should remain upright during and after the impact and there should not be any loose elements which can penetrate the vehicle.
- After impact, the final stopping position of the errant vehicle should intrude only minimally into the adjacent traffic lanes.
- It should provide good visual guide to the road users.

- It should not entail heavy maintenance expenditure.
- It should involve reasonably low initial cost, maintenance cost and accident cost to the motorist.
- It should have an aesthetically pleasing experience.
- There should be documented evidence of barrier's performance in the field.

31.5.3. Types of safety barriers

Broadly, there are only two types of traffic safety barriers viz. Road Edge Barriers and Median Barriers (Fig. 31.10), placed beyond the road edge and within the median, respectively. These are briefly described below:

31.5.3.1. Road edge barriers

A road edge barrier or roadside barrier is a longitudinal system used to shield vehicles from hazards on the edge of a road, and is provided at the following locations :

– Embankments with high fills and steep slopes.
– Near roadside obstacles.
– Bridge rail ends.

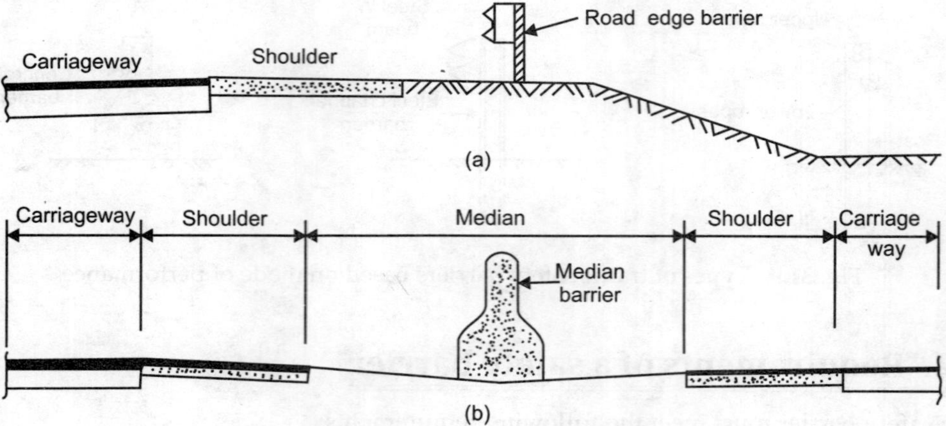

Fig. 31.10. (*a*) Road edge barrier, (*b*) Median barrier.

– At specific locations for ensuring safety of bystanders, pedestrians and cyclists.
– Dangerous ditches. – Steep grades.

The types of road edge barriers commonly used abroad are :

1. "W" beam type steel barrier. **2.** Three beam type steel barrier.
3. Wire rope barrier.

The above three types are shown in Figs. 31.11, 31.12 and 31.13. Ordinarily a road edge barrier should be placed beyond 2.5 m of the travelled way. In case of embankments, a minimum distance of 60 cm should be maintained between the barrier and the start of embankment slope or a hazard (Fig. 31.14) to prevent the wheels dropping over the edge.

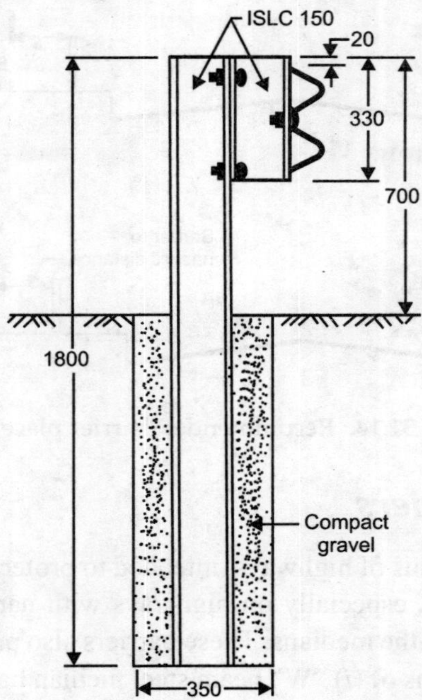

Fig. 31.11. A 'W' beam type steel barrier

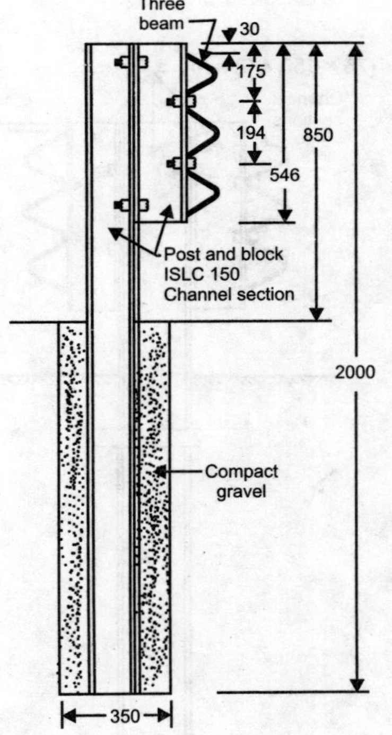

Fig. 31.12. A Three beam type steel barrier

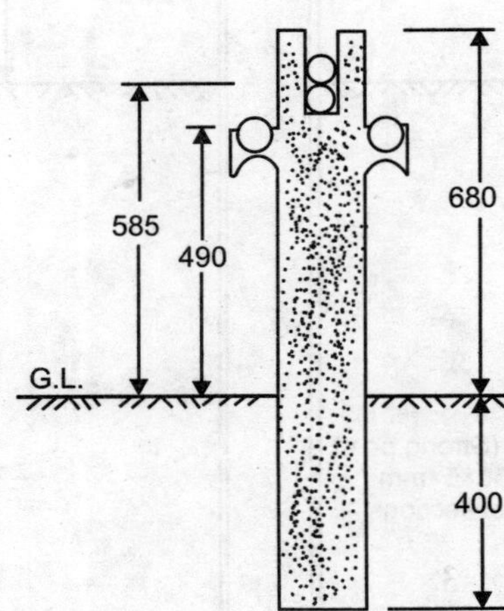

Fig. 31.13. General arrangement of ropes and post in wire rope barrier

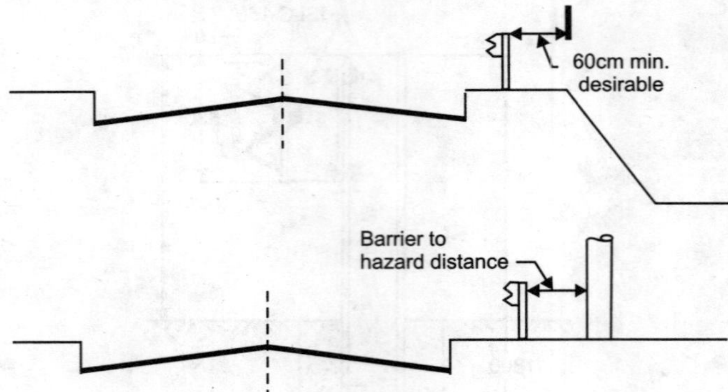

Fig. 31.14. Recommended barrier placement

31.5.3.2. Median barriers

These are provided in the medians of highways, intended to protect the traffic on both carriageways and prevent head-on collisions, especially on highways with narrow medians, caused by out-of-control vehicles jumping across the medians. These barriers also protect fixed objects on the median from traffic flow. Typical designs of (*i*) "W" beam steel median barrier and (*ii*) Three beam median barrier are shown in Figs. 31.15 and 31.16.

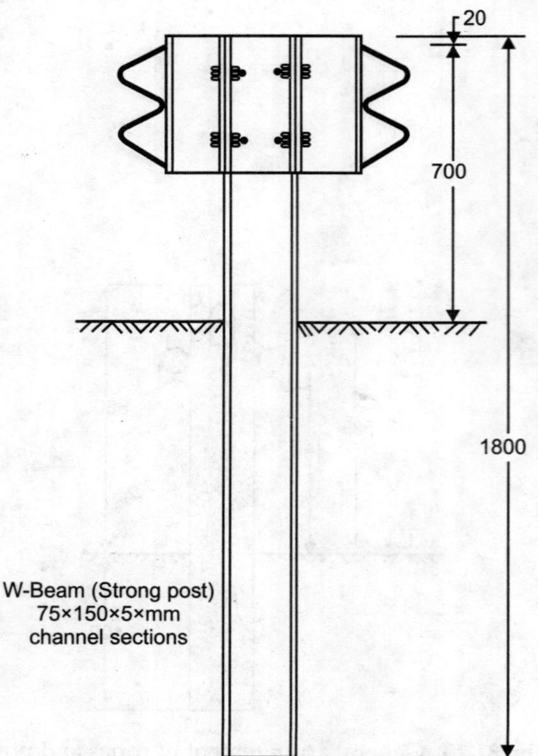

Fig. 31.15. Typical details of 'W' beam steel median barrier

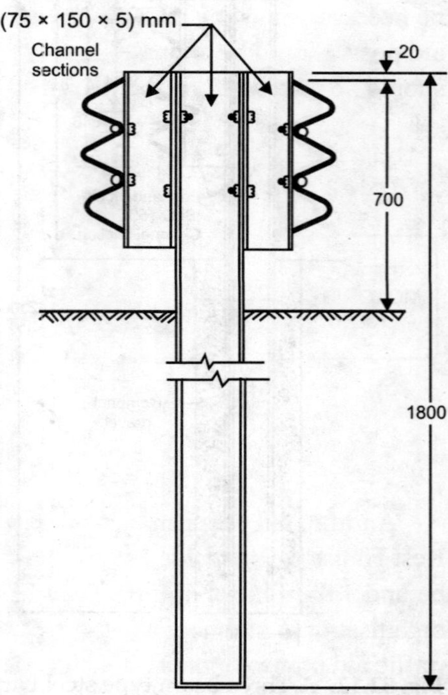

Fig. 31.16. Typical details of three beam median barrier

31.6. TRAFFIC AID POSTS

31.6.1. Background

On rural sections of busy National Highways and on Expressways, when an accident takes place, the accident victims and damaged vehicles sometimes have to lie unattended for long periods of time due to a variety of factors. These factors include lack of adequate communication facilities; long distances from centres where medical facilities are available and non-availability of transport vehicles. Besides road accidents, motorists may sometimes need immediate help in case of any mechanical problems suddenly arising in their vehicles etc. In order to overcome such problems, it is necessary for the highway authorities that adequate facilities be provided for the accident victims or drivers of damaged/disabled vehicles, depending on the type of highway facility. For example, on Expressways, a modern system comprising of closely spaced telephone call boxes, a central monitoring station and a system of quick dispatch of needed help will be necessary, while on National Highways, a simplified system of mobile traffic patrol and stationary traffic aid posts will be required.

31.6.2. Traffic aid posts for national highways

The basic criterion for locating traffic aid posts is to ensure that not more than half an hour or so elapses before any accident victim is provided with first aid and taken to a medical centre as may be necessary and any damaged vehicle towed away. Following this criterion, the spacing of traffic aid posts should be about 40 Km in plain terrain and 25 Km in hilly terrain. The traffic aid system should be structured on the lines of the chart shown in Fig. 31.17

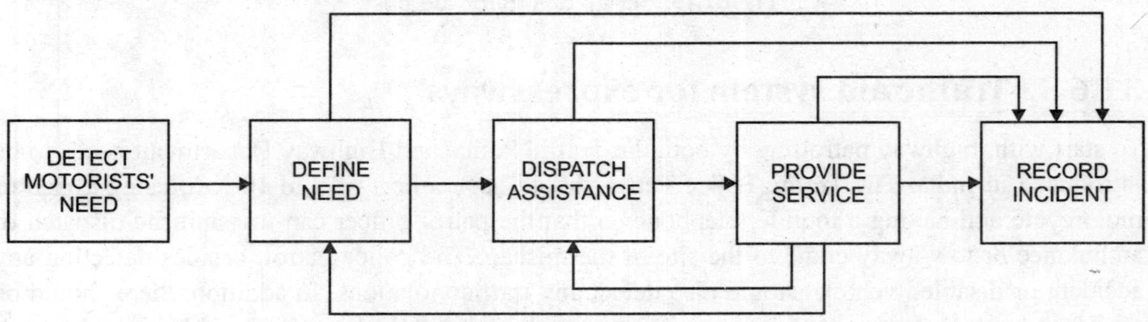

Fig. 31.17. Motorists' aid system

A traffic aid post can be conveniently located within the premises of the PWD offices or PWD Rest House or Roadside Maintenance Units or Roadside Rest Area. Such a traffic aid post should be under the control of the concerned officer of the PWD maintaining the National Highway. The organisational structure maybe as indicated in Fig. 31.18. It should be the responsibility of the traffic aid post to organise constant patrolling through patrolling squads to regulate traffic to ensure free flow of traffic on highways and flash news of any accident as and when any mishap occurs. It should be the responsibility of the traffic aid post to render immediate first aid medical assistance to the victims of accidents and remove the vehicle involved in the accident from the highway. An ambulance van, a tow-away crane, a jeep, atleast two motor cycles, 2 wireless sets, radar speed meter, camera, a dozen each of traffic safety cones and barricades and at least six emergency traffic warning lights should always be available with the traffic aid post.

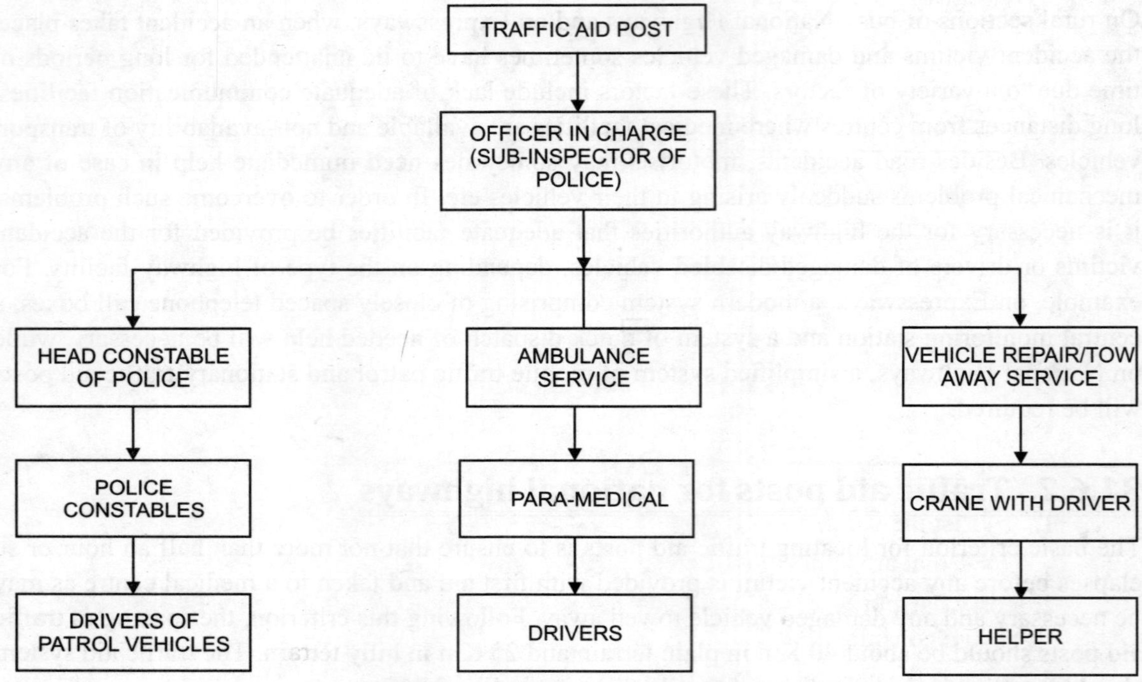

Fig. 31.18. Set-up of a traffic aid post.

31.6.3. Traffic aid system for expressways

To start with, highway patrolling by both the Traffic Police and Highway Department needs to be introduced in India. The Traffic Police Patrol should have a beat of 5 to 10 Km, using a jeep or motorcycle and having a mobile telephone so that the patrol officer can transmit the dispatch of ambulance or tow-away crane to the site of the mishap. The police patrol, besides detecting any accident or disabled vehicle should also detect any traffic violations. In addition, there should be a highway patrol, preferably in a van equipped with mobile telephone, first aid kit, tool box etc. with a beat of about 25 Km, mainly to assist the motorists needing help. Along Expressways, it is also desirable, if not necessary, to have a call box system with call boxes located at an interval of 1 Km in pairs (one on each side of the carriageway) as shown Fig. 31.19. The traffic aid posts should be provided, adopting the same criterion as for National Highways viz not more than 30 minutes should elapse before any accident victim is administered first aid or a damaged vehicle is towed away from the scene of the accident. In view of higher design speeds, the traffic aid posts on Expressways may be spaced, 50 Km in plain terrain and 35 Km in hilly terrain. In addition to the equipment required in a traffic aid post for National Highways, a fire-fighting vehicle should also be provided in the traffic aid posts along Expressways.

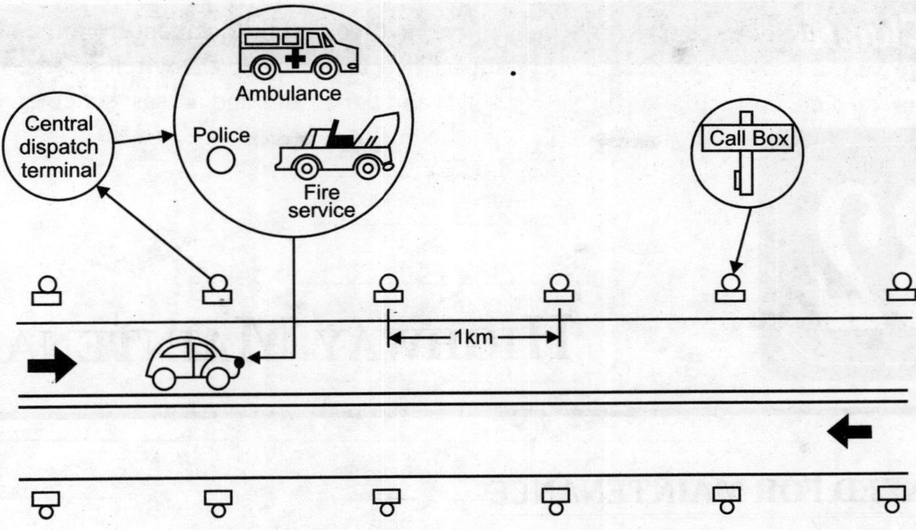

Fig. 31.19. Call box system for motorists' aid

QUESTIONS

1. (*a*) List the various safety appurtenances relevant to Indian roads.
   (*b*) What are delineators ? Describe briefly the different types of delineators.
2. For what purpose Traffic impact attenuators are provided? Give their main types and design considerations.
3. What is the need of roadside rest areas? Draw a typical layout of the area showing various facilities provided in it.
4. What is the necessity of providing safety barriers? List the requirements of a safety barrier
5. With the help of sketches, explain the following .
   (a) Road edge barriers
   (*b*) Median barriers.
6. (*a*) State the background of traffic aid posts.
   (*b*) Explain the criteria for locating traffic aid posts for National Highways and Expressways.

Chapter

HIGHWAY MAINTENANCE

32.1. NEED FOR MAINTENANCE

A highway facility deteriorates in its level of serviceability due to various causes. These are:

1. **Traffic Factors**

   The traffic operating on the facility can cause different types of distress like ravelling, rutting, corrugations, cracking, loss of material, loss of skid-resistance and structural deformation. The extent of deterioration depends upon the intensity of traffic, especially the wheel load and its repetitions. Solid-wheeled cart traffic can be significant in the case of water-bound macadam roads and earthen roads.

2. **Environmental Factors**

   The external influence of environmental factors such as rainfall, snowfall, temperature variations and atmospheric conditions can also cause deterioration of the pavement. Rainfall causes erosion of shoulders and slopes and ingress of water into the pavement structure and subgrade and affects the performance of drainage structures. Snowfall can cause ingress of moisture into the pavement structure and subgrade and result in frost action. It can also disrupt traffic. Temperature variations can adversely affect the performance of bituminous surfaces and cement concrete pavements. Atmospheric action can oxidise the binder and cause deterioration.

In addition to the above, the extent of deterioration and its rate are governed by the standards to which a facility was designed initially. If a facility is designed to higher standards initially, its maintenance needs will be lower than for the facility designed to lower standards initially.

The economic benefits of a well-planned maintenance policy are :

1. Reduction in road user costs, such as vehicle operating costs, travel time savings and accident costs.
2. Reduction in the level of future maintenance and rehabilitation costs (remember : a stitch in time saves nine).
3. Reduction or prevention of the economic loss due to road closures.

From the above, it is clear that a good policy of highway maintenance should be one of the aims of any highway department.

32.2. ASSESSING MAINTENANCE NEEDS

Till recently, the assessment of maintenance needs was done by intuition and past experience of the highway engineer. He used to travel on roads, visually inspecting the condition of the surface and its extent of deterioration. The travel in inspection vehicles used to enable him to judge the riding quality. His accumulated experience used to guide him in determining the periodicity and specifications for resurfacing and resealing.

While in India the above system is still in vogue, in the advanced countries great strides have been recently made in assessing maintenance needs on a more scientific and exact basis. The steps involved in such a system are :

1. Evaluation of the pavement characteristics in quantitative terms by various methods.
2. Development of minimum standards for road characteristics like roughness and skid resistance for different categories of roads.
3. Provision of the needed maintenance inputs based on (1) and (2) above at appropriate periodicity.

The various methods, visual and instrument-based, for pavement evaluation have been described in Chapter 33.

The selection of minimum standards for maintenance can be scientifically done if research is conducted on the deterioration of road characteristics over a period of time under traffic and the road user costs under different levels of road characteristics. The most common road characteristic used for this purpose is roughness. It is generally observed that a road deteriorates in its roughness over time under traffic, due to cracking, ravelling, rutting, deformation etc. When the deterioration reaches such a level that the road user costs mount up, it is advisable to resurface the road and restore it to its original condition.

This principle is illustrated in Fig. 32.1. The curves pertaining to road deterioration as a function of age (or equivalent standard axles) can be constructed after careful research.

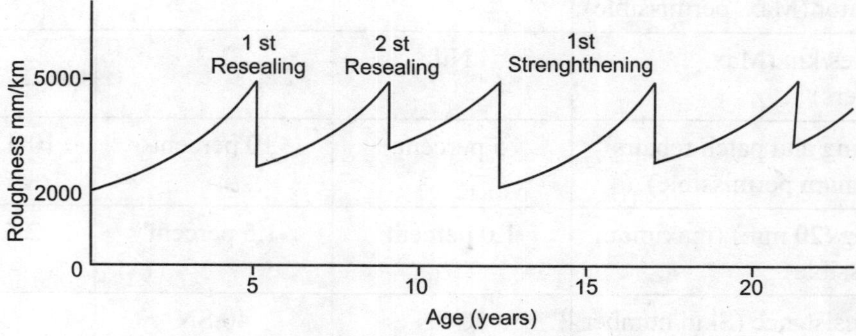

Fig. 32.1. Typical road deterioration-vs-age curve.

Some organisations have fixed standards for road characteristics for purposes of a good maintenance management policy. Standards for road roughness suggested for Indian conditions are given in Table 32.1.

Table 32.1. Recommended roughness values for roads in India (mm/km measured on towed fifth wheel bump integrator)

Surface type		Road condition			
		Good	Average	Poor	Very Poor
1.	Asphaltic concrete	2000–2500	2500–3500	3500–4000	Over 4000
2.	Premix open textured bituminous carpet	2500–4500	4500–5500	5500–6500	Over 6500
3.	Surface dressing	4000–5000	5000–6500	6500–7500	Over 7500
4.	Water-bound macadam or gravel	8000–9000	9000–10000	10000–12000	Over 12000

Resurfacing can be done when the roughness values reach the lower values under the column "poor" in the above Table.

For the National Highways and State Highways in India, the Ministry of Road Transport and Highways has proposed to divide the maintenance programme into three maintenance levels : 1, 2 and 3. Level 1 provides for the highest level of comfort, convenience and safety. Level 2 signifies the level to which the highway deteriorates after a couple of years of service before fresh maintenance is implemented. Level 3 represents the minimum level necessary to protect the investment and provide reasonable levels of safety. The proposed levels based on various serviceability indicators are shown in Table 32.2. For the lower categories of roads *i.e.* MDR/ODR/VR, the concept of level of service is envisaged to be adopted after full connectivity by all-weather road is accomplished.

Table 32.2. Serviceability indicators for Indian highways

S. No.	Serviceability indicator	Level 1 (Good)	Level 2 (Average)	Level 3 (Acceptable)
1.	Roughness by bump integrator (Max. permissible)	2000 mm/km	3000 mm/km	4000 mm/km
2.	Potholes/km (Max. Numbers)	Nil	2-3	4-8
3.	Cracking and patch repairs (maximum permissible)	5 percent	10 percent	10-15 percent
4.	Rutting (20 mm) (maximum permissible)	1.0 percent	1.5 percent	2.5 percent
5.	Skid resistance (Skid number by ASTM-274) Minimum desirable	50 SN	40 SN	35 SN

6.	Defective bridge deck area and bump at approach (maximum permissible)	Nil	10 percent	15 percent
7.	User information	All road signs, km stones and road markings in good condition	Only major road signs, km stones, some road markings in good condition	Signs only for major destinations and km stones in fair condition.

32.3. MAINTENANCE OF EARTH ROADS

Earth roads form a major percentage of rural roads in India and hence their efficient maintenance is of great importance. Because of the low specifications (inadequate embankment height, small roadway width and low cost drainage arrangements), good maintenance can preserve the assets and prolong their life.

The principal maintenance operation consists of maintaining the cross-section by grading and dragging.

32.3.1. Grading

The shaping and sectioning of an earth road is best done by blading with a grader or motor grader. A grader of about 110 HP is suitable for the purpose. In India, it is most unlikely that mechanical graders would be available for routine maintenance operations for low volume roads. Manual methods should include making up of ruts and deformations by additional soil from borrow pits and restoring the camber. If solid-wheeled traffic is heavy and ruts are formed, the ruts can be filled by quarry run, gravel or other local materials.

32.3.2. Dragging

Dragging rectifies to an extent, the formation of corrugations. A typical drag, which can be towed by animal power, by men or by motor grader, is illustrated in Fig. 32.2. Dragging does not restore the cross-section of the road.

32.3.3. Rolling

The earth surface should be rolled and compacted after grading and dragging. A light sprinkling of water needs to be done if rolling is done in dry season.

32.3.4. Filling of rain-cuts

Rain-cuts in the embankment slopes should be filled up after the rainy season. Turfing prevents soil erosion.

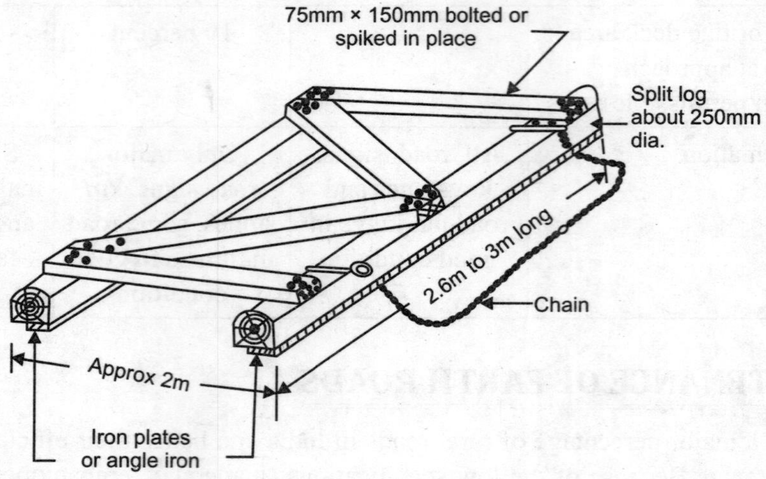

Fig. 32.2. Drag.

32.4. MAINTENANCE OF GRAVEL ROADS

Gravel roads (also known as moorum roads) are very common in India. The maintenance operations involved are filling local depressions, grading, dragging, rolling and regravelling.

32.4.1. Filling local depressions

Local depressions and longitudinal ruts should be filled up by adding fresh materials of the same specifications as the original material. Light sprinkling and tamping with hand rammers will help in compacting the material.

32.4.2. Grading

Grading is an operation intended to restore the camber and shape of the gravel surface. A motor grader is ideal for this purpose. For gravel roads, heavy grading is inadvisable without the provision of additional surfacing material if the remaining thickness of gravel is less than 75 mm

32.4.3. Dragging

One of the common defects that develops in a gravel road is corrugations. Dragging can rectify the formation of corrugations. The drag shown in Fig. 32.2 can be used for this purpose.

32.4.4. Regravelling

Regravelling is necessary to make up the loss in material caused by the combined action of traffic, rain and wind. The loss per year is about 25 mm thickness. Regravelling is done once in 2–5 years. Additional gravel, 25–75 mm loose thickness is spread after scarifying the old surface. Water is sprinkled to facilitate compaction which is done preferably at optimum moisture content. The layer is rolled by a power roller.

32.5. MAINTENANCE OF WATER-BOUND MACADAM ROADS

Untreated water-bound macadam is a common specification in India. Water-bound macadam surface deteriorates in the following typical ways :

1. Formation of ruts
2. Formation of potholes
3. Formation of corrugations
4. Ravelling
5. Damaged edges.

Formation of ruts is caused by excessive camber and preponderance of iron-tyred traffic. Filling up the rutted portion consists of (*i*) cleaning and watering the rut, (*ii*) scarifying and removing the stones to an approximately rectangular section with flat bottom and vertical sides, (*iii*) filling the section with salvaged metal and fresh metal, (*iv*) rolling with the addition of screenings, gravel and watering, (*v*) finally spreading 6 mm sand layer.

Pot-holes are formed due to lack of binding properties in the binder material, poor quality of stones, local subgrade failure or defects in consolidation. Pot-holes are remedied by patch repairs. The area to be patched is cut to a rectangular or square shape with vertical sides. The sequence of operations for laying patch material is the same as in filling up the rutted portion. Hand rammers may be used for patch repairs instead of power rollers.

Corrugations result in a wavy surface, causing discomfort to travel. Corrugations are caused by a variety of factors such as :

1. Defective rolling	Lack of control in rolling can result in corrugations. Especially jerks and non-uniform rolling speed can cause corrugations.
2. Vibrations set up pneumatic tyres	Pneumatic tyred vehicles running at speed cause vibrations to be set up in the spring loaded axles. These vibrations are harmonic in nature and cause waves.
3. Vibrations set up due to braking	At locations where frequent braking of vehicles is involved, such as a bus stop, vibrations are set up in the vehicle These eventually cause corrugations.
4. Use of excessive quantity of blindage material	Excessive blindage material at the surface tends to get deposited in regular waves, depending upon the spring action in the wheels.
5. Transverse picking of WBM surface	If during the laying of a thin renewal coat of WBM, transverse picking has been done, corrugations are bound to result. Picking should therefore be randomised or longitudinal.

When corrugations have been formed, the excess blindage material that has got deposited in ripples should be immediately removed by dragging or brooming. If corrugations have developed in the WBM course itself, a renewal layer is needed. This should be laid after careful scarifying of the corrugated surface.

Ravelling is the phenomenon under which stones get loosened and are freely scattered on the surface. Ravelling is due to (*i*) lack of binding properties in the binding material (*ii*) inadequate consolidation (*iii*) use of too plastic a binder (*iv*) excess quantity of binder and (*v*) evaporation of moisture in the hot weather. Ravelling can be detected early by the presence of tiny haircracks. The tendency can be remedied by blinding with a good binder and watering the surface.

Damaged edges are caused by lack of shoulder support. Unless prompt action is taken, the damage can progressively travel to the inner portions of the carriageway. The damage portion is removed and renewed with fresh material. Rolling the edge and the shoulder should be done simultaneously. Reverse camber in the shoulders should be remedied by grading.

Water bound macadam surface gets worn out due to traffic and needs periodic renewal. The periodicity of renewal varies from 2–6 years, depending upon traffic, presence of solid wheeled vehicles, quality of aggregates etc. Renewal is done in layers of 50 to 75 mm loose. Renewal consists of the following sequence of operations :

1. Cleaning the surface of all dust and caked mud by wire brushes and brooms.
2. Picking up and scarifying the surface after moistening the surface.
3. Screening the salvaged materials.
4. Forming stable shoulders by additional earthwork.
5. Spreading the salvaged materials and additional material.
6. Dry rolling with a power roller.
7. Wet rolling.
8. Application of screenings.
9. Spreading binding material and rolling.
10. Spreading a 6 mm layer of coarse sand.
11. Curing by light sprinkling of water. Traffic can be allowed after 2-3 days.

32.6. MAINTENANCE OF BITUMINOUS SURFACES

32.6.1. Defects, symptoms, causes and remedies

A bituminous surface wears out due to (*i*) traffic (*ii*) weather, such as ingress of water, loss of volatiles in the binder and oxidation of binder (*iii*) inadequacies in the initial specifications and construction standards. Table 32.3 lists out for each type of distress, symptoms, probable causes and possible types of treatments.

Table 32.3. Symptoms, causes, and treatment of defects In bituminous surfacings

	Types of distress 1	Symptoms 2	Probable causes 3	Possible types of treatment 4
A.	**Surface defects**			
1.	Fatty surface	Collection of binder on the surface	Excessive binder in premix, spray or tack coat ; loss of cover aggregates, excessively heavy axle load.	Sand-blinding ; open-graded premix ; liquid seal coat ; burning of excess binder ; removal of affected area.
2.	Smooth surface	Slippery	Polishing of aggregates under traffic, or excessive binder.	Resurfacing with surface dressing or premix carpet.
3.	Streaking	Presence of alternate lean and heavy lines of bitumen	Non-uniform application of bitumen, or at a low temperature	Application of a new surface.
4.	`	Loss of aggregates or presence of fine cracks	Use of less bitumen en or absorptive aggregates	Slurry seal or seal.
B.	**Cracks**			
1.	Hair-line cracks	Short and fine cracks at close intervals on the surface	Insufficient bitumen, excessive filler or improper compaction	The treatment will depend on whether pavement is structurally sound or unsound.
2.	Inter-connected cracks forming series of small blocks	Inter-connected cracks forming series of small blocks	Weak pavement, unstable conditions of subgrade or lower layers, excessive overloads or brittleness of binder.	The treatment will depend on whether pavement is structurally sound or unsound. Where the pavement is structurally sound, the cracks should be filled with a low viscosity binder or a slurry seal or fog seal depending on the width of cracks.
3.	Longitudinal cracks	Cracks on a straight line along the road	Poor drainage, shoulder settlement, weak joint between adjoining spreads of pavement layers or differential frost heave	Unsound cracked pavements will need strengthening or rehabilitation treatment.
4.	Edge cracks	Cracks near and parallel to pavement edge	Lack of support from shoulder, poor drainage, frost heave, or inadequate pavement width	
5.	Shrinkage cracks	Cracks in transverse direction or inter-connected cracks forming a series of large blocks	Shrinkage of bituminous layer with age	

6.	Reflection cracks	Sympathetic cracks over joints and cracks in the pavement underneath	Due to joints and cracks in the pavement layer underneath	
C.	**Deformation**			
1.	Slippage	Formation of crescent shaped cracks pointing in the direction of the thrust of wheels	Unusual thrust of wheels in a direction, lack or failure of bond between surface and lower pavement courses	Removal of the surface layer in the affected area and replacement with fresh material.
2.	Rutting	Longitudinal depression in the wheel tracks	Heavy channelised traffic, inadequate compaction of pavement layers, poor stability of pavement material or heavy bullockcart traffic	Filling the depressions with premix material.
3.	Corrugations	Formation of regular undulations	Lack of stability in the mix, oscillations set up by vehicle springs, or faulty laying of surface course	Scarification and relaying of surfacing, or cutting of high spots and filling of low spots.
4.	Shoving	Localised bulging of pavement surface alongwith crescent-shaped cracks	Unstable mix, lack of bond between layers, or heavy start-stop type movements and those involving negota-tions of curves and gradients	Removing the material to firm base and relaying a stable mix.
5.	Shallow	Localised shallow depressions	Presence of indequately compacted pockets	Filling with premix materials.
6.	Settlement and upheaval	Large deformation of pavement	Poor compaction of fills, poor drainage, inadequate pavement or frost heave	Where fill is weak the defective fill should be excavated and redone. Where inadequate pavement is the cause, the pavement should be strengthened.
D.	**Disintegration**			
1.	Stripping	Separation of bitumen from aggregates in the presence of moisture	Use of hydrophilic aggregate, inadequate mix composition, continuous contact with water, poor bond between aggregate and bitumen at the time of construction, etc.	Spreading and compacting heated sand over the affected area in the case of surface dressing ; replacement with fresh bituminous mix with added anti-stripping agent in other cases.

2.	Loss of aggregate	Rough surface with loss of aggregate in some portions	Ageing and hardening of binder, stripping, poor bond between binder and aggregate, poor compaction etc.	Application of liquid seal, fog seal or slurry seal depending on the extent of damage.
3.	Ravelling	Failure of binder to hold the aggregates shown up by pock marks of eroded areas on the surface.	Poor compaction, poor bond between binder and aggregate, insufficient binder, brittleness of binder etc.	Application of cutback covered with coarse sand, or slurry seal, or a premix renewal coat.
4.	Pot-hole	Appearance of bowl-shaped holes, usually after rain	Ingress of water into the pavement, lack of bond between the surfacing and WBM base, insufficient bitumen content etc.	Filling pot-holes with premix material, or penetration patching.
5.	Edge-breaking	Irregular breakage of pavement edges.	Water infiltration, poor lateral support from shoulders, inadequate	Cutting the affected area to regular sections and rebuilding with simultaneous attention paid to the proper construction of shoulders.

32.6.2. Pot-hole repair (patch repair)

The amount of patching needed to make up pot-holes and localised failures may vary from 0 to 25 per cent of the surface area annually. Patching prolongs the surface life until a time will come when it will be more economical and desirable to renew the surface entirely.

Patching can be done by (*i*) sand premix, (*ii*) open-graded premix (*iii*) dense-graded premix (*iv*) penetration patching or (*v*) surface dressing. Dense-graded premix patch is rarely used, only where the existing surface itself is dense-graded asphaltic concrete. Surface dressing (one or two coats) can be done for existing surfaces with a similar specification and where the traffic is not too heavy.

Patching consists of the following sequence of operations :

1. Cleaning the area by brooming.
2. Trimming the sides vertically and the shape to a rectangle or square and making the bottom level.
3. Painting the sides and bottom of the hole with a tack coat if a premixed material is used.
4. Following the regular specifications of the treatment.
5. Rolling or hand tamping and checking the profile with straight edge.

Sealing the surface is resorted to rectify hungry surface, repair cracks, and arrest loss of aggregates. Sealing can take the form of the following treatments :

1. Liquid seal 2. Fog seal 3. Slurry seal.

Liquid seal is an application of a binder (penetration grade or emulsion) at 9.8 kg/10 sq. m followed up with a spread of cover aggregates, 6.3 mm nominal size, at a rate of 0.09 cu m/10 sq m and rolling in position.

Fog seal is a spray of slow-setting emulsion diluted with equal amount of water at a rate 0.5–1 litre/sq m. Traffic is allowed after the seal sets in. It is provided over a hungry surface, a cracked surface, a surface where there is loss of aggregates and over a surface exhibiting ravelling.

A slurry seal is an application of a slurry composed of slow-setting emulsion, water and aggregates to a thickness of 5–10 mm. The emulsion and water are 18–20 per cent and 10–12 per cent respectively of the weight of aggregates. The slurry is spread at the rate of 200 sq. m per tonne. No rolling is needed. Slurry seal is provided over a hungry surface, cracked surface, a surface where there is loss of aggregates and over a surface exhibiting ravelling. Because of low viscosity of the binder, the specification results in sealing voids and cracks.

When patching becomes too high, it is more economical to renew the surfaces with suitable treatments which include : Metal Renewal 75 mm (MR-I) ; Single-coat or Double-coat Surface Dressing (SD-I or SD-II); Premix Chipping Carpet with Seal Coat (PC + SC) ; Mix Seal Surfacing (MSS) ; Semi-Dense Bituminous Concrete (SDBC) or Bituminous Concrete (BC).

Such renewals are part of preventive maintenance and prolong the life of a pavement. These renewals result in a better riding quality when the surface has deteriorated.

The type and periodicity of such renewals are given in Table 32.4 (*I*) for National/State Highways and in Table 32.4 (II) for MDR/ODR/VR

Table 32.4. Recommended type and periodicity of renewals

(*i*) Life cycle (in years) For National Highways/State Highways

Type of treatment/ category of road	Traffic intensity in CVD	MR-I	SD-II	PC+SC	20 mm MSS	25 mm SDBC	25 mm BC	40 mm BC
NH/SHs (Normal)	>4500	–	–	–	–	–	5/4*	5/4*
	1500–4500	–	–	–	–	5/4*	5/4*	–
	450–1500	–	–	–	5/4*	5/4*	–	–
	<450	–	–	5/4*	5/4*	–	–	–
NH/SHs (Urban)	>4500	–	–	–	–	–	4/3*	4/3*
	1500–4500	–	–	–	–	4/3*	4/3*	–
	450–1500	–	–	4/3*	4/3*	4/3*	–	–
	<450	–	–	4/3*	4/3*	–	–	–
NH/SHs (Hills)	> 1500	–	–	–	–	4/3*+	4/3*	4/3*+
	450–1500	–	–	–	4/3+	4/3*+	4/3*+	–
	<450	–	–	5/4*	5/4*+	–	–	–

(ii) Life Cycle (in years) for MDR/ODR/VR

Type of treatment/ category of road	Traffic intensity in CVD	MR-I	**SD-I/ SD-II	PC + SC	20 mm MSS	25 mm SDBC	25 mm BC
MDR/ ODR/VR (Normal)	>1500	–	–	–	–	5/4*	5/4*
	450–1500	–	–	–	5/4*	5/4*	–
	150–450	–	–	–	5/4*	5/4*	–
	> 150	5/4*	5/4*	5/4*	–	–	–
MDR/ODR/ VR (Urban)	> 1500	–	–	–	–	4/3*	4/3*
	450–1500	–	–	–	4/3*	4/3*	–
	150–450	–	3	4/3*	4/3*	–	–
	< 150	3	4/3*	5/4*	5/4*	–	–
MDR/ODR/ VR (Hills)	> 1500	–	–	–	–	4/3*+	4/3*+
	450–1500	–	–	–	–	5/4*+	5/4*+
	150–450	–	3	5/4*	5/4*+	–	–
	< 150	4/3*	4/3*	5/4*	–	–	–

Note : A. * Indicates reduced life of treatment due to high rainfall *i.e.* > 3000 mm.

B. + Indicates reduced life due to higher altitude *i.e.* > 2000 Mts.

C. ** This treatment of SD-I is to be used under condition of severe resource crunch only.

D. The treatment symbols have been explained in para 32.6.4.

32.7. MAINTENANCE OF CEMENT CONCRETE SURFACE

A well designed and properly constructed cement concrete pavement needs hardly any maintenance. In fact, this is one of the strong points of this specification. However, defects do appear due to the following reasons :

1. Ingress of water to the subgrade causing uneven settlement especially through joints.
2. Inadequate design and faulty workmanship.

32.7.1. Cracks

A common defect noticed in a cement concrete slab is the appearance of cracks. Cracks can be shrinkage cracks, structural cracks, contraction cracks, corner cracks and warping cracks. They can be of varying width. Usually hair cracks are not dangerous since they do not admit water to the subgrade. Medium and wide cracks are harmful since they can cause progressive destruction of the subgrade support by allowing water to percolate. Such medium and wide cracks are filled up by liquefiable substances such as low viscosity epoxy. Before the cracks are sealed, they are cleaned of dust and foreign matter. Compressed air jets and nozzles are useful to achieve this. The dry joints are then filled with appropriate liquid epoxy poured by cans. The poured material is topped up with sand or fine chips to prevent the removal of binder under traffic.

32.7.2. Joints

The maintenance of joints consists in examining whether the joints are properly sealed and, if not, to immediately seal them. If the preformed joint filler has rotted and deteriorated, it should be removed and substituted by a fresh compressible filling material. The sealing material is then poured.

32.7.3. Patching of slabs

A variety of defects, such as scaling, spalling, depressions, irregularities and failures, can occur locally in a slab. In such cases, it is necessary to patch up the defective portions immediately to arrest further deterioration. Bituminous premix materials though very widely used for this purpose are not very satisfactory. The best materials are concrete, and epoxy/ polyester resin mortars. Such patches are of regular geometrical shapes, without acute-angled corners. The sides are first trimmed and made vertical and fresh concrete is laid and tamped.

32.7.4. Mud-pumping and blowing

When the subgrade becomes moist with free accumulation of water, heavy axle loads passing over the slab will eject water and mud through the joints, cracks and pavement edges. This phenomenon is known as mud-pumping and blowing. When a pavement exhibits this phenomenon, the joints and cracks should be inspected and defective ones refilled and sealed. A bituminous underseal can be pumped underneath the slab to prevent recurrence of the defect. This is accomplished through drilled holes in the slab. A viscous binder is preferred. This fills voids in-between the slab and the subgrade.

32.7.5. Loss of texture

If the surface becomes smooth and slippery, texture can be restored by cutting grooves by diamond grooving machines or diamond grinding machines. Acid etching used to be an earlier solution, but now rarely used.

32.7.6. Partial depth crack repair

If the cracks extend upto one-third the depth of slab, partial depth crack repair technique can be adopted. The area is chipped off, the exposed surface is air blasted and a patching mix consisting of epoxy mortar is filled and compacted.

32.7.7. Full depth crack repair

Cross-stiching with inclined tie bars (Fig. 32.3) or stapling with U-bars (Fig. 32.4) or replacement of the cracked area with fresh concrete with dowel bars inserted to secure adequate load transfer are some of the techniques.

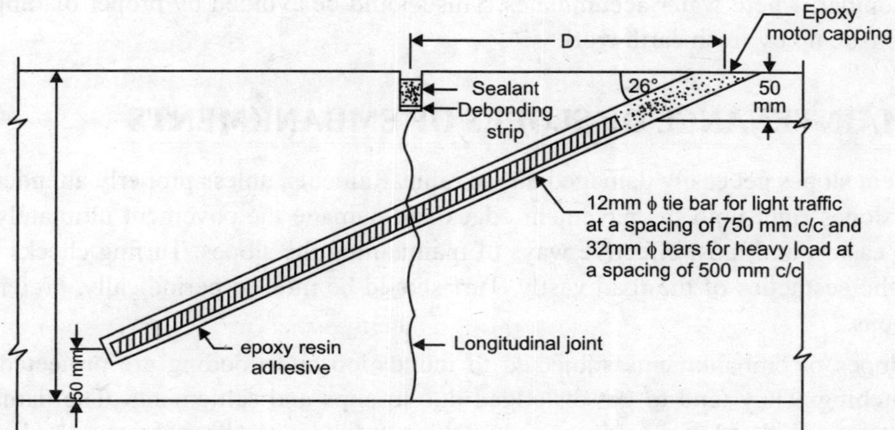

Fig. 32.3. Cross-stitching a crack.

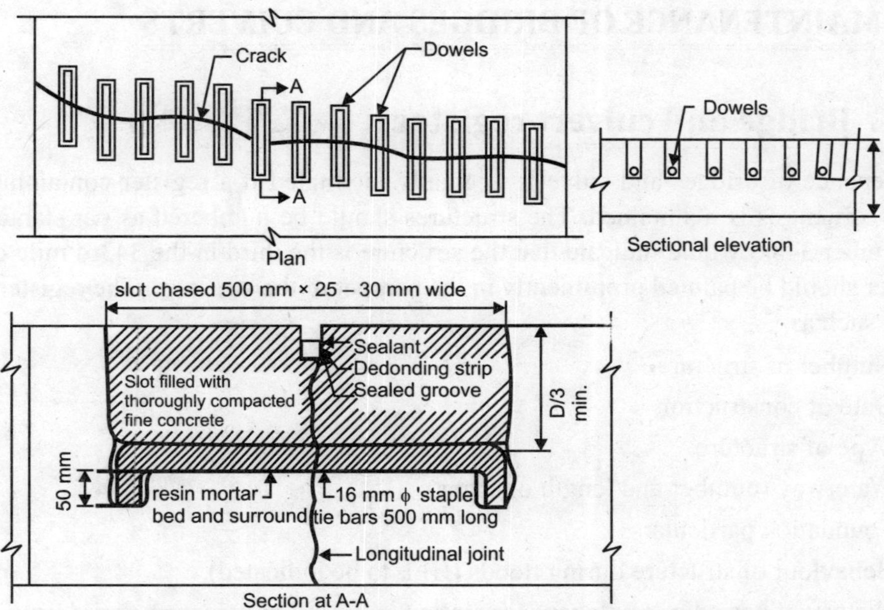

Fig. 32.4. Stapling a cracked longitudinal joint

32.8. MAINTENANCE OF SHOULDERS

Shoulders give lateral support to the pavement and provide room for vehicles when crossing and overtaking on narrow pavements. They are also used for parking of stalled vehicles.

Shoulders constructed of gravel, WBM or bituminous specifications are maintained in the same manner as the pavement of such specifications.

Shoulders of earth or gravel need periodic attention. Proper maintenance of cross-section and camber are the key to successful pavement performance. Shoulders should never be allowed to be depressed below the pavement level. Reverse camber causes a ditch at the junction of the pavement and the shoulder, where water accumulates. This should be avoided by proper blading. Rain cuts should be made up by fresh earth work.

32.9. MAINTENANCE OF SLOPES OF EMBANKMENTS

Embankment slopes get easily damaged due to rains. Raincuts, unless properly attended to in time, erode the slopes right upto the pavement edges and damage the pavement ultimately. Turfing is one of the easiest and most effective ways of maintaining the slopes. Turfing checks erosion and improves the aesthetics of the road vastly. Turf should be mowed periodically, preferably before the monsoons.

The slopes of embankments subjected to inundation and flooding are protected often with boulder pitching. They tend to get dislodged due to slips and settlements. The damaged stones should be removed, the slopes made up and pitching redone with adequate granular bedding.

32.10. MAINTENANCE OF BRIDGES AND CULVERTS

32.10.1. Bridge and culvert register

The maintenance of bridges and culverts is greatly facilitated if a register containing the salient features of structures is maintained. The structures should be numbered as per standard practice. Thus, a number 343/3 would indicate that the structure is the third in the 343rd mile or kilometre. The number should be painted prominently in the parapet of the structure. The register should give particulars such as :

1. Number of structure
2. Date of construction
3. Type of structure
4. Waterway (number and length of spans)
5. Foundation particulars
6. Behaviour of structure during floods (HFL to be indicated).
7. History of periodic maintenance (painting, pointing of masonry, regirdering etc.).

32.10.2. Periodic inspection

The structures should be periodically inspected at least once in a year by (1) Junior Engineers in case of culverts of waterway upto 6 m (2) Assistant Engineers in case of minor bridges (length 6–30 m), Executive Engineers in case of medium bridges (length 30–150 m) and Superintending

Engineers in case of major bridges (length greater than 150 m). The following points should be noted during inspection :

1. Condition of foundations, and any signs of scour
2. Condition of substructure and any signs of damage
3. Condition of floor protection works, and signs of scour and dislodgement
4. Bearings : greasing, tilts, signs of corrosion
5. Superstructure : signs of cracks, corrosion
6. Condition of painting of steel girders
7. Signs of settlement
8. Condition of wing walls
9. Condition of guide bunds
10. Condition of approaches
11. Condition of wearing coat, hand rails, approach slabs curbs, drainage spots and guard stones.
12. H.F.L. reached
13. Condition of river channel and its banks
14. Adequacy of the opening
15. Condition of pipes in a pipe culvert and minimum cushion.

32.10.3. Painting of steel bridges

Steel members get corroded unless protected by painting. Painting should be done once in 6 to 12 years, depending on location and nearness to sea.

32.10.4. Maintenance of masonry

Repointing of joints of brick and stone masonry should be done if deterioration is noticed. All vegetable growth should be cleared. Roots of trees which are likely to cause disruption of the masonry of abutments and wing walls should be cleared. Weepholes of abutments and wing walls should not be allowed to clog.

32.10.5. Scour

Control of scour of foundation can be obtained by dumping boulders, construction of spurs and dumping a garland of concrete blocks or stone sausages around piers. Scour meters provide good information on the extent of scour.

32.10.6. Bearings

Metallic bearings deserve special attention. They should be cleaned and greased with a natural graphite grease.

32.10.7. Expansion joints

Expansion joints get dislodged and loosened frequently. They should be immediately restored.

32.10.8. Weak and narrow structures

Rating of bridges and culverts should be done periodically and the safe load displayed prominently on the structures.

32.11. SPECIAL PROBLEMS OF HILL ROAD MAINTENANCE

Some of the special problems of hill road maintenance are :

1. Snow clearance
2. Slips and landslides
3. Drainage.

32.11.1. Snow clearance

The roads in the Himalayan hill region get covered with snow. Snow clearance can be done manually or by special equipment (dozers, snow masters and snow blasts).

32.11.2. Landslides and slips

Landslides and slips are a common feature of hill roads. The causes for these and remedial measures to prevent them are discussed elsewhere. Road maintenance in areas subjected to landslides and slips poses severe problems. The greatest of them is quick removal of debris and restoration of traffic. Mechanical equipment like dozers become very necessary.

32.11.3. Drainage maintenance

Maintenance of drainage structures is an important task in hill road maintenance. The quick and efficient removal of water prevents slips and landslides and road deterioration. Drainage arrangements such as catchwater drains, cross-drainage structures, side drains and subsurface drains should be inspected before the monsoons and cleared of all obstructions.

32.12. MAINTENANCE PRACTICE IN INDIA

32.12.1. Organisation

Roads are maintained in India through Public Works Departments, Zilla Parishads, Rural Engineering Organisations and the Border Road Organisation. National Highways are maintained by the State PWDs who are agents to the Central Government, who provide necessary funds for the same. The basic unit of maintenance organisation is a Division under an Executive Engineer.

A Division generally looks after 800–1500 kms of roads. The maintenance is generally done by labour intensive methods. Routine maintenance has generally been attended to by road gangmen under a gangmate. The norms for the strength of gang labour areas are given in Table 32.5.

Table 32.5. Norms for strength of gang-labour

S. No.	Road category	Labour per km
1.	National Highway/State Highway	
	Single lane	0.3
	Double lane	0.5
2.	Other Highways	
	Black-topped	0.3
	Water-Bound Macadam	0.4
	Earth Road	0.5

Currently, the percentage of labour, material and machinery costs of ordinary repairs and periodic renewals for different categories of roads are given in Table 32.6

Table 32.6. Percentage of labour, material and machinery

Category	Ordinary repairs			Periodical renewal		
	Labour	Material	Machinery	Labour	Material	Machinery
NH/SH	28	66	6	5	70	25
MDR/ODR/VR	30	67	3	10	72	18

32.12.2. Types of maintenance operations

In India, maintenance operations fall under the following groups

1. **Routine repairs,** including patching, earthwork, shoulders, drainage, road furniture, road signs, arboriculture.
2. **Periodic repairs,** including renewal of surface at specified intervals.
3. **Special repairs,** such as flood damage restoration, major painting of steel girders, etc.

32.12.3. Norms for maintenance

A summary of maintenance and repair costs per year for double lane roads with reference to traffic and terrain for six price zones (I to VIA) are given in Table 32.7 for National/State Highways and in Table 29.8 for Major District Roads/Other District Roads and Village Roads. These costs have been given for different ranges of traffic in terms of commercial vehicles plying per day. The pricing zones I to VIA are based on the cost of stone chips, as under :

Price zones	I	II	III	IV	V	VI	VIA
Range of cost of stone chips in Rs. per cum	280 to 350	350 to 450	400 to 500	500 to 580	580 to 660	660 to 740	740 to 860

Table 32.7. Maintenance and repair costs per year in Rs. per km double lane (7.0 m) with 1.5 m paved shoulder of National/State Highways with reference to traffic and terrain

Traffic in CVD*

	Less than 450 CVD **B.T. Surface	450-1500 CVD **B.T. Surface	More than 4500 CVD **B.T. Surface	More than 4500 CVD Cement concrete	WBM
Zone-I					
Normal	328724	406429	455807	621503	138165
Urban	476673	529271	588326	793250	186395
Zone-II					
Normal	337750	418039	468363	640415	148110
Urban	489565	543270	603362	816128	196340
Zone-III					
Normal	344567	427218	478575	655801	157665
Urban	500249	554332	615555	834685	205895
Zone-IV					
Normal	353974	434688	493886	679047	167220
Urban	515951	563127	633946	862858	215450
Zone-V					
Normal	362021	452518	506876	698719	170405
Urban	529465	584931	649518	886661	218635
Hills	689633	754145	842058	927986	—
Zone-VI					
Normal	369591	463545	519402	717647	176775
Urban	542424	598232	664509	909535	225005
Hills	707634	772550	862628	952619	–

Zone-VI A					
Hills	735087	801916	890511	988264	–

*CVD : Commercial vehicles per day

**BT : Black topped.

#WBM: Water bound macadam

Table 32.8. Summery of M and R cost for MDR/ODR/VR for Two Lane (7.0 m)/km Zone Wise

Zone	Category	Traffic in CVD*				
		< 150		150–450	450–1500	> 1500
		BT**	WBM#	BT**	BT**	BT**
(1)	(2)	(3)	(4)	(5)	(6)	(7)
I	Normal	160,927	132,687	170,383	202,997	227,095
	Urban	208,324	220,290	226,119	266,069	297,100
II	Normal	164,520	145,388	175,859	208,824	233,057
	Urban	213,094	239,365	232,765	273,132	302,449
III	Normal	167,762	158,931	180,653	214,291	238,749
	Urban	216,882	259,508	238,616	279,807	309,389
IV	Normal	174,095	181,794	188,333	224,441	244,866
	Urban	224,280	83,708	247,931	292,178	316,683
V	Normal	178,486	190,168	194,281	231,792	257,119
	Urban	229,387	307,182	255,138	301,113	331,718
	Hills	216,170	248,702	252,549	278,229	368,871
VI	Normal	182,777	205,051	199,704	238,820	264,426
	Urban	234,374	329,759	261,689	309,645	340,571
	Hills	221,688	267,128	259,732	285,691	378,553
VI-A	Hills	229,812	284,595	271,935	297,814	391,254

*CVD–Commercial vehicles per day

**BT–Black-topped

#WBM–Water bound macadam.

32.12.4. Mechanised road maintenance

Though road maintenance in India is mostly labour oriented, it is desirable to introduce mechanised road maintenance. This can be done with the small size equipment already being manufactured in India .

32.13. MAINTENANCE MANAGEMENT SYSTEM (MMS)

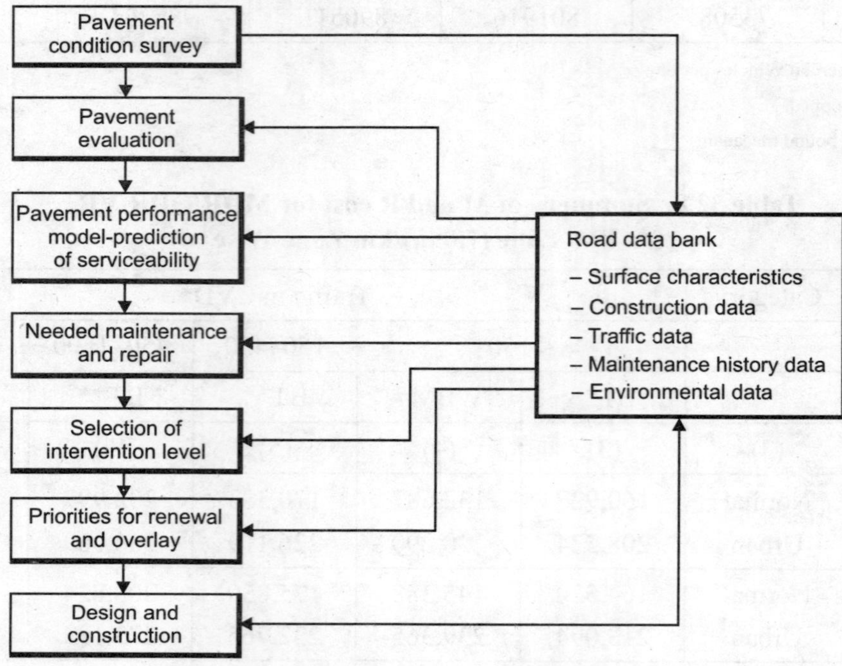

Fig. 32.3. Flow chart of pavement maintenance management system.

A Maintenance Management System (MMS), also known as Pavement Management System (PMS), is a computer package which facilitates maintenance planning and optimal allocation of resources. Its main elements are (*i*) a basic road data bank (*ii*) a pavement performance model (*iii*) selection of intervention levels and (*iv*) listing out priorities for maintenance (renewal and overlay) for a given budget. Many countries have developed and implemented their own MMS. A typical flow chart of pavement maintenance management system is shown in Fig. 32.3.

QUESTIONS

1. Describe why highway maintenance is needed.
2. How are maintenance needs assessed ?
3. Describe how an earth road is maintained.
4. Describe how a gravel road is maintained.
5. Write short note on :

   Corrugations in pavements
6. Describe how a water-bound-macadam road is maintained.
7. What are the common defects, symptoms, causes and remedies for bituminous surfaces ?
8. Describe the operations of sealing of bituminous surfaces.
9. What is the current practice for periodic renewal of surfaces of National Highways ?
10. Describe the maintenance of cement concrete roads.

Chapter

PAVEMENT EVALUATION

33.1. INTRODUCTION

Pavement evaluation is a technique of assessing the condition of a pavement, both structurally and from the point of view of surface characteristics. It is also known as pavement condition survey and rating of pavement.

Pavement evaluation is a handy tool in the hands of a highway engineer and serves a variety of purposes, such as:

1. To research on the performance of pavements of different specifications over a period of time.
2. To assess maintenance needs such as patch repairs, renewals and resealing.
3. To assess the need for structural overlays on distressed pavements.

33.2. METHODS OF PAVEMENT EVALUATION

The methods available for pavement evaluation are:

1. Visual rating
2. Present serviceability index concept
3. Roughness measurements
4. Benkelman Beam deflection method.

33.3. VISUAL RATING

Visual rating is a simple method of inspecting the pavement surface for detecting and assessing the amount and severity of various types of damage. The usual manifestation ,of distress or damage occurs in the form of:

1. rutting
2. corrugations
3. ravelling

4. flushing
5. alligator cracking
6. pot holing
7. longitudinal cracking
8. transverse cracking.

There are various methods of visual rating in use by different organisations the world over. One of the most widespread methods was initially developed at the Texas *A* and *M* University and is commonly known as the Deduct Value or Deduct Point method. In this method, certain deduct points are associated with specific values of various distress factors. The deduct points, indicate the relative importance of the distress type. These deduct points are then subtracted from an established "perfect" score (usually 100) to arrive at the overall rating score of the pavement. Table 33.1 gives the deduct points for various levels of distress.

There is as yet no rating system developed for Indian conditions though there is a great need for developing one.

Table 33.1. Deduct values for flexible pavement

Type of distress	Degree of distress	Extent or amount of distress		
		(1)	(2)	(3)
Rutting	Slight	0	2	5
	Moderate	5	7	10
	Severe	10	12	15
Ravelling	Slight	5	8	10
	Moderate	10	12	15
	Severe	15	18	20
Flushing	Slight	5	8	10
	Moderate	10	12	15
	Severe	15	18	20
Corrugations	Slight	5	8	10
	Moderate	10	12	15
	Severe	15	18	20
Alligator cracking	Slight	5	10	15
	Moderate	10	15	20
	Severe	15	20	25

Patching	Good	0	2	5
	Fair	5	7	10
	Poor	7	15	20
Failures	–	20	30	40

Deduct points for cracking

	Sealed			Partially sealed			Not sealed		
	(1)	**(2)**	**(3)**	**(1)**	**(2)**	**(3)**	**(1)**	**(2)**	**(3)**
Longitudinal cracking									
Slight	2	5	8	3	7	12	5	10	15
Moderate	5	8	10	7	12	15	10	15	20
Severe	8	10	15	12	15	20	15	20	25
Transverse cracking									
Slight	2	5	8	3	7	10	3	7	12
Moderate	5	8	10	7	10	15	7	12	15
Severe	8	10	15	10	15	20	12	15	20

33.4. PRESENT SERVICEABILITY INDEX (PSI)

One of the major contributions of the AASHO Road Test was the development of a rating system involving the measurement of permanent deformation, riding quality and the extent of cracking and patching. The rating is well-known by the term Present Serviceability Index (*PSI*) and is probably the most widely used pavement rating measure in existence today. The following equations give the value of *PSI* for flexible and rigid pavements:

Flexible Pavements:

$$PSI = 5.03 - 1.91 \log (1 + SV) - 1.38 (RD)^2 - 0.01 \sqrt{C + P} \quad (33.1)$$

Rigid Pavements:

$$PSI = 5.41 - 1.80 \log (1 + SV) - 0.09 \sqrt{C + P} \quad (33.2)$$

In the above equations:

PSI = Present serviceability index

SV = Slope variance giving an index of the longitudinal profile

RD = Rut depth under a 4 ft. (1.2 m) straight edge

C = Cracking distress in terms of lineal feet of cracks per 1000 sq. ft area

P = Patched area in ft^2 per 1000 ft^2 area

In the AASHO Test, the longitudinal profile was monitored by the CHOLE Profilometer.

33.5. ROUGHNESS MEASUREMENTS

The riding quality of a pavement is determined to a large extent by its structural adequacy, the traffic load repetitions it has been subjected to, the specifications adopted for the surfacing initially and the maintenance inputs. Hence a measure of the pavement performance can be obtained by monitoring its roughness. In view of the importance of the subject, a detailed account of roughness measurement is given below.

33.5.1. Importance of smooth riding surface

For the fast motor traffic, one of the desirable characteristics is a reasonably smooth riding quality. A smooth surface brings about many advantages. Some of them are:

1. **Higher speeds**

   The smoother the road surface, the higher is the speed at which vehicles can drive.

2. **Less fuel consumption**

   The fuel consumed in a vehicle is proportional to the work required to be done to overcome various resistances (air resistance, frictional resistance and grade resistance). The frictional resistance is a function of the coefficient of friction at tyre-road interface. Smoother surfaces thus bring about considerable fuel economy.

3. **Less wear and tear of tyres**

   The life of a tyre depends to a large extent on the smoothness of the road surfaces. Rough roads shorten tyre life.

4. **Less consumption of spare parts**

   Important spare parts of a vehicle such as the suspension system, springs, shock absorbers, body and chassis get damaged more on rough roads than on smooth roads.

5. **Riding comfort**

   Smooth roads result in high riding comfort. On the other hand, rough roads result in jerky motion and poor riding comfort.

6. **Safety**

   Roads which are full of potholes, ruts and corrugations endanger the safety of travel since the possibility of sudden failure of vital parts increases on bad roads. However, extremely smooth roads increase the chances of skidding. Thus, the smoothness to be aimed at has to be a compromise between the conflicting needs of anti-skid properties and economic and comfortable travel.

33.5.2. Need for roughness measurements

Since roughness of a road is a major determinant of the safety, cost, comfort and speed of travel, its accurate measurement is of vital importance to a highway engineer. The important uses of roughness measurements are:

1. To assess maintenance needs

The riding quality of a road surface is an indirect measure of its structural behaviour (pavement distress and structural capacity) and a direct measure of the need for maintenance inputs such as patch repairs, pot-hole filling, resurfacing, provision of overlays etc.

2. To assess quality of construction

Since different types of specifications give different values of roughness measurements, it is possible to assess the quality of new constructions by running roughness measurements over them and comparing the readings obtained with the specified values.

3. For research purposes

Since road roughness is closely related to safety, fuel consumption, speed and vehicle operating costs, any research programme involving any of these parameters, requires accurate measurement of roughness.

33.5.3. What constitutes road roughness?

Road roughness is a very difficult term to define and still more difficult a characteristic to measure. It can be defined as "deviations of a travelled surface from a true planar surface with characteristic dimensions that affect riding quality, vehicle dynamics, dynamic pavement loads and pavement damage" The deviations from the true planar surface could be due to the surface texture, lack of care in finishing operations, tolerances specified during construction, potholes, cracks, corrugations, rutting and pavement deformations.

33.5.4. Measurement of road roughness

The methods of measuring roughness can be broadly grouped under two categories:

1. Direct measurement of the longitudinal profile.
2. Response-type instrument methods.

The direct measurement of the longitudinal profile is an ideal and accurate method since it theoretically gives scaled reproductions of the pavement profile along a straight line. In practice, however, the range and resolution of the profiling devices are limited, but within these limits, the measurements may be called absolute.

The simplest profiling method is by running levels along a desired straight line at intervals of 50 cm. Though a tedius method, this gives an absolute measurement.

Another simple device is a straight edge, mounted on wheels. The device is moved at a low speed. A typical example is the CHOLE profilometer.

A modern profiling equipment is the GMR (General Motors Research) Profilometer. In this instrument, the road profile is accurately measured and the data stored on a digital magnetic recorder for processing. The instrument is very costly and is generally used for research purposes.

A recently developed instrument for measuring the longitudinal profile directly is the TRRL Beam (commonly known as the Abay Beam, after Abaynayaka, its developer). The beam is 3 metre long, along which an instrumented sliding fixture is made to travel. The slider contains transducers which detect the vertical profile and the longitudinal position along the beam. The beam has been found to be very versatile and is expected to be a popular and handy tool in future.

In the response-type systems, the dynamic response of mechanical systems travelling over the rough road is recorded. Thus, this system gives only a relative measurement of roughness, and depends on the characteristics of the mechanical system and the speed of travel.

The well-known response-type of instruments are:

1. Bureau of Public Roads (BPR) Roughometer, developed as early as 1925. It is a single -wheel trailer which measures the uni-directional vertical movements of the damped, leaf sprung wheel by a mechanical integrator unit. The results are recorded in cm/km. The standard speed of measurement is 32 KMPH.
2. The British Towed Fifth Wheel Bump Integrator, which is basically a BPR Roughometer type instrument that has undergone considerable development at the T.R.R.L. (U.K.). A detailed description of this device is given later.
3. APL trailer, developed by the French Bridge and Pavement Laboratory (LCPC).
4. Mays Meter, which is a car-mounted device, giving a paper plot, whose length at the end of a test is the raw roughness numeric for that test. The instrument can be operated at various speeds.
5. Car-mounted integrator unit of the TRRL Towed Fifth Wheel.

This is described in detail later.

The common disadvantages of all response-type roughness measurement systems are:

1. Absence of an internationally accepted conversion method to relate readings from different instruments.
2. Variability of readings depending upon speed of measurement.
3. Need for calibrating the instruments at periodic intervals.

33.5.5. Towed fifth wheel bump integrator

The towed fifth wheel bump integrator is the standard roughness measuring device in U.K. An indigenously manufactured unit is available in India.

The instrument (Fig. 33.1) consists of a trailer of a single wheel mounted within a rectangular steel frame. The wheel is of pneumatic type of a standardised size. The wheel itself supports the chassis through two leaf springs. The rectangular steel frame is tongue-shaped at the forward end and carries a hitch (fastener) for connection to the towing vehicle. The vertical movement of the frame is dampened by means of piston rod, operating in a dash-pot filled with standard fluid. The uni-directional movement between the axle of the wheel and the chassis is recorded by a recorder unit. The unit performs the function of closing a pair of contacts inserted in the circuit of an electro-magnetic counter, once for every 2.5 cm of integrated uni-directional movement. The distance travelled by the unit is measured by counting the number of revolutions of the wheel. The roughness measurement and the distance travelled are recorded on electromagnetic counters placed in the electric circuit operation on 12 V battery of the towing vehicle. A jeep is used for towing. A speed of 32 KMPH is the standard for measurement. The instrument gives measurements in terms of mm/km.

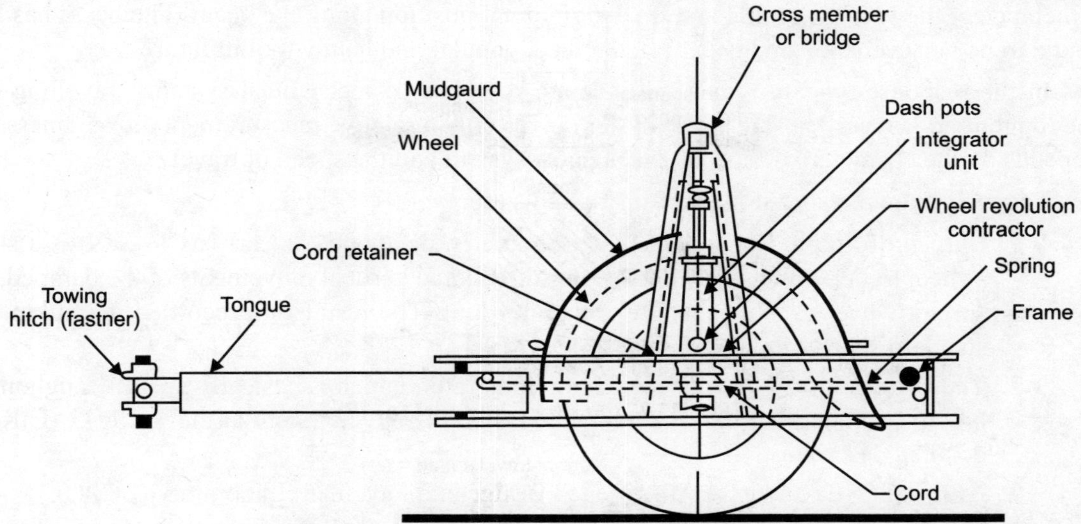

Fig. 33.1. Towed fifth wheal bump integrator.

Precautions to be observed in its use include:

1. Checking radius of tyre.
2. Ensuring correct level of dash-pot fluid.
3. Measurement at standardised speed.

33.5.6. CAR-mounted roughness measuring device

For quick and routine measurements of roughness, a car-mounted unit can be used. The integrator unit is the same as that provided in the towed fifth wheel bump integrator. The unit is placed directly above the rear axle of a car and a cable taken from the axle is wound around a pulley in the integrating unit. The unit records the roughness in mm. The distance can be recorded separately by a distance measuring device taking off from the speedocable. The system needs to be periodically calibrated against the towed fifth wheel integrator; equations are available to convert car-mounted roughness readings to standard towed fifth wheel readings.

33.5.7. International roughness index (IRI)

The purpose of developing an International Roughness Index (IRI) was to prepare guidelines for measuring roughness on a standard scale. It is related to measurements obtained by various devices installed on vehicles or trailers as discussed above, and is infact a mathematical model applied to a measured profile. It is calculated as the cumulative movement of the suspension of a standard 'Quarter Car System' (Fig. 33.2), travelling at a constant speed of 80 km/hr, divided by the travelled distance.

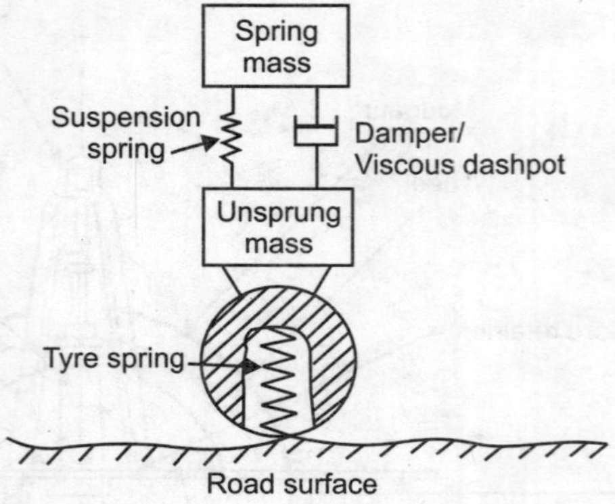

Fig. 33.2. Quarter car model for IRI.

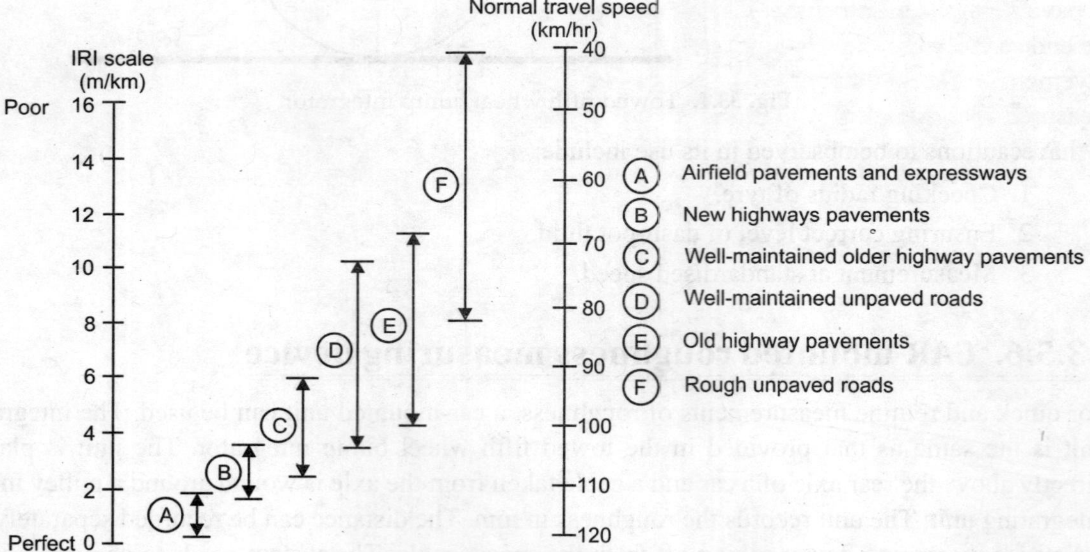

Fig. 33.3. The IRI Roughness scale.

33.6. NON-DESTRUCTIVE PAVEMENT DEFLECTION TESTING

Pavements which have been subjected to traffic, deform elastically under load, the elastic deflection depending on various factors such as:

- Subgrade soil type
- Moisture content and compaction level of subgrade
- Pavement thickness, composition, quality and condition
- Drainage conditions
- +Pavement surface temperature
- Wheel load.

There are a number of different types of pavement deflection testing equipment being used all over the world. The more common types are described below.

33.6.1. Benkelman Beam

The Benkelman beam measures the deflections under standard wheel load conditions. Two kinds of deflection measurements are possible:

1. Rebound deflection, which is the recoverable deflection or the elastic deflection. In a well-designed road, the deflection is entirely elastic and recoverable.
2. Residual deflection, which is the non-recoverable deflection. As a pavement ages, it loses a portion of its elastic properties and a permanent deflection takes place.

The Benkelman beam is a handy instrument which is most widely used for measuring deflection of pavements. The instrument is illustrated in Fig. 33.4 It consists of a long pivoted 2.44 m from the end carrying the contact point which rests on the surface of the pavement. The deflection of the pavement surface produced by the test load is transmitted to the other end of the beam where it is measured by a dial gauge or recorder. The movement at the dial gauge end of the beam is one-half of that at the contact point end. The load on the dual wheel can be in the range 2.7 to 4.1 Tonnes.

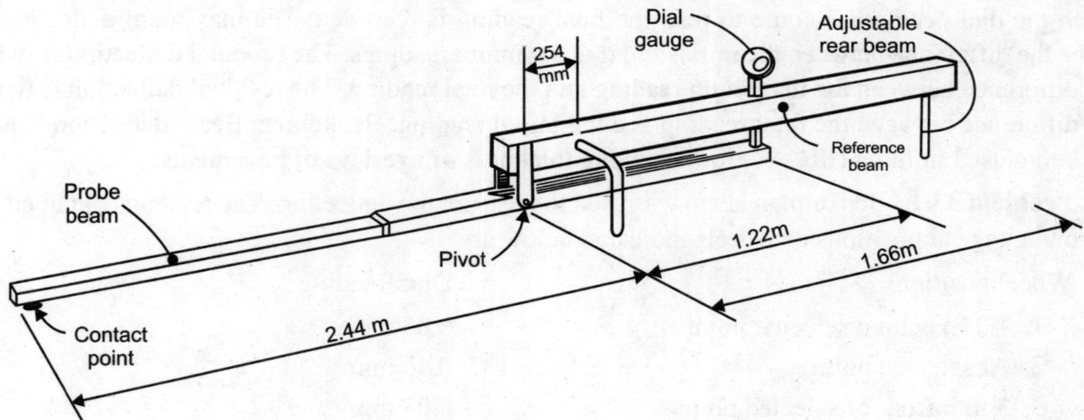

Fig. 33.4. Benkelman Beam.

The procedure of measuring the rebound deflection is as follows

1. Select 10 points along the outer wheel path (*i.e.*, 60 cm from the pavement edge) for each lane.
2. Bring the rear dual wheel assembly of the truck over the marked point and insert the probe of the beam between the dual wheels so that the probe is placed exactly over the point where the deflection is to be measured.
3. A standard wheel load of 4085 kg is used for the test, the tyre pressure being 5.6 kg/cm^2 (or 560 kN/m^2).
4. The dial gauge reading is noted initially (D_0) in the position described under 2 above.
5. The truck is driven forward at a slow speed and dial gauge readings (D_1 and D_2) are taken

when the truck stops at 2.7 m and 9 m from the measuring point, and when the rate of recovery is equal to 0.025 mm per minute or less.

6. Pavement temperature is recorded.
7. If $D_1 - D_2 \leq 0.025$ mm, the actual rebound deflection is $2(D_0 - D_2)$. If, however, $D_1 - D_2 > 0.025$ mm, correction is needed for the vertical movement of the front legs ; the true deflection is obtained by the formula:

$$X_T = X_A + 2.91\, Y$$

where X_T = True pavement deflection

X_A = Apparent pavement deflection

Y = Vertical movement of the front legs *i.e.*, twice the difference between the final and intermediate dial readings.

The WASHO procedure, known as creep load method, is similar to the above, except that the truck rear is initially located 1.2 m behind the selected point. The probe arm is located 1.2 m in front of the wheel. The initial reading is noted and the truck is moved forward at a creeping speed of 2 km/hr to at least 3 m past the tip of the beam. The maximum dial reading will occur when the wheels are in the line with the probe arm. This value is noted; After a reasonable length of time or when the dial needle has come to rest, the final reading is recorded. The maximum deflection is twice the difference between the initial and the maximum readings. The rebound deflection is twice the difference between the maximum reading and the final reading. The residual deflection is twice the difference between the final reading and the initial reading. Benkelman Beam deflection values are being used in India (IRC 81) for designing thickness of overlays of pavements.

Problem 33.1. Benkelman beam was used to measure the deflection. The readings obtained on the dial gauge at positions of wheels indicated below are:

Wheel positions	Dial Reading
1. 1.2 m behind selected point	0.06 mm
2. At selected point	0.46 mm
3. 3 m infront of selected point	0.08 mm

Calculate the (1) maximum deflection (2) rebound deflection (3) residual deflection.

Solution. Maximum deflection

$$= 2 \times (\text{Maximum deflection} - \text{Initial deflection})$$

$$= 2 \times (0.46 - 0.06) = 0.80 \text{ mm.}$$

Rebound deflection $= 2 \times (\text{Maximum deflection} - \text{Final deflection})$

$$= 2 \times (0.46 - 0.08) = 0.76 \text{ mm.}$$

Residual deflection $= 2 \times (\text{Final deflection} - \text{Initial deflection})$

$$= 2 \times (0.08 - 0.06) = 0.04 \text{ mm.}$$

33.6.2. Lacroix Deflectograph

A Lacroix Deflectograph consists of a truck with a rear-wheel assembly, an arrangement for carrying the deflection measuring beam on the truck itself and an arrangement for advancing it intermittently. The truck advances at a constant speed (3 km/hour) while the measuring system, consisting of a reference beam and sensor rods, advances intermittently. Each measuring cycle consists of the following sequence:

1. The reference beam is placed in front of the rear axle of the vehicle in such a manner that the; sensor rods are situated in the path of a twin rear wheel, outside the zone subjected to the deflection.
2. The twin wheels advance towards the sensor rods, entering into the deformed zone of the road. The deflection is recorded when the rear axle has passed the extremity of the sensor rod. The deflection recorded is maximum in this position.
3. The reference beam is now pulled forward (4 to 6 metres) in front of the rear wheels to its initial position.
4. The process is again repeated.

The system is capable of giving an output of 15–20 km/day (600 measurements per km) The results can be recorded on tapes and analysed on a computer.

33.6.3. Dynaflect

The Dynaflect is another truck-mounted device for quick measurement of dynamic deflections.. The Dynaflect carries contra-rotating masses in a small trailer and applies a maximum dynamic force of 2270 N at a fixed frequency of 8 Hz superimposed on a static load k of 725 kg., and the same is transmitted to the road surface through two small, rigid wheels 0.5 m apart. The maximum deflection midway between the wheels and four other points along the centre line between the vehicle is recorded by velocity sensitive transducers. The equipment can be trailer mounted or carried on the front of a light vehicle. About 300 measurements at a single frequency are possible in a working day.

The Dynaflect measures the shape of the deflected pavement near the point of maximum deflection. This space is sensitive to changes in upper pavement layers and is relatively unaffected by the subgrade. Thus, the profile of the deflected shape provides sufficient information to characterise the stiffness of the pavement.

33.6.4. Falling weight deflectometer

Another simple instrument to measure the dynamic deflection is the Falling Weight Deflectometer (FWD), which drops a weight of 150 kg from a variable height on to a spring system. This in turn transmits a load pulse of 28 ms duration to the road surface through a circular plate. A maximum peak load of 60 kN develops a deflected dish that can be recorded by upto five velocity-sensitive transducers, arrayed radially from the loaded area. The equipment is carried on a single axle trailer. About 200 measurements can be taken daily. Fig. 33.5 gives the arrangement of the device.

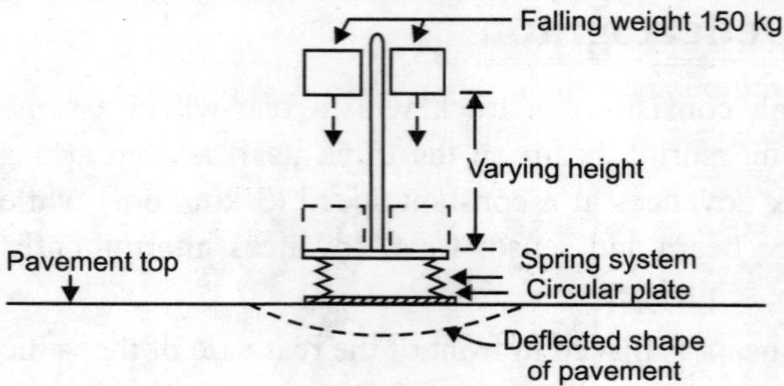

Fig. 33.5. Diagrammatic arrangement of falling weight deflectometer.

33.7. DYNAMIC CONE PENETROMETER (DCP) TEST

The Dynamic Cone Penetrometer is a simple device developed in UK for rapid in situ strength evaluation of subgrade and other unbound pavement layers. Basically, a DCP measures the penetration of a standard cone when driven by a standard force, the reported DCP value being in terms of the penetration of the standard cone, in mm per blow of the standard hammer. Fig. 33.6 shows a typical DCP. The standard steel cone with an angle of 60° has a diameter of 20 mm. The standard 8 kg. drop hammer slides over a 16 mm dia. steel rod with a fall height of 575 mm.

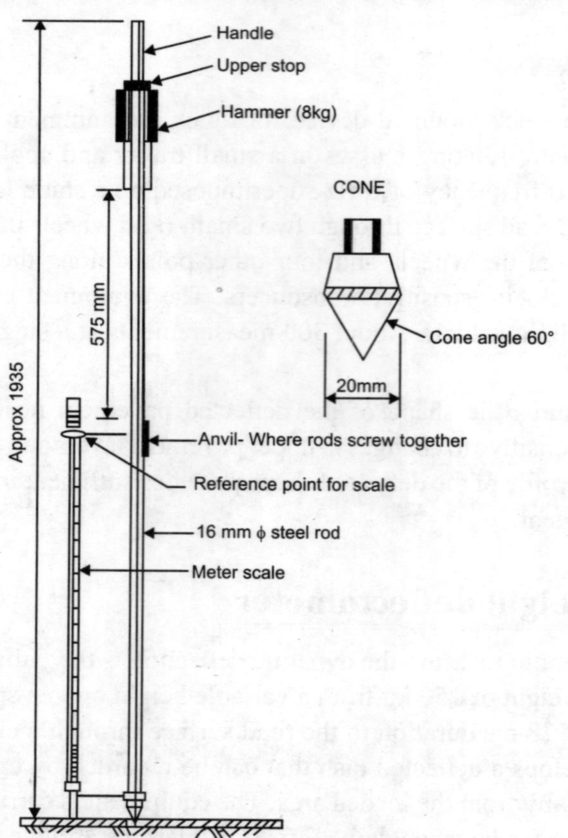

Fig. 33.6 A Typical dynamic cone penetrometer (DCP).

The penetration of the cone can be measured on a graduated metre scale. The DCP test is specially useful for bituminous pavement rehabilitation design and is now being used extensively in several countries including India. A number of reliable correlations between DCP test values and the CBR values have been developed.

33.8. SKID RESISTANCE SURVEYS

A smooth surface is dangerous to traffic, especially when the surface is wet and the vehicles move fast. While care is normally taken to construct reasonably skid-resistant surfaces, the passage of vehicles polish the aggregates. Excess bitumen tends to fatten the surface and render it slippery. An evaluation of the skid-resistance of the surface at periodic intervals is needed to ensure that the roughness level has not fallen to dangerously low levels. This is accomplished by measurement of skid-resistance periodically.

33.9. PAVEMENT DETERIORATION RESEARCH

Pavements deteriorate due to traffic and environmental factors. The extent of deterioration is also a function of the initial pavement thickness and composition. The exact way in which a pavement deteriorates is of great importance to a maintenance engineer to work out the maintenance strategy and to a highway planner to work out the economic evaluation of schemes. Interest in this field is, therefore, increasing the world over. The Kenya study and the Brazilian sutdy have determined the pavement deterioration models. A similar study in India is now being implemented.

A recent development in the field of pavement research and evaluation is the Heavy Vehicle Simulator (HVS). This machine is capable of applying wheel loads of upto 100 kN through a dual or single wheel assembly on a pavement and can apply upto 1400 repetitions of load per hour. Thus, upto half a million repetitions of load can be applied to a pavement in 20 to 30 days. This equipment can, therefore, save considerable time in testing a pavement under actual traffic.

QUESTIONS

1. What are the methods of pavement evaluation ?
2. What is the importance of pavement roughness?
3. How is pavement roughness measured ?
4. What is the International Roughness Index (IRI)?
5. Describe the Benkelman Beam and its use.
6. Write short rotes on:
   (*a*) Lacroix deflectograph
   (*b*) Dynaflect
   (*c*) Falling weight deflectometer
   (*d*) Dynamic cone penetrometer test

Chapter

OVERLAY DESIGN AND CONSTRUCTION

34.1. NEED FOR OVERLAYS

Pavements, which have been in service, deteriorate due to a variety of factors. A part of such deterioration can be made good by patching and periodic renewals. When the extent of deterioration is beyond such simple maintenance solutions, the pavement needs an additional overlay. Strengthening with such an overlay will overcome the structural inadequacy caused by traffic that has used the pavement so far and will enable the strengthened pavement to withstand the expected traffic in the design period.

34.2. OVERLAY DESIGN FOR FLEXIBLE PAVEMENTS

34.2.1. Principles of design

An overlay design differs from the design of a new pavement in that in the former, the strength of the existing pavement is to be evaluated, whereas in the latter, the strength of the subgrade on which the new pavement has to be constructed, is evaluated.

Thus, overlay designs involve the following steps :

1. Estimation of the traffic to be carried by the overlaid pavement
2. Measurement and estimation of the strength of the existing pavement
3. Determination of the thickness and type of the overlay.

The exact design of overlays by analytical methods is rather difficult. Most of the design methods are empirical. The estimation of future traffic is done on the same lines as indicated in Chapter 11 in terms of standard axle loads.

34.3. OVERLAY DESIGN METHODS FOR FLEXIBLE PAVEMENTS

34.3.1. Measurement of pavement strength

After estimation of the future traffic, the next information needed for design is the strength of the existing pavement. The measurement of deflection of a flexible pavement is one of the indirect methods for assessing its strength. Most of the methods currently in use by various organisations round the world use deflection criterion as the basis of design. The appeal of this method is the ease and speed with which deflections can be measured without disturbing the pavement structure. The method is based on the assumption that there is a strong correlation between surface deflection and the stresses and strains developed in the layered system of the flexible pavement. It must, however, be recognised that surface deflection is not uniquely related to pavement strength.

34.3.2. I.R.C. Guidelines

The Indian Roads Congress have formulated guidelines for design of overlays flexible pavements. The method is based on measurement of pavement deflections by the Benkelman beam. Two methods of deflection measurements are possible, viz., (*i*) C.G.R.A, static load test and (*ii*) WASHO creep load test.

Deflection measurements are very sensitive to temperatures. The IRC guidelines recommend that deflection measurements are related to a standard temperature of 35°C. Readings are taken when the temperature is not always 35°C. The correction of deflection measurements is 0.0065 mm for each degree of temperature, greater or less than 35°C. The correction will be positive for temperatures lower than and negative for greater than 35°C. For higher altitude areas, an ambient temperature of 20°C is recommended with no corrections.

Deflection measurements vary with the season. They should be taken when the sub-grade is moist, which is after the monsoons. If measurements are taken in dry season, a correction factor is to be applied. For clayey soils, the correction factor may be taken as 2, whereas for sandy subgrade it may be taken as 1.2 to 1.3. For intermediate soils, values may be interpolated.

A large number of measurements of deflections (n) are taken at locations away from the pavement edges, typically along wheel paths. The mean ($\bar{x}$) and standard deviation (σ) are calculated. The characteristic deflection Δ_0 is then taken as the mean plus one standard deviation.

Recent research carried out in India however, has led to the conclusion that the characteristic deflection should be the mean plus two standard deviations.

Thus,

$$\bar{x} = \frac{\Sigma x}{n} \tag{34.1}$$

$$\sigma = \sqrt{\frac{\Sigma(x-\bar{x})^2}{n-1}} \tag{34.2}$$

$$\Delta_0 = \bar{x} + 2\sigma \tag{34.3}$$

The 1984 IRC guidelines recommend the allowable deflection given in Table 34.1.

Table 34.1. Allowable deflections as per IRC guidelines (1984)

Design traffic in the 10 year design period (commercial veh./day)	Allowable Deflection	
	CGRA Method	WASHO Method
150–450	1.50 mm	1.40 mm
450–1500	1.25 mm	1.10 mm
1500–4500	1.00 mm	0.80 mm

The allowable deflections as per Ministry of Surface Transport research findings are given in Table 34.2.

Table 34.2. Allowable deflections as per MOST research project R-6

Design traffic (Milion standard axles)	Allowable deflection (mm)
Upto 2	1.00
2—10	0.80
10—30	0.75
>30	0.70

If the characteristic deflection is greater than the allowable deflection, the thickness of the overlay is then determined by the following formula :

$$h = R \log_{10} \frac{\Delta_0}{\Delta} \tag{34.4}$$

where h = thickness of granular overlay (WBM) in mm

Δ_0 = characteristics deflection,

Δ = allowable deflection

R = constant, whose value may be taken as 550.

The latest IRC guidelines (1997), does away with the above equation (34.4) and, on the other hand, gives a set of curves for determining the overlay thickness. These curves are given in Fig. 34.1. The thickness is in terms of bituminous macadam construction. In case other materials are used, the following conversion factors are adopted :

1 cm of bituminous macadam = 1.5 cm of WBM/WMM/BUSG

= 0.7 cm of dense bituminous macadam/ Bituminous concrete/semi-dense bituminous concrete.

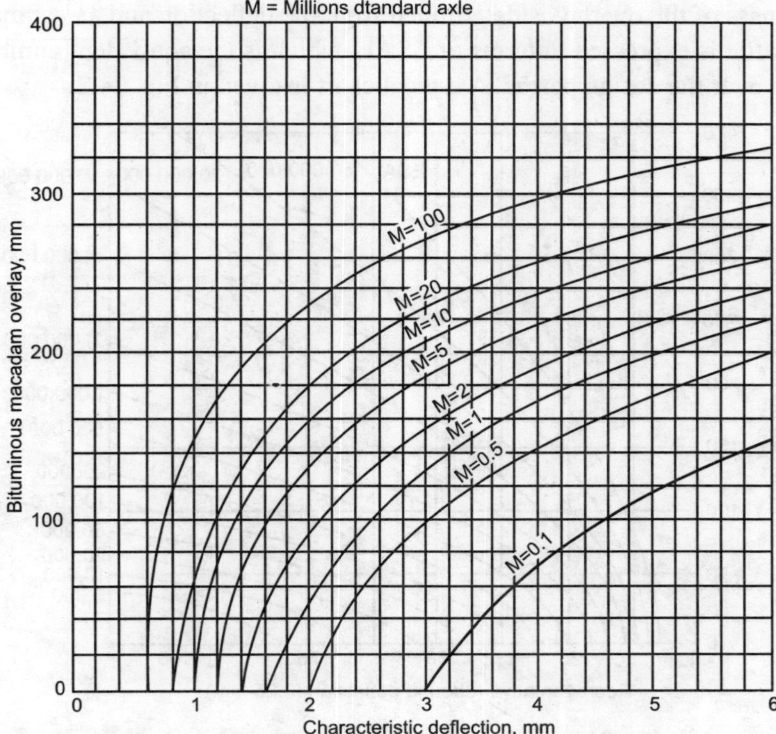

Fig. 34.1. Overlay thickness design curves (IRC).

34.3.3. TRRL procedure

A detailed procedure for overlay design has been developed by TRRL. The method is based on extensive measurements of surface deflection and their relationship with performance obtained by TRRL during many years' work on road experiments. The following basic types of design charts are given :

1. Relationship between the existing deflection and the amount of traffic carried since construction, enabling an assessment of the time when it may require structural strengthening.

2. Chart showing the thickness of rolled asphalt overlay to reduce the present deflection to a value consistent with the satisfactory performance under traffic which is forecast for the future.

With these charts, it is possible to conclude whether any given section of pavement needs an overlay now or later, and if now, what should be the thickness of the overlay.

34.3.4. Asphalt Institute method

In the U.S.A., the design procedure for asphalt overlays published by the Asphalt Institute is followed. The existing pavement is evaluated by measuring the surface deflection by the Benkelman beam. The thickness of the overlay is determined from the deflection and an estimate of the traffic to be carried. Traffic is expressed in terms of ESAL, which is the equivalent number of 8.2 Tonne single axle loads over the design period. A typical chart is given in Fig. 34.2.

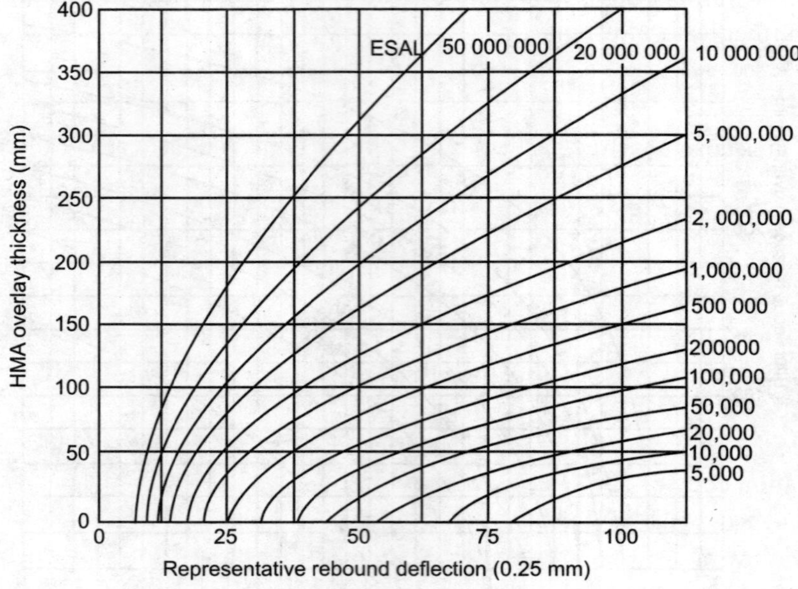

Fig. 34.2. Asphalt Institute overlay design chart.

34.3.5. Extension of design of new pavements

A simple method of determining the overlay thickness of flexible pavement is to use the normal design method for new pavements. Knowing the soil strength (say in terms of CBR) and the estimated future traffic volume, the thickness of the pavement for a selected design period is assessed as for a new pavement. The difference between the thickness so calculated and the present thickness gives the overlay thickness needed. Refinements can be made in the method by converting the pavement thicknesses of different layers into a standard value using coefficients, developed from past experience and research.

34.3.6. Analytical methods

Just as a flexible pavement can be designed both by empirical methods and analytical methods, so also it is possible to design overlay thickness by analytical methods in addition to the empirical methods described so far. The methods are based on sound theoretical formulation of pavement response using multilayer elastic systems. These have the advantage that they are much less dependent on local conditions and can be applied universally. They, however, suffer from the disadvantage that they are very complex, involving elaborate computer analysis.

34.4. OVERLAY DESIGN METHODS FOR RIGID PAVEMENT

Concrete pavements develop structural cracks if they are under-designed or if they have been subjected to heavy traffic. These slabs can be rehabilitated with a rigid or a flexible overlay, thus giving them a further lease of useful life.

34.4.1. Types of rigid overlays

Rigid overlays are those constructed with a cement concrete slab. There are three types of rigid overlays, viz:

1. Bonded, or monolithic overlays, in which the thin overlay slab is bonded on to the existing slab after specially preparing the existing surface through acid-etching or scarifying, using cold milling machines equipped with silicon carbide teeth and mortar coating. Such overlays act monolithically with the original slab and hence need the minimum thickness.
2. Partially bonded overlay, in which the overlay slab is placed directly over the existing slab after cleaning the surface.
3. Unbonded overlay (also called over-slabbing) consisting of a thick slab laid over a separation course. The separation course is laid over the existing slab. The overlay slab acts independently of the underlying concrete slab. Fig. 34.3 gives the three types.

Thin bonded overlays (minimum thickness 25 mm) are not recommended where the existing slab has severely failed. Partially bonded overlays (minimum thickness 120 mm) are also not recommended if the existing slab has severely failed. Unbonded overlays of a minimum thickness of 150 mm can be provided over badly failed slabs.

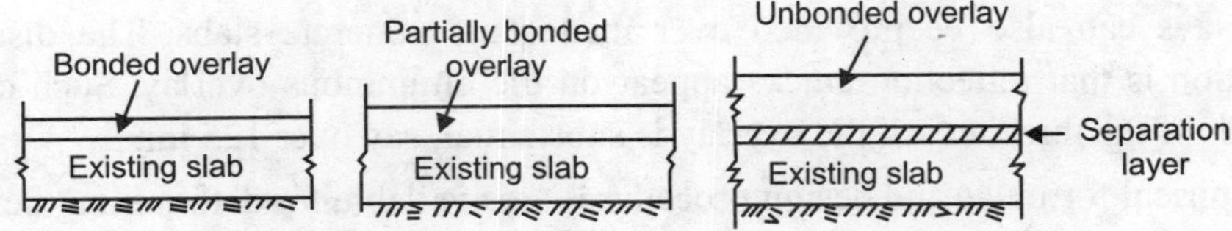

Fig. 34.3. Types of rigid overlays.

Reflection cracks can be expected in thin bonded overlays. They are also usually expected in partially bonded overlays, but are not normally expected in unbonded overlays.

Steel reinforcement is not normally used in thin bonded overlays. In partially bonded overlays, steel requirement is independent of the steel in the existing pavement. In unbonded overlays, steel requirement is entirely independent of steel in the existing pavement.

34.4.2. Design of rigid overlays

Various organisations have evolved empirical formulae for design of rigid overlays. The formulae of the Corps of Engineers and the Federal Aviation Agency are popular. These are given below:

Unbonded overlay

$$h_0 = (h_m^2 - Ch_e^2)^{1/2} \quad (34.5)$$

(Corps of Engineers and F.A.A. Formula)

Partially bonded overlay

$$h_0 = (h_m^{\ 1.87} - Ch_e^{\ 2})^{1/2} \tag{34.6}$$

(F.A.A. Formula)

$$h_0 = (h_m^{\ 1.4} - Ch_e^{\ 1.4})^{1/1.4} \tag{34.7}$$

(Corps of Engineers Formula)

Fully bonded overlay

$$h_0 = h_m - h_e \tag{34.8}$$

In the above formulae,

h_0 = overlay thickness required (inches)

h_m = thickness of monolithic slab required (inches)

h_e = thickness of existing pavement slab (inches)

C = Pavement condition factor

= 1.00 when the existing pavement is in good condition

= 0.75 when the existing pavement shows initial cracking.

= 0.35 when the existing pavement is badly cracked.

34.4.3. Flexible overlays Over rigid slabs

Flexible overlays can also be provided over inadequate concrete slabs. The disadvantage of this specification is that reflection cracks appear on the bituminous overlay. Such cracks can be eliminated only if the thickness of the overlay is substantial, say over 125 mm.

Many empirical formulae and design procedures are available for determining flexible overlay thickness.

IRC recommendations for flexible overlays are summarised below :

(A) For areas of medium rainfall (40–125 cm per annum), favourable drainage conditions and low plasticity of soil (PI not exceeding 14).

(*i*) For very heavy traffic (exceeding 1500 commercial vehicles per day):

(*a*) 7.5 cm Bituminous Macadam (BM) under 4 cm Asphaltic Concrete (AC).

Or

(*b*) 15 cm granular layer under 4 cm AC

(*ii*) For heavy and medium heavy traffic (151–1500 commercial vehicles per day):

(*a*) 7.5 cm BM under 2 cm premix carpet with seal coat

Or

(b) 7.5 cm Built-up-spray-grout (BUSG) under 2 cm premix with seal coat (for medium heavy traffic only).

Or

(*c*) 15 cm granular layer under 2 cm premix carpet with seal coat.

Or

(*d*) 7.5 cm granular layer under 4 cm AC.

(*B*) For areas with high intensity of rainfall (exceeding 125 cm and upto 200 cm per annum), unfavourable drainage conditions and subgrade of high plasticity (P.I. 20 and above).

(*i*) For very heavy traffic (exceeding 1500 commercial vehicles per day): 10 cm B.M. under 4 cm A.C.

(*ii*) For heavy and medium heavy traffic (151 – 1500 commercial vehicles per day): 7.5 cm B.M. under 4 cm A.C.

(*C*) For areas of very heavy traffic in areas where the rainfall is more than 200 cm per annum, unfavourable drainage conditions and subgrade of high plasticity :

(*i*) 5 cm coated macadam with 40 mm single size aggregate mixed with 2.5–3.0 per cent bitumen under 7.5 cm B.M. with 4.0 cm A.C. as wearing course.

Or

(*ii*) 11 cm B.M. with 4 cm A.C. as wearing course.

34.5. RIGID OVERLAYS FOR FLEXIBLE PAVEMENTS–WHITE-TOPPING

The growing scarcity of cement globally has generated interest in providing cement concrete overlays for black-topped pavements. This technique is known as "White-topping". The thickness of the overlay is determined assuming the flexible pavement as a base. The Modulus of Reaction (K value) of the pavement is determined and the overlay thickness calculated in the normal way. It is customary to restrict the value of K to around 14 kg/cm^3. Before laying the concrete overlay, all irregularities in the profile of the existing pavement are made good by a levelling course of lean concrete or bituminous mix.

QUESTIONS

1. Describe the method of design of overlays for flexible pavement as per IRC guidelines.
2. A number of deflection readings were taken on a pavement. The mean and standard deviation were 1.50 and 0.20 mm respectively. The allowable deflection is 1.00 mm. Determine the overlay thickness. **[Ans.** $\Delta_0 = 1.7$, $h = 127$ mm]
3. What are bonded, partially bonded and unbonded overlays ?
4. How is the thickness of overlay on cement concrete pavement determined ?

Chapter

USE OF GEOSYNTHETICS

35.1. TERMS RELATED TO GEOSYNTHETICS

Geosynthetics cover a wide range of applications in Highway Engineering as under:

1. Geotextiles, which are synthetic woven or non-woven fabrics;
2. Geogrids, which are net like tensile elements with apertures of sufficient size to allow interlocking with soil and aggregates;
3. Geomembrane, which is an impermeable membrane acting as a barrier or liner;
4. Geonets;
5. Geocomposite
6. Geocell, used for erosion control of slopes
7. Paving Fabric, used for strengthening bituminous pavements and acting as stress relieving membranes and water barrier; and
8. Glass grids used for strengthening bituminous pavements.

35.2. GEOTEXTILE AS SEPARATION LAYER

The geotextile acts to prevent intermixing of two materials. For example, if provided below a granular layer and above the earth subgrade, it prevents the soil mixing with the granular layer. It can also be used in railway track between embankment and ballast.

Geotextile can be laid over soft soils over which the pavement is built so as to improve the load bearing property. Geotextile and geogrids are used as the reinforcing element in Reinforced Earth Walls.

35.3. GEOSYNTHETICS IN DRAINAGE

Geosynthetics can be used in subsurface drainage of roads. A Fin Drain consisting of a geonet sandwiched between two layers of geotextile and Prefabricated Vertical Drains are used to drain water in roads constructed over swampy or marshy soil. (Fefer to Chapter 22)

35.4. GEOSYNTHETICS FOR EROSION CONTROL

Geosynthetics are used to resist erosion of embankment slopes and allowing vegetation to grow.

QUESTION

Discuss the use of Geosynthetics in Highway Engineering

Chapter

ENVIRONMENT, LANDSCAPING AND ARBORICULTURE

36.1. EFFECT OF HIGHWAYS ON ENVIRONMENT

Highway and traffic have earned the dubious distinction of being the worst defilers of the environment. They have a direct impact on social and community values, the environment and the ecology. In view of the growing awareness of the community and the government to preserve and enhance the environmental values, highway engineers have to plan, construct and maintain highways with this special requirement in mind.

Before any scheme is prepared by the highway engineer, he should prepare an Environmental Impact Statement (EIS), in which he should discuss the effects of the scheme on various components of the environment. It is possible to assign monetary values to the effects of some, but not all of the factors affecting the environment. Wherever it is possible to do so, it should be considered in the economic evaluation of the scheme. The highway engineer should consider the "do nothing" alternative as one of the solutions, as only then will the beneficial or adverse effects of the proposed scheme on the environment become apparent.

The effect of the highways and traffic on the environment will be of the following types :

1. Noise pollution
2. Air pollution
3. Visual intrusion and degrading the aesthetics
4. Community impact.

36.2. ROAD ALIGNMENT AND THE ENVIRONMENT

Basically, roads should be so aligned and located that disturbance of the ecological balance is minimum, *i.e.*, removal of vegetative cover, upsetting of the natural drainage pattern and instability of slopes etc. are minimised. It must be recognised that removal of vegetation not only leads to excessive surface erosion but also increases the tendency of water to seep through and render the slopes unstable, especially in hilly areas. As a result of road development, if any change is brought about in the natural drainage system, appropriate remedial measures should be taken by suitably designing the surface, subsurface and cross-drainage works. In hilly terrain, before fixing an alignment, a landslide susceptibility study should be carefully carried out, since the Himalayas are considered to be geologically very young mountains and blasting may trigger landslides in the vicinity. Erosion at the toe of the slope where a river or stream might be taking a sharp turn, has also been known to cause landslides. It must be ensured that the road alignment is taken sufficiently above the zone of influence of toe erosion.

36.3. EFFECT OF ROAD CONSTRUCTION AND MAINTENANCE OPERATIONS ON THE ENVIRONMENT

All works of road construction and maintenance should be carried out in conformity with the statutory and regulatory environmental requirements. Where construction operations involve removal of vegetation, suitable compensatory measures need to be taken. Any spoil, debris, waste or any deleterious substance from the site, deposited on any adjacent land should be immediately removed and the affected area restored to its original state. All water courses, waterways, ditches, canals, drains, lakes etc. should be protected from pollution arising out of the construction operations. During construction operations, any dust, gaseous or other airborne emissions must be minimised so that air quality is not adversely affected. Effective water sprays should be used during delivery, manufacture, processing and handling of materials when dust is likely to be created. Stockpiles of friable materials should be covered with clean tarpaulins, with application of sprayed water during dry and windy weather. Mixing plant at the site should be equipped with a dust collector. Where gravel or earth roads have been made for haulage to and from the site of construction works, spraying over the gravel or earth road with water should be carried out at regular intervals. Any waste products like surplus aggregates, gravels, cement, bitumen, fuel and engine oils etc. should be disposed of conforming to local regulations. Rural roads, which do not have a proper surfacing, tend to become dusty in dry and windy weather. On such roads, low cost techniques like periodic spraying of water, use of dust palliatives like calcium chloride etc. are necessary to prevent the dust nuisance. On all categories of roads, turfing the earthen shoulders can go a long way in reducing the dust nuisance.

36.4. NOISE POLLUTION

Noise is unwanted sound. The growth of traffic on highways and streets has increased the noise levels, intruding upon the quiet life and privacy of the dwellers nearby. Noise causes annoyance and discomfort, can interfere with sleep or enjoyment of radio or television and causes distraction at schools, libraries and hospitals.

Noise on roads is caused by the engine of the vehicles, its exhaust, brakes, horn, tyres and other parts. Noise is measured in 'A' weighted decibel, dB (A). Acceptable noise levels recommended by the Wilson Committee are given in Table 36.1.

Table 36.1. Acceptable Noise Levels Recommended by Wilson Committee

Situation	Noise Levels dB (A)	
	Day	Night
Country areas	40	30
Suburban areas	45	35
Busy urban areas	50	35

A highway engineer can help in noise abatement in the following ways :

1. Providing smooth surface for the roads.
2. Providing bypasses and ring roads to siphon off the through traffic entering the towns.
3. Designing wide streets, so that the canyon effect is eliminated.
4. Provide noise screens such as trees and shrubs.
5. Through noise-sensitive areas, special sound barriers along the ground level or elevated highways may be provided ; these barriers may be made of concrete, wood, metal or masonry walls. Earth berms to the desired height, where possible or lower earth berms in combination with a wall or screen could be considered.
6. Depressing a highway section, wherever possible has the same effect as providing noise barriers.

36.5. AIR POLLUTION

Air pollution takes the form of poisonous fumes and smell caused by the emissions from the engine exhaust. The situation deserves special notice in busy streets and in tunnels.

The major pollutants are carbon monoxide, oxides of nitrogen, lead particles and smoke (carbon particles). Smoke m combination with fog can cause smog which is hazardous to driving.

The major thrust area of reducing pollution is the vehicle itself. Suitable legislative measures defining the maximum emission from vehicles are urgently needed. The highway engineer can also help in reducing air pollution by constructing bypasses and ring-roads.

36.6. VISUAL INTRUSION AND DEGRADING THE AESTHETICS

Highway and traffic give rise to visual intrusion and a general degradation of aesthetics. The following are the typical ways :

1. Ribbon development along highways giving rise to clutter (crowded and untidy) of repair shops, eating places and fuelling stations.
2. Ugly bill-boards marring scenic beauty and distracting driver's attention.
3. Intrusion of peace and tranquility of environmentally protected areas such as national parks and game sanctuaries and monuments of architectural and historical importance.

The highway engineer has to be very cautious to avoid the above situations.

36.7. COMMUNITY IMPACT

The construction of a highway has adverse impact on the social conditions of a community. A particular route location may cause severance of land and uproot a settled community. It may entail land consumption, including precious agricultural land. Such impact should be evaluated carefully before a scheme is implemented.

36.8. ROADSIDE AMENITIES

Adequate provision of well-designed roadside amenities brings about orderliness in road traffic and enhances the safety and convenience of travel. It prevents the springing up of ugly shops along the highway catering to the eating and resting needs of drivers and passengers. Roadside amenities comprise of rest areas for relaxation of long-distance travellers, truck parks and service stations. Rest areas should be planned sufficiently away from the main highway, preferably close to places of natural scenic beauty. Sufficient provision of toilet facilities, water supply, overnight stay and food arrangements should be made. Truck parks are intended for convenience of truck-drivers. They should have ample space for parking, dormitories for overnight stay, food arrangements, water-supply, toilet facilities and truck repair and servicing facilities.

36.9. LANDSCAPING

Highway landscaping covers a wide area of activities dealing with planning and designing a highway to fit into the adjoining landscape and enhance its beauty. Some of the principles of highway landscaping are enumerated below :

1. The highway design should help to preserve pleasing and aesthetically valuable features, whether artificial or manmade, along the road. Examples are : wooded areas, scenic vistas, rock outcrops, streams, ponds, waterfalls, monuments, architectural masterpieces etc.
2. The road should follow the natural terrain and harmoniously blend with it.
3. Ugly scars due to cutting the existing slopes should be made good by vegetation.
4. Flatter slopes of embankments, closely blending with the natural ground, are pleasing to the eyes, besides being conducive to slope protection.
5. Turfing of shoulders and slopes should be carried out to obtain a lush green appearance. It also prevents soil erosion.
6. Medians and rotary islands should be provided with flowering shrubs and plants.
7. Road-side arboriculture, with shady and flowery trees, beautifies the road.
8. Borrow areas should be treated so that they do not leave an adverse impact on the landscape.
9. The architecture of bridges and structures should be given careful thought to make the appearance pleasing and blending with the natural setting.

36.10. ROADSIDE ARBORICULTURE

Trees afford shade, provide colour if they are of flowering variety, yield fruits, screen unsightly views such as slums, junk yards, storage depots etc. and prevent rain erosion and wind erosion.

Roadside planting can be of three types :

1. Avenue plantation, having one or two lines of trees on either side.
2. Group plantation, consisting of a cluster of trees at intervals.
3. Mixed planting, having different varieties of trees.

Avenue trees are spaced 10-15 m depending upon the tree variety. They are generally planted in a staggered way. The distance of the trees from the edge of the roadway (keeping future developments in view) should be 4-6 m. The variety of trees which are evergreen should be preferred to those which shed leaves in a particular season. Trees with long life are to be preferred. Suitable species for different regional conditions in India should be preferred.

36.11. GREEN HIGHWAYS

Green highways are those that are environmentally friendly, pollution-free, aesthetically pleasing and ecologically sustainable. Some of the guidelines for developing green highways are :

- promote growth of native species of trees and vegetation and control population of invasive species.
- protect hydrology and restore natural drainage paths.
- promote porous pavements for better management of stormwater.
- promote wildlife corridors and passages in areas identified through wildlife conservation plans.
- use recycled materials.
- reduce energy consumption in building roads.
- promote water-harvesting.

36.12. SUSTAINABILITY IN HIGHWAY CONSTRUCTION

The resources available for highway construction are limited. This is particularly true of stone aggregates. India consumes 175 million cubic metres of stone aggregates in the roads sector annually. Over-exploitation of stone quarries is not a sustainable option. Solutions like soil-stabilised sub-bases and bases and concrete pavements in preference to flexible pavements should be explored. Bitumen, a petroleum crude derived binder, is going to be scarcer in the years to come. The use of hydraulic binders (cement, lime, slag, flyash) is a better option.

QUESTIONS

1. (*a*) What are the effects of highways on environment?
   (*b*) What environmental considerations will you keep in mind while selecting a road alignment?
2. Give the effects of road construction and maintenance operation on the environment.
3. How can a highway Engineer help in keeping noise and air pollution within limits?
4. (*a*) State the principles of highway landscaping.
   (*b*) Briefly discuss road side arboriculture.
5. What are green highways? State the guidelines for developing green highways.

Chapter

AIRPORT ENGINEERING

37.1. HISTORY OF GROWTH OF AIR TRANSPORTATION

The first power driven plane was flown by Wright brothers in 1903 at Kitty Hawk, North Carolina, U.S.A. The early planes were driven by propellers, which were later replaced by turbo-props and turbo-jets.

The First World War and the Second World War used aeroplanes for the war effort. In 1927, Lindburg made the first solo flight between New York and Paris. In India, in 1911, Henri Piquet, a Frenchman flew the first flight from Allahabad to Naini. J.R.D. Tata made a solo flight in 1932 from Karachi to Bombay in a single engine Puss Moth.

In the last century, air transport has made dramatic strides. Whereas the propellers driven planes fly at speeds of 250-300 km per hour, modern jet aircraft can travel at a speed of 1000 km per hour. Supersonic aircraft, for example Concord, can fly at a speed of 2300 km per hour which is greater than the speed of sound. Modern jets can carry more than 300 passengers.

37.2. ADVANTAGES AND DISADVANTAGES OF AIR TRANSPORTATION

37.2.1. Advantages

(*i*) ***Speed***: The biggest advantage of air transport is its high speed. Speeds of 1000 km per hour are quite common.

(*ii*) ***Encourages tourism***: International and domestic travel by air has encouraged tourism industry.

(*iii*) ***Promotion of business and trade***: High value goods and perishables are transported quickly by air, giving a boost to business.

(*iv*) ***Military use***: Air transport is invaluable for the defence of the country. It can transport military personnel and defence equipment quickly.

(*v*) ***Relief and rescue operations***: In times of natural and man-made calamities, air transport can provide relief to effected people and rescue them to safety quickly. The recent floods in Uttarakhand and Gujarate have amply proved the benefits of air transport in this regard.

(*vi*) ***Accessibility to Remote Areas***: For example, Ladakh in India is difficult of excess by road. The airport at Leh provides good connectivity to Ladakh.

(*vii*) ***Safety***: Air transport is very safe compared to road transport and railways. Modern technology and rigid regulations for licensing of planes and pilots make air travel very safe.

37.2.2. Disadvantages

(*i*) ***High Consumption of liquid fuel***: The only fuel for aircraft now is petroleum crude based aviation petrol. As availability of crude worldwide is shrinking, this is a serious disadvantage.

(*ii*) ***High noise pollution***: Residential areas near airports and the path of landing and take-off are subjected to high noise levels.

(*iii*) ***High cost***: Compared to road transport and rail transport, air travel is very costly and only the rich can avail of it.

37.3. AIR TRANSPORT IN INDIA

Till a few years ago, air transport was the monopoly of the government sector. Indian Airlines, Air India and Vayudoot are all owned by the government. But, now private airlines also operate, both on domestic routes and international routes. The private sector is also involved in developing airports. India now has the ninth largest civil aviation market in the world. The passenger handling capacity has risen three-fold from 72 millions in 2006 to 220 million in 2011. The cargo handling capacity has risen six fold from 0.5 million tonnes in 2006 to 3.3 million tonnes in 2011. During the Twelfth Five Year Plan, domestic air travel is expected to grow at an average annual growth rate of 12 per cent, and the international air travel is expected to grow at 8 per cent per annum. The cargo traffic is expected to grow annually at 12 per cent and 8 per cent in the domestic and international sectors respectively. Several new airports in Tier II and Tier III towns will have to be built soon. To sum up, air transportation is expected to grow at a healthy rate in the coming years.

37.4. AIRCRAFT CHARACTERISTICS

37.4.1. Speed

The airport infrastructure facilities must match the speed of aircraft. Modern wide-bodied jets can travel at speeds of 1000 km per hour.

37.4.2. Size and dimensions

The size and dimensions of the aircraft expected to use the airport determine the layout of the airport. Fig. 37.1 gives a typical sketch of an aircraft showing the controlling dimensions

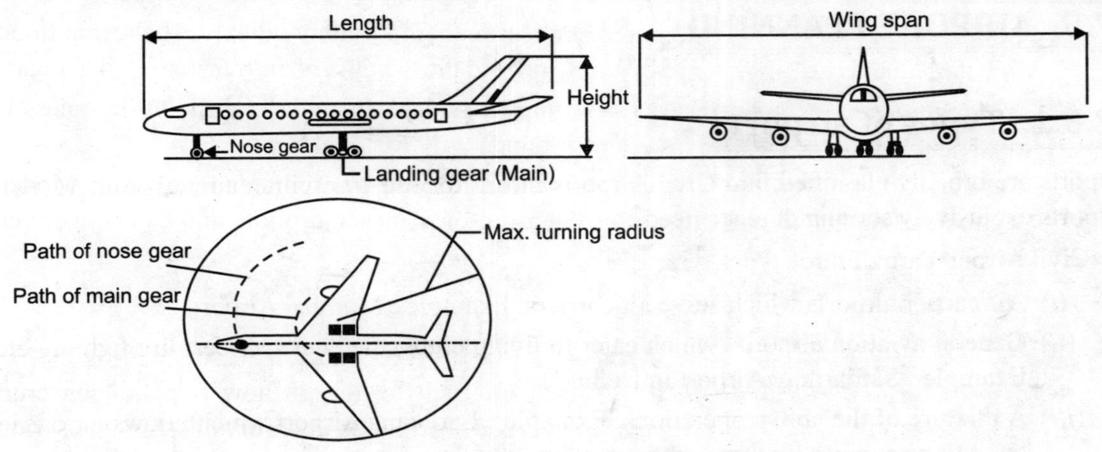

Fig. 37.1: Aircraft dimensions

The dimensions of a few common aircraft types are given in Table 37.1.

Table 37.1. Principal dimensions and characteristics of aircraft

Aircraft type	Wing span (m)	Length (m)	Maximum weight at take off (kg)	Wheel track (m)	Wheel base (m)	Height (m)	Cruising speed (km/h)
Boeing 747	59.6	70.5	351,500	11.0	25.6	19.3	940
Boeing 737	28.3	30.5	45,600	5.3	11.4	11.3	850
Airbus 300	44.8	53.8	152,000	9.6	18.6	16.6	900

37.4.3. Weight, landing gear and tyre pressure

The maximum weight of an aircraft is when it is loaded with fuel and is at take-off. Table 37.1 gives the maximum weight of some aircrafts. The load of an aircraft is transmitted to the airport pavement by wheels conforming to various configurations. This is dealt with in para 37.7.

Tyre pressures are high in aircraft. Typical values are as under:

Aircraft type	Tyre pressure (MPa)
Boeing 747	1.43
Boeing 737	1.04
Airbus	1.18

The main tyres are in the vicinity of 50 × 16 size. The nose wheels are of the 40 × 15 size.

37.5. AIRPORT PLANNING

37.5.1. Types of airports

Airports are broadly classified into Civil Airports meant for use by civilian aircrafts and Military Airports exclusively serving defence needs.

Civil Airports are of three types, viz.

(*i*) Air carrier airports which serve air carriers. Example : Mumbai Airport

(*ii*) General aviation airports, which cater to flying clubs, medical services, fire fighting etc. Example : Safdarjung Airport in Delhi

(*iii*) A mixture of the above operations. Example : Ludhiana Airport which allows air carrier operations and use by flying club.

Airports can also be (*i*) International, which have facilities for customs and immigration or (*ii*) Domestic, meant only for use by domestic services.

The types of landing that are possible are:

(*i*) **CTOL:** Conventional Take-off and Landing

(*ii*) **STOL:** Short Take-off and Landing, with runways of less than 600 m length.

(*iii*) **VTOL:** Vertical Take-off and Landing - Example : Heliport

(*iv*) **RTOL:** Reduced Take-off and Landing with a runway 1000 – 1200 m long.

37.5.2. Demand forecast

The planning for an airport is dictated by the future demand it is expected to cater for. The forecast is generally done for a period of 20 years. For new airports it must be recognised that traffic forecast is based on trends in growth of national air travel, growth of the economy and the growth of business and industry in the region. If an airport already exists, and the planning is for its expansion and modernisation, the historical growth of air travel provides a good basis, combined with forecast of the economy, industry and business in the region.

37.5.3. Site selection

The site selection is governed by the following engineering and physical factors:

(*i*) The site should have fairly level area so that the earthwork cost is minimal.

(*ii*) The land required for a small airport is about 20 to 40 hectares, but for large airports it can be as large as 6,000 to 15,000 hectares.

(*iii*) The site should permit future expansion.

(*iv*) Airport approaches must be clear of obstructions like tall buildings, towers, hills, and transmission lines.

(*v*) The site should be easily accessible to all users, such as passengers, industry and business.

(*vi*) The area should be well drained, free from flooding and have a low water table.

(*vii*) A firm soil can reduce the cost of pavement of runways, taxiways and apron.

(*viii*) The site should not create noise nuisance to residential areas and hospitals.

(*ix*) Services like electricity, water, gas, sewerage and telephone should be easily available.

(*x*) Bird hits cause serious damages to the aircraft and endanger the safety. The site should therefore be away from slaughter houses and refuse dumps that attract birds.

(*xi*) Good quality construction materials should be available nearby.

Meteorological Factors

(*i*) Wind velocity, direction, frequency, turbulence and gusts influence aircraft operation and these should be favourable. Wind direction and velocity influence the orientation of the runway.

(*ii*) Visibility conditions like fog, dust-storms influence safe landing and take-off. These should be favourable.

37.6. AIRPORT LAYOUT

37.6.1. Principal components of an airport

The airport has three main components:

(*i*) Airfield

(*ii*) Terminal area

(*iii*) Flight support facilities

37.6.2. Airfield

The airfield consists of:

(*i*) landing strip consisting of runways (one or more), shoulders and stopways.

(*ii*) Taxiways

(*iii*) Apron.

A runway is the place where the aircraft lands and takes off. It has a strong paved surface with shoulders on either side for serving as safety zones in the event an aircraft goes off sideways. Stop-ways are provided at the ends of the runway to accommodate an aircraft that overshoots a runway during landing or has an aborted take-off. The taxi-way is a paved strip connecting the runway with the apron or connecting runways with one another. The apron is the hard-standing area where aircraft is parked for embarking/disembarking the passengers or where cargo is loaded/ unloaded. Fueling and catering services are also carried out here. A holding apron (also called a run-up area) is that portion which is provided near the ends of the runway for waiting aircraft to take-off.

37.6.3. Terminal Area

The terminal area of an airport consists of:

(*i*) Terminal building

(*ii*) Gates

(*iii*) Aircraft service facilities.

The terminal buildings is where the passengers are rendered services and cargo and mail are handled.

Gates are the openings in the terminal building where the embarking passengers are allowed to proceed to the aircraft. The transfer of passengers from the gate to the aircraft and vice versa is accomplished by buses or through aerobridges.

The aircraft servicing facilities include hangars which are covered area intended for repairing and servicing of aircraft.

37.6.4. Flight support facilities

These include structures for air traffic control, navigational aids and fuelling the aircraft.

37.6.5. Illustrative layout

An illustrative layout of an airport is given in Fig. 37.2.

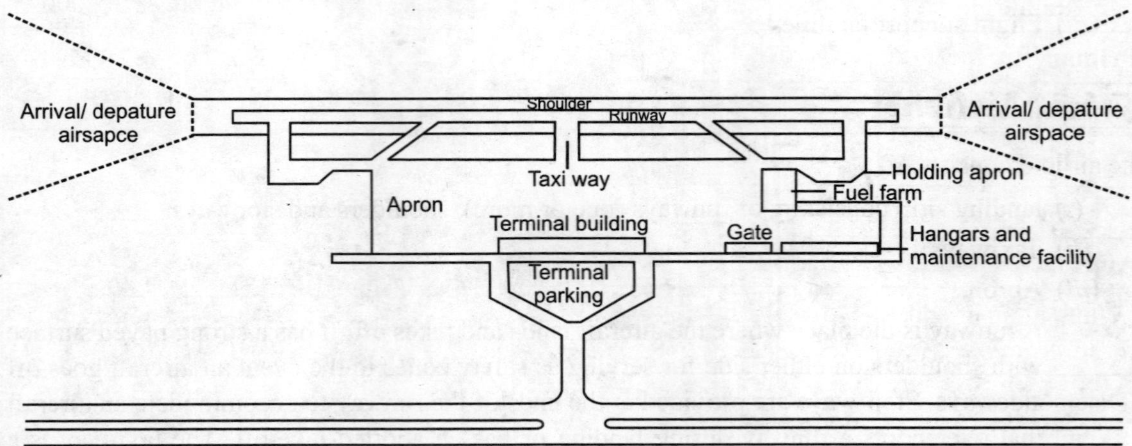

Fig. 37.2: Illustrative layout of an airport

37.6.6. Obstruction criteria

Landing and take-off of aircrafts require that the zones at either end of the runway be clear of obstructions. In order to achieve this, the dimensions and shapes of obstacle surfaces have been standardised. The International Civil Aviation Organisation (ICAO) recommendations for approach runaways and take-off runways are given in Tables 37.2 and 37.3 respectively. As per ICAO classification, airports may be categorised in A, B, C, D and E categories, the highest among them being category A. The primary surface is one which is longitudinally centred on a runway and which extends 60m beyond each end of the runway.

Table 37.2 : ICAO Standards for approach runways

Obstruction clearance surface	Approach conditions and runway code designations							
	Visual runways				Non-precision instrumentation runways		Precision instrumentation runways	
							Category I	Category II
	A, B	C	D	E	A, B	C	A,B,C	A,B,C
Conical surface:								
Slope, %	5	5	5	5	5	5	5	5
Height, m	100	75	55	35	100	75	100	100
Horizontal surface:								
Height, m	45	45	45	45	45	45	45	45
Radius, m	4000	4000	2500	2000	4000	4000	4000	4000
Primary surface:								
Width, m	- -	- -	- -	- -	- -	- -	- -	- -
Length, m	- -	- -	- -	- -	- -	- -	- -	- -
Slope, %	- -	- -	- -	- -	- -	- -	- -	- -
Approach surface:								
Width at inner end, m	150	150	80	60	300	300	300	300
Distance from inner end to runway threshold, m	60	60	30	30	60	60	60	60
Uniform flare, %	10	10	10	10	15	15	15	15
First sector:								
Length, m	3000	3000	2500	1600	3000	3000	3000	3000
Slope, %	2.5	3.33	4	5	2	2	2	2

Second sector:								
Length, m	--	--	--	--	3600	3600	3600	3600
Slope, %	--	--	--	--	2.5	2.5	2.5	2.5
Horizontal sector:								
Length, m	--	--	--	--	8400	5400	8400	8400
Total length of sector, m	3000	3000	2500	1600	1500	1500	1500	
Transitional surface:								
Slope, %	14.3	14.3	20	20	14.3	14.3	14.3	14.3
Internal transition surface:								
Slope, %	--	--	--	--	--	--	--	3.33

The horizontal surface is a horizontal plane 45 m above the established airport elevation and is circular whose centre is at the airport reference point (geographic centre of the airport) and whose radius varies from 2000 to 4000 m.

Table 37.3 : ICAO Standards for take-off runways

Obstruction clearance surface	**Main runways**		**Other runways**	
	Category I	**Category II**		
	A, B,C	**A,B,C**	**D**	**E**
Take-off climb surface:				
Width at inner end, m	180	180	80	60
Distance from inner end to runway threshold, m	60	60	30	30
Uniform flare, %	13	13	10	10
Width at outer end, m	1200	1200	580	580
Length, m	15000	12000	2500	1600
Slope, %	2	2.5	4	5

The conical surface extends outwards and upwards from the periphery of the horizontal surface at a slope of 20 to 1.

The approach surface is longitudinally centred on extended runway centre line. The approach surface slopes at an angle to the horizontal, which is known as the approach angle. A slope of 20:1 is commonly adopted for visual runway.

Fig. 37.3 explains the above terms.

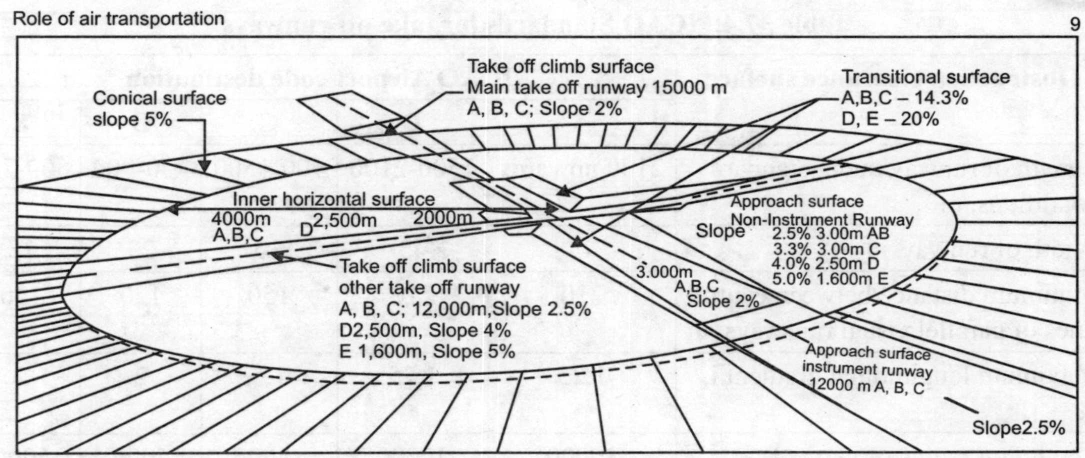

Fig. 37.3: Perspective view of surfaces connected with airport

37.6.7. Configuration of Runways

Various configurations of runway are shown in Fig. 37.4.

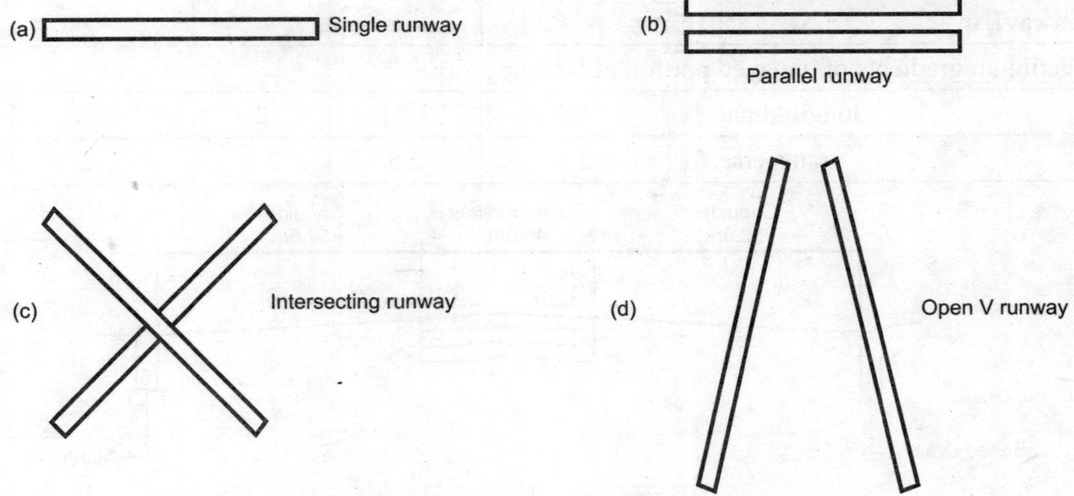

Fig. 37.4 : Configurations of runway

The configurations are dictated by the prevailing wind directions. Designs (*a*) and (*b*) give the configurations parallel to the prevailing wind directions. Designs (*c*) and (*d*) are adopted when the winds are in more than one direction.

37.6.8. Runway and taxiway dimensions

Tables 37.4 and 37.5 give the dimensions of runway and taxiway recommended by ICAO.

Table 37.4: ICAO Standards for take-off runways

Obstruction clearance surface	ICAO Airport code destination				
	A	B	C	D	E
Length of runway under standard conditions, m	2100 upwards	1500-2100	900-1500	750-900	600-750
Width of runway, m	45	45	30	23	18
Minimum distance between centre lines of parallel visual run-ways, m	210	210	150	120	120
Maximum longitudinal gradient i, %	1.25	1.25	1.5	2	2
Maximum radius of vertical curves, m	30000	30000	15000	5000	5000
Maximum traverse gradient, %	1.5	1.5	1.5	2	2
Width of landing strip (precision instrument runway), m	300	300	300		
Width of landing strip (visual runway), m	150	150	150	80	60
Maximum gradients of unpaved portion of landing strip:					
longitudinal, %	1.5	1.75	2	2	2
transverse, %	2.5	2.5	2.5	3	3

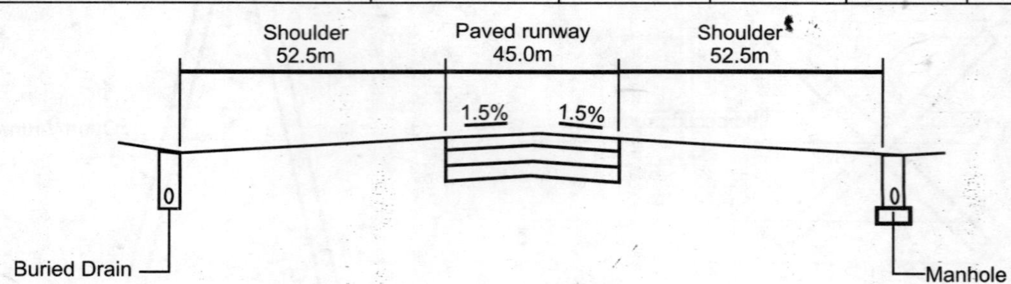

Fig. 37.5 : Runway cross-section

The longitudinal gradient of runway affects the performance of the aircraft. Rising gradients demand extra power for take-off, and downward gradients need more braking distance. For large airports intended for commercial use, a maximum gradient of 1.5% is used. For utility airports, the gradients can be slightly increased to 2%.

Longitudinal grade changes should be avoided. When they are unavoidable, they should be serviced by suitably designed vertical curves. The minimum length of the vertical curve of each 1% grade change is 300 m for carrier airports and 90 m for utility airports. Transverse gradients should not exceed 1.5%.

Since aircraft travels at a much reduced speed on taxiways, their dimensional requirements are lower than for runways. ICAO standards for taxiway dimensions are given in Table 37.5

Table 37.5: Taxiway dimensions and details

Items	ICAO Airport destination				
	A	B	C	D	E
Width of taxiway (m)	23	23	15	10	7.5
Maximum transverse gradient taxiway (%)	1.5	1.5	1.5	2	2
Maximum longitudinal gradient of taxiway (%)	1.5	1.5	3	3	3
Width of taxiway margins (m)	10.5	10.5	7.5		
Width of taxiway strip (taxiway + paved or un-paved margins) (m)	44	44	30	25	22

The exit taxiway from a runway can be either right angled or at some other angle. An angle of about 30° permits to clear the runway quickly at speeds 100 km per hour.

Such exit taxiways can be conveniently located at about 800 m, 1200 m, 1800 m, and 2000 m from the runway threshold such that a variety of aircraft can be served. Taxiway exit details are given . in Figs. 37.6 and 37.7.

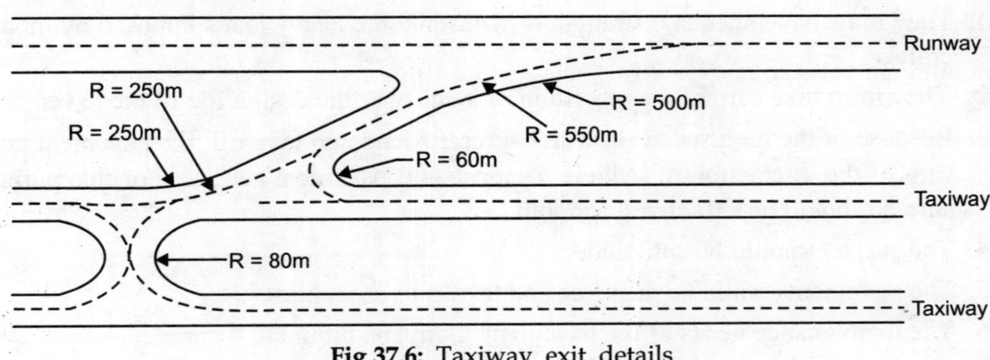

Fig 37.6: Taxiway exit details

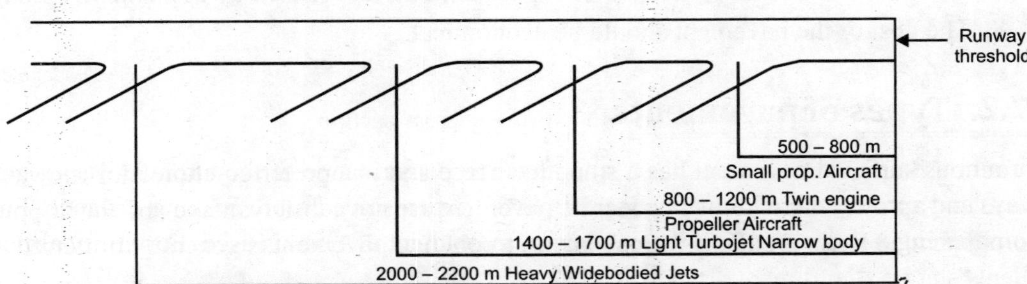

Fig. 37.6 : Taxiway exit details for various types of aircraft.

For safety of operations during taxiway exit, separation clearance is needed. Table 37.6 gives the clearances for various airport designations.

Table 37.6 : Separation clearances

Items	ICAO Airport destination				
	A	B	C	D	E
Minimum distance between taxiway and runway edges (m):					
• Precision instrument runway	150	150	150		
• Non-precision instrument runway	75	73	73	36	29
Minimum distance between taxiway edges (m)	62	52	43	27	23
Minimum distance from edge of taxiway to fixed obstruction outside taxiway (m)	38	30	26	18	16

37.7. AIRPORT PAVEMENT

37.7.1. Runway and taxiway dimensions

Airport pavements must meet the following requirements:

1. They must have adequate strength to withstand the heavy loads imposed by modern aircrafts.
2. They must take care of the repetition of loads over the design life of the pavement.
3. Because of the high speeds at which aircrafts land and take-off, the pavement must take care of the interaction of vehicle response and pavement surface. For this purpose, the surface should be sufficiently smooth.
4. The surface should be anti-skid.
5. The pavement should be dust free and take care of jet blast.
6. The maintenance needs of the pavement should be minimal.
7. The pavement should withstand the detrimental effect caused by fuel and oil spillage.
8. The cost of the pavement should be economical.

37.7.2. Types of pavements

A bituminous surfaced pavement has a smooth surface and is a preferred choice for runways. For taxiways and aprons where joints in concrete pavement are not a disadvantage and the drippings of oil from the engines do not affect the pavement, a concrete pavement is generally preferred.

37.7.3. Design of pavement thickness for flexible pavements

U.S. Corps of Engineering Method: The U.S. Corps of Engineers, who were involved in the construction of many military airfields during the World War II, have developed their own design methodology. The soil strength is determined in terms of CBR values. The early curves developed by them are given in Fig. 37.8. Analysing the performance of various airfield pavements, the following mathematical expression was evolved:

$$t = \sqrt{\frac{P}{8.1(CBR)} - \frac{A}{\square}}$$

where t = pavement thickness, in inches

P = Single wheel load, in lb.

A = Tyre contact area, in sq. inch.

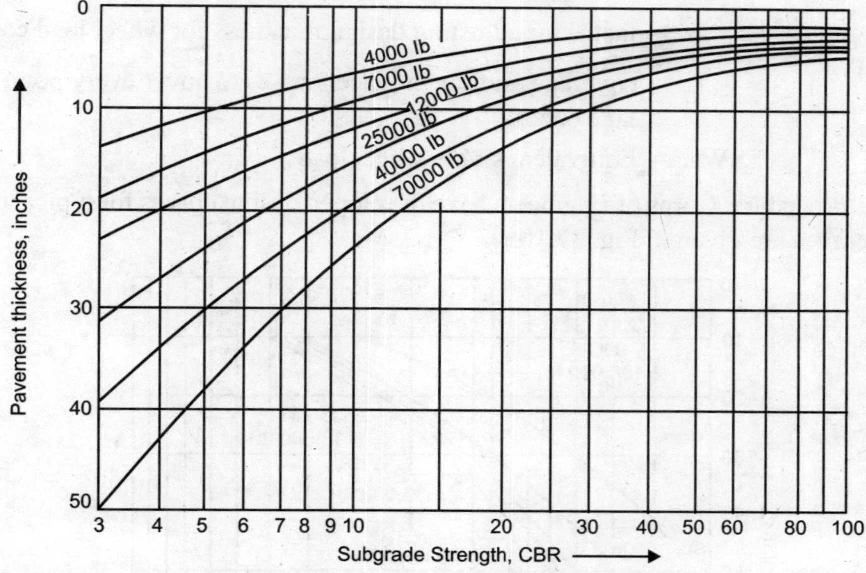

Fig. 37.8 : U.S. Corps of Engineers curves for airport pavement design for different wheel loads

A Set of curves for 200 PSI tyre pressure is given in Fig. 37.9.

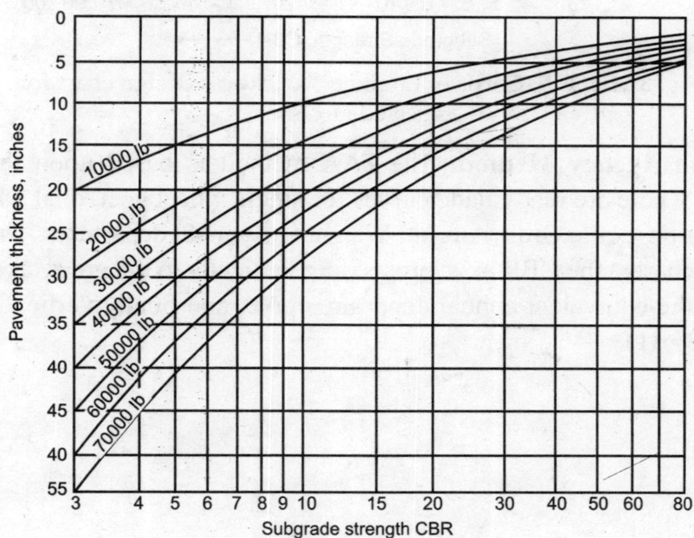

Fig. 37.9: U.S. Corps of Engineers curves for airport pavement design for different wheel loads for 200 PSI tyre pressure

Since different aircrafts had different wheel configurations, the concept of Equivalent Single Wheel Load (ESWL) was developed. ESWL for a particular wheel assembly is that single load which produces the same strain in the foundation as the multiple wheel loads. ESWL can be computed by using Boussinesq's theory.

If ESWL is considered, Equation above gets modified as under:

$$t = f\sqrt{\frac{ESWL}{8.1(CBR)} - \frac{A}{\square}}$$

where f = factor for adjusting design thickness for wheel load coverage

C = coverage (sufficient wheel passes to cover every point of traffic lane once)

ESWL = Equivalent single wheel load

Using this procedure, Corps of Engineers have developed design curves for typical commercial aircraft. The curves are given in Fig. 37.10.

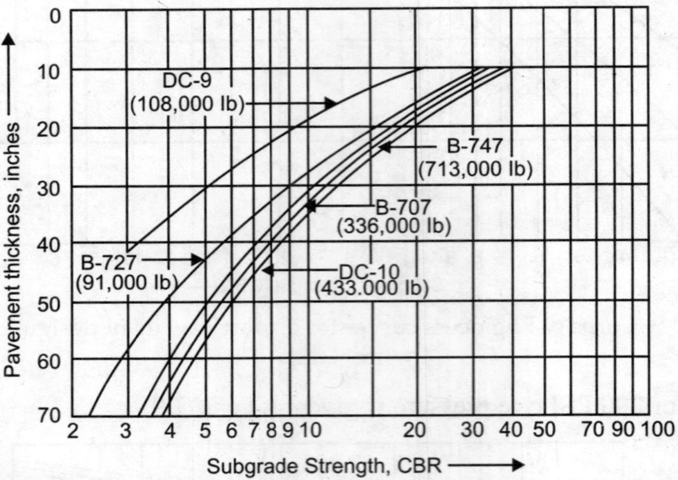

Fig. 37.10: U.S. Corps of Engineers thickness design chart for selected aircraft

Federal Aviation Agency Method: The FAA method is based upon the soil strength as determined by CBR. There are readymade curves for single wheel gear, dual wheel gear, and dual tandem gear. The chart uses the Gross Aircraft Weight and annual departures. The design charts are used by entering the chart at the CBR axis, proceeding vertically to the gross take-off weight curve, then horizontally to the equivalent annual departure curve, and finally vertically to the pavement thickness axis (Fig. 37.11).

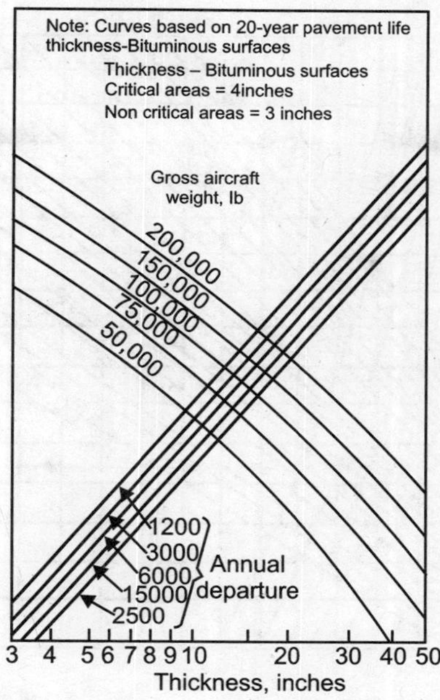

Fig. 37.11 FAA chart for flexible pavement design

37.7.4. Design of rigid pavement for airports

U.S. Corps of Engineering method: The U.S. Corps of Engineers have evolved charts for the design of concrete slabs on Westergaard's formulae presented in Chapter 12. The edge loading condition is taken as the critical one. It is assumed that when the load is placed at an edge of slab, one quarter of the load is transferred to the adjacent slab. The remaining load (0.75P) is increased by 30 percent to take care of curling stresses, load repetitions and other contingencies. The net load ($= 0.75\ P \times 1.30 = 0.89P$) is used for design.

Federal aviation agency method: The typical FAA rigid pavement design chart is given in Fig. 37.12.

The chart is for dual wheel gear aircraft and uses inputs such as:

1. Aircraft weight
2. Annual departures
3. Modulus of subgrage reaction
4. Concrete flexural strength.

General features of rigid airport pavements: Transverse contraction joints are provided at longer spacing than in highway. Typical range is 4 to 8 m. Transverse expansion joints are not generally used except at intersections. Dowels are used where load transfer occurs. A sub-base is generally provided to give a firm support. Reinforcement is desirable for controlling cracks.

LCN Method of design: The Load Classification Number (LCN) is a method of aircraft evaluation of pavement design first used by the Air Ministry, UK. The method expresses the supporting capacity of a pavement the terms of LCN. The wheel load of the aircraft can also be expressed in terms of LCN.

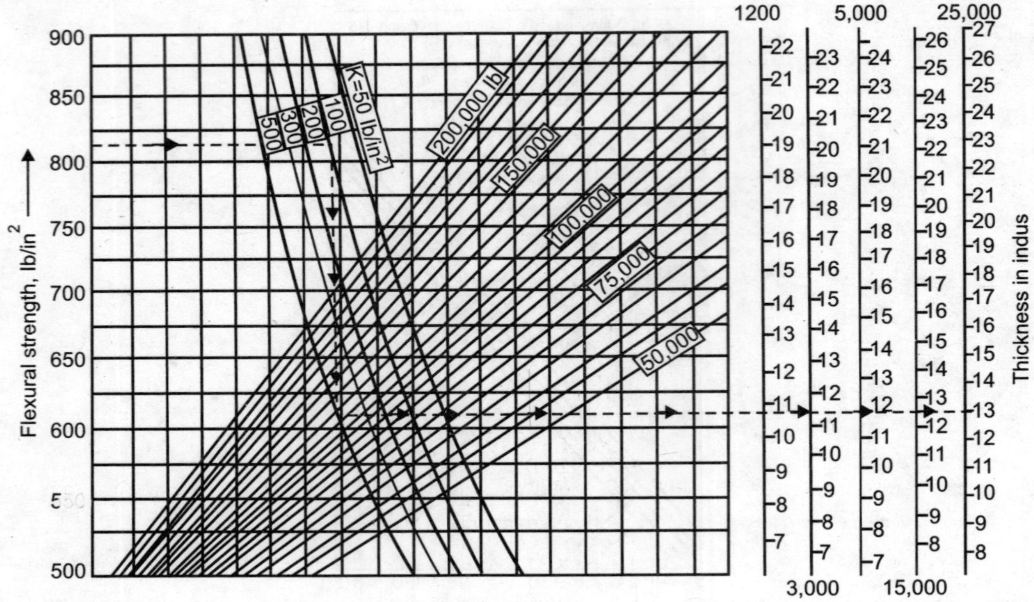

Fig. 37.12 : Rigid pavement design chart for critical areas, dual wheel gear

From the known values of the wheel loads and the contact areas of the aircraft in use, the LCN method was devised and a smooth curve was drawn (Fig. 37.13).

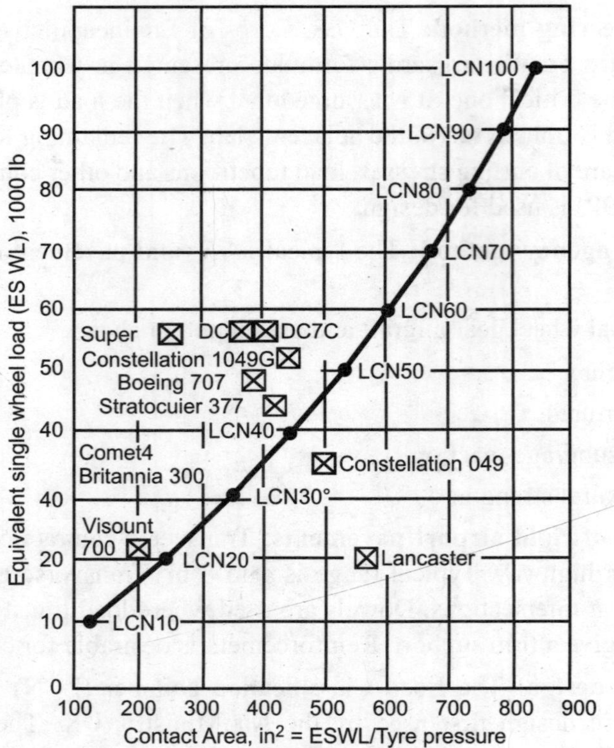

Fig. 37.13: LCN Curve

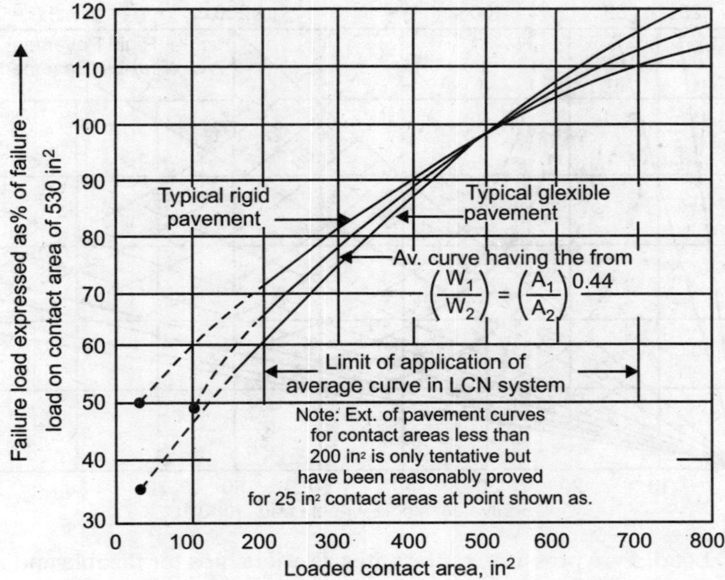

Fig. 37.14 : Failure, loaded/contact area relationship

Various types of pavements, rigid and flexible, were tested with bearing plates having a range of contact areas (20 to 70 sq. in), which represented the aircraft then in use. Over this range of contact areas, both rigid and flexible pavements showed similar load deflection behaviour. From these studies, carried out till failure, the following relationship was established:

$$\frac{W_1}{W_2} = \left(\frac{A_1}{A_2}\right)^{0.44}$$

where W_1, W_2 = failure loads, lb

A_1, A_2 = load contact areas, sq. in

The curve that developed is shown in Fig. 37.14.

In this curve the ordinate (failure load) is expressed as a percentage of failure load on a contact area of 530 sq. in (plate diameter 26 in).

The LCN curve and failure load curve were then combined to produce a chart given in 37.15 using the following procedure:

1. The tyre contact area lines were drawn from the relationship.

$$\text{Contact area} = \frac{\text{Load}}{\text{Tyre pressure}}$$

2. One point on each LCN line comes directly from the standard load classification curve.
3. Other points were calculated using the equation given above.
4. Dotted lines are tentative extensions of the LCN system to accommodate contact areas less than 20 sq.in.

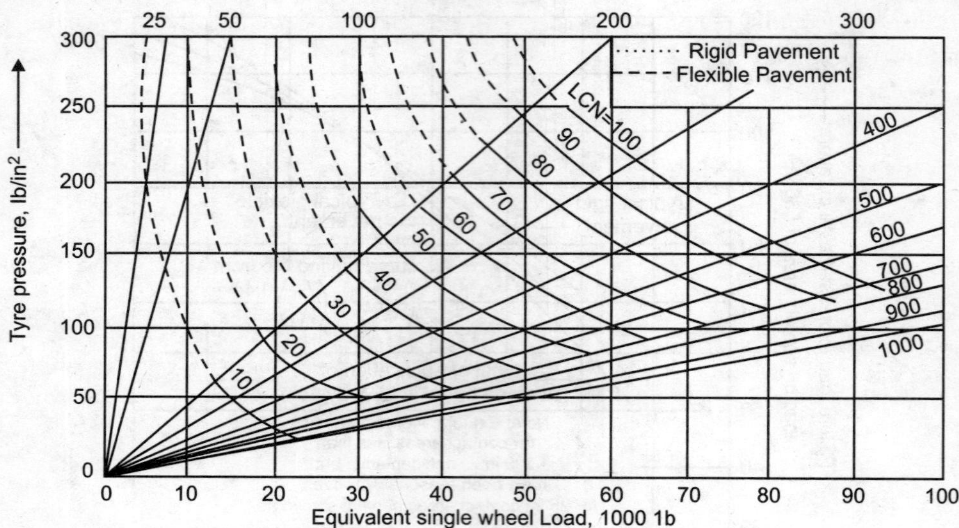

Fig. 37.15: LCN, Load, Tyre pressure, contact area relationships for flexible and rigid pavements

ACN-PCN System: The ICAO has developed an aircraft pavement classification system for reporting pavement strength. The system, known as ACN-PCN, reports a pavement classification number (PCN) which indicates that an aircraft with an aircraft classification number (ACN) equal to or less than the PCN can operate on the pavement subject to any limitations on the tyre pressure. A given aircraft will have a different ACN dependent upon the pavement on which it operates, whether flexible (*F*) or rigid (*R*) and the subgrade strength.

ICAO permits some overloading of the pavement by aircraft with ACNs slightly larger than the reported PCN.

37.7.5. Overlays

When an existing pavement deteriorates structurally and functionally, an overlay is provided. Overlays can be rigid overlays or flexible overlays.

The following equation is recommended by US Corps of Engineers for a flexible overlay on a rigid pavement.

$$t = 2.5\,(Fh_d - C_b h)$$

where t = overlay (flexible) thickness, (inches)

h_d = thickness of cement concrete pavement assuming that the existing pavement did not exist, (inches)

h = thickness of existing pavement, (inches)

F = a factor having values in the range of 0.6 to 1.0, which can be determined from a chart if the Modulus of Subgrade Reaction and the number of annual departures are known.

C_b = a condition factor for exiting pavement, having values in the range of 0.75 to 1.0

The following formula is recommended by Corps of Engineers and Federal Aviation Agency for determining the thickness of concrete overlays.

$$h_c = \sqrt[1.4]{h^{1.4} - Ch_e^{\ 1.4}}$$

or

$$h_c = \sqrt{h^2 - Ch_e^{\ 2}}$$

Where h_c = thickness of overlay (inches)

h_e = thickness of existing slab (inches)

h = thickness of equivalent single slab placed directly on the subgrade with a working stress equal to that of the overlay slab, (inches)

C = a coefficient having the following values:

= 1.0, if the existing pavement is good

= 0.75, if the slab has initial corner cracks, but no progressive cracks

= 0.35, if the slab is badly cracked.

37.8. TERMINAL BUILDING

37.8.1. Function

The terminal building receives passengers intending to travel, processes them and their baggage and transfers them to the aircraft. It also receives the passengers disembarking, handles their baggage and enables them to continue their journey to the city by taxies, buses or Metro lines.

37.8.2. Facilities provided and design requirements

The facilities to be provided and their design requirements can be dealt with under the following heads:

(*i*) Passenger needs

(*ii*) Airlines needs

(*iii*) Airport management requirements

(*iv*) Community requirements

Passenger facilities needed

(*i*) Passengers and their baggage should be handled within a short time. There should be adequate number of check-in counters.

(*ii*) There should be facilities for seating, lounges for passengers drinking water, refreshments, shopping, toilets, bookstalls, postal services, telephone, banking (ATMs) and money exchange.

(*iii*) There should be good passenger information system, displaying the arrival/ departure schedules, a good public address system and pre-paid taxi countres.

(*iv*) Arrival and departure zones should be separated to avoid conflict and confusion.

(*v*) Parking facilities should be adequate and should be close to the terminal building.

(*vi*) Baggage trolleys and wheel-chair facilities should be provided.

(*vii*) Baggage should be handled by means of conveyor belts.

Airlines requirements

(*i*) There should be space for ticketing, passenger check-in for the various airlines operating.

(*ii*) Immigration and Customs clearance facilities for international passengers should be available.

(*iii*) Security checking facilities should be available.

Airport management requirements

(*i*) Provision should be made for their office space.

(*ii*) Facilities for cleaning the airport spaces shall be available.

Community requirements

(*i*) The passengers should be able to access the airport through their own cars, taxies, buses and Metro Lines.

(*ii*) The airport should be close to the city it serves so that a minimum time is lost in transit.

37.9. AIR TRAFFIC CONTROL

Air traffic control aids are located in a separate space, accommodating the administrative staff, control tower, weather stations, and a radio communication centre.

37.10. ILLUMINATION

The lighting provided at the airport should be adequate. The edges of the runway and the taxiways and their centre should be clearly visible.

37.11. Radio Equipment

For controlling air traffic, use is made of radio and radar. The radio equipment is used for air surveillance, weather observation, monitoring take-offs and landings, guidance to aircraft to reach the terminal area and facilitating instrument aided landing.

37.12. HELIPORTS

37.12.1. Helicopters and their role

Helicopter find extensive use in military operations, rescue operations, police patrolling and for civilian use for short hauls, such as Guwahati to Arunachal Pradesh. India has a public sector helicopter operation entity known as Pavan Hans. They can take-off or land in a nearly vertical position in heliports.

37.12.2. Characteristics of helicopters

Helicopters of weight 3 to 5 tonnes can carry 1 to 3 passengers. Those of weight 5 to 10 tonnes can carry 10 to 30 passengers. Very large ones which carry upto 50 passengers weight about 15 tonnes. Their speed is in the range of 150 to 200 km/hr. The rotors which are above the helicopter body have a diameter of 10 to 20 m. The overall length is 10 to 25 m and the height is 2.5 to 7.5 m.

3712.3. Site selection for heliports

The considerations for selecting a site for a heliport are:-

(*i*) A clear approach and departure area free from obstructions with good visibility.

(*ii*) Avoidance of conflict with other air traffic.

(*iii*) Good potential for air traffic.

(*iv*) Good connectivity with other modes of transport.

(*v*) Avoidance of residential area so as not to cause noise pollution.

37.12.4. Design standards

The layout and dimensions are given in Figs. 37.16 and 37.17 and Table 37.7.

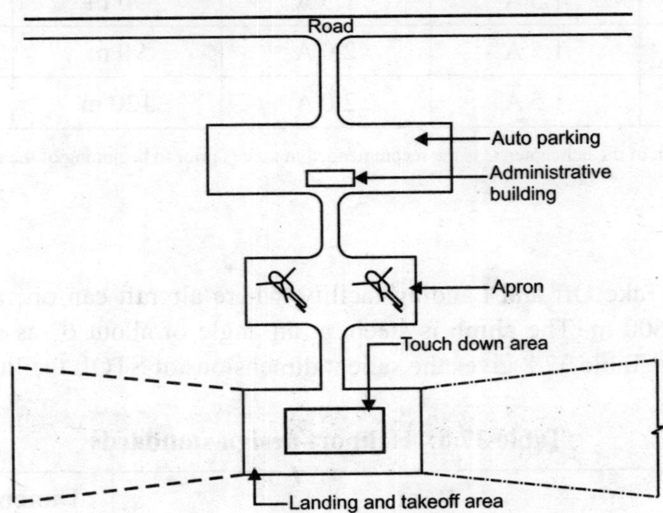

Fig. 37.16 : Heliport layout

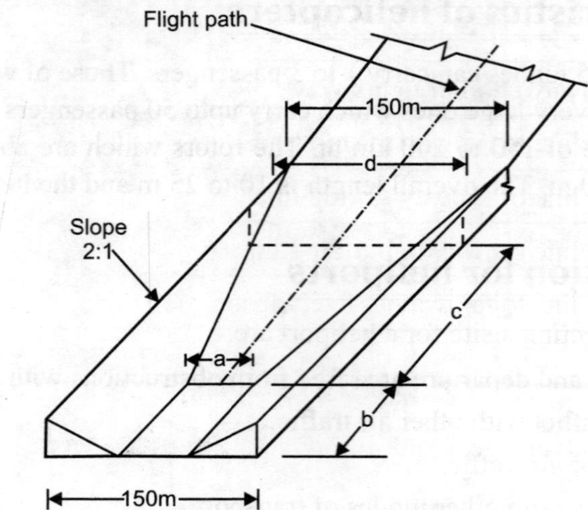

Fig. 37.17: Heliport dimensions

Items	Dimensions			
	A	b	c	d
Private	1.5 A	1.5 A	90 m	60 m
Small public	1.5 A	2.0 A	90 m	90 m
Large public	1.5 A	2.0 A	120 m	90 m

[**Note:** A is the overall length of the helicopter. C is the recommended in metres prior to beginning of the curved path

37.13. STOL

STOL means Short Take Off and Landing facility, where aircraft can operate on short runways of length less than 600 m. The climb is steep, at an angle of about 6° as against 3° needed by conventional aircraft. Table 37.8 gives the salient dimensions of STOL facility.

Table 37.8: Heliport design standards

	Item	Dimension (m)
1.	Length	600
2.	Width of runway	30
3.	Length of Stop way	30-120
4.	Width of shoulder	30-60
5.	Width of taxiway	18

6.	Distances	
	Between runway parallel to taxiway	60
	From centre line to runway to stop line	75
	From centre line of runway to stop line	30
	From centre line runway to building line	90
	From centre line to taxiway to fixed obstruction	30

QUESTIONS

1. What are the advantages and disadvantages of air transportation vis–a–vis land transportation?
2. What are the three levels of airlines and the government airlines in India?
3. Compare the tyre pressure in a modern aircraft with that of a commercial vehicle plying on road.
4. What are the main considerations in selection of site for an airport and in airport planning?
5. How is the flight route of an aircraft controlled?
6. What do you understand by STOL, VTOL, CTOL, RTOL, PDI and ASR?
7. Define : (*i*) air field (*ii*) terminal area (*iii*) flight support area (*iv*) runway (*v*) taxiway (*vi*) apron (*vii*) holding apron (*viii*) terminal building and (*ix*) hangars.
9. What are the common configurations possible for runways ? How is the runway orientation decided ? What are the factors that govern the dimensions of a runway?
9. Give a typical dimensioned cross-section of a runway and a taxiway.
10. What is the difference in the design approaches of US Corps of Engineers method and the Federal Aviation Administration method of design of flexible airfield pavement?
11. What do you understand by LCN, ACN and PCN?
12. What are the facilities needed for : aircraft maintenance, fuelling and cargo handling at an airport?
13. What do you understand by 'Air Traffic Control'?
14. Describe the salient features of airport lighting.
15. What is the role of heliports in air transport? What are the components of a heliport? Give a typical layout and various requirements for the suitability of a site for a heliport.
16. What are the characteristics of STOL aircraft? Give the main dimensions of a STOL port.

Elements of Mechanical Engineering

THIRD EDITION

Dr. D.S. Bedi
Formerly
Professor Emeritus
Department of Mechanical Engg.
Institute of Engineering and Technology
Bhaddal, Dist. Ropar Punjab

Prof. M.P. Poonia
Vice-Chairman
All India Council for Technical Education (AICTE)

KHANNA BOOK PUBLISHING CO. (P) LTD.

Publisher, of Science, Technology and Engineering Books

4C/4344, Ansari Road, Darya Ganj, New Delhi-110002
Phone: 011-23244447-48 **Mobile:** +91-99109 09320
E-mail: contact@khannabooks.com, sales@khannapublishers.com
Website: www.khannabooks.com

Price : ₹ 450.00

ELEMENTS OF MECHANICAL ENGINEERING

Dr. D.S. Bedi, Prof. M.P. Poonia

ISBN: 978-93-86173-06-5

First Edition: 2010

Reprint: 2019

Published by:

KHANNA BOOK PUBLISHING CO. (P) LTD.

4C/4344, Ansari Road, Darya Ganj, New Delhi-110 002

Phone: 011-23244447-48 **Mobile:** +91-9910909320

E-mail: contact@khannabooks.com

Printed in India by:

India Book Printers & Binders, Delhi

Preface to The Third Edition

In the scheme of course for all the engineering dispense for B.Tech students has included a course intitled "Elements of Mechanical Engineering" right during the first year of their studies. This course covers a variety of fields, for example Basic Thermodynamics, Strength of Materials and Theory of Machines.

During the teaching career of the author, spreading to fifty years not only in India but in USA, the author has taught these subjects to B.Tech students. Based on the personal contacts with the students, he is well versed with the details of their requirements of not only those students who fall in the top ranking group but of those students also who are relatively weaker in studies. Keeping these aspects in view, this book is written. Each chapter contains a large number of solved problems so that every student can easily grasp the basic principles involved.

To take into account the additional topics included in the syllabi of various universities courses, chapter 13 has been added, which deals with the materials including the ones, which are used in the electronics industries as well as automobile, airplane and other industries.

The author is induced to his students on whose express demand this text has been written.

In otherhand the author extends his sincere thanks to a large number of eminent personalities in the field of technical education for their valuable suggestions. To mention a few, they are Dr. K.D. Mannan of Baba Bhag Singh Institute of Engg. & Technology, Prof. S.S. Jolly of Sukhmani Institute of Engg. & Technology, Prof. Amit Arora and Prof. Hemant Kumar, Institute of Engg. & Technology Bhoddal, Prof. Beant Singh, Punjab College of Engg. & Technology, Lalru, Mandi, Dr. J.K. Sharma Ambala College of Engg. & Technology Mithapur, Prof. S.S. Sengal of Indo-Canadian Institute of Technology, Mandi, Gobingarh, Dr. B.L. Sethi, Environment Engg. College, Mullana and above all large number of students.

Special Thanks to Dr. Buta Singh, Dean, Planning & Development, Maharaja Ranjit Singh, Punjab Technical University, Bhatinda for completely reviewing the final manuscript.

–Authors

Brief Table of Contents

Detailed Table of Contents

CHAPTER 1

BASIC CONCEPTS OF THERMODYNAMICS

1.1 INTRODUCTION

Thermo-dynamics, as any other science has a unique vocabulary associated with it, precise definition of basic concepts forms the firm formulation for the development of any science keeping thus in view, in this chapter the unit system in use will be reviewed and the basic concepts of thermo-dynamics such as energy, system state cycle, process pressure and temperature will be discussed.

1.2 THERMODYNAMICS AND ENERGY

Thermodynamics in defined as the science of energy transfer. Although everyone has a feeling of what energy is, it is very difficult to give a precise definition for it. Energy can be defined as the capacity to do work or the ability to cause changes.

One of the laws of natures is the "Law of conservation of Energy." It states that during an interaction, energy can change form from one form to another but the total amount of energy remains constant In other words energy can neither we created nor destroyed. For example a stone falling from a tower picks up speed as a result of its potential energy being converted into kinetic energy. This law equally applies to the diet industry: a person who has a greater energy input in the form of food than energy output in the form of exercise will gain weight by storing energy in the form of fat and on the contrary a person who has a smaller energy input (diet) than output (exercise) will lose weight.

Thermodynamics deals with conversion of energy from one form to another. In addition it also deals with various properties of substances and the changes in these properties due to energy transformation. Thermodynamics, like all other sciences, is based on experimental observations. These findings, over a long period of time, have been consolidated into what is known as basic laws of thermodynamics. There is no mathematical proof of the three laws of thermodynamics namely zeroth, first and second laws of thermo dynamics. These laws have been universally accepted as valid since as not in a single case these laws have been violated. Zeroth law relates to thermal equilibrium and deals with the concept of equality of temperature and leads to the means for measuring temperature. The first law of thermodynamics is simply an expression of the conservation of energy. Likewise the second law of thermodynamics asserts that the processes occur in a certain direction but not in the reverse direction. A cup of tea left on a table in an office, for example, eventually cools but a cup of tea on the same table never gets hot by itself. Similarly the third law of thermodynamics defines the absolute zero entropy.

Although the principles of thermodynaincs have been in existence since the creation of the universe, thermo-dynamics did not emerge as a science until about 1700 when the first attempt to make a steam engine was made by **Savory and Newcommen.** It was in 1849 Lord Kelvin for first time used the term thermo-dynamics in one of his publications The first text book on thermo-dynamics was written by *W.* Rankine, in 1859 the greatest progress in this science was made in early 1900 *s*.

1.3 CLASSICAL AND STATISTICAL THERMODYNAMICS

It is universally accepted fact that a substance consists of a large number of particles called molecules. The properties of the substance depend on the behavior of these molecules. For example, the pressure of a gas in a container is the result of momentum transfer between the molecules and the walls of the container. To know the pressure of the gas simply a pressure gauge is to the attached to the container. This macroscopic approach to the study of thermo-dynamics which does not require the knowledge of the behavior of individual molecules is called classical thermodynamics this gives an easy and direct way to the solution of engineering problems. On the other hand a more elaborate approach based on the average behavior of large groups of individual particles is called statistical thermodynamics. This microscopic approach is **rather** involved and will not form part of the present text.

1.4 AREAS OF APPLICATION OF THERMODYNAMICS:

All the engineering activities involve interaction between matter and energy which in turn relate to thermodynamic. Therefore it is very essential to possess good understanding of thermodynamic principles and this is the reason why thermodynamics has become essential part of engineering education.

In order to see various application areas of thermodynamics one does not need to go anywhere but see these areas where one lives. Many ordinary household utensils and appliances are designed in whole or part by using thermodynamic principles. Some examples include refrigerator, air conditioning systems, the pressure cooker, the water heater, the ironing press, microwave and even the computer, the TV and VCR etc. On the industry side thermodynamics plays a major role in the design and analysis of automotive engines, jet engines, nuclear power plants. These are a few examples to mention. Above all one need not feel surprised if it is said that human body is also an interesting application area of thermodynamics.

1.5 DIMENSIONS AND UNITS

Any physical quantity can he characterised by dimensions. Arbitrary magnitudes assigned to dimensions are called units. Some basic dimensions like mass m, length *L*, time *t* and temperature T are known as primary dimensions. While others such as veloeity V, volume V and energy E etc. are expressed in terms of primary dimensions and are called secondary dimensions or derived dimensions.

Despite great efforts in the scientific community to unify the world with a single unit system, two sets of units are still in common use today. But the most popular and the one being followed in India and other most of the countries including England is the metric S.I. (from Le system International *d'* units) which is also known as the System International. The other known as English System which is also known as United States Customary system is prevalent only in United States. Even in U.S.A, the S.I. system is slowly being adopted in order to the capture the export market of other countries.

The S.I. is based on decimal relationship between units. The prefixes like tera, T (10^{12}), Giga, G (10^9), Mega, M (10^6), kilo, k (10^3), centi, c (10^{-2}), mili, m (10^{-3}), micro, m (10^{-6}), nano, n (10^{-9}) and pico, p (10^{-12}) are used to express the multiples of units.

The S.I. units of mass, length and time are kilogram (kg), meter (m) and second (s) respectively. The unit of force is Newton (N) and is defined as the force required to accelerate a mass of 1 kg at a rate of 1 m/sec^2. The term weight is usually incorrectly used to express mass specially by the "weight watchers". Unlike mass, weight W is actually a force. It is the gravitational force acting on the body and is estimated from Newtons second law, and this force is directed towards the centre of earth

$$W = mg$$

Where m is the mass of the body and g the local grairtational accleration (g = 9.81 m/s^2 at sea level and 45° latitude)

Similarly the weight of a unit volume of a substance is called the specific weight w and is calculated from

$$w = \rho g$$

where ρ is the density in kg/m^3.

The mass of a body will remain same irrespective of its location any where in the universe but its weight will change with change of g from place to place. The body will weigh less on the top of a hill as g is less there. Like wise on the surface of moon its weight will be one-sixth of its weight on earth.

Work, which is a form of energy, can simply the defined as force times the distance moved. Therefore units of work is Newton-meter" (N-m) which is called a joule (J)

$$1\text{J} = 1 \text{ N-m}$$

Since joule is a very small unit, more common unit of energy in S.I. is the kilo joule (1kJ = 10^3J)

1.5.1 Dimensional Homogeneity

We all know from our school days that mangoes and bananas do not add. But some how, sometimes we do manage to do it (by mistake of course). In engineering all equations must be dimensionally homogeneous *i.e.* every term in equation must have the same units. While making calculations at some stage we notice addition of two quantities having different units, we must get alarmed that an error has been committed at an earlier stage. Thus checking units can serve as a tool to locate errors.

It is very important to keep in mind that a formula which is not dimensionally homogeneous is definitely wrong but a dimensionally homogeneous formula may not necessarily be right.

1.6 CLOSED AND OPEN SYSTEMS

A thermodynamic system or simply a system is defined as quantity of matter or a region in space chosen for study. The region outside the system is known as surroundings. Further the imaginary or real surface which separates the system from its surroundings is called the boundary. The system boundary can be fixed or movable.

Depending on whether a fixed mass or a fixed volume in space is chosen for study the system is considered to be closed or open. A closed system (also known as a control mass) comprises of a fixed amount of mass, and no mass can cross its boundary. In other words no mass can enter or leave a closed system. Fig 1.1 is an illustration of system, surroundings and boundary

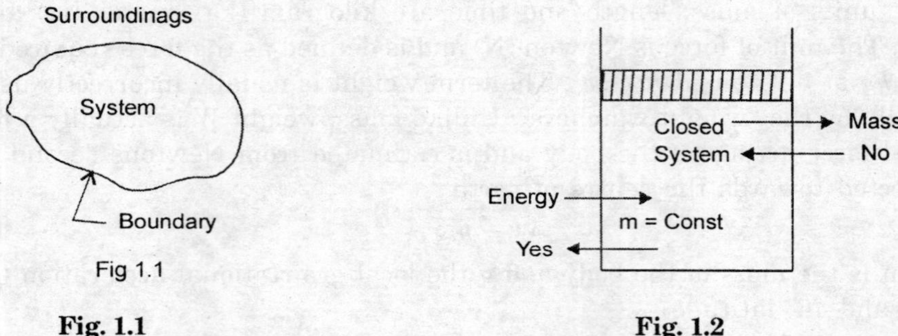

Fig. 1.1 **Fig. 1.2**

Similarly Fig 1.2 shows a closed system where no mass can enter or leave the system. But energy in the form of heat or work can cross the boundary and the volume of the closed system is not fixed. Further let us consider a special case where even energy is not allowed to errors the boundary of the system then that system is called an Isolated system.

Take the example of a piston-cylinder device shows in Fig. 1.3. Let us assume that

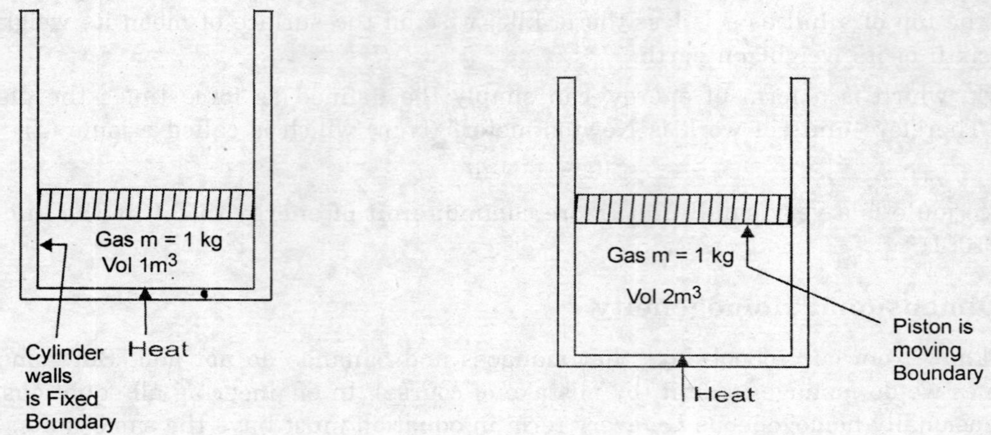

Fig. 1.3 A closed system with moving boundary

in the cylinder there is 1 *kg* of gas at any instant and volume is 1 m^3. If attention is focused then this on the mass of the gas mass of gas will be a closed system. The inner walls of cylinder form a fixed boundary and the inner surface of the piston forms a moving boundary of this closed system. Here no mass is crossing the boundary. However energy in the form of heat may cross the boundary and the part of the boundary in the inner surface of the piston, moves when energy in the form of heat enters the system thereby increasing the volume of the same mass of gas. Every thing outside the gas, including the cylinder and piston, is surroundings.

An "open system" which is also commonly called a "control volume", is a properly selected region in space. It often encloses a device which involves mass flow such as a turbine, compressor or nozzle, water heater etc. Flow through devices is usually studied by selecting a region within the device as the control volume. Both mass and energy can cross the boundary of the control volume The boundary of the control volume is also known as "control surface". This is illustrated in Fig. 1.4.

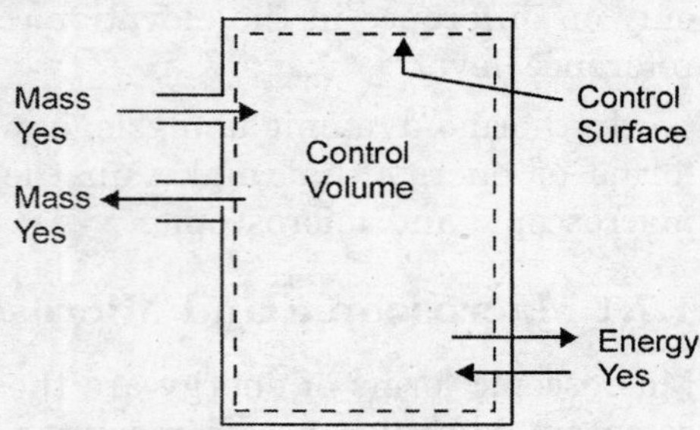

Fig. 1.4 Both mass and energy can cross the boundaries of control volume

The case of water heater is a typical example of "OPEN SYSTEM". It is desired to know the quantity of heat energy to be added to water so that we get the steady supply of hot water stream. Since hot water leaves the heater tank and cold water enters, therefore it is not convenient to choose a fixed mass as the system for analysis. It is therefore, convenient to take into consideration for analysis purpose, the volume formed by internal surface of the tank and concentrate our attention on the cold water and hot water streams entering and leaving the control volume respectively. While analysing the system the hot and cold water streams are taken as mass leaving and enterning the control volume. Evidently the interior surface of the tank of the heater forms the control surface for this case and mass is crossing the control surface at two locations Thus the thermodynamic relations that are applicable to the open and closed systems are different. Therefore it is of utmost importance to establish whether the system under study is an open or closed system. A careful and proper choice of the system can greatly simplify the thermodynamic analysis of even seemingly involved problems.

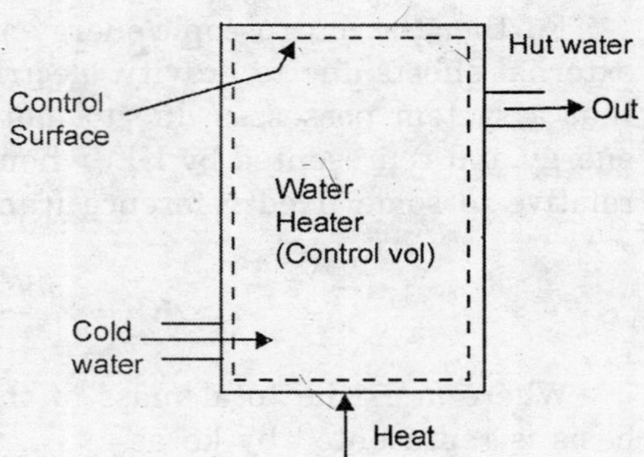

Fig. 1.5 Open system (control volume)

1.7 FORMS OF ENERGY

Energy exists in numerous forms, namely mechanical, kinetic, potential, internal, magnetic, chemical, nuclear, electrical and their sum is called TOTAL ENERGY and is usually represented by the symbol E for a particular system. The total energy of the system per unit mass is denoted by e and defined as

$$e = \frac{E}{m} \text{ kJ/kg}$$

Thermo-dynamics gives no information about the absolute value of the total energy. It only deals with the change of the total energy of the system and this is what is not only of importance but it matters in engineering problems. Thus the total value of the energy of a system can arbitrarily be assigned zero value at source, convenient reference point.

That $E = 0$ at source chosen reference point it must be realised that the change in total energy of the system is NOT DEPENDENT. On the selection of the reference point. For example the decrease of potential energy of a body falling from the top of a cliff will depend only on difference in the elevation of the top of cliff and the ground and not on the chosen reference level.

In thermo-dynamic analysis, however, it is very useful to take into consideration various forms of energy that make up the total energy of the system into two groups namely macroscopic and microscopic.

1.7.1 Macroscopic and Microscopic forms

Macroscopic forms of energy are those which a system possesses as a whole with respect to some outside reference frame, such as kinetic and potential energies. On the other hand, microscopic forms of energy are those which are related to molecular structure of the system and the degree of molecular activity and are independent of the outside reference frame. The sum total of all these energies is called the INTERNAL ENERGY of the system and is represented by U.

Further the macroscopic energy of the system is related to motion and influence of some external effects due to gravity, electric field, magnetic field and surface tension. The energy that a system possesses due to motion relative to some reference frame is called kinetie energy and represented by KE, when all the parts of the system move with same velocity V, relative to some fixed reference frame.

$$KE = \frac{mV^2}{2} \text{ (KJ)} \qquad \text{...(1.1)}$$

Where m is the total mass of the system. The kinetic energy expressed on unit mass basis is represented by ke as

$$k.e. = \frac{V^2}{2} \text{ (kJ/kg)} \qquad \text{...(1.2)}$$

The energy that a system possesses due to its elevation in the gravitational field is called POTENTIAL ENERGY and is represented by P.E as

$$PE = mgz \text{ (kJ)} \qquad \text{...(1.3)}$$

Where g is the acceleration due to gravity and z is the elevation of the centre of gravity of the system with respect to the chosen reference frame Again the potential energy on unit mass basis is expressed as

$$p.e. = gz \text{ (kJ/kg)} \qquad \text{...(1.4)}$$

Even though energies due to electric field, magnetic field and surface tension are important for consideration in certain specified cases only but will not be considered in this text. Thus in the alesence of these three energies the total energy of a system will then consist of potential energy, kinetic energy and internal energy. Therefore from equations (1.1) through (1.4) use have

$$E = U + KE + PE$$

or

$$E = U + \frac{mV^2}{2} + mgz \text{ (kJ)} \qquad ...(1.5)$$

And on the basis of unit mass of system it is expressed as

$$e = u + ke + pe$$

$$= u + \frac{v^2}{2} + mz \text{ (KJ/kg)} \qquad ...(1.6)$$

Whenever a closed system remains stationary during a process it experience no change in its kinetic and potential energies. Then from equations (1.5) and (1.6)

$$E = U \qquad ...(1.7a)$$

or on unit mass basies

$$e = u \qquad ...(1.7b)$$

Thus a closed system when velocity and elevation of the centre of gravity of the system with respect to some chosen reference plane is constant during a process is commonly referred to as "STATIONARY SYSTEM" and change of total energy ΔE of a stationary system will be equal to the change of internal energy ΔU is

$$\Delta E = \Delta U \qquad ...(1.8)$$

Note: In this text a closed system will be assumed to be statiunary unless it is specifically specified otherwise

1.8 FURTHER INSIGHT TO INTERNAL ENERGY (*U*)

In article (1.7) the internal energy has been defined as the sum total of all the microscopic forms of energy of a system. It is dependant on the molecular structure and the degree of molecular activity and it can be taken as the sum of kinetic and potential energies of the molecules

Consider a system at molecular level. In general the individual molecules of a system will move around with some velocity, vibrate about each other and will also rotate about certain axis during their random motion. Thus associated with these motions are translational, rotational and vibrational kinetic energies. Sum of these kinetic energies is called the "KINETIC ENERGY OF THE MOLECULES" this portion of the internal energy of the system is called "SENSIBLE ENERGY" this is illustrated in Fig (1.6)

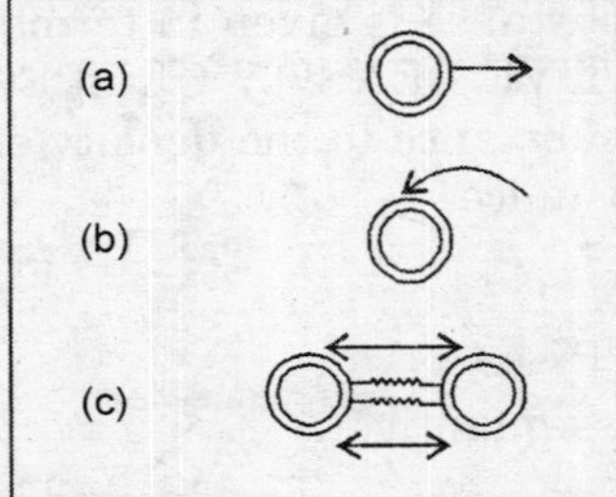

(*a*) Translational knetic energy
(*b*) Rotational kinetic energy
(*c*) Vibrtional kinetic energy

Fig. 1.6 Different forms of moleculargy that make up sensible internal energy

Further the average velocity and the degree of molecular activity are proportional to the temperature of the gas. i.e. higher the temperature higher will be kinetic energies which in turn will result in higher internal energy of the system.

The internal energy is also associated with the inter molecular forces between the molecules of the system. These forces bind the molecules of the system to each other. These forces are strongest in solids and weakest in gases. Thus if sufficient energy is added to the molecules of solids they will overcome the inter molecular forces and break away to turn solid into liquid similarly further addition of energy will turn liquid into a gas without changing the chemical composition of the system. This process is known PHASE CHANGE PROCESS. Evidently a system in gas-phase is at higher internal energy level than that in liquid -phase or than that in solid phase. The internal energy associated with the phase of the system is called the LATENT ENERGY. The most important point to keep in mind is that the chemical composition of the system remains the same, it is only the change in the phase from solid to liquid or liquid to gas. Here no attention is paid to the internal energy associated with the atomic bonds in the molecules of the substance, which is called the CHEMICAL ENERGY OR BOND ENERGY. During chemical reaction like combustion process, more chemical bonds are broken while others formed thereby causing changes in the internal energy thus in the process where combustion will be involved the internal energy will also contain the "Bond energy" component which will otherwise be absent in most of the thermodynamic problems.

It may further be mentioned that tremendous amount of internal energy is associated with the bonds with in the nucleus of the atom itself. This part of internal energy is called that NUCLEAR ENERGY and is released during a nuclear reaction. In the present text we will not deal with this part of the internal energy.

1.9 SYSTEM PROPERTIES

Any system will have its own characteristies and any particular characteristic is called a PROPERTY. For example the pressure P, volume V, temperature T, mass m and density ρ are the characteristics of a system. To this list can be added viscosity, coefficient of thermal expansion, modules of elasticity, electric resistivity. Even its velocity and elevation with respect to certain reference surface can be added to the list.

However it may be noted that all these characteristics are not independent. Since properties are derived from others. For example density is defined as mass per unit volume

i.e.
$$\rho = \frac{m}{V} \text{ (kg/m}^3\text{)} \quad \text{...(1.9)}$$

Sometimes the density of a substance is given in terms of the density of a better known substance. Then it is called RELATIVE DENSITY OR SPECIFIC GRAVITY which is defined as the ratio of the density of the substance to the density of water at 4° C which is 1000 kg/m^3 that is specific Gravity of substance

$$\rho_s = \frac{\rho}{\rho_{water}} \quad \text{...(1.10)}$$

Evidently ρ_s is a dimensionless quantity.

Similarly SPECIFIC VOLUME is more frequently used property in thermo-dynamics. This is the reciprocal of density and is defined as volume per unit mass. It is represented by v as

$$v = \frac{V}{m} = \frac{1}{\rho} \text{ (m}^3\text{/kg)} \quad \text{...(1.11)}$$

An IMPORTANT POINT to note is that in the classical thermodynamics the atomic structure of the substance and the associated spaces between and within the molecules are neglected and the substance is considered to be continuous, homogeneous matter with no microscopic holes. This substance is then known as CONTINUUM. This assumption is valid only as long as we deal with volumes, areas and lengths which are large as compared to the inter molecular spacings.

1.10 INTENSIVE AND EXTENSIVE PROPERTIES

Intensive properties of a system are those which are independent of the size of the system.

Extensive properties are those which vary directly with the size or extent of the system. For example mass, m, volume V and total energy E of the system. Extensive properties per unit mass are called SPECIFIC PROPERTIES. For example specific volume v, specific total energy e and specific internal energy u are defined as follows

$$v = \frac{V}{m} \quad ...(1.12)$$

$$e = E/m \quad ...(1.13)$$

and

$$u = \frac{U}{m} \quad ...(1.14)$$

1.11 STATE AND EQUILIBRIUM

Let there be a system which does not undergo any change. Then at that point all the properties of the system can be measured or calculated through out the system which results in a set of properties that completely define the condition or STATE of the system. In other words all the properties of the system have fixed values and even if the value of one property changes the STATE of the system will change to a new one. For example consider a closed system shown in Fig (1.7 a) and (1.11 b). In Fig (1.7 a) the mass of gas in cylinder is 1 kg, Temperature is 20° C and volume is 1 m^3. Which is state 1 (say). In Fig (1.7 b) the same mass of gas i.e. 1 kg is occuphing a volume of 2 m^3 at temperature of 20° C which is state 2 (say). Evidently state 1 and state 2 are different

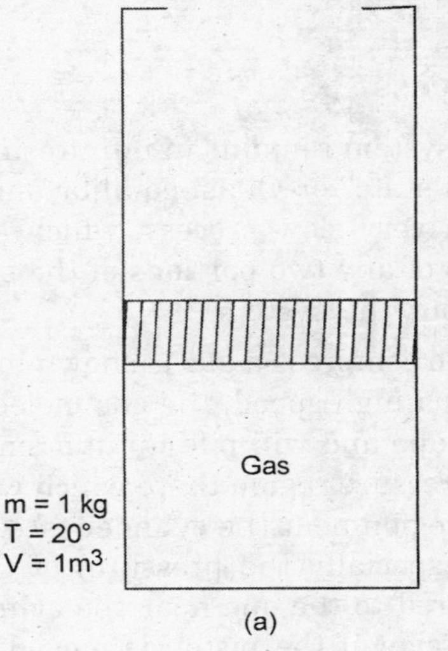

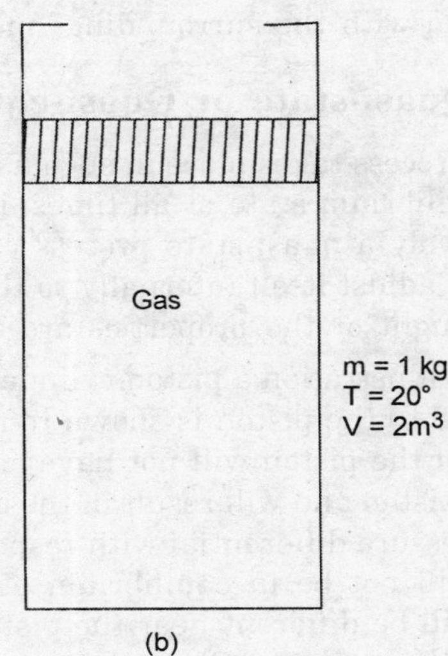

Fig. 1.7

Further thermodynamics deals with equilibrium states. In an equilibrium state there are no unbalanced driving forces within the system. Thus a system which is in equilibrium will experience no change when it is isolated from its surroundings.

There are many types of thermodynamic equilibriums and the system is not in thermodynamic equilibrium until and unless the conditions of all the types of equilibrium are satisfied. Following are the different types of thermo-dynamic equilibrium

(*a*) **Thermal equilibrium:–** If the temperature is same through out the system, it is known to be in thermal equilibrium. In other words the system involves no temperature differential which will be the driving force for heat flow.

(*b*) **Mechanical equilibrium:–** Mechanical equilibrium pertains to pressure. A system is in mechanical equilibrium if there is no change in pressure at any point of the system with time. However the pressure may vary within the system with elevation as a result of gravitational effects. The higher pressure at the bottom layer is balanced by the extra weight it must carry. Thus there is no inbalance of forces. However in most of the thermo-dynamic systems the variation of pressure as a result of gravity is relatively very small and usually neglected.

(*c*) **Phase equilibrium:–** If a system cousists of two phases then the situation of phase-equilibrium is said to prevail when the mass of each phase reaches an equilibrium level and stays there.

(*d*) **Chemical equilibrium:–** A system is said to be in chemical equilibrium when its chemical composition does not change with time. In other words no chemical reaction occurs within the system.

Finally to sum up a system will be in thermodynamic equilibrium when all the above mentioned four equilibrium criteria are satisfied.

1.12 PROCESSES AND CYCLES

When a system undergoes any change from one equilibrium state to another equilibrium state, this change is called a process. Further the series of states through which a system passes during a process is called the PATH of the process. In order to define a process completely its initial and final states along with the path to be followed by it as well as its interaction with the surroundings most be specified.

1.12.1 Quasi-state or Quasi-equilibrium process

When a process takes place in such a manner that the system remains in infinitesimally slow to an equilibrium state at all times, it is called "Quasi-state" or Quasi-equilibrium" process. Alternatively a quasi-state process can be taken as a very slow process. which allows the system to adjust itself internally so that the properties of any two portions of the system are same thought or the properties are uniform throughout the system.

As an illustration a piston-cylinder assembly used to compress a gas is shown in Fig (1.8). In Fig (1.8 *a*) the piston is shown to move suddenly at a high speed, the gas molecules near the face of the piston will not have enough time to escape and will pile up in a small region near the piston and will result in the creation of high pressure region there which will in turn create pressure differential with respect the left most region near the cylinder face. Thus the system will not be in equilibrium as the properties (specially the pressure) of the gas or system will be different near the piston face as compared to the one near the cylinder head face and the process will not be quasi-static one. However if the piston is moved at a very slow speed, the gas molecules will have ample time to redistribute themselves and there will be no piling up of the molecules near the piston face. As a consequence the pressure in the cylinder will always be uniform and will rise at all points within the system. In other words, as the equilibrium is maintained at all times during the process, this is a quasi-state equilibrium process.

Fig. 1.8

It may be pointed out that quasi-static process is an ideal process and is not a true representation of an actual process. But many engineering process are appraximated to be quasi-static condition. This has two advantages : firstly the calculations for the process analysis are easy and secondly the work producing devices deliver the most work when they are assumed to operate under quasi-static equilibrium condition. Therefore quasi-static equilibrium processes are usually taken an standard while analysing the actual process.

1.12.2 Process Diagrams

In order to visualize analysis a thermodynamic process, the process diagrams are drawn by taking some common thermodynamic properties like pressure *P*, volume *V* (or specific volume *v*) and temperature *T*) as co-ordinates For example in Fig (1.9) is shown a compression process with *P* and *V* as co-ordinates

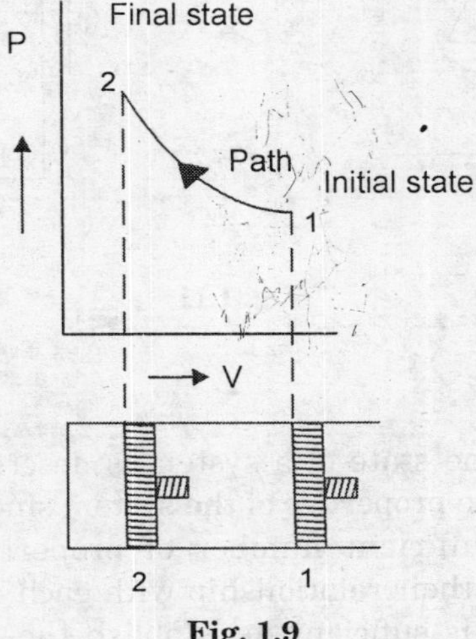

Fig. 1.9

As already pointed out earlier the process path is series of equilibrium states through which the system passes and has significance only for quasi-static equilibrium process. For non-quasi static processes the states through which the system passes can not be specified. However it is universally accepted practice to denote non-quasi-equilibrium process by a dashed line between the initial and final states. It is illustrated in Fig (1.10) and we cannot specify the process path.

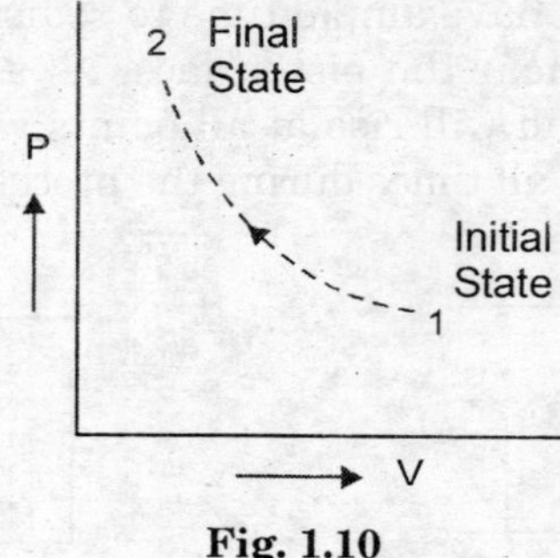

Fig. 1.10

Further it is accepted practice to use a prefix "iso" to designate a process for which a particular property remains crustant. For example an ISOTHERMAL PROCESS in one during which the temperature T, remains constant, an 'ISOBARIC PROCESS" is a process during which the pressure P remains constant and "ISOCHORIC (or ISOMETRIC) PROCESS" is process during which the SPECIFIC VOLUME remains constant.

1.12.3 Cycle

A system is said to undergo a cycle if it comes back to its initial state at the end of a process, alternatively the initial and final states are same for a cycle. In Fig (1.11 *a*) is shown a cycle which involves two processes In Fig (1.11 *b*) is illustrated a cycle which consists of four processes

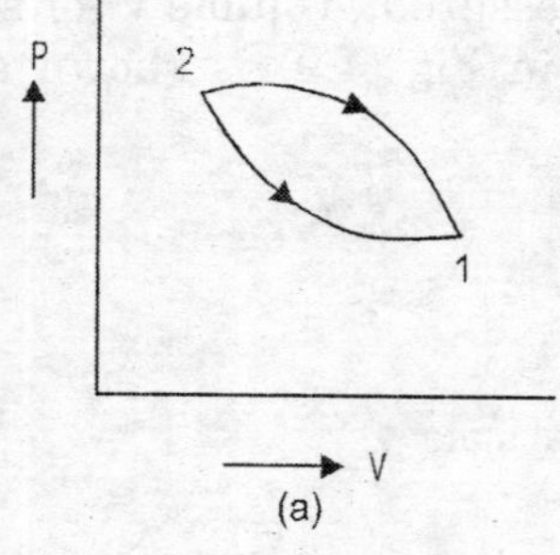

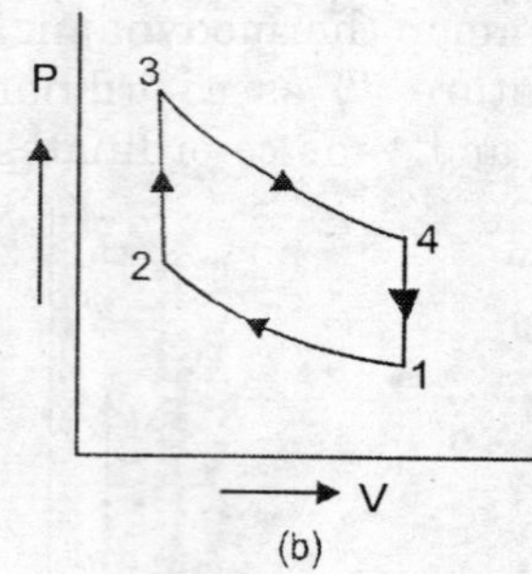

Fig. 1.11

1.13 STATE POSTULATE

As pointed out in article 1.11 the state of a system is described by its properties. However we may not need to specify all the properties of the system since some of these are interrelated to each other. Thus once a sufficient number of properties are specified the rest are automatically fixed by virtue of their relationship with each other. In other words specifying a certain number of properties is suficient to establish the state of a system.

The number of properties required to fix the state of a system is given by the state postulate as given below

The state of a simple compressible system is completely specified by two independent "Intensive Properties". Further a "Simple compressible system" is one which is free from the electrical, gravitational, magnetic, motion and surface tension effects. These effects are due to external force fields and are negligible in most of the engineering problems. Otherwise if the effect of one or more of the above mentioned external force fields is significant then additional. corresponding property needs to be specified over and above the two properties needed to fix the state.

1.13.1 Further Discussion on State Postulate

State postulate requires that two properties which fix the state must be independent which means that one property can be varied while the other is kept fixed. For example temperature and specific volume are always independent properties and they can certainly fix the state. However temperature and pressure are two independent properties for a single-phase system but are dependent properties for a multiphase system. As an example at sea level water boils at 100° C where pressure (P = 1 atmosphere) while at hill top water boils at lower temperature because the pressure at hill top is lower that is $T = f(p)$ during phase change process. Thus temperature and pressure are not sufficient properties to fix the state of a two-phase system.

1.14 PRESSURE

Pressure is the term associated with fluids (liquids and gases). In a fluid contained in a container and at rest the molecules are moving at random in all possible directions. One third of the particles can be assumed to be moving along each of the three arbitrarily chosen axes of reference. The force exerted per unit area due to the impact of molecules on the walls of the container due to molecules moving normal to it, is known as pressure. However at a point within the fluid where there is no wall, an imaginary plane can be considered to be passing through this point then the force of impact of the moving molecules on this plane and normal to it per unit area will be the pressure at that point. Since the imaginary plane can be rotated to change its orientation, the impact force per unit area on the opposite side of this plane and normal to it will be pressure at that point and it will be same in all directions This is precisely in conformity with the Pascals law.

The pressure in the fluid contained in a container and at rest increases with depth as a result of the weight of fluid above the point under consideration. This is due to the fact that fluid at lower level will have to carry more weight than the fluid at the upper level. Thus the pressure in the fluid increases in the vertical downward direction as a result of gravitational effects. But there is no variation of pressure in the horizontal direction. However in the case of gas contained in a container or tank may be considered uniform since the weight of the gas is too small to make a significant difference.

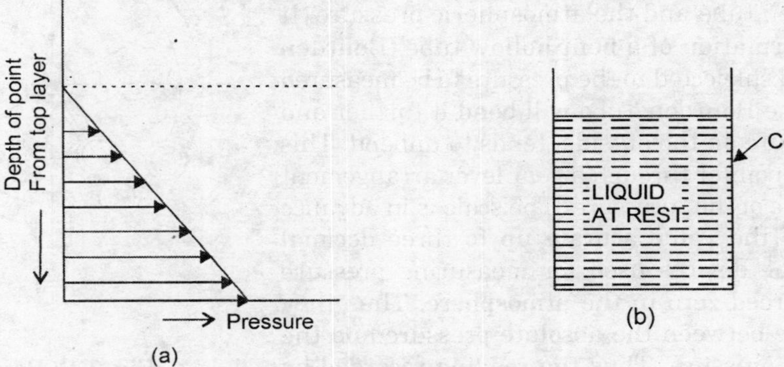

Fig. 1.12

Since the pressure is the force her unit area its units are N/m^2 which is called Pascal

i.e., $$1\ Pa = 1\ \text{N/m}^2 \quad ...(1.15)$$

As unit pascal is too small as compared to the pressures encountered in actual practice, its multiples are used.

For example

$$\text{kilo paseal } KPa = 10^3\ Pa \quad ...(1.16a)$$
$$\text{Mega Paseal } MPa = 10^6\ Pa \quad ...(1.16b)$$
$$\text{Gage paseal } GPa = 10^9\ Pa \quad ...(1.16c)$$

Internationality accepted, two other pressure units are "bar" and "standard atmosphere" which are defined below

$$1\ bar = 10^5\ Pa$$
$$= 0.1\ MPa = 100\ KPa \quad ...(1.17)$$

1.14.1 Absolute pressure and gauge pressure

The actual pressure at a point in the fluid in ABSOLUTE PRESSURE. However most of the instruments used to measure pressure indicate it relative to the atmospheric pressure, where as the pressure actually is above zero or relative to perfect vacuum. The pressure relative to atmospheric pressure which is indicated or read on the gange is known as GAUGE PRESSURE and the pressure relative to prefect vacuum is known as absolute pressure. Similarly the pressures below atmospheric pressure are called vacuum pressure and are measured by vacuum gauges which indicate the difference between the atmospheric pressure and the absolute pressure. Gauge, vacuum and absolute pressures are positive quantities and related to each other by the following equation

$$P_{gauge} = P_{abs} - P_{atm} \quad ...(1.18)$$
$$P_{vac} = P_{atm} - P_{abs} \quad ...(1.19)$$

Usually P_g and P_a are abriviations for Gange pressure and absolute pressure respectively. In the thermo-dynamic tables absolute pressure is invariably used. However in thermo-dynamic calculation the absolute pressure is denoted by P and in the text also P has been used to represent absolute pressure unless it is otherwise specified.

1.14.2 Pressure Measuring Devices

Figure (1.13) shows BOURDON Pressure gauge which measures the difference between the system pressure inside the Bourden tube and the atmospheric pressure. It relies on the deformation of a bent hollow tube (Bourden Tube) which when sulejected to the pressure to be measured on the inside of the Bourden Tube will bend it further and atmosphere pressure on the outside, tends to unbend. This in turn moves the pointer through a gear lever arrangement and gives the value of the "pressure" The scale is in advance calibrated to give the value acurate up to three decimal places. Most of the devices used for measuring pressure are calibrated to reed zero in the atmosphere. Thus they give the difference between the absolute pressure and the local atmospheric pressure. Thus the reading recorded by them is the gauge pressure.

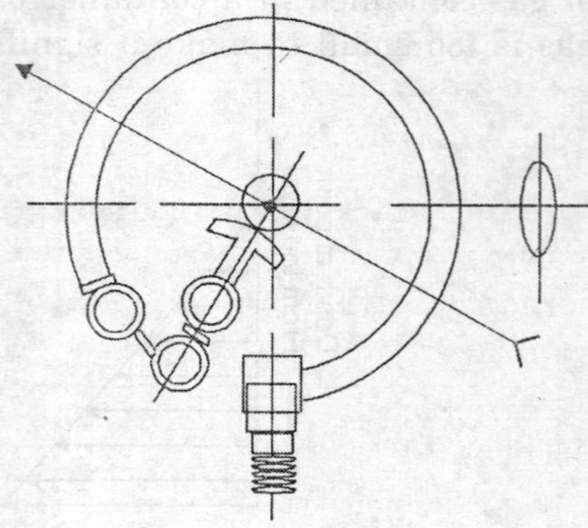

Fig. 1.13 Bourden tube pressure gauge

Similarly the pressure below the atmosphere pressure is called vacuum pressure and is measured by vacuum gauge which gives the difference between the absolute pressure and the local atmospheric pressure. Absolute, gauge and vacuum pressures are related to each other by following relationships.

$$P_{gauge} = P_{abs} - P_{atm} \quad ...(1.20)$$

$$P_{vac} = P_{atm} - P_{abs} \quad ...(1.21)]$$

Where P_{gauge} = Gauge pressure

P_{abs} = Absolute pressure

P_{atm} = Atmosphere pressure

P_{vac} = Vacuum pressure

Note:– In thermo-dynamic relations and tables absolute pressure is almost always used and is represented by '*P*'. Similarly P_a and P_g are the symbols for Atmospheric Pressure, and gauge pressure respectively.

1.14.3 MANOMETER

Manometer is a simple device with which small and medium pressure difference is measured. It consists of a glass or plastic U tube containing fluid like water, alcohol, oil and even mercury. In order to keep the size of the manometer at a manageable level, high density fluid like mercury is used if large pressure differences are expected to the measured.

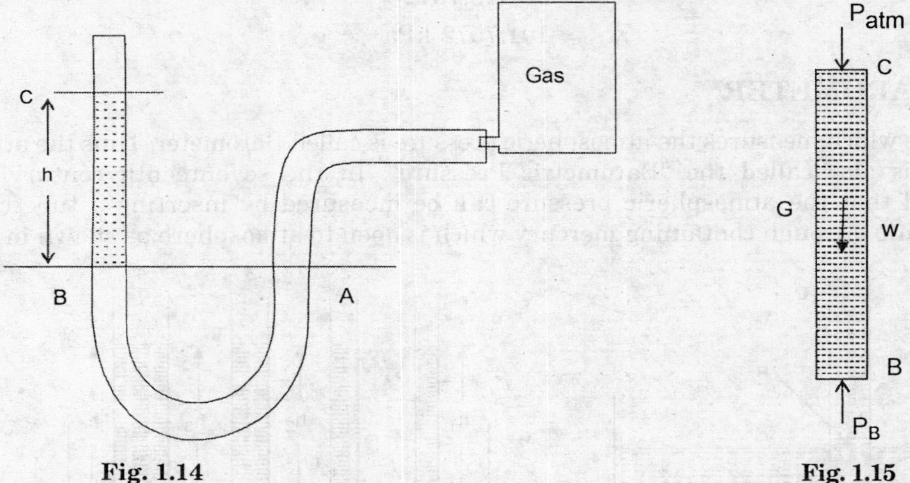

Fig. 1.14 **Fig. 1.15**

In Fig. 1.14 is shown a *U*-tube containing some fluid. It is used to measure the pressure of gas in a tank. As the gravitational effects of gases are negligible therefore the pressure any where in the tank as well as at point A will be same. Further since the pressure in the fluid remains constant in the horizontal direction the pressure at points A and B within the fluid will be same.

That is $$P_A = P_B \quad ...(1.22)$$

Considering the equilibrium of the column *BC* of the fluid for which the free-body diagram is shown in Fig. 1.15 where *W* is the weight of fluid column *BC* of height-*h* metre. Assuming the area of cross-section of the fluid column as *A* and the desity of the fluid as *P* then equating the algebraic sum of verticle forces to zero, we get

$$AP_{atm} + W = AP_B$$

$$= A\,P_A \qquad \text{(from eqs 1.22)}$$

or $$AP_{atm} + (Ah\rho)g = AP_A$$

or $$P_A = P_{atm} + pgh\ (k\ Pa) \qquad ...(1.23)$$

Where P_{atm} is the local atmospheric pressure in $k\ Pa$. It may be noted from eqn (1.23) that the cross-sectional area of the U-tube has no effect on the differential height of the fluid column.

Example 1.1. *A manoneter containing oil (density ρ = 850 kg/m³) is attached to a tank filled with air under pressure. If the oil level difference between two columns is 450 mm and the atmospheric pressure is 98 k Pa, find the absolute pressure of the air in the tank. Taken g = 9.807 m/s²*

Solution: It P_A is the absolute pressure of air in the tank then from equation (1.23) we have

$$P_A = P_{atm} + \rho gh$$

Now $$\rho gh = 850 \times 9.807 \times \frac{450}{1000}$$

$$= 3757.2 \quad P_a$$

$$= 3.7512 \quad P_a$$

$$P_A = P_{atm} + 3.7572$$

$$= 98 + 3.7572$$

$$= 101.7572 \text{ kPa}$$

1.14.4 BAROMETER

The device which measures the atmospheric pressure is called Barometer. Thus the atmospheric pressure is also called the "Barometric Pressure". In the seventeenth century Ferricelli discovered that the atmospheric pressure can be measured by inserting a tube containing mercury into a trough containing mercury which is open to atmosphere as shown in Fig (1.16)

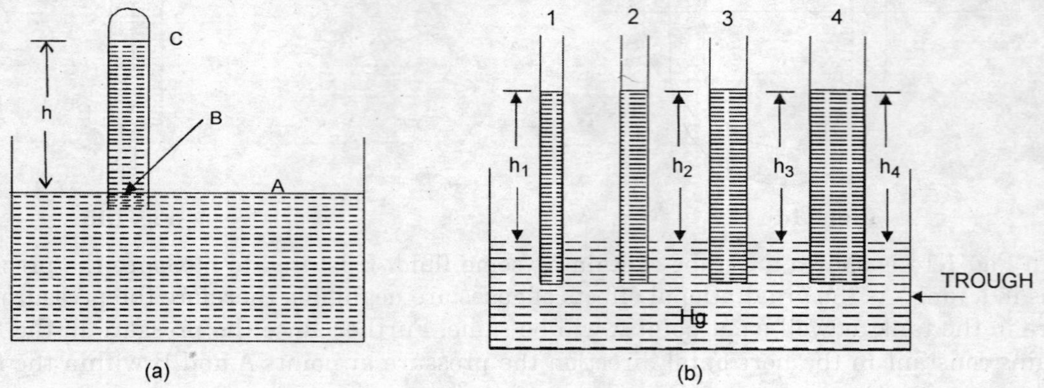

Fig. 1.16

The pressure at the point A is equal to atmospheric pressure. Further since A and B points are lying in the same horizontal plane

$$P_A = P_B = P_{atm} \qquad ...(1.24)$$

The pressure at the point *C* ca y mercury vapour above the point *C* in the tube

$P_C =$ (...(1.25)

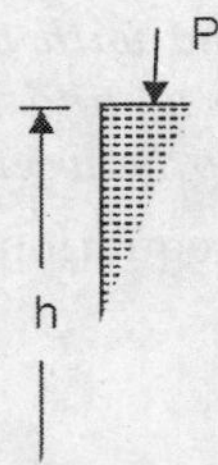

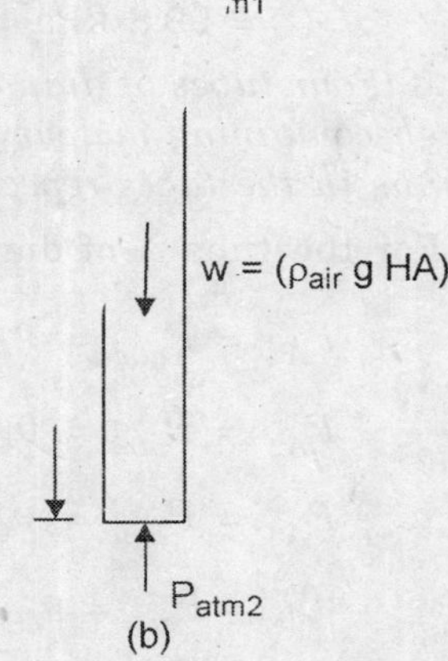

Fig (1.17 the mercury column *BC* in Fig. (1.16), which is in ng the algebric sum of the vertical forces and equati

from (1.25)

or ...(11.26)

ıercury

n due to gravity

mercury column above this free surface in the trough.

of tube and cross sectional area of tube have no effect of the barometer.

e

ntly "STANDARD ATMOSPHERE", which is defined as the ı of mercury 760 *mm* in height at 0° *C* (ρ of mercury = 13.595) acceleration $g = 9.807$ m/s^2.

ote is that the atmospheric pressure at any location is the weight at that location. Since the weight of air column will change with on and the weather condition the barometric pressure will change ion as well as the weather condition.

t a certain location the reading on the barometer in 750 mm of the density of mercury at 10° C is 13.58 and acceleration due to find the pressure at the location.

Solution : $P_{atm} = \rho gh$

$$= \left(13.58\frac{\text{kg}}{\text{m}^3}\right)(9.8\ \text{m/sec}^2) \times \frac{750}{1000}\left(\frac{1\text{kPa}}{1000\text{N/m}^2}\right)$$

$= 99.8$ kPa

Example 1.3. *Four tubes of diameters d_1, d_2, d_3 and d_4 filled with mercury are monted into an open trough containing mercury simultaneously. If h_1, h_2, h_3 and h_4 are the height of the mercury columns in the tubes repectively, find the relationship between h_1, h_2, h_3 and h_4.*

Solution : For the tube A of diamether d, we have from equation. (1.26)

$$P_{A1} = P_{atm} = \rho gh_1 \quad ...(1.27)$$

Similarly

$$P_{A2} = P_{atm} = \rho gh_2 \quad ...(1.28)$$

$$P_{A3} = P_{atm} = \rho gh_3 \quad ...(1.29)$$

$$P_{A_4} = P_{atm} = \rho gh_4 \quad ...(1.30)$$

Thus from equations (1.27) through (1.30) we get

$$h_1 = h_2 = h_3 = h_4$$

Example 1.4. *A pressure gauge connected to a tank reads 3.5 bars at a location where the reading on the barometer is 750 mm of mercury. Determine the absolute pressure in the tank. Take density of Hg = 13590 kg/m^3 and g = 9.807 m/sec^2.*

Solution : The atmospheric pressure can be calculated from equation (1.26)

$P_{atm} = \rho gh$

$$= \left(13590\frac{\text{kg}}{\text{m}^3}\right)\left(9.807\frac{\text{m}}{\text{s}^2}\right)(0.75\ \text{m})\left(\frac{1N}{1\ \text{kg}\cdot\text{m/s}^2}\right)\left(\frac{1\ \text{bar}}{10^5\ \text{N/m}^2}\right)$$

$= 1.00$ bar

Thus $P_{atm} = P_{gauge} + P_{atm}$

$= 3.5 + 1.0 = 4.5$ bars

Example 1.5. *A pressure gauge connected to a tank reads 500 kPa at a location where atmospheric pressure is 97.8 kPa. What is the absolute pressure in the tank?*

Solution : $P_{abs} = P_{gauge} + P_{atm}$

$= 500 + 97.8 = 597.8$ kPa

Example 1.6. *A vacuum gauge connected to a tank reads 25 kPa at a location where the barometric reading is 760 mm of Hg. Determine the absolute pressure in the tank. Take ρ_{Hg} = 13590 kg/m^3 and g = 9.807 m/s^2.*

Solution : $P_{atm} = \rho gh$

$$= (13590)\left(\text{kg/m}^3\right)\left(9.807\ \text{m/s}^2\right)(.760\ \text{m})\left(\frac{1N}{1\ \text{kg m/s}^2}\right)\left(\frac{1\ \text{kPa}}{1000\ \text{N/m}}\right)$$

$= 101.266$ kPa

Then from equation (1.21) we have

$$P_{abs} = P_{atm} - P_{vac} = 101.266 - 25 = 76.266\ \text{kPa}$$

Example 1.7. *Barometer is used to determine the altitude of an aeroplane. The ground control room reports a barometric reading of 750 mm Hg while the reading taken by the pilot is 687 mm Hg Determine the altitude of the plane. Take the average density of air as 1.2 kg/m³ and g = 9.8 m/s² and Density of Hg = 13, 600 kg/m³.*

Solution : Let the atmospheric pressure at the plane level be = P_{atm_1} and that at ground level = P_{atm_2}.

Then $P_{atm_1} = \rho g h_1$

$$= \left(13600\ \text{kg/m}^3\right)\left(9.8\ \text{m/s}^2\right)(0.687\ \text{m})\left(\frac{1N}{1\ \text{kg}\cdot\text{m/s}^2}\right)\left(\frac{1\text{kPa}}{1000\ \text{N/m}^2}\right)$$

$$= 91.56\ \text{kPa}$$

Similarly $P_{atm_2} = \rho g h_2 = \left(13600\ \text{kg/m}^3\right)(9.8\ \text{m/s})(0.750\ \text{m})\left(\frac{1\text{N}}{1\ \text{kg}\cdot\text{m/s}^2}\right)\left(\frac{1\text{kPa}}{1000\ \text{N/m}^2}\right)$

$$= 99.96\ \text{kPa}$$

Now consider an air column of unit area of cross-section between the ground and plane and drawing its free body diagram and considering the equilibrium of this air column and resolving the forces along the vertical axis. (Refer to Fig. 1.17 b)

Let the W = weight of air column of Height

$= \rho_{air}\, g(HA)$

where A = Area of cross-section of air column

$= 1\ \text{m}^2$

H = Height of aeroplane

Therefore $W = (1.2\ \text{kg/m}^3)\ (9.8\ \text{m/s}^2)\ (\text{Hm})(1\text{m}^2)\left(\frac{1N}{1\ \text{kg}\cdot\text{m/s}^2}\right)\left(\frac{1\text{kPa}}{100\ \text{N/m}^2}\right)$

$= 0.01176\ \text{H kg}$

For the equation of air column, resolving the forces along steal direction we have

$$AP_{at_{m1}} + W = P_{at_{m2}} A$$

or $W = \left(P_{at_{m2}} - P_{at_{m1}}\right) A$

$= (99.96 - 91.56) \times 1\ \text{kg} = 8.4\ \text{kg}$

$\therefore$ $0.1176\, H = 8.4$

or $H = \dfrac{8.4}{0.1176} = 714.28\ \text{m}$

Example 1.8. *A diver is at a depth of 30 m below the free surface of sea. If the atmospheric pressure is 101 kPa find the pressure exerted on the diver. Take specific gravity of sea water as 1.03 and g = 9.807 m/s² and density of water as 1000 kg/m³.*

Solution : Let ρ = density of sea water

= (specific gravity of sea water) × (density of water)

$= 1.03 \times 1000 = 1030\ \text{kg/m}^3$

P = Pressure exerted on diver

Then $P = P_{atm} + \rho g h$

$$= 101 \text{ kPa} + (1030 \text{ kg/m}^3)(9.807 \text{ m/s}^2)\ (30 \text{ m})\left(\frac{1 \text{ kPa}}{1000 \text{ N/m}^2}\right)$$

$$= 404.4 \text{ kPa}$$

Example 1.9. *Lower half of cylindrical tank 8 m high is filled with water whose density is 1000 kg/m³, upper half contains oil of specific gravity 0.8. If g = 9.807 m/s² find the pressure difference in kPa between the top and bottom points of the tank. Fig. 1.18*

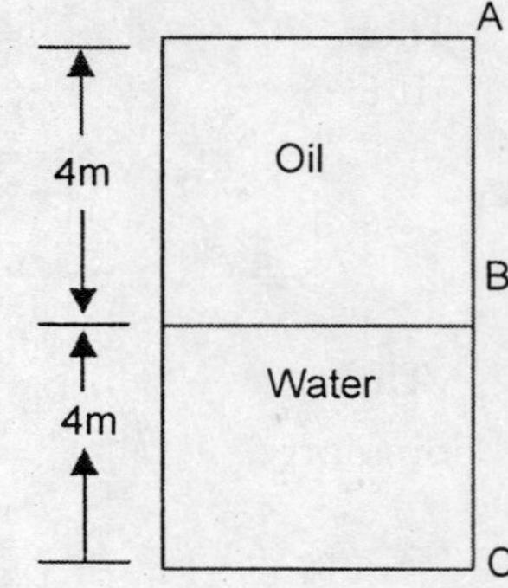

Fig. 1.18

Solution : Specific gravity of oil = 0.8

Therefore $\rho_{oil} = 0.8 \times 1000 = 800 \text{ kg/m}^3$

$$\Delta P_{AB} = \rho_{oil}\, g h_{oil}$$

$$= (800 \text{ kg/m}^3)(9.807 \text{ m/s}^2)\ (4.0 \text{ m})\left(\frac{1.0 \text{ kPa}}{1000 \text{ N/m}^2}\right)$$

$$= 31.38 \text{ kPa}$$

$$\Delta P_{BC} = \rho_{water}\, g h_{water}$$

$$= (1000 \text{ kg/m}^3)(9.807 \text{ m/s}^2)\ (4.00 \text{ m})\left(\frac{1 \text{ kPa}}{100 \text{ N/m}^2}\right)$$

$$= 39.228 \text{ kPa}$$

$$\Delta P_{AC} = \Delta\rho_{AB} + \Delta\rho_{BC}$$

$$= 31.38 + 39.228$$

$$= 71.608 \text{ kPa}$$

Example 1.10. *A vertical piston cylinder device contains gas at a pressure of 300 kPa. The atmospheric pressure is 10⁵ Pa. If the piston area is 200 mm² find the mass of the piston. Take g = 9.807 m/s² and assume the piston cylinder assembly is frictionless.*

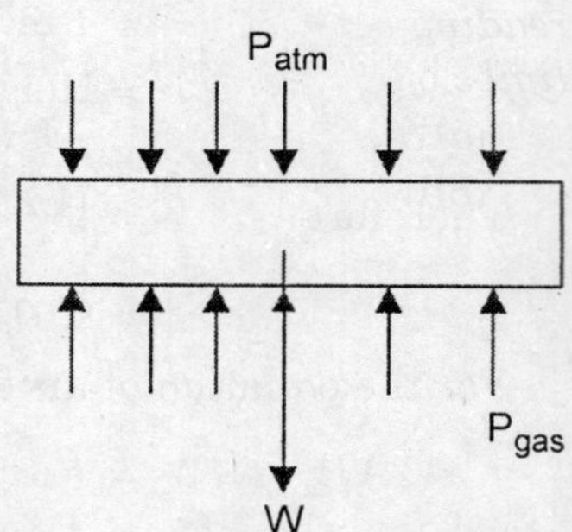

Fig. 1.19 (a)

Solution : In Fig. 1.19 (*a*) is shown the free body diagram of the piston of weight. If A is the area of the piston and m its mass then for the equilibrium of the piston and resolving the forces along the vertical direction we have

$$A\,P_{atm} + mg = A\,P_{gas}$$

or $$mg = (P_{gas} - P_{atm})\,A \qquad \text{...(1.31)}$$

Given $A = 2000 \text{ mm}^2 = 2000 \times 10^{-6} \text{ m}^2$

$P_{atm} = 10^5\ Pa = 100 \text{ kPa}$

$P_{gas} = 300 \text{ kPa}$

Therefore from equation (1.31)

$$(m \text{ kg})\ (9.807 \text{ m/s}^2) = (300 - 100)\ (10^3\ Pa) \times 2000 \times 10^{-6} \text{ m}^2$$

or $$m = \frac{200 \times 10^3 \times 2000 \times 10^{-6}}{9.807} \text{ kg}$$

$$= 40.7 \text{ kg}$$

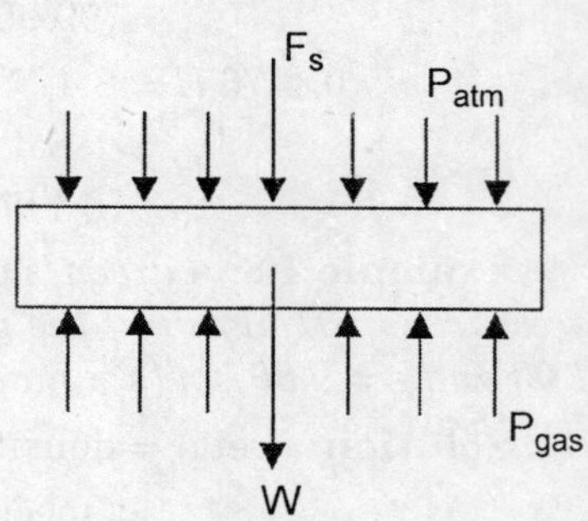

Fig. 1.19 (b)

Example 1.11. *A vertical frictionless piston cylinder device contains a gas. The mass of the piston is 4.5 kg and cross-section area 2450 mm². A compressed spring exerts a force of 70 N on the piston. If the atmospheric pressure is 100 kPa, find the pressure of the gas inside the cylinder. Take g = 9.807 m/s².*

Solution : In Fig. 1.19 (*b*) is shown Free Body diagram of the piston of weight *W*, and cross-section are *A* and F_s = spring force.

For the equilibrium of the piston and balancing the forces in the vertical direction

$$F_s + (P_{atm})A + mg = A\,P_{gas} \qquad \text{...(1.32)}$$

It is given
$$P_{atm} = 100\ kP_a$$
$$A = 2450\ \text{mm}^2$$
$$= 2450 \times \text{mm}^2$$
$$= 2450 \times 10^{-6}\ \text{m}^2$$
$$F_s = 70\ \text{N}$$

Therefore substituting values in equation (1.32)

$$P_{gas} = \frac{1}{A}\left[AP_{atm} + F_s + mg\right]$$
$$= P_{atm} + \frac{F_s}{A} + \frac{mg}{A}$$
$$= 100\ \text{kPa} + \left[\frac{70}{1000 \times 2450 \times 10^{-6}}\right] + \left[\frac{4.5 \times 9.807}{2450 \times 10^{-6}}\right]\frac{1}{1000}$$
$$= 100\ \text{kPa} + 28.57\ \text{kPa} + 18.01\ \text{kPa}$$
$$= 146.58\ \text{kPa}$$

Example 1.12. *To a gas tank both a pressure gauge and a manometer are attached. The reading on the pressure gauge is 80 kPa. What will be the level difference of the fluid in the two limbs of the manometer in the following two cases?*

(a) When the fluid is water (density 1000 kg/m³)

(b) When the fluid is mercury (density 13600 kg/m³)

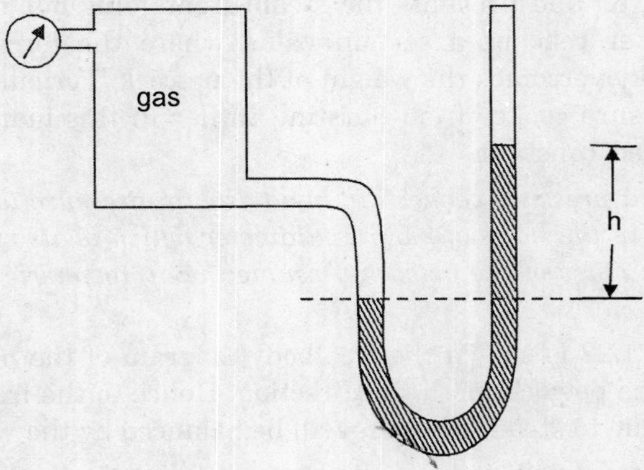

Fig. 1.24

Solution : $P_{gauge} = \rho g h$

or $$h = \frac{P_{gauge}}{\rho_{Hg} g}$$

(a) For water $h = \dfrac{P_{\text{gauge}}}{P_w\, g}$

$$= \frac{80 \text{ kPa}}{(1000 \text{ kg/m}^3)\,(9.807 \text{ m/s}^2)} \times \left(\frac{1000 \text{ kg/ms}^2}{1 \text{ kPa}}\right)$$

$= 8.16$ m

(b) For mercury $h = \dfrac{P_{\text{gauge}}}{P_{Hg}\, g} = \dfrac{80}{13600 \times 9.807} \times 1000 = 0.6$ m

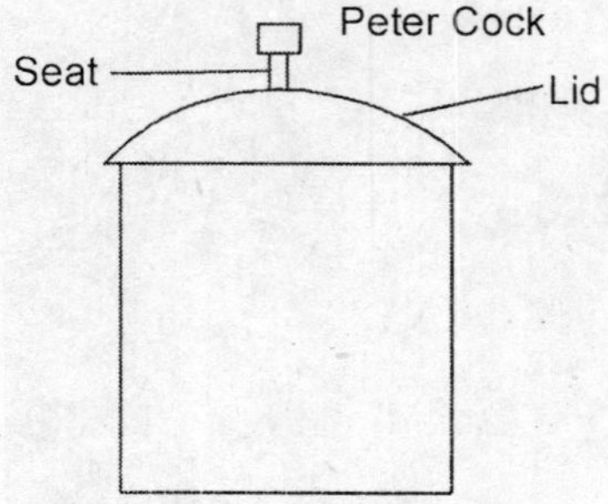

Fig. 121 Pressure cokker

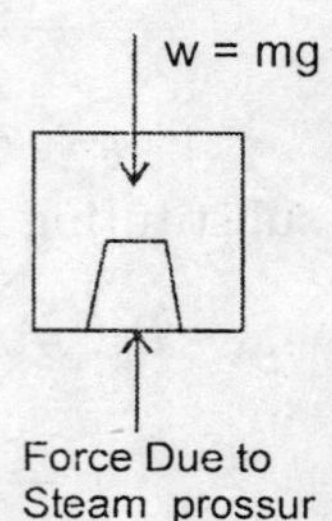

Fig. 1.22 Free body dagram of dead weight

1.14.5 Domestic Pressure Cooker

Domestic pressure cooker cooks food at a much faster rate than ordinary pan because it is maintained at higher pressure and temperature. The lid of the cooker is well sealed and the steam from the cooker can escape only through an opening in the middle of the lid. On the seat of the opening is made to sit a certain mass, which is called "PETERCOCK" or more popularly as "PETCOCK" and prevents the steam from going out until the pressure of the steam inside the cooker reaches a certain value where the force exerted by the steam pressure on the petcock overcomes the weight of the petcock. Periodic escape of steam in this way prevents the pressure cooker from bursting rather in this manner the pressure inside the cooker a maintained constant.

Example 1.12. *In a pressure cooker the inside steam pressure is maintained at 100 kPa gauge and the opening in the lid is of 2.5 mm diameter which is also the size of the opening in the seat. Determine the mass of the petcock. Assume the atmospheric pressure to be 98.7 kPa. Take* $g = 9.81\ m/s^2$.

Solution : In Fig. 1.22 is shown the free body diagram of the petcock. The atmospheric pressure is acting on the petcock in all the direction. Hence in the free body diagram it is not considered. The force due to steam pressure will be balanced by the weight of the petcock.

If m is the mass of petcock, then the balance of forces in the vertical direction, on the petcock gives

$$mg = P_{\text{gauge}} \text{ (area of opening)}$$

or

$$m = \frac{P_{\text{gauge}} \times A}{g}$$

$$= \frac{(100 \text{ kPa})\left(\frac{\pi}{4} \times 2.5^2 \times 10^{-6} \text{ m}^2\right)}{(9.81 \text{ m/s}^2)}\left(\frac{1000 \text{ kg.m/s}^2}{1 \text{ kPa}}\right)$$

$$= 0.05 \text{ kg}$$

1.15 TEMPERATURE AND THE ZEROTH LAW OF THERMODYNAMIC'S

Despite the fact that everybody is familiar with the TEMPERATURE as a measure of "coldness" or "hotness", it is very difficult to exactly define this term. However based on physiological sensations the level of temperature is quantitatively expressed commonly by terms like "freezing cold, "cold", "warm", "hot", "steaming hot" and "red hot". But are cornet assign any numerical values the temperatures corresponding to the terms mentioned above. Rather on senses may be misleading sometimes. The metallic heads of a spade, for example, will feel much colder than its wooden handie even though both are at the same temperature.

Over a period of time, scientists have established after long experimentation that there are several properties of materials which change with temperature in a repeatable and predictable manner. This fact has led to the establishment of the basis for the accurate measurement of temperature. For example the commonly used mercury-in-glass thermometer is based on the uniform expansion of mercury with temperature. Similarly temperature is also measured by using several other temperature-dependent properties.

In day to day life it is the common experience of even a layman that a cold drink left on a table will slowly work up to room temperature and on the contrary a cup of hot tea left on the same table will eventually cool down to room temperature. These facts lead us to conclude that when two bodies, which are at different temperatures are brought into contact then the heat is transferred from the body at higher temperature to the one at lower temperature until both the bodies attain the same temperature. At that point the transfer of heat from one body to the other ceases and the two bodies are said to have reached a stage which is called "Thermal Equilibrium" Thus for thermal equilibrium". The only requirement is the "Equality of temperature".

The Zeroth Law of thermodynamics" states that if two bodies A and B are in thermal equilibrium with a third body C then they (A and B) are also in thermal equilibrium with each other. Such an obvious fact was taken for granted when the First and second Laws of thermodynamics were enunciated. However it was later on realised that this fact formed the basics law of thermodynamics and was called Zeroth Law of thermodynamics and its main contribution is to establish basis for the validity of temperature measurement. In case we treat the thermometer as a third body the "Zeroth Law of Thermodynamics" can the restated as " Two bodies A and B are in thermal equilibrium if both have same temperature even though they are not in contact.

1.15.1 Temperature Scales

Several scales have been devised by scientists. The scales are based on some easily reproducible states such as freezing and boiling points of water. Alternatively these are also called the "ice point" and "steam point" respectively. A mixture of water and ice which is in equilibrium with air saturated with water vapor at 1-atmospheric pressure is known as the "ice point" and the mixture of liquid water and water vapor (having no air) in equilibrium at 1-atmosphere pressure is said to be the "steam point".

Formerly two t ıt. Firstly the English system called the centigrade scale ar named after *G*. Fahrenheit who was a German Instrumen system has been adopted the temperature scale and is called Astronomer, A Celicus who devised it. In this scale "the ice" ed the values of 0°*C* and 100°*C* respectively while the corresp le are 32°*F* and 212°*F*. Both these scales are known as "tw rature values are assigned at two different points. However used in thermodynamics is the "absolute temperature sca ve temperatures and the lowest attainable temperature is lute temperature scale is called "KELVIN SCALE" named ature unit on this scale is kelvin which is designated by degree sign.

A sealed ri essure can also be used as a thermometer to measure absol called a "constant volume gas thermometer". It is based on ume the absolute temperature of a gas at low pressure is p Thus when the absolute pressure is halved so is the absolu n it can be determined that as the absolute pressure in ter approached zero the absolute temperature will also app nding reading on the celcius scale will approach –273.15°*C*. re of –273.15°C corresponds to 0.0 K on absolute temperatur onship between celcius scale and kelvin scale is given by

73.15 ...(1.33)

How s internationaly accepted, is to round the constant in equati

Duri s and measures held in 1954 the celcius scale was redefine d the absolute temperature scale. The designated single p ter which is the state at which all the three phases of wate int is assigned the value of 0.01°C. The magnitude of the emperature scale and as before the boiling point of water sure. Evidently the new celcius scale evolved in 1954 is esse

It agnitude of 1k and 1°C are identical.

T °C) ...(1.34)

I e changes on both the scales are same. In some ther ature *T* is involved and generally a question arises whe e temperature difference is involved either one can be use). However when in the relations *T* is involved then it is ill not lead to erroneous results.

process the temperature of water is raised by 20°C. Find

th ı scale.

ΔT (°C)

20°C

Thus ΔT (K) = 20

PROBLEMS

1. Explain why does a cyclist pick up speed on a down hill road even though he does not peddle. Does this violate the principle of conservation of energy?

2. A hot cup of tea at 60°C is placed on a table in on office where the room temperature is maintained at 20°C. Explain what will be the final temperature of tea if it is left on the table for more than an hour.

3. A cyclist claimed that he picked up speed on a level road even though he was not pedaling. Explain whether his claim is valid or not. Does it violate any thermodynamic law?

4. The radiator of a car is used to reject heat to the air through circulating water. Explain for the purpose of analysis the radiator is to be treated as a closed system or open system. (**Ans.** Open system)

5. A bottle of soft drink is kept in a refrigerator to cool it. Explain whether the bottle of soft-drink is an open system or closed system. (**Ans.** Closed System)

6. State whether following statements are true or false

(*a*) Mass can cross the boundaries of closed system.

(*b*) Energy can cross the boundaries of closed system. (**Ans.** False, True)

7. A convector heater is used to heat a small room in winter. Explain the process of energy transformation during this process.

(**Ans.** Electric energy is transformed into sensible internal energy)

8. Water kept in a saucepan is heated by placing it an electric hot plate. What energy transformations take place in this process?

(**Ans.** Electric energy is converted to sensible internal energy of water)

9. What are the energy transformations involved in the burning of a candle placed in a room?

(**Ans.** During the burning of candle chemical energy is converted into sensible internal energy of the air in the room)

10. State whether following are true or false :

(*a*) Macroscopic forms of energy are those which a system possesses as a whole with respect to some outside reference frame. (**Ans.** True)

(*b*) The microscopic forms of energy are those which are related to the molecular structure of the system and the degree of molecular activity, and are independent of the outside reference frame. (**Ans.** True)

(*c*) Sum of all forms of energy a system possesses is called the total energy.

(**Ans.** True)

(*d*) In the absence of magnetic, electrical and surface tension effects total energy of a system is equal to the sum of kinetic, potential and internal energies.

(**Ans.** True)

(*e*) Internal energy of a system is equal to the sum of sensible, latent, chemical and nuclear energies. (**Ans.** True)

(*f*) The sensible internal energy is due to translational, rotational and vibrational effects of the molecules. (**Ans.** True)

(*g*) Intensive properties do not depend on the size of the system but extensive properties do. (**Ans.** True) (*h*) For a system to be in thermodynamic equilibrium, the temperature has to be same throughout but pressure does not. However there should be no unbalanced pressure forces present. (**Ans.** True)

11. What is a quasi-equilibrium process? What is its importance in engineering?

12. What are isothermal, isobaric and isochoric processes?

13. What is the state postulate?

14. Explain whether the state of air in an isolated room is completely specified by the pressure and temperature. (**Ans.** Yes)

15. A pressure gauge connected to a gas tank reads 4 bars at a place where the barometric reading is 750 mm Hg. What the absolute pressure of the gas in the tank? Take P_{Hg} = 13590 kg/m^3. (**Ans.** 5 bars)

16. A vacuum gauge connected to a tank reads 35 kPa at a location where the barometric pressure reading is 760 mm Hg. Find the absolute pressure in the tank. Take specific gravity of Hg = 13.59 and ρ_{Hg} = 9.807 m/s^2. (**Ans.** 66.266 kPa)

17. The barometer connected in the cabin of the pilot of an aeroplane shows the pressure as 690 mm Hg. However the barometric reading at the ground control room is 753 mm kg. If the density of air is 102 kg/m^3 and g = 9.8 m/s^2, find the height of the plane from ground level. (**Ans.** 714 m)

18. A submarine is 100 m below the free surface of sea of the barometric pressure is 100 kPa and the density of sea water is 1030 kg/m^3, determine the pressure exerted on the surface of submarine take g = 9.807 m/s^2. (**Ans.** 111.1 kPa)

19. A vertical piston cylinder device contains gas at a pressure of 400 kPa. The atmospheric pressure is 1 bar. It piston area is 2000 mm^2 find the mass of the piston. Take g = 9.807 m/s^2 and assume the piston cylinder assembly frictionless. (**Ans.** 61.05 kg)

20. A vertical frictionless piston-cylinder device contains a gas. The cross-sectional area of the piston is 2500 mm^2 and its mass is 5 kg. A compressed spring exerts a force of 70 N on the piston. If the atmospheric pressure is 0.98 bar find the pressure of gas inside the cylinder. (**Ans.** 147.55 kPa)

21. A manometer containing oil is connected to a tank filled with air under pressure. If the difference between the oil levels in the two columns is 450 mm and the density of oil is 850 kg/m^3, find the absolute pressure of air in the tank. Take atmospheric pressure as 98 kPa and g = 9.807 m/s^2. (**Ans.** 101.75 kPa)

22. A manometer containing mercury is connected to an air duct to measure pressure of air in the duct at a certain point. If the level difference between the two columns is 15 mm and atmospheric pressure is 1 bar determine the absolute pressure in the duct. Take specific gravity of Hg as 13.6 and g = 9.807 m/s^2. The mercury level in monometer on the duct side is lower. (**Ans.** 102 kPa)

23. In a pressure cooker the inside steam pressure is maintained at 101 kPa gauge. The opening in the lid is of 2.5 mm diameter which is also the diameter of the opening in the seat of petcock. Determine the mass of the petcock. Take atmospheric pressure as 1 bar and g = 9.81 m/s^2. (**Ans.** 0.0505 kg)

24. Normal temperature of a healthy person is 37°C. What is its value in kelvin? (**Ans.** 310 k)

25. Soft drink bottle at 20°C is placed in a refrigerator. After sometime it is noticed that its temperature has dropped by 10°C. Express this temperature drop in kelvin. (**Ans.** 10 k)

❑❑❑

CHAPTER 2

PROPERTIES OF PURE SUBSTANCES

2.1 PURE SUBSTANCES

A substance having a fixed chemical composition throughout is called a "Pure Substance" water, Oxygen, Nitrogen, Helium, carbon dia-oxide are few examples of pure substances. A pure susbtance need not be a single chemical element or compound. A mixture of various chemical elements or compounds can also form a pure substance provided the mixture is homogeneous for example, air which is mixture of various gases is often considered a pure substance as it has uniform chemical composition. However a mixture of water and oil is not a pure substance as oil is not soluble in water rather it will collect over water to form the district chemical regious with different chemical compositions.

A mixture of two or more phases of a pure substance will be a pure substance as long as the chemical composition of all the phases is same. For example mixture of liquid water and ice will be a pure substance since both have same chemical composition. On the contrary mixture of liquid air and gaseous air is not a pure substance. Since the composition of liquid air is different from that of gaseous air. This is due to the fact that different constituents of air have different condensation temperatures at a specified pressure.

2.1.1 Phases of Pure Substance

From experience we know that substances exist in different phases. For example at room temperature Aluminium is solid, mercury is liquid and oxygen is a gas. Under different conditions these may exist in different phases. Even though there are three "PRINCIPAL PHASES", i.e. solid, liquid and gas, a substance may have more than one phases within a principal phase each with a different molecular structure. For example carbon may exist as diamond or graphite in solid phase. Similarly liquid Helium has two liquid phases while iron has three solid phases. Thus a phase is identified as possessing a distinct molecular arrangement which is homogeneous throughout and is separated from the others by identifiable boundary surfaces. Two phases of water, viz ice in liquid water is a good example of this.

In the scientific world it is believed that molecular bonds are strongest in solids and weaktest in gases the reason being that in solids the molecules are closely packed together while in gases they are separated by great distances. In solids the molecules are arranged in three-dimensional pattern called the lattice which is repeated throughout solid. Since the distance between the molecules in a solid is small the forces of attraction of the molecules on each other are large which keep the molecules at fixed positions within the solid. However the attractive forces between molecules turn to repulsive forces as the distance between the molecules approaches zero, thus the piling up molecules on top of each other is prevented. Even though in a solid the molecules can not move relative to each other but they do oscillate about their equilibrium position. Further the velocity of molecules during their oscillation depends upon the temperature. At sufficiently high temperatures the velocity and in turn the momentum of the molecules may reach a stage when the intermolecular forces are partially overcome and a group of molecules break away. This is what is known as the beginning of the melting process.

In the liquid phase the molecular spacing is almost of the same order as in solid phase. However the molecules are no longer at fixed position relative to each other rather groups of molecules float about each other. Further the molecules do maintain an orderly structure within each group or chunk and retain their original positions with respect to each other. However the distance between molecules slightly increases as the solid turns into liquid with water being a rare exception.

In the gaseous phase, the molecules are at a large distance apart from each other and there exists no molecular order. The gas molecules move about at random in all directions colliding with each other and the walls of the container they are in. Particularly at low densities the inter-molecular forces are very small and the collision is the only mode of interaction between the molecules. Molecules in gaseous phase are relatively at much higher energy level than those in liquid or solid phases. Thus the gas will release a large amount of energy before it can condense or freeze.

2.1.2 Phase-change Processes of Pure Substances

In Engineering field there are many practical situations where two phases of pure substances co-eixst in equilibrium. Water co-exists as liquid and vapor in a boiler. Similarly the refrigerant changes its phase from liquid to vapor in the freezer of a refrigrator. In order to explain the basic principles involved in the process of phase change of a substance we take the example of water, which is the most commonly used pure substance, in the following paragraphs. It must however, he noted that all pure substances show the same general characteristics as is exhibited by water.

(*i*) Compressed Liquid and Saturated Liquid

Consider a frectionless piston-cylinder device (assuming piston to be weightless) containing water at room temperaturative of 20°C at pressure = 1 atm, as shown in Fig. 2.1. Let this be called state 1 of the water. Evidently under these conditions water exists in liquid phase and it is called "compressed liquid" or "subcooled liquid" thereby meaning that it is not about to evaporate.

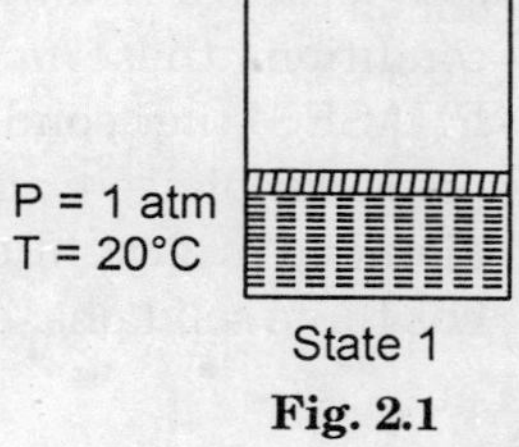

P = 1 atm
T = 20°C

State 1

Fig. 2.1

Heat is now added to water and its temperature rises to say 50°C. Due to temperature rise the liquid water will slightly expand and its specific volume will increase. As a result of this the piston will slightly move up to accomodate the expansion of water. The pressure will still continue to be 1 atm, water will still be "Compressed Liquid" and will not be at the point of evaporating.

However if we continue adding more heat the temperature of liquid water will continue to rise till it reaches 100°C. At this point the water is still in liquid phase and at 1 atm pressure. With additional heat, which may be very small, the liquid water will start evaporating. Let this state he designated as state 2 as shown in Fig. 2.2. At this state a phase change from liquid to vapor is about to start. A liquid which is about to evaporate is called "Saturated Liquid". In other words state 2 is a "Saturated liquid" state.

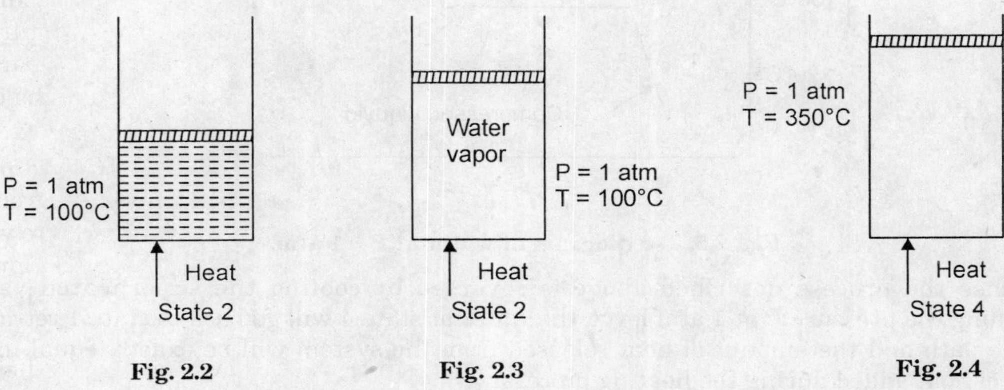

Fig. 2.2 **Fig. 2.3** **Fig. 2.4**

(*ii*) Compressed Vapour and Superheated Vapour

In case more heat is added to the water which is at state 2, the water will start boiling and the temperature will stop rising until the saturated liquid is completely evaporated. In other words the temperature will remain constant during the phase change from liquid to vapor of the entire liquid. During the evaporation of the liquid there will be large increase in the volume, if the pressure on the piston is maintained constant at 1 atm. Side by side there will be continuous decline in the liquid water level in the cylinder. Thus at any stage the cylinder will contain liquid water at the bottom upto certain height and water vapor alove it and ithe liquid and vapor phases will co-exist in equilibrium at the same temperature. As the heat is continued to be added the vaporisation of the liquid will continue until the last drop of liquid has vaporised. Then the entire liquid would have changed the phase from liquid to gaseous. During the phase change process, at constant pressure of 1 atm., there will be increase in volume as the specific volume will increase. As pointed earlier during this change of phase, the temperature will remain constant at 100°C. At this point (say state 3) the entire cylinder is with water vapor which is on the border line of liquid phase and any heat loss from the vapor, may be very small, will cause some of the water vapor to condense. A vapor which is first on the point of condensing is called "SATURATED VAPOR". Thus the state-3, shown in Fig. 2.3, is a saturated vapor state. Evidently any state between state 2 and 3 is generally "SATURATED LIQUID-STURATED VAPOURE MIXTURE" since the liquid and vapor phases co-eixst in equilibrium. However state-3 is a single phase having only vapor phase.

In case the heat is further added to the system it will result in increase of both the temperature and specific volume even though the pressure is maintained constant at 1 atm. Let us assume at a certain stage (called state 4) Fig. 2.4 the temperature is 350°C and pressure is 1 atm, there is some heat loss and the temperature falls to 300°C, then the vapor will not condense, since its temperature is alove 100°C. This state of vapor is called the "SUPERHEATED VAPOR". This vapor which is not about to condense is called a superheated vapor.

Above stated phase-change process at constant pressure in represented on T–v diagram as in Fig. 2.5.

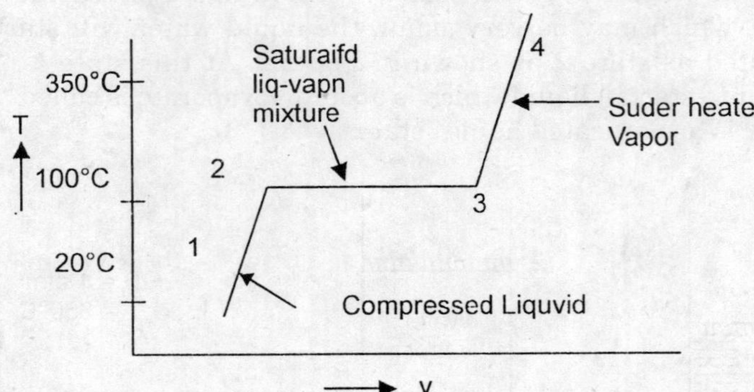

Fig. 2.5. *T–v* diagram of water at *P* = 1 atm.

In case the process described above is reversed by cooling the superheated vapor, maintaining the pressure P at 1 atm level, the water at state 4 will go back to state 1 retracing the same path and the amount of heat released from the system will be exactly equal to the amount of heat added during the heating process.

(*iii*) Saturation Temperature and Saturation Pressure

In the above discussion mentioned in para (*ii*) for which the T–v diagram is shown in Fig. 2.5, the pressure in side the cylinder was maintained constant at 1 atm (101.35 *kPa*) level. Suppose the pressure inside the cylinder were raised to 313 *kPa* by adding weights on the top of the piston the water would start boiling at 135°C. In other words the temperature at which water starts boiling is dependent on pressure. Thus at a fixed value of pressure P, the boiling point is also fixed. At a given pressure the temperature at which a pure substance starts boiling is called "Saturation Temperature" represented by T_{sat}. Likewise at a given temperature, the pressure at which pure substance starts boiling is known as "SATURATION PRESSURE" represented by P_{sat}. Thus at P = 101.35 *kPa*, T_{sat} = 100°C and at a temperature = 100°C, P_{sat} = 101.35 *kPa*.

Evidently during a phase-change process Temperature and Pressure are dependent properties and in mathematical terms, there is a definite relation between them for a pure substance of the form.

$$T_{sat} = f(P_{sat}) \quad \text{...(2.1)}$$

For water the relationship of equation (1.35) is shown in Fig. 2.6. This curve is known as "LIQUID-VAPOR SATURATION CURVE" for water.

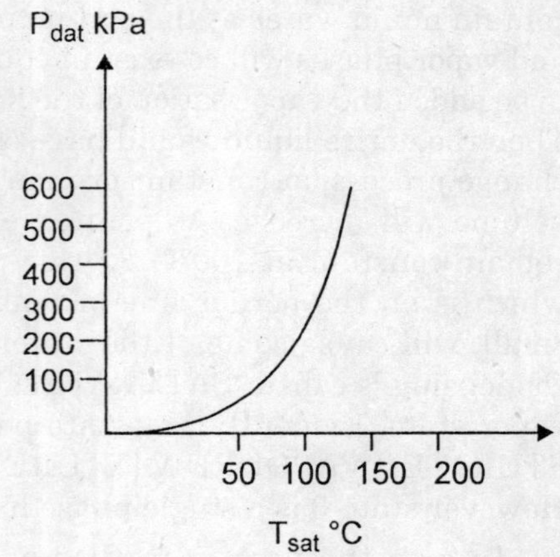

Fig. 2.6. $P_{sat} - T_{sat}$ curve for water.

In general the shape for other pure substance is similar to the one shown in Fig. 2.6.

2.1.3 Property Diagrams for Phase-change Processes

It has been disucssed above that with the change in pressure the boiling point of the liquid changes and variation of P_{sat} with T_{sat} has been shown in Fig. 2.6. However even at one constant pressure the specific volume changes with temperature. Thus in order to have better understanding of the variations in properties of a pure-substance following three sets of diagrams are drawn.

(*i*) *T–v* Diagram

Phase-change process of water has been described above and T.v. diagram at 1 atm was drawn and shown in Fig. 2.5. Let us now add weights on the top of piston so as to increase the pressure in the cylinder and repeat the process of adding heat and drawing the T-v curve at a new but increased pressure. Again the third, then 4th, 5th and so on curves can be drawn by increasing the weight on the piston. These curves are shown in Fig. 2.7 with every increase in the weight on the piston, the process follows the path which looks very much like the path followed at 1 atm. In Fig. 2.7 are shown the path curves at various paths corresponding to various pressures ranging from 1 atm (= .01 *mPa*) to 25 *mPa*.

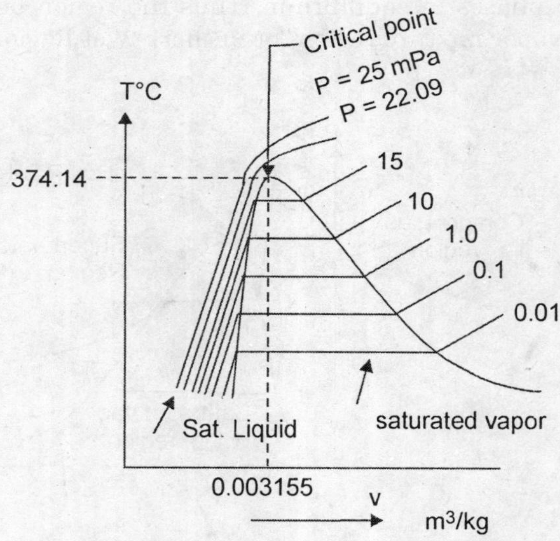

Fig. 2.7. *T–v* diagram at just *P* for water (Phase change process)

Following differences are noticed in various curves as the pressure increases from 0.01 mPa to 25 mPa.

(*a*) The water will start boiling at much higher temperature as the pressure is increased.

(*b*) At higher pressure the specific volume of the saturated liquid is larger and specific volume of saturated vapor is smaller than the corresponding values at 1 atm pressure . In other words the horizontal line which connects the saturated liquid and saturated vapor states goes on becoming much shorter as air pressure increases and ultimately it becomes a point when the pressure reaches 22.09 MPa in the case of water. This point in called the CRITICAL POINT and it is defined as the point at which the saturated liquid and saturated vapor are identical. Then the temperature, pressure and specific volume at the critical point for a pure substance are respectively known as “CRITICAL

TEMPERATURE", T_{cr}, "CRITICAL PRESSURE", P_{cr} and "CRITICAL SPECIFIC VOLUME", v_{cr}. These values for water are T_{cr} = 374.14°C, P_{cr} = 22.09 mPa and v_{cr} = 0.003155 m^3/kg.

(*c*) At pressures above critical pressure, there will be no distinct phase-change process rather there will be only one phase and the specific volume will continuously increase. It will ultimately resemble vapor and it will not be possible to tell when the change has occured. Above the critical state there is no line that separates the compressed liquid region and the super heated vapor region. However, it is internationally accepted notation to refer to a substance, which is above the critical temperature, as superheated vapor and as compressed liquid if the temperature is below critical temperature.

(*d*) The saturated liquid states in Fig. 2.7 are connected by a smooth line which is called the "SATURATED LIQUID LINE". Similarly Saturated Vapour States are connected by another line which is called the "SATURATED VAPOR LINE". Further these two lines meet each other at the "critical point", forming a dome. All the compressed liquid states lie to the left of the saturated liquid line and this region is known as "COMPRESSED LIQUID REGION". Similarly all the super heated vapor states lie on the right of the saturated vapor line and this region is known as "SUPERHEATED VAPOR REGION". In all the states which we represented by points lying under the dome, involve two phases in equilibrium. Thus the region below the dome is called "Saturated-liquid-vapor mixture region" or in short "Wet Region".

(*ii*) The *P-v* Diagram

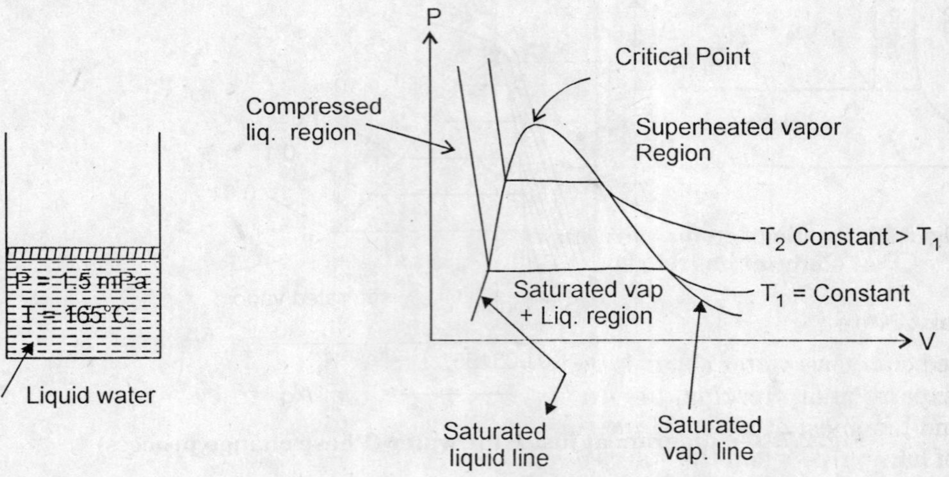

Fig. 2.8

Fig. 2.9. *P-v* diagram

In Fig. 2.8 is shown a piston-cylinder device containing liquid water at pressure of 1.5 MPa and 165°C. Water at this state will exist as compressed liquid. Let this state be designated as state 1. Let the weights on the top of the piston be removed one by one gradually, there by resulting in gradual decrease in pressure. Further water is allowed to exchange heat with the surrounding so that its temperature remains constant at 165°C. Then as the pressure decreases the specific volume of liquid water will slightly increase. When the pressure reaches the saturation pressure value corresponding to 165°C as saturation temperature (From steam tables at T = 165°C the value of P_{sat} is = 0.7 MPa the liquid water will start-boiling. Let this state be represented by point 2. During the process of vaporisation, both the temperature and pressure will remain constant as during this period no weight from the top of piston is removed but the specific volume will increase. This will continue till the last drop of water everporated which is represent by state 3. After reaching the state 3 we remove the weights from the top of the piston so that the pressure decreases gradually. This will result in gradual increase of the specific volume and side by side the temperature will also drop since from equation (2.1), $T_{sat} = f(P_{sat})$ and the process will no longer be isothermal.

Again when the process is repeated for other temperatures similar curves will be obtained for the phase-change process. Points corresponding to saturated liquid and saturated vapor states are connected with a smooth curve to give again a dome shaped boundary line in the case of *P-v*. diagram as well.

(*iii*) Extending Diagrams to Include Solid Phase

T.v. and P.v. diagrams shown in Fig. 2.7 and 2.9 can be extended to include the solid phase as well as solid-liquid and solid vapor saturation regions. Basic fundamentals which apply to liquid-vapor change processes discussed earlier apply to solid-liquid and solid-vapor phase changes also. Most of the pure solids contract during solidification or freezing but water expands as it freezes. The *P-v* diagrams for both types of substances are given in Fig. 2.10 and 2.11.

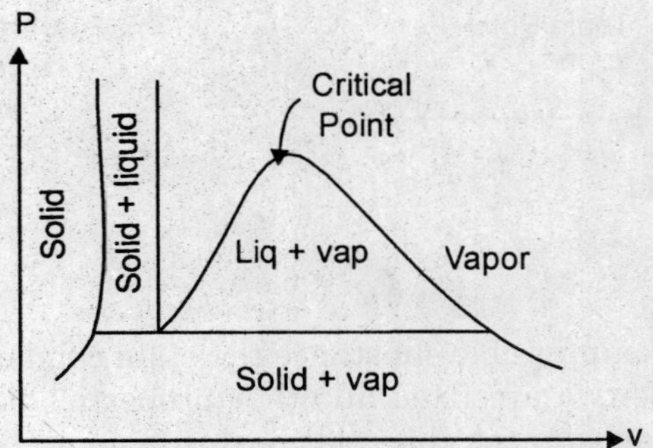

Fig. 2.10. P.v. diagram for substances that contract on freezing.

Fig 2.11. P-v diagram for substances that expand while freezing (e.g. water)

Important Note

As pointed out above water expands on freezing. This has far reaching consequences in nature. Due to expansion on freezing, the ice formed floats on the water surface. Had it been other way round like most other substance had ice contracted on freezing it will settle down at the bottom of lakes, rivers and the sun rays would not reach it to melt it. Rather with every winter season passing the ice would have accumulated at the bottom and then seriously affectining the marine life. Thus this exception in the case of water is in a way gift to the mankind.

In the above paragraphs the existence of two phases in equilibrium for a pure substance has been discussed, but under certain conditions all the three phases of a pure substance co-exist in equilibrium. (shown in Fig. 2.11) "TRIPLE LINE". On P.v. diagrams the triple-phase states form a line which is called the "TRIPLE LINE". The states corresponding to the points on the triple line have the same temperature and pressure but different specific volumes. However on P.T. diagrams (Fig. 2.12) the triple line appears as a point and, there, it is called the "TRIPLE POINT". For water the Temperature and Pressure, at triple point are 0.01°C and 0.6113 kPa respectively. In other words the three phases of water co-exist in equilibrium only at this temperature and pressure. If pressure is lower than the triple point pressure no pure substance will exist in liquid phase. Similarly no pure substance will exist in liquid phase if the Temperature is lower than the triple point temperature. In general there are two ways the substance is made to change phase from solid to vapor. Firstly it melts into liquid and then liquid evaporates to vapor phase. Secondly it may directly evaporate from solid to vapor state. The second process in known as SUBLIMATION and is possible only if the pressure is below the Triple Point pressure since the pure substance will exist in the liquid phase at this pressure. The triple point pressure for solid CO_2 is higher than atmospheric pressure. Therefore at atmospheric pressure solid CO_2 (also known as dry ice) will evaporate to vapor phase directly by sublimation.

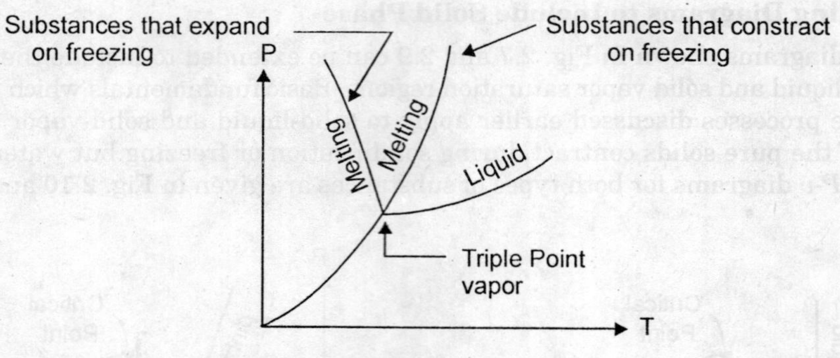

Fig. 2.12

2.2 PROPERTY TABLES

In article 1.13 it was indicated that for a simple compressible substance the state can be fixed by two independent intensive properties. If these two are fixed all other properties become dependent properties. However for majority of substances the relationship among various thermodynamic properties are too complex and cannot be expressed by simple equations. Therefore the common practice is to present these properties in the form of tables. Some of the thermodynamic properties can be easily measured, but others which cannot be measured are calculated with the help of relations between these properties and measurable properties. In this way the property tables are prepared using a convenient-to-use format. For water these tables are known as steam tables.

2.2.1 Enthalpy

In the property tables values of various variables pertaining to superheated vapor region, compressed liquid region and saturated mixture region are given. In addition two new properties viz Enthalpy *'h'* and entropy *'s'* are also given. Entropy is the property which is associated with second law of Thermodynamics and will be defined and discussed later. However Enthalpy which is a combination property will be taken up now.

In most of the thermodynamic calculations encountered in the power plant analysis as well as as in the field of refrigeration one has very frequently deal with the combination properties U + PV. Thus for the sake of convenience and simplicity this combination is introduced which a called the Enthalpy which is represented by H and is defined as

$$H = U + PV \quad \text{(kJ)} \qquad \text{...(2.2)}$$

and per unit mass

$$h = u + Pv \quad \text{(kJ/kg)} \qquad \text{...(2.3)}$$

where H is the total Enthalpy and 'h' is the specific enthalpy and u is the specific internal energy. In practice in some tables the value of internal energy u is not given but it can be very easily calculated form the following equation.

$$u = h - Pv \quad \text{...(2.4)}$$

In addition another subscript, which is commonly used is fg which denotes the difference between the saturated vapor and saturated liquid values of the same property. For example

v_f = specific volume of saturated liquid

v_g = specific volume of saturated vapor

v_{fg} = difference betwene v_g and v_f

or $= v_g - v_f$ ...(2.5)

The quantity h_{fg} is called the "Enthalpy of vaporisation (or latent heat of vaporisation). It is equal to the amount of energy needed to vaporise one kg of saturated liquid at a given pressure or temperature. It decreases as the temperature or pressure increases. It becomes zero at the critical point.

Example 2.1. *A tank contains 60 kg of saturated liquid water at 95°. Find the pressure in the tank and the volume of the tank.*

Solution : Since the saturation conditions exist in the tank, the pressure must also be the saturation pressure at 95°C. From the steam tables.

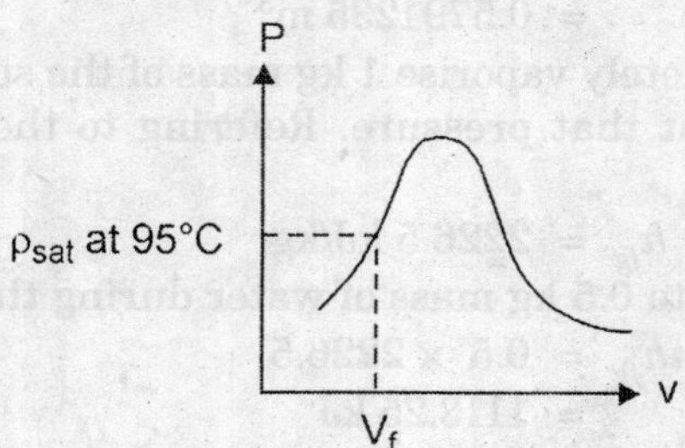

Fig. 2.13

At $T_{sat} = 95°$, we have

$$P = P_{sat} \text{ at } 95°C = 84.55 \text{ kPa}$$

Also at $T_{sat} = 95°C$, the specific volume v_f the saturated liquid at 95°C is read from the steam tables as

$$v = v_f \text{ at } 95°C = 0.001040 \text{ m}^3/\text{kg}$$

Therefore total volume of tank which is equal to the total volume of saturated liquid is

$$V = mv$$
$$= 60 \times 0.001040$$
$$= 0.0624 \text{ m}^3$$

Example 2.2. *A mean 0.5 kg of saturated liquid water is completely evaporated at a constant, pressure of 150 kPa. Find the volume change and the amount of energy added to the water, during the process.*

Solution : (*a*) In Fig. 2.14 points A and B respectively represent the state of water. The volume change per unit mass will evidently be the difference of v_g and v_f at pressure 150 kPa.

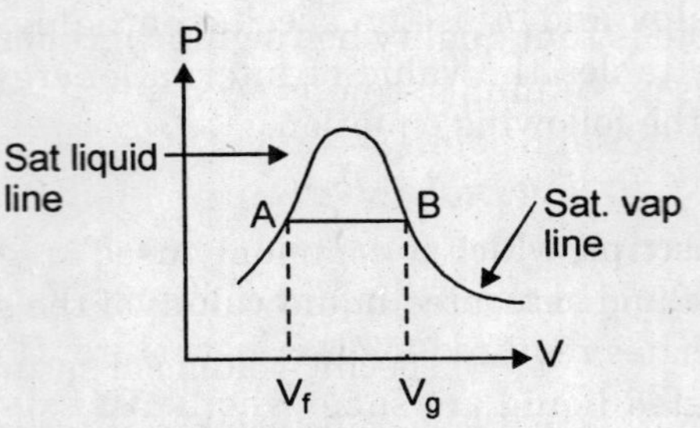

Fig. 2.14

That is $v_{fg} = v_g - v_f$

Reading the values of v_g and v_f from steam tables corresponding to pressure of 150 kPa we have

$$v_f = 0.001053 \text{ m}^3/\text{kg}$$
$$v_g = 1.1593 \text{ m}^3\text{kg}$$

Therefore substituting these values in equation for v_{fg}

$$v_{fg} = 1.1593 - .001053$$
$$= 1.158247 \text{ m}^3/\text{kg}$$

Total volume change for 0.5 kg water

$$\Delta V = 1.158247 \times 0.5$$
$$= 0.5791235 \text{ m}^3$$

(*b*) The energy added to completely vaporise 1 kg mass of the substance at a since pressure is the enthalpy of vaporisation at that pressure. Refering to the steam tables, at 150 kPa pressure h_{fg} can be read, which is

or $h_{fg} = 2226.5$ kJ/kg

Therefore total energy added to 0.5 kg mass of water during the process

$$mh_{fg} = 0.5 \times 2226.5$$
$$= 1113.25 \text{ kJ}$$

2.3 SATURATED LIQUID-VAPOR MIXTURE AND QUALITY OF MIXTURE

Let the point *A* represent the state of saturated liquid at pressure *P* and the point *B* correspond to the state of saturated vapor at the same pressure *P*. During the vaporisation process at any state represented by point *C* the substance exists as part saturated liquid and part as saturated vapor. In order to analyse the mixture the proprotions of liquid and vapor phases in the mixture must be known. In this connection a new property called the "QUALITY", *x*, is introduced. Quality is defined as the ratio of mass of vapor to the total mass of the mixture.

i.e., $$x = \frac{m_{\text{vapor}}}{m_{\text{total}}} \quad ...(2.6)$$

Where m_{vapor} = mass of vapor

$$m_{\text{total}} = \text{mass of liquid + mass of vapor}$$
$$= m_{\text{liquid}} + m_{\text{vapor}}$$
$$= m_f + m_g$$

Therefore $$x = \frac{m_g}{m_f + m_g} = \frac{m_g}{m_t} \quad ...(2.7)$$

Where $$m_t = m_f + m_g = \text{total mass} \quad ...(2.8)$$

An important point to note is that quality has significance for saturated mixtures only and it has no meaning for regions pertaining to compressed liquids or superheated vapor. In other words if the state point lies within points A and B it will have significance. At the state corresponding to point A, $m_g = 0$ and thus the value of quality x from equation (1.41) will be zero (or zero percent). Similarly for the state represented by point B, $m_f = 0$ and then the value of quality $x = 1$ or (100%). In the saturated mixtures, quality serves as one of the two independent intensive properties required to describe a state. However, it is important to note that the properties of saturated liquid are same whether it exists alone or in a mixture with saturated vapor. During the evaporation process only the amount of saturated liquid changes and not its properties. Same argument applies to saturated vapor.

In the saturated mixture, the mass of each phase is usually not known. In order to deal with such cases it is assumed that both liquid and vapor phase are mixed very well so that the total mass of mixture of the two phases is homogeneous and its average specific volume is assume as v_{av}.

The total value V is given in terms of V_f and V_g which are the total volumes of liquid and vapor phases respectively.

Then $$V = V_f + V_g \quad ...(2.9)$$

It $$m_t = \text{Total mass}$$
$$= m_f + m_g \quad ...(2.10)$$

where m_f and m_g are the respective masses of fluid and vapor present in the mixture.

Then $$V = m_t\, v_{av} \quad ...(2.11)$$
$$V_f = m_f\, v_f \quad ...(2.12)$$

From equations (2.10) and (2.11) we get

$$V_f = (m_f - m_g)\, v_f \quad ...(2.13)$$

or $$V_g = m_g v_g \quad ...(2.14)$$

From equations, (2.11), (213) and (2.14)

$$m_t v_{av} = (m_t - m_g)\, v_f + m_g v_g$$

Divide by m_t

or $$v_{av} = \left(1 - \frac{m_g}{m_t}\right) v_g + \frac{m_g}{m_t}\, v_g \quad ...(2.15)$$

Then from equations (2.6),

$$v_{av} = (1 - x)\, v_f + x\, v_g$$

or $$v_{av} = v_f + x\,(v_g - v_f)$$
$$v_{av} = v_f + x v_{fg} \quad ...(2.16)$$

where $$v_{fg} = (v_g - v_f) \quad ...(2.17)$$

Then from equation (2.16)

$$x = \frac{v_{av} - v_f}{v_{fg}} \quad ...(2.18)$$

Let us now refer is the $P - v$ diagram shown in Fig. 2.15. Let the point C represent the state of the mixture of saturated liquid and saturated vapor the specific volume at state A is v_f and that at state B it is v_g.

Then to some scale the line AB represents the vector $v_g - v_f$

i.e., $$\overrightarrow{AB} = v_g - v_f = v_{fg} \quad \text{...(2.19)}$$

Also $$\overrightarrow{AC} = v_C - v_A$$

$$= (v_{av} - v_f) \quad \text{...(2.20)}$$

From equation (2.18), (2.19) and (2.20) we get

$$x = \frac{\overrightarrow{AC}}{\overrightarrow{AB}} \quad \text{...(2.21)}$$

If x is know v_{av} can be evaluated from equation (2.16)

i.e., $$v_{av} = v_f + xv_g \quad \text{...(2.22)}$$

Proceeding on similar lines and following similar equations for the Internal Energy and Enthalpy of a saturated mixture can be derived.

$$u_{av} = u_f + xu_{fg} \quad \text{...(2.23)}$$

$$h_{av} = h_f + xh_{fg} \quad \text{...(2.24)}$$

Values of the average properties that is specific volume v, or internal energy u, or enthalpy h of the mixture are always between the values of saturated liquid and saturated vapor *i.e.,*

$$v_f < v_{av} < v_g \quad \text{...(2.25}a\text{)}$$

$$u_f < u_{av} < u_g \quad \text{...(2.25}b\text{)}$$

$$h_f < h_{av} < h_g \quad \text{...(2.25}c\text{)}$$

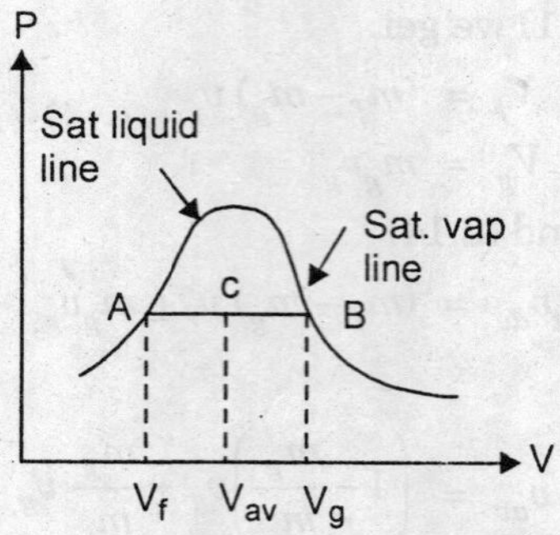

Fig. 2.15

Evidently a reference to Fig. 2.15 shows that the point representing saturated mixtures will be under the saturation curve. In other words if we know the quality and the Saturation Pressure or Saturation Temperature the values of the appropriate property of the saturated liquid and saturated vapor can be read from the table and then calculate the value of the corresponding property of the saturated mixture.

Example 2.3. *A tank contains 12 kg of water at 85°C. If 9 kg of water is in liquid form and rest in vapor form, find (a) the pressure in the tank and (b) the volume of the tank.*

Solution : (*a*) Since the two phases viz liquid and vapor are in equilibrium, the mixture must he saturated mixture. Accordingly the pressure will be the saturation pressure at the given temperature *i.e.*, 85°C. The state of the mixture is represented by point *C* in the *T–V* diagram (Fig. 2.16).

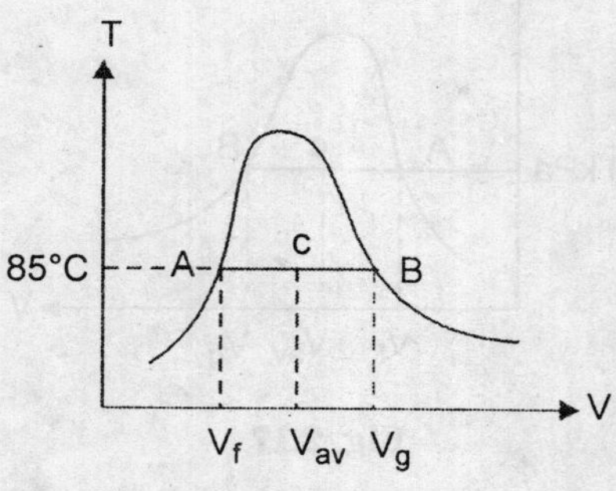

Fig. 2.16

From steam tables 4 the saturation pressure corresponding to saturation temperature of 85°C is

$$P = P_{\text{sat – at 85°C}} = 57.83 \text{ kPa}$$

(*b*) At 85°C we see from tables

$$v_f = 0.001033 \text{ m}^3/\text{kg}$$

$$v_g = 2.828 \text{ m}^3/\text{kg}$$

Method 1 : Therefore,

$$\begin{aligned} V &= V_f + V_g \\ &= v_f m_f + v_g m_g \\ &= (0.001033) \times 9 + (2.828) \times 3 \\ &= 0.009297 + 8.784 \text{ m}^3 \\ V &= 8.493297 \text{ m}^3 \end{aligned}$$

Method 2 : In this method the quality of the mixture is asertained first and in gensal this method will be useful specially in those cases where the mass of the two phases in the mixture is not known.

From equation (2.7)

$$x = \frac{mg}{mf} = \frac{3}{12} = 0.25$$

$$\begin{aligned} v_{av} &= v_f + xv_{fg} \\ &= 0.001033 + 0.25 \times (v_g - v_f) \\ &= .001033 + 0.25 \times (2.828 - .001033) \\ &= 0.70777475 \text{ m}^3/\text{kg} \end{aligned}$$

Therefore

$$\begin{aligned} V &= m_f v_{av} \\ &= 12 \times .70777475 \\ &= 8.493297 \text{ m}^3 \end{aligned}$$

Example 2.4. *A rigid tank of 2.5 m³ volume, contains 5 kg of saturated liquid-vapor mixture water at a pressure of 40 kPa. Determine (a) the temperature of the water (b) the quality (c) the enthalpy of the water and (d) the volume occupied by the vapor phase.*

Solution : (*a*) The state of mixture is represented by point *C* on the ρ–*v* diagram shown in Fig. 2.17. The two phases are in equilibrium.

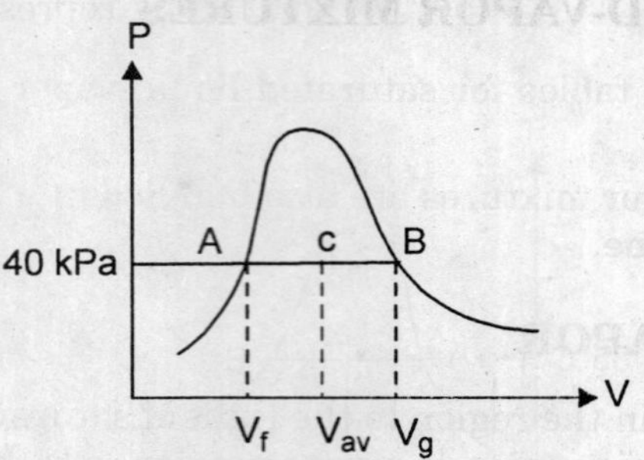

Fig. 2.17

From Steam Table we have at pressure of 40 kPa

$$v_f = 0.001027 \text{ m}^3/\text{kg}$$

$$v_g = 3.993 \text{ m}^3/\text{kg}$$

$$v_{fg} = v_g - v_f = 3.993 - 0.001027 = 3.98273 \text{ m}^3/\text{kg}$$

And
$$v_{av} = \frac{V}{m_t} = \frac{2.5}{5} = 0.5 \text{ m}^3/\text{kg}$$

Evidently $v_f < v_{av} < v_g$. Thus the mixture is a saturated mixture. Therefore the temperature will be the saturation temperature at pressure 40 kPa. From steam tables, the saturation temperature at this pressure is

$$T = T_{\text{sat - at}} \; 40 \text{ kPa}$$

$$= 75.87°\text{C}$$

(*b*) From equation (2.18)
$$x = \frac{v_{av} - v_f}{v_{fg}}$$

$$= \frac{0.5 - 0.001027}{3.98273} = 0.125295$$

(*c*) From steam table, the value of enthalpy, at pressure of 40 kPa can be read for the states corresponding to the points *A* and *B* are

$$h_f = 317.58 \text{ kJ/kg}$$

$$h_g = 2636.8 \text{ kJ/kg}$$

$$h_{fg} = h_g - h_f = 36.360.317.58 = 2319122$$

Therefore from equation (2.24)

No
$$h_{av} = h_f + x h_{fg}$$

$$= 3175.8 + 0.125295 \times 2319.22$$

$$= 608.166 \text{ (kJ/kg)}$$

(*d*) From equation (2.7)
$$x = \frac{m_g}{m_t}$$

or
$$m_g = x m_f = (0.125295)(5) = 0.626475 \text{ kg}$$

Therefore
$$V_g = m_g v_g = 0.626475 \times 3.993 = 2.5015 \text{ m}^3$$

2.4 SATURATED SOLID-VAPOR MIXTURES

In addition to the property tables for saturated liquid-vapor mixtures, tables for solid-vapor mixtures are also available.

The saturated solid-vapor mixtures are also handled in a manner similar to the saturated liquid-vapor mixture are done.

2.5 SUPERHEATED VAPOR

As already pointed earlier, in the region to the right of saturated vapor line a substance exists as superheated vapor. As this region is a vapor region, which is a single phase, temperature and pressure are no longer dependent variables. Rather they can be used as two independent properties. This is exactly what is done for preparing the property tables for this region.

Superheated vapor will prevail under any of the following conditions :

(*a*) Higher Temperature ($T > T_{sat}$ at given P)

(*b*) Lower pressure ($P < P_{sat}$ at given T)

(*c*) Higher specific volume ($v > v_g$) at given P or T)

(*d*) Higher manual energies ($u > u_g$ at given P or T)

(*e*) Higher enthalpies ($h > h_g$ at given T or P)

Example 2.5. *Find the internal energy and Enthalpy of water at 0.4 MPa and 300°C.*

Solution : From steam tables it is noted that at pressure of 0.4 MPa the saturation temperature is = 143.63°C. Therefore at the given pressure, 300°C is greater than saturation temperature.

i.e., $$T > T_{sat}$$

In otherwards the water vapor is in superheated region. Again from the steam tables, we note that at 300°C.

$$v = 2804.8 \text{ kJ/kg}$$

and $$h = 3066.8 \text{ kJ/kg}$$

Example 2.6. *At certain state water has following parameters*

$$\textit{Pressure} = 1.0\ \textit{MPa}$$

$$h = 3100\ kJ/kg$$

What is its temperature?

Solution : From steam table we notice that at pressure of 1.0 MPa the enthalpy at saturation point is 2778.1.

Since $h > h_{sat}$ the water is in superheated state. Again refering the steam table we notice following at P = 1.00 MPa.

Temp.	h
300°C	3051.2
350°C	3157.7

From these figures we infer that the temperature will be somewhere between 300°C and 350°C. The exact value will be found by interpolation

$$h_{350} - h_{300} = 3157.7 - 3051.2 = 106.5 \text{ kJ/kg}$$

$$h - h_{300} = 3100 - 3051.2 = 48.8 \text{ kJ/kg}$$

Therefore required temperature T can be calculated as follows. Calculate the change of temperature ΔT over 300°C so that the corresponding increase in h over 3051.2 is 48.8.

i.e.,

$$\Delta T = \frac{106.5}{1350 - 3001} \times 48.8$$

$$= 0.976°C$$

Therefore

$$T = 300 + \Delta T = 300 + 0.976 = 300.976 \text{ kJ/kg}$$

2.6 COMPRESSED LIQUID

Very little data is available for compressed liquids in literature. The main reason is that the properties of compressed liquids are relatively independent of pressure. The variation in the properties is very small due to change in pressure. The only property which is affected to some extant is the enthalpy.

It is the accepted practice to treat compressed liquid as saturated liquid at the given temperature. This primarily is due to the fact that the properties of compressed liquids depend more on temperature than pressure.

Generally a compressed liquid is characterised by the following :

(*a*) Lower temperature ($T < T_{sat}$ at a given P)

(*b*) Higher pressure ($P > P_{sat}$ at a given T)

(*c*) Lower specific volume ($v < v_f$ at a given P or T)

(*d*) Lower internal energies ($v < u_v$ at a given P or T)

(*e*) Lower enthalpies ($h < h_{sat}$ at a given P or T)

Example 2.7. *A rigid tank with a volume 2.5 m³ contains 5 kg of saturated liquid-vapor mixture of water at 75°C. Water is then slowly heated. Find the temperature at which the liquid in the tank is completely vaporised. Also show, by drawing free hand sketch, the process on T–V diagram with respect to saturation line.*

Solution : Since the tank is rigid, it is a constant volume process.

That is

$$v = \frac{V}{m} = \text{constant}$$

$$= \frac{2.5(\text{m}^3)}{5(\text{kg})} = 0.5 \text{ m}^3/\text{kg}$$

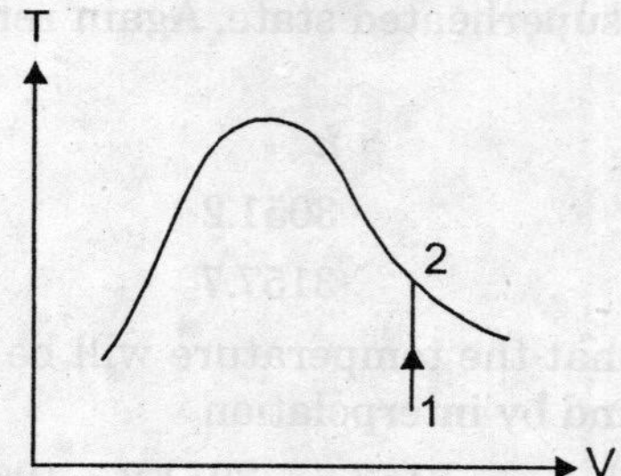

Fig. 2.18

At the final stage the tank will contain saturated vapor only thus $v_2 = v_g = 0.5$ m^3/kg (because v is constant). From steam tables the values are as follows :

Temp	v_g
140°C	0.5089
145°C	0.4463

The value of T corresponding to $v_g = 0.5$ can be calculated by interpolation

$$T - 140 = \Delta T \text{ corresponding } v_g = 0.5$$

For $\Delta T = 145 - 140 = 5$

$$(\Delta v_g)_1 = 0.5089 - .4463 = 0.00626$$

$$\therefore \quad (\Delta v_g)_2 \text{ for } \Delta T = .5089 - .5 = .0089$$

That is change in v_g, is 0.0626 for = 5° change the T

Therefore corresponding to 0.0089 change in v_g, the

$$\Delta T = \frac{5}{.0626} \times .0089 = 0.7108$$

$$T = 140 + \Delta T = 140 + .7108 = 140.7108°\text{C}$$ **Ans.**

Example 2.8. *Find the internal energy of compressed water at 100°C and pressure 10.0 MPa by the following two methods and then find the error involved.*

(a) Using the data from tables

(b) By treating it as saturated liquid and using table for the same

Solution : (*a*) From the steam tables at 100°C the saturation pressure of water is 0.10135 MPa which is less than the given pressure of 10.0 MPa. Thus the liquid water is in compressed liquid state.

Then from property tables

At P = 10 MPa and T = 100°C, the value of

$$u = 416.12 \text{ kJ/kg}$$

(*b*) If the liquid is treated as saturated liquid, then from table we have

$$u = u_f \text{ at } 100°\text{C} = 418.94 \text{ kJ/kg}$$

(*c*) Error involved

$$= \frac{918.94 - 416.12}{416.12} \times 100 = 0.6777\%$$

2.7 THE IDEAL GAS EQUATION OF STATE

In the preceding paras the properties of the pure substances have been discussed and property tables based on experimental observations have been brought out. It is one way of reporting the property data. This data, though very accurate and reliable, but is very bulky.

Accordingly another way which is simple and easy to use is to have simple equation which gives the relationship between the pressure, temperature and specific volume. Any equation which gives the relationship between Pressure, Temperature and volume of a substance is called the "EQUATION OF STATE". Several equations have been put forth. The simplest and most popular equation of state for gas phase is known as "IDEAL GAS EQUATION OF STATE". This equation gives the relationship between P, v and T and predicts the properties of the gas quite accurately within certain regions.

Gas and vapor are usually taken synonymous with each other. Vapor phase of a substance is called gas if it is above the critical temperature. On the contrary vapor generally is a gas which is very near to the state of condensation.

Robert Boyle described his famous law in 1662 known as Boyle's Law, which states that the pressure of gases is inversely propertional to their volume. Later on in earlier part of 18th century J. Charles observed on the basis of extensive experiments that at low pressures the volume of a gas is proportional to its temperature.

If these two laws are grouped together, then we get

$$P = R\left(\frac{T}{v}\right)$$

or

$$Pv = RT \qquad \text{...(2.26)}$$

where P is the absolute pressure

T is the absolute temperature

v is the specific volume

R is the constant of proportionality and called 'GAS CONSTANT'

Equation (2.6) is called the "IDEAL GAS EQUATION OF STATE" or in short "IDEAL-GAS EQUATION". A gas which obeys this relationship is called an "IDEAL GAS".

The gas constant R is different for each gas and it is determined from the relationship

$$R = \frac{R_u}{M} \text{ [kJ/(kg}\cdot\text{K) or kPa} - \text{m}^3\text{/(kg}\cdot\text{K)]} \qquad \text{...(2.27)}$$

Where R_u is Universal Gas constant

M is molar mass of the gas also known as molecular weight

Further R_u is same for all substances and its value is

$$R_u = 8.314 \text{ kJ/(k mol} - \text{K) or kPa} - \text{m}^3\text{/(k mol} - \text{K)} \qquad \text{...(2.28)}$$

or

$$= 0.08314 \text{ bar m}^3\text{/(k mol} - \text{K)}$$

The molar mass M is defined as the mass of one mole if the substance in grams, which is also called GRAM-MOL. the abbreviation for this is g mol. Similarly mass of one kilogram mole of the substance is called kilogram mole which is abbereviated on kg mol.

Evidently if m is the mass of the system and the member of moles is N and M is the molar mass, then

$$m = MN \text{ (kg)} \qquad \text{...(2.39)}$$

The values of R and M for various substances are available in the form of tables.

Further ideal gas equation (2.6) can also be written in following different forms.

$$Pv = RT$$

or

$$P\left(\frac{V}{m}\right) = RT$$

or

$$Pv = mRT \qquad \text{...(2.30)}$$

Again

$$mR = (MN)R$$
$$= N(MR)$$
$$= MR$$

Therefore from equation (2.10)

$$Pv = NR_uT \qquad ...(2.31)$$

Also is $\bar{v}$ is the molar specific volume (volume per unit mole in m^{3}k mol, then

$$V = N\bar{v}$$

Thus equation (2.31) takes the from

$$PN\bar{v} = NR_uT$$
$$P\bar{v} = R_uT \qquad ...(2.32)$$

The relationship between properties of an ideal gas at two different states designated by sufixes 1 and 2 respectively can easily be done by writing the equation (2.30) twice for the same mass of the substance.

$$m = \frac{P_1V_1}{RT_1} = \frac{P_2V_2}{RT_2}$$

or

$$\frac{P_1V_1}{T_1} = \frac{P_2V_2}{T_2} \qquad ...(2.33)$$

As stated above an ideal gas is one which obeys the relationship given in equation (2.26). Evidently it is an imaginary substance. However, it has been experimentally observed that in the case of real gases the relationship of equation (2.24) holds good approximately at low densities. Since at low pressure and high temperatures the density of a gas decreases, the real gas will behave like an ideal gas under low pressures and high temperatures. A question arises as what constitutes the low pressure and high temperature quantitatively. In the case of water vapor it has been experimentally observed that at "pressures below 100 kPa" water vapor can be treated as an ideal gas regeradless of its temperature with error less than 0.1 percent. On the other hand at higher pressures of water vapor is assumed as an ideal gas it will result in very erroneous results and the error will be still more in the vicinity of critical point and the saturated vapor line. Therefore where pressures involved are very low, for example in air conditioning applications the water vapor can be treated as an ideal gas without significant error. However in the case of applications in steam power plants, the pressures involved are very high, then the assumption of water vapor as ideal gas will lead to eroneous results and as much it must not be used. Instead the use of steam tables, data which is derived from experimental results is recommended.

Similarly within practical range of applications, many gases such as Oxygen, Nitrogen, Helium, Argon, Neon, Air, Krypton and even heavier gases like carbon dioxide are treated as ideal gases which entail an error less than one percent.

Example 2.8. *Determine the mass of air in a class room of size 8 m × 16 m × 6 m at 100 kPa pressure and 20°C temperature.*

Solution : Volume of air contained in class room

$$= 8 \times 16 \times 6 = 768 \text{ m}^3$$

From property tables for air $R = 0.287$ kPa – m^3/(kg · K) and $$T = 20 + 273 = 293 \text{ K}$$

From equation (2.30) $$PV = mRT$$

or $$m = \frac{PV}{RT} = \frac{(100 \text{ kPa})\,(768 \text{ m}^{3})}{0.287\,(\text{k} \cdot \text{Pa} - \text{m}^3/\text{kg} \cdot \text{K})(293 \text{ K})}$$

$$= 913.296 \text{ kg}$$

2.7.1 Compressibility Factor

As pointed out earlier ideal gas equation is simple and convenient to use. However, real gases devite significantly from ideal gas behaviour near the critical point and the saturation region. This deviation from the ideal gas behaviour at a given pressure and temperature, can be accurated by inserting a correction factor z which is called the compressibility factor and defined as

$$z = \frac{Pv}{RT} \quad \text{...(2.34)}$$

or $$Pv = zRT \quad \text{...(2.35a)}$$

or $$v = \frac{zRT}{P} \quad \text{...(2.35b)}$$

However, in the case of ideal gas

$$v_{\text{ideal}} = \frac{RT}{P} \quad \text{...(2.36)}$$

This if we replace v with v actual for which equation (2.35b) applies, we get by dividing equation (2.35b) with equation (2.36).

$$z = \frac{v_{\text{actual}}}{v_{\text{ideal}}} \quad \text{...(2.37)}$$

Thus equation (2.37) is an alternative expression for z and it, therefore, leads us to (2.37) call z as the "compressibility factor". Obviously for ideal gas.

$$z = 1$$

For real gases z can be greater than or less than unity. The more is the deviation of z from unity the more will be deviation in the behaviour of real gas from that of the ideal gas. It has been stated above that most gases obey the ideal gas equation at low pressure and high temperatures. But the question arises as to what these values are quantitatively? Answer to this question is that the pressure or temperature of a substance is high or low relative to its critical temperature or pressure.

It has been experimentally observed that gases behave differently at a given pressure and temperature but they behave very much alike at "normalised" temperatures and pressures with respect to their critical temperatures and pressures. The normalisation is done as under

$$P_R = \frac{P}{P_r} \quad \text{...(2.38)}$$

$$T_R = \frac{T}{T_r} \quad \text{...(2.39)}$$

where P_R is called "REDUCED PRESSURE" and T_R the "REDUCED TEMPERATURE". Experimentally it has been observed that compressibility factor z for all gases is approximately same at the same reduced pressure and temperature. This phenomenon is called the "PRINCIPLE OF CORRESPONDING STATES". In this connection one "GENERALISED COMPRESSIBILITY CHART" has been prepared and is available for all gases. The use of this chart requires the knowledge of critical point data for repective gases the results so obtained are correct within a few percent. When P and v, or T and v are given instead of P and T, the compresibility chart is still used to determine the third property but this involves tedious that and error. Therefore in order to simplify the calculations one more reduced-property called "PSEUDO-REDUCED SPECIFIC VOLUME", v_R, is introduced and it is defined as follows

$$v_R = \frac{v_{\text{actual}}}{RT_{cr}/P_{cr}} \quad ...(2.40)$$

Note : It is worth noting that v_R is defined in equation (2.40) which is different from P_{cr} and T_{cr}. It is related to T_{cr} and P_{cr} instead of V_{cr}. For the convenience of user lines of constant v_R are also added to the compressibility charts so as to determine T and P directly.

Example 2.9. *Determine the specific volume of the Refrigerant – 12 at 0.8 MPa and 60°C, using following methods.*

(a) The Refrigerant tables

(b) The ideal gas equation of state

(c) The generalised compressibility chart

Also calculate the error involved in the methods given at (b) and (c)

Solution : (*a*) The gas constant, the critical pressure and the critical temperature of refrigerant are read from the property tables of Refrigerant-12 and these are

$$R = 0.0688\ Pa - \text{m}^3\ (\text{kg} - K)$$

$$P_{cr} = 4.01\ \text{MPa}$$

$$T_{cr} = 384.7\text{K}$$

Similarly the specific volume of refrigerant-12 at the given specific pressure and temperature can be determined from Property tables which is as follows

$$\left.\begin{array}{r} P = 0.8MPa \\ T = 60°C \end{array}\right\} \; v = 0.02525\ \text{m}^3/\text{kg}$$

Since the tables give values which have been experimentally determined, this value of v will be most accurate.

(*b*) The specific volume of the refrigerant 12 can be calculate by treating it an ideal gas and using the ideal gas equation we have

$$v = \frac{RT}{P} = \frac{(0.0688\ \text{kPa} - \text{m}^3/(\text{kg} - \text{K})\ [333\ \text{K})}{0.8 \times 1000\ (\text{kPa})}$$

$$= 0.028638\ \text{m}^3/\text{kg}$$

Therefore the error involved in treating the refrigerant as an ideal gas will be equal to

$$= \frac{.02838 - 0.2525}{0.02525} \times 100 = 2.525\%$$

(*c*) In order to determine the correction factor z by using compressibility chart, the reduced pressure and reduced temperature can be elaborated as:

$$P_R = \frac{P}{P_{cr}} = \frac{0.8 \text{ mPa}}{4.01 \text{ mPa}} = 0.19950$$

$$T_R = \frac{T}{T_{cr}} = \frac{333 \text{ k}}{384.7 \text{ k}} = 0.86561$$

Then from compressibility chart at these values of P_R and T_R

$$z = 0.88$$

Thus $v = zv_{\text{ideal}} = (0.88)\,(0.028638) = 0.02520$

% error in this result with respect to the experimental value

$$= \frac{.02525 - .02520}{.02525} \times 100$$

$$= 0.198\ \%$$

Example 2.10. *Determine the specific volume of superheated water vapor at 10 MPa and 400°C using (a) ideal gas equation, the generalised compressibility chart and (c) the steam tables. What is the error involved in first two cases.*

Solution : Refering to steam tables the values of gas constant R, critical pressure P_r and critical temperature T_r are

$$R = 0.4615 \text{ kPa} \cdot \text{m}^3/(\text{kg} \cdot \text{K})$$

$$T_{cr} = 647.3 \text{ K}$$

$$P_{cr} = 22.09 \text{ MPa}$$

Also $T = 400 + 273 = 673$ K and $P = 10$ MPa $= 10000$ kPa

(*a*) Therefore from ideal gas equation

$$v = \frac{RT}{P} = \frac{0.4615 \text{ kPa m}^3/(\text{kg} \cdot \text{K}) \times (673 \text{ K})}{10000 \text{ (kPa)}}$$

$$= 0.03106 \text{ m}^3/\text{kg} \qquad \text{...(2.41)}$$

(*b*)

$$P_R = \frac{P}{P_{cr}} = \frac{10 \text{ MPa}}{22.09 \text{ MPa}} = 0.453$$

$$T_R = \frac{T}{T_{cr}} = \frac{673 \text{ k}}{647.3 \text{ k}} = 1.04$$

Therefore, P_R lies between 0 and 1 is $0 < P_R < 1.0$

Thus refering to compressibility Chart for this range of P_R we notice for $P_R = 0.453$ and $T_R = 104$, the value of

$$z = 0.84$$

Therefore $v = (v_{\text{ideal}}) \times z$

$= (.03106) \times .84$ from (equation 2.41)

$= 0.2609 \text{ m}^3/\text{kg}$

(*c*) Refering to superheated steam tables

At $P = 10$ MPa

$T = 400°C$

The value of $v = 0.02641$ m³/kg

Therefore error in part (*a*) $= \dfrac{0.03106 - 0.02641}{0.02641} \times 100$

$= 17.61\%$

And error in part (*b*) $= \dfrac{0.02641 - 0.02609}{.02641} \times 100$

$= 11.21\%$

Example 2.11. *Steam at 400°C has specific volume of 0.02 m³/kg. Find the pressure of the steam based on (a) ideal gas equation (b) the generalised compressibility chart and (c) the steam tables. What are the errors in parts (a) and (b).*

Solution : From steam tables, the values of R, T_{cr} and P_{cr} for water are

$R = 0.4615 \text{ kPa} \cdot \text{m}^3/(\text{kg} \cdot \text{K})$

$T_{cr} = 647.3 \text{ K}$

$P_{cr} = 22.09 \text{ MPa}$

Also $T = 400 + 273 = 673$

$v = 0.02 \text{ m}^3/\text{kg}$

(*a*) From the ideal gas equation

$$P = \frac{RT}{v} = \frac{0.4615 \text{ kPa} \cdot \text{m}^3/(\text{kg} \cdot \text{K}) \times 673K}{0.02(\text{m}^3/\text{kg})}$$

$= 15529$ kPa

(*b*) $T_R = \dfrac{T}{T_{cr}} = \dfrac{673 \text{ K}}{647.3 \text{ K}} = 1.04$

And $V_R = \dfrac{v_{\text{actual}}}{RT_{cr}/P_{cr}}$

$$= \frac{(0.02 \text{ m}^2/\text{kg})\ (22090 \text{ kPa})}{[(0.4615 \text{ kPa} \cdot \text{m}^3/(\text{kg} - \text{K})]\ (647.3 \text{ K})}$$

$= 1.48$

Corresponding to $T_{cr} = 1.04$ and $v_R = 1.48$, we read from chart the value P_R which is

$P_R = 0.57$

Thus $P = P_R \cdot P_{cr} = (0.57)\ (22090)$

$= 12591$ kPa

(*c*) From superheated steam tables

At $T = 400°C$

$v = 0.02 \text{ m}^3/\text{kg}$

Corresponding $P = 12500$ kPa

Therefore error in part (*a*) $= \dfrac{15529 - 12500}{12500} \times 100 = 24.23\%$

And % error in part (*b*) $= \dfrac{12591 - 12500}{12500} \times 100 = 0.728\%$

Example 2.12. *Find the specific volume of Nitrogen gas at 10 MPa and 150 K using (a) the ideal gas equation, (b) the generalised compressibility chart, and (c) using tables where experimental values are given. Also calculate the error involved in the first two cases.*

Solution : The values of gas constant, the critical pressure and critical temperature of Nitrogen gas can be taken from the Nilwgen gas tables.

$$R = 0.2968 \text{ kPa} \cdot \text{m}^3/\text{kg} - \text{K}$$

$$T_{cr} = 126.2 \text{ K}$$

$$P_{cr} = 3.39 \text{ mPa}$$

Also $T = 150$ K and $P = 10$ MPa $= 10000$ kPa

(*a*) From ideal gas equation $v = \dfrac{RT}{P} = \dfrac{0.2966(\text{kPa} \cdot \text{m}^3/\text{kg} - \text{K})(150 \text{ K})}{10000 \text{ (kPa)}}$

$$= 0.004452 \text{ m}^3/\text{kg}$$

(*b*) From the compressibility chart

$$P_R = \frac{P}{(P_{cr})} = \frac{10 \text{ MPa}}{3.39 \text{(MPa)}} = 2.95$$

$$T_R = \frac{T}{T_{cr}} = \frac{150 \text{ k}}{126.2 \text{ k}} = 1.19$$

Therefore from compressibility chart corresponding to these values of P_R and T_R, $z = 0.54$

Therefore $v = (z)(v_{ideal})$

Now v_{ideal} as calculated in part (*a*) $= 0.004452$ m³/kg

Therefore $v = .54 \times 0.004452$

$$= 0.002404 \text{ m}^3/\text{kg}$$

(*c*) From superheated Nitrogen tables

For $P = 10$ MPa

and $T = 150$ K

Corresponding value of $v = 0.002388$ m³/kg

(*d*) Error in part (*a*) $= \dfrac{0.004452 - 0.002388}{0.002388} \times 100 = 86.43\%$

And Error in part (*b*) $= \dfrac{0.002404 - 0.002388}{0.002388} \times 100 = 0.67\%$

Example 2.13. *Determine the compressibility factor of Oxygen at pressure of 2 MPa and temperature 140 K and hence find the value of error involved in treating oxygen gas an ideal gas at this pressure and temperature.*

Solution : The critical pressure and critical temperature of oxygen from Table are

$$P_{cr} = 5.08 \text{ MPa and } T_{cr} = 154.8 \text{ K}$$

Then

$$P_R = \frac{P}{P_r} = \frac{2 \text{ mPa}}{5.08 \text{ mPa}} = 0.394$$

and

$$T_R = \frac{T}{T_{cr}} = \frac{140 \text{ k}}{154.8 \text{ k}} = 0.904$$

For these values of P_R and T_R the corresponding value of z from the compressibility chart

$\frac{T}{T_{cr}} = \frac{140 \text{ k}}{154.8 \text{ k}} = 0.904$ is read as

$$z = 0.79$$

Now error

$$= \frac{v - v_{\text{ideal}}}{v}$$

$$= 1 - \frac{v_{\text{ideal}}}{v}$$

$$= 1 - \frac{1}{z} = 1 - \frac{1}{0.79} = -.266$$

or

$$= -26.6\%$$

Example 2.14. *Determine the error involved in treating carbon dioxide at 3 MPa and 10°C as ideal gas.*

Solution : From Table the critical pressure and critical temperature for carbon dioxide are

$$P_{cr} = 7.39 \text{ MPa}$$

$$T_{cr} = 304.2 \text{ K}$$

Also

$$T = 273 + 10 = 283 \text{ K}$$

Then

$$P_R = \frac{P}{P_{cr}} = \frac{3 \text{ MPa}}{7.39 \text{ MPa}} = 0.406$$

and

$$T_R = \frac{T}{T_{cr}} = \frac{283 \text{ k}}{304.2 \text{ k}} = 0.93\text{S}$$

Then from compressibility chart the corresponding value of z is read as

$$z = 0.80$$

Error

$$= \frac{v - v_{\text{ideal}}}{v} = 1 - \frac{v_{\text{ideal}}}{v}$$

$$= 1 - \frac{1}{z} = 1 - \frac{1}{0.8} = 0.25 \text{ or+-}25.0\%$$

2.8 OTHER EQUATIONS OF STATE

Although the ideal gas equation is very simple but its range of applicability is very limited as has been demonstrated in the above examples. With a view to find equation of state which represents the P-v-T behaviour of substances accurately over a larger region with no limitations, people have put forth many equations of state which are more complicated than the ideal gas equation.

2.8.1 Van-der Waals Equation of State

Ideal gas equation can be derived from the Clerk Maxwell's postilate of kinetic theory of gases with two important assumptions that there is no attraction between the molecules of the gas and the volume occupied by the molecules themselves is negligibly small compared to the volume of the gas. When pressure is small and temperature is very high, the intermolecular attraction and the volume of the molecules compared to total volume of the gas, are not of much significance and thus the real gas obeys the ideal gas equation. But as the pressure increases the inter molecular attraction or repulsion increases and also the volume of the molecules becomes appreciable as compared to the volume of gas. Under these conditions the real gases derviate considerably from the ideal gas equation. This deviation is predominant in the region near the critical point. Van-der Waals, by applying the laws of mechani's to individual molecules introduced two terms in the ideal gas equation in 1873. The equation proposed is given below.

$$\left(P+\frac{a}{v^2}\right)(v-b) = RT \qquad \text{...(2.42)}$$

The constant 'a' accounts for the existance of mutual attraction between the molecules and term $\left(\frac{a}{v^2}\right)$ is called the FORCE OF COHESION. The constant 'b' had been introduced to account for the volume of the molecules and is called as "CO-VOLUME" and represents the volume occupied by the gas molecules per unit volume. These constant are determined from the behaviour of the substance at the critical point. It has been observed that the ciritical isotherm on P–v diagram has a horizontal inflexion point at the critical point as shown in Fig. (2.19). In mathematical terms the first and second derivatives of P with respect to v at the critical point will be zero that is

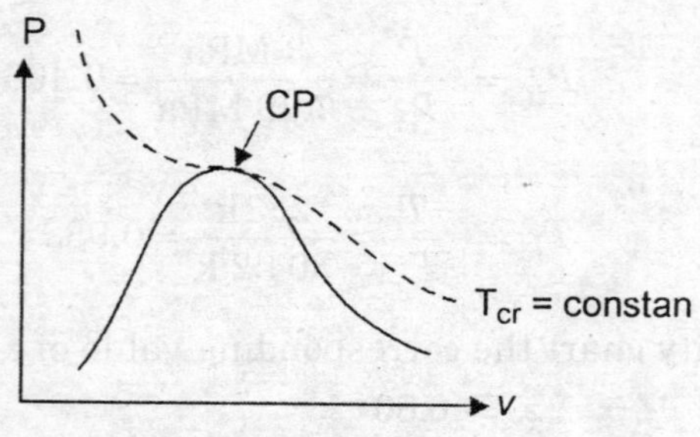

Fig. 2.19

$$\left(\frac{\partial P}{\partial v}\right)_{T=T_{cr}=\text{const}} = 0 \qquad \text{...(2.43)}$$

and $$\left(\frac{\partial^2 P}{\partial v^2}\right)_{T=T_{cr}=\text{const}} = 0 \qquad ...(2.44)$$

Equation (2.42) can also be written in the form given below

$$P = \frac{RT}{(v-b)} - \frac{a}{v^2} \qquad ...(2.45\ a)$$

or $$Pv^3 - (Pb + RT)\,v^2 + av - b = 0 \qquad ...(2.45\ b)$$

Equations (2.45) are cubic in v and for given values of P and T it has three roots out of which only one need to be real. For low temperatures, three positive real roots exist for a certain range of pressure. As the temperature increases the three roots approach one another and they become one at the critical point.

Thus from equations (2.45), taking partial decorative with respect to v we get

$$\left(\frac{\partial P}{\partial v}\right) = -\frac{RT}{(v-b)^2} + \frac{2a}{v^3} \qquad ...(2.46)$$

$$\left(\frac{\partial^2 P}{\partial v^2}\right) = \frac{2RT}{(v-b)^3} - \frac{6a}{v^4} \qquad ...(2.47)$$

Therefore from equations (2.43), (2.44), (2.46) and (2.47) at critical point

where $T = T_{cr}$, $P = P_{cr}$ and $v = v_{cr}$ we get

$$\left(\frac{\partial P}{\partial v_{cr}}\right)_{T=T_{cr}} = -\frac{RT_c}{(v_{cr}-b)^2} + \frac{2a}{v_{cr}^3} = 0 \qquad ...(2.48\ a)$$

or $$\frac{RT_c}{(v_{cr} \cdot b)^2} = \frac{2a}{v_{cr}^3} \qquad ...(2.48\ b)$$

and $$\left(\frac{\partial^2 P}{\partial v_{cr}^2}\right)_{T=T_{cr}} = \frac{2RT_{cr}}{(v_{cr}-b)^3} - \frac{6^a}{(v_{cr})^4} = 0 \qquad ...(2.49\ a)$$

or $$\frac{RT_{cr}}{(v_{cr}-b)^3} = \frac{3a}{(v_{cr})^4} \qquad ...(2.49\ b)$$

Dividing equation (2.48 b) with equation (2.49 b) we have

$$(v_{cr} - b) = \frac{2}{3}v_{cr}$$

or $$\frac{1}{3}v_{cr} = b$$

or $$b = \frac{v_{cr}}{3} \qquad ...(2.50)$$

Then substituting the value of 'b' from equation (2.51) in equation (2.48 b) we have at critical point

$$\frac{RT_{cr}}{\left(\frac{2}{3}v_{cr}\right)^2} = \frac{2a}{v_{cr}^{\ 3}}$$

or
$$R = \frac{8a}{9T_{cr}\, v_{cr}} \qquad \text{...(2.51)}$$

Then substituting the values of '*b*' and '*R*' from equation (2.50) and (2.51) into equation (2.42) at critical point

$$\left(P_{cr} + \frac{a}{v_{cr}^{\ 2}}\right)\left(v_{cr} - \frac{v_{cr}}{3}\right) = \frac{8aT_{cr}}{9T_{cr}v_{cr}}$$

or
$$P_{cr} = \frac{8aT_{cr}}{9T_{cr}\, v_{cr}}\,\frac{3}{2v_{cr}} - \frac{a}{v_{cr}^{\ 2}}$$

$$= \frac{4}{3}\frac{a}{v_{cr}^{\ 2}} - \frac{a}{v_{cr}^{\ 2}} = 0$$

i.e.,
$$P_{cr} = \frac{a}{3v_{cr}^{\ 2}}$$

or
$$a = 3P_{cr}v_{cr}^2 \qquad \text{...(2.52)}$$

Therefore from equations (2.51) and (2.52)

$$R = \frac{8}{3}\left(\frac{P_{cr}\, v_{cr}}{T_{cr}}\right) \qquad \text{...(2.53)}$$

Substituting the values of the constant '*a*', '*b*' and '*R*' in equation (2.42) the Van-der-Walls equation reduces to

$$\left(P + \frac{3P_{cr}\, v_{cr}^{\ 2}}{v^2}\right)\left(v - \frac{v_{cr}}{3}\right) = \frac{8}{3}\frac{P_{cr}\, v_{cr}}{T_{cr}}\,T \qquad \text{...(2.54)}$$

Yet another form of the van-der Waals equation can be derived from equation (2.51) and (2.52) where the equation does not contain v_{cr}

$$R = \left(\frac{8}{9T_{cr}\, v_{cr}}\right)(3P_{cr}\, v_{cr}^{\ 2})$$

or
$$R = \frac{8}{3}\frac{P_{cr}\, V_{cr}}{T_{cr}}$$

Thus
$$v_{cr} = \frac{3}{8}\frac{RT_{cr}}{P_{cr}} \qquad \text{...(2.55)}$$

Then from equation (2.50), (2.51) and (2.55)

$$a = 3P_{cr}\left(\frac{3}{8}\frac{RT_{cr}}{P_{cr}}\right)^2$$

or $$a = \frac{27}{64}\frac{R^2T_{cr}^2}{P_{cr}} \quad ...(2.56)$$

and $$b = \frac{1}{3}v_{cr} = \frac{1}{3}\cdot\frac{3}{8}\frac{RT_{cr}}{P_{cr}}$$

or $$b = \frac{RT_{cr}}{8P_{cr}} \quad ...(2.57)$$

Then the Van-der Waals equation takes the form

$$\left(P + \frac{27}{65}\frac{R^2T_{cr}^2}{P_{cr}}\frac{1}{v^2}\right)\left(v - \frac{RT_{cr}}{8P_{cr}}\right) = RT \quad ...(2.58)$$

Van der Waals equation written in terms of reduced pressure reduced temperature and reduced specific volume can be written by process of normalisation where

$$P_R = \frac{P}{P_{cr}} \quad ...(2.59)$$

$$T_R = \frac{T}{T_{cr}} \quad ...(2.60)$$

and $$v_R = \frac{v}{v_{cr}} \quad ...(2.61)$$

In the light of equation (2.59), (2.60) and (2.61) equation (2.54) when divided by $(P_{cr} \cdot v_{cr})$ takes the form

$$\left(\frac{P}{P_{cr}} + \frac{3v_{cr}^2}{v^2}\right)\left(\frac{v}{v_{cr}} - \frac{1}{3}\right) = \frac{8}{3}\frac{T}{T_{cr}}$$

or $$\left(P_R + \frac{3}{v_R^2}\right)(3v_R - 1) = 8T_R \quad ...(2.62\ a)$$

or $$\left(P_R v_R + \frac{3}{v_R}\right)(3v_{R-1}) = 8T_R v_R \quad ...(2.62\ b)$$

Equations (2.62 *a*) and (2.62 *b*) represent the reduced equation of state in which the constants *a*, *b* and *R* for a particular gas do not appear. Thus equation (2.62) is Van der Waals law of corresponding states because it reduces the properties of all gases to one formula. It is a "law" to the extent that real gases obey Var-der Waals equation. Two different substances are considered to be in "CORRESPONDING STATES" if their pressures, temperatures and specific volumes fractions of the respective critical pressure, temperatures and specific volumes of the two substances.

Note : For the purpose of making calculations with the help of Van-der Waals equation the constants *'a'* and *'b'* are determined for any gas from the critical-point data given in respective gas tables.

The accuracy of Van-der Waals equation of state is often found inadequate but it can be improved by using values of *'a'* and *'b'* that are based on actual behaviour of gas over wider range instead of a single point. Van-der-Waals equation of state can also be expressed on Mole basis by replacing the v in the equation (2.45) by $\bar{v}$ and R by R_u.

Thus it reduces to

$$\left[P + \frac{a}{(\bar{v})^2}\right](\bar{v} - b) = R_u T \quad ...(2.63)$$

2.8.2 Beattie-Bridgeman Equation of State

Real gases conform closely with Van-der Waals equation of state than the ideal gas equation of state particularly at higher pressures. But it is not obeyed in all ranges of temperatures and pressures. In this connection many more equations were introduced later and notable among these are given below :

(*i*) Van-der Waals equation — Has 2 constants and is accurate over limited range

(*ii*) Beattie-Bridgeman equation — Has 5 constants which one determined experimentally. It is accurate for $\rho \le 0.8\,\rho_{cr}$

(*iii*) Stro bridge — Has 16 constants. More suitable for Computer calculations

(*iv*) Benediel-Webb-Rubex equation — Has 8 constants. Accurate for $\rho \le 2.5\,\rho_{cr}$

(*v*) Virial equation — Number of constants may vary. Accuracy dependent on the number of terms used

Beattie-Bridgeman Equation of state was proposed in 1928 and is based on five experimentally determined constants. It is given below

$$P = \frac{R_u T}{(\bar{v})^2}\left(1 - \frac{c}{\bar{v}T^3}\right)(\bar{v} + B) - \frac{A}{(\bar{v})^2} \quad ...(2.64)$$

where

$$A = A_0\left(1 - \frac{a}{\bar{v}}\right) \quad ...(2.65)$$

$$B = B_0\left(1 - \frac{b}{\bar{v}}\right) \quad ...(2.66)$$

and $\bar{v}$ is unit number specific volume. These constants are given in the respective tables for various substances. This equation is reasonably accurate for densities upto $0.8\,\rho_{cr}$ where ρ_{cr} is the density of the substance at the critical point.

2.8.3 Benedict-Webb-Rubin Equation of State

Benedict-Webbs and Rubin extended the equation suggested by Beattie-Bridgeman in 1940 by raising the number of constant to eight. It is given below

$$P = \frac{R_u T}{\bar{v}} + \left(B_0 R_u T - A_0 - \frac{C_0}{T^2}\right)\frac{1}{(\bar{v})^2} + \frac{bR_u T - a}{(\bar{v})^3} + \frac{a\alpha}{(\bar{v})^6} + \frac{C}{(\bar{v})^3 T^2}\left[1 + \frac{\gamma}{(\bar{v})^2}\right]_e - \gamma/(\bar{v})^2 \quad ...(2.67)$$

The values of the constants are given in table A-29*b* of the Appendix. This equation give accorate results for $\rho \le 2.5\,\rho_{cr}$.

2.8.4 Virial Equation of State or Virial Expansion

The relationship between P and v can also be expressed is series from as

$$P = \frac{RT}{v} + \frac{a(T)}{v^2} + \frac{b(T)}{v^3} + \frac{c(T)}{v^4} + \frac{d(T)}{v^5} + ... \qquad ...(2.68)$$

This equation is called virial expansion or virial equation of state where $a(T)$, $b(T)$, $c(T)$ ... etc. are functions of temperature alone and are called virial coefficients. These coefficients can be determined experimentally or even theoretically from statistical mechanics. Evidently as the pressure approaches zero all the virial coefficients will vanish and the equation (2.68) will reduce to ideal gas equation of state. By taking sufficient number of terms, the equation (2.68) represents the behaviour of the real gas very accurately over a wider range.

It may be noted that above equations are applicable to gas phase of the substance only and are not applicable to liquid or liquid-vapor mixtures.

Example 2.15. *Determine the pressure of Nitrogen gas which has a specific volume of 0.041884 m^3/kg at 150 K by using (a) the ideal gas equation, μ the Beattie-Bridgeman equation and (c) the Nitrogen Tables.*

Solution : (a) From ideal gas equation

$$P = \frac{RT}{v}$$

For Nitrogen $R = 0.2968$ kPa · m³/kg · K

Therefore $$P = \frac{(0.2968 \text{ kPa} \cdot \text{m}^3/\text{kg} \cdot \text{K})\ (150 \text{ K})}{0.041884 \text{ m}^3/\text{kg})} = 1063 \text{ kPa}$$

(b) The evaluating the Beattie-Bridgeman constants refer to table for Nitrogen

$$A = A_0\left(1 - \frac{a}{\bar{v}}\right)$$

$$B = B_0\left(1 - \frac{b}{\bar{v}}\right)$$

$$C = b,\ 2 \times 10^4 \text{ m}^3 \text{ k}^3/\text{k mol}$$

where $\bar{v} = Mv$ where v is specific volume is m³/kg and

M = mass per k mol

Therefore $\bar{v}$ = [(28.013(v_3/k mol] (0.041884 m³/kg)

= 1.1733 m³/kmol

Thus $$A = 136.2315\left[1 - \frac{0.02617}{1.1733}\right]$$

$$= 133.193$$

And $$B = 0.05046\left[1 - \frac{-0.00691}{1.1733}\right]$$

$$= 0.05076$$

Substituting these values in Beattie-Bridgeman equation

$$P = \frac{R_u T}{(\bar{v})^2}\left[1 - \frac{C}{\bar{v}T^3}\right][\bar{v} + B) - \frac{A}{(\bar{v})^2}$$

$$= \frac{8.314 \times 150}{(1.1733)}\left[1 - \frac{4.2 \times 10^4}{1.1733 \times (150)^3}\right][1.1733 + 0.5076] - \frac{133.193}{(1.1733)^2}$$

$$= 1000.4 \text{ kPa}$$

(*c*) From superheated Nitrogen table : At T = 150 K and v = .041884 m³/kg the value of P = 1000 kPa

2.9 MIXTURE OF GASES

Let us take a vessel containing a homogeneous, mixture of large number (say c) of ideal gases at a temperature T, pressure P and volume V. Suppose there are N_1 moles of gas A_1, N_2 moles of gas A_2, ..., N_c moles of gas A_c present in the vessel. Gases being inert there will be no chemical reaction, the mixture will be in a state of equilibrium and then the governing equation of state will be

$$PV = (N_1 + N_2 + N_3 + \ldots + N_c)\, R_u T \qquad \ldots(2.68)$$

where $R_u = 8.3143$ kJ/mol K

Then $$p = \frac{N_1 R_u T}{V} + \frac{N_2 R_u T}{V} + \frac{N_3 R_u T}{V} + \ldots + \frac{N_c R_u T}{V} \qquad \ldots(2.69)$$

where $$p_1 = \frac{N_1 R_u T}{V},\ P_2 = \frac{N_2 R_u T}{V}, \ldots P_c = \frac{N_c R_u T}{V} \qquad \ldots(2.70)$$

The expression $\frac{N_k R_u T}{V}$ where k = 1, 2, ... c, represents the pressure that the kth, gas would exert if it occupies the volume v alone at temperature T. This is known as the partial pressure of the kth gas and is represented by p_k.

Then form equation (2.69) we get

$$p = p_1 + p_2 + p_3 + \ldots + p_c \qquad \ldots(2.71)$$

This is called the Dalton's law of partial pressures which states that the total pressure of mixture of ideal gases is equal to the sum of partial pressures.

Again of N_t is the total of moles in the gas mixture, then

$$N_t = N_1 + N_2 + \ldots + N_c \qquad \ldots(2.72)$$

Then $$pV = N_t + R_u T$$

or $$V = \frac{N_t R_u T}{p} \qquad \ldots(2.73)$$

Also for the kth gas, then partial pressure P_R is

$$p_k = \frac{N_k R_u T}{V}$$

Substituting value of v for (2.73)

$$p_k = \frac{N_k R_u T}{N_t R_u T} P$$

$$= \frac{N_k}{N_t} p$$

But $$N_t = \sum_{k=1}^{t} N_k = N_1 + N_2 + N_3 + \dots + N_e$$

Then $$p_k = \frac{N_k}{\Sigma N_k} p \quad \text{...(2.74 a)}$$

$$= x_k p \quad \text{...(2.74 b)}$$

where $x_k = \dfrac{N_k}{\Sigma N_k}$ is called the mole fraction of the kth gas such that

$$x_1 = \frac{N_1}{\Sigma N_k}, x_2 = \frac{N_2}{\Sigma N_k}, \dots x_c = \frac{N_c}{\Sigma N_k}$$

Then from (2.74) $$p_1 = x_1 p, p_2, x_2 p, \dots p_c = x_c p \quad \text{...(2.75 a)}$$

or $$p_k = x_k p \quad \text{...(2.75 b)}$$

Also $$\sum_{k=1}^{t} x_k = x_1 + x_2 + \dots + x_c$$

$$= \frac{N_1}{\Sigma N_k} + \frac{N_2}{\Sigma N_k} + \dots + \frac{N_c}{\Sigma N_k}$$

$$= \frac{1}{\Sigma N_k} [N_1 + N_2 + \dots + N_c]$$

$$= \frac{\Sigma N_k}{\Sigma N_k} = 1 \quad \text{...(2.76)}$$

Thus in a mixture of gases if all the mole fractions except one are known, the unknown mole fraction can be found with the help of equation (2.76).

It must be reiterated that in Daltons law it is implicit that the temperature of individual gases is same, that is

$$T = T_1 = T_2 = T_3 \dots = T_c \quad \text{...(2.77)}$$

Similarly since each gas is occupying the same volume V

$$V = V_1 = V_2 = V_3 = \dots = V_c \quad \text{...(2.78)}$$

Therefore in terms of masses we have

$$p_1 V = m_1 R_1 T \quad \text{...(2.79 a)}$$

$$p_2 V = m_2 R_2 T \quad \text{...(2.79 b)}$$

$$\vdots$$

$$p_c V_c = m_c R_C T \quad \text{...(2.79 c)}$$

where R_1, R_2, ... R_c are the respective gas constants

Adding equation (2.79 *a*), (2.79 *b*) ... we have

$$V(p_1 + p_2 + p_3 ... + p_c) = (m_1R_1 + m_2R_2 + m_cR_c)\, T$$

or

$$pV = (m_1R_1 + m_2R_2 ... m_cR_c)\, T \quad \text{...(2.80)}$$

Let

$$m_1R_1 + m_2R_2 + ... + m_cR_c = \sum_{k=1}^{c} m_k R_m$$

$$= (m_1 + m_2 ... m_c)\, R_m$$

or

$$R_m = \frac{m_1R_1 + m_2R_2 + ... + m_cR_c}{m_1 + m_2 + ... + m_c} \quad \text{...(2.81)}$$

Further if

$$m = \sum_{k=1}^{c} m_k = m_1 + m_2 + ... + m_c$$

$$= \text{Total mass of gas mixture} \quad \text{...(2.82)}$$

Then equation (2.80) reduces as

$$pV = mR_mT \quad \text{...(2.83)}$$

Where R_m is called the gas constant of the mixture of gases.

Equation (2.78) can be written in terms of specific volumes as

$$mv = m_1v_1 = m_2v_2 ... = m_cv_c \quad \text{...(2.83)}$$

and

$$m = m_1 + m_2 + ... m_c$$

or

$$= \frac{mv}{v_1} + \frac{mv}{v_2} + ... + \frac{mv}{v_c} \quad \text{...(2.84)}$$

$$= mv\left[\frac{1}{v_1} + \frac{1}{v_2} + ... + \frac{1}{v_c}\right]$$

or

$$\frac{m}{mv} = \frac{1}{v_1} + \frac{1}{v_2} + ... \frac{1}{v_c}$$

or

$$\frac{1}{v} = \frac{1}{v_1} + \frac{1}{v_2} + ... \frac{1}{v_c} \quad \text{...(2.85)}$$

Further density

$$\rho = \frac{1}{v}$$

Therefore

$$\rho = \rho_1 + \rho_2 + \rho_3 + \rho_c \quad \text{...(2.86)}$$

Which means the density of the mixture is equal to the sum of densities of the component gases.

2.9.1 Mixture of Gases : Mass, Volume and Molar Fractions

Sometimes the components of gas mixture are specific by mass (mass fraction) or volume (volume fraction). the mass fraction of each component of a mixture is defined as the ratio of the mass of the given component to the mass of the whole mixture.

If m is the mass of the mixture and $m_1, m_2 \ldots m_c$ are the masses of c components, then

$$m = m_1 + m_2 + \ldots + m_c \qquad \ldots(2.87)$$

Dividing by m through out

$$1 = \frac{m_1}{m} + \frac{m_2}{m} + \ldots + \frac{m_c}{m}$$

$$= g_1 + g_2 + \ldots + g_c \qquad \ldots(2.88)$$

where $g_1, g_2, \ldots g_c$ are mass fractions obviously from (2.88) we see that sum of the parts by mass fractions is equal to unity. This is known as GRAVIMETRIC ANALYSIS

Similarly volume fraction is defined as the ratio of partial (or reduced) volume of each component of the gas mixture the total volume of the gas mixture, where the partial (or reduced) volume the gas component is the volume that this gas component would occupy if the temperature and pressure were equal to those of the gas mixtures respectively. Thus if p and T are the pressure and temperature of the gas mixture respectively then from the ideal gas equation

$$pV_1 = m_1R_1T \qquad \ldots(2.89)$$

$$pV_2 = m_2R_2T$$

$$\vdots$$

$$pV_c = m_cR_cT$$

Summation of equations (2.89) gives

$$p(V_1 + V_2 + V_3 + \ldots + V_c) = (m_1R_1 + m_2R_2 + \ldots m_cR_c)\,T$$

$$= p_1V + p_2V + \ldots + p_cV$$

$$= (p_1 + p_2 + \ldots + p_c)\,V \qquad \ldots(2.90)$$

But from Daltan's law of partial pressures

$$p_1 + p_2 + \ldots p_c = P$$

Therefore (2.90) reduces to

$$p(V_1 + V_2 \ldots + V_c) = pV$$

or

$$V_1 + V_2 + \ldots + V_c = V \qquad \ldots(2.91)$$

From equation (2.91) we conclude that the total volume of the mixture of gases is equal to the sum of partial (reduced) volumes of the constituent gases. This is called Amagat Leduc law of partial volumes.

Again dividing (2.91) by V, we have

$$1 = \frac{V_1}{V} + \frac{V_2}{V} + \ldots + \frac{V_c}{V}$$

or

$$= r_1 + r_2 + \ldots + r_c \qquad \ldots(2.92)$$

where $r_1, r_2 \ldots r_c$ are called the volume fractions which are defined as

$$r_1 = \frac{V_1}{V},\ r_1 = \frac{V_2}{V}, \ldots r_c = \frac{V_c}{V} \qquad \ldots(2.93)$$

Further the sum of volume fractions is equal to unity. When the gas mixture analysis is made an volume basis, it is called volumetric analysis.

Similarly the mole fraction is defined as the ratio of kilomoles of each constituent gas to the total number of kilomoles contained in the gas mixture.

Thus if $N_1, N_2 \ldots N_c$ are the kilo moles of first gas second gas and so on ... and N are the total number of kilomoles contained in the mixture.

Then the mole fractions $\mu_1, \mu_2 \ldots$ are given by

$$\mu_1 = \frac{N_1}{N}, \mu_2 = \frac{N_2}{N} \ldots, \mu_c = \frac{N_C}{N} \quad \ldots(2.94)$$

Therefore we have $N = N_1 + N_2 + \ldots N_c$

or

$$1 = \frac{N}{N} + \frac{N_2}{N} \ldots \frac{N_c}{N}$$

$$= \mu_1 + \mu_2 \ldots + \mu_c \quad \ldots(2.95)$$

Again the sum of mole fractions is unity. But applying the ideal gas equation the constituents of gas mixture and the gas mixture itself we get for the kth gas where $k = 1, 2 \ldots c$ in turm.

$$V_k = \frac{N_k R_u T}{p} \quad \ldots(2.96)$$

$$V = \frac{N R_u T}{p} \quad \ldots(2.97)$$

Dividing (2.96) with (2.97)

$$\frac{V_k}{V} = \frac{N_k}{N} \quad \ldots(2.98\ a)$$

But from equation (2.97)

$$\frac{V_k}{V} = r_k \quad \ldots(2.99)$$

and from equation (2.97)

$$\frac{N_k}{N} = \mu_k \quad \ldots(2.100)$$

Therefore from (2.98) through (2.100) we conclude

$$r_k = \mu_k \quad \ldots(2.101)$$

In other words for mixture of ideal gases, the mole fraction and volume fraction are equal.

2.9.2 Relation Between Volumetric and Gravimetric Analysis of Ideal Gas Mixture

Let M_k be the molecular mass of constituent kth gas in the mixture where $k = 1, 2 \ldots c$ and m_k and N_v be the mass and kilo moles of the respective gas constituents. Then

$$m_1 = N_1 M_1, m_2 = N_2 M_2 \ldots m_c = N_c M_c \quad \ldots(2.102)$$

Then $$(m_1 + m_2 + \ldots m_c) = N_1 M_1 + N_2 M_2 + \ldots + N_c M_c \quad \ldots(2.103)$$

Then from (2.102) and (2.103) we get by dividing

$$\frac{m_1}{(m_1 + m_2 + ...m_c)} = \frac{N_1 M_1}{(N_1 M_1 + N_2 M_2 + ... + N_c M_c)}$$

Dividing by N, the numerator and denominator of RHS yieds

$$\frac{m_1}{m_1 + m_2 + ...m_c} = \frac{(N_1/N) M_1}{\left(\frac{N_1}{N} M_1 + \frac{N_2}{N} M_2 + ... + \frac{N_c}{N} M_c\right)}$$

Substituting values for equation (2.100)

$$= \frac{\mu_1 M_1}{(\mu_1 M_1 + \mu_2 M_2 + ... + \mu_c M_c)}$$

Further $\mu_1 = r_1$ (Eqation 2.101)

Therefore $$\frac{m_1}{m_1 + m_2 + ... + m_c} = \frac{(V_1/V) M_1}{\left(\frac{V_1}{V}\right) M_1 + \left(\frac{V_2}{V}\right) M_2 + ... + \left(\frac{V_c}{V}\right) M_c}$$

or $$\frac{M_1}{m} = \frac{(V_1/V) M_1}{\Sigma\left[\left(\frac{V_1}{V}\right) M_1 + ...\right]} \quad ...(2.104)$$

Generalising this equation, for any constituent gas

$$\frac{m_k}{m} = \frac{(V_k/V) M_k}{\Sigma\left(\frac{V_k}{V} \cdot M_k\right)} \quad ...(2.105)$$

With the help of equation (2.105) one can solve the problems, sometimes very conveniently) by changing the mode from mass fraction to volume fraction.

2.9.3.1 Determination of Partial Pressures (Gas Mixture Specified in Terms of Mass Fraction)

In case the gas mixture is specified in terms of mass fractions, we have from equation (2.59), for each constituent gas

For $K = 1, 2.... P_K V = m_k R_k T$ ...(2.106)

And for the entire gas mixture $pV = m R_{\text{mix}} T$ ...(2.107)

Dividing (2.106) with (2.107) $$\frac{p_R}{p} = \frac{m_v}{m} \frac{R_k}{R_{\text{mix}}} \quad ...(2.108)$$

or $$p_k = p\, g_k \frac{R_k}{R_{\text{mix}}}$$

Thus the partial pressure of any constituent gas is a mixture multiplied by the mass fraction of that particular gas and the ratio of the gas constants of that gas and of the gas mixture.

Again writing the gas equations for each constituent gas and for the mixture in terms of number of kilomoles we have

$$p_K V = N_k R_u T \quad ...(2.109)$$

and

$$pV = NR_u T \quad ...(2.110)$$

Dividing (2.109) by (2.110)

$$\frac{p_K}{p} = \frac{N_k}{N} \quad ...(2.111)$$

From equation (2.111) it is concluded that the ratio of the partial pressure of kth constituent gas to the total pressure at the same temperature and volume is equal to the mole fraction of the gas. Alternatively the partial pressure of any constituent gas in a mixture of gases is directly proportional to the number of moles of that gas

or

$$\frac{p}{N} = \frac{p_1}{N_1} = \frac{p_2}{N_2} = ... = \frac{p_k}{N_k} \quad ...(2.112)$$

2.9.3.2 Determination of Partial Pressures (Gas Mixtures Specified in Terms of Volume Fraction)

In case the gas mixture is specified in terms of volume fraction, then Boyle's law will apply because the gases change from one state to another state at constant temperature.

That is

$$p_k V = pV_k$$

or

$$\frac{p_k}{p} = \frac{V_u}{V}$$

or

$$p_k = pr_k \quad ...(2.113)$$

In other words the partial pressure of any constituent gas in a mixture is equal to the pressure of the gas mixture multiplied by the volume fraction of that gas.

Example 2.16. *A closed rigid vessel contained 1.5 kg of Nitrogen at 3 MPa pressure. Then Oxygen was added into the vessel such that the pressure rises to 4 MPa, temperature remaining same. Dtermine the mass of the oxygen added.*

Solution : Here $m_n = 1.5$ kg

Also $M_n = 28$

Therefore kg mols of Nitrogen $= N_n = \frac{m_n}{M_n}$

$$= \frac{1.5}{28} = 0.05357$$

Now $p_n = 3$ MPa

Therefore $P_0 = 4 - 3 = 1.0$ MPa

We have from equation (2.113)

$$\frac{p_0}{N_0} = \frac{p_n}{N_n}$$

$$\therefore \qquad N_0 = p_0 \frac{N_n}{p_n} = \frac{1.0}{3.0} \times 0.05357$$

$$= 0.178566\, k \text{ mol}$$

Therefore $\quad m_0 = N_0 M_0$

$$= 0.0178566 \times 32$$

$$= 0.5714 \text{ kg}$$

Eample 2.17. *An engine uses air fuel mixture cousisting of 0.1 kg of fuel (moler mass 118) and 1.0 kg of air. At entry to the cylinder the pressure and temperature have been recorded as 100 kPa and 17°C respectively. Determine the volume of the mixture and partial pressure of air in the mixture. Take molecular mass of air M_a as 29.*

Solution : Let N_a be no. of moles of air and N_f the number of moles of fuel.

Then $\quad N_a = \frac{1.0}{29} = 0.3448$

$$N_f = \frac{0.1}{118} = 0.00084745$$

Therefore total number of moles in mxiture

$$N = N_a + N_f$$

$$= .03448 + .00084745$$

Now $\quad T = 273 + 17 = 290\,K$

From the gas equation as applied to mixture

$$V = \frac{N R_u T}{p}$$

Where R_u = universal gas constant

$$V = \frac{0.035335 \times 8314 \times 290}{100 \times 10^3}$$

$$= 0.8519 \text{ m}^3$$

(*b*) Let partial pressure of air in the mixture be p_a)

Then $\quad p_a = p\frac{N_a}{N} = 100 \times \frac{0.03448}{0.35335}$

$$= 97.5804 \text{ kPa Ans.}$$

Example 2.18. *A 5.5 m^3 capacity rigid tank contain a mixture of Oxygen, Nitrogen and carbon dia oxide at 27°C and pressure of 10 bar. Determine the mass of each constituent gas in the mixture. Take R_u = 8314 kJ / (k mol · k),*

Moler mass of O_2 = 32

Moler mass of N_2 = 28.016

Moler mass of CO_2 = 44

Solution : $\quad T = 273 + 27 = 300$ K

$$P = 10 \text{ bar} = 10 \times 10^5 \text{ Pa} = 10^6 \text{ Pa}$$

Gas constant of oxygen $\quad R_0$ = is given as

$$R_0 = \frac{R_u}{\text{Moler mass of } O_2} = \frac{R_u}{M_0}$$

$$= \frac{8314}{32} = 259.81 \text{ J/(kg} \cdot K)$$

For N_2 $$R_N = \frac{R_u}{M_N} = \frac{8314}{28.016} = 296.759 \text{ J/(kg} \cdot K)$$

For CO_2 $$R_{CO_2} = \frac{R_u}{M_{CO_2}} = \frac{8314}{44} = 188.954 \text{ J/(kg} \cdot K)$$

From gas equation we then get

$$pV = mRT$$

or $$\frac{pV}{R_0 T} = \frac{10^6 \times 5.5}{259.81 \times 300} = 70.564 \text{ kg}$$

Mass of N_2 $$M_0 = \frac{PV}{R_N T} = \frac{10^6 \times 5.5}{296.759 \times 300} = 61.7785 \text{ kg}$$

And for CO_2 $$M_{CO_2} = \frac{PV}{R_{CO_2} T} = \frac{10^6 \times 5.5}{188.954 \times 300}$$

$$= 97.02537 \text{ kg}$$

Example 2.19. *A rigid vessel of capacity 0.5 m^3 contained air at 85 kPa and 27°C. Hydrogen was then added in the vessel thereby raising the pressure to 90 kPa, the temperature remains constant. Find the final masses of oxygen, Nitrogen and Hydrogen in the vessel along with their respective partial pressures. Take gas constant for air as 287 J/kg K and for hydrogen 4124 J/kg K. Assume that air contains 77% of Nitrogen on mass basis.*

Given $T = 273 + 27 = 300 \text{ K}$

Solution : From gas equation as applied to air, we have

$$m_a = \frac{pV}{R_a T} = \frac{(85 \times 10^3)(0.5)}{287 \times 300} = 0.4936 \text{ kg}$$

The mass of Nitrogen $= 0.77 \times .4936$

$= 0.3800$ kg

And mass of Oxygen $= 0.4936 - .3800$

$= 0.1136$ kg

Now partial pressure of Hydrogen

$= 90 - 85 = 5$ kPa

Therefore from gas equation we get the mass of hydrogen as

$$m_h = \frac{pV}{R_h T} = \frac{(5 \times 10^3)(.5)}{4124 \times 300}$$

$$= 0.002021 \text{ kg}$$

Partial pressure of Oxygen, p_0 is again obtained by applying gas equation

$$p_0 = \frac{m_0 R_0 T}{V} = \frac{0.1136 \times 287 \times 300}{0.5}$$

$$= 19561.92 \text{ Pa}$$

$$= 191619 \text{ kPa}$$

Example 2.20. *A gas mixture consists of 12 mass fractions of air and 1.5 mass fraction of natural gas. If the density of natural gas ρ_n = 0.5 kg/m^3 at 273 K temperature and 101325 Pa pressure, find the gas constant, partial pressures and desnity of the components of the gas mixture. Take gas constant for air 287 J/kg K.*

Solution : $R_a = 287$ J/kg K

Now for natural gas $R_n = \frac{pV}{T} = \frac{p}{\rho_n T}$

or $R_n = \frac{101325}{0.5 \times 273} = 724.3 \text{ J/kg} \cdot K$

Also gas constant for the mixture

$$R_{\text{mix}} = g_a R_a + g_n R_n$$

$$= \frac{12}{13.5} \times 287 + \frac{1.5}{13.5} \times 724.3$$

$$= 255.1111 + 80.47777$$

$$= 335.58887 \text{ kJ/(kg} - \text{k)}$$

Again from gas equation as applied to mixture

$$\rho_{\text{mix}} = \frac{p}{R_{\text{mix}} T} = \frac{101325}{335.58887 \times 273}$$

$$= 1.105978 \text{ kg/m}^3$$

Partial pressure of air $p_a = p g_a \frac{R_a}{R_{\text{mix}}}$

$$= 101325 \times \frac{12}{13.5} \times \frac{287}{335.58887}$$

$$= 77026.1938 \text{ Pa}$$

$$= 77.0261938 \text{ kPa}$$

Partial pressure of neutral gas (From Dalton's law)

$$p_n = 101325 - 77026 \text{ Pa}$$

$$= 24299 \text{ Pa}$$

Alternatively $p_n = p g_n \frac{R_n}{R_{\text{mix}}} = 101325 \times \frac{1.5}{13.5} \times \frac{724.3}{335.58887}$

$$= 24299 \text{ Pa}$$

Example 2.21. *A rigid vessel contains 8 kg of Oxygen, 10 kg of Nitrogen and 20 kg of CO_2 at 100°C temperature and 250 kPa pressure. If the universal gas constant is $R_u = 8314$ J/kg · K and Molar mass of O_2, N_2 and CO_2 are 32, 28 and 44 respectively, calculate the capacity of the vessel and the partial pressure of each gas in the vessel and the total pressure when the temperature is raised to 175°C.*

Solution : For O_2, $R_1 = \dfrac{R_u}{M_0} = \dfrac{8314}{32}$

For N_2, $R_2 = \dfrac{R_u}{M_n} = \dfrac{8314}{28}$

for CO_2, $R_3 = \dfrac{R_u}{M_{CO_2}} = \dfrac{8314}{44}$

$$R_{mix} = g_1R_1 + g_2R_2 + g_3R_3$$

$$= \frac{8}{38} \times \frac{8314}{32} + \frac{10}{38} \times \frac{8314}{28} + \frac{20}{38} \times \frac{8314}{44}$$

$$= 54.69737 + 78.13991 + 99.44976$$

$$= 232.2862 \text{ J/kg} \cdot \text{K}$$

$$V_{mix} = \frac{m\ R_{mix}T}{p} = \frac{38 \times 232.2862 \times 373}{250 \times 10^3}$$

$$= 13.1697 \text{ m}^3$$

(*b*) For heating the mixture at constant volume we have

$$\frac{p_1}{T_1} = \frac{p_2}{T_2}$$

Therefore $p_2 = p_1 \dfrac{T_2}{T_1}$

$$= 250 \times \frac{448}{373} = 300.268 \text{ kPa}$$

(*c*) Partial pressures of constituents are

For O_2, $P_1 = \dfrac{p_1g_1R_1}{R_{mix}} = 250 \times \dfrac{8}{38} \times \left(\dfrac{8314}{32}\right) \times \dfrac{1}{232.2862}$

$$= 58.8685 \text{ kPa}$$

For N_2 $P_2 = \dfrac{p_2g_2R_2}{R_{mix}} = 250 \times \dfrac{10}{38} \times \dfrac{8314}{28} \times \dfrac{1}{232.2862}$

$$= 84.09787 \text{ kPa}$$

For CO_2 $P_3 = \dfrac{p_3g_3R_3}{R_{mix}}$

$$= 250 \times \frac{20}{38} \times \frac{8314}{44} \times \frac{1}{232.2862}$$

$$= 107.03365 \text{ kPa}$$

PROBLEMS

1. Define a pure substance.
2. Is iced water a pure substance ? Why? (**Ans.** Yes as it has same chemical composition)
3. Is sweetened coffee a pure substance?

   (**Ans.** Yes as it has same chemical composition throught)
4. Explain what you understand by triple point. Give the temperature and pressure of water at triple point. (**Ans.** 0.01°C, 0.6113 kPa)
5. At what phase the energy level of a substance is highest? (**Ans.** Gas phase)
6. Differentiate between saturated vapor and superheated vapor.
7. Are the properties of saturated vapor at a given temperature and of a saturated mixture at the same given temperature, are same? (**Ans.** Yes)
8. Will the water boil at higher temperature at higher pressure? (**Ans.** Yes)
9. During boiling process the pressure of water is increased. Will the temperature also increase or will it remain constant? (**Ans.** Yes, the temperature will also increase)
10. Explain the difference between critical point and triple point.

    (**Ans.** At critical point the saturated liquid and saturated vapor states are identical. At triple point the trhree phases of a pure substance co-exist in equilibrium)
11. Can liquid water eixst at 0.4 kPa pressure?

    (**Ans.** No. as the pressure is below that corresponding to triple point)
12. Can water vapor exist at –10°C?

    (**Ans.** Yes because water expands on freezing and for such substances at temperatures below triple point temperature the vapor phase eixsts)
13. A rigid tank is partially filled with water at 25°C and rest is air at atmospheric pressure. Can the water be made to boil with out raising the temperature?

    (**Ans.** Yes, by lowering the pressure in the tank by sucking air).
14. A sealed rigid tank is half filled with water at atmospheric pressure and the 4est is air The tank is heated so that water starts boiling. Will the temperature remain constant during this process?

    (**Ans.** No. since the tank is sealed the pressure will start rising when it boils. And with the rising pressure the saturation temperature will also start-rising)
15. Is it possible to determine the enthalpy of a substance at a given state from the knowledge of u, P and v *data.*

    (**Ans.** Yes with the help of equation $h = u + Pv$)
16. State the physical significance of h_{fg} Can it be estimated if h_f and h_g are known?

    (**Ans.** h_{fg} represents the amount of energy required to vaporise a unit mass of saturated liquid at a given pressure or temperature. it is estimated by using the equation $h_{fg} = h_g - h_f$)

17. It the value of h_{fg} for water at 100°C more than at 125°C. **(Ans.** Yes)

18. Will the energy required to completly vaporise 1 kg of saturated liquid at 1 bar pressure will be more that required to vaporise 1 kg of saturated liquid at 10 bar pressure?

(**Ans.** It will be more at 1 bar pressure than at 10 bar pressure because at higher pressure h_{fg} is lower main that at lower pressure)

19. A tank contains 50 kg of saturated liquid water at 90°C. Find the pressure in the tank and the volume of the tank. Show the state of saturated liquid water on T.v. diagram.

(**Ans.** 70.14 kPa, 0.0518 m^3)

20. A mass of 400 gm of saturated liquid water is completely evaporated at a constant pressure of 150 kPa. Find the volume change and the amount of energy given to the water during this process. (**Ans.** 0.4633 m^3, 890.6 kJ)

21. A tank contains 12 kg of water at 85°C. If 10 kg of water is in liquid form and rest in vapor form, find (*a*) the pressure in the tank and (*b*) the volume of the tank.

(**Ans.** 57.83 kPa; 5.6663 m^3)

22. A rigid tank with a volume of 3m^3 contains 6 kg of saturated-vapor mixture of water at a pressure of 40 kPa. Find (*a*) the temperature of the water, (*b*) the quality (*c*) the enthalpy of the water and (*d*) the volume occupied by the vapor phase.

(**Ans.** (*a*) 75.87°C, (*b*) 0.125295, (*c*) 608.166 (kJ/kg), (*d*) 3.0018 m^3)

23. A cooking pan whose inner diameter is 200 mm is filled with water and covered with a lid of mass 0.5 kg. It the atmospheric pressure is 101 kPa, find the temperature at which the water, when heated, will start boiling. (**Ans.** 99.9°C)

24. A vertical piston cylinder device contains water. The piston has a mass of 20 kg and the cross-sectional area is 0.01 m^2. Determine the temperature at which the water if heated, will start boiling? Take atmospheric pressure as 100 kPa. (**Ans.** 10.4.6°C)

25. A rigid tank of 0.2 m^3 volume obtains 5 kg of water at 150 kPa. Find (*a*) the temperature (*b*) the total enthalpy and (*c*) the mass of each phase.

(**Ans.** $T = T_{sat}$ at 150 kPa = 111.37°C; H = 3709.5 kJ; (*c*) m_g = .168 kg and m_f = 41832 kg)

26. A piston cylinder device contains 0.9 m^3 water vapor and 0.1 m^3 of liquid water at 500 kPa pressure. Heat is added at constant pressure to the water till the temperature reaches 200°C. Find the total mass of water and final volume.

Show the process on P.v. diagram with respect to saturation lines.

(**Ans.** m_f = 91.5 kg, m_g = 2.40 kg, m_{total} = 93.9 kg; v_2 = .4249 m^3/kg and $V_2 = m_t\ v_2$ = 39.9 m^3)

27. With the help of steam tables find the enthalpy of steam when the pressure is 2 MPa and specific volume is 0.09 m^3/kg. (**Ans.** 2618.79 kJ/kg)

28. A vessel of volume 0.04 m^3 contains a mixture of saturated water vapour and liquid water at 250°C. If the mass of the liquid water is 9 kg find the pressure, the quality of steam, total mass of water in the vessel, specific volume and the enthalpy.

(**Ans.** p_{sat} = 3.973 MPa; 0.06, 9.575 kg; v = 0.00418 m^3/kg; 1188.32 kJ/kg)

29. A rigid vessel contains steam at pressure of 0.3 MPa and temperature of 250°C. Steam is cooled at constant volume. Find (*a*) at what temperature will the steam become saturated vapour? (*b*) What will be the quality at 80°C? (**Ans.** 123.9°C t_{sat}, (*b*) 0.234)

30. A rigid tank contains superheated water vapour at 1 MPa pressure and 300°C. It is now allowed to cool to 150°C. Show this process on T.v. diagram with respect ot saturation lines and at the final state determine the pressure, the quality and the enthalpy. (**Ans.** 475.8 kPa; 0.656; 2030.5 kJ/kg)

31. A piston cylinder device contains 0.05 m^3 Of liquid water at 300 kPa pressure and 250°C temperature. Heat is added to water at constant pressure and entire water evaporates. State whether at the initial state water is in compressed liquid state or not and find the mass of water. What is the temperature at the final state. Also determine the change in enthalpy. Show the process on T.v. diagram. (**Ans.** 49.85 kg; 133.56°C, 130.63 kJ)

32. A mixture of 3 kg of Nitrogen and 5 kg of Carbon diaoxide at 20°C has a pressure of 300 kPa. Find (*a*) the mole fraction of each constituent gas (*b*) the equivalent molecular weight of the mixture (*c*) Equivalent gas constant of the mixture (*d*) Partial pressures of N_2 and CO_2 Take R_u = 8.3143 kJ/kg · K.

(**Ans.** (*a*) = 0.485; x_{CO_2} = 0.515, (*b*) 36.25 kg/k mol, (*c*) 0.229 kJ/kg · K,
(*d*) p_{N2}= 145.5 kPa, p_{CO_2} = 154.5 kPa)

33. A rigid vessel contains 1.25 kg Nitrogen at 37°C and 2.5 MPa pressure. Certain quantity of Oxygen is added in the vessel at constant temperature such that the pressure becomes 3.5 MPa. Calculate the mass of Oxygen added. (**Ans.** 0.5714 kg)

34. A rigid vessel of capacity 0.4 m^3 contained air at 80 kPa pressure and temperature Of 27°C. Hydrogen was added at constant temperature such that the pressure was raised to 82 kPa. Assuming air contains 77% nitrogen by mass, find the final masses of Hydrogen, Nitrogen and Oxygen in the vessel and their respective partial pressures. Take R_{air} = 287 J/kg · K and R_{H_2} = 4124 J/kg· K.

(**Ans.** m_{1+2} = 0.0006466 kg, m_{N_2} = 0.2862 kg; m_{O_2} = 0.0855 kg; p_{H_2} = 2 kPa;
p_{N_2} = 61.596 kPa; p_{O_2} = 18.404 kPa)

35. A mixture of 10 kg of O_2, 8 kg of N_2 and 25 kg of CO_2 is constained in a rigid vessel, the temperature being 102°C and pressure equal to 250 kPa. Calculate the capacity of the vessel and the partial pressure of each onstituent gas

Take R_u = 8314 kJ/kg · k

(**Ans.** 14.546 m^3, p_{O_2} = 66.98 kPa, p_{N_2} = 61.24 kPa; p_{CO_2} = 121.78 kPa)

□□□

CHAPTER 3

FIRST LAW OF THERMODYNAMIC-CLOSED SYSTEMS

3.1 INTRODUCTION TO FIRST LAW OF THERMODYNAMICS

As explained in chapter 1 that energy can neither be created nor destroyed, it can change forms only. This principle is based on experimental observations and is called the "First Law of Thermodynamics" or "The principle of conservation of energy". The first law of thermodynamics can be stated as follows :

"During an interaction between a system and its surroundings, the amount of energy gained (or lost) by the system must be exactly equal to the amount of energy lost (or gained) by the surroundings. It may be noted that the energy can cross the boundary of the closed system in the form of heat or work both ways.

It is common experience that a bottle of cold drink left on a table in a room will slowly warm up and on the other hand a hot-cup of tea left on the table will cool down to room temperature. In other words we conclude that when a body is left in some medium which is at different temperature, energy trnasfer takes place between the body and the surrounding medium until thermal equilibrium is established, *i.e.*, the body and the surrounding medium reach the same temperature. The direction of energy transfer is always from higher temperature body to lower temperature one. In the case of cold drink bottle the energy in the form of heat was leaving the surrounding medium and entering the cold-drink-bottle until both the bottle and surroundings were at the same temperature. Similarly in the case of hot cup of tea the energy in the form of heat was leaving the tea cup until it cooled down to room temperature. In other words it can be said that once the equality of temperature is achievd the energy transfer stops. Heat is defined as the form of energy that is tranferred between two systems or a system and its surroundings by virtue of temperature difference. In other words, an energy interaction is HEAT only if it takes place due to temperature difference. Therefore it can be concluded that there cannot be heat transfer between two sysems if they are at the same temperature. Alternatively 'Heat' is energy in transition. It is recognised only when it crosses the boundaries of a system. Again referring to the hot cup of tea lying on a table. The tea contained energy but this energy is called Heat only as it passes through the boundaries of the cup to reach air in the room. Once in the surroundings this heat becomes the part of internal energy of the surroundings. Therefore in thermodynamics the term "HEAT" means "HEAT TRANSFER". The transfer of heat is into the system is called "HEAT ADDITION" and similarly transfer of "Heat" out of the system is called "HEAT REJECTION".

3.1.1 Adiabltic Process

A process during which there is no Heat Transfer is called "ADIABATIC PROCESS". There are two ways in which it can happen. Firstly the system is well insulated so that no heat can cross the boundary of the system. secondly both system and the surroundings are at the same temperature and therefore there is no heat transfer. It is very important to note that "ADIABATIC PROCESS" must not be confused with "ISOTHORMAL PROCESS" during which the temperature remains constant. Even though there is no heat tranfer across the boundary of the system during an adiabatic process, still the energy content and as a result of this the temperature of the system can be changed by other means such as work.

Heat being a form of energy has energy units kJ. The amount of heat transfer during a process between two states is denoted by 'Q'. Heat transfer per unit mass of the system is denoted by 'q' such that

$$q = \frac{Q}{m} \text{ kJ/kg} \qquad ...(3.1)$$

Similarly heat tranfer rate is denoted by Q or $\frac{dQ}{dt}$ and is defined as the amount of Heat transfer per unit time and its units are kJ/s which is equivalent to kW.

3.1.2 Sign Convention

Universally accepted sign convention for heat transfer is as follows.

"Heat transfer to a system is positive and heat tranfer from a system is negative". It means any heat transfer which increases the energy of the system is positive and any heat transfer which decreases the energy of the system is negative.

3.2 WORK

Work, like heat, is an energy interaction between a system and its surroundings. As pointed out earlier energy can cross the boundary of a system in the form of heat or work. Therefore if the energy crossing the boundary of a system is not heat (as in the adiabatic process), it must be work. Alternatively an energy interaction which is not caused by a temperature difference between a system and its surroundings is work. To elucidate this further following three examples are cited.

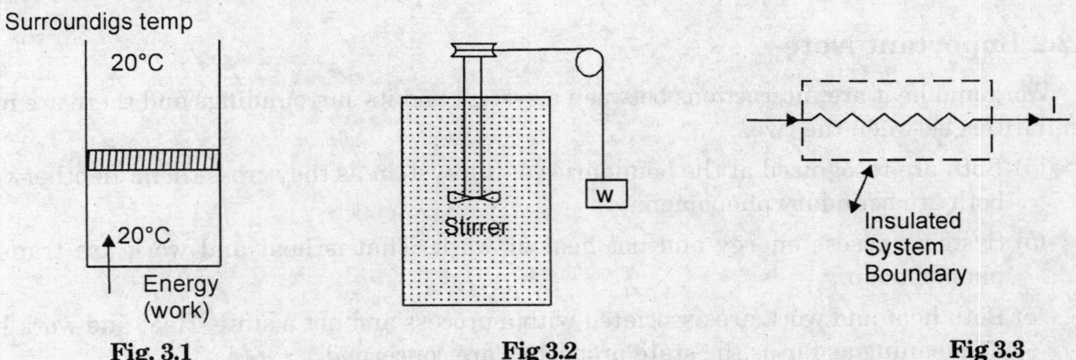

Fig. 3.1 Fig 3.2 Fig 3.3

Fig. 3.1 shows a piston-cylinder assembly in which the system temperature and the surrounding temperature is same. Thus the energy input in the system will be in the form of work, which will result in the upward movement of piston. In Fig. 3.2, the loss of potential energy in the falling weight will be transfered to water through a stirser in the form of work as the vessel is insulated. Similarly Fig. 3.3 is showing a resistance wire through which electric current flows and the energy transfer takes place in the form of work as the boundary is insulated.

Work too is a form of energy and therefore its units are also kJ. If the work done during a process between states 1 and 2 is denoted by W then work done per unit mass of the system, denoted by w is given by

$$w = \frac{W}{m} \quad ...(3.2)$$

The work done per unit time is called 'Power' and denoted by W or $\frac{dW}{dt}$ and its units are kJ/s or kilowatt (kw).

3.2.1 Sign Convention

In case the system produces work it is considered a positive effect on the other-had if the system consumes work it is taken as negative effect. Thus the sign convention for the work will be as under.

"Work done by the system is positive and work done on the system is negative". Thus according to this sign convention the work produced by a tractor engine, or by steam or gas turbine will be taken as positive and the work consumed by pumps, mixers and compresors

It may further be pointed out that the energy of a system decreases as it does work and increases when work is done on the system.

Sometimes in place of positive or negative signs safixes 'in' and 'out' are used. For example Heat transfer to a system is represented by Q_{in} and heat transfer from a system by Q_{out}, that is a heat loss of 10 kJ from a system can be expressed on $Q = -10$ kJ or $Q_{out} = 10$ kJ. Similarly 10 kJ work out can be expressed either as $W = 10$ kJ or $W_{out} = 10$ kJ.

Special Note : When one exlusively, is dealing with work consuming devices like pumps and compressors, the negative sign associated with the work term can conveniently be avoided by speaking of work or power input instead of work done (for example $W_{in} = 2$ kW in place of $W = -2$ kW).

3.2.2 Important Note

Work and heat are interactions between a system and its surroundings and there are many similarities between the two.

(*a*) Both are recognised at the boundaries of the system as they cross them. In otherwords both are boundary phenomena.

(*b*) Systems posess energy and not heat or work. That is heat and work are transient phenomenon.

(*c*) Both heat and work are associated with a process and not a state. Heat and work have no meaning as far as the state properties are concerned.

(*d*) Both are path functions. That is their magnitudes depend on the path followed during a process as well as the end states.

(*e*) Path functions have "inexact differentials" denoted by symbol δ. Therefore, a differential amount of work or heat is represented by δW or δQ respectively and not by dW or dQ. On the contrary properties are point functions *i.e.*, they depend on state only and not how the system reaches that state and they have exact differentials denoted by the symbol *d*. For example a small change in volume is represented by dV and the total volume charge during a process from state 1 and state 2 will be

$$\int_1^2 dV = V_2 - V_1 = \Delta V \quad ...(3.3)$$

That is the volume change during the process from state 1 to state 2 will always be the volume at 2 minus the volume at state 1 irrespective of the path followed. In Figure 3.4 is shown the states 1 and 2. Starting from state 1 two different processes represented by paths *A* and *B* have been used to reach the state 2.

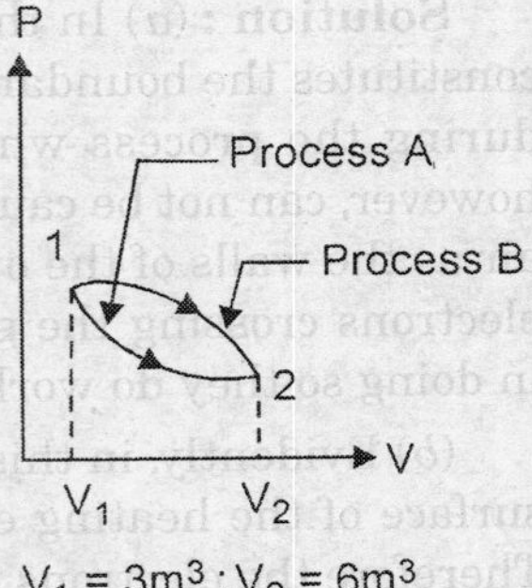

Fig. 3.4

However total work done during the process 1 – 2 is

$$\int_1^2 \delta W = W_{12} \quad ...(3.4)$$

In other words total work is obtained by taking with account the process path and adding the differential amounts of work (δw) done along the path *A* or *B* as the case may be. Evidently

$$\int_1^2 \delta W \text{ is not } (W_2 - W_1).$$

i.e., $\delta W \neq$ (work at state 2 – work at state 1)

This is so because work is not a property and systems do not posses work at a state.

Example 3.1. *A candle is burning in an insulated room. Considering the room including the candle and air, as a system, determine (a) if there is any heat transfer during the burning of candle process and (b) if there is any change in the internal energy.*

Solution : (*a*) In this case system boundary is the inner surface of the room. Since heat is recognised if it crosses the boundary and in this case the room is insulated, there will be no heat transfer across the boundary.

Thus $Q = 0$ for the process.

(*b*) Internal energy is the sum total of energies that exist in the system, in various forms viz sensible, chemical, nuclear etc. In the given problem part of chemical energy is converted to sensible energy. That is part of internal energy of the system is changed from one form to another form and there is no increase or decrease in the total amount of internal energy of system.

Therefore $\Delta U = 0$ for this process

Example 3.2. *A piece of sweet potato which is initially at room tempeature of 28°C is kept in an oven for baking. The temperature of the oven is maintained at 190°C and its walls are insulated from the atmosphere. Discuss whether there will be heat transfer in the following two cases*

(*i*) *If the sweet patato is taken as system*

(*b*) *It the oven including the sweet potato is taken as the system.*

Solution : (*a*) In this case the skin of the sweet potato will be the boundary of the system. Part of the energy in the oven will pass through the skin of the sweet potato because of the temperature difference. Thus this will be heat transfer process.

(*b*) In this case the oven including the sweet potato is the system and its boundary is the innder periphery of the over. This boundary is insulated also. Thus there will be no crossing of energy across the boundary *i.e.*, $\Delta Q = 0$ or there will be no heat transfer.

Example 3.3. *An electric oven having well insulated walls is being heated with the help of an electric heating element. It the oven along with the heating element is considered as one system, discuss whether this is a heat or work internaction. Also discuss the case where the system is taken as only air in the oven without the heating element.*

Solution : (*a*) In this problem the inner periphery of oven including the heating element constitutes the boundary which is well insulated. The energy content of the oven is increasing during the process which is evidenced by the rise in temperature. This energy transfer, however, can not be caused by the temperature difference between the oven and surroundings air as the walls of the oven are insulated. However the energy transfer is being caused by the electrons crossing the system boundary, when they flow through the heating element. Thus in doing so they do work. Therefore it is a work interaction.

(*b*) Evidently, in this case the system boundary will be the outer surface of the heating element as shown in Fig. 3.5 by dotted line. Therefore the electrons flowing in the heating element will not cross the system boundary at any point. The energy generated inside the heating element will be transfered to the surrounding air as a result of temperature difference between the heating element and the surrounding air. Therefore it will be a heat transfer process.

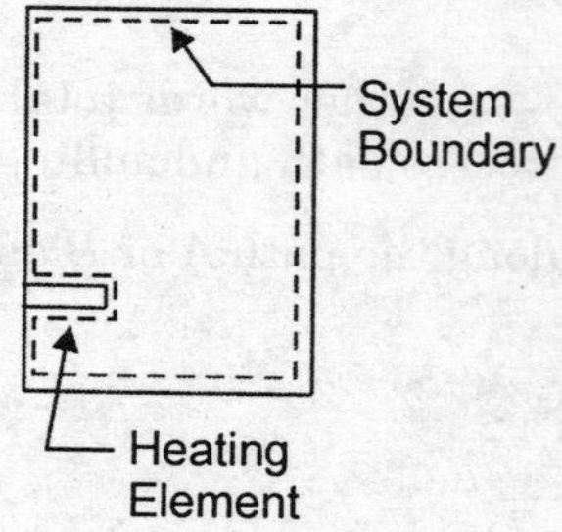

Fig. 3.5

A point worth noting is that in both the cases the energy transfer to the air is same. Thus depending on how the system is demarcated the process could be work interaction or heat transfer.

3.3 ELECTRICAL WORK

It was pointed out in example 3.3 that electrons crossing the system boundary do work on the system. However in an electric field electrons move in the wire or element due to potential difference. Thus when N electrons move through a potential difference V the electrical work done is given by

$$W_e = VN \text{ kJ} \qquad \text{...(3.5)}$$

Equation (3.5) when expressed in rate form

$$W_e = V\frac{dN}{dt}$$

$$= VI \text{ (kw)} \qquad \text{...(3.6)}$$

Where W_e = eletrical power in kw

V = Number of electrons moving per unit time

i.e., I = The current

In general both V and I may change with time and the electrical work done during the time internal ΔT is expressed as

$$W_e = \int_{t_1}^{t_2} VI\,dt \qquad ...(3.7)$$

In case both V and I remain constant during the time internal ΔT equation (3.7) reduces to

$$W_e = VI\,\Delta T \qquad \text{(kJ)} \qquad ...(3.8)$$

Example 3.3. *(a) A small tank, B containing iced water at 0°C is placed in a well insulated bigger tank, A filled with oil as shown in Fig. (3.6). the oil is also at 0°C temperature. An electric heater in the oil is then switched on for some time so that 15 kJ of electrical work is done on the oil. Even some of the ice in the tank B has melted but after some time gap it is noticed that the temperature of the entire system is again at 0°C. Considering the oil as system A and water as system B discuss whether the interaction between the system A and system B will be heat transfer or work.*

(b) What type of interaction will prevail if the oil and iced water are considered as a combined system.

Solution : *(a)* Mark the boundaries of the tanks A and B by dotted lines as shown in Fig. 3.6.

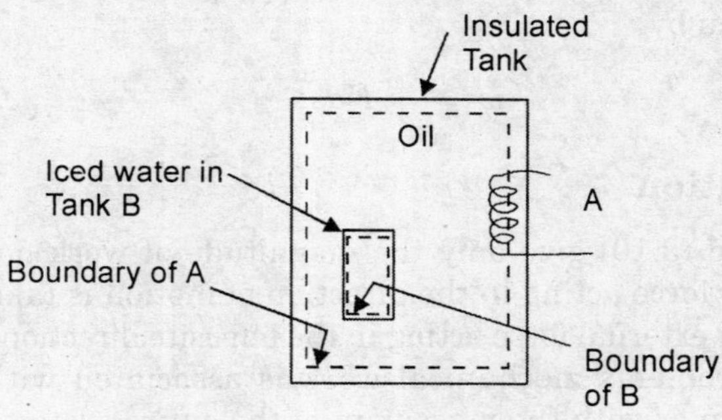

Fig. 3.6

When the heater is turned on electrons will cross the boundary of system A. Therefore work will be done. Thus work is being done on the system this work which is called the electrical work, will be done on the system A.

The work is done on the system, then as per the sign convention.

$$W_A = 15 \text{ kJ}$$

Due to this energy which is added to the system A, it will raise its temperature. Due to this temperature differential the temperature of oil will rise thereby creating a temperature gradient between oil and water which in turn will result in heat flow process from oil to water through their common boundary. Since the final temperature of oil again is 0°C the energy lost by system A as heat must be equal to energy gained as work

There $\qquad Q_A = -15 \text{ kJ}$

or $\qquad Q_{A,\,out} = 15 \text{ kJ}$

System *B* The only energy interaction at the boundary of system B is the heat flow from system A. All the heat lost by system A *i.e.*, oil is gained by the system B *i.e.*, iced water. there is no work interaction at the boundary of system B.

Thus $\qquad W_B = 0 \qquad$ and $\qquad Q_B = +\,15$ KJ

(b) Oil and Iced Water as one Combined System

In this case the outer boundary of system A will form the entire boundary of the combined system. As explained in part (a), the only energy interaction of this boundary is the electrical work. Further the tank is insulated therefore no heat will cross this boundary.

Thus $$W_{combined} = -15 \text{ kJ}$$

and $$Q_{combined} = 0$$

It may be added here that the heat flow from oil to iced water will be an internal process for the combined system and it is simply the re-distribution of interval energy of the combined system.

3.4 MECHANICAL FORMS OF WORK

Mechanical work can be done in different ways. However each in some way is the outcome of a force acting through a distance. It is a well known equation of mechanics that work done by a constant force F acting on a body which is displaced by a distance s is given by

$$W = Fs \text{ kJ} \quad ...(3.9)$$

In case F is not constant and varies with s, then for moving the body from position 1 to position 2 work is given by

$$w = \int_1^2 F\,ds \quad ...(3.10)$$

3.4.1 Sign Convention

Equation (3.9) and (3.10) give only the magnitude of work done. The work done on the system by an external force acting in the direction of motion is taken negative and work done by a system against an external force acting in the opposite direction to motion is taken positive. In thermodynamic problems mechanical work is associated with either movement of the boundary of the system or with the movement of the entire system.

3.4.2 Moving Boundary Work

In thermodynamic problems frequently encountered mechanical work is associated with compression or expansion of gas in a piston-cylinder device. During this process part of the boundary *i.e.*, the inner surface of the piston moves up and down. Therefore the expansion and compression work is often called "MOVING BOUNDARY WORK" or sometimes "BOUNDARY WORK". Some authors call it $P \cdot dV$ work as they argue that the moving boundary work is the primary form of work involved in automobile engines. During their expansion the combustion gases force the piston move and this in turn forces the crank shaft to rotate.

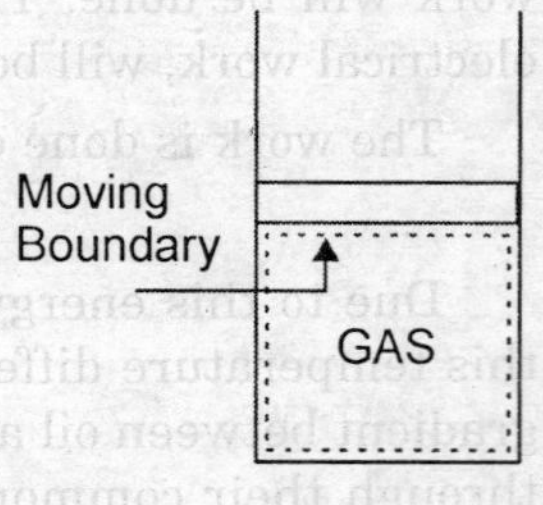

Fig. 3.7

However the moving boundary work in real engines and compressors cannot be determined exactly with the help of thermodyanmic analysis alone because the piston usually moves at a very high speed which makes it very difficult for the gas inside to maintain equilirbium and in addition there is friction between the piston and cylinder. As a result of high speed of the piston the states that the system passes through can not be specific and the process path can not be drawn. Thus work being a path function, cannot be determined analytically in the absence of the knowledge of the path. Therefore in real engines or compressors the boundary work is determined by direct measurements experimentally. Still in order to make thermodynamic analysis an assumption of quasi-equilibrium process is made. A quasi-equilibrium process is one in which the system remains in equilibrium at all times. This process is closely approximated by real engines, especially when the piston speed is very low. Under identical conditions, the work out put of the engines is found to be a maximum (and the work input to the compressor to be minimum) when quasi-equilibrium processes are used in place of non-quasi-equilibrium processes.

In order to calculate the moving boundary work for a quasi-equilibrium process, consider the gas contained in a piston-cylinder device shown in Fig. 3.8. Let the initial pressure of the gas be P, total volume U and the area of cross-section of the piston be A. Let the piston be allowed to move a distance ds in a quasi-equilibrium manner, then the differential work done in this process is given by

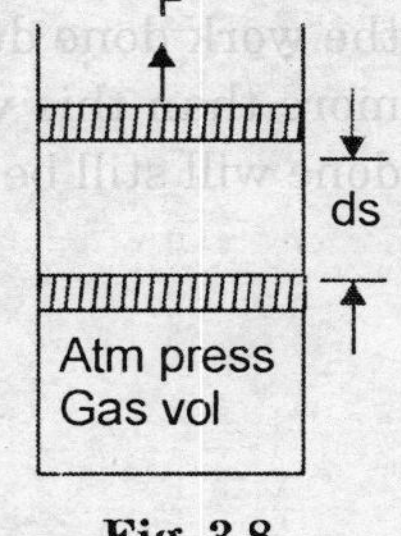

Fig. 3.8

$$\delta W_b = F\,ds$$

$$= PA\,ds$$

or

$$\delta W_b = P\,dV \text{ (kJ)} \quad ...(3.11)$$

where δW_b = Differential moving boundary work

W_b = Moving boundary work

dV = Differential change in volume

P = Absolute pressure of the gas at the instant when the volume is V

Thus from equation (3.11) we conclude that the boundary work in the differential form is equal to the product of absolute pressure and the differential change in volume. Because of the expression, some of the authors prefer to call this as $P \cdot dV$ work as pointed above.

P being the absolute pressure, will always be positive. However volume change, dV, will be positive during the expansion process but negative during the compression process. Thus during expansion process boundary work will be positive and during compression process it will be negative.

Integrating equation (3.11) from the initial state to the final state we get the total boundary work during the entire process as

$$W_b = \int_1^2 P\,dV \quad \text{(kJ)} \quad ...(3.12)$$

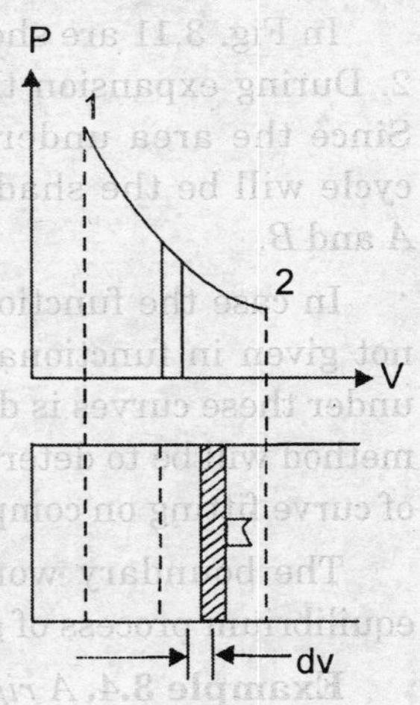

Fig. 3.9

In order to evaluate the integral functional relationship between P and V during the process must be known. That is $P = f(V)$ should be known or simply the equation of the process path on P–V space should be available. Quasi-equilirbium expansion process described above is shown in P.V. diagram in Fig. 3.9

Elemental area under the process path dA is given by

$$dA = P\,(dV)$$

The total area under the process curve from states 1 to 2 is then

$$A = \int_1^2 P\,dV \quad ...(3.13)$$

Comparasion of equation (3.13) with equation (3.12) reveals that the area under the process curve on P–V diagram is equal in magnitude to the work done during quasi-equilibrium expansion or compression process of a closed system. Similarly on P.v. diagram it will represent the boundary work done per unit mass.

A gas can, in general, follow different paths as it expands from state 1 to state 2. The area under one path curve will be different from the area under other path curve. In other words the work done during expansion from state 1 to state 2 if path A is followed. (Fig 3.10) will be more than this work done if path B is followed. Similarly if the path C is followed the work done will still be smaller

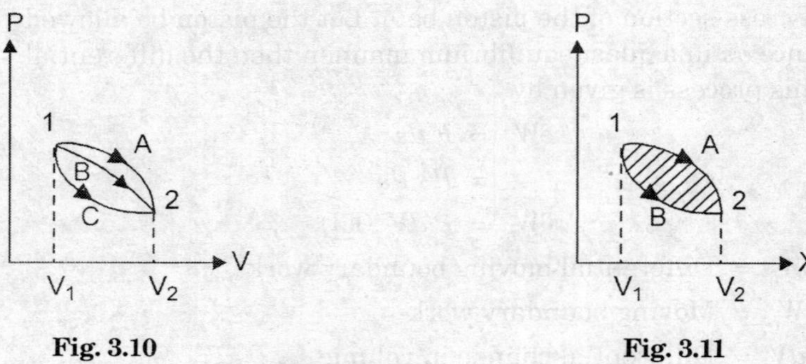

Fig. 3.10 **Fig. 3.11**

i.e.,

$$W_A > W_B > W_C \qquad ...(3.14)$$

Note : If work were not a path function, no cycle device (car engines, or power plants) could operate to genrate power because the work output during expansion would then get consumed during another path.

In Fig. 3.11 are shown two quasi-equilibrium process paths between the state 1 and state 2. During expansion the path A is followed and during compression the path B is followed. Since the area under the curve A is more than that under curve B, the net out put in cycle will be the shaded area which is the difference between the areas under the curve A and B.

In case the functional relationship between P and V during expansion or compression is not given in functional form, rather it is given in the form of experimental data the areas under these curves is determined graphically to calculate the net work out put per cycle. Other method will be to determine the relationship between P and V for both the paths by the method of curve fitting on computer for which software packages are available.

The boundary work relationship given in equation (3.12) is not limited to the quasi-equilibrium process of gases only but it can also be applied to liquids and solids.

Example 3.4. *A rigid pressure vessel contains air at 400 kPa and 140°C. In the absence of any insulation heat is transfered to the surroundings and the pressure and temperature inside the pressure vessel drop to 300 kPa and 60°C respectively. Calculate the boundary work during the process.*

Solution : Since the pressure vessel is rigid and not insulated heat is allowed to cross the boundary of the vessel but there will be no boundary work. Alternatively assuming the process to be quasi equilibrium, we can evaluate the boundary work by using equation (3.12).

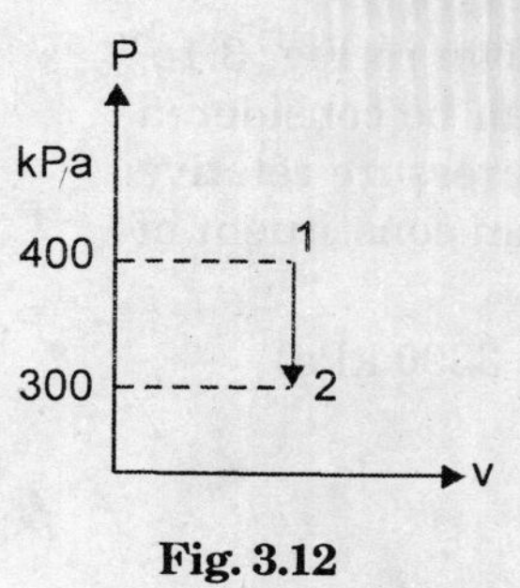

Fig. 3.12

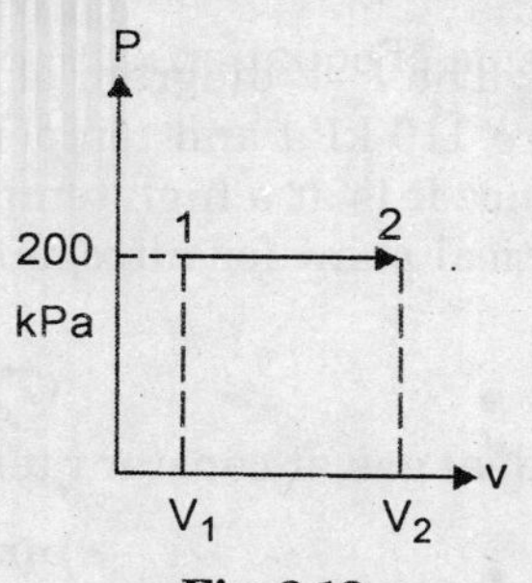

Fig. 3.13

i.e., $$W_b = \int_1^2 P\,dV \qquad \text{...(3.15a)}$$

Since the pressure vessel is rigid it volume will remain constant and the process path on P.v. diagram will be as shown in Fig. 3.12. It is evident from the Fig. 3.12 that

$$dV = 0$$

Therefore the boundary work W_b, then, from equation (3.15) will be

$$W_b = 0$$

Thus there is no boundary work done during this process. From this we conclude that boundary work done during a constant-volume process is always zero.

From Fig. 3.12, where P.v. diagam for constant volume process is shown the area under the path curve which represents the boundary work is zero.

Example 3.5. *A frictionless piston cylinder device contains 3 kg of saturated water vapor at 200 kPa. Heat is now added to water vapor until the temperature reaches 400°C. Calculate the work done by the steam during the process.*

Solution : The P.v. diagram for the constant pressure process is shown in Fig. 3.13. At state (1) the water vapor is saturated. Therefore its specific volume can be read from the steam table. Corresponding to pressure of 200 *kPa*, it value is given as v_g = 0.8857.

Since the piston is moving the work done is due to moving boundary. Moreover the pressure remains constant at 200 kPa level.

Therefore $$W_b = \int_1^2 P\,dV$$

$$= P\int_1^2 dV = P(V_2 - V_1) = P_m(V_2 - V_1) \qquad \text{...(3.15b)}$$

Also form steam table for superheated steam corresponding to pressure of 200 kPa and Temperature 400°C the specific volume is 1.5493.

Therefore $$W_b = 200(\text{kPa})\,(3\text{ kg})\,(1.5493 - .8857)\,(\text{m}^3/\text{kg})$$
$$= 398 \text{ kJ}$$

Note : The positive sign W_b indicates that the work is done by the system. That is the steam used 398 kJ of its energy to do this work. The area under the process curve (Fig. 3.13) would also represent this work done by steam.

Example 3.6. *An air tight piston-cylinder device contains 0.6 m³ of air at 110 kPa and 85°C. The air is now compressed to 0.15 m³ slowly so that the temperature inside the cylinder remains constant. Find the work done during the process.*

Solution : The P–V diagram of the process is drawn in Fig. 3.14. At the pressure 110 kPa and temperature 85°C air can be considered as ideal gas since it is at a high temperature and low pressure relative to those at critical point for nitrogen, which is the main constituent of air

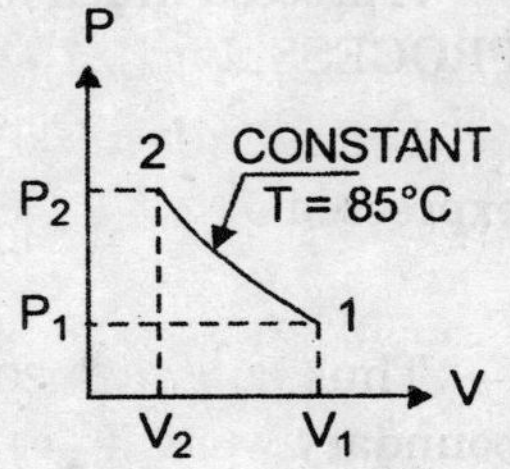

Fig. 3.14

$$(T_{cr} = -147°C, P_{cr} = 3390 \text{ kPa})$$

Thus for ideal gas at constant temperature T

$$\frac{PV}{T} = mR$$

or

$$PV = mRT = C$$

where T and C are constants

or

$$P = \frac{C}{V} \qquad \text{...(3.16)}$$

thus substituting value of P from this equation in equation (3.12)

$$W_b = \int_1^2 P\,dV$$

$$= \int_1^2 \frac{C}{V}\,dV$$

$$= C[\ln V]_1^2$$

$$= C\ln\left(\frac{V_2}{V_1}\right)$$

or

$$W_b = PV\,ln\left(\frac{V_2}{V_1}\right) \qquad \text{...(3.17)}$$

In equation (3.17) $\quad PV = P_1V_1 = P_2V_2 = \text{mRT}$

Therefore substituting numerical values in equation (3.17), we have

$$W_b = (110 \text{ kPa})\,(0.6 \text{ m}^3)\ln\left(\frac{0.15}{0.6}\right)\left(\frac{1 \text{ kJ}}{1 \text{ kPa} - \text{m}^3}\right)$$

$$= 66\, l_n\,(0.25) \text{ kJ} = -91.495 \text{ kJ}$$

Note : Negative sign indicates that work is done on the system which is always the case for compression process.

3.4.3 Work Done During Polytropic Process

During expansion and compression process of real gases pressure and volume mostly follow the following relationship

$$PV^n = C \qquad \text{...(3.18)}$$

where n and C are constants

A process which follows the equation (3.18) is called a POLYTROPIC PROCESS. Fig. 3.15 shows the P.v. diagram of such process.

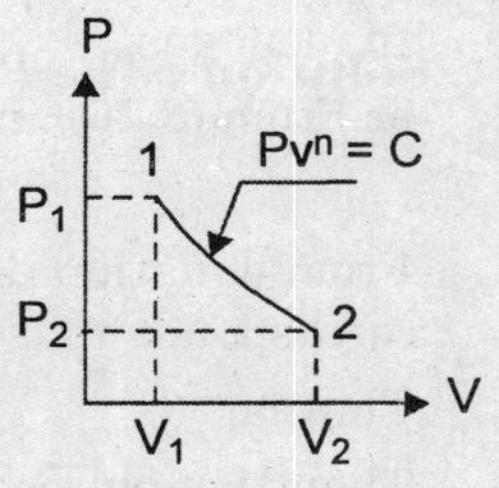

Fig. 3.15

From equation (3.18) the pressure during the process can be expressed as

$$P = C V^{-n} \quad ...(3.19)$$

Thus substituting this value of P in equation (3.12), the moving boundary work W_b is given by

$$W_b = \int_1^2 P\, dV$$

$$= \int_1^2 C V^{-n}\, dV$$

$$= C\left[\frac{V^{-n+1}}{-n+1}\right]_1^2$$

$$W_b = C\left[V_2^{-n+1} - V_1^{-n+1}\right] \quad ...(3.20)$$

Also, from equation (3.18)

$$C = P_1V_1^n = P_2V_2^n \quad ...(3.21)$$

From equation (3.20) and (3.21)

$$W_b = \left(\frac{1}{1-n}\right)\left[P_2V_2^nV_2^{-n+1} - P_1V_1^nV_1^{-n+1}\right]$$

$$= \left(\frac{1}{1-n}\right)[P_2V_2 - P_1V_1]$$

or

$$W_b = \frac{P_2V_2 - P_1V_1}{(1-n)} \quad ...(3.22)$$

Also for ideal gas $\quad P_1V_1 = mRT_1 \quad ...(3.23a)$

and $\quad P_2V_2 = mRT_2 \quad ...(3.23b)$

Therefore from equations (3.22) and (3.23), we get

$$W_b = \frac{mRT_2 - mRT_1}{1-n}$$

$$W_b = \frac{mR(T_2 - T_1)}{(1-n)} \quad ...(3.24)$$

Thus the final expression for moving boundary work for polytropic process with respect to ideal gases is

$$W_b = \frac{P_2V_2 - P_1V_1}{1-n} = \frac{mR(T_2 - T_1)}{1-n} \quad ...(3.25)$$

Important Note : For a special case in equation (3.18) if $n = 1$, then

$$PV = C \quad ...(3.26)$$

In ideal gas $$PV = mRT \quad \text{...(3.27)}$$

or From (3.26) and (3.27) we have

$$PV = mRT = C \quad \text{...(3.28)}$$

From equation (3.28) we infer that when $n = 1$ than T must be constant. In other words $n = 1$ corresponds to isothermal process.

That is $$T_2 = T_1$$

Then the expression for W_b given in equation (3.24) fails as it leads to an in determinate solution for W_b.

Thus for an ideal gas undergoing an isothermal process W_b is found as follows

$$PV = mRT_0 = \text{constant} = C \text{ (say)}$$

or $$PV = C$$

or $$P = \frac{C}{V}$$

Then from equation (3.12) $$W_b = \int_1^2 P\,dV = \int_1^2 \frac{C}{V} dV = C[ln(V)]_1^2$$

or $$W_b = C\,ln\left(\frac{V_2}{V_1}\right)$$

$$= PV\,ln\left(\frac{V_2}{V_1}\right)$$

or $$W_b = P_1V_1\,ln\left(\frac{V_2}{V_1}\right) = P_2V_2\,ln\left(\frac{V_2}{V_1}\right) = mRT\,lr\left(\frac{V_2}{V_1}\right)$$

Example 3.7. *A frictionless piston-cylinder device contains 7 kg of superheated steam at 500 kPa and 300°C. Steam is now cooled at constant pressure until 60% of it by mass condenses. Find the work done during this process.*

Solution : The cooling process is shown in Fig. (3.16). During the process the pressure remains constant. The specific volumes at the states 1 and 2 are read from the steam tables.

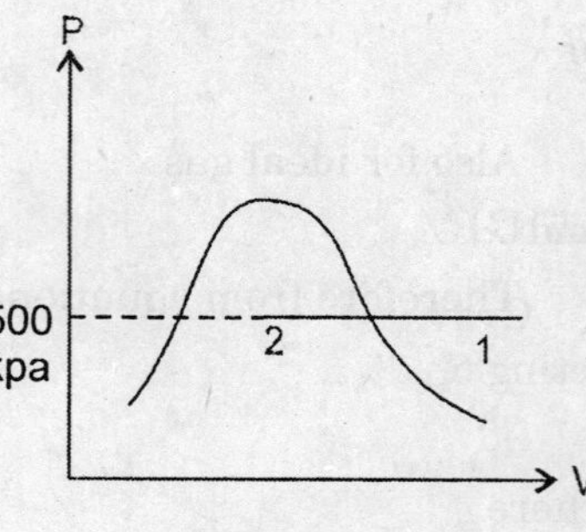

Fig. 3.16

At state 1 $$\left.\begin{array}{l} P_1 = 500\,kPa \\ T_1 = 300°C \end{array}\right]$$

Corresponding to these, the specific volume $v_1 = 0.522$ m³/kg.

At state 2
$$P_2 = 500 \text{ kPa}$$
$$x_2 = 1 - 0.6 = 0.4$$
$$\therefore \quad v_2 = v_f + x_2\,v_{fg}$$
$$= 0.001093 + 0.4\,(.3749 - 0.001093)$$
$$= 0.001093 + 0.1495228$$
$$= 0.1506 \text{ m}^3\text{/kg}$$

Thus $$W_b = \int_1^2 PdV = P(V_2 - V_1) = P_m(v_2 - v_1)$$

$$= (500\,kPa)\,(7\text{ kg})(0.1506 - 0.5226)\,\text{m}^3/\text{kg}\left[\frac{1\text{ kg}}{1\text{ kPa}\cdot\text{m}^3}\right]$$

$$= 1302\text{ kJ}$$

Example 3.8. *A frictionless piston-cylinder device contains 0.05 m³ gas at 100 kPa pressure. If this gas expands to the final volume of 0.3 m³ following the equation $PV^{1.2}$ = constant, find the work done*

Solution : The P–V diagram for this polytropic process is shown in Fig 3.15 where n = 1.2, V_1 = 0.05 m³ and V_2 = 0.3 m³ The pressure at state 2 can be calculated from the given relationship.

$$p_1 V_1^{1.2} = P_2 V_2^{1.2} = C$$

or $$P_2 = P_1\left(\frac{V_1}{V_2}\right)^{1.2}$$

$$= (100)\left[\frac{0.05\,\text{m}^3}{0.3\,\text{m}^3}\right]^{1.2}$$

$$= 11.65\text{ kPa}$$

From equation (3.5) the moving boundary work is given as

$$W_b = \frac{P_2V_2 - P_1V_1}{1-n}$$

$$= \frac{11.65\times 0.3 - 100\times .05}{1-1.2}\left(\text{kPa}\cdot\text{m}^3\right)\left[\frac{1\text{ kJ}}{\text{kPa}\cdot\text{m}^3}\right]$$

$$= 7.53\text{ kJ}$$

3.5 GRAVITATIONAL WORK

Gravitational work is defined as the work done by or against gravitational field. The force acting on a body of mass m is given by

$$F = mg \qquad ...(3.30)$$

where g is the acceleration due to gravity and is constant.

The work required to raise the body from a level z_1 to z_2 is given by

$$W_g = \int_1^2 Fdz = \int_1^2 mg\,dz$$

$$= mg(z_2 - z_1) \qquad ...(3.31)$$

Where $(z_2 - z_1)$ is the vertical distance moved by the body. Evidently the expression on the right hand side of equation (3.31) represents the change in potential energy. Moreover W_g is independent of the path followed but it does depend on the end states and the work done is equal in magnitude to the change of potential energy.

3.5.1 Sign Convention

Gravitation work done by the system is taken as positive. Similarly it is taken negative if work is done on the system.

Example 3.9. *Find the work done by a person to lift 20 kg log by 0.5 m from the ground.*

Solution :

$$W_g = mg(z_2 - z_1)$$
$$= 20 \times 9.807\,(0.5)$$
$$= 98.07 \text{ kJ}$$

3.6 ACCELERATION WORK

Whenever the velocity of a system changes, it accelerates or retards (negative acceleration). The work associated with the change in velocity is dfined as "Acceleration" work.

The acceleration work required to accelerate a body of mass m from initial velocity V_1 to final velocity V_2 is determined by invoking by Newton's second law of motion. Force required to generate an acceleration 'a' in a body of mass m is given by

$$F = ma$$

$$= m\frac{dV}{dt} \qquad \text{...(3.32)}$$

Also differential displacement ds occuring in time dt we have

$$\frac{ds}{dt} = V$$

or

$$ds = V\,dt \qquad \text{...(3.33)}$$

Further acceleration work done in time dt

$$W_a = F\,ds \qquad \text{...(3.34)}$$

Therefore from equation (3.32), (3.33) and (3.34), we get

$$W_a = \int_1^2 F\,ds = \int_1^2 \left(m\frac{dV}{dt}\right)(V\,dt)$$

$$= m\int_1^2 V\,dv = \frac{1}{2}m\left(V_2^2 - V_1^2\right) \quad \text{(kJ)} \qquad \text{...(3.35)}$$

Clearly the work done to accelerate a body is independent of path. From equation (3.35) it is evident that the acceleration work is same as the change in Kinetic Energy.

3.6.1 Sign Convention

If the work is done by the system, the Acceleration work is taken positive, Similarly the acceleration work is taken negative (*–ve*) if work is done on the system.

Example 3.10. *Find the power required to accelerate a car from rest to a velocity of 100 km/hr in 30 seconds on a level road? Take the mass of the car as 1000 kg.*

Solution : The acceleration work W_a is calculated from equation 3.35

$$W_a = \frac{1}{2}m(V_2^2 - V_1^2)$$

$$= \frac{1}{2}\times(1000)\left[\frac{(1000\times 1000\text{ m}^2)}{3600\text{ s}} - 0\right]\left(\frac{1\text{ kJ}}{1000\text{ kg}\cdot\text{m}^3/\text{s}^2}\right)$$

$$= 500\left[\frac{1000}{36}\right]^2\frac{1}{1000}\text{ kJ}$$

$$= 385.8\text{ kJ}$$

Average power $\dot{W}_a = \frac{W_a}{\Delta t} = \frac{385.8}{30}$

$$= 12.86\text{ kW}$$

3.7 SHAFT WORK

In Engineering pratice, energy transmission with a rotating shaft is very common. Usually the torque, T, transmitted through a shaft is constant which is normally due to a constant force F acting with a fixed moment arm r.

Thus, $T = F\cdot r$

or $$F = \frac{T}{r} \quad \text{...(3.36)}$$

If the rotational speed of the shaft in r.p.m is N, then in one revolution the work done by force F is obtained by multiplying the distance, through which force acts, which is equal to 2π r.

Thus shaft work in N revolution

$$W_{sh} = F(2\pi rN) \quad \text{...(3.37)}$$

From equation (3.36) and (3.37)

$$W_{sh} = 2\pi NT\text{ Nm / min}$$

The power transmitted through the shaft per sec will then be

$$W_{sh} = P_{sh} = \frac{2\pi NT}{60}\text{ Nm/sec}$$

3.7.1 Sign Convention

The sign of the shaft work in taken positive if done by the system and negative if done on the system

Example 3.11. *Find the power transmitted through the input shaft of a speed reducer when the torque applied is 100 Nm and the shaft rotates at 2000 rpm*

Solution : $\dot{W}_{sh} = P_{sh} = \frac{2\pi NT}{60}$

$$= \frac{2\pi(2000)(100)}{60}\text{ J /s or watts}$$

$$= \frac{2\pi(2000)(100)}{60}\times\frac{1}{1000}\text{ kW}$$

$$= 20.94\text{ kW}$$

3.8 SPRING WORK

When an axial force is applied on a spring it deflects. The deflection is directly proportional to the force and the deflection x is given by the relationship

$$F = kx \quad ...(3.40)$$

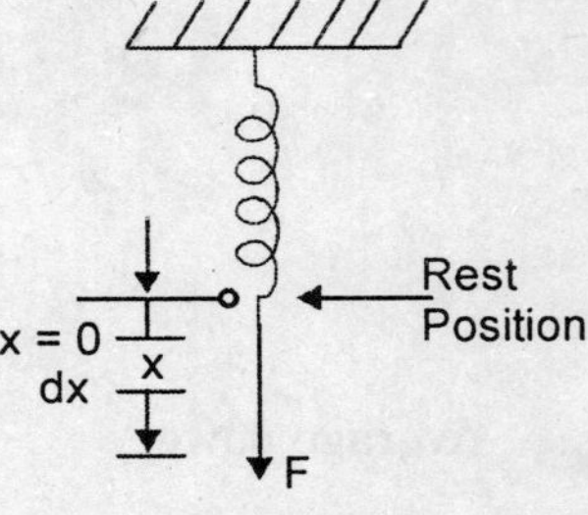

Fig. 3.17

where k is the spring constant also known as stiffness of spring which is defined as the force required to cause unit deflection *i.e.*, (kN/m) The deflection or displacement x of the spring is measured from the position of rest or undisturbed position when the force is zero (*i.e.*, when $F = 0$, $x = 0$).

For elemental deflection dx, corresponding to force F, the elemental spring work is then

$$\delta W_{\text{spring}} = F\,dx$$

$$= kx\,dx \quad \text{(From 3.40)}$$

Integrating this for the range x_1 to x_2 where x_1 and x_2 are the initial and final displacements of the spring both measured from the position of rest, we get

$$W_{\text{spring}} = \int_{x_1}^{x_2} kx\,dx$$

or

$$W_{\text{spring}} = \frac{1}{2}k(x_2^2 - x_1^2)$$

Example 3.12. *A frictionless piston-cylinder device initially contains 1 m³ Hydrogen at 100 kPa. At this state a linear spring ($F \propto x$) with a stiffness of 200 kN/m is touching the piston but exerts no force on it. The cross-sectional area of the pistion is 0.8 m². Heat is transferred to Hydrogen, causing it to expand until its volume doubles. Determine (a) the final pressure, (b) the total work done by the hydrogen and (c) the fraction of this work done against the spring show the process on P.v. diagram.*

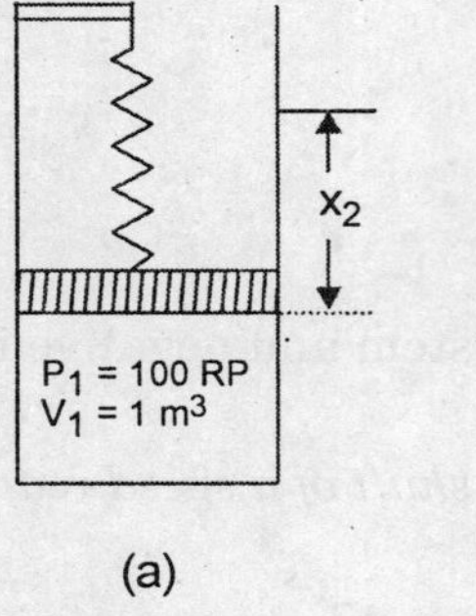

(a)

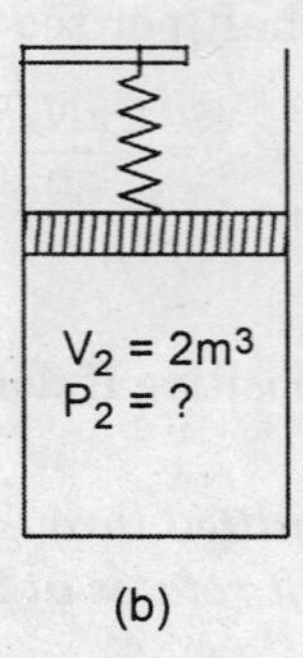

(b)

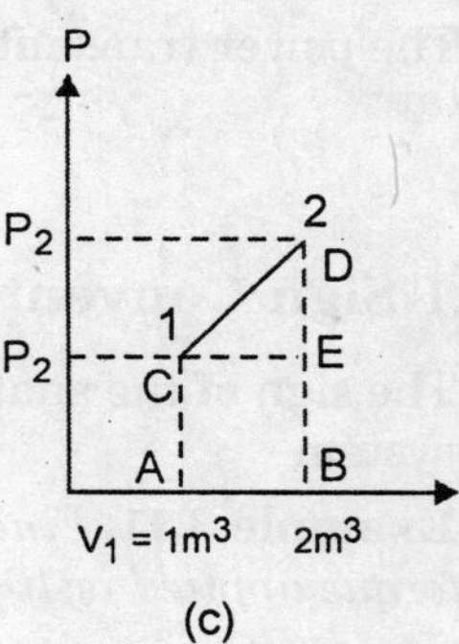

(c)

Fig. 3.18

Solution : Area of piston $= A = 0.8\text{m}^2 \qquad V_1 = 1\text{ m}^3 \qquad V_2 = 2\text{m}^2$

$$k = 200 \text{ kN/m}$$

Then
$$x = \frac{\text{change in volume}}{A} = \frac{V_2 - V_1}{A} = \frac{(2-1)}{(.8)} = \frac{1}{0.8}\text{ m}$$

Then
$$F_2 = \text{spring force at the position D}$$

New at position D the deflection of spsing is x_2

$$F_2 = kx_2 = 200\left(\frac{\text{kN}}{\text{m}}\right)\frac{1}{0.8}(\text{m}) = 250 \text{ kN}$$

Therefore
$$P_2 = P_1 + \frac{F_2}{A}$$

$$= 100(\text{kPa}) + \frac{250 \text{ kN}}{(0.8\text{m}^2)}$$

$$= 412.5 \text{ kPa}$$

(*b*) The pressure of Hydrogen will change linearly with volume during the process. Therefore the P.v. diagram will be a straight line as shown in Fig. (3.18 *c*). Since the piston has moved there will be boundary work which will be equal to the area under the process curve.

Thus
$$W_b = \text{Area of Trapezium CABD}$$

$$= \frac{1}{2}(AC + BD) \times AB$$

$$= \frac{1}{2}(P_1 + P_2) \times (V_2 - V_1)$$

$$= \left[\frac{1}{2}(100 + 412.5) \text{ kPa}\right] [(2\text{-}1)\text{m}^3]\left[\frac{1 \text{ kJ}}{1 \text{ kPa} \cdot \text{m}^3}\right]$$

$$= 256.25 \text{ kJ}$$

(*c*) Had there been no spring, we would have got a constant pressure process represented by the line *CE* (Fig. 3.18 *c*). then the work done during the process would have been (at constant *P*).

$$W_B, \text{ no spring} = \int_0^2 P_1 W = \text{Area } ABEC$$

$$= AC \times AB$$

$$= P_1(2-1) = (100 \text{ kPa})(1 \text{ m}^3)\left[\frac{1 \text{ kJ}}{1 \text{ kPa} \cdot \text{m}^3}\right]$$

$$= 100 \text{ kJ}$$

Thus spring work
$$= W_b - W_{b,\text{ spring}}$$

$$= 256.25 - 100 = 156.25 \text{ kJ}$$

Example 3.13. *A piston-cylinder device constains 50 kg of water at 150 kPa and 250°C. The cross-sectional area of the piston is 0.1 m^2. Heat is now transferred to water, causing part mass of it to evaporate and expand. When the volume reaches 0.2 m^3, the piston reaches a linear spring whose spring constant is 100 kN/m. More heat is transferred to water until the piston moves 0.2 m more. Find (a) the final pressure and temperature and (b) the work done during the process. Show the process on P.V. diagram.*

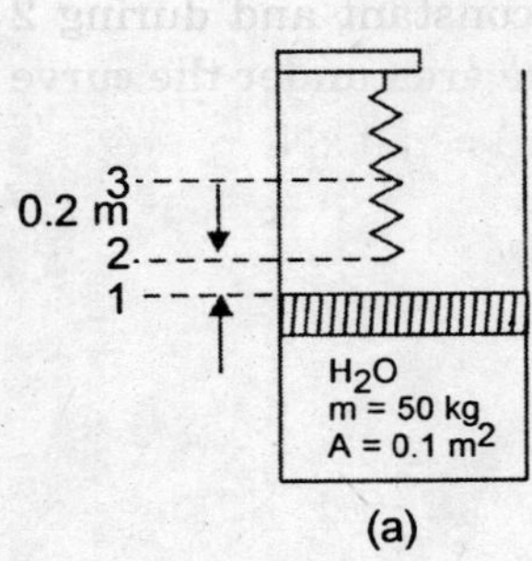

(a)

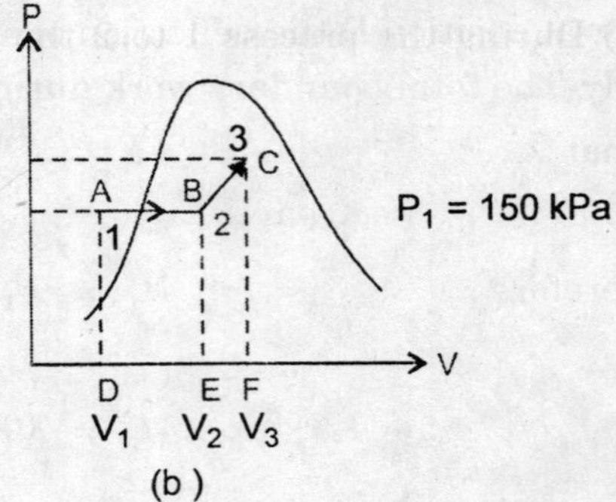

(b)

Fig. 3.19

Solution : When heat is added to water, it starts evaporating and the pressure in the cylinder remains constant till the piston touches the lower end of the spring (Fig. 3.19 *a*) that is the process 1 – 2 is a constant pressure process. After that due to the presence of the spring the pressure will rise linearly (Fig. 3.19 *b*) such that

$$P_3 = P_2 + \frac{F_s}{A}$$

$$= P_2 + \frac{kx}{A}$$

or

$$P_3 = 150(\text{kPa}) + \frac{(100\ \text{kN/m})(0.2\ \text{m})}{0.1\ \text{m}^2}\left[\frac{1\ \text{kPa}}{1\ \text{kN/m}^2}\right]$$

$$= 350\ \text{kPa}$$

The specific and total volumes at the three states are ;

At $T_1 = 25°C$

$P_1 = 150$ kPa

$v_1 \simeq v_f$ at 25°C because the point lies in the liquid state

$\approx .001003$ m³/kg (from steam tables)

Therefore $V_1 = mv_1 = 50 \times .001003 = 0.05\ \text{m}^3$

At State 2 : $V_2 = 0.2\ \text{m}^3$ (given)

At State 3 : $v_3 = V_2$ + Area of piston × (displacement from points 2 to 3)

$= (0.2)\ \text{m}^3 + (0.1\ \text{m}^2)(0.2\ \text{m}) = 0.22\ \text{m}^3$

Then $$V_3 = \frac{V_3}{m} = \frac{0.22\ \text{m}^3}{50\ \text{kg}} = 0.0044\ \text{m}^3/\text{kg}$$

Now from steam table at $P_3 = 350$ kPa = 0.350 MPa

We have $v_f = .001079$ and $v_g = 0.5243$

Thus we see that $v_f < v_3 > v_g$ at $P_3 = 350$ kPa from which we infer that the state 3 is saturated mixture and hence the final temperature T_3 is equal to the saturation temperature of water at 350 kPa which again from steam table can be read as 138.88°C.

i.e., $$T_3 = 138.88°C$$

(*b*) During the process 1 to 2 the pressure remains constant and during 2 to 3 it rises linearly. The total boundary work during this process is the area under the curve *ABC*.

That is W_b = Area *ABC FED*

= Area *ABED* + Area *BCFE*

$$= P_1(V_2 - V_1) + \left(\frac{P_3 + P_2}{2}\right)(V_3 - V_2)$$

$$= \left[(150 \text{ kPa})(0.2 - 0.05)\,\text{m}^3 + \frac{1}{2}(150 + 350)\text{ kPa}(.22 - .2)\text{ m}^3\right] \times \left[\frac{1 \text{ kJ}}{1 \text{ kPa}\cdot\text{m}^3}\right]$$

= 27.5 kJ Ans.

Example 3.14. *A frictionless piston cylinder device contains a mass of 10 kg of liquid – vapour mixture of water at 100 kPa. To start with, 4 kg of water is in liquid phase and rest 6 kg in vapour phase. Heat is now transferred to water, and the piston, which is resting on a set of stops, starts moving. when the pressure inside reaches 200 kPa. Heat transfer to the water is continued until the total volume increases by 20 percent Determine (a) the initial and final temperatures (b) mass of liquid water when the piston first starts moving and (c) the work done during this process. Show the process on P.v. diagram.*

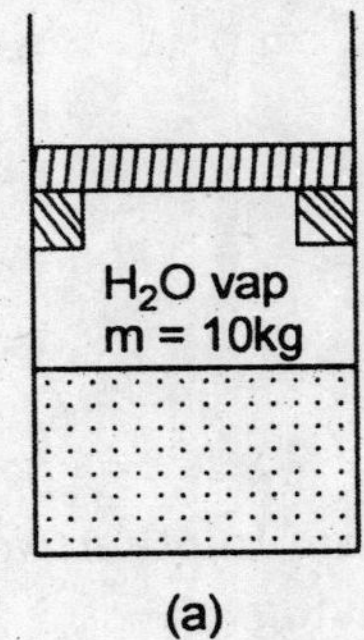

(a)

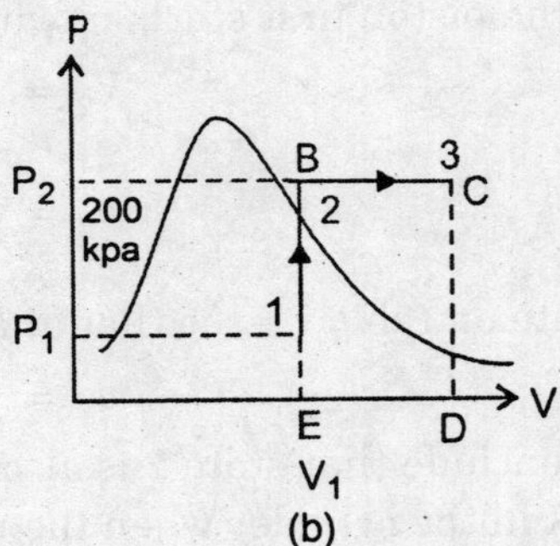

(b)

Fig. 3.20

Solution : (*a*) The P.V. diagram for this process is shown in Fig. (3.20 b) When heat is transfered to the system which is initially a saturated mixture of water at 100 kPa at state 1, water will continue evaporating at constant volume till the piston starts moving when its pressure will reach 200 kPa. When heat is further transfered and piston starts moving the pressure will remain constant till the final volume V_3 is 20% more than the initial volume V_1. Thus the process 1 – 2 will be represented by the vertical line *AB* and the process 2 – 3 will be represented by horizontal line *BC* on the P.V. diagram.

At the state 1 $\quad T_1 = T_{sat}$ at 100 kPa

$= 99.63°C$ (from steam table)

Also $\quad V_1 = m_f v_f + m_g v_g$

Again from steam tables we have at P_1 = 100 kPa, v_f = 0.001043 and v_g = 1.6940

Then $\quad V_1 = 4 \times .001043 + 6 \times 1.6940 = 10.168 \text{ m}^3$

At the final state $\quad V_3 = 1.2\, V_1$ (given)

$= 1.2 \times 10.168 = 12.20 \text{ m}^3$

then $$v_3 = \frac{V_3}{m} = \frac{12.20}{10} = 1.22 \text{ m}^3/\text{kg}$$

Also $$P_3 = P_2 = 200 \text{ kPa (given)}$$

Now from steam table at $P_3 = 200$ kPa, $v_g = .8857$ and $v_f = .001061$, *i.e.*, $v_g < v_f < v_3$ thus the state 3 is in superheated region.

From steam table, we see that

At $T = 250°\text{C}, v = 1.1988$

At $T = 300°\text{C}, v = 1.3162$

Therefore by interpolation at $v_3 = 1.22 \text{ m}^3/\text{kg}$, we get

$$T_3 = \frac{(1.22 - 1.1988)(300 - 250)}{(1.3162 - 1.1988)} \times 250$$

$$= \frac{0.0212}{.1172} \times 50 + 250$$

$$= 259.044°\text{C}$$

(*b*) When the piston first starts moving $P_2 = 200$ kPa

and $$V_2 = V_1 = 10.168 \text{ m}^3$$

Then $$v_2 = \frac{V_2}{m} = \frac{10.168}{10} = 1.0168$$

But from steam table we see that at $P_2 = 200$ kPa

$$v_g = 0.8857 \text{ m}^3/\text{kg} < v_2$$

Thus we conclude that state 2 is also in superheated region. In other words it means that no liquid is left in the cylinder when the piston first starts moving.

(*c*) The boundary work done is the area under the process curve since the process 1 – 2 is represented by vertical line their will be no work done in this part of the process.

During the process 2 – 3 the pressure remains constant. The boundary work will be represented by the area *BCDE*, which will also be the total boundary work for the entire process 1 – 2 – 3.

Therefore $$W_b = \int_2^3 P_2 dV = P\int_2^3 dV = P_2(V_3 - V_2)$$

$$= (200)(\text{kPa})[12.2 - 10.168](\text{m}^3)\left[\frac{1 \text{ kJ}}{1 \text{ kPa} \cdot \text{m}^3}\right]$$

$$= 406.4 \text{ kJ}$$

Example 3.15. *A piston-cylinder device with a set of stops contains 10 kg of saturated liquid-vapour mixture of water at 100 kPa pressure. Initially 9 kg of the water is in liquid phase and rest* 1.0 *kg in the vapour phase. Heat is now transferred to the water until the piston hits the stops, at which point the volume is 10 m³. Determine (a) the temperature when the piston first hits the stops and (b) the work done during the process. Also show the process on P.v. diagram.*

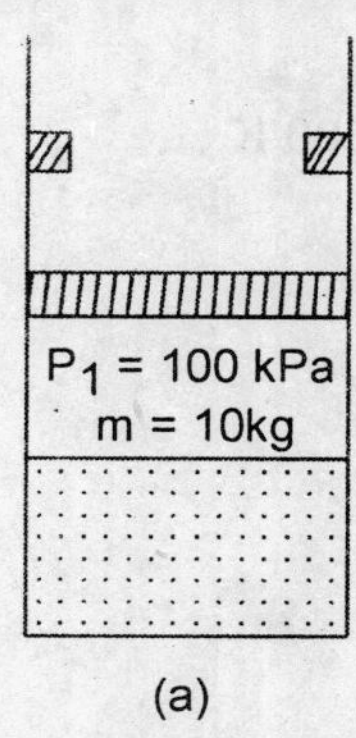

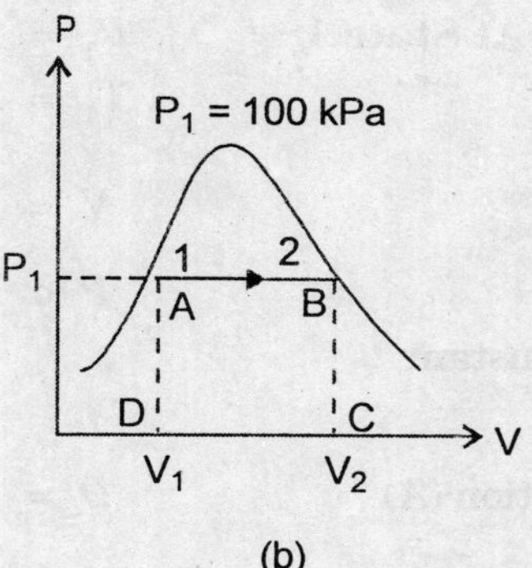

Fig. 3.21

Solution : At state 1 $T_1 = T_{sat}$ at 100 kPa

$= 99.63°C$ (from steam table)

Also at $P_1 = 100$ kPa

$v_f = 0.001043$ and $v_g = 1.6940$ (From steam tables)

and $m_g = 9$ kg; $m_f = 1$ kg

Therefore $V_1 = m_f v_f + m_g v_g$

$= 9 \times .001043 + 1 \times 1.6940$

$= 1.70687$ m^3

When the piston moves up to touch the stops the pressure inside remains consant at 100 kPa

At state 2 $V_2 = 10$ m^3 (given)

and $P_2 = 100$ kPa

Therefore $v_2 = \frac{10}{10} = 1.00 \text{ m}^3$

Thus at state 2 since $v_f < v_2 < v_f$ the mixture is in saturated state. Therefore on the P.v. diagram the process 1 – 2 is represented by a horizontal line *AB* as shown in Fig. 3.21 (*b*). Since the process 1 – 2 is in the saturated region and the temperature will remain constant at T_{sat}

Therefore $T_2 = T_{sat}$ at 100 kPa pressure

$= 99.63°C$

(*b*) The boundary work in this process 1 – 2 is represented by the area under the process curve *AB* is the area *ABCD*.

Therefore W_b = Area *ABCD*

$= P_1(V_2 - V_1)$

$= (100 \text{ kpa}) (10 - 1.70687) \text{ m}^3$

$= 829.313$ kJ **Ans.**

Example 3.16. *A spherical balloon contains 5 kg air at 200 kPa and 227°C. The baloon is made of such a material that the inside pressure is always proportional to the square of the baloon diameter. Find the work done when the inside volume of the baloon becomes twice of its initial volume due to heat transfer.*

Solution : At State 1 $P_1 = 200 \text{ kPa}$

and $T_1 = 227°C = 227 + 273 = 500 \text{ K}$

Now $$V = \frac{\pi D^3}{6} \quad ...(A)$$

and $P \; \alpha \; D^2$ or $P = aD^2$ ...(*B*)

where a is a constant

From equation (*A*) $$D = \left(\frac{6V}{\pi}\right)^{\frac{1}{3}}$$

Therefore *P* from equation (*B*) $$P = a\left(\frac{6V}{\pi}\right)^{2/3} \quad ...(C)$$

or $$P(V)^{-\frac{2}{3}} = a\left(\frac{6}{\pi}\right)^{2/3} = b$$

where $b = a\left(\frac{6}{\pi}\right)^{2/3}$ another constant thus the product $PV^{-2/3}$ is constant.

Therefore $$P_1 V_1^{-2/3} = P_2 V_2^{-2/3}$$

or $$\left(\frac{V_2}{V_1}\right)^{2/3} = \frac{P_2}{P_1}$$

That is $$\frac{P_2}{P_1} = \left(2\frac{V_1}{V_1}\right)^{2/3} \quad \text{(Since } V_2 = 2V_1\text{)}$$

$$= (2)^{2/3} = 1.587$$

Assuming air to be an ideal gas

$$\frac{P_1 V_1}{T_1} = \frac{P_2 V_2}{T_2}$$

$$T_2 = \left(\frac{P_2 V_2}{P_1 V_1}\right) T_1 = \frac{P_2}{P_1} \times \frac{V_2}{V_1} \times T_1$$

$$= 1.587 \times 2 \times 500$$

$$= 1.587 \text{ K}$$

Then $$W_b = \int_1^2 P\,dV = \int_1^2 a\left(\frac{6V}{\pi}\right)^{2/3} dV \quad \text{(From equation } C\text{)}$$

or $$W_b = a\left(\frac{6}{\pi}\right)^{2/3} \int_1^2 (V)^{2/3} dV$$

$$= a\left(\frac{6}{\pi}\right)^{2/3} \left(\frac{3}{5}\right)\left[V_2^{5/3} - V_1^{5/3}\right]$$

$$= \frac{3}{5}\left[\left\{a\left(\frac{6}{\pi}\right)^{2/3} V_2^{2/3}\right\}V_2 - \left\{a\left(\frac{6}{\pi}\right)^{2/3} V_1^{2/3}\right\}V_1\right]$$

or $$W_b = \frac{3}{5}[P_2V_2 - P_1V_1] \quad \text{(from equation } C)$$

Now for ideal gas

$$PV = mRT$$

i.e., $$m = \frac{P_1V_1}{RT_1} = \frac{P_2V_2}{RT_2}$$

Therefore $$W_b = \frac{3}{5}mR(T_2 - T_1)$$

But for air $$R = 0.287\frac{\text{kJ}}{\text{kg}\cdot\text{k}}$$

Then $$W_b = \frac{3}{5}\times(5\text{ kg})(0.28)\left(\frac{\text{kJ}}{\text{kg}\cdot\text{k}}\right)(1587-500)\text{ k}$$

$$= 936\text{ kJ}$$

3.9 WORK DONE ON ELASTIC SOLID BARS

According to well known Hooke's Law the load versis deformation (extension or compression) of a solid bar curve is a straight line as long as the stress generated by the load in the bar is within elastic limit.

Therefore if x is the stretch due to a gradually applied load F, then

$$x \propto F \qquad ...(3.42)$$

If l is the original length and A is the original area of cross-section of the bar then the strain ε and the stress σ are defined as

$$\varepsilon = \frac{x}{l} \text{ and } \sigma = \frac{F}{A}$$

Evidently the original length and area of cross-section are constants. Then

$$\sigma \propto \varepsilon$$

or $$\sigma = E\varepsilon \qquad ...(3.43)$$

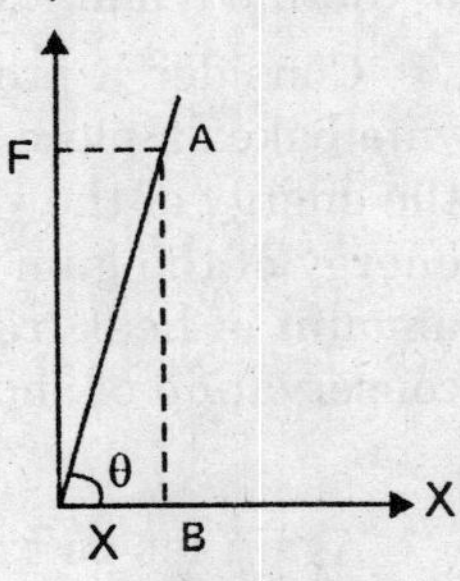

Fig. 3.22

Where E is a constant for the material and is known as youngs moduless or moduless of elasticity and it will be the slope of the load-extension curve upto elastic limit point (Fig. 3.22)

That is $$\tan\theta = E \qquad ...(3.44)$$

If the stretch corresponding to load F is x then the work done on the solid bar will be the area under the curve *i.e.,* area *OAB*.

Thus $$W_b = \text{Area of triangle } OAB$$

i.e., $$W_b = \frac{1}{2}(A\sigma)(l\varepsilon) \qquad ...(3.45)$$

$$= \frac{1}{2}(A\sigma)(l\varepsilon)$$

$$= \frac{1}{2}(\sigma\varepsilon)(Al)$$

or
$$W_b = \frac{1}{2}\sigma \times \text{ volume}$$

or
$$W_b/\text{volume} = \frac{1}{2}\sigma\varepsilon \quad ...(3.46)$$

In the case of solid bar, when the load F is removed the solid body regains back its original shape and dimensions provided the load had not crossed the elastic limit point of the material. This behaviour is exactly similar to that of a spring. Therefore the equations derived for a linear spring can also be applied in the case of elastic solid bars.

3.10 THE FIRST LAW OF THERMODYNAMICS

In the previous articles various forms of energy such as heat Q, work W and total energy have been discussed individually. First law of thermodyanmic, which is also knwon as "The Principle of Conservation of Energy" provides firm foundation for studying the relationship among various forms of energy. Right at the outset it is to be emphasized that there is no mathematical proof of this law.

Based on experimental observations, "The First Law of Thermodynamcis" states that energy can neither be created nor destroyed; it can only change from one firm to other firm. It is implied in this statement that every bit of energy must be accounted for during a process. As pointed out earlier there is no mathematical proof of this law but no process in nature is known to have violated this law. Thus it is accepted as the sufficient proof of the statement of this law of thermodyanmics.

As an example let us consider a stone which is at some elevation with respect to the sea level, which is taken as the reference surface. This stone possesses some potential energy, and some part of this potential energy is converted to kinetic energy as the stone falls. Experimental observations confirm that the decrease in potential energy (= mgZ) is exactly equal to the increase in kinetic energy $\left[\frac{1}{2}(mV_2^2 - mV_1^2\right]$ where V_1 and V_2 are initial and final velocities of the stone after the fall through height Z, provided air resistance in negligible. Thus it confirms the principle of conservation of energy. Similarly the validity of the first-law of thermodynamics for closed systems is examined in the following paragraphs.

Consider a process which involves only heat transfer and no work interaction. A house wife bakes a piece of m (sweet protato) in a microwave oven. As a result of heat trasnfer to it the energy of the yum will increase. If we neglect the loss of moisture from the yum, the total energy of the gum will be equal to the amount of heat transfer from the oven thus if Q is the amount of heat transfer and ΔE, the increase in the energy of the yum then the principle of conservation of energy or first law of thermodynamics is expressed as

$$Q = \Delta E$$

Again let us take another example where water in a sauce pan is being heated on a gas stove by a house wife. If 20 kJ of heat is transferred to the water from the burning gas and 5 kJ is lost to the surrounding air, then the increase in energy of the water will be equal to the net amount of heat transferred to water, which will be 15 kJ.

In other words $$Q = Q_{net} = \Delta E$$

From the above examples, we conclude as under : "In the absence of any work interaction between a system and its surroundings the amount of net heat transfer is equal to the change in energy of the closed system which is expressed mathematically as

$$Q = \Delta E \quad \text{...(3.47)}$$

Let us take a case where there is "No Heat Interaction between the System and Surroundings but Work is Involved". Consider well insulated room (*i.e.*, adiabatic) which is being heated with an electric heater When an electric current is allowed to pass through the heater element, electrical work is done on the closed system resulting in the increase in energy of the system. Since the process is adiabatic, as being well insulated there can not be heat interaction between the system and surroundings. (*i.e.*, $Q = 0$) However the energy of the system will increase due to the electrical work. As per the principle of conservation of energy, the electrical work done on the system must be equal to the increase in energy of the system. That is

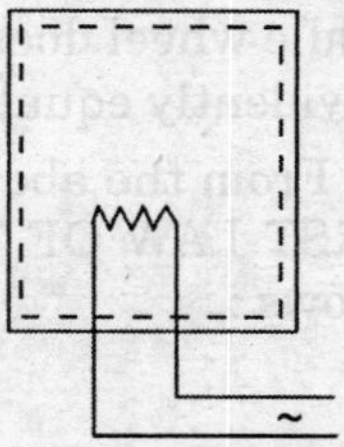

Electricity mains

Fig. 3.23

$$-W_e = \Delta E \quad \text{...(3.48)}$$

Note : Negative sign assigned to We in equation (3.48) is due to the sign convention being followed that is work done on the system is taken as negative and work done by the system as positive. Similarly if the electric heater is replaced with a paddle wheel, then the work done by the paddle wheel as a result of stirring process Fig. (3.24) the energy of the system will increase. But due to the fact that the room is well insulated there will be no heat interaction between the system and surroundings ($Q = 0$), that is the paddle wheel work will result in the increase in energy of the system.

W_{pw}

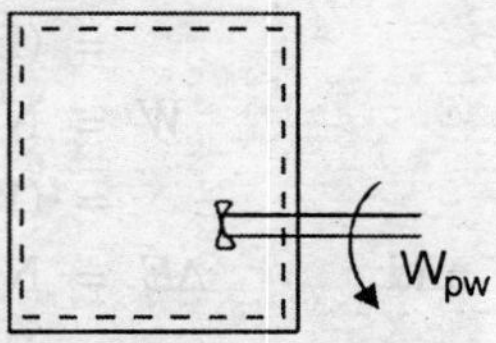

Fig. 3.24

$$-W_{pedak} = \Delta E \quad \text{...(3.49)}$$

Still further, let us consider the case of air being compressed in a piston cylinder assembly (Fig. 3.25) in which the cylinder and piston surface are supposed to be well insulated. That is the air in the cylinder is being compressed under adiabatic conditions. When air is compressed its temperature will increase because energy is added to the air in the form of boundary work. Since the process is adiabatic there is no heat transfer (*i.e.*, $Q = 0$) the entire boundary work will be stored in the air and will form part of its total energy. Thus as per the principle of conservation of energy the increase in energy, ΔE, of the system will be equal to the bounday work W_b.

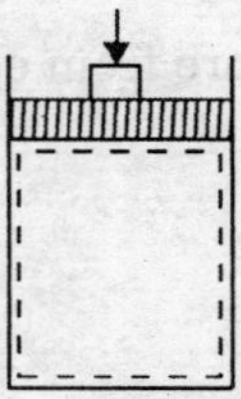

Insulated Surfaces

Adiabatic

Fig. 3.25

i.e., $$-W_b = \Delta E \ (\because Q = 0) \quad \text{...(3.50)}$$

Once again note that negative sign of W_b in equation (3.49) indicates that work is done an the system and as per the sign convention it is taken as negative.

From the equation (3.48) through (3.50) which refer to different modes of doing work on a system under adiabatic conditions we conclude as under.

For adiabatic processes the amount of work done is equal to the change in the energy of the closed system "which is expressed in the mathematical form as

$$-W = \Delta E \quad \text{...(3.51)}$$

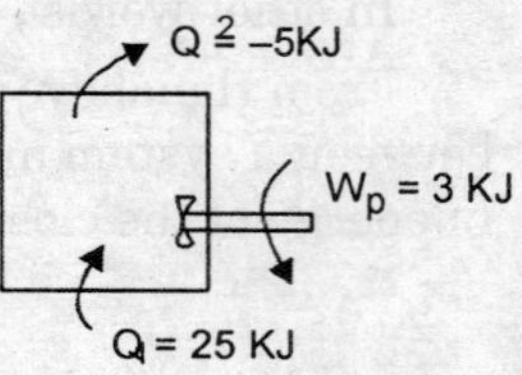

Fig. 3.26

Now after having discussed the cases of closed systems where either heat or work interactions alone were involved, we take up the case of the system where there is simultaneous heat and work interactions. When a system involves both heat and work interactions during a process their contributions are added. As an example if a system receives 25 kJ of heat and 5 kJ of heat is rejected and in addition a paddle wheel does 3 kJ work on the system, the net increase in energy, is evidently equal to (25 – 5) kJ + 3 kJ = 23 kJ.

From the above discussion we conclude that the principle of conservation of energy or the FIRST LAW OF THERMODYNAMICS for a closed system or a fixed mass, is expressed as follows :

$$\begin{bmatrix}\text{Net energy transfer to} \\ \text{(or from) the system as} \\ \text{heat and work}\end{bmatrix} = \begin{bmatrix}\text{Net increase (or decrease)} \\ \text{in the total energy} \\ \text{of the system}\end{bmatrix}$$

or

$$Q - W = \Delta E \quad ...(3.52)$$

Where Q = Net heat transfer across system boundaries

$= (\Sigma Q_{in} - \Sigma Q_{out})$

W = Net work done in all forms

$= (W_{out} - W_{in})$

and ΔE = Net change in total energy of the system

$= E_2 - E_1$

As already discussed earlier in chapter 1 the total energy of a system consists of there components. Firstly, the internal energy, secondly the kinetic energy and thirdly the potential energy.

i.e.,

$$\Delta E = \Delta U + \Delta KE + \Delta PE \text{ (kJ)} \quad ...(3.53)$$

Therefore from equations (3.52) and (3.53)

We get

$$Q - W = \Delta U + \Delta KE + \Delta PE \quad ...(3.54)$$

where

$$\Delta U = m(u_2 - u_1) \quad ...(3.55a)$$

$$\Delta KE = \frac{1}{2} m (V_2^2 - V_1^2) \quad ...(3.55b)$$

$$\Delta PE = mgh = mg(Z_2 - Z_1) \quad ...(3.55c)$$

3.10.1 First Law Applied to Stationary Closed System

In practice sometimes stationary closed system are encountered, *i.e.*, they do not involve any velocity changes or even change in their elevation in their center of gravity. Thus for a closed stationary system

$$\Delta(KE) = \frac{1}{2} m (V_2^2 - V_1^2) = 0 \quad ...(3.56a)$$

and

$$\Delta(PE) = mg (Z_2 - Z_1) = 0 \quad ...(3.56b)$$

The first law relationship (3.53) reduces to

$$Q - W = \Delta U \text{ kJ} \quad ...(3.57)$$

In case the initial and final states of the system are specified, then the internal specifice energies u_1 and u_2 can be easily determined with the help of property tables or else by using some thermodynamic relations.

Sometimes it is convenient to consider the work term W, *in* equation (3.57) in two parts viz W_b and W_{others} where W_b is the boundary work and W_{others} represents all forms of work other than boundary work. Than the equation (3.57) reduces to

$$Q - W_{others} - W_b = \Delta E \text{ kJ} \quad \text{...(3.58)}$$

Important Notes

1. The importance of splitting W into two parts will be realised when we discuss the second law of thermodynamics in the forth coming chapters.
2. **Sign Convention of Heat and Work Interactions** : It is once again emphasized that the sign convention for the heat and work interactions be thoroughly understood and stirctly adhered to. These are reproduced below.
   (*i*) Heat flow to the system is taken positive.
   (*ii*) Work done by the system is taken positive.
   (*iii*) Heat flow from the system is taken negative.
   (*iv*) Work done on the system is taken negative.
3. The system may involve more than one form of work during a process. One need not brother about the sign of W_b. The boundary work calculation by using equation (3.12) will always give the correct sign. However, proper sign is to be assigned to other forms of work as per the sign convention.

3.10.2 Other forms of First Law Relation for Closed System (Non Flow Processes)

Given below are other forms of First Low of Thermodynamic for closed system. Dividing equation (3.52) by the mass of the system, gives the first law equation on a "Unit-mass" basis as

$$q - w = \Delta e \text{ (kJ/kg)} \quad \text{...(3.59)}$$

Similarly dividing equation 13.52, by the time interval Δt and taking the limit $\Delta t \to 0$ we get

$$\frac{dQ}{dt} - \frac{dW}{dt} = \frac{dE}{dt} \text{ (kW)} \quad \text{...(3.60)}$$

or

$$\dot{Q} - \dot{W} = \frac{dE}{dt} \text{ (kW)} \quad \text{...(3.61)}$$

Where $\dot{Q}$ is the rate of heat transfer, $\dot{W}$ is the power and $\frac{dE}{dt}$ is the rate of change of total energy.

Yet another form in which equation (3.52) can be expressed in the **"differential form"** which is

$$\delta Q - \delta W = dE \text{ (kJ)} \quad \text{...(3.62)}$$

or

$$\delta q - \delta w = de \text{ (kJ/kg)} \quad \text{...(3.63)}$$

3.10.3 First Law Relation for a Closed System Undergoing Cyclic Process

For a cyclic process the initial and final states one identical

There
$$\Delta E = E_2 - E_1 = 0$$

Then the first law relation equation (3.52) for the cyclic process reduces to

$$Q - W = 0 \quad \text{kJ} \qquad \text{...(3.64}a\text{)}$$

From equation (3.64) it is evident that for a cylic process, the net heat transfer and the net work done during a cyclic process must be equal

i.e.
$$Q = W \qquad \text{...(3.64}b\text{)}$$

Fig. 3.27

Important Note

Heat and work being energy quantities are not much different but still we have been differentiating these two as two separate entities. From the proceeding discussion we observe that the change in the energy content of the system is equal to the amount of energy that crosses the boundary as "heat or work" and it makes no difference whether energy crosses the boundary as heat and work. One may, then, be tempted to suggest that the first-law relations could be made simpler if we had just one quantity which we could call "Energy Interaction" to represent both heat and work. This could do well as far as 'First-law' is concerned but from the second law point of view heat and work are different (as discussed later in the relevant chapter).

Example 3.17. *Water is being heated in a closed pan on top of a gas stove. In addition it is being stirred by a paddle wheel. During the process 35 kJ of heat is added to the water while 7 kJ is lost to the surrounding air. The paddle wheel work is equivalent to 600* N-m. *If the initial energy of the system is 10 kJ, find the final energy of the system.*

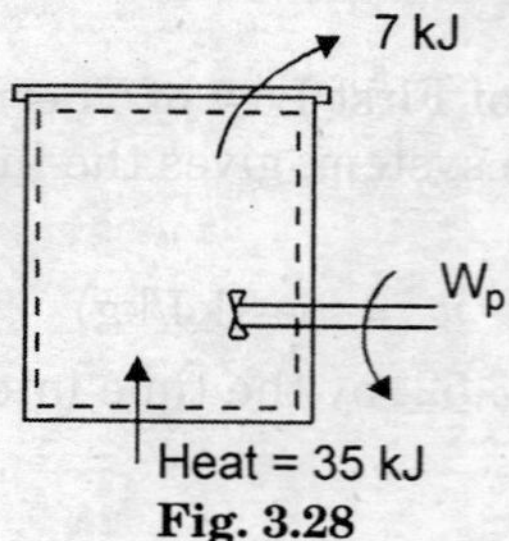

Fig. 3.28

Solution : Let the water in the pan be our system since no mass enters or leaves the pan, it is a closed system. Then from first law we have

$$\Delta E = Q - W$$

or
$$E_2 - E_1 = Q - W$$

or
$$E_2 = E_1 + Q - W$$

or
$$E_2 - E_1 = Q - W$$

$$E_2 - 10 = (35 - 7) - \left(-\frac{600}{1000}\right) \text{ kJ}$$

or
$$E_2 = 10 + 28 + 0.6$$
$$= \mathbf{38.6\ kJ}$$

Example 3.18. *A closed system undergoes a cycle consisting of two process. During the first process 50 kJ of heat is transfered to the system while the system does 70 kJ of work. During the second process 30 kJ of work is done on the ystem. Determine (a) the heat transfer during the second process. (b) Calculate the net work and net heat transfer for the cycle.*

Solution : (*a*) The closed system on ρ.v. diagram is shown in Fig. 3.29. The arrow heads show the cyclic processes 1 and 2 as from states *A* to *B* and *B* to *A* respectively where

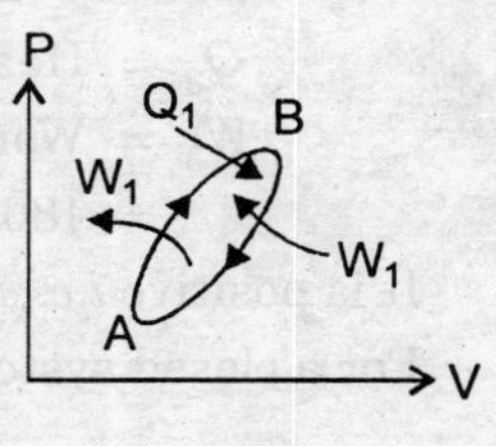

Fig. 3.29

Q_1 = 50 kJ of heat transferred to closed system

W_1 = 70 kJ of work done by the system during first process

W_2 = –30 kJ of work done on the system during second process (Note: it is –ve as work is done on the system)

Let Q_2 = Heat transfer to the system during the second process

Then from the conservation of energy equation or First Law equation for closed system undergoing cyclic process *i.e.*, equation (3.64*a*) we have

$$Q - W = 0$$

Here $Q = Q_1 + Q_2$

and $W = W_1 + W_2$

or $(Q_1 + Q_2) - (W_1 + W_2) = 0$

or $50 + Q_2 - (70 - 30) = 0$

or $10 + Q_2 = 0$

i.e., $Q_2 = -10$ kJ

(Negative sign of Q_2 shows that heat is being transferred from the system)

(*b*) Net heat transfer during the cycle

$$Q_{net} = Q_1 + Q_2 = 50 - 10 = 40 \text{ kJ}$$

(It being positive 40 kJ of heat is being transferred to the system)

Net work done during the system

$$W_{net} = W_1 + W_2$$
$$= 70 - 30 = 40 \text{ kJ}$$

(It being positive 40 kJ is the work done by the system)

Example 3.19. *A closed system undergoes a cycle of three processes. The first process is adiabatic and consists of 100 kJ of work done on the system. During the second process 40 kJ of heat is transferred to the system and with no work interaction. During the third process the system does 180 kJ of work to return to the initial state. Find (a) the heat transfer during the third process and (b) the net work done during the cycle.*

Solution : The cycle of three processes from states *A* to *B*, B to *C* and *C* to *A* is shown in Fig. (3.30)

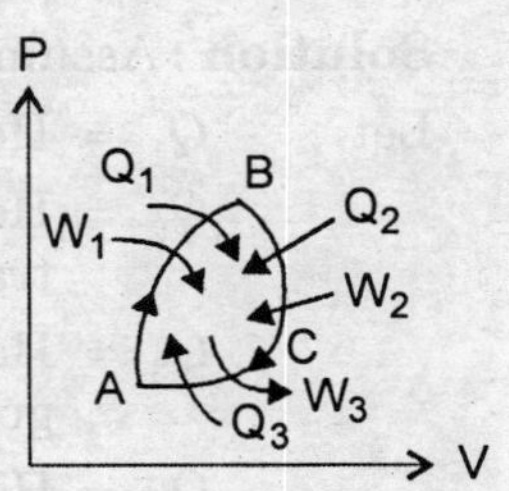

Fig. 3.30

Q_1 = Heat transfer to the closed system during the 1st process

= 0 (as this process being adiabatic there is no heat interaction at the boundary)

W_1 = Work done on the system

= –100 kJ

Q_2 = Heat transfer to the closed system during second process
= 400 kJ
W_2 = Work done on the system during second process
= 0 (as there is no work interaction)
Q_3 = Heat transfer to the system during third process
W_3 = Work done by the system
= 180 kJ

It is positive *i.e.*, work is doneby the system

For a closed system undergoing process in cycle we have from equation (3.64 *a*)

$$Q - W = 0 \quad ...(3.64a)$$

where

$$Q = Q_1 + Q_2 + Q_3$$
$$= 0 + 400 \text{ kJ} + Q_3$$
$$= 400 + Q_3$$

and

$$W = W_1 + W_2 + W_3$$
$$= (-100) + 0 + (180)$$
$$= 80$$

Therefore from equation (3.64*a*)

$$(400 + Q_3) - 80 = 0$$

or

$$Q_3 = -400 + 80$$
$$= -320 \text{ kJ}$$

(Negative sign indicates that heat is being transferred from the system during the third process)

(*b*) Net work done during the cycle

$$W_{net} = W_1 + W_2 + W_3$$
$$= (-100) + 0 + 180 = 80 \text{ kJ}$$

(As it is positive *i.e.*, work is done by the system)

Example 3.20. *A seminar room where normally 100 people are there during a seminar, is to be air conditioned by installing window air conditioning units of 4 kW rating each. The total load of the light sources and other electrical gadgets is 4 kW. A person at rest and present in the room is assumed to dissipate heat at the rate of 360 kJ/hour The room temperature is to be maintained at 20°C. During the summer months it is estimated that rate of heat transfer from the atmosphere through the walls, roof and windows is 36000 kJ/hour for the room temperature of 20°C. If the seminar room is to be maintained at constant temperature of 20°C find the number of window air conditioning units required.*

Solution : Assuming the seminar room as the closed system

Let Q_1 = Heat transfer from the atmosphere to the system
Q_2 = Heat dissipation by the 100 people which is being transferred to the closed system.
Q_3 = Heat transfer to the closed system by the electrical gudgets and light sources.
Q_4 = Heat to be transferred by the air conditioning units from the system to the surroundings.

Fig. 3.31

Since $Q_4 = Q_1 + Q_2 + Q_3$

Now $Q_1 = 36000$ kJ/m

$$= \frac{36000}{3600} = 10 \text{ kW}$$

$$Q_2 = (360 \times 100) \text{ kJ/m}$$

$$= \frac{36000}{3600} = 10 \text{ kW}$$

$$Q_3 = 4 \text{ kW}$$

Thus

$$Q_4 = Q_1 + Q_2 + Q_3$$
$$= 10 \text{ kw} + 10 \text{ kw} + 4 \text{ kW}$$
$$= 24 \text{ kw}$$

Since each air conditioning units is of rating 4 kw, the number of units required is

$$= \frac{24}{4} = 6 \text{ units}$$

3.10.4 First Law Relation for A closed System Undergoing constant-pressure Quasi-Equilibrium Process

Neglecting the changes in kinetic and potential energies we have from equation (3.54)

$$Q - W = \Delta U + \Delta KE + \Delta PE$$
$$= \Delta U + 0 + 0$$

Also if we express the work as sum of boundary work and other forms of work, then the above equation reduces to

$$Q - (W_b + W_{other}) = \Delta U$$

or

$$Q - W_{other} = W_b + U_2 - U_1 \qquad \text{...(3.65)}$$

where U_1 and U_2 are the internal energies at the initial and final states repectively.

Since the system undergoes constant pressure process

$$P_2 = P_1 = P_0 \text{ (say)}$$

and the corresponding boundary work is then given by equation (3.15 *b*)

$$W_b = P_0(V_2 - V_1) = P_2V_2 - P_1V_1 \qquad \text{...(3.66)}$$

Therefore from equation (3.65) and (3.66)

$$Q - W_{others} = (P_2V_2 - P_1V_1) + (U_2 - U_1)$$
$$= (P_2V_2 + U_2) - (P_1V_1 + U_1)$$
$$= H_2 - H_1 \qquad \text{...(3.67)}$$

where

$$H = U + PV \qquad \text{...(3.68)}$$

and H is known as the enthalpy of the system which is defined as the sum of internal energy and the boundary work. Thus the final form of the First Law relation for a closed system undergoing constant pressure quasi-equilibrium process is given by equation (3.67)

$$Q - W_{others} = H_2 - H_1 \text{ kJ} \qquad \text{...(3.677)}$$

Note : Equation (3.67) is very convenient to use for the closed system undergoing constnat pressure quasi-equilibrium process as we need not calculate the boundary work seperately as it is included in the enthalpy. Usually the enthalpy of the system is available in the property tables of various substances.

Example 3.21. *A piston cylinder device contains water. Sixty kJ heat is transferred to water and 10 kJ of it is lost to the atmosphere through the walls of the cylinder. Due to this heating the piston rises as a result of evaporation and a boundary work of 6 kJ is done. Calculate the change in energy of the water in this process.*

Solution : Since the mass of water remains constant it is a closed system. From equation 3.52 we have

$$Q - W = \Delta E$$

Here $Q = 60 - 10 = 50$ kJ

and $W = W_b = 6$ (as work done by the system it is positive)

Therefore $\Delta E = 50 - 6 = 44$ kJ

Example 3.22. *In the following table certain data is given in kJ for four different processes of a closed system between states 1 and 2. For each process find the missing terms.*

S.No.	*Q*	*W*	*E₁*	*E₂*	*ΔE*
(1)	20	–4		37	
(2)	–8			6	–13
(3)		14	5		34
(4)	27		16		12

Solution :

(*i*) As $Q - W = \Delta E$

Therefore $\Delta E = 20 - (-4) = 24$ kJ

Also $E_2 - E_1 = \Delta E$

Therefore $E_1 = E_2 - \Delta E = 37 - 24 = 13$ kJ

(*ii*) Since $Q - W = \Delta E$

or $W = Q - \Delta E = -8 - (-13) = 5$ kJ

Also $E_2 - E_1 = \Delta E$

or $E_1 = E_2 - \Delta E = 6 - (-13) = 19$ kJ

(*iii*) Since $Q - W = \Delta E$

or $Q = W + \Delta E = 14 + 34 = 48$ kJ

Also $E_2 - E_1 = \Delta E$

or $E_2 = E_1 + \Delta E = 5 + 34 = 39$ kJ

(*iv*) Since $Q - W = \Delta E$

$W = Q - \Delta E = 27 - 12 = 15$ kJ

Also $E_2 - E_1 = \Delta E$

or $E_2 = E_1 + \Delta E = 16 + 12 = 28$ kJ

Example 3.23. *The volume of the radiator of steam heating system has a volume of 15 Litre and at a certain point of time is filled with superheated water vapour at 200 kPa and 200°C. At this instant both the intet and output values of the radiator are closed. Find the* *amount of heat that will be transferred to the room when the steam pressure drops to 100 kPa. Draw the P–V Diagram.*

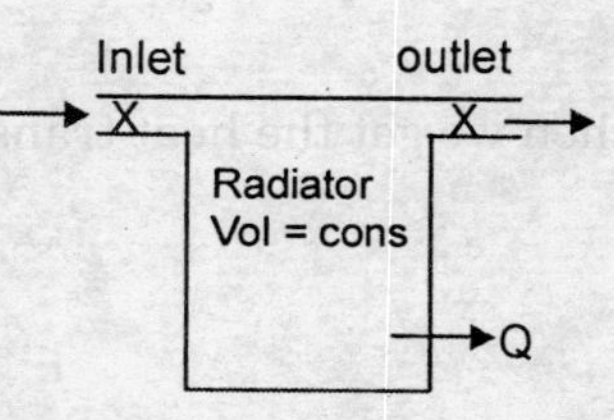

Fig. 3.32

Solution : Take the water in the radiator as the system. Since both the inlet and outlet valves are closed, no mass leaves or enters the radiator. Thus it is a closed system. Then from the first-law equation (3.54) we have

$$Q - W = \Delta U + \Delta KE + \Delta PE \qquad ...(3.54)$$

As the two valves are closed, there is no change in the position of centre of gravity of the system *i.e.*, $\Delta PE = 0$ similarly no flow is involved $\Delta KE = 0$. In addition the boundary remains stationary as volume of the system is constant *i.e.*, $W = 0$

Therefore equation (3.54) reduces to

$$Q = \Delta U$$

$$= m(u_2 - u_1)$$

At the initial state

$$P_1 = 200 \text{ kPa}$$

$$T_1 = 200°\text{C}$$

$$V_1 = 15 \text{ litre} = 0.015 \text{ m}^3$$

From steam tables

$$u_1 = 2654.4 \quad \text{kJ/kg}$$

$$v_1 = 1.0803 \quad \text{m}^3/\text{kg}$$

$$\therefore \quad m = \frac{v}{v_1} = \frac{0.015 \text{ m}^3}{1.0803 \text{ m}^3/\text{kg}} = 0.0139 \text{ kg}$$

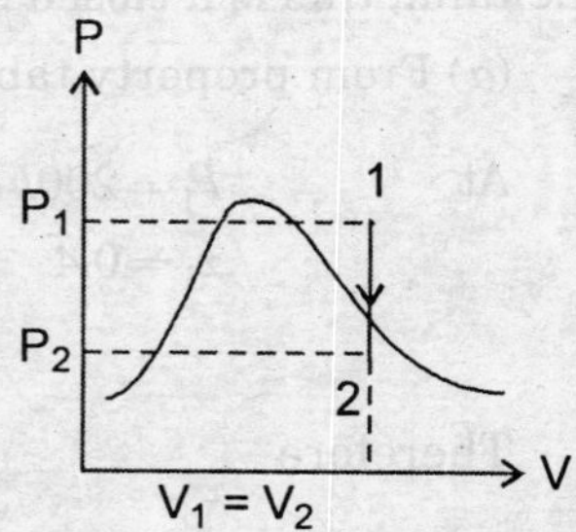

Fig. 3.33

At the final state

$$P_2 = 100 \text{ kPa}$$

$$v_2 = v_1$$

(Since volume is constant and being closed system mass is also constant

$$\therefore \quad v_2 = v_1 = 1.0803$$

Also at

$$P_2 = 100 \text{ kPa}$$

$$v_f = 0.001043 \text{ m}^3/\text{kg}$$

$$v_g = 1.6940 \text{ m}^3/\text{kg}$$

and

$$u_f = 417.36 \text{ kJ/kg}$$

$$u_{fg} = 2088.7 \text{ kJ/kg}$$

At these values,. since $v_f < v_2 < v_g$, the state 2 is a saturated state and quality x_2 is found as given below

$$x_2 = \frac{v_2 - v_f}{v_{fg}}$$

or

$$x_2 = \frac{1.0803 - .001043}{1.6940 - .001043}$$

$$= .06375$$

Therefore
$$u_2 = u_f + x_2 u_{fg}$$
$$= 417.36 + .6375 \times 2088.7$$
$$= 1748.9 \text{ kJ/kg}$$

then we get the heat transfer by substituting These values in following equation
$$Q = m\,(u_2 - u_1)$$
$$= 0.0139\,(1748.9 - 2654.4) \text{ kJ/kg}$$
$$= -12.58 \text{ kJ}$$

Note : Negative sign indicates that the heat is flowing out of the system (*i.e.*, steam in the radiator)

Example 3.24. *A rigid tank whose volume is 0.5 m³, contains refrigerant–12 at 200 kPa and 40% quality. Heat is then transferred to the refrigerant until the pressure reaches 800 kPa. Find*

(a) The mass of the refrigerant in the tank

(b) The amount of heat tranferred to the refrigerant

Show the process on the P – v diagram with respect to the saturation lines.

Solution : Take the refrigerant in the tank as system. Since no mass enters or leaves the tank, this is a closed system.

(*a*) From property table for $R - 12$ we have at $P_1 = 200$ kPa

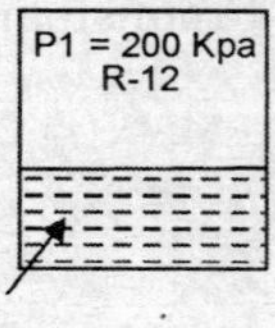

P1 = 200 Kpa
R-12
Q

Fig. 3.34

At
$$\left.\begin{aligned} P_1 &= 200kPa \\ x_1 &= 0.4 \end{aligned}\right\} \rightarrow \begin{aligned} v_f &= 0.0006962 \text{ m}^3\text{/kg} \\ v_g &= 0.08354 \text{ m}^3\text{/kg} \\ u_f &= 24.43 \text{ kJ/kg} \\ v_g &= 165.36 \text{ kJ/kg} \end{aligned}$$

Therefore
$$v_1 = v_f + x_1 v_{fg}$$
$$= v_f + x_1(v_g - v_f)$$
$$= 0.0006962 + 0.4(0.08354 - 0.0006962)$$
$$= 0.03383 \text{ m}^3\text{/kg}$$
$$u_1 = u_f + x_1 u_{fg}$$
$$= 24.43 + 0.4\,(u_g - u_f)$$
$$= 24.43 + 0.4\,(165.36 - 24.43)$$
$$= 80.80 \text{ kJ/kg}$$

Mass of the refrigernat
$$m = \frac{V_1}{v_1}$$
$$= \frac{0.5 \text{ m}^3}{0.03383 \text{ m}^3\text{/kg}} = 14.78 \text{ kg}$$

(*b*) From first law of thermodynamics
$$Q - W_b - W_c = \Delta U + \Delta PE + \Delta KE$$

Since the tank is rigid and no work is involved in this process and the volume remains constant the process is shown in Fig. 3.35.

Now during the process the volume remains constant. therefore $v_2 = v_1 = 0.03383$ m³/kg.

At stage 2 $P_2 = 800$ kP$_G$ we have following data from property tales of R – 12

$$\begin{array}{lll} \text{At} & v = 0.03359 \rightarrow & u = 251.39 \\ & v_2 = 0.03383 & u = ? \text{ (to be estimated by interpolation)} \\ \text{At} & v = 0.03552 \rightarrow & u = 264.41 \end{array}$$

∴ By interpolation
$$u_2 = 251.39 + \frac{(264.41 - 251.39)}{(0.03552 - .03359)} \times (.03383 - .03359)$$
$$= 251.39 + \frac{13.02}{.00193} \times .00024$$
$$= 251.39 + 1.6191$$
$$= 253.01 \quad \text{kJ/kg}$$

Therefore
$$Q = m(u_2 - u_1)$$
$$= 14.78(253.01 - 80 - 80)$$
$$= 2545 \quad \text{kJ}$$

Fig. 3.35

Example 3.25. *An insulated rigid tank cortains 4 kg of saturated liquid-vapour mixture of water at 100 kPa. In the beginning 3 kg of the mass is in liquid phase and 1 kg in vapor phase. An electric resistor is placed in the tank and is connected to a 230 V mains through a switch. When the switch is turned on a current of 15A flows through the resistor. Find how long will it take to turn the entire liquid in the tank into saturated vapour. Show the Process on T.V. diagram with respect to saturation lines.*

Solution : Take the water in the insulated rigid tank as the system. Since no mass enters or leaves the tank, this is a closed system and the volume remains constant.

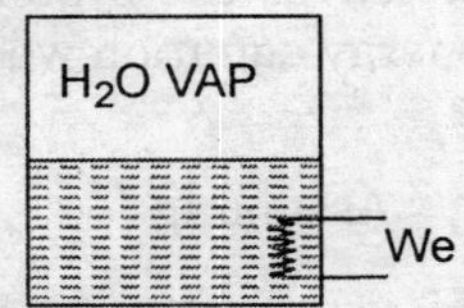

Fig. 3.36

At the initial state

$P_1 = 100$ kPa and since 3 kg is m liquid state and 1 kg in vapors state, we conclude that $\frac{1}{4}$th of the total mass is vapor

Therefore $x_1 = 0.25$

Also from steam tables

At pressure 100 kPa,
$$P_1 = 100 \text{ kPa}$$
$$v_f = 0.001043 \text{ m}^3/\text{kg}, \; v_g = 1.6940 \text{ m}^3/\text{kg}$$
$$u_f = 417.36 \text{ kJ/kg and } u_{fg} = 2088.7 \text{ kJ/kg}$$

Therefore
$$v_1 = v_f + x_1 v_{fg}$$
$$= 0.001043 + 0.25\,(v_g - v_f)$$
$$= 0.001043 + 0.25\,(1.6940 - 0.001043)$$
$$= 0.42428 \text{ m}^3/\text{kg}$$

and
$$u_1 = u_f + x_1 u_{fg}$$
$$= 417.36 + 0.25 \times 2088.7$$
$$= 939.5 \quad \text{kJ/kg}$$

Now from first law of thermodynamics
$$Q - W_b = \Delta U + \Delta PE + \Delta KE$$

or $W_e = Q$

$\because$ [Because $W_b = 0, \quad \Delta PE = 0, \quad \Delta KE = 0$]

$$Q = \Delta U = m(u_2 - u_1)$$

Now $W_e = -VI\ \Delta t$

where V = Line voltage

I = Current in Amps flowing through the resistor

Δt = time taken

Note : We is taken as negative, since the electric work is being done on the system.

Thus $$-(-230 \times 15 \times \Delta t) = (4\text{ kg})\ [(2556.75 - 939.5)\text{ kJ/kg}]\left(\frac{1000\text{ VA}}{1\text{ kJ/s}}\right)$$

or $$\Delta t = \frac{4 \times 1617.25 \times 1000}{230 \times 15}\text{ sec}$$

$$= 1875\text{ sec}$$

$$= 31.25\text{ minutes}$$

Example 3.26. *An insulated rigid tank is divided into two equal parts with the help of a partition. One part of the tank contains 1kg of compressed liquid water at 600 kPa pressure and 60°C temperature while the other part is empty. The partition is now removed to allow the water to expand to fill the entire tank. If the final pressure is 10 kPa find the final temperature of the water and the volume of the tank.*

Solution : Let the entire tank be the system. Since no mass enters or leaves the system, it is a closed system. Then from conservation of energy equation we have

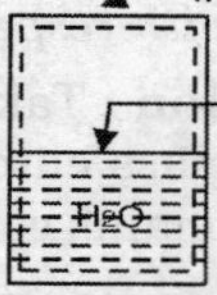

$$Q - W = \Delta U + \Delta KE + \Delta PE$$

Assuming $\Delta KE = 0$

$\Delta PE = 0$

Insulated

Partition

H_2O

Fig. 3.37

Also as the boundary of the system is insualted, then

$$Q = 0$$

Further no work is done $W = 0$

Therefore substituting these values in the above equation we get

$$0 - 0 = \Delta U + 0 + 0$$

i.e., $\Delta U = (u_2 - u_1) = 0$

or $u_2 = u_1$

Now at initial state $P_1 = 600\text{ kPa}$

$T_1 = 60°\text{C}$

From the property tables at this state for saturated water

$$v_1 \simeq v_f \text{ at } 60°\text{C} = 0.001017\text{ m}^3\text{/kg}$$

and $u_1 = u_f$ at $60°\text{C} = 251.11$ kJ/kg

We assume that at the final state in the tank is saturated liquid-vapour mixture of quantity to be calculated. (In case the value of the quality comes out to the absurd then our assumption of the final state being saturated liquid vapour mixture will be wrong). If the value of quality comes at any thing between 0 and 1 it will be all right.

At State 2: $P_2 = 10$ kPa

and $u_2 = u_1 = 251.11$ kJ/kg (found earlier)

At this pressure from table for saturated liquid vapour mixture

$$v_f = 0.00101 \text{ m}^3/\text{kg}$$
$$v_g = 14.67 \text{ m}^3/\text{kg}$$
$$u_f = 191.82 \text{ kJ/kg}$$
$$u_{fg} = 2246.1 \text{ kJ/kg}$$

Then $u_2 = u_f + x_2 u_{fg}$

or
$$x_2 = \frac{u_2 - u_f}{u_{fg}} = \frac{251.11 - 191.82}{2246.1}$$
$$= 0.0264$$

And $T_2 = T_{sat}$ at P_2 of 10 kPa

$= 4581°\text{C}$ (from saturate water pressure table) **Ans.**

(*b*) Volume of tank $V = mv_2$

$= (1.0 \text{ kg})\,(v_2)$

But $v_2 = v_f + x_2 v_{fg}$ At 10 kPa

$$= .00101 + .0264 \times (v_g - v_f)$$
$$= .00101 + .0264\,(14.67 - .00101)$$
$$= 0.388 \text{ m}^3/\text{kg}$$

$\therefore$ $V = 1 \times 3.88 = \mathbf{0.388\ m^3}$

Example 3.27. *A rigid tank is divided into two equal parts with the help of a partition. Initially one side is evacuated and other side contains 10 kg of water at 200 kPa and 25°C. The partition is now removed and the water expands to fill the entire space of the tank. The water is then allowed to exchange heat with the surroundings until its temperature returns to the initial value of 25°C. Find (a) The volume of the tank (b) The final pressure (c) The heat transfer for this process.*

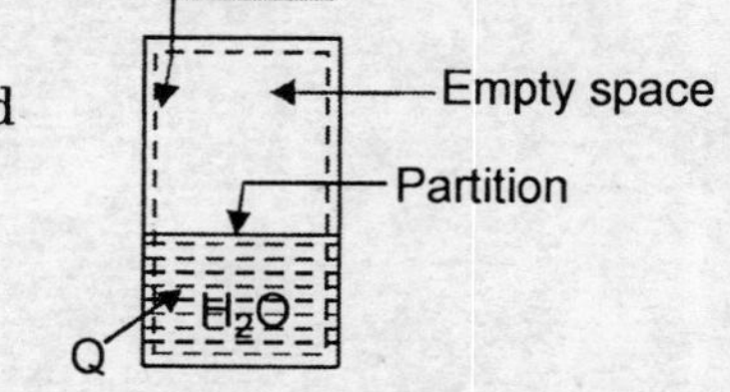

Fig. 3.38

Solution : Let us take the water in the tank as our system

At State 1 : Initially the water pressure is 200 kPa and temperature 25°*C*

i.e., $P_1 = 200$ kPa

$T_1 = 25$°C

From steam tables at saturation temperature of 25°C the saturation pressure is = 3.169 kPa.

But the given pressure is 200 kPa which is higher than 3.169 kPa. Thus we conclude that initially the water is in compressed state. However it is internationally accepted norm to approximate the compressed liquid as saturated liquid at the given temperature.

At $T_1 = 25°$ $v_1 \approx v_f$ at 25°C

And $v_g = 0.001003$ m³/kg (Table 4)

$= 43.36$ m³/kg

Then $$V_1 = mv_1$$
$$= 10 \text{ kg} \times 1001003 \text{ m}^3\text{/kg}$$
$$= 0.01003 \text{ m}^3 = \text{volume of half portion of tank}$$

But the total volume of the tank is then

$$V_{tank} = 2V_1 = \mathbf{0.02006 \text{ m}^3} = V_2$$

At Final State 2 : $$v_2 = \frac{V_2}{m} = \frac{0.02006}{10} = 0.002006 \text{ m}^3\text{/kg}$$

After heat exchange the final temperature is 25 *i.e.*, $T_2 = 25°C$ and at this temperature the corresponding values of

$$v_{f_2} = .001003 \text{ m}^3\text{/kg}$$

$$v_{g_2} = 43.36$$

Evidently $$v_{f_2} < v_2 < v_{g_2}$$

From this we conclude that at final state the water is a saturated liquid-vapour mixture to which first law of Thermodynamics well be applied. Again choosing the water in whole of tank as our system. Neglecting the change in *PE* and *KE*, we get from the first law keeping in mind that $W = 0$

$$Q - W = \Delta U + \Delta KE + \Delta PE$$

or $$Q - 0 = \Delta U + 0 + 0$$

or $$Q = m(u_2 - u_1)$$

Note : Even though the water is expanding during this process, the system choosen has the fixed boundary. Therefore W_b is also zero.

Initially $$u_1 = u_f \text{ at } 25°C = 104.89 \text{ kJ/kg}$$

Let x_2 be the quality of water vapour mixture at state 2,

Then $$v_2 = v_{f_2} + x_2\, v_{f_{g2}}$$

or $$x_2 = \frac{v_2 - v_{f_2}}{v_{f_{g2}}}$$

$$= \frac{.002006 - .001003}{v_{g_2} - v_{f_2}}$$

$$= \frac{0.001003}{43.36 - 0.001003}$$

$$= \frac{0.001003}{43.3589} = 2.3 \times 10^{-5}$$

Then $$u_2 = u_{f_2} + x_2\, u_{f_{g2}}$$

$$= 104.88 + (2.3 \times 10^{-5})(2304.9)$$

$$= 104.93 \text{ kJ/kg}$$

Substituting $Q = 10(u_2 - u_1)$

$= 10(104.93 - 104.88)$

$= 10 \times .05 = \mathbf{0.5\ kJ}$

Example 3.28. *A pistion cylinder device as shown in Fig. 3.39 contains 0.1 m³ of steam at 200 kPa and 150°C. Lower end of coil spring AB having stiffness as x_1 is just touching the piston while its upper end A is fixed (At this point the spring exerts no face on the piston). Heat is now transferred to the steam so as to raise the pressure and volume to 500 kPa and 0.12 m³ respectively. Find (a) the final temperature, (b) the work done by the steam and (c) the total heat transferred.*

Show the process and P.V. diagram with respect to solution lines.

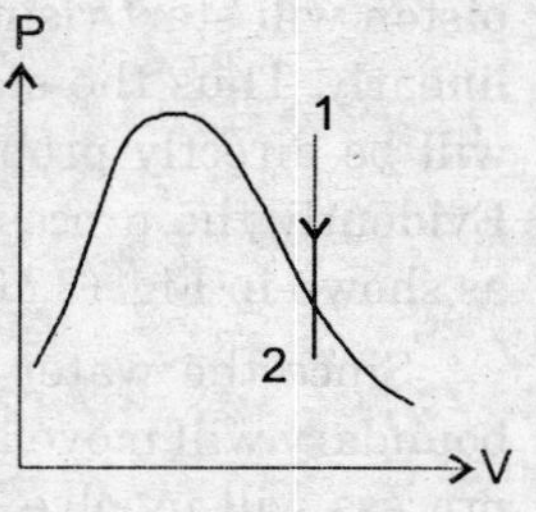

Fig. 3.39

Solution : Take the steam in the piston cylinder device as the system. This is a closed system as no mass learns or enters the device. From steam tables we have

At State 1 :

$$\left.\begin{matrix} P_1 = 200kPa \\ T_1 = 150°C \end{matrix}\right\} \rightarrow \begin{matrix} v_1 = 0.9596 \text{ m}^3/\text{kg} \\ u_1 = 2576.9 \text{ kJ/kg} \end{matrix}$$

and

Since $V_1 = 0.1$ m³, therefore $m_1 = \dfrac{V_1}{v_1}$

$$= \frac{0.1}{0.9559} = 0.104 \text{ kg}$$

At State 2 : $V_2 = 0.12$ m³/kg (given)

$P_2 = 500$ kPa

Therefore $v_2 = \dfrac{V_2}{m} = \dfrac{0.12}{0.104} = 1.154 \text{ m}^3/\text{kg}$

Therefore from steam tables at

$P_2 = 500$ kPa

and $v_2 = 1.154$ m³/kg

The data available from the table for superheated water is

$P_2 = 500$ kPa At $v_2 = 1.0822$, $T = 900$ °C, $u_2 = 3853.6$

$v_2 = 1.1747$, $T = 1000$ °C, $u_2 = 4051.8$

Therefore by interpolation

At $v_1 = 1.154$ $T = 900 + \dfrac{(1000 - 900)}{(1.1747 - 1.08221)}(1.154 - 1.0822)$

$$= 900 + \frac{100 \times 0.718}{0.0925} = 977.62°\text{C}$$

Similarly at $v_1 = 1.154$ $u_2 = 3853.6 + \dfrac{(4051.8 - 3853.6)}{(1.1747 - 1.0822)}(1.154 - 1.0822)$

$$= 3853.6 + \frac{198.2}{925} \times .0719$$

$$= 3853.6 + 153.85$$
$$= 4007.45 \text{ kJ/kg}$$

(*b*) The spring does not exert any force at the begining. As the piston will start rising the force exerted by the spring will increase linearly. Thus the steam pressure during the time the piston rises will be directly proportional to the distance moved by the piston. Evidently the process curve on P.v. diagram will be a straight line as shown in Fig. (3.40).

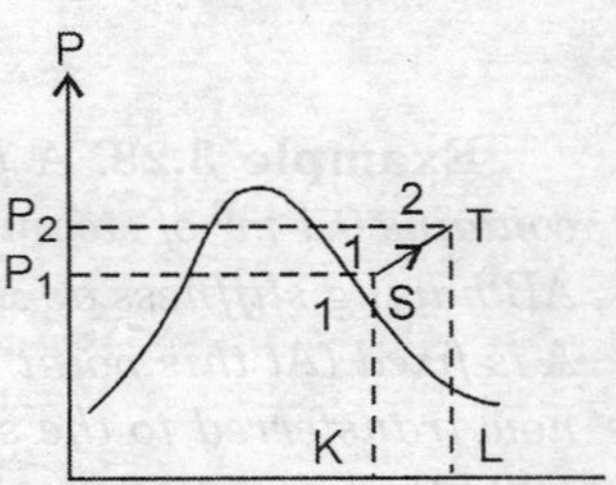

Fig. 3.40

Since the water in the device constitutes the system and the boundary will move along with the inner surface of the piston, this process will involve boundary work which will be the area under the process curve. In this case this curve will be a straight line and the area will be a trapezium STLK.

Thus W_b = area of trapezium STLk

$$= \left(\frac{P_1 + P_2}{2}\right)(V_2 - V_1)$$

$$= \left[\left(\frac{200+100}{2}\right)\text{kPa}\right][(0.12 - 0.1)\text{ m}^3)]\left[\frac{1\text{ kJ}}{1\text{ kPa}\cdot\text{m}^3}\right]$$

$$= 7 \text{ kJ}$$

(*c*) From conservation of energy equation

$$Q - W_{\text{others}} - W_b = \Delta U + \Delta KE + \Delta PE$$

Since $W_{\text{others}} = \Delta KE = \Delta PE = 0$, we get

$$Q = W_b + \Delta U$$
$$= W_b + m(u_2 - u_1)$$
$$= 7 \text{ kJ} + (0.104 \text{ kg})(4007.45 - 2576.9)$$
$$= 155.8 \text{ kJ}$$

3.11 SPECIFIC HEATS

From day to day experience we know that difference amounts of energy are required to raise the temperature of identical masses of different substances by one degree. Therefore, it is desirable to have a property which will enable us to compare the energy storage capabilities of various substances. This property is called the specific heat.

The specific heat is "defined as the energy required to raise the temperature of a unit mass of a substance by one degree". However in general, this energy will defined on how the process is exented. In thermodynamics two kinds of specific heats viz specific heat at constant volume C_v and specific heat at constant pressure C_p are encountered.

The specific heat at constant volume C_v is defined as the energy required to raise the temperature of the unit man of a substance by one degree as the volume is maintained constant. Similarly the energy required to raise the temperature of the unit mass of the substance by one degree as the pressure is maintained constant is called the specific heat at constant pressure C_p, this is illustrated in Fig. 3.41. Figure 3.41 (*a*) is shows a piston cylinder assembly which

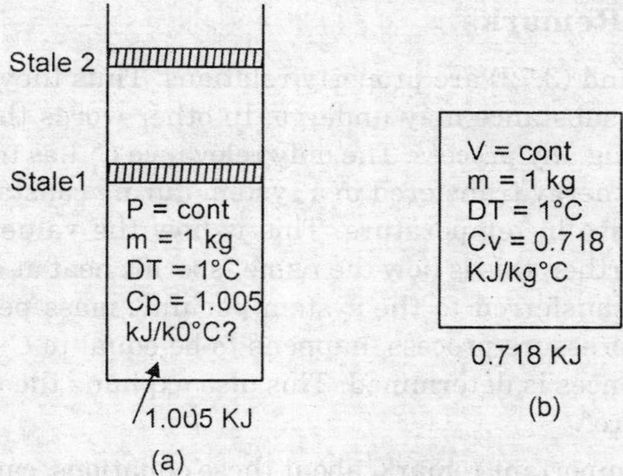

Fig. 3.41

contains 1 kg of air, 1.005 kJ of energy is transferred to the system and the piston moves up thus maintaining the pressure constant and the temperature is raised by 1°C. In Fig. 3.41 (*b*) the volume of 1 kg of air is kept constant, the temperature is being raised by 1°C by supplying 0.718 kJ of energy to this system. Evidently under constant pressure conditions the amount of energy required to raise the temperature of 1 kg of air will more th n that in the case of constant volume case because at constant pressure the system is allowed to expand and the energy required for the expension work must also be supplied to the system there by rendering C_p to be greater than C_v.

3.11.1 Relationship of Specidic Heats and Other Thermodynamic Properties

Consider a stationary closed system undergoing a constant volume process. Evidently for this process the boundary work will be zero. Then the first law relationship for this process in differential from is as under.

$$\delta_q - \delta W_{\text{others}} = du \qquad \text{...(3.69)}$$

The left hand side of the equation (3.69) represents the amount of energy transferred to the system in the form of heat and/or work. From the definition of C_v, this energy must be equal to $C_v\, dT$ where dT is the differential change in temperature.

Therefore $$C_v(dT) = du \qquad \text{...(3.70 } a\text{)}$$

or $$C_v = \left(\frac{\partial u}{\partial T}\right)_v \qquad \text{...(3.70 } b\text{)}$$

Similarly consider a constant pressure process for which the mathematical (differential form) expression for first law is

$$w_b + \Delta u = \Delta h \qquad \text{...(3.71)}$$

The left hand side of equation (3.71) represents the amount of energy transferred to the system at constant pressure and is equal to $C_p\, dT$ where ΔT is the differential change in temperature.

Therefore $$C_p(\Delta T) = \Delta h \qquad \text{...(3.72 } a\text{)}$$

or $$C_p = \left(\frac{\partial h}{\partial T}\right)_p \qquad \text{...(3.76 } b\text{)}$$

3.11.1.1 Important Remarks

1. Equation (3.71) and (3.72) are property relations. Thus they do not depend on the type of processes which any substance may undergo. In other words these equations are valid for any substance undergoing any process. The only relevance C_v has to a constant volume process is that C_v is amount of energy transfered to a system during constant volume process per unit mass per unit degree rise in temperature. This is how the values of C_v are determined for different substances. Further, this is how the name "specific heat at constant volume" orginated. Similarly, the energy transferred to the system per unit mass per unit degree temperature rise during a constant pressure process, happens to be equal to C_p. This is the way the value of C_p for various substances is determined. This also explains the origin of the name "specific heat at constant pressure".

2. Another equally important remark, about these equations, can be made that C_v is related to "the changes in internal energy of the substance and C_p to changes in enthalpy.

In fact it would be more proper to define C_v as the change in the "specific internal energy of a substance per unit change in temperature at constant volume. Likewise, C_p can be defined as the change in specific enthalpy of a substance per unit change in temperature at constant pressure. In otherwords, C_v is the measure of variation of internal energy of a substance with temperature and C_p is the measure of variation of enthalpy of a substance with temperature.

3. Another important point to note is that both the internal energy and enthalpy of a substance can be changed by the transfer of energy, in any form (Heat is just one of them). Therefore, the term "specific energy" is more appropriate then the terms "specific heat", which implies that energy is transferred or stored in form of heat.

3.11.1.2 Units of Specific Heats

Common unit for both specific heats C_v and C_p is kJ/(kg°c) or (kJ/kg ·K) and these units are identical since.

$$\Delta T(°\text{C}) = \Delta T\,(K)$$

3.11.1.3 Molar Specific Heats

Specific heats for some substances, are sometimes given in respective tables on "MOLAR BASIS" and symbols for these are $\bar{C}_v$ and $\bar{C}_p$ respectively and have units as kJ/(k mol · °C) or kJ/(k mol · K).

3.12 INTERNAL ENERGY, ENTHALPY AND SPECIFIC HEATS OF IDEAL GASES

In chapter 2, the ideal gas has been defined as a substance whose temperature, pressure and specific volume are related by the equation.

$$Pv = RT \quad \text{...(3.73)}$$

Joule in 1843, concluded on the basis of intensive experimentation that for an ideal gas the internal energy is a function of temperature only.

Thus

$$u = u(T) \quad \text{...(3.74)}$$

In the experimental set up used by Joule two tanks connected with pipe and valve were submerged in a bigger water tank fitted with a thermometer. The tank A was filled with air under pressure while the tank B was evacuated. The valve was closed initially and when the thermal equilibrium was obtained i.e when both the tanks A and B were at the same temperature as that of water in bigger tank, the valve fitted in the connecting pipe was opened to let air pass from tank A to tank B. until the pressure equalised. Joule observed no change in the temperature of the water bath and sssumed that no heat was transferred to or form air. Further since there was also no work done, he concluded that the internal energy of the air did not change even though the volume and pressure had changed. Therefore, he argued, the internal energy is a function of temperature only and not a function of pressure and specific volume. Later on Joule showed experimentally that for gases which deviate significantly from the ideal gas behaviour, the internal energy is not a function of temperature alone.

Also the enthalpy has earlier been defined in article 2.21 and is given as

$$h = u + Pv \quad ...(3.75)$$

Also as in equation (3.73)

$$Pv = RT \quad ...(3.73)$$

Therefore from equation (3.73) and (3.75), we get

$$h = u + RT \quad ...(3.76)$$

Further since R is constant and internal energy u is a function of temperature only that is $u = u(T)$, we conclude from equation (3.76) that enthalpy of an ideal gas is also a function of temperature only i.e.

$$h = h(T) \quad ...(3.77)$$

From the above discussion we conclude that since both 'u' and 'h' depend on temperature only, for an ideal gas, the specific heats C_v and C_p also depend, on temperature only. Therefore at a given temperature u, h, C_v and C_p of an ideal gas will have fixed values regardless of the pressure and specific volume. Thus the partial derivative in equations (3.70 a) and (3.72 a) can be replaced by ordinary derivatives.

Thus the differential changes in internal energy and enthalpy of an ideal gas is expressed as

$$du = C_v(T)\, dT \quad ...(3.78)$$

and

$$dh = C_p(T)\, dT \quad ...(3.79)$$

The change in internal energy and enthalpy for an ideal gas during a process from state 1 to state 2 are found by integrating equation (3.78) and (3.79).

$$\Delta u = u_2 - u_1 = \int_1^2 C_v(T)\, dT \quad \text{(kJ/kg)} \quad ...(3.80)$$

and

$$\Delta h = h_2 - h_1 = \int_1^2 C_p(T)\, dT \quad \text{(kJ/kg)} \quad ...(3.81)$$

In order to evaluate the change in internal energy and the enthalpy the functions, $C_v(T)$ and $C_p(T)$ must be known then only integration will possible.

At low pressure, all real gases behave like ideal gases. Thus their specific heats depend upon temperature only. The specific heats of real gases at low pressures are called "IDEAL GAS SPECIFIC HEATS" or "zero pressure specific heats" and are denoted by C_{v0} and C_{p0} respectively. Analytical expressions for ideal gas specific heats, based on direct measurement are available in the form of tables and are given as a third degree polynomial.

The use of ideal gas specific heat data is limited to low pressures, but these data can also be used at moderately high pressure with reasonable accuracy provided gas does not deviate from ideal gas behaviour significantly. The '*i*' and '*h*' data given in these tables for number of gases have been obtained by choosing states on reference state which coincides with zero kelvin. The choice of reference state has no effect in ΔU and Δh.

Further the specific heats of gases having complex molecule (molecules with two or more atoms) are higher and increase with temperature. Further the variation of specific heats with temperature is smooth and is usually approximated as linear over small temperature intervals (a few hundred degrees or less) then the specific heat functions $C_v(T)$ or $C_p(T)$ in equation (3.80) and (3.81) can be replaces by constant average specific heat values. Thus equation (3.80) and (3.81) respectively yield.

$$u_2 - u_1 = C_{v,\,av}\,(T_2 - T_1)\ \text{(kJ/kg)} \quad ...(3.82)$$

$$h_2 - h_1 = C_{p,\,av}\,(T_2 - T_1)\ \text{(kJ/kg)} \quad ...(3.83)$$

The specific heat values for common gases are given as function of temperature. The average specific heats $C_{p,\,av}$ and $C_{v,\,av}$ are evaluated from the table average temperature $\left(=\dfrac{T_1 + T_2}{2}\right)$. It the final temperature T_2 is not known, the specific heats may be evaluated at T_1 or at an anticipated average temperature. T_2, then can be calculated by using these specific heat values. Then the temperature T_2 is defined by first finding the specific heats at the new average temperature.

Another method of determining specific heats is to evaluate them at T_1 and T_2 and then take average. Usually both the methods give reasonably good results and one is not necessarily better than the other.

Fig. 3.42

Accurate analytical expressions for specific heats of ideal gases based on experiments are available. The variation of $\bar{C}_{p0}$ for some common gases is given in Fig. (3.43). It may be noted from Fig. 3.42 that ideal gas specific heats of monotonic gases such as argen, helium, neon, krypton remain constant over entire temperature range. Thus Δu and Δh of monotonic gases can easily be determined sum equations (3.82) and (3.83).

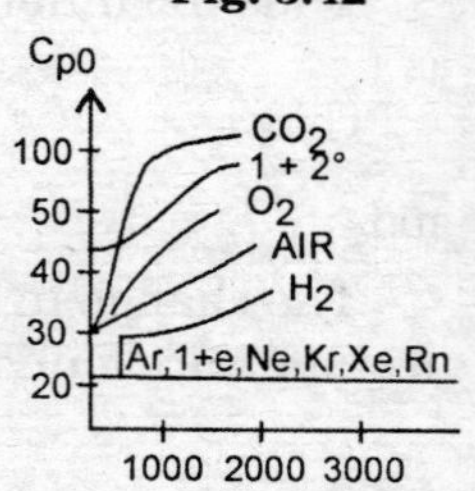

Fig. 3.43

Another important point to keep in mind is that the relations for Δu and Δh are not restricted to any kind of process rather they are valid for all the processes. The presence of constant volume specific heat C_v in equation (3.83) should not lead one to believe that this equation is valid for constant volume process only. On the countrary the relation $\Delta u = C_{v,\,av}\,\Delta T$ is valid for any ideal gas undergoing any process. A similar argument can be given for the relation $\Delta h = C_{p,\,av}\,\Delta T$.

3.13 RELATIONSHIP OF SPECIFIC HEATS OF IDEAL GASSES

In article (3.12) the equation giving the relationship between h and u for ideal gases has been derived as

$$h = u + RT \quad ...(3.76)$$

Differentiating equation (3.76) w.r.t. T we get

$$dh = du + RdT \quad ...(3.84)$$

Also for ideal gases we have from equations (3.78) and (3.79)

$$du = C_v\,dT \quad ...(3.78)$$

and

$$dh = C_p\,dT \quad ...(3.79)$$

Therefore dividing equation (3.84) by dT we get

$$\frac{dh}{dT} = \frac{du}{dT} + R$$

or $$C_p = C_v + R \text{ kJ/(kg}\cdot\text{K)} \quad ...(3.85)$$

When the specific heats are given on molar basis, R in equation (3.81) is replaced by universal gas constant R_u that is

$$\bar{C}_p = \bar{C}_v + R_u \text{ kJ/(k mol}\cdot\text{K)} \quad ...(3.86)$$

3.14 SPECIFIC HEAT RATIO

After having defined C_p and C_v, another ideal gas property can be introduced this property is called the specific heat ratio, σ which is defined as

$$\gamma = \frac{C_p}{C_v} \quad ...(3.87)$$

Since both C_p and C_v vary with temperature the specific heat ratio γ also varies with temperature but the variation is very small. For monotonic gases it is essentially constant and it value is 1.667. Many diatomic gases, including air, have the ratio $\gamma = 1.4$ at room temperature. The specific heat ratio γ is also called adiabatic index or adiabatic exponent.

Following important relations between R, γ and specific heats for ideal gases can be derived

We have from equation (3.85)

$$C_p - C_v = R \quad ...(3.85)$$

and $$\frac{C_p}{C_v} = \gamma \quad ...(3.87)$$

$$C_v\gamma - C_v = R$$

That is $$C_v = \frac{R}{\gamma - 1} \quad ...(3.88)$$

Similarly again from equations (3.85) and (3.87)

$$C_p - \frac{C_p}{\gamma} = R$$

or $$\frac{(\gamma-1)C_p}{\gamma} = R$$

That is $$C_p = \frac{\gamma R}{\gamma - 1} \quad ...(3.89)$$

3.14.2 Molar Specific Heats of Gases

If M is the molar mass of the gas, then the molar specific heats at constant pressure and constant value for the gas are given by $\bar{C}_p$ and $\bar{C}_v$ respectively as

$$\bar{C}_p = MC_p \quad ...(3.90)$$

$$\bar{C}_v = MC_v \quad ...(3.91)$$

That is molar specific heat at constant pressure, $\bar{C}_p$, is defined as the quality of heat required to raise the temperature of one mole (22.4 m^3) of the gas by 1 degree at constant pressure. Similarly molar specific heat at constant valume, $\bar{C}_v$, is defined as the quantity of heat required to raise the temperature of one mole of the gas by one degree when the volume is kept constant.

Again multiplying both sides of equation (3.85) with M we get

$$MC_p - MC_v = MR$$
$$= R_u$$

where R_u is the universal gas constant and its value is 8.314 kJ/(kg mol · K).

Therefore $\bar{C}_p - \bar{C}_v = R_u = 8.314$ kJ/(kg mol · K)

3.14.3 Adibatic Process

A change in state of a system where there is no exchange of heat with the surroundings is called an adiabatic process. This process can be carried out in piston cylinder assembly by expansion or compresion of the gas in the cylinder whose walls are insulated.

If unit mass is taken as the base, the conservation of energy for non flow process is

$$dq = dw + du$$

But for adiabatic process $dq = 0$ and $du = C_p\, dT$

Also $dw = pdv$

Therefore substituting these values in upper equation we have

$$0 = p\,dv + C_p\,dT \qquad ...(3.90)$$

Also for ideal gas $Pv = RT$

Differentiating gas equation

$$RdT = pdv + vdp$$

or
$$dT = \frac{pdv + vdp}{R} \qquad ...(3.91)$$

From equations (3.90) and (3.91)

$$C_v \frac{(pdv + vdp)}{R} + pdv = 0$$

or
$$C_v\,(pdv + vdp) + Rpdv = 0$$

$$C_v\,(pdv + vdp) + p(C_p - C_v)\,dv = 0$$

or
$$C_p\,pd_v + C_v\,vdp = 0$$

Dividing by $C_v pv$
$$\frac{C_p}{C_v}\frac{d_v}{v} + \frac{d_p}{p} = 0$$

or
$$\gamma\frac{dv}{v} + \frac{d_p}{p} = 0$$

Integrating we get $\gamma ln\,(v) + ln\,(p) = ln\,[constant]$

or
$$pv^{\gamma} = e^{(\text{constant})} \qquad ...(3.92)$$
$$= \text{constant} = C$$

Thus the mathematical expression for adiabatic process is

$$pv^{\gamma} = \text{contant}$$

Expression for work done is then given by

$$W_{12} = \int dv$$

Most $$pv^{\gamma} = \text{const and } p = \frac{C}{v^{\gamma}}$$

therefore $$W_{1-2} = C\int \frac{dv}{v^{\gamma}} = \frac{C}{(1-\gamma)}\left[v_2^{1-\gamma} - v_1^{1-\gamma}\right]$$

Also $$C = p_1 v_1 v_1^{\gamma} = p_2 v_2^{\gamma}$$

$\therefore$ $$W_{1-2} = \left(\frac{1}{1-\gamma}\right)\left[Cv_2^{1-\gamma} - Cv_1^{1-\gamma}\right]$$

Also

$$= \left(\frac{1}{1-\gamma}\right)\left[p_2 v_2^{\gamma}\left(v_2^{1-\gamma}\right) - p_1 v_1^{\gamma}(v_1)^{1-\gamma}\right]$$

$$= \frac{1}{1-r}\left[p_2 v_2 - p_1 v_1\right]$$

$$u\ W_{1-2} = \frac{p_1 v_1 - p_2 v_2}{\gamma - 1} \quad ...(3.93)$$

Since for adabatic process $\delta q = 0$, the energy equation becomes for this process

$$\delta w + du = 0$$

or $$U_1 - U_2 = \frac{pV_1^{\gamma} - p_2 V_2^{\gamma}}{\gamma - 1} \quad ...(3.94)$$

Further since $$\frac{p_1 V_1}{T_1} = \frac{p_2 V_2}{T_2} \text{ and } p_1 V_1^{\gamma} = p_2 V_2^{\gamma}$$

from these relations we get $$\frac{T_2}{T_1} = \frac{p_2}{p_1}\frac{V_2}{V_1}$$

$$= \left(\frac{V_1^{\gamma}}{V_2^{\gamma}}\right)\frac{V_2}{V_1} = \left(\frac{V_1}{V_2}\right)^{\gamma-1} \quad ...(3.95)$$

3.14.3 Constant Volume or Isochoric Process

When a system undergoes a change in the state at constant volume, it is called isochoric process. As an example isochoric process will result when a gas is heated or cooled in an enclosed space. In this process the volume will remain constant but both the temperature and pressure will change. The pressure and temperature will increase when heat is supplied to the system and decrease when the heat is rejected by the system. On the P–V space, the two states designated by points 1 and 2 will lie on vertical line as shown is Fig 3.44

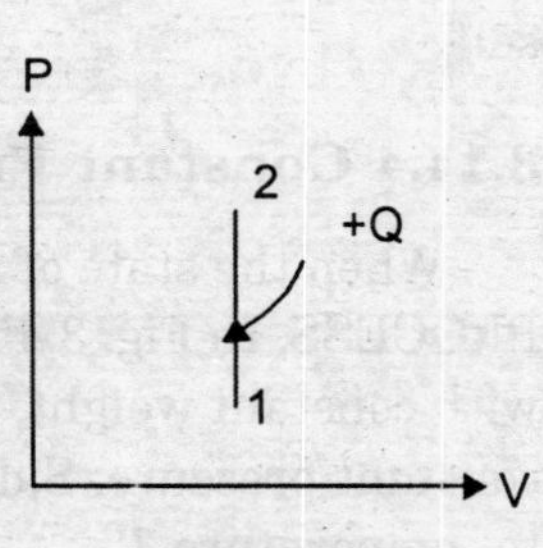

Fig. 3.44

At state 1

$$p_1V_1 = m\,RT_1 \qquad ...(3.96)$$

As state 2

$$p_2V_2 = m\,RT_2 \qquad ...(3.97)$$

dividing equation (3.96) and (3.97)

$$\frac{p_1}{p_2} = \frac{T_1}{T_2} \quad (\because V_1 = V_2)$$

or

$$\frac{p}{T} = \text{Constant} \qquad ...(3.98)$$

This is known as Charles Law

Further as this volume remains constant, $dV = 0$. Thus no mechanical work is done on or by the system.

or

$$dw = 0 \qquad ...(3.99)$$

And from the non flow energy equation

$$\delta Q = \delta w + \delta U$$

$$= 0 + \delta U \quad \text{(from 3.99)}$$

or

$$\delta Q = \delta U \qquad ...(3.100)$$

It follows for the isochoric process all the heat is utilised to change the internal energy. Further since the heat is added during constant volume process

$$\delta Q = m\,C_v\,(T_2 - T_1)$$

or

$$\delta U = \delta Q$$

$$= mC_v(T_2 - T_1) \qquad ...(3.101)$$

Alternatively for unit mass of the gas

$$du = \delta q = C_v\,dT$$

or

$$C_v = \frac{du}{dT} \qquad ...(3.102)$$

With the help of equation (3.102) the specific heat at constant volume is defined as the rate of change of internal energy with respect to the change of temperature when the process is carried out at constant volume.

As C_v has been defined in terms of state properties, in turn it is a state property it self.

3.14.4 Constant Pressure or Isobaric Process

When the state of a system undergoes a change at constant pressure it is called ISOBARIC PROCESS. In Fig. 3.45 perfect gas contained in piston cylinder assembly where piston is loaded with constant weight, will expand due to heat Q supplied to it from volume V_1 to volume V_2 at constant pressure. Side by side the temperature of this gas will also rise from initial value T_1 to temperature T_2.

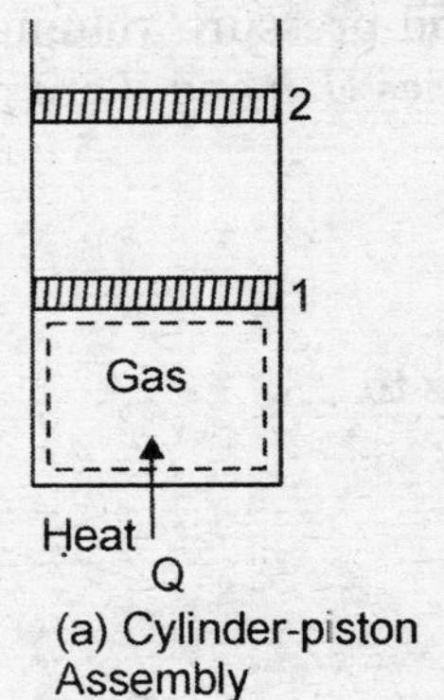

(a) Cylinder-piston Assembly

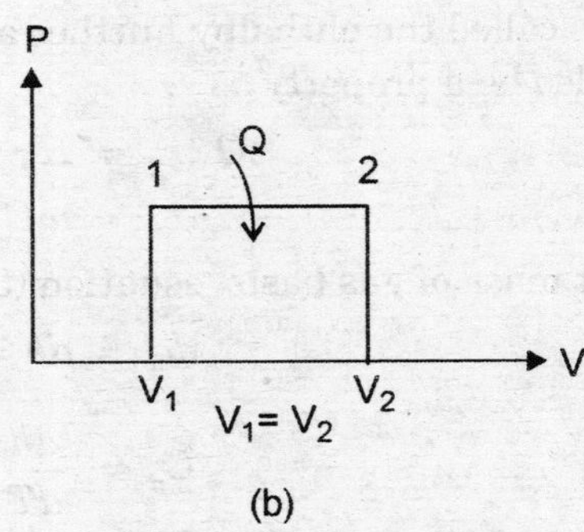

(b)

Fig. 3.45

Neglecting friction between piston and cylinder, this piston will move slowly upwards During Isoboric process when heat is supplied to the system both temperature and volume will increase on the contrary when the system rejests heat then the temperature and volume both will reduce.

Thus from gas law

At initial state

$$p_1V_1 = m\,RT_1 \qquad ...(3.103)$$

At final state

$$p_2V_2 = m\,RT_2 \quad (\therefore p_1 = p_2) \qquad ...(3.104)$$

Dividing expressions (3.103) and (3.104),

$$\frac{V_1}{V_2} = \frac{T_1}{T_2} \qquad [\therefore p_1 = p_2]$$

or

$$\frac{V}{T} = \text{Constant} \qquad ...(3.105)$$

i.e., in an isobaric process volume of a given mass of perfect gas is directly proportional to its absolute temperature. This is called the Charl's Law.

It may be seen that when heat energy is added to the gas in cylinder-piston assembly, the boundary of this control volume moves up. Thus there will be involved the boundary work.

In mathematical terms it is represented as follows.

Work done

$$W_{1-2} = \int_1^2 pdV$$

or

$$W_{1-2} = p(V_2 - V_1) \qquad ...(3106)$$

Also for a non-flow process

$$\delta Q = \delta W + \delta U$$

or

$$Q_{1-2} = p\,(V_2 - V_1) + (U_2 - U_1)$$

$$= (U_2 + p_2\,V_2) - (U_1 + p_1\,V_1) \qquad ...(3.107)$$

But as defined earlier the sum of internal energy and pressure volume product *i.e.*, the term $(U + pV)$ is called the enthalpy Further all the quantities U, p and V are property, therefore enthalpy will also be a property

Your
$$Q_{1-2} = H_2 - H_1$$
$$= mC_p (T_2 - T_1) \quad ...(3.108)$$

On the unit mass of gas basis esuation (3.108) reduces to

$$dq = dh = C_p\, dT$$

or
$$C_p = \frac{dh}{dT} \quad ...(3.109)$$

With the help of equation (3.109) C_p, specific heat at constant pressure is defined as the rate of change of specific enthalpy with respect the change of temperature

During isobaric process

specific work done
$$w_{1-2} = \int_1^2 p\, dv$$
$$= p\,(v_2 - v_1) \quad ...(3.110)$$

Also
$$p = p_1 = p_2 = \text{constant}$$

$$p_1 v_1 = RT_1 \text{ or } \quad v_1 = \frac{RT_1}{p_1}$$

and
$$p_2 V_2 = RT_2 \text{ or } \quad v_2 = \frac{RT_2}{P_2}$$

Therefore
$$w_{1-2} = p\left(R\frac{T_2}{p_2} - \frac{RT_1}{p_1}\right)$$

$$= p\frac{R}{p}(T_2 - T_1) \quad ...(3.111)$$

or
$$w_{1-2} = \mathrm{R}(\mathrm{T}_2 - \mathrm{T}_1)$$

is
$$R = \frac{w_{1-2}}{(T_2 - T_1)} \quad ...(3.112)$$

With the help of equation (3.112) The gas constant, R, is defined as the work done on one Kilogram of ideal gas in an isobaric process when the temperature changes by 1 K.

3.14.5 Isothermal Process (Constant Temperature Process)

A change in the state of a system at constant Temperature is called an isothermal process. In this process the pressure and volume will undergo change.

For an example an ideal gas is contained in a piston cylinder assembly which itself is enclosed in a constant temperature jacket. Evidently when the volume decreases the presume is raised in the gas and vice versa.

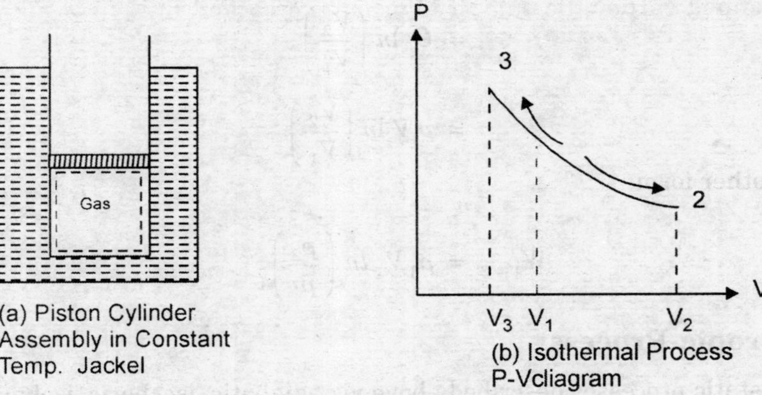

(a) Piston Cylinder Assembly in Constant Temp. Jackel

(b) Isothermal Process P-Vcliagram

Fig. 3.46

Since the gas in the cylinder and the surrounding jacket (reservoir) are at the same temperature, there will be thermal equilibrium between the ideal gas and the reservoir. Thus there will be no heat interaction between the gas and the surroundings across the system boundary. Therefore when the piston is forced to move slowly downwards the gas will get compressed thereby raising the pressure and decreasing the volume. There will be slight increase in temperature dT. Heat will then flow from gas to reservoir till again the temperature comes back to original value T of the reservoir.

At state 1 $$p_1 V_1 = mRT_1 \quad ...(3.113)$$

At state 2 $$p_2 V_2 = mRT_2 \quad ...(3.114)$$

or $$\frac{p_1 V_1}{p_2 V_2} = \frac{mRT_1}{mRT_2}$$

or $$\frac{p_1}{p_2} = \frac{V_2}{V_1} \quad (\therefore T_1 = T_2)$$

or $$p_1 V_1 = p_2 V_2 = pV = \text{constant} \quad ...(3.115)$$

From Equation (3.115) it is concluded that volume is inversely proportional to pressure. It is known as Boyles Law.

In Figure (3.46) is shown this process in $P - V$ chart.

Work done from state 1 to state 2

$$W_{1-2} = \int_1^2 pdV \quad ...(3.116)$$

Since $$pV = \text{constant} = C \text{ say}$$

Then $$P = \frac{C}{V} \quad ....(3.116)$$

Where C = constant

From equation (3.115) and (3.116 We have equations

$$W_{1-2} = \int_1^2 \frac{C}{V}(dV) = C[\ln(V)]_1^2$$

$$= C \ln\left(\frac{V_2}{V_1}\right)$$

$$= p\, V \ln\left(\frac{V_2}{V_1}\right) \qquad ....(3.117)$$

Also in another form

$$W_{1-2} = p_1 V_1\, ln\left(\frac{p_2}{p_1}\right) \qquad ...(3.118)$$

3.14.6 Polytropic Process

The quasi-static processes described above viz adiabatic, isochoric, isobaric and isothermal can all be described by a single equation.

$$pv^n = C \qquad ...(3.119)$$

When n is a positive number and remains constant during any process.

(i) When $n = \gamma$ it will represent adiabatic process

(ii) When $n = 1$ it will represent isothermal process

(iii) When $n = 0$ it will represent constant pressure or isobaric process

(iv) When $n = \infty$ Then from (3.119) we have

$$(p)^{\frac{1}{n}} V = C'$$

Where C' is another constant

when $n \to \infty$

$$p^{\circ} V = C'$$

$$V = C' \text{ constant}$$

The general equation (3.119) is known as Polytropic Law

Various processes for different values of n and $P - V$ diagrams are shown is Fig (3.47)

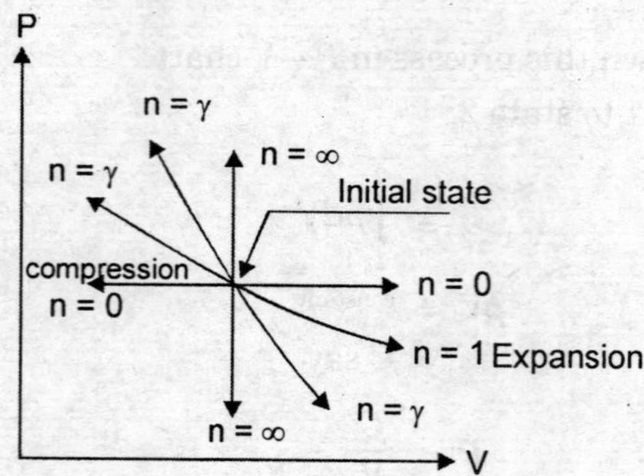

Fig. 3.47

If the origin O represents the initial state, the compression process curves will lie in second quadrant and the expansion curve in the fourth Quadrant.

Again from equation (3.119) on differentiation we get

$$V^n\, dp + pn\, V^{n-1}\, dV = 0$$

or $$\frac{dp}{dV} = -n\frac{p}{V} \qquad ...(3.120)$$

From equation (3.120) we observe that the slope of the process curve increases in the negative direction with the increase in value of n

3.14.6.1 Polytropic Index

The value of the index n can be evaluated during the change of state from 1 to 2 as explained below

The polytropic law

$$pV^n = C$$

or $$p_1V_1^{\,n} = p_2V_2^{\,n}$$

or $$\left(\frac{p_2}{p_1}\right) = \left(\frac{V_1}{V_2}\right)^n$$

Taking natural log

in $$\ln\left(\frac{p_2}{p_1}\right) = n\ln\left(\frac{V_1}{V_2}\right)$$

or $$n = \frac{\ln\left(\dfrac{p_2}{p_1}\right)}{\ln\left(\dfrac{V_1}{V_2}\right)} \qquad ...(3.121)$$

Again from gas equation

we $$\frac{p_2V_2}{p_1V_1} = \frac{T_2}{T_1}$$

or $$\frac{p_2}{p_1} = \frac{V_1}{V_2}\cdot\frac{T_2}{T_1} \qquad ...(3.122)$$

But from equations (3.119)

$$\frac{p_2}{p_1} = \left(\frac{V_1}{V_2}\right)^n \qquad ...(3.123)$$

∴ Equating 3.123 and (3.122) we get

$$\left(\frac{T_2}{T_1}\right)\left(\frac{V_1}{V_2}\right) = \left(\frac{V_1}{V_2}\right)^n$$

or $$\frac{T_2}{T_1} = \left(\frac{V_1}{V_2}\right)^{n-1} \qquad ...(3.124)$$

Again from (3.119) $$\left(\frac{V_1}{V_2}\right) = \left(\frac{p_2}{p_1}\right)^{\frac{1}{n}} \quad ...(3.125)$$

and from gas law $$\frac{V_1}{V_2} = \frac{p_2}{p_1}\frac{T_1}{T_2} \quad ...(3.126)$$

Equation (3.125) and (3.126), give

$$\left(\frac{p_2}{p_1}\right)^{\frac{1}{n}} = \left(\frac{p_2}{p_1}\right)\left(\frac{T_1}{T_2}\right)$$

or $$\left(\frac{p_2}{p_1}\right)^{\frac{1}{n}-1} = \frac{T_1}{T_2}$$

or $$\frac{T_2}{T_1} = \left(\frac{p_2}{p_1}\right)^{1-\frac{1}{n}} = \left(\frac{p_2}{p_1}\right)^{\frac{n-1}{n}} \quad ...(3.127)$$

3.15 POLYTROPIC PROCESS

3.15.1 Work Done during Polytropic Process

During a polytropic process, let a system changes its state from 1 to 2

then $$WD_{1-2} = \int_1^2 p dV \quad ...(3.128)$$

But $$pV^n = C \text{ or } p = CV^{-n}$$

Then the substituting value of p we have

$$WD_{1-2} = \int_1^2 C V^{-n}\, dV$$

$$= C\left[\frac{V^{1-n}}{1-n}\right]_1^2$$

$$= \frac{C}{(1-n)}\left[V_2^{1-n} - V_1^{1-n}\right] \quad ...(3.129)$$

Again $$C = p_1V_1^n = p_2V_2^n$$

$$CV_1^{-n} = p_1 \text{ and } CV_2^{-n} = p_2$$

Therefore $$WD_{1-2} = \frac{1}{1-n}\left[CV_2^{1-n} - CV_1^{1-n}\right]$$

$$= \left(\frac{1}{n-1}\right)\left[CV_2^{-n}V_2 - CV_1^{-n}V_1\right]$$

$$= \left(\frac{1}{n-1}\right)\left[p_2V_2 - p_1V_1\right] \quad ...(3.130)$$

But from gas equation

$$p_1V_1 = mRT_1, \text{ and } p_2V_2 = mRT_2,$$

Thus equation (3.130) reduces to

$$WD_{1-2} = \frac{p_2V_2 - p_1V_1}{1-n} = \frac{p_1V_1 - p_2V_2}{n-1}$$

$$= \frac{mR(T_1 - T_2)}{n-1} \quad ...(3.131\ a)$$

or work done per unit mass $w_{1-2} = \frac{p_1v_1 - p_2v_2}{n-1} = R\frac{(T_1 - T_2)}{n-1}$ ...(3.131 *b*)

3.15.2 Change in Internal Energy

Change in internal energy is given as

$$\delta U_{1-2} = U_2 - U_1$$

$$= mC_v(T_2 - T_1) \quad ...(31.32\ a)$$

On Unit mass basis $\delta u_{1-2} = u_2 - u_1 = C_v(T_2 - T_1)$

3.15.3 Heat Transfer during Polytropic Process

(For a non flow Process)

From non-flow energy equation the heat transfer during polytropic process on unit mass basis is given by

$$q_{1-2} = w_{1-2} + (u_2 - u_1)$$

$$= \frac{p_1v_1 - p_2v_2}{n-1} + C_v(T_2 - T_1) \quad ...(3.133)$$

But from equation (3.88) we have

$$C_v = \frac{R}{(\gamma - 1)}$$

Therefore equation (3.133) reduces to

$$q_{12} = \frac{p_1v_1 - p_2v_2}{n-1} + \left(\frac{R}{\gamma-1}\right)(T_2 - T_1)$$

$$= \left(\frac{p_1v_1 - p_2v_2}{n-1}\right) + \left[\frac{p_2V_2 - p_1V_1}{\gamma - 1}\right]$$

$$= \left(\frac{p_1v_1 - p_2v_2}{n-1}\right) + \left[\frac{p_2v_2 - p_1v_1}{\gamma - 1}\right]$$

$$= \frac{p_1v_1\left[(\gamma-1)-(n-1)\right]-p_2v_2\left[(\gamma-1)-(n-1)\right]}{(n-1)(\gamma-1)}$$

$$= \frac{p_1v_1(\gamma-n)-p_2v_2(\gamma-n)}{(n-1)(\gamma-1)}$$

$$= \frac{(\gamma-n)}{(n-1)(\gamma-1)}\left[p_1v_1-p_2v_2\right]$$

$$q_{12} = \frac{(\gamma-n)}{(\gamma-1)}\left[\frac{p_1v_1-p_2v_2}{n-1}\right]$$

substituting equation (3.131b)...in (3.134a)

or $$q_{12} = \frac{(\gamma-n)}{(\gamma-1)}w_{12} \qquad (3.134h)$$

Evidently from equation (3.134 h) we conclude that the signs of work done and heat interaction are same because $\frac{(\gamma-n)}{(\gamma-1)}$ is positive Thus in the case of polytropic expansion process work is done by the working fluid as well as the heat is absorbed.

3.15.4 Polytropic Specific heat

We have from equation (3.133)$q_{12} = \frac{p_1v_1-p_2v_2}{(n-1)}+C_v(T_2-T_1)$

Again we have from gas equation

$$p_1v_1 = RT_1 \text{ and } p_2v_2=RT_2$$

Therefore $$q_{12} = \frac{R(T_1-T_2)}{(n-1)}+C_v(T_2-T_1)$$

$$= (T_2-T_1)\left[-\frac{R}{n-1}+C_v\right]$$

or $$q_{12} = (T_2-T_1)\left[-\frac{(C_p-C_v)}{n-1}+C_v\right]$$

$$= (T_2-T_1)\left[\frac{-C_p+C_v+nC_v-C_v}{(n-1)}\right]$$

or $$q_{12} = (T_2-T_1)\left[\frac{-C_p+nC_v}{n-1}\right]$$

$$q_{12} = (T_2-T_1)\left[\frac{C_p-nC_v}{1-n}\right] \qquad ...(3.135a)$$

or

$$q_{12} = (T_2 - T_1)C_n \qquad ...(3.135b)$$

where

$$C_n = \left[\frac{C_p - nC_v}{1-n}\right] \qquad ...(3.136)$$

Equation (3.135 *b*) defines the polytropic specific heat as given is equation (3.136). It is denoted by C_n

Further since $\frac{c_p}{c_v} = \gamma$ we get from equation (3.136)

$$C_n = \left[\frac{\gamma C_v - nC_v}{(1-n)}\right]$$

$$= \left(\frac{\gamma - n}{1-n}\right)C_v$$

$$= -\left[\frac{(\gamma - n)}{(n-1)}\right]C_v \qquad ...(3.137)$$

If $\gamma > n > 1$, then the value of C_n will be negative

Example 3.29. *Two kg of an ideal gas was heated under constant pressure conditions and it required 2300 kJ of heat to raise its temperature from 293 K to 373 K When the same gas was heated under constant volume conditions it required 1650 kJ of heat for the same temperature range. Find the values of C_p, C_v, adiabatic index, the gas constant R and molar mass of the gas.*

Solution : (*a*) During the process at constant Pressure

$$Q = m\,C_p\,(T_2 - T_1)$$

$$2300 = (2.0)\,C_p\,(373 - 293)$$

(*b*)

$$C_p = \frac{2300}{2\times 80} = 14.375 \text{ KJ/Kg.K}$$

(*b*) During the precess at Constant Volume

$$Q = mc_v(T_2 - T_1)$$

$$1650 = 2C_v\,(373 - 293)$$

$$C_v = \frac{1650}{2\times 80} = 10.312\ kJ/kg.K$$

(*c*) Adiabatic index

$$\gamma = \frac{C_p}{C_v}$$

or

$$\gamma = \frac{14.375}{10.312} = 1.394$$

(*d*) Gas constant R $\quad R = C_p - C_v$

$= 14.375 - 10.312 = 4.063$ kJ/(kg.K)

(*e*) Molar mass $\quad M = \dfrac{R_u}{R}$

Since for ideal gases $\quad R_u = 8.314$

Therefore molor mass $\quad = \dfrac{8.314}{4.063} = 2.046$

Example 3.30. *Find the enthalpy change Δh in kJ/kg of nitrogen as it is heated from 400 to 600 K using (a) the empirical data for h from Nitrogen table*

(h) C_p value at the average temperature by using relavant table

Also determine the percentage error involved in each case.

Solution : (*a*) At the given temperature Nitrogen can be treated as an ideal gas and its enthalpy depends on temperature only. Taking data from Table

$$h_1 = \text{h}_{\text{at 400 k}} = 11640$$

$$\bar{h}_2 = \bar{h}_{at600K} = 17563\text{kJ/k mol}$$

Thus if M is the molar mass of Nitrogen, then M for Nitrogen = 28.013 kg/k mol.

And $$\Delta h = \frac{\bar{h}_2 - \bar{h}_1}{M} = \frac{(17563 - 11640)}{M}\text{ kJ/mol}$$

$$= \frac{5923}{28.013} = 211.4 kJ/kg$$

(*b*) Average temperature $= \dfrac{T_1 + T_2}{2} = \dfrac{400 + 600}{2} = 500K$ From table we note that value of C_p of Nitrogen at temperature $500K$ is

$$Cp, av = C_{p\text{ at 500 k}} = 1.056\text{ kJ/kg.K}$$

$$Now\ \Delta h = C_p, av\,(T_2 - T_1)$$

$$= [1056\text{ kJ/kg.K}]\,[600 - 400]\text{K}$$

$$= 211.2\text{ kJ/kg}$$

$$\%\text{ Error} = \frac{(211.4 - 211.2)^{100}}{211.4} = 0.095\%$$

Example 3.31. *Two kg of Nitrogen gas in contained in piston cylinder assembly and the weight of the piston is such that the pressure is 6 bar and the temperature is 300 K. Heat is now added and gas expands to three times its original volume under Coustant pressure Conditions (i.e. isobaric expansion) find the (a) initial volume (b) final temperature (c) Work done by gas, (d) Heat added and (e) the change in internal energy (du).*

Take for Nitrogen $C_p = 1.05\ kJ/kG$ and $R = 295\ kJ/kg.K$

Solution : (*a*) From gas equation

$$V_1 = \frac{mRT_1}{P_1} = \frac{2 \times 295 \times 300}{6 \times 10^5} = 0.295\text{ m}^3$$

(*b*) Final Volume $\quad V_2 = 3V_1 = 3 \times .295$

$= 0.885\text{ m}^3$

Since it is constant pressure expansion

We have $P_2 = P_1$

Therefore $P_1V_1 = mRT_1$

and $P_2V_2 = mRT_2$

Dividing these $\frac{V_2}{V_1} = \frac{T_2}{T_1}$

i.e., $T_2 = \frac{V_2}{V_1} \times T_1 = \frac{3V_1}{V_1} \times 300 = 900\text{K}$

(*c*) Work done $\delta W = P(V_2 - V_1)$

or $\delta W = (6 \times 10^5)(0.885 - 0.295)$

$= 3.54 \times 10^5$ Nm or J

$= 354$ kJ

(*d*) Heat added $\delta Q = mC_p(T_2 - T_1)$

or $\delta Q = 2 \times 1.05\,(900 - 300)$

$= 1260$ kJ or closed

(*e*) From first Law for non flow System, we have

$dU = \delta Q - \delta W$

$= 1260 - 354 = 906$ kJ (increase)

Example 3.32. *Certain mass of air was compressed in a cylinder-piston assembly to 50% of its original volume and pressure remaining constant at 0.15 MPa. If the original volume is 0.5 m³ and 50 kJ of heat is transferred from the air to the surroundings during the process, calculate the change in internal energy of air*

Solution : For non flow process, we have

$$Q - W_b = dU$$

But $(Q) = -50$ *kJ*(– *ve* as heat is lost by the air to its surroundings)

And $W_b = \int PdV = P(V_2 - V_1)$

$= (0.15 \times 10^6)\,(0.25 - .5)$

$= -0.15 \times 10^6\,(0.25)$

$= -37.5 \times 10^3 \text{ J} = -37.5\ kJ$

Negative sign shows that work is done on the gas

Substituting these values in the energy equation, we get

$$-(50) - (-37.5) = dU$$

i.e., $dU = -50 + 37.5 = -12.5$ kJ

Negative sign indicates that the internal energy decrease by 12.5 kJ

Example 3.33. *(a) Calculate the enthalpy change dh of oxygen in kJ/kg, when it is heated from 500 K to 800 K using the empirical data for h from oxygen tables.*

(b) C_p valve of oxygen at the average temperature with the help of data Given in oxygen table.

Solution : (*a*) At the temperatures which are specified oxygen can be treated as an ideal gas. Thus the enthalpy will depend on Temperature From oxygen table

$$\bar{h}_1 \text{ at } 500\text{K} = 14770 \text{ kJ/k mol}$$

and

$$\bar{h}_2 \text{ at } 800\text{K} = 24532 \text{ kJ/k mol}$$

Therefore

$$\delta h = \frac{\bar{h}_2 - \bar{h}_1}{M}$$

where M mass per k mol Oxygen $= 31.999$ kg/k mol

i.e.

$$\delta h = \frac{24523 - 14770}{31.999} = 211.4 \text{ kJ/kg}$$

(*b*)

$$\delta h = C_p (T_2 - T_v)$$

$\therefore$

$$C_p = \frac{\delta h}{T_2 - T_k} = \frac{211.4}{800 - 500} = \frac{211.4}{300} = 0.705$$

Example 3.34. *Find the change in internal energy Δu of hydrogen, in kj/kg, when it is heated from 400 K* to *1000 K using (a) the empirical data for u from hydrogen table (b) the C_v Valve at average temperature (c) the C_v Valve at room temperature of* 300 K *What is the percentage errors in each case.*

Solution : (*a*) At the specified temperature hydrogen can be treated as an ideal gas and its internal energy will depend on temperature only therefore from hydrogen tables on k mol,

$$\bar{u}_{\text{at}}\ 400\,K = 8100 \text{ kJ/k mol}$$

$$\bar{u}_2 = \bar{u}_{\text{at}}\ 1000\,K = 20839 \text{ kJ/k mol}$$

and

$$M = 2.016 \text{ kg/k mol}$$

Therefore

$$\Delta u = \frac{u_2 - u_1}{M}$$

$$= \frac{20839 - 8100}{2.016} = 6318.9 \text{ kJ/kg}$$

(*b*) Average Temperature

$$= \frac{400 + 1000}{2} = 700 \text{ K}$$

Thus from hydrogen table at average temperature of 700 K

$$C_v, av = C_v \text{ at } 700\,K = 10.48 \text{ (kJ/kg·K)}$$

$$\text{Now } \Delta u = C_{v,\,av}\,(T_2 - T_1)$$

$$= 10.48\,(1000 - 400) = 6288 \text{ kJ/kg}$$

% age error

$$= \frac{6318.9 - 6288}{6318.9} \times 100 = 0.489\ \%$$

(*c*) Again from hydrogen table at 300 K

$$C_{v,}av = 10.183 \text{ kJ/(kg.K)}$$

Then

$$\Delta u = (C_{v\,av} \text{ at } 300\,K)\,(T_2 - T_1)$$

$$= 10.183\,(1000 - 400) = 6100 \text{ kJ/kg}$$

$\therefore$ % age error

$$= \frac{6318.9 - 6110}{6318.9} \times 100 = 3.31\%$$

Example 3.35. *A closed system undergoes a cycle of four processes. The heat transfer in the first and fourth processes are 450 kJ and 300 kJ, while during the second and third processes they are – 411 kJ and – 250 kJ respectively. However during this first and third processes there are also work transfers of 150 kJ and W kJ respectively and there are no work transfers during second and fourth processes. Determine the value of W.*

Solution : During the cycle of four prosess net heat transfer

$$= Q = 450 - 411 - 250 + 300$$

$$= 89 \text{ kJ}$$

Similarly net work transfer $= W_{1, 2, 3, 4}$

$$= 150 + 0 + W + 0$$

$$= 150 + W$$

Therefore from first Law of thermodynamics

$$Q_{1, 2, 3, 4} = W_{1, 2, 3, 4}$$

or

$$89 = 150 + W$$

$$\therefore \quad W = -150 + 89 = -61 \text{ kJ}$$

(Since it is negative the work is being done on the system)

Example 3.36. *The working fluid in an engine undergoes a cycle of four process. In the cycle two work interaction are involved viz 15 kJ to the working fluid and 40 kJ from the fluid. The working cycle also includes three heat transters, Viz 100 kJ to the working fluid and 50 kJ from the working fluid and x respectively. Determine the magnitude and direction of the third heat interaction x.*

Solution : During the cycle of 4 processes net work transfer

$$W = -15 + 40 - 25 \text{ kJ}$$

Similarly net heat transfer,

$$Q = 100 - 50 + x$$

$$= 50 + x$$

Where x is the third heat interaction

From first law of thermodynamics for cyclic process

$$Q = W$$

$$50 + x = 25$$

$$x = -25 \text{ kJ}$$

Negative sign indicates that there is heat flow out of the system.

Example 3.37. *A closed rigid vessel containing 5 kg of air at 100 kPa and 27 °C is heated until its pressure doubles. Calculate (a) the volume of the tank (b) the final temperature (c) change in internal energy(d) change in enthalpy and (e) The amount of heat transfer*

Take Molar mass of air = 28.97 kg/k mol and

$$R_u = 8.314 \text{ kJ/(kmol.K)}$$

$$C_v = 0.718 \text{ kJ/(kg.K)}$$

Solution : (*a*) At the specified conditions this air can be treated as an ideal gas. Then from this ideal gas relation.

$$P_1 V_1 = mRT_1$$

Where V_1 = Volume of the tank

$T_1 = 273 + 27 = 300$ K

$P_1 = 100$ kPa

and
$$R = \frac{R_u}{molar\ mass} = \frac{R_u}{M}$$

$$= \frac{8.314}{28.97} = \frac{\text{kJ/(k mol} \cdot \text{K)}}{\text{kg/}(k \cdot \text{mol})} = 0.287 \text{ kJ/(kg} \cdot K)$$

Substituting values in gas equation

$$V_1 = \frac{mRT_1}{P_1} = \frac{5 \times 0.287 \times 300}{100} = 4.305m^3$$

(*b*) As the vessel is rigid the volume remains constant that is

$$V_2 = V_1$$

Also $P_2 = 2P_1$

Again from gas equation
$$\frac{P_1V_1}{T_1} = \frac{P_2V_2}{T_2}$$

$$T_2 = \left(\frac{T_1P_2}{P_1}\right)\left(\frac{V_2}{V_1}\right) = T_1\frac{P_2}{P_1}$$

$$= 300 \times \frac{2P_1}{P_1} = 600K$$

(*c*) Change in internal energy

As the vessel is rigid and no mass leaves or enters the vessel, it is isochoric process or constant volume non flow process

Further as it is ideal gas $u = u\,(T)$

and
$$dU = m\,C_v\,(T_2 - T_1)$$
$$= (5 \text{ kg})\,(0.718 \text{ kJ/kg} \cdot \text{K})\,(600 - 300) \text{ K}$$
$$= 1077 \text{ kJ}$$

(*d*) Change in enthalpy $dH = mc_p\,(T_2 - T_2)$

and $C_p - C_v = R$

or
$$C_p = R + C_v$$
$$= 0.287 \text{ kJ/(kg} \cdot \text{K)} + 0.718 \text{ kJ/(kg} \cdot \text{K)}$$
$$= 1.005 \text{ k}J/(\text{kg} \cdot \text{K})$$

Therefore
$$dH = (5 \text{ kg})\,[1.005 \text{ kJ/(kg} \cdot \text{K)}]\,(600 - 300)\text{K}$$
$$= 1507.5 \text{ kJ}$$

(*c*) Since it is a closed system and the process is isochoric (contant volume), there will be no boundary work

ie $W = 0$

Therefore from first low of therewdyngnis we have

$$Q - W = dU$$

$$Q - 0 = dU$$

ie $$Q = 0 + 1077 = 1077 \text{ kJ}$$

Example 3.38. *A rigid tank contains 0.5 m^3 hydrogen at 500 kPa pressure and 500 K temperature. It is now cooled until its temperature drops to 300 K. Determine*

(*a*) the final pressure (*b*) change is enthalpy (*d*) the amount of heat transfer

Take $$R_u = 8.314 \text{ kJ/(k mol} \cdot \text{K)}$$

molar mass of hydrogen $$= M = 2.016 \text{ kg/k mol}$$

$$C_v = 10.183 \text{ kJ/(kg} \cdot \text{K)}$$

Solution :

(*a*) At this Specified conditions hydrogen can be treated as an ideal gas. Also the tank is rigid therefore process is isochoric (or constant volume process)

That is $$V_2 = V_1 = 0.5\ m^3$$

and $$R = \frac{Ru}{M} = \frac{8.314 \text{ kJ/(k mol. K)}}{2.016 \text{ kg/(k mol. K)}}$$

$$= 4.124 \text{ kJ/(kg} \cdot \text{K)}$$

Then from gas equation $$\frac{P_1V_1}{T_1} = \frac{P_2V_2}{T_2}$$

$$P_2 = \frac{P_1T_2}{T_1}\left(\frac{V_1}{V_2}\right) = \frac{P_1T_2}{T_1} = \frac{500 \times 300}{500} \qquad (\because V_1 = V_2)$$

$$= 300 \text{ kPa}$$

(*b*) Change in internal energy

Since it is a non flow (closed system) constant volume process

$$dU = mC_v (R_2 - T_1)$$

now $$m = \frac{P_1V_1}{RT_1} = \frac{500(\text{kP}_a) \times 0.5\text{m}^3}{4.124(\text{kJ/kg} \cdot \text{K}) \times 500\text{k}}$$

$$= 0.1212 \text{ kg}$$

Therefore $$dU = (.1212 \text{ kg})\ (10.183 \text{ kJ/kg} \cdot \text{k})\ (300 \times 500)$$

$$= -246.84 \text{ kJ}$$

(Negative sign indicates that heat is going out)

(*c*) Change in enthalpy

$$dH = m\,C_p\,(T_2 - T_1)$$

now $$C_p = C_v + R$$

$$= 10.183 + 4.124$$

$$= 14.307 \text{ kJ/(kg} \cdot \text{K)}$$

Therefore $dH = (0.1212)(14.307)(300-500)$

$= -346.8$ kJ

(*d*) Amount of heat transfer Q

$$Q - W = dU$$

Since the volume remains constant boundary work $W = o$

Therefore $Q = dU$

$= -246.8$ kJ

(Heat flows out of the system)

Example 3.39. *Two students living in a hostel room 6 m × 6 m × 4 m high, leave the 100 W fan on before leaving the room for their college on some summer day in the hope that they will find the room relatively cooler when they come back in the evening after attending classes for 8 hours. Assuming that all the windows and doors were tightly closed and there was no heat transfer from the walls, windows and roof, determine the temperature in the room when they come back after attending the college. The morning temperature in be room be assumed as 15 °C at 100 kPa.*

Take $C_v = 0.718\ kJ/kg\ K$

$R_u = 8.314\ kJ/(k\ mol \cdot K)$

Molar mass $= M = 28.97\ kg/k \cdot mol$

Solution : At the specified conditions air can be treated as ideal gas. Let the air in the room be taken as our system. Since the doors and windows were tightly closed no air leaked out nor it leaked in. Thus this is a constant volume closed non flow system.

Therefore the conservation of energy equation for this is

$$Q - W_e - W_b = \Delta U + \Delta KE + \Delta PE$$

Where Q = Heat transfer

W_e = Electric work done by the fan

ΔU = Change in internal energy

ΔKE = Change in Kinetic energy

ΔPE = Change in Potential energy

Here, neither there is transfer of heat nor boundary work, nor change in potential or kinetic energy.

We have $Q = 0,\ W_b = 0,\ \Delta KE = 0,\ \Delta PE = 0$

Substituting these values in energy equation we get

$$-W_e = \Delta U = m(u_2 - u_1)$$

$$= m\,C_v\,(T_2 - T_1)$$

Now $m = \dfrac{P_1 V_1}{R T_1}$

where V_1 = Volume of the air in the room

$= 6 \times 6 \times 4 = 144\ m^3$

$T_1 = 15 + 273 = 288$ K

$$R = \frac{R_u}{M} = \frac{8.314}{28.97} = 0.287 /(\text{Kg/K})$$

$$\therefore \quad m = \frac{100(\text{kP}_\text{a}) \times (144\,\text{m}^3)}{0.287\,\text{kJ}/(\text{kg}\cdot\text{K}) \times (288\,\text{K})}$$

$$= 174.2\ \text{kg}$$

Also $W_e = \text{W}_\text{e}\Delta\text{t}$

Fan wattage $= 100\ W = \dfrac{100}{1000}\ \text{kJ/Sec}$

$$= 0.1\ \text{kg/sec}$$

$$\Delta t = 8\ \text{hours}$$

$$= 8 \times 3600\ \text{sec} = 28800\ \text{sec}$$

$$\therefore \quad W_e = 0.1\ (\text{kJ/s})\ (28800\ \text{s})$$

$$= 2880\ \text{kJ}$$

We have earlier found that

$$-W_e = mC_v\,(T_2 - T_1)$$

Substituting this values of W_e, m, C_v, T_1 we get

$$-(-2800) = (174.2\ \text{kg})\ (0.718\ \text{kJ/kg}\cdot\text{K})\ (T_2 - 288)\text{K}$$

or $$(T_2 - 288) = \frac{2880}{174.2 \times 0.718} = 23\text{K}$$

or $$T_2 = 288 + 23$$

$$= 311\ \text{K or } 38°\text{C}$$

Example 3.40. *A rigid tank contains 0.3 m³ Oxygen initially at 100 kPa and 27° C. A paddle wheel fitted in the tank is rotated until the pressure inside the tank rises to 150 kPa. In this process 2 kJ of heat is lost to the surroundings. Find the paddle wheel work done neglecting the energy stored in the paddle wheel. Take* R_u *= 8.314 kJ/k mol.K M= 31.999 kg/mol and* C_v *= 0.658 kJ/kg.K*

Solution : Under this specified conditions oxygen can be treated as perfect gas. Take the oxygen in the tank as the System. Since no mass leaves or enters the tank, it is non flow or closed system more over it is a constant volume prosess.

Therefore W_b = Boundary work = 0

Also $\Delta KE = 0;\ \Delta PE = 0$

Then the first law equation reduces to

$$Q - W_{pw} - W_b = \Delta U + \Delta KE + \Delta PE$$

Substituting the values it reduces to

$$Q - W_{pw} = m(u_2 - u_1)$$

$$= mC_v\,(T_2 - T_1)$$

Now $Q = -2\ \text{kJ}$

$$R = \frac{8.314}{31.999} = 0.2598 \text{ kJ/(kg} \cdot \text{K)}$$

$$V_1 = 0.3\text{m}^3$$

$$T_1 = 273 + 27 = 300 \text{ K}$$

$$P_1 = 100\, kP_a$$

$$P_2 = 150\, kPa$$

Therefore from gas equation

$$P_1V_1 = mRT_1$$

$$M = \frac{P_1V_1}{R_1T_1} = \frac{(100\,\text{kP}_a)(0.3\,\text{m}^3)}{0.2598(k\text{J/kg} \cdot \text{K})(300\,\text{K})}$$

$$= 0.3849 \text{ kg}$$

$$V_2 = V_1$$

$$\frac{P_1V_1}{T_1} = \frac{P_2V_2}{T_2}$$

$$T_2 = \frac{P_2V_2}{P_1V_1}T_1$$

$$= \frac{150 \times 300}{100} \times \left(\frac{V_2}{V_1}\right) = 450 \text{ K}$$

Therefore substituting these values in the above equation we get

$$(-2 \text{ kJ}) - W_{pw} = (0.3849 \text{ kg})\,(0.658 \text{ kJ/kg} \cdot \text{K})\,(450\text{-}300)$$

$$-2 - W_{pw} = 0.3849 \times 0.658 \times 150$$

$$= 37.9896$$

that is $W_{pw} = -39.9896$ kJ

(Negative sign indicates that work is done on the system

Example 3.41. *An insulated rigid vessel is divided into two equal chambers by a partition. One chamber is evacuated while the other contains 3 kg of an ideal gas at 450 kPa and 80°C. The partition is then removed and the gas expands to occupy the entire space in the vessel. Find the final temperature and pressure.*

Solution : Take the entire vessel as the system. Then this will be a constant volume closed system as no mass leaves or enters the system. Then from the conservation of energy equation for an ideal gas

A Vaccum	B gas

Fig. 3.48

$$Q - W_b = dU + \Delta KE + \Delta P\varepsilon$$

Since $\Delta KE = \Delta PE = 0$

Also $W_b = 0$ (as there is no work done for a constant volume Systin)

and $Q = 0$ (the vessel is insulated and their is no heat transfer)

Then the equation reduces to

$$dU = 0$$

or $$mC_{v,\,av}(T_2 - T_1) = 0$$

i.e., $$T_2 = T_1 \text{ and } T_1 = 80°\text{C (given)}$$

$\therefore$ $$T_2 = 80°\text{C}$$

For finding the pressure we apply the gas equation

$$\frac{P_1V_1}{T_1} = \frac{P_2V_2}{T_2}$$

Where P_1, V_1 and T_1 are pressure, volume and temperature of gas in chamber before the removal of partition. Similarly P_2, V_2 and T_2 are respectively the pressure, volume and temperature of the same gas when it occupies the entire vessel after removal of the partition.

Evidently $$V_2 = 2V_1$$

and $$T_2 = T_1 \text{ (as proved above)}$$

Thus from gas equation

$$P_2 = P_1\left(\frac{V_1}{V_2}\right)\left(\frac{T_2}{T_1}\right)$$

$$= P\left(\frac{1}{2}\right)(1) = \frac{P}{2}$$

$$= \frac{450}{2} = \mathbf{225\ kPa}$$

Example 3.42. *A piston cylinder device contains 0.4 m³ of Nitrogen gas at 300 kPa and 327°C. The gas is allowed to cool at constant pressure till the temperature drops to 77°C. Using the data from nitrogen tables calculate the amount of heat transfer.*

Solution : At the specified conditions Nitrogen gas can be treated as an ideal gas. If we take gas as our system it will then be a closed non flow system as there is no mass leaving or entering the system but it will be a constant pressure process.

Then from conservation of energy equation we have

$$Q - W_{\text{others}} - W_b = \Delta U + \Delta KE + \Delta PE$$

Here $$\Delta KE = \Delta PE = 0$$

and $$W_{\text{others}} = 0$$

then this equation reduces to

$$Q - W_b = \Delta U$$

or $$Q = dU + W_b$$

$$= dH \text{ (as it is constant pressure process)}$$

$$= m(h_2 - h_1)$$

$$= m(\bar{h}_2 - \bar{h}_1)$$

where
$$m = \text{mass of gas}$$
$$M = \text{molar mass}$$
$$\bar{h}_1, \bar{h}_2 = \text{Enthalpy on molar mass basis}$$

From the ideal gas equation we have

$$P_1V_1 = MR_uT_1$$

or $$M = \frac{P_1V_1}{R_uT_1}$$

Also $$T_1 = 273 + 327 = 600 \text{ K}$$

From nitrogen gas tables

$$R_u = 8.314 \text{ kJ/(k mol} \cdot \text{K)}$$

and $$V_1 = 0.4 \text{ m}^3$$

Therefore we get $$M = \frac{P_1V_1}{R_uT_1} = \frac{300(\text{kPa}) \times 0.4 \text{ m}^3}{8.314 \text{ kJ/(k mol} \cdot \text{K)} \times (600 \text{ K})}$$

Again from nitrogen tables the enthalpy at two states are (on mol basis)

$$\bar{h}_1 = h_{\text{at 600 k}} = 17563 \text{ kJ/k mol}$$

$$\bar{h}_2 = \bar{h}_{\text{at 350°k}} = 10180 \text{ kJ/k mol}$$

Substituting these values in the above equation

$$Q = M(\bar{h}_2 - \bar{h}_1)$$

$$= (0.0241 \text{ k mol})/(10180 - 17563)$$

$$= 177.6 \text{ kJ}$$

Example 3.43. *A piston cylinder device contains 15 kg of air. By passing current through a resistance heater inside the cylinder the air is heated from 27°C to 87°C, keeping the pressure inside constant at 200 kPa during the process. It the heat loss is 10 kJ, determine the electric energy consumed in kWh.*

Solution : At the specified conditions air can be heated as an ideal gas. Take the air in the cylinder as our system. It is closed system (non flow system) as no mass enters or leaves the cylinder. From first law of thermodynamics we have

$$Q - W_e - W_b = \Delta U + \Delta KE + \Delta PE$$

Here $$\Delta KE = 0 \text{ and } \Delta PE = 0$$

Therefore $$Q - W_e - W_b = \Delta U$$

or $$Q - W_e = \Delta U + W_b$$

$$= m(h_2 - h_1)$$

$$W_e = -m(h_2 - h_1) + Q$$

From the air tables, we have

At $$T_1 = 273 + 27 = 300 \text{ K}, \quad h_1 = 300.19 \text{ kJ/kg}$$

and at $$T_2 = 273 + 87 = 360 \text{ K}, \quad h_2 = 360.58 \text{ kJ/kg}$$

Substituting these values in the above equation

$$W_e = -15 \text{ kg } (360.58 - 300.19) \text{ kJ/kg} + (-10 \text{ kJ})$$

$$= -901.35 - 10 = -911.35 \text{ kJ}$$

$$= (911.35 \text{ kJ}) = -0.2532 \text{ kWh}$$

Negative sign of W_e indicates that electrical work is done on the system.

Example 3.44. *An insulated piston cylinder device initially contains 0.5 m^3 air at 200 kpa and 27°C. Electric current is made to pass through a resistance heater fitted inside the cylinder for 15 minutes and the volume is doubled the pressure inside the cylinder remaining constant.*

If the line voltage is 230 volts determine the current in amperes that has passed through the heater element.

Solution : The air at the specified conditions can be treated as an ideal gas. Take the air in the cylinder as our system. This is a closed or non flow system as no mass enters or leaves the system. Then from the conservation of energy principle we have.

$$Q - W_e - W_b = \Delta U + \Delta KE + \Delta PE$$

Here $Q = 0$ as the device is insulated

$$\Delta KE = 0$$

and $\Delta PE = 0$

Then the equation reduces to

$$Q - W_e = \Delta U + W_b$$

$$= m(h_2 - h_1)$$

$$W_e = N(\bar{h}_2 - \bar{h}_1)$$

Where $\bar{h}_1$ and $\bar{h}_2$ are enthalpies on mol basis and N is the number of moles of air present.

From the ideal gas equation

$$\frac{P_1V_1}{T_1} = \frac{P_2V_2}{T_2}$$

or

$$T_2 = T_1\left(\frac{P_2}{P_1}\right)\left(\frac{C_2}{V_1}\right) = (273 + 27)\ (1)(2) = 600 \text{ K}$$

Again $P_1V_1 = NR_uT_1$

or

$$N = \frac{P_1V_1}{R_uT_1} = \frac{(200 \text{ kPa})(0.5 \text{ m}^3)}{(8.314 \text{ kPa}/_{\text{k mol}\cdot\text{k}})(300 \text{ K})}$$

$$= 0.04 \text{ k mol}$$

Non from air takes

At 300 K $\bar{h}_1 = 8736$ kJ/k mol

At 600 K $\bar{h}_2 = 17929$ kJ/k mol

Substituting these values in the above equation

$$W_e = -(0.04 \text{ k mol})\ (17929 - 8736) \text{ kJ/k mol}$$

$$= -367.72 \text{ kJ}$$

It E is the live voltage and I the current and ΔT the time, the current flows through the resistance heater, then

$$|W_e| = EI\Delta T$$

Then

$$I = \frac{|W_e|}{E(\Delta T)} = \frac{367.72 \text{ kJ}}{(230 \text{ k})(15 \times 60 \text{ s})}\left(\frac{1000 \text{ VA}}{1 \text{ kJ/s}}\right)$$

$$= 1.776 \text{ A}$$

Example 3.45: *A piston cylinder device contains 0.5 kg of nitrogen at 100 kPa and 300 K. It is slowly compressed in a polytropic process $PV^{1.3}$ = constant until its volume is reduced to* *half its original volume. Find the work done and heat transfer for this process. For Nitrogen take C_v = 0.743 kJ / (kg · K) and R = 0.2968 kJ / (kg · K).*

Solution : (*a*) Nitrogen at the specified conditions can be treated as an ideal gas. Then the ideal gas equation will apply in addition to the process equation.

For the polytropic process $P_1V_1^{1.3} = P_2V_2^{1.3}$

$$P_2 = P_1\left(\frac{V_1}{V_2}\right)^{1.3} = (100 \text{ kPa})\,(2)^{1.3}$$

$$= 246.2 \text{ kPa}$$

Then from ideal gas equation $\dfrac{P_1V_1}{T_1} = \dfrac{P_2V_2}{T_2}$

$$T_2 = T_1\left(\frac{P_2}{P_1}\right)\left(\frac{V_2}{V_1}\right) = 300\left(\frac{246.2}{100}\right)\left(\frac{1}{2}\right)$$

$$= 369.3 \text{ K}$$

The boundary work for this polytropic process from equation (3.25)

$$W_b = \frac{P_2V_2 - P_1V_1}{1-n} = \frac{mR\,(T_2 - T_1)}{1-n} \quad ...(3.25)$$

$$= \frac{(0.5 \text{ kg})(2.2968 \text{ kJ/kg}\cdot\text{K})\ (369.3 - 3.00)\text{ K}}{1 - 1.3}$$

$$= -34.28 \text{ kJ}$$

(*b*) Take the Nitrogen in the cylinder as our system since no mass leaves or enters the system, it is a closed or nonflow system. Then from the first law of thermodynamics one have

$$Q - W_{other} - W_b = \Delta U + \Delta KE + \Delta PE$$

here $\qquad W_{others} = 0;\ \Delta KE = 0;\ \Delta PE = 0$

There $\qquad Q = W_b + \Delta U$

$$= W_b + m\,C_v(T_2 - T_1)$$

$$= (-34.28) + 0.5 \times 0.743\ (369.3 - 300)$$

$$= -34.28 + 25.74$$

$$= -8.54 \text{ kJ}$$

Negative sign indicates that the heat is lost to the surroundings.

Example 3.46. *A piston cylinder device contains 0.5 m³ helium gas at 150 kPa and 20°C. It is then compressed in a polytropic process $PV^n = C$ to 400 kPa and 140°C. Find the value of the index n and calculate the heat transfer for the process. For helium take R = 2.0769 kJ / (kg · K.) and C_v = 3.1156 kJ / (kg · K).*

Solution : At the specified condition helium gas can be treated as an ideal gas. Then from gas equation.

$$P_1V_1 = mRT_1$$

or $\qquad m = \dfrac{P_1V_1}{RT_1} = \dfrac{(150 \text{ kPa})(0.5 \text{ m}^3)}{(2.0769 \text{ kPa m}^3/\text{kg}\cdot K)\ (293\ K)}$

$$= 0.123 \text{ kg}$$

Again from ideal gas equation

$$\frac{P_1V_1}{T_1} = \frac{P_2V_2}{T_2}$$

$$V_2 = \frac{T_2}{T_1}\frac{P_1V_1}{P_2} = \frac{(140+273)}{(293)}\left(\frac{150}{400}\right)\times(.5)$$

$$= \frac{413}{293}\times\frac{150}{400}\times.5 = .264 \text{ m}^3$$

Since $P_1V_1^{\,n} = P_2V_2^{\,n}$

or $$\left(\frac{V_1}{V_2}\right)^n = \left(\frac{P_2}{P_1}\right) = \left(\frac{400}{150}\right)$$

or $$\left(\frac{0.5}{0.264}\right)^n = \left(\frac{400}{150}\right)$$

or $$(1.894)^n = 2.6666$$

$$n\, ln(1.894) = ln(2.6666)$$

$$n \times 0.6387 = 0.9808$$

or $$n = \frac{0.9808}{0.6387} = 1.536$$

(*b*) Boundary work from equation (3.25)

$$W_b = \frac{P_2V_2 - P_1V_1}{1-n} = \frac{mR(T_2 - T_1)}{1-n}$$

$$= \frac{(0.123 \text{ kg})\ (2.0769 \text{ kJ/(kg}\cdot\text{k)})\ (413 - 293) \text{ k}}{1-1.536}$$

$$= -57.2 \text{ kJ}$$

Taking helium as the system. This system is closed system as no mass enters the system or leaves the system. Then from conservation of energy equation.

$$Q - W_{other} - W_b = \Delta U + \Delta PE + \Delta KE$$

Here $W_{others} = 0$, $\Delta PE = 0$; and $\Delta KE = 0$

Therefore $$Q = W_b + \Delta U = W_b + mC_v(T_2 - T_1)$$

$$= -57.2 + 0.123 \times 3.1156\ (413 - 293)$$

$$= -11.2 \text{ kJ (Heat leaves the system)}$$

Example 3.47. *On a winter day a room is heated with an electric heater. Assuming the heat loss from the room is 18000 kJ/h, find rating of the heater so that the room is kept at constant temperature. Assume that the room is air tight and air to be an ideal gas.*

Solution : We take the air in the room, as our system. Since the room is air tight, no mass enters the room nor any mass leaves, it is a closed or non flow system. Then from conservation of energy principle we have for this isothermal process.

$$Q - W_b - W_e = \Delta U + \Delta PE + \Delta KE$$

Since the volume is constant

$$W_b = 0$$

Further the temperature remains constant

Then $\Delta U = m\,C_v(T_2 - T_1) = 0$

Also here $\Delta PE = 0$

$\Delta KE = 0$

Thus $Q = W_e$

$$\therefore \quad W_e = (1800 \text{ kJ/h})\left(\frac{1 \text{ kW}}{3600 \text{ kJ/h}}\right)$$

$= 5$ kW which is the required rating of the heater

Example 3.48. *A piston cylinder device contains 0.5m³ of air at 300 kPa and 60°C When 40 kJ heat is trnsferred to the air, it expands isothermally. Find the boundary work during the process.*

Solution : At the specified condition the air can be treated as an ideal gas. Take the air in the piston cylinder assembly as our system. It is a non flow isothermal process, since neither any mass enters nor it leaves the system.

Therefore from the conservation of energy principle we have

$$Q - W_b - W_{\text{others}} = \Delta U + \Delta PE + \Delta KE$$

Here $\Delta PE = 0, \Delta KE = 0$ and $W_{\text{others}} = 0$

At $\Delta U = mC_v\,\Delta T = 0$ (Being isothermal process)

Therefore substituting these values in the above equation we get

$$Q = W_b$$

or $W_b = Q = 40$ kJ

Example 3.49. *A piston cylinder device contains 3 kg of carbon dioxide at 250 kPa and 27°C. During a quasi-equilibrium isothermal expansion process 15 kJ of boundary work is done by the system and 5 kJ of paddle-wheel work is done on the system.Determine the heat transfer for the process.*

Solution : At the specified condition, carbon dioxide gas can be treated as an ideal gas. Take the gas in the piston-cylinder device as the system. Since no gas either enters or leaves the system it is non flow or closed system. Further being an isothermal expansion the temperature will remain constant.

i.e. $\Delta T = 0$

or $dU = m\,C_v\,\Delta T = 0$

Therefore from the conservation of energy principle we have for this process

$$Q - W_{pw} - W_b = \Delta U + \Delta KE + \Delta PE$$

Here $\Delta U = 0, \Delta KE = 0$ and $\Delta PE = 0$

Therefore $Q = W_{pw} + W_b$

$= (-5) + 15 = 15$ kJ

(Positive sign indicates that the heat is entering the system)

Example 3.50. *A rigid vessel and a frictionless piston cylinder device initially contain 12 kg of air each and both have the same volume, temperature and pressure. Now-heat is supplied* *to both the systems such that the temperature of air in both of them rises by 15°C. Calculate the extra amount of heat supplied to the air contained in the piston-cylinder device than that contained in the rigid cylinder. Take the molar mass of air as 28.97 kg/k mol and R_u = 8.314 kJ(k mol · k). Assume that air behaves like an ideal gas at the given conditions.*

Solution : At the given conditions air is to be treated as an ideal gas. Further in both the systems no mass enters the system nor it leaves the system Both of Them are closed or non flow system. However in the case of rigid vessel it will be a constant volume process while in the case of piston cylinder assembly, the gas will expand when heat is supplied to it and the piston (whose weight is fixed) will start rising and the volume will increase but the pressure will reamin constant. Thus in the case of piston cylinder where pressure remains constant will require more heat than that where volume remains constant because in the former process heat will be consumed to do boundary work.

Thus for the rigid vessel case

$$Q_1 - W_b = \Delta U + \Delta PE + \Delta KE$$

Here $\Delta PE = 0, \Delta KE = 0 \text{ and } W_b = 0$

$$\therefore \quad Q_1 = \Delta U + W_b = \Delta U$$

For the piston cylinder assembly the have

$$Q_2 - W_b = \Delta U + \Delta PE + \Delta KE$$

Here $\Delta KE = 0$

i.e., $Q_2 = \Delta U + W_b + \Delta KE + \Delta PE$

Here $\Delta PE = 0 \text{ and } \Delta KE = 0$

$$Q_2 = \Delta U + W_b = \Delta H$$

That is $= mC_p \Delta T$

Thus from These two equations

Extra heat $Q_2 - Q_1 = \Delta H - \Delta U$

$$= mC_p \Delta T - mC_v \Delta T$$

or extra heat required $= Q_2 - Q_1$

$$= m(C_p - C_v) \Delta T$$

$$= mR \Delta T$$

Now $R = \dfrac{R_u}{M} = \dfrac{8.34 \text{ kJ/(k mol} \cdot K)}{28.97 \text{ kg(k mol)}}$

$$= 0.287 \text{ kJ/(kg} \cdot \text{K)}$$

Thus $Q_2 - Q_1 = (12 \text{ kg}) [0.287 \text{ kgJ/kg.K} (15 \text{ K}]$

$$= 51.66 \text{ kJ}$$

Example 3.51. *A piston cylinder device, in which the piston is resting on a set of stops (Fig. 3.45), contains 4 kg of air at 250 kPa and 300 K. The mass of the piston is such that a pressure of 450 kPa is required to move it. Heat is now added to the air until its volume is doubled. Find the work done by the air and the total amount of heat added to air. Show the process on P-V diagram. Take C_v = 0.718 kJ/(kg · K), molar mass = 28.97 kg/k mol and R_u = 8.314 kJ/(kg mol · K).*

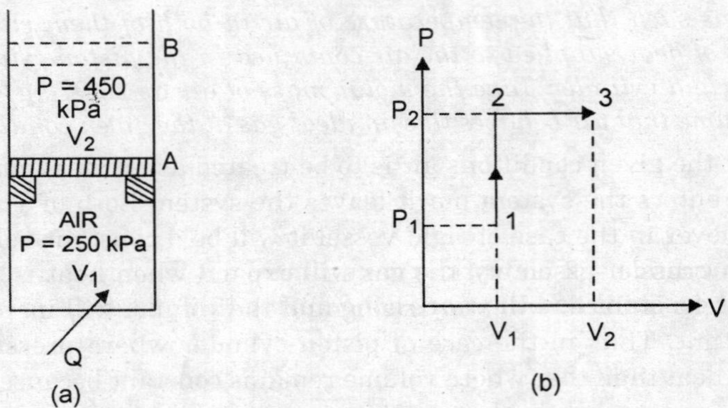

Fig. 3.49

Solution : The complete process consists of two parts. During the first part when the heat is added to air its pressure will start increasing but the volume will remain constant at V_1 till the pressure becomes 450 kPa. Then the piston will start rising. During this second part the pressure will remain constant at 450 kPa and piston will start moving upward till the volume at position B will be twice that at the initial position A. Thus the P.V. diagram in the first part of the process will be a vertical line such that volume remains constant and during the second part it will be a horizontal line *i.e.*, pressure is constant. It is shown in Fig. 3.49 (*b*).

We take the air contained in the device as our system which is a closed or non flow system as no mass cross the boundary of the system. Under the specified conditions the air can be treated as a perfect gas and the temperature and volumes at various points are calculated as given below :

$$R = \frac{R_u}{M} = \frac{8.314}{28.97} = 0.287 \qquad \text{kJ/(kg}\cdot\text{K)}$$

$$\therefore \quad V_1 = \frac{mRT_1}{P_1} = \frac{(4.0 \text{ kg})\,(0.287 \text{ kJ/kg}\cdot K)\,(300 \text{ K})}{250 \text{ (kPa)}}$$

$$= 1.3776 \text{ m}^3$$

At the point 2, the temperature T_2 is again found by using gas equation

$$\frac{P_1V_1}{T_1} = \frac{P_2V_2}{T_2}$$

or

$$T_2 = T_1\left(\frac{P_2}{P_1}\right)\left(\frac{V_2}{V_1}\right)$$

$$= (300)\left(\frac{450}{250}\right)(1)$$

$$= 540 \text{ K}$$

At the point 3, if T_3 is the temperature and V_3 the volume

$$V_3 = 2V_1 = 2 \times 1.3776 = 2.7552 \text{ m}^3$$

Then
$$\frac{P_1V_1}{T_1} = \frac{P_3V_3}{T_3}$$

$$T_3 = \frac{P_3}{P_1}\frac{V_3}{V_1}\cdot T_1 = \left(\frac{450}{250}\right)(2)300$$
$$= 1080 \text{ K}$$

During the process 1 – 2 no work is done since $V_1 = V_2$. and during the process 2 – 3 the pressure remains constant and the work done during this process is

$$W_b = \int_2^3 PdV = P_2(V_3 - V_2) = 450(2.7552 - 1.3776)$$
$$= 619.92 \text{ kJ}$$

Now from the first law of thermodynamics for the closed system

$$Q - W_b = \Delta U + \Delta KE + \Delta PE$$

Here $\Delta KE = 0$ and $\Delta PE = 0$

Therefore
$$Q = \Delta U + W_b$$
$$= mCv(T_3 - T_1) + W_b$$
$$= (4 \text{ kg})\,[0.718 \text{ kJ/(kg}\cdot\text{K)}]\,[1080 - 300] \text{ K} + 619.92 \text{ kJ}$$
$$= 2240.16 + 619.92$$
$$= 2860.08 \text{ kJ}$$

Example 3.52. *A pistion cylinder device has a set of stops at the top. It contains 4 kg of air at 250 kPa and 300 K. Heat is transferred to the air until the piston rises and hits the stops. At this stage the volume of the air is twice the initial volume. More heat is added to air so that the pressure is twice the value when the piston was rising. Show the process on P–V chart and calculate the work done and the amount of heat transfer. Take C_v = 0.718 kJ(kg · K) molar mass of air = 28.97 kg/k mol and R_u = 8.314 kJ/(k mol ·K).*

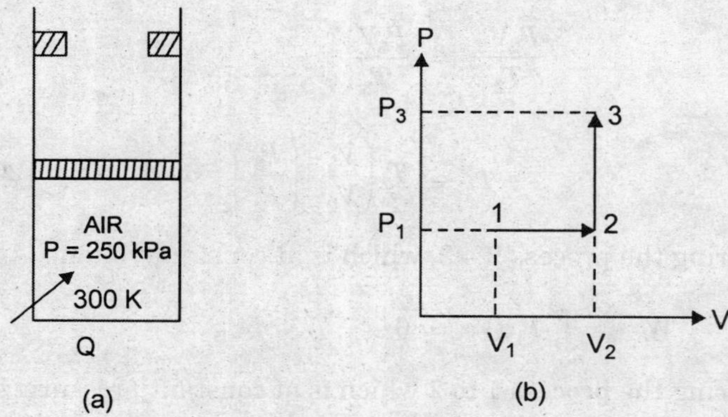

Fig. 3.50

Solution : The complete process is in two stages. During the first stage the piston will start rising when heat is added at constant pressure till it hits the stops at the top. During this stage the volume will increase and pressure will remain constant at the initial value of 250 kPa and this part of process is represented by horizontal line 1 – 2 on P.v chart. During the second stage with the addition of more heat the piston will stay there because of stops and the pressure will start increasing and the volume will remain constant at value V_2. This part of process is represented by the vertical line 2 – 3 on P.v chart Fig. (3.50 *b*).

We take the air contained in the device as our system. It is a non flow or closed system as no mass crosses the boundary of the system. Further at the specified conditions the air can be treated as perfact gas and the temperature, volume and pressure at various points is calculated with the help of gas equation.

$$R = \frac{R_u}{M} = \frac{8.314}{28.97} = 0.287 \text{ kJ/(kg} \cdot K)$$

$$V_1 = \frac{mRT_1}{P_1} = \frac{(4.0 \text{ kg})\ [0.287 \text{ kJ/(kg} \cdot K)]\ (300 \text{ K})}{(250 \text{ kPa})}$$

$$= 1.3776 \text{ m}^3$$

At the point 2,

$$V_2 = 2V_1 = 2 \times 1.3776 = 2.7552 \text{ m}^3$$

$$= V_3$$

$$P_2 = P_1 = 250 \text{ kPa}$$

From gas equation for points 1 and 2

$$\frac{P_1V_1}{T_1} = \frac{P_2V_2}{T_2}$$

$$T_2 = T_1\left(\frac{V_2}{V_1}\right)\left(\frac{P_2}{P_1}\right) = 300 \times 2 \times 1 = 600 \text{ K}$$

At the point 3,

$$V_3 = V_2 = 2.7552 \text{ m}^2$$

$$P_3 = 2P_2 = 2 \times 250 = 500 \text{ kPa}$$

From gas equation for points 2 and 3

$$\frac{P_2V_2}{T_2} = \frac{P_3V_3}{T_3}$$

$$T_3 = T_2\left(\frac{V_3}{V_2}\right)\left(\frac{P_3}{P_2}\right) = 600 \times 1 \times 2 = 1200 \text{ K}$$

Work done during the process 2 – 3, which is at constant volume is

$$W_{b_{1-2}} = \int_1^2 P_1 dV = 0 \qquad (\because dV = 0)$$

Work done during the process 1 to 2 which is at constant pressure

$$W_{b_{1-2}} = \int_1^2 P_1 dV = P_1(V_2 - V_1)$$

$$= (250 \text{ kPa})(2.7552 - 1.3776) \text{ m}^3 \left(\frac{1 \text{ kJ}}{1 \text{ kPa} \cdot \text{m}^3}\right)$$

$$= 344.4 \text{ kJ}$$

Now from the first law of thermodynamics for a closed system 1 – 2 – 3.

$$Q - W_{b_{1-2}} - W_{b_{2-3}} = \Delta U + \Delta PE + \Delta KE$$

Since here $W_{b_{2-3}} = 0, \Delta PE = 0$ and $\Delta KE = 0$

We get

$$\begin{aligned} Q &= W_{b_{1-2}} + \Delta U \\ &= W_{b_{1-2}} + mC_v(T_3 - T_1) \\ &= 344.4 \text{ kJ} + (4 \text{ kg})/(0.718 \text{ kJ/(kg} \cdot \text{K))} \times (1200 - 300) \text{ K} \\ &= 344.4 + 4246.2 \\ &= 4590.6 \text{ kJ} \end{aligned}$$

Example 3.53. *A piston cylinder device contains 0.6 m³ oxygen gas initially at 100 kPa and 27°C. At this position lower and of a linear spring just touches the piston and exerts no force on the piston. Heat is then transferred to oxygen until both the pressure and volume become twice the respective initial values. Find the work done and the amount of heat transfer for this process. Show the process on P.v. chart. Take $C_v = 0.658$ kJ(kg · K), $R_u = 8.314$ kJ(k mol · K) and molar mass of oxygen $M = 31.999$ kg / k mol.*

Solution : (*a*) As the heat is transferred to oxygen due to the presence of linear spring the pressure in the device will increase linearly during the process. The process curve is shown in Fig. (3.51 *b*) the work done during the process will be area under the curve.

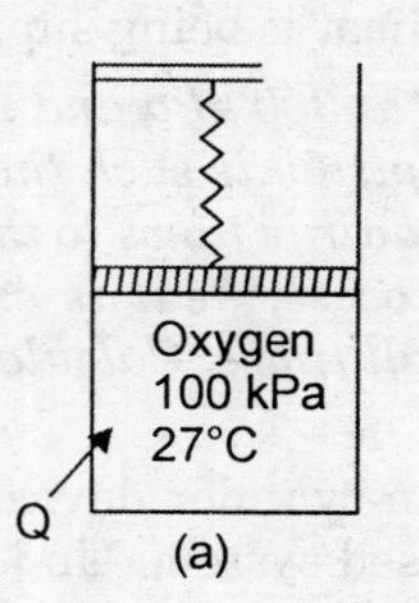

(a)

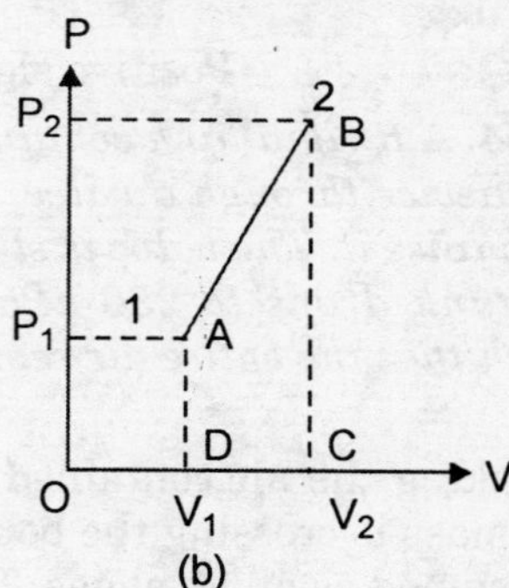

(b)

Fig. 3.51

We take the oxygen gas contained in the cylinder as our system. Since during the process no mass crosses the boundary it is a closed system. At the specified conditions oxygen can be treated as an ideal gas.

The pressure, volume and temperature at the initial state are given and at the final state temperature can be found by using gas equation as the volume and pressure at state 2 are known.

For oxygen $R = \dfrac{R_u}{M} = \dfrac{8.134}{31.999} = 0.2598 \text{ kJ(kg} \cdot K)$

At state1 $P_1 = 100$ kPa, $V_1 = 0.6$ m³ and $T_1 = 300$ K

At state 2 $P_2 = 2P_1 = 200$ kPa, $V_2 = 2V_1 = 1.2$ m³

$$\frac{P_1V_1}{T_1} = \frac{P_{2d}V_2}{T_2}$$

$$\therefore \quad T_2 = T_1\left(\frac{P_2}{P_1}\right)\left(\frac{V_2}{V_1}\right) = 300\times 2\times 2 = 1200 \text{ K}$$

$$W_b = \text{area } ABCD$$

$$= \left(\frac{P_1+P_2}{2}\right)(V_2 - V_1) = \left(\frac{100+200}{2}\right)(1.2-0.6)$$

$$= 90 \text{ kJ}$$

(*b*) Total heat transfer Q is found by using consvervation of energy equation for a closed system

$$Q - W_b = \Delta U + \Delta KE + \Delta PE$$

Here $\Delta KE = 0; \Delta PE = 0$

Therefore $Q = W_b + \Delta U$

$$= W_b + mC_v(T_2 - T_1)$$

Non $m = \dfrac{P_1V_1}{RT_1} = \dfrac{100\times 0.6}{(0.2598)(300)} = 0.7698 \text{ kg}$

Thus substituting the values in the above equation

$$Q = 90 \text{ kJ} + (0.7698 \text{ kg})\,[0.658 \text{ kJ(kg}\cdot\text{K)}]\,(1200 - 300] \text{ K}$$

$$= 90 + 455.87 = 545.87 \text{ kJ}$$

(Positive sign indicates that heat is being supplied to the system)

Example 3.54. *A R rigid tank contaning 0.6 m³ of air at 350 kPa and 27°C is connected to a piston cylinder device through a valve. The mass of the piston is such that a pressure of 200 kPa is required to raise it. The valve is slightly opened and air is flows to the cylinder until the pressure in the tank drops to 250 kPa. During the process, heat is exchanged with the surroundings such that the entire air remains at 27°C at all times. Calculate the heat transfer for this process.*

Solution : Taking the air contained within the piston-cylinder device as system. In this system, since no mass is crossing the boundary, it is a closed system. Under the given system the air may be assumed as an ideal gas. Then the final volume is calculated from gas equation.

$$\frac{P_1V_1}{T_1} = \frac{P_2V_2}{T_2}$$

Also $T_1 = T_2 = 27 + 273 = 300$ k

or $V_2 = \dfrac{P_1}{P_2}\dfrac{T_2}{T_1}V_1 = \dfrac{350 \text{ kPa}}{250 \text{ kPa}}\left(\dfrac{T_2}{T_1}\right)\times 0.6$

$$= 0.84 \text{ m}^3$$

At the piston face the pressure will be constant and it will be equal to 200 kPa corresponding to the weight of piston. Thus the boundary work done during the process is

$$W_b = \int_1^2 PdV = P_1(V_2 - V_1)$$

$$= (200 \text{ kPa})\,(0.84 - 0.6)$$

$$= 48 \text{ kJ}$$

Then from the energy equation for the closed system for the change from the initial to final states, we have

$$Q - W_b = \Delta U + \Delta KE + \Delta PE$$

But $$\Delta U = mC_v\,\Delta T = mC_v(0) = 0$$

(∴ Temperature is constant)

$$\Delta T = 0$$

Also $$\Delta KE = 0 \text{ and } \Delta PE = 0$$

∴ $$Q - W_b = 0$$

or $$Q = W_b = 48 \text{ kJ}$$

Example 3.55. *The cross-sectional area of the piston of a frictionless piston-cylinder assembly is 0.1 m^2. It contains 50 kg of water at 150 kPa and 25°C. Heat is now added to the water a portion of which evaporates and expands. When the volume of the contents reaches 0.2 m^3 the piston starts touching a linear spring whose stuffness is 100 kN/m. More heat is then added to water till the piston rises further by 0.2 m. Find (a) the final pressure and temperature (b) the work done during the process show the process on P–V diagram.*

Solution : When heat is transferred to water in piston-cylinder assembly the pressure will remain 150 kPa level even through the piston will be moving up till it touches the lower end of the spring. After that when it rises further the spring will start exerting force on the piston which will be proportional to the displacement or compression of the spring. At the end of the process the compression in spring will be 0.2 m.

Thus the force exerted at this stage will be given by

$$F = kx$$

$$= 0.2 \times k \quad \text{(where k is the stiffness of spring)}$$

Therefore additional pressure in the cylinder due to spring force will be

$$\Delta P = \frac{F}{A} = \frac{0.2 \times 100}{0.1} = 200 \text{ kPa}$$

Thus at this stage

$$P_3 = 150 + 200 = 350 \text{ kPa}$$

(*b*) The process is shown as P.V. chart. At stage 1 it is liquid water. During the path 1 – 2, the pressure remains constant. For the process 2 – 3 the path will be straight line as the pressure will rise linearly due to spring deflection.

From steam tables the specific volumes at these states are given below

$T_1 = 25°C$ $$v_1 = v_f \text{ at } 25°C = 0.001003 \text{ m}^3/\text{kg}$$

$P_1 = 150$ kPa

∴ $$V_1 = mv_1 = 50 \times 0.001003 = 0.05 \text{ m}^3$$

$$V_2 = 0.2 \text{ m}^3$$

(when the piston touches the lower end of spring)

$$V_3 = V_2 + x_{2-3}A = 0.2 + 0.2 \times .1 = 0.22 \text{ m}^3$$

∴ $$v_3 = \frac{V_3}{m} = \frac{0.22}{50} = 0.0044 \text{ m}^3/\text{kg}$$

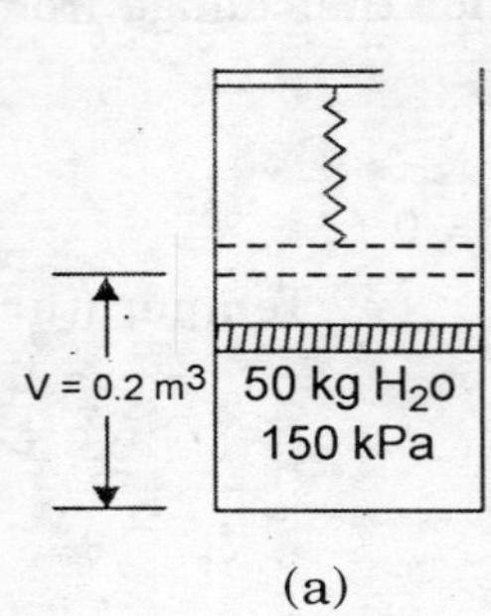

(a)

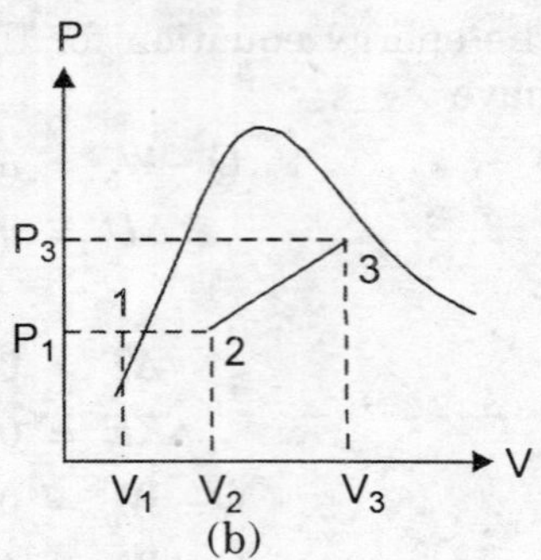

(b)

Fig. 3.52

Again from steam tables

At $\qquad P_3 = 350\text{ kPa}, v_f = 0.0010\text{ m}^3/\text{kg}$

and $\qquad v_g = 0.5243\text{ m}^3/\text{kg}$

Here since $v_f < v_3 < v_g$, the final stage at point 3 will be saturated mixture. Thus the final temperature will be

$$T_3 = T_{\text{out at}}\ 350\text{ kPa} = \mathbf{138.88°C}$$

(*c*) The pressure remains constant during the process 1 – 2 and clanges linearly during (2 – 3). Then the boundary work during the entire process will be the area under the curve

$$\therefore \quad W_b = \text{Area under the curve}$$

$$= P_1(V_2 - V_1) + \frac{1}{2}(P_2 + P_3)(V_3 - V_2)$$

$$= 150(0.2 - 0.05) + \frac{1}{2}(150 + 350)(0.22 - 0.2)$$

$$= 27.5\text{ kJ}$$

Example 3.56. *An aluminium block at 85°C is dropped into an insulated tank containing 150°C of water at 35°C and at atmospheric pressure. If the final temperature at equilibrium is 40°C, find the mass of aluminium block. Take density of water as 1000 kg/m³ and specific heat of aluminium and water as 0.9 kJ/kg °C and 4.184 kJ/kg °C respectively.*

Solution : Take the Aluminium block and water together as our system. As there is no transfer of mass across the boundary it is a closed system. Then from the conservation of energy equation we have

$$Q - W = \Delta U + \Delta KE + \Delta PE$$

Since the tank is insulated, there is no heat transfer across the boundary

Therefore $\qquad Q = 0$

In addition there is no boundary work

i.e., $\qquad W_b = 0$

Further $\qquad \Delta KE = 0 \text{ and } \Delta PE = 0$

Thus the energy equation reduces to

$$0 - 0 = \Delta U + 0 + 0$$

or $\qquad \Delta U = 0$

or $\qquad \Delta U_{Al} - \Delta U_{\text{water}} = 0$

Now $\Delta U_{Al} = m_{Al}\, C(T_2 - T_1)$

or $\Delta U_{Al} = m_{Al} \times (0.9 \text{ kJ/kg °C})\,(40 - 85)$

$= -40.5\, m_{Al}$

and $\Delta U_{\text{water}} = m_{\text{water}} (4.184 \text{ kJ/kg°C})\,(40 - 35)$

But $m_{water} = \text{vol} \times \text{density}$

$= (0.150) \times 1000$

$= 1500 \text{ kg}$

$\therefore$ $\Delta U_{water} = (150 \text{ kg})\,(4.184 \text{ kJ/}_{\text{kg°C}})\,(40 - 35)$

$= 3138 \text{ kJ}$

Therefore from energy equation

$$-\,40.5\, m_{al} + 3138 = 0$$

or $$m_{Al} = \frac{3138}{40.5} = 77.48 \text{ kg}$$

Example 3.57. *A copper piece weighing 20 kg at 85°C is dropped into an insulated tank containing 125 kg of water at 25°C. Determine the final equilibrium temperature in tank.*

Take $C_{water} = 4.184\ kJ/kg\ °C$

$C_{cu} = 0.386\ (kJ/kg°C)$

Solution : Take Copper and water as system. Then conservation of energy equation gives

$$Q - W = \Delta U + \Delta KE + \Delta PE$$

Here $W = \Delta KE - \Delta PE = 0$

$\therefore$ $\Delta U = 0$

But $\Delta U_{cu} + \Delta U_{cv} = 0$

Non $\Delta U_{cu} = m_{cu} C_{cu} (T_2 - T_1)_{cw}$

$= 20 \times 0.386\,(T_2 - 85)$

$\Delta Uw = m_w * C_w\,(T_2 - 25)$

$\therefore$ From energy equation

$$20 \times 3.86\,(T_2 - 85) + 125 \times 4.184\,(T_2 - 25) = 0$$

$$7.72\, T_2 - 656.2 + 523\, T_2 - 13075 = 0$$

$$530.72\, T_2 = 13731.2$$

or $$T_2 = \frac{13731.2}{530.72} = 28.873\text{°C}$$

Example 3.58. *A 20 kg mass of aluminium at 200°C is brought in contact with 20 kg mass of iron at 100°C in an insulated chamber. Find the final equilibrium temperature of the combined system. Take* $C_{iron} = 0.45\ kJ/kg°C$ *and* $C_{Al} = 0.9\ kg/kg°C$.

Solution : Take aluminium and iron blocks as system. Then from conservation of energy equation we get

$$Q - W = \Delta U + \Delta KE + \Delta PE$$

Since $Q = 0,\ W = 0,\ \Delta KE = 0$ and $\Delta PE = 0$

We have $\Delta U = 0$

or $\Delta U_{Al} + \Delta U_{\text{iron}} = 0$

$$m_{Al}\, C_{Al}(T_2 - T_1)_{Al} + m_i c_i (T_2 - T_1) = 0$$

$$20 \times 0.9\,(T_2 - 200) + 20 \times .45\,(T_2 - 100) = 0$$

$$2(T_2 - 200) + (T_2 - 100) = 0$$

$$3T_2 = 500$$

$$T_2 = 167°C$$

Example 3.59. *A 40 kg mass of iron at 80°C and 60 kg mass of copper at 70°C are dropped into a tank containing 60 L of water at 20°C. It is estimated that 20 kJ of that is lost to the surroundings during the process. Taking C_{cu} = 3.86 kJ/kg°C C_{iron} = 0.45 kJ/kg°C and C_{water} = 4.184 kJ/kg°C, determine the final equlibrium temperature.Assume density of water 1000 kg/m^3.*

Solution : Take copper iron and water as system then from conservation of energy equation we have

$$Q - W = \Delta U + \Delta KE + \Delta PE$$

Here $\quad \Delta KE = 0, \Delta PE = 0 \text{ and } W = 0$

Then $\quad Q = \Delta U$

$$= \Delta U_w + \Delta \text{U}_{\text{iron}} + \Delta U_{\text{water}}$$

Now $\quad Q = -20 \text{ kJ}$

$$\Delta U_{cu} = m_w\, C_{cu}(T_2 - T_1)_{cu}$$

$$= 60 \times .386\,(T_2 - 70)$$

$$= 23.16\,(T_2 - 70)$$

$$\Delta U_{\text{iron}} = m_{\text{iron}}\, C_{\text{iron}}(T_2 - T_1)_{\text{iron}}$$

$$= 40 \times 0.45\,(T_2 - 80)$$

$$= 18(T_2 - 80)$$

$$\Delta U_{\text{water}} = m_w c_w (T_2 - T_1)_w$$

$$= (600 \times 10^{-3} \times 1000)\,(4.184)\,(T_2 - 20)$$

$$= 2510.4\,(T_2 - 200)$$

∴ Energy conservation equation gives

$$-20 = 23.16(T_2 - 70) + 18(T_2 - 80) + 2510.4(T_2 - 20)$$

$$= 2551.56T_2 - 1621.2 - 1440 - 50208$$

or $\quad T_2 = \dfrac{53249.2}{2551.56} = 20.869\ °C$

Example 3.60. *A piece of iron at 85°C is dropped into an insulated tank which contains 200 L of water at 20°C. Simultaneously a paddle wheel driven by 200 watt motor is activated to stir the water. Thermal equlibrium is achieved after 20 min with final temperature of 24°C. Take water density 1000 kg/m^3. Taking C_{iron} = 0.45 kJ/kg°C, C_{water} = 4.184 kJ/kg°C find the mass of iron peice.*

Solution : Taking iron block and water as system. Then from energy conservation equation

$$Q - W_{pw} = \Delta U + \Delta KE + \Delta PE$$

Since $\quad W_{pw} = 0, \Delta KE = 0 \text{ and } \Delta PE = 0$

We get $\quad Q - W_{pw} = \Delta U_{\text{iron}} + \Delta U_{\text{water}}$

Also $\quad Q = 0 \text{ (As tank insulated)}$

Therefore from energy equation we have

$$-W_{pw} = \Delta U_{iron} + \Delta U_{water}$$

Now
$$\Delta U_{iron} = m_i c_i (T_2 - T_1)_{iron}$$
$$= m_i (0.45)(24 - 85) = -27.45\, m_i$$
$$\Delta U_{water} = m_w C_w (T_2 - T_1)_w$$

But
$$m_w = \rho \text{ vol}$$
$$= (1000 \text{ kg/m}^3)\left(\frac{200}{1000}\right)$$

$\therefore$
$$\Delta U_{water} = 200 \times 4.148\,(24 - 20)$$
$$= 3347.2 \text{ kJ}$$
$$W_{pw} = W_{pw} \times \Delta T$$
$$= (0.2 \text{ kJ/sec}) \times (20 \times 605) = 240 \text{ kJ}$$

Therefore from energy equation

$$-(-240 \text{ kJ}) = -27.45\, m_i + 3347.2$$
$$-3107.2 = -27.45\, m_i$$

$\therefore$
$$m_i = 113.19 \text{ kg}$$

REVIEW QUESTIONS

1. Is the process of burning coal in thermal plant boiler an energy creation or energy transformation process?

(**Ans.** It is an energy transformation process. Chemical energy is converted into heat energy)

2. In how many forms can energy cross boundaries of a system? Name them

(**Ans.** Two; Heat and work)

3. When the energy crossing the boundaries is heat, what is it due to?

(**Ans.** Heat crosses the boundary due to temperature difference)

4. In what form the energy crosses the boundaries if there is no temperature diference?

(**Ans.** All forms of work)

5. Define adiabiatic process.

6. Specify in sign conventions for work and heat.

(**Ans.** Work done by the system is positive and work done on the system is negative. Heat transfer to the system is positive and heat transfer from the system is – *ve*.

7. On a cold winter day a room is heated by plugging in a resistance type room heater. Is this a heat or work interaction? Take the entire room including the heater as a system.

(**Ans.** It is a work interaction as the electrons are crossing the system boundary doing electrical work)

8. A room is heated as a result of solar radiations coming through the windows. Is this a heat or work interaction?

(**Ans.** It is heat interaction as it is the result of temperature difference between the sum and room)

9. During winter an insulated room is heated by burning charcoal. Is this a heat or work interaction? Take the entire room including the charcoal as a system.

(**Ans.** It is neither work nor heat interaction as no energy crosses the system boundary. It is simply the conversion of chemical energy to another from *i.e.*, sensible energy)

10. Explain point and path functions. Give some example.

(**Ans.** Point functions depend on state only on the other hand path functions depend on the path followed during a process.

Properties are point functions. Heat and work are path functions.

11. What does area under the process curve, on p.v. diagram, represent?

(**Ans.** It represents the boundary work for quasi-equilibrium processes)

12. In a piston-cylinder device, explain why the work output is greater, when the process takes place in a quasi equilibrium manner, than the actual process.

(**Ans.** In actual processes the pressure at the piston face is lower than the pressure in other parts of the system. During a quasi-equilibrium process the system is always at a state of equilibrium, and thus the pressure at the piston face is, almost identical to the pressure in other parts of the system, thereby maximising the work output)

13. Is the boundary work associated with constant-volume system, always zero? (**Ans.** Yes)

14. An ideal gas at a given state expands to a fixed final volume in two different ways; firstly at constant pressure and secondly at constant temperature. For which case the work done will be greater?

(**Ans.** The process curve on constant pressure is *AB* while at constant temperature is *AC*. The area under the curve *AB* is more than that under *AC* (Fig 353) Therefore the work done in the constant pressure case will be more).

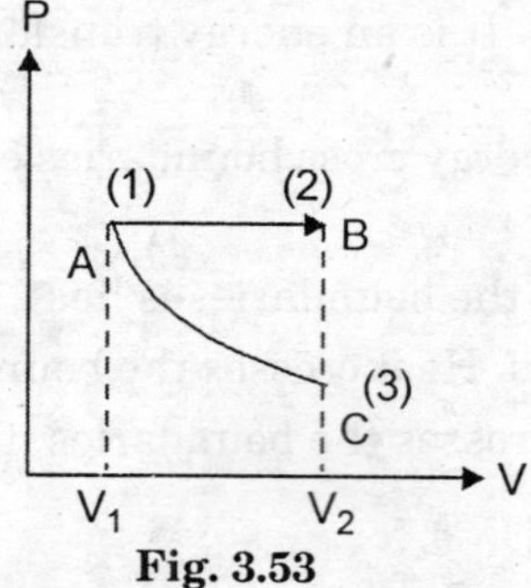

Fig. 3.53

15. Prove taht 1kJ = 1 kPa m^3. (**Ans.** 1 kJ = 1 kN · m = 1(kN/m^2) m^3 = 1 kPa · m^3)

PROBLEMS

1. One kg of saturated water vapor at 200 kPa is heated at constant pressure till the temperature reaches 400°C. Determine the work done by the steam during the process. (**Ans.** 132.733 kJ

2. A frictionless piston cylinder device contains 4 kg of superheated water vapor at 500 kPa and 300°C. Steam is now cooled at constant pressure untill 70% of it by mass, condenses. Calculate the work done during this process. (**Ans.** –819 kJ)

3. An air tight frictionless piston cylinder device contains 0.4 kg of air at 100 kPa pressure and temperature of 298 K. The air is now compressed to final pressure of 500 kPa.

During this process heat is transferred from air such that the temperature inside the cylinder remains constant. Calculate the work done during the process taking R_{air} = 0.287 kJ/kg · K. **(Ans.** $W_b = -55.05$ kJ)

4. Nitrogen at an initial state of 300 K, 130 kPa and 0.4 m^3 is compressed slowly under isothermal conditions to a final pressure of 700 kPa. Determine the work done during the process. **(Ans.** –87.34438 kJ)

5. Certain gas is compressed from initial volume of 0.38 m^3 to final volume of 0.1 m^3 under quasi equilibrium conditions where the pressure changes with volume as per the relationship $P = aV + b$ where $a = -1200$ kPa/m^3 and $b = 600$ kPa. Show the process on P.v. chart.

Calculate the work done during this process. (*a*) by integration (*b*) by finding the once under the process curve.

[Hint : (*a*) $W_b = \int_1^2 P\,dV = \int_1^2 (aV + b)\,dV$

(*b*) $W_b = \frac{P_1 + P_2}{2}(V_1 - V_2)$] **(Ans.** –87.4 kJ)

6. A gas is compressed from 100 kPa to 900 kPa according to relation $P = aV + b$ where a = 1 mPA/m^3 and b = a constant. If the initial volume of the gas is 0.2 m^3 find the final volume and work done during the process. **(Ans.** $V_2 = 1$ m^3; $W_b = 400$ kJ)

7. A gas expands slowly from a state of 100 kPa and 0.05 m^3 to a final volume of 0.3 m^3. Assuming the expansion process to be a polytropic one given by the relationship $PV^{1.2}$ = constant. Find the work done in the process. $\left(\textbf{Ans. } W_b = \frac{P_1V_1 - P_2V}{1-n} = 7.53 \text{ kJ}\right)$

8. A mass of 2 kg of Nitrogen is initially at a state 100 kPa pressure and temperature 300 K. It is now compressed slowly as per the relationship $PV^{1.4}$ = constat until it reaches final temperature of 360 K. Find the work done during the process. Take R_{N_2} = 0.2968 kJ/kg · K. $\left(\textbf{Ans. } W_b = \frac{P_2V_2 - P_1V_1}{1-n} = \frac{mR(T_2 - T_1}{1-n} = -89.0 \text{ kJ}\right)$

9. The equation of state of a certain gas is given below :

$$\bar{v}\left[P + \frac{10}{(\bar{v})^2}\right] = R_uT$$

where P is in bars and $\bar{v}$ in m^3/k · mol and R_u = 8.314 kJ/k · mol.

Now 0.5 k · mol of this gas is expanded in a quasi-equilibrium manner from 2.5 m^3 to 5 m^3 at constant temperature 27°C. Find the work done during the isothermal expension.

[Hint : $P = \frac{R_uT}{\bar{v}} - \frac{10}{(\bar{v})^2}$

$$W_b = \int_1^2 P\,dV = \int_1^2 P\left[\frac{R_uT}{\bar{v}} - \frac{10}{(\bar{v})^2}\right]dv$$

Now $\bar{v} = \frac{V}{N}$

Where V = Volume

N = number of k mol of gas

= 0.5 given

Therefore $W_b = \int_1^2 P\left[\frac{NR_uT}{V} - \frac{10N^2}{(V)^2}\right]dV$ **(Ans.** 863.8 kJ)

10. A piston cylinder device contains 0.6 m^3 carbon dioxide. It is compressed to 0.2 m^3. During the process the pressure and volume are related by $P = a\bar{v}^{-2}$ where $a = 8$ kPa · m^6. Calculate the work done on carbon dioxide. (**Ans.** –266667 kJ)

11. A frictionless piston cylinder device contains air at 125 kPa and 0.2 m^3. At this stage a linear spring is just touching the piston without exerting any force on it. The air is then heated so that at the final state its volume is 0.4 m^3 and pressure 500 kPa. Find (*a*) the total work done by air (*b*) work done if there were no spring (*c*) the work done against the spring. Show the process on P–V. diagram.

(**Ans.** (*a*) 6.25 kJ, (*b*) 25 kJ, (*c*) 37.5 kJ)

12. A firctionless piston-cylinder device contains 2 m^3 hydrogen at 150 kPa. At this stage a linear spring with stiffness of 200 kN/m is just touching the piston but exerts no force on it. The area of cross-section of the piston is 0.9 m^2. Heat is then transferred to hydrogen due to which the hydrogen expands and its final volume becomes 3 m^3. Calculate the (*a*) final pressure (*b*) the total work done by hydrogen (*c*) work done against the spring.

(**Ans.** (*a*) 396.9135 kPa, (*b*) 273.456 kJ, (*c*) 123.456 kJ)

13. A piston-cylinder device contains a 10 kg of saturated liquid-vapor mixture of water at 100 kPa in which 4 kg is in liquid phase. The piston rests on a set of stops. Heat is non tranferred to water till the piston starts moving and at that time the pressure inside water is 200 kPa. Heat is further transferred to water till the volume increases by 20%. Find (*a*) the initial and final temperatures. (*b*) The mass of liquid water when the piston fusi starts moving and (*c*) the work done during this process.

(**Ans.** (*a*) $T_1 = 99.63°C$, $T_3 = 252.6°C$, (*b*), no liquid left (*c*) 402 kJ)

14. Water contained in a vertical piston-cylinder device is being heated. During this process 50 kJ of heat is transferred to water while from the walls of the cylinder there is heat loss of 8 kJ. The piston rises due to evaporation and 5 kJ of boundary work is done. Find the change in energy of the water in this process. (**Ans.** 37 kJ)

15. Fill in the missing data for each of the following processes between states 1 and 2 (the data in the kJ).

	Q	W	E_1	E_2	ΔE
(*a*)	18	–6		35	
(*b*)	–10			4	–15
(*c*)		12	3		32
(*d*)	25		14		10

16. Find the missing data in the following table the data is no kJ and it pertains to closed systems between stats 1 and 2.

	Q	W	E_1	E_2	ΔE
(*a*)	–9			12	–15
(*b*)	5		20		35
(*c*)		18	6		20
(*d*)	25	–10		40	

17. A closed system is subjected to a cycle consesting of two processes. During the first process 30 kJ of heat is transferred to the system while the system does 50 kJ of work. During the second process 35 kJ of work is done on the system (*a*) Calculate the heat transfer during the second process (*b*) Also determine the network and net heat transfer for the cycle. [**Hint :** For the cycle $Q - W = 0$]

(**Ans.** –15 kJ; $Q_{net} = 15$ kJ and $W_{net} = W_1 + W_2 = 50 - 35 = -15 \cdot$ kJ)

❑❑❑

CHAPTER 4

FIRST LAW OF THERMODYNAMICS APPLIED TO FLOW PROCESSES

4.1 CONTROL VOLUME

In Chapter 3, the energy interaction between a system and its surroundings, and the conservation of energy principle for closed system (*i.e.* non flow systems) are discussed. In this chapter the analysis is extended to systems that involve mass flow across their boundaries *i.e* control volumes. When there is mass transfer across the system boundary, the system is called an Open System. Most of the engineering devices are open systems involving mass flow in and out of the system and, therefore, are modelled as Control Volumes. A water heater, a turbine and a compressor all involve mass flow and should be analyzed as control volume (or open system) problems. In general, any arbitrary region in space can be taken as control volume. There is no set rule for its selection, but the proper choice makes the analysis easier. For example if the flow of air through nozzle were to be analyzed, a good choice for the control volume would be the region within the nozzle. (Fig. 4.1).

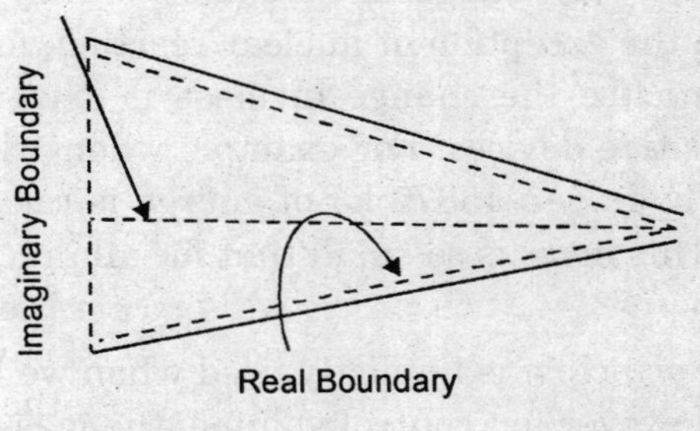

(a) Nozzzle

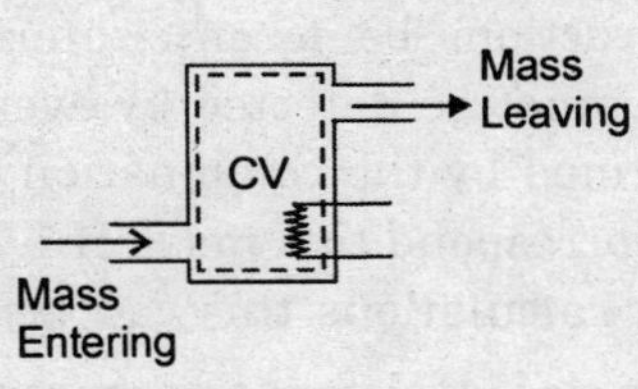

(b) Water Heater control vol with moving Boundaries

Fig. 4.1

The boundaries of control volume are called "control surface" and they can be real or imaginary or both. It may also involve a moving boundary. Fig. 4.1 (b).

A control volume can be fixed in size and shape Fig. 4.1 (a) as in the case of nozzle or it may involve a moving boundary, shown in Fig. (4.1b). In case the control volume has fixed boundaries, they do not involve any moving boundary work (Fig. 4.1 (a).

4.2 STEADY AND UNSTEADY FLOW PROCESS

A flow process is said to be steady when the fluid parameters at any point of the control volume remain constant and do not change with time, the parameters may, however be different at different cross-section of the flow passage. From this we conclude that in a steady flow process, quantities like, velocity, pressure, temperature etc. are functions only of location and do not vary with time.

If any of these parameters, is P (say), then mathematically a steady flow is defined as

$$\frac{\partial P}{\partial t} = 0 \qquad ...(4.1)$$

i.e. time rate of change of P at a position is zero.

When the conditions at a particular point in control volume vary with time, the flow process is Unsteady. That is, Unsteady Process refers to changing parameters with passage of time at a particular point in control volume. Mathematically unsteady flow process is defined by as unequally.

$$\frac{\partial P}{\partial t} \neq 0 \qquad ...(4.2)$$

4.3 CONSERVATION OF MASS PRINCIPLE

Conservation of mass is the most fundamental principle in nature. Mass, like energy, is conserved property *i.e.* it cannot be created or destroyed. However mass 'm' and energy 'E' are interconvertible according to the famous Einstein formula

$$E = mC^2 \qquad ...(4.3)$$

where C is the velocity of light evidently equation (4.3) suggests that mass of a system will change when its energy changes. However with the exception of nuclear reaction, for all energy interactions being encountered in actual practice the change in mass is extremely small and cannot be detected by even the most sensitive devices. For example when 1 kg of water in formed by the combination of oxygen and hydrogen 15879 kJ of energy is released which will correspond to a mass of 1.76×10^{-10} kg. This mass is so small that for all practical engineering calculations this can be neglected.

Thus for closed system the conservation of mass principle is implictly used when we keep the mass of the system constant during a process. However for, control volume, the mass can cross the boundaries and so we must keep track of the amount of mass entering and leaving the control volume.

Therefore the conservation of mass principle for a control volume (C.V.) undergoing a process can be expressed by the following equation.

$$\begin{bmatrix}\text{Total mass entering}\\ C_V\end{bmatrix} - \begin{bmatrix}\text{Total mass leaving}\\ C_V\end{bmatrix} = \begin{bmatrix}\text{Net change in mass in}\\ C_V\end{bmatrix}$$

or $$\Sigma m_i - \Sigma m_e = (\Delta m)_{CV} \quad ...(4.4)$$

where the subscript i,e and C_V refer to inlet, exit and control volume respectively.

The conservation of mass principle could also be written in the rate form by expressing the corresponding quantities, per unit time as

$$\Sigma \dot{m}_i - \Sigma \dot{m}_e = (\Delta \dot{m} C_V) \quad ...(4.4)$$

The conservation of mass equation (4.4) is usually referred to as the Continuity Equation in fluid mechanics.

4.3.1 Mass and Volume Rate

Mass flow rate through a cross-section is defined as the amount of mass flowing through the cross-section per unit time and is denoted by $\dot{m}$. The mass flow rate $\dot{m}$ of fluid flowing in and out of a control volume through pipe or ducts is proportional to the cross-sectional area 'A' of the pipe or duct, the density 'ρ' and the velocity 'V' of the fluid. Then the mass flow, rate through a differential area 'dA' will be

$$d\dot{m} = \int \rho V_n dA \quad ...(4.5)$$

where V_n is velocity component normal to dA. Then the mass flow rate in through the entire cross-sectional area of pipe or duct will be

$$\dot{m} = \rho V_n \, dA \quad ...(4.6)$$

the density ρ of the fluid flowing in the pipe or duct usually remains constant over any cross-section but the velocity V_n does not. Due to viscous effects the velocity, V_n, will vary from zero at the walls to maximum at the centre. For most of the purposes equivalent average, velocity, V_{av}, is taken for the entire cross-section.

Then the equation (4.6) referred to

$$\dot{m} = \rho V_{av} A \quad ...(4.7)$$

where ρ = density, kg/m^3 = $\left(\frac{1}{v}\right)$

v = specific volume

V_{av} = Average fluid velocity normal to A, m/s

A = Area of cross-section, m^2

Similarly volume flow rate $\dot{V}ol$ is defined as the volume of the fluid flowing through a cross-section per unit time and is given by

$$\dot{V}ol = \int_A V_n \, dA = V_{av} A \quad ...(4.8)$$

From equation (4.7) and (4.8) we get

$$\dot{m} = \dot{V}ol = \frac{\dot{V}ol}{v} \quad ...(4.9)$$

i.e. mass flow rate $= \frac{1}{v}$ [Vol. flow rate]

where v = is specific volume.

4.4 CONSERVATION OF ENERGY PRINCIPLE AS APPLIED TO CONTROL VOLUME

In Chapter 3, the conservation of energy system principle *in* relation to closed system was discussed. The energy of the closed system could be changed by heat or work interaction only and the change in energy of the closed system during a process is equal to the net heat or work transfer across the boundary.

This is given by the equation for closed system

$$Q - W = \Delta E \qquad \text{...(4.10)}$$

For the control volume, however, the energy of the system can be affected by the mass flow in and out of the control volume. When mass enters the control volume, the energy of the control volume increases because the entering mass carries some energy along with it. Similarly when some mass leaves the control volume, the energy content of the control volume decreases as the mass leaving the control volume takes out some energy along with it.

$$\begin{bmatrix}\text{(Total energy}\\ \text{crossing boundary}\\ \text{as heat and mass)}\end{bmatrix} + \begin{bmatrix}\text{(Total energy of}\\ \text{mass entering}\\ \text{C.V.)}\end{bmatrix} - \begin{bmatrix}\text{(Total energy of}\\ \text{mass leaving}\\ \text{C.V.)}\end{bmatrix} = \begin{bmatrix}\text{(Net change in}\\ \text{energies of}\\ \text{C.V.)}\end{bmatrix}$$

or $$(Q - W) + \Sigma E_{in} - \Sigma E_{out} = - \Delta E_{c.v} \qquad \text{...(4.11)}$$

Evidently when no mass is entering or leaving the control volume the second and third terms is ΣE_{in} and ΣE_{out} in equation (4.11) will drop out and the resultant equation will reduce to equation (4.10) for the closed system.

Thus equation (4.11) is applicable to any control volume undergoing any process. Equation (4.11) can also be expressed in the rate from as

$$\dot{Q} - \dot{W} + \Sigma \dot{E}_{in} - \Sigma \dot{E}_{out} = \Delta \dot{E}_{eV} \qquad \text{...(4.12)}$$

4.4.1 Important Points to Remember

(*i*) Heat transfer to and from control volume must not be confused with the energy transported with the mass in and out of control volume. Further heat is in the form of energy transferred as a result of temperature difference between the control volume and surroundings.

(*ii*) A control volume, like a closed system, may involve one or more forms of work at the same time.

In the example shown in Fig. (4.2) the work term will involve boundary work shaft work and electrical work for simple compressible system.

In case the boundary of the control volume is stationary the moving boundary work W_b will be zero.

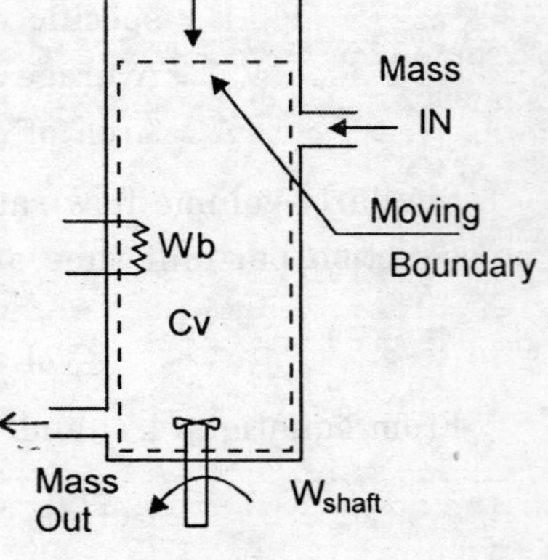

Fig. 4.2

Further when the control volume is insulated, the heat transfer term becomes zero.

(*iii*) The energy required to push the fluid into or out of the control volume is called the flow work or flow energy and will form part of the energy transported with the fluid and the flow energy or work is discussed in the following paragraph.

4.5 FLOW WORK

As explained above the control volumes involve mass flow across their boundaries and some work is required to push the mass into or out of the control volume. This work is called the "Flow Work" or "Flow Energy" and is essential for maintaining a continuous flow through the control volume.

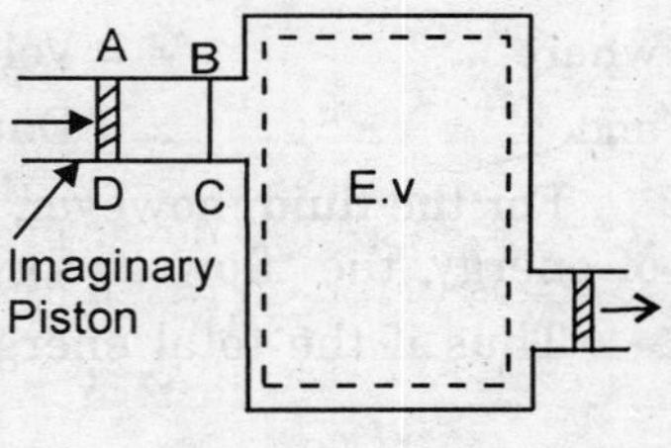

Fig. 4.3

In Fig. 4.3 is shown a fluid element. *ABCD*, immediately upstream the control volume, of volume V, having mass equal to m and the pressure is P. If the area of cross-section of this element is A, then the force applied on this fluid element by the imaginary piston will be F, given by

$$F = PA \quad ...(4.13)$$

In order to push the entire fluid element into the control volume, the imaginary piston will have to move through the distance 'l' thus the work done in pushing the fluid element across the boundary.

$$W_{flow} = F \times l$$

$$= PAl = PV \text{ (kJ)} \quad ...(4.14)$$

The flow work per unit mass will be

$$W_{flow} = \frac{W_{flow}}{m}$$

$$= \frac{PV}{m} = Pv \quad \text{(kJ/kg)} \quad ...(4.15)$$

The flow work per unit mass is the product of pressure and specific volume and the relationship as given in equation (4.15), will be same whether the fluid is pushed into or out of the control volume.

4.5.1 Important Discussion on Equation (4.15)

From equation (4.15) we see that flow work per unit mass is expressed in terms of the properties of the fluid *i.e.* it is the product of pressure and specific volume. For this reason some people view the flow work as a "combination property" and refer to it as "Flow Energy" or "Convected Energy" or "Transport Energy" instead of flow work.

According to another school of thought, however, people argue that the product Pv represents energy for flowing fluids only and it does not represent any form of energy for non flow or closed system. Therefore according to this group of people it should be treated as work.

This controversy is not likely to end until and unless an International Conference is held to resource this. But it is comforting to note that both arguments yield the same result for the energy equation.

In the following paragraph the flow energy will be considered to be part of the energy of the flowing fluid since this greatly simplifies the derivation of energy equation for the control volume.

4.5.2 Total Energy of Flowing Fluid

In Chapter 1, the total energy of a simple compressible system was considered and the total energy per unit mass was expressed as

$$e = u + (ke) + (pe) = u + \frac{V^2}{2} + gz \qquad ...(4.16)$$

where V = Velocity

and z = Datum height of the system relative to some external reference point

For the fluid, however, entering a control volume or leaving it possess an additional form of energy, the "flow energy" (= Pv).

Thus if the total energy per unit mass of the flowing fluid is θ, then

$$\theta = Pv + e$$

$$= Pv + u + (ke) + (pe)$$

Substituting equation (4.16) in it we get

or $$\theta = Pv + u + \frac{V^2}{2} + gz \qquad ...(4.17)$$

Put in Chapter 1, the combination ($P_v + u$) has been defined as enthalpy h

i.e. $$h = Pv + u \qquad ...(4.18)$$

Therefore equations (4.17) and (4.18)

$$\theta = h + \frac{V^2}{2} + gz \quad \text{(kJ/kg)} \qquad ...(4.19)$$

It is emphasized that by using "enthalpy" instead of internal energy to represent the energy of a flowing fluid, one has not to be concerned about the flow work. The energy associated with pushing the fluid into and out of the control volume is automatically taken care of by enthalpy.

As a matter of fact this is the main reason for defining the property enthalpy. Equation (4.19) represents the energy of the fluid stream entering into or out of control volume. Thus the work term W in the control volume energy equation will represent all forms of work including boundary, shaft electrical etc. except the flow work.

4.3 THE STEADY FLOW PROCESS

A good number of devices like turbines, compressors, diffusers, nozzles operate for long period of time under the same conditions and they are classified as "steady-flow devices". The "processes" involving "steady-flow devices" can be represented by a some what idealized process, called the Steady-Flow-Process.

A steady flow process is defined as the process during which a fluid flows through a control volume steadly. In other words the fluid properties can change from point to point with in the control volume, but at any fixed point within the control volume, remain same during the entire process. evidently the steady flow process is characterized by the following.

(*i*) No properties, including extensive or intensive within the control volume change. That is the mass m, volume V and the total energy content E of the control volume remain constant during a steady flow process. Evidently, it results into zero, boundary work for steady flow system became V_{CV} = constant. Further total mass or energy entering the control volume must be equal to the total mass or energy leaving the control volume since m_{CV} = constant and E_{CV} = constant.

Alternatively in the mathematical from these can be written as follows for steady flow process.

$$V_{CV} = \text{constant} \quad ...(4.20)$$

$$m_{CV} = \text{constant} \quad ...(4.21)$$

$$E_{CV} = \text{constant} \quad ...(4.22)$$

$$W_b = 0 \quad ...(4.23)$$

Fig. 4.4 represents this aspect for steady flow.

Mass out
CV
mcv = const
CV
mcv = const
Mass In

Fig. 4.4

(*ii*) Since no properties change at the boundaries of the control volume with time, the fluid properties at an inlet or an outlet will remain the same during the process. The properties may, however, be different at different (opening inlets and exits). The properties may even vary over the cross-section of an inlet or an exit. But it is emphasized that all properties including the velocity, elevation must remain constant with time at some fixed point. From this we conclude that the mass flow rate of the fluid at an opening must remain constant during a steady flow process (Fig. 4.5).

Further as an additional simplification, the fluid properties at an opening are usually assumed uniform at some average value over the cross-section. Thus the fluid properties at inlet or at outlet can be specified by average single values.

(*iii*) The work and heat interaction between the steady flow system and its surroundings do not change with time. In other words the power delivered by the system and the rate of heat transfer to or from the system remain constant during the steady-flow process.

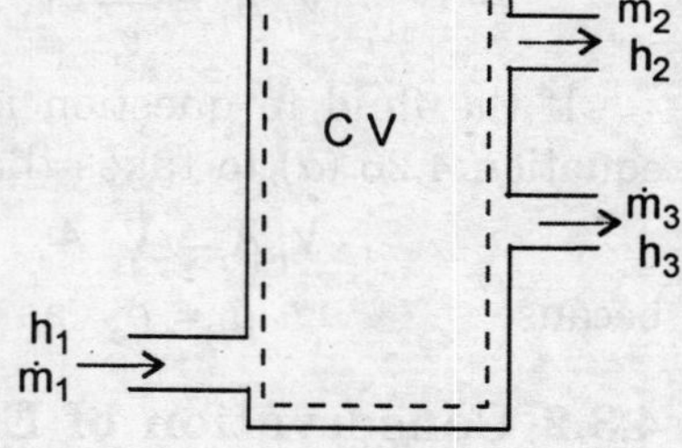

Fig. 4.5

(*iv*) It is pertinent to mention here that some cyclic devices such as reciprocating pumps and compressors do not satisfy the above stated conditions because the flow at the inlets and outlets will be steady but it will be pulsating. However, this fluid properties vary with time in a periodic manner and the flow through these devices can still be analyzed as a steady flow process by using time averaged values of the properties and the heat flow rates through boundaries.

4.3.1 Conservation of Mass

As explained above during a steady flow process total amount of mass contained in a control volume does not change with time. That is

$$m_{CV} = \text{Constant}$$

Evidently, the conservation of mass principle requires that total amount of mass entering control volume must be equal to the total amount of mass leaving it over a period of time. Alternatively if we take the mass flowing per unit time then mass flow rate in and mass flow rate out must be equal. In the case of multiple inlets and exits we have

$$\Sigma \dot{m}_i = \Sigma \dot{m}_e \quad ...(4.24)$$

where the suffix '*i*' stands for inlet and '*e*' for exit.

In case there is only one inlet and one exit as in the case of nozzle diffurers, turbines, compressors and pumps the equation (4.24) reduces to

$$\dot{m}_i = \dot{m}_e \qquad \text{...(4.25)}$$

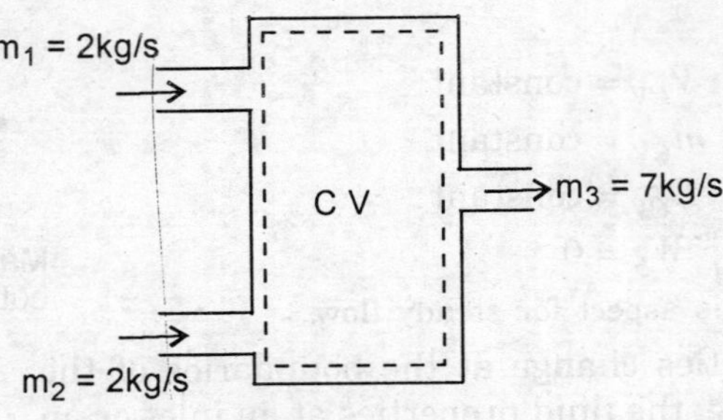

Fig. 4.6

If V = Average flow velocity in the flow direction m/s

A = Area of cross-section normal to the direction of flow m^2

ρ = Density of fluid, kg/m^3

v = Specific volume m^3/kg

Then equation (4.25), reduces to

$$\rho_i V_i A_i = \rho_e V_e A_e \qquad \text{...(4.26 } a\text{)}$$

$$\frac{1}{v_i} V_i A_i = \frac{1}{v_e} V_e A_e \qquad \text{...(4.26 } b\text{)}$$

If the fluid in question is a liquid, which is considered to be incompressible then the equation 4.26 (*a*) to takes the form

$$V_i A_i = V_e A_e \qquad \text{...(4.27)}$$

because $\rho_i = \rho_e$ as it is incompressible.

4.3.2 Conservation of Energy

It was pointed out earlier that for a steady flow process the total energy content of a control volume remains constant. That is change in the total energy of a control volume during a steady flow process is zero

That $\Delta E_{CV} = 0$

Thus the total amount of energy entering a control volume in all forms (heat, works and mass transfer) must be equal to the total amount of energy leaving it.

In the mathematical from it is written as

$$\begin{bmatrix} \text{(Total energy} \\ \text{crossing bundary} \\ \text{as heat or work per} \\ \text{unit time} \end{bmatrix} = \begin{bmatrix} \text{(Total energy transported} \\ \text{out of C.V. with mass per} \\ \text{unit time} \end{bmatrix} - \begin{bmatrix} \text{(Total energy transported} \\ \text{into C.V. with mass per} \\ \text{unit time)} \end{bmatrix}$$

For multiple openings

or $$\dot{Q} - \dot{W} = \Sigma \dot{m}_e \,\theta_e - \Sigma \dot{m}_i \,\theta_i \qquad \text{...(4.28)}$$

where θ is the total energy of the flowing fluid, including the flow work, per unit mass. Equation (4.28) is rewritten in another from by substituting equation (14.19) into it

$$\dot{Q} - \dot{W} = \sum_{\substack{\text{for each}\\ \text{exit}}} \dot{m}_e \left(h_e + \frac{V_e^2}{2} + g\, z_e \right) - \sum_{\substack{\text{for each}\\ \text{inlet}}} \dot{m}_i \left(h_i + \frac{V_i^2}{2} + g\, z_i \right) \quad ...(4.29)$$

Equation (4.29) is the general form of First Law Relationship for Steady Flow Processes.

In case there is a single stream system (i.e. one inlet and one exit from *C.V.*) than *i* will be replaced by suffix '1' and *e* by suffix '2' in equation 4.29 to give conservation of energy equation single-steam steady flow system as

$$\dot{Q} - \dot{W} = \dot{m} \left[(h_2 - h_1) + \frac{V_2^2}{2} + g\,(z_2 - z_1) \right] \quad ...(4.30)$$

Dividing equation (4.30) by $\dot{m}$, it yields the First Law Relation on a unit mass basis for single steam steady flow system

as $$q - w = (h_2 - h_1) + \frac{V_2^3 - V_1^2}{2} + g\,(z_2 - z_1) \text{ kJ/kg} \quad ...(4.31)$$

or $$q - w = \Delta h + \Delta\,(K.E.) + g\,(z_2 - z_1) \quad ...(4.32)$$

where $$q = \frac{\dot{Q}}{\dot{m}} \text{ (Heat transfer per unit mass kJ/kg)} \quad ...(4.33)$$

and $$w = \frac{\dot{W}}{\dot{m}} \text{ (Workdone per unit mass kJ/kg)} \quad ...(4.34)$$

If the fluid does not experience any appreciable change in Kinetic energy and potential energy.

$$\Delta K.E. = 0 \text{ and } \Delta P.E. = 0$$

as it flows through the control volume then the energy equation (4.32) for a single steam steady flow is further simplified to

$$q - w = \Delta h \text{ (kJ/kg)} \quad ...(4.35a)$$

4.3.3 Important Points to Remember Regarding Various Terms of Equation (4.30)

(*i*) $\dot{Q}$ which is the rate of Heat Transfer between the control volume and surroundings, will be negative if the control volume is losing heat. As an example in the water heater the control volume will be losing heat to surroundings if it is not insulated.

(*ii*) Secondly if the control volume is well insulated (i.e. adiabatic), then

$$\dot{Q} = 0$$

(*iii*) $\dot{W}$ = Power and for steady flow devices, the volume of the control volume is constant, thus there is no boundary work involved. The work required to push mass into and out of

C.V. is also taken care of by using enthalpies for the energies of the fluids streams instead of internal energies. Thus $\dot{W}$ represents the remaining forms of work done per unit time. In the case of steady flow devices, like pumps, turbines and compressors where power is transmitted through a shaft, $\dot{W}$ simply becomes equal to shaft power. In case the control surface is crossed by electric wires, as in the case of electric heater $\dot{W}$ will represent electrical work done per unit time. If neither shaft work nor electric work are crossing C.V., then

$$\dot{W} = 0$$

(*iv*) **Change in Enthalpy.** $\Delta h = h_2 - h_1$

The enthalpy change of the fluid can easily he calculated by reading the enthalpy values at the two states (exit and inlet of *C.V.*) from the tables. However in the case of ideal gas it can be calculated as under

$$\Delta h = C_{p,\,av}\,(T_2 - T_1) \qquad \text{...(4.35 } b\text{)}$$

Further it may be noted that

(kg/s) (kJ/kg) = kW

(*v*) $$\Delta KE = \frac{1}{2}\,(V_2^{\,2} - V_1^{\,2})$$

It may be noted that units of kinetic energy per unit mass are $\frac{m^2}{s^2}$ and it is equivalent to J/kg

Because $$\text{J/kg} = \text{N.m/kg} = (\text{kg . m/s}^2)\ (\text{m/kg}) = \frac{m^2}{s^2}$$

Further the units of enthalpy are usually kJ/kg as read from tables. Therefore the kinetic energy term is divided by 1000 to convert the units to kJ/kg.

It will not be out of place to mention here that the velocity at which kinetic energy is 1kJ/kg

then $$1\ \text{kJ/kg} = \frac{V^2}{2} = \frac{m^2}{s^2}$$

or $$V = \sqrt{2000} = 44.72\ \text{m/s}$$

Thus a velocity of 44.72 m/s will correspond to a kinetic energy of 1 kJ/kg which is usually very small as compared to enthalpy values found in practice. That is Kinetic energy terms contributes very little to the total energy and it could be neglected being small. Similarly if

$$V_2 \cong V_1$$

then $$\Delta\ KE \approx 0$$

(*vi*) $$\Delta\ PE = g\,(z_2 - z_1)$$

Here potential energy change of 1 kJ/g will correspond to

$$(z_2 - z_1) = \frac{1\ \text{kJ/kg}}{g\,\dfrac{m}{s^2}}$$

$$= \frac{1000 \frac{(\text{N.m})}{\text{kg}}}{9.81 \frac{m}{s^2}}$$

$$= 102 \frac{\text{N.m}}{\left(\frac{m}{s^2}\right)} \cdot \frac{m}{N s^2} = 102 \text{ m}$$

In practice the difference in elevation of inlet and exit of various industrial devies like turbines, compressors, heat exchanges etc. is far below this value. Thus the potential energy term is neglected in these devices. this term is only considered when the process deals with pumping of fluid to a very high attitude. This will be true for systems involving negligible heat transfer.

4.4 ENGINEERING APPLICATIONS OF STEADY FLOW ENERGY EQUATION

Large number of engineering device operate under same conditions for long periods of time. For example turbines, compressors, heat exchangers which are components of steam power plant, operate non-stop for months together before the systems is shut down for maintenance. Therefore these devices are analysed as steady flow devices.

In the following paragraphs some common steady flow devices are described and the thermodynamic aspects of the flow through them is analysed.

4.4.1 Nozzles and Diffusers

A nozzle is a device that increases the velocity of the fluid at the expense of pressure. On the contrary diffuser is a device that increases the pressure of the fluid by slowing it down The cross-sectional area of the nozzle decreases in the flow direction (of course for subsonic flows) and for diffusers reverse is true. In Fig. 4.7 are shown nozzle and diffuser schematically.

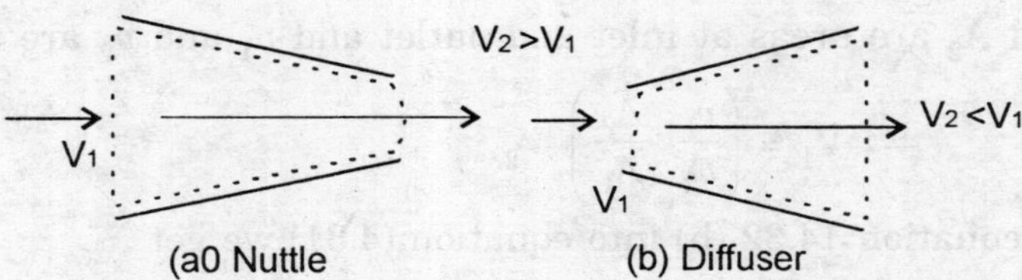

Fig. 4.7. Nozzle and Diffuser.

The energy equation for nozzles and diffusers will be the equation (4.30) as there is one stream of fluid, that is

i.e.
$$\dot{Q} - \dot{W} = \dot{m}\left[(h_2 - h_1) + \frac{V_2^2 - V_1^2}{2} + g(z_2 - z_1)\right] \quad ...(4.30)$$

It is important to note the relative contribution of various terms appearing in equation (4.30) in the case of nozzles and diffusers.

(*i*) The rate of heat transfer between the fluid flowing through the nozzle or diffuser and the surroundings is usually very small, even though these devices are not insulated because the residence time of the fluid within the nozzle or diffuser is very small due to prevalent high velocities. During this very short span of time no significant heat-transfer takes place. Therefore in the absence of any heat transfer data, the flow through nozzles and diffuser may be assumed adiabatic for which

$$\dot{Q} \approx 0$$

(*ii*) The work term $\dot{W}$ for nozzles and diffusers is zero because these devices are properly shaped ducts and they do not involve any shaft or electrical work that is

$$\dot{W} = 0$$

(*iii*) Nozzles and diffusers involve very high velocities. More over there are subsantial changes in velocities during the flow of fluid through these devices. Therefore Kinetic energy change must be taken into account while applying the equation (4.30).

(*iv*) There is practically no or very little change in the elevation when the fluid flows through nozzles or diffusers. Therefore potential energy term is neglected. That is

$$\Delta PE = 0$$

In the light of above discussion, the equation for nozzles reduces to

$$0 - 0 = \dot{m}\left[(h_2 - h_1) + \frac{V_2^2 - V_1^2}{2}\right]$$

or

$$\frac{V_2^2 - V_1^2}{2} = h_1 - h_2 \qquad \text{...(4.31)}$$

or Increase in K.E. = Decrease in enthalpy.

Also from continuity of mass flow

$$\dot{m} = \frac{A_1 V_1}{v_1} = \frac{A_2 V_2}{v_2} \qquad \text{...(4.32 } a\text{)}$$

where A_1 and A_2 are areas at inlet and outlet and v_1 and v_2 are specific volumes

i.e.

$$V_1 = \left(\frac{v_1}{v_2} \cdot \frac{A_2}{A_1}\right) V_2 \qquad \text{...(4.32 } b\text{)}$$

Substituting equation 14.32 (b) into equation (4.31) we get

$$V_2^2 - V_2^2\left(\frac{v_1}{v_2} \cdot \frac{A_2}{A_1}\right) = 2\,(h_1 - h_2)$$

or

$$V_2 = \left[\frac{1}{1 - \left(\dfrac{v_1 A_2}{v_2 A_1}\right)^2}\right]^{\frac{1}{2}} \sqrt{2(h_1 - h_2)}$$

$$= \left[\frac{v_1^2 A_1^2}{(v_2 A_1)^2 - (v_1 A_2)^2}\right]^{\frac{1}{2}} \sqrt{2\,(h_1 - h_2)}$$

$$V_2 = \frac{v_1 A_1 \sqrt{2(h_1 - h_2)}}{\sqrt{(v_2 A_1 + v_1 A_2)(v_2 A_1 - v_1 A_2)}} \quad ...(4.33)$$

In case inlet velocity or velocity of approach V_1 is small compared to V_2, the exit velocity then equation (4.31) reduces to

$$V_2 = \sqrt{2(h_1 - h_2)} \quad \text{when } V_1 << V_2 \quad ...(4.34)$$

If the working fluid is a perfect gas

then $pv = RT$

and $\Delta h = C_{p,\,av}\,(T_2 - T_1)$ ...(4.35)

Also $\left(\dfrac{T_2}{T_1}\right) = \left(\dfrac{p_2}{p_1}\right)^{\frac{\gamma-1}{\gamma}}$ ...(4.36)

Therefore substituting values from equation (4.35) and (4.36) in equation (4.31), we get

$$\frac{V_2^2 - V_1^2}{2} = (h_1 - h_2)$$

$$= C_{p,\,av}\,(T_1 - T_2)$$

$$= C_{p,\,av}\left[T_1 - T_1\left\{\frac{p_2}{p_1}\right\}^{\frac{\gamma-1}{\gamma}}\right]$$

$$= C_{p,\,av}\,T_1\left[1 - \left(\frac{p_2}{p_1}\right)^{\frac{\gamma-1}{\gamma}}\right] \quad ...(4.37)$$

Examples on Nozzles and Diffusers

Example 4.1. *Air enters an adiabatic nozzle steadily at 500 kPa, 200 °C and 30 m/sec and leaves at 100 kPa and 180 m/sec. If nozzle inlet area is 0.012 m², find*

(a) the mass flow rate through the nozzle

(b) the exit temperature of air and

(c) exit area of the nozzle.

Take $R = 0.287$ *kJ/kg-K and* $\gamma = 1.4$

Solution. (*a*) Treating air as an ideal gas we have at inlet

$$v_1 = \frac{RT_1}{P_1}$$

and $R = 0.287$ kPa m³/kg-K

$T_1 = 200 + 273 = 473$ K

$P_1 = 500$ kPa

$$\therefore \quad v_1 = \frac{(0.287\,\text{kPa m}^3/\text{kg.K})\,(473\,\text{K})}{500\,\text{kPa}}$$

$$= 0.2715\ \text{m}^3/\text{kg}$$

Therefore mass flow rate

$$\dot{m} = \frac{A_1 V_1}{v_1}$$

$$= \frac{(0.012 m^3)(30 m/s)}{0.2715 m^3/kg}$$

$$= 1.326 \text{ kg/s}$$

(*b*) As it is an adiabatic nozzle

$$\gamma = \frac{C_p}{C_v} = 1.4 \text{ For air}$$

Also $\quad C_p - C_v = \text{R}$

or $\quad C_p - \dfrac{C_p}{1.40} = 0.287 \text{ kPa m}^3\text{/k-K}$

or $\quad 0.40\ C_p = 0.287$

$\therefore \quad C_p = \dfrac{0.287}{0.40} = \text{kJ/kg-K}$

$$= 0.7175 \text{ kJ/kg-K}$$

Now for steady flow energy equation

$$q - w = (h_2 - h_1) + \left(\frac{V_2^2 - V_1^2}{2}\right)(m^2/s^2)\left(\frac{1 \text{ kJ/kg}}{1000 \text{ m}^2/\text{s}^2}\right)$$

But here $\quad q = 0$ and $w = 0$

or $\quad 0 = C_p\ (T_2 - T_1) + \left(\dfrac{V_2^2 - V_1^2}{2}\right)\left(\dfrac{1}{1000}\right)$

i.e. $\quad 0 = 0.7175 \text{ (kJ/kg-K)} (T_2 - 200\ °\text{C}) + \left(\dfrac{180^2 - 30^2}{2}\right) + \dfrac{1}{1000}$

or $\quad (T_2 - 200) = -\left(\dfrac{100 \times 210}{2000}\right) \times \dfrac{1}{0.7175}$

or $\quad T_2 - 200 = 21.95$

or $\quad T_2 = 200 - 21.95$

$$= 178.05\ °\text{C}$$

$$= 273 + 178.04 = 451.05 \text{ K}$$

(c) The specific volume of air at exit is

$$v_2 = \frac{R\,T_2}{P_2} = \frac{(0.287 \text{ kPa m}^3\text{/kg- K})(451.05 \text{ K})}{(100 \text{ kPa})}$$

$$= 1.2945 \text{ m}^3\text{/kg}$$

But mass flow rate = $\dot{m}$ = 1.326 kg/sec

But $$\dot{m} = \frac{A_2 V_2}{v_2}$$

$$\therefore \quad A_2 = \frac{\dot{m}\, v_2}{V_2} = \frac{1.326 \times 1.2945}{180}$$

$$= .009536 \text{ m}^2$$

$$= 95.36 \text{ cm}^2$$

Example 4.2. *Carbon dioxide enters an adiabatic nozzle steadily at 1 MPa and 500 °C. The mass flow rate is 6000 kg/hr. It leaves the nozzle at 100 MPa with a velocity of 450 m/s. If the inlet area of the nozzle is 4 × 10^{-3} m^2 find the inlet velocity and outlet temperature. Assume that CO_2 is an ideal gas and R for CO_2 = 0.1889 (kPa m^3/kg-K) and γ = 1.289.*

Solution. Treating the CO_2 as an ideal gas

$$v_1 = \frac{RT_1}{P_1}$$

$$= \frac{(0.1889 \text{ kPa m}^3\text{kg- K})\,(500 + 273)}{(1000 \text{ kPa})}$$

$$= 0.146 \text{ m}^3/\text{kg}$$

But mass flow rate = $\dot{m} = \dfrac{A_1 V_1}{v_1}$

i.e. $$\left(\frac{6000}{3600}\text{kg/s}\right) = \frac{A_1 V_1}{v_1}$$

$$= \frac{(4 \times 10^{-3}\ m^2) \times (V_1 \text{ m/s})}{(0.146)\text{m}^3/\text{kg}}$$

Therefore $$V_1 = \frac{6000}{3600} \times \frac{0.146}{4 \times 10^{-3}}$$

$$= 60.8 \text{ m/s}$$

(b) For outlet temperature, we apply the steady flow energy relation

$$q - w = (h_2 - h_1) + \frac{V_2^2 - V_1^2}{2} \times \frac{1}{1000}$$

$$= C_p\,(T_2 - T_1) + \frac{V_2^2 - V_1^2}{2} \times \frac{1}{1000}$$

But here $q = 0$

and $w = 0$

$V_2 = 450$ m/s

and $V_1 = 60.8$ m/s

$$\therefore \quad C_p\,(T_2 - T_1) = -\frac{(450)^2 - (608)^2}{2} \times \frac{1}{1000}$$

$$= -\frac{202500 - 368.64}{2 \times 1000}$$

$$= -99.4016$$

Now $\dfrac{C_p}{C_v} = \gamma = 1.289$ (given)

and $C_p - C_v = R$ kJ/kg-K

or $C_p - \dfrac{C_p}{1.285} = 0.1889$

Therefore $C_p = \dfrac{0.1889}{0.289}$

$= 0.6536$ kJ/kg-K

Also $T_1 = 500 + 273 = 773$ K

Therefore

$$0.6536\,(T_2 - 773) = -99.40168$$

or $T_2 - 773 = -\dfrac{99.40168}{.6536}$

$$T_2 = 773 - 152.08$$

$$= 620.92\ K$$

Example 4.3. *Air at 300 kPa and 350 K) enters an adiabatic nozzles with a velocity 50 m/s and leaves at 100 kPa and 320 m/s. If the heat loss from this nozzle is 3.2 kJ/kg of air flowing determine the temperature of air at exit and the exit area of nozzle; the inlet area being 0.01 m*^{2}.

Take $R = 0.287$ *kJ/kg K*

and $\gamma = 1.4$

Solution. From the steady flow energy equation we have

$$q - w = (h_2 - h_1) + \frac{V_2^2 - V_1^2}{2} + \Delta PE$$

Here $q = -3.2$ kJ/kg (It is negative as it is loss of heat from air)

and $w = 0$

also $\Delta PE = 0$

Therefore $-3.2 \text{ kJ/kg} = C_p\,(T_2 - T_1) + \dfrac{(320\,\text{m/s})^2 - (500\,\text{m/s})^2}{2}\left(\dfrac{1\,\text{kJ/kg}}{1000\,\text{m}^2/\text{s}^2}\right)$

Now $\gamma = \dfrac{C_p}{C_V} = 1.4$

and $C_p - C_v = R$

$$C_p - \frac{C_p}{\gamma} = R$$

or $$-\left(\frac{1-\gamma}{\gamma}\right) C_p = R$$

or $$C_p = \left(\frac{\gamma}{\gamma-1}\right) R$$

$$= \left(\frac{1.4}{0.4}\right) 0.287 \text{ kJ/kg}$$

$$= 1.9945$$

Therefore from energy equation we get

$$-C_p (T_2 - T_1) = 3.2 + \frac{(320+50)(320-50)}{2\times 1000}$$

or $$-1.0045 (T_2 - 350) = 3.2 + 49.95$$

$$= 53.15$$

or $$-T_2 = \frac{53.15 - 350\times 1.0045}{1.0045}$$

$$= -297.088$$

or $$T_2 = 297.088\text{K}$$

(b) For the conservation of mass, we have

$$\dot{m} = \frac{A_2 V_2}{v_2} = \frac{A_1 V_1}{v_1}$$

Now $$v_1 = \frac{RT_1}{P_1}$$

and $$v_2 = \frac{RT_2}{P_2}$$

$\therefore$ $$A_2 = \left(\frac{A_1 V_1}{v_1}\right)\cdot\left(\frac{v_2}{V_2}\right)$$

$$= A_1 \left(\frac{V_1}{V_2}\right)\cdot\left(\frac{v_2}{v_1}\right)$$

$$= A_1 \left(\frac{V_1}{V_2}\right)\left(\frac{RT_2}{P_2}\cdot\frac{P_1}{RT_1}\right)$$

$$= A_1 \frac{V_1}{V_2}\frac{P_1}{P_2}\cdot\frac{T_2}{T_1}$$

$$= (0.01)\left(\frac{50}{320}\right)\left(\frac{300}{100}\right)\left(\frac{297.088}{350}\right)$$

$$= 0.00397 \text{ m}^2$$

$$= 3970 \text{ mm}^2$$

Example 4.4. *Steam at 3 MPa and 400 °C enters an adiabatic nozzle steadily with a* velocity of 40 m */s and leaves at 2.5 MPa and 300 m/s. Find the exit temperature and the ratio of inlet area to outlet area.*

Solution. From steam cables

At inlet

$$P_1 = 3 \text{ MPa}$$
$$T_1 = 400 \text{ °C}$$

Then from steam tables we get values of v_1 and h_1

$$v_1 = 0.09936 \text{ m}^3\text{/kg}$$
$$h_1 = 3230.9 \text{ kJ/kg}$$
$$V_1 = 40 \text{ m/s}$$
$$V_2 = 300 \text{ m/s}$$

Thus from steady flow energy equation

$$q - w = (h_2 - h_1) + \frac{V_2^2 - V_1^2}{2} + \Delta PE$$

Here
$$q = 0$$
$$w = 0$$
$$\Delta PE = 0$$

Therefore
$$h_2 = h_1 - \frac{V_2^2 - V_1^2}{2}$$
$$= 3230.9 - \frac{(300 \text{ m/s})^2 - (40 \text{ m/s})^2}{2 \times 1000} \text{ kJ/kg}$$
$$= 3230.9 - \frac{340 \times 260}{2 \times 1000}$$
$$= 3186.7 \text{ kJ/kg}$$

Thus from steam cables

By interpolation

For $P_2 = 2.5 mPa$ and $h_2 = 3186.7 kJ/kg$ $\Big]$ $T_2 = 376.7\text{°C}$, $v_2 = 0.1153\text{m}^3\text{/kg}$

Also from conservation of mass equation

$$\frac{A_1 V_1}{v_1} = \frac{A_2 V_2}{v_2}$$

Therefore
$$\left(\frac{A_1}{A_2}\right) = \frac{v_1}{v_2} \cdot \frac{V_2}{V_1}$$
$$= \frac{(0.09936 \text{ m}^3\text{/kg})}{(0.1153 \text{ m}^3\text{/kg})} \cdot \frac{(300 \text{ m/s})}{(40 \text{ m/s})}$$
$$= 6.46$$

Example 4.5. *Refrigerant – 12 at pressure 800 kPa and 80 °C enters an adiabatic nozzle steadily at a velocity of 15 m/s and leaves at 300 kPa and 30°C. Find the exit velocity and the ratio of* $\frac{A_1}{A_2}$.

Solution. From this R – 12 Tables

At Inlet for given P_1 and T_1

$$\left.\begin{aligned} P_1 &= 800\,kPa \\ T_1 &= 80°C \end{aligned}\right] \qquad \begin{aligned} v_1 &= 0.02748 \text{ m}^3/\text{kg} \\ h_1 &= 325.11 \text{ kJ/kg} \end{aligned}$$

And at exit for

$$\left.\begin{aligned} P_2 &= 300\,kPa \\ T_2 &= 30°C \end{aligned}\right] \qquad \begin{aligned} v_2 &= 0.06573 \text{ m}^3/\text{kg} \\ h_2 &= 207.12 \text{ kJ/kg} \end{aligned}$$

Then from steady flow energy equation

$$q - w = (h_2 - h_1) + \frac{V_2^2 - V_1^2}{2} + \Delta PE$$

Here

$$q = 0,$$
$$w = 0,$$
$$\Delta PE = 0$$

$$\therefore \qquad -(h_2 - h_1) = \frac{V_2^2 - V_1^2}{2}$$

$$\frac{V_2^2 - V_1^2}{2} = -h_2 + h_1$$
$$= -207.12 + 235.11$$
$$= 27.99$$

$$\frac{V_2^2 - V_1^2}{2 \times 1000} = 27.99 \qquad \text{As } \frac{1\,\text{m}^2}{\text{s}^2} = \frac{1\,\text{kJ/kg}}{1000\,\text{m/s}}$$

$$V_2^2 = V_1^2 + 2000 \times 27.99$$
$$= (15)^2 + 55980$$
$$= 56205$$

$$\therefore \qquad V_2 = 237.07 \text{ m/s}$$

(*b*) For mass conservation equation

$$\frac{A_2 V_2}{v_2} = \frac{A_1 V_1}{v_1}$$

$$\therefore \qquad \frac{A_1}{A_2} = \frac{V_2}{V_1} \cdot \frac{v_1}{v_2}$$

$$= \left(\frac{273.01}{15}\right)\left(\frac{0.02748}{0.06573}\right)$$

$$= 6.607$$

Example 4.6. *Air at 600 kPa and 500 K enters an adiabatic nozzle with a velocity of 120 m/s and leaves with a velocity of 380 m/s. If $A_1 : A_2$ ratio is 2 : 1 find the exit temperature and exit pressure. For air R = 0.287 kJ/g.K and γ = 1.4.*

Solution.
$$C_p = \left(\frac{\gamma}{\gamma-1}\right) R$$
$$= \frac{1.4}{1.4-1} \times 0.287$$
$$= 1.0045 \text{ kJ/kg}$$

From steady flow energy equation

$$q - w = (h_2 - h_1) + \frac{V_2^2 - V_1^2}{2} + \Delta PE$$

Here
$$q = 0$$
$$w = 0$$
$$\Delta PE = 0$$

$$\therefore \quad 0 = C_p(T_2 - T_1) + \frac{(380^2 - 120^2)\,\text{m}^2/\text{s}^2}{2}\left(\frac{1 \text{ kJ/kg}}{1000\,\frac{m^2}{s^2}}\right)$$

i.e.
$$1.0045\ (T_2 - 500) = -\frac{(380+120)(380-120)}{2\times 1000}$$
$$1.0045\ T_2 = 500 \times 1.0045 - 65$$
$$= 437.25 \text{ K}$$
$$T_2 = 435.29 \text{ K}$$

(*b*) **For P_2**

We have from mass conservation equation

$$\frac{A_1 V_1}{v_1} = \frac{A_2 V_2}{v_2}$$

or
$$\frac{v_2}{v_1} = \frac{A_2 V_2}{A_1 V_1}$$

and
$$P_1 v_1 = RT_1$$
$$P_2 v_2 = RT_2$$
$$\frac{P_2}{P_1} \cdot \frac{v_2}{v_1} = \frac{T_2}{T_1}$$

Substituting the value of $\frac{v_2}{v_1}$ we get

$$\frac{P_2}{P_1} \cdot \frac{A_2 V_2}{A_1 V_1} = \frac{T_2}{T_1}$$

$$\therefore \quad P_2 = \frac{T_2}{T_1} \cdot \frac{A_1 V_1}{A_2 V_2}$$

$$P_1 = \frac{435.29}{500} \times \frac{2}{1} \times \frac{120}{380} \times 600 \text{ kPa}$$

$$= 329.9 \text{ kPa}$$

Example 4.7. *Air at 100 kPa and 400 K enters an adiabatic diffuser steadily at a rate 8000 kg/h and leaves at 120 kPa. The velocity at inlet is 230 m/s and at exit 30 m/s. Find the exit temperature and areas at exit and entry to the diffuser.*

Take $R = 0.287$ *kJ/kg. K*

and $\gamma = 1.4.$

Solution. From steady flow energy equation

$$q - w = (h_2 - h_1) + \frac{V_2^2 - V_1^2}{2} + \Delta PE$$

Here $q = 0; \quad w = 0; \quad$ and $\quad \Delta PE = 0$

Therefore $$0 = C_p (T_2 - T_1) + \left(\frac{V_2^2 - V_1^2}{2}\right) (\text{m/s})^2 \left(\frac{1 \text{ kJ/kg}}{1000 \text{ m}^2/\text{s}^2}\right) \quad ...(A)$$

Now $$C_p = \left(\frac{\gamma}{\gamma - 1}\right) R$$

$$= \frac{1.4}{0.4} \times .287$$

$$= 1.0045 \text{ kJ/kg}$$

Now $$P_1 v_1 = RT,$$

or $$v_1 = \frac{RT,}{P,} = \frac{R \;\; T \;,}{P \;\;,} = 1.148 \text{ m}^3/\text{kg}$$

But air flow rate is $\dot{m} = \frac{8000}{3600}$ kg/sec (given)

Also $$\dot{m} = \frac{A_1 V_1}{v_1}$$

or $$A_1 = \dot{m}\frac{v_1}{V_1} = \left(\frac{8000}{3600}\right) \times \frac{1.148}{230}$$

$$= 0.01109 \text{ m}^2$$

From energy equation (A)

$$C_p (T_2 - T_1) = -\frac{V_2^2 - V_1^2}{2000}$$

or $$1.0045\, T_2 = 1.0045 \times 400 - \left(\frac{30^2 - 230^2}{2000}\right)$$

$$= 1.0045 \times 400 + 26$$

$$= 427.80$$

$\therefore$ $$T_2 = \frac{427.80}{1.0045} = 425.88K$$

Again from gas equation

$$P_1\, v_1 = R\, T_1$$

and $$P_2\, v_2 = R\, T_2$$

By dividing there we have

$$\left(\frac{P_1}{P_2}\right)\frac{v_1}{v_2} = \frac{T_1}{T_2} \quad \text{...(B)}$$

Also $$\dot{m} = \frac{A_1 V_1}{V_1} = \frac{A_2 V_2}{V_2}$$

or $$\frac{V_1}{V_2} = \frac{A_1}{A_2} \cdot \frac{V_1}{V_2} \quad \text{...}(C)$$

Therefore from equation (B) and (C)

$$\frac{P_1}{P_2} \cdot \frac{A_1}{A_2} \cdot \frac{V_1}{V_2} = \frac{T_1}{T_2}$$

$\therefore$ $$A_2 = \left(\frac{P_1}{P_2}\right)\left(\frac{A_1}{A_2}\right)\left(\frac{V_1}{V_2}\right)A,$$

$$= \left(\frac{100}{120}\right)\left(\frac{230}{30}\right)\left(\frac{425.88}{400}\right)A,$$

$$= 6.802\text{ A},$$

or $$A_2 = 6.802 \times 0.01109 = 0.0754\text{m}^2$$

Example 4.8. *Air at 110 kPa and – 5°C enters an adiabatic diffuser steadily at a velocity of 220 m/sec and leaves it at a very low velocity (such that $V_2 << V_1$) at a pressure 90 kPa. If the ratio $\frac{A_2}{A_1}$ is 7 find the temperature and velocity of air at exit.*

Take $$R = 0.287 \text{ kJ/kg}$$

$$\gamma = 1.4$$

Solution. Treating air as an ideal gas and applying the steady flow energy equation we have

$$q - w = (h_2 - h_1) + \frac{V_2{}^2 - V_1{}^2}{2} + \Delta\, PE$$

But here $$q = 0$$

$$w = 0$$

and $$\Delta PE = 0$$

$$\therefore \quad 0 = C_p (T_2 - T_1) + \frac{V_2^2 - V_1^2}{2}$$

Now $$C_p = \frac{\gamma}{\gamma - 1} R$$

$$= \frac{1.4}{0.4} \times 0.287$$

$$= 1.0045 \text{ kJ/kg}$$

Therefore $$C_p (T_2 - T_1) = -\frac{V_2^2 - V_1^2}{2}$$

Now $$V_2 << V_1$$

or $$C_p (T_2 - T_1) = -\frac{0 - V_1^2}{2} \times \left(\frac{1 \text{ kJ/kg}}{1000 \text{ m}^2/\text{s}^2}\right)$$

or $$1.0045 (T_2 - T_1) = \frac{V_1^2}{2000}$$

or $$T_2 = T_1 + \frac{(220)^2}{2000} = T_1 + 24.2$$

Now $$T_1 = 273 - 5 = 268\text{K}$$

or $$T_2 = (273 - 5) + 24.2$$

$$= 292.2 \text{ K}$$

(*b*) From conservation of mass relation we have

$$\frac{A_1 V_1}{v_1} = \frac{A_2 V_2}{v_2}$$

or $$\frac{v_1}{v_2} = \frac{A_1 V_1}{A_2 V_2}$$

Also $$\frac{P_1 v_1}{T_1} = \frac{P_2 v_2}{T_2}$$

i.e. $$\frac{v_1}{v_2} = \frac{P_2}{P_1} \cdot \frac{T_1}{T_2}$$

Therefore $$\frac{P_2}{P_1} \frac{T_1}{T_2} = \frac{A_1 V_1}{A_2 V_2}$$

or $$V_2 = \frac{A_1}{A_2} \cdot \frac{T_2}{T_1} \cdot \frac{P_1}{P_2} V_1$$

$$= \frac{1}{7} \times \frac{292.2}{268} \cdot \frac{110}{90} \times 220$$

$$= 41.85 \text{ m/s.}$$

Example 4.9. *Air at 85 kPa and 300 K enters a diffuser with a velocity 225 m/s and the rate of flow is 2.5 kg/s. It leaves the diffuser at 313K. The exit area of diffuser is 0.04 m^2. If the heat lost by air to the surroundings is 18.5 kJ/s during the flow through diffuser find the exit velocity and exit pressure of the air.*

Take $R = 0.287 \text{ kJ/kg} - K$

and $\gamma = 1.4$.

Solution. Treating the air as an ideal gas and applying the energy equation

$$\dot{Q} - \dot{W} = \dot{m}\left[(h_2 - h_1) + \frac{V_2^2 - V_1^2}{2} + \Delta PE\right]$$

Now $\dot{W} = 0$ and $\Delta PE = 0$

and $\dot{Q} = -18.5$ kJ/s

and $\dot{m} = 2.5$ kg/s

($\dot{Q}$ is – ve as heat is leaving the system).

Therefore $$-18.5 \text{ (kg/s)} = (2.5 \text{ kg/s})\left[C_p(T_2 - T_1) + \frac{V_2^2 - V_1^2}{2000}\right]$$

Now $$C_p = \left(\frac{\gamma}{\gamma - 1}\right) R$$

$$= \frac{1.4}{0.4} \times 0.287$$

$$= 1.0045 \text{ kJ/kg}$$

Therefore $$-18.5 = 2.5\left[1.0045(313 - 300) + \frac{V_2^2 - (225)^2}{2 \times 1000}\right]$$

$$= 2.5\left[13.0585 + \frac{V_2^2}{2000} - 25.31\right]$$

or $$\frac{-18.5}{2.5} = -12.25 + \frac{V_2^2}{2000}$$

or $$V_2^2 = 2000(12.25 - 7.4)$$

or $$V_2 = 98.489$$

Now $$\dot{m} = \frac{A_2 V_2}{v_2} \text{ or } v_2 = \frac{A_2 V_2}{\dot{m}} = \frac{0.04 \times 98.489}{2.5} = 1.5758 \text{ m}^3/\text{kg}$$

Again $$P_2 = \frac{RT_2}{v_2} = \frac{0.287 \times 313}{1.5758} = 57 \text{ } kPa$$

Example 4.10. *Nitrogen gas at 80 kPa and 7°C enters an adiabatic diffuser steadily with a velocity of 200 m/s andleaves at 105 kPa and 22°C. Find the axit velocity and the ratio of inlet to exit area $A_1 : A_2$.*

Take R for nitrogen = 0.2968 kJ/kg-K

and $\gamma = 1.4$

Solution:

Assuming Nitrogen to be an ideal gas

Applying steady flow energy equation

$$q - w = (h_2 - h_1) + \frac{1}{2}(V_2^2 - V_1^2) + \Delta PE$$

But here $q = 0$, $w = 0$ and $\Delta PE = 0$

$$\therefore \quad 0 = C_P(T_2 - T_1) + \frac{1}{2}(V_2^2 - V_1^2)\left(\frac{m}{s}\right)^2\left(\frac{1kJ/kg}{1000m^2/s^2}\right)$$

or
$$0 = C_P(T_2 - T_1) + \frac{V_2^2 - V_1^2}{2000} \quad \text{...(A)}$$

Now
$$C_P = \left(\frac{\gamma}{\gamma - 1}\right)R = 0.2968 \times \frac{1.4}{0.4}$$
$$= 1.0388$$

Now it is given that
$$T_2 = 22 + 273 = 295K$$
$$T_1 = 7 + 273 = 280\ K$$

and
$$V_1 = 200 \text{ m/s}$$

Substituting values in equation (A)

$$0 = 1.0388\ (295 - 280) + \frac{V_2^2 - (200)^2}{2000}$$

or
$$V_2^2 = (200)^2 - 2000 \times 1.0388 \times 15$$

or
$$V_2 = 94.0 \text{ m/s}$$

Now for ideal gas

$$\frac{P_1 v_1}{T_1} = \frac{P_2 v_2}{T_2}$$

or
$$\frac{v_1}{v_2} = \frac{P_2}{P_1} \cdot \frac{T_1}{T_2} \quad \text{...(B)}$$

Also for conservation of mass

$$m = \frac{A_1 V_1}{v_1} = \frac{A_2 V_2}{v_2}$$

or
$$\frac{v_1}{v_2} = \frac{A_1}{A_2} \cdot \frac{V_1}{V_2} \quad \text{...(C)}$$

From equation (B) and (C)

or
$$\frac{A_1}{A_2}\left(\frac{V_1}{V_2}\right) = \frac{P_2}{P_1} \cdot \frac{T_1}{T_2}$$

or $$\frac{A_1}{A_2} = \frac{V_2}{V_1} \cdot \frac{P_2}{P_1} \cdot \frac{T_1}{T_2}$$

$$= \left(\frac{94.0}{200}\right)\left(\frac{105}{50}\right) \cdot \left(\frac{280}{295}\right)$$

$$= 0.5855$$

Example 4.11. *Steam at 9.1 MPa and 150 °C enters a diffuser with a velocity of 180 m/s and leaves as saturated vapor at 120 °C with a velocity of 50 m/s. The exit area of the diffuser is 0.08 m². Find (a) the mass flow rate of steam (b) the rate of heat transfer (c) inlet area of diffuser.*

Solution :

From steam tables

At inlet $$\left.\begin{matrix} P_1 = 0.1 MPa \\ T_1 = 150°C \end{matrix}\right] \rightarrow \begin{matrix} v_1 = 1.9364\, m^3/kg \\ h_1 = 2776.4\, kJ/kg \end{matrix}$$

At exit $$\left.\begin{matrix} Salurated\ vapor \\ T_2 = 120°C \end{matrix}\right] \rightarrow \begin{matrix} v_2 = 0.8912\, m^3/kg \\ h_2 = 2706.3\, kJ/kg \end{matrix}$$

Also it is given A_2 = 0.08 m² and V_2 = 50 m/s

Therefore mass flow rate

$$m = \frac{V_2 A_2}{v_2} = \frac{(50\, m/s)(0.08\, m^2)}{0.8912\,(m^3/Kg)}$$

$$= 4.488 \text{ kg/s}$$

(*b*) Heat Transfer rate

We have from steady flow evergy equation

$$\dot{Q} - \dot{W} = \dot{m}\left[(h_2 - h_1) + \frac{1}{2}\left(V_2^2 - V_1^2\right) + \Delta PE\right]$$

Here $\dot{W} = 0$ and $\Delta PE = 0$

Therefore . $$\dot{Q} - 0 = (4.488\, kg/s)\left|(2706.3 - 2776.4)\, kJ/kg\right.$$

$$+\left(\frac{50^2 - 150^2}{2}\right)\left(\frac{m}{s}\right)^2\left(\frac{1kJ/kg}{1000\, m^2/s^2}\right)$$

or $$\dot{Q} = 4.488\left[-70.1 - \frac{32400 - 2500}{2000}\right]$$

$$= 4.488\,(-70.1 - 14.95) = -381.7\ kJ/s$$

(*c*) *We have* $$\dot{m} = \frac{A_1 V_1}{v_1}$$

$$A_1 = \frac{\dot{m} v_1}{V_1} = \frac{4.488 \times 1.9364}{180}$$

$$= 0.048\ m^2$$

Example 4.12. *Refrigerant 12 enters a diffuser steadily as saturated vapor at 700 kPa with a velocity of 120 m/s. It leaves at 500 kPa and 40°C the refrigerant gains heat at a rate of 12kW while passing through the diffuser. If A_2 = 1.3 A, find (a) the exit velocity (b) the mass flow rate of R–12.*

Solution:
From R–12 tables we have
At inlet

$$\left.\begin{array}{r} P_1 = 700\,kPa \\ Saturated\ vapor \end{array}\right] \rightarrow \begin{array}{l} v_1 = 0.02501\,m^3/kg \\ h_1 = 198.74\,kJ/kg \end{array}$$

At exit

$$\left.\begin{array}{r} P_1 = 800\,kPa \\ T_2 = 40°C \end{array}\right] \rightarrow \begin{array}{l} v_2 = 0.02283\,m^3/kg \\ h_2 = 206.07\,kJ/kg \end{array}$$

Applying the conservation of mass flow equation, we have

$$\dot{m} = \frac{A_1 V_1}{v_1} = \frac{A_2 V_2}{v_2}$$

or
$$V_2 = \frac{A_1}{A_2} \cdot \frac{v_2}{v_1} \cdot V_1$$

$$= \left(\frac{1}{1.3}\right)\left(\frac{0.02283}{0.02507}\right)(120)$$

$$= 84.3 \text{ m/s.}$$

(*b*) Again we have from steady flow energy equation

$$\dot{Q} - \dot{W} = \dot{m}\left[(h_2 - h_1) + \frac{V_2^2 - V_1^2}{2} + \Delta PE\right]$$

But here $\dot{W} = 0$

and $\Delta PE = 0$

and $\dot{Q} = 12 \text{ kW} = 12 \text{ kJ/s}$

Therefore
$$12 - 0 = \dot{m}\left[(206.07 - 198.74) + \frac{(84.3)^2 - (120)^2}{2000}\right]$$

$$12 = \dot{m}\left[7.33 + \frac{7106.49 - 144.0}{2000}\right]$$

$$= \dot{m}\ [7.33 - 3.647]$$

$$= 3.683\ \dot{m}$$

or
$$\dot{m} = 3.258 \text{ kg/s}$$

PROBLEMS OF TURBINES AND COMPRESSORS

Example 4.13. *Steam enters a turbine steadily at 10 MPa and 550 °C with a velocity 60 m/s and leaves at 25 kPa with quality of 95%. During the process there is a heat loss of 30 kJ/kg. The inlet area of turbine is 0.0150 m² and outlet area is 0.14 m². Determine the mass flow rate of steam, the exit velocity and power output.*

Solution. From steam tables

At inlet Corresponding to given volues of P_1 and T_1, values of v_1 and h_1 are read from steam tables, by resosting to interpolation where ever necessary.

$$\left.\begin{array}{l} P_1 = 10MPa \\ T_1 = 550°C \end{array}\right] \quad \begin{array}{l} v_1 = 0.03564\ m^3 kg \\ h_1 = 3500.9\ kJ/kg \end{array}$$

$$A_1 = 0.015\ \text{m}^2$$

$$A_2 = 0.14\ \text{m}^2$$

At outlet

$$P_2 = 25\ \text{kPa}$$

$$x_2 = 0.95$$

$$h_2 = h_f + x_2 h_{fg} = 271.93 + 0.95 \times 2346.3$$

$$= 2500.9\ \text{kJ/kg}$$

$$v_2 = v_f + x_2 v_{fg} = 0.00102 + .95 \times 6.203$$

$$= 5.854\ \text{m}^3/\text{kg}$$

Then $$\dot{m} = \frac{A_1 V_1}{v_1} = \frac{(0.015)(60)}{0.03564} = 25.3\ \text{kg/s}$$

Again $$\dot{m} = \frac{A_2 V_2}{v_2}$$

$\therefore$ $$V_2 = \frac{\dot{m} v_2}{A_2} = \frac{(25.3)(5.894)}{0.14} = 106.5\ \text{m/s}$$

(*c*) **Power output**

From steady flow energy equation

$$\dot{Q} - \dot{W} = \dot{m}\left[(h_2 - h_1) + \frac{V_2^2 - V_1^2}{2} + \Delta PE\right]$$

Here $$\dot{Q} = -(30\ \text{kJ/kg}) \quad \dot{m} = -30 \times 25.3$$

$$= -759\ \text{kJ/s}\ (\text{–ve as it is heat loss.})$$

$\therefore$ $$\dot{W} = \dot{Q} - \dot{m}\left[(h_2 - h_1) + \frac{V_2^2 - V_1^2}{2} \times \left(\frac{1\ \text{kJ/kg}}{1000\ \text{m}^2/\text{s}^2}\right)\right]$$

$$= -759 - (25.3)\left[(2500.9 - 3500.9) + \frac{(1065)^2 - (60)^2}{2000}\right]$$

$$= -759 - (25.3)[-1000 + 565.3]$$

$$= -759 - 25.3\ (-434.7)$$

$$= -759 + 10997.9$$

$$= 10238.9\ \text{kJ/s}$$

$$= 10238.9\ \text{kW}$$

Example 4.14. *Steam enters steadily a turbine at a rate of 25000 kg/hr at 8 MPa pressure and 450°C Temperature and it leaves at 30 kPa as saturated vapour. If the power generated is 4 MW find the rate of heat loss.*

Solution. From steam tables

At inlet corresponding to given P and the value of h is

$$P_1 = 8 \text{ MPa } h_1 = 3272.0 \text{ kJ/kg}$$

$$T_1 = 450°\text{C}$$

$$h_1 = 3272.0 \text{ kJ/kg}$$

At exit similarly for the known properties

$$\left.\begin{array}{l} P_2 = 30kPa \\ \text{and sat. vap.} \end{array}\right] \quad h_2 = 2625.3 \text{ kJ/kg}$$

Saturated vapour

$$h_2 = 2625.3 \text{ kJ/kg}$$

then from steady flow energy equation

$$\dot{Q} - \dot{W} = \dot{m}\left[(h_2 - h_1) + \frac{V_2^{\,2} - V_1^{\,2}}{2} + \Delta PE\right]$$

Here

$$\dot{W} = 4 \text{ MW}$$

$$= 4000 \text{ kW} = 4000 \text{ kJ/s}$$

$$\Delta KE = 0$$

and

$$\Delta PE = \frac{V_2^{\,2} - V_1^{\,2}}{2} = 0$$

$\therefore$

$$\dot{Q} = \dot{W} + \dot{m}\ [(h_2 - h_1)$$

$$= 4000 \text{ kJ/s} + \left(\frac{25000}{3600}\right) [2625.3 - 3272]$$

$$= 4000 - 4490.0$$

$$= -\ 490.9 \text{ kJ/s}$$

Example 4.15. *Air enters steadily an adiabatic turbine at 1 MPa, 500°C and 120 m/s and leaves at 150 kPa, 150 °C and 250 m/s. If the inlet area is 0.08 m² find the mass flow rate and the power output. Assume air an ideal gas and*

Take $R = 0.287\ kJ/kg\text{-}K$

and $\gamma = 1.4$

Solution. Give $T_1 = 500 + 273 = 773$ K

$$P_1 = 1000 \text{ kPa}$$

$$P_1\, v_1 = RT_1$$

$$v_1 = \frac{RT_1}{P_1} = \frac{0.287 \times 773}{1000} = 0.2219 \text{ m}^3\text{/kg}$$

That is

$$\dot{m} = \frac{A_1 V_1}{v_1} = \frac{0.008 \times 120}{0.2219} = 4.326 \text{ kg/s}$$

Now $$C_p = \left(\frac{\gamma}{\gamma - 1}\right) R$$

$$= \frac{1.4}{0.4} \times 0.287 = 1.0045$$

Then from steady flow energy equation

$$\dot{Q} - \dot{W} = \dot{m}\ [(h_2 - h_1) + \Delta\ KE + \Delta\ PE]$$

Here $\dot{Q} = 0$

$$\Delta\ PE = 0$$

Therefore $$\dot{W} = \dot{Q} - \dot{m}\left[(h_2 - h_1) + \frac{V_2^2 - V_1^2}{2}\right]$$

$$= -\ (4.326\ \text{kg/s})\left[C_p\ (T_2 - T_1) + \frac{250^2 - 120^2}{2} \times \left(\frac{1\ \text{kJ/kg}}{1000\ \text{m}^2/\text{s}^2}\right)\right]$$

$$= -\ 4.326\left[1.0045\,(100 - 500) + \frac{62500 - 14400}{2000}\right]$$

$$= 4.326\ [-\ 351.575 + 24.05]$$

$$= 1416.87\ \text{kJ/s}$$

$$= 1416.87\ \text{kW}$$

Example 4.16. *Air enters a compressor of a gas turbine power plant at 96 kPa and 25° C with very low velocity and leaves the compressor at 1 MPa and 327 °C with a velocity of 120 m/s. The compressor is cooled at a rate of 25 kJ/s. If the power input is 300 kW find the mass flow rate of air.*

Take $R = -\ 0.287\ \ kJ/kg\text{–}K$

and $\gamma = 1.4.$

Solution. $$C_p = \left(\frac{\gamma}{\gamma - 1}\right) R$$

$$= \frac{1.4}{0.4} \times 0.287 = 1.0045\ \text{kJ/kg}$$

From steady flow energy equation

$$\dot{Q} - \dot{W} = \dot{m}\left[(h_2 - h_1) + \frac{V_2^2 - V_1^2}{2} + \Delta\ PE\right]$$

Here $\Delta\ PE = 0$

$$\dot{Q} = -\ 25\ \text{kJ/s}\ (-\ \text{ve being loss})$$

$$\dot{W} = -\ 300\ \text{kJ/s}$$

Therefore $$-25-(300)=\dot{m}\left[C_p\,(T_2-T_1+\frac{V_2^{\,2}}{2}\times\frac{1}{1000}\right]$$

or $$275=\dot{m}\left[1.004\,(327-25)+\frac{120^2}{2000}\right]$$

$$=\dot{m}\,[1.9945\times 302+72]$$

$$275=375.35\,\dot{m}$$

$$\dot{m}=\frac{275}{375.35}=0.732\text{ kg/s}$$

Example 4.17. *Air is compressed from 100 kPa and 22 °C to a pressure of 1 MPa. It is being cooled at an intermediate stage at the rate of 16 kJ/s. The volume flow rate of air at inlet conditions is 2.5 m³/s. The power input to compressor is 500 kW. Find mass flow rate of air and temperature at the exit of compressor.*

Take $R = 0.287$ kJ/kg–K

and $\gamma = 1.4$.

Solution. Taking air as ideal gas

$$T_1=273+22=295\text{ K}$$

We have $$C_p=\left(\frac{\gamma}{\gamma-1}\right)R$$

$$=\frac{1.4}{0.4}\times .287$$

$$=1.0045\text{ kJ/kg-K}$$

Also $$v_1=\frac{RT_1}{P_1}$$

$$=\frac{0.287\times 295}{100}=0.847\text{ m}^3\text{/kg}$$

Given V_1 = Vol flow rate = 2.5 m³/s

Therefore $$\dot{m}=\frac{V_1}{v_1}=\frac{(2.5\,\text{m}^3/\text{s})}{0.847\,\text{m}^3/\text{kg}}$$

$$=\frac{2.5}{0.847}=2.95\text{ kg/s}$$

(*b*) From steady flow energy equation

$$\dot{Q}-\dot{W}=\dot{m}\;\dot{m}\,[(h_2-h_1)+\Delta\,KE+\Sigma\,PE]$$

Here $\Delta\,KE = 0$

and $\Delta\,PE = 0$

$\therefore$ $$\dot{Q}-\dot{W}=\dot{m}\;C_p\,(T_2-T_1)$$

$$(-\,\dot{m}\times 16\text{ kJ/s})-(-\,500\text{ kJ/s})=2.95\,[1.0045\,(T_2-295)]$$

$$2.95 \times 16 - (-500 \text{ kJ/s}) = 2.95[1.0045\ (T_2 - 295)]$$

$$-47.2 + 500 = 2.95 \times 1.0045\ (T_2 - 295)$$

$$\frac{452.8}{2.95 \times 1.0045} + 298 = T_2 \text{ K}$$

$$1528 + 295 = T_2$$

$$T_2 = 470.8 \text{ K}$$

$$= 197.8\ °\text{C}$$

PROBLEMS ON THROTTLING VALVES

Example 4.18. *Refrigerant-12 is throttled from saturated liquid state at 800 kPa to a pressure of 140 kPa. Determine the temperature drop during the process and find the final specific volume of the refrigerant.*

Solution. From refrigerant-12 tables

Sat liquid At $P_1 = 0.8\text{MPa}$ $T_1 = T_{Sat} = 32.74°\text{C}$

Sat. liquid $h_1 = h_{f_1} = 67.3$ kJ/kg

After throttling $P_2 = 0.14$ MPa $h_{f_2} = 16.09$ kJ/kg

$$T_{sat} = -21.91°\text{C}$$

$$h_{g_2} = 177.8 \text{ kJ/kg}$$

And since $h_2 = h_1$, we have $h_2 = 67.3$ kJ/kg

Evidently $h_{f_2} < h_2 < h_{g_2}$. Thus the refrigerant exists as a saturated mixture at the exit state. Therefore

$$T_2 = T_{sat} \text{ at } 0.14 \text{ MPa or } 140 \text{ kPa}$$

$$= -21.91\ °\text{C}$$

Therefore temperature drop

$$= T_2 - T_1$$

$$= -21.91 - 32.74$$

$$= -54.65\ °\text{C}$$

For the quality at exit

$$h_2 = h_{f_2} + x_2\, h_{fg2}$$

or

$$x_2 = \frac{h_2 - h_{f_2}}{h_{fg2}}$$

$$= \frac{h_1 - h_{f_2}}{h_{fg2}} \qquad (\because\ h_2 = h_1)$$

$$= \frac{67.3 - 16.09}{h_{g_2} - h_{f_2}} = \frac{67.3 - 16.09}{177.87 - 16.09}$$

$$= \frac{51.21}{161.78} = 0.3165$$

Therefore for tables

$$v_2 = v_f + x\ v_{fg}$$
$$= 0.0006828 + 0.3165 \times 0.11611$$
$$= 0.1375\ \text{m}^3/\text{kg}$$

Example 4.19. *An insulated valve is used to throttle steam from 10 MPa and 550 °C to 8 MPa. Find the final temperature.*

Solution. At inlet from steam tables

$$\left.\begin{matrix} P_1 = 10 MPa \\ T_1 = 6550 \end{matrix}\right] \quad h_1 = 3500.9\ \text{kJ/kg}$$

At outlet from steam tables

$$\left.\begin{matrix} P_2 = 8 MPa \\ h_2 = h_1 \end{matrix}\right] \quad T_2 = 541.8°C$$

(As during throttling enthalpy remains constant)

Example 4.20. *Air at 1 MPa and 25 °C is throttled to the atmospheric pressure of 100 kPa. Determine the final temperature of the air.*

Solution. At the specified conditions air can be treated as ideal gas. Thus the enthalpy of air will be function of temperature only.

Further during throttling process

$$h_2 = h_1$$

Such h_2 is function of T_2 and h1 is same function of T_1

and $\quad h_2 = h_f$

Therefore $\quad T_2 = T_1 = 25\ °C$

PROBLEMS ON MIXING CHAMBERS AND HEAT EXCHANGERS

Example 4.21. *A stream of hot water at 70 °C enters a mixing chamber with a mass flow rate of 0.6 kg/s where it is mixed with a stream of cold water at 20 °C. It is desired that the mixture leaves the chamber at 42°C. Determine the mass flow rate of cold water stream. Assume that all the stream are at pressure of 300 kPa.*

T_3 = 42°C
P_3 = 300 kPa
m_2 = ?
R_2 = 20°C
P_2 = 300kPa

Fig. 4.8

Solution. From conservation of mass flow

$$\Sigma \dot{m}_i = \Sigma m_{out}$$

i.e. $\quad \dot{m}_1 + \dot{m}_2 = \dot{m}_3$ ...(A)

Also from conservation of energy equations

$$\dot{Q} - \dot{W} = \Sigma \dot{m}_0\ h_0 - \Sigma \dot{m}_i\ h_i + \Delta KE + \Delta PE$$

Since $\quad \dot{Q} = 0$

$$\dot{W} = 0$$
$$\Delta PE = 0$$
$$\Delta KE = 0$$

we get

$$\dot{m}_1 h_1 + \dot{m}_2 h_2 = \dot{m}_3 h_3 \qquad \text{...(B)}$$

$$= (\dot{m}_1 + \dot{m}_2)\, h_3$$

From equation (A)

or $$\dot{m}_2\, h_2 = \dot{m}_1\, h_3 + \dot{m}_2\, h_3 - \dot{m}_3\, h_3$$

$$= \dot{m}_1\, (h_3 - h_1) + \dot{m}_2\, h_3$$

$$\dot{m}_2\, (h_2 - h_3) = \dot{m}_1\, (h_3 - h_1)$$

or $$\dot{m}_2 = \left(\frac{h_1 - h_3}{h_3 - h_2}\right) \dot{m}_1 \qquad \text{...(C)}$$

From water tables we see that at 300 kPa pressure

$$T_{sat} = 135.55\ °C$$

In other words the temperature of all the three individual steam is less than T_{sat}. Therefore it is concluded that these streams are all compressed liquids and can be approximated as saturated liquids.

Then from tables

$$h_1 = h_f \text{ at } 70\ °C = 292.88 \text{ kJ/kg}$$

$$h_2 = h_f \text{ at } 20\ °C = 83.96 \text{ kJ/kg}$$

$$h_3 = h_f \text{ at } 42\ °C = 175.92 \text{ kJ/kg}$$

Therefore substituting these values is equation (C) we get

$$\dot{m}_2 = \left(\frac{292.88 - 175.92}{175.92 - 83.96}\right) (0.6)$$

$$= \frac{116.96}{91.96} \times 0.6 = 0.763 \text{ kg/s}$$

Example 4.22. *Superheated steam at 200 kPa and 300 °C is used to heat water which is at 200 kPa and 20 °C by mixing it in a chamber. The flow rate of cold water in the chamber is 2.5 kg/s. If the mixture leaving the chamber is at 60 °C find the mass flow rate of steam entering the chamber. Neglect kinetic and potential energies of the streams.*

Solution. From the mass flow conservation equation

$$\Sigma m_i = \Sigma m_0$$

or $$\dot{m}_1 + \dot{m}_2 = \dot{m}_3 \qquad \text{...(A)}$$

where $\dot{m}_1$ = mass flow rate of incoming water whose enthalpy = h_1

at $T_1 = 20\ °C$

$\dot{m}_2$ = mass flow rate of incoming superheated steam, having enthalpy $= h_2$ at $T_2 = 300\ °C$

$\dot{m}_3$ = mass flow rate of mixture going out of mixing chamber having enthalpy h_3 at

$$T_3 = 60\ °C$$

For the conservation of energy equation

$$\dot{Q} - \dot{W} = \Sigma(\dot{m}_0 h_0) - \Sigma(\dot{m}_i h_i)$$

But $\dot{Q} = 0$

and $\dot{W} = 0$

$\therefore \quad 0 = -\dot{m}_3 h_3 + (\dot{m}_1 h_1 + \dot{m}_2 h_2)$

or $\dot{m}_1 h_1 + \dot{m}_2 h_2 = (\dot{m}_1 + \dot{m}_2) h_3$ ...(B)

or $\dot{m}_1 h_1 - \dot{m}_1 h_3 = \dot{m}_2 h_3 - \dot{m}_2 h_2$

or $\dot{m}_1 (h_1 - h_3) = \dot{m}_2 (h_3 - h_2)$

or $\dot{m}_2 = \left(\dfrac{h_1 - h_3}{h_3 - h_2}\right) \dot{m}_1$ ...(3)

From tables we note that at 200 kPa

$T_{sat} = 120.23$ °C

Further it is given that

$T_1 = 20$ °C

and $T_3 = 60$ °C

Since $T_1 < T_{sat}$ and $T_3 < T_{sat}$, it is concluded that his incoming water steams and out-going mixture steam exist as compressed liquids and they can be approximated as saturated liquids therefore from tables.

At 20 °C $\quad h_1 = h_f$ at 20 °C = 83.96 kJ/kg

At 60 °C $\quad h_3 = h_g$ at 60 °C = 251.13 kJ/kg

For Steam at $\quad \left.\begin{matrix} P_2 = 200 \text{kPa} \\ T_2 = 300°C \end{matrix}\right] \rightarrow h_2 = 3071.8\ kJ/kg$

and

Substituting these values in equation (C)

$$\dot{m}_2 = \frac{(83.96 - 251.13)}{(251.13 - 3071.80)} \times 2.5$$

$$= \frac{(-167.17)}{(-2820.67)} \times 2.5 = 0.148 \text{ kg/s}$$

Example 4.23. *In a feed water heater attached to a steam pressure power plant, water at 600 kPa and 50 °C is to be heated by mixing it with superheated steam at 200 °C and 600 kPa pressure. The mixture leaves the feed water heated as saturated liquid at 600 kPa. Determine the ratio of mass flow rates of feed water and superheated steam. Neglect the kinetic and potential energies.*

Solution. Let $\dot{m}_1$ = mass flow rate of feed water at 50 °C and P_1 = 600 kPa with h_1 as enthalpy

$\dot{m}_2$ = mass flow rate of superheated steam at 200 °C and 600 kPa with enthalpy = h_2

$\dot{m}_3$ = mass flow rate of mixture at pressure 600 kPa.

From conservation of mass flow equation

$$\Sigma \dot{m}_i = \Sigma \dot{m}_e$$

$$\dot{m}_1 + \dot{m}_2 = \dot{m}_3 \quad \text{...(A)}$$

From conservation of energy equation

$$\dot{Q} - \dot{W} = \Sigma \dot{m}_i h_i - \Sigma m_0 h_0$$

or $$0 = (\dot{m}_1 h_1 + \dot{m}_2 h_2) - \dot{m}_3 h_3 \quad \text{...(B)}$$

or $$\dot{m}_1 h_1 + \dot{m}_2 h_2 = \dot{m}_3 h_3$$

$$= (\dot{m}_1 + \dot{m}_2) h_3$$

or $$\dot{m}_1 (h_1 - h_3) = \dot{m}_2 (h_3 - h_2)$$

or $$\dot{m}_2 = \left(\frac{h_1 - h_3}{h_3 - h_2}\right) \dot{m}_1$$

Let $$r = \frac{\dot{m}_1}{\dot{m}_2} = \left(\frac{h_3 - h_2}{h_1 - h_3}\right) \quad \text{...(C)}$$

From steam tables we note that at 600 kPa

$$T_{sat} = 158.85 \text{ °C}$$

Since T_1 = 50 °C and the mixture is saturated liquid at 600 kPa, we conclude that the feed water and the mixture are in the state of compressed liquids as $T_1 < T_{sat}$. Thus treating them as saturated liquids.

At $$T_1 = 50 \text{ °C}$$

$$h_1 = h_f \text{ at } 50 \text{ °C} = 209.33 \text{ kJ/kg}$$

at exit $$h_3 = h_f \text{ at } 600 \text{ kPa} = 670.56 \text{ kJ/kg}$$

For superheated steam

at $P_2 = 600kPa$

and $T_2 = 200°C$ $\Big]$ $h_2 = 2850.1$ kJ/kg

Therefore substituting these values in equation (C), we get

$$r = \frac{\dot{m}_1}{\dot{m}_2} = \frac{(2850.1 - 670.56)}{(670.56 - 209.33)}$$

$$= \frac{2179.54}{461.23}$$

$$= 4.7255$$

Example 4.24. *Water at 25 °C and 300 kPa is heated in a chamber by mixing it with saturated water vapour at 300 kPa. The mass flow rates of both the streams entering the chamber is same. Find the temperature and quality of mixture leaving the chamber. Neglect the kinetic and potential energies.*

Solution. Let $\dot{m}_1$ = mass flow rate of water

$\dot{m}_2$ = mass flow rate of steam

$\dot{m}_3$ = mass flow rate of mixture.

Also since $\dot{m}_1 = \dot{m}_2$

From the conservation of mass flow rate equation

$$\Sigma \dot{m}_i = \Sigma \dot{m}_0$$

i.e. $$\dot{m}_1 + \dot{m}_2 = \dot{m}_3$$

or $$\dot{m}_3 = 2\ \dot{m}_1 \quad \text{...(A)}$$

Also from energy conservation equation

$$\dot{Q} - \dot{W} = \Sigma \dot{m}_i\ h_i - \Sigma \dot{m}_0\ h_0$$

or $$0 = (\dot{m}_1\ h_1 + \dot{m}_2\ h_2) - \dot{m}_3\ h_3$$

or $$\dot{m}_1\ h_1 + \dot{m}_1\ h_2 = \dot{m}_3\ h_3$$

$$= 2\ \dot{m}_1\ h_3$$

or $$h_1 + h_2 = 2\ h_3$$

or $$h_3 = \frac{h_1 + h_2}{2} \quad \text{...(B)}$$

From steam tables

At 25 °C

$$h_1 = h_f \text{ at } 25\ °C = 104.89 \text{ kJ/kg}$$

For saturated water vapour at 300 kPa

$$h_2 = h_g \text{ at } 300 \text{ kPa} = 2725.3 \text{ kJ/kg}$$

Then from (B)

$$h_3 = \frac{104.89 + 2725.3}{2}$$

$$= 1415.1 \text{ kJ/kg}$$

Again from steam tables

At 300 kPa $\rightarrow$ $h_f = 561.47$ kJ/kg

$$h_{fg} = 2725.3 \text{ kJ/kg}$$

$$h_{fg} = 2725.3 - 561.47$$

$$= 2163.83 \text{ kJ/kg}$$

Therefore the volue of h_3 lies between h_f and h_g

or $$h_f < h_3 < h_g$$

From this we conclude that the exit steam is a saturatid mixture. Let its quality be x_3 and the correspuding saturation temperature at pressure 300 *kPa* is

$$T_{sat} = 133.55°C$$

and $$x_3 = \frac{h_3 - h_f}{h_{fg}} = \frac{1415.1 - 561.47}{2163.83}$$

$$= 0.394 \textbf{ Ans.}$$

Example 4.25. *A stream of refrigerant -12 at 8°C and 1000 kPa pressure is mixed with a stream of R-12 at same pressure but at 60°C. If the mass flow rate of first sheam is twice that of the second, find this temperature and quality of the exit stream.*

Solution: Let $\dot{m}$, be the mass flow rate of first stream at 10^3 kPa(or 1MPa) and 8°C and $\dot{m}_2$ be that of the second stream. Then from the mass conservation equation.

$$\sum \dot{m}_i = \sum \dot{m}_0$$

or $$\dot{m}_1 + \dot{m}_2 = \dot{m}_3$$

But $$\dot{m}_1 = 2\dot{m}_2$$

Therefore $$2\dot{m}_2 + \dot{m}_2 = \dot{m}_3$$

or $$\dot{m}_3 = 3\dot{m}_2 \quad \text{...(A)}$$

Negleating ΔPE and ΔKE, we have from the energy conservation equation.

$$\dot{Q} - \dot{W} = \sum \dot{m}_0 \dot{h}_0 - \sum \dot{m}_i h_i$$

But heve $$\dot{Q} = 0 \text{ and } \dot{W} = 0$$

Therefore $$0 = \dot{m}_3 h_3 - (\dot{m}_1 h_1 + \dot{m}_2 g_2)$$

or $$\dot{m}_3 h_3 = \dot{m}_1 h_1 + \dot{m}_2 g_2$$

$$= 2\dot{m}_2 h_1 + \dot{m}_2 h_2$$

or $$(3\dot{m}_2) h_3 = \dot{m}_2 (2h_1 + h_2)$$

$\therefore$ $$3g_3 = 2h_1 + h_2 \quad \text{...(B)}$$

From $R - 12$ tables

At 10^3 kPa(= 1mPa) and 8°C $h_1 = h_f$ at 8°C

$= 43.50$ kJ/kg

At 1 MPa and 60°C $= h$ at 1 MPa 60°C

$= 217.97$ kJ/kg

Therefore

$$3h_3 = 2 \times 43.50 + 217.97$$

$$= 304.97$$

$\therefore$ $$h_3 = \frac{304.97}{3} = 101.656 \text{ kJ/kg}$$

Again at 1 MPa $h_f = 76.26$ kJ/kg

and $h_g = 203.76$ kJ/kg

Evidently $h_f < h_3 < h_g$. Therefore we conclude that exit stream is a saturated mixture.

Therefore $$T_3 = T_{sat} \text{ at 1 MPa}$$

$$= 41.64\text{ °C}$$

and the quality

$$x_3 = \frac{h_3 - h_f}{h_{fg}}$$

$$= \frac{101.656 - 76.26}{127.5}$$

$$= 0.199$$

Example 4.26. *In an air conditioner condenser refrigerant-12 at 1 MPa and 80° C is to be cooled to 1 MPa and 30 °C by an air stream. The air enters at 100 kPa and 27 °C with a volume flow rate 800 m³/min and leaves at 95 kPa and 60 °C. Calculate the mass flow rate of the refrigerant-12.*

Neglect kinetic and potential energies.

Take $R_{air} = 0.287$ kJ/kg – K

and $\gamma_{air} = 1.4$

Assume air to be perfect gas.

Solution.

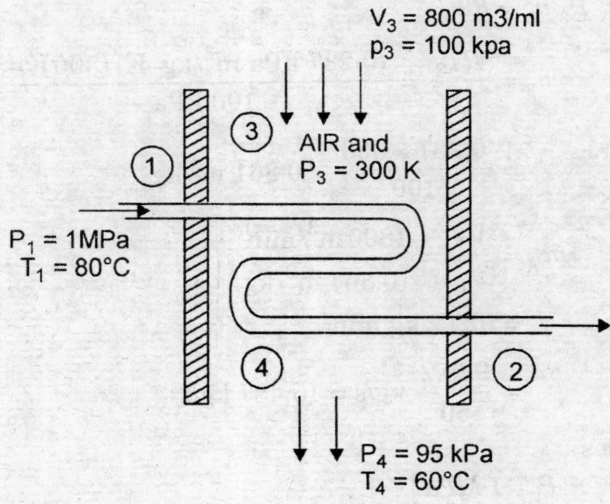

Fig. 4.9

From mass conservation equations (Refer to fig 4.9)

$$\dot{m}_1 = \dot{m}_2 = \dot{m}_R \quad \text{...(A)}$$

where $\dot{m}_R$ = mass flow rate of $R - 12$

and

$$\dot{m}_3 = \dot{m}_4 = \dot{m}_A \quad \text{...(B)}$$

$\dot{m}_A$ = mass flow rate of air

From energy conservation equation

$$\dot{Q} - \dot{W} = \Sigma \dot{m}_0 h_0 - \Sigma \dot{m}_i h_i$$

But $\dot{Q} = 0$

and $\dot{W} = 0$

$\therefore \qquad 0 = (\dot{m}_1 h_1 + \dot{m}_3 h_3) - (\dot{m}_2 h_2 + \dot{m}_4 h_4)$

or $\qquad \dot{m}_1 h_1 + \dot{m}_3 h_3 = \dot{m}_2 h_2 + \dot{m}_4 h_4$

$$\dot{m}_1 h_1 - \dot{m}_2 h_2 = \dot{m}_4 h_4 - \dot{m}_3 h_3 \qquad \text{...(C)}$$

From equations (A), (B) and (C)

$$\dot{m}_R (h_1 - h_2) = \dot{m}_A (h_4 - h_3)$$

Therefore $\qquad \dot{m}_R = \left(\frac{h_4 - h_3}{h_1 - h_2}\right) m_A \qquad \text{...(D)}$

Now for air

$$h_4 - h_3 = C_p (T_4 - T_3)$$

But $\qquad C_p = \left(\frac{\gamma}{\gamma - 1}\right) R$

$$= \frac{1.4}{0.4} \times 0.287 = 1.0045 \text{ kJ/kg-K}$$

Also from gas equation for air

$$P_3 v_3 = RT_3$$

or $\qquad v_3 = \frac{RT_3}{P_3} = \frac{(0.287 \text{ kPa m}^3/\text{kg- K})(300\text{ K})}{100 \text{ kPa}}$

$$= \frac{0.287 \times 300}{100} = 0.861 \text{ m}^3/\text{kg}$$

But $\qquad \dot{m}_A = \frac{\dot{V}_3}{v_3} = \frac{(800\text{ m}^3/\text{min})}{(0.861 \text{ m}^3/\text{kg})}$

$$= 929.2 \text{ kg/min}$$

$$= \frac{929.2}{60} \text{ kg/s} = 15.487 \text{ kg/s}$$

Now from R-12 tables

$$\left.\begin{array}{l} P_1 = 1\, MPa \\ T_1 = 80°\, C \end{array}\right] \qquad h_1 = 232.91 \text{ kJ/kg}$$

and at

$$\left.\begin{array}{l} P_2 = 1\, MPa \\ T_2 = 30°\, C \end{array}\right] \qquad h_2 = h_f \text{ at } 30° \text{ C} = 64.59 \text{ kJ/kg}$$

$$h_2 = h_f \text{ at } 30° \text{ C}$$

$\therefore \qquad h_1 - h_2 = 232.91 - 64.59 = 168.32 \text{ kJ/kg}$

Therefore

$$(h_4 - h_3) = C_p (T_4 - T_3)$$

$$= 1.0045\,(60 - 27)$$

$$= 1.0045 \times 33$$

$$= 33.1485 \text{ kJ/kg}$$

Thus substituting values in equation (D)

$$\dot{m}_R = \frac{33.1485}{168.32} \times 15.487 \text{ kg/s}$$
$$= 3.05 \text{ kg/s}$$
$$= 182.99 \text{ kg/min.}$$

Example 4.27. *Air enters the evaporator section of a window air conditioner at 100 kPa and 27 °C with volume flow rate of 12 m³/min. Refrigerant-12 at 140 kPa with quality of 0.3 enters the evaporator at the rate of 2 kg/min and leaves as saturated vapour at the same pressure. Determine the exit temperature of air and the rate of heat transfer from the air. Treat air as ideal gas and take*

$$R = 0.287 \; kJ/kg - K$$

and $\gamma = 1.4$ *for air.*

Solution.

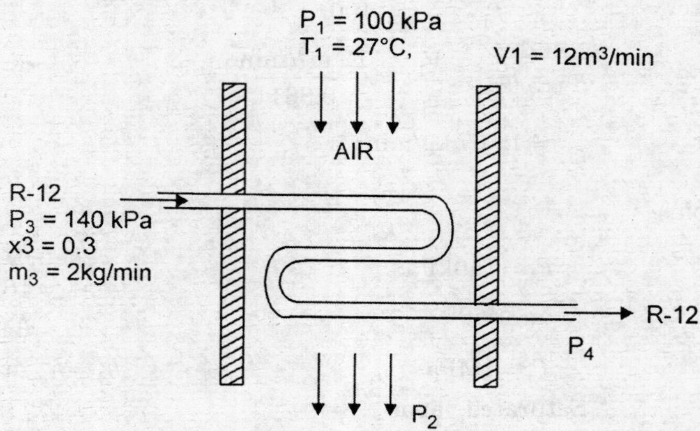

Fig. 4.10

For air $\quad C_p = \left(\frac{\gamma}{\gamma - 1}\right) R$

$$= \frac{1.4}{0.4} \times .287 = 1.0045 \text{ kJ/kg-K}$$

Thus conservation of mass flow rate of air

$$\Sigma \dot{m}_i = \Sigma \dot{m}_0$$

$$\dot{m}_1 = \dot{m}_2 = \dot{m}_a \quad \text{...(1)}$$

For R-12 $\quad \dot{m}_3 = \dot{m}_4 = \dot{m}_R \quad \text{...(2)}$

From conservation of energy equation for the heat exchanger

$$\dot{Q} - \dot{W} = \Sigma \dot{m}_0 h_0 - \Sigma \dot{m}_i h_i$$

Here $\quad \dot{Q} = 0$

$$\dot{W} = 0$$

$$0 = (\dot{m}_4 h_4 + \dot{m}_2 h_2) - (\dot{m}_1 h_1 + \dot{m}_3 h_3)$$

or
$$0 = (\dot{m}_4 h_4 - \dot{m}_3 h_3) + (\dot{m}_2 h_2 - \dot{m}_1 h_1)$$
$$= \dot{m}_R (h_4 - h_3) + \dot{m}_a (h_2 - h_1)$$

or
$$\dot{m}_R (h_3 - h_4) = \dot{m}_a (h_2 - h_1)$$
$$= \dot{m}_a C_p (T_2 - T_1)$$

i.e.
$$(T_2 - T_1) = \frac{\dot{m}_R}{\dot{m}_a C_p} (h_3 - h_4)$$

or
$$T_2 = T_1 + \frac{\dot{m}_R}{\dot{m}_a C_p} (h_3 - h_4) \qquad \text{...(C)}$$

Now for air

$$v_1 = \frac{RT_1}{P_1} = \frac{0.287 \times 300}{100} = 0.861 \text{ m}^3/\text{kg}$$

$$\therefore \quad \dot{m}_1 = \dot{m}_a = \frac{\dot{V}_1}{v_1} = \frac{12 \text{ m}^3/\text{min}}{0.861}$$
$$= 19.67 \text{ kg/min}$$

For R-12

From *R*-12 tables

At $\left.\begin{array}{l} P_3 = 140\text{kPa} \\ x_3 = 0.3 \end{array}\right]$ $\quad h_3 = h_f + x_3 h_{fg}$
$$= 16.09 + 0.3 \times 161.78$$
$$= 64.62 \text{ kJ/kg}$$

At $\left.\begin{array}{l} P_4 = 1\text{MPa} \\ \text{Saturated vapor} \end{array}\right]$ $\quad h_4 = h_g \text{ at} + 1\text{MPa}$
$$= 177.87 \text{ kJ/kg}$$

Substituting these values in equation (C)

$$T_2 = 27 + \frac{(2 \text{ kg/min})(64.62 - 177.87) \text{ kJ/kg}}{(19.67)(\text{kg/min})(1.0045 \text{ kJ/kg})}$$

$$= 27 + \frac{2 \times (-113.25)}{19.67 \times 1.005}$$

$$= 27 - 11.46 = 15.54\ °\text{C}$$

(*b*) For steady flow energy equation for air we have

$$\dot{Q} - \dot{W} = \dot{m}_a (h_2 - h_1) + \Delta KE + \Delta PE$$

Here
$$\Delta KE = 0$$
$$\Delta PE = 0$$

and
$$\dot{W} = 0$$

$$\therefore \quad \dot{Q} = \dot{m}_a C_p (T_2 - T_1)$$
$$= (19.67 \text{ kg/min})(1.0045 \text{ kJ/kg-K})(15.54 - 27)$$
$$= -518.2 \text{ kJ/min}$$

(– ve sign means heat loss from air).

Example 4.28. *R-12 at 800 kPa pressure and 70 °C temperature flowing at 6 kg/min rate is cooled by water in a condenser until it is turned into a saturated liquid at the same pressure. The cooling water enters the condenser at 300 kPa and 15 °C and leaves at 25 °C at the same pressure. Determine the required mass flow rate of cooling water to cool the refrigerant.*

Solution. From mass flow rate conservation principle

$$\Sigma \dot{m}_i = \Sigma \dot{m}_0$$

i.e. $$\dot{m}_1 = \dot{m}_2 = \dot{m}_w \quad ...(1)$$

and $$\dot{m}_3 = \dot{m}_4 = \dot{m}_R \quad ...(2)$$

From conservation of energy equation we have, taking

$$\Delta KE = \Delta PE = 0$$

and $$\dot{W} = 0, \ \dot{Q} = 0$$

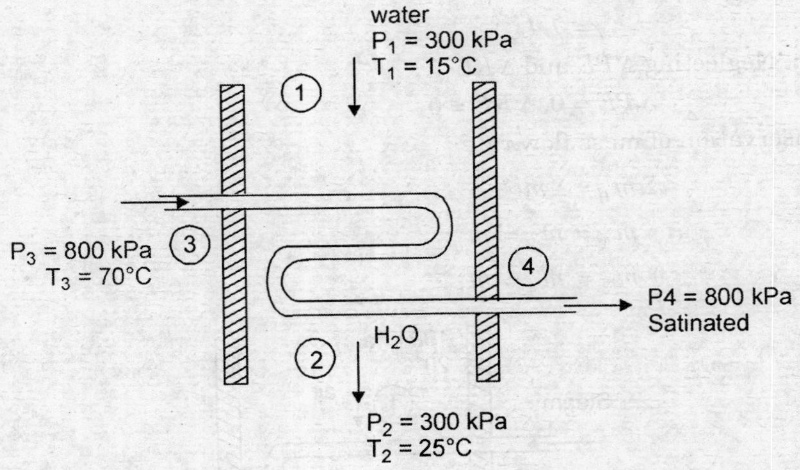

Fig. 4.11

$$\dot{Q} - \dot{W} = \Sigma \dot{m}_0 h_0 - \Sigma \dot{m}_i h_i$$

$$0 = (\dot{m}_2 h_2 + \dot{m}_4 h_4) - (\dot{m}_3 h_3 + \dot{m}_1 h_1)$$

$$= \dot{m}_1 (h_2 - h_1) + \dot{m}_3 (h_4 - h_3)$$

i.e. $$\dot{m}_w (h_2 - h_1) = \dot{m}_R (h_3 - h_4)$$

or $$\dot{m}_w = \left(\frac{h_3 - h_4}{h_2 - h_1}\right) \dot{m}_R$$

From R-12 tables

At inlet

$$\left.\begin{array}{l} P_3 = 800 \text{ kPa} \\ T_3 = 70°C \end{array}\right] \qquad h_3 = 272.915 \text{ kJ/kg}$$

At outlet

$$\left.\begin{array}{l} P_4 = 800 \text{ kPa} \\ \text{Sat. Liquid} \end{array}\right] \qquad h_4 = h_f \text{ at } 800\text{kPa} = 67.3 \text{ kJ/kg}$$

From steam tables at 15 °C and 75 °C water is at 300 kPa is in compressed liquid state.

Therefore at 15 °C

$$h_1 = h_f \text{ at } 15\ °C = 62.99 \text{ kJ/kg}$$

At 25 °C

$$h_2 = h_f \text{ at } 25\ °C = 104.89 \text{ kJ/kg}$$

Therefore

$$\dot{m}_w = \frac{(227.915 - 67.3)}{(104.85 - 62.99)} \times 6 \text{ (kg/min)}$$

$$= 23.0 \text{ kg/min}$$

Example 4.29 *In a steam type room heater, air is heated in winter by passing over a set of tubes which form a heat exchanger. Through the tubes flows steadily steam which enters at 200 kPa and at 200 °C at a rate 8 kg/min and leaves the heat exchanger at 180 kPa and 100 °C. Air enters the heat exchanger at 100 kPa and 20 °C and leaves at 47 °C. Find the flow rate of air at inlet.*

Take for air $R = 0.287\ kJ/kg - K$

and $\gamma = 1.4.$

Solution. Neglecting ΔPE and ΔKE

i.e $\Delta PE = 0, \Delta KE = 0$

From conservation of mass flow rate

$$\Sigma \dot{m}_0 = \Sigma \dot{m}_i$$

We have $\dot{m}_1 = \dot{m}_2 = \dot{m}_a$ ...(A)

and $\dot{m}_3 = \dot{m}_4 = \dot{m}_s$ ...(B)

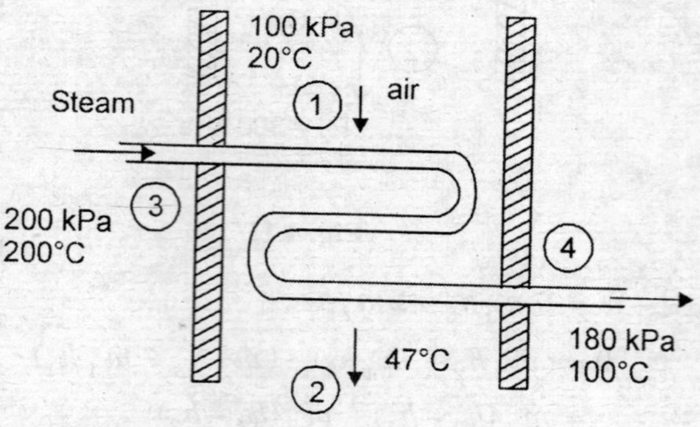

Fig. 4.12

Again from energy conservation equation

$$\dot{Q} - \dot{W} = \Sigma \dot{m}_0 h_0 - \Sigma \dot{m}_i h_i + \Delta KE + \Sigma PE$$

$$0 = (\dot{m}_2 h_2 + \dot{m}_4 h_4) - (\dot{m}_1 h_1 + \dot{m}_3 h_3)$$

In equation (1) and (2), it reduces to

$$0 = \dot{m}_2 (h_2 - h_1) - \dot{m}_3 (h_3 - h_1)$$

or

$$\dot{m}_a = \frac{h_3 - h_4}{h_4 - h_1}\ \dot{m}_s \qquad ...(C)$$

In air

$$h_2 - h_1 = C_p (T_2 - T_1)$$

But $$C_p = \frac{\gamma}{\gamma - 1} R$$

$$= \frac{1.4}{0.4} \times .287 = 1.0045$$

$\therefore$ $$(h_2 - h_1) = C_p (T_2 - T_1)$$

$$= 1.0045\ (47 - 20)$$

$$= 1.0045 \times 27 = 27.122 \text{ kJ/kg}$$

Also from steam tables

At inlet $$\left.\begin{array}{l} P_3 = 200\ kPa \\ T_3 = 200°\ C \end{array}\right] \rightarrow h_3 = 2870.5 \text{ kJ/kg}$$

At exit $$\left.\begin{array}{l} P_4 = 180\ kPa \\ T_4 = 100°\ C \end{array}\right] \rightarrow h_4 = h_f \text{ at } 100°\text{ C} = 419.04 \text{ kJ/kg}$$

Substituting values in equation (C)

$$\dot{m}_a = m_s \frac{(2870.5 - 419.4)}{27.122} = 90.373 \times 8$$

$$= 722.984 \text{ kg/min}$$

$$= \frac{722.984}{60} = 12.0497 \text{ kg/s}$$

Air can be treated as perfect gas

From gas equation we have

$$\frac{P_1 v_1}{T_1} = R$$

$$v_1 = \frac{RT_1}{P_1} = \frac{0.287 \times 293}{100} = 0.8409 \text{ m}^3\text{/kg}$$

Now $$\dot{V}_1 = \dot{m}_a v_1$$

$$= 12.0497 \times 0.8409$$

$$= 10.133 \text{ m/sec}$$

Example 4.30. *In the condenser of a steam power plant water taken from a near by river is circulated through the tubes of the condenser and the steam flows over the tubes. In such a condenser steam enters the unit at 20 kPa pressure and quality of 95% at a flow rate of 2000 kg/h. Steam is to be cooled so as to leave the condenser as saturated liquid at 20 kPa. At the same time the river water is not allowed to experience a temperature rise above 10 °C. Determine the mass flow rate of cooling water required. Take its specific heat = 4.18 kJ/kg °C.*

Solution. From conservation of mass flow rate we have

$$\Sigma \dot{m}_0 = \Sigma \dot{m}_i$$

$$\dot{m}_1 = \dot{m}_2 = \dot{m}_w \quad \text{...(1)}$$

and $$\dot{m}_3 = \dot{m}_4 = \dot{m}_s \quad \text{...(2)}$$

Also from energy conservation equation

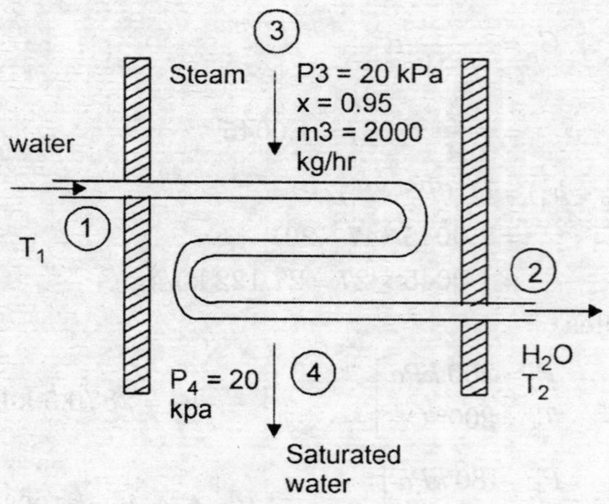

Fig. 4.13

$$\dot{Q} - \dot{W} = \Sigma \dot{m}_0 h_0 - \Sigma \dot{m}_i h_i + \Delta KE + \Delta PE$$

Here $\dot{Q} = 0$

$\dot{W} = 0$ and neglect $\Delta KE + \Delta PE$

We get
$$0 = (\dot{m}_2 h_2 + \dot{m}_4 h_4) - (\dot{m}_1 h_1 + \dot{m}_3 h_3)$$
$$0 = \dot{m}_w (h_2 - h_1) + \dot{m}_s (h_4 - h_3)$$

or
$$\dot{m}_w = \frac{(h_3 - h_4)}{(h_2 - h_1)} \dot{m}_s \qquad \text{...(C)}$$

From steam tables

At inlet

$$\left.\begin{aligned} P_3 &= 20 \text{ kPa} \\ x_3 &= 0.95 \end{aligned}\right] \rightarrow \begin{aligned} h_3 &= h_g + 0.95 h_{fg} \\ &= 251.4 + 0.95 \times 2358.3 \\ &= 2491.8 \text{ kJ/kg} \end{aligned}$$

At exit

$$\left.\begin{aligned} P_4 &= 20 \text{ kPa} \\ \text{Saturated liquid} \end{aligned}\right] \rightarrow \begin{aligned} h_4 &= h_f at 20 \text{ kPa} \\ &= 251.4 \text{ kJ/kg} \end{aligned}$$

Water at atmospheric presume can be treated as compressed liquid at both the temperature Taking its specific heat as 4.18 kJ/kg °C

Therefore
$$\begin{aligned} h_2 - h_1 &= C(\Delta T) \\ &= 4.18 \times 10 \\ &= 41.8 \text{ kJ/kg °C} \end{aligned}$$

Substituting values in equation (C)

$$\dot{m}_w = \left(\frac{249.18 - 251.4}{41.8}\right) \frac{20000}{3600} \text{ kg/sec}$$

$$= 298 \text{ kg/sec}$$

Example 4.31. *In commercial gas turbine power plants air, before entering the combustion chamber, is always preheated by the exhaust gases in a heat exchanger which is called the Regenerator. In such a regenerator air entire at 1 MPa and 277 °C at a mass flow rate of 950 kg / min. Heat is transferred to the air at the rate of 3550 kJ / sec. Exhaust gases enter the regenerator at 150 kPa and 527 °C and leave at 130 kPa and 327 °C. If the exhaust gases are treated as air determine (a) the exit temperation of air (b) the mass flow rate of exhaust gases.*

Assume $C_p = 1.005$

Solution. $\dot{Q} = 3550$ kJ/s

Air can be treated as ideal gas. From steady flow energy equation for air gives

$$\dot{Q} - \dot{W} = \dot{m}_a (h_2 - h_1) + \Delta KE + \Delta PE$$

Here $\dot{Q} = 3550$ kJ/sec

$$\dot{m}_a = \frac{950}{60} \text{ kg/sec} = 15.833 \text{ kg/sec}$$

$$T_1 = 273 + 277 = 550$$

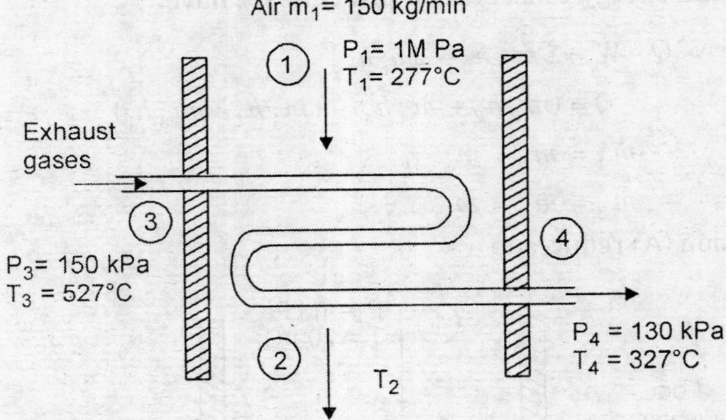

Fig. 4.14

$$3550 - 0 = \frac{950}{60} (T_2 - T_1) C_p + 0 + 0$$

$$3550 = \frac{950}{60} \times 1.005 (T_2 - 550)$$

$$T_2 = 550 + \frac{3550 \times 60}{1.005 \times 950}$$

$$= 550 + 223.09$$

$$= 773.09 \text{ K}$$

or $$= (773.09 - 273) = 500.09 \text{ °C}$$

(*b*) Again treating exhaust gases as ideal air to applying the energy conservation equation to exhaust gases we have and taking

$$C_p = 1.005$$

$$\dot{Q} - \dot{W} = \dot{m}_{eg} (h_4 - h_3) + \Delta KE + \Delta PE$$

$$-3550 - 0 = \dot{m}_{eg} (T_4 - T_3) C_p$$

Here $\dot{Q}$ is negative as from these gases heat flows out

$$\therefore \quad 3550 = \dot{m}_{eg} (327 - 527) \times 1.005$$

or

$$\dot{m}_{eg} = \frac{3550}{200 \times 1.005} = 17.66 \text{ kg/s}$$

Example 4.32. In sugar industry, steam power plants the feed water is mostly heated by using steam which is extracted from steam turbine at some stage (known as the process of bleeding). In such a feed water heater steam enters at 1 MPa and 200 °C and leaves as saturated liquid at the same pressure. The feed water enters the heater at 2.5 MPa and 50°C and leaves at 10° below the exit temperature of the steam. Find the ratio of mass flow rates of the extracted steam and feed water.

Solution. Neglecting ΔKE and ΔPE *In addition* $\dot{W}$ and $\dot{Q}$ are also zero.

Therefore from the energy conservation principle we have

$$\dot{Q} - \dot{W} = \Sigma \dot{m}_0 h_0 - \Sigma m_i h_i$$

$$0 = (\dot{m}_2 h_2 + \dot{m}_4 h_4) - (\dot{m}_1 h_1 + \dot{m}_3 h_3) \quad \text{...(A)}$$

Here

$$\dot{m}_1 = \dot{m}_2 = \dot{m}_s$$

and

$$\dot{m}_3 = \dot{m}_4 = \dot{m}_w$$

Therefore equation (A) reduces to

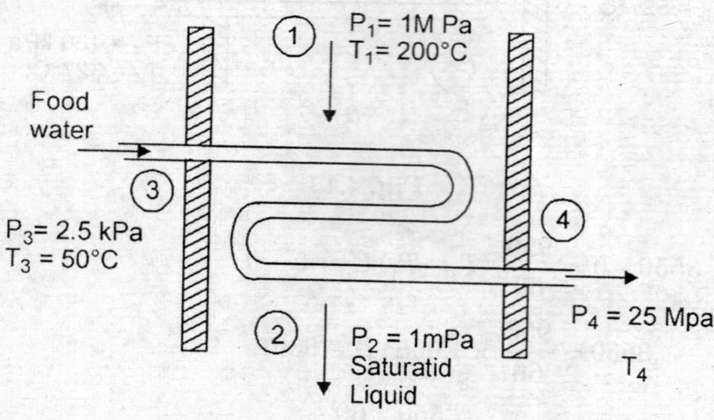

Fig. 4.15

$$0 = \dot{m}_s (h_2 - h_1) + \dot{m}_w (h_2 - h_3)$$

or

$$\frac{\dot{m}_s}{\dot{m}_w} = \left(\frac{h_3 - h_4}{h_2 - h_1}\right) \quad \text{...(2)}$$

From Steam tables

For Steam at

Inlet $\left.\begin{matrix} P_1 = 1\ MPa \\ T_1 = 200°\ C \end{matrix}\right] \rightarrow h_1 = 2827.9\ kJ/kg$

Outlet At P_2 = 1 MPa Saturated Temperature

$h_2 = h_f$ at 1 MPa = 762.81 kJ/kg

And Saturated temperature = T_2 = 179.9 °C

From compressed water tables

At entry

$$\left.\begin{matrix} P_3 = 2.5\ MPa \\ T_3 = 50°\ C \end{matrix}\right] \rightarrow h_3 = h_f \text{ at } 50°\ C\ kJ/kg$$

At exit

$$\left.\begin{matrix} P_4 = 205\ MPa \\ T_4 = T_2 - 10 \\ = 179.91 - 10 \\ = 169.91°\ C \end{matrix}\right] \rightarrow h_3 = h_f \text{ at } 50°\ C\ kJ/kg$$

Therefore substituting these value in equation (2)

$$\frac{\dot{m}_s}{\dot{m}_w} = \frac{h_3 - h_4}{h_2 - h_1} = \frac{209.33 - 71901}{762.81 - 2927.9}$$

$$= \frac{-509.77}{-2065.09} = 0.2468$$

PROBLEMS OF PIPE AND DUCT FLOW

Example 4.33. *A conference room 8 m × 6 m × 5 m is to be heated by a resistance type electric heater placed in short duct in the room. On some cold day the room temperature is 15 °C and atmospheric persons 98 kPa. A 200 W fan is used to circulate the air steadily through the duct and the electric heater at a mass flow rate of 50 kg/min. The duct can be assumed to be adiabatic. If it takes 15 minutes to raise the temperature of the room by 10 °C when the heat lost from the room to the surroundings is at a steady rate of 300 kJ / min, find (a) the power rating of the electric heater (b) the temperature rise which the air experiences each time it passes through the heater. Assume air as ideal gas and for air Take*

$$R = 0.287 \text{ kJ/kg} - K$$

and $\gamma = 1.4.$

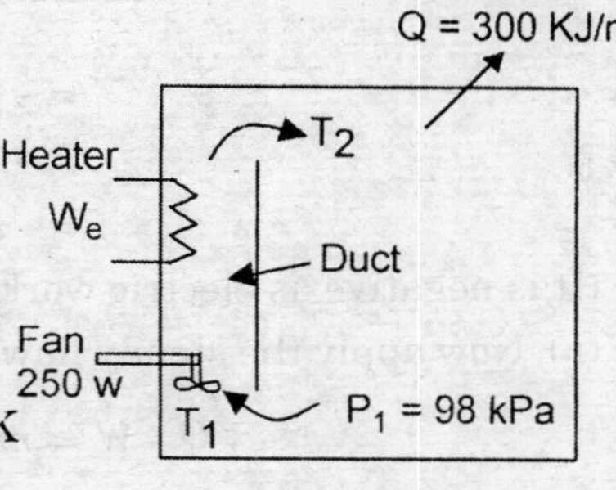

Fig. 4.16

Solution. Taking air an ideal gas we have

$$C_p = \left(\frac{\gamma}{\gamma - 1}\right) R$$

$$= \frac{1.4}{0.4} \times 0.287 = 1.0045 \text{ kJ/kg-K}$$

Then $C_V = C_p - R$

$= 1.0045 - 0.287$

$= 0.71875$ kJ/kg-K

$$\text{Volume}, V \text{ of air} = 8 \times 6 \times 5\ 250 \text{ m}^3$$

$$\rho_1 = 98 \text{ kPa}$$

and $T = 273 + 15 = 288\ K$

Now $\dfrac{P_1 V}{RT} = m$

i.e. $m = \dfrac{(98 \text{ kPa})(240\ m^3)}{0.287 \text{ (kJ/kg–K)} \times 288 \text{ K}}$

$= 284.55 \text{ kg}$

(a) Taking entire room as the system and applying the conservation of energy principle to the closed system having constant volume of 240 m^3.

$$Q - W_e - W_{\text{fan}} - W_b = \Delta U + \Delta KE + \Delta PE \qquad ...(1)$$

Here $\dot{Q} = \Delta t \times Q$

$= \Delta t \left(\dfrac{-300}{60} \text{ kJ/s} = -5 \text{ kJ/s} \right)$

$W_e = \dot{W}_e \times \Delta t$

$W_{\text{fan}} = \dot{W}_{\text{fan}} \times \Delta t$

$= - \dfrac{-200}{1000} \text{ kW} = 0.2 \text{ kJ/s}$

W_b = Boundary work = 0

Also $\Delta KE = 0$

and $\Delta PE = 0$

Take $\Delta t = 1$ sec

Then equation (1) reduces to

$$(\dot{Q} - \dot{W}_e - \dot{W}_{\text{fan}}) = \frac{m\, C_v (T_2 - T_1)}{\Delta t} \text{ kJ/sec}$$

or $\dot{W}_e = (-5 \text{ kJ/s}) - (-0.2 \text{ kJ/s}) - (284.5 \text{ kg}) \times \dfrac{0.7175}{(15 \times 60)} \text{ (kJ/kg-°C) (10)}$

$= -5 + .2 - 2.2685$

$= -7.0685$ kJ/sec

$= -7.0685$ kW

(It is negative as electric work is done on air)

(b) Now apply the steady flow energy equation to the duct

$$\dot{Q} - \dot{W} = \dot{m}(h_2 - h_1)$$

Here $\dot{Q} = 0$ as it is adiabatic

$\dot{W} = -\dot{W}_{fan} - \dot{W}_e = -.2 - 7.2860$ kJ/s

Therefore $0 - (-7.2685) = \dot{m} C_p (\Delta T)$

where $\dot{m}$ = mass flow rate of due to fan = $\frac{50}{60}$ kJ/s

ΔT = Temperature rise in the pass through duct

Therefore $$7.2685 = \Delta T \left(\frac{50}{60} \text{ kg/s}\right) (1.0045 \text{ kJ/ks})$$

or $$\Delta T = \frac{7.2685 \times 60}{50 \times 1.0045}$$

$$= 8.683\ °C$$

Example 4.34. *A 400 m³ volume room is to be heated with the help of a resistance type 30 kW heater placed in a duct. Initially the air in the room is at 95 kPa atmospheric pressure and it is intended to raise its temperature by 10 °C by forcing the air through the duct with the help of a 250 W fan so as to give a steady flow of air through the duct and raises the temperature of air by 5 °C every time it flows through the duct. The duct may be assumed adiabatic. If air is treated as an ideal gas with*

$$R = 0.287 \text{ kJ/kg} - K$$

and $$\gamma = 1.4$$

If heat loss to surroundings is 450 kJ/m then find (a) how long will it take to raise the temperature of the room by 10 °C ? (b) Mass flow rate of air through the duct.

Solution. Treating air as ideal gas

$$C_p = \left(\frac{\gamma}{\gamma - 1}\right) R$$

$$= \frac{1.4}{0.4} \times .287$$

$$= 1.0045 \text{ kJ/kg °C}$$

$$C_V = C_p - R$$

$$= 1.0045 - 0.287$$

$$= 0.7\,175 \text{ kJ/kg °C}$$

Fig. 4.17

Volume, $V = 400 \text{ m}^3$

and $T = 15 + 273 = 288$

$$\therefore \quad m = \frac{P_1 V}{RT} = \frac{95 \times 400}{0.287 \times 2.88}$$

$$= 459.74 \text{ kg}$$

Take room as closed system and applying energy conservation equation

$$Q - W_e - W_{\text{fan}} - W_b = \Delta U + \Delta KE + \Delta PE$$

Here $$W_b = 0$$

$$\Delta KE = 0$$

$$\Delta PE = 0$$

or $$\Delta t\,[\dot{Q} - \dot{W}_e - \dot{W}_{fan}] = \Delta U$$

$$= m\,C_V\,(\Delta T) \qquad ...(A)$$

Now $$\dot{Q} = -\frac{450}{60}\text{ kJ/s} = -7.5\text{ kJ/s}$$

$$\dot{W}_e = -30\text{ kJ/s} = -30\text{ kJ/s}$$

$$\dot{W}_{fan} = \frac{280}{1000}\text{ kJ/s} = -0.25\text{ kJ/s}$$

Therefore from A

$$\Delta t\,[-7.5 - (-30) - (-25)] = (459.74)(0.7175))(10)$$

$$= (459.74)(0.7175)(10)$$

$\therefore$ $$\Delta t\,(30.25 - 7.5) = 3556.9$$

$\therefore$ $$\Delta t = \frac{3556.9}{22.75} = 156.31$$

(*b*) **Applying steady flow energy equation to adiabatic duct if in the average mass flow rate through duct.**

$$\dot{Q} - \dot{W} = \dot{m}\,(h_2 - h_1)$$

$$= \dot{m}\,C_p\,(T_2 - T_1)$$

But $$\dot{Q} = 0$$

$$\dot{W} = \dot{W}_e + \dot{W}_{fan}$$

$$= (-30) + \left(-\frac{250}{1000}\right)$$

$$= -30.25\text{ kJ/s}$$

$\therefore$ $\therefore$ From eqn. (B) $0 - (30.25) = \dot{m} \times 1.0045 \times 5$

$\therefore$ $$\dot{m} = \frac{30.25}{1.0045 \times 5} = 6.02\text{ kg/s}$$

Example 4.35. *In a hair dryer a small fan pulls air and forces it to flow over electric resistance wire where the air is heated. Air enters the hair dryer at 100 kPa and 200 °C and leaves at 45 °C. The rating of the resistors is 1000 W and the cross-section of the outlet of the hair dryer is 0.0025 m² and the fan is of 25 W rating determine the volume flow rate of air at inlet and velocity of air at outlet. Assume air to be an ideal gas and*

$$R = 0.287\ kJ/kg-K$$

and $$\gamma = 1.4$$

and this is no loss of heat through the walls of the hair dryer.

Solution. $$C_p = \left(\frac{\gamma}{\gamma - 1}\right) R$$

$$= \left(\frac{1.4}{1.4 - 1}\right) R$$

$$= 1.0045\text{ kJ/kg °C}$$

From the energy conservation equation for the steady flow we have

$$\dot{Q} - \dot{W}_e - \dot{W}_{fan} = \dot{m}_1 (h_2 - h_1) + \Delta KE + \Delta PE \quad \text{...(A)}$$

Here $\dot{Q} = 0$

and $\Delta KE = 0 \quad \Delta PE = 0$

and $$\dot{W}_e = -\frac{1000}{1000} = -1 \text{ kW} = -1 \text{ kJ/s}$$

or $$\dot{W}_{fan} = -\frac{40}{1000} = -0.04 \text{ kJ/s}$$

Therefore from equation (A) we get

$$0 - (-1) - (-0.04) = \dot{m} C_p (T_2 - T_1)$$

or $$1.04 = \dot{m} \times 1.0045 (45 - 20)$$

$$= \dot{m}_1 \times 1.0045 \times 25$$

or $$\dot{m}_1 = \frac{1.04}{1.0045 \times 25}$$

$$= 0.0414 \text{ kg/s}$$

For volume flow rate we have from gas equation

$$v_1 = \frac{RT_1}{P_1} = \frac{0.287 \times (273 + 20)}{100}$$

$$= 0.8409 \text{ m}^3\text{/kg}$$

Then $$\dot{V}_1 = \dot{m} v_1 = (0.0414 \text{ kg/s}) (0.8409 \text{ m}^3\text{/kg})$$

$$= 0.0348 \text{ m}^3\text{/s}$$

(*b*) From the mass conservation equation

$$\dot{m}_1 = \dot{m}_2 = \frac{A_2 V_2}{v_2} \quad \text{...(B)}$$

Now $$v_2 = \frac{RT_2}{P_2} = \frac{0.287 \times (273 + 45)}{100}$$

$$= \frac{0.287 \times 318}{100} = 0.91266 \text{ m}^3\text{/kg}$$

Therefore from equation (B)

$$0.0414 = \frac{A_2 V_2}{0.91266}$$

or $$V_2 = \frac{0.0414 \times 0.91266}{(0.0025)}$$

$$= 15.11365 \text{ m/s}$$

Example 4.36. *Air enters the duct of an air conditioning plant at 100 kPa and 15 °C at a flow rate of 160 m^3/hr. The area of cross-section of the duct is 0.00032 m^2 and there is transfer of heat to the air at the rate of 2 kJ/s from the surroundings. If air is assumed to be an ideal gas having*

$$R = .287\ kJ/kg\text{-}K$$

and $$C_p = 1.005$$

determine the velocity of air at the inlet of the duct and temperature of air at the outlet of duct.

Solution. At entry point

Let inlet velocity be V_1

Then $$V_1 = \frac{\dot{V}_{vol}}{A_1}$$

where $\dot{V}$vol = Volume flow rate

$$= 160 \times \frac{1}{3600}\ \text{m}^3/\text{s} = 4.4444 \times 10^{-2}\ \text{m}^{-2}$$

$$\therefore \quad = \frac{4.4444 \times 10^{-2}}{0.00032}$$

$$= 0.13889\ m/s$$

Given $$T_1 = 15 + 273 = 288\ K$$

$$P_1 = 100\ \text{kPa}$$

Also $$v_1 = \frac{RT_1}{P_1} = \frac{0.287 \times 288}{100}$$

$$= 0.82656\ \text{m}^3/\text{kg}$$

Now $$\dot{m}_1 = \frac{\dot{V}_{vol}}{v_1} = (4.444 \times 10^{-2}\ m^3/s) \times \frac{1}{v_1}\ (\text{m}^3/\text{kg})$$

$$= \frac{4.4444 \times 10^{-2} m^3 \backslash s}{0.82656\ \text{m}^3/\text{kg}}$$

$$= 5.3769 \times 10\ \text{kg/s}$$

(*b*) At exit $$\dot{m}_2 = \dot{m}_1$$

Therefore from the steady flow energy equation

$$\dot{Q} - \dot{W} = \dot{m}\ (h_2 - h_1) + \Delta KE + \Delta PE$$

Here $$\dot{Q} = 2\ \text{kJ/s}$$

$$\dot{W} = 0$$

Therefore $$2 - 0 = 5.3769 \times 10^{-2} \times C_p\ (T_2 - T_1)$$

$$= 5.3769 \times 1.005\ (T_2 - T_1)$$

or $$T_2 = T_1 + \frac{2}{5.3769 \times 10^{-2} \times 1.005}$$

$$= 15 + \frac{2}{0.3226 \times 1.005}$$

$$= 15 + 37.01$$

$$= 52.01°C$$

Example 4.37. *A domestic water heater consists of an insulated pipe of 25 mm diameter; Cold water enters the pipe at 20 °C and leaves at 45 °C while flowing steadily at the rate 20 L/min. Determine the power rating of the resistance type heater which is fitted inside the pipe and the average velocity of water in pipe. Take density of water as 1000 kg / m3 and specific heat*

C = 4.184 kJ/kg °C.

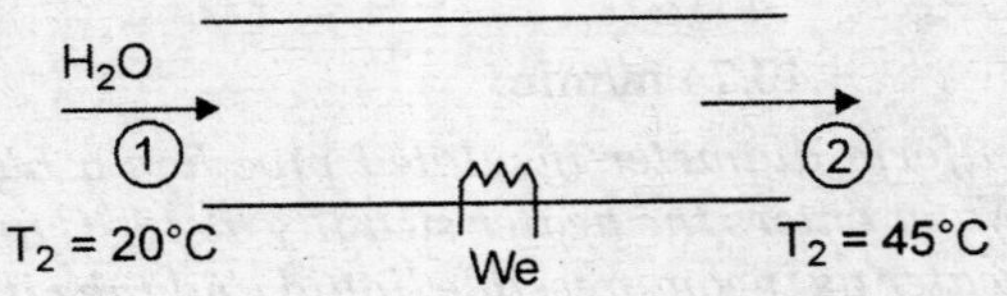

H$_2$O
①
②
T_2 = 20°C
We
T_2 = 45°C

Fig. 4.18

T_1 = 20 °C

T2 = 45 °C

Solution. Water is assumed to be incompressible fluid. Volume flow rate of water

$$\dot{V}_{vol} = 20 \text{ L/min}$$

$$= \frac{20}{60} = \frac{1}{3} \text{ L/s}$$

$$= \frac{1}{3} \times .001 \text{ m}^3\text{/s}$$

Therefore $$\dot{m} = \rho\ \dot{V}_1 = (1000 \text{ kg/m}^3) \times \left(\frac{.001}{3} \text{ m}^3\text{/s}\right)$$

$$= \frac{1000 \times 0.001}{3} \text{ kg/s}$$

$$= \frac{1}{3} = 0.3333 \text{ kg/s}$$

Then from the steady flow energy equation we have

$$\dot{Q} - \dot{W}_e = \dot{m}\ (h_2 - h_1) + \Delta KE + \Delta PE$$

Since $$\dot{Q} = 0$$

$$\Delta KE = 0$$

and $$\Delta PE = 0$$

$$-\ \dot{W}_e = \left(\frac{1}{3} \text{ kg/s}\right) C\ (T_2 - T_1)$$

$$= \frac{1}{3} \times 4.184\,(45 - 20)$$

$$= 34.8667 \text{ kJ/s}$$

or $$\dot{W}_e = -34.6867 \text{ kW}$$

(*b*) If average velocity of water is V_1

Then $$V_1 = \frac{\text{Volume flow rate of water}}{\text{(Area of cross-section) of pipe}}$$

$$= \frac{\left(\frac{1}{3} \times .001 \text{ m}^3/\text{s}\right)}{\frac{\pi}{4} \times (.025)^2 \; m^2} = 0.679 \text{ m/sec}$$

$$= 40.74 \text{ m/min.}$$

Example 4.38. *A uniform diameter insulated pipe has a built-in electric resistance type heater of 15 kW rating. Water enters the heater steadily at 15 °C and leaves at 88 °C. Determine the mass flow rate treat water as incompressible liquid and take its specific heat*

$$C = 4.184 \; kJ/kg \; °C$$

Solution. Neglect ΔKE and ΔPE

$$T_1 = 15 \text{ °C}$$

$$T_2 = 88 \text{ °C}$$

Applying the steady flow energy equation

$$\dot{Q} - \dot{W}_e = \dot{m}\,(h_2 - h_1) + \Delta KE + \Delta PE$$

Here $$\dot{Q} = 0$$

$$\Delta KE = 0,$$

$$\Delta PE = 0$$

and $$\dot{W}_e = -15 \text{ kJ/s}$$

Therefore from equation (1)

$$0 - (-15) = \dot{m}\,C\,(88 - 15)$$

or $$15 = \dot{m} \times 4.184\,(73)$$

or $$\dot{m} = \frac{15}{4.184 \times 73} = 0.0491 \text{ kg/s}$$

Example 4.39. *Steam flows through a long horizontal un-insulated pipe 110 mm diameter at entry with a velocity of 5 m/s. At entry point the pressure and temperature are respectively 1 MPa and 300 °C. At exit its diameter is 90 mm and its pressure is 800 kPa and temperature 250 °C. Find the mass flow rate of the steam and the rate of heat loss to the surroundings.*

Solution. From steam tables

At entry velocity = 5 m/s

and $$D_1 = 110 \text{ mm}$$

$$D_2 = 90 \text{ mm}$$

or $$A_1 = \frac{\pi}{4} \times (110)^2 \text{ mm}^2$$

$$= \frac{\pi}{4}(0.11)^2 \text{ m}^2$$

From steam tables

At $\left.\begin{matrix} P_1 = 1\ MPa \\ T_1 = 300°\ C \end{matrix}\right] \rightarrow \begin{matrix} v_1 = 0.2579\ m^3/kg \\ h_1 = 3051.2\ kJ/kg \end{matrix}$ And $A_2 = \frac{\pi}{4}(0.09)^2 = 0.0063\ m^2$

At exit point $\left.\begin{matrix} P_2 = 800 \text{ MPa} \\ T_2 = 250°C \end{matrix}\right] \rightarrow \begin{matrix} h_2 = 2950.2 \text{ kJ/kg} \\ v_2 = 0.2931 \text{ m}^2\text{/kg} \end{matrix}$

Therefore $$\dot{m}_1 = \frac{A_1 \text{ (Velocity)}}{v_1}$$

$$= \frac{\pi}{4} \times \frac{(0.11)^2}{0.2579} \times 5 = 0.1842 \text{ kg/s}$$

Now for conservation of mass we have

$$\dot{m}_2 = \dot{m}_1$$

or $$\frac{A_2 V_2}{v_2} = 0.1842$$

or $$V_2 = \frac{(0.1842)(0.2931)}{0.006362} = 8.4862 \text{ m/s}$$

(*b*) Now applying the steady flow energy equation we have

$$\dot{Q} - \dot{W} = \dot{m}\ (h_2 - h_1) + \Delta\ KE + \Delta\ PE$$

Here $$\dot{W} = 0 \text{ and } \Delta KE = \frac{\dot{m}}{2}[v_2^2 - v_1^2] = \frac{\dot{m}}{2}(8.4862^2 - 5^2)$$

and neglecting $\Delta\ PE$ we get

$$\dot{Q} = \dot{m}\ (2950 - 3051.2) + \dot{m}(23.5078)$$

$$= 0.1842\,(-77.6922)$$

$$= -14.3109 \text{ kJ/s}$$

(Negative sign indicates heat loss)

PROBLEMS ON PUMPS

Example 4.40. *Water level in a wall is 20 m below the ground level and water is to be pumped steadily at a point 25 m above ground level at flow rate of 2 m³/min. Neglecting KE changes frictional losses and heat transfer, calculate the power input to the pump.*

Take density $= 1000\ kg/m^3$.

Solution. $$\dot{m} = \rho \dot{V}$$

$$= (1000 \text{ kg/m}^3)(2 \text{ m}^3\text{/min})$$

$$= 2000 \text{ kg/min}$$

From steady flow energy equation

$$\dot{Q} - \dot{W} = \dot{m}(C\Delta T + v\ \Delta P + \Delta\ KE + \Delta\ PE)$$

Here $\dot{Q} = 0$

$\Delta T = 0$

$\Delta P = 0$

$\Delta KE = 0$

Therefore $\dot{W} = \Delta PE$

$= \dot{m} g (Z_2 - Z_1)$

Here $Z_2 = 25$ m

and $Z_1 = -20$

$$-\dot{W} = (2000 \text{ kg/min}) g [25 - (-20)]$$

$$= \left(\frac{2000}{60} \text{ kg/s}\right)\left(\frac{9.81 \text{ m}}{s^2}\right)(45 \text{ m}) \times \left[\frac{1 \text{ kJ/kg}}{1000 \text{ m}^2/\text{s}^2}\right]$$

$$= \frac{200}{60} \times \frac{9.81 \times 45}{1000}$$

i.e. $\dot{W} = -14.715$ kW

PROBLEMS OF UNSTEADY FLOW PROCESSES

Example 4.41. *An insulated rigid tank in evacuated. The value is then opened to allow the atmospheric air at 100 kPa and 25 °C enter the tank until the pressure inside tank become 100 kPa. Then the value is closed calculate the temperature inside the tank assuming constant C_p and C_V. Treat air as an ideal gas*

$R = 0.287$ kJ/kg – K

and $\gamma = 1.4.$

Fig. 4.19

Solution. $C_p = \left(\frac{\gamma}{\gamma - 1}\right) R$

$$= \left(\frac{1.4}{1.4 - 1}\right) 0.287 = 1.0045$$

$$C_V = C_p - R = 1.0045 - 0.287 = 0.7175$$

This is unsteady flow process and can be analysed as uniform flow process. Then from conservation of mass and energy equations as applied to control volume which is the volume of Tank.

For mass conservation

$$\Sigma m_i - \Sigma m_e = (m_2 - m_1)_{CV} \quad \text{...(1)}$$

Since $\Sigma m_e = 0$

and $m_1 = 0$

as it was evacuated ∴ $m_1 = m_2$ ...(2)

For energy conservation for C_V

$$Q - W = (\Sigma m_e h_e - \Sigma m_i h_i) + (m_2 u_2 - m_1 u_1)_{c.v} \quad \text{...(3)}$$

Now $Q = 0$ as tank is insulated

$W = 0$

$m_e = 0$

and $m_1 = 0$

Therefore equation (3) reduces to

$$0 = 0 - m_i h_1 + m_2 u_2$$

or $m_i h_1 = m_2 u_2$

But from equation (2) $m_i = m_2$

Therefore $u_2 = h_1$

or $C_V T_2 = C_p T_1$

$$T_2 = T_1 \left(\frac{C_p}{C_V}\right) = \gamma T_1$$

$$= 1.4 \times (273 + 25)$$

$$= 1.4 \times 298 = 417 \text{ K}$$

$$= 417 - 273 = 144 \text{ °C}$$

Example 4.42. *A 2.25 m³ capacity rigid tank initially contains air 100 kPa and 22 °C. It is connected through a value to the supply line in which air is flowing at 600 kPa and 22 °C. By opening the value the air is allowed to entire this tanks until the pressure reaches 600 kPa when the value is closed. It is found that the temperature of air inside tank is 77 °C. Calculate the mass of air which has entered the tank and this amount of heat transfer. Treat air as local gas and take*

$$R = 0.287 \text{ kJ/kg–K}$$

and $\gamma = 1.4.$

Use air tables

Solution.

$$C_p = \left(\frac{\gamma}{\gamma - 1}\right) R$$

$$= \left(\frac{1.4}{1.4 - 1}\right) \times .287 = 1.0045 \text{ kJ/kg}$$

Then

$$C_V = \frac{C_p}{\gamma} = \frac{1.0045}{1.4} = 0.7175 \text{ kJ/kg °C}$$

Let m_1 be the mass of air in tank at initial conditions

i.e. $P_1 = 100$ kPa

$V_1 = 2.25$ m³

$R = 0.287$ kJ/kg.K

$T_1 = 72 \text{ °C} = 273 + 22 = 295$ K

$$m_1 = \frac{P_1 V_1}{R T_1} = \frac{100 \times 2.25}{0.287 \times 295} = 2.675 \text{ kg}$$

If m_2 is mass at the end of filling

$P_2 = 600$ kPa

$V = 2.25$ m³

$$T_2 = 273 + 77 = 350 \text{ K}$$

i.e.
$$m_2 = \frac{P_2 V}{RT_2} = \frac{600 \times 2.25}{0.287 \times 350}$$
$$= 13.4395 \text{ kg}$$

Therefor for conservation of mass equation for the control volume (= Tank Vol.)

$$\Sigma m_i - \Sigma m_e = (m_2 - m_1)_{CV}$$
$$m_i = m_2 - m_1$$
$$= 134.395 - 2.6575$$
$$m_i = m_2 - m_1 = 10.782 \text{ kg}$$

(*b*) Heat transfer is found by applying the uniform flow energy equation to control volume

$$Q - W = (\Sigma m_e h_e - \Sigma m_i h_i) + (m_2 u_2 - m_1 u_1)_{CV} + \Delta KE + \Delta PE$$

Here
$$W = 0$$
$$\Delta KE = 0$$
$$\Delta PE = 0$$

Now
$$m_e = 0$$

Therefore equation reduces to

$$Q - W = 0 - \Sigma m_i h_i + (m_2 u_2 - m_1 u_1) C_V$$

or
$$Q = - m_i h_i + m_2 u_2 - m_1 u_1$$

From air tables we have

At
$$T_1 = 295 \text{ K}$$
$$h_1 = 295.17 \text{ kJ/kg}$$
$$u_1 = 210.49 \text{ kJ/kg}$$

At
$$T_2 = 350 \text{ K}$$
$$u_2 = 250.02 \text{ kJ/kg}$$

This substituting these values in energy equation we get

$$Q = -10.782\,(295.17) + 13.4395\,(750.02) - 2.6575\,(210.49)$$
$$= -3187.52 + 3774.02 - 559.1377$$
$$= -367.877 \text{ kJ}$$

(*i.e.* heat lost to surroundings)

Example 4.43. *A rigid tank of capacity 0.2 m³ contains initially refrigerant 12 at 8 °C. At this state 40% of mass is in liquid phase and 60% in vapour phase. It is connected through a value to a supply line in which R-12 is flowing steadily at 1.2 MPa and 100 °C. Refrigerant is allowed to flow in the tank by opening the value. When the pressure in the tank reaches 800 kPa the entire R-12 in the tank is in vapour phase. Then the value is closed. Determine the temperature and mass of R-12 which entered in tank and heat transfer between R-12 and surrounding.*

Solution. This is unsteady flow process. Taking this as uniform flow process and the C.V. as the volume of tank.

From R-12 tables

At

$$\left.\begin{array}{l} T_1 = 8°C \\ x_1 = 0.6 \end{array}\right] \quad \begin{aligned} v_1 &= v_f + x_1 v_{fg} \\ &= 0.0007297 + 0.6[0.0434 - 0.0007297] \\ &= 0.02633 m^3/kg \\ u_1 &= u_f + x_1 u_{fg} = 43.21 + 0.6[173.63 - 43.21] \\ &= 121.46 kJ/kg \end{aligned}$$

At $P_2 = 800$ kPa Saturated vapour] $v_2 = v_g$ at 800 kPa = 0.02188 m^3/kg

$u_2 = u_g$ at 800 kPa = 183.13 kJ/kg

At $P_i = 1.2$ MPa $\quad h_i = 245.70$ kJ/kg

$T_i = 100$ °C

The tank contains m_1 kg at the initial stage

$$T_1 = 8 °C$$

and

$$x_1 = 0.6$$

i.e.

$$m_1 = \frac{V}{v_1} = \frac{0.2 m^3}{(0.02633 m^3/kg)} = 7.6 \text{ kg}$$

(*a*) It contains at the final stage say m_2 kg (at 800 kPa and saturated vapour)

$$T_2 = T_{sat} \text{ at } 800 \text{ kPa}$$

Therefore

$$T_2 = T_{sat} \text{ at } 800 \text{ kPa} = 32.34 °C$$

and

$$m_2 = \frac{V}{v_i} = \frac{0.2 m^3}{(0.02188 m^3/kg)} = 9.14 \text{ kg}$$

(*b*) Then from conservation of mass equation

$$\Sigma m_i - \Sigma m_e = (m_2 - m_1)_{CV}$$

Now

$$\Sigma m_e = (m_2 - m_1)_{CV}$$

Now

$$\Sigma m_e = 0$$

$\therefore$

$$m_i = m_2 - m_1 = 9.14 - 7.6 = 1.54 \text{ kg}$$

(*c*) For heat transfer we have from the uniform flow energy equation

$$Q - W = \Sigma m_e - h_e \Sigma m_i h_i + (m_2 - u_2 - m_1 u_1)_{CV}$$

Since

$$W = 0$$

$$m_e = 0$$

$\therefore$

$$Q = -\Sigma m_i h_i + (m_2 u_2 - m_1 u_1)_{CV}$$

or

$$Q = -m_i h_i + (m_2 u_2 - m_1 u_1)_{CV}$$

$$= -1.54 \times 245.70 + 9.14 (183.13) \; 7.6 (121.46)$$

$$= 372.3 \text{ kJ}$$

Example 4.44. *A 0.2 m^3 capacity rigid tank initially contains saturated water vapour at 110 °C. The tank is connected through a value to supply line which carries steam at 1 MPa and 250 °C. Steam is allowed to enter the tank by opening the value. During this process heat is lost to the surroundings such that the temperature in the tank remains at 110 °C when half of the tank volume is occupied by liquid water the value is closed. Determine the pressure in the tank after the value is closed (b) the amount of steam that has entered the tank (c) heat transfer to the surroundings.*

Solution. The given unsteady flow can be analysed as a uniform flow process.

From the steam tables

At $T_1 = 110°C$ Saturated vapour] $v_1 = v_g$ at 110°C = 1.2102 m³/kg

$u_1 = u_g$ at 110°C = 2518.1kJ/kg

$v_f = 0.001052$, $v_g = 1.2102$ m³/kg

$T_2 = 110°C$] $u_f = u_g$ at 110°C, $u_g = 2518.1$ kJ/kg

$P_i = 1$ MPa

$T_i = 250$ °C $h_i = 2942.6$ kJ/kg

(*a*) The tank contains saturated mixture at the first state at 110 °C. Therefore the pressure will be equal to saturation pressure at 110 °C. For steam tables at 110 °C.

$$P_2 = \text{Saturated pressure at } 110\ °C = 0.14327 \text{ MPa}$$

(*b*) Initial and final masses m_1 and m_2

$$m_1 = \frac{V}{v_1} = \frac{0.2 \text{ m}^3}{1.2102 \text{ m}^3/\text{kg}} = 0.16526 \text{ kg}$$

$$m_2 = m_f + m_g = \frac{V_f}{v_f} + \frac{V_g}{v_g} \text{ (at 110 °C)}$$

$$= \frac{0.1 \text{m}^3}{0.001052 \text{ m}^3/\text{kg}} + \frac{0.1 \text{m}^3}{1.2102 \text{ m}^3/\text{kg}}$$

$$= 95.07 + 0.0526 = 95.1996 \text{ kg}$$

Then from conservation of mass equation for *C.V*

$$\Sigma m_i - \Sigma m_e = (m_2 - m_1)_{CV}$$

Since $m_e = 0$

$\therefore$ $m_i = m_2 - m_1$

$$= 95.1396 - 0.16526$$

$$= 94.974 \text{ kg}$$

(*c*) For heat transfer we have from uniform flow energy equation

$$Q - W = (\Sigma m_e\, h_e - m_i\, h_i) + (m_2\, u_2 - m_1\, u_1)_{CV}$$

But $W = 0$

and $m_e = 0$

$\therefore$ $Q = -\Sigma m_i\, h_i + (m_2\, u_2 - m_1\, u_1)_{CV}$

$$= -m_1\, h_i + m_2\, u_2 - m_1\, u_1$$

Now $m_2\, u_2 = m_{2_f}\, u_f + m_{2_g}\, u_g$ at 110 °C

$$= \left(\frac{0.1}{0.001052}\right) \times 461.14 + \frac{0.1}{1.2102} \times 2518.1$$

$$= 43834.6 + 208.06$$

$$= 44042.66$$

and $m_1 h_1 = 194.9747\,(2942.6) = -279470.49$

and $m_1 u_1 = 0.16526 \times 2518.1$

$= 416.14$

Therefore $Q = -279470.49 + 44042.66 - 416.14$

$= -235843.97$ kJ

Example 4.45. *A balloon contains 70 m³ helium at atmospheric condition of 100 kPa pressure and temperature 22 °C. Through a value it is connected to a large tank of helium where the pressure is 150 kPa and temperature 25 °C. Helium is allowed to flow into the balloon by opening the value till the pressure in the balloon also becomes 150 kPa. Then the value is closed. Assuming that the material of the balloon is such that its volume increases linerarily with pressure find the final temperature of helium in the balloon if no heat transfer takes place during the process*

Take $R = 2.0769\ kJ/kg\text{–}K$

$C_V = 3.1156\ kJ/kg\text{–}K$

and $C_p = 5.1925\ kJ/kg\text{-}K$

Solution. This is an unsteady flow process. Assuming this to be a uniforms flow process and treating helium as an ideal gas. At initial state

$$T_1 = 273 + 22 = 295$$

$$m_1 = \frac{P_1 V_1}{RT_1} = \frac{100 \times 70}{2.0769 \times 295}$$

$$= 116.221 \text{ kg}$$

At the final state let volume = V_2 since volume is linearly proportional to pressure

Hence $$\frac{P_2}{P_1} = \frac{V_2}{V_1}$$

or $$V_2 = V_1 \frac{P_2}{P_1}$$

$$= 70 \times \frac{150}{100}$$

$$= 105 \text{ m}^3$$

Therefore $$m_2 = \frac{P_2 V_2}{RT_2} = \frac{150 \times 105}{2.0769 \times T_2}$$

$$= \frac{7583.42}{T_2} \text{ kg}$$

From the conservation of mass equation

$$\Sigma m_i - \Sigma m_0 = (m_2 - m_1)_{CV}$$

But $$\Sigma m_0 = 0$$

Therefore $$\Sigma m_i = (m_2 - m_1)_{CV}$$

$$m_i = \frac{7583.42}{T_2} - 116.221$$

Again from the uniform flow energy equation

$$Q - W = (\Sigma m_0 h_0 - \Sigma m_i h_i) + (m_2 u_2 - m_1 u_1)_{CV} \quad ...(1)$$

Now $Q = 0$

$W = W_b$

and $m_0 = 0$

Then $0 - W_b = 0 - m_i C_p T_2 - m_2 C_V T_2 - m_1 C_V T_1 \quad ...(2)$

Now $W_b = \text{(Average Pressure)} (V_2 - V_1)$

$$= \left(\frac{P_1 + P_2}{2}\right)(105 - 70)$$

$$= \left(\frac{100 + 150}{2}\right)(35)$$

$$= 125 \times 35 = 4375.00 \text{ kJ}$$

Therefore substituting values in equation (2) we get

$$-4375 = -\left(\frac{7583.42}{T_2} - 116.221\right) 5.1925 \times T_2$$

$$= \frac{7583.42}{T_2}(3.1156\, T_2)$$

$$= 116.221 \times 3.1156 \times 295$$

or $$4375 = \frac{7583.42}{T_2} \times 5.1925 \times T_2 - 116.227 \times 5.1925\, T_2$$

$$+ \frac{7583.42}{T_2} \times 3.1156\, T_2 + 116.22 \times 3.1156 \times 295$$

$$= 7583.42\,(5.1925 + 3.1156) - 116.225 \times 5.1925\, T_2 + 106818.95351$$

$$= 63003.8117 - 116.221 \times 5.1925\, T_2 + 106818.95317$$

$$= 169822.7645 - 603.4775\, T_2$$

Therefore $$T_2 = \frac{169822.7645 - 4375}{603.4775}$$

$$= \frac{165447.76}{603.4775} = 274.157 \text{ K}$$

$$= 1.16\,°\text{C}$$

Example 4.46. *An insulated piston-cylinder device contains 5 kg of water 80% of which is in vapour state. The mass of the piston is such that the pressure inside the cylinder remains constant at 300 kPa. The cylinder is connected through a value to a steam supply line carrying steam at 1 MPa and 400 °C. The value is opened and the steam from the supply line enters the cylinder. The value is closed when the liquid in the cylinder has just evaporated. Find (a) the final temperature in the cylinder (b) the mass of the addilional steam that has entered the cylinder.*

Solution. This is an unsteady flow process and can be analysed as uniform flow process. At final stage the cylinder contains saturated steam at 300 kPa

Therefore at 300 kPa from steam tables

$$T_2 = T_{sat} \text{ at } 300 \text{ kPa} = 133.6\,°\text{C}$$

(*b*) From the mass conservation equation for the control volume

$$\Sigma m_i - \Sigma m_e = (m_2 - m_1)_{CV}$$

Since $\Sigma m_0 = 0$

$\therefore$ $$m_i = m_2 - m_1 \quad ...(1)$$

Again from the energy conservation equation

$$Q - W = (\Sigma m_e\, h_e - \Sigma m_i\, h_i) + (m_2\, u_2 - m_1\, u_1)_{CV} \quad ...(2)$$

Now $Q = 0$

and $W = W_b$

and $m_e = 0$

Therefore substituting (1) in equation (2)

$$-W_b = -m_i\, h_i + (m_2\, u_2 - m_1\, u_1)$$

$$= -(m_2 - m_1)\, h_1 + (m_2\, u_2 - m_1\, u_1)$$

or $$0 = [W_b + m_2\,(h_2\ u_1) - (m_2 - m_1)\, h_1 \quad ...(3)$$

But $$W_b + m_2\,(u_2 - u_1) = H_2 - H_1 = m_2\, h_2 - m_1\, h_1 \quad ...(4)$$

$\therefore$ From (3), and (4)

$$0 = (m_2\, h_2 - m_1\, h_1) - m_2\, h_i + m_1\, h_i$$

$$m_2\,(h_i - h_2) = m_1\,(h_i - h_1)$$

$$m_2 = \left(\frac{h_i - h_1}{h_i - h_2}\right) m_1 \quad ...(5)$$

Again from steam tables

At $\left.\begin{matrix} P_1 = 300 kPa \\ x_1 = 0.8 \end{matrix}\right] \rightarrow \begin{aligned} h_1 &= h_f + 0.8\, h_{fg} \\ &= 561.47 + 0.8 \times 2163.8 \\ &= 2295.51\ kJ/kg \end{aligned}$

At $P_2 = 300 kPa$ and Saturated vapour$] \rightarrow \begin{aligned} h_2 &= h_g (300 kPa \\ &= 2725.3\ kJ/kg \end{aligned}$

At $\left.\begin{matrix} P_i = 1\ MPa \\ T_1 = 400°\ C \end{matrix}\right] \rightarrow h_i = 3263.9\ kJ/kg$

Therefore substituting values in equation (5), we have

$$m_2 = \left(\frac{h_i - h_1}{h_i - h_2}\right) m_1$$

$$= \left(\frac{3263.5 - 2292.51}{3263.9 - 2725.3}\right) \times 5$$

$$= 9.02 \text{ kg}$$

Then $$m_i = m_2 - m_1$$

$$= 9.02 - 5 = 4.02 \text{ kg}$$

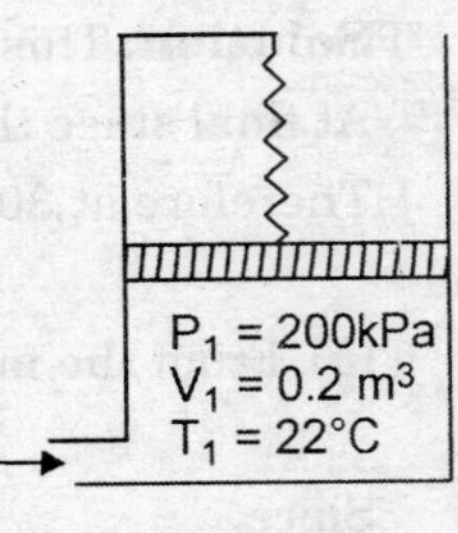

Fig. 4.20

Example 4.47. *An insulated piston-cylinder device initially contains 0.2 m^3 of air at 200 kPa pressure and temperature equal to 22 °C. At this state a linear spring just touches the piston and no force is exerted on the piston (Fig. 4.20). The cylinder is connected through a value to an air supply line in which the air pressure is 800 kPa and temperature 22 °C.*

Now the value is opened and the air is allowed to enter the cylinder till the pressure reaches 600 kPa and then it is closed. It is observed that the volume of air in the cylinder doubles during this process. Find the mass of additional air that has entered the cylinder and the final temperature inside the cylinder. Assume air an ideal gas and take C_P = 1.005 kJ / kg -K, C_v = 0.718 kJ / kg-K

and $\quad R = 0.287$ *kJ/kg.K*

$C_p = 1.005$

and $\quad C_V = 0.718$ *kJ/kg.K*

Solution. At initial state

$$P_1 = 200 \text{ kPa}$$

$$T_1 = 273 + 22 = 295$$

$$m_1 = \frac{P_1 V_1}{RT_1} = \frac{200 \times 0.2}{(0.287)(295)} = 0.472 \text{ kg}$$

Similarly at final state

$$P_2 = 600 \text{ kPa}$$

$$V_2 = 0.4 \text{ m}^3$$

Therefore $\quad m_2 = \dfrac{P_2 V_2}{RT_2} = \dfrac{600 \times 0.4}{(0.287)\, T_2}$

$$= \frac{836.2}{T_2} \qquad \text{...(1)}$$

Therefore from conservation of mass equation we have

$$\Sigma m_i - \Sigma m_e = (m_2 - m_1)_{CV}$$

But $\quad \Sigma m_e = 0$

Therefore $\quad m_i = (m_2 - m_1) \qquad \text{...(2)}$

$$= \frac{836.2}{T_2} - 0.412$$

Since the spring is linear, pressure variation will also be linear and the P – V diagram will be as shown in Fig. (4.21).

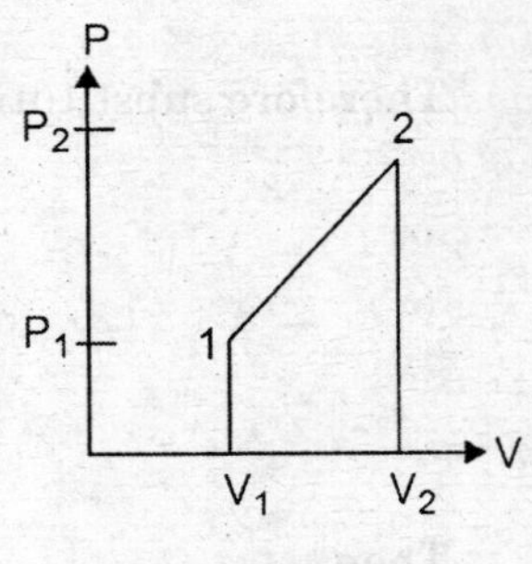

Fig. 4.21

Then $\quad W_b$ = Work done

= Area under the curve

$$= \left(\frac{P_1 + P_2}{2}\right)(V_2 - V_1)$$

$$= \left(\frac{200 + 600}{2}\right)(0.4 - 0.2) = 80 \text{ kJ}$$

Then further conservation of energy for uniform flow, equation we have

$$Q - W = (\Sigma m_e h_e - \Sigma m_i h_i) + (m_2 u_2 - m_1 u_1)_{CV}$$

Here $Q = 0$

and $m_e = 0$ ...(3)

Therefore equation (3), reduces to

$$0 - 80 = - m_i h_1 + (m_2 u_2 - m_1 u_1)$$

or $$80 = - (m_2 - m_1) h_1 + (m_2 u_2 - m_1 u_1)$$

$$= - (m_2 - m_1) C_p T_1 + m_2 C_V T_2 - m_1 C_V T_1$$

or $$- 80 = - \left[\frac{836.2}{T_2} - 0.472\right] (1.005)(295) + \left[\frac{836.2}{T_2}\right] (0.718) T_2$$

$$- (0.472)(.718)(295)$$

$$- 80 = - \frac{247912.4}{T_2} + 139.936 + 600.39 - 99.9743$$

or $$- 80 = - \frac{247912.4}{T_2} + 640.3517$$

or $$T_2 = \frac{247912.4}{720.3517}$$

$$= 344.15 \text{ K}$$

Therefore $$m_2 = \frac{836.2}{T_2} = \frac{836.2}{344.15}$$

$$= 2.43 \text{ kg}$$

Then $$m_i = m_2 - m_1$$

$$= 2.43 - 0.472$$

$$= 1.958 \text{ kg}$$

DISCHARGING PROCESS

Example 4.48. *A rigid tank of capacity 0.4 m³ is filled with saturated liquid water at 200 °C. Through a valume fitted at the bottom of the tank liquid is withdrawn from the tank when simultaneously heat is being transferred to the tank so that the temperature inside remains constant When half of the total mass of water present in the tank is withdrawn how much heat has been transferred to the tank.*

Solution. This is an unsteady flow process and can be analysed as uniform flow process.

For steam tables

At the initial state

$T_1 = 200$ °C Saturated liquid $\quad v_1 = v_f$ at 200 °C = 0.001157 m³/kg

$u_1 = u_f$ at 200 °C = 850 kJ/kg

At the end state $\quad T_e = 200$ °C Saturated liquid $\quad h_e = h_f$ at 200 °C = 852.45 kJ/kg

At initial state

$$m_1 = \frac{V}{v_1} = \frac{0.4\ \text{m}^3}{0.001157\ \text{m}^3/\text{kg}} = 345.77\ \text{kg}$$

Therefore $$m_2 = \frac{1}{2}\ m_1 = \frac{345.72}{2} = 172.86\ \text{kg}$$

From the conservation of mass equation we have

$$\Sigma m_i - \Sigma m_e = (m_2 - m_1)_{CV}$$

But $$\Sigma m_i = 0$$

Therefore $$0 - m_e = (m_2 - m_1)$$

$$= (172.86 - 345.72)$$

or $$m_e = 172.86\ \text{kg}$$

At final state

Therefore $$v_2 = \frac{V}{m_2} = \frac{0.4}{172.86} = 0.002314\ \text{m}^3/\text{kg}$$

Again for steam tables we have at saturated temperature = 200 °C

$$v_f = 0.00157$$

$$v_g = 0.001157$$

Therefore at Saturated temperature of 200 °C $$v_{fg} = v_g - v_f = 0.12736 - 0.001157$$

$$= 0.1762$$

But $$v_2 = 0.002314$$

Therefore $$x_2 = \frac{v_2 - v_f}{v_{fg}} = \_ = .00115$$

$$= 0.00917$$

Thus at state 2 from steam tables

At $$\left.\begin{array}{l} T_2 = 200°\ C \\ x_2 = 0.00917 \end{array}\right] \rightarrow \begin{array}{l} u_2 = u_f + x_3\ u_{fg} \\ = 850.65 + 0.00917 \times 1744.7 \\ = 866.65\ \text{kJ/kg} \end{array}$$

Again for conservation of energy equation for uniform flow

$$Q - W = (\Sigma m_e\ h_e - \Sigma m_i\ h_i) + (m_2\ u_2 - m_1\ u_1)_{CV}$$

Here $$W = 0$$

$$\Sigma m_i = 0$$

$$\therefore \quad Q = (m_e\ h_e - 0) + (\text{m}_2\ u_2 - m_1\ u_1)_{CV}$$

or $$Q = m_e\ h_e + m_1\ u_1 - m_1\ u_1$$

Therefore substituting values in it.

$$Q = (172.86)(852.45) + (72.86)(866.65) - (345.86)(850.65)$$

$$= 2958\ \text{kJ}$$

Example 4.49. *A domestic pressure cooker of a 5-L capacity has an operating pressure of 200 kPa. Initially 80% volume is occupied by vapour and 20% by liquid. It is placed on a cooking range which supplies heat to water at a rate 400 W. Determine the time to be taken for be liquid to disappear so that in the cooker is left only saturated vapour.*

Solution. Working pressure = 200 kPa constant

At initial state

$$P_1 = 200 \text{ kPa}$$
$$V_f = 20\% \text{ of } 5\text{L}$$
$$= 1\text{L} = 0.001 \text{ m}^3$$

and
$$V_g = 80\% \text{ of } 5\text{L}$$
$$= 4\text{L} = 0.004 \text{ m}^3 \quad ...(1)$$

Now from steam tables

At $P_1 = 200$ kPa

$v_f = 0.0010605 \text{ m}^3$

$V_g = 0.8851$ m3/kg

$u_{f1} = 504.49$ kJ/kg

$u_{g1} = 2529.5$ kJ/kg

Therefore
$$m_1 = m_{f1} + m_{g1}$$
$$= \frac{V_{f_1}}{v_{f_1}} + \frac{V_{g_1}}{v_{g_1}} = \frac{0.001 \text{ m}^3}{0.0010605} + \frac{0.004 \text{ m}^3}{0.885 \text{ m}^3}$$
$$= 0.9430 + 0.0045 = 0.9475 \text{ kg} \quad ...(3)$$

Also
$$U_1 = m_1 u_1 = m_{f_1} u_{f_1} + m_{g_1} u_{g_1}$$
$$= 0.9430 \times 504.49 + 0.0045 \times 2529.5 \quad ...(4)$$
$$= 487.12 \text{ kJ}$$

Now the given process is an unsteady flow process and can be dealt with an uniform flow process and taking the conservation of mass and energy equations

$$\Sigma m_i - \Sigma m_e = (m_2 - m_1)_{CV}$$

Now
$$\Sigma m_i = 0$$

As there is outflow of saturated steam only

$\therefore$
$$-m_e = m_2 - m_1$$

or
$$m_e = m_1 - m_2 \quad ...(5)$$
$$= 0.9475 - m_2$$

At final stage there is only saturated steam at 200 kPa is left.

For steam tables

And at 200 kPa Saturated vapour

$$v_2 = v_g \text{ at } 200 \text{ kPa} = 0.8857 \text{ m}^2/\text{kg}$$
$$u_2 = v_g \text{ at } 200 \text{ kPa}$$
$$= 2529.5 \text{ kJ/kg}$$

Therefore
$$m_2 = \frac{V}{v_2} = \frac{0.005 \text{ m}^3}{0.8857 \text{ m}^3/\text{kg}}$$

$$= +\frac{0.005 \text{ m}^3}{0.8857 \text{ m}^3/\text{kg}} = 0.0057 \text{ kg}$$

Then mass conservation equation gives

$$m_e = m_1 - m_2 = 0.9475 - 0.0057 = 0.9418$$

Thus substituting values in energy equation

$$Q - W = (\Sigma m_e h_e - \Sigma m_i h_i) + (m_2 u_2 - m_1 u_1)_{CV} \quad ...(6)$$

Here $W = 0$

and $m_i = 0$

$$Q - 0 = \Sigma (m_e h_e) - 0 + (m_2 u_2 - m_1 u_1)$$

$$= m_e h_e + m_2 u_2 - m_1 u_1$$

or $(0.4 \text{ kJ/s}) \Sigma\Delta T = m_e h_e + m_2 u_2 - m_1 u_1 \quad [\because Q = 0.4 \, kW = 0.4 \, kJ/s] \quad ...(7)$

Now m_e = final mass of Saturated vapour left in cooker at 200 kPa

At 200 kPa Saturated vapour

$$u_2 = 2529.5 \text{ kJ}$$

and h_e = 200 kPa = 2706.7 kJ/kg

Therefore inserting these values in equation (2) we get

$$(0.4 \text{ kJ/sec}) \Delta T = (0.9418)(2706.7) + (0.0056 \text{ kg})(2529.5 \text{ kJ/kg}) - 487.12 \text{ kJ}$$

$$= 2529.17 + 14.165 - 487.12$$

$$= 2076.215$$

$$\therefore \quad \Delta T = \frac{2076.215}{0.4}$$

$$= 5190.5 \text{ sec}$$

$$= 1.44 \text{ hrs}$$

Example 4.50. *A pressure cooker of 4 litre capacity works at a pressure of 175 kPa. Initially it contains liquid water having volume of 2 litre and remaining in gaseous state. It is required to determine the heat transfer rate to water inside pressure cooker so that the lost drop of water to evaporated after one hour*

Solution. Initially volume of liquid water = 2 litre

$$= 2 \times 0.001 \text{ m}^3$$

Volume of saturated steam at 175 kPa

$$= 2 \times .001 = 0.002 \text{ m}^3$$

From steam tables

At P_1 = 175 kPa Saturated vapour] → $u_f = 486.8$ kJ/kg, $v_f = 0.001057 \text{ m}^3/\text{kg}$, $v_g = 1.0036 \text{ m}^3/\text{kg}$, $u_g = 2524.9$ kJ/kg

Therefore $m_1 = m_f + m_g$

$$= \frac{0.002 \text{ m}^3}{(0.001057 \text{ m}^3/\text{kg})} + \frac{0.002 \text{ m}^3}{1.0036 \text{ m}^3/\text{kg}}$$

$$= 1.892 + 0.001993$$

$$= 1.893993 \text{ kg}$$

Also $U_1 = m_1 u_1 = m_f u_f + m_g u_g \quad ...(1)$

$$= 1.892 \times 486.8 + 0.001993 \times 2524.9$$

$$= 921.0256 + 5.032 = 926.0576 \quad ...(2)$$

At the final state i.e after one hour

P_2 = 175 kPa Saturated vapour]

$h_e = h_{ge}$ @175 kPa saturated = 2700.6 kJ/k

$v_2 = v_g = 1.0036\ m^3/kg$

$u_2 = u_g$ saturated vapour at 175 kPa

= 2524.9 kJ/kg

Therefore $m_2 = \dfrac{V}{v_2} = \dfrac{0.004}{1.0036} = 0.003986$ kg

Then we have from conservation of mass equation

$$\Sigma m_i - \Sigma m_e = (m_2 - m_1)_{CV}$$

But $m_i = 0$

Therefore $-m_e = (m_2 - m_1)$

or $m_e = m_1 - m_2$

$= 1.893993 - 0.003986$

$= 1.89$ kg ...(4)

Now from the uniform flow conservation of energy equation

$$Q - W = (\Sigma m_e h_e - \Sigma m_i h_i) + (m_2 u_2 - m_1 u_1)_{CV}$$

Here $W = 0$

$m_i = 0$

Substituting values from (1), (2), (3) and (4)

Therefore $Q = m_e h_e + m_2 u_2 - m_1 u_1$

$= (1.89)(2700.6) + (0.003986)(2524.9) - 926.0576$

$= 4187.0576$

$$\therefore \quad \dot{Q} = \frac{Q}{\Delta T} = \frac{4187.0576}{3600} \text{ (kJ/sec)} = 1.163 \text{ kW}$$

Example 4.51. *A piston cylinder device is kept vertical and the mass of the piston is such that in it the air pressure remains constant at 300 kPa and initially the volume of air is 0.2 m³ and temperature is 27 °C. A value which is connected to the device is then opened so that the air from the cylinder escapes to the atmosphere till the volume of air in the cylinder is reduced to half. Side by side heat transfer takes place during the process so that the temperature inside remains constant. Determine the mass of air left inside the cylinder and the amount of heat transfer. Assume air to be an ideal gas Take*

$$R = 0.287\ kJ/kg\text{-}K$$

and $C_p = 1.0045\ kJ/kg\text{-}K$

and $C_V = 0.7175\ kJ/kg\text{-}K$

Solution Initial state

$T_1 = 273 + 27 = 300$ K

$P_1 = 300$ kPa

$V_1 = 0.2$ m^3

Then $$m_1 = \frac{P_1 V_1}{RT_1} = \frac{(300 \text{ kPa}) \times 0.2 \text{ m}^3}{(300 \text{ K})(0.28 \text{ kPa m}^3/\text{kg- K})}$$

$= 0.69686$ kg

At final state

$$P_1 = 300 \text{ kPa}$$
$$T_2 = 300 \text{ K}$$
$$V_2 = 0.1 \text{ m}^3$$

$$\therefore \quad m_2 = \frac{P_2 V_2}{RT_2} = \frac{(300 \text{ kPa})\times(0.1\text{m}^3)}{(0.287 \text{ kPa m}^3/\text{kg-K})(300\text{K})} = 0.34843 \text{ kg}$$

The process is an unsteady flow process and can be assumed as uniform flow. Then from the conservation of mass equation we have

$$\Sigma m_i - \Sigma m_e = (m_2 - m_1)_{CV} \quad \text{...(1 } a\text{)}$$

But here $m_i = 0$

Therefore from (1) we get

$$-m_e = m_2 - m_1 \quad \text{...(1}b\text{)}$$

or

$$m_e = m_1 - m_2$$
$$= 0.69686 - 0.34843$$
$$= 0.34843 \text{ kg}$$

Mass of air left inside

$$m_2 = 0.34843 \text{ kg}$$

(*b*) Again from the energy conservation equation

$$Q - W = (\Sigma m_e h_e - \Sigma m_i h_i) + (m_2 u_2 - m_1 u_1) \quad \text{...(2)}$$

Here $m_i = 0$

and as the pressure of air during the process remains constant this W which is boundary work W_b can be combined with Δu terms to give enthalpy Δh terms

Thus from equation (2)

$$Q = m_e h_e + W_b + m_2 u_2 - m_1 u_1$$
$$= m_e h_e + m_2 u_2 - m_1 u_1 \quad \text{...(3)}$$

Thereafter the temperature also remains constant during the process

then $h_1 = h_2 = h \text{ (say)} = h_i$

Therefore equation (2) reduces to

$$Q = m_e h + m_2 h - m_1 h$$
$$= h (m_e + m_2 - m_1)$$

But from equation (1*b*)

$$m_e = m_1 - m_2$$

Therefore $Q = h [(m_1 - m_2) + (m_2 - m_1)] = 0$

Example 4.52. *A balloon initially contains 10 m3 helium at 110 kPa pressure and 27 °C temperature. The value is opened so as to allow helium to escape till the pressure drops to 100 kPa at which point the value is closed. During the process the volume of balloon decreases by 20%. Assuming the balloon material is such that its volume changes linearly with pressure and there is no heat transfer during the process find the mass of helium that escaped and the final temperature of helium in the balloon. Treat helium as an ideal gas and take*

$$C_p = 5.1926 \text{ kJ/kg-K}$$

and

$$C_V = 3.1156 \text{ kJ/kg-K}$$

Solution.

$$R = C_p - C_V$$
$$= 5.1926 - 3.1156$$
$$= 2.0770 \text{ kJ/kg - } K$$

At initial state

$$P_1 = 110 \text{ kPa}$$
$$T_1 = 273 + 27 = 300 \text{ K}$$
$$V_1 = 10 \text{ m}^3$$

Therefore

$$m_1 = \frac{P_1 V_1}{RT_1} = \frac{(110 \text{ kPa}) \times (10 \text{ m}^3)}{(2.077 \text{ kPa m}^3)/\text{kg- K})(300 \text{ K})}$$
$$= 1.76537 \text{ kg}$$

At final state

$$V_2 = 0.8\, V_1$$

or

$$V_2 = 0.8 \times 10 = 8 \text{ m}^3$$
$$P_2 = 100 \text{ kPa}$$

Therefore

$$m_2 = \frac{P_2 V_2}{RT_2} = \frac{(100 \text{ kPa}) \times (8 \text{ m}^3)}{(2.077 \text{ kPa m}^3/\text{kg- K})\, T_2}$$

$$= \frac{385.171}{T_2} \text{ kg} \qquad \text{...(1)}$$

From the conservation of mass equation

$$\Sigma m_i - \Sigma m_e = (m_2 - m_1)_{CV} \qquad \text{...(2)}$$

Here

$$m_i = 0$$

Therefore

$$-m_e = m_2 - m_1$$

or

$$m_e = m_1 - m_2$$

From the conservation of energy relationship for the uniform flow we have

$$Q - W = (\Sigma m_e\, h_e - \Sigma m_i\, h_i) + (m_2\, u_2 - m_1\, u_1)_{CV}$$
$$Q = 0 \text{ and}$$

Here

$$W = W_b = \frac{P_1 + P_2}{2}\,(V_2 - V_1)(\because V \propto P)$$

or

$$W_b = \left(\frac{110 + 100}{2}\right)(V_2 - V_1)$$
$$= (105)\,(0.8\, V_1 - V_2)$$
$$= (105)\,(8 - 10)$$
$$= -210 \text{ kJ}$$

Therefore substituting values in equation (3)

$$0 - (-210) = m_e\, h_e + (m_2\, u_2 - m_1\, u_1)$$

or

$$210 = (m_1 - m_2)\, h_e + m_2\, C_V\, T_2 - m_1 C_V T_1$$

$$= \left(1.76537 - \frac{385.171}{T_2}\right)\left(\frac{T_1 + T_2}{2}\right) C_P + C_V\,(m_2\, T_2 - m_1\, T_1) \qquad \text{...(3)}$$

Sulestituting the values of C_p, m_1, m_2 and T_1 in equation (3), it reduces to

$$T_2^2 = 216.4\, T_2 - 21822.8 = 0 \qquad \text{...(4)}$$

The solution of equation (4) gives

$$= 291.3\,°\text{C}$$

$$\therefore \qquad m_2 = \frac{385.17}{291.3} = 1.32 kg$$

Then $\qquad m_e = m_1 - m_2 = 1.76537 - 1.32 = 0.44537\ kg$

Example 4.52. *An empty household gas cylinder is to be filled with gas from a gas supply line where the enthalpy of the gas is 620 kJ / kg ;When filled with gas the cylinder contains 15 kg of gas at 10^3 kPa. The enthalpy of gas at 10^3 kPa is 750 kJ/kg and its specific volume is 0.0488 m³/ kg. Determine the heat interaction of the cylinder with surroundings during the gas filling process.*

Solution. This is an unsteady flow process and can be analysed as uniform flow process.

For the conservation of mass and energy equations we have

For mass

$$\Sigma m_i - \Sigma m_e = (m_2 - m_1)_{CV}$$

Here $\qquad m_e = 0$

and $\qquad m_1 = 0$

Therefore $\qquad m_i = m_2 \qquad \text{...(1)}$

For Energy

$$Q - W = \Sigma m_e\, h_e - \Sigma m_i\, h_i + (m_2\, u_2 - m_1\, u_1)_{CV}$$

Here $\qquad Q = 0$

$$W = 0$$

$$m_e = 0$$

and $\qquad m_1 = 0$

Therefore $\qquad Q = -\, m_i\, h_i + (m_2\, u_2 - 0 \times u_1)_{CV}$

$$R = -\, m_i\, h_i + (m_2\, u_2)_{CV}$$

or $\qquad Q + m_i\, h_i = m_2\, u_2 \qquad \text{...(2)}$

But from equation (1)

$$m_i = m_2$$

Thus equation (2) reduces to

$$Q + m_2\, h_i = m_2\, u_2$$

or $\qquad Q = m_2\, u_2 - m_2\, h_i$

$$= m_2\, [u_2 - h_i]$$

Here $\qquad m_2 = 15 \text{ kg}$

$$h_i = h_1 = 620 \text{ kJ/kg}$$

$$h_2 = 750 \text{ kJ/kg}$$

But $\qquad u_2 = h_2 - p_2 v_2$

$$= 750 - (10^3) \times 0.0487$$

$$= 750 - 48.7 = 701.3 \text{ kJ/kg}$$

Therefore $Q = 15\,[u_2 - h_1]$
$= 15\,[701.3 - 620]$
$= 1219.5$ kJ/kg

Since it is positive, heat is received by the cylinder from its surroundings.

4.5 SUMMARY

In this chapter conservation of mass and conservation of energy principles for control volumes were discussed. When mass enters or leaves a control volume the energy content will also change as the mass carries energy with it.

(*i*) Thus the mass flow rate through a cross-section per unit time is called the mass flow rate

$$\dot{m} = \rho\, V_{av}\, A \quad ...(4.38)$$

where ρ = density kg/m³ $\left(= \frac{1}{v}\right)$

v = specific volume or volume/kg

V_{av} = average velocity of fluid normal to area of cross-sectional are, m/s

A = area of cross-section m²

(*ii*) Volume flow rate is the fluid volume flowing through the cross-section per unit time

$$\dot{V} = \int V_n\, dA$$
$$= V_{av}\, A \text{ m}^3/\text{s}$$

(*iii*) Then from equation (4.38) and (4.39(we have

$$\dot{m} = \rho \frac{\dot{V}}{v} = \frac{\dot{V}}{v} \quad ...(4.40)$$

(*iv*) During a steady flow process through a control volume, the fluid experiences no change with time at a fixed position and the mass and the energy content of the control volume remains constant. The mass and energy conservation equation are as follows

$$\Sigma m_i = \Sigma m_e \quad ...(4.41)$$

$$\dot{Q} - \dot{W} = \dot{m}_e \left(h_e + \frac{V_e^2}{2} + g\, z_e \right) - \Sigma \dot{m}_i \left(h_i + \frac{V_i^2}{2} + g\, z_i \right) \quad ...(4.42)$$

where subscript i stands for inlet and e for exit.

In case there is only one inlet and one outlet these subscripts are changed 1 and 2 respectively that is

$$\dot{m}_1 = \dot{m}_2 \quad ...(4.43)$$

or

$$\frac{V_1\, A_1}{v_1} = \frac{V_2\, A_2}{v_2} \quad ...(4.44)$$

Similarly equation (4.42) reduces to

$$\dot{Q} - \dot{W} = \dot{m} \left[(h_2 - h_1) + \frac{V_2^2 - V_1^2}{2} + g\,(z_2 - z_1) \right] \text{kW} \quad ...(4.45\,a)$$

or

$$q - w = (h_2 - h_1) + \frac{V_2^2 - V_1^2}{2} + g\,(z_2 - z_1) \text{ (kJ/kg)} \quad ...(4.45\,b)$$

or $$q - w = \Delta h + \Delta\, KE + \Delta\, PE \text{ kJ/kg} \quad ...(4.45\,c)$$

where $$q = \frac{\dot{Q}}{\dot{m}} \text{ heat transfer per unit mass kJ/kg}$$

and $$w = \frac{\dot{W}}{\dot{m}} \text{ work down per unit mass kJ/kg}$$

(*v*) Unsteady flow processes are very difficult to analyse. However, these are approximated to uniform flow processes such that during uniform flow process the state of the control volume may change with time but it does so uniformly and the fluid properties at inlets and exits are assumed to remain constant during the entire process.

The controlling equations their reduce to

$$Q - W = \Sigma m_e \left(h_e + \frac{V_e^2}{2} + g\, z_e \right) - \Sigma m_i \left(h_i + \frac{V_i^2}{2} + g\, z_i \right) + (m_2\, u_2 - m_1\, u_1)_{CV} \text{ kJ} \quad ...(4.46)$$

when $\Delta\, KE$ and $\Delta\, PE$ are negligibility small it as reduces to

$$Q - W = (\Sigma m_e\, h_e - m_i\, h_i) + (m_2\, u_2 - m_1\, u_1)_{CV} \text{ kJ} \quad ...(4.47)$$

REVIEW QUESTIONS

1. Explain the difference between "control volume" and "closed system".
2. Explain the steady flow process. Stating the assumptions made, write down the energy conservation equation for the same.
3. Explain the principle of conservation of mass for a control volume.
4. Differentiate between mass flow rate and volume flow rate.
5. Write down conservation of energy equation for steady flow when a single steam enters a control volume and a single steam leaves the control volume.
6. Explain the concept of flow work.
7. Explain the difference in the terms "*pdv*" and "*vdp*".
8. Explain "Throttling valves". What function do they perform ?

   How do they affect the flow process of a fluid passing through them ?

   How do you justify them to be called 'Isenthalpic Devices'.
9. Write down the energy equation for (a) "filling a tank (b) emptying a tank.

PROBLEMS

1. Atmospheric air at pressure 100 kPa and 17 °C is compressed to pressure to 250 kPa pressure with the help of a compressor The temperature of air at the outlet of compressor is 127 °C. Assuming the steady flow rate of air through compressor is 5 m^3/s find the power rating of the compressor. Take the heat loss during the process as 15 kJ/s. Take

   $$R = 0.287 \text{ kJ/kg-K}$$

   and $$\gamma = 1.4.$$

   Neglect changes is kinetic and potential energies. **(Ans.** 679.07 kW)

2. Determine the amount of heat required per unit mass to raise the temperature of air from 17 °C to 97 °C when heated (a) at constant pressur (b) at constant volume. Assume air to be an ideal gas

Take $\gamma = 1.4$

$R = 0.287$ kJ/kg-K (**Ans.** (a) 80.36 kJ/kg (b) 57.4 kJ/kg)

3. A rigid cylinder contains 5 kg of oxygen at 17 °. Heat is supplied to the oxygen until the pressure inside the cylinder rises to twice that of the initial value. Calculate the final temperature, change in internal energy, change in enthalpy and the amount of heat interaction across the boundary between the oxygen and the surroundings. Treat oxygen as an ideal gas. Take

$R = 0.2589$ kJ/kg-K

and $C_V = 0.65$. (**Ans.** 307 °C, 942.5 kJ, 1319.2 kJ, 942.5 kJ)

4. A piston cylinder assembly contains 1.5 kg of Nitrogen at 600 kPa pressure and 27 °C temperature. The mass of the piston is such that the pressure inside the cylinder remains constant. Heat is now supplied to Nitrogen such that Nitrogen expands in the cylinder so that the final volume is three times the original volume. Calculate (a) the final temperature (b) the work done by Nitrogen (c) heat added and (d) changes in internal energy Take

R = 0.295 kJ/kg-K

and $C_p = 1.05$ kJ/kg-K

(**Ans.** (a) 627 °C (b) 265.2 kJ (c) 945 kJ (d) 679.8 kJ (increase))

5. A rigid cylinder contains 5 kg of oxygen at 17 °C. The cylinder is supplied heat until the final pressure becomes twice that of original pressure. Determine the final temperature, change in internal energy and enthalpy and the heat interaction across the boundary. Assume oxygen as an ideal gas. Take

$R = 259.85$ kJ/kg

and $C_V = 0.65$ kJ/kg

(**Ans.** 307 °C, $dU = 942.5$ kJ, $dH = 1319.2$ kJ, $Q = 942.5$ kJ)

6. In an insulated piston cylinder assembly containing CO_2, the piston diameter is 250 mm and is loaded with a 1200 N weight. The piston is 300 mm above the cylinder bottom to start with and its temperature is 290 K. Heat is then supplied to CO_2 with 1 watt electric heater for one hour. Find the distance through which the piston will rise after one hour. Assume that CO_2 behaves like an ideal gas Take

$R = 189$ kJ/kg-K

$C_p = 820$ kJ/kg-K

and atmospheric pressure as 10^5 N/m^2. (**Ans.** 0.728 m)

7. A vertical piston cylinder assembly contains a combustible mixture at 27 °C and weight of piston is such that the pressure of the mixture is 305×10^5 N/m^2. The mixture is now ignited and as the piston moves up slowly there is a heat loss to the surroundings When the reaction is complete and the contents have come to original temperature it is observed that the piston has moved up by 80 mm and the heat lost to surroundings 4 kJ. If the diameter of the piston is 180 mm, calculate the work done by the system and the change in its internal energy. (**Ans.** 0.7125 kJ, 4.7125 kJ is decrease)

□□□

CHAPTER 5

SECOND LAW OF THERMODYNAMICS

5.1 INTRODUCTION

In the previous chapters the First Law of Thermodynamics and its application to processes involving closed and open systems have been discussed. The first law is nothing but the conservation of energy principle that is energy is neither created nor destroyed, its form can only be changed. However, as explained below, satisfying the first law alone does not ensure that the process will actually take place.

Consider a resistance type electric room heater used to heat a room by passing electric current through it. The first law dictates that the amount of electric energy applied to the resistance wire of the heater will be equal to the energy transferred to the air in the room as heat. However, if we intend to reverse the process by heating the resistance wire it will not generate an equal amount of electrical energy in the wire.

Similarly a cup of hot tea left in a cooler room will cool down to room temperature. This process will conform to first law of thermodynamics as the amount of heat energy lost by tea will be equal to the energy gained by the surrounding air. However, if we try to reverse the process, that is hot tea getting hotter in the relatively cooler atmosphere as a result of heat transfer from surrounding air to tea will not happen.

Again consider two bodies A and B at different temperatures T_A and T_B (where $T_A > T_B$) respectively when are brought in contact with each other heat will flow from body A to body B because of temperature difference such that after some time both the bodies will attain a common temperature. The heat lost by A will be equal to heat gained by B and first law will not be violated. But if we intened to reverse the process that is the body A gets hotter due to transfer of heat from body B to body A, it will not happen.

Thirdly as an additional example consider a peddal wheel mechanism which is operated by a falling mass will stir a fluid contained in an insulated container. As a result the loss of potential energy of the falling mass will decrease and in turn increase the internal energy of the fluid. The first law of thermodynamics will not be violated as the potential energy lost by the falling mass will be equal to the energy gained by the fluid in the form of heat. However the reverse process that is raising the mass by transferming heat from the fluid to the peddal does not occur in nature.

From the above examples we observe that the processes proceed in a certain direction and not in reverse direction. The first law does not impose any restriction on the direction of the process but at the same time satisfying the first law does not ensure that the process will actually occur.

Thus in the light of above discussion we conclude the following limitations in the first law of thermodynamics.

(*i*) It does not indicate the direction of heat transfer.

(*ii*) It does not specify the conditions under which heat can be transformed into work.

(*iii*) It does not indicate as to why the whole of heat energy can not be converted to mechanical work continuously.

This inadequacy of the first law to indentify whether the process can take place is remedied by introducing another general principle called the Second Law of Thermodynamics. However, the use of second law of thermodyanmics is not only limited to the direction of processes but it also dictates the concept that the energy has "QUALITY" as well as "QUANTITY". The first law is concerned only with the quantity of energy and its transformation from one form to another form without bothering about the quality but on the contrary the major concern of the second law is to provide the necessary means to determine the quality as well as the degree of degradation of energy during a process; for example more of high temperature energy can be converted to work than the same amount of energy at low temperature. Thus energy at higher temperature has higher quality than the same amount of energy at lower temperature. It is also used in finding the theroretical limits for the performance of common engineering systems such as heat engines and refrigrators as well as the degree of completion of chemical reactors.

5.2 THERMAL ENERGY RESERVOIRS

Before we proceed to develop the second law of thermodynamics we will define a hypothetical body with a relatively very large thermal energy capacity (that is mass × specific heat) that is in a position to supply or absorb finite amount of energy as heat without undergoing any charge in temperature. Such a body in called "THERMAL ENERGY RESERVOIR" or in short a "Reservoir". In actual practice large bodies of water such as ocean, lakes and rivers, atmospheric air can be considered as thermal energy reservoirs. For example lage quantities of waste energy (to the tune of millions of Jouls) dumped in large rivers by thermal Power Plants do not significantly alter the temperature of water similarly the heat losses from the residentical buildings in the winter months to the atmosphere does not warm up the air. Further a two-phase system can also be modeled as a reservoir because it can absorb or release large quantities of heat while remaining at constant temperature. Similarly an Industrial Furnace can also be considered as a thermal energy reservoir since its temperature is carefully controlled and is capable of supplying large quantities of thermal energy at constant temperature.

A body need not be very large to be considered as a reservoir. Any physical body whose thermal energy capacity is large in relation to the amount of energy it suplies or absorbs can be treated as a reservoir. Air in a room, for example, can be considered as a reservoir while analysing the heat dissipation from a T.V. playing in the room since the amount of heat transfer from T.V. set is not large enough to have a noticeable effect on the temperature of the air in room.

A reservoir that supplies energy in the form of heat is called the "SOURCE" and that receives or absorbs energy in the form of heat is called a "SINK".

Thermal reservoirs are commonly known as Heat Reservoirs beccause they absorb or supply energy in the form of heat.

Heat Engines

As pointed out earlier wrok can easily be converted into other forms of energy but converting other forms of energy into work is not that easy. As a matter of fact converting heat to work requires some sepcial device. These devices are called "HEAT ENGINES". There are various type of heat engines but in general their working principle is characterised by the following :

1. They receive heat Q, from a high temeprature T_1, heat source (*e.g.*, oil furnace, nuclear reactor, solar energy).
2. They convert part of the heat to work, W (usually in the form of rotating shaft).
3. They reject the remaining waste heat, Q_2, to a low temperature (T_2) sink (*e.g.*, river, ocean, lake, atmosphere).
4. They operate in a cycle.

Heat engines and other cycle devices involve a fluid to and from which heat is transferred which undergoing a cycle. The fluid is called the "WORKING FLUID".

The term "Heat Engine" is used in a broader sense to include work producing devices that do not strictly speaking operate in a Thermodynamic cycle. The engines that involve internal combustion *e.g.*, automobile engine, gas turbines fall in this catagory. These engines operate in a "Mechanical Cycle" and not in a Thermodynamic cycle because the working fluid which consists of combustion gases do not undergo complete cycle as instead of being cooled to initial temperature they are thrown out as exhaust gases and fresh air-fuel mixture is sucked in as their replacement.

The work producing device that best fits into the difinition of Heat Engines is the "Steam Power Plant" Fig. 5.1 briefly describes the "steam power plant" construction. Pump is used to pump water into the Boiler where the energy source is high temperature furnace supplies the heat Q, to generate high pressure steam. The steam while expanding in turbine produces work in the form rotating shaft. The steam rejected by the Turbine is condensed into water and waste heat is rejected to the sink. Evidently net work out put of this power plant is the difference between the total work out put and the work input *i.e.*,

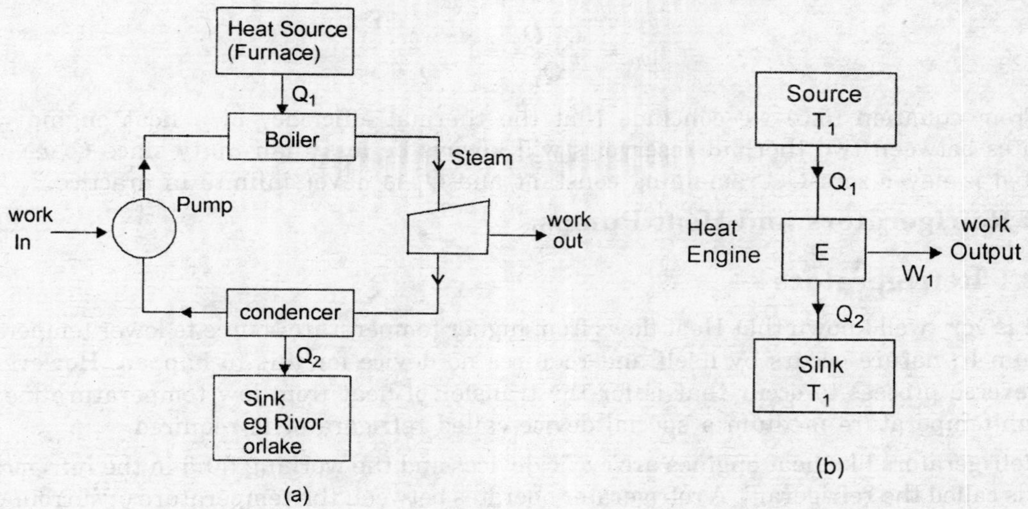

Fig. 5.1

$$\text{work}_{\text{net out}} = \text{work}_{\text{out}} - \text{work}_{\text{in}} \quad ...(5.1)$$

Altenatively net work out put can also be determined from the Heat transfer data as well. The steam power plant are involved the mass flow in and mass flow out. Therefore from this angle it should be treated as an open system. However these components (*i.e.*, in flowing water and outgoing condensate) along with the connecting pipes is considered. Then since these pipes always contains the same fluid (neglecting the leakage of steam if any), then no mass enters or leaves this combination system. Thus it can be analysed as a closed system. And further for a closed system undergoing a cycle, the change in internal energy ΔU is zero and therefore the net work output of the system is also equal to the net heat transfer to the system.

Thus
$$W = W_{\text{net out}} = Q_1 - Q_2 \text{ (kJ)} \quad ...(5.2)$$

5.3.1 Thermal Efficiency

In Equation (5.2), Q_2 is the magnitude of heat energy wasted in order to complete the cycle and Q_2 is never zero. From this we conclude that the work out put of the engine is never equal to the amount of heat input in other words part of the heat input is converted to work. The ratio of the net work output to the heat input is called the Thermal Efficiency.

In general Efficiency or Performance is defined as

$$\text{Performance} = \frac{\text{Derived output}}{\text{Required input}} \quad ...(5.3)$$

For Heat Engines

$$\text{Thermal efficiency} = \frac{\text{Net work output}}{\text{Total Heat input}}$$

or
$$\eta_{th} = \frac{W}{Q_1} \quad ...(5.4)$$

From equation (5.2) and (5.4)

$$\eta_{th} = \frac{Q_1 - Q_2}{Q_1} = 1 - \frac{Q_2}{Q_1} \quad ...(5.5)$$

From equation (5.5) we conclude that the thermal efficiency of a heat engine which operates between two thermal reservoirs will always be less than unity since Q_2 *i.e.*, heat rejected is never zero Q_1, remaining constant and Q_1 is never infinite in practice.

5.3.2 Refrigerators and Heat Pumps.

5.3.2.1 Refrigerators

It is very well known that Heat flows from higher temperature source to lower temperature medium In nature occurs by itself and requires no device for this to happen. However, for the reverse process to occur that is for the transfer of heat from low temperature medium to high-temperature medium a special device called refrigerator is required.

Refrigerators like heat engines are cyclic devices and the working fluid in the refrigerators cycle is called the refrigerant. A refrigerator operates between the temperature of surroundings and a temperature lower than that of surroundings. Thus its main propose is to remove heat from a medium at lower temperature to transfer it to another medium at higher temperature. Most commonly used refrigerator cycle is vapor compression refrigerator cycle and consists of four components a compressor, a condenser, an expansion value and an evaporator (Fig. 5.2).

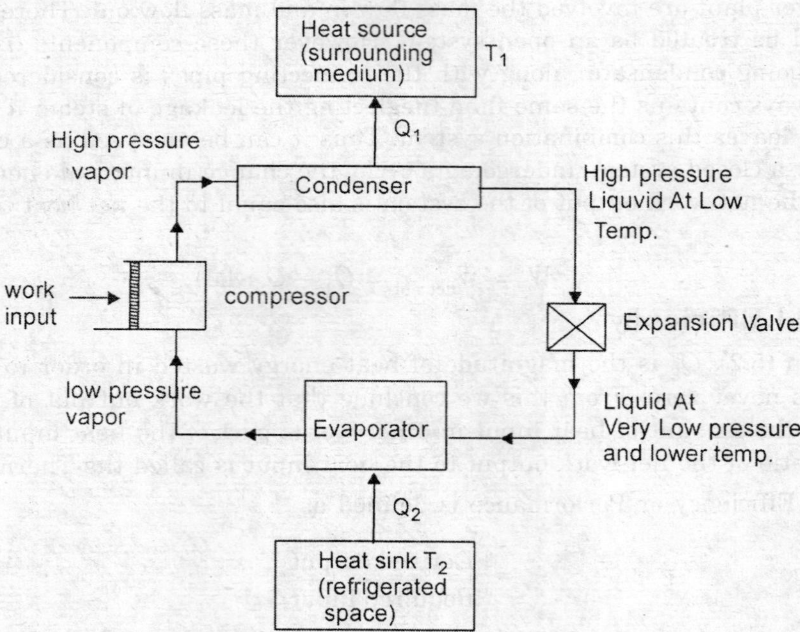

Fig. 5.2 System diagram of refrigerator

Refrigerant in the form of vapor enters the compressor where it is compressed to condenser pressure and leaves the compressor at a relatively high temperature while passing through the condensor it cools down and condenses by rejecting the heat to the surrounding medium which is mostly the kitchen area. It then enters the capillary tube where pressure and temperature drop drastically due to throttling effect. The low temperature refirgerant then enters the evaporator. Here it absorbs heat from the sink which is the refrigerated area and gets evaporated to enter the compressor.

If Q_2 is the heat removed from the refrigerated space at temp T_2 and Q_1 magnitude of heat rejected to the warm enviornment at temp T'_1 and W is the work input to the refrigerator, the efficiency of the refrigerator is expressed in the form of coefficient of performance (COP) as the ratio of the (desired output) which is the amount of heat removed from the refrigerated space to the (required input) in the form of net work in put W.

Then
$$\text{COP} = \frac{\text{Desired output}}{\text{Required input}}$$

$$= \frac{Q_2}{W} \qquad \text{...(5.6)}$$

Further from the conservation of energy principle for cyclic device we have

$$W = Q_1 - Q_2 \qquad \text{...(5.7)}$$

Therefore from equations (5.6) and (5.7)

$$(\text{C.O.P.}) = \frac{Q_2}{Q_1 - Q_2}$$

$$= \frac{1}{\left(\dfrac{Q_1}{Q_2} - 1\right)} \qquad \text{...(5.8)}$$

5.3.2.2 HEAT PUMPS

Heat pumps are devices that transfer heat from a low temperature medium to a higher temperature medium shown schematically in Fig. 5.3.

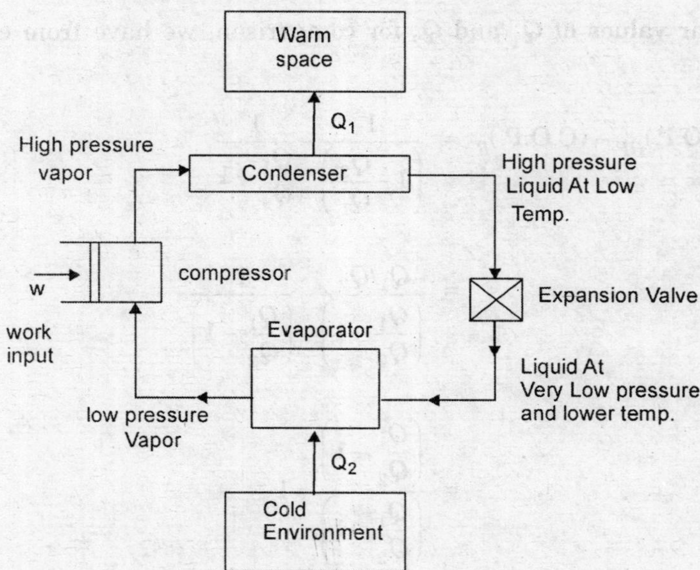

Fig. 5.3 System Diagram of Heat Pump

Both the heat pump and refrigerator operate on the same cycle even though their objectives are slightly different. In the refrigerator its inner space is kept at low temperature by removing heat from it and discharging it to higher temperature surrounding atmosphere Evidently this surronding atmosphere becomes a necessary part of the system and it is actually not the purpose. However in the case of Heat pump it has to maintain a heated space at higher temperature by extracting heat from a low temperature source such as cold air in winter season outside the house and supplying it to the high temperature medium that is air inside the house during winter season.

The measure of the performance of Heat pump is also expressed is terms of C.O.P. (coefficient of performance).

$$(\text{C.O.P.}) = \frac{\text{Desired output}}{\text{Required input}}$$

$$= \frac{Q_1}{W} \qquad \text{...(5.9)}$$

Also from conservation of energy principle

$$Q_1 - Q_2 = W \qquad \text{...(5.10)}$$

From equations (5.9) and (5.10)

$$(\text{C.O.P.})_{HP} = \frac{Q_1}{Q_1 - Q_2}$$

or

$$(\text{C.O.P.})_{HP} = \frac{1}{\left(1 - \frac{Q_2}{Q_1}\right)} \qquad \text{...(5.11)}$$

Thus for similar values of Q_1 and Q_2 for comparison, we have from equation (5.8) and (5.11).

$$(\text{C.O.P.})_{HP} - (\text{C.O.P.})_R = \frac{1}{\left(1 - \frac{Q_2}{Q_1}\right)} - \frac{1}{\frac{Q_1}{Q_2} - 1}$$

$$= \frac{Q_1/Q_2}{\left(\frac{Q_1}{Q_2} - 1\right)} - \frac{1}{\left(\frac{Q_1}{Q_2} - 1\right)}$$

$$= \frac{\left(\frac{Q_1}{Q_2} - 1\right)}{\left(\frac{Q_1}{Q_2} - 1\right)} = 1$$

i.e.,

$$(\text{C.O.P.})_{HP} = (\text{C.O.P})_R + 1 \qquad \text{...(5.12)}$$

Thus for fixed values of Q_1 and Q_2 we infer from equation (5.12) that COP of heat pump is always greater than unity since $(\text{COP})_R$ is a positive quantity.

5.3.2.3 Air Conditioners

Air conditioners are primarily refrigerators in which the refrigerated space is a room in place of food chamber. A window type air conditioner cools the room by absorbing heat from the room air and throwing it into the surroundings air outside the room. However the same window air conditioner can be used as heat pump in winter by installing it in the reversed position. Thus it will pick up heat from the outside air and discharge it into the room.

Example 5.1. *Heat is transferred from a furnace to a heat engine at the rate of 90 MW. The corresponding heat rejection to a near by river is 60 MW. Determine the net power output and thermal efficiency of the heat engine.*

Solution : $\dot{Q}_1 = 90 \text{ MW}$

$\dot{Q}_2 = 60 \text{ MW}$

Therefore neglecting losses

$$W_{\text{output}} = Q_1 - Q_2$$
$$= 90 - 60 = 30 \text{ MW}$$

Then thermal efficiency $\eta_{th} = \dfrac{\text{Work output rate}}{\dot{Q}_1}$

$$= \frac{30\,\text{MW}}{Q_1} = \frac{30}{90} = \frac{1}{3}$$

$$= \mathbf{0.333 \text{ or } 33.3\%.}$$

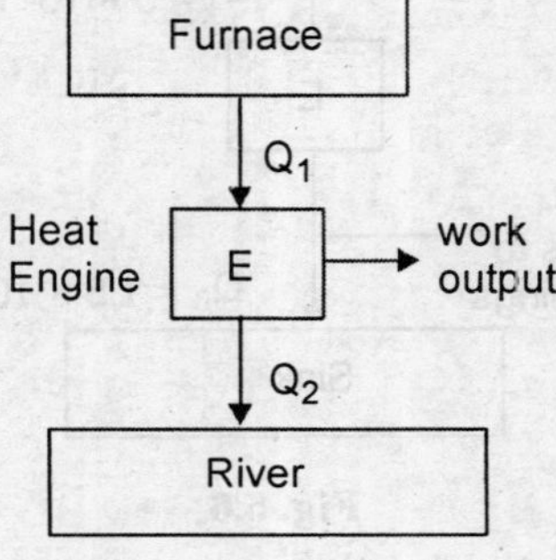

Fig. 5.4

Example 5.2. *The thermal efficiency of a 500 MW steam power plant is 35%. Find the rate of waste heat transfer to a nearly river water.*

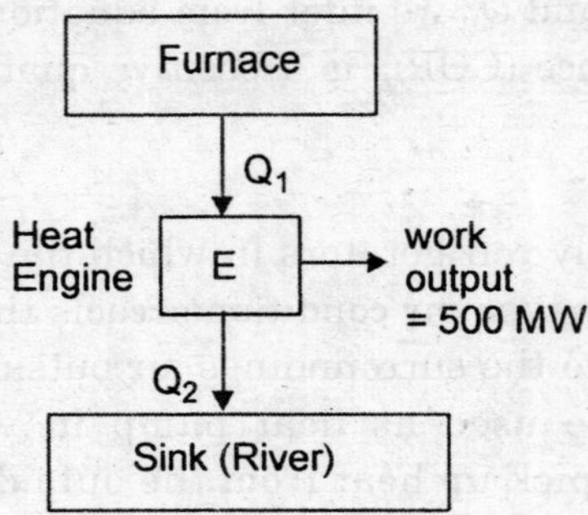

Fig. 5.5

Solution : Thermal efficiency $= \eta_{th}$

$$= \frac{W_{output}}{\dot{Q}_1}$$

or $$\dot{Q}_1 = \frac{500}{0.35} = 1928.57 \text{ MW}$$

Rate of heat transfer to river water

$$\dot{Q}_2 = \dot{Q}_1 - W_{out}$$
$$= 1428.57 - 500$$
$$= 928.57 \text{ MW}$$

Example 5.3. *A steam power plant receives heat from a furnace at the rate of 300 GJ/h. The loss of heat to the surroundings when the steam passes through the pipes etc is 10 GJ/h. If the waste heat transferred to river water is 150 GJ/h calculate the net power output and the thermal efficiency of the power plant.*

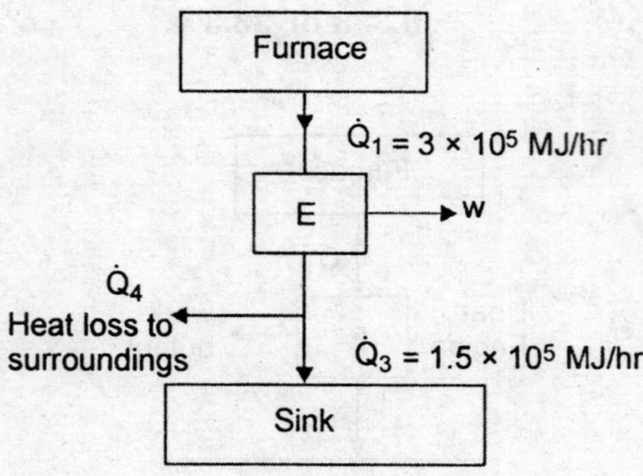

Fig. 5.6

Solution : (*a*) Total heat rejected by the power plant

$$\dot{Q}_2 = \dot{Q}_3 + \dot{Q}_4 \ (\dot{Q}_3 = \text{Loss to river}, \ \dot{Q}_4 = \text{Loss through pipes})$$
$$= 150 + 10 \text{ GJ/h}$$
$$= 160 \text{ GJ/h}$$

There $\qquad W_{out} = \dot{Q}_1 - \dot{Q}_2$

$= 300 \text{ GJ/h} - 160 \text{ GJ/h}$

$= 140 \text{ GJ/h}$

$$= \frac{140 \times 100^3}{3600} \text{ MJ/s} = 38.8 \text{ MW}$$

(*b*) Thermal efficiency $\qquad = \eta_{th} = \dfrac{W_{out}}{\dot{Q}_1}$

$$= \frac{140 \text{ GJ/h}}{300 \text{ GJ/h}} = 0.466 \text{ } i.e., 46.6\%$$

Example 5.4. *An automobile engine has a power output of 75 KW. It is thermal efficiency is 20%, determine the rate of fuel consumption if the energy content of fuel is 4500 kJ/kg.*

Solution : $\qquad \eta_{th} = \dfrac{W_{out}}{\dot{Q}_1}$

Here $\qquad W_{out} = 75 \, kW$ and $\eta_{th} = 0.2$

$$\therefore \qquad \dot{Q}_1 = \frac{75 \text{ kJ/s}}{0.20} = 375 \text{ kJ/s}$$

For producing 375 kJ/s the fuel consumption will be say $\dot{m}$

$$\text{Then } \dot{m} = \frac{\dot{Q}_1 \text{ kJ/s}}{45000 \text{ kJ/kg}} = \frac{375}{45000} \text{kg/s} = 8.3 \times 10^{-3} \text{kg/s} = 8.3 \text{ g/s}$$

Example 5.5. *A gas turbine having 16% efficiency develop 6400 kW output power. If the energy content of the fuel is 46000 kJ/kg find the fuel consumption rate in litres/min. Take the desity of fuel 0.8 g/c.c*

Solution : $\qquad \eta_{Th} = \dfrac{W_{out}}{\dot{Q}_1} = \dfrac{6400}{\dot{Q}_1} \text{ kW}$

Therefore $\qquad \dot{Q}_1 = \dfrac{6400 \text{ kW}}{0.16} = 40000 \text{ kJ/s}$

Let $\qquad \dot{m}$ = rate of fuel consumption

Therefore $\qquad \dot{m} = \dfrac{\dot{Q}_1 \text{ kJ/s}}{46000 \text{ kJ/kg}} = \dfrac{40000}{46000} = 0.8696 \text{ kg/s}$

The vol. flow rate of fuel $\qquad = \dfrac{\dot{m}}{\rho} = \dfrac{0.8696}{\rho} \text{ g/sec}$

Now $\qquad \rho = 0.8 \text{ g/c.c} = 0.8 \times 10^3 \text{ g/litre}$

$= 0.8 \text{ kg/litre}$

$\therefore$ vol flow rate $= \dfrac{0.8696}{0.8} \dfrac{\text{kg/sec}}{\text{kg/litre}}$

$= 1.087$ litre/sec

Example 5.6. *An automobile engine consumes fuel at the rate of 25 litre/h and delivers 75 kW power at the wheels. If the energy content of the fuel is 45000 kJ/kg and its density 0.8 kg/L, determine the efficiency of the engine.*

Solution : Mass consumption rate of fuel

$$= \dot{m} = \rho\dot{V}$$

$$= \left(0.8\frac{\text{kg}}{\text{L}}\right)\left(\frac{25\ \text{L}}{\text{h}}\right) = 20\ \text{kg/h}$$

Therefore rate of heat supply to engine

$$= \dot{Q}_1 = (20\ \text{kg/h})\ (45000\ \text{kJ/kg})$$

$$= 90000\frac{\text{kJ}}{\text{h}}$$

$$= \frac{30000}{3600}\ \text{W or kJ/s}$$

$$= 250\ \text{kW}$$

Now thermal efficiency $= \dfrac{\dot{W}_{out}}{\dot{Q}_1} = \dfrac{75\ \text{kW}}{250\ \text{kW}}$

$$= 0.3 \text{ or } 30\%$$

Example 5.7. *A house hold refrigerator having COP = 1.6 removes heat from the food chamber of the refrigerator at the rate 64 kJ/min. Calculate the electric power being consumed by the refrigerator and the rate of heat transfer to the kitchen.*

Solution : $\dot{Q}_2 = 64$ kJ/min

Now $\text{COP} = \dfrac{\dot{Q}_2}{\dot{W}}$

or $1.6 = \dfrac{\dot{Q}_2}{\dot{W}}$

or $\dot{W} = \dfrac{\dot{Q}_2}{1.6}$

$$= \frac{(64\ \text{kJ/min})}{1.6}$$

$$= 40\ \text{kJ/min}$$

Fig. 5.7

$$= \frac{40}{60} \text{ kJ/s} = 0.666 \text{ kW}$$

Heat transfer rate to kitchen

$$\dot{Q}_1 = \dot{Q}_2 + \dot{W} = 64 \text{ kJ/min} + 40 \text{ kJ/min}$$

$$= 104 \text{ kJ/min}$$

$$= \frac{104}{60} \text{ kJ/s} = 1.733 \text{ kJ/sec}$$

Example 5.8. *An air conditioner removes heat steadly from a house at the rate of 625 kJ/min and consumes electric power at the rate of 5 kW. Find its COP and the rate of heat transfer to the outside air.*

Solution : $\text{COP} = \frac{\dot{Q}_2}{\dot{W}}$

$$= \left(\frac{625 \text{ kJ/min}}{5 \text{ kW}}\right)\left(\frac{1 \text{ kW}}{60 \text{ kJ/min}}\right)$$

$$= 2.08$$

Rate of Heat transfer to outside air

$$\dot{Q}_1 = \dot{Q}_2 + \dot{W}$$

$$= (625 \text{ kJ/min}) + (5 \text{ kW})\left(\frac{60 \text{ kJ/min}}{1 \text{ kW}}\right)$$

$$= 625 + 300 = 925 \text{ kJ/min}$$

Fig. 5.8

Example 5.9. *A house hold refrigerator runs one fourth of the time The average heat removal rate from the food chamber is 780 kJ/h. If the COP of the refrigerator is 2.0, find the power that it draws when running.*

Solution : As the refrigerator runs $\frac{1}{4}$th of the time the actual heat removal rate when it is running

$$= 4 \times (\text{Average heat removal rate})$$

$$= 4 \times 780 \text{ kJ/h}$$

$$= 3120 \text{ kJ/h} = \dot{Q}_2$$

$$\text{COP} = \frac{\dot{Q}_2}{\dot{W}}$$

or $$\dot{W} = \frac{\dot{Q}_2}{\text{COP}} = \frac{3120}{2} \text{ kJ/h}$$

$$= 1560 \text{ kJ/h}$$

$$= \left(1560 \frac{\text{kJ}}{\text{h}}\right) \cdot \left(\frac{1 \text{ kW}}{3600 \text{ kJ/h}}\right)$$

$$= \frac{1560}{3600} = 0.433 \text{ kW}$$

Example 5.10. *An air conditioning system is designed to maintain the temperature of a house at 20°C (Fig 5.8). The house is estimated to gain heat from outside at the rate of 18000 kJ/h. In addition it will gather heat from within the house due to the presence of electrical gadgets, kitchen cooking range and people living inside, amounting to 7000 kJ/m. If the COP = 2.2, determine power required to run the air conditioning unit.*

Solution : Total cooling load

$$\dot{Q}_1 = \text{Heat gain from outside + Heat generated from within the house}$$

$$= 18000 + 7000 \text{ kJ/h}$$

$$= 25000 \text{ kJ/h}$$

Now $$\text{COP} = \frac{\dot{Q}_1}{\dot{W}} \text{ kJ/h}$$

$$2.2 = \frac{25000 \text{ kJ/h}}{\dot{W}}$$

$$\dot{W} \text{ kW} = \left(\frac{2500 \text{ kJ}}{2.2 \quad \text{h}}\right)\left(\frac{1 \text{ kW}}{3600 \text{ kJ/h}}\right)$$

$$= 3.16 \text{ kW}$$

Example 5.11. *A house-owner uses resistance type electric heater and in a particular winter month the consumption of electric energy was 1500 kWh. He has been advised to employ a heat pump instead of resistance type heaters. If the average COP of the heat pump is 2.5 determine the saving of the house owner with this switch over. Tariff of electricity is Rs 3 per kWh of energy.*

Solution : Assuming the amount of heat supplied by the resistance heaters to be equal to the amount of electricity they consume. Thus the heat pump must also supply 1500 kWh of energy.

Then for the heat pump $\dot{Q}_1 = 1500$ kWh

$$\text{COP} = \frac{\dot{Q}_1}{\dot{W}}$$

$$\dot{W} = \left(\frac{1500 \text{ kWh}}{2.5}\right)$$

$$= 600 \text{ kWh}$$

Thus when using heat pump he will save

$$= 1500 - 600$$

$$= 900 \text{ kWh}$$

Cost of 900 kWh of electricity $= 3 \times 900 = \text{Rs. } 2700$

Example 5.12. *A heat pump is installed to heat a house in the winter. However it runs one third of time. If the heat loss in house is 16000 kJ/h find the power requirement of the heat pump. Take COP as 2.8.*

Solution : As the heat pump runs one third of the time and supplies 16000 kJ/h, the heat pmup, must supply heat when running

$$= \dot{Q}_1$$

$$\dot{Q}_1 = 3 \times (16000) \text{ kJ/h}$$

$$= 48000 \text{ kJh}$$

Thus

$$\dot{W} = \frac{\dot{Q}_1}{\text{COP}} = \left(\frac{48000}{2.8} \text{ kJ/h}\right)\left(\frac{1 \text{ kW}}{3600 \text{ kJ/h}}\right)$$

$$= 4.76 \text{ kW}$$

5.3.2.3 THERMAL EFFICIENCY AND WASTE HEAT

The thermal efficiency of heat engine is very low and large quantity of heat is thrown into sink an waste heat.

The question arises whether it is possible to get rid of the condenser and save all the waste heat. The answer is firm 'No' because without the cooling process in the condenser the cycle can not be completed and the engine or power plant can not run continuously unless the cycle is completed.

5.4 THE SECOND LAW OF THERMODYNAMICS

5.4.1 Kelvin-Plank Statement

In the preceding article it has been concluded that a heat engine must reject some heat to a low temperature sink, then only the cycle will be completed. In other words no heat engine converts the entire heat to be received from the high temperature reservoir into useful work.

That is the thermal efficiency can not be 100%. On this limitation the thermal efficiency is based on the Kelvin-Plank Statement of the Second Law of Thermodynamics which is given below.

"It is impossible for any device working on a cyclic process to receive heat energy from a single source and convert the entire heat energy into equivalent amount of work".

In other words it implies that heat engine must exchange heat with low temperature sink as well as "have high temperature source. The Kelvin-Plnak statement can also be expressed as follows "No heat-engine can have 100% thermal efficiency and two thermal reservoirs (one source and one sink) are required for the working of engine in a cycle."

5.4.2 The Second Law of Thermodynamics Clausius Statement

It is impossible to construct a device that operates in a cycle and produces no effect other than the transfer of heat from a low temperature system to a system at high temperature.

From the statement it is implied that heat cannot flow by itself from a low temperature system to a system at high temperature. In Fig. (5.9 *a*) is shown the schematic arrangement that is not permitted by clausius statement. Thus the

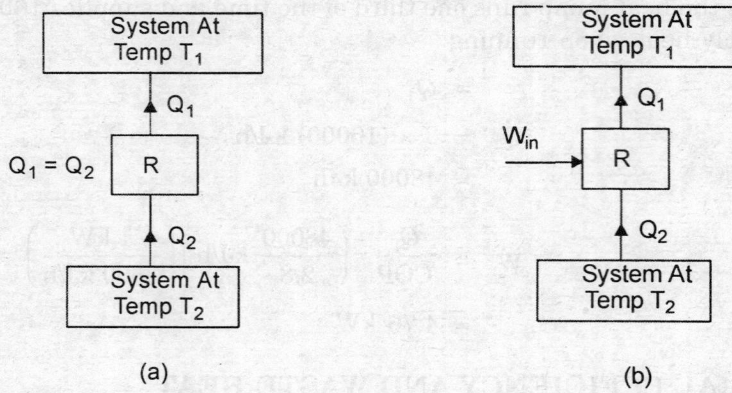

Fig. 5.9

only possible way for the transfer of heat from low temperature system to the system at higher temperature is to supply external work as shown in Fig. (5.9 *b*)

Evidently the Kelvin-Plank statement is applied to heat engines while the Clausius Statement deals with refrigerators and heat pumps. Both these statements are negative statements and do not have mathematical proofs. The second law is based on experimental observations and to date no experiment has been conducted that contradicts the second law.

5.4.3 Equivalence of the Two Statements

Even though the Kelvin-Plank and Clausius statements have been applied to heat engines and refrigerators respectively but in their consequences can be used as the expression of the second law of Thermodynamics. It is impossible to have a system that satisfies one and violates the other and vice-versa. It is proved below that if a device violates one statement it also violates the other.

5.4.3.1 Violation of Kelvin-Plank Statement Leading to the violation of Clausius Statement

Consider an arrangement shown in Fig. 5.10 the Heat engine and refrigerator combination, operating between the same reservoirs (*i.e.*, same source and same sink). Let the Heat engine be in violation of Kelvin-Plank statement *i.e.*, its thermal efficiency is 100%. In otherwords it converts the entire heat energy Q_1 that it draws from the reservoir into work W.

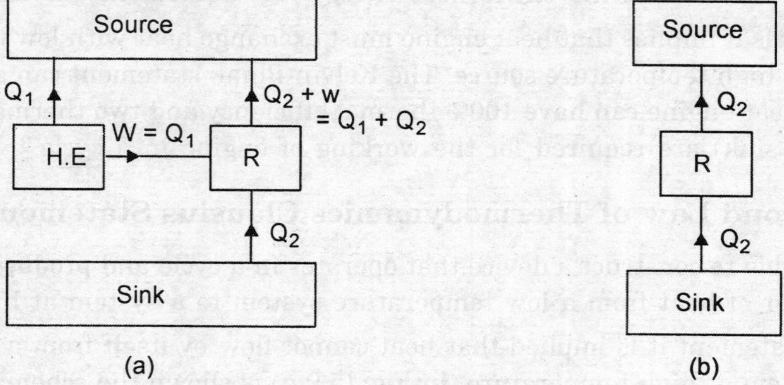

Fig. 5.10

Along with it is a refrigerator which extracts heat Q_2 units from low temperature reservoir called sink and is supplied with Q_1 units of work from an external agency (surroundings) and it supplies $(Q_1 + Q_2)$ units heat to high temperature reservoir. The refrigerator is operating in confirmity of Clausisus statement. However the Heat engine and refrigerator are coupled togther *ie* the Heat engine runs the refrigerator then $W = Q_1$. This composite system will be such that it draws no work from outside agency and it draws Q_2 units heat from sink and delivers $(Q_1 + Q_2) - (Q_1) = Q_2$ units to high temperature source without any work input. That is it is equivalent to a system as shown in Fig. 5.11*b* which is in vlolation of Clausius statement.

5.4.3.2 Violation of Clausius Statement Leads to Violation of Kelvin-Plank Statement

Consider an arrangement shown in Fig. 5.11. The heat engine is supplied Q_1 units of heat by the High temperature thermal source and rejects Q_2 units of heat to sink at low temperature T_2. The remaining heat $Q_1 - Q_2$ is converted into work, W is in conformity with Kelvin-Plank statement. However, the heat pump transfers heat Q_2 from heat source (sink) at low temperature T_2 to thermal source at temperature T_1 without absorbing easy work in violation of the Clausius statement. Now combining the heat engine and the

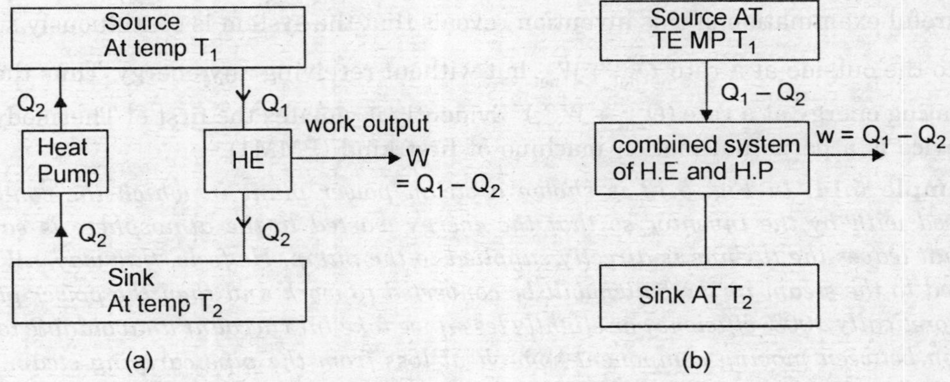

Fig. 5.11

heat pump, we can eleminate the Heat sink because then is no heat interaction as Q_2 units heat is rejected by the engine to it and Q_2 units of heat is taken by heat pump. Thus the combined system of heat pump and heat engine are exchanging heat with only one reservoir and converting it into work. This violates the Kelvin Plank Statement. Therefore we conclude that violation of Clasius statement leads to the violation of Kelvin Plank statement.

5.5 PERPETUAL MOTION MACHINES

As stated above for any process to take place in nature it must satisfy both the First are Second Law of Thermodynamics. A device that violates either law is called a perpetual motion machine. a device which violate the first law of Thermodynamics (*i.e.*, it creats energy is called perpetual motion machine of First kind (PMM1) and the device that violates the second law of Thermodynamics is called the perpetual motion machine of second kind (PMM2).

Example 5.13. *In Fig. 5.12 is shown a steam power plant. An inventor proposes to heat the steam with resistance heaters placed in side the boiler instead of heating with coal. Part of the electricity generated by the plant is to be used to power these electric heaters and the pump. The rest of an electric energy to be supplied to be electric grid as net work output. The inventor claims that once the system is started the power plant will produce electricity idefinitely without requiring any energy input from outside. Discuss the claim of the inventor whether it is True or False.*

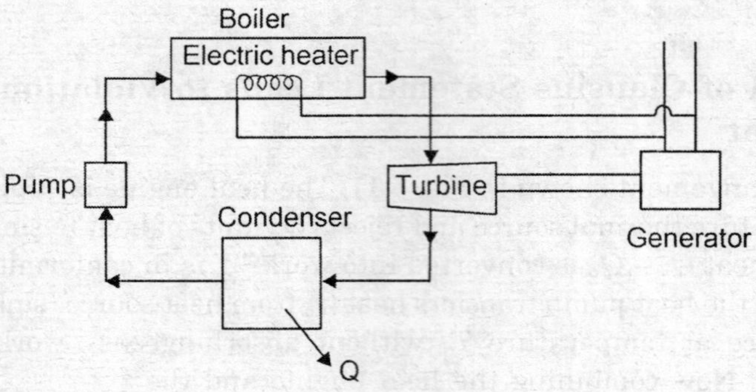

Fig. 5.12

A careful examination of this invention reveals that the system is continuously supplying energy to the outside at a rate $\dot{Q}_{out} + \dot{W}_{net}$ but without receiving any energy. Thus the system is generating energy at a rate (Q_{out} + W_{net}). Evidently it violates the first of Thermodynamics. Thus device is a perpetual motion machine of first kind (PMM1).

Example 5.14. *In Fig. 5.13 is shown a steam power plant in which the condenser is dispenssed with by the inventor so that the energy wasted to the atmosphere is saved and steam that leaves the turbine is directly supplied to the pump. He feels, this way, all the heat transfered to the steam in the boiler will be converted to work and then the power plant will have theoretically 100% efficiency or slightly less if we take into account unavoidable losses due to friction between moving component some heat loss from the pipe carrying steam. That is he expects the efficiency of the order of 80% which will be much higher then the actual plant efficiency of about 30 to 40%.*

Discuss his claim whether it is true or false.

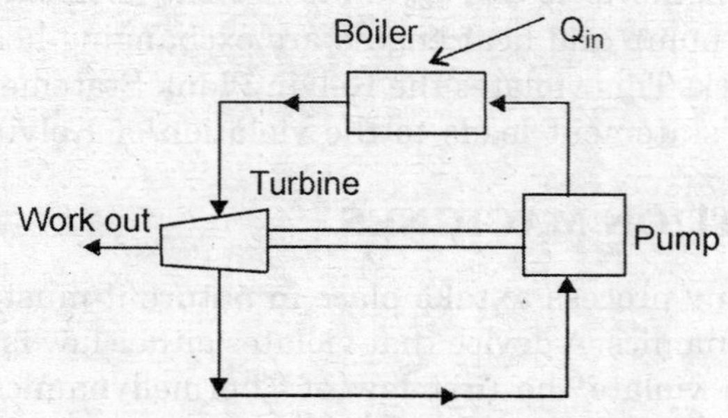

Fig. 5.13

Solution : As per the inventors idea the device works on a cycle and produces net amount of work while exchanging heat with a single reservoir which is the furnace here. Evidently it satisfies the First Law of Thermodynamics but the second law is violated. Therefore it will not work and the device is a perpetual motion machine of the second kind (PMM2).

5.6 REVERSIBLE AND IRREVERSIBLE PROCESSES

A Thermodynamic reversible process is defined as the process which can be reversed without leaving any trace on the surroundings. In other words both the system and the surroundings can be restored to their respective states by reversing the direction of the process. this is possible only if the net heat and work exchange between the system and the surroundings is zero for the combined original and reserve processes. Processes which are not reversible are known as Irreversible processes.

It may be mentioned here that a system can always be restored to its initial states following a process, irrespective of the fact that it may be reversible or irreversible. However for the reversible process, the restoration is made without leaving any net change on the surroundings, whereas for irreversible process the surroundings usually do some work on the system and therefore will not return to their original state. A reversible process is a quasi-static process that is a process carried out at infinitaly slow rate with infinitesimal gredient with the system passing through a continuous series of equilibrium states. A reversible process can proceed in either direction without violating the second law of thermodynamics. In nature reversible processes do not actually occur because the processes take place at finite speeds through finite discontinuities between the system and its surroundings rendering them irreversible. Reversible process can be approximated by actual devices but can never be achieved. This is due to the degradation of energy by such factors on friction, diffusion, inelasticity turbulance and electrical resistance. These factors can be reduced but cannot be completely eliminated.

Even though the reversible process can never be achieved, still these fictitious processes are studied because of two reasons. First, they are easy to analyse, since the system passes through a series of equilibrium states during a reversible process, second they surve as idealised models to which actual processes can be compared. Engineers, specially, are intersted in reversible process because work producing devices such as car-engines, gas or steam turbine deliver the most-work and work consuming devices such as fans compressors, and pumps require least work when reversible processes are used intead of irreversible one. The reversible processes can be viewed as theoretical limits for the correspondings irreversible ones.

Thus the concept of reversible processes leads to the definition of second law of Thermodynamics for actual processes, which is the degree of approximation to the corresponding reversible processes. This way we can compare the performance of different devices which have been designed to do the same work on the basis of their efficiencies. For better design, the irreversibilities will be lower and second law efficiency will be higher.

5.7 IRREVERSIBILITIES

The factors which are responsible for rendering a process irreversible are called IRREVERSIBILITIES. These include function, unrestrained expansion, mixing of two gases, electric resistance, Heat transfer across a finite temperature difference chemical reactions, inelastic deformation of solids. Whenever any one of these present in a process it will render it irreversible.

5.7.1 Friction

When two bodies in contact are made to move relative to each other, friction force which opposes the motion is developed at the interface of the bodies and work is needed to overcome the friction force. The energy consumed as work is eventualy converted into heat during the process and it is transferred to the bodies in contact which results in the rise of temperature in the interface. When the direction of motion is reversed and the bodies are brought to their original position the interface will not cool and the heat will not be converted back to work. On the contrary more heat will be generated while overcoming the friction forces that are opposing the reverse motion. Evidently the system of two bodies and the surroundings connot return to original states, the process is irreversible. Thus presence of friction renders the process irreversible. It may be mentioned that the friction may be involved between fluid and solid body or even between adjoining layers of a fluid moving with different velocities.

5.7.2 Non quasi-equilirbium Expasion or Compression

A quasi-equilibrium process is one in which the system remains infinitesimally close to the state of equilirbium at all times. For example take a frictionless, adiabatic piston-cylinder device that contains a gas. If the piston is pushed to compress the gas with a low velocity, the pressure and temperature of the gas will increase uniformally thought out the entire gas and the system is maintained at a state close to equilibrium which is known as quasi-equilibrium process. Similarly if the piston is slowly moved in the reverse direction the pressure in the gas will slightly decrease and the gas will expand, this will then be quasi-equilibrium expansion of the gas. When the pistion has returned to its original position, the boundary work done on the gas during compression process is returned to the surroundings during expansion process. Thus the net work during the combined process is zero. Thus it can be termed as reversible process since there is no heat transfer involvement and both the system and the surroundings return to their respective original states.

Again we repeat the adiabatic process but this time at a very rapid speed. In so doing the molecules near the piston face will not have sufficient time to escape & thus will tend to pile up in front of the pistion. The net result will be to raise the pressure near the piston face and there will be pressure gradient developed that is the pressure will not be uniform. This is known as Non-quasi-equilibrium process. Similar type of situation will exist during the return expansion stroke when the pressure near the piston face will be lower than that at other points.

5.7.2.1 Heat Transfer

Another familiar irreversibility is that of heat transfer through a finite temperature difference Consider the example of a glass of cold water placed in warm room. Heat will flow form the room air to cold water. The only way the process can be reversed so that water is restored to its original temperature is to employ refrigeration to restore the original temperature of the water. To do this we need same work to go into he refrigeration process. This way the glass of water will be retored to its original temperature but the surroundings will not, however side by side internal energy of the surroundings will increase by an amount equal to the work supplied to the refrigerator. The surroundings can be brought back to their initial state only by converting this excess internal energy completely to work, which is impossible to do so without violating the second laaw of thermodynamics. Therfore it is not possible to have a reversible heat transfer process. However, a heat transfer process becomes less and less irreversible as the temperature difference between the two bodies approaches zero. It is therefore customary to consider a heat transfer through a differential temperature dT as reversible. Theoretically as dT approaches zero the process will be reversible and in that event no refrigerator will be required to reverse the direction of the process. Once again it is retreeted that this is only theoretical and practically it cannot be achieved.

5.7.2.2 Internally and Externally Reversible Processes

A process is called internally reversible if no irreversibilities occur within the boundaries of the system during the process. In other words during on internally reversible process, the system proceeds through a series of equilibrium states and during the reverssal, the system passes through exactly the same equilibrium states.

A system is called externally reversible of no irreversibles occur outside the system boundaries during the process.

Likewise a system is called TOTALLY REVERSIBLE (or simply reversible) if no irreversibilities within it or outside surroundings occur. Fig. 5.14 (*a*) and (*b*) show a phase change phenomenon taking place at constant pressure and constact temperature. However, in Fig. 5.14 (*a*) the thermal energy reservoir is 25°.000001°C whereas in Fig. 5.14 (*b*) the thermal energy reservoir is at 30°C but the gaseous phase in both is at 25°C. Evidently in Fig 5.14 *a* the process is totally reversible but in Fig. 5.14 *b* it is internally reversible only.

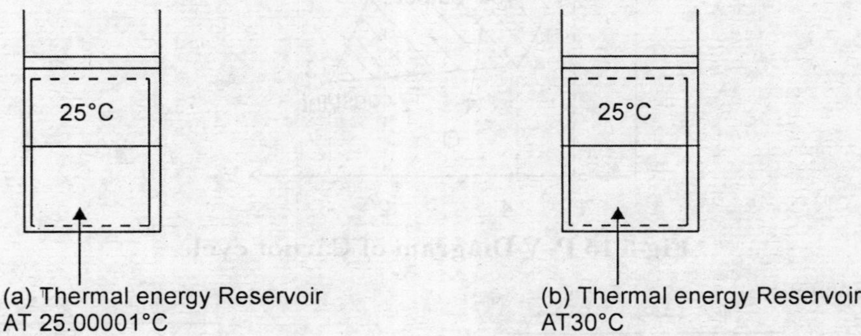

Fig. 5.14

5.8 THE CARNOT CYCLE

In article 5.3 Heat Engines had been introduced as cyclic devices in which the working fluid returns to its initial state at the end of each cycle. Work is done by the working fluid during one part of the cycle and on the working fluid in the other part. The difference between these two is the net work delivered by the engine. Thus the efficiency of the heat-engine will depend upon the execution of individual processes during each constituent part of the cycle. Thus the cycle efficiency will be highest in the process when the fluid does work is such that it is maximum and similarly while line work is being done on the fluid it is the least. This can be achieved by using process that require least amount of work and deliver the most by using reversible processes. Thus we conclude that most efficient cycles are reversible cycles. But unfortunately this can not be achieved practically because the presence of ir-reversibilities in such processes. However as pointed out earlier the assumption of using reversible cycles gives the upper limits on the performance of real cycles.

It was in 1824 that a French engineer Sadi-Carnot proposed a reversible cycle for the heat engines known after the name as Carnot heat engine.

The Carnot cycle consists of four reversible processes, two isothermal and two adiabatic. Since all the constitutent processes are reversible the entire Carnot Cycle is also reversible. Following assumptions are made in the operation of Carnot Engine.

1. The working fluid is assumed to be a perfect gas.
2. Two heat reservoirs, the hot reservoir or heat source at temperature T_1 and cold reservoir (heat sink) at temperature T_2.
3. Piston cylinder arrangement for getting net work out of the cycle. The piston and the cylinder walls and piston head are taken as perfect insulators The cylinder head is assumed to be such that it provides alternatively diathermic cover (perfect heat conductor) and an diabatic cover (perfect heat insulator). There is no friction between the piston and cylinder walls.

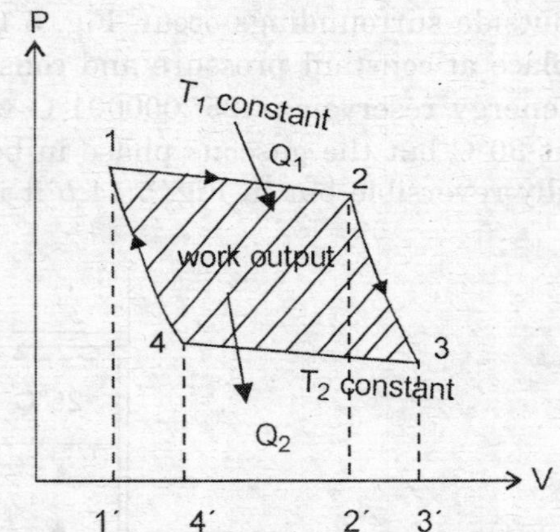

Fig 5.15 P–V Diagram of Carnot cycle

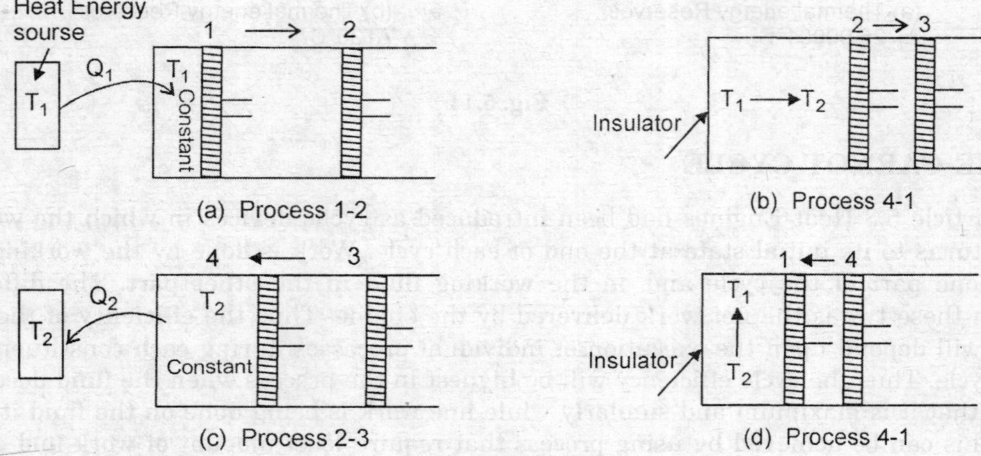

Fig. 5.16

The four reversible processes constitutting the carnot cycle are explained below :

(*i*) Reversible Isothermal Expansion (1 – 2)

To start with at state 1 the gas is at temperature T_1 and the cylinder head is in close contact with the energy source at temperature T_1. The gas is allowed to expand slowly doing work on the surroundings. In so doing the temperature of the gas tends to fall. But the movment the temperature falls by an infinitesimal amount dT heat starts flowing from the energy source at T_1, raising the temperature back to T_1. In this way the temperature is kept constant at T_1. As the temperature difference between the temperature of the gas and the heat energy source is less than dT, it can be taken as reversible isothermal expansion. Total heat transferred from source at T_1 to gas during this processes will be Q_1.

Then
$$Q_1 = W_{1-2} = P_1V_1 \ln\left(\frac{V_2}{V_1}\right)$$

$$= mRT_1 \ln\left(\frac{V_2}{V_1}\right) \qquad ...(5.13)$$

(*ii*) Reversible Adiabatic Expansion (Process 2 – 3)

Temperature drops for T_1 to T_2. During this process the cylinder head kept during the process 1 – 2 is replaced by one which is insulated so that the system becomes adiabatic. The gas continues to expand slowly doing work on the surroundings until the temperature drops to T_2. Here again it is assumed that the piston is frictionless and the process to be quasi-equilibrium thus the process is rendered reversible adiabatic.

The work done by the working fluid *i.e.*, gas is at the expense of internal energy and that is the why the temperature falls from T_2 to T_3 at the point 3 (Fig. 5.15 and 5.16).

(*iii*) Reversible Isothermal Compression (Process 3 – 4) T_2

At the state 3 insulating cylinder head is removed and the cylinder is brought into contact with the sink at temperature T_2. The piston is made to move inwards with the help of an external force, doing work on the gas so as to compress it. As the gas is compressed, its temperature tends to rise. But as soon as it rises by an infinitisional amount dT, heat flows from the gas to sink, thereby causing the gas temperature to drop to T_2. This is how the gas temperature is made to remain T_2. The temperature difference between the gas and the sink never exceeds dT. That is this heat transfer process 3 to 4 is a reversible heat transfer process. If the amount of heat rejected from the gas to the sink is Q_2 and it is given by

$$Q_2 = W_{3-4} = P_3V_3 \ln\left(\frac{V_4}{V_3}\right)$$

$$= mRT_2 ln\left(\frac{V_4}{V_3}\right) \qquad ...(5.14)$$

(*iv*) Reversible Adiabatic Compression (Process 4 – 1)

When the low temper sink is removed and the insulation is put back on the cylinder head and the gas is compressed in a reversible manner, the gas returns to its initial state (1) from where we started. The temperature will rise for T_2 to T_1 during the reversible adiabatic compression. Evidently work is done on the working fluid (that is the gas) and the internal energy of the gas increases.

The P-V. diagram of the cycle is shown in Fig. (5.15). It may be noted that area under the process curve represents the boundary work for quasi-equilibrium (reversible internally) the area $1-2-3-3^1-4^1-1^{1-1}$ is the work done by the gas during the expansion stroke. similarly area under the curve 3 – 4 – 1 *i.e.,* the area $1-1^1-4^1-3^1-3-1$ is the work done on the gas. Therefore the difference between these areas is the area 1 – 2 – 3 – 4– 1, represents the net work done during the cycle. Alternatively, since there is no heat interaction along the reversible adiabatic processes 2 – 3 and 4 – 1, we get with the application of first law of thermodynamics for the complete cycle.

$$W_{net} = |Q_1| - |Q_2|$$

$$= mRT_1 \ln\left(\frac{V_2}{V_1}\right) - mRT_2 \ln\left(\frac{V_3}{V_4}\right) \quad ...(5.15)$$

Therefore efficiency

$$\eta = \frac{|Q_1| - |Q_2|}{|Q_1|} \quad ...(5.16a)$$

or

$$\eta = \frac{mRT_1 h_0\left(\frac{V_2}{V_1}\right) - mRT_2 \ln\left(\frac{V_3}{V_4}\right)}{mR_1 \ln\left(\frac{V_2}{V_1}\right)}$$

$$= 1 - \frac{T_2}{T_1}\frac{\ln(V_3/V_4)}{\ln(V_2/V_1)} \quad ...(5.16b)$$

Further the working fluid is assumed an ideal gas. Therefore for the reversible adiabadtic process the temperature and volume are related by the relationship.

$$TV^{\gamma-1} = \text{constant}$$

Therefore for the adiabatic processes 2 – 3

$$T_1 V_2^{\gamma-1} = T_2 V_3^{\gamma-1}$$

or

$$\frac{T_1}{T_2} = \left(\frac{V_3}{V_2}\right)^{\gamma-1} \quad ...(15.17)$$

Similarly for the adiabatic processes (4–1)

$$T_2 (V_4)^{\gamma-1} = T_1 V_1^{\gamma-1}$$

or

$$\frac{T_1}{T_2} = \left(\frac{V_4}{V_1}\right)^{\gamma-1} \quad ...(15.18)$$

Therefore from equations (15.17) and (15.18) we have

$$\frac{T_1}{T_2} = \left(\frac{V_3}{V_2}\right)^{\gamma-1} = \left(\frac{V_4}{V_1}\right)^{\gamma-1}$$

Therefore from equations (15.17) and (15.18) we have

$$\frac{T_1}{T_2} = \left(\frac{V_3}{V_2}\right)^{\gamma-1} = \left(\frac{V_4}{V_1}\right)^{\gamma-1}$$

or
$$\frac{V_3}{V_2} = \frac{V_4}{V_1} \qquad ...(15.19)$$

or
$$\ln\left(\frac{V_3}{V_2}\right) = \ln\left(\frac{V_4}{V_1}\right) \qquad ...(15.20)$$

Therefore from equation (15.16) and (15.20) we get

$$\eta = 1 - \frac{T_2}{T_1}$$

$$= \frac{T_1 - T_2}{T_1} \qquad ...(15.21)$$

And for equation (15.16*a*) and (15.21)

$$\eta = \frac{Q_1 - Q_2}{Q_1}$$

$\therefore$
$$\eta = \frac{T_1 - T_2}{T_1} = \frac{Q_1 - Q_2}{Q_1}$$

From this following conclusions are made

or
$$1 - \frac{T_2}{T_1} = 1 - \frac{Q_2}{Q_1}$$

or
$$\frac{T_2}{T_1} = \frac{Q_2}{Q_1} \qquad ...(15.21b)$$

(*i*) Efficiency is independent of the working fluid. It only depends upon the temperatures of source and sink.

(*ii*) If $T_2 = 0$ then $\eta = 100\%$. But this is not possible as per second law of thermodynamics there must be a heat sink. Therefore when T_2 is nearer to zero, the efficiency will be maximum.

(*iii*) Higher the difference between the temperatures of source and sink higher will be the efficiency. In other words increase T_1 and reduce T_2.

(*iv*) At $T_1 = T_2$ no work will be done and efficiency will be zero.

5.8.1 Impracticability of Carnot Cylce

Theoretically we have been able to show that an engine working on Carnot Cycle has the maximum efficiency but practically it is impossible to make such an engine due to following reasons :

(i) Since all the four processes have to be reversible which implies that the working fluid must have no internal friction between the fluid particles. In addition there should not be any friction between the piston and cylinder walls which is not possible.

(ii) The heat rejection or absorption have to take place with infinitesimal temperature difference. This means that every transfer will be at a very slow rate and the engine will deliver only infinitesimal power. Further it is impossible to transfer heat without finite difference in temperature.

(iii) During one part of the expansion stroke the piston is required to move very slowly to achieve isothermal expansion but during the later part it must move very fast to get adiabatic expansion. This variation in speed is impossible to achieve.

5.8.2 Reversed Carnot Cycle

As pointed out above carnot Heat Engine cycle is totally reversible. Thus all the processes of the Carnot Heat Engine when reversed it becomes the carnot refrigeration cycle or Carnot Heat pump cycle. When these mechanical processes are reversed and operated backwards all the energy transfers associated with these individual processes get reversed in direction but their magnitudes remain same Fig. 5.17 show the P-V. plot of such a Carnot Heat Pump or (Carnot Refrigerator) *i.e.*, reversed (Carnot Heat Engine). The operational sequence of these processes are as follows :

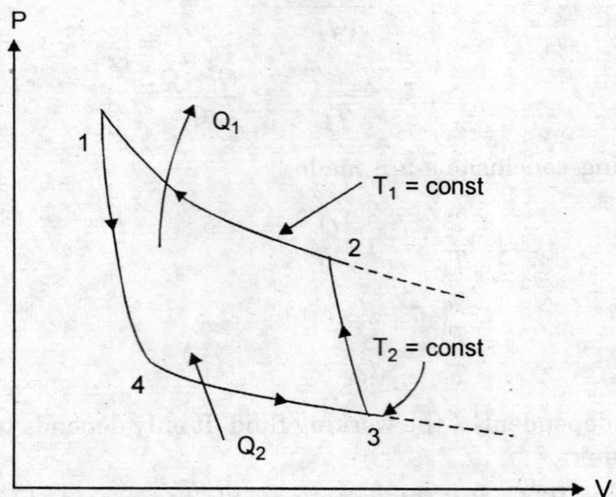

Fig. 5.17 Reversed Carnot Cycle P–V Diagram

(i) Reversible adiabatic expansion (1 – 4) (also known as isentropic expansion) of the working fluid in the clearance space of the cylinder.

(ii) Isothermal expansion during which heat Q_2 is absorbed at temperature T_2 from the space being cooled. (4 – 3)

(iii) Isentropic compression (Reversible adiabatic compression) of fluid at a very slow speed the temperature rises from T_2 to T_1. (3 – 2)

(iv) Isothermal compression of working fluid during which heat Q_1 is rejected to a system at higher temperature. (2 – 1)

As pointed out above

$$Q_1 = mRT_1 \ln\left(\frac{V_2}{V_1}\right)$$

$$Q_2 = mRT_2 \ln\left(\frac{V_3}{V_4}\right)$$

(*a*) Therefore for a Heat Pump

$$\text{C.O.P.} = \frac{Q_1}{Q_1 - Q_2}$$

$$= \frac{mRT_1 \ln\left(\frac{V_2}{V_1}\right)}{mRT_1 \ln\left(\frac{V_2}{V_1}\right) - mRT_2 \ln\left(\frac{V_3}{V_4}\right)} \quad ...(15.22)$$

As proved in equation (15.19) here also

$$\frac{V_2}{V_1} = \frac{V_3}{V_4} \quad ...(15.23)$$

Therefore from equations (15.22) and (15.23)

$$(\text{COP})_{\text{Heat Pump}} = \frac{T_1}{T_1 - T_2}$$

(*b*) Again for Carnot Refrigerator

$$(\text{COP})_{\text{Ret}} = \frac{Q_2}{Q_1 - Q_2}$$

$$= \frac{mRT_2 \ln\left(\frac{V_3}{V_2}\right)}{mRT_1 \ln\left(\frac{V_4}{V_1}\right) - mRT_2 \ln\left(\frac{V_3}{V_2}\right)}$$

Here again $\frac{V_4}{V_1} = \frac{V_3}{V_2}$

$$\therefore \quad (\text{COP})_{\text{Ref}} = \frac{\text{T}_2}{\text{T}_1 - \text{T}_2} \quad ...(15.24)$$

Example 5.15. *A carnot heat engine operates between a source at 1250K and sink at 300 K. If the heat engine consumes fuel which produces 820 kJ/min of heat determine the (a) thermal efficiency and (b) the power output of the engine.*

Solution : Carnot efficiency $= 1-\frac{T_2}{T_1}$

where T_2 = Sink temperature

T_1 = Source temperature

$$\therefore \qquad \eta = 1-\frac{300}{1250} = 1-.29 = 0.76 \text{ or } 76\%$$

Also thermal efficiency $\eta = \frac{\dot{W}_{output}}{\dot{Q}_1}$

or

$$\dot{W}_{output} = \dot{Q}\eta$$
$$= 0.76 \times 820 \text{ kJ/min}$$
$$= 623.2 \text{ kJ/min}$$
$$= \frac{623.2}{60} \text{ (kJ/s)}$$
$$= 10.38 \text{ kW}$$

Example 5.16. *A heat engine working on carnot cycle converts 25% of the heat input into work. However when the temperature of the sink is reduced by 90°C its efficiency becomes twice the earlier value Calculate the temperatures of source and sink.*

Solution : the Carnot engine $\eta = 1-\frac{T_2}{T_1} = 0.25$

$$T_1 - T_2 = 0.25\, T_1$$

or

$$0.75\, T_1 = T_2$$

or

$$T_1 = \frac{4}{3}T_2 \qquad ...(A)$$

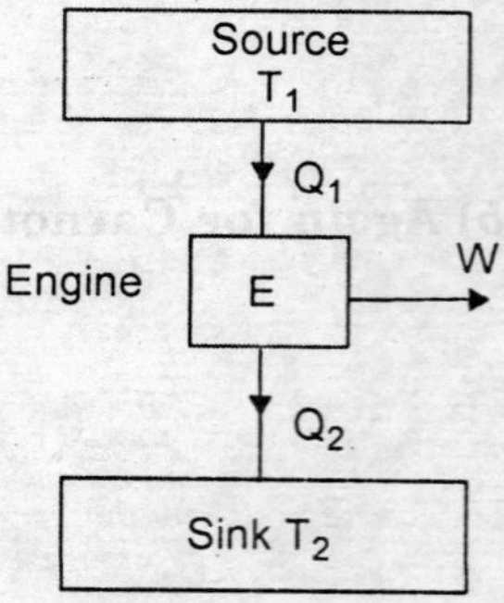

Source T_1

Q_1

Engine E W

Q_2

Sink T_2

Fig. 5.18

In the second case $\eta = 0.5 = 1-\frac{T_2^1}{T_1}$

or

$$0.5\, T_1 = T_1 - T_2^1$$
$$T_2^1 = 0.5\, T_1$$

But

$$T_2^1 = T_2 - 90$$

i.e.,

$$T_2 - 90 = 0.5\, T_1$$
$$= 0.5\times\frac{4}{3}T_2$$

or

$$3T_2 - 270 = 2T_2$$
$$T_2 = 270\, K$$

Then the (*A*) $T = \frac{4}{3}\times 270 = 360K$

Example 5.17. *A Carnot Engine receives 480 kJ of heat from a source at T_1°C and rejects 180 kJ of it into a sink at 20°C. Find the value of source temperature T_1 and the thermal efficiency.*

Solution : Carnot engine works on a reversible cycle and for a reversible cycle from equation (15.21 *b*)

$$\frac{T_1}{T_2} = \frac{Q_1}{Q_2}$$

$$\therefore \quad T_1 = T_2 \cdot \frac{Q_1}{Q_2}$$

$$= \frac{(20+273)\times 480}{180}$$

$$= \frac{293\times 480}{180} = 787.3 \text{ K}$$

Then

$$\eta = 1-\frac{T_2}{T_1} = \frac{T_1-T_2}{T_1}$$

$$= \frac{781.3-293}{781.3} = 0.62 \text{ } i.e., \text{ } 62\%$$

Exaple 15.18. *A heat engine working on Carnot Cycle has thermal efficiency of 54%. The waste heat from the engine is rejected to a river at 17°C at a rate of 850 kJ/min Calculate the engine output in kW and temperature of the source.*

Solution : For carnot engine $\eta = 1-\frac{\dot{Q}_2}{\dot{Q}_1}$

$$0.54 = 1-\frac{850}{\dot{Q}_1}$$

or

$$\frac{850}{\dot{Q}_1} = 1 - 0.54 = 0.46$$

$$\dot{Q}_1 = \frac{850}{0.46} = 1847.83 \text{ kJ/min}$$

Now again

$$\frac{\dot{W}_{out}}{\dot{Q}_1} = \eta$$

or

$$\dot{W}_{out} = 0.54 \times 1847.83 \text{ kJ/min}$$

$$= 997.8 \text{ kJ/min}$$

$$= \frac{997.8}{60} = 16.63 \text{ kW}$$

Also for reversible Carnot Engine

$$\frac{T_1}{T_2} = \frac{\dot{Q}_1}{\dot{Q}_2}$$

i.e.,

$$T_1 = T_2 \times \frac{\dot{Q}_1}{\dot{Q}_2} = \frac{(17+273)\times 1847.83 \text{ kJ/m}}{8.5 \text{ kJ/m}}$$

$$= \frac{290 \times 1847.83}{850} = 630.4 \text{ K}$$

Example 5.19. *A heat engine working on carnot cycle has wroking source and sink temperatures T_1 and T_2 respectively. It is desired to double the thermal efficiency of the engine by raising the temperature of the source. Determine the new temperature of the source if the temperature of the sink is kept the same.*

Solution : Let the new temperature of source be $T_1{}^1$.

If the first case

$$\text{Efficiency} = \eta = 1 - \frac{T_2}{T_1} \quad ...(A)$$

In the second case

$$\text{Efficiency=} \quad \eta^1 = 1 - \frac{T_2}{T_1^1} \quad ...(B)$$

Since

$$\eta^1 = 2\eta$$

$$\left(1 - \frac{T_2}{T_1^1}\right) = 2\left(1 - \frac{T_2}{T_1}\right)$$

$$1 - \frac{T_2}{T_1^1} = 2 - \frac{2T_2}{T_1}$$

or

$$\frac{2T_2}{T_1} = 2 - \frac{2T_2}{T_1}$$

or

$$\frac{2T_2 - T_1}{T_1} = \frac{T_2}{T_1^1}$$

or

$$T_1^1 = \frac{T_1 T_2}{2T_2 - T_1}$$

5.9 POWER GENERATION FROM SOLAR ENERGY

Solar energy has been used to generate power by constructing deep artificial lakes which are called SOLAR PONDS. Solar energy is absorbed by all parts of the water in the lake and the water temperature rises every where. However the upper layers of water lose much of the heat it absorbs to the atmosphere. As a result the temperature of the upper layers is lesser than those at the bottom. Moreover salt is added at the bottom of the pond which helps to prevent the rise of hot water to the top. The difference in the temperatures of bottom and top layers is used to run a power plant in which the bottom and top layers of water are treated as source and sink respectively. Generally the efficiency of such power plants is very low. For example if the temperature of the layers is 30°C and that of top bottom layers 75°C.

$$\eta = 1-\frac{T_2}{T_1}=1-\frac{(30+273)}{(75+273)}=1-\frac{293}{345}=0.1293 \text{ or } 12.93\%$$

Example 5.20. *A refrigerator working on cornot cycle consumes 2 kW power when operating and the corresponding temperature of the food compartment is maintained at 4°C. If the room temperature is 22°C, determine the rate of heat removal from the food chamber.*

Solution : Here $T_1 = 22 + 273 = 295$ K

$T_2 = 4 + 273 = 277$ K

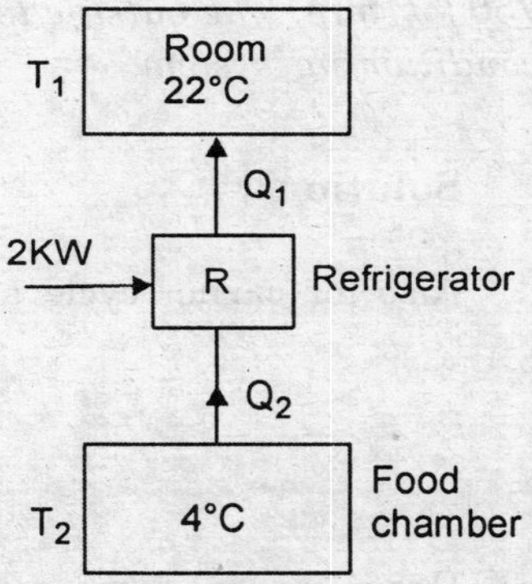

$$\text{COP} = \frac{1}{\frac{T_1}{T_2}-1}=\frac{1}{\frac{295}{277}-1}=\frac{277}{18}=15.4$$

Also $$\text{COP} = \frac{\dot{Q}_2}{\dot{W}_{in}}$$

$\therefore$ $$\dot{Q}_2 = 15.4\times \dot{W}_{in}$$

$= 15.4 \times 2$ kW

$= 30.8$ kW

$= 30.8 \times 60 = 1848.0$ kJ/min

Fig. 5.19

Example 5.21. *A refrigerator removes heat from the cooled space which is kept at a temperature –4°C for which it extracts heat from this space at the rate of 145 kJ/min. The room temperature is 22°C. Determine the minimum power in put required for the refrigerator.*

Solution : Since the minimum power input is to be estimated, it must their be running on a reversible cycle, COP for such a reversible refrigerator is given by

$$\text{COP} = \frac{T_2}{T_1-T_2}=\frac{1}{\frac{T_1}{T_2}-1}$$

Here $T_1 = 22 + 273 = 295$ K

$T_2 = -4 + 273 = 269$ K

$\therefore$ $$\text{COP} = \frac{1}{\frac{295}{269}-1}=\frac{269}{265-269}$$

$$= \frac{269}{26}=10.346$$

Now $\text{COP} = \dfrac{\dot{Q}_1}{\dot{W}}$

$\therefore$ $\dot{W} = \dfrac{\dot{Q}_1}{\text{COP}} = \dfrac{145 \text{ kJ/min}}{10.346}$

$= 14.015 \text{ kJ/min}$

$= \dfrac{14.015}{60} = 0.2335 \text{ kW}$

Example 5.22. *An air conditioning system operating on reversed carnot cycle is used to maintain a house at 20°C for which it is required to extract heat from the house at a rate of 750 kJ/min. The outside temperature is 35°C. Calculate the power requirement to run the air conditioning system.*

Solution : $\text{COP} = \dfrac{\dot{Q}_1}{\dot{W}}$

Also for carnot cycle air conditioner

$$\text{COP} = \frac{1}{\dfrac{T_1}{T_2} - 1}$$

Here $T_1 = 35 + 273 = 308$

$T_2 = 20 + 273 = 293$

$\therefore$ $\text{COP} = \dfrac{1}{\dfrac{308}{293} - 1} = 19.5$

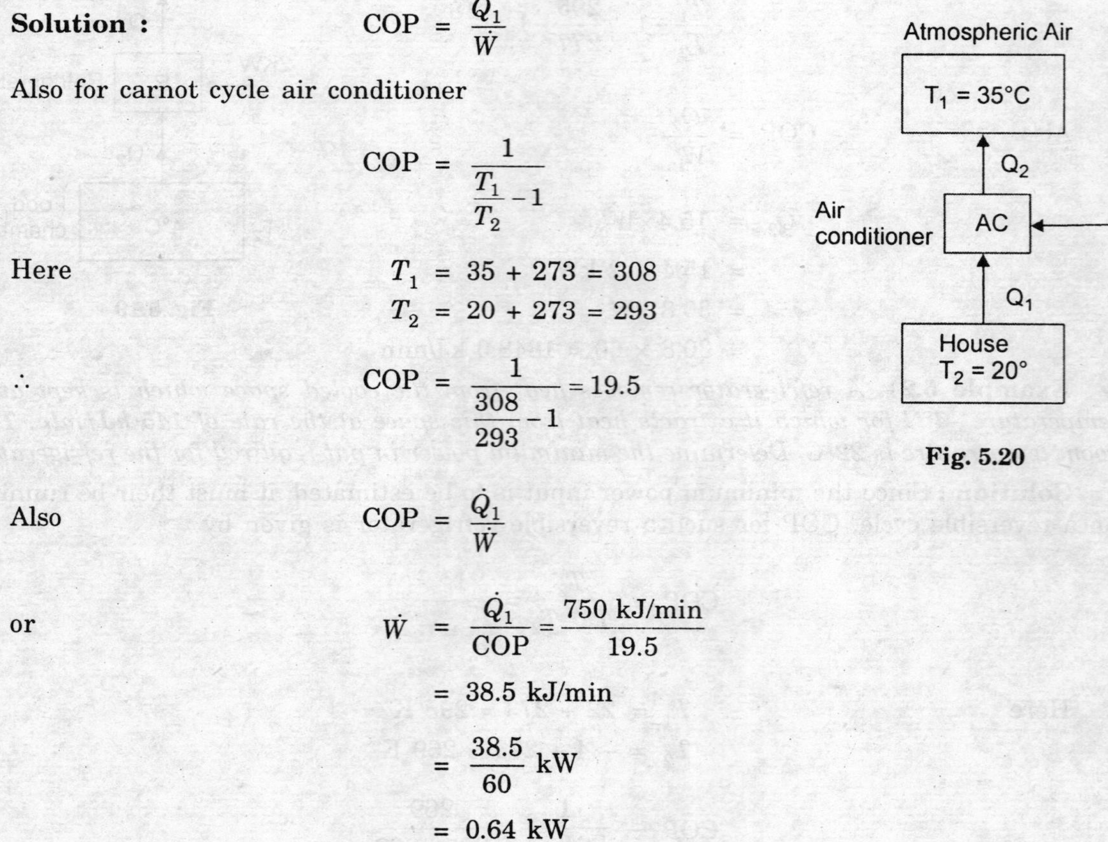

Fig. 5.20

Also $\text{COP} = \dfrac{\dot{Q}_1}{\dot{W}}$

or $\dot{W} = \dfrac{\dot{Q}_1}{\text{COP}} = \dfrac{750 \text{ kJ/min}}{19.5}$

$= 38.5 \text{ kJ/min}$

$= \dfrac{38.5}{60} \text{ kW}$

$= 0.64 \text{ kW}$

Example 5.23. *An air conditioning system is used to keep the temperature inside a house at 22°C when the outside temperature is 33°C. The house gains heat though its walls and windows at a rate of 750 kJ/min and the heat generation rate from within the house due to electric fittings and the people present there amounts to 150 kJ/min. Determine the minimum power requirement for this air condition unit.*

Solution : Since the power requirement is to be least, the air conditioning system must be designed on the basis of reversible cycle.

For such refrigerator $$COP = \frac{1}{\frac{T_1}{T_2} - 1}$$

Where $T_1 = 33 + 273 = 306$ K

$T_2 = 22 + 273 = 295$ K

Therefore $$COP = \frac{1}{\frac{306}{295} - 1} = 26.8$$

Also $$COP = \frac{\dot{Q}_1}{\dot{W}}$$

Now $$\dot{Q}_1 = 750 + 150 = 900 \text{ kJ/min}$$

$$\therefore \quad 26.8 = \frac{900 \text{ kJ/min}}{\dot{w}}$$

or $$\dot{W} = \frac{900}{26.8} \text{ kJ/min}$$

$$= 33.6 \text{ kJ/min}$$

$$= \frac{33.6}{60} \text{ kw}$$

$$= 0.56 \text{ kw}$$

Example 5.24. *A heat pump working on carnot cycle is used to heat and maintain a house at 20°C in winter The house loses heat from windows etc. at the rate 2400 kJ/hr per degree C temperature difference between the window and outside of the house. If the outside temperature is 5°C find (a) the COP (b) power input to the heat pump.*

Solution : $$COP = \frac{1}{1 - \frac{T_2}{T_1}}$$

$$= \frac{1}{1 - \frac{278}{293}} = \frac{293}{15} = 19.53$$

Also $$COP = \frac{\dot{Q}_1}{\dot{W}}$$

or $$\dot{W} = \frac{\dot{Q}_1}{COP}$$

Fig. 5.21

But $\dot{Q}_1$ = 2400 kJ/h × (Temperature Difference between window outdoor)

$$= 2400 \times \left(\frac{\text{kJ}}{\text{h}}\right) \times (20-5)$$

$$= 2400 \times 15 \text{ kJ/h}$$

$$\therefore \quad \dot{W} = \frac{2400 \times 15}{19.53} = 1843.32 \text{ kJ/h}$$

$$= \frac{1843.32}{3600} = 0.512 kW$$

Example 5.25. *A carnot heat pump which requires 8 kw power input is available for installing it in the house the temperature of which is to be maintained at 22°C. The heat loss through the windows etc. has been estimated to be 3000 kJ/hr per degree C difference between the indoor and outdoor temperatures. Determine the lowest outdoor temperature for which this pump can meet the requirement.*

$$\dot{W} = 8 \; kW$$

Solution : Let the otuside temperature be = T_2 K

Inside temperature T_1 = 22 + 273 = 295 K

Temperature difference = $(295 - T_2)$ K

Since it is working on carnot cycle $\text{COP} = \dfrac{1}{1-\dfrac{T_2}{T_1}} = \dfrac{295}{(295-T_2)}$ ...(*A*)

Also $\text{COP} = \dfrac{\dot{Q}_1}{\dot{W}}$ ...(*B*)

Here $\dot{Q}_1$ = Heat lost to surroundings at T_1

$$= \frac{3000 \times (T_1 - T_2)}{\dot{W}}$$

But $\dot{W}$ = 8 kW

= 8 × 3600 kJ/h

Therefore substituting values in equation (*B*) and solving equation (*A*) and (*B*)

$$\frac{295}{(295-T_2)} = \frac{3000(295-T_2)}{8 \times 3600}$$

$$(295 - T_2)^2 = \frac{295 \times 8 \times 3600}{3000}$$

$$= 2832.0$$

or $(295 - T_2) = \sqrt{2832.0} = 53.21$

or $$T_2 = 295 - 53.21$$
$$= 241.78 \text{ K}$$
$$= 241.78 - 273$$
$$= -31.2°\text{C}$$

Example 5.26. *A carnot heat pump is intended to be used for heating and maintaining the indoor temperature of a house at 20°C during winter when the outside average temperature is 2°C. The estimated heat loss from the windows etc. is 70000 kJ/hr. If the heat pump consumes 8 kW of power to run it, determine the (a) total time pump ran or a particular day (b) the total cost to be incurred to run the pump if electricity costs Rs. 3.0 per kWh (c) The Cost to be incurred if resistance type heater is used.*

Solution : Carnot cycle Heat pump

Therefore $$\text{COP} = \frac{1}{1-\frac{T_2}{T_1}}$$

$$T_1 = 20 + 273 = 293 \text{ K}$$
$$T_2 = 2 \times 273 = 275 \text{ K}$$

$$\text{COP} = \frac{1}{1-\frac{275}{293}} = \frac{293}{18} = 16.28$$

Heat lost during the day

$$Q_1 = \dot{Q}_1(1 \text{ day}) = 7000\left(\frac{\text{kJ}}{\text{h}}\right)\times(24 \text{ hr}) = 1680000 \text{ kJ}$$

Now $$\text{COP} = \frac{Q_1}{W_{net}}$$

$$W_{net} = \frac{1680000}{\text{COP}} = \frac{1680000}{16.28} \text{ kJ}$$
$$= 103194.1 \text{ kJ}$$

If the pump ran for t sec on that day then

$$\frac{W_{net}}{\dot{W}} = t$$

Where $\dot{W}$ = rate of input per sec
= 8 kW
= 8 kJ/s

$$\therefore \quad t = \frac{103194.1(\text{kJ})}{8 \text{ (kJ/s)}}$$
$$= 12899.3 \text{ sec} = \mathbf{3.58 \text{ hrs}}$$

(*b*) Total cost to run the pump

$$= (8 \text{ kw})(3.58 \text{ hr}) \times \frac{\text{Rs. } 3}{\text{kW hr}}$$

$$= \text{Rs. } \mathbf{85.995}$$

(*c*) If the resistance type heater is used

$$\text{Total cost} = Q_1 \times \text{Price} = (168000 \text{ kJ}) \left(\frac{1 \text{ kWh}}{3600 \text{ kJ}}\right) \times (3 \text{ Rs/kWh})$$

$$= \frac{1680000 \times 3}{3600} = 1400 \text{ rupees}$$

Example 5.27. *A carnot heat engine receives heat from a thermal source at 450°C and rejects the waste heat to the enviornment at 27°C. The work output of this engine is utilised to drive a carnot refrigerator which removes heat from the food compartment to be kept at –10°C at the rate of 375 kJ/min (Fig 5.22) The refrigerator also rejects the heat to same environment at 27°C. Determine the rate of heat supplied to the heat engine (b) total rate of heat rejection to the atmosphere.*

Solution : $T_1 = 450 + 273 = 723$ K

For Refrigerator $Q_2, R = 375$ kJ/min

$$(\text{COP})_R = \frac{1}{\frac{T_1}{T_2} - 1}$$

$$= \frac{1}{\frac{300}{263} - 1}$$

$$= \frac{263}{37} = 7.1$$

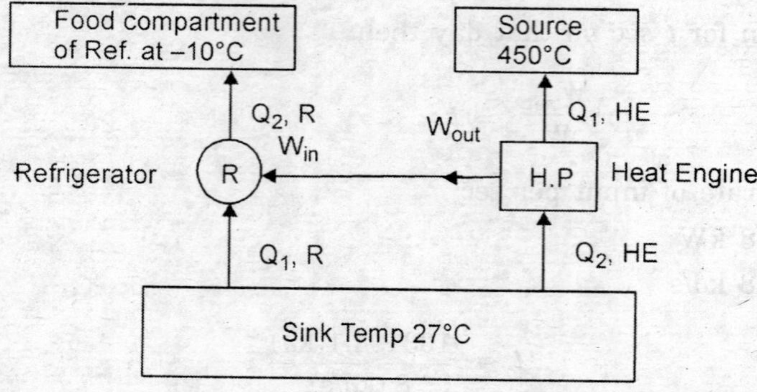

Fig. 5.22

Also $$\text{COP} = \frac{\dot{Q}_{2,R}}{\dot{W}_{in,R}}$$

$\therefore$ $$\dot{W}_{\text{in},R} = \frac{\dot{Q}_{2,R}}{\text{COP}} = \frac{375 \text{ kJ/min}}{7.1}$$

$$= 52.82 \text{ kJ/min}$$

$$= \dot{W}_{\text{out}}, HE$$

Now for *HE AT ENGINE* $$T_2 = 27 + 273 = 300 \text{ K}$$

Also $$T_{1,HE} = 450 + 273 = 723 \text{ K}$$

$$\eta_{HE} = 1 - \frac{T_{2,HE}}{T_{1,HE}} = 1 - \frac{300}{723}$$

$$= 1 - \frac{300}{723} = \frac{423}{723} = 0.585$$

But $$\eta = \frac{\dot{W}_{\text{net},HE}}{Q_{1,HE}} = \frac{\dot{W}_{\text{in},R}}{Q_{1,HE}}$$

i.e. $$Q_{1,HE} = \frac{\dot{W}_{\text{in},R}}{\eta_{HE}}$$

$$= \frac{52.82}{0.585} \text{ kJ/min}$$

$$= 90.29 \text{ kJ/min} \quad \textbf{Ans.}$$

Therefore Heat rejected by *HE* to environment

$$= Q_{1,HE} - \dot{W}_{\text{out},HE}$$

$$= 90.29 - 52.82 = 37.47 \text{ kJ/min}$$

Heat rejected by refrigerator to environment

$$Q_{1,R} = Q_{2,R} + W_{\text{in},R}$$

$$= 375 + 52.82 = 427.82 \text{ kJ/min}$$

Total heat rejected to environment

$$= 37.47 + 427.82 = 465.29 \text{ kJ/min} \quad \textbf{Ans.}$$

$$= 465.29 \text{ kJ/min} \quad \textbf{Ans.}$$

Example 5.28. *A heat engine operates between two reservoirs at 777 and 20°C. Half of the work output of this engine is used to drive a heat pump working on carnot cycle. This heat pump in turn maintains the temperature of the house at 22°C by extracting heat from the surroundings which are at 5°C and transferring it to the house. House at the same time loses heat at the rate of 75000 kJ/h. Determine the minimum rate of heat supply to the heat engine to run this unit.*

Solution : For the carnot ***HP***

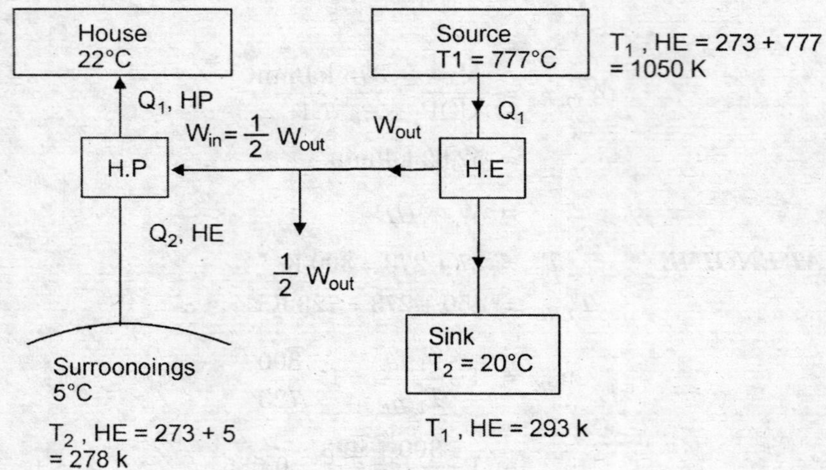

Fig. 5.23 HE → Heat Engine

HP → Heat Pump

H.P

$$T_2, HP = 5 + 273 = 275K$$

$$T_1, HP = 22 + 273 = 295 \text{ K}$$

$$T_1, HE = 777 + 273$$

$$= 1050\ K$$

H.E

$$T_2, HE = 20 + 273$$

$$= 293\ K$$

$$\text{COP} = \frac{1}{1 - \frac{T_{2,\,HP}}{T_{1,\,HP}}}$$

$$= \frac{1}{1 - \frac{278}{295}}$$

$$= \frac{295}{17} = 17.35$$

Heat loss form House = 75000 kJ/h

= Heat supplied by ***HP***

But $\text{COP} = \frac{\dot{Q}_1}{W_{in,\,HP}}$

$$\therefore \quad \dot{W}_{in, HP} = \frac{\dot{Q}_1}{COP}$$

$$= \frac{75000}{17.35} \text{ kJ/h}$$

$$= 4322.76 \text{ kJ/h}$$

= This input work *HP* is half of the output work of the *HE*

$$\therefore \quad (\dot{W}_{out})_{HE} = 2(\dot{W}_{in})_{HP}$$

$$= 2 \times 4322.76 \text{ kJ/h}$$

$$= 8645.52 \text{ kJ/h}$$

Since the minimum rate of heat supply is to be determined, this *HE* heat engine must be assumed to be working on carnot cycle.

Then $$\eta_{HE} = 1 - \frac{T_{2, HE}}{T_{1,HE}}$$

$$= 1 - \frac{293}{1050} = \frac{751}{1050} = 0.7209$$

Also $$\dot{Q}_{2,HE} = \frac{\dot{W}_{out, HE}}{\eta_{HE}} = \frac{8645.52}{0.7209} \text{ kJ/h}$$

$$= 11992.67 \text{ kJ/h}$$

Example 5.29. *The efficiency of an engine mounted on a ship is 75% of that of an engine working on carnot cycle. The heat source for the engine is sea water at 315 K and the sink is the atmospheric air at 295 k. The engine work output is utilised to operate on agitator to heat 400 kg of sea water to 350 K. If the specific heat of sea water C_p is 4.186 kJ/kg K determine the quantity of heat energy that must be extracted from the sea water so as to achieve the required objective.*

Solution : Thermal efficiency of the given engine = 75% of carnot engine efficiency working between the given heat source at temperature T_1 and sink (temperature T_2)

$$\eta = 0.75 \times \text{carnot efficiency}$$

$$= 0.75 \times \frac{T_1 - T_2}{T_1} = 0.75 \times \frac{315 - 295}{315}$$

$$= 0.75 \times \frac{20}{315} = 0.0476$$

Now work required for heating 400 kg of

Sea water $= 400 \times C_p\,(350 - 315)$

$= 400 \times 4.186\,(35)$

$= 58604$ kJ

Also $\eta = \dfrac{W}{Q_1} = \dfrac{58604}{Q_1}$

Where Q_1 is the heat requirement

But $\eta = 0.0476$

$\therefore \quad 0.0476 = \dfrac{58604}{Q_1}$

or $Q_1 = \dfrac{58604}{0.0476}$

$= 1231176$ kJ **Ans.**

Example 5.30. *A source S_A can supply heat energy at a rate of 13000 kJ/min at 580 K. There is another source S_B which can supply heat energy at a rate 13000 kJ/min at 360 K. A reversible heat engine which has to produce large amount of power is to be installed to extract heat from either of these sources and waste heat to be rejected to surroundings at 300 K. Which source should be selected for this propose justify your answer with proper calculations.*

Solution : In the case of source S_A

$(\dot{Q}_1)_A$ = Rate of supply of energy

= 13000 kJ/min

$$\eta_A = \frac{(T_1)_A - T_2}{(T_1)_A}$$

$$= \frac{580 - 300}{580} = \frac{280}{580} = 0.482$$

Therefore $\eta_A = \dfrac{\dot{W}_A}{(\dot{Q}_1)_A}$

or $W_A = 0.482 \times 13000$

$= 6266$ kJ/min ...(A)

Similarly in the case of source S_B

$(\dot{Q}_1)_B$ = Rate of supply of energy

= 130000 kJ/min

$$\eta_B = \frac{(T_1)_B - T_2}{(T_1)_B}$$

$$= \frac{360 - 300}{360} = \frac{60}{360} = 0.1667$$

$$= \frac{\dot{W}_B}{(\dot{Q}_1)_B}$$

or

$$\dot{W}_B = \eta_B \times (\dot{Q}_1)_B$$

$$= 0.1667 \times 130000$$

$$= 21671 \text{ kJ/min} \qquad ...(B)$$

From (A) and (B) we conclude that the source S_A will be chosen as its output is more then that of source S_B despite the fact that the efficiency may be lower than that in the case of S_A because the criterion for selection is the larger output.

Example 5.31. *A refrigeration plant working on carnot cycle is required to be designed for maintaining the temperature of food store at –5°C. The ambient temperature is 27°C and the expected waste heat rate is 5 kW. Determine the COP of the plant and the power required to operate to.*

Solution : Refer to Fig 5.19 $T_1 = 273 + 27 = 300 \text{ K}$

$$T_2 = -5 + 273 = 268 \text{ K}$$

$$\text{COP} = \frac{T_2}{T_1 - T_2} = \frac{268}{300 - 268}$$

$$= \frac{268}{32} = 8.375$$

Also

$$\text{COP} = \frac{Q_2}{Q_1 - Q_2}$$

Where Q_2 = waste heat rate

= 5 kW

or

$$8.375 = \frac{5 \text{ kW}}{Q_1 - 5 \text{ kW}}$$

is

$$(Q_1 - 5) = \frac{5}{8.375} = 0.5714$$

or $Q_1 = 5.5714$ kw

Also $Q_1 = W + Q_2$

i.e., $W = Q_1 - Q_2$

$$= 5.5714 - 5$$

$$= 0.5714 \text{ kW}$$

Where W is the minimum power required to run the refrigerator

Example 5.32. *A carnot engine absorbs Q_1 units to heat from source at uniform temperature T_1 and rejects Q_2 units of heat to a sink at temperature T_2. However due to inperfact heat* transfer the temperatures attained by the working fluid are respectively θ_1 *and* θ_2 *given by the following relations.*

$$\theta_1 = T_1 - kQ_1$$

and

$$\theta_2 = T_2 + kQ_2$$

Where k is a constant show that the efficiency of the engine is given by

$$\eta = 1 - \frac{T_2}{T_1 - 2kQ_1}$$

Solution : Fig. (5.24) shows the arrangement of the system. since the engine works on carnot cycle therefore

$$\eta = \frac{Q_1 - Q_2}{Q_1} = \frac{\theta_1 - \theta_2}{\theta_1}$$

or $$\frac{\theta_2}{\theta_1} = \frac{Q_2}{Q_1} \quad ...(B)$$

But it is given that $$\theta_1 = T_1 - kQ_1$$

That is $$\frac{\theta_1 - T_1}{k} = -Q_1 \quad (C)$$

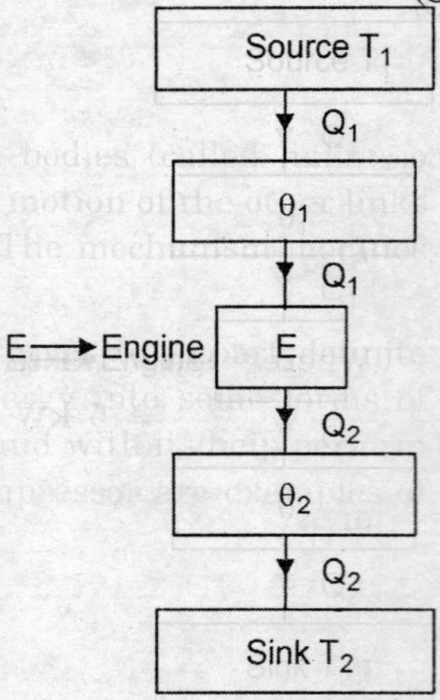

and $$\frac{\theta_2 - T_2}{k} = Q_2 \quad ...(D)$$

Therefore dividing (*C*) and

we get $$\frac{\theta_1 - T_1}{\theta_2 - T_2} = -\frac{Q_1}{Q_2}$$

$$= -\frac{\theta_1}{\theta_2} \text{ (From B)}$$

Source T_1 → Q_1 → θ_1 → Q_1 → E⟶Engine E → Q_2 → θ_2 → Q_2 → Sink T_2

Fig. 5.24

By cross-multiplication

$$\theta_1\theta_2 - T_1\theta_2 = -\theta_1\theta_2 + \theta_1 T_2$$

or $$2\theta_1\theta_2 - T_1\theta_2 = \theta_1 T_2$$

or $$\theta_2(2\theta_1 - T_1) = \theta_1 T_2$$

or $$\theta_2 = \frac{\theta_1 T_2}{2\theta_1 - T_1} \quad ...(E)$$

But from given relationship

$$\theta_1 = (T_1 - kQ_1) \quad ...(F)$$

Substituting this in equation (*E*)

$$\theta_2 = \frac{\theta_1 T_2}{2(T_1 - kQ_1)} - T_1$$

$$= \frac{\theta_1 T_2}{T_1 - 2kQ_1} \quad ...(G)$$

Therefore substituting the values of θ_1 and θ_2 in the expression for efficiency we get .

$$\eta = \frac{\theta_1 - \theta_2}{\theta_1}$$

$$= 1 - \frac{\theta_2}{\theta_1}$$

$$= 1 - \frac{\theta_1 T_2}{\theta_1 (T_1 - 2kQ_1)}$$

or $$\eta = 1 - \frac{T_2}{(T_1 - 2kQ_1)} \quad \text{Hence Proved}$$

Example 5.33. *A refrigerator working on carnot principle operates between temperatures –5°C and 25°C, (a) find its COP (b) If the higher operating temperature is raised by an amount T°C and that the lower one is reduced by T°C. and the new COP is 10, determine the value of T.*

Solution : (*a*)

$$T_1 = 273 + 25 = 298 \text{ K}$$

$$T_2 = 273 - 5 = 268 \text{ K}$$

$$\text{COP} = \frac{T_2}{T_1 - T_2} = \frac{268}{298 - 268}$$

$$= \frac{268}{30} = 8.933$$

(*b*) Here new working temperature are

$$T_1 = 298 + T$$

$$T_2 = 268 - T$$

$$\text{COP} = \frac{T_2}{T_1 - T_2} = \frac{268 - T}{(298 + T)} - (268 - T)$$

$$= \frac{268 - T}{30 + 2T} = 10 \text{ (given)}$$

Therefore $$268 - T = 300 + 20\,T$$

or $$(268 - 300) = 21\,T$$

or $$T = \frac{-32}{21} = -1.524°\text{C}$$

Thus the range of working temperature becomes

$$T_1 = 298 - 1.524 = 296.76 \text{ K}$$

$$T_2 = 268 - (-1.524) = 269.524 \text{ K}$$

Example 5.34. *A carnot heat engine has the working source and sink temperatures as T_1 and T_2 respectively. In order to increase the efficiency there are two proposals viz one to increase the source temperature by say T°C and keeping the sink temperature same and the second proposal is to lower the sink temperature by T°C while keeping the source temperature constant at T_1 Show by calculation which of the proposals is better.*

Solution : In the first case working temperatures are $(T_1 + T)$ and T_2

$$\eta_1 = \frac{(T_1 + T) - T_2}{(T_1 + T)} \quad ...(A)$$

However in the second case

$$\eta_2 = \frac{T_1 - (T_2 - T)}{T_1}$$

$$= \frac{(T_1 + T) - T_2}{T_1} \quad ...(B)$$

We have from A and B

$$(T_1 + T) - T_2 = \eta_1(T_1 + T) = \eta_2 T_1$$

or

$$\frac{\eta_2}{\eta_1} = \frac{T_1 + T}{T_1}$$

$$= 1 + \frac{T}{T_1}$$

Evidently $\eta_2 > \eta_1$

Thus we conclude that the second proposal is better as it will result in greater efficiency.

Example 5.35. *A cornot Heat engine draws equal amounts of heat from two heat sources at temperatures T_A and T_B as shown an Fig. 5.25 and rejects heat to sink at temperature T_C. If the efficiency of the engine is 'a' times the efficiency of a reversible engine operating between the reservoirs A and C find the relationship between T_A/T_B and T_A/T_C.*

Solution : When operating between the reservoirs A and C

$$\eta = \frac{T_A - T_C}{T_A}$$

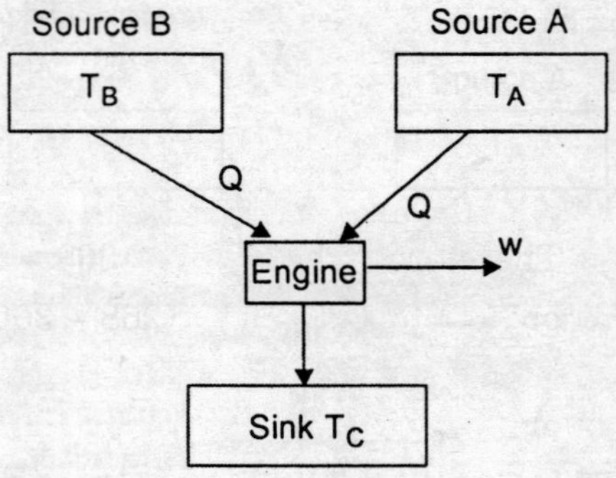

Then corresponding work output

$$\frac{W_{AC}}{\text{Input}} = \eta$$

If input = Q,

$$W_{AC} = Q \times \frac{T_A - T_C}{T_A} \quad ...(A)$$

Fig. 5.25

Similarly when operating between the reservoirs B and C

$$W_{BC} = Q \times \frac{T_B - T_C}{T_B} \quad ...(B)$$

Therefore total work output corresponding to the total heat in take of

$$Q + Q = 2Q$$

$$W = W_{AC} + W_{BC}$$

$$= Q\left(\frac{T_A - T_C}{T_A}\right) + Q\left(\frac{T_B - T_C}{T_B}\right)$$

then $$\eta_{net} = \frac{W}{2Q}$$

$$= \frac{1}{2Q}\left[Q\left(\frac{T_A - T_C}{T_A}\right) + Q\frac{T_B - T_C}{T_B}\right]$$

or $$\eta_{net} = \frac{1}{2}\left[\left(1 - \frac{T_C}{T_A}\right) + \left(1 - \frac{T_C}{T_B}\right)\right] \quad ...(C)$$

$$= \frac{1}{2}\left[2 - \frac{T_C}{T_A} - \frac{T_C}{T_B}\right]$$

Now the efficiency of reversible engine working between reservoirs A and C

is $$\eta_2 = \frac{T_A - T_C}{T_A}$$

$$= \left(1 - \frac{T_C}{T_A}\right) \quad ...(D)$$

It is given that $\eta = a\,(\eta_2)$

i.e. from equations (C) and (D)

$$\frac{1}{2}\left[2 - \frac{T_C}{T_A} - \frac{T_C}{T_A}\right] = a\left[1 - \frac{T_C}{T_A}\right]$$

$$1 - \frac{1}{2}\frac{T_C}{T_A} - \frac{1}{2}\frac{T_C}{T_B} = a - \frac{aT_C}{T_A}$$

or $$2 - \frac{T_C}{T_B} - \frac{T_C}{T_B} = 2a - \frac{2aT_C}{T_A}$$

$$\left((2a-1)\frac{T_C}{T_A} + 2(1-a)\right) = \frac{T_C}{T_B}$$

or $$(2a-1) + 2(1-a)\frac{T_A}{T_C} = \frac{T_A}{T_B}$$

Example 5.36. *A reversible heat engine mounted on a·satalite in the space operates between a high temperature reservoir at T_1 and low temperature solar panel at temperature T_2. The radiation heat transfer from the panel is proportional to fourth power of the absolute temperature. For a given work output and value of T_1 show that the area of the panel will be minimum when $\frac{T_2}{T_1} = \frac{3}{4}$ and find the corresponding area.*

Solution :

$$\eta = \frac{W}{Q_1} = \frac{Q_1 - Q_2}{Q_1} = \frac{T_1 - T_2}{T_1} \quad ...(A)$$

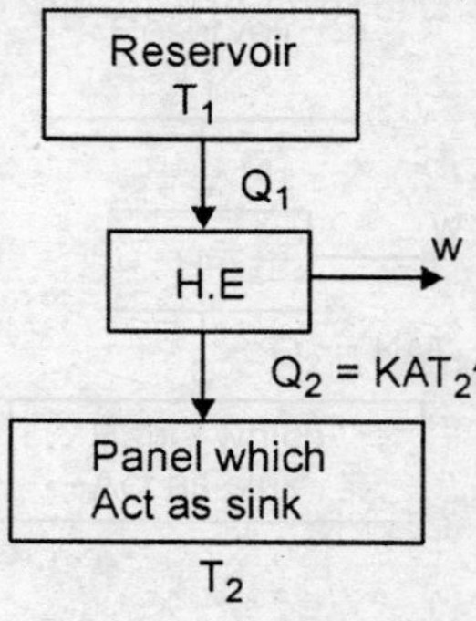

Reservoir T_1

Q_1

H.E

W

$Q_2 = KAT_2^4$

Panel which Act as sink

T_2

Fig. 5.26

or $$1 - \frac{Q_2}{Q_1} = 1 - \frac{T_2}{T_1}$$

or $$\frac{Q_2}{Q_1} = \frac{T_2}{T_1}$$

or $$\frac{Q_1}{T_1} = \frac{Q_2}{T_2} \quad ...(B)$$

Again from (*A*) $$\frac{W}{Q_1} = \frac{T_1 - T_2}{T_1} \quad ...(C)$$

or $$\frac{W}{T_1 - T_2} = \frac{Q_1}{T_1}$$

$$= \frac{Q_2}{T_2} \quad \text{[From B]} \quad ...(D)$$

Now it is given $$Q_2 \propto AT_2^4 \quad ...(E)$$

$$= KAT_2^4$$

when K is constant

∴ From equations D and E

$$\frac{W}{T_1 - T_2} = \frac{KAT_2^4}{T_2}$$

or $$A = \frac{W}{(T_1 - T_2)\, K T_2^3} \quad ...(F)$$

Thus for given values of W and T_1 area A will be minimum when the denominator of RHS of equation F is maximum. For which

$$\frac{d}{dT_2}\left[(T_1 - T_2) K T_2^3\right] = 0$$

$$\text{or} \qquad 3T_1KT_2^2 - 4KT_2^3 = 0$$

$$\text{or} \qquad 3T_1K - 4KT_2 = 0$$

$$\text{i.e.,} \qquad T_2 = \frac{3}{4}T_1 \qquad \text{...(G)}$$

Substituting the value of T_2 from equation (G) in equation F we get the minimum area as

$$A_{min} = \frac{W}{(T_1 - 0.75T_1)K(0.75T_1)^3}$$

$$= \frac{W}{K(0.25T_1)(0.75T_1)^3}$$

$$= \frac{256W}{27KT_1^4}$$

$$\text{or} \qquad A_{min} = \frac{9.48W}{KT_1^4}$$

5.10 CARNOT THEOREM

Statement

No heat engine working in a cycle between two given thermal reservoirs of constant temperatures can be more efficient than a reversible engine operating between the same two reservoirs.

Proof

Let there be a reversible engine E_A working between two heat sources at temperatures T_1 and T_2 as shown in Fig. (5.27*a*) (where $T_1 > T_2$). Let an irreversible engine E_B be aslo operating between the same heat sources at T_1 and T_2 K on which the reversible engine E_A is working. Let both the engines receive the same quantity of heat, Q kJ/unit line. At this stage we assume that the heat engine 'E_B' is more efficient than, E_A.

That is if

$$\eta_B > \eta_A$$

$$\text{or} \qquad \frac{W_B}{Q} > \frac{W_A}{Q}$$

$$\text{or} \qquad W_B > W_A \qquad \text{...(5.25)}$$

$$\text{and then} \qquad Q - W_A < (Q - W_B) \qquad \text{...(5.26)}$$

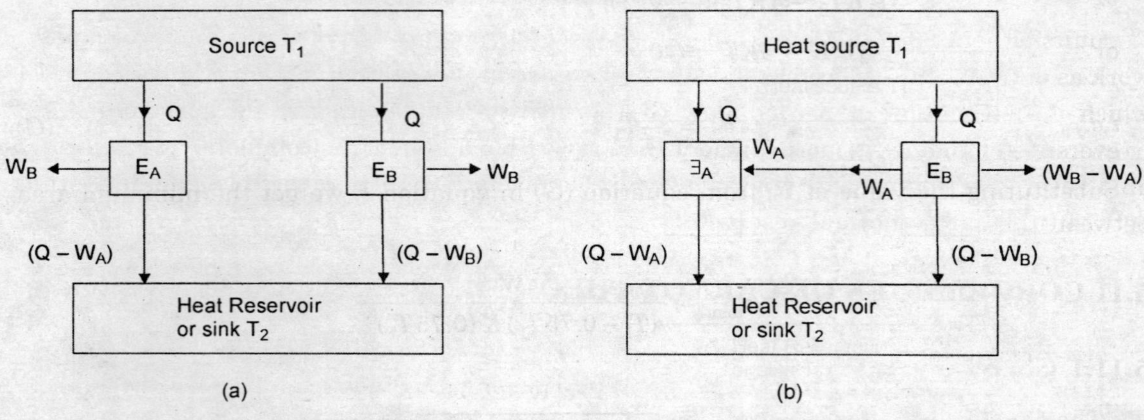

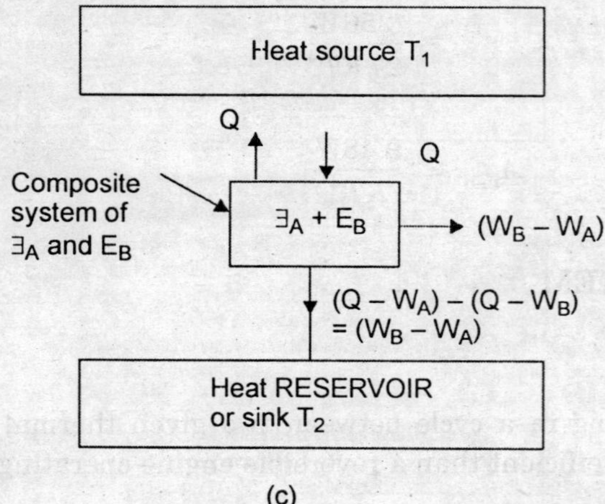

Fig. 5.27

From equation (5.25) we conclude that the output of irreversible engine E_B is more than that of reversible engine E_A.

Let the reversible engine E_A be made to operate as refrigerator or heat pump and the irreversible engine E_B continues to operate as engine between the given heat reservoirs as shown in Fig. (5.27 *b*).

As the engine E_A is reversible the heat and work interactions will be same as before but their directions will be reversed. The work now required to operate the refrigerator/heat pump E_A will be W_A which can be drawn from the irreversible Engine E_B, whose output is W_B, by directly coupling the two engines which will then form a composite system comprising of reversible engine E_A (now working as heat pump/refrigerator) and the irreversible engine E_B. From the Fig. (5.27*b*) it is evident that the net heat interaction with the high temperature reservoir is zero as it supplies an amount equal to Q and receives back Q units of heat. Secondly the composite system withdraws.

$$(Q - W_A) - (Q - W_B) = (W_B - W_A)$$

units of heat form the low temperature reservoir and converts it into $(W_B - W_A)$ units of work as output. Thus the composite system behaves as a perpetual motion of the second kind which is in violation of Second Law of Thermodynamics. Evidently the assumption that irreversible engine E_B is more efficient then reversible engine E_A is completely wrong. Hence an irreversible engine can not have higher efficiency than a reversible engine working between the same thermal reservoirs.

5.11 COROLLARIES OF CARNOT THEOREM

5.11.1 Corollary 1

The efficiency of all reversible heat engines operating between two constant temperature thermal reservoirs, have the same efficiency.

Proof

Let both the engines E_A and E_B shown in Fig. (5.28 (*a*)) be reversible and both are working between the heat reservoirs of constant temperatures T_1 and T_2 ($T_1 > T_2$). Further assume that both the engines receive the same quantity of heat Q from the higher temperature heat source. Let the efficiency of heat engine E_A be more than that of Engine E_B. If the output of these engines is W_A and W_B respectively, then

$$\eta_A > \eta_B$$

or
$$\frac{W_A}{Q} > \frac{W_B}{Q}$$

or
$$W_A > W_B \qquad \text{...(5.27)}$$

If now the engine E_B is reversed to act as heat pump and engine

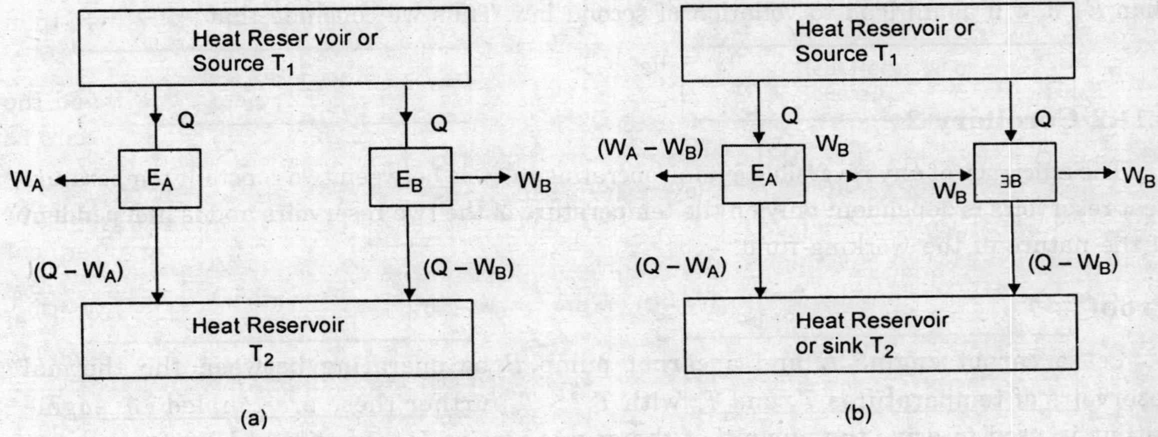

Fig. 5.28

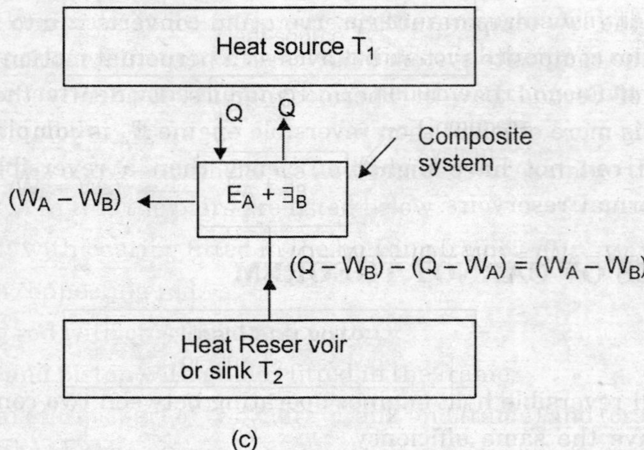

(c)

Fig. 5.28

E_A continuous work as engine between the same reservoirs. Since the engine E_B is reversible, heat and work interactions will be same as before but their directions will be reversed. The work now required to operate the engine E_B as heat pump will be W_B which being lower than W_A can be drawn from the output of heat engine W_A by simply coupling the heat pump E_B with the heat engine E_A to form a composite system as shown in Fig. (5.32 *b*). From the Fig. (5.28 *b*) it is evident that the net heat interaction with the high temperature reservoir in zero but the composite system withdraws

$$(Q - W_B) - (Q - W_A) = (W_A - W_B)$$

amount of heat from low temperature reservoir and converts it into $(W_A - W_B)$ units of work as net output. Thus this composite system behaves as a perpetual motion machine of second kind in voilation of Second Law of Thermodynamics. Thus we conclude that the assumption $\eta_A > \eta_B$ is wrong. Similarly if we assume that efficiency of heat engine E_B is more than E_A it will again lead to voilation of second law. Thus we conclude that

$$\eta_A = \eta_B$$

5.11.2 Corollary 2

The efficiency of any reversible engine operating in cycle between two constant temperature heat reservoirs is dependent only on the temperature of the two reservoirs and is independent of the nature of the working fluid.

Proof

Let a carnot engine E and a carnot pump P be operating between the thermal reservoirs of temperatures T_1 and T_2 with $T_1 > T_2$ further these are coupled *i.e.* engine output is used to drive the pump. Let the engine receive Q_1 quantity of heat from

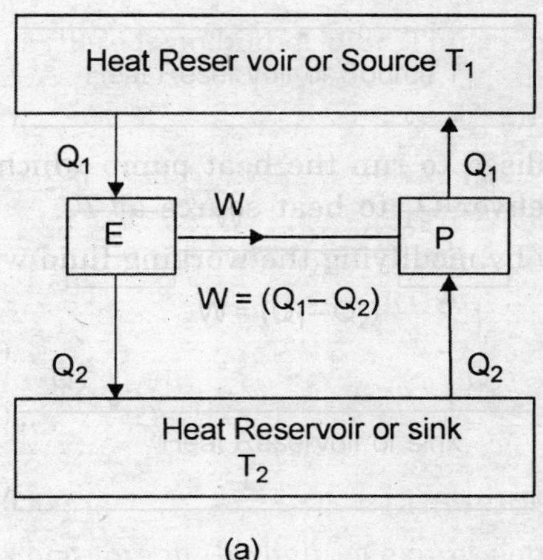

(a)

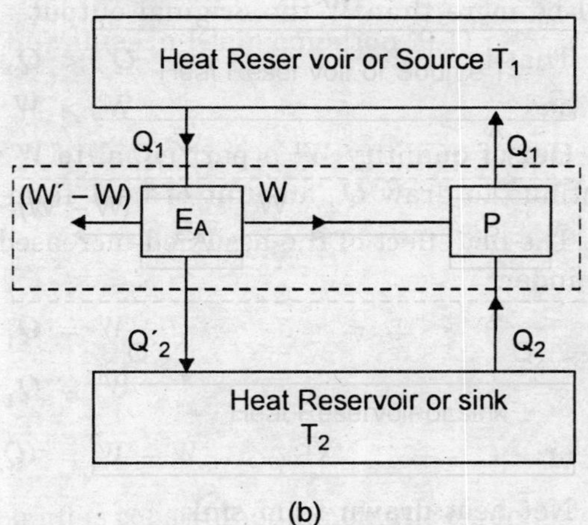

(b)

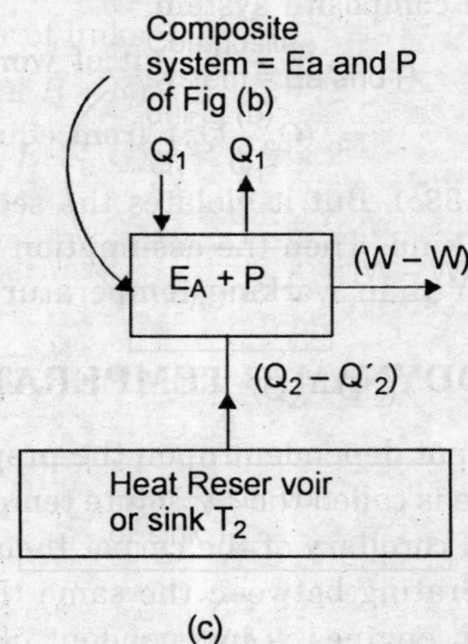

(c)

Fig. 5.29

reservoir at T_1 and reject Q_2 quantity of heat to sink at T_2. As both the engine and pump work on carnot cycle and both operate between the same heat reservoirs their efficiency will be same and dependent upon the temperatures of the reseviors. Further since carnot pump is a reversed carnot engine, the work and heat transfers will be same but in the reversed direction. Thus the heat pump will draw Q_2 amount of heat from same at T_2 and deliver Q_1 amount of heat to heat source T_1 while absorbing work W equivalent to the output of Engine E as shown in Fig. (5.29*a*). Net effect of this composite system on the surrounding will be zero.

Let it now be assumed that factors other than the temperatures of the two reservoirs, like working fluid affects the efficiency of the machines. Let us assume due to the working fluid effect the efficiency of the engine E only is increased by modifying the working fluid and the pump continuous to be same. Then in that case let Q_2' be the amount of heat rejected to the sink at T_2 will be less than Q_2 and the corresponding work output W' of the engine will be more than W the original output.

That is $Q_2' < Q_2$

and $W' > W$

Out of quantity W^1 a part equal to W will be utilised to run the heat pump which will continue to draw Q_2 amount of heat from sink to deliver Q_1 to heat source at T_1.

The net effect of the assumed increased efficiency by modifying the working fluid will be as under.

$$W = Q_1 - Q_2 \quad ...(5.28)$$

$$W' = Q_1 - Q_2' \quad ...(5.29)$$

or

$$W' - W = (Q_2 - Q_2') \quad ...(5.30)$$

Net heat drawn from sink

$$= Q_2 - Q_2'$$

and net work output of the composite system

$$= W' - W \text{ unit of work}$$

$$= (Q_2 - Q_2^1) \text{ from equation (5.30)}$$

These are shown in Fig. (5.33*c*). But it violates the second law of thermodynamic as it is a perpetual motion of second kind. Then the assumption that the efficiency of a reversible engine depends on factors other than working temperature of reservoirs is wrong.

5.12 ABSOLUTE THERMODYNAMIC TEMPERATURE SCALE

Temperature scale which is not dependent upon the properties of the substances that are used to measure the temperature is called the "Absolute temperature scale". For the derivation of such a scale the use of second corollary of the carnot theorem is invoked which states that all reversible engines when operating between the same thermal reservoirs have the same efficiency. The efficiency of such engines is independent of the working fluid employed, its properties, the way the cycle is executed or even the type of the reversible engine used.

Because the thermal energies of the two reservoirs is characterised by their respective temperatures, the thermal efficiency of the heat engines will be function of reservoir temperatures say $g(T_1, T_2)$

Where T_1 and T_2 are temperatures of heat source and heat sink reservoirs respectively and $g(T_1, T_2)$ is any function of (T_1, T_2).

Further since

$$\eta_{thermal} = 1 - \frac{Q_2}{Q_1} \quad ...(5.31)$$

But as stated above Q is same fuction of T

We have

$$\frac{Q_1}{Q_2} = f(T_1, T_2) \quad ...(5.32)$$

Where $f(T_1, T_2)$ is another function, the form of the function $f(T_1, T_2)$ can be found with the help of their reversible heat engines arranged to work.

(*i*) Engine E_A receives Heat Q_1 at temperature T_1 and rejects heat Q_2 at temperature T_2 and gives W_A work output.

(*ii*) Engine E_B receives heat Q_2 at temperature T_2 as rejected by engine E_A. Engine E_B rejects heat Q_3 to the sink at temperature T_3 its output is W_B.

(*iii*) Engine E_C receives heat Q_1 at temperature T_1 and rejects heat Q_3 to the sink at temperature T_3 and its work output is W_C.

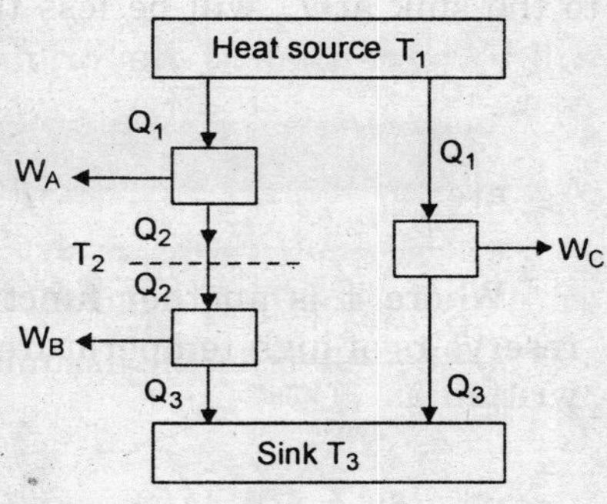

Fig. 5.30

Between two heat reservoirs of temperatures T_1 and T_3 as shown in Fig. (5.30). The engines *A* and *C* both draw Q_1 amount of heat from the higher temperature heat source at T_1. The engine *C* rejects Q_3 amount of heat energy to the sink at temperature T_3 and generates W_C work output. While the engine *A* gives a work output of W_A and rejects Q_2 amount of heat energy to engine *B*. The engine *B* which receives Q_2 units of heat form engine A_1 gives work out W_B and rejects Q_3 units of heat to the sink at T_3 the sink is same as that which receives Q_3 units of heat from engine *C*.

Now applying equation (5.32) to those three reversible engines we have

$$\frac{Q_1}{Q_2} = f(T_1, T_2) \quad ...(5.33)$$

$$\frac{Q_1}{Q_3} = f(T_1, T_3) \quad ...(5.34)$$

and

$$\frac{Q_2}{Q_3} = f(T_2, T_3) \quad ...(5.35)$$

From equation (5.33) and (5.35) we get on multiplication

$$\frac{Q_1}{Q_2} \cdot \frac{Q_2}{Q_3} = f(T_1, T_2) \cdot f(T_2, T_3)$$

or

$$\frac{Q_1}{Q_3} = f(T_1, T_2) \cdot f(T_2, T_3) \quad ...(5.36)$$

Again from equation (5.34) and equation (5.36) we have

$$f(T_1, T_3) = f(T_1, T_2) \cdot f(T_2, T_3) \quad ...(5.37)$$

From equation (5.37) we see that left hand side is function of T_1, T_2 which leads us to conclude that the right hand side of the equation must also be a function of T_1 and T_2. In other words the right hand side of the equation is independent of temperature T_2. This condition will be satisfied only if the function *f* is the form given below

$$f(T_1, T_2) = \frac{\psi(T_1)}{\psi(T_2)} \qquad ...(5.38a)$$

are $$f(T_2, T_3) = \frac{\varphi(T_2)}{\varphi(T_3)} \qquad ...(5.38b)$$

Where ψ is another function. Then for a reversible heat engine working between two reservoirs of high temperature T_1 and low temperature T_3 equation (5.38*a*) and (5.32) can be written as

$$\frac{Q_1}{Q_2} = \frac{\psi(T_1)}{\psi(T_2)} \qquad ...(5.39a)$$

$$\frac{Q_1}{Q_3} = \frac{\psi(T_1)}{\psi(T_3)} \qquad ...(5.39b)$$

and $$\frac{Q_2}{Q_3} = \frac{\psi(T_2)}{\psi(T_3)} \qquad ...(5.39c)$$

Various forms of the function ψ can be found which will satisfy the equation (5.39). However, Lord Kelvin proposed that from $\psi(T) = T$ in order to define a thermodynamic temperature scale. Thus as per this proposal.

$$\frac{Q_1}{Q_2} = \frac{T_1}{T_2}, \frac{Q_2}{Q_3} = \frac{T_2}{T_3} \text{ and } \frac{Q_1}{Q_3} = \frac{T_1}{T_3} \qquad ...(5.40)$$

This temperature scale is known as Kelvin scale and the temperature on this scale in called the absolute temperature. Equation (5.40) dictates that the temperature ratios depend in the ratios of heat transfer between a reversible engine and the heat reservoirs and this ratio is independent of the physical properties of the working fluid. On this scale the temperature varies between zero and infinity.

Equation (5.40) gives us the ratio of absolute temperature on Kelvin scale is equal to the ratio of heat absorbed to heat rejected by a carnot engine operating between these temperatures. Thus the relation leads to the conclusion that if one of the two temperatures is fixed and assigned a numerical value the other temperature can be determined. Thus if such a temperature is taken as T_0 as the reference temperature the value of the other temperature T will be uniquely determined by the relation.

$$T = T_0\left(\frac{Q}{Q_0}\right) \qquad ...(5.41)$$

The ratio (Q/Q_0) is fixed by virture of second law of thermodynamics and is only a function only of the temperature of reservoirs. During the discussions which took place at the International conference on weights and measures held in 1954 it was agreed to assign the value of 273.16 K to the TRIPPLE POINT of water which refers to the state at which the three phases of water exist in equilibrium. Then the carnot engine operating between thermal reservoirs of temperature T and the triple point temperature of water T_t, we then have

$$T = T_t\left(\frac{Q}{Q_t}\right) = 273.16\,\frac{Q}{Q_t} \qquad ...(5.42)$$

Evidently in the Kelvin scale, the amount of heat supplied plays the role of thermodynamic property. The amount of heat supplied changes with temperature. Equation (5.42) shows that if $Q = 0$ then $T = 0$. Thus the thermodynamic scale gives the idea of absolute zero.

Further consider a series of carnot engines arranged between a heat source at Temperature T_1 and to lower temperature (as shown in Fig. 5.31) such that (*a*) all the carnot engines operate between only two reservoirs (*b*) each sink is the source of the following engine *i.e.*, heat rejected by one engine is the heat supply to the next engine (*c*) the temperature of the thermal reservoirs are so selected that the work done by each engine is same.

Since the work done by each engine is same we have

$$W = Q_1 - Q_2 = Q_2 - Q_3 = Q_3 - Q_4$$
$$= Q_n - Q_{n+1} \text{ so on} \qquad \ldots(5.43)$$

Again for the thermodynamic or Kelvin scale

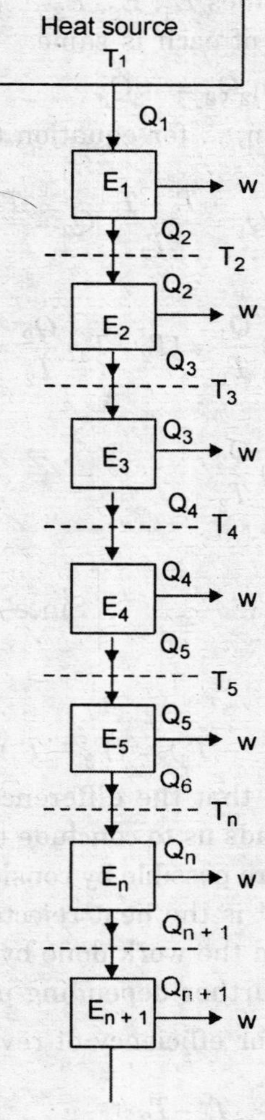

Fig. 5.31. Carnot engine operating (in series)

$$\frac{Q_1}{Q_2} = \frac{T_1}{T_2} \quad ...(5.44a)$$

$$\frac{Q_2}{Q_3} = \frac{T_2}{T_3} \quad ...(5.44b)$$

$$\frac{Q_3}{Q_4} = \frac{T_3}{T_4} \text{ and so on} \quad ...(5.44c)$$

From equation (5.44) we get

$$\frac{Q_1}{T_1} = \frac{Q_2}{T_2} = \frac{Q_3}{T_3} = \frac{Q_4}{T_4} \text{ so on} \quad ...(5.45)$$

Further let the efficiencies of engines E_1, E_2, E_3 ... are respectively η_1, η_2, η_3, η. Then for individual engines the work done of each is same.

$$W = \eta_1 Q_1 = \eta_2 Q_2 = \eta_3 Q_3 \quad ...(5.46)$$

Substituting the values of η_1, η_2, η, ... for equation (5.21) in equation (5.46)

$$W = \frac{T_1 - T_2}{T_1} Q_1 = \frac{T_2 - T_3}{T_2} Q_2 = \frac{T_3 - T_4}{T_3} Q_3$$

or
$$W = (T_1 - T_2)\frac{Q_1}{T_1} = (T_2 - T_3)\frac{Q_2}{T_2} = (T_3 - T_4)\frac{Q_3}{T_3} \text{ and son on} \quad ...(5.47)$$

i.e.,
$$(T_1 - T_2)\frac{Q_1}{T_1} = (T_2 - T_3)\frac{Q_2}{T_2}$$

or
$$(T_1 - T_2) = (T_2 - T_3) \qquad \left[\text{Since } \frac{Q_1}{T_1} = \frac{Q_2}{T_2} \text{ from eqn. 5.45}\right]$$

Similarly it can be shown that

$$(T_1 - T_2) = (T_2 - T_3) = (T_3 - T_4) \quad ...(5.48)$$

From equation (5.48) we conclude that the difference between the temperatures of the successive reservoirs is same which leads us to conclude that equal temperature intervals on the absolute or thermodynamic scale are possible by considering a series of reversible engines in the manner as indicated above that is the heat rejected by one engine becomes the heat supply of the next engine and provided the work done by each such engine is same. Further the intervals can be made larger or further depending upon the requirements of the scale.

Again refering to the expression for efficiency of revesible engine equation (5.21)

$$\eta = \frac{T_1 - T_2}{T_1}$$

We observe that as the sink temperature T_2 is reduced the efficiency will increase. Since the efficiency cannot be more than one, the lowest possible (theoretical) temperature on thermodynamic scale is zero but that carnot be achieved as it will then violate the second law of thermodynamic.

A scale having one hundred equal intervals between ice point and steam point can be achieved by arranging 100 carnot engines in series as indicated in Fig. 5.31. Such a scale will be independent of the working shortance.

Let T_s be the absolute temperature corresponding to steam point and T_i that corresponding to ice point.

Then $$T_s - T_i = 100$$

and $$\eta = \frac{T_s - T_i}{T_s}$$

Further during the discussion held at the International conference on weights and measures in the year 1954 it was agreed to assign the value of 273.16 K the tripple point of water which refers to the state at which the three phases of water exist in equilibrium. The magnitude of the Kelvin degree is defined as $\frac{1}{273.16}$ of temperature interval between absolute zero and the triple point temperature of water. Thus the magnitudes of the temperature units on Kelvin and Calsius scales are identical.

i.e., $$1 \text{ K} = 1°\text{C}$$

The temperature on these scales differ by 273.16

$$T(°C) = T(K) - 273.16 \qquad ...(5.49)$$

Then if T_i(°C) corresponding to the temperature at ice point on Celsius scale the on Kelvin scale its value is

$$T_i = 273.16 \text{ K}$$

Then $$T_s = 100 + T_i$$
$$= 373.16 \text{ K}$$

Therefore efficiency of carnot engine working between reservoirs at T_s and T_i will be

$$\eta = \frac{T_s - T_i}{T_s} = \frac{100}{373.16} = 0.26 \textit{ i.e.}, 26.8\%$$

Even though the Internationally agreed relationship between Kelvin degree and Celisus degree is given in equation (5.49) but for pratical purposes following relationship is taken.

$$T(°C) = T(K) - 273 \qquad ...(5.50)$$

Example 5.37. *A heat engine having an efficiency of 20% derives a heat pump whose COP is 4.5. The heat rejected by the heat engine and the heat delivered by the heat pump are both used to heat the water circulating through the heat radiator used for heating the building during winter season. Calculate the ratio of heat transferred to the circulating water to the heat taken by the engine from the heat source.*

Solution : Fig. 5.32 shows the arrangement of heat source, engine and heat pump.

Let Q_1 be the amount of heat supplied to the engine E which rejects heat Q_2 to the water.

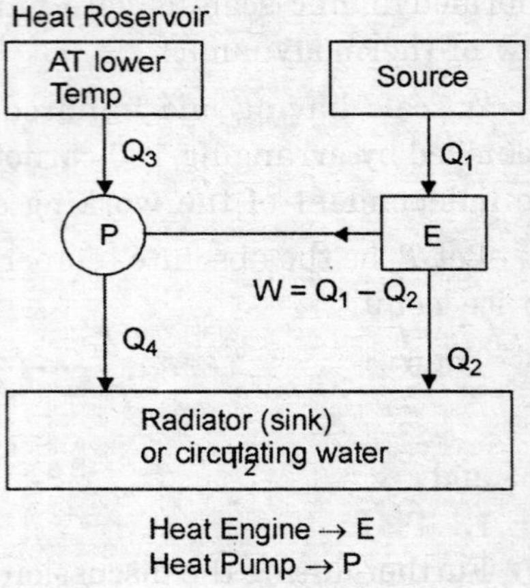

Heat Rosevoir
AT lower Temp
Source
Q_3
Q_1
P
E
$W = Q_1 - Q_2$
Q_4
Q_2
Radiator (sink) or circqlating water
Heat Engine → E
Heat Pump → P

Fig. 5.32

Then work output of engine W

$$W = (Q_1 - Q_2)$$

Also efficiency of the engine

$$\eta_e = \frac{Q_1 - Q_2}{Q_1}$$

$$0.2 = 1 - \frac{Q_2}{Q_1} \quad (\text{As } \eta_e = 0.2)$$

or $$\frac{Q_2}{Q_1} = 0.8 \; i.e., \; Q_2 = 0.8\, Q_1 \quad ...(A)$$

Further let Q_3 be the amount of heat drawn by the heat pump P from low temperature surroundings and Q_4 the amount of heat delivered to the circulating water in the radiator while the entire work no is being used by the heat pump P.

Then COP of pump $$= \frac{Q_4}{Q_4 - Q_3} = 4.5 \quad ...(B)$$

Also $$W = \text{Work out of the engine}$$
$$= (Q_1 - Q_2)$$
$$= \text{Work output to the pump}$$

But for the pump $$\text{COP} = \frac{\text{Heat output or delivered}}{\text{Work input}}$$

or $$\text{COP} = \frac{Q_4}{W}$$

$$4.5 = \frac{Q_4}{Q_1 - Q_2}$$

or $$Q_4 = 4.5\,(Q_1 - Q_2) \quad ...(C)$$

Also work input to pump $= (Q_4 - Q_3)$

That $$Q_1 - Q_2 = Q_4 - Q_3$$

Now heat transfer to the circulating water

$$= Q_2 + Q_4$$

Also the heat taken by the engine = Q_1

Therefore the ratio $$= \frac{Q_2 + Q_4}{Q_1}$$

$$= \frac{Q_2 + 4.5(Q_1 - Q_2)}{Q_1} \quad \text{(From C)}$$

$$= \frac{4.5Q_1 - 3.5Q_2}{Q_1}$$

$$= 4.5 - 3.5\frac{Q_2}{Q_1}$$

$$= 4.5 - 3.5 \times 0.8$$

$$= 4.5 - 2.8$$

$$= 1.70 \textbf{ Ans.}$$

5.13 THE CLAUSIUS INEQUALITY

In article 5.11, it has been proved that an irreversible heat engine is less efficient than a reversible operating between the same two thermal energy reservoirs. Similarly an irreversible heat pump or refrigerator has lower COP than that of reversible one operating between the same temperature limits. Thus it leads to the existance of one important inequality known as Clausius Inequality which is mathematically expressed as

$$\oint \frac{\delta Q}{T} \leq 0 \quad \text{...(5.51)}$$

and which states the cyclic integral of $\frac{\delta Q}{T}$ is always less than or equal to zero. This inequality is valid for all cycles whether reversible or irreversible. Any heat flow to or from a system can be considered to consist of small differential ammounts of heat. Then the cyclic integral $\oint \frac{\delta Q}{T}$ can be viewed as the sum of all the differential amounts of heat divided by the boundary temperature. This is illustrated in the Fig. 5.33.

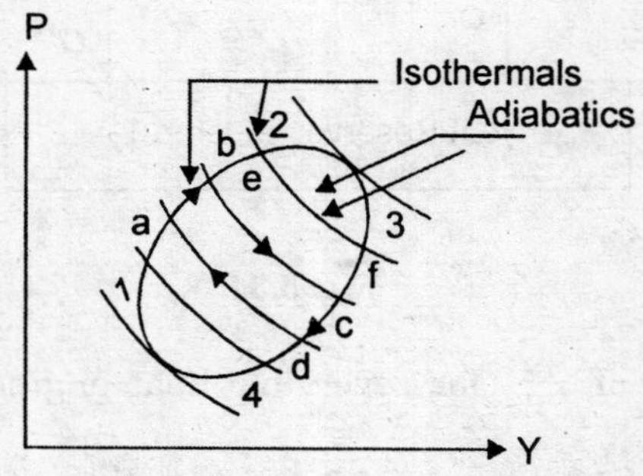

Fig. 5.33

Fig. 5.33 shows the P-V diagram of a system performing reversible cyclic process 1 – 2 – 3 – 4. The cycle is subdivided by closely spaced adiabiatic lines similar to *ad, bc,* etc. The elementary cycle *abcd* then consists of two adiabiatic *bc* and *da* and two elementary paths *ab* and *ed* Since the adiabatics are very close to each other the distance *ab* or *ed* tends to be zero. As a result of which the temperatures at points *'a'* and *'b'* can be taken as same. Similarly, the temperatures at points *'c'* and *'d'* is also taken to be same but essentially it will be different from that of the points *'a'* and *'b'*. Thus the elementary cycle *ab* and *cd* conform to isothermal conditions. Then the cycle *abcd* constitutes an elementary carnot cycle. Similarly *befc*, will be another elementary carnot cycle. As a result of this discussion we conclude that the given cycle 1 – 2 – 3 – 4 consists of a large number of carnot cycles.

For the carnot cycle *abcd* the process *bc* represents adiabatic expansion and for the adjoining cycle *befc*. it represents an equal amount of adiabatic compression. Therefore for the composite reversible cycle 1 – 2 – 3 – 4 the adiabatics contribute neither towards work nor towards heat interaction. Consequently the work output from the engine is net heat transfer during the elementary isothermal processes. Then the quantity $\oint \frac{\delta Q}{T}$ *i.e.*, cyclic integral of $\frac{\delta Q}{T}$ can be viewed as the sum of all these differential amounts of heat divided by the absolute temperature at the boundary.

The validity of the Clausius Inequality can be proved with the help of two heat engines one reversible E_1 and the other irreversible E_2 both operating between high temperature reservoir at T_1 and low temperature reservoir at T_2 as shown in Fig. 5.34. First consider the reversible heat engine E_1 receives Q_1 amount of heat during the cycle and rejects Q_2 amount of heat to sink at T_2 and the net output is W_1. It is further given that Q_1 and Q_2 amounts of heat are transfered at contant temperatures T_1 and T_2 respectively.

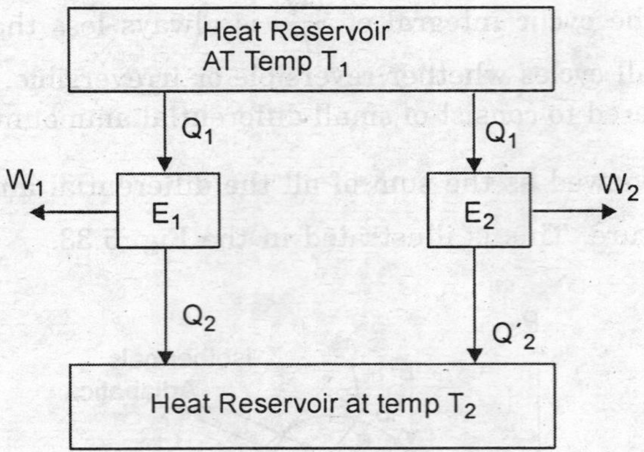

Fig. 5.34

Then the cyclic integral of $\frac{\delta Q}{T}$ for reversible heat engine E_1 becomes

$$\oint \left(\frac{\delta Q}{T}\right)_{rw} = \oint \frac{\delta Q_1}{T_1} - \oint \frac{\delta Q_2}{T_2}$$

$$= \frac{1}{T_1}, \oint \delta Q_1 - \frac{1}{T_2} \oint \delta Q_2$$

$$\text{or} \qquad \oint\left(\frac{\delta Q}{T}\right)_{rw} = \frac{Q_1}{T_1} - \frac{Q_2}{T_2}$$

$$= 0 \text{ [because for evensible heat engine as in equatis 15.16)]} \qquad ...(5.52a)$$

$$\frac{Q_1}{T_1} = \frac{Q_2}{T_2} \qquad ...(5.52b)$$

Thus for a reversible heat engine

$$\oint\left(\frac{\delta Q}{T}\right)_{rw} = 0 \qquad ...(5.53)$$

Even through equation (5.53) helds good for totally reversible heat engine, but it is equally valid for only internally reversible engines where T_1 and T_2 are the temperature of the working fluid at locations where it receives and rejects heats respectively.

Let us now consider the irreversible heat engine E_2 operating between the same thermal energy reservoirs as the reversible engine E_1 receiving the same amount of heat Q_1 during the cycle. Then as per the first carnot theorem (article 5.10), the irreversible heat engine E_2 will deliver less work or in otherwords it will reject more heat to the sink at temperature T_2.

$$\text{Thus} \qquad {Q_2}^1 > Q_2 \qquad ...(5.54)$$

$$\text{or} \qquad {Q_2}^1 = Q_2 + Q_d$$

Where Q_d = Difference between the heat rejected by real (irreversible) engine and that rejected by reversible one. It is a positive quantity.

Thus taking the cyclic integral of $\frac{\delta Q}{T}$ for the irreversible engine we get

$$\oint_{in} \frac{\delta Q}{T} = \frac{Q_1}{T_1} - \frac{Q_2^1}{T_2} < 0 \qquad ...(5.55)$$

$$\text{and} \qquad \oint_{rw} \frac{\delta Q}{T} = \frac{Q_1}{T_1} - \frac{Q_2}{T_2} = 0 \qquad ...(5.56)$$

$$\text{Thus if} \qquad \oint \frac{\delta Q}{T} = \left[\frac{Q_1}{T_1} - \frac{Q_2}{T_2}\right] \text{ is zero} \qquad ...(5.57)$$

Then the engine is reversible now and for irreversible engine

$$\oint \frac{\delta Q}{T} = \frac{Q_1}{T_1} - \frac{Q_2^1}{T_2}$$

$$\text{or} \qquad \oint \frac{\delta Q}{T_2} = \frac{Q_1}{T_1} - \left(\frac{Q_2 + Q_d}{T_2}\right)$$

$$= \frac{Q_1}{T_1} - \frac{Q_2}{T_2} - \frac{Q_d}{T_2}$$

$$= -\frac{Q_d}{T_2} \qquad ...(5.58)$$

Thus combining expressions (5.56) and (5.58) we have the relationship

$$\oint \frac{\delta Q}{T} \leq 0 \qquad ...(5.59)$$

Expression (5.59) is known as Clausius Inequality and the result can be expressed as

When $\oint \frac{\delta Q}{T} = \frac{Q_1}{T_1} - \frac{Q_2}{T_2}$

= 0 Then the cycle in reversible

When $\oint \frac{\delta Q}{T} < 0$

$$= -I \qquad ...(5.60)$$

Then the engine is irreversible and then I is called the Irreversibility of the engine.

5.13.1 Mathematical form of Clausius Inequality for a Refrigerator or Heat pump Cycle

Let R_1 (Reversible) and R_2 (Irreversible) be two referigerators operating between High Temperature reservoir at T_1 and Low Temperature reservoir at T_2. Refering Fig 5.35 Both the referigerators remove same amount of heat say Q_2. The reversible refrigerator rejects heat of amount Q_1 to high temperature surroundings at T_1 while the irreversible Refrigerator being less efficient will absorb more power *i.e.*, $(W_2 > W_1)$. Thus Q_1' the heat rejected by R_2 to high temperature surroundings will be more than Q_1

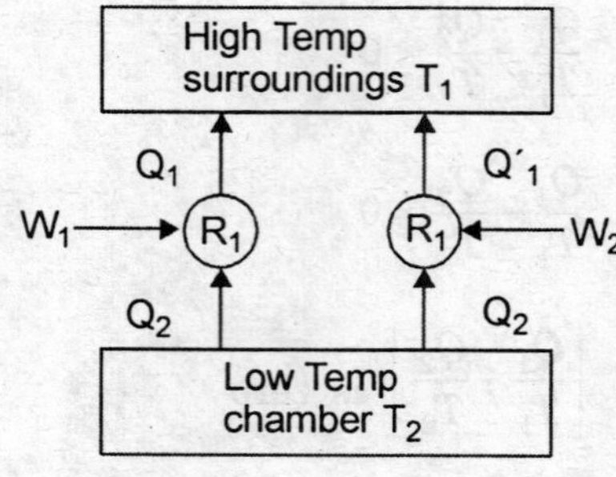

where
R_1 = Reversible refigerator
R_2 = Ir. reversible refrigerator
$Q_d = (Q'_1 - Q_1)$

Fig. 5.35

Since $\quad Q_1' = Q_2 + W_2 \qquad ...(5.61)$

and $\quad Q_1 = Q_2 + W_1 \qquad ...(5.62)$

Thus $\quad Q_2' = Q_2 + Q_d \qquad ...(5.63)$

Where Q_d = The positive quantity

$= Q_2' - Q_2$

The cyclic integral of $\frac{\delta Q}{T}$ is therefore given by

$$\oint_{rw} \frac{\delta Q}{T} = \oint \frac{\delta Q_2}{T_2} - \oint \frac{\delta Q_1}{T_1} \quad ...(5.64)$$

$$= \frac{1}{T_2}\int \delta Q_2 - \frac{1}{T_1}\int \delta Q_1$$

$$= \frac{Q_2}{T_2} - \frac{Q_1}{T_1} = 0 \qquad \left[\because \text{For reversible} \frac{Q_1}{T_1} = \frac{Q_2}{T_2}\right]$$

Again for Irreversible refrigerator

$$\oint_{in} \frac{\delta Q}{T} = \frac{Q_2}{T_2} - \frac{Q_1^1}{T_1} \quad ...(5.66)$$

$$= \frac{Q_2}{T_2} - \frac{Q_1 + Q_d}{T_1}$$

$$= \frac{Q_2}{T_2} - \frac{Q_1}{T_1} - \frac{Q_d}{T_1}$$

$$= -\frac{Q_d}{T_1} < 0$$

Thus for refrigerator cycle also

$$\oint \frac{\delta Q}{T} < 0 \quad ...(5.66)$$

An important point to note is that when ever $\oint \frac{\delta Q}{T} > 0$, then the second Law of Thermodynamics will be violated and the process in question will be invalid

The relationship (5.66) can also be written as

$$\oint \frac{\delta Q}{T} + I = 0 \quad ...(5.67a)$$

Where I = Positive quantity and is known as IRREVERSIBILITY

or

$$I = -\oint \frac{\delta Q}{T} \quad ...(5.67b)$$

Example 5.38. *An engine receives 790 kJ of heat from a high temperature source at 1090 K and rejects 490 kJ to low temperature sink at 300 K. Investigate whether this heat engines violates the second low of the thermodynamics on the basis of (a) Clausius Inequality (b) Carnot Principle.*

Solution : (*a*) On the basis of Clausius Inequality

$$\oint \frac{\delta Q}{T} = \frac{Q_1}{T_1} - \frac{Q_2}{T_2}$$

Now $Q_1 = 790$ kJ, $Q_2 = 490$ kJ

$T_1 = 1090$ K and $T_2 = 300$K

$$\therefore \quad \oint \frac{\delta Q}{T} = \frac{790}{1090} - \frac{490}{300}$$

$$= 0.7248 - 1.6333$$

$$= -0.9085 < 0$$

Therefore Clausius Inequality is satisfied.

(*b*) Thermal efficiency of carnot engine is

$$\eta_{rw} = 1 - \frac{T_2}{T_1} = 1 - \frac{300}{1090}$$

$$= \frac{790}{1090} = 0.724$$

Thermal efficiency of the given engine

$$\eta = 1 - \frac{Q_2}{Q_1} = 1 - \frac{490}{790}$$

$$= 1 - \frac{300}{790} = 0.38$$

Since $\eta_{thermal} < \eta_{rw}$, the heat engines satisfies the Carnot Principle.

Example 5.39. *(a) A refrigerator absorbs 48000 kJ/h heat at 5°C and rejects heat to the surroundings at 44°C and 3 W motor drivers the compressor. Examine whether the clausius in equally is satisfied. If so then find the irreversibility.*

(b) Also determine the heat rejection rate.

Solution : Refer to Fig. 5.19 Q_2 = Heat absorbed = 48000 kJ/h

$T_2 = 5 + 273 = 278$ K

$\dot{Q}_1$ = Heat rejection at higher temperature surroundings

$= \dot{Q}_2 + W$

And $T_1 = 44 + 273 = 317$ K

Where W = 3 KW = 3 kJ/sec

= 3 × 3600 kJ/h

= 10800 kJ/h

Therefore $\dot{Q}_1 = \dot{Q}_2 + W$

$= 48000 \text{ kJ/h} + 10800 \text{ kJ/h}$

$= 58800 \text{ kJ/h}$

Thus the heat rejection rate

$= 58800 \text{ kJ/h}$

Now for refrigerator, we have from

$$\oint \frac{\delta Q}{T} = \frac{Q_2}{T_2} - \frac{\dot{Q}_1}{T_1}$$

$$= \frac{48000}{278} - \frac{58800}{317}$$

$$= 172.66 - 185.49$$

$$= -12.83 \text{ kJ/h}$$

$$< 0$$

This being *–ve*, the Clasius inequality is satisfied

Then Irreversibility $= I = -\oint \frac{\delta Q}{T}$

$$= -(-12.83)$$

$$= 12.83 \text{ kJ/h}$$

Example 5.40. *During winter a heat pump supplies heat to a house, which is maintained at 21°C, at a rate of 72000 kJ/h while consuming 7 W of power. If the outside temperature is –3°C find whether if the heat pump satisfies the clausius inequality.*

Solution : Refer to Fig. 5.21 $W = 7 \text{ kJ/s} = 7 \times 3600 \text{ kJ/h} = 25200 \text{ kJ/h}$

$$\dot{Q}_1 = 72000 \text{ kJ/h}$$

$$T_1 = 21 + 273$$

$$= 294 \text{ K}$$

Also $\dot{Q}_2 = \dot{Q}_1 - W$

$$= 72000 - 25200$$

$$= 46800 \text{ kJ/h}$$

$$T_2 = -3 + 273$$

$$= 270 \text{ k}$$

From Clausius inequality $\oint \frac{\delta Q}{T} = \frac{\dot{Q}_2}{T_2} - \frac{\dot{Q}_1}{T_1}$

$$= \frac{46800}{270} - \frac{72000}{294}$$

$= 173.33 - 244.9$

$= -71.57$

< 0

Thus the Clausius inequality is satisfied

Example 5.41. *a carnot engine E_1 receives heat from a reservoir at temperature T_1 and the heat rejected by it at temperature T_2 is received by an other carnot Engine E_2 which rejects the waste heat to a sink at temperature T_3. [Fig 5.36] Determine the ratio of the work outputs of the two engines.*

Solution : Let η_1 and η_2 be the efficiencies of given two carnot engines respectively

Then $$\eta_1 = \frac{W_1}{Q_1} = 1 - \frac{T_2}{T_1} \quad ...(A)$$

Also $$Q_1 = Q_2 + W_1 \quad ...(B)$$

Substituting value of Q_1 from (B) into (A)

$$\frac{W_1}{Q_2 + W_1} = 1 - \frac{T_2}{T_1}$$

or $$\frac{T_2}{T_1} = 1 - \frac{W_1}{Q_2 + W_1}$$

$$= \frac{Q_2 + W_1 - W_1}{(Q_2 + W_1)}$$

or $$\frac{T_2}{T_1} = \frac{Q_2}{Q_2 + W_1}$$

or $$T_2Q_2 + T_2W_1 = Q_2T_1$$

or $$T_2W_1 = Q_2(T_1 - T_2)$$

or $$Q_2 = W_1\left(\frac{T_2}{T_1 - T_2}\right) \quad ...(C)$$

Fig. 5.36

Similarly for the engine E_2

$$\eta_2 = \frac{W_2}{Q_2} = 1 - \frac{T_3}{T_2} \quad ...(D)$$

and $$Q_2 = Q_3 + W_2 \quad ...(E)$$

From (D) and (E)

$$\frac{W_2}{Q_3 + W_2} = 1 - \frac{T_3}{T_2}$$

or $$\frac{T_3}{T_2} = 1 - \frac{W_2}{Q_3 + W_2}$$

$$= \frac{Q_3}{Q_3 + W_2}$$

or $$T_3(Q_3 + W_2) = T_2Q_3$$

or $$Q_3 = W_2\left(\frac{T_3}{T_2 - T_3}\right) \quad \text{...(F)}$$

Substituting the values of Q_2 and Q_3 for equations (*C*) and (*F*) into equation (*E*) we get

$$W_1\left(\frac{T_2}{T_1 - T_2}\right) = W_2\left(\frac{T_3}{T_2 - T_3}\right) - W_2$$

$$= W_2\left[\frac{T_3 + T_2 - T_3}{T_2 - T_3}\right]$$

or $$W_1\left(\frac{T_2}{T_1 - T_2}\right) = \frac{W_2T_2}{(T_2 - T_3)}$$

or $$\frac{W_1}{W_2} = \left(\frac{T_1 - T_2}{T_2 - T_3}\right) \quad \text{...(G)}$$

Example 5.42. *A designer claims that an engine designed by him is capable of delivering 0.2 kWh of work for 1250 kJ of heat received in one hour from a heat source at temperature of 460 K and rejecting the waste heat to the sink at 290 K. Make suitable calculations to check whether the claim of the designer is false or true.*

Solution : Work out put $= 0.2$ kWh

$= 0.2 \times 3600$ kJ

$= 720$ kJ

Heat input $Q_1 = 1250$ kJ

Therefore $$\eta_{th} = \frac{720}{1250} = 0.576 \text{ or } 56.7\%$$

Now the thermal efficiency of a reversible engine working between the source temperature $T_1 = 460$ K and sink temperature $T_2 = 290$ K is given by

$$\eta_{rw} = \frac{T_1 - T_2}{T_1}$$

$$= \frac{460 - 290}{460} = \frac{170}{460}$$

$= 0.3696$ or 36.97 %

As the efficiency of real irreversible engine can not be more than that of reversible engine, the claim of the designer is false.

Example 5.43. *An engine manufacturing company claims that they have developed an engine with the following specifications*

Temperature of heat source	= *1100 K*
Temperature of sink	= *500 K*
Power output	= *80 kW*
Fuel consumption	= *5.2 kg/hr*
Heating value of fuel	= *75000 kJ/kg*

Varify making appropriate calculations, whether the claim of the company is correct.

Solution : W = work output = 80 kW

= 80 × 3600 kJ/h

= 288000 kJ/h

Q_1 = Heat supplied from source at temperature T_1

= Quantity of fuel kg/hr × heating value of fuel

= (5.2 kg/h) × (75000 kJ/kg)

= 390000 kJ/h

Therefore
$$\eta_{irr} = \frac{W}{Q_1}$$

$$= \frac{288000}{390000} = 0.7385 \text{ or } 73.85\%$$

Now the thermal efficiency of reversible engine working between the same source at T_1 and sink T_2 is given by

$$\eta_{in} = \frac{T_1 - T_2}{T_1} = \frac{1100-500}{1100}$$

$$= \frac{500}{1100} = 0.5455 \text{ or } 54.55\%$$

Since the calculations show that the efficiency of irreversible engine in more than that of reversible engine, the claim of the company is false.

Example 5.44. *Two heat sources at constant temperatures 950 K and 650 K respectively supply heat to a reversible engine which rejects waste heat to the sink at 310 K. The engine develops 125 kW while rejecting 3650 kJ/min of heat to the sink. Determine the heat supplied by each source per minute and the thermal efficiency of the engine.*

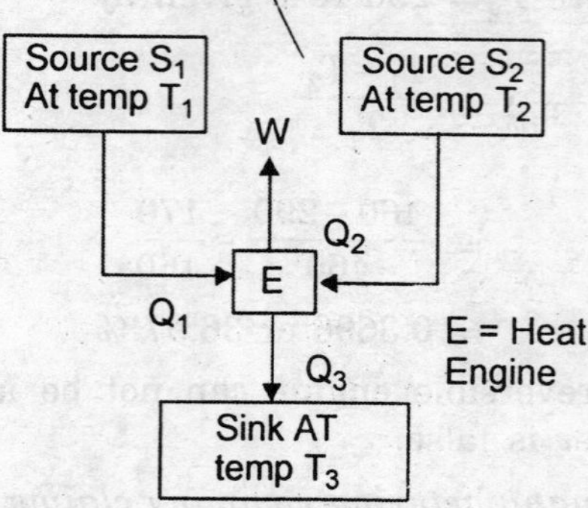

Fig. 5.37

Solution : $T_1 = 950$ K

$T_2 = 650$ K

$T_3 = 310$ K

$W = 125$ kW

$= 125$ kJ/s

$= 125 \times 60 = 7500$ kJ/min

$Q_3 = 3650$ kJ/min

Heat supplied to the engine

$$Q_1 + Q_2 = \text{Heat rejected + work output}$$

$$= Q_3 + W$$

or $$Q_1 + Q_2 = 3650 + 7500$$

$$= 11150 \text{ kJ/min}$$

Therefore $$Q_2 = 11150 - Q_1 \quad \text{...(A)}$$

The engine being reversible, Clausius theorem will give

$$\oint \frac{\delta Q}{T} = 0$$

or $$\frac{Q_1}{T_1} + \frac{Q_2}{T_2} - \frac{Q_3}{T_3} = 0$$

or $$\frac{Q_1}{T_1} + \frac{11150 - Q_1}{T_2} - \frac{3650}{T_3} = 0$$

or $$\frac{Q_1}{950} + \frac{11150 - Q_1}{650} - \frac{3650}{310} = 0$$

$$\frac{Q_1}{95} + \frac{11150 - Q_1}{65} - \frac{3650}{31} = 0$$

$$0.0105\, Q_1 + 171.538 - 0.015385\, Q_1 - 117.742 = 0$$

or $$53.796 = 0.004885\, Q_1$$

or $$Q_1 = \frac{53.796}{0.004884} = 11014.74 \text{ kJ/min}$$

Then $$Q_2 = 11150 - 11014.74 = 135.258 \text{ kJ/min}$$

Efficiency of Engine $$\eta = \frac{\text{Work Output}}{\text{Total Heat Supplied}}$$

$$= \frac{W}{Q_1 + Q_2}$$

$$= \frac{7500}{11150}$$

$$= 0.673 \text{ } i.e., \text{ } 67.3\%$$

Example 5.45. *A reversible heat engine receives 1500 kJ/min from a reservoir at temperature 500 K and rejects heat to two reservoirs which are at constant temperatures 400 K and 300 K respectively. It develops 400 kJ/min of work. Determine the heat transactions of the two sinks.*

Solution : Given

$$W = 400 \text{ kJ/min}$$
$$Q_1 = 1500 \text{ kJ/min}$$
$$T_1 = 500 \text{ K}$$
$$T_2 = 400 \text{ K}$$
$$T_3 = 300 \text{ K}$$

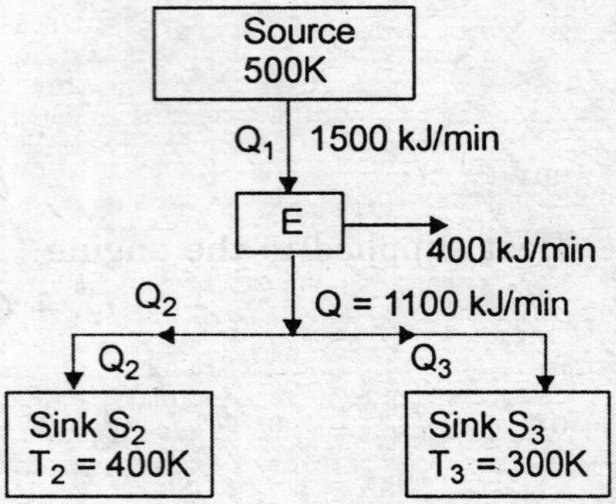

Fig. 5.38

Let the heat rejected by the engine

$$= Q \text{ kJ/min}$$
$$Q_1 = Q + W$$
$$150 \text{ kJ/min} = Q + 400 \text{ kJ/min}$$

Therefore $Q = 1100$ kJ/min

Let Q_2 be the heat rejected to sink at temperature 400 K = T_2.

Let the heat rejected to sink at temperature

T_3, 300 k = Q_3

be = Q_3

Then $Q_3 = Q - Q_2$

$$= 1100 - Q_2$$

Since the engine is reversible we have from clausius equation.

$$\oint \frac{\delta Q}{T} = 0$$

or
$$\frac{Q_1}{T_1} - \frac{Q_2}{T_2} - \frac{Q_3}{T_3} = 0$$

or
$$\frac{1500}{500} - \frac{Q_2}{400} - \frac{1100 - Q_2}{300} = 0$$

$$\frac{1500}{5} - \frac{Q_2}{4} - \frac{1100 - Q_2}{3} = 0$$

or multiplying by 60 through out we get

$$18000 - 15\,Q_2 - 20\,(1100 - Q_2) = 0$$

or $18000 - 15Q_2 - 22000 + 20Q_2 = 0$

or $5Q_2 = 4000$

or
$$Q_2 = \frac{400}{5} = 800 \text{ kJ/min}$$

Then
$$Q_3 = 1100 - 800$$
$$= 300 \text{ kJ/min}$$

Example 5.46. *A carnot engine working between a source temperature T_1 and sink temperature T_2 has efficiency of 25%. If the sink temperature is reduced by 20°C the efficiency is increased to 30%. Find the source and sink temperature.* *(P.T.U. Dec. 2004)*

Solution : $\eta_{carnot} = \frac{T_1 - T_2}{T_1} = 0.25$ (Given)

or $T_1 - T_2 = 0.25\ T_1$

or $0.75\ T_1 = T_2$ ...(*A*)

carnot efficiency with reduced sink temperature

$= (T_2 - 20)$

$$\eta'_{carnot} = \frac{T_1 - (T_2 - 20)}{T_1} = 0.3$$

or $T_1 - T_2 + 20 = 0.3\ T_1$

or $0.7\ T_1 - T_2 = -20$ ...(*B*)

$\therefore$ From (*A*) to (*B*) $0.7 \times T_1 = 0.75\ T_1 = -20$

or $T_1 = \frac{(20)}{0.05} = 400K$

and from (*A*) $T_2 = 0.75 \times 400$ K

$= 300$ k

Example 5.47. *Cold storage plant requires 7000 kJ/kg (20 tonnes) of refrigerant. Determine power required for the following conditions.*

Evaporator temperature = *–40°C*

Ambient temperature = *50°C*

$(COP)_{refrigerator}$ = *25% of ideal cycle (COP)* *(P.T.U)*

Solution : T_1 = Temperature of surroundings or ambient temperature

$= 50 + 273 = 323$ K

Evaporate temperature $= -40 + 273$

$= 233$ K

Therefore $(COP)_{ideal} = \frac{T_2}{T_1 - T_2}$

$$= \frac{233}{323 - 233} = 2.58$$

Therefore $(COP)_{Ref\ plant} = 0.25 \times 2.58$

$= 0.645$

Also $(COP)_{Ref} = \frac{\text{Refrigerant effect}}{\text{Work done}}$

or $\qquad 0.645 = \dfrac{7000 \text{ kJ/kg}}{\text{work done}}$

or work Done $\qquad = \dfrac{7000}{0.645} = 10852.7 \text{ kJ/kg}$

But it is given that mass of refrigerant = 20 Tonnes = 20000 kg

$\therefore$ Total work Done $\qquad = 10852.7 \times 20000$

$\qquad = 21.705 \times 11^{6}$ kJ (**Ans.**)

Example 5.48. *Three real heat engines have the same thermal efficiency and are connected in series as shown in Fig 5.39 The first engine absorbs 2400 kJ of heat from thermal reservoir at 1250 k and the third engine rejects the waste heat of 300 kJ to a sink at 150 K. Determine the work output from each engine.*

Solution : $\qquad Q_1 = 2400$ kJ

$\qquad Q_4 = 300$ kJ

$\qquad T_1 = 1250$ K

AND $\qquad T_4 = 150$ K

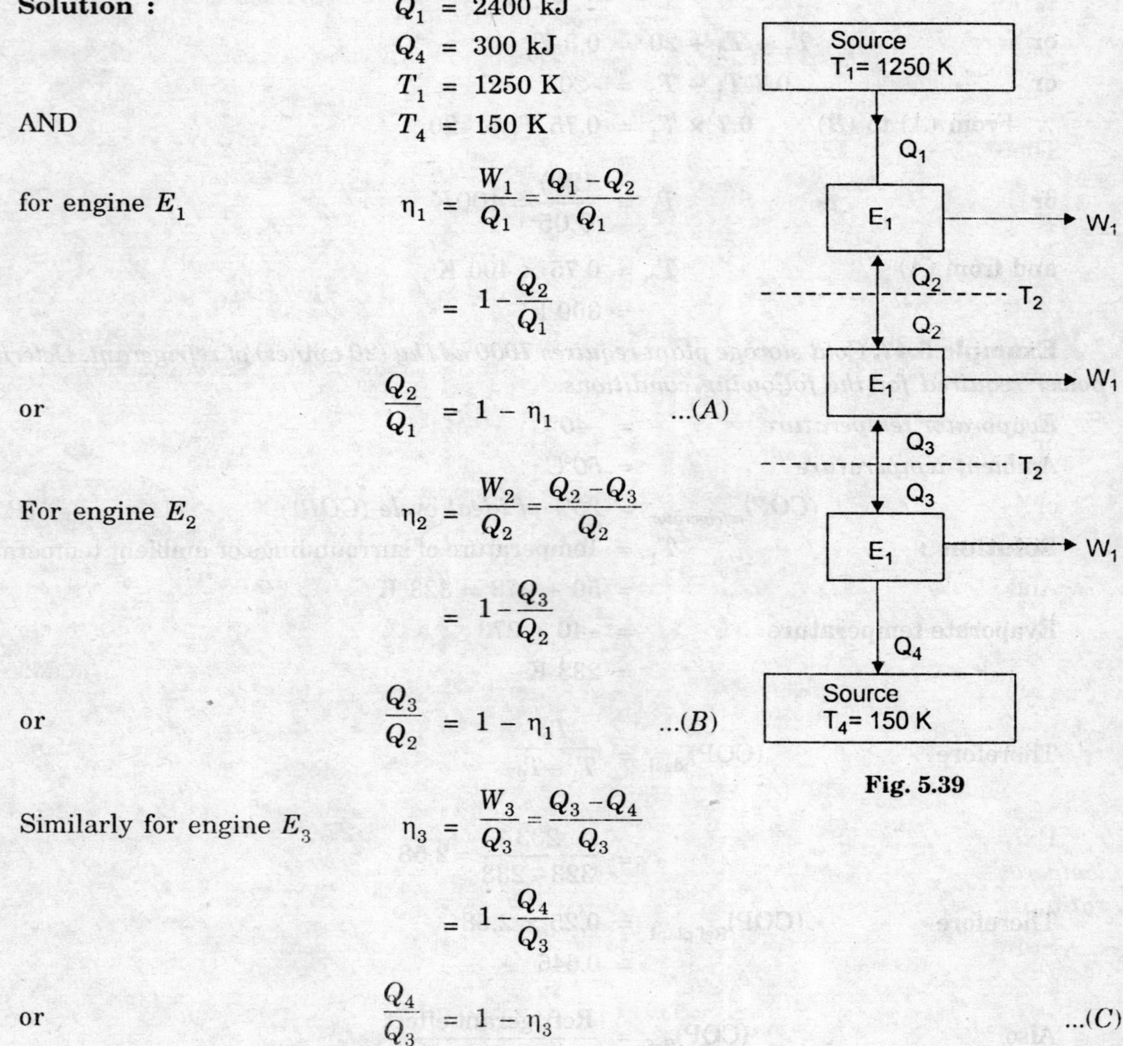

Fig. 5.39

for engine E_1 $\qquad \eta_1 = \dfrac{W_1}{Q_1} = \dfrac{Q_1 - Q_2}{Q_1}$

$\qquad = 1 - \dfrac{Q_2}{Q_1}$

or $\qquad \dfrac{Q_2}{Q_1} = 1 - \eta_1 \qquad ...(A)$

For engine E_2 $\qquad \eta_2 = \dfrac{W_2}{Q_2} = \dfrac{Q_2 - Q_3}{Q_2}$

$\qquad = 1 - \dfrac{Q_3}{Q_2}$

or $\qquad \dfrac{Q_3}{Q_2} = 1 - \eta_1 \qquad ...(B)$

Similarly for engine E_3 $\qquad \eta_3 = \dfrac{W_3}{Q_3} = \dfrac{Q_3 - Q_4}{Q_3}$

$\qquad = 1 - \dfrac{Q_4}{Q_3}$

or $\qquad \dfrac{Q_4}{Q_3} = 1 - \eta_3 \qquad ...(C)$

Multiplying (A), (B) and (C)

$$\frac{Q_2}{Q_1}\cdot\frac{Q_3}{Q_2}\cdot\frac{Q_4}{Q_3} = (1 - \eta_1)(1 - \eta_2)(1 - \eta_3)$$

or $$\frac{Q_4}{Q_1} = (1 - \eta)^3 \text{ (since } \eta_1 = \eta_2 = \eta_3 = \eta)$$

i.e. $$(1 - \eta)^3 = \frac{300}{2400} = \frac{1}{8}$$

or $$1 - \eta = \frac{1}{2}$$

i.e. $$\eta = \frac{1}{2} \text{ or } 50\%$$

Therefore from (A) $$\frac{Q_2}{Q_1} = 1 - \eta_1 = 1 - 0.5 = 0.5$$

or $$Q_2 = 0.5\, Q_1 = 0.5 + 2400 = 1200 \text{ kJ}$$

$\therefore$ $$W = Q_1 - Q_2 = 2400 - 1200 = 1200 \text{ kJ } \textbf{Ans.}$$

Similarly from (B) $$\frac{Q_3}{Q_2} = 1 - \eta_2 = 1 - 0.5 = 0.5$$

or $$Q_3 = \frac{1}{2}Q_2 = \frac{1}{2}\times 1200 = 600$$

or $$W_2 = Q_2 - Q_3 = 1200 - 600 = 600 \text{ kJ}$$

And from (C) $$\frac{Q_4}{Q_3} = 1 - \eta_3 = 1 - 0.5 = 0.5$$

$$Q_4 = \frac{1}{2}Q_3 = \frac{1}{2}\times 600 = 300$$

$\therefore$ $$W_3 = Q_3 - Q_4$$

$$= 600 - 300 = 300 \text{ kJ } \textbf{Ans.}$$

Example 5.49. *Three carnot engines E_1, E_2 and E_3 work in series (Fig 5.40) between the temperature limits of 1000 K and 300 K. The amount of work produced by the engines in the ratio 5:4:3. Determine the intermediate temperatures.*

Solution : The efficiencies of engines E_1, E_2 and E_3 are given by

$$\eta_1 = \frac{W_1}{Q_1} = \frac{T_1 - T_2}{T_1}$$

or $$\frac{W_1}{Q_1} = \frac{1000 - T_2}{1000} \quad \text{...(A)}$$

$$\eta_2 = \frac{W_2}{Q_2} = \frac{T_2 - T_3}{T_2} \quad \text{...(B)}$$

and $$\eta_3 = \frac{W_3}{Q_3} = \frac{T_3 - T_4}{T_3}$$

$$= \frac{T_3 - 300}{T_3} \quad \text{...(C)}$$

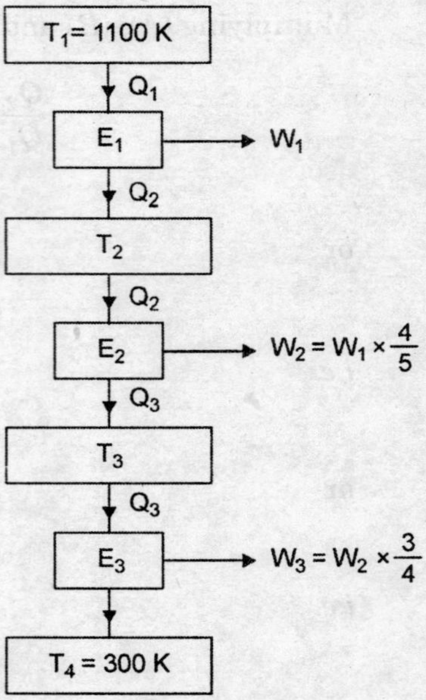

Also from Ist Law of Thermodynamics Fig. 5.40

$$Q_1 = Q_2 + W_1 \quad \text{...(D)}$$

Then from equation (*A*) and (*D*)

$$\frac{W_1}{W_1 + Q_2} = \frac{1000 - T_2}{1000}$$

or $$W_1\left(\frac{1000}{1000 - T_2}\right) = W_1 + Q_2$$

or $$W_1\left(\frac{1000}{1000 - T_2}\right) - W_1 = Q_2$$

or $$Q_2 = W_1\left[\frac{1000}{1000 - T_2} - 1\right]$$

or $$Q_2 = W_1\left[\frac{T_2}{1000 - T_2}\right] \quad \text{...(E)}$$

Fig. 5.40

Substituting the value of Q_2 from equation (*E*) into equation (*B*) we get

$$\frac{W_2}{W_1}\left(\frac{1000 - T_2}{T_2}\right) = \left(\frac{T_2 - T_3}{T_2}\right)$$

But it is given that $\frac{W_1}{W_2} = \frac{5}{4}$ therefore above equation reduces to

$$\frac{4}{5}\left(\frac{1000 - T_2}{T_2}\right) = \left(\frac{T_2 - T_3}{T_2}\right)$$

or $$4000 - 4T_2 = 5T_2 - 5T_3$$

$$4000 = 9T_2 - 5T_3$$

or $$9T_2 = 4000 + 5T_3$$

or $$T_2 = \frac{5}{9}(T_3 + 800) \qquad ...(F)$$

Also for engine E_2

$$Q_2 = W_2 + Q_3 \qquad ...(G)$$

Therefore from equations (*B*) and (*G*)

$$\frac{W_2}{W_2 + Q_3} = \frac{T_2 - T_3}{T_2}$$

or $$W_2\left(\frac{T_2}{T_2 - T_3}\right) = W_2 + Q_3$$

or $$W_2\left[\frac{T_2}{T_2 - T_3} - 1\right] = Q_3$$

or $$W_2\left[\frac{T_2 - T_2 + T_3}{T_2 - T_3}\right] = Q_3$$

or $$Q_3 = W_2\left[\frac{T_3}{T_2 - T_3}\right] \qquad ...(H)$$

Then inserting this value of Q_3 in equation (*C*)

$$\frac{W_3}{W_2}\left(\frac{T_2 - T_3}{T_3}\right) = \frac{T_3 - 300}{T_3}$$

or $$\frac{W_3}{W_2}(T_2 - T_3) = (T_3 - 300)$$

But $$\frac{W_3}{W_2} = \frac{3}{4} \text{ given}$$

or $$\frac{3}{4}(T_2 - T_3) = (T_3 - 300)$$

$$3T_2 - 3T_3 = 4T_3 - 1200$$

or $$3T_3 - 7T_3 + 1200 = 0 \qquad ...(J)$$

Substituting the value T_2 for equation (*F*) into (*J*)

$$3 \times \frac{5}{7}(T_3 + 800) = 7T_3 - 1200$$

or $$15T_3 + 12000 = 63T_3 - 10800$$

$$48T_3 = 22800$$

$$T_3 = \frac{22800}{48} = 475 \text{ K } \textbf{Ans.}$$

Then from (*F*) $$T_2 = \frac{5}{9}(475+800)$$

$$= \frac{5}{9} \times 1275 = 708.3 \text{ K } \textbf{Ans.}$$

Example 5.50. *Two carnot refrigerators R_1 and R_2 having individual coefficient of performance C_1 and C_2 respectively are arranged in series an shown in Fig. 5.41 show that the coefficient of performance C, of the composite system is*

$$C = \frac{C_1 C_2}{1+C_1+C_2}$$

Solution : COP of $R_1 = C_1 = \dfrac{Q_3}{W_1}$

Also $$Q_2 = Q_3 + W_1$$

or $$W_1 = Q_2 - Q_3$$

$\therefore$ $$C_1 = \frac{Q_3}{Q_2 - Q_3}$$

or $$C_1 Q_2 = Q_3(1 + C_1)$$

or $$Q_2 = \frac{(1+C_1)Q_3}{C_1} \quad \text{...(A)}$$

Similarly COP of $R_2 = C_2$

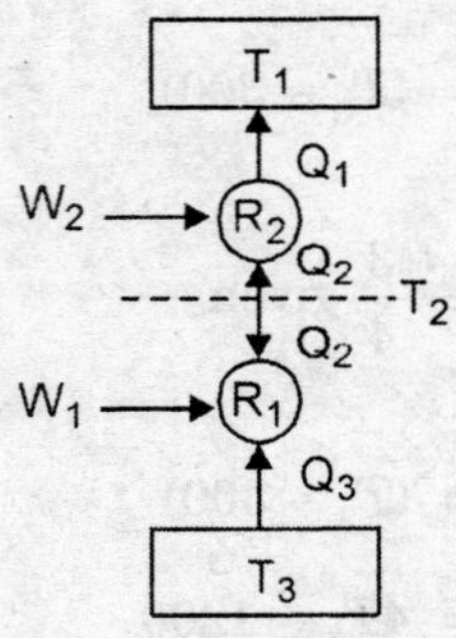

Fig. 5.41

$$= \frac{Q_2}{W_2}$$

But $$Q_2 + W_2 = Q_1$$

or $$W_2 = Q_1 - Q_2$$

$$\therefore \qquad C_2 = \frac{Q_2}{Q_1 - Q_2}$$

or $$C_2Q_1 = Q_2 + C_2Q_2$$

$$= Q_2\,(1 + C_2)$$

or $$Q_1 = \frac{Q_2(1+C_2)}{C_2} \quad ...(B)$$

Multiplying (A) and (B) $$Q_2Q_1 = \frac{(1+C_1)Q_3}{C_1} \times \frac{Q_2(1+C_2)}{C_2}$$

or $$\frac{Q_1}{Q_3} = \frac{(1+C_1)\,(1+C_2)}{C_1C_2} \quad ...(C)$$

Now if C_3 is the COP of the composite system which is working between the reservoirs at T_3 and T_1 then

$$C_3 = \frac{Q_3}{W_1 + W_2} \quad ...(C\text{-}2)$$

But $$W_1 + Q_3 = Q_2 \quad ...(D)$$

and $$W_2 + Q_2 = Q_1$$

or $$Q_2 = Q_1 - W_2$$

Therefore equation (D) reduces to

$$W_1 + Q_3 = Q_1 - W_2$$

or $$W_1 + W_2 = Q_1 - Q_3 \quad ...(E)$$

Substituting equation (E) into equation (C)

$$C_3 = \frac{Q_3}{Q_1 - Q_3}$$

$$C_3 = \frac{1}{\dfrac{Q_1}{Q_3} - 1}$$

or $$C_3\left(\frac{Q_1}{Q_3} - 1\right) = 1$$

or $$\frac{Q_1}{Q_3} = 1 + \frac{1}{C_3} \quad ...(F)$$

From equations (C) and (F)

$$1 + \frac{1}{C_3} = \frac{(1+C_1)\,(1+C_2)}{C_1C_2}$$

or $$\frac{1}{C_3} = \frac{(1+C_1)(1+C_2)-C_1C_2}{C_1C_2}$$

or $$C_3 = \frac{C_1C_2}{(1+C_1)(1+C_2)-C_1C_2}$$

$$= \frac{C_1C_2}{1+C_1+C_2+C_1C_2-C_1C_2}$$

or $$C_3 = \frac{C_1C_2}{1+C_1+C_2}$$

Example 5.48. *A refrigerator working on carnot cycle absorbs 240 kJ of heat from a reservoir at 145 K and rejects heat to a reservoir at 425 K. This rejected heat is then drawn by another carnot refrigerator operating between the temperatures 225 K and 1350 K (Fig. 5.42). Determine the COP of both the refrigerators and that of the composite system.*

Solution : $$\text{COP of } R_1 = C_1 = \frac{Q_3}{Q_2-Q_3}$$

$$= \frac{T_3}{T_2-T_3}$$

as it is a carnot refrigerator $$\frac{Q_3}{Q_2} = \frac{T_3}{T_2}$$

$\therefore$ $$Q_2 = Q_3\frac{T_2}{T_3}$$

refering to Fig. 5.42

Here $$Q_3 = 240 \text{ kJ}, T_2 = 425 \text{ K}$$

$$T_3 = 145 \text{ K}$$

$\therefore$ $$Q_2 = \frac{240\times 425}{145}$$

$$= 703.448 \text{ kJ}$$

$\therefore$ $$C_1 = \frac{Q_3}{Q_2-Q_3} = \frac{240}{703.448-240}$$

$$= \frac{240}{463.448} = 0.5178 \text{ or } 51.78\%$$

For the refrigerator R_2

$$\text{COP} = C_2 = \frac{Q_2}{Q_1-Q_2} = \frac{T_2}{T_1-T_2}$$

Again for carnot refrigerator

$$\frac{T_2}{T_1} = \frac{Q_2}{Q_1}$$

or $$Q_1 = Q_2 \frac{T_1}{T_2} = \frac{703.448 \times 1350}{145}$$

$$= 6549.34$$

$\therefore$ $$C_2 = \frac{Q_2}{Q_1 - Q_2} = \frac{703.448}{6549.34 - 703.48} = 0.12 \text{ or} 12\%$$

For the composite system $$\text{COP} = Q_3 = \frac{Q_3}{Q_1 - Q_3}$$

$$= \frac{240}{6549.34 - 240}$$

$$= \frac{240}{6309.74} = 0.038 \text{ } i.e., 3.8\%$$

Example 5.49. *A carnot heat engine operates between the reservoirs at temperatures 607°C and 47°C. Part of the work produced by the engine is used to drive a reversible refrigerator which works between reservoirs at temperatures 47°C and –13°C. If the heat transfer to the heat engine is 2050 kJ and the net work output of the combined engine and refrigerator plant is 350 kJ determine (a) the heat transfer to the refrigerant and the net heat transfer to the reservoir at 47°C. (b) Further if the actual efficiency of the heat engine is 40% of the maximum efficiency and the COP of the refrigerator is 40% that of the maximum possible value, then calculate the heat transfer to the refrigerant and the net heat transfer to the reservoir at 47°C.*

Solution :

(*a*) Maximum efficiency of engine $= 1 - \dfrac{T_2}{T_1}$

$$\eta_{max} = 1 - \frac{320}{880}$$

$$= \frac{560}{880} = 0.6364$$

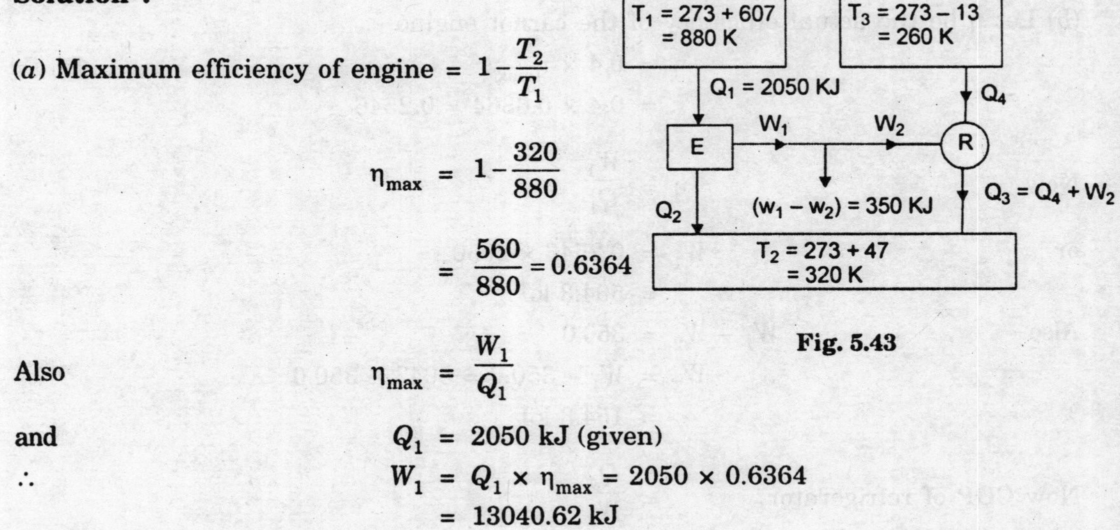

Fig. 5.43

Also $$\eta_{max} = \frac{W_1}{Q_1}$$

and $$Q_1 = 2050 \text{ kJ (given)}$$

$\therefore$ $$W_1 = Q_1 \times \eta_{max} = 2050 \times 0.6364$$

$$= 13040.62 \text{ kJ}$$

For the reversible refrigerator

$$\text{COP} = \frac{T_3}{T_2 - T_3} = \frac{260}{320 - 260}$$

$$= \frac{260}{60} = 4.32$$

Also $\text{COP} = \frac{Q_4}{W_2} = 4.32$

But $W_1 - W_2 = 350$ kJ (given)

or $W_2 = W_1 - 350$

$= 1304.62 - 350$

$= 954.62$ kJ

Therefore $Q_4 = W_2 \times 4.32$

$= 954.62 \times 4.32$

$= 4123.96$ kJ

Therefore $Q_3 = Q_4 + W_2$

$= 4123.96 + 954.62$

$= 5078.58$ kJ

Then $Q_2 = Q_1 - W_1$

$= 2050 - 1304.62$

$= 745.38$ kJ

Therefore net heat rejected to reservoir at 47°C

$= Q_2 + Q_3 = 745.38 + 5078.58$

$= 5823.96$ kJ **Ans.**

(*b*) Let η be the actual efficiency of the carnot engine

$\eta = 0.4 \times \eta_{max}$

$= 0.4 \times 0.6364 = 0.2546$

Now $\eta = \frac{W_1}{Q_1}$

or $W_1 = 0.2546 \times 2050$

$= 504.3$ kJ

Also $W_1 - W_2 = 350.0$

$W_2 = W_1 - 350.0 = 5043 - 350.0$

$= 154.3$ kJ

Now COP of refrigerator $= \frac{Q_4}{W_2}$

But actual $\text{COP} = 0.4 \times (\text{COP})_{max}$

$= 0.4 \times 4.32$

$= 1.732$

$\therefore \quad Q_4 = W_2 \times 1.732$

$= 154.3 \times 1.732$

$= 267.25$ kJ

Further $\quad Q_3 = Q_4 + W_2$

$= 267.25 + 154.3$

$= 421.55$ kJ

And $\quad Q_2 = Q_1 - W_1$

$= 2050 - 504.3$

$= 1545.3$ kJ

Heat rejected to reservoir at 47°C

$= Q_2 + Q_3$

$= 1545.7 + 421.55$

$= 1967.25$ kJ **Ans.**

Example 5.50. *The efficiency of a carnot engine can be increased in two ways viz (1) by increasing the source reservoir temperature T_1 and keeping the sink temperature constant and (2) by decreasing the sink temperature T_2 but keeping the source temperature T_1. Show that the in the second option method depicted is better than that given in the of option no 1.*

Solution : Let η_1 he the efficiency of the carnot engine where T_2 is decreased by a small amount ΔT_2 and T_1 source temperature kept at T_1.

Then
$$\eta_1 = 1 - \frac{T_2 - \Delta T}{T_1} \quad \text{...(A)}$$

Similarly in the second case T_2 is kept fixed and T_1 in increased by ΔT, then

$$\eta_2 = 1 - \frac{T_2}{T_1 + \Delta T} \quad \text{...(B)}$$

Subtract (*B*) from (*A*)

$$\eta_1 - \eta_2 = \frac{T_2}{T_1 + \Delta T} - \frac{T_2 - \Delta T}{T_1}$$

$$= \frac{T_1 T_2 - (T_1 + \Delta T)(T_2 - \Delta T)}{T_1 (T_1 + \Delta T)}$$

$$= T_1 T_2 - \left[\frac{T_1 T_2 + (T_2 - T_1)\Delta T - \Delta T^2}{T_1 (T_1 + \Delta T)} \right]$$

$$\eta_1 - \eta_2 = \frac{(T_1 - T_2)\,\Delta T + \Delta T^2}{T(T_1 + \Delta T)} \quad \text{...(C)}$$

Since $T_1 > T_2$ the Right hand side of equation (C) is positive,

$$\eta_1 > \eta_2 \quad ...(D)$$

The conclusion of (D) is that better method will be the first option.

Example 5.51. *A heat engine receives heat Q_1 from a source at temperature T_1 and drives a heat pump. The heat rejected by the engine and the heat delivered by the heat pump are both used to heat the water circulating through the radiator used for heating the room in winter. If the efficiency of the heat engine is 25% and the C.O.P. of heat pump is 3.8, find the ratio of the heat transferred to the circulating water to heat Q_1 received by the engine.*

Solution : Let efficiency of heat engine = η

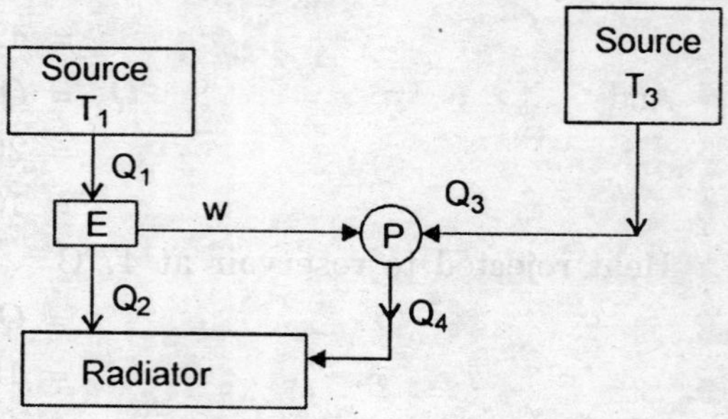

$$\eta = \frac{Q_1 - Q_2}{Q_1}$$

or

$$0.25 = 1 - \frac{Q_2}{Q_1}$$

i.e.,

$$\frac{Q_2}{Q_1} = 0.75$$

Fig. 5.42

or

$$Q_2 = 0.75\ Q_1 \quad ...(A)$$

Now COP of heat pump

$$\text{COP} = 3.8 = \frac{Q_4}{W} = \frac{Q_4}{Q_4 - Q_3} \quad ...(B)$$

But for the conservation of energy for the heat engine

$$W = Q_1 - Q_2$$

Therefore

$$Q_1 - Q_2 = Q_4 - Q_3$$

or

$$Q_4 - Q_3 = Q_1 - .75\ Q_1 \text{ (from } A)$$

$$= 0.25\ Q_1$$

Thus from equation (B)

$$3.8 = \frac{Q_4}{0.25 Q_1}$$

or

$$Q_4 = 3.8 \times 0.25\ Q_1$$

$$= 0.95\ Q_1$$

Total amount of heat transferred to circulating water

$$= Q_2 + Q_4$$

$$= 0.75\ Q_1 + 0.95\ Q_1$$

$$= 1.70\ Q_1$$

Therefore required ratio

$$= \frac{1.70 Q_1}{Q_1} = 1.70$$

REVEIW QUESTIONS

1. State the limitations of first law of thermodynamics.
2. Explain the following terms :
   (a) Thermal energy reservoir
   (b) Heat source
   (c) Heat sink
3. What is a cyclic heat engine?
4. Define the thermal efficiency of a heat engine. Can it be 100%?
5. Explain the difference between a reversible and reversed engine.
6. What is a heat pump? How does it differ from a refrigerator?
7. Can same plant be used as a refrigerator in summer and heat pump in winte? Explain.
8. Show that the C.O.P. of a pump is greater than the COP of a refrigerator by unity.
9. State the Kelvin-Plank and Clausius statements of the second law of thermodynamics and show that they are equivalent to each other.
10. What is a reversible process? Give example of reversible and irreversible processes.
11. What is a perpertial motion machine of a second kind? Can such a machine be constructed in actual practice?
12. Explain the carnot cycle. What are the main conclusions that can be drown from the efficiency of carnot cycle.
13. Can a heat engine operate without rejecting any heat to a low temperature reservoir? Explain.
14. What is a reversible process? Explain how is reversible process only a limiting process, never to be attained in practice?
15. Prove that no heat engine operating in a cycle between the constant temperature heat reservoirs can be more efficient than a reversible engine operating between the same two reservoirs.
16. Prove that all reversible engines opeating between the same two thermal reservoirs have the same efficiency.
17. Explain the establishment of thermodynamics temperature scale. Why this scale is called the absolute temperature scale?
18. How many fixed points of reference are used in establishing the thermodynamic temperature scale and what are these fixed points?
19. What is the difference between the thermodynamic and emperical scales of temperature?
20. State clausius inequatity?
21. Explain the importance of second law of thermodynamics.
22. What are the reasons of irreversibility of a process?
23. Explain how all spoutaneous processes are irreversible?

24. Explain, with the help of second law of thermodynamics, how friction makes the process irreversible.

25. State the four processes that constitute a carnot cycle.

26. Show that the efficiency of a reversible engine is independent of the working fluid rather it depends upon the temperature of the heat reservoirs between which it is operates.

27. Explain how thermodynamic temperature is independent of the working substance.

PROBLEMS

1. An inventor claims to have developed an engine that draws 100 MJ of heat energy from a source at 127°C and rejects 40 MJ to a sink at 27°C and develops 15 kW/h. Show by calculations whether his claim is correct or false.

2. The output of an engine having a thermal efficiency of 30%, is used to drive the motor of a refrigerator having COP of five. Find the heat input to the engine for every MJ of heat extracted by the refrigerator from its food chamber (*b*) If the system of engine and refrigerator is used as a heat pump, then for each MJ of input to the engine how much heat energy will be available for heating the from. (**Ans.** (*a*) 666.67 kJ (*b*) 1.8 MJ)

3. A carnot heat engine receives during a cycle, 500 kJ heat from a heat source at 652°C and rejects heat to a sink at 30°C. Calculate the thermal efficiency of the engine and the quantity of heat rejected to the sink. (**Ans.** 67.2%; 163.75 kJ)

4. During winter a house in Shimla is to be maintained at 21°C with the help of a heat pump. The temperature out side the house is –5°C. It is estimated that the loss of heat through windows etc. is 2250 kJ/min. Find the minimum power required to drive the heat pump. (**Ans.** 3.318 kW)

5. A domestic refrigerator runs one-fifth of the time and removes heat from the food compartment at the rate of 640 kJ/hr on average basis. If the COP of the refrigerator is 2 calculate the power the refrigerator consumes when running. (**Ans.** 1600 kJ/h or 0.444 kW)

6. An ice making machine is fed water at 15°C and converts at the rate of 8kg/h and it converts it into ice at –5°C. The COP of the machine is 2.2. If 384 kJ of energy is to be removed from 1 kg of water at 15°C to convert it into ice at –5°C, determine the required power input to the machine. (**Ans.** 0.3878 kw)

7. Power input for a household refrigerator having COP of 2.5, is 450 W. House wife keeps 4 kg of mangoes which were initially at 20°C in the food chamber and intends to cool the mangoes to 8°C. Assuming the specific heat of mangoes as 4.2 kJ/kg °C, estimate the time it will take to cool the mangoes to 8°C. (**Ans.** 224 sec)

8. Before the air conditioner is switched on the temperature of the air in the well sealed committee room attached to office of the chief executive of a company is 32°C on a summer day. When the air conditioner is switched on the office temperature is brought down to 20°C in 15 minutes. The COP of the air conditioner is 2.5 and it is estimated that the mass of air in committee room is 800 kg. Determine the power drawn by the air conditioner.

Take for air $C_v = 0.72$ kJ/kg °C

and $C_p = 1.0$ kJ/kg°C (**Ans.** 3.072 kw)

9. A heat engine working on carnot cycle receives heat from a source at a temperature 1173 k at a rate of 700 kJ/min and rejects the waste heat to air at 300 K. The output of the engine is used to drive a refrigerator which removes heat from the food chamber at 288 k and transfers to the atmospheric air at 300 K. Calculate (*a*) the maximum heat removal rate from the food chamber (*b*) total rate of heat rejected to the atmosphere air. (**Ans.** 4359.1 kJ/min, 5059 kJ/min)

10. A heat engine receivers heat from a reservoir at 1000 K and rejects waste heat to sink at 300 K. Half of the output of the heat engine is utilized to drive a carnot heat pump which extracts heat from the cold surroundings at 275 k and supplies it to a house whose temperature is maintained at 295 k. If the house is losing heat at a rate of 80,000 kJ/hr, calculate the minimum rate of heat supply to the heat engine so that the house could be kept at 295 K. (**Ans.** 15279 kJ/h)

□□□

CHAPTER 6

ENTROPY

6.1 INTRODUCTION

The Clausius inequality discussed in chapter 5 forms the basis for introducing a new property called the "Entropy".

In order to develop a relationship which will lead to the definition of Entropy we refer to the expression (5.59) which is the mathematical statement of Clausius inequality. That is the cycle integral of the quantity $\frac{\partial Q}{T}$ for a reversible cycle process is zero and for an irreversible one it less than zero, we re-write this expression for further discussion.

$$\oint \frac{\partial Q}{T} \leq 0 \qquad \text{...(6.1)}$$

That for Irreversible process it is

$$\oint_{\text{irr}} \frac{\delta Q}{T} < 0 \qquad \text{...(6.2}a\text{)}$$

and for reversible process it is

$$\oint_{\text{rev}} \frac{\delta Q}{T} = 0 \qquad \text{...(6.2}b\text{)}$$

Let us discuss the equation (6.2*b*) in more detail. that is we have the quantity $\frac{\delta Q}{T}$ whose cyclic integral is zero. As discussed in earlier chapters that neither the cyclic integral of work nor that of heat is zero. However, if we fix our attention on volume of the gas occupied in a piston-cylinder assembly undergoing a cycle we come to the conclusion that when the piston returns to its initial position the volume of the gas also returns to its original value. Thus the net change in volume during the cycle is zero.

or

$$\oint dV = 0$$

In other words the cyclic integral of volume is zero and further the volume of a gas depends on the state. Similarly a quantity whose cyclic integral is zero depends on the state only and not on the process path, is the property. Therefore if leads us to state that the quantity $\left(\frac{\delta Q}{T}\right)_{rev}$ must represent a property in differential form this argument will equally apply to the intenally reversible processes.

6.1.1 Entropy Change in a Reversible Process

In figure 6.1 is shown a cycle that consist of two the initially reversible process A and B. For these states applying equation (6.2b).

$$\oint_{rev} \frac{\delta Q}{T} = \oint_2^1 \left(\frac{\delta Q}{T}\right)_A + \int_2^1 \left(\frac{\delta Q}{T}\right)_B = 0$$

or

$$\oint_1^2 \left(\frac{\delta Q}{T}\right)_A = -\int_2^1 \left(\frac{\delta Q}{T}\right)_B = \int_1^2 \left(\frac{\delta Q}{T}\right)_B \quad ...(6.3)$$

Paths A and B are both internally reversible

Fig. 6.1 Paths A and B are internally rev paths.

Since A and B are any two internally reversible paths between the states 1 and 2 the value of this integral depends on the end states and not on the path. Therefore the quantity $\oint \frac{\delta Q}{T}$ depends in the end states only 1 and then it represents the change of the property. This property is called the ENTROPY. It is represented by S and is defined as

$$dS = \left(\frac{\delta Q}{T}\right)_{rev} \quad ...(6.4)$$

or in words

$$\text{Change in Entropy} = \frac{\text{Heat Supplied or Rejected } \delta Q}{\text{Absolute Temperature } T} \text{ kJ/K}$$

Entropy is an extensive property of a system and sometimes it is called Total Entropy. However Entropy per unit mass is represented by 's' and is called the specific entropy.

Thus the change in entropy during a process between two states

$$\Delta S = S_2 - S_1 = \int_1^2 \left(\frac{\delta Q}{T}\right)_{rev} \text{ kJ/K} \quad ...(6.5)$$

At this stage we have been able to define the change in entropy. However the absolute values of entropy will be defined later which will be defined on the basis of Third law of thermodynamics. An important point to note is that the entropy is a property, and like any other property, it has fixed values at fixed states. Therefore the change in entropy between two specified states is same and is not dependent on the path, reversible or irreversible, that is followed during the process.

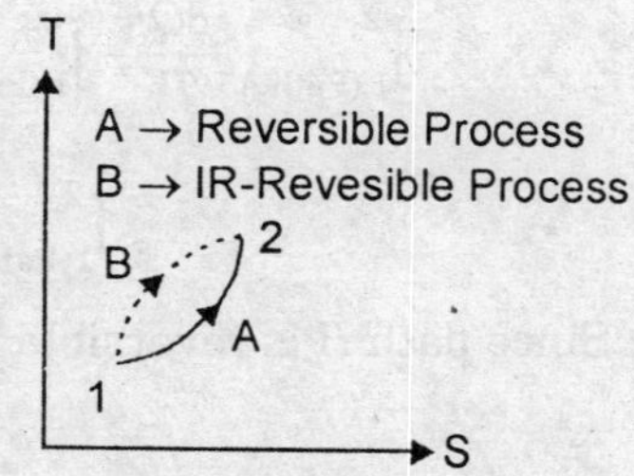

Fig. 6.2

Another important point to note is that $\oint \left(\frac{\delta Q}{T}\right)_{rev}$ will give the value of entropy change between two states only if the integration is carried out along an internally reversible path between the two given sates. Integral of $\left(\frac{\delta Q}{T}\right)$ along the irreversible path is not A PROPERTY and we will get different values if integration is done along different irreversible paths. Therefore we conclude that even when we have irreversible processes, the entropy change between the states must be determined by carrying out the integration along some convenient imaginary but certainly reversible path between the two given states.

6.2 ENTROPY CHANGE IN AN IR-REVERSIBLE PROCESS

Let a system undergo a change from state 1 to state 2 (Fig. 6.3) following an ir-reversible path 1-A-2. Further the system follows the reversible path 2-B-1 to come to its initial state 1.

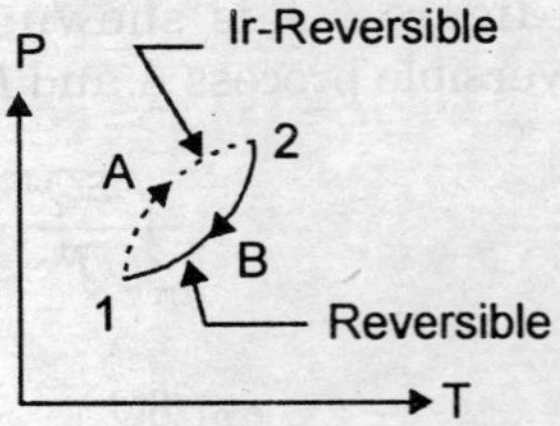

Fig. 6.3

The two process 1-A-2 and 2-B-1 together form an ir-reversible cycle. Then we have from Clausius inequality.

$$\oint \frac{\delta Q}{T} < 0$$

or $$\int_{1\,(\text{Path } A)}^{2} \frac{\delta Q}{T} + \int_{2\,(\text{Path } B)}^{1} \frac{\delta Q}{T} < 0 \qquad \text{...(6.6)}$$

Since path B is reversible then The integral of $\left(\frac{\delta Q}{T}\right)$ will give change of entropy.

i.e., $$\int_{2\,(\text{Path } B)}^{1} \frac{\delta Q}{T} = -\int_{2(\text{Path } B)}^{1} ds$$

Now for the reversible path the limits can be reversed

i.e., $$\int_{2(\text{Path } B)}^{1} \frac{\delta Q}{T} = -\int_{2\,(\text{Path } B)}^{1} \frac{\delta Q}{T}$$

or $$\int_{2\,(\text{Path } B)}^{1} \frac{\delta Q}{T} = -\int_{1\,(\text{Path } B)}^{2} \frac{\delta Q}{T}$$

$$= -\int_{2(\text{Path } B)}^{1} ds \qquad \text{...(6.7)}$$

Therefore in the light of equation (6.7), inequality (6.6) reduces to

$$\int_{1\,(\text{Path } A)}^{2} \frac{\delta Q}{T} - \int_{1\,(\text{Path} B)}^{2} ds < 0$$

or $$\int_{1(\text{Path} A)}^{2} \frac{\delta Q}{T} < \int_{1\,(\text{Path } B)}^{2} ds$$

Since path B is reversible it gives the change in entropy

$$\int_{1}^{2} ds = S_2 - S_1$$

Then $$\int_{1\,(Path A)}^{2} \left(\frac{\delta Q}{T}\right) < (S_2 - S_1)$$

or
$$S_2 - S_1 \geq \int_1^2 \frac{\delta Q}{T} \quad \text{...(6.8)}$$
(irreversible path)

In the expression (6.8), the equality size holds good for reversible process and inequality for irreversible process. Further from expression (6.8) we conclude that entropy increases during a reversible process and remains constant (or it is conserved) during a reversible process.

Another important point to note is that for an adiabatic process in which $\delta Q = 0$ we get from (6.8)

$$ds \geq 0$$

That if the process is reversible adiabatic

$$ds = 0$$

and is case it is irreversible adiabatic

$$ds > 0$$

6.3 ENTROPY PRINCIPLE

Statement

Entropy of an isolated system can never decrease. It increases for an irreversible process and remains constant for reversible process.

Proof

An isolated system is one that does not undergo any work or heat interaction with the surroundings and the total energy of all the states remain constant.

Therefore $$(\delta Q)_{rev} = 0 \quad \text{...(6.9)}$$

Further change of entropy $$ds = \frac{\delta Q}{T}$$

Since entropy is a thermodynamic property and the cyclic integral of thermodynamic property is zero.

$$\oint ds = 0 \quad \text{...(6.10)}$$

But from clausius inequality one have

$$\oint \frac{\delta Q}{T} \leq 0 \quad \text{...(6.11)}$$

∴ From (6.10) and (6.11)

$$\oint \frac{\delta Q}{T} \leq \oint ds$$

or
$$\frac{\delta Q}{T} \leq ds$$

or
$$ds \geq \frac{\delta Q}{T} \quad \text{...(6.12)}$$

Again from (6.9) and (6.12) for isolated system

$$(ds)_{iso} \geq 0$$

However for reversible process

$$(ds)_{iso} = 0 \qquad \text{or} \qquad S = \text{constant}$$

For irreversible process $(ds)_{iso} > 0$

6.3.1 Universe

Combination of an isolated system and the surroundings within a single boundary is known as universe.

In Fig. 6.4 the combination of system and surroundings is shown. Let there be heat transfer δQ from the isolated system at temperature T_{system} to its surroundings at temperature T_{sur}

Then $$(ds)_{sys} = -\frac{\delta Q}{T_{sys}}$$

and $$(ds)_{sur} = \frac{\delta Q}{T_{sur}}$$

Fig. 6.4

Net change in entropy of the universe

$$(ds)_{universe} = (ds)_{sys} + (ds)_{sur}$$

$$= -\frac{\delta Q}{T_{sys}} + \frac{\delta Q}{T_{sur}}$$

$$= \delta Q\left[\frac{1}{T_{sur}} - \frac{1}{T_{sys}}\right] \qquad \text{...(6.13)}$$

since $T_{sur} < T_{sys}$ (second low of thermo of thermal)

The right hand side of (6.13) is greater than zero

$$\therefore \quad (ds) = \delta Q\left[\frac{1}{T_{sur}} - \frac{-1}{T_{sys}}\right]$$

$$> 0$$

Thus the entropy of an isolated system along with its surroundings or for the universe increases if the system undergoes an irreversible process.

However in case $T_{sur} = T_{sys}$

the process woud occur reversibly and then

$$ds = \delta Q\left[\frac{1}{T_{sur}} - \frac{1}{T_{sys}}\right] \qquad \text{...(6.14)}$$

$$= 0$$

Thus in general we may write

$$ds \geq 0 \text{ for an isolated system}$$

In the physical sense it means that if the entropy decreases at region within the isolated system it must increase by an equal amount some where else within the system. The net result for an irreversible system would be an increase in the entropy of the whole system.

Further we know that all real processes in the universe are irreversible and universe can be taken as an isolated system the entropy of the universe is progressively increasing.

6.4 TEMPERATURE-ENTROPY DIAGRAM

Diagrams which depict the properties serve as great visual aids while analysing the thermodynamic processes. In the earlier chapters the use of P-V. and T-V. diagram was extensively employed, similarly entropy-temperature diagrams in which one of the co-ordinates is entropy are very useful. There are two sets of diagrams which are extensively used in the second law analysis viz Temperature-Entropy and Enthelpy-Entropy diagrams.

6.4.1 The T–S Diagram

For a reversible process or for an internally reversible system being changed from initial state 1 to final state 2 we have from equation (6.4)

$$ds = \left(\frac{dQ}{T}\right)_{rev} \quad \text{...(6.15a)}$$

or

$$Q_{rev} = \int_1^2 T\,ds \quad \text{...(6.15b)}$$

From equation (6.15) it is infered that the area under the curve plotted for a reversible process, on T–S space, represents the quantity of heat interaction during the process (Fig. 6.5a)

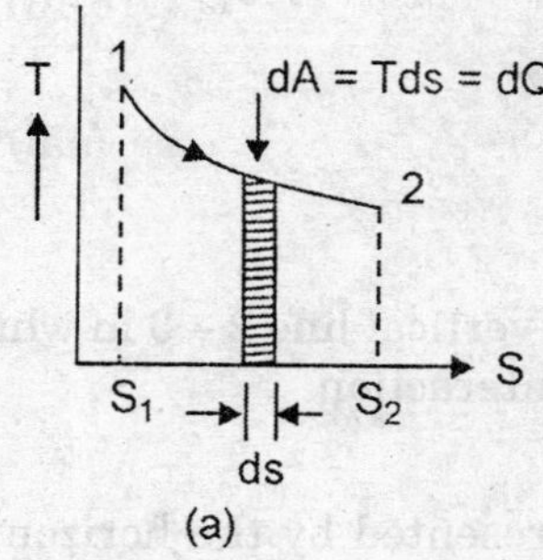

(a)

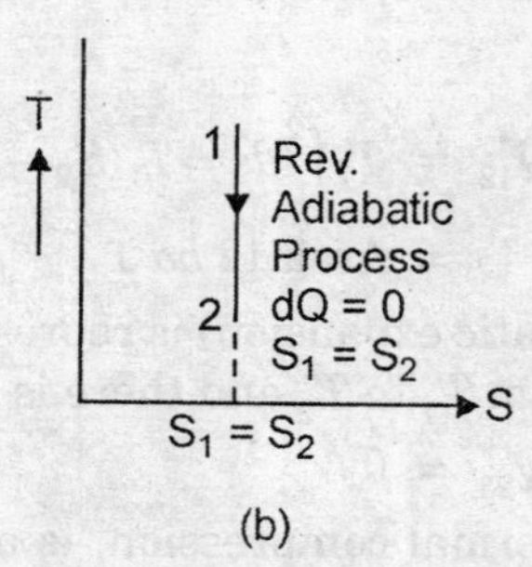

(b)

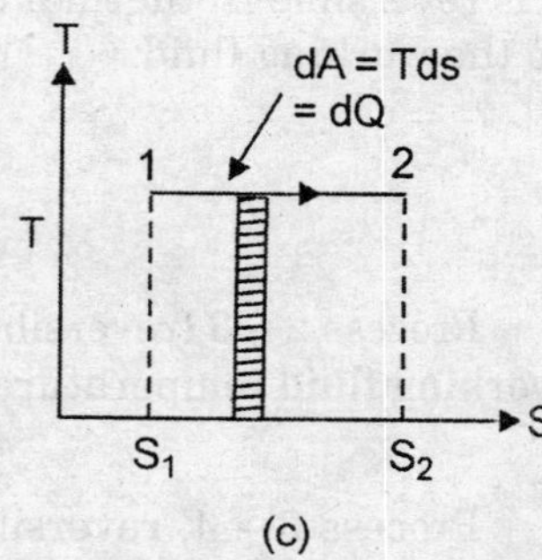

(c)

Fig. 6.5

6.4.2 Reversible Adiabiatic Process

For reversible adiabiatic process we have

$$dQ = 0$$

Therefore from (6.15a) $\quad ds = 0 \quad$ or $\quad S = \int ds = \text{constant}$

Thus the reversible adiabatic process will be represented by a vertical line is (constant entropy), on $T - S$ diagram Fig. 6.5(b)

6.4.3 Reversible Isothermal Process

In this process as the temperature remains constant equation (6.15a) reduces to

$$Q_{rev} = T_0\int_1^2 ds = T_0(S_2 - S_1) \quad \text{...(6.16)}$$

Thus the reversible isothermal process is represented by a horizontal line at (constant temperature T_0) (Fig. 6.5 *c*).

6.4.4 T–S, Diagram for Carnot Cycle

Since Carnot cycle consists of two reversible isothermals and was reversible adiabatics, it will be represented by a rectangle on T–S. diagram as shown in Fig. (6.6).

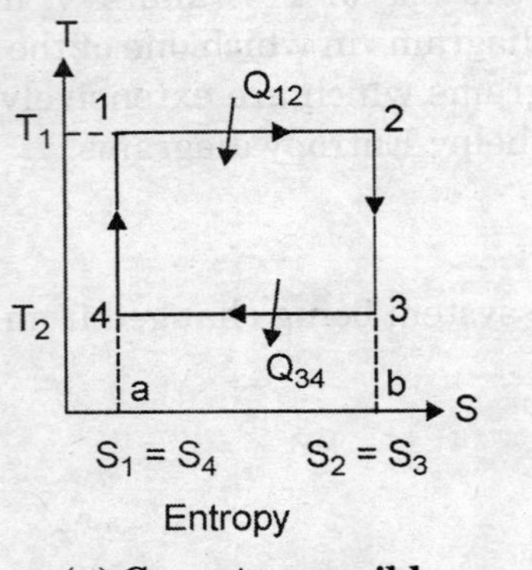

(*a*) Carnot reversible heat engine cycle

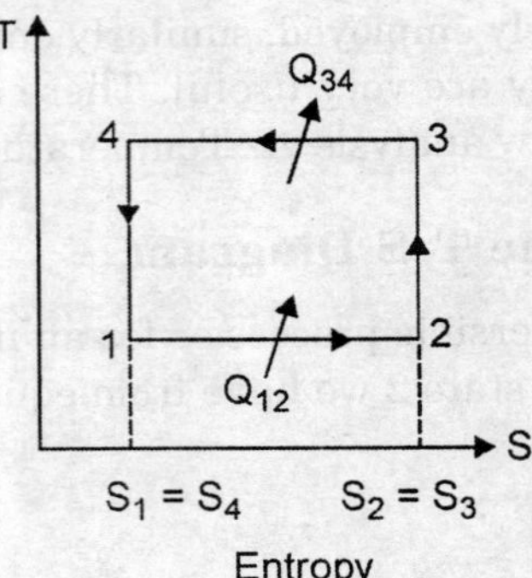

(*b*) Carnot reversible refrigerator cycle

Fig. 6.6

If Fig. (6.6*a*) is shown T–S. diagram for a reversible Carnot Cycle heat engine Process 1 – 2 is reversible isothermal expansion at constant Temperature *T*, in which heat, Q_{12} is supplied to the working fluid.

$$Q_{12} = T_1\int_1^2 ds = T_1(S_2 - S_1)$$
$$= \text{Area } 12\, ba\, 1$$

Process 2 – 3 (reversible adiabatic expansion) is represented by vertical line 2 – 3 in which working fluid temperature falls from T_1 to T_2 and there is no heat interaction

$$Q_{23} = 0$$

Process 3 – 4, reversible isothermal compression, is again represented by the horizontal line 3 – 4 in which the heat is transferred from the working fluid to the low temperature sink at temperature T_2.

and
$$Q_{34} = \int_3^4 T_2 ds = T_2\int_3^4 ds = T_2(S_4 - S_3)$$
$$= \text{Area } 34\, ab$$

Process 4 – 1 is reversible adiabiatic compression during which there is no heat interaction

i.e., $Q_{4\text{-}1} = 0$ and $S_4 = S_1$

However there is increase in temperature of fluid but entropy remains constant.

During the cyclic process there is no change in the entropy and $S_4 = S_1$. It is evident that increase in entropy during the process 1 – 2, (as the heat is supplied to the working fluid) is equal to the decrease in entropy during the process 3 – 4 (as the heat is rejected by the fluid).

or
$$S_2 - S_1 = S_4 - S_3$$

Now during the cycle the heat input is Q_{12} and the heat rejected is Q_{3-4}.

Thus the equivalent work output

$$= Q_{1-2} - Q_{3-4}$$

Therefore efficiency $$\eta = \frac{Q_{12} - Q_{34}}{Q_{12}}$$

$$= \frac{\text{Area } 12\,ba1 - \text{Area } 34\,ab3}{\text{Area } 12\,ba1}$$

$$= \frac{\text{Area } 12341}{\text{Area } 12\,ba\,1} \quad ...(6.17)$$

Also net work output $= \text{Area } 12341$ ...(6.18)

Similarly the reversible carnot refrigerator cycle or heat pump which is reversed of that for heat engine is shown in Fig. (6.6 *b*). Here the entropy increase is at lower temperature as the heat input happens at T_2 and the rejection of heat is at higher temperature T_1.

6.5 ENTROPY CHANGE DURING VARIOUS THERMODYNAMIC PROCESSES

In equation (6.15) entropy change was evaluated by integrating $\frac{\delta Q}{T}$ along some imaginary internally reversible path between the actual states. For internally reversible isothermal process, the integration is simple as T is constant. However, when the temperature varies during a process we have to have a relation between δQ and T so that the integration can be performed. The differential form of the conservation of energy equation for a closed stationary system (or for a fixed mass) containing a simple compressible substance we have for an internally reversible process (chapter 3, equation 3.57).

$$\delta Q_{\text{int-rev}} - \delta W_{\text{int-rev}} = dU \quad ...(3.57)$$

But $$\delta Q_{\text{int-rev}} = T\,dS \quad (6.15)$$

and $$\delta W_{\text{int-rev}} = P\,dV \quad (3.11)$$

Therefore substituting the values of $\delta Q_{\text{int-rev}}$ and $\delta W_{\text{int-rev}}$ in equation (3.57) we get

$$T\,dS - \rho\,dV = dU$$

or $$T\,dS = dU + P\,dV \text{ kJ} \quad ...(6.19\,a)$$

And on unit mass basis $$Tds = du + P\,dv \text{ kJ/kg} \quad ...(6.19\,b)$$

This equation is known as FIRST $T - ds$ equation or GIBBS equation

An important point to be noted is that the only type of work interaction a simple compressible system may involve during an internally reversible process is the quasi-equilibrium boundary work.

Further we know from the definition of enthalpy

$$h = u + Pv \quad ...(3.75)$$

or $$dh = du + Pdv + vd\rho \quad ...(6.20)$$

But from (6.19) $$du + \rho dv = Tds$$

Therefore from (16.20) $$dh = Tds + vdP$$

or $$Tds = dh - vdP \quad ...(6.21)$$

This equation is known as "Gibbs second Tds equation".

Gibbs equations are very useful since it relate entropy changes of a system to the changes in other properties. Even though we have stated above that these processes are internally reversible but these equations give property relations and therefore they are independent of the type of the process whether reversible or irreversible. Further these are relations between the properties of a unit mass of a simple compressible system as it undergoes a change of state, and these are applicable whether the changes occur in a closed or an open system.

For finding the differential changes in entropy the Gibbs equations (6.19) and (6.21) are re-written in the following form.

$$ds = \frac{du}{T} + \frac{P\,dv}{T} \quad ...(6.22)$$

and

$$ds = \frac{dh}{T} - \frac{v\,dP}{T} \quad ...(6.23)$$

6.5.1 Important Note On the six equations as obtained from the First and Second Laws of thermodynamcis.

(*a*) $dQ = dE + dW$

This equation holds good for any process reversible or irreversible and for any system.

(*b*) $dQ = dU + dW$

This equation holds good for any process taking place by a closed stationary system.

(*c*) $dQ = dU + PdV$

This equation holds good for a closed system when only pdv work is present and further it is true only for a reversible (quasi-static) process.

(*d*) $dQ = Tds$

This equation is true only for reversible process.

(*e*) $Tds = dU + Pdv$

This equation holds good for any process reversible or irreversible, undergone by a closed system as it is a relation among properties and it is dependent upon the path of the process.

(*f*) $T\,dS = dH - V\,dP$

This equation too relates to the properties of the closed system. Therefore it is valid for both reversible and irreversible process undergone by a closed system.

6.6 CHANGE OF ENTROPY FOR DIFFERENT PROCESSES

6.6.1 Constant Volume Process (Isochoric process)

Let a certain mass of a perfect gas be heated at constant volume from temperature T_1 to T_2 and the corresponding pressure changes from P_1 to P_2 Fig. 6.7 depicts the two states in T–S. digram and this process is represented by the curve 12.

Heat supplied for a small change of temperature dT is

in $$\delta Q = m\,C_v\,dT \qquad ...(6.22)$$

When C_v = specific heat at constant volume

Dividing equation (6.22) with T

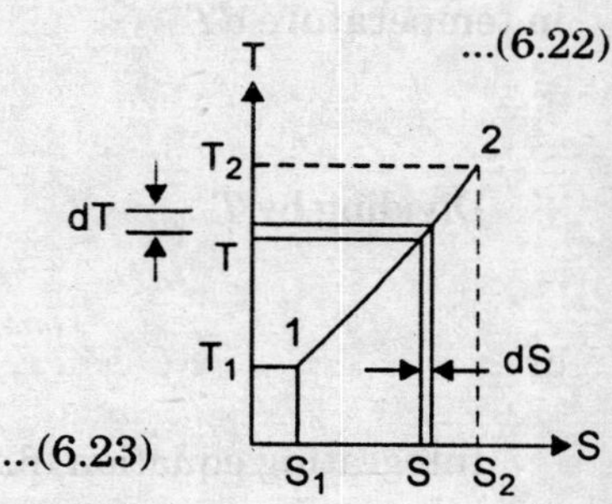

Fig. 6.7

$$\frac{\delta Q}{T} = mC_v\,\frac{dT}{T}$$

or $$dS = mC_v\,\frac{dT}{T} \qquad ...(6.23)$$

Integrating equation (6.23) for change from state 1 to state 2

$$\int_1^2 dS = \int_{T_1}^{T_2} mC_v\,\frac{dT}{T}$$

or $$S_2 - S_1 = mC_v \ln\left(\frac{T_2}{T_1}\right) \qquad ...(6.24)$$

Further for a perfect gas for constant volume process

$$\frac{P_1V_1}{T_1} = \frac{P_2V_2}{T_2} \qquad (\because V_1 = V_2)$$

$$\frac{P_1}{T_2} = \frac{P_2}{T_2} \text{ or } \frac{T_2}{T_1} = \frac{P_2}{P_1}$$

Therefore equation (6.25) reduces to

$$S_2 - S_1 = mC_v \ln\left(\frac{P_2}{P_1}\right) = mC_v \ln\left(\frac{T_2}{T_1}\right) \qquad ...(6.25a)$$

And for unit mass it is written as

$$s_2 - s_1 = C_v \ln\left(\frac{P_2}{P_1}\right) = C_v \ln\left(\frac{T_2}{T_1}\right) \qquad ...(6.25b)$$

The equation (6.23), the slop of the curve

$$\frac{dT}{dS} = \frac{T}{mC_V} \qquad ..(6.25c)$$

and for unit mass $$\frac{dT}{ds} = \frac{T}{C_v}$$

6.6.2 Constant Pressure Process (Isobaric Process)

Let a certain mass of a perfect gas be heated from temperature T_1 to temperature T_2 at constant pressure as shown in Fig. 6.8

Heat supplied at temperature T for a small elemental change in temperature dT

$$dQ = mC_p\, dT$$

Dividing by T

$$\frac{dQ}{T} = mC_p \frac{dT}{T}$$

or

$$ds = mC_p \frac{dT}{T} \qquad ...(6.26)$$

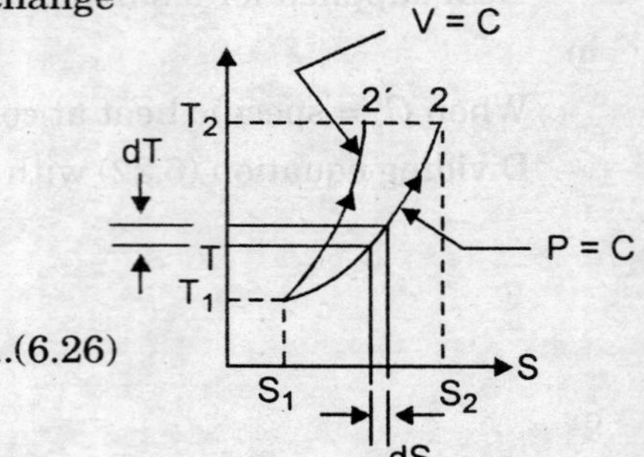

Fig. 6.8. Constant pressure and constant volume processes

Integrating equation (6.26)

$$\int_1^2 ds = \int_{T_1}^{T_2} mC_p \frac{dT}{T}$$

or

$$S_2 - S_1 = mC_p \ln\left(\frac{T_2}{T_1}\right) \qquad ...(6.27)$$

For ideal gas from gas equation one have

$$\frac{P_1V_1}{T_1} = \frac{P_2V_2}{T_2}$$

or

$$\frac{V_1}{V_1} = \frac{V_2}{T_2} \qquad (\text{as } P_1 = P_2)$$

or

$$\frac{T_2}{T_1} = \frac{V_2}{V_1}$$

Thus equation (6.26) reduces to

$$S_2 - S_1 = mC_p \ln\left(\frac{T_2}{T_1}\right) = mC_p \ln\left(\frac{V_2}{V_1}\right) \qquad ...(6.27a)$$

For unit mass it is rewritten as

$$s_2 - s_1 = C_p \ln\left(\frac{T_2}{T_1}\right) = C_p \ln\left(\frac{V_2}{V_1}\right) \qquad ...(6.27b)$$

From equation (6.26), the slope of this curve is

$$\frac{dT}{dS} = \frac{T}{mC_p}$$

and an unit mass basis

$$\frac{dT}{ds} = \frac{T}{C_p} \qquad ...(6.27c)$$

Note : Comparison of slope equations (6.25*c*) and (6.27*c*) shows that the slope of the constant pressure curve $\left(\frac{dT}{dS}\right)_{const\ p}$ is smaller than the curve for constant volume $\left(\frac{dT}{dS}\right)_{const\ v}$ because $c_p > c_v$

Therefore $$\frac{1}{c_p} < \frac{1}{c_v}$$

or $$\frac{T}{c_v} > \frac{T}{c_p} \qquad ...(6.28)$$

Thus of these two process curves are plotted on the same $T - S$ space (Fig. 6.8) they will look like the one as shown in Fig. 6.8.

6.6.3 Constant Temperature Process (Isothermal Process)

Let a certain mass of a perfect gas be heated at a constant temperature T from initial pressure and volume P_1, V_1 to final state where pressure and volume are P_2, V_2. The process on T–S. diagam is represented by a straight horizontal line as shown in Fig. 6.9.

Work done by the gas will be equal to heat supplied to it because during isothermal process there is no change in internal energies

i.e., $$dU = 0$$

Therefore from $$\delta Q = dU + PdV$$

$$= PdV$$

or $$\delta Q = \frac{mRT}{V} dv \quad (\because PV = mRT)$$

$$= mRT \frac{dV}{V}$$

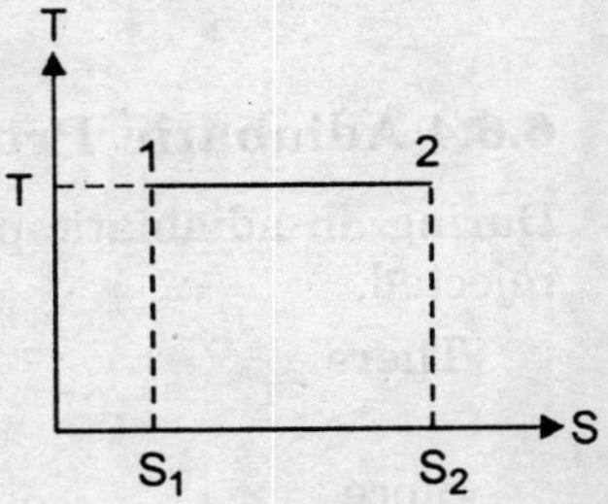

or $$\frac{\delta Q}{T} = mR \frac{dV}{V}$$

or $$dS = mR \frac{dV}{V} \qquad ...(6.29)$$

T, 1, 2, T, S, S_1, S_2

Fig. 6.9. T.S. diagram isothermal Process

Integrating (6.29) we get

$$\int_1^2 dS = \int_1^2 mR \frac{dV}{V}$$

or $$S_2 - S_1 = mR \int_1^2 \frac{dV}{V}$$

$$= mR \ln\left(\frac{V_2}{V_1}\right) \qquad ...(6.30a)$$

$$= m(C_p - C_v) \ln\left(\frac{V_2}{V_1}\right) \qquad ...(6.30b)$$

Also for perfect gas $$\frac{P_v V_1}{T_1} = \frac{P_2 V_2}{T_2}$$

or $$P_1 V_1 = P_2 V_2 \qquad (\because T_1 = T_2 \text{ as isothermal process})$$

or
$$\frac{P_1}{P_2} = \frac{V_2}{V_1}$$

Therefore equation (6.30*b*) has another form

$$S_2 - S_1 = m(C_p - C_v)\ln\left(\frac{P_1}{P_2}\right) \quad ...(6.30c)$$

Equation (6.30) is re-written as

$$S_2 - S_1 = mR\ln\left(\frac{V_2}{V_1}\right) = mR\ln\left(\frac{P_1}{P_2}\right)$$

$$= m(C_p - C_v)\ln\left(\frac{V_2}{V_1}\right) = m(C_p - C_v)\ln\left(\frac{P_1}{P_2}\right) \quad ...(6.30)$$

Again an unit mass basis $s_2 - s_1 = (C_p - C_v)\ln\left(\frac{v_2}{v_1}\right) = (C_p - C_v)\ln\left(\frac{P_1}{P_2}\right)$ ...(6.31)

6.6.4 Adiabatic Process (Isentropic Process)

During an adiabiatic process there is no heat interaction *i.e.*, neither heat is supplied nor rejected.

There $\delta Q = 0$

Since
$$\delta S = \frac{\delta Q}{T}$$
$$= 0$$

That is there is no change in entropy and it remains constant thus its name ISENTROPIC PROCESS (it is functionless adiabatic process). The reversible adiabatic (Frictionless process is shown on T–S. diagram as in Fig. 6.10 by a vertical line 1 – 2.

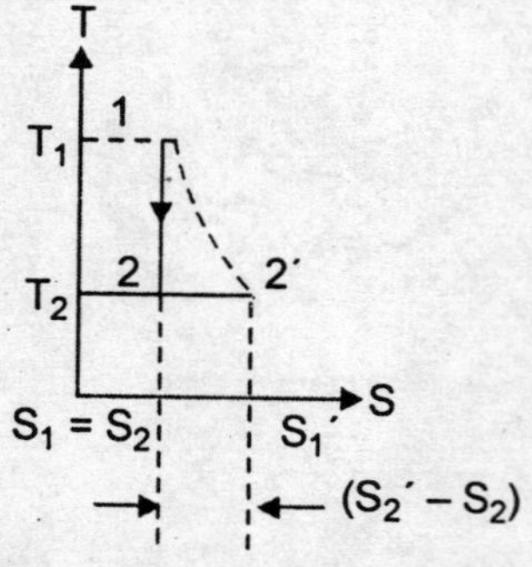

***(a)* (*i*) Entropy constant 1 – 2 as it is reversible adiabatic expansion (frictionless)**
(*ii*) Entropy increase 2 – 2′ is due to friction or irreversibility

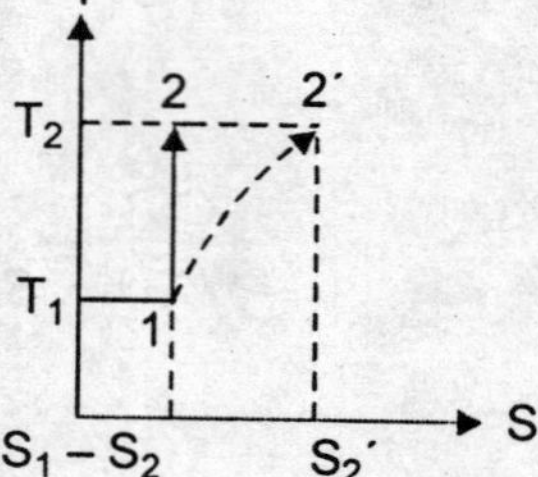

***(b)* (*i*) Entropy constant 1 – 2 as it is reversible reversible adiabatic compression (frictionless)**
(*ii*) Entropy increase 2 – 2′ is due to friction or irreversible adiabatic

Fig. 6.10

However, when the effect of friction is taken into account the internal friction during expansion or compression will raise the internal energy even though the temperature range may be T_1 to T_2. In this way additional quantity of heat (say = $\delta Q'$) will be there so that this corresponding entropy change will be equal to $\frac{\delta Q'}{T}$. Thus during irreversible adiabatic process.

$$ds > 0 \quad ...(6.32)$$

6.6.5 Polytropic Process (PV^n = Constant)

Consider a certain mass m of a perfect gas is heated by polytropic process (PV^n = constant). Let P_1, V_1 and T_1 be the initial pressure, volume and temperature respectively of the gas and at the final stage these are P_2, V_2 and T_2

$$PV^n = C \quad ...(6.33)$$

Work done

$$W_{1-2} = \int_1^2 PdV$$

$$= \int_1^2 \left(\frac{C}{V^n}\right) dV$$

$$= \frac{C}{1-n}\left[V^{1-n}\right]_1^2$$

$$= \left(\frac{C}{1-n}\right)\left[V_2^{1-n} - V_1^{1-n}\right] \quad ...(6.34)$$

But also $\rho_1 V_1^n = \rho_2 V_2^n = C$

Therefore equation (6.34) reduces to

$$W_{1-2} = \left(\frac{1}{1-n}\right)\left[P_2 V_2^n V_2^{1-n} - P_1 V_1^n V_1^{1-n}\right]$$

$$= \frac{P_2 V_2 - P_1 V_1}{1-n} = \frac{mR(T_2 - T_1)}{1-n}$$

or

$$W_{1-2} = \frac{P_1 V_1 - P_2 V_2}{n-1} = \frac{mR(T_1 - T_2)}{n-1} \quad ...(6.35a)$$

On unit mass basis it reduces to

$$w_{1-2} = \frac{\rho_1 v_1 - \rho_2 v_2}{n-1} = \frac{R(T_1 - T_2)}{n-1} \quad ...(6.35b)$$

Further the heat interaction during polytropic process can be determined with the help of first law of thermodynamics.

i.e.,

$$\delta Q = \delta W + \delta U \quad ...(6.36a)$$

$$\delta q = \delta w + \delta u \quad ...(6.36b)$$

Where δU = Change in internal energy for constant specific heat

or

$$U_2 - U_1 = mC_v(T_2 - T_1) \quad ...(6.37a)$$

and on unit mass basis $\delta u = C_v(T_2 - T_1)$

or

$$u_2 - u_1 = C_v(T_2 - T_1) \quad ...(6.37b)$$

Then equation (6.36) reduces on unit mass basis

$$q_{1-2} = w_{1-2} + (u_2 - u_1)$$

$$= \frac{p_1v_1 - p_2v_2}{n-1} + C_v(T_2 - T_1)$$

$$= \left(\frac{p_1v_1 - p_2v_2}{n-1}\right) + \frac{R}{\gamma - 1}(T_2 - T_1)$$

or

$$q_{1-2} = \frac{p_1v_1 - p_2v_2}{n-1} + \left(\frac{p_2v_2 - p_1v_1}{\gamma - 1}\right)R$$

$$= (p_1v_1 - p_2v_2)\left[\frac{1}{n-1} - \frac{1}{\gamma-1}\right]$$

$$= \frac{p_1v_1 - p_2v_2}{n-1}\left[1 - \frac{n-1}{\gamma-1}\right]$$

$$= \frac{p_1v_1 - p_2v_2}{(n-1)}\left(\frac{\gamma - n}{\gamma - 1}\right)$$

$$= \left(\frac{\gamma - n}{\gamma - 1}\right) \text{ [polytropic work done]}$$

Thus

$$\delta Q = \left(\frac{\gamma - n}{\gamma - 1}\right)\delta W \quad ...(6.38)$$

$$= \left(\frac{\gamma - n}{\gamma - 1}\right)PdV \quad ...(6.39)$$

or dividing by T

$$\frac{\delta Q}{T} = \left(\frac{\gamma - n}{\gamma - 1}\right)\frac{PdV}{T}$$

or

$$dS = \left(\frac{\gamma - n}{\gamma - 1}\right)\frac{mR}{V}dV \qquad \left(\because \frac{P}{T} = \frac{mR}{V}\right)$$

Integrating between the initial and final state

$$\int_1^2 ds = \left(\frac{\gamma - n}{\gamma - 1}\right)mR\int_1^2 \frac{dV}{V}$$

or

$$S_2 - S_1 = \left(\frac{\gamma - n}{\gamma - 1}\right)mR\ln\left(\frac{V_2}{V_1}\right) \quad ...(6.40)$$

$$= (\gamma - n)\, m\left(\frac{R}{\gamma - 1}\right)\ln\left(\frac{V_2}{V_1}\right)$$

$$= (\gamma - n)\, m C_V \ln\left(\frac{V_2}{V_1}\right)$$

$$= (\gamma C_v - n C_V)\, m \ln\left(\frac{V_2}{V_1}\right)$$

But $$C_p/C_v = \gamma$$

$\therefore$ $$S_2 - S_1 = m(C_p - nC_v) \ln\left(\frac{V_2}{V_1}\right) \qquad \text{...(6.40b)}$$

It is emphasized that C_p and C_v are assumed to be constant specific heats

Equations (6.40) can also be expressed in terms of temperature T_1 and T_2. Since for polytropic process

$$P_1 V_1^n = P_2 V_2^n$$

or $$\frac{P_1}{P_2} = \left(\frac{V_2}{V_1}\right)^n \qquad \text{...(6.41)}$$

and $$\frac{P_1 V_1}{T_1} = \frac{P_2 V_2}{T_2}$$

or $$\frac{P_1}{P_2} = \left(\frac{T_1}{T_2}\right)\left(\frac{V_2}{V_1}\right) \qquad \text{...(6.42)}$$

For (6.41) and (6.42)

$$\left(\frac{V_2}{V_1}\right)^n = \frac{T_1}{T_2}\left(\frac{V_2}{V_1}\right)$$

$$\frac{T_1}{T_2} = \left(\frac{V_2}{V_1}\right)^{n-1}$$

or $$\left(\frac{V_2}{V_1}\right) = \left(\frac{T_1}{T_2}\right)^{\frac{1}{n-1}} \qquad \text{...(6.43)}$$

Thus substituting equation (6.43) into equation (6.40)

$$S_2 - S_1 = (mR)\left(\frac{\gamma - n}{\gamma - 1}\right) \ln\left(\frac{T_1}{T_2}\right)^{\frac{1}{n-1}}$$

$$= \left(\frac{\gamma - n}{\gamma - 1}\right) \frac{mR}{n-1} \ln\left(\frac{T_1}{T_2}\right)$$

Now $$C_v = \frac{R}{\gamma - 1}$$

or $$R = C_v(\gamma - 1)$$

$\therefore$ or $$S_2 - S_1 = \left(\frac{\gamma - n}{\gamma - 1}\right)\frac{m}{n-1}C_v(\gamma - 1)\ln\left(\frac{T_1}{T_2}\right)$$

$$= m\left(\frac{\gamma - n}{n-1}\right)C_v \ln\left(\frac{T_1}{T_2}\right) \quad \text{...(6.44a)}$$

$$= \frac{m(\gamma C_v - nC_v)}{(n-1)}\ln\left(\frac{T_1}{T_2}\right)$$

$$= \frac{m(C_p - nC_v)}{n-1}\ln\left(\frac{T_1}{T_2}\right) \quad \text{...(6.44b)}$$

Yet there is another form of the equation (6.40), in terms of p_1 and p_2. We have from equations (6.40*a*) and (6.41).

$$S_2 - S_1 = \left(\frac{\gamma - n}{\gamma - 1}\right)m\,R\ln\left(\frac{p_1}{p_2}\right)^{\frac{1}{n}}$$

$$= m\left(\frac{\gamma - n}{\gamma - 1}\right)\frac{R}{n}\ln\left(\frac{p_1}{p_2}\right)$$

$$= m\left(\frac{\gamma - n}{\gamma - 1}\right)[(c_v(\gamma - 1)]\frac{1}{n}\ln\left(\frac{p_1}{p_2}\right)$$

$$= mc_v\left(\frac{\gamma - n}{n}\right)\ln\left(\frac{p_1}{p_2}\right)$$

$$= \frac{m(c_p - nc_v)}{n}\ln\left(\frac{p_1}{p_2}\right) \qquad \because \gamma = \frac{C_p}{C_v} \quad \text{...(6.45)}$$

Equations (6.40*a*), (6.40*b*), (6.44) and (6.45) are valid for both reversible and irreversible polytropic processes.

6.6.6 Alternative Forms of Expression for Entropy Change for A Polytropic Process for Perfect Gases (With specific heats constants)

With constant specific (heats) we have from equation (6.36*b*) on unit mass basis

$$\delta_q = \delta u + \delta w \quad \text{...(6.36b)}$$

But $$\delta u = c_v dt$$

and $$\delta w = p\,dv$$

Therefore dividing equation (6.36*b*) and substituting the values of δu and δw we get.

$$\frac{\delta_q}{T} = c_v\frac{dT}{T} + \frac{p}{T}dv$$

or $$ds = c_v\frac{dT}{T} + \frac{R}{v}dv \qquad (\because pv = RT)$$

Integrating between the initial and final states we get

$$s_2 - s_1 = c_v \int_1^2 \frac{dT}{T} + R\int_1^2 \frac{dv}{v}$$

$$= c_v \ln\left(\frac{T_2}{T_1}\right) + R \ln\left(\frac{v_2}{v_1}\right) \quad \text{...(6.46)}$$

Equation (6.46) gives the entropy change in line of specific volume and absolute temperature.

Equation 4.46 can be re-written in another from in terms of specific volumes and pressues as follows :

$$s_2 - s_1 = c_v \ln\left(\frac{p_2 v_2}{p_1 v_1}\right) + R \ln\left(\frac{v_2}{v_1}\right) \quad \left(\because \frac{p_1 v_1}{T_1} = \frac{p_2 v_2}{T_2}\right)$$

or

$$s_2 - s_1 = c_v \ln\left(\frac{p_2}{p_1}\right) + c_v \ln\left(\frac{v_2}{v_1}\right) + R \ln\left(\frac{v_2}{v_1}\right)$$

or

$$s_2 - s_1 = c_v \ln\left(\frac{p_2}{p_1}\right) + (R + c_v) \ln\left(\frac{v_2}{v_1}\right)$$

$$= c_v \ln\left(\frac{p_2}{p_1}\right) + c_p \ln\left(\frac{v_2}{v_1}\right) \quad \text{...(6.47)}$$

Equation (6.46) can also be re-written in terms of pressures and absolute temperature as follows

$$s_2 - s_1 = c_v \ln\left(\frac{T_2}{T_1}\right) + R \ln\left(\frac{v_2}{v_1}\right)$$

$$= c_v \ln\left(\frac{T_2}{T_1}\right) + R \ln\left(\frac{T_2 p_1}{T_1 p_2}\right) \quad \left[\because \frac{p_1 v_1}{T_1} = \frac{p_2 v_2}{T_2}\right]$$

or

$$s_2 - s_1 = c_v \ln\left(\frac{T_2}{T_1}\right) + R \ln\left(\frac{T_2}{T_1}\right) + R \ln\left(\frac{p_1}{p_2}\right)$$

$$= (c_v + R) \ln\left(\frac{T_2}{T_1}\right) + R \ln\left(\frac{p_1}{p_2}\right)$$

$$= c_p \ln\left(\frac{T_2}{T_1}\right) + R \ln\left(\frac{p_1}{p_2}\right)$$

$\therefore$

$$s_2 - s_1 = c_p \ln\left(\frac{T_2}{T_1}\right) - R \ln\left(\frac{p_2}{p_1}\right) \quad \text{...(6.48)}$$

Thus we notice that the entropy change during a polytropic process for a perfect gas can be found by using any of the six expressions derived above vide equation (6.40), (6.44), (6.41), (6.46), (6.47).

6.6.7 Isentropic Processes of Ideal Gases (Constant Specific Heats)

Normally specific heats are dependent on absolute temperature. However in equations (6.40), through (6.47) it has been assumed that c_p and c_v are constant or in otherwards now average values are taken.

For an isentropic process the entropy will remain constant or $dS = 0$. Accordingly equating $(s_2 - s_2) = 0$ we get from (6.46).

$$(s_2 - s_1) = c_v \ln\left(\frac{T_2}{T_1}\right) + R \ln\left(\frac{v_2}{v_1}\right)$$

or $$\ln\left(\frac{T_2}{T_1}\right) = \frac{-R}{c_v} \ln\left(\frac{v_1}{v_2}\right)$$

But $$R = C_p - C_v$$

or $$\frac{R}{C_v} = \frac{C_p}{C_v} - 1$$

$$= (\gamma - 1)$$

Therefore $$\ln\left(\frac{T_2}{T_1}\right) = (\gamma - 1) \ln\left(\frac{v_1}{v_2}\right)$$

$$= \ln\left[\frac{v_1}{v_2}\right]^{(\gamma-1)}$$

or $$\left(\frac{T_2}{T_1}\right) = \left(\frac{v_1}{v_2}\right)^{(\gamma-1)} \qquad \text{...(6.49}a\text{)}$$

Equation (6.49a) is the first isentropic relation for ideal gases with constant specific heats. Similarly for constant specific heat assumption (6.46) for isentropic process, we get

$$(s_2 - s_1) = 0 = c_p \ln\left(\frac{T_2}{T_1}\right) - R \ln\left(\frac{p_2}{p_1}\right)$$

or $$c_p \ln\left(\frac{T_2}{T_1}\right) = R \ln\left(\frac{p_2}{p_1}\right)$$

$$\ln\left(\frac{T_2}{T_1}\right) = \frac{R}{c_p} \ln\left(\frac{p_2}{p_1}\right)$$

Now $$c_p - c_v = R$$

$$1 - \frac{c_v}{c_p} = \frac{R}{c_p}$$

$$1-\frac{1}{\gamma} = \frac{R}{c_p}$$

Therefore $$\ln\left(\frac{T_2}{T_1}\right) = \left(\frac{\gamma-1}{\gamma}\right)\ln\left(\frac{p_2}{p_1}\right)$$

or $$\ln\left(\frac{T_2}{T_1}\right) = \ln\left(\frac{p_2}{p_1}\right)^{\left(\frac{\gamma-1}{\gamma}\right)}$$

or $$\left(\frac{T_2}{T_1}\right) = \left(\frac{p_2}{p_1}\right)^{\left(\frac{\gamma-1}{\gamma}\right)} \quad ...(6.49b)$$

Equation (6.49*b*) is the second isentropic relation for ideal gases with constant specific heats. The third isentropic relation for ideal gases with constant specific heats assumption is obtained from equation (6.47).

$$(s_2-s_1)=0 = c_v\ \ln\left(\frac{p_2}{p_1}\right)+c_p\ \ln\left(\frac{v_2}{v_1}\right)$$

or $$c_v\ \ln\left(\frac{p_2}{p_1}\right) = -c_p\ \ln\left(\frac{v_2}{v_1}\right)$$

or $$\ln\left(\frac{p_2}{p_1}\right) = -\frac{c_p}{c_v}\ln\left(\frac{v_2}{v_1}\right)$$

or $$\ln\left(\frac{p_2}{p_1}\right) = -v\ \ln\left(\frac{v_2}{v_1}\right)$$

$$= \ln\left(\frac{v_1}{v_2}\right)^{\gamma}$$

or $$\left(\frac{p_2}{p_1}\right) = \left(\frac{v_1}{v_2}\right)^{\gamma} \quad ...(6.49c)$$

Equation (6.49*c*) is known as the third isentropic relation for ideal gases with assumption of constant specific heats. Equations (6.49*a*), (6.49*b*) and (6.49*c*) can also be expressed in compact form as

$$T(v)^{\gamma-1} = \text{constant} \quad ...(6.50a)$$

$$T(p)^{\frac{\gamma-1}{\gamma}} = \text{constant} \quad ...(6.50b)$$

$$p(v)^{\gamma} = \text{constant} \quad ...(6.50c)$$

Once again it is re-iterated that specific heat C_v varies with temperature and so do the above three relations. Therefore in these relations an average value of C_v for the full range of temperature should be used.

Note : The relations (6.50*a*) through (6.50*c*) are stirictly valid for isentropic processes only when the specific heats assumed constant.

6.7 ISENTROPIC PROCESSES OF IDEAL GASES (VARIABLE SPECIFIC HEATS)

In those processes during which the temperature change is large and the specific heats of the gas very non-linearly within the range of the temperature variation, the assumption of constant specific heats leads to very erroneous results while making calculations for entropy change. For such cases the exact function as how the specific heats (c_v and c_p) very with temperature must be known and then the integration is performed.

6.7.1 Increase in Entropy Principle for Closed Systems

Since the closed system involves no mass flow across the boundaries its entropy change is simply the difference between the final and initial entropies of the system. The entropy of the surroundings of the closed system will be affected by the process if the system exchanges heat with the surroundings during the process. In such cases the surroundings are usually treated as thermal energy reservoir at temperature T_{sum} and its entropy change will be related to the heat transfer. Thus the increase in entropy principle for a closed system will be expressed mathematically as

$$S_{generated} = \Delta S_{total}$$

$$= \Delta S_{system} + \Delta S_{sur} \geq 0$$

where

$$\Delta S_{system} = S_2 - S_1$$

$$= m(s_2 - s_1) \quad \text{...(6.51)}$$

and

$$\Delta S_{sur} = \frac{Q_{sum}}{T_{sur}} \quad \text{...(6.52)}$$

In case a closed system exchanges heat with more than one system or reservoir, the total entropy changes of all the systems and reservoirs involved will be added up. The increase in entropy principle for such a closed system will be as under

$$\text{Sgenerated} = \Delta S_{total} = m(s_2 - s_1) \geq 0 \quad \text{...(6.53)}$$

Note : The entropy of a closed system will never decrease duing an adiabatic process.

6.7.2 The Increase-in-Entropy Principle (Control Volumes)

During the process mass flow across the boundaries of the control volume takes place, the mass that enters or leaves the control volume, carries along will it some energy and some entropy as well. Therefore this mass transfer will affect the entropy content of both the control volume as well as the surroundings. The mathematical expression for *c.v* is then

$$(\Delta S)_{cv} = (S_2 - S_1)_{cv}$$

$$(\Delta S)_{sur} = \frac{Q_{sur}}{T_{sur}} + (S_{exit} - S_{inlet}) \quad \text{...(6.54)}$$

Where the subscripts 1 and 2 refer to initial and final states respectively. Also *i* and *e* refer to inlet and exist states entropy value leaving the control volume.

and S_i = Total entropy transported into the control volume (from the surroundings) with the mass entering the *c.v.*

S_e = Total entropy transported out of the control volume (into the surroundings) with the mass leaving the control volume

6.8 REVERSIBLE STEADY-FLOW WORK

As discussed earlier work and heat are path functions. Thus during a process the heat transfer or work done are also dependent on the path as well as on the properties at the end states. The reversible (quasi-equilibrium) moving boundary work associated with closed system as expressed in terms of fluid properties is given by

$$W_b = \int_1^2 p\,dv$$

As discussed earlier quasi-equilibrium work interaction leads to maximum work output for work producing devices and minimum work input for work consuming devices.

Now we proceed to express work associated with steady flow devices in terms of fluid properties. We have for internally reversible steady flow process (in differential form).

$$\delta q_{\text{rev}} - \delta w_{\text{rev}} = dh + dke + dpe \quad \text{...(6.55)}$$

But

$$\delta q_{\text{rev}} = Tds \quad \text{...(Eqn. 6.4)}$$

and

$$Tds = dh - v\,dp \quad \text{...(Eqn. 6.21)}$$

Substituting value of *Tds* in equation (6.21) we get

$$\delta q_{\text{rev}} = dh - vdp$$

or

$$dh = \delta q_{\text{rev}} + vdp \quad \text{...(6.56)}$$

From equations (6.55) and (6.56)

$$\delta q_{\text{rev}} - \delta w_{\text{rev}} = \delta q_{\text{rev}} + vdp + dke + dpe$$

$$-\delta w_{\text{rev}} = vdp + dke + dpe \quad \text{...(6.57)}$$

Integrating equation (6.57)

$$w_{\text{rev}} = -\int_1^2 vdp - \Delta ke - \Delta pe \quad \text{...(6.58)}$$

In case the change in potential and kinetic energies are negligible the equation (6.58) reduces to

$$w_{\text{rev}} = -\int_1^2 v\,dp \quad \text{...(6.59)}$$

In order to evaluate the reversible steady flow work we need to have v the specific volume, as function of p.

6.8.1 In Case the Working Fluid is Incompressible Fluid, Then v is Constant

or

$$v = \text{constant}$$

Then equation (6.59) reduces to

$$w_{\text{rev}} = -\int_1^2 vdp$$

$$= -v(p_2 - p_1)$$

$$w_{\text{rev}} = v(p_1 - p_2) \quad \text{...(6.60)}$$

Thus the equation (6.58) reduces to

$$w_{rev} = v(p_1 - p_2) - \Delta ke - \Delta pe \quad ...(6.61)$$

But for steady flow of a liquid through a device which involves no work interaction such as high flow nozzles and pipes the work term is zero, then of equation (6.61) reduces to

$$v(P_2 - P_1) + \frac{V_2^2 - V_1^2}{2} + g(z_2 - z_1) = 0 \quad ...(6.62)$$

which is the famous Bernoulli equation for flow of liquids.

6.9 WHEN THE PROCESS IS REVERSIBLE, THE STEADY FLOW DEVICES DELIVER THE MOST AND CONSUME THE LEAST WORK

Consider two steady flow devices, one reversible and one actual device which is not reversible. Both these are operating between the same inlet and outlet states.

The conservation of energy equations for devices are

$$\delta q_{act} - \delta w_{act} = dh + dke + dpe \quad ...(6.63)$$

or

$$\delta q_{rev} - \delta w_{rev} = dh + dke + dpe \quad ...(6.64)$$

Since the right hand sides of these equations is same

$$\delta q_{act} - \delta w_{act} = \delta q_{rev} - \delta w_{rev}$$

or

$$\delta w_{rev} - \delta w_{act} = \delta q_{rev} - \delta q_{act}$$

$$= Tds - \delta q_{act}$$

or

$$\frac{\delta w_{rev} - \delta w_{act}}{T} = ds - \frac{\delta q_{act}}{T} \quad ...(6.65)$$

But according to increase in entropy principle the right hand side of equation (6.65) is positive

i.e.,

$$ds - \frac{\delta q_{act}}{T} \geq 0 \quad ...(6.66)$$

Thus for (6.65) and (6.66) we conclude

$$\frac{\delta w_{rev} - \delta w_{act}}{T} \geq 0 \quad ...(6.67)$$

Further since $\delta w_{rev} > \delta w_{act}$ and T's being absolute temperature is always positive

$$\delta w_{rev} - \delta w_{act} \geq 0$$

or

$$\delta w_{rev} \geq \delta w_{act} \quad ...(6.68)$$

From equation (6.68) we conclude that work producing devices delivers more work such as turbines.

6.10 THE COMPRESSOR WORK, ITS MINIMIZATION

In the previous article it has been shown that when the changes in kinetic and potential energies are negligible the work input to the compressor is minimized when the compression process is executed as an internally reversible manner, then the compression work is given by

$$w_{rev} = -\int_1^2 v\,dp \quad ...(6.59)$$

From equation (6.59) it is implied the irreversibilites like friction, turbulance and non-quasi-equilibrium compression manner be reduced to the minimum possible level. More practical and cheaper way be to keep the function v, specific volume as small as possible. Since the specific volume of the gas is proportional to temperature and temperature tends to increase during the compression process. In order to achieve this goal the gas is cooled as it is compressed. Further the work input to the compressor will also depend upon kind of the process whether it is iso-thermal process, polytropic compression or else it is isentropic compression process. In order to compare the work input requirement corresponding to each of these processes these expressions for keeping the same end pressure levels.

Assuming that the ideal gas is being compressed from pressure p_1 to pressure p_2. Then compression work corresponding to these processes are

(*i*) Isentropic Process (pv^{γ} = constant)

$$w_{12} = -\int_1^2 p\,dv$$

or

$$w_{12} = \int_1^2 cv^{-\gamma}\,dv \qquad (\therefore pv^{\gamma} = c)$$

$$= \left(-\frac{\gamma c}{1-\gamma}\right)[v^{1-\gamma}]_1^2$$

$$= \left(-\frac{\gamma c}{1-\gamma}\right)\left(v_2^{1-\gamma} - v_1^{1-\gamma}\right)$$

$$= \frac{\gamma}{1-\gamma}\left(cv_2^{1-\gamma} - cv_1^{1-\gamma}\right)$$

$$= \frac{-\gamma}{1-\gamma}\left[p_2\,v_2^{\gamma}\,v_2^{1-\gamma} - p_1\,p_1^{\gamma}\,v_1^{1-\gamma}\right]$$

$$= \left(\frac{\gamma}{1-\gamma}\right)[p_1v_1 - p_2v_2]$$

$$= \frac{\gamma p_1v_1}{(1-\gamma)}\left[1 - \frac{p_2v_2}{p_1v_1}\right]$$

$$= \frac{rR_1T_1}{(1-v)}\left[1 - \frac{p_2}{p_1}\left(\frac{v_2}{v_1}\right)\right] \qquad ...(6.69)$$

But

$$p_1v_1^{\gamma} = p_2v_2^{\gamma}$$

or

$$\left(\frac{v_2}{v_1}\right)^{\gamma} = \left(\frac{p_1}{p_2}\right)$$

or

$$\left(\frac{v_2}{v_1}\right) = \left(\frac{p_1}{p_2}\right)^{\frac{1}{\gamma}} \qquad ...(6.70)$$

From equations (6.69) and (6.70) we get

$$w_{12} = \frac{\gamma RT_1}{(1-\gamma)}\left[1-\left(\frac{p_2}{p_1}\right)\left(\frac{p_1}{p_2}\right)^{\frac{1}{\gamma}}\right]$$

$$= \frac{\gamma RT_1}{(1-\gamma)}\left[1-\left(\frac{p_2}{p_1}\right)^{\frac{\gamma-1}{\gamma}}\right] \qquad ...(6.71)$$

(*ii*) Similarly for polytropic process where

$$p_1 v_1^n = \text{constant (where } 1 < n < \gamma)$$

The expression for the work done will be obtained by replacing γ with n in equation (6.71)

That is
$$w_{12} = \frac{nRT_1}{(1-n)}\left[1-\left(\frac{p_2}{p_1}\right)^{\frac{n-1}{n}}\right]$$

(*iii*) For isothermal process when pv = constant we have pv > constant = c(say)

$$w_{12} = \int_1^2 v\,dp$$

$$= \int_1^2 \frac{c}{p}dp = c[\ln p]_1^2 \qquad (\therefore pv = c)$$

$$= -p_1 v_1 \ln\left(\frac{p_2}{p_1}\right) \qquad ...(6.73a)$$

$$= -RT_1 \ln\left(\frac{p_2}{p_1}\right) \qquad ...(6.73b)$$

These three processes, for comparison sake are plotted in Fig. 6.11 corresponding to the same inlet and exist pressures p_1 and p_2 on $p - v$ space. The area to the left of process curve represents that compressor work for each process. Evidently the work requirement for the isentropic compression is the maximum and that corresponding isothermal process is minimum and work requirement corresponding to polytropic case is in between these two cases and as n decreases the work requirement decreases and it is minimum when n approaches unity or when it is isothermal. From this discussion we conclude that if the gas is cooled during compression by providing cooling jacket-arround the cyclinder it will lower the work requirement. However, in practice it is very costly affair to employ. In the absence of effective cooling through the compressor easing therefore, the technique of multistage compression with inter cooling is adopted. In this method the gas is compressed in stages and it is cooled between each stage by passing it through a heat exchanger which is given a special name as INTERCOOLER. Idealy at each stage the cooling process takes place at constant pressure and is cooled to the initial temperature T_1 at which the gas at low pressure is available.

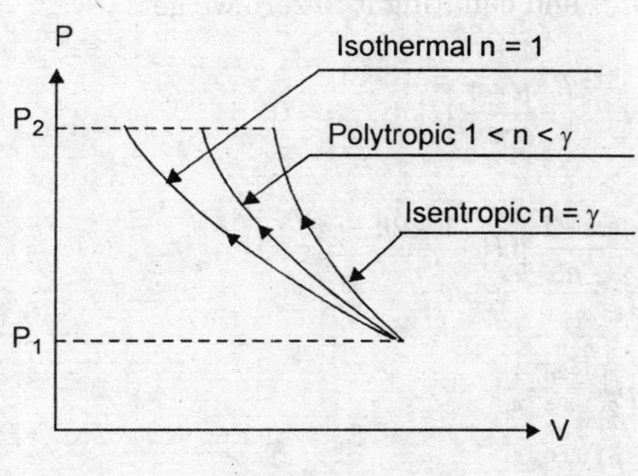

Fig. 6.11

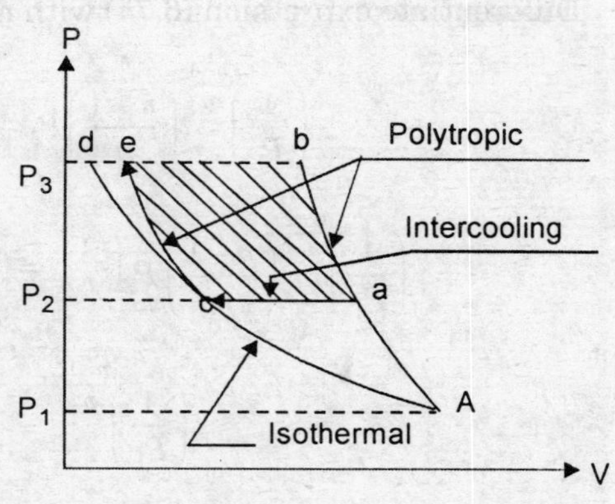

Fig. 6.12

In Fig. 6.12 is shown the process of the low stage compression with intercooling. The gas initially is available at pressure P_1 and temperature T_1. This state is represented by point *A* in the P–V. diagram. It is compressed to pressure P_2 which lies between the initial pressure P_1 and P_3 the final desired pressure P_3. Had it been compressed polytropically to pressure P_3 it would have consumed work proportional to the area on the left of the curve *Aab*. But in the present case it is compressed upto pressure P_2 and intercooled temperature T_1 at constant pressure P_2. As to followed by another spell of polytropic compression to reach the pressure P_3. This portion of the path is represented by the curve *ce*. Had there beeen isothermal compression through out it would have followed the path *Acd*. It is evident from the diagram (Fig. 6.12) that due to intercooling the work saved is represented by the shaded area. The size of the shaded area which represents the proportional saving in compressive work values with the value of the intermediate pressure P_2. It will be of interest to determine the optimum value of P_2 so as to maximise the saving is the work done or minimise the total work input required to compress the gas. For this we proceed as under.

Total work input in a two stage, compression with intercooling

$$w_{13} = w_{12} + w_{23}$$

where w_{13} = Represents the work input for polytropic compression from pressure P_1 to pressure P_2 with T_1 as initial gas temperature

and w_{23} = Work input for polytropic compression from pressure P_2 to pressure P_3 with initial temperature T_1

Therefore we have with the help of equation (6.72)

$$w_{13} = \frac{nRT_1}{(n-1)}\left[1\left(\frac{P_2}{P_1}\right)^{\frac{n-1}{n}}\right] + \frac{nRT_1}{(n-1)}\left[1-\left(\frac{P_3}{P_2}\right)^{\frac{n-1}{n}}\right] \quad ...(6.74)$$

In the equation (6.74) the only variable is P_2. In order to minimise w_{13} we equate the $\frac{\partial w_{13}}{\partial p_2}$ to zero and solve for P_2.

Thus $$\frac{\partial w_{13}}{\partial p_2} = 0$$

Differentiate expression (6.74) with respect to P_2 and equating it to zero we get

$$-\left(\frac{1}{P_1}\right)^{\frac{n-1}{n}}\left(\frac{n-1}{n}\right)P_2^{\left(\frac{n-1}{n}-1\right)}-(P_3)^{\frac{n-1}{n}}\left(-\frac{n-1}{n}\right)P_2^{\left(-\frac{n-1}{n}-1\right)}=0$$

or
$$\left(\frac{1}{P_1}\right)^{\frac{n-1}{n}}\left(\frac{n-1}{n}\right)P_2^{\left(-\frac{1}{n}\right)} = (P_3)^{\frac{n-1}{n}}\left(-\frac{n-1}{n}\right)(P_2)\frac{1-2n}{n}$$

or
$$\left(\frac{1}{P_1}\right)^{\frac{n-1}{n}} = (P_3)^{\frac{n-1}{n}}\ P_2^{\left(\frac{1-2n}{n}+\frac{1}{n}\right)}$$

or
$$\left(\frac{1}{P_1}\right)^{\frac{n-1}{n}} = (P_3)^{\frac{n-1}{n}}\ P_2^{\left(\frac{2-2n}{n}\right)}$$

$$\left(\frac{1}{P_1}\right)^{\frac{n-1}{n}} = P_3^{\left(\frac{n-1}{n}\right)}\ P_2^{2\left(\frac{1-n}{n}\right)}$$

or
$$P_2^{2\left(\frac{n-1}{n}\right)} = (P_3)^{\frac{n-1}{n}}(P_1)^{\frac{n-1}{n}}$$

or
$$(P_2^2)^{\frac{n-1}{n}} = (P_1P_3)^{\frac{n-1}{n}}$$

or
$$P_2^2 = P_1P_3$$

or
$$P_2 = \sqrt{P_1P_3} \quad \text{...(6.75a)}$$

or
$$\frac{P_2}{P_1} = \frac{P_3}{P_2} \quad \text{...(6.75b)}$$

From equations (6.75) we infer that for a two stage compression with intercooling to the original temperature level the intermediate pressure should be the geometric mean of P_1 and P_3 or else in general and more appropriate terms the pressure ratio across each stage (for a multistage compression with intercooling). to original temperature T_1, must be same. That is in general.

$$\frac{P_2}{P_1} = \frac{P_3}{P_2}=\frac{P_4}{P_3}=\ldots\frac{P_n}{P_{n-1}} \quad \text{...(6.76)}$$

When the above mentioned condition is satisfied, the compression work at each stage becomes identical. That is mathematically expressed as follows

$$w_{12} = w_{23}=w_{34} \quad \text{...(6.77)}$$

6.11 THE ENTROPY CHANGE OF PURE SUBSTANCES

The $T - ds$ relations derived above are not limited to a particular substance in a particular phase. They are valid for all pure substances at any phase or combination of phases. The applicability of these relations, however, depends on the availability of relations between T

and du or dh and the $P - v - T$ behaviour of the substance. It is to be noted that for pure substance these relations are too complicated which in turn make them too involved to obtain simple relationship for entropy changes. Therefore the values of entropy 's' are determined from measureable property data and are tabulated in exactly the same manner as the properties v, v and h. The entropy values in the property tables are given relative to an arbitrary reference state. For example, in the steam tables the entropy of saturated liquid, s_f, at 0.01°C is assigned the value of zero. Similarly for refrigerant-12, the zero value is assigned to saturated liquid at –40°C and below this reference temperature, the entropy values will be negative. Likewise in the compressed liquid or vapour regions, the entropy values can be obtained directly from the tables at any specified state. However in the saturated mixture region it is determined from the relation.

$$s = s_f + xs_{fg} \quad \text{...(6.78)}$$

Where x is the quality and s_f and s_{fg} are the entropy values listed in the saturation tables. However, in the absence of compressed liquid data, its entropy can be approximated by the entropy of saturated liquid at the given temperature.

$$s_{\text{at } p,\, T} = s_{f \text{ at } T} \quad \text{...(6.79)}$$

Thus the entropy change of a pure substance during a process is the difference between the entropy values at final and initial states.

$$\Delta S = m(s_2 - s_1) \text{ kJ/K} \quad \text{...(6.80a)}$$

or

$$\Delta S = (s_2 - s_1) \text{ kJ/(kg – K)} \quad \text{...(6.80b)}$$

6.12 ENTROPY CHANGE OF SOLIDS AND LIQUIDS

A substance whose specific volume (or density) remains constant is called incompressible substance. The specific volume of solids and liquids essentially remains constant during an actual process. Therefore it is customary to approximate solids and liquids as incompressible substances without sacrificing much in accuracy. The constant volume assumption, then implies that energy associated with volume change, such as boundary work is negligible as compared with other forms of energy.

The incompressible property can be expressed in mathematical form as

$$dv = 0 \text{ (for solids and liquids)}$$

The equation (6.22) applicable to solids and liquids, then reduces to

$$ds = \frac{du}{T}$$

$$= \frac{CdT}{T} \quad \text{...(6.81)}$$

and

$$du = CdT \quad \text{...(6.82)}$$

because $c_p = c_v = c$ (say) where c is constant for incompressible solids and liquids and it depends on temperature that is C is function of temperature. Then equation (6.81) reduces to

$$\int_1^2 ds = \int_1^2 [C(T)]\frac{dT}{T} \quad \text{...(6.83)}$$

Specific heat must be known as function of C for proper integration of equation (6.83). In the absence of proper expression of variation of C with temperature the specific heats c_p and

c_v are assumed as constant having average value C_{av} depending upon the temperature range of the process. In that event equation (6.83) reduces to

$$s_2 - s_1 = C_{av}\int_1^2 \frac{dT}{T} = C_{av} \ln\left(\frac{T_2}{T_1}\right) \quad ...(6.84)$$

Evidently the entropy change of incompressible substances depends upon temperature only.

6.13 THIRD LAW OF THERMODYNAMICS (NERNST LAW)

The third law of thermodynamics states that the entropy of any pure substance in thermodynamic equilibrium tends to become zero as the absolute temperature approaches zero mathematically it is expressed as

$$\underset{T\to 0}{Lt}\ S = 0 \quad (6.85)$$

From equation (6.85) it is implied that at absolute zero temperature the entropy ceases to be a function of state. Actually it approaches a constant value independent of parameters of state. However, for various substances like alloys, amorphous bodies, chemical compounds like CO, CO_2, NO etc. the entropy does not tend to be zero when T approaches zero but it has a finite positive value becasue these substances are not found in equilibrium states. As this law was enunciated by Nernst, it is also known as Nernst Law of thermodynamics.

6.13.1 Importance of Nernst Law

(*i*) Third law of thermodynamics provides an absolute base from which the entropy of each substance can be measured and then the entropy is referred to as absolute entropy.

(*ii*) It is helpful in determining the chemical forces of reacting substances.

(*iii*) It helps in explaning the behaviour of solids at very law temperature.

(*iv*) Similarly it is helpful in analysing the chemical and phase equilibriums.

Example 6.1. *A thermal energy source at 850 K looses 2150 kJ of heat to a sink at 550 K. Determine (a) the total entropy change for the heat transfer process between the two reservoirs.*

(a) If a carnot heat engine operating between the above thermal sources were used, then what would be the total entropy change.

Solution : (*a*) For the heat transfer of 2150 kJ heat energy from source to sink we have

$$Q_{\text{source}} = Q_{\text{sink}} = 2150 \text{ kJ}$$

Therefore
$$\Delta S_{\text{source}} = \frac{-2150}{T_1} \text{ (It is negative as heat is being rejected)}$$

$$= \frac{2150}{850} = -2.5294 \text{ kJ/K}$$

$$\Delta S_{\text{sink}} = \frac{Q_{\text{sink}}}{T_2} = \frac{2150}{550} = +3.9091 \text{ kJ/K}$$

Therefore total entropy change

$$\Delta S_{\text{total}} = \Delta S_{\text{source}} + \Delta S_{\text{sink}}$$
$$= -2.5294 + 3.9091$$
$$= 1.3777 \text{ kJ/K}$$

(*b*) when carnot heat engine is used to transfer 2150 kJ of heat from source at 850 K to sink at 550 K.

$$\Delta S_{total} = \Delta S_{source} + \Delta S_{sink}$$

$$= -\int_1^2 \frac{\delta Q}{T} + \int_1^2 \frac{\delta Q}{T}$$

Since $\delta Q_{source} = \delta Q_{sink}$

$$\Delta S_{total} = -\delta Q \ln\left(\frac{T_2}{T_1}\right) + \delta Q \ln\left(\frac{T_2}{T_1}\right)$$

$$= 0$$

Example 6.2. *A heat engine receives 960 kJ of heat from a thermal source at 1100 K and rejects 600 kJ of heat to sink at 300 K. Fig (6.13) By calculations show whether this engine violates the second law of thermodynamics on the basis of (a) clausius inequality and (b) on the basis of carnot principle.*

Solution : (*a*) We have from clausius inequality

$$\oint \frac{\delta Q}{T} < 0$$

$$\int \frac{\delta Q}{T} = \frac{Q_1}{T_1} - \frac{Q_2}{T_2}$$

$$= \frac{960}{1100} - \frac{600}{300}$$

$$= 0.8727 - 2$$

$$= -0.1273 \text{ kJ/K}$$

$$< 0$$

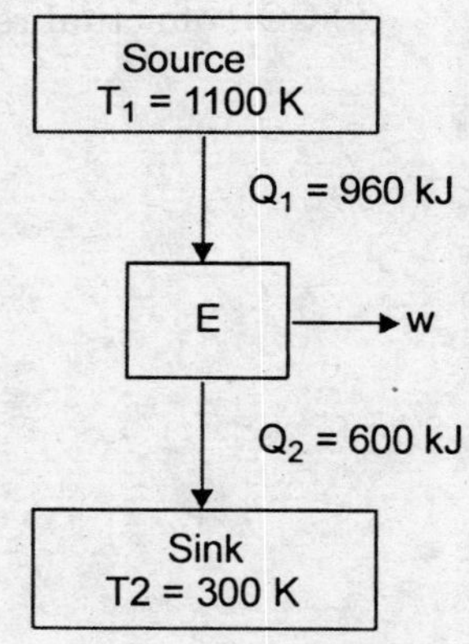

Fig. 6.13

Thus clausius inequality is satisfied.

(*b*) Thermal efficiency of the actual heat engine is given by

$$\eta_{th} = 1 - \frac{Q_2}{Q_1} = 1 - \frac{600}{960} = 1 - 0.625$$

$$= .375 \text{ or } 37.5\%$$

Thermal efficiency of a reversible carnot engine

$$\eta_{in,\,rev} = 1 - \frac{T_2}{T_1} = 1 - \frac{300}{1100}$$

$$= 1 - 0.2727$$

$$= 0.7273 \text{ i.e., } 72.73\%$$

Since $\eta_{th} < \eta_{th-rev}$

The heat engine satisfies the carnot principle

Example 6.3. *A refrigerator removes heat from the cooling chamber which is at 2°C, at a rate 295 kJ/min and rejects heat to surroundings air at 25°C at a rate 340 kJ/min, Fig (6.14)*

Show by calculations whether this refrigerator violates the second law of thermodynamics on the basis of (a) the clausius inequality and (b) the Carnot principle.

Solution : (*a*) Applying the clausius inequality to the refrigerator.

Here $T_2 = 273 + 2 = 275 \text{ K}, T_1 = 273 + 25 = 298 \text{ K}$

and $\delta\dot{Q}_1 = 295 \text{ kJ/min and } \delta\dot{Q}_2 = 340 \text{ kJ/min}$

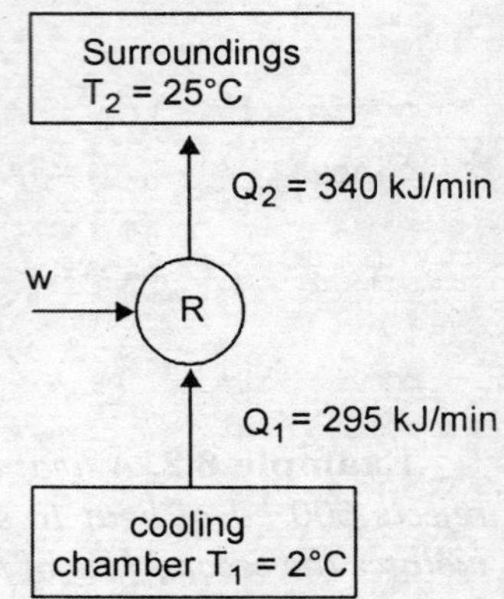

Fig. 6.14

$$\oint \frac{\delta\dot{Q}}{T} = \frac{\delta\dot{Q}_1}{T_1} - \frac{\delta\dot{Q}_2}{T_2}$$

$$= \frac{295}{275} - \frac{340}{298}$$

$$= 1.0727 - 1.1409$$

$$= -0.0682 \text{ kJ/min K}$$

It is negative thus the clausius inequality is satisfied

(*b*) COP of actual refrigerator is

$$\text{COP}_{\text{actual}} = \frac{\delta\dot{Q}_1}{\delta\dot{Q}_2 - \delta\dot{Q}_1}$$

$$= \frac{295}{340 - 295} = \frac{295}{45} = 6.5555$$

$$(\text{COP})_{\text{rev}} = \frac{T_1}{T_2 - T_1} = \frac{275}{298 - 275}$$

$$= \frac{275}{23} = 11.9$$

Evidently $\text{COP}_{\text{actual ref}} < \text{COP}_{\text{rev, Ref}}$

Thus it satisfies the cannot-principle

Example 6.4. *A heat pump Fig (6.15) while consuming 8 kW power, supplies heat to a house during a winter night at the rate of 75000 kJ/h and keeps maintaining the temperature inside the house at 21°C when the outside temperature is –2°C. Determine if this heat pump violates the second law of thermodynamics on the bases of clausius inequality.*

Solution : From conservation of energy equation for heat pump

$$Q_2 = Q_1 - W$$

$$= 75000 \text{ kJ/h} - 8 \text{ kW}$$

But $Q_1 = \frac{75000}{3600} \text{ kJ/s} = 20.83 \text{ kW}$

$\therefore$ $Q_2 = Q_1 - w$

$$= 20.83 - 8 = 12.83 \text{ kW}$$

Also $T_1 = 21 + 273 = 294 \text{ K}$

and $T_2 = 273 - 2 = 271 \text{ K}$

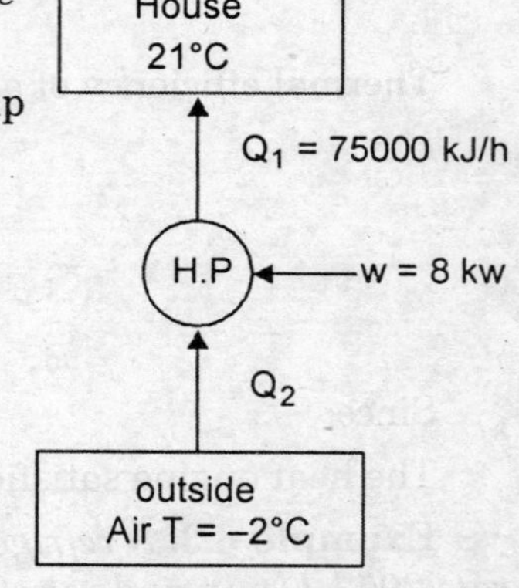

Fig. 6.15

But from clausius inequality

$$\int \frac{\delta \dot{Q}}{T} = \frac{\delta \dot{Q}_2}{T_2} - \frac{\delta \dot{Q}_1}{T_1}$$

$$= \frac{12.83 \text{ kW}}{271 \text{K}} - \frac{20.83}{294} \text{ kW/K}$$

$$= -0.0235 \text{ kW/K}$$

$$< 0$$

Thus the clausius inequality is satisfied

Example 6.5. *Heat flows from a reservoir at 1000 K to another reservoir at 500 K. If the entropy change of the hot reservoir is –6.5 kJ/K find the entropy change of the other reservoir.*

Solution : Heat lost of by hot reservoir

= Heat gained by the other reservoir

Also for the hot reservoir $(ds)_h = \frac{Q}{T}$

or $(ds)_h = \frac{Q_h}{1000}$ kJ/K

or $Q_h = T_h (ds)_h$

$= 1000 (-6.5)$ kJ

$= -6.500$ kJ

But $Q_c = Q_h$

$= 6500 = (ds)_c T_c$

$$(ds)_c = \frac{6500}{500}$$

$= 13.0$ kJ/K

Example 6.6. *Heat flows from the hot reservoir at 1000 K to another reservoir at 500 K. If the entropy change of the overall process is 7.5 kJ/K determine the amount of heat flowing out of the hot reservoir.*

Solution : Heat lost from hot reservoir

= Heat gained by cold reservoir

= Q (say)

Total entropy change ds of system = $(ds)_{\text{total}} = (ds)_h + (ds)_c$

$$(ds)_f = -\frac{Q}{T_h} + \frac{Q}{T_c}$$

or $$7.5 = -\frac{Q}{1000} + \frac{Q}{500}$$

$$= -\frac{Q}{1000}(1-2) = \frac{Q}{1000}$$

or $Q = 7500$ kJ

Example 6.7. *The specific entropy of a fluid system changes reversibly from initial state of 6.5 kJ/K at 400 K to final state of 8 kJ/K at 480 K. Determine the heat flow involved. Assume the variation of specific entropy with temperature as linear.*

Solution : The initial state is represented by the point 'a' and final state by 'b' in Fig. 6.16

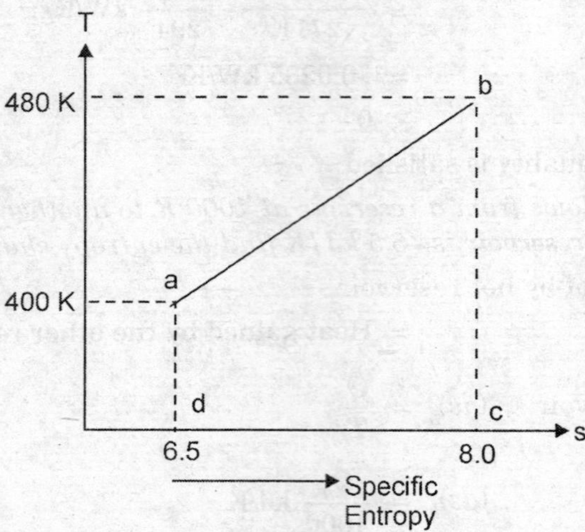

Fig. 6.16

$$\text{Heat flow} = \text{area of trapezium } abcd$$

$$= \frac{1}{2}(ab + bc)(dc)$$

or

$$Q = \frac{1}{2}(400 + 480)(8 - 65)$$

$$= \frac{880}{2} \times 1.5 = 660 \text{ KJ}$$

Example 6.8. *A bucket-full of hot water left in the open lost 1000 kJ of heat to the surrounding atmospheric air at 295 K. If the change of entropy of the water is –2.5 kJ/K find the total entropy change. Discuss the result whether it is possible, reversible or irreversible.*

Solution :

$$(ds)_{\text{total}} = (ds)_{\text{water}} + (ds)_{\text{air}}$$

$$= (-2.5) + \frac{1000}{295}$$

$$= -2.5 + 3.38998$$

$$= 0.8898 \text{ kJ/K}$$

Thus as per clausius inequality the process is possible and it is irreversible

Example 6.9. *H heat engine cycle is represented by a circle of 5.5 cm diameter at T– S diagram whose horizontal diameter is along 490 K temperature line. The temperature scale is 1 cm = 100 K and entropy scale is 1 cm = 0.1 kJ/K. Engine is running at 1200 rpm and one engine cycle is completed in 2 revolutions. Determine the power developed by the engine, heat supplied per cycle to the engine and its efficiency.*

Solution : Scale of temperature axis

$$1 \text{ cm} = 100 \text{ K}$$

scale of entropy axis $\quad 1 \text{ cm} = 0.1 \text{ kJ/K}$

Therefore 1cm² area of T – S. diagram

$$= 100 \times 0.1 = 10 \text{ kJ}$$

Area of T– S. diagram gives the heat transfer and as per the first law of thermodynamics

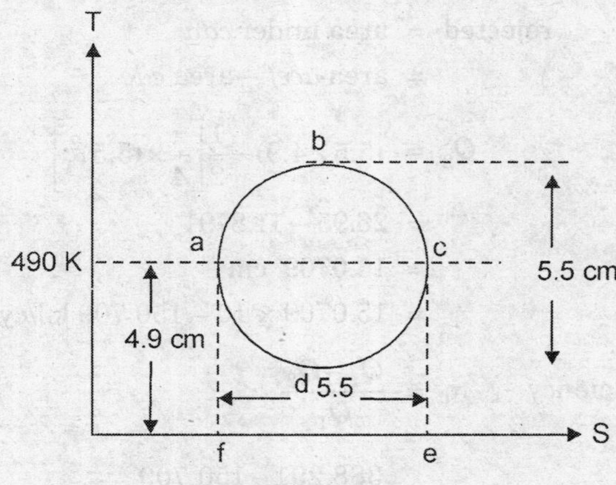

Fig. 6.17

$$\oint \delta Q = \int \delta W$$

i.e., area of T– S diagram = work output

or work output in one cycle $= \frac{\pi}{4}(5.1)^2 \times \text{area scale}$

$$= \frac{\pi}{4}(5.5)^2 \times 10 \text{ kJ}$$

$$= 237.5829 \text{ kJ}$$

the speed of engine is 1200 rpm and one cycle is completed in 2 revolutions

Therefore no of cycles per moments

$$= \frac{1200}{2} = 600$$

Number of cycles per sec. $= \frac{600}{60} = 10$

∴ Power developed = work done in 1 sec.

= work done in 10 cycles

= 237.5829 × 10

= 2375.829 kJ/s

= 2375.829 kw

Now heat supplied Q_1 = Area under the curve a *bc*

= Area *abc* + Area *acef*

$$= \frac{1}{2}\left[\frac{\pi}{4}(5.5)^2\right] + [(5.5) \times 4.9]$$

$= 11.879 + 26.95$

$= 38.8291 \text{ cm}^2$

$= 38.8291 \times 10$ kJ/cycle

$= 388.291$

Heat rejected = area under *cda*

= area *acef* – area *cde*

or

$$Q_2 = (5.5 \times 4.9) - \frac{1}{2}\left[\frac{\pi}{4} \times (5.5)^2\right]$$

$= 26.95 - 11.8791$

$= 15.0709 \text{ cm}^2$

$= 15.0709 \times 10 - 150.709$ kJ/cycle

Therefore engine efficiency

$$\eta = \frac{Q_1 - Q_2}{Q_1}$$

$$= \frac{388.291 - 150.709}{388.291}$$

$= 1 - 0.3881 = 0.6119$ or 61.19 %

Example 6.10. *Hot gases at 575 K enter a heat exchanger in which 70 kg of water is heated from 325 K to 450 K per minute. If the flow of the gases is 110 kg/min calculate the net change in entropy. Take c_p (water) is 186 kJ/kg and c_p gas = 1 kJ/kg k. Assume no loss of heat to the surroundings.*

Solution : Initial temperature of water $T_1 = 325$ K

Final temperature to water $T_2 = 450$ K

Initial temperature of hot gases $T_3 = 575$ K

Let final temperature of gas $= T_4$

From the energy conservation principle we have Heat lost by gases = Heat gained by water

$$(m\, c_p\, dT_g)_{gas} = (m\, c_p\, dT)_{water}$$

110 kg/min × 1 kJ/kg k × $(575 - T_4)$ k

= (70 kJ/min) × (4.186 kJ/kg k) (450 – 325) K

$$\sigma(575 - T_4) = \frac{70 \times 4.186 \times 125}{110 \times 1}$$

$= 332.9773$

or

$T_4 = 575 - 332.9773$

$= 242.0229$ K

Therefore

$$(ds)_{gas} = m c_p \ln\left(\frac{T_4}{T_3}\right)$$

$$= 110 \times 1 \times \ln \times \left(\frac{242.0229}{450}\right)$$

$$= 110 \ln (0.5378)$$

$$= 110 \times (-.62027)$$

$$= -68.2297 \text{ kJ/K}$$

$$(ds)_{water} = m\, c_p \ln \left(\frac{T_2}{T_1}\right)$$

$$= 70 \times 4.186 \ln \left(\frac{450}{325}\right)$$

$$= 70 \times 4.186 \times \ln (1.3846)$$

$$= 70 \times 4.186 \times 0.3254 = 95.34$$

Therefore net change in entropy

$$(ds)_{net} = (ds)_{gas} + (ds)_{water}$$

$$= -68.2297 + 95.34$$

$$= 27.1103 \text{ kJ/k increase}$$

Example 6.11. *Calculate the change in entropy of the universe owing each of the frollowing processes.*

(a) 0.5 kg of copper block at 100°C is placed in water reservoir at 10°C.

(b) Same block at 100°C is dropped from the height of 100 m into the reservoir and comes to rest without any loss of heat of the block.

(c) Two blocks at 100°C and 0°C are joined together.

Take $C_{cu} = 0.393 \text{ kJ/kg K at } 100°C$

and $= 0.381 \text{ kJ/kg K at } 0°C$ *(PTU 2007 Dec)*

Solution : (*a*) Assume water reservoir to be of very large size so that there is no appericiable change in temperature when the copper block is placed in it.

$$\Delta S_{universe} = (\Delta S)_{cu} + (\Delta S)_{water}$$

Here $(\Delta S)_{water} = 0$

as the temperature remains constant at 10°C.

and $$(\Delta S)_{cu} = m\, c_p \ln \left(\frac{T_2}{T_1}\right)$$

Where T_1 = Initial temperature of *cu*-block

T_2 = Final temperature of *cu*-block

here $T_1 = 100°C = 100 + 273 = 373 \text{ K}$

$T_2 = 10°C = 10 + 273 = 283 \text{ K}$

$$\therefore \quad (\Delta S)_{cu} = 0.5 \times 0.393 \ln \left(\frac{283}{373}\right)$$

$$= 0.1965 (\ln 283 - \ln 373)$$

$$= 0.1965\,(5.6464 - 5.9216)$$
$$= -0.1965 \times 0.2762$$
$$= -0.05427 \text{ kJ/K (being negative it is decrease)}$$

There $\Delta S_{\text{universe}} = -0.05427 + 0$
$$= -0.5427 \text{ kJ/K}$$

(*b*) When the block falls from 100 m height the loss in potential energy will result in rise in temperature of the block. If ΔT is the corresponding increase in temperature due to fall.

Then $mgh = m\,c_p\,\Delta T$

$$\text{or } \Delta T = \frac{gh}{c_p} = \frac{9.81 \times 100}{0.393}$$
$$= 2.496 \text{ K}$$

Therefore temperature of block just before it enters the water reservoir

$$T_f = 373 + 2.496 = 375.496 \text{ K}$$

Initial temperature of the block $= 100 + 273 = 373$

$$(\Delta S)_{cu} \text{ due to fall} = m\,c_p \ln\left(\frac{T_t}{T_i}\right)$$
$$= 0.5 \times 0.393 \ln\left(\frac{375.496}{373}\right)$$
$$= 0.1965 \ln (1.00669)$$
$$= 0.1965 \times 0.006677$$
$$= 0.0013102 \text{ kJ/K}$$

and $(\Delta S)_{\text{water}} = 0$

Thus $(\Delta S)_{\text{universe}} = 0.0013102 + 0 = 0.0013102$ kJ/K

(*c*) When two blocks are joined together, let the final equilibrium temperature be T_f.

Then Heat lost by hot block = Heat gained by cold block

$$m(c_p)_n\,(100 - T_f) = m(c_p)_c\,(T_f - 0)$$
$$0.393(100 - T_f) = 0.381\,(T_f - 0)$$

or $39.3 - .393\,T_f = 0.381\,T_f$

$$39.3 = 0.774\,T_f$$

or $T_f = \dfrac{39.3}{0.774} = 50.7752°\text{C}$

$$= 273 + 50.7752$$
$$= 323.7752 \text{ K}$$

Initial temperature of hot block $= 100\text{ °C} + 273 = 373$ K

$$(\Delta S)_{\text{hot block}} = m\,c_p \ln\left(\frac{323.7752}{373}\right)$$
$$= 0.5 \times 0.393\,[\ln 32.37752 - \ln 373]$$

$$= 0.1965\,[5.78005 - 5.92158]$$
$$= -0.1965 \times 0.14153$$
$$= -0.02781 \text{ kJ/K}$$

$$(\Delta S)_{\text{cold block}} = m\,c_p \ln\left(\frac{T_t}{T_c}\right)$$
$$= 0.5 \times 0.381 \ln\left(\frac{323.7752}{273}\right)$$
$$= 0.1905\,[\ln 323.7752 - \ln 273]$$
$$= .1905\,[5.78005 - 5.6095]$$
$$= .1905 \times 0.1706$$
$$= 0.0325 \text{ kJ/k}$$

There
$$(\Delta S)_{\text{universe}} = (\Delta S)_{\text{hot block}} + (\Delta S)_{\text{cold block}}$$
$$= -0.02781 + .0325$$
$$= \mathbf{0.0469} \text{ kJ/K (increase)}$$

Example 6.12. *An iron forging of mass 5 kg, initially at 400°C is quenched in an insulated tank which contains 150 kg of water at 20°C. Assuming the water that vaporises during the process condenses back in the tank, calculate the change in entropy for this process. Take the average specific heats of iron and liquid water as 0.45 kJ/kg°C and 4.184 kJ/kg °C respectively.*

Solution : Let T_f he the final temperature at equilibrium. Taking the iron forging and the water as our system, we have from the energy conservation equation.

Heat lost by iron forging = Heat gained by water

$$[mc(T_i - T_f)]_{\text{iron}} = [\int mc(T_j - T_i)]_{\text{water}}$$
$$5 \times 0.45\,(400 - T_f) = 150 \times 4.184\,(T_f - 20)$$
$$2.25\,(400 - T_f) = 627.6\,(T_f - 20)$$

or
$$900 - 2.25\,T_f = 627.6\,T_f - 12552$$

or
$$13452 = 629.85\,T_f$$

or
$$T_f = \frac{13452}{629.85}$$
$$= 21.356\ °\text{C}$$
$$= 273 + 21.356 = 294.356 \text{ K}$$
$$(T_i)_{\text{iron}} = 400 + 273 = 673 \text{ K}$$
$$(T_i)_{\text{water}} = 20 + 273 = 293$$
$$(\Delta S)_{\text{Total}} = (\Delta S)_{\text{iron}} + (\Delta S)_{\text{water}}$$
$$= \left[m\,c_{cv} \ln\left(\frac{T_t}{T_i}\right)\right]_{\text{iron}} + \left[m c_v \ln \frac{T_f}{T_i}\right]_{\text{water}}$$
$$= 5 \times 0.45 \ln\left(\frac{294.356}{673}\right) + 150 \times 4.184 \ln\left(\frac{294.356}{293}\right)$$

$$= 2.25\ [\ln 294.356 - \ln 673 + 627.6\ [\ln 294.356 - \ln 293]$$

$$= 2.25[5.6848 - 6.5117] + 627.6\ [5.6848 - 5.6802]$$

$$= -1.8605 + 2.88696$$

$$= 1.02646 \text{ kJ/K (increase)}$$

Example 6.13. *An iron block of mass 20 kg at 100°C is brought into contact of an aluminium block of same mass but initially at 200°C in an insulated enclosure. Assuming the average specific heats of iorn and aluminium as 0.45 kJ/kg °C and 0.973 kJ/kg °C respectively, determine the final equilibrium temperature and the total change in entropy for this process.*

Solution : Taking the iron and aluminium blocks as our system. Then from the conservation of energy equation we have

$$Q - W = DU + DKE + DPE$$

Here $Q = 0, W = 0, \Delta KE = 0\ \Delta PE = 0$

$\therefore$ $\Delta U = 0$

or $[mc_{cu}(T_f - T_i)]_{\text{iron}} + [m\,c_{an}(T_f - T_i)]_{Al} = 0$

or $20 \times 0.45\ (T_J - 100) + 20 \times 0.973\ (T_f - 200) = 0$

or $9.0\ (T_f - 100) + 19.46\ (T_f - 200) = 0$

or $28.47\ T_f = 4792$

or
$$T_f = \frac{4792}{28.47}$$

$$= 168.32°\text{C} = 168.32 + 273$$

$$= 441.32 \text{ K}$$

$$(T_i)_{\text{iron}} = 100°\text{C} = 100 + 273 = 373 \text{ K}$$

$$(T_0)_{Al} = 200°\text{C} = 200 + 273 = 473 \text{ K}$$

$$(\Delta S)_{\text{Total}} = (\Delta S)_{\text{iron}} + (\Delta S)_{Al}$$

$$= \left[m\,c_{av} \ln\left(\frac{T_f}{T_i}\right)\right]_{\text{iron}} + \left[m\ c_{av} \ln\left(\frac{T_f}{T_i}\right)\right]_{Al}$$

$$= 20 \times 0.45 \ln\left(\frac{441.32}{373}\right) + 20 \times 0.973 \ln\left(\frac{441.32}{473}\right)$$

$$= 9.0\ [\ln 441.32 - \ln 373] + 19.46\ [\ln 441.32 - \ln 473]$$

$$= 9.0\ [6.0898 - 5.9296] + 19.46\ [6.0898 - 6.1591]$$

$$= 1.5138 - 1.3486$$

$$= 0.1652 \text{ kJ/K}$$

Example 6.14. *A copper block of mass 20 kg and an iron block of mass 30 kg, both initially at 80°C are simultaneously dropped into a lake which is at 20°C. As a result of heat transfer between the lake water, thermal equilibrium was established after a while. Determine the total entropy change for the process. Assume average values of specific heats for iron and copper as 0.45 kJ/kg K and 0.386 kJ/kg K respectively.*

Solution : Since the thermal energy capacity of the lake is very large, the temperature of both the iorn and copper blocks will drop to lake temperature of 20°C when the thermal equilibrium is established.

Taking the iron, copper system, the heat transfer during this process is determined for this closed system energy equation

$$Q - W = \Delta U + \Delta KE + \Delta PE$$

here $W = 0, \Delta KE = 0, \Delta PE = 0$

$$\therefore \quad Q = (\Delta U)_{iron} + (\Delta U)_{cu}$$
$$= 300\,(\text{m}\, c_{av}\,(T_f - T_i)_{iron} + [m\, c_{av}\,(T_f - T_i)]$$

or
$$Q = 30 \times 0.45\,(285 - 353) + 20\,(0.356)\,(285 - 353)$$
$$= -[135 \times 68 + 7.72 \times 68]$$
$$= -(21.22 \times 68) = -1442.96 \text{ kJ}$$

Negative sign indicates heat loss for the hot bodies

Heat gained by lake water $= Q = 1442.96$ kJ

$$\therefore \quad (\Delta S)_{Total} = (\Delta S)_{iron} + (\Delta S)_{cv} + (\Delta S)_{lable}$$
$$= \left[m\, c_{av}\,\ln\left(\frac{T_f}{T_i}\right)\right]_{iron} + \left[m\, c_{av}\,\ln\left(\frac{T_t}{T_i}\right)\right]_{cv} + \frac{Q}{T_{lake}}$$
$$= 30 + 0.45\,\ln\left(\frac{293}{353}\right) + 20 + 0.386\,\ln\left(\frac{293}{353}\right) + \frac{1442.96}{293}$$
$$= (13.5 + 7.72)\,(\ln 293 - \ln 353) + 4.9215$$
$$= 21.22\,(5.6802 - 5.8665) + 4.9215$$

or
$$(\Delta S)_{Total} = -3.9533 + 4.9215$$
$$= 0.9682 \text{ kJ/K}$$

Example 6.15. *Seven kg of water at 25°C is mixed with 1.5 kg of ice at 0°C. The mixing process is adiabatic and system is open to atmosphere. Calculate the final temperature of the mixture and the change in entropy for the process.*

Take $(c_p)_{water} = 4.187\ kJ/kgK$

Solution : Water temperature $= 25°\text{C} = 273 + 25 = 298$ K

Temperature of ice $= 0°\text{C} = 273$ K

Let T = temperature of mixture after melting of ice

Then from the conservation of energy principle, if L = Latest heat of ice

Heat lost by water = Heat gained by ice

$$[m\, c_p\,(298 - T)]_{water} = [mL + m\, c_p (T - 273)]_{ice}$$
$$7 \times 4.187\,(298 - T) = 1.5 \times 335 + 1.5 \times 4.187\,T$$
$$8734.08 - 29.309\,T = 902.5 + 6.2805\,T$$

or
$$9946.16 = 1714.58 - 35.5895\,T$$

$$\therefore \quad T = 279.47$$
$$= 279.47 \text{ K}$$

or
$$= 6.47\ °\text{C}$$

(*b*) $(\Delta S)_{water} = m\, c_p \ln\left(\frac{T}{298}\right)$

$$= 7 \times 4.187 \ln\left(\frac{279.47}{298}\right)$$

$$= 29.309\ (\ln 279.47 - \ln 298)$$

$$= 29.309\ (5.633 - 5.6971)$$

$$= -29.309 \times 0.0641$$

$$= -1.8787 \text{ kJ/K}$$

$(\Delta S)_{ice}$ = Change in entropy due to melting ice at constant temperature + Change in entropy due to heating of water from 0°C to 6.47°C

$$= \frac{mL}{T_{ice}} + m c_p \ln\left(\frac{279.47}{273}\right)$$

$$= \frac{1.5 \times 335}{273} + 1.5 + 4.187\ [\ln\ (279.47 - \ln 273)]$$

$$= 1.8407 + 6.2805\ [5.6329 - 5.6095]$$

$$= 1.8407 + 6.2805 \times 0.0234$$

$$= 1.8407 + 0.1496 = 1.98766 \text{ kJ/K}$$

Therefore $(\Delta S)_{system} = (\Delta S)_{water} + (\Delta S)_{ice}$

$$= -1.8787 + 1.98766$$

$$= 0.10896 \text{ kJ/K (increase)}$$

Example 6.16. *Nitrogen gas is compressed from an initial state of 90 kPa and 17°C to a state of 450 kPa and 57°C. Determine the change in entropy during this process assuming the* average value of c_p *for nitrogen to be 1.0394 kJ/kgK and R = 0.295 kJ/kg · K. Treat Nitrogen as perfect gas.*

Solution : Initial state $P_1 = 90$ kPa $\quad T_1 = 17 + 273 = 290$ K

Final sate $P_2 = 450$ kPa $\quad T_2 = 57 + 273 = 330$ K

The entropy change from equation (6.48) is given by

$$\Delta S = s_2 - s_1 = c_p \ln\left(\frac{T_2}{T_1}\right) = R \ln\left(\frac{P_2}{P_1}\right)$$

$$= 1.0394 \ln\left(\frac{330}{290}\right) - 0.295 \ln\left(\frac{450}{90}\right)$$

$$= 1.0394 \ln (1.1379) - .295 \ln (5)$$

$$= 0.1343 - 0.4748$$

$$= -0.3405 \text{ kJ/(kg} \cdot \text{K)}$$

Example 6.17. *Air is compressed with the help of a 5-kW compressor from pressure P_1 to P_2. The air temperature is kept constant at 27°C despite the heat transfer to the surroundings which is at 12°C calculate the rate of entropy change of air and of the surroundings. State whether this process is reversible, irreversible or impossible. Assume air to be an ideal gas.*

Solution : As the temperature is maintained constant it is an isothermal compression process. The rate of heat transfer during this process can be found from the study-flow energy equation.

$$\dot{Q} - \dot{W} = \dot{m}(\Delta h + \Delta KE + \Delta PE)$$

Here $\dot{W} = -5 \text{ kW}, \Delta KE = 0, \Delta PE = 0$

and $\Delta h = c_p \Delta T = 0 \text{ as } \Delta T = 0$

Therefore from the above equation

$$\dot{Q} - \dot{W} = 0$$

or $\dot{Q} = \dot{W} = -5 \text{ kW}$

Assuming that the process is internally reversible then

$$\Delta \dot{S}_{\text{air}} = \frac{\dot{Q}_{av}}{T_{\text{air}}} = \frac{-5 \text{ kW}}{273+27} = \frac{-5 \text{ kW}}{300 \text{ K}}$$

$$= -0.01667 \text{ kW/K}$$

and $\Delta \dot{S}_{\text{surroundings}} = \dfrac{-\dot{Q}}{T_{\text{sur}}} + \dfrac{5 \text{ kW}}{T_{\text{sur}}} = \dfrac{+5 \text{ kW}}{285} = 0.01754 \text{ KW/k}$

Therefore total $\dot{s}_{\text{total}} = (\Delta \dot{s})_{\text{sur}} + (\Delta \dot{s})_{\text{surronding}}$

$$= -0.01667 + 0.01754$$

$$= 0.00087 \text{ kw/k}$$

$$> 0$$

Since $\dot{s}_{\text{total}} > 0$ it is irreversible process

Example 6.18. *(a) m_1 kg of water at temperature T_1 is isobarically and adiabtically mixed with m_2 kg of water at temperature T_2 where $T_1 > T_2$. Find the expression for change of entropy of the mixture.*

(b) In case m_1 were equal to m_2 then derive the modifed expression for the entropy change and discuss the validity of this expression.

Solution : Let T be the final temperature of the mixture and c_p the specific heat of water. We have from the conservation of energy principle.

Heat lost by m_1 kg of water = Heat gained by the m_2 kg water

$$m_1 c_p(T_1 - T) = m_2 c_p(T - T_2)$$

or $m_1 T_1 - m_1 T = m_2 T - m_2 T_2$

or $m_1, T_1 - m_2 T_2 = m_1 T + m_2 T$

or $T = \dfrac{m_1 T_1 + m_2 T_2}{(m_1 + m_2)}$

Change in entropy of m_1 kg of water when its temperature falls from T_1 to T.

$$dS_1 = m_1 c_p \ln\left(\frac{T}{T_1}\right)$$

Similarly change in entropy of m_2 kg water when its temperature rises from T_2 to T

$$dS_2 = m_2 c_p \ln\left(\frac{T}{T_2}\right)$$

Therefore total change in entropy of the mixture

$$(dS)_{\text{universe}} = dS_1 + dS_2$$

$$= m_1 c_p \ln\left(\frac{T}{T_1}\right) + m_2 c_p \ln\left(\frac{T}{T_2}\right)$$

$$= m_1 c_p \ln\left(\frac{m_1T_1 + m_2T_2}{(m_1+m_2)T_1}\right) + m_2 c_p m\left[\frac{m_1T_1 + m_2T_2}{(m_1+m_2)T_2}\right]$$

(*b*) when $m_1 = m_2 = m$

the above expression (*A*) reduces to

$$(dS)_{\text{universe}} = m c_p \ln\left[\frac{m(T_1+T_2)}{2mT_1}\right] + m c_p \ln\left[\frac{m(T_1+T_2)}{2m\,T_2}\right]$$

$$= m c_p \ln\left(\frac{T_1+T_2}{2T_1}\right) + m\,c_p \ln\left[\frac{T_1 \times T_2}{2T_2}\right]$$

$$= m c_p \ln\left[\left(\frac{T_1+T_2}{2T_1}\right)\left(\frac{T_1+T_2}{2T_2}\right)\right]$$

$$= m c_p \ln\left[\frac{(T_1+T_2)^2}{4T_1\,T_2}\right]$$

$$= m c_p \ln\left[\frac{T_1+T_2}{2\sqrt{T_1\,T_2}}\right]^2$$

$$= 2m c_p \ln\left[\frac{T_1+T_2}{2\sqrt{T_1+T_2}}\right]$$

or $$(dS)_{\text{universe}} = 2m\,c_p\,h\left[\frac{\left(\frac{T_1 \times T_2}{2}\right)}{\sqrt{T_1T_2}}\right] \quad ...(B)$$

Non $\dfrac{T_1+T_2}{2}$ = Arithmatic mean of T_1 an T_2

and $\sqrt{T_1T_2}$ = Geometric mean of T_1 and T_2

Since Arithmatic mean > Geometric mean. The right hand side of expression B is positive. In otherwords the entropy of the universe is increasing and as per the entropy principle it is a valid process which is irreversible in water.

Example 6.19. *A heat engine working between two thermal reservoirs with source at T_1 and sink at T_2. The masses of both the source and sink are equal and the specific heats are c_p, prove that the maximum possible work output is given by*

$$W_{max} = c\,m\left(\sqrt{T_1} - \sqrt{T_2}\right)^2$$

Solution : For maximum output efficiency of the heat engine will be maximum which occurs when the engine operates on a reversible cycle for which (Ref. Fig. (5.1 b))

$$\oint \frac{\delta Q}{T} = 0$$

in other words the total entropy change must be Zero. Let the common temperature when the engine stops working be T

Then $$(dS)_{total} = (dS)_{Source} + (dS)_{Sink} = 0$$

or $$mc\ln\left(\frac{T}{T_1}\right) + mc\ln\left(\frac{T}{T_2}\right) = 0$$

or $$mc\ln\left[\left(\frac{T}{T_1}\right)\left(\frac{T}{T_2}\right)\right] = 0$$

or $$mc\ln\left(\frac{T^2}{T_1T_2}\right) = 0$$

Taking antilog $$\ln\frac{T^2}{T_1T_2} = 0 = \ln(1)$$

or $$\frac{T^2}{T_1T_2} = 1$$

or $$T = \sqrt{T_1T_2}$$

Then the workoutput $$W_{max} = \text{Heat supplied} - \text{Heat rejected}$$

$$= mc\,(T_1 - T) - mc\,(T - T_2)$$

$$W_{max} = mc\left[T_1 - \sqrt{T_1T_2}\right] - m\left[\sqrt{T_1T_2} - T_2\right]$$

$$= mc\left[T_1 + T_2 - 2\sqrt{T_1T_2}\right]$$

$$= mc\left(\sqrt{T_1} - T_2\right)^2$$

Example 6.20. *An ideal gas at temperature T_1 is heated keeping the volume constant, to temperature T_2. The gas is then subjected to expansion process to bring it back to initial* *temperature T_1 according to the law PV^n = constant. If the entropy changes in the two processes ore equal find the relationship between n and the adiabatic index γ.*

Solution : Entropy change during compression while keeping volume constant, we have from equation (6.24).

$$(\Delta S)_{\text{constant vol}} = mC_v \ln\left(\frac{T_2}{T_1}\right) \quad ...(A)$$

Further we have from equation (6.14*a*) keeping in mind that it is expansion process and initial temperature in T_2 at the start of process which ends at temperature T_1 (final temperature during the expansion process.

$$(\Delta S)_{\text{exp}} = mc_v\left(\frac{\gamma - n}{n-1}\right)\ln\left(\frac{T_2}{T_1}\right) \quad ...(B)$$

Equating equation (*A*) & (*B*) we get

$$mc_v \ln\left(\frac{T_2}{T_1}\right) = mc_v\left(\frac{\gamma - n}{n-1}\right)\ln\left(\frac{T_2}{T_1}\right)$$

or
$$1 = \frac{\gamma - n}{n-1}$$

or
$$n - 1 = \gamma - n$$

or
$$2n = (\gamma - 1)$$

or
$$n = \left(\frac{\gamma - 1}{2}\right)$$

Example 6.21. *An ideal gas is contained is a rigid tank. The gas is being stirred by a paddle wheel and does 200 kJ of work on the gas. As a result of heat transfer between the system and surrounding (which are at 298 K), the temperature of the ideal gas in the tank remains at 323 K. Determine the entropy change of the ideal gas and of suroundings.*

(***b***) Also state whether the increase-in-entropy principle is satisfied

Solution :

(*a*) The initial and final states of the gas are same,

Therefore
$$S_2 = S_1$$

∴
$$(\Delta s)_{\text{gas}}\; S_2 - S_1 = 0$$

(*b*) From the conservation of energy equation we have

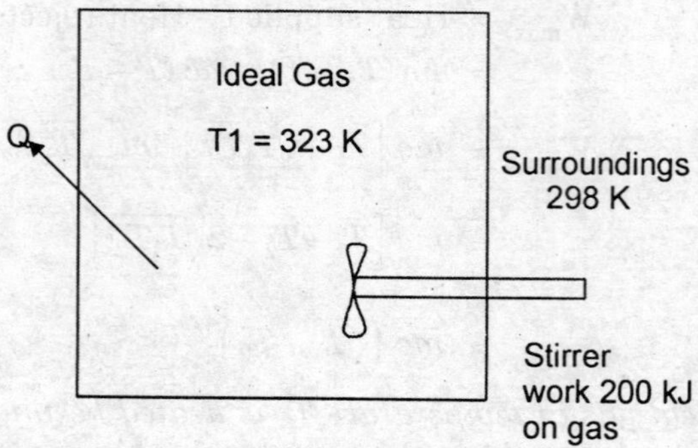

Fig. 6.18

$$Q - W = \Delta U + \Delta PE + \Delta KE$$

Now $\Delta KE = \Delta PE = 0$

Therefore $Q = W + \Delta U$

$$= W + mc_v \, \Delta T$$

or $Q = -200 \text{ kJ} + 0 = -200 \text{ kJ (as } \Delta T = 0)$

Then $$\Delta S_{sur} = \frac{Q_{sur}}{T_{sur}}$$

$$= \frac{-Q}{T_{sur}} = \frac{200 \text{ kJ}}{298 \text{ k}}$$

$$= 0.671 \text{ kJ/k}$$

(*c*) Total entropy change $= \Delta S_{total}$

$$= \Delta S_{gas} + \Delta S_{sur}$$

$$= 0 + 0.671 = 0.671 \text{ kJ/K}$$

(*d*) Since ΔS_{total} is positive, it satisfies the increase-in-entropy principle.

Example 6.22. *Saturated liquid water at 250 kPa pressure is contained in a frictionless piston cylinder device. When 450 kJ of heat is supplied to the water from a heat source at 525 °C, part of water vaporises at constant pressure. Find the total entropy change for this process. Is this process reversible, irreversible or impossible?*

Solution : Since the pressure of the saturated steam is maintained constant, the temperature of the steams will reamin constant at saturated steam point corresponding to this pressure. From steam tables

At 250 kPa pressure T_{sat} = 127.44°C. Further since the device is functionless, it is assumed that the process involves no ir-reversibilities. Then the process will be isothermal and internally reversible.

$$T_{system} = 127.44 + 273$$

$$= 400.44 \text{ K}$$

and $T_{source} = 525 + 273 = 798 \text{ K}$

$$\Delta S_{sys} = \frac{Q_{sys}}{T_{sys}} = \frac{450}{400.44} = 1.1238 \text{ kJ/K}$$

The entropy change of the heat source

$$\Delta S_{source} = \frac{Q_{source}}{T_{source}} = \frac{-450}{798} = -.5639 \text{ kJ/K}$$

Total entropy change $= \Delta S_{sys} + \Delta S_{source}$

$$= 1.1238 - .5639$$

$$= 0.5599 \text{ kJ/K} > 0$$

Since the total entropy change is positive the process is irreversible.

Example 6.23. (*a*) *Refrigerant-12 in the form of saturated liquid-vapour mixture enters the evaporator of a refrigerator at 180 kPa. The refrigerant absorbs 65 kJ of heat from the cooling chamber of the refrigerator which is maintained at –2°C and leaves the device as* *saturated vapor at the same pressure. Find (a) the entropy change of the refrigerant (b) the entropy change of the cooling chamber (c) the total entropy change for the process.*

Solution : Since the pressure of the *R*-12 refrigerant remains constant, the refrigerant remains at constant temperature which is equal to the saturation temperature at pressure of 180 kPa.

From *R-12* properties tables we have

at P = 180 kPa

$$T_{sat} = -15.38\ °C$$
$$= 273 - 15.38$$
$$= 257.62\ K$$

Assuming that their are no internal irreversibilities, then for this isothermal process.

$$\Delta S_{sys} = \frac{Q_{sys}}{T_{sys}} = \frac{Q_{sys}}{T_{sat}}$$

$$= \frac{65\ kJ}{257.62\ K}$$

$$= 0.25231\ kJ/K$$

(*b*) Entropy change of the cooling chamber space

$$T_{space} = 273 - 2$$
$$= 271\ K$$

$$\Delta S_{space} = \frac{Q_{space}}{T_{space}}$$

$$= \frac{-65\ kJ}{271\ K}$$

$$= -0.2398\ kJ/K$$

(*c*)

$$\Delta S_{total} = \Delta S_{sys} + \Delta S_{space}$$
$$= 0.25231 - .2398$$
$$= 0.01251\ kJ/K$$

Example 6.24. *Radiator of steam heating system has a volume of 1.5 Litre and at any point of time is filled with superheated water vapour at 200 kPa and 150°C. At this stage both the inlet and outlet values of the radiator are closed. After sometime the temperature of the water vapour in the radiator drops to 75°C due to transfer of heat to the room air. Find the entropy charge of the steam during this process.*

Solution : From steam tables

$$\left.\begin{array}{l} P_1 = 200\ \text{kPa} \\ T_1 = 150°\text{C} \end{array}\right\} \begin{array}{l} v_1 = 0.9596 m^3/kg \\ s_1 = 7.2795 kJ/kgK \end{array}$$

Since both the inlet and outlet valves are closed

At $\left.\begin{array}{l} v_2 = v_1 \\ T_2 = 25°EC \end{array}\right\}$ we have from steam tables

$$v_f = 0.001026 \text{ m}^3/\text{kg}$$
$$v_g = 4.131 \text{ m}^2/\text{kg}$$
$$s_f = 1.0155 \text{ kJ/kg} \cdot \text{K}$$
$$sf_g = 6.6669 \text{ kJ/kg} \cdot \text{K}$$
$$s_g = 7.6824 \text{ kJ/kg} \cdot \text{K}$$

It is given $v_2 = v_1 = 0.9596$

Since at the final state the steam will be saturated mixture. Let its quality be x_2. Then

Then
$$x_2 = \frac{v_2 - v_f}{v_{fg}} = \frac{v_2 - v_f}{v_g - v_f}$$
$$= \frac{0.9596 - 0.001026}{4.131 - 0.001026}$$
$$= \frac{0.9586}{4.12997} = 0.2321$$

Therefore
$$s_2 = s_f + x_2 s_{fg}$$
$$= 1.0155 + 0.2321 \times 6.6669$$
$$= 1.0155 + 1.5474$$
$$= 2.5629 \text{ kJ/kg} \cdot \text{K}$$

Now the mass of steam
$$m = \frac{V}{v_1}$$

Here
$$V = 1.5 \text{ L}$$
$$= 0.015 \text{ m}^3$$
$$= \frac{0.015 \text{ m}^3}{0.9596 \text{ m}^3/\text{kg}} = 0.1563 \text{ kg}$$

Therefore entropy change of the steam
$$\Delta S = m(s_2 - s_1)$$
$$= 0.1563\,(2.5629 - 7.2795)$$
$$= -0.1563 \times 4.7166$$
$$= -0.7372 \text{ kJ/K}$$

Example 6.25. *A rigid tank 0.5 m^3 capacity contains refrigerant –12 initially at 200 kPa and quality 0.4. Now heat is added from a heat source at 35°C until the pressure rises to 320 kPa. Find*

(a) The entropy change of the refrigerant

(b) The entropy change of the source

(c) The total entropy change for the process

Solution : (*a*) From the refrigerant 12 table

At $P_1 = 200$ kPa $\quad v_1 = v_f + x_1 v_{fg} = 24.43 + 0.4\,(165.36 - 24.4)$

$= 80.8$ kJ/kg

$x_1 = 0.4 \quad s_1 = s_f + x_1 s_{fg} = 0.0992 + .4\,(0.7035 - 0.099)$

$$= 0.3409 \text{ kJ/kg}$$

$$v_1 = v_f + x_1 v_{fg} = 0.0006962 + .4\,(0.08354 - 0.0006964)$$

$$= 0.03383 \text{ m}^3/\text{kg}$$

As the tank is rigid $v_1 = v_2$. Again from R-12 tables at 320 kPa

$P_2 = 320$ kPa Then $x_2 = \dfrac{v_2 - v_f}{v_{fg}} = \dfrac{v_1 - v_f}{v_{fg}} = \dfrac{v_1 - v_f}{v_g - v_f}$

$v_2 = v_1 = 0.03383$ $= \dfrac{0.03383 - 0.0007177}{0.05351 - 0.0007177}$

$v_f = 0.0007177 \text{ m}^3/\text{kg}$ $= \dfrac{0.03311}{0.05279} = 0.6272$

$v_g = 0.05351 \text{ m}^3/\text{kg}$

$s_f = 0.1457 \text{ kJ/kg} \cdot \text{K}$

$s_f = 0.6960 \text{ kJ/kg} \cdot \text{K}$

Mass of the refrigerant $= \dfrac{V}{v_1} = \dfrac{0.5 \text{ m}^3}{0.3383} \text{ m}^3/\text{kg}$

$= 14.78$ kg

Therefore entropy change of the refrigerant

$$\Delta S = m(s_2 - s_1)$$

But $s_2 = s_f + x_2 s_{fg}$

$$= 0.1457 + 0.6272\,(0.6560 - 0.1457)$$

$$= 0.1457 + .6272 \times 0.5503$$

$$= 0.4908 \text{ kJ/k}$$

and $u_2 = u_f + x_2\, u_{fg}$

$$= 36.85 + 0.6272\,(u_g - u_f)$$

$$= 36.85 + .6272\,(170.88 - 36.85)$$

$$= 36.85 + .6272 + 134.03$$

$$= 120.91 \text{ kJ/kg}$$

Therefore Entropy change of refrigerant

$$= m\,(s_2 - s_1)$$

$$= 14.78\,(0.4908 - 0.3409)$$

$$= 14.78 \times .1499 = 2.2155 \text{ kJ/K}$$

(*b*) For finding the entropy change of the heat source, we need to calculate the heat transfer. Since it is closed system for which energy conservation equation gives

$$Q - W = \Delta U + \Delta KE + \Delta PE$$

Here $W = 0, \Delta KE = 0 \text{ and } \Delta PE = 0$

$\therefore$ $Q = \Delta U$

$$= m(u_2 - u_1) = 14.78\,(120.91 - 80.8)$$

$$= 14.78 \times 40.11 = 592.8258 \text{ kJ}$$

Therefore $$\Delta S_{source} = \frac{Q_{source}}{T_{source}} = -\frac{Q}{T_{source}}$$

$$= \frac{-592.8258}{T_{source}}$$

Now $$T_{source} = 35 + 273 = 308$$

$$\Delta S_{source} = \frac{-552.8258}{308}$$

$$= -1.9248 \text{ kJ/K}$$

(*c*) Total entropy change for the process

$$\Delta S_{total} = \Delta S_{ref} + \Delta S_{source}$$

$$= 2.2155 - 1.9248$$

$$= 0.2607 \text{ kg/K}$$

6.14 CONCLUDING REMARKS ABOUT ENTROPY CHANGE AND ENTROPY

From the discussions during the course of the present chapter coupled with the variety of examples it is clear that two factors cause the entropy of the system to change (*a*) Heat transfer and ir-reversibilities.

(*a*) Heat transfer to the system increases the entropy of that system and heat transfer from the system decreases it. Entropy change due to reversible heat transfer is called the Entropy Flow and does not involve any entropy generation.

(*b*) Inversibilities such as friction, high speed compression or expansion and heat transfer through a finite temperature difference always result in increase of entropy. In thermodynamic terms the entropy of a system will increase during an adiabatic process. Mathematically for adiabatic process it is expressed as

(*i*) $$S_{generated} > 0 \text{ (for irrversible process)}$$

and $$S_{gas} = 0 \text{ (for reversible process)}$$

The adiabatic process with no irreversibilities within the system or "internally reversible adiabatic" process is also known as ISENTROPIC process for which $S_{gas} = 0$.

(*c*) In view of the discussions made in this chapter following inferences are drawn.

(*i*) Process can occur in a certain direction only and "not in any direction". The direction of the process in which it proceeds, is decided in accordance with the "Increase in Entropy Principle" *i.e* $S_{total} \geq 0$. A process that violates this principle is not possible to take place.

(*ii*) In the presence of irreversibilities entropy is non conserved property. It must increase in such cases. However, during the idealised reversible process it is conserved or $\Delta S = 0$. But in actual practice due to the presence of ir-reversibilities, it must increase. Therefore we conclude that the entropy of the universe is increasing continuously.

For any engineering system, the entropy generation is then the measure of the irreversibilities present during the process. This property is therefore used to determine the eriteria for the performance of engineering devices.

Example 6.26. *Air is compressed in a piston cylinder assembly in a reversible-adiabatic manner from 110 kPa and 290 k to 880 kPa. Assuming constant specific heat as 0.727 kJ/kg · k find the final temperature. Take γ = 1.4. Also find the work done during the process.*

Solution : Since this process is reversible adiabatic (*i.e.*, isentropic) we have, from equation (6.49 *b*)

$$\frac{T_2}{T_1} = \left(\frac{P_2}{P_1}\right)^{\frac{\gamma-1}{\gamma}}$$

or
$$T_2 = T_1\left(\frac{P_2}{P_1}\right)^{\frac{\gamma-1}{\gamma}} = (290)\left[\frac{880}{110}\right]^{\frac{0.4}{1.4}}$$

$$= (290)\,(8)^{\frac{0.4}{1.4}} = 525.304 \text{ K}$$

(*b*) the work done during the process is determined from the closed system energy equation

$$q - w = \Delta u + \Delta KE + \Delta PE$$

Here
$$q = 0,\ \Delta KE = \Delta PE = 0$$

i.e.
$$-w = \Delta u$$

or
$$w = -(u_2 - u_1)$$
$$= -c_v(T_2 - T_1)$$
$$= -0.727\,(525.304 - 290)$$
$$= -171.1 \text{ kJ/kg}$$

Example 6.27. *An insulated rigid tank is divided into two equal volume chambers with a partition wall Initially chamber certains 3 kmol of an ideal gas at 500 kPa and 343 K and the other side is evacuated. Now the partition is removed and the gas fills the entire tank. Determine the final temperature and the entropy change of the system during this process. Take R_u = 8.314 kJ/k mol · K. Also find the entropy change of the suaroundings.*

Solution : Taking the entire rigid tank as the system and applying the energy equation for the given closed system we get

$$Q - W = \Delta U + \Delta KE + \Delta PE$$

Here
$$Q = 0,\ W = 0,\ \Delta KE = \Delta PE = 0$$

Therefore
$$\Delta U = 0$$

or
$$mc_v(T_2 - T_1) = 0$$

Since m and c_v are constants

$$T_2 - T_1 = 0$$

or
$$T_2 = T_1$$
$$= 343 \text{ K } \textbf{Ans.}$$

(*b*) From equation (6.46) we have the entropy change in terms of *N* k mol.

$$(S_2 - S_1)_{\text{kmol}} = N\left[c_v \ln \frac{T_2}{T_1} + R_u \ln \frac{V_2}{V_1}\right]$$

or
$$\Delta S = N\left[c_v \ln\left(\frac{T_2}{T_1}\right) + R_u \ln \frac{V_2}{V_1}\right]$$

Here $\quad T_2 = T_1$

and $\quad V_2 = 2V_1$

and $\quad N = 3$ k mol

$$\therefore \quad \Delta S = 3[c_v \ln(1) + R_u \ln(2)]$$
$$= 3(0 + 8.314 \ln(2)]$$
$$= 24.942 \ln(2) = 17.288 \text{ kJ/K}$$

(*b*) Since the rigid tank is insulated there is no heat transfer to the surroundings. Thus the entropy change of surroundings is zero.

Example 6.28. *(a) An ideal gas is compressed from an initial state of 100 kPa and 27°C to 250 kPa in a reversible isothermal manner. Determine the specific entropy change of the gas and (b) the work done per unit volume. Take R = 0.287 kJ/kg · K and c_p = .287 kJ/kg.*

Solution : (*a*)We have from equation (6.48)

$$\Delta s_{gas} = c_p \ln \frac{T_2}{T_1} - R \ln \frac{P_2}{P_1}$$

Since $\quad T_2 = T_1 = 300, \ \ln \frac{T_2}{T_1} = 0$

$$\therefore \quad \Delta S_{gas} = 0 - (0.287) \ln \frac{250}{100}$$
$$= -0.263 \text{ kJ/kg} \cdot \text{K}$$

(*b*) Work done per unit volume $q = T \Delta s$

$$= (300)(-.263)$$
$$= -78\text{-}9 \text{ kJ}$$

Example 6.29. *Helium gas is compressed from 110 kPa and 303 k to 550 kPa in a reversible adiabatic process. Find the final temperature and work done if the process take place in.*

(a) Piston-cylinder assembly (a closed system).

(b) In rotory compressor (i.e., steady flow process)

Take $\quad \gamma = 1.667$ *and c_v average = 3.1156 kJ/kg · K*

Solution : (*a*) Since the specific heats average value is given, it is implied that we assume it constant. Moreover the process is reversible adiabatic or isentropic. Therefore we have from equation (6.49*b*).

$$\frac{T_2}{T_1} = \left(\frac{P_2}{P_1}\right)^{\frac{\gamma-1}{\gamma}}$$

Here $\quad T_1 = 303$ K

$\quad P_2 = 550$ kPa

$\quad P_1 = 110$ kPa

$$\therefore \quad T_2 = (303)\left[\frac{550}{110}\right]^{\frac{1.667-1}{1.607}}$$
$$= 576.9 \text{ K}$$

In the first case it is a closed mass system for which we have from the closed system energy equation.

$$q - w = \Delta u + \Delta KE + \Delta PE$$

Since here $q = 0, \Delta KE = \Delta PE = 0$

Therefore $q - w = \Delta u$

or
$$\begin{aligned} w &= -c_v(T_2 - T_1) \\ &= -3.1156\,(576.9 - 303) \\ &= -835.4 \text{ kJ/kg} \end{aligned}$$

(*b*) In this steady flow device, the steady flow energy equation is

$$q - w = \Delta h + \Delta KE + \Delta PE$$

Here $q = 0 \text{ and } \Delta KE - \Delta PE = 0$

$\therefore$
$$\begin{aligned} w &= -\Delta h \\ &= -c_p(T_2 - T_1) \end{aligned}$$

But
$$\begin{aligned} c_p &= c_v\,\gamma \\ &= 3.1156 \times 1.667 \\ &= 5.1937 \end{aligned}$$

Therefore
$$\begin{aligned} w &= -5.1937\,(T_2 - T_1) \\ &= -5.1937\,(576.9 - 303) \\ &= -1422.3 \text{ kJ/kg} \end{aligned}$$

Example 6.30. *Five kg of Helium is contained in an insulated rigid tank at 630 kPa and 303 K. A value is now opened and helium is allowed to escape until the pressure in the tank drops to 210 kPa. Assuming the helium remaining in the tank undergoes a reversible adiabatic process, find the final mass of helium left in the tank and its temperature. Treat helium as ideal gas and take its* $\gamma = 1.667$.

Solution : As the process is reversible adiabatic *i.e.*, isentropic we have from equation (6.49*b*)

$$\frac{T_2}{T_1} = \left(\frac{P_2}{P_1}\right)^{\frac{\gamma-1}{\gamma}}$$

Here
$$\begin{aligned} T_1 &= 303 \text{ k} \\ P_1 &= 630 \text{ kPa} \\ P_2 &= 210 \text{ kPa} \end{aligned}$$

Therefore
$$\begin{aligned} T_2 &= 303\left[\frac{210}{630}\right]^{\frac{1.667-1}{1.667}} \\ &= 303\left[\frac{1}{3}\right]^{\frac{0.667}{1.667}} \\ &= \mathbf{195\ K\ Ans.} \end{aligned}$$

(*b*) From the ideal gas equation

$$P_1V_1 = m_1RT_1$$

and $$P_2V_2 = m_2RT_2$$

Dividing these $$\frac{P_1V_1}{P_2V_2} = \frac{m_1RT_1}{m_2RT_2}$$

Since the tank is rigid $$V_1 = V_2$$

$$\therefore \quad \frac{P_1}{P_2} = \frac{m_1T_1}{m_2T_2}$$

or $$m_2 = m_1\left(\frac{T_1}{T_2}\right)\left(\frac{P_2}{P_1}\right)$$

$$= 5\left(\frac{303}{195}\right)\left(\frac{210}{630}\right)$$

$$= 2.5897 \text{ kg}$$ **Ans.**

Example 6.31. *Helium is compressed steadily by a compressor from 100 kPa and 290 K to 700 kPa at a rate of 2 kg/min. Find the minimum power input required if the process is (a) adiabatic and (b) isothermal. Assume that Helium is a perfect gas and has constant specific heats c_p = 5.1973 kJ/kg · K. Also neglect the changes in kinetic and potential energies. Take γ = 1.667*

Solution : (*a*) For minimum power consumption the process is to be reversible. Since the process is reversible adiabatic (it is isentropic). Therefore from the relation (6.49 *b*) we have

$$\frac{T_2}{T_1} = \left(\frac{P_2}{P_1}\right)^{\frac{\gamma-1}{\gamma}} = \left(\frac{700}{100}\right)^{\frac{0.667}{0.667}}$$

or $$T_2 = 290(7)^{0.4001}$$

$$= 631.72 \text{ K}$$

Now we have from the steady flow equation

$$\dot{Q} - \dot{W} = \dot{m}(\Delta h + \Delta KE + \Delta PE)$$

Since $$\dot{Q} = 0, \Delta KE = 0, \Delta PE = 0$$

We get $$\dot{w} = -\dot{m}\ (\Delta h)$$

$$= -\dot{m}(h_2 - h_1)$$

$$= -\dot{m}c_p(T_2 - T_1)$$

Here $$\dot{m} = 2 \text{ kg/min}$$

$$= \frac{2}{60} = \frac{1}{30} \text{ kg/sec}$$

and $$c_p = 5.1973$$

$$\therefore \quad \dot{w} = -\left(\frac{1}{30}\right)5.1973\ (631.72 - 290)$$

$$= -\frac{1}{30} \times 5.1973 \times 341.72$$

$$= -59.2 \text{ kW}$$

(*b*) In this case it is reversible isothermal process

$$T_2 = T_1$$

and $$W_{12} = P_1 V_1 \ln\left(\frac{P_2}{p_1}\right) \text{ (equation 6.73)}$$

Also $$P_1 V_1 = RT_1$$

But $$R = C_p - C_v$$

$$= c_p - \frac{c_p}{\gamma}$$

$$= 5.1973\left(1 - \frac{-1}{\gamma}\right)$$

$$= 5.1973\left(\frac{1.667 - 1}{1.667}\right)$$

$$= \frac{5.1973 \times .667}{1.667}$$

$$= 2.0795$$

$\therefore$ $$v_1 = \frac{RT}{P_1} = \frac{2.0795 \times 290}{100}$$

$$= 6.031 \text{ m}^3/\text{kg}$$

$\therefore$ $$w_{12} = P_1 V_1 \ln\left(\frac{P_2}{P_1}\right)$$

$$= RT_1 \ln\left(\frac{P_2}{P_1}\right)$$

$$= 2.0795 \times 290 \ln\left(\frac{700}{100}\right)$$

$$= 2.0795 \times 290 \times 1.7859$$

$$= 1173.2 \text{ kJ/m}$$

Then power required $$= \dot{m} w_{12}$$

$$= \frac{2}{60} \times 1173.2$$

$$= 39.1 \text{ kw } \textbf{Ans.}$$

Example 6.32. *Nitrogen is compressed steadily by a reversible compressor from an inlet state of 110 kPa and 27°C to an exit pressure 990 kPa. Calculate the compressor work required per unit mass of gas for the following cases.*

(a) Isentropic compression with $\gamma = 1.4$

(b) Polytropic compression with $n = 1.3$

(c) Isothermal compression

Assume $R = 0.287\ kJ/kg \cdot K$

Solution : (*a*) For isentropic compressure we have from equation (6.71), the work required per unit mass we have $T_1 = 273 + 27 = 300$ K.

$$w_{12} = -\frac{\gamma R T_1}{(\gamma - 1)}\left[1 - \left(\frac{P_2}{P_1}\right)^{\frac{\gamma-1}{\gamma}}\right]$$

$$= -\frac{(1.4)(.287)(300)}{(1.4-1)}\left[1 - \left(\frac{990}{110}\right)^{\frac{0.4}{1.4}}\right]$$

$$= -301.35\,[1 - (9)^{0.2857}]$$

$$= -263.2 \text{ kJ/kg}$$

(*b*) For polytropic compression we have from equation (6.72)

$$w_{12} = \frac{nRT_1}{(1-n)}\left[1 - \left(\frac{P_2}{P_1}\right)^{\frac{n-1}{n}}\right]$$

Here $n = 1.3$

$$w_{12} = \frac{1.3\,(.287)\,(300)}{1-1.3}\left[1 - \left(\frac{990}{110}\right)^{\frac{1.3-1}{1.3}}\right]$$

$$= -373.1\,[1 - (9)^{0.2307}]$$

$$= -246.4 \text{ kJ/kg}$$

(*c*) For isothermal compression we have from equation (6.73)

$$w_{12} = -RT_1 \ln\left(\frac{P_2}{P_1}\right)$$

$$= -0.287 \times 300 In\left(\frac{990}{110}\right)$$

$$= -189.2 \text{ kJ/kg}$$

Example 6.33. *Air is compressed in a two stage ideal compressor with intercooling from the initial inlet state of 100 kPa 27°C to final exist pressure 900 kPa. The process is polytropic with $n = 1.3$. Determine the compressor work required per unit masses.*

Take $R = 0.287\ kJ/kg \cdot K$

Solution : In case of ideal gas two stage compression with intercooling consuming the minimum work for polytropic (n = 1.31) process we have from equation (6.35).

$$\frac{P_2}{P_1} = \frac{P_3}{P_2}$$

where P_1 = initial stage pressure at inlet

P_2 = Intermediate pressure at the and of first stage

P_3 = Final pressure at outlet

i.e.,
$$P_2 = \sqrt{P_1 P_3}$$
$$= \sqrt{100 \times 900} = 300 \text{ kPa}$$

Further the compressor work in each stage will also be same. Thus the total work input will be twice the compression work of end stage.

$\therefore$
$$W_{13} = \text{Total compressor work}$$
$$= 2W_{12}$$
$$= 2 \times (\text{compressor work of 1st stage})$$
$$= 2 \times \frac{nRT_1}{(n-1)}\left[1 - \left(\frac{P_2}{P_1}\right)^{\frac{n-1}{n}}\right]$$

Here
$$T_1 = 273 + 27 = 300 \text{ K}$$

or
$$W_{13} = \frac{2 \times 1.3 \times 0.287 \times 300}{(1.3-1)}\left[1 - \left(\frac{300}{100}\right)^{\frac{0.3}{1.3}}\right]$$
$$= 748.2\,[1 - (3)^{.2308}]$$
$$= 748.2\,[1 - 1.2886]$$
$$= -215.35 \text{ kJ/kg}$$

PROBLEMS

1. Using a reversible and an irreversible refrigerator operating between the same heat reservoirs, show that clausius inequality is justified.

2. In a cyclic process $\oint \delta Q > 0$. Does this process violate the clausius inequality? Give reason.

3. A heat engine receives heat from thermal source S_1 and rejects to the sink S_2. The temperatures of S_1 and S_2 are 827°C and 37°C respectively. If the engine receives 800 kJ of heat and rejects 500 kJ to the sink, then show whether the engine violates the clausius inequality or not. (**Ans.** Violates)

4. Show whether the heat engine of the problem 3 violates the Carnot Principle or not. (**Ans.** Violates)

5. A thermal source at 1027°C supplies 75000 kJ heat to a heat engine per hour which rejects heat to river water at 17°C to give 8 kw power output. Determine whether this heat engine violates the clausius inequality or not. (**Ans.** Satisfies)
6. A house in Simla during the winter season is heated with the help of an air based pump. The house is maintained at 21°C while the outside temperature is –2°C. The house requires heat at the rate of 75 MJ/h to maintain the desired temperature inside, for which the heat pump consumes 8 kw of power. Show by calculations whether the pump violates the clausius inequality. (**Ans.** Satisfies)
7. Can the cyclic integral $\oint \delta Q$ be zero for a system explain? (**Ans.** No)
8. Is the value of the integral $\int_1^2 \frac{\delta Q}{T}$ the same for all processes between the state 1 and 2? (**Ans.** No)
9. In order to find the entropy change for an irreversible process between states 1 and 2 should the integral $\int_1^2 \frac{\delta Q}{T}$ be performed along the actual path or an imaginary reversible path? Explain (**Ans.** Along imaginary reversible path)
10. The entropy of a hot baked cake decreases as it cools. Is this a violation of the increase-in-entropy principle? Explain (**Ans.** No the increase in entropy of surroundings his also to be considered)
11. A piston-cylinder device contains air. During a reversible adiabatic process will the entropy of air increase or decrease or remain constant. (**Ans.** No change)
12. During a reversible process, the magnitudes of entropy changes of the system and surroundings are always equal. State whether it is true or false. (**Ans.** True)
13. Can the entropy of an ideal gas change during an iso-thermal process? Explain (**Ans.** Yes)
14. An ideal gas undergoes a process between two temperature T_1 and T_2 in two different ways firstly at constant volume and secondly at constant pressure. For which case the entropy change will be larger? Explain why (**Ans.** Larger in 2nd case)
15. Oyxgen gas is compressed in a piston cylinder assembly from initial state of 0.125 m^3/kg and 298 k to final state of 1.0 m^3/kg 560 K. Assuming constant average specific heat c_v as 0.69 kJ/kg · K and R = 0.2598 kJ/kg find the entropy change during the process. (**Ans.** –0.105 kJ/kg · k)
16. An insulated rigid tank contains 0.6 kg of carbon dioxide at 120 kPa pressure. Pedal work is done on the gas to raise the pressure in the tank to 144 kPa. Assuming constant average specific heat c_v as 0.657 kJ/kg · K and R = 0.286 kJ/kg · K. Find the entropy change in CO_2 during the process. (**Ans.** 0.07185 kJ/k)
17. An insulated piston-cylinder device contains air at 120 kPa and 290 K. Initially its volume is 0.3 m^3. It is now supplied heat with the help of a resistance heater placed in it for 15 minutes. The rating of the heater is 200 W. As a result of this the frictionless piston moves so that the volume changes but the pressure remain constant. Assuming the air to be an ideal gas having R = 0.287 kPa · m^3/kg · K and constant specific heat c_p = 1.02 kJ/kg · k make calculation to find out the final temperature of air and then calculate the change in entropy of the system. (**Ans.** 698 k, 0.187 kJ/K)

18. 0.8 kg of nitrogen gas is contained in a frictionless piston-cylinder device at 100 kPa and 300 K. It is now compressed slowly corresponding to the polytropic equation pv^n = constant (where n = 1.3) to an extent that its volume reduces to half of the original volume. If the specific heat is $c_{v,\ av}$ = 0.744 kJ/kg · and R = 0.297 kJ/kg · k find the change in entropy of N_2 in the process and the final temperature.

(Ans. T_2 = 369.3 k, ΔS_{N_2} = 0.331) m^2

19. A mass of 1.5 kg of helium is subjected to a certain process such that from its initial state 3 m^3/kg and 288 k to final state of 0.5 m^3/kg 353 k. Find the entropy change assuming (*a*) the process is reversible (*b*) the process is in reversible.

(Ans. (*a*) –4.6305 kg/k,
(*b*) some as in (*a*))

20. Two kg air is contained in a frinctionless piston-cylinder assembly from initial state of 100 kPa and 300 K to 250 kPa, the compression process being reversible isothermal. Find the entropy change of the air and to the work done. Take $c_{p,\ av}$ = 1.005 kJ/kg · k and R = 0.287 kJ/kg · k. **(Ans.** –147.8 kJ/k, –147.8 kJ/k)

❑❑❑

CHAPTER 7

GAS POWER CYCLES

7.1 INTRODUCTION

There are two important cases where the knowledge of thermodynamic is applied namely : (*i*) power generation and (*ii*) Refrigeration. The power generation and refrigeration systems operate on thermodynamic cycles which are called the Power Cycles and Refrigeration Cycles. In this chapter we shall deal with power cycles.

7.2 BASIC DEFINITION

7.2.1 Engines and Power Cycles

The devices used to produce net power output are called the 'Engines' and the thermodynamic cycles they operate on, are called the power cycles. These thermodynamic power cycles are further classified on the basis of phase of the working fluid that circulate through the cyclic device. If the working fluid remains in the gaseous phase throught the entire cycle it is called the Gas Power Cycle. Similarly if the working fluid exists in vapor form in one part of the cycle and liquid phase during other point, it is called Vapor Cycle.

7.2.2 Closed and Open Cycles

Gas power cycles are further catagorised into two catagories namely closed cycle and open cycle.

In the closed cycle, the working fluid is returned to the initial state at the and of the cycle as is the case of steam power plant. However in the open cycle the working fluid is renewed at the and of each cycle and it is not recirculated. In automobile engines, for example, the combustion gases are throw out at the end of the cycle and fresh air fuel mixture is sucked in during the next cycle. The engine operates in a mechanical cycle and the working fluid in such devices does not a gas through the complete thermodynamic cycle.

Most popular way to categorise the Heat Engines is as Internal Combustion or External Combustion Engines depending on how the heat is supplied to the working fluid. In the internal combustion engines heat is supplied by burning the fluid within the system (*e.g.*, automobiles engines). In the external combustion engines heat is supplied to the working fluid from out side external source (as in steam power plants). In such systems the heat is suplied by the furnance to the working fluid.

Internal combustion engines are further classified on the basis of the method used for igniting the fuel.

(*i*) Spark Iginition (Petrol) Engines

In such engines the combustible air-fuel mixture is drawn into he cylinder and compressed. After the compression an electric spark ignites the mixture and heat energy is liberated by the spontaneous exothermic chemical reaction and the temperature is raised to a very high value. The gases produced after the chemical reaction expand and in doing so work is done on the piston it moves within the cylinder which in. turn rotates the engine crank shaft. Then after doing work the gases are expelled from the cylinder to the atmosphere.

(*ii*) Compression Ignition (Diesel) Engines

In compression ignition engines air is drawn into the cylinder and compressed to a high pressure which results in raising the temperature of air above the ignition temperature of the fuel. Thus near the end of the compression process fuel is injected into the cylinder to be ignited quickly. The combustion products in the form of gases expand and work is done and in the final stroke the exhaust takes place so as to throw the exhaust gases into the surrounding atmosphere. Evidently the working fluid in both the spark and compression iginition engines, is thrown out to the atmosphere and never restored to the initial state rather fresh charge is sucked in during the suction stroke of the next cycle. Thus these engines execute only mechanical cycles and not thermodynamic cycle because in the true thermodynamic cycle the working fluid has eventually to return to its initial state.

7.3 BASIC ASSUMPTIONS IN THE ANALYSIS OF POWER CYCLES

As explanied above, most power producing devices operate on mechanical cycles and not on true thermodynamic cycles. The actual cycles encountered in these devices are very difficult to analyse because of the presence of friction and absence of sufficient time for the establishment of equilibrium conditions during the cycle. But still in order to analyse the actual cycle analytically some idealisation assumptions are made so that the internal irreversibilities and complexities are removed and the resulting cycle is approximately equivalent to the actual cycle but is entirely internally reversible and the working fluid is taken as air. Such a cycle is called an ideal cycle or standard cycle.

This simple idealised model is used to study the effects of major parameters which dominate the working of the various cycles.

7.3.1 Basic Definitions

(*i*) Cycle

A cycle is defined as a repeated set of operations in a particular order so that the process is repeated again and again in the same order.

(ii) Ideal Cycle

Ideal cycle is the cycle of an imaginary ideal engine in which there is no heat loss, no internal irreversibilities. The working substance is an ideal gas which comes back to its initial state at the end of the cycle. The ideal cycle is closed cycle.

(iii) Actual Cycle

The cycle of an actual engine is called the actual cycle. The working substance does not go through a complete cycle and it does not come back to initial state at the end of cycle rather it is thrown out and fresh change is sucked in at the start of next cycle. Actual cycle is an open cycle.

(iv) Air Standard Cycle

When air is assumed to be the working substance in internal combustion engines then these cycles are called Air-Standard Cycles. Following four are the most common air standard cycles.

(*a*) Otto cycle

(*b*) Diesel Cycle.

(*c*) Dual combustion cycled

(*d*) Bryton cycle.

In addition to these four cycles there are following few more gas cycles which have also been used but are not that popular.

(*i*) Stirling cycle

(*ii*) Erricson cycle

(*iii*) Atkinson cycle

(*iv*) Lenoir cycle

7.3.2 Air Standard Efficiency and Associated Assumptions

The analysis of the mechanical cycle of *I.C.*, engines (Petrol and Diesel) is based on the under mentioned assumptions :

1. The working fluid is air and it is taken as perfect gas.
2. The engine operates in a closed cycle and the working fluid is used repeatedly.
3. The properties of the working fluid at any state can be found by applying the characteristic gas equation. Further the working fluid is hemogeneous at all times and no chemical action takes place.
4. The compression and expansion processes are truely adiabatic and during these processes no heat is lost or gained.
5. All these processes are internally reversible and there is no frictional or mechanical loss.
6. The specific heats C_p, C_v as well as their ratio γ is constant.
7. The heat content of the fuel due to its combustion in the cylinder is replaced by the same quantity of heat supply from a heat source, brought in contact with the cylinder head at the designated time and the cylinder head behaves as perfect conductor during this operation. Thus, despite the fact that it is an internal combustion engine but with in air as working substance the heat added from an external imaginary source and not due to combustion of fuel inside the cylinder.
8. The exhaust process, is similarly air replaced by an equivalent amount of heat rejection to heat sink so that the working fluid (*i.e.*, air) is restored back to its orginal state.
9. Both, the heat source and heat sink are of infinitely large capacity so that due to heat transfer to or from them their temperature remains constant.

7.3.3 Efficiency

The efficiency of the engine with working substance as perfect gas coupled with the assumption mentioned above is given by the following expression.

$$\text{Efficiency} \quad \eta = \frac{\text{Workdone}}{\text{Heat supplied}} \qquad ...(7.1)$$

Let Q_1 = Heat supplied kJ

Q_2 = Heat rejected kJ

and work done = (Heat supplied-Heat rejected). Then expression (7.1) reduces to

$$\eta = \frac{Q_1 - Q_2}{Q_1} \qquad ...(7.2)$$

The efficiency given by the expression (7.2) is known as the thermal efficiency of the cycle also known as Air Standard Efficiency under assumed ideal conditions certainly the air standard efficiency of the cycle will be higher than the actual efficiency of the engine. Accordingly the concept of relative efficiency is introduced wihile studying the relative performance of engines. It is defined as

$$\text{Relative efficiency} = \frac{\text{Actual Thermal Efficiency}}{\text{Air Standard Efficiency}} \qquad ...(7.3)$$

7.4 NOMENCLATURE FOR RECIPROCATING ENGINES

Reciprocating engines are basically piston-cylinder devices and have wide range of applications in majority of automobiles, trucks, light aircraft, and even in electric power generators and so many other such devices.

The basic components of a vertical reciprocating engine are shown in Fig. 7.1. The piston reciprocates in the cylinder between two fixed pistions TDC and BDC, called the Top Dead Centre and Bottom Dead Centre in the vertical engine. (However in a Horizontal Engine these fixed positions are ODC and IDC called the Outer Dead Centre and Inner Dead Centre).

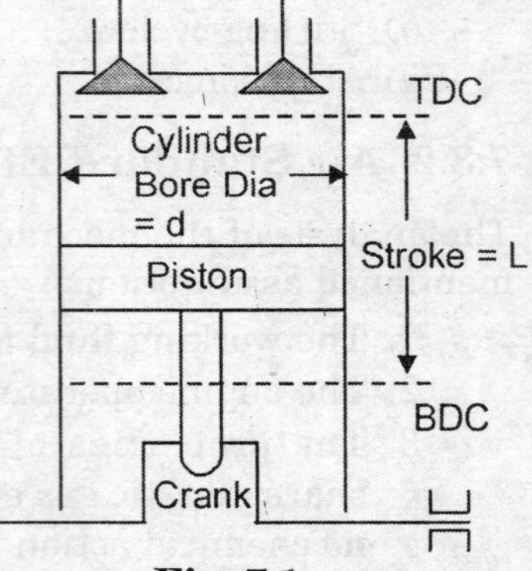

Fig. 7.1

In the case of vertical engine the piston will be at BDC it will form the largest volume in the cylinder while when at TDC it will form the smallest volume in the cylinder. The corresponding positions in the horizontal engine will be IDC and ODC). The distance between the TDC and BDC is the largest distance the piston can move in the cylinder and it is called the Stroke of the engine. The stroke is equal to twice the throw of the crank or twice the crank radius. Alternatively the crank radius is half the stroke length. Other basic definitions are as follows :

(*i*) Piston Displacement or Swept Volume

Swept volume is the volume swept by the piston in one stroke. It is equal to the volume swept by the piston when it moves from TDC to BDC and mathematically it is given as

$$\text{Swept volume by} \quad = \frac{\pi}{4} d^2 L \qquad ...(7.4)$$

Where d = Position or cylinder diameter

L = Strike length

(ii) Clearance Volume

The volume between the piston and cylinder head when the piston is in TDC position is called the clearance volume. This is generally expressed as the percentage of swept volume.

(iii) Compression Ratio

The ratio of the total volume to the clearance volume is called the compression ratio. Alternatively it is also defined as the ratio of the volume when the piston is at BDC to the volume when it is at TDC. That is, if '*r*' compression ratio

$$r = \frac{\text{Total volume}}{\text{Clearance volume}}$$

$$= \frac{\text{Swept volume + Clearance volume}}{\text{Clearance volume}}$$

$$r = \frac{V_S + V_C}{V_C} = 1 + \frac{V_S}{V_C} \quad \text{...(7.5)}$$

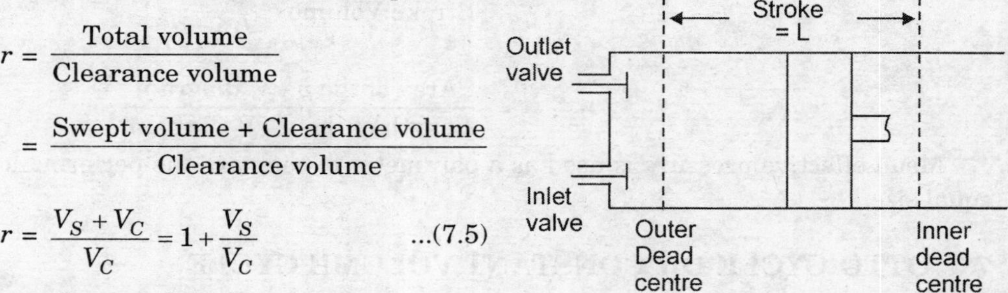

Fig. 7.2

Where V_S = Swept volume

V_C = Clearance volume

It will be seen later in the chapter that higher the compression ratio, letter will be the efficiency of the engine.

(iv) Piston Speed

When the crank speed is uniform the linear speed of the piston will be varying. It will be zero when it is at TDC or BDC. Its speed will be maximum when it will be at the middle of the stroke. Even through the piston speed is variable, as explained above the average piston speed is defined as follows :

Piston speed = Distance moved by piston in one minute

= (Distance travelled by piston in one revolution of the crank) × (Angular Speed of Crank in revolutions per minute r.p.m.)

= (Distance travelled from BDC to TDC) + Distance travelled from TDC to BDC) × (r.p.m of Engine)

= $(L + L)\,N$

$$v_{\text{piston}} = 2LN \quad \text{...(7.6)}$$

(v) Mean Effective Pressure

Another term which is extensively used with the reciprocating engines is the mean effective pressure. It is defined as the fictitious pressure which if acted on the piston during the entire power stroke would produce the same amount of work as that produced during the actual cycle.

Mathematically,

$$W_{\text{net}} = p_m(\text{Displacement Volume})$$

Where W_{net} = The net amount of work produced during the cycle

p_m = Mean effective pressure

or

$$W_{\text{net}} = p_m\,(\text{displacement Volume})$$

$$= p_m \times V_S$$

or $$p_m = \frac{W_{net}}{V_S} \qquad [\because V_S = V_{total} - V_C]$$

$$= \frac{W_{net}}{V_{total} - V_C} \qquad ...(7.7a)$$

Also p_m can be expressed in another form as

$$p_m = \frac{W_{net}}{\text{Stroke Volume}} = \frac{W_{net}}{L \times \frac{\pi}{4} d^2}$$

$$= \frac{\text{Area of the } p - V \text{ diagram}}{\text{Length of the } p - V \text{ diagram}} \qquad ...(7.7b)$$

Mean effective pressure is used as a parameter to compare the performance of engines of equal size.

7.5 OTTO CYCLE OR CONSTANT VOLUME CYCLE

The Otto cycle is an ideal cycle for spark ignition reciprocating engine. It is named after Nikolous A. Otto who built a successful four stroke internal combustion engine in 1876 in Germany.

The Otto cycle consists of two adiabatic and two constant volume processes. Fig 7.3 shows Otto cycles on P-V. and T- S. space respectively. Sequence of processes is as follows :

To strat with both the intake and exhaust values are closed and the piston is at BDC position in vertical engines (and at IDc in the chase of horizontal engines) The ideal gas (air fuel mixture) is at state 1 When the crank rotates the piston from BDC to TDC and the

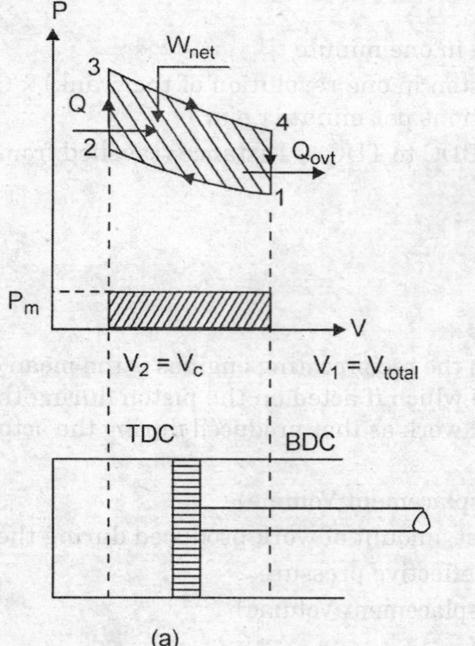

(a)

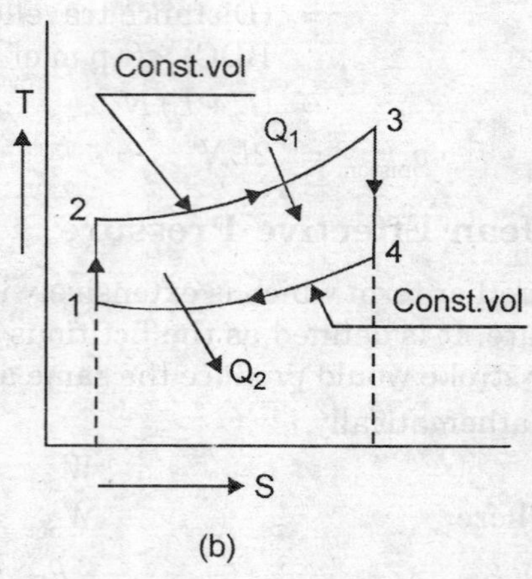

(b)

Fig. 7.3

gas is compressed Shortly before the piston reaches the highest point (TDC) the spark plug fires and the mixture ignites increasing the pressure and temperature of the system. The high pressure gases force the piston down which in turn forces the crank shaft to roate thereby producing useful work output during the stroke 2-3 which is called the power stroke or expansion stroke at the end of the stroke the piston is at its lowermost position BDC and the first mechanical cycle is complete. At this stage the combustion gas is still in the cylinder. Then the piston moves up one more time thereby forcing the combustion gas out through the outlet value. This stroke is called the exhaust stroke. It then moves down second time thereby drawing in fresh-air-fuel mixture though the inlet value to commence the second cycle and so on.

Above mentioned mechanical cycle is replaced by the imaginary thermodynamic internally reversible cycle with the purpose of simplifying the analyses by utilising air-standard assumption. Thus the Otto Cycle is assumed to consist of following four internally reversible process.

(i) **Process 1-2 :** Isentropic compression.

(ii) **Process 2-3 :** Shortly before the piston reaches the highest position (TDC) the spark plug forces the mixture to ignite which in turn raises the pressure and temperature of system instantly. Since near the TDC the velocity of the piston is very low, it can be assumed that heat is added at constant volume.

Thus during the process 2-3 heat is added at constant volume and the pressure rises form p_2 to p_3.

(iii) **Process 3-4 :** The increased pressure exerts greater force on the piston and pushes it down to BDC and the working fluid expands ISENTROPICALLY and V_4/V_3 is called the isentropic volumetric expansion.

(iv) **Process 4-1 :** Here again the piston is momentarily at rest at BDC and heat is rejected to external sink (here atmospheric air). During this stroke which is called the exhaust stroke during which the burnt gases are thrown out. Next cycle then starts by sucking the fuel air mixture & then the compression process.

7.5.1 Derivation of Expression for Air Standard Thermal Efficiency of Otto Cycle

Process 1-2 is Isentropic compression. Therefore during this process there is no heat interaction Further it is assumed that Otto Cycle is executed in a closed system. Thus from first law of thermodynamics we have on unit mass basis.

$$q - w = \Delta u \text{ (kJ/kg)} \quad \text{...(7.8)}$$

Heat is supplied during process 2-3 and rejected during the process 4-1. Both of these processes are at constant volume there will be no work involved *i.e.*, $w = 0$ during these two processes.

Similarly during the process 1-2 and process 3-4 being isentropic processes there is no heat interaction but only work is spent during process 1-2 and gained during process 3-4.

Now heat supplied during the process 2-3.

$$Q_1 = C_v(T_3 - T_2) \quad \text{...(7.9)}$$

And heat rejected during the process 4-1

$$Q_2 = C_v(T_4 - T_1) \quad \text{...(7.10)}$$

Therefore workdone

$$= \text{(Heat supplied)} - \text{(Heat rejected)}$$

$$= C_v(T_3 - T_2) - C_v(T_4 - T_1) \quad \text{...(7.11)}$$

Therefore thermal efficiency

$$\eta = \frac{\text{Work done}}{\text{Heat supplied}}$$

$$= \frac{C_v(T_3 - T_2) - C_v(T_4 - T_1)}{C_v(T_3 - T_2)} \quad \text{...(7.12)}$$

Assuming constant specific heat equation (7.12) reduces to

$$\eta = \frac{(T_3 - T_2) - (T_4 - T_1)}{(T_3 - T_2)}$$

or $$\eta = 1 - \left(\frac{T_4 - T_1}{T_3 - T_2}\right) \quad \text{...(7.13a)}$$

$$= 1 - \frac{T_1\left(\frac{T_4}{T_1} - 1\right)}{T_2\left(\frac{T_3}{T_2} - 1\right)} \quad \text{...(7.13b)}$$

The expression (7.13) can be re-written in terms of the compression ratio $r_c = \frac{V_1}{V_2}$ and expression ration $r_e = \frac{V_4}{V_3}$ But in Otto cycle we have at point 2.

$$\frac{V_1}{V_2} = \frac{V_4}{V_3}\, r(\text{say})$$

i.e., $$r_c = r_e = r \quad \text{...(7.14)}$$

Now for the Isentropic process 1-2

$$P_1 V_1^{\gamma} = P_2 V_2^{\gamma}$$

or $$P_2 = P_1\left(\frac{V_1}{V_2}\right)^{\gamma}$$

$$= P_1 r^{\gamma} \quad \text{...(7.15)}$$

$$\frac{P_1 V_1}{T_1} = \frac{P_2 V_2}{T_2}$$

i.e., $$\frac{T_1}{T_2} = \frac{P_1 V_1}{P_2 V_2}$$

or $$\left(\frac{T_1}{T_2}\right) = \left(\frac{V_1}{V_2}\right)\left(\frac{1}{r}\right)^{\gamma}$$

$$= \frac{r}{(r)^{\gamma}} = r^{(1-\gamma)}$$

or $$T_2 = T_1 (r)^{\gamma - 1} \quad \text{...(7.16)}$$

Again at Point 3

$$\frac{V_3}{V_2} = 1 \quad \text{or} \quad V_3 = V_2 = \frac{V_1}{r} \quad \text{...(7.17)}$$

Similarly for the expension process 3-4 we have for isentropic expension (by replacing T_2) with T_4 and T_1 with T_3 in equation (7.16), we get

$$\left(\frac{T_3}{T_4}\right) = \left(\frac{V_4}{V_3}\right)^{(\gamma-1)}$$

or $$T_3 = T_4(r)^{(\gamma-1)} \quad ...(7.18)$$

And $$\frac{P_4}{P_3} = \left(\frac{V_3}{V_4}\right)^{\gamma}$$

or $$P_4 = P_3(r)^{\gamma} \quad ...(7.19)$$

Therefore substituting values from equation (7.16) and (7.18) into equation (7.1)

$$\eta = 1 - \left(\frac{T_4 - T_1}{T_3 - T_2}\right)$$

$$= 1 - \left[\frac{T_4 - T_1}{T_4(r)^{\gamma-1} - T_1(r)^{\gamma-1}}\right]$$

$$= 1 - \frac{(T_4 - T_1)}{(r^{\gamma-1})\,(T_4 - T_1)}$$

Thus air standard efficiency of Otto Cycle is

$$\eta_{otto} = 1 - \frac{1}{(r)^{\gamma-1}} \quad ...(7.20)$$

Equation (7.20) gives the value of efficiency which is known as AIR STANDARD EFFICIENCY OF OTTO CYCLE. Equation (7.20) shows that based on the air standard assumptions the thermal efficiency of otto cycle is dependent on compression ratio and the specific heat ratio $\gamma(= C_p/C_v)$.

The thermal efficiency of ideal Otto Cycle increases with both r and γ. However for a given set of values of r and γ the thermal efficiency of actual spark-ignition engine will be less than the ideal Otto Cycle because of irreversibilities such as friction and other factors like incomplete combustion and variation of γ with temperature.

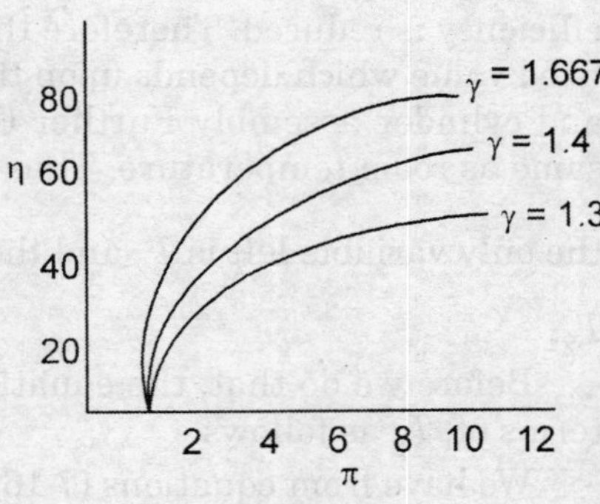

Fig. 7.4

From the curves shown in Fig. 7.4 that the thermal efficiency curve is steep at low values of r and it gets flatter after the compression ratio value of about 8. From this we infer that these is no pronounced increase in efficiency at higher than 8 compression ratios. Further when high compression ratios are used the temperature of airfuel mixture rises above the auto-ignition temperature of the fuel (the temperature at which the fuel ignites without the help of spark) during the compression stroke causing an early and quick burning of the fuel. This premature ignition of the fuel is called the AUTOIGNITION which produces an audible noise which is called the ENGINE KNOCK. This in turn badly effects not only the performance of the engine but can cause permanent damage to the engine. Thus this in its own turn puts an upper limit on the compression ratio that can be used in spark ignition internal combustion engines. However the problem of auto ignition was overcome by blending the petrol with tetra ethyl lead and this practice continued till the end of sixth decade of the last century when the people realised that the exhaust gases from the petrol engines were injurious to health and in most of the western countries the lead based additives were banned The loss of efficiency due to non increase of compression ratio was compensated by advanced research in the field of aero-dynamic body design and body weight reduction by employing new light weight but equally strong materials.

Now coming to the effect of γ, the specific heat ratio on the thermal efficiency for a given compression ratio. The value of γ will depend upon the working fluid. If it is mono-automic gas $\gamma = 1.667$. In case the working fluid is air $v = 1.4$ at room temperature and for CO_2 it is 1.3 and for ethane its value is 1.2. Since for larger molecules the value is lower, the efficiency will be lower. In addition v also decreases with the increase in temperature. In actual engine the working fluid is a mixture of gases and that two at higher temperature which results lower thermal efficiencies of otto cycle. The ratio of actual thermal efficiency to the air standard thermal efficiency is known as Relative Efficiency or Efficiency ratio.

$$ie., \text{ Relative efficienty} = \frac{\text{Actual Efficiency}}{\text{Air Standard Efficiency}}$$

7.5.3 Condition for Maximum Work

From equation (7.11) we have the (7.20*b*) expression for work done

$$W = C_v(T_3 - T_2) - C_v(T_4 - T_1) \qquad ...(7.11)$$

In actual engines working an Otto Cycle the compression ratio is between 5 to 8, corresponding to working fluid as air $\gamma = 1.4$.As for as possible it is kept near the higher limit depending upon the fuel and other design factors. If we intend to raise its value above 8, it results in higher temperature T_3 which in turn lowers the actual value of γ and thus the efficiency is reduced. Therefore the temperature T_3 after ignition of the mixture is a maximum fixed value which depends upon the characteristics of the material used for fabrication of piston and cylinder assembly Further the initial temperature T_1 of the air-fuel mixture is also the same as room temperature. Thus in equation (7.11) which is the expression for the work done, the only variable left is T_2 and the work done is maximised by setting $\frac{dW}{dT_2} = 0$ and solving for T_2.

Before we do that, the equation (7.11) is re-written by substituting values of T_1 and T_4 in terms of, T_2 as follows.

We have from equations (7.16) and (7.18)

$$T_2 = T_1(r)^{(\gamma - 1)} \qquad ...(7.17)$$

$$T_3 = T_4(r)^{(\gamma - 1)} \qquad ...(7.18)$$

Dividing equation (7.18) by (7.16)

$$\frac{T_3}{T_2} = \frac{T_4}{T_1}$$

or $$T_4 = \frac{T_1 T_3}{T_2} \qquad ...(7.19a)$$

or $$T_2 = \frac{T_1 T_3}{T_4} \qquad ...(7.19b)$$

Substituting value of T_4 from equation (7.19*a*) with equation (7.11) we get

$$\frac{W}{C_v} = (T_3 - T_2) - \left[\frac{T_1 T_3}{T_2} - T_1\right]$$

$$= (T_3 + T_1) - \left[T_2 + \frac{T_1 T_3}{T_2}\right] \qquad ...(7.20)$$

Differentiating equation (7.20) w.r.t. T_2 and equating it to zero, we have for maximum work

$$\frac{1}{C_v}\frac{dW}{dT_2} = 0 - 1 + \frac{T_1 T_3}{T_2^2} = 0$$

or $$T_1 T_3 = T_2^2$$

i.e., $$T_2 = \sqrt{T_1 T_3} \qquad ...(7.21)$$

The equation (7.11) can also be written in yet another form by taking the value of T_2 as in equation (7.19*b*)

i.e., $$\frac{W}{C_v} = (T_3 - T_2) - [T_4 - T_1]$$

$$= \left(T_3 - \frac{T_1 T_3}{T_4}\right) - (T_4 - T_1)$$

$$= (T_3 + T_1) - \left(T_4 + \frac{T_1 T_3}{T_4}\right) \qquad ...(7.22)$$

Differentiating equation (7.22) w.r.t. T_4 and equating it to zero for maximum work we get

$$\frac{1}{C_v}\frac{dw}{dT_4} = 0 - 1 + \frac{T_1 T_3}{T_4} = 0$$

or $$T_4^2 = T_1 T_3$$

or $$T_4 = \sqrt{T_1 T_3} \qquad ...(7.23a)$$

From equation (7.21) and (7.23) we see

$$T_2 = T_4 = \sqrt{T_1 T_3} \qquad ...(7.23b)$$

In other words the intermediate temperature T_2 and T_4 must be equal and the value should be geometric mean of T_1 and T_3. Thus substituting these values of T_1 and T_4 into equation (7.17) we have

$$W_{max} = C_v(T_3 - T_2 - T_4 + T_1)$$

$$= C_v(T_3 - 2\sqrt{T_1 T_3} + T_1)$$

Also from (7.16), (7.18), (7.19), (7.21), (7.22)

$$T_2 = T_4 = \sqrt{T_1 T_3}$$

or $$T_2^2 = T_1\, T_3 \qquad ...(7.24)$$

Substituting value of T_2 from (7.17)

$$\left[T_1 (r)^{\gamma - 1}\right]^2 = T_1 T_3$$

or $$T_1^2 (r)^{2(\gamma-1)} = T_1 T_3$$

or $$\frac{T_3}{T_1} = r^{2(\gamma - 1)}$$

or $$r = \left(\frac{T_3}{T_1}\right)^{\frac{1}{2(\gamma-1)}} \quad ...(7.25)$$

7.5.4 Mean Effective Pressure

As per the definition, the expression for the mean effective pressure p_m is

$$p_m = \frac{\text{Work done per cycle}}{\text{Swept volume}}$$

$$= \frac{C_v (T_3 - T_2) - C_v (T_4 - T_1)}{(V_1 - V_2)} \quad \text{[Refer eq. 7.11]} \quad ...(7.26)$$

Now swept volume $= V_1 - V_2$

$$= V_1\left(1 - \frac{V_2}{V_1}\right) = \frac{RT_1}{p_1}\left(1 - \frac{1}{r}\right)$$

where r = compression ratio $= \frac{V_1}{V_2}$

or swept volume $$= \frac{RT_1}{p_1 r}(r - 1) \quad ...(7.27)$$

Therefore from (7.25) and (7.26)

$$p_m = \frac{C_v[(T_3 - T_2) - (T_4 - T_1)]}{\frac{RT_1}{p_1 r}(r-1)}$$

$$= \frac{C_v p_1 r}{RT_1 (r-1)}[(T_3 - T_2) - (T_4 - T_1)] \quad ...(7.28)$$

Let constant volume pressure ratio *i.e.*, $\frac{p_3}{p_2}$ be represented by r_p

i.e., $$r_p = \frac{p_3}{p_2} \quad ...(7.29)$$

Then the expression (7.28) for p_m can be expressed in terms of v, r and r_p as follows :
We have from equation (7.16) and (7.18)

$$T_2 = T_1 (r)^{(\gamma - 1)} \quad ...(7.16)$$
$$T_3 = T_4 (r)^{(\gamma - 1)}$$

or $$T_4 = \frac{T_3}{(r)^{\gamma-1}} \quad ...(7.18)$$

multiplying (7.16) and (7.18) we get

$$T_2T_4 = T_1T_3$$

or $$T_4 = \frac{T_1}{T_2}T_3 \quad ...(7.30)$$

Again for the constant volume process 2-3 we have

$$\frac{p_2}{T_2} = \frac{p_3}{T_3}$$

or $$\frac{T_3}{T_2} = \frac{p_3}{p_2} = r_p$$

or $$T_3 = r_p T_2$$

or $$\frac{T_3}{T_2} = r_p \quad ...(7.31)$$

Therefore from (7.30) and (7.31*a*) we get

$$T_4 = T_1 r_p \quad ...(7.32)$$

Again from (7.18) and (7.32) we have

$$T_3 = T_4(r)^{\gamma-1}$$
$$= T_1 r_p (r)^{\gamma-1} \quad ...(7.33)$$

Thus substituting the values of T_2, T_3 and T_4 from equations (7.16), (7.23) and (7.31) respectively in terms of T_1 the equation (7.28) reduces to

$$p_m = \frac{C_v p_1 r}{RT_1(r-1)}\left[T_1 r_p (r)^{\gamma-1} - T_1(r)^{\gamma-1} - T_1 r_p + T_1\right]$$

$$= \frac{C_v p_1 r}{R(r-1)}\left[(r)^{(\gamma-1)}(r_p - 1) - (r_p - 1)\right]$$

or $$p_m = \frac{C_v p_1, r}{R(r-1)}[(r_p - 1)(r^{\gamma-1} - 1) \quad ...(7.34)$$

Now we have $$C_p - C_v = R$$

or $$\frac{C_p}{C_v} - 1 = \frac{R}{C_v}$$

or $$\gamma - 1 = \frac{R}{C_v}$$

i.e., $$\frac{C_v}{R} = \left(\frac{1}{\gamma-1}\right)$$

Therefore equation (7.34) reduces to

$$p_m = \frac{p_1 r}{(\gamma-1)(r-1)}[(r_p - 1)(r^{\gamma-1} - 1)] \quad ...(7.35)$$

Expression (7.35) is an alternative form for mean effective pressure in terms of r, v and r_p.

Example 7.1. *An engine working on otto cycle has compression ratio 8. It $\gamma = 1.4$, what is the thermal efficiency of the engines.*

Solution : Given $\gamma = 1.4 \qquad r = 8$

$$\eta = 1-\frac{1}{r^{\upsilon-1}} = 1-\frac{1}{8^{1.4-1}}$$

$$= 1-\frac{1}{8^{0.4}} = 1-\frac{1}{2.2974} = 0.5647 \text{ or } 56.47\%$$

Example 7.2. *An engine working on Otto cycle has the following data :*

Cylinder diameter = 1400 mm
Stroke = 1280 mm
Clearance volume = 0.28 m³

Calculate the compression ratio and the air standard efficiency of the engine if $\gamma = 1.4$.

Solution : (*a*) Given d = 1400 mm, L = 1280, V_c = 0.28 m³

Swept volume = $$V_3 = \frac{\pi}{4}d^2 l$$

$$= \frac{\pi}{4}\times(1400)^2\times 1280$$

$$= 19.7 \times 10^8 \text{ mm}^3$$

$$= 1.97 \text{ m}^3$$

Now $$r = \frac{V_S + V_C}{V_C} = \frac{1.97+.28}{0.28}$$

$$= 8.03 \textbf{ Ans.}$$

(*b*) $$\eta = 1-\frac{1}{r^{\gamma-1}}$$

$$= 1-\frac{1}{2.3008} = 0.5654$$

i.e., $$= 56.54\%$$

Example 7.3. *In an engine working an Otto cycle, energy generated during the expansion stroke is twice that of rejected during the exhaust stroke. If γ = 1.4, calculate the thermal efficiency of the engine and its compression ratio.*

Solution : (*a*) $$\eta = \frac{Q_1 - Q_2}{Q_1}$$

But it is given $$Q_1 = 2Q_2$$

$\therefore$ $$\eta = \frac{2Q_2 - Q_2}{2Q_2} = \frac{Q_2}{2Q_2} = 0.5 \quad \textit{i.e.}, 50\%$$

(*b*) Now $$\eta = 1-\frac{1}{r^{\gamma-1}}$$

$$0.5 = 1-\frac{1}{(r)^{0.4}}$$

$$\frac{1}{(r)^{0.4}} = 1 - 0.5 = 0.5$$

$$\therefore \quad r^{0.4} = \frac{1}{0.5} = 2$$

$$\text{or} \quad r = (2)^{\frac{1}{0.4}} = (2)^{2.5} = 5.66$$

Example 7.4. *An Otto cycle engine has a compression ratio 7.5. At the start of compression stroke the temperature and pressure are 293 K and 100 kPa. The maximum pressure is limited to 4500 kPa. Determine the temperature, pressure and volume of air at the four point 1, 2, 3, 4 of the P–V. diagram of the cycle on the basis of 1 kg of air if*

$$R = 0.287\ kJ/kg\ K \text{ and } v = 1.6$$

Also find the ratio of heat supplied to heat rejected.

Solution : P–V. diagram of the Otto cycle for the problem is shown in Fig. 7.5

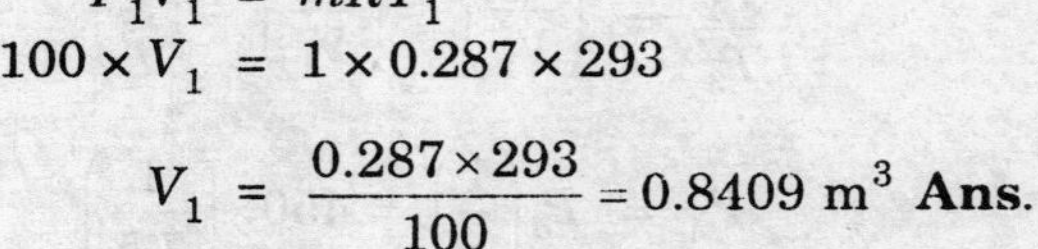

Fig. 7.5

Given Data :

$$P_1 = 100 \text{ kPa}$$
$$T_1 = 293 \text{ K}$$
$$r = 7.5$$
$$P_3 = 4500 \text{ kPa}$$
$$m = 1 \text{ kg}$$

At the Point 1 : We have from gas equation

$$P_1V_1 = mRT_1$$
$$100 \times V_1 = 1 \times 0.287 \times 293$$

$$\therefore \quad V_1 = \frac{0.287 \times 293}{100} = 0.8409 \text{ m}^3 \textbf{ Ans.}$$

At Point 2 : The process 1 – 2 is an isentropic compression

$$P_1V_1^{\gamma} = P_2V_2^{\gamma}$$

$$\frac{P_2}{P_1} = \left(\frac{V_1}{V_2}\right)^{\gamma} = (r)^{\gamma}$$

$$P_2 = P_1(r)^{\gamma} = 100 \times (7.5)^{1.4}$$
$$= 1679.1 \text{ kPa } \textbf{Ans.}$$

$$\text{Also} \quad \frac{P_1V_1}{T_1} = \frac{P_2V_2}{T_2}$$

$$\text{or} \quad \frac{T_2}{T_1} = \frac{P_2V_2}{V_1}$$

$$\text{or} \quad \frac{T_2}{T_1} = \left(\frac{P_2}{P_1}\right)\left(\frac{V_2}{V_1}\right)$$

$$= (r)^{\gamma}\left(\frac{1}{r}\right) = (r)^{\gamma-1}$$

$$T_2 = T_1\, r^{(\gamma-1)} = 293 \times (7.5)^{0.4} = 656 \text{ K } \textbf{Ans.}$$

$$\text{and} \quad V_2 = \left(\frac{1}{r}\right)V_1 = \left[\frac{0.8409}{7.5}\right] = 0.1121 \text{ m}^3 \textbf{ Ans.}$$

Process 2-3 is constant volume process

At Point 3 :

$$P_3 = 4500 \text{ kPa}$$

$$V_3 = V_2 = 0.1121 \text{ m}^3$$

Now
$$\frac{P_2V_2}{T_2} = \frac{P_3V_3}{T_3}$$

or
$$\frac{T_3}{T_2} = \frac{P_3}{P_2}$$

or
$$T_3 = T_3\frac{P_3}{P_2}$$

$$= T_2\left(\frac{4500}{1679.1}\right)$$

$$= 2.68\ T_2 = 2.68 \times 656 = \mathbf{1758\ K}\ \textit{\textbf{Ans.}}$$

Process 3-4 is isentropic expansion

$$P_3V_3^{\gamma} = P_4V_4^{\gamma}$$

$$= P_4V_1^{\gamma} \qquad \because V_4 - V_1$$

$$\because \quad \frac{P_4}{P_3} = \left(\frac{V_3}{V_1}\right)^{\gamma} = \left(\frac{V_2}{V_1}\right)^{\gamma} = \left(\frac{1}{r}\right)^{\gamma} \qquad \because V_3 = V_2$$

$$\therefore \quad P_4 = P_3\left(\frac{1}{r}\right)^{\gamma} = 450 \times \left(\frac{1}{7.5}\right)^{1.4}$$

$$= \frac{450}{16.7913} = 26.8 \text{ kPa}$$

Also
$$\frac{P_3V_3}{T_3} = \frac{P_4V_4}{T_4} \quad \text{and} \quad P_3V_3^{\gamma} = P_4V_4^{\gamma}$$

$$\because \quad \frac{T_4}{T_3} = \frac{P_4}{P_3}\frac{V_4}{V_3} \quad \text{or} \quad \frac{P_4}{P_3} = \left(\frac{V_3}{V_4}\right)^{\gamma}$$

$$= \left(\frac{V_3}{V_4}\right)^{\gamma}\left(\frac{V_4}{V_3}\right) = \left(\frac{V_3}{V_4}\right)^{\gamma-1}$$

$$= \left(\frac{V_2}{V_1}\right)^{\gamma-1} = \left(\frac{1}{r}\right)^{\gamma-1}$$

$$T_4 = T_3\left(\frac{1}{r}\right)^{\gamma-1} = 1758 \times \left(\frac{1}{7.5}\right)^{0.4}$$

$$= \frac{1758}{2.2358} = \mathbf{785.24}\ K\ \textbf{Ans.}$$

(*b*) Heat supplied
$$Q_1 = C_v\,(T_3 - T_2)$$

Now

$$C_p - C_v = R$$

$$\frac{C_p}{C_v} - 1 = \frac{R}{C_v}$$

$$\gamma - 1 = \frac{R}{C_v}$$

i.e.,

$$C_v = \frac{R}{\gamma - 1} = \frac{0.287}{1.4 - 1} = \frac{0.287}{0.4}$$

$$= 0.7175 \text{ kJ/kg K}$$

∴ Heat supplied

$$Q_1 = C_v(T_3 - T_2)$$

$$= 0.7175\,(1758 - 656)$$

$$= 790.685 \text{ kJ}$$

Heat rejected

$$Q_2 = C_v(T_4 - T_1) = 0.7175\,(785.24 - 293)$$

$$= 353.18 \text{ kJ}$$

Therefore ratio

$$\frac{\text{Heat supplied}}{\text{Heat rejected}} = \frac{790\text{-}685}{353.18}$$

$$= 2.239$$

Example 7.5. *(a) An engine working on Otto cycle has air as its working fluid. Its compression ratio is 7.5. It takes in air at pressure of 100 kPa and temperature of 17°C. The energy released to air in each cycle is 1800 kJ/kg.*

Determine its efficiency and the net work output per kg of air take $\gamma = 1.4$

(b) It is intended to increase the net work output 10% by raising the compression ratio. Estimate the new compression ratio and discuss the result.

Solution : (*a*) Given data

$$r = 7.5,\ \gamma = 1.4$$

$$T_1 = 273 + 17 = 290 \text{ K}$$

$$P_1 = 100 \text{ kPa}$$

$$\eta = 1 - \frac{1}{r^{(\gamma - 1)}} = 1 - \frac{1}{(7.5)^{1.4-1}}$$

$$= 1 - \frac{1}{(7.5)^{0.4}}$$

or

$$\eta = 1 - \frac{1}{2.2388} = 0.5533 \quad i.e., \quad 55.33\%$$

Also

$$\eta = \frac{\text{Work output}}{\text{Heat supplied}}$$

∴ Work out put

$$W = \eta \times \text{Heat supplied}$$

$$= 0.5533 \times 1800$$

$$= 995.94 \text{ kJ/kg}$$

(*b*) Desired new work out put

$$W' = 995.94 \times 1.1$$

$$= 1095.534 \text{ kJ/kg}$$

$\therefore$ new $\qquad \eta_2 = \dfrac{1095.534}{1800} = 0.6086$

If r_2 = new compression ratio

Then $\qquad 0.6056 = 1 - \dfrac{1}{r_2^{\gamma-1}}$

or $\qquad \left(\dfrac{1}{r_2}\right)^{\gamma-1} = 1 - 0.6086 = 0.3914$

$$\left(\frac{1}{r_2}\right) = (0.3914)^{\frac{1}{0.4}} = 0.09584$$

$$r_2 = 10.434$$

Therefore percentage increased in r

$$= \frac{r_2 - r_1}{r_1} = \frac{10.434 - 7.5}{7.5}$$

$$= 0.3912 \text{ or } 39.12\ \%$$

The result shows that for a mere 10% increase in net work output the compression ratio should be increased by 39.12%. This aspect has already been discussed in article 7.5.1 in which with reference to Fig. (7.4) it was seen that near the value of $r = 8$ the η curve becomes flatter there by indicating that for a small increase in η the relative increase in compression ratio is very large.

Example 7.6. *An air standard Otto cycle has been designed for maximum cycle pressure of 4500 kPa and maximum cycle temperature of 2200 K. The ambient pressure and temperature are 100 kPa and 27°C.*

Calculate the designed net work output per unit mass of air and the thermal efficiency. Also find the mean effective pressure.

Take for air $C_v = 0.718\ kJ/kg \cdot K$ and $R = 0.287\ kJ/kg \cdot K$.

Solution : Refer to Fig 7.5

Given data $\qquad P_3 = 4500$ kPa, $\qquad T_3 = 2000$ K

$\qquad P_1 = 100$ kPa $\qquad T_1 = 27 + 273 = 300$ K

Compression ratio $\qquad r = \dfrac{V_1}{V_2}$

But $\qquad \dfrac{P_1 V_1}{T_1} = \dfrac{P_2 V_2}{T_2}$

$$\frac{V_1}{V_2} = \frac{P_2}{P_1}\frac{T_1}{T_2}$$

$$\therefore \qquad r = \left(\frac{P_2}{P_1}\right)\left(\frac{T_1}{T_2}\right) = \left(\frac{P_2}{T_2}\right)\left(\frac{T_1}{P_1}\right) \qquad \text{...(A)}$$

Also for constant volume process 2-3

$$\frac{P_2}{T_2} = \frac{P_3}{T_3} \qquad \text{...(B)}$$

Therefore from (A) and (B)

$$r = \left(\frac{P_3}{T_3}\right)\left(\frac{T_1}{P_1}\right)$$

$$= \left(\frac{4500}{2000}\right)\left(\frac{300}{100}\right)$$

$$= 6.75$$

Then $$\eta = 1 - \frac{1}{r^{(\gamma-1)}}$$

$$= 1 - \frac{1}{(6.75)^{0.4}}$$

$$= 1 - \frac{1}{2.1465} = 0.5341 \quad \textit{i.e.,}\ 53.41\%$$

For adiabatic process 1-2

$$T_1 = T_1\left(\frac{V_1}{V_2}\right)^{\gamma-1}$$

$$= 300\,(r)^{\gamma-1} = 300 \times (6.75)^{0.4}$$

$$= 643.94 \text{ K}$$

Heat supplied per unit mass of air

$$Q_1 = C_v - (T_3 - T_2)$$

$$= 0.718\,(2000 - 673.94)$$

$$= 973.65 \text{ kJ/kg}$$

Net work Done $$= \eta Q_1$$

$$= 0.5341 \times 973.65$$

$$= 520.03 \text{ kJ/kg}$$

(b) Mean effective pressure $$= \frac{\text{Work done per cycle}}{\text{Swept volume}}$$

$$\text{m.e.p} = \frac{W_{\text{net}} \text{ per unit mass of air}}{(v_1 - v_2)} \quad ...(C)$$

Now at point 1 $$P_1 v_1 = RT_1$$

or $$v_1 = \frac{RT_1}{P_1} = \frac{0.287 \times 300}{100} - 1.461 \text{ m}^3/\text{kg}$$

$$v_2 = \frac{RT_2}{P_2}$$

But for adiabatic process 1.2

$$P_1 v_1^{\gamma} = P_2 v_2^{\gamma}$$

$$P_2 = P_1\left(\frac{v_1}{v_2}\right)^{\gamma}$$

$$= P_1(r)^{\gamma} = 100 \times (6.75)^{1.4}$$
$$= 1448.86 \text{ kPa}$$

Therefore $$v_2 = \frac{0.287 \times 643.94}{1448 \cdot 86}$$
$$= 0.1276 \text{ m}^3/\text{kg}$$

Thus resubstituting values in eqation C we get

$$\text{Mean effective pressure (m. e. p)} = \frac{520 \cdot 03}{(1.761 - 0.1276)}$$

$$= \frac{520 \cdot 03}{1.334} = 390.00 \text{ kPa}$$

Example 7.7. *(a) In an air standard Otto cycle the minimum cycle temperature is the ambient room temperature of 25°C. Similarly the maximum temperature is decided on the considerations of metallurgical properties of the materials of piston and cylinder which is 1400 K. Keeping in view these limiting temperatures of the Otto cycle it is required to estimate the compression ratio of the cycle so as to have the maximum possible work output.*

(b) Under these conditions also estimate the temperatures at the end of compression and power strokes and then find the power output of the engine if air flow rate is 0.4 kg/min. Take for air $\gamma = 1.4$ and $C_{\gamma} = 0.718$ kJ/kg · K.

Solution : Refer to Fig. (7.5) for the Otto cycle given data $T_3 = 1400$ K; $T_1 = 273 + 25$ $\gamma = 1.4$ and $C_v = 0.718$ kJ/kg · K. For maximum work output of Otto cycle we have from equation (7.25) and (7.23).

$$r = \left(\frac{T_3}{T_1}\right)^{\left[\frac{1}{2(\gamma-1)}\right]} \quad \text{...(7.25)}$$

$$T_2 = T_4 = \sqrt{T_1 T_3} \quad \text{...(7.23}b\text{)}$$

Therefore $$r = \left(\frac{1400}{298}\right)^{\frac{1}{2(1.4-1)}}$$

$$= (4.698)^{\frac{1}{0.8}} = (4.698)^{1.25}$$
$$= 6.9166$$

$$T_2 = T_4 = \sqrt{1400 \times 298}$$
$$= \sqrt{417200}$$
$$= 645.91 \text{ K}$$

(*b*) Workdone per kg of air flow = (Heat supplied)

$$= C_v(T_3 - T_2) - C_v(T_4 - T_1)$$
$$= 0.718(1400 - 645.91) - 0.718\,(645.91 - 298)$$
$$= (0.718)\,(1698 - 2 \times 645.91)$$
$$= (0.718)\,(406.18)$$
$$= 291.637 \text{ kJ/kg}$$

$$\therefore \text{ Maximum power } = 291.637 \times \frac{0.4}{60}$$

$$= 1.944 \text{ kW}$$

Example 7.8. *(a) The compression process of a gas engine working on Otto cycle follow the polytropic law $PV^{1.3}$ = constant. During the compression stroke the measured pressures at 30% and 80% stroke length were 120 kPa and 310 kPa respectively. Determine (a) the compression ratio of the gas engine.*

(b) If the relative efficiency of the engine is 0.4 estimate the gas consumption rate per kW · h of the output. Take the calorific value of gas as 18000 kJ/m³. For air take γ = 1.4.

Solution : Fig (7.6) shows the P–V plot of the compression stroke of the cycle. Curve 1 AB 2 represents the P–V curve of the polytropic compression process and

points *A* and *B* represent the position of the piston at 30% and 80% of stroke

Thus if V_c = Clearance volume

V_s = Stroke volume

V_1 = Total volume of air when piston is at position 1

V_a = Volume at *A*

$= V_c + V_s - 0.3\ V_s$

and $V_b = V_c + V_s - 0.8\ V_s$

or $V_a = V_c + 0.7\ V_s$

and $V_b = V_c + 0.2\ V_s$

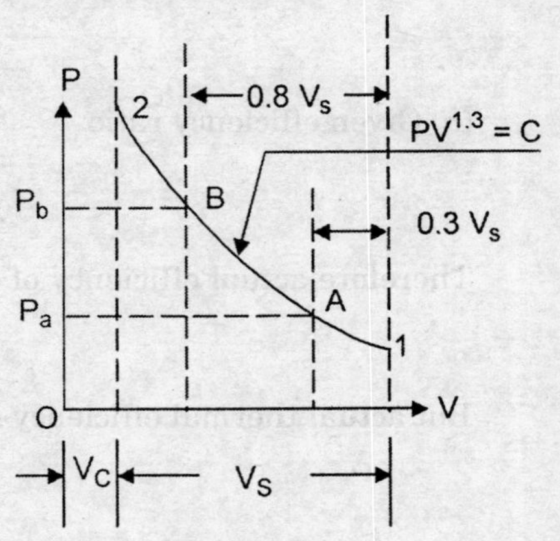

Fig. 7.6

Since this process is following polytropic law

Therefore we have $P_a V_a^{1.3} = P_b V_b^{1.3}$

or $$\left(\frac{P_a}{P_b}\right) = \left(\frac{V_b}{V_a}\right)^{1.3}$$

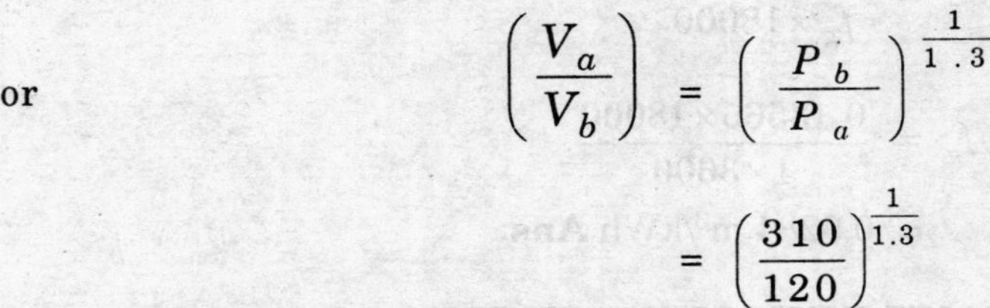

or $$\left(\frac{V_a}{V_b}\right) = \left(\frac{P_b}{P_a}\right)^{\frac{1}{1.3}}$$

$$= \left(\frac{310}{120}\right)^{\frac{1}{1.3}}$$

or $$\left(\frac{V_a}{V_b}\right) = \left(\frac{310}{120}\right)^{\frac{1}{1.3}}$$

$$= (2.5833)^{0.7692}$$

$$= 2.075$$

or $$\frac{V_c + 0.7\ V_s}{V_c + 0.2\ V_s} = 2.075$$

or $$V_c + 0.7\ V_s = 2.075\ V_c + 0.4150\ V_s$$

or $$0.285\ V_s = 1.075\ V_c$$

or $$\frac{V_s}{V_c} = \frac{1.075}{0.285} = 3.77$$

now $$r = \frac{V_1}{V_2} = \frac{V_c + V_s}{V_c}$$

$$= 1 + \frac{V_s}{V_c}$$

$$= 1 + 3.77 = 4.77$$

Therefore air standard efficiency, had air been the working fluid

$$\eta_{\text{air standard}} = 1 - \frac{1}{(r)^{(\gamma-1)}}$$

$$= 1 - \frac{1}{(4.77)^{0.4}} \quad \textit{i.e.}, 46.47\%$$

(*b*) Given efficiency ratio $= 0.4$

$$= \frac{\text{Actual Efficiency of Engine}}{\text{Air Standard Efficiency}}$$

Therefore actual efficiency of engine

$$= 0.4 \times 0.7647$$

$$= 0.18568 \quad \textit{i.e.}, 18.568\%$$

But actual thermal efficiency of engine

$$= \frac{\text{Output Power in kw} \times 3600}{f_g \times \text{Calorific value of Gas}}$$

where f_g = gas consumption per kWh

Therefore $$0.18568 = \frac{1\,\text{kw} \times 3600}{f_g \times 18000}$$

or $$f_g = \frac{0.18565 \times 18000}{1 \times 3600}$$

$$= 0.9284 \text{ m}^3/\text{kWh}$$ **Ans.**

7.6 DIESEL CYCLE (CONSTANT PRESSURE CYCLE)

The Diesel Cycle is the ideal cycle for (C.I.) compression-ignition reciprocating engines. Compression ignition engine was first developed by Rudolf Diesel in 1893. It is very similar to spark ignition (S.I.) engine discussed earlier in this chapter, the only difference is the initiation of the combustion process : In the S.I. engine combustion is initiated by firing a spark plug which ignites the compressed air-fuel mixture while in C.I. engine the combustion starts when the injected fuel comes into contact with the hot compressed air.

The fuel injection process in diesel engine starts when the piston reaches TDC and continues to some part of the power stroke. Therefore the combustion process is over a stretched period of time contrary to the S.I. engines where the combustion is instantaneious at the appearance of spark. Therefore the combustion process in diesel cycle is approximated as constant pressure. heat addition process. it is the power stroke only where the diesel and Otto cyles differ, the other strokes are similar in the respective processes.

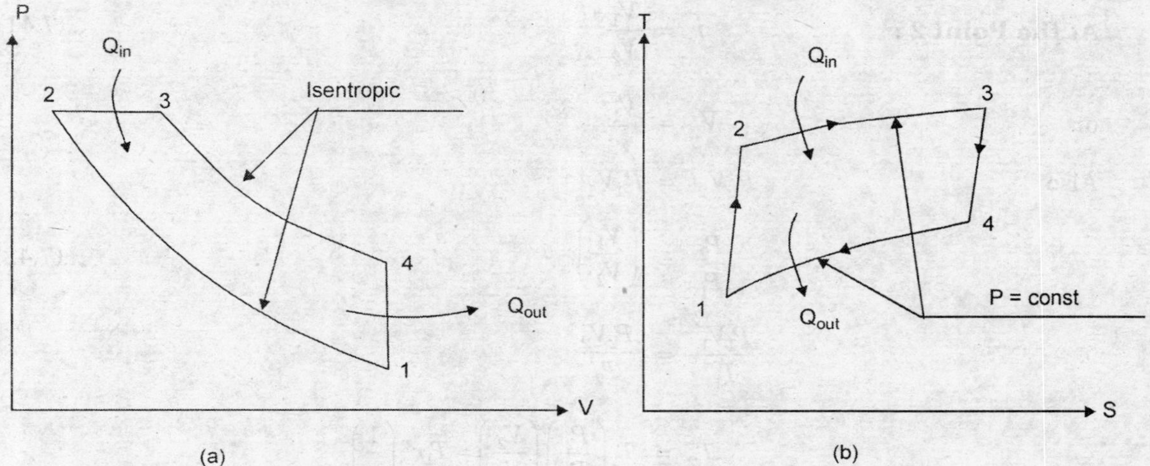

Fig. 7.7

In Fig. 7.7 (*a*) the diesel cycles is shown on P.V. space and in Fig. 7.7 (*b*) it is shown on T–s. space, process 1-2 is isentropic compression, 2-3 is the constant pressure heat addition, 3-4 is the isentropic expansion and 4-1 is the constant volume heat rejction. We now proceed below to develop an expression for the thermal efficiency of a diesel cycle by taking air as the system along with the associated assumptions. Under these circumstances the thermal efficiency is also known as Air standard Efficiency of the diesel cycle.

Consider a unit mass of air undergoing the cyclic change. Heat supplied during the constant pressure process 2-3.

$$Q_1 = C_p(T_3 - T_2) \quad \text{...(7.36)}$$

Heat rejected during the process 4-1 at constant volume

$$Q_2 = C_v(T_4 - T_1) \quad \text{...(7.37)}$$

Work done $= \text{Heat supplied} - \text{Heat rejected}$

$$= C_p(T_3 - T_2) - C_v(T_4 - T_1) \quad \text{...(7.38)}$$

Therefore thermal efficiency

$$\eta = \frac{\text{Work done}}{\text{Heat supplied}}$$

$$= \frac{C_p(T_3 - T_2) - C_v(T_4 - T_1)}{C_p(T_3 - T_2)}$$

or

$$\eta = 1 - \frac{1}{\gamma}\left(\frac{T_4 - T_1}{T_3 - T_2}\right) \quad \text{...(7.39)}$$

The ratio of the volumes after and before the combustion process is called the CUTOFF RATIO and is represented by r_c where

$$r_c = \frac{V_3}{V_2} = \frac{v_3}{v_2} \quad \text{...(7.40)}$$

After having defined the cut off ratio we now proceed to develop an alternative expression for the thermal efficiency of diesel cycle in terms of compression ratio r, cut off ratio r_c and adiabatic exponent γ.

At the Point 2 : $$r = \frac{V_1}{V_2} \quad ...(7.41)$$

or $$V_2 = \frac{V_1}{r}$$

Also $$P_2 V_2^{\gamma} = P_1 V_1^{\gamma}$$

$$\frac{P_2}{P_1} = \left(\frac{V_1}{V_2}\right) = r^{\gamma} \quad ...(7.42)$$

$$\frac{P_1 V_1}{T_1} = \frac{P_2 V_2}{T_2}$$

$$T_2 = T_1\left(\frac{P_2}{P_1}\right)\left(\frac{V_2}{V_1}\right) = T_1 r^{v}\left(\frac{1}{r}\right)$$

$$= T_1\, r^{(\gamma - 1)} \quad ...(7.43)$$

At Point 3 : $$P_3 = P_2 = P_1\, r^{\gamma} \quad ...(7.44)$$

Also from (7.40)

$$V_3 = V_2\, r_c \quad ...(7.40)$$

Also at constant pressure process 2-3

$$\frac{V_2}{T_2} = \frac{V_3}{T_3}$$

or $$T_3 = T_2\left(\frac{V_3}{V_2}\right) = T_2 r_C \quad \text{[From 7.40]}$$

$$= T_1\, r^{(\gamma - 1)}\, r_c \quad ...(7.45)$$

At the end of process 3-4, which is isentropic expansion

$$P_3 V_3^{\gamma} = P_4 V_4^{\gamma}$$

$$P_4 = P_3\left(\frac{V_3}{V_4}\right)^{\gamma}$$

$$= P_3\left(\frac{V_3}{V_1}\right)^{\gamma} \quad [\because V_4 = V_1]$$

$$= P_3\left[\frac{V_3}{V_2}\cdot\frac{V_2}{V_1}\right]^{\gamma}$$

or $$P_4 = P_3\left[r_c \cdot \frac{1}{r}\right]^{\gamma} \quad \text{[From (7.40) and (7.41)]} \quad ...(7.46)$$

Also $$\frac{P_3V_3}{T_3} = \frac{P_4V_4}{T_4}$$

$$T_4 = T_3\left(\frac{P_4}{P_3}\right)\left(\frac{V_4}{V_3}\right)$$

$$= T_3\left(\frac{r_c}{r}\right)^{\gamma}\left[\frac{V_4}{V_3}\right] \quad ...(7.47)$$

But $$\frac{V_4}{V_3} = \left(\frac{V_4}{V_2}\cdot\frac{V_2}{V_3}\right)$$

or $$\frac{V_4}{V_3} = \left(\frac{V_1}{V_2}\cdot\frac{V_2}{V_3}\right) \quad [\because V_4 = V_1]$$

$$= (r)\left(\frac{1}{r_c}\right) \quad \text{[From (7.41) and (7.40)]}$$

or $$\frac{V_4}{V_3} = \frac{r}{r_c} \quad ...(7.48)$$

Therefore from equations (7.47) and (7.45) we get

$$T_4 = T_3\left(\frac{r_c}{r}\right)^{\gamma}\left(\frac{r}{r_c}\right)$$

$$= T_3\left(\frac{r_c}{r}\right)^{\gamma-1} \quad ...(7.49)$$

Again substituting the value of T_3 from equation (7.45) into equation (7.45)

$$T_4 = T_1(r)^{\gamma-1}\,r_c\cdot\left(\frac{r_c}{r}\right)^{\gamma-1}$$

$$= T_1 r_c^{\gamma} \quad ...(7.50)$$

Thus substituting the values of T_2, T_3 and T_4 from equation (7.43), (7.45) and (7.50) into equation (7.39) we get the alternative form of expression for air cycle thermal efficiency of diesel cycle as

$$\eta = 1-\frac{1}{\gamma}\left[\frac{T_1 r^{\gamma} - T_1}{T_1 r^{\gamma-1} r_c - T_1 r^{(\gamma-1)}}\right]$$

$$= 1-\frac{1}{\gamma}\left[\frac{1}{r^{(\gamma-1)}}\right]\left[\frac{(r_c^{\gamma}-1)}{(r_c-1)}\right]$$

$$\eta_{\text{diesal}} = 1-\frac{1}{r^{(\gamma-1)}}\left[\frac{(r_c^{\gamma}-1)}{\gamma(r_c-1)}\right] \quad ...(7.51)$$

7.6.1 General Discussion on Equation 7.51

A perusal of the equations (7.20) and (7.51) which refer to the air standard efficiencies of Otto and Diesel cycles respectively shows that the efficiency of Diesel cycles differs from that of Otto cycle by the quantity given in the square bracket (Equn. 7.51). The quantity in the square bracket is always greater than unity.

Therefore $$\eta_{Otto} > \eta_{diesel} \quad ...(7.52)$$

Provided both the cycles operate with the same compression ratio.

Further as the value of the cut off ratio r_c decreases the efficiency of the diesel cycle increases as is clear from the Fig. (7.8) where the efficiency verses compression ratio curves have been drawn as per equation (7.51) for different value of r_c.

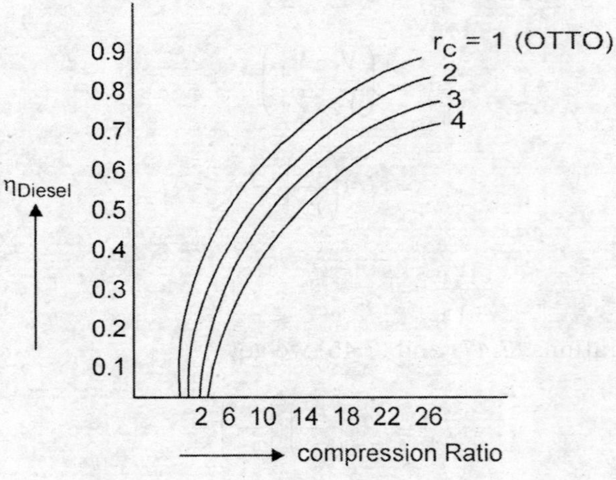

Fig. 7.8

In the limiting case when $r_c = 1$, the quantity in square brackets (eqn. 7.51) becomes unity and the efficiencies of Diesel and Otto cycles become identical. However the diesel engines operate at much higher values of compression ratio thus are usually more efficient than S.I (Petrol) engines as petrol engines at higher compression ratios give rise to auto-ignition and they cannot operate at higher values of r_c. In addition in the diesel engines the fuel burms more completely as these engines work at lower speeds than S.I. engines. A detailed discussion about the relative efficiencies of Otto, Diesel or Dual cycles is given in article 7.9. Due to higher efficiency in certain cases and lower fuel cost the diesel engines are more widely used in industry specially where demand for power is larger.

7.6.2 Mean Effective Pressure of Diesel Cycle

With reference to Fig. 7.7 the mean effective pressure

$$p_m = \frac{\text{Work done per cycle}}{\text{Swept volume}} \quad ...(7.53)$$

Now work done $= \text{Heat Supplied} - \text{Heat Rejected}$

$$= C_p(T_3 - T_2) - C_v (T_4 - T_1) \quad ...(7.54)$$

And swept volume $V_s = (V_1 - V_2)$

$$= (V_1 - V_2)$$

$$= V_1\left(1-\frac{V_2}{V_1}\right)$$

$$= \frac{RT_1}{P_1}\left(1-\frac{1}{r}\right) \qquad [\because V_1 - V_2]$$

$$= \frac{RT_1}{rP_1}(r-1) \qquad ...(7.55)$$

Therefore from (7.53), (7.54) and (7.55)

$$p_m = \frac{C_p(T_3-T_2)-C_v(T_4-T_1)}{\frac{RT_1}{rP_1}(r-1)} \qquad ...(7.56)$$

But $\quad C_p - C_v = R$

or $\quad \left(\frac{C_p}{C_v}-1\right) = \frac{R}{C_v}$

or $\quad (\gamma-1) = \frac{R}{C_v}$

or $\quad R = C_v(\gamma-1)$

Substituting the value of R in (7.56)

$$p_m = \frac{C_p(T_3-T_2)-C_v(T_4-T_1)}{\frac{C_v(\gamma-1)}{rp_1}(r-1)T_1}$$

$$= \frac{r\rho_1}{T_1}\left[\frac{C_p}{C_v}(T_3-T_2)-(T_4-T_1)\right]\frac{1}{(\gamma-1)(r-1)}$$

or $$p_m = \frac{r\rho_1}{(\gamma-1)(r-1)T}[\gamma(T_3-T_2)-(T_4-T_1)] \qquad ...(7.57)$$

Substituting the values of T_4, T_3 and T_2 from equations (7.50), (7.45) and (7.43) respectively in eqn. (7.57) we get

$$p_m = \frac{r\rho_1}{(v-1)(r-1)T_1}\left[\gamma\{T_1 r^{(\gamma-1)}r_c - T_1 r^{(\gamma-1)}\} - \{T_1 r_c^{\gamma} - T_1\}\right]$$

$$= \frac{r\rho_1}{(\gamma-1)(r-1)T_1}\left[T_1\gamma r^{(\gamma-1)}(r_c-1) - T_1(r_c^{\gamma}-1)\right]$$

$$p_m = \frac{r\rho_1}{(\gamma-1)(r-1)}[\gamma_r{}^{(\gamma-1)}(r_c-1)-(r_c{}^{\gamma}-1)] \qquad ...(7.58)$$

Example 7.9. *A four stroke diesel engine working on Diesel cycle has the following specifications.*

Cylinder diameter = *350 mm*

Stroke length = *200 mm*

Clearance volume = *0.00125 m^3*

The cut off takes place at 6% of the stroke. Find (a) the compression ratio (b) cut off ratio (c) Air standard efficiency.

Take for air $\gamma = 1.4$

Solution : Refer to Fig. 7.7

Stroke length l = 200 mm

Cylinder d = 350 mm

Clearance volume = V_c = 0.00125 m^3

Therefore swept volume or stroke volume

$$V_s = \frac{\pi}{4} \times (350)^2 \times 200 \text{ mm}^3 = 0.02037 m^3$$

$$V_2 = V_c = 0.00125 \text{ m}^3$$

$$V_1 = V_C + \text{stroke volume}$$

$$= 0.00125 + 0.02037$$

$$= 0.0216\ m^3$$

∴ Compression ratio $r = \frac{V_1}{V_2} = \frac{0.0216}{0.00125} = 17.28$ **Ans.**

$$V_3 = \text{volume at point 3} = V_2 + 6\ \%\text{ of } V_s$$

$$= V_2 + \frac{6}{100} \times .02037 \text{ m}^3$$

$$= 0.00125 + 0.0012192$$

$$= 0.002492 \text{ m}^3$$

Cut off ratio $r_c = \frac{V_3}{V_2} = \frac{0.002492}{0.00125}$

$= 1.9936$ **Ans.**

Air standard efficiency $= 1 - \frac{1}{r^{\gamma-1}}\left[\frac{r_c^{\gamma} - 1}{\gamma(r_c - 1)}\right]$

or $\eta = 1 - \frac{1}{(17.28)^{1.4-1}}\left[\frac{(1.9936)^{1.4} - 1}{(1.4)\,(1.9936 - 1)}\right]$

$$= 1 - \frac{1 \times 1.6272}{3.126 \times 1.4 \times .9936}$$

$\eta = 1 - 0.374 = 0.626$ *i.e.,* 62.6%

Example 7.10. *The compression ratio for an air standard Diesel cycle is 14.5. The temperature and pressure at the begining of the cycle are 17°C and 100 kPa. If the maximum temperature of the cycle is 2600°C, calculate air standard efficiency of the cycle take* $\gamma = 1.4$.

Solution : Refer to Fig. 7.7

Given $r = \frac{V_1}{V_2} = 14.5,\ P_1 = 100 \text{ kPa},\ T_1 = 273 + 17 = 290 \text{ K}$

and $T_3 = 2600 + 273 = 2873 \text{ K}$

For the adiabatic process 1 – 2

$$P_1V_1^{1.4} = P_2V_2^{1.4}$$

or
$$P_2 = P_1\left(\frac{V_1}{V_2}\right)^{1.4} = P_1(14.5)^{1.4}$$

$$P_2 = P_1 \times 42.2586$$

or
$$P_2 = 100 \times 42.2586 = 4225.86 \text{ kPa}$$

Also
$$\frac{P_1V_1}{T_1} = \frac{P_2V_2}{T_2}$$

$$T_2 = T_1\left(\frac{P_2}{P_1}\right)\cdot\left(\frac{V_2}{V_1}\right)$$

$$= 290 \times 42.2586 \times \frac{1}{14.5}$$

$$= 845.172 \text{ K}$$

For the constant pressure process 2-3

$$\frac{V_2}{T_2} = \frac{V_3}{T_3}$$

or
$$\frac{V_3}{V_2} = \frac{T_3}{T_2} = \frac{2873}{845.172} = 3.4$$

Therefore cut off ratio
$$r_c = \frac{V_3}{V_2} = 3.4$$

Then air standard efficiency
$$\eta = 1 - \frac{1}{r^{(\gamma-1)}}\left[\frac{r_c^{\gamma} - 1}{\gamma(r_c - 1)}\right]$$

or
$$\eta = 1 - \frac{1}{(14.5)^{0.4}}\left[\frac{(3.4)^{1.4} - 1}{(1.4)(3.4 - 1)}\right]$$

$$= 1 - \frac{1 \times 4.5472}{2.9144 \times (1.4) \times 2.4}$$

$$= 1 - 0.464 - 0.5356 \text{ or } 53.56\%$$

Example 7.11. *An engine working on diesel cycle drawn air which is at 100 kPa pressure and 20°C temperature. Find the air standard efficiency of the engine if the maximum temperature and pressure are 2100 K and 6000 kPa respectively Take for air $\gamma = 1.4$*

Solution : Refer to Figure 7.7

Higher
$$T_1 = 273 = 20 = 293 \text{ K}$$
$$P_1 = 100 \text{ kPa} \qquad P_2 = P_3 = 6000 \text{ kPa}$$
$$T_3 = 2100 \text{ K} \qquad \text{and} \qquad P_3 = 6000 \text{ kPa}$$

For isentropic process 1.2

$$\left(\frac{V_1}{V_2}\right)^{\gamma} = \frac{P_2}{P_1} = \frac{6000}{100} = 60$$

$$\therefore \quad \left(\frac{V_1}{V_2}\right) = (60)^{\frac{1}{1.4}} = (60)^{0.71429}$$

$$= 18.6258$$

Also $\frac{P_1 V_1}{T_1} = \frac{P_2 V_2}{T_2}$ or $T_2 = T_1\left(\frac{P_2}{P_1}\right)\left(\frac{V_2}{V_1}\right) = 293 \times \frac{6000}{100} \times \frac{1}{18.6258}$

$$= 943.85 \text{ K}$$

Also for constant pressure process 2–3

$$\frac{V_3}{T_3} = \frac{V_2}{T_2}$$

$$\therefore \quad \frac{V_3}{V_2} = \frac{T_3}{T_2} = \frac{2100}{943.85} = 2.2249 = r_c$$

Now

$$\eta = 1 - \frac{1}{r^{(\gamma-1)}}\left[\frac{r_c^{\gamma} - 1}{\gamma(r_c - 1)}\right]$$

$$= 1 - \frac{1}{(18.6258)^{0.4}}\left[\frac{(2.2249)^{1.4} - 1}{(1.4)\,(2.2249 - 1)}\right]$$

$$= 1 - \frac{1}{3.2214}\left[\frac{2.0636}{1.4 \times 2.2249}\right]$$

$$= 1 - 0.3735$$

$$= 0.6265 \text{ or } 62.65\ \%$$

Example 7.12. *The compression ratio of a diesel engine is 18. It draws air during the Suction stroke at 100 kPa pressure and 20°C temperature. If the maximum temperature during the cycle is 1800 K find (a) the cut off ratio and percentage of stroke at cutt off. (b) the cycle efficiency and (c) work output per kg of air and (d) mean effective pressure. Take* $\gamma = 1.4$ *and* $R = 0.287\ kJ/kg \cdot K$.

Solution : Refer to Fig. 7.7

Given data

$$r = 18 = V_1/V_2$$

$$T_1 = 273 + 20 = 293 \text{ K}$$

$$P_1 = 100 \text{ kPa}$$

$$T_3 = 1800 \text{ k}, \qquad \upsilon = 1.4$$

Process 1.2 (Isentropic compression) $\gamma = 1.4$

$$\frac{V_1}{V_2} = r = 18 \text{ (given)}$$

$$P_1 V_1^{\gamma} = P_2 V_2^{\gamma}$$

or $$\left(\frac{V_1}{V_2}\right)^\gamma = \frac{P_2}{P_1} \quad i.e., \quad (r)^\gamma = \frac{P_2}{P_1}$$

$$\therefore \quad P_2 = P_1(r)^\gamma = 100 \times (18)^{1.4}$$
$$= 57.2$$

Also $$\frac{P_1V_1}{T_1} = \frac{P_2V_2}{T_2}$$

$$T_2 = T_1 \cdot \left(\frac{P_2}{P_1}\right)\frac{V_2}{V_1} = 293 \times (r)^\gamma \times \frac{1}{r}$$
$$= 293 \times (r)^{\gamma-1}$$
$$= 293 \times (18)^{.4} = 293 \times 3.1777$$
$$= 931.07$$

For constant pressure 2 – 3

$$\frac{V_2}{T_2} = \frac{V_3}{T_3} \quad \text{or} \quad \frac{V_3}{V_2} = \frac{T_3}{T_2}$$

$$= \frac{1800}{931.07} = 1.9933$$

$$= r_c \textbf{ Ans.}$$

(*b*) cycle efficiency $$\eta = 1 - \frac{1}{r^{\gamma-1}}\left[\frac{(r_c^\gamma - 1)}{\gamma(r_c - 1)}\right]$$

$$= 1 - \frac{1}{(18)^{0.4}}\left[\frac{(1.9933)^{1.4} - 1}{(1.4)\,(1.9933 - 1)}\right]$$

$$= 1 - \frac{1}{3.1776} \times \frac{1.6266}{1.4 \times 0.9933}$$

$$= 1 - 0.368 = 0.632 \quad i.e., \quad 63.2\%$$

(*c*) % of stroke at cut off

$$= \frac{V_3 - V_2}{V_1 - V_2}$$

Per Unit Mass of Air

Now $$V_1 = \frac{RT_1}{P_1} = \frac{0.287 \times 293}{100} = 0.841 \text{ m}^3/\text{kg}$$

$$\frac{V_2}{V_1} = \frac{1}{r} = \frac{1}{18}$$

$$\therefore \quad V_2 = \frac{V_1}{18} = \frac{0.841}{18} = 0.0467 \text{ m}^3/\text{kg}$$

Now $r_c = \frac{V_3}{V_2} = \frac{V_3}{0.0467} = 1.9933$

$\therefore$ $V_3 = 0.0467 \times 1.9933$

$= 0.0931 \text{ m}^3/\text{kg}$

(*d*) Therefore work out put per cycle per kg of air

$= \text{Heat supplied} - \text{Heat rejected}$

Heat supplied $Q_1 = C_p(T_3 - T_2)$

But $C_p - C_v = R \text{ and } \frac{C_p}{C_v} = \gamma = 1.4$

$$C_p - \frac{C_p}{1.4} = 0.287$$

or $C_p\left(\frac{0.4}{1.4}\right) = 0.287$

or $C_p = \frac{1.4 \times 0.287}{0.4} = 1.0045$

And $T_3 = 1800 \text{ K and } T_2 = 931.07 \text{ K}$

$\therefore$ $Q_1 = 1800 - 931.07$

$= 868.93 \text{ kJ/kg}$

and $\eta = \frac{\text{Work output}}{\text{Input}}$

$\therefore$ Work output $= Q_1 \eta$

$= 868 \cdot 93 \times 0.6265$

$= 544.38$ kJ/kg **Ans.**

(*d*) Mean effective pressure $p_m = \frac{\text{Work done}}{\text{Swept volume}}$

$$= \frac{\text{Work done}}{V_1 - V_2}$$

$$= \frac{544.38}{0.841 - 0.0467} = \frac{544.38}{0.7943}$$

$= 685.36 \text{ kPa}$

Example 7.13. *An oil engine with 20 cm cylinder diameter and 30 cm stroke works on diesel cycle. The initial pressure and temperatue of air used are 1 bar and 30°C. The cut off takes place at 10% of stroke. Find*

(a) Thermal efficiency.

(b) Mean effective pressure.

(c) The power developed by the engine if the working strokes per minute are 400. Assume compression ratio of 16. *[PTU 2005 Dec.]*

Solution : Refer to Fig 7.7

Given data Cylinder diameter $d = 20 \text{ cm} = 0.2 \text{ m}$

Stroke length $l = 30 \text{ cm} = 0.3 \text{ m}$

Compression ratio $r = 16$

$$P_1 = 1 \text{ bar} = 10^5 \text{ Pa} = 100 \text{ kPa}$$

$$V_s = \text{Swept volume} = \frac{\pi}{4} d^2 l = \frac{\pi}{4} (.2)^2 (.3) = 0.0094 \text{ m}^3$$

Let V_c = Clearance volume

Then
$$r = 1 + \frac{V_S}{V_C}$$

or
$$16 = 1 + \frac{0.0094}{V_C}$$

$$\therefore \quad V_C = \frac{.0094}{15} = 0.0006267 \text{ m}^3$$

From Fig. 7.7
$$V_2 = V_C = 0.0006267 \text{ m}^3$$

Now
$$V_3 = V_2 + 10\% \text{ stroke volume} = V_2 + \frac{10}{100} V_S$$

$$= .0006267 + \frac{10}{100} \times 0.0094$$

$$= 0.00157 \text{ m}^3$$

Therefore cut off ratio
$$r_c = \frac{V_3}{V_2} = \frac{0.00157}{0.0006267} = 2.499 = 2.5$$

Now thermal efficiency
$$\eta_{\text{Diesel}} = 1 - \frac{1}{r^{\gamma-1}} \left[\frac{v_c^{\gamma} - 1}{\gamma(r_c - 1)} \right]$$

$$= 1 - \frac{1}{10^{(1.4-1)}} \left[\frac{2.5^{1.4} - 1}{1.4(2.5 - 1)} \right]$$

$$= 1 - 0.4098 = 0.59 \text{ or } 59\%$$

(*b*) mean effective pressure = p_m from eqn. (7.58) we know that

$$p_m = \frac{r p_1}{(\gamma - 1)(r - 1)} \left[\gamma r^{(\gamma-1)} (r_c - 1) - (r_c^{\gamma} - 1) \right]$$

or
$$p_m = \frac{16 \times 100}{(1.4 - 1)(16 - 1)} [(1.4) 16^{0.4} (2.5 - 1) - (2.5^{1.4} - 1)]$$

$$= 1010 \text{ kPa}$$

(c) Power Development

We have from equation (7.58)

$$p_m = \frac{\text{Work output per cycle}}{\text{Stroke volume}}$$

$\therefore$ work done per cycle $= p_m \times$ stroke volume

$= (1010 \text{ kPa})(0.0094 \text{ m}^3)$

$= 9.494$ kJ

Therefore power developed $=$ W.D. per cycle $\times$ (cycle/sec)

$$= 9.494 \times \frac{400}{60}$$

$= 63.3$ kJ/sec or 63.3 kW

7.7 EFFECT OF VARIATION IN C_V ON EFFICIENCIES OF OTTO AND DIESEL CYCLES

In the preceeding sections it has been assumed that the specific heat C_v remains constant. But in actuality it depends upon the temperature of the gas (in the present case air). As the temperature no longer remains constant, it does affect the efficiency of the cycle. We take the cases of Otto and Diesel cycles separately in the following paragraphs.

7.7.1 Effect on Otto Cycle Efficiency

The efficiency of Otto cycle is given in equation 7.20

$$\eta_{\text{Otto}} = 1 - \frac{1}{r^{(\gamma-1)}} \qquad \text{...(7.20)}$$

Now

$$C_p - C_v = R$$

$$\frac{C_p}{C_v} - 1 = \frac{R}{C_v}$$

or

$$(\gamma - 1) = \frac{R}{C_v}$$

Therefore equation (7.20) reduces to

$$(1 - \eta) = \frac{1}{r^{(R/C_v)}}$$

or

$$(1 - \eta) = r^{-(R/C_v)}$$

Taking log on both sides

$$ln(1 - \eta) = -\frac{R}{C_v} ln(r)$$

Differentiating w.r.t C_v

$$\left(\frac{1}{1-\eta}\right)(-1)\frac{d\eta}{d_{C_v}} = -[ln(r)\,(R)\left(\frac{-1}{C_v^{\,2}}\right) \qquad \text{...(7.58)}$$

$$\frac{d\eta}{1-\eta} = -[R]\,(lnr)\,\frac{d_{C_v}}{C_v^{\,2}} \qquad \text{...(7.58)}$$

But $\frac{R}{C_v} = \gamma - 1$

or $R = C_v(\gamma - 1)$

$\therefore$ Equation (7.58) reduces to

$$\frac{d\eta}{(1-\eta)} = C_v(\gamma-1)\, ln(r) \frac{d_{C_v}}{C_v^{\,2}}$$

or
$$d\eta = \frac{(\gamma-1)\,(1-\eta)\, ln(r)}{C_v} d_{C_v}$$

During by η
$$\frac{d\eta}{\eta} = \left(\frac{1-\eta}{\eta}\right)(\gamma-1)\, ln(r)\left(\frac{d_{C_v}}{C_V}\right) \quad ...(7.59)$$

Therefore if % age change in C_v or $\frac{d_{C_v}}{C_v}$ is known corresponding % age variation of η can be calculated from equation (7.59).

7.7.2 Effect on Diesel Cycle Efficiency

Equation (7.51) gives the efficiency of Diesel cycle based on the assumption that C_v is constant *i.e.*,

$$\eta = 1 - \frac{1}{r^{(\gamma-1)}}\left[\frac{(r_c^\gamma - 1)}{\gamma(r_c - 1)}\right] \quad ...(7.51)$$

or
$$(1-\eta) = \frac{1}{r^{(\gamma-1)}} \cdot \frac{1}{\gamma} \cdot \frac{1}{(r_c - 1)} \cdot (r_c^\gamma - 1)$$

Taking logarithm on both sides

$$ln(1-\eta) = -ln(r^{\gamma-1}) - ln(\gamma) - ln(r_c - 1) + ln(r_c^{\,\gamma} - 1)$$

But
$$\gamma = 1 + \frac{R}{C_v}$$

$$ln(1-\eta) = -ln(r)^{\frac{R}{C_v}} - ln\left(\frac{R}{C_v} + 1\right) - ln(r_c - 1) + ln\left(r_c^{\left(\frac{R}{C_v}+1\right)} - 1\right)$$

or
$$ln(1-\eta) = -\frac{R}{C_v} ln(r) - ln\left(\frac{R}{C_v} + 1\right) - ln(r_c - 1) + ln\left[r_c^{\left(\frac{R}{C_v}+1\right)} - 1\right]$$

Differentiate w.r.t C_v

$$\frac{d\eta}{1-\eta} = \left[\frac{R}{C_v^{\,2}} ln(r)\, dC_v\right] - \left[\frac{1}{\left(\frac{R}{C_v}+1\right)}\left(-\frac{R}{C_v^2}\, d_{C_v}\right)\right] + \left[\frac{1}{r_c^{\left(\frac{R}{C_v}+1\right)} - 1}\right]\left[r_c^{\left(\frac{R}{C_v}+1\right)} \cdot ln(r_c) \cdot \left(-\frac{R}{C_v^{\,2}} \cdot dC_v\right)\right]$$

Again $\frac{R}{C_v} = \gamma - 1$

$$\therefore \quad -\frac{d\eta}{1-\eta} = \left[\left(\frac{\gamma-1}{C_v}\right) ln(r)\, dC_v\right] + \left[\frac{1}{\gamma}\left(\frac{\gamma-1}{C_v}\right) dC_v\right] - \left[\frac{1}{(r^\gamma-1)}\right]\left[r_c^\gamma\, ln(r_c)\left(\frac{\gamma-1}{C_v}\right) dC_v\right]$$

$$= \left(\frac{\gamma-1}{C_v}\right)\left[ln(r) dC_v + \frac{1}{\gamma} dC_v - \frac{r_c^\gamma ln(r_c)}{(r^\gamma-1)} dC_v\right]$$

i.e.
$$d\eta = (1-\eta)(\gamma-1)\frac{dC_v}{C_v}\left[ln(r) + \frac{1}{\gamma} - \frac{r_c^\gamma ln(r)}{(r^\gamma-1)}\right]$$

or
$$\frac{d\eta}{\eta} = \left(\frac{1-\eta}{\eta}\right)(\gamma-1)\frac{dC_v}{C_v}\left[\frac{r_c^\gamma\, ln(r_c)}{(r^\gamma-1)} - ln(r) - \frac{1}{\gamma}\right] \quad ...(7.59)$$

Example 7.14. *The compression ratio of an engine working on Otto cycle is 8. If the specific heat C_v increases by 1% what will be the corresponding change in the air standard efficiency of the engine.*

Take $\gamma = 1.4$

Solution : Given $r = 8, \quad \gamma = 1.4$

$$\frac{d\eta}{\eta} = \left(\frac{1-\eta}{\eta}\right)(\gamma-1)\, ln(r)\left(\frac{dC_v}{C_v}\right)$$

and for Otto cycle
$$\eta = 1 - \frac{1}{r^{\gamma-1}}$$

$$= 1 - \frac{1}{(8)^{0.4}} = 1 - \frac{1}{2.2974}$$

$$= 0.5647$$

Also it is given
$$\frac{dC_v}{C_v} = \frac{1}{100} = 0.01$$

$$\therefore \quad \frac{d\eta}{\eta} = \left(\frac{1-.564}{0.564}\right)(1.4-1) ln(8) \times .01$$

$$= \frac{0.4353 \times 0.4 \times 2.0794 \times .01}{.5647}$$

$$= 0.006419$$

i.e., 0.6419% decrease **Ans.**

Example 7.15. *An engine working on Diesel cycle has compression ratio 18 and cut off ratio 2. If the specific heat C_v changes by 2% what will be the %age change in the efficiency. Take $\gamma = 1.4$.*

Solution : $r = 18 \quad r_c = 2.0 \quad \gamma = 1.4$

$$\eta = 1 - \frac{1}{r^{\gamma-1}}\left[\frac{r_c^\gamma - 1}{\gamma(r_c-1)}\right]$$

$$= 1 - \frac{1}{(18)^{0.4}}\left[\frac{2^{1.4}-1}{1.4 \times 1.0}\right]$$

$$= 1 - \frac{1}{3.1777}\left[\frac{2.639 - 1}{1.4 \times 1}\right]$$

$$= 1 - 0.3075 = 0.6325 \quad i.e., 63.25\%$$

We have from equation (7.59)

$$\frac{d\eta}{\eta} = \left(\frac{1-\eta}{\eta}\right)(\gamma - 1)\left[\frac{r_c^{\gamma}\, ln(r_c)}{(r^{\gamma}-1)} - ln(r) - \frac{1}{\gamma}\right]\frac{dC_v}{C_v}$$

$$= \frac{(1-.6325)}{.6325}(0.4)\left[\frac{2^{1.4}\, ln(2)}{(18^{1.4}-1)} - ln(18) - \frac{1}{1.4}\right](0.02)$$

$$= \frac{0.3675 \times 0.4}{0.6325}\left[\frac{2^{1.4}\ \ln(2)}{(18^{1.4}-1)} - 2.89.4 - 0.7143\right] \times (0.02)$$

$$= \frac{0.3675 \times 0.4}{0.6325}[0.03255 - 2.8904 - 0.7143]\,(.02)$$

$$= 0.2324\ [-3.57215]\ (.02)$$

$$= -0.01660 \quad i.e., \ 0.16\% \text{ decrease } \textbf{Ans.}$$

7.8 DUAL CYCLE (MIXED OR COMPOSITE CYCLE)

In the preceeding sections the combustion process in internal combustion engines at constant volume (as in Otto cycle) or constant pressure (as in diesel cycle) heat addition processes is actually over simplified approximation which is far from reality. Probably a better but slightly more complex approach would be to model the combustion process in both the petrol and diesel engines as a combination of two heat transfer processes *viz* one occurring at constant volume and the other at constant pressure. Based on this concept the ideal air cycle is called the Dual Cycle (or mixed or composite cycle) the P–V. diagram for such a cycle is given in Fig. 7.9.

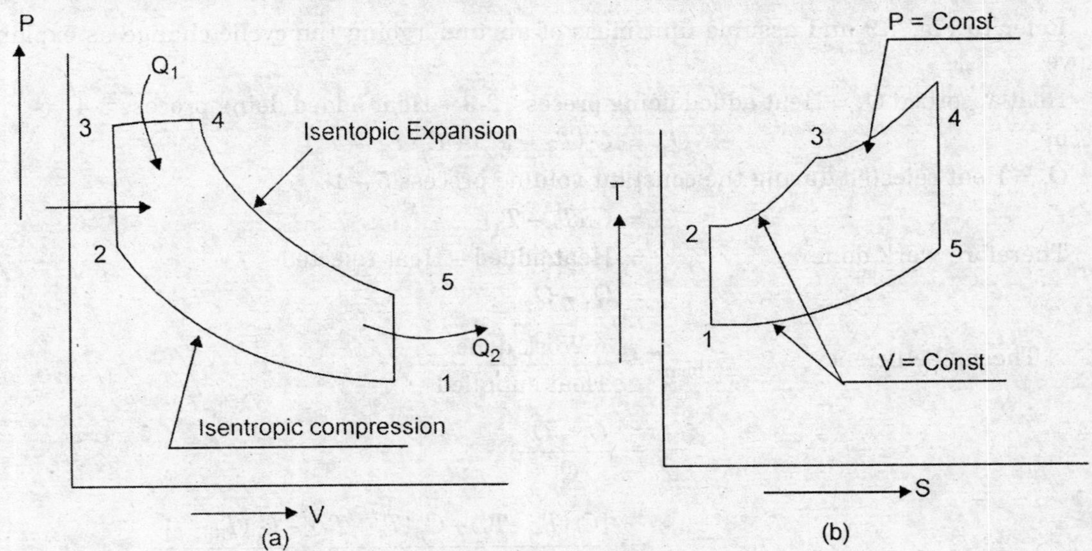

Fig. 7.9

Process 1-2 : Piston moves from crank end dead centre to cover end dead centre (Top dead centre). The air of state 1 is compressed isentropically. If compression ratio is

$$r = \frac{V_1}{V_2}$$

Process 2-3 : For a very small interval of time when the piston is at TDC position heat is added to air at constant volume from an external heat source. The pressure, as a result of the same, will increase. If r_p is the explosion ratio then

$$r_p = \frac{P_3}{P_2} \quad ...(7.60)$$

Process 3-4 : More heat is added at constant pressure during the early portion of the explosion stroke. The heat supply is cut off at point 4, which is called the cut off point. The ratio V_4/V_3 is called the cut off ratio and represented by r_c *i.e.*,

$$r_c = \frac{V_4}{V_3} \quad ...(7.61)$$

Process 4-5 : During this process the expansion of air takesplace isentropically and work is done by the systems. The ratio V_5/V_4 is known is isentropic expansion ratio.

Process 5-1 : The piston is momentarily at rest at bottom dead centre (BDC) and heat is rejected to the external sink by bringing it in contact with the cylinder head and heat rejection takes place at constant volume. The cycle is thus completed and initial parameters are restored at the completion of the cycle.

7.8.1 Thermal Efficiency of Dual Cycle

As explained above the compression process 1-2 and expansion process 4-5 are both isentropic. Therefore there is no heat interaction during these processes. Heat is absorbed during the process 2-3 at constant volume and process 3-4 at constant pressure while during the process 5-1 heat is rejected at constant volume.

Refer to Fig. 7.9 and assume unit mass of air undergoing the cyclic change as explaned above.

Heat absorbed Q_1 = Heat added doing process, 2-3 + Heat added doing process 3-4

or $$Q_1 = C_v(T_3 - T_2) + C_p(T_4 - T_3)$$

Q_2 = Heat rejected during the constant volume process 5 – 1

$$= C_v(T_5 - T_1)$$

Therefore work done = Heat added – Heat rejected

$$= Q_1 - Q_2$$

∴ Thermal efficiency $$\eta_{max} = \frac{\text{Work done}}{\text{Heat supplied}}$$

$$= \frac{Q_1 - Q_2}{Q_1}$$

$$= \frac{[C_v(T_3 - T_2) + C_p(T_4 - T_3)] - C_v(T_5 - T_1)}{[C_v(T_3 - T_2) + C_p(T_4 - T_3)]}$$

$$= 1 - \frac{C_v(T_5 - T_1)}{[C_v(T_3 - T_2) + C_p(T_4 - T_3)]}$$

or $$\eta_{max} = 1 - \frac{(T_5 - T_1)}{(T_3 - T_2) + \gamma(T_4 - T_3)} \quad ...(7.62)$$

Where $$\gamma = \frac{C_p}{C_v}$$

The expression (7.62) for efficiency can be represented in another form in terms of explosion ratio r_p, cut off ratio r_c, compression rate r and adiabiatic exponent γ as follows.

For process 1-2 : $$\frac{V_1}{V_2} = r \quad \textit{i.e.,} \quad V_2 = \frac{V_1}{r} \quad (7.63)$$

For isentropic compression

$$P_1V_1^{\gamma} = P_2V^{\gamma}$$

$$\frac{P_2}{P_1} = \left(\frac{V_1}{V_2}\right)^{\gamma} \quad \text{or} \quad P_2 = P_1r^{\gamma} \quad ...(7.64)$$

and $$\frac{P_1V_1}{T_1} = \frac{P_2V_2}{T_2}$$

or $$T_2 = T_1\left(\frac{P_2}{P_1}\right)\cdot\left(\frac{V_2}{V_1}\right)$$

$$= T_1\left(\frac{V_1}{V_2}\right)^{\gamma}\left(\frac{V_2}{V_1}\right)$$

$$= T_1\left(\frac{V_1}{V_2}\right)^{\gamma-1} = T_1r^{(\gamma-1)} \quad ...(7.65)$$

For Process 2-3 : Constant volume process

$$V_3 = V_2 = \frac{V_1}{r} \quad \text{(from 7.63)}$$

$$\frac{P_3}{P_2} = r_p \quad ...(7.66)$$

Then inserting value from equation (7.64) $P_3 = P_2r_p = P_1r^{\gamma}\, r_p$ ...(7.67)

Also for constant volume process 2-3 (7.67)

$$\frac{P_2}{T_2} = \frac{P_3}{T_3}$$

i.e., $$T_3 = T_2\left(\frac{P_3}{P_2}\right) = T_2\, r_p \quad \text{[from (7.66)]}$$

or $$T_3 = T_1r^{(\gamma-1)}\, r_p \quad \text{[from (7.63)]} \quad ...(7.68)$$

For the Constant Process 3-4

$$P_3 = P_4$$

$$\therefore \quad \frac{V_3}{T_3} = \frac{V_4}{T_4}$$

or $$T_4 = T_3\left(\frac{V_4}{V_3}\right) = r_c T_3$$

Substituting value of T_3 from (7.68)

$$T_4 = r_c r_p (r)^{\gamma-1} T_1 \quad \text{...(7.69)}$$

For the isentropic expansion 4-5

$$P_4 V_4^{\gamma} = P_5 V_5^{\upsilon}$$

or $$P_5 = P_4\left(\frac{V_4}{V_5}\right)^{\gamma} \quad \text{...(7.70)}$$

But $$\frac{V_4}{V_5} = \frac{V_4}{V_1} \quad (\because V_5 - V_1)$$

$$= \left(\frac{V_4}{V_3} \cdot \frac{V_3}{V_1}\right)$$

$$= \left(\frac{V_4}{V_3} \cdot \frac{V_2}{V_1}\right) \quad (\because V_2 - V_3)$$

i.e. $$\frac{V_4}{V_5} = \frac{r_c}{r} \quad \text{...(7.71)}$$

$\therefore$ From (7.70) and (7.71)

$$P_5 = P_4\left(\frac{r_c}{r}\right)^{\gamma} \quad \text{...(7.72)}$$

Also $$\frac{P_4 V_4}{T_4} = \frac{P_5 V_5}{T_5}$$

or $$\frac{T_5}{T_3} = \left(\frac{P_5}{\rho_4}\right) \cdot \left(\frac{V_5}{V_4}\right) \quad \text{...(7.73)}$$

Therefore from (7.71), (7.72) and (7.73) we have

$$\frac{T_5}{T_4} = \left(\frac{r_c}{r}\right)^{\gamma}\left(\frac{r}{r_c}\right)$$

$$= \left(\frac{r_c}{r}\right)^{\gamma-1}$$

or $$T_5 = T_4\left(\frac{r_c}{r}\right)^{\gamma-1}$$

Substituting value of T_4 from eqn. (7.69)

$$T_5 = T_1 r^{\gamma-1} r_p r_c \left(\frac{r_c}{r}\right)^{\gamma-1}$$

$$= T_1 r_p r_c^{\gamma} \qquad ...(7.74)$$

Now substituting the values of T_2, T_3, T_4 and T_5 from equations (7.65), (7.68), (7.69) and (7.74) respectively into equation (7.62) we get

$$\eta_{max} = 1 - \frac{(T_1 r_p r_c^{\gamma} - T_1)}{(T_1 r^{\gamma-1} r_p - T_1 r^{\gamma-1}) + \gamma\left[r_c r_p r^{\gamma-1} T_1 - T_1 r^{\gamma-1} r_p\right]}$$

$$= 1 - \frac{(r_p r_c^{\gamma} - 1)}{r^{(\gamma-1)}(r_p - 1) + \gamma r_p r^{\gamma-1}(r_c - 1)}$$

or

$$\eta_{max} = 1 - \frac{1}{r^{(\gamma-1)}}\left[\frac{(r_p r_c^{\gamma} - 1)}{(r_p - 1) + \gamma r_p (r_c - 1)}\right] \qquad ...(7.75)$$

7.8.2 Mean Effective Pressure of Dual Cycle

As per definition of mean effective pressure.

$$p_m = \frac{\text{Work done per cycle}}{\text{Swept volume}} \qquad ...(7.76)$$

But work done per cycle = Heat Supplied – Heat Rejected

$$= [C_v(T_3 - T_2) + C_p(T_4 - T_3)] - [C_v(T_5 - T_4)] \qquad ...(7.77)$$

And swept volume $V_s = V_1 - V_2$

$$= V_1\left(1 - \frac{V_2}{V_1}\right)$$

Now $P_1 V_1 = RT_1$

i.e. $V_1 = \dfrac{RT_1}{P_1}$

and $r = \dfrac{V_1}{V_2}$

Therefore swept volume $V_s = V_1\left(1 - \dfrac{V_2}{V_1}\right) = V_1\left(1 - \dfrac{1}{r}\right)$

$$= \frac{RT_1}{rP_1}(r - 1) \qquad ..(7.78)$$

Therefore from equations (7.76), (7.77) and (7.78)

$$p_m = \frac{[C_v(T_3 - T_2) + C_p(T_4 - T_3)] - C_V(T_5 - T_4)}{\frac{RT_1}{rP_1}(r-1)}$$

or
$$p_m = \frac{rp_1 C_v}{R(r-1)}\left[\frac{(T_3 - T_2) + \frac{C_p}{C_r}(T_4 - T_3) - (T_5 - T_4)}{T_1}\right]$$

$$= \frac{C_v rP_1}{R(r-1)}\left[\frac{(T_3 - T_2) + \gamma(T_4 - T_3) - (T_5 - T_6)}{T_1}\right] \quad ...(7.79)$$

Substituting the values of T_2, T_3, T_4 and T_5 in terms of T_1, from equations (7.65), (7.68), (7.69) and (7.74) respectively in equation (7.75) we get

$$p_m = \frac{C_v rP_1}{T_1 R(r-1)}[\{T_1 r^{(\gamma-1)} r_p - T_1 r^{(\gamma-1)}\} + \gamma\{r_c r_p r^{(\gamma-1)} T_1 - T_1 r^{(\gamma-1)}\} - \{r_p r_c^\gamma T_1 - T_1\}]$$

$$= \frac{C_v rP_1}{R(r-1)}\left[r^{(\gamma-1)}(r_p - 1) + \gamma r_p r^{(\gamma-1)}(r_c - 1) - (r_p r_c^\gamma - 1)\right]$$

But
$$\frac{C_v}{R} = \frac{1}{\gamma - 1}$$

Therefore
$$p_m = \frac{P_1 r}{(r-1)(\gamma-1)}[(r_p - 1)r^{(\gamma-1)} - \gamma r_p r^{(\gamma-1)}(r_c - 1) - (r_p r_c^\gamma - 1)] \quad ...(7.80)$$

7.8.3 Effect of Cut-off and Pressure ratios (r_c and r_p) on Thermal Efficiency and p_m of Dual Cycle

Effect of r_c : Equation (7.75) gives the value of thermal efficiency of dual cycle which is re-written below

$$\eta_{max} = 1 - \frac{1}{r^{(\gamma-1)}}\left[\frac{r_p r_c^\gamma - 1}{(r_p - 1) + \gamma r_p (r_c - 1)}\right] \quad ...(7.75)$$

When $r_c = 1$ *i.e.*, when $(V_4 - V_3)$, then from (7.75)

$$\eta_{max} = 1 - \frac{1}{(r)^{(\gamma-1)}}\left[\frac{(r_p - 1)}{(r_p - 1) + \gamma r_p (1 - 1)}\right]$$

$$= 1 - \frac{1}{r^{(\gamma-1)}}\left[\frac{(r_p - 1)}{(r_p - 1)}\right]$$

$$= 1 - \frac{1}{r^{(\gamma-1)}}$$

i.e., $\eta_{max} = \eta_{otto}$

Secondly when r_p : 1 *i.e.,* when $p_3 - p_2$ or when the points 2 and 3 concide, then equation (7.75) reduces to

$$\eta_{max} = 1 - \frac{1}{r^{(\gamma-1)}}\left[\frac{r_c^\gamma - 1}{\gamma(r_c - 1)}\right]$$

$$= \eta_{Diesel}$$

Thus from the above we conclude that otto cycle and diesel cycle are the special cases of dual cycle.

Further a perusal of the right hand side of equation (7.75) shows that the thermal efficiency of dual cycle will increase with the increase in value of r_c. That is increased supply of heat energy at constant pressure and smaller value of r_p or smaller duration of heat supply at constant volume.

Similarly in the expression (7.80) for p_m, if cut off ratio $r_c = 1$, when $V_4 = V_3$ the equation (7.80) reduces to

$$(p_m)_{Dual} = \frac{P_1 r}{(r-1)(\gamma-1)}[(r_p - 1)r^{\gamma-1} - (r_p - 1)]$$

$$= \frac{P_1 r}{(r-1)(v-1)}[(r_p - 1)(r^{\gamma-1} - 1)]$$

$$= (p_m)_{otto} \qquad \text{[from eqn. (7.35)]}$$

Again in equation (7.80) if the pressure ratio $r_p = 1$ *i.e.,* when $P_3 = P_2$, the value of p_m becomes.

$$(p_m)_{Dual} = \frac{P_1 r}{(r-1)(\gamma-1)}[\gamma r_p r^{\gamma-1}(r_c - 1) - (r_p r_c^\gamma - 1)]$$

$$= \frac{P_1 r}{(r-1)(\gamma-1)}[\gamma r^{(\gamma-1)}(r_c - 1) - (r_c^\gamma - 1)]$$

$$= (p_m)_{Diesel} \qquad \text{[from eqn. (7.58)]}$$

7.9 COMPARISON OF OTTO, DIESEL AND DUAL COMBUSTION CYCLES

Various factors which affect the performance of different air cycles are given below :

(*i*) Compression ratio
(*ii*) Maximum pressure
(*iii*) Heat supplied
(*iv*) Heat rejected
(*v*) Work output
(*vi*) Maximum temperature

Therefore depending upon the factors which are considered of prime importance for certain requirements, the performance of one of there cycles may be better than the others. Therefore we proceed below to compare their performance under different set of conditions.

7.9.1 For Same Compression Ratio and Same Quantity of Heat Input

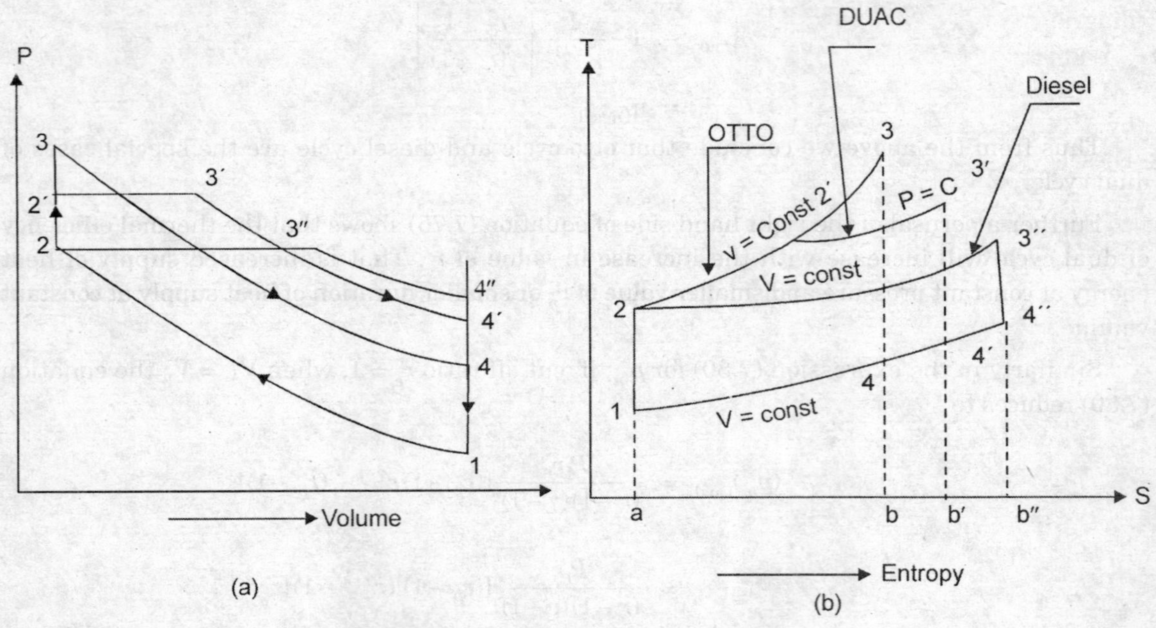

Fig. 7.10

The thermal efficiency η is given by the expression

$$\eta = \frac{\text{Work output}}{\text{Heat Input}}$$

if Q_1 = Heat Input

Q_2 = heat rejected

Then work output $= Q_1 - Q_2$

$$\therefore \quad \eta = \frac{Q_1 - Q_2}{Q_1} = 1 - \frac{Q_2}{Q_1}$$

If the Heat input Q_1 is same then the efficiency of any cycle will be greater if Q_2 smaller. In the present case it is desired that comparison of efficiency of Otto, Diesel and Dual cycles be made when the compression ratio is same and heat in put is same.

In Fig. (7.10 *a*) are drawn the pressure volume diagrams of the three cycles having the same heat input and all of them have the same initial parameters corresponding to state point 1 with isentropic compression 1-2 and all the three have the same compression ratio $r = \frac{V_1}{V_2}$.

Addition of heat is under varying condition for different cycles. Similarly Fig. (7.10*b*) shows these cycles on T.s. chart.

The otto cycle is represented, on these two figures, by 12341.

Dual cycle by 1 2 2′ 3′ 4′ 1 and

Diesel cycle by 1 2 3″ 4″ 1

Since the heat supplied is same for all the three cycles, the corresponding areas of T.S. diagrams for these three cycles are equal.

Further the isentropic expansion for each cycle continues upto the constant volume lines 1 – 4, 1 – 4 and 1 – 4″ respectively. That is on T–s. chart the heat rejected by Otto cycle is represented by *a* 14 *ba*, Dual cycle by *a* 14′ *b*′ *a* and Diesel cycle by *a* 14″ *b*″ *a*. Evidently heat rejected by Diesel cycle is the longest while by otto cycles is lowest. Mathematically it is represented by

$$(Q_2)_{otto} < (Q_2)_{Dual} < (Q_2)_{Diesel}$$

Thus as per the above discussion

$$\eta_{otto} > \eta_{max} > \eta_{Diesel} \quad ...(7.81)$$

7.9.2 For Same Heat input and Same Maximum Pressure and Temperature

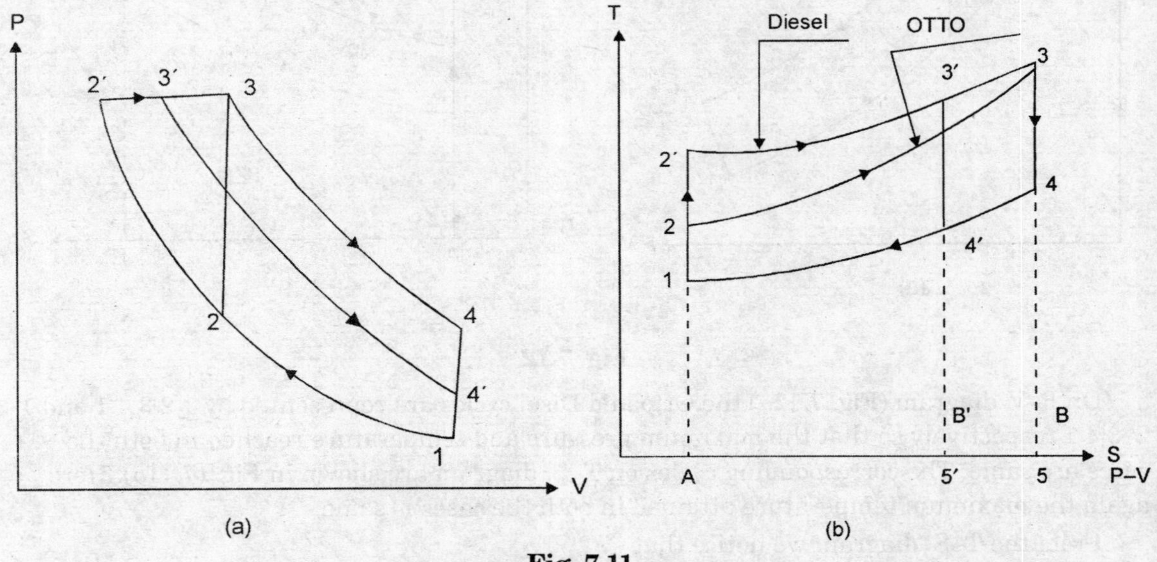

Fig. 7.11

On P–V. diagram Fig. (7.11*a*) the otto and Diesel cycles are represented by 1 2 3 4 1 and 12′ 3′ 4′ 1 respectively. The corresponding areas on T–S. diagram are 1 2 3 4 and 1*v* 3′ 4′. Since the maximum pressure is to be same in both the cycles the points 3 and 3′ on P–V. diagram must be on the same horizontal line corresponding to the maximum pressure value. Similarly as the heat input is same in the two cycles the areas *A* 2 3 *B A* and *A*2′ 3′ *B*′ *A* on T–S. diagram must be same.

From the T–S. diagram it is evident that the area A′ 4′ B′ A, which is the heat rejected by the diesel cycle, is less than the curve *A* 1 4 *B A* which is the heat rejected by the otto cycle. Thus the since work output

$$= Q_1 - Q_2$$

work out put of Diesel cycle > work out put of otto cycle.

is $\eta_{Diesel} > \eta_{otto}$

The efficiency of dual cycle will fall in between there two

i.e., $\eta_{Diesel} > \eta_{Dual} > \eta_{Otto}$

Further in the present case, it is clear from P–V. diagram that

Compression ratio of Diesel cycle > Compression ratio of otto cycle

7.9.3 Comparison for the Same Maximum Pressure and Maximum Temperature

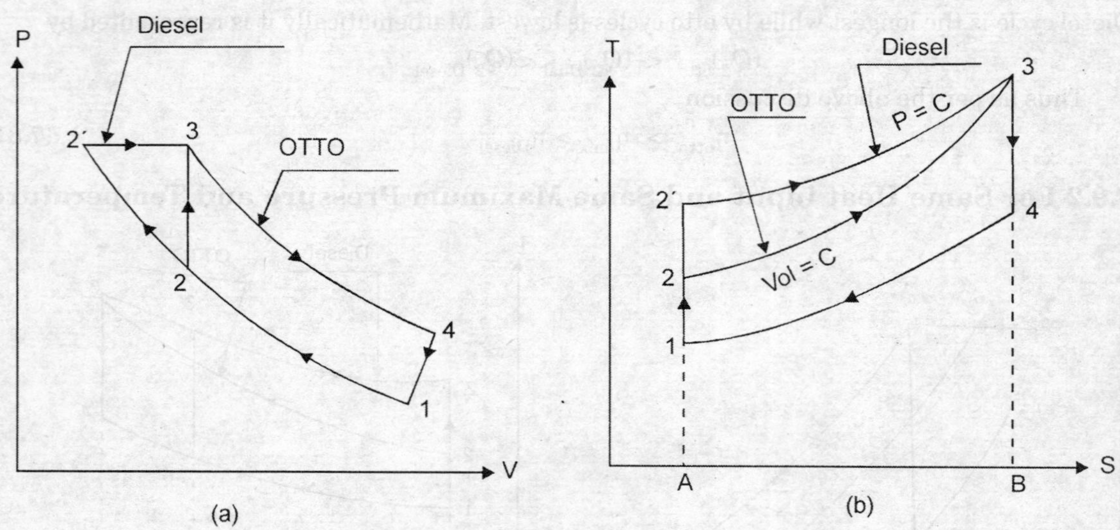

Fig. 7.12

On P–V. diagram (Fig. 7.12*a*) the Otto and Disel cycles are represented by 1 2 3 4 1 and 1 2′ 3 4 1 respectively so that the maximum pressure and temperature reached in both the cases are same. The corresponding cycles on T–S. diagram are shown in Fig. (7.11*b*). Here again the maximum temperature attained in both the cases is same.

From the T–S. diagram we notice that

(*i*) For Otto cycle

$(Q_1)_{Otto}$ = Heat supplied is represented by area 2 3 *B A* 2

$(Q_2)_{Otto}$ = Heat rejected is the area 4 1 *A B* 4.

(i) For Diesel Cycle

$(Q_1)_{Diesel}$ = Heat supplied = Area 2′ 3 *B A* 2′

$(Q_2)_{Diesel}$ = Heat repeated = Area 4 1 *A B* 4

From the above it is clear that

$$(Q_2)_{Otto} = (Q_2)_{Diesel} \quad (7.82a)$$

and $$(Q_1)_{Diesel} > (Q_1)_{Otto} \quad ...(7.82b)$$

Now output is given by

$$(W)_{Diesel} = (Q_1)_{Diesel} - (Q_2)_{Diesel} \quad ...(7.83a)$$

and $$(W)_{Otto} = (Q_1)_{Otto} - (Q_2)_{Otto} \quad ...(7.83b)$$

In the light of equations (7.82) and (7.83) we conclude that.

$$W_{Diesel} > W_{Otto} \quad ...(7.84a)$$

Which is term means that for the given conditions of maximum pressure and maximum temperature in Diesel, Otto cycle, Diesel cycle is more efficient than otto cycle. The efficiency of Dual cycle will fall in between these.

$$\eta_{Diesel} > \eta_{Dual} > \eta_{Otto} \quad ...(7.84b)$$

Example 7.16. *In an air standard diesel cycle the compression ratio is 15 and fluid properties at the beginning of compression are 100 kPa and 300 K.*

For the peak temperature of 1600 K, calculate the efficiency of the engine and work output per kg of air supply. *(PTU 2006)*

Solution : Refer to Fig. 7.7

Given data $$r = 15 = \frac{V_1}{V_2} \quad i.e., \quad V_1 = 15V_2$$

$$P_1 = 100 \text{ kPa}, \quad T_1 = 300 \text{ K}, \quad T_3 = 1600 \text{ K}$$

For Process 1-2 : Isentropic compression

$$P_1V_1^\gamma = P_2V_2^\gamma$$

i.e., $$P_2 = P_1\left(\frac{V_1}{V_2}\right)^\gamma = 100 \times (15)^{1.4} = 4431.26 \text{ kl}$$

Now $$\frac{P_1V_1}{T_1} = \frac{P_2V_2}{T_2}$$

$$T_2 = T_1\left(\frac{P_2}{P_1}\right)\left(\frac{V_2}{V_1}\right) = T_1\left(\frac{V_1}{V_2}\right)^\gamma\left(\frac{V_2}{V_1}\right)$$

$$= T_1\left(\frac{V_1}{V_2}\right)^{\gamma-1}$$

$$= 300 \times (15)^{0.4} = 886.25 \text{ K}$$

Process 2-3 : (Constant pressure)

$$\frac{V_2}{T_2} = \frac{V_3}{T_3}$$

$$r_c = \frac{V_3}{V_2} = \frac{T_3}{T_2} = \frac{1600}{886.25} = 1.8$$

$$\eta_{Diesel} = 1 - \frac{1}{r^{\gamma-1}}\left[\frac{r_c^\gamma - 1}{\gamma(r_c - 1)}\right]$$

$$= 1 - \frac{1}{(15)^{.4}}\left[\frac{(1.8)^{1.4} - 1}{(1.4)\,(1.8 - 1)}\right]$$

$$= 1 - 0.387 = 0.613 \text{ or } 61.3\%$$

Heat supplied $Q_1 = C_p(T_3 - T_2)$ per 1 kg air

Now
$$C_p - C_v = R$$
$$= 0.287$$
$$\frac{C_p}{C_v} = \gamma = 1.4$$
$$C_p - \frac{C_p}{1.4} = 0.287$$

$$\therefore \quad C_p = \frac{0.281 \times 1.4}{0.4} = 1.0045 \text{ kJ/kg} \cdot \text{K}$$

$$\therefore \quad Q_1 = 1.0045\,(1600 - 886.25)$$
$$= 716.96 \text{ kJ/kg}$$

Now
$$\eta = \frac{\text{Work done}}{\text{Heat input}}$$

$$\therefore \quad \text{Work done} = \eta \times \text{Heat input}$$
$$= 0.613 \times Q_1 = 0.613 \times 716.96$$
$$= 439.5 \text{ kJ/kg}$$

Example 7.17. *An engine working on dual combustion cycle has compression ratio 16 and explosion ratio 1.5. If the cut off takes place at 5.5% of the stroke, find the air standard efficiency of the cycle.*

Solution : Refer to Fig. 7.8

Given data
$$r = 16 = \frac{V_1}{V_2} \quad \text{or} \quad V_1 = 16V_2 = 16V_3 \text{ Becuase } \quad V_2 = V_3$$

$$r_p = 1.5 = \frac{P_3}{P_2} \quad \textit{i.e.,} \quad p_3 = 1.5\,P_2$$

$$V_4 = V_3 + 5.5\% \text{ of stroke volume}$$

Now stroke volume
$$= V_1 - V_2$$
$$= 16V_2 - V_2 = 15V_2$$

$$\therefore \quad V_2 = V_3 \times \frac{55}{100} \times 15V_2$$
$$= V_2 \times .825\,V_2 = 1.825\,V_2$$

$$r_c = \frac{V_4}{V_3} = \frac{V_4}{V_2} = \frac{1.825\,V_2}{V_2} = 1.825$$

From equation (7.25)

$$\eta_{\text{Dual}} = 1 - \frac{1}{r^{\gamma-1}} \left[\frac{(r_p r_c^{\gamma} - 1)}{(r_p - 1) + \gamma r_p (r_c - 1)} \right]$$

$$= 1 - \frac{1}{(16)^{1.4}} \left[\frac{1.5 \times (1.825)^{1.4} - 1}{(1.5 - 1) + (1.4)\,1.5\,(1.825 - 1)} \right]$$

$$= 1-\frac{1}{3.03}\left[\frac{2.4822}{2.2325}\right]$$

$$= 1-.3669$$

$$= 0.633 \quad \text{or} \quad 63.3\ \%$$

Example 7.18. *An engine working on dual combustion cycle has the following specifications.*

Compression ratio	= *16*
Stroke volume	= *0.009 m^3*
Pressure and temperature at the end of section stroke	= *1 bar and 300 K*
Maximum pressure in the cycle	= *68 bar*
Air fuel ratio	= *21:1*
Colorific value of fuel	= *43800 kJ/kg*

Find the ideal air standard efficiency and the power developed by the engine if there are 180 working cycles per minute. Take for air $\gamma = 1.4$ an $R = 0.287$ kJ/kg.K

Solution : Refer to Fig. 7.8

$$P_1 = 1\text{ bar} = 100\text{ kPa}$$

$$T_1 = 300\text{ K}$$

$$r = 16 = \frac{V_1}{V_2} \qquad \text{or} \qquad V_1 = 16V_2$$

$$V_S = 0.009\text{ m}^3$$

$$P_3 = P_4 = 68\text{ bar} = 6800\text{ kPa}$$

Now $V_1 = V_2 + \text{stroke volume}$

$$16V_2 = V_2 + .009$$

$$15V_2 = .009$$

$$V_2 = \frac{.009}{15} = 0.0006\text{ m}^3$$

$\therefore$ $V_1 = 16 \times .0006 = .0096\text{ m}^3$

$$\gamma = 1.4 \quad R = 0.287$$

$$C_p - C_v = R$$

$$C_p - \frac{C_p}{1.4} = R \qquad i.e., \qquad C_p = \frac{1.4R}{0.4} = \frac{1.4\times 287}{0.4} = 1.0045$$

and $C_v = C_p - R$

$$= 1.0045 - .287$$

$$= 0.7175$$

Isentropic Process 1-2 : $P_1V_1^{\gamma} = P_2V_2^{\gamma}$

or $P_2 = P_1\left(\frac{V_1}{V_2}\right)^{\gamma} = 100\times(16)^{1.4}$

$$= 4850.3\text{ kPa}$$

$$\frac{T_2}{T_1} = \left(\frac{V_1}{V_2}\right)^{v-1}$$

$$T_2 = T_1 \times (16)^{0.4} = 300 \times 3.0314 = 909.43 \text{ K}$$

Constant Volume Process 2-3

$$\frac{P_2}{T_2} = \frac{P_3}{T_3}$$

$$T_3 = T_2 \frac{P_3}{P_2} = 909.43 \times \frac{6800}{4850.3}$$

$$= 1275 \text{ K}$$

For Mass of Air at Point 1

$$P_1V_1 = mRT_1$$

$$m = \frac{P_1V_1}{T_1} = \frac{100 \times .0036}{.287 \times 300} = 0.001115 \text{ kg}$$

Heat Added During Constant Volume Process 2-3

$$= mC_v(T_3 - T_2)$$

$$= 0.01115 \times 0.7175\,(1275 - 909.43)$$

$$= 2.9246 \text{ kJ}$$

Therefore mass of fuel added during this process 2-3 (constant volume)

$$= \frac{2.9246}{\text{Calorific volume of fuel}}$$

$$= \frac{2.9246}{43800} = 6.6772 \times 10^{-5} \text{ kg}$$

Mass of total air at point 1 $=$ **0.01115 kg**

Since air fuel ratio is 21:1

Total fuel added in the cycle $= \frac{0.01115}{21} = 5.3095 \times 10^{-4}$ kg

Mass of fuel added during the process 3-4

$$= 5.3095 \times 10^{-4} - 6.6772 \times 10^{-5}$$

$$= 46.4178 \times 10^{-5} \text{ kg}$$

∴ Heat added during constant pressure process 3-4

$$= 46.4178 \times 10^{-5} \times 43800 = 20.331 \text{ kJ}$$

But heat added during constant pressure process 3-4 is also

$$= (\text{Total mass of air + Total mass of fuel}) \times C_p(T_4 - T_3)$$

$$= (0.01115 + 5.3095 \times 10^{-4}) \times 1.0045\,(T_4 - 1275)$$

$$= 0.011681 \times 1.0045\,(T_4 - 1275)$$

i.e., $20.331 = 0.0117335\,(T_4 - 1275)$

$$\text{or} \qquad T_4 - 1275 = \frac{20.331}{.0117335}$$

$$= 1732.73$$

$$\therefore \qquad T_4 = 1732.73 + 1275$$

$$= 3007.73 \text{ K}$$

For constant pressure 3 process 3-4

$$\frac{V_3}{T_3} = \frac{V_4}{T_4}$$

$$\text{Also} \qquad V_3 = V_2 = .0006$$

$$\therefore \qquad V_4 = V_3\left(\frac{T_4}{T_3}\right) = 0.0006 \times \frac{3007.73}{1275}$$

$$= 0.001415^4 \text{ m}^3$$

For isentropic expansion process 4.5

$$\frac{T_4}{T_5} = \left(\frac{V_5}{V_4}\right)^{\gamma-1} = \left(\frac{V_1}{V_4}\right)^{\gamma-1} \qquad [\because V_1 = V_5]$$

$$\text{or} \qquad \frac{T_4}{T_5} = \left(\frac{0.0096}{0.0014154}\right)^{0.4}$$

$$= (6.7825)^{0.4} = 2.1506$$

$$\because \qquad T_5 = \frac{T_4}{2.1506} = \frac{3007.73}{2.1506} = 1398.55 \text{ K}$$

Therefore heat rejected during constant volume process 5-1

$$= m_t C_v (T_5 - T_1)$$

where m_t = Total mass of air + Total mass of fuel

$$= (0.0115 + 5.3095 \times 10^{-4})$$

$$= 0.011681 \text{ kg}$$

$\therefore$ Heat repeated during 5-1 $= 0.011681 \times .7175\,(T_5 - T_1)$

$$= 0.008381\,((1388.55 - 300)$$

$= \mathbf{9.207\ kJ}$

Total heat added = Heat added during constant volume process 2-3 + Heat added during constant pressure process 3-4

$$= 2.9246 + 20.331$$

$$= 23.2556 \text{ kJ}$$

Therefore work done = Heat supplied – heat Reject

$$= 23.2556 \text{ kJ} - 9.207$$

$$= 14.0486 \text{ kJ}$$

$$\therefore \qquad \eta = \frac{\text{W.D.}}{\text{Heat supplied}}$$

$$= \frac{14.0486}{23.2556} = 0.6041 \quad \text{or} \quad 60.41\%$$

Power of Engine $= \text{W.D. per sec}$

$= \text{W.D./cycle} \times \text{cycles/sec}$

$$= 14.0486 \times \frac{180}{60}$$

$= 42.15 \text{ kW}$

Example 7.19. *A diesel engine working or Dual combustion cycle has compression ratio of 14. The pressure and temperature at the end of section stroke are 100 kPa and 350 K respectively. The maximum pressure in the cycle is 6000 kPa. During the power stroke heat at constant pressure is added during 0.03 of the stroke and rest of heat added at constant volume. Find the thermal efficiency take for air v = 1.4 and R = 0.287 kJ/kg · k.*

Solution : Refer to Fig. 7.8

Given data

$$r = 14 - \frac{V_1}{V_2} \quad \text{or} \quad V_1 = 14V_2$$

$$P_1 = 100 \text{ kPa}, T_1 \ 350 \text{ K}$$

$$P_3 = P_4 = 6000 \text{ kPa}$$

$$V_4 = V_3 + 0.03 \text{ stroke vol}$$

$$= V_3 + 0.03\,(V_1 - V_2) \qquad (\because V_3 = V_2)$$

$$= V_2 + 0.03\,(14V_2 - V_2)$$

i.e.,

$$V_4 = V_2(1 + 0.39) = \mathbf{1.39}\ V_2$$

And

$$C_p - C_v = R, C_p/C_v = 1.4$$

$$C_p - \frac{C_p}{1.4} = 0.287$$

$$C_p = \frac{0.287 \times 1.4}{0.4} = 1.0045$$

$$C_v = C_p = .287 = 1.0045 - .287 = 0.7175$$

The isentropic compression process 1-2

$$\frac{T_2}{T_1} = \left(\frac{V_1}{V_2}\right)^{v-1} = r^{v-1} = (14)^{0.4} = 2.8738$$

or

$$T_2 = T_1 \times 2.8738 = 350 \times 2.8738$$

$$= 1005.83 \text{ K}$$

and

$$P_1V_1^{\gamma} = P_2V_2^{\gamma}$$

$$\frac{P_2}{P_1} = \left(\frac{V_1}{V_2}\right)^{\gamma} = (14)^{1.4} = 40.2327$$

$$P_2 = 100 \times 40.2327 \text{ kPa}$$

$$= 4023.27 \text{ kPa}$$

For constant volume process 2-3

$$\frac{P_2}{T_2} = \frac{P_3}{T_3}$$

$$T_3 = T_2\left(\frac{P_3}{P_2}\right) = (1005.83)\left(\frac{6000}{4023.27}\right)$$

$$= 1005.83 \times 1.4913$$

$$= 1499.999 \text{ or } 1500 \text{ K}$$

At constant pressure process 3-4

$$\frac{V_3}{T_3} = \frac{V_4}{T_4} \text{ or } T_4 = \frac{V_4}{Vl3} \times T_3 = \frac{V_4}{V_2} T_3 = 1.39 \times 1500 = 2085$$

$$P_4(V_4)^{1.4} = P_5 V_5^{1.4} \quad \text{or} \quad \frac{P_5}{P_4} = \left(\frac{V_4}{V_5}\right)^{1.4}$$

or

$$\frac{P_4 V_4}{T_4} = \frac{P_4 V_4}{T_4}$$

$$= \left(\frac{V_4}{V_5}\right)^{1.4} \left(\frac{V_5}{V_4}\right)$$

$$= \left(\frac{V_4}{V_5}\right)^{0.4}$$

$$T_4 = T_4\left(\frac{V_4}{V_5}\right)^{0.4} = T_4\left(\frac{V_4}{V_1}\right)^{0.4} \qquad (\because V_5 = V_1 -$$

or

$$\frac{T_5}{T_4} = \left(\frac{1.3942}{V_1}\right)^{0.4} \qquad (\because V_4 = 1.39\, V_2)$$

$$= \left(\frac{1.39}{14}\right)^{0.4} = (0.09928)^{.4}$$

$$= 0.3969$$

$$T_5 = T_4 \times .3969 = 2085 \times .3969 = 827.54 \text{ K}$$

For unit mass of air Heat supplied

$$= C_v(T_3 - T_2) + C_p(T_4 - T_3)$$

$$= 0.7175\,(1500 - 1005.83) + 1.0045\,(2085 - 1500)$$

$$= 354.567 + 587.6325$$

$$= 942.1995 = 942.2 \text{ kJ/kg}$$

Heat rejected

$$= C_v(T_2 - T_1)$$

$$= 0.7175\,(827.54 - 350)$$

$$= 477.54 \text{ kJ/kg}$$

$$\text{W.D.} = 942.200 - 477.54 = 464.60$$

$$\therefore \quad \eta = \frac{464.66}{942.2}$$

$$= 0.4647 \quad \text{or} \quad 46.47\%$$

7.10 THE CARNOT CYCLE VS POWER CYCLES

The carnot cycle which has been introduced in earlier chapter is composed of four totally reversible processes *viz.* isothermal heat addition, isentropic expansion isothermal heat rejection and isentropic compression.

Evidently the carnot cycle can be executed in a closed system a slidly flow system (By using two turbines and two compressors) and either a gas or vapor can be utilised as its working fluid Further the carnot cycle is the most efficient cycle that can be executed between two thermal sources one at high temperature T_H and other called sink at low temperature T_L and its efficiency is given by

$$\eta_{\text{connot}} = 1 - \frac{T_L}{T_H} \qquad \text{...(7.85)}$$

However, reversible isothermal heat transfer is very difficult to achieve in actual practice because it will require very large heat exchanger and it would take infinitely large time which is contrary to the power cycles in practice as they take a fraction of a second to complete the cycle. Thus it is not possible to design and fabricate an engine that will work not only on carnot cycle but not even very close it.

But still the importance of carnot cycle is maintained for it being a standard against which an actual or ideal cycle can be compared. As shown above (eqn. 7.85) carnot efficiency is a function of temperature of source and sink. The inference that one can draw from equation 7.85 is that the thermal efficiency increases with increase in average temperature at which heat is added to the system or with the decrease in the average temperature at which the heat is rejected from the system.

In practice there is no limit that can be set for the temperature of source and sink. But certainly there is a limit to the heighst temperature in the cycle which is the maximum temperature that the components of the engine such as piston, gudgeon pin or even the turbine blades can withstand without expanding beyond certain limit so that it does not impair the smooth working of that particular component. Similarly the lowest temperature is limited by the cooling properties of the cooling medium being utilised in the engine.

Keeping in view the above facts following assumptions one made which are called the air-standard assumption for various cycles.

(*i*) The working fluid is air which continuously circulates in a closed loop and always behaves like an ideal gas.

PROBLEMS

1. How does gas power cycle differ from actual power cycle?

2. Why is the carnot cycle not suitable as an ideal cycle for all power cycles?

3. State whether the thermal efficiency of an actual cycle is less or more than that of carnot cycle operating between the same temperature limits. (Less)

4. What does the area enclosed by a cycle represent on P.V. diagram? (work)

5. What does the area enclosed by a cycle represent an T.S. diagram. (Work)

6. The carnot cycle efficiency expression *viz*, $\eta = 1 - \frac{T_L}{T_H}$ be of any help in improving the thermal efficiency of the actual cycle?

(**Ans.** Thermal η can be increased by raining T_H or lowering T_L)

7. State the air stand assumptions.

8. Do the internal combustion engines operate on closed or open cycle? Explain why?

(**Ans.** Open because at the end of exhaust stroke fresh air fuel mixture is sucked in the next cycle)

9. Define the following terms encountered in the reciprocating engines : (*i*) Bore (*ii*) stroke volume, (*iii*) Swept volume (*iv*) TDC (*v*) BDC (*vi*) clearance volume

10. State whether true or false "The mean effective pressure is always more than the atmospheric pressure." (False)

11. An engine working an closed cycle consists of the following four processes :

(*i*) Isentropic compression from 100 kPa and 300 K to 800 kPa.

(*ii*) Heat addition at constant volume to 1800 K.

(*iii*) Isentropic expansion to 100 kPa.

(*iv*) Heat rejection to initial state at constant pressure.

Draw the cycle or P.V. diagram. Find the temperatures at the end of compression and expansion strokes and also calculate the net work output of the cycle per unit mass of air and the thermal efficiency take $\gamma = 1.4$ and $R = 0.297$ kJ/kg · K.

(**Ans.** T_2 = 543.44 k, T_4 = 705.7 K, 494.08 kJ/kg, 54.8%)

12. An air standard cycle is executed in a closed system and consists of following four processes.

(*i*) Heat is added at constant volume from initial state of 100 kPa end 300 K to 300 kPa.

(*ii*) Heat is added at constant pressure to 1300 K

(*iii*) Isentropic expansion to 100 kPa.

(*iv*) Heat rejection to initial state at constant pressure.

Show that cycle as P.V. and T.s. diagrams. Find the temperatures at the end of first and third processes. Then calculate the total heat input and heat rejected per unit mass of air and the work done and efficiency of the cycle.

Take $\gamma = 1.4$ and $R = 0.287$ kJ/kg treat specific heats as constant.

(**Ans.** T_2 = 900 k, T_4 = 949.78 k, 832.3 kJ, 652.7 kJ, 179.6 kJ, 21.57%)

13. What is the relationship between the rpm of an actual 4-stroke petrol engine with the number of thermodynamic cycles?

(**Ans.** RP is twice the thermodyanamic cycles)

14. Show on P.V. diagram as how the thermal efficiency of otto cycle engine change with compression ratio and specific heat ratio of the working fluid.

15. Why in spark ignition engines high compression ratio is not used.

16. One otto cycle engine using air as working fluid and the other having same compression ratio is using Argon gas which engine will give higher efficiency $\gamma_{air} = 1.4, \gamma_{argon} = 1.667$ Explain why? (with Argon)

17. The compression ratio of an air standard otto cycle is 9.5 Prior to the isentropic compression process the air is at 100 kPa, 290 K and 600 cm^3. The temperature at the end of sentropic expansion process is 800 K. Assuming constant specific heats of air with $\gamma = 1.4$ and $R = 0.287$ kJ/kg determine (*a*) the highest temperature and pressure in the cycle (*b*) the amount of heat added (*c*) thermal efficiency (*d*) mean effective pressure.

(**Ans.** (*a*)1968.7 k, 6449.2 kPa; (*b*) 0.65 kJ, (*c*) 59.4 %, (*d*) 719 kPa)

18. How does a diesel engine differ from a petrol engine?

19. Define cut off ratio. How does it affect the thermal efficiency of Diesel cycle.

20. What is a dual cycle? Show how can the otto and diesel cycles be considered as special cases of dual cycle.

21. An air standard Diesel cycle has a compression ratio of 18.2. At the begining of compression process air is at 300 k and 100 kPa and 200 K at the end of heat addition process. Assuming the specific heat of air remains constant determine (*a*) the cut off ratio (*b*) heat rejection per unit mass of air and (*c*) the thermal efficiency. Take for air $\gamma = 1.4$ and $R = 0.287$ kJ/kg · K. (**Ans.** 2.089, 388.4 kJ/kg; 62.9%)

22. An oil engine with 20 cm cylinder diameter and 30 cm stroke works on diesel cycle. The initial pressure and temperature of the air used are 1 bar and 30°C. The cut off takes place 10% of the stroke find.

(*a*) Thermal efficiency.

(*b*) Mean effective pressure.

(*c*) The power developed by the engine if the working strokes per minute are 400.

Take compression ratio as 16 and $\gamma = 1.4$ and R = 0.287 kJ/kg · K. (UPTU 2005)

(**Ans.** 59%, 10.10 bar; 63.46 kw)

23. An ideal diesel engine has compression ratio of 20 and uses air as working fluid. At the beginning of compression process the pressure and temperature are 95 kPa and 20°C Maximum temperature in the cycle is limited to 2200 k. Determine the thermal efficiency and mean effective pressure. Assume constant specific heat for air and $\gamma = 1.4$, $R = 0.287$ kJ/kg · K.

(**Ans.** 63.5%, 933 kPa)

24. In an air standard diesel cycle the compression ratio is 15 and fluid properties at the beginning of the compression are 100 kPa and 300 K.

For a peak temperature of 1600 k, calculate the efficiency of the engine and the work output per kg of air supply. Assuming constant specific heat and take $\gamma = 1.4$ are $C_p = 1$ kJ/kg · K.

(**Ans.** 63.3%, 413.53 kJ/kg)

25. An oil engine working on ideal dual cycle has compression ratio of 12 and uses air as working fluid. At the beginning of the compression process air is at 100 kPa, 30°C and occupies a volume of 1.2L During the heat addition process 0.3 kJ of heat is added to air at constant volume and 1.1 kJ at constant pressure. Assuming specific heat remains constant determine the thermal efficiency Take $\gamma = 1.4$ and $R = 0.287$ kJ/kg · K.

(**Ans.** 59.4%)

□□□

CHAPTER 8

INTERNAL COMBUSTION ENGINES

8.1 INTRODUCTION

In general a device which converts, the heat energy into mechanical work is called a heat engine. The heat engines are classified as per the combustion of fuel whether it takes place outside the engine or within the engine itself. There are two types of heat engines.

1. External combustion engine
2. Internal combustion engines.

In the external combustion engines the fuel which may be coal, crude oil, bagasse (which is left over after crushing the sugar cane in the sugar mills) is burnt in the fire grate of the work and the heat energy so produced is used to convert water into stream. This high pressure stream is then fed to stream engine turbine when the mechanical work is produced. On the contrary in the internal combustion engines the fuel is burnt inside the cylinder. The gaseous products of the combustion then from the working substance which is responsible for moving the piston and produce mechanical work at the crank shaft. Evidently in the former case the steam which is the working substances, is produced outside the engine, it is known as external combustion engine while in the later case the fuel is burnt inside the engine cylinder is called the internal combustion engine.

8.1.1 Relative Merits of I.C. Engines Over External Combustion Engines

(*i*) I.C. Engines are compact and smaller in size for the same power output. They have lower, weight to power ratio.

(*ii*) Higher overall efficiency. Their efficiency ranges from 30 to 35% while that of external combustion engines is in the order of 15-20%.

(*iii*) They are easy and quick to start as compared to stream engines.

(*iv*) Maintenance and running cost low.

(*v*) Vehicles for defence purposes can only he run on I.C. Engines. Steam engines can not be used for this.

8.1.2 Important Applications of I.C. Engines

1. Automobiles, locomotives, ships, aircraft, defence vehicles. Earth moving machinery, public road transport vehicles.
2. Stand bye, domestic as well as industrial generations at times of power cut, failure of electric supply.
3. Farm machinery like tractors combines, thrashers etc.

8.2 CLASSIFICATION OF I.C. ENGINES

I.C. Engines are classified in number of ways, depending upon the design, fuel and systems on which their working depends.

8.2.1 Cycle of Operation

(*a*) Otto cycle engines

(*b*) Diesel cycle engines

(*c*) Dual cycle engines.

8.2.2 Fuel Used

(*a*) Petrol engines

(*b*) Diesel engines

(*c*) Light diesel engines.

8.2.3 Number of Cylinder

(*a*) Single cylinder engines

(*b*) Multi cylinder engines.

8.2.4 Number of Strokes per cycle

(*a*) Four stroke engines

(*b*) Two stroke engines.

8.2.5 Method of Ignition

(*a*) Spark ignition engines

(*b*) Compression ignition engines.

8.2.6 Method of Cooling as

(*a*) Air cooled engines

(*b*) Water cooled engines.

8.2.7 Method of Governing

(*a*) Quantity governing
(*b*) Quality governing
(*c*) Hit and Miss governing.

8.2.8 On the Basis of their Use

(*a*) Automobile engines
(*b*) Aircraft engines
(*c*) Marine engines
(*d*) Locomotive engines.

8.2.9 Valve Mechanism

(*a*) Over head valve engines.
(*b*) Side value engines.

8.2.10 Cylinder Arrangement

(*a*) In line cylinder engines
(*b*) V-Type engine
(*c*) Radial engine
(*d*) Opposed piston engines.

8.2.11 Fuel Injection Method

(*a*) Carburettor injection engine
(*b*) Air injection engine.

8.2.12 Lubrication System

(*a*) Splash lubrication system for light dirty engines.
(*b*) Pressure lubrication for heavy dirty engines.

8.2.13 Speed

(*a*) High speed engines where the crank shaft speed is more than 900 rpm.
(*b*) Low speed engines where the speed is less than 400 rpm.

8.3 CONSTRUCTIONAL FEATURES OF I.C. ENGINES

In Fig. 8.1 are shown various principal parts of an I.C. Engine. Most of the commonly used I.C. Engines are either Petrol engines working on Otto cycle or diesel engines working on diesel cycle. Basic constructional details are mostly similar and both have parts which are essentially common. Therefore to avoid repetition the common parts are grouped together as the sub-section 'A' below 1.

(*A*) Parts common to both petrol and diesel engines.

(*i*) **Cylinder and cylinder Head. Cylinder** is the main part of the engine where in the piston reciprocates, due to the explosion pressure generated by the combustion of air fuel mixture present in the cylinder above the piston. Cylinder is stationary and is generally made of cast-iron or alloy steel and has a cylinder linear mode of nickel chrome steel inserted in it for easy maintenance or replacement when it is worm out. Cylinder head is used to close the top end of the cylinder and is bolted to the cylinder with a gasket in between to make leak proof. The cylinder hard are housed the value sects and ports and supports in the value actuating mechanism. Usually it is made of C.I. but where weight is the consideration it is of aluminium alloy.

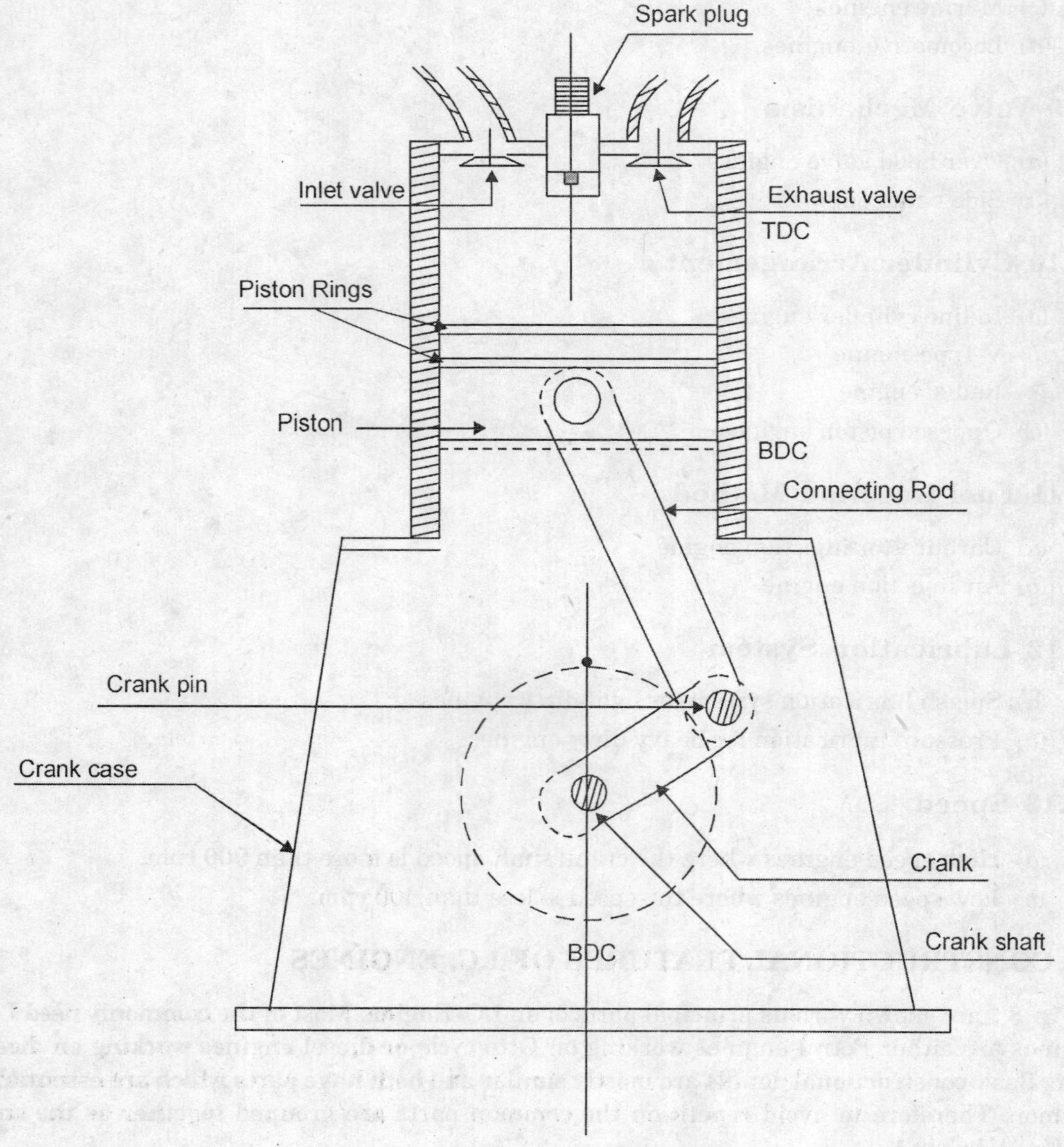

Fig. 8.1. Nomenclature of engine parts

(ii) **Piston and Piston Rings :** Piston is a tumbler shaped cylinder with its crown facing the combustion space. On the outer side of the walls of the piston, grooves are provided to accommodate the piston rings. The function of the piston and the piston rings is to confine the gases in the combustion space so that they can exert full force on the piston during the expansion stroke. The force exerted by the gases on the piston is transmitted through the connecting rod to the crank shaft as a result of which it starts rotating. The piston is usually made of grey C.I. or Aluminium alloy for high speed engines.

Due to the running clearance between the piston and the cylinder the gases under pressure tend to leak over the piston. Then leakage is checked by the presence of piston rings. They are rings with an inclined cuts that they can be assembled into the grooves cut on the piston. There are two sets of rings. One set which is put on the upper side is called the compression rings, they are the ones to check the leakage of gases past the piston. The other set which is put on the lower side called the oil rings check the lubricating oil go to the gases chamber side, and also to provide uniform lubrication for the smooth reciprocating movement of piston in the cylinder.

(iii) **Connecting Rod :** The force of combustion gases acting on the piston is transmitted to the crank shaft through the connecting rod and also it converts the reciprocating motion of piston into rotary motion of crank shaft. The connecting rod is pin-jointed to the piston with the help of what is called the gudgeon pin which is forced fit into the housing specially made in the piston. This end of the connecting rod which is connected to the piston through pin-joint is called small end of this connecting rod. The other end of the connecting rod has a split bush bearing arround the crank shaft. It is essentially split type bush bearing so that the assembly of this connecting rod end and the crank can be achieved without interfering in the complete rotation of the rank shaft.

The connecting rod is made of T-section forged steel so as to have maximum strength to weight ratio and is tapered to become smaller towards the piston and so that it could be assembled with case on the piston side.

(iv) **Crank and Crankshaft :** The crank is part and parcel of crank shaft. The crank shaft consists of as many cranks as are the number of cylinders. The annular position of the cranks is so adjusted that it is in consonance with the firing order of the gases in cylinder for smooth transmission of the motion from piston to crank shaft and the components are properly balanced to avoid vibrations.

The power required to operate other accessories like cam shaft, oil pumps distributor etc. is drawn from the crank shaft with proper gear or belt drives.

Both the crank and crank shafts are steel forgings as one unit along with the machined portion for mounting it in bearings. Fig. 8.2 shows a typical crank shaft for 4 cylinder engine.

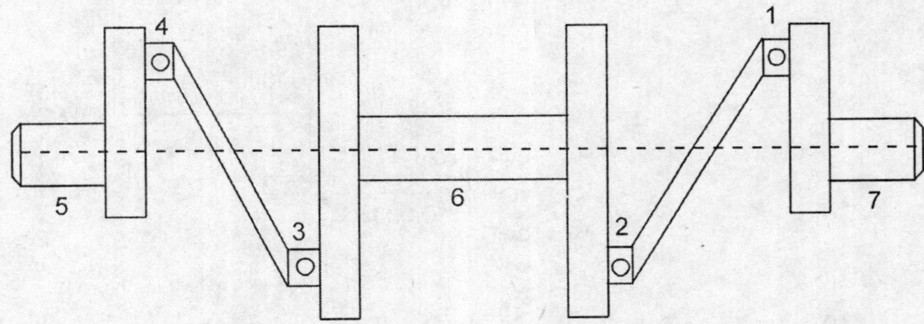

Fig. 8.2

***(v)* Crank Case :** The engine cylinders, and crank shaft are mounted in a housing which is called crank case. It also serves as an oil sum for the storage of lubricating oil.

***(vi)* Cam Shaft and Valve Mechanism :** In Fig. 8.3 is shown the cam shaft and valve mechanism. The inlet and outlet valves which are fitted in the cylinder head are operated by the cam shaft through cam, cam followers, push rods and rocker arms. For an engine working on four stroke cycle the inlet and exhaust values operate once per cycle. In other words it takes two revolutions of the crank shaft to complete one cycle of operation. As a result of this the cam shaft speed is exactly half that of the crank shaft so that the gear train through which the cam shaft gets motion from or crank shaft has a speed reduction of 2. The valves are mushroom shaped and are known as poppet valves. They have conical surfaces which are ground and lapped on the conical seats fitted on the cylinder head. When cam lifts the push rod through the cam follower which in turn actuates one end of the rocker arm that turns around the central pin to press it valve down against the spring force so as to open it, once the valve opens. The valve returns to its seating due to spring force after the cam has rotated into dwell state.

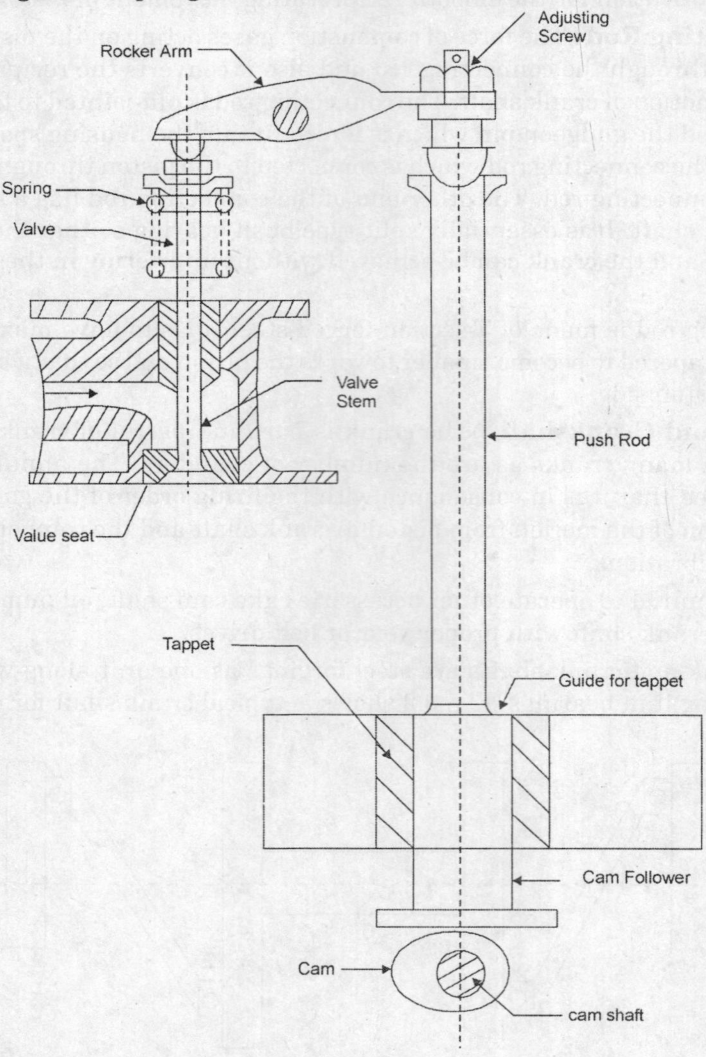

Fig 8.3

(*vii*) Flywheel is a heavy perfectly balanced wheel fitted on the crank shaft. Its purpose to smoothen the output power of the engine despite the fact-that out of the 4 strokes there is power output only during one stroke. It absorbs the extra energy during the power stroke and during the rest of three strokes it supplies the energy to the crank shaft. thus it reduces to cyclic variation of speed of the crank shaft.

(*viii*) Governor : Its purpose is to adjust the supply of the fuel so that engine power output matches with the load and the crank shaft of engine rotate, at almost constant speed.

(*B*) Components Which Are Not Common on Petrol and Diesel Engines

(*i*) Carburettor and Spark Plug : Main function of the carburettor is to prepare and supply air fuel mixture in an atomised form to the cylinder of the petrol engine so that it is easily and quickly burnt when the spark plug gives spark due to instantaneous supply of electric current passing through it.

(*ii*) Fuel Pump and Injector : In diesel engines fuel pump is used to deliver fuel with in the combustion chamber through a nozzle so that it is atomised in spray form. The atomised form ensures that sufficient quantity of air (which surrounds the atomised particles of fuel) is there for proper instant burning of the fuel.

8.4 WORKING OF FOUR STROKE PETROL ENGINE

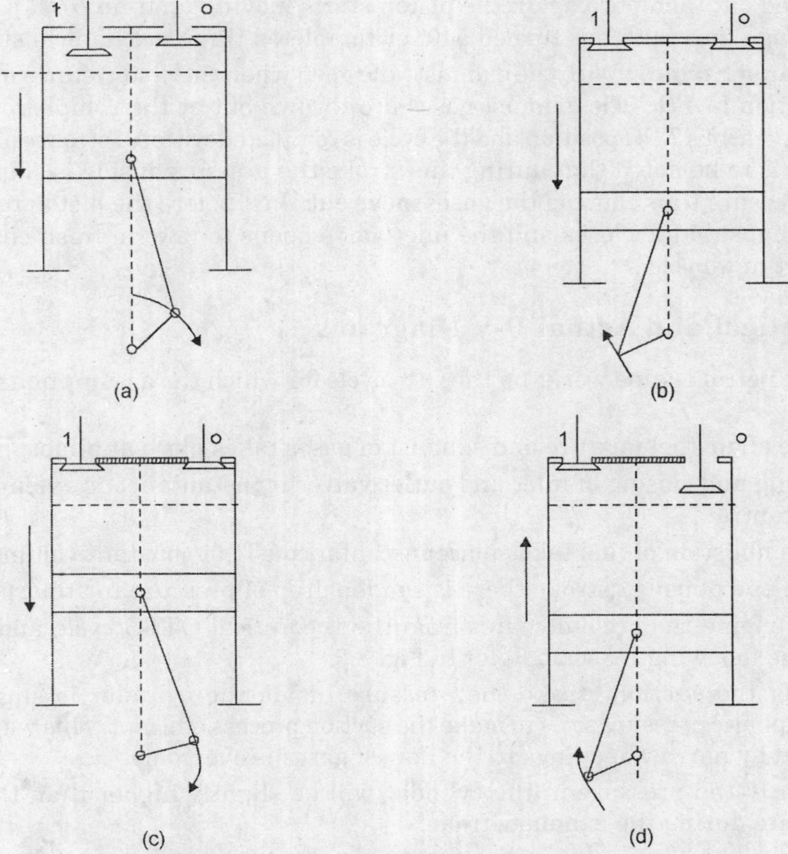

Fig. 8.4

Four stroke petrol engine works on OTTO cycle in which piston executes four complete strokes (or two mechanical cycles) within the cylinder and the crank shaft completes two revolutions for each thermodynamic cycle.

***(i)* Suction Stroke :** Initially the piston is at *TDC* position, the inlet value is open and outlet valve is closed. The piston moves downwards with the rotation of the crank and the pressure in the cylinder is reduced below the atmospheric pressure. Thus air fuel mixture is sucked in to fill the space vacated by the movement of the piston and it continues till the piston reaches *BDC* position thereby completing one stroke. The inlet valve closes and in crank has rotated through 180°.

***(ii)* Compression Stroke :** With both the valves closed the piston moves upwards from *BDC* towards *TDC* thereby compressing the charge inside the cylinder to the clearance volume. The pressure continuously increases and along with it the temperature of the air mixture also rises.

***(iii)* Power or E xpansion storoke :** Just before the piston reaches *TDC* position the crank has completed one revolution (or 360° rotation of crank) the spark plug fires still both inlet and outlet valves closed, the mixture ignites instantly. Then the pressure and temperature of the system increases. The high pressure gases force the piston down towards *BDC* which in turn force the crank shaft rotate producing the useful work output. This stroke is known as the power or expansion stroke. Just before the end of this stroke the exhaust valve opens and the gases are pushed out then once again the piston starts moving up from *BDC* position towards *TDC*. At this stage the crank has turned 540° or completed three mechanical strokes.

***(iv)* Exhaust Stroke :** With the exhaust vale open when the piston starts moving upwards from *BDC* position to *TDC* the exhaust gases are thrown out at the completion of 4th stroke when the piston reaches *TDC* position and the cycle is completed within 4 strokes or two revolution of the crank. It is to be noted that during this stroke the pressure inside is slightly more than atmospheric pressure thus causing the gases move out. Just before the piston reaches the *TDC* position the exhaust value closes and the inlet valve opens to take in fresh charge of air fuel mixture to start new cycle.

8.4.1 Theoretical and Actual P-V Diagrams

As stated above petrol engine works on the Otto cycle for which the assumption made are listed below.

- *(i)* Intake of air-fuel mixture and exhaust of gases takes place at atmospheric pressure.
- *(ii)* Opening and closing of inlet and outlet valves is instantaneous when the piston is at dead centre.
- *(iii)* The combustion of fuel takes place instantaneously at constant volume.
- *(iv)* At the end of power stroke there is sudden drop of pressure to atmosphere pressure.

In actual engine the cycle deviates from the theoretical OTTO cycle and so do the *P-V* diagrams for the following reasons. Refer to Fig. 8.5.

- *(i)* During the suction stroke the pressure inside the cylinder is slightly less than atmospheric pressure so as to make the suction process efficient. This way the resistance offered by narrow openings to the flow of gases is overcome.
- *(ii)* Similarly the pressure in the cylinder will be slightly higher than the atmospheric pressure during the exhaust stroke.
- *(iii)* The compression and expansion have been assumed isentropic *i.e.* internally reversible adiabatic. But friction is always present to render irreversibility to the process. In

addition there is considerable interaction of heat between the gases and the cylinder walls In other words it is not truly adiabatic.

(*iv*) Due to inertia the opening and closing of the valves will never be instantaneous as assumed.

(*v*) Opening of valves and ignition of charge is also not at dead centres. These events actually occur at some degrees of crank rotation on either side of dead centres to get better charging and scavenging (throwing out of burnt gases) effects.

(*vi*) At high temperatures the specific heats of exhaust gases, which contain CO_2 and water vapour, the will decrease and not remain constant at room temperature value. In other words γ will decrease thereby resulting in lowering of pressure and temperature during the actual cycle.

The cummulative effects of these rotates will modify the *P-V* diagrams. The theoretical and actual P-V diagrams for the cycle are shown in Fig. 8.5.

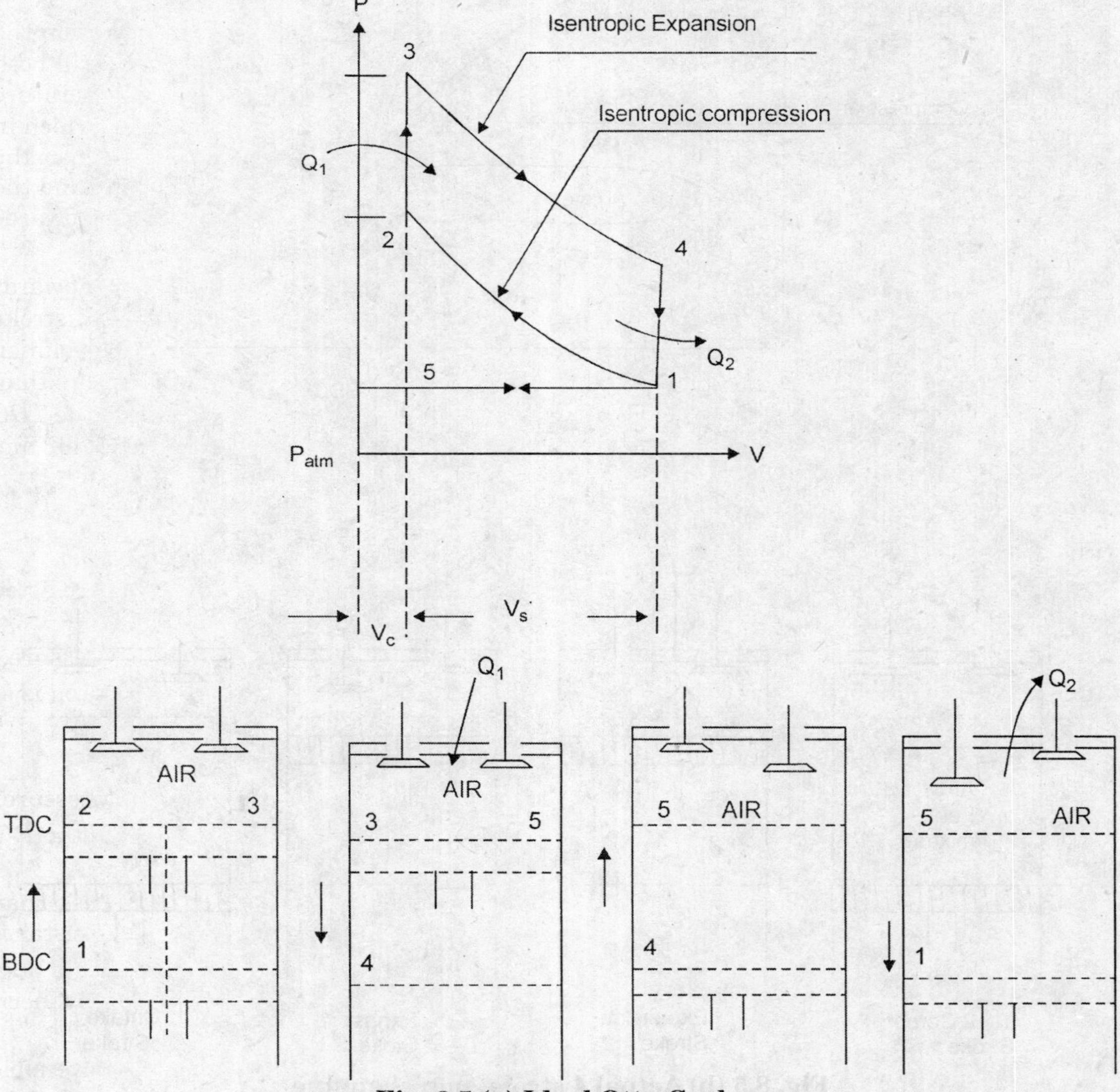

Fig. 8.5 (a) Ideal Otto Cycle

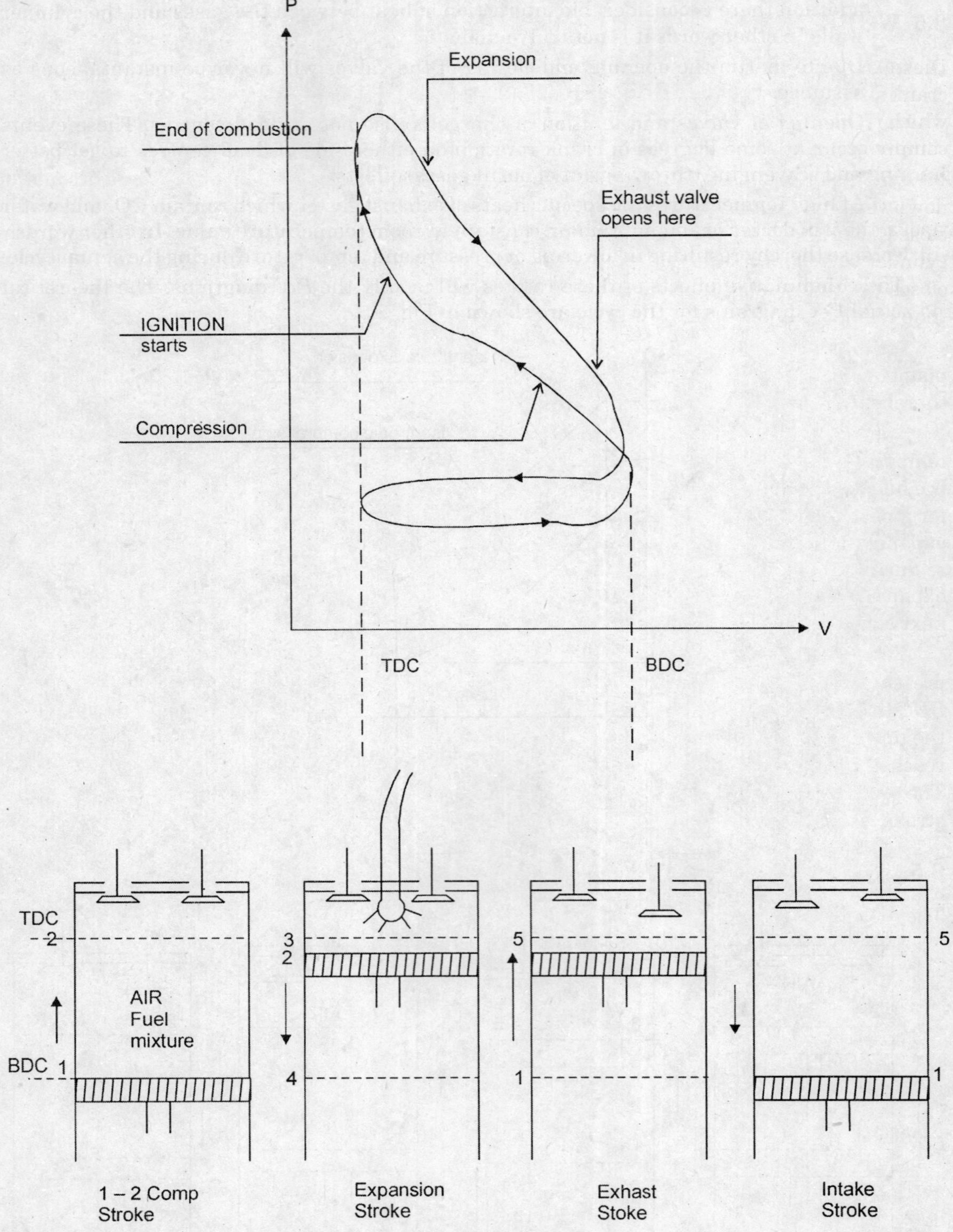

Fig. 8.5 (b) Actual 4 stroke petrol engine

8.5 WORKING OF FOUR STROKE DIESEL ENGINE

Diesel engine works an diesel cycle which is completed in four strokes or in two revolutions of crank shaft. Each stroke is completed in 180° of crank rotation. In the case of diesel engine, fuel which is diesel oil is injected with the help of an injector fitted on the cylinder head into highly compressed air. Details of the four strokes are given below.

***(i)* Suction Stroke :** The suction stroke starts when the piston is at *TDC* and its motion downward towards *BDC* starts. At this instant the inlet value is open and the exhaust value is closed. Due to vacuum created by the movement of piston, air continues to be sucked in the cylinder through the inlet value till the piston reaches the *BDC* position when the inlet value closes to bring an end to the suction stroke to be followed by the compression stroke. By this time crank has turned 180°.

***(ii)* Compression Stroke :** Corresponding to the next 180° of the crank rotation, the piston moves up for the *BDC* towards the *TDC* when both the inlet and outlet values are closed, thereby compressing the air sucked in the cylinder during the suction stroke.

(iii) **Power stroke (Expansion Stroke) :** As a result of compression the pressure and temperature of air rises considerably which ignites the fuel that is sprayed in the cylinder with the help of injector just before the piston reaches the *TDC*. The fuel burns at constant pressure for a part of the next stroke when the fuel supply is cut off and both valves are closed. Due to expansion of higher pressure gases in the cylinder the piston moves down towards the *BDC*. By so during mechanical power is obtained at the crank shaft. Thus it is known as power stroke or expansion strokes. When the piston reaches near the *BDC*, the exhaust value opens and the next stroke ensues which is called the exhaust stroke.

***(iv)* Exhaust Stroke :** After the expansion of the gases during the power stroke the pressure and temperature of the burnt gases decreases and when the piston reaches near the *BDC* the pressure falls to atmospheric pressure value and the burnt gases are pushed out due to the upward movement of the piston. This stroke is thus called exhaust stroke when the piston reaches near the *TDC* the exhaust valve closes and the inlet valve opens to start the new cycle. The pressure in the cylinder is slightly higher than the atmospheric pressure during the exhaust stroke which makes the burnt gases flow out the exhaust value.

8.5.1 Theoretical and Actual P-V Diagrams

The diesel engine works on diesel cycle for which following assumptions have been made.

(*i*) The suction and exhaust takes place at atmospheric pressure.

(*ii*) Both the inlet and exhaust values open instantaneously.

(*iii*) Compression and expansion process are reversible adiabatic or isentropic.

(*iv*) There is sudden drop in pressure to atmospheric pressure at the end of expansion stroke.

In actual engine the cycle deviates from the theoretical diesel cycle and so do the P-V diagrams due to following reasons. Refer to Fig. 8.6.

(*a*) During suction stroke the pressure inside the cylinder is slightly less than the atmospheric pressure so as to make the suction process efficient. This way the resistance offered by narrow openings to the flow of gases is overcome.

(*b*) Similarly the pressure in the cylinder will be slightly higher than the atmospheric pressure during the exhaust stroke.

(*c*) The compression and expansion have been assumed isentropic is internally adiabatic. But friction is always present to render irreversibility to the process. In addition there is considerable heat interaction between the gases and cylinder walls. In other words it is not truly adiabatic.

(*d*) Due to inertia the opening and closing of valves will never be instantaneous as assumed but take some finite time. Thus these events start taking place at some degrees of crank rotation on either side of the dead centre.

(*e*) At high temperature, the specific heats of exhaust gases, which contain CO_2 and water vapour, will decrease and not remain constant at room temperature value. Thus the values of γ will decrease thereby resulting in lowering of pressure and temperature during the actual cycle.

The cummulative effect of these factors will modify the P-V diagram. The theoretical and actual P-V diagrams for the cycle are shown in Fig. 8.6.

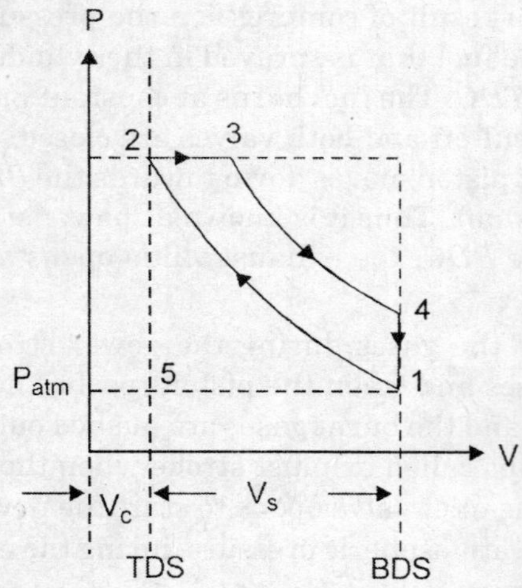

(a) Theoretical P-V diagram of diesel engine

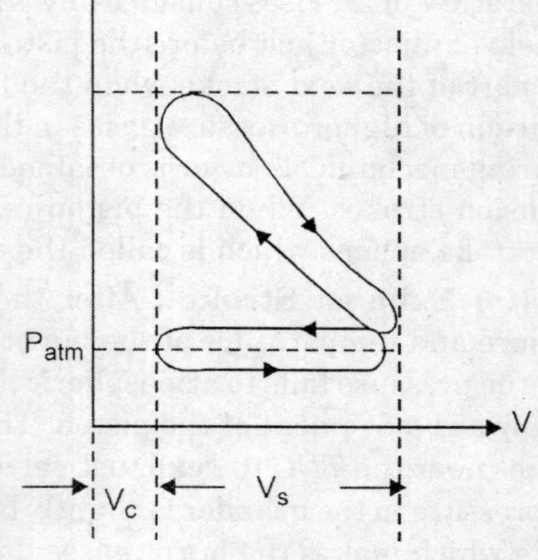

(b) Actual P-V diagram or diesel engine

Fig. 8.6

8.6 COMPARISON BETWEEN 4-STROKE PETROL AND DIESEL ENGINE

S. No.	Petrol Engine (S.I. Engine)	Diesel Engine (C.I. Engine)
1.	Working cycle is Otto cycle.	1. Working cycle diesel cycle.
2.	Fuel used is petrol.	2. Fuel used is diesel.
3.	During suction stroke the engine draws a mixture of petrol vapour and air.	3. Diesel engines draws, during suction stroke, atmospheric air only.
4.	carburator is used to mix air and petrol in the desired	4. The injector is used to inject fuel in the form of spray in the high pressure and

	proportion and to supply it to the engine during suction stroke.		high temperature air at the end of compression stroke.
5.	Compression ratio during compression stroke is between 6 to 8.	5.	Compression ratio is between 15 to 20.
6.	Pressure at the end of compression stroke is about 1000 kPa.	6.	Pressure at the end of compression stroke is 3500 kPa.
7.	The charge is ignited with the help of a spark plug.	7.	The high temperature of air gained due to high compression ratio is responsible for the ignition of the fuel spray.
8.	For same power petrol engine is lighter and cheaper due to smaller compression ratio.	8.	For same power diesel engine is heavier as compression ratio is higher consequently cost is higher.
9.	Starting is easy as compression ratio is low.	9.	Starting is more difficult as compression ratio is higher.
10.	For same compression ratio efficiency is higher.	10.	For same compression ratio efficiency is lower. But high compression ratio is a must otherwise the fuel will not get ignited as the temperature at the end of compression stroke will not be sufficiently high to ignite the fuel.
11.	Maintenance cost is low.	11.	Maintenance cost is high.
12.	Running cost high due to costlier fuel.	12.	Running cost low as diesel oil is cheaper.
13.	Maximum thermal efficiency is about 26%.	12.	Maximum thermal efficiency is about 40%.
14.	There are high speed engine with speed values from (3000 to 5000 rpm).	13.	Low speed engines equally 1500-3000 rpm.
15.	There are quantity control engines. Carburettor prepares the air fuel mixture. To control the load air fuel ratio is increased or decreased as per requirement.	14.	There are quality control engines. The ratio of fuel/air is changed by changing the proportion of fuel being injected by the injector.
16.	Risk of over heating more as the efficiency is lower the losses will be higher.	16.	Over heating problem is less acute because efficiency is higher and loss of heat is lesser.

8.7 TWO STROKE ENGINES PETROL AND DIESEL

In two stroke engines the working cycle is completed in two strokes or in one revolution of the crank shaft. The suction and exhaust strokes are eliminated. As a matter of fact the suction and exhaust are combined with the compression and expansion strokes. In these engines three ports replace the two valves. These three ports are cut in the cylinder wall. They are called the inlet port, transfer port and the exhaust port. The opening and closing of these ports is controlled by the movement of the piston itself.

There are two different methods to achieve the desired objectives. First one is to employ a separate pump outside the engine cylinder to compress the charge (air-fuel mixture from carburetor in case of petrol engine or air alone in the case diesel engine) before forcing it in the cylinder. This method is used in the case of multi cylinder engines.

The second method is meant for single cylinder engines. This is known as crank-case compression system in which with the downward (from *TDC* towards *BDC*) movement of the piston the charge in the crank case gets compressed on the under side of the piston. The working of this is explained below. Refer Fig. (8.7).

8.7.1 Working of two stroke engine

The crank case is filled by the charge through the inlet port. The charge consists of mixture of air and petrol prepared by the carburettor in the case of petrol engine or only fresh air from atmosphere in the case of diesel engine. The piston is near the TDC position, during compression stroke, and the charge is compressed into the clearance volume above the piston. At this position the charge is ignited by spark plug in the case of petrol engine (in the cse of diesel engine the fuel is injected in the form of spray with injector). Due to the burning of the fuel the temperature and pressure in the cylinder above piston rise to push the piston down. The piston is connected to the crank shaft through connecting rod so that the power is transmitted to crank shaft. When the piston moves down it opens the exhaust port, Theres by making the burnt gases to escape through the exhaust port and simultaneously compressing the charge within the crank case. When the piston moves downward a bit more it uncovers the transfer port so that the (compressed) high pressure charge present in the crank case is transferred to the cylinder on the upper side of piston. The compressed charge will have a tendency to escape directly along with the exhaust gases (through the exhaust port). To check the charge is deflected towards the spark plug (or injector in the case of diesel engine) with the help of deflector which is an integrated part of the crown of the piston. The deflector also helps to push out the burnt gases through the exhaust port. The throughing out of this burnt gases with the help of fresh charge is called the scavenging which continues till the piston reaches *BDC* position. Then the piston moves up and it opens the inlet port and covers the transfer port and then the exhaust port. Due to upward movement of the piston the pressure in the crank case starts falling to come to atmospheric pressure value. Then the fresh charge enters the crank case through the inlet port. Simultaneously the charge above the piston goes and getting compressed till the piston reaches near the TDC and turns the cycle process continues. Evidently it takes only one revolution (or the strokes) of the crank shaft to complete the cycle.

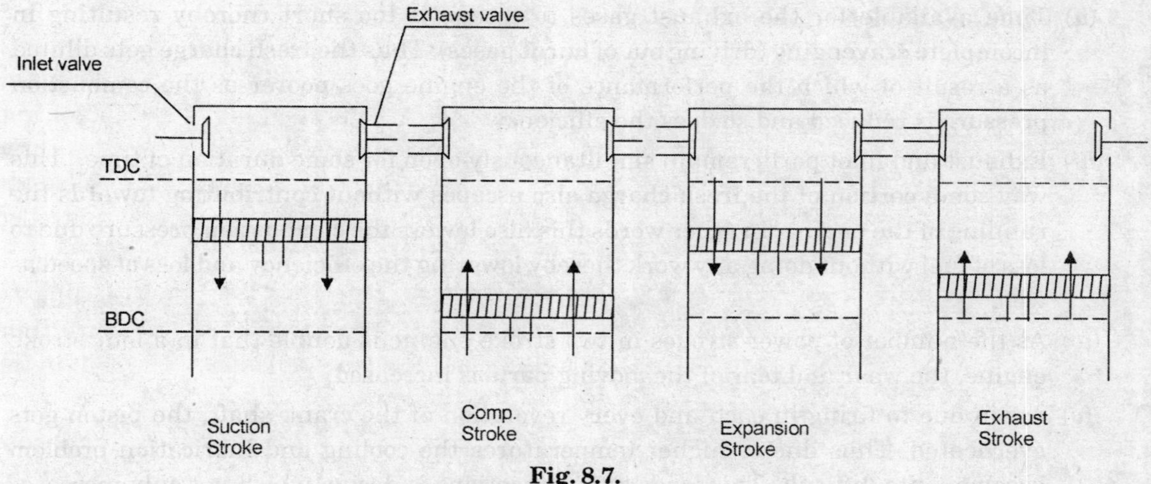

Fig. 8.7.

8.8 RELATIVE MERITS AND DEMERITS OF TWO STROKE AND FOUR STROKE ENGINES

8.8.1 (*a*) Merits of Two Stroke Engines

(*i*) For same power rating two stroke engine is simpler in design and having less components. Consequently they are not only easy to manufacture but easy to operate.

(*ii*) Comparatively lighter in weight for the same power output. Hence they are widely used on scooters, motor cycles and autorickshaw.

(*iii*) They have no valve and valve actuating mechanisms. Thus not only makes them simple in design and efficient in working. The ports are opened and closed by the piston itself consequently they are cheaper and user friendly.

(*iv*) Easy to start. Low maintenance cost.

(*v*) Simple construction, no valve actuating mechanism, four ports less space required light weight.

(*vi*) The number of working strokes are twice than that in four stroke engine. Which results in more uniform torque output and then these engines need very light flywheel.

(*vii*) Less friction due to absence of suction and exhaust strokes, this results in higher efficiency.

(*viii*) In this absence of value actuating mechanism, there is no cam shaft cam, tappet, piston rod, rocker arm, springs etc. This also conributes towards higher efficiency and also the cost of two stroke engine is much less than that of 4 stroke engine of the same power rating.

(*ix*) Much easier to start.

8.8.2 Demerits of Two Stroke Engines

(*i*) For the same stroke and clearance volumes the compression ratio is much smaller due to large volume of crank case. The net effect is to lower the combustion pressure, poor efficiency.

(*ii*) Time available for the exhaust gases to escape is too short thereby resulting in incomplete scavenging (driving out of burnt gases). Thus the fresh charge gets diluted as a result of which the performance of the engine goes poorer as the combustion pressure is reduced and so does the efficiency.

(*iii*) Exhaust and inlet ports remain simultaneously open for some duration of time. This way some portion of the fresh charge also escapes without contributing towards the running of the engine. In other words this also lowers the combustion pressure due to loss of fuel without doing any work thereby lowering the efficiency and loss of speed as well.

(*iv*) As the number of power strokes in two stroke engine is double that in a four stroke engine, the wear and tear of the moving parts is increased.

(*v*) Again due to firing in each and every revolution of the crank shaft, the piston gets overheated. Thus due to higher temperatures the cooling and lubrication problem become quite difficult. This leads to modifications is design which not only increases the manufacturing cost but also the operating cost as the consumption of the lubricant also increases due to higher temperature.

(*vi*) Two stroke engine is more noisy due to sudden escape of burnt gases.

8.9 APPLICATION OF TWO STROKE ENGINES

Two stroke petrol engines find their use in those situations where light weight and compactness are the primary considerations along with low initial cost. Thus they find their use in scooters, motorcycles and mopeds.

Similarly two stroke opposed piston diesel engines find their use in the marine engines because of the space limitations.

PROBLEMS

1. What do you understand by external combustion engine and internal combustion engine ? Give example of each of them.

2. What do the terms S.I. and I.C. Engine represent ?

3. Enumerate different ways of classifying I.C. Engines.

4. Give the merits and demerits of I.C. Engines and external combustion engines.

6. Give various applications of I.C. Engines.

7. Explain the function of following parts of I.C. Engines.

(*a*) Piston (*b*) Connecting rod

(*c*) Gudgeon pin (*d*) Crank pin

(*e*) Crank (*f*) Crank shaft

(*g*) Cam shaft (*h*) Cams

(*i*) Rocker arm (*j*) Fly wheel.

8. What is the relationship between the speeds of crank shaft and cam shaft of a 4-stroke petrol engine ?

9. With the help of neat sketches explain the working of a 4-stroke petrol engine.
10. Give the sketches of theoretical *P-V* diagram and the actual *P-V* diagram of a 4-stroke petrol engine. Give reasons for deviations in them.
11. Give the comparison between 4-stroke petrol and diesel engines.
12. Differentiate between 4-stroke and two stroke petrol engines.
13. Why the effective compression ratio in a 2-stroke petrol engine is less than that in 4-stroke engine for the same clearance and stroke volumes ?
14. Explain at to why there is no cam shaft in a two stroke engine.
15. Why does the 2-stroke engine produce more noise ?

CHAPTER 9

MECHANISMS AND MACHINES

9.1 INTRODUCTION AND DEFINITIONS

***(i)* Mechanism :** A mechanism is a combination of resistant bodies (called links) so arranged that the motion of one link causes predictable and definite motion of the other links according to the law depending on the nature of the combination. The mechanism modifies the motion. Spring toys, watches are examples of mechanisms.

***(ii)* Machine :** Machine is combination of mechanisms that serves to impart definite motion to the parts and transmit and/or modify the available energy into some forms of desired work. The machine receives energy in some available form and with its help perform some particular form of work. Petrol and diesel engines, lactic, compressor are examples of machines.

9.2 THEORY OF MACHINES

Theory of machines is a branch of engineering which deals with the motion and forces between various components of machine. This subject is divided into two parts viz. Kinematics of machine are dynamics of machines.

9.2.1 Kinematics of Machines

Within the purview of kinematics of machines we deal with the study of motion of various components (called kinematic links) of which the machine is built without the consideration of forces acting on them or being transmitted by them.

9.2.2 Dynamics of Machines

This branch of theory of machines deals with the steady of forces acting on various machines elements.

9.3 RIGID AND RESISTANT BODIES

A rigid body is one which does not suffer any change in shape or change in dimension when acted upon α by a set of forces. In other words the distance between any two points on a rigid body does not change due to the application of external force on this body. In actual practice such a rigid body does not exist. All the known solids deform under the action of external forces due to the inherent elasticity present in them but the deformation will depend on the elastic properties of the material and the size of the body. In case the deformation of the body under the action of forces is not appreciable so that it does not bring in the noticeable changes in the configuration of the machine to affect the working and performance of the machine.

Therefore a body which behaves as a rigid body for the purpose it has to serve under given loading condition is known as Resistant Body. The resistant body will transmit the required forces, torgue and power without any appreciable deformation. However belts and ropes, own through are flexible act a resistant bodies when transmitting power. These links transmit only tensile load. Even fluid which compressed in hydro active process, hydro active jacks and brakes act as resistant bodies.

9.4 STRUCTURE AND MACHINE

A structure is combination of resistant bodies capable of transmitting forces or carrying loads but having no relative motion between parts of the machine. There is only straining of the parts due to this forces acting on union. The frame of a machine is a structure which supports the moving parts and regulates the path, or kind of motion, of stone of the parts.

Machine receives energy in some available form and with its help perform some particular form of work. On the other hand there is no energy transformation with useful work in the case of structure.

The links or elements which comprise the machine transmit with motion and power. However the structure machines only sustain the loads and do not transmit any motion or power.

Bicycle is a machine, however, the frame of bicycle is a structure. Railway locomotive is a machine but railway linkage is a structure.

9.5 KINEMATIC LINK OR ELEMENT

As pointed out earlier in Article 9.2.1, the subject of machine kinematics deals with the analysis of motion of various components or members of the machine, with which the machine is built or assembled are called the kinematic links or elements as they from part of the study of kinematics of the machine.

The kinematic link or element which in short is referred to as 'link' is element which has motion relative to some other part. The link is a resistant body or an assembly of resistant bodies connected to each other in such a way that these assembled parts have no ralative motion among themselves. For example, the piston, piston rod and cross head of steam engine though being separate parts, are connected with each other rigidly form one link.

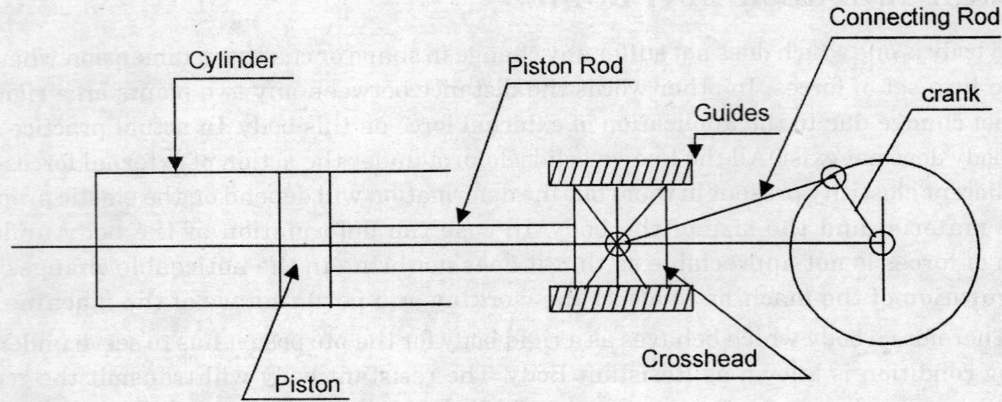

Fig. 9.1.

In Fig. 9.1 is shown a slider crank mechanism for converting reciprocating motion of this piston into rotary motion of the cranke shaft of a steam engine. The details of the four links are given below :

1. Link 1 is the fixed frame, along with the cylinder and crank shaft, bearing housing
2. Link 2 crank shaft, fly wheel etc.
3. Link 3 connecting rod.
4. Link 4 a piston, piston rod, cross head, as one link.

9.6 TYPES OF LINKS

There are three types of links which are as follows :

***(i)* Rigid Link.** A link is said to be rigid if it does not deform in any form while transmitting motion. In other words the distance between any two points on the link remains constant. In actuality such a link is purely a hypothetical. However in real links, the deformation is very insignificaut and it does not alter the characteristics of the motion which it is required to transmit.

***(ii)* Flexible Link.** A link which gets partly deformed while transmitting motion is called a flexible link *e.g.* belts, ropes, springs.

***(iii)* Fluid Link.** In case the motion is transmitted through the fluid by pressure or compression it is called fluid link *e.g.* Hydraulic jacks, hydraulic brakes, hyudraulic presses.

9.7 KINEMATIC PAIR

Combination of two links or elements which are joined together in such a way that their relative motion is completely or successfully constrained is known as a Kinematic Pair.

In the slider crank mechanisms shown in Fig. 9.1, following Kinematic pairs are there :

(*i*) Crank and connecting rod.

(ii) Connecting rod and piston rod.

(*iii*) Piston and cylinder.

(*iv*) Crank shaft and frame along with bearing housing.

It may be seen that in each of these pairs, the two links which are connected together have completely successfully constrained relative motion between them.

Types of constrained motions are described below;

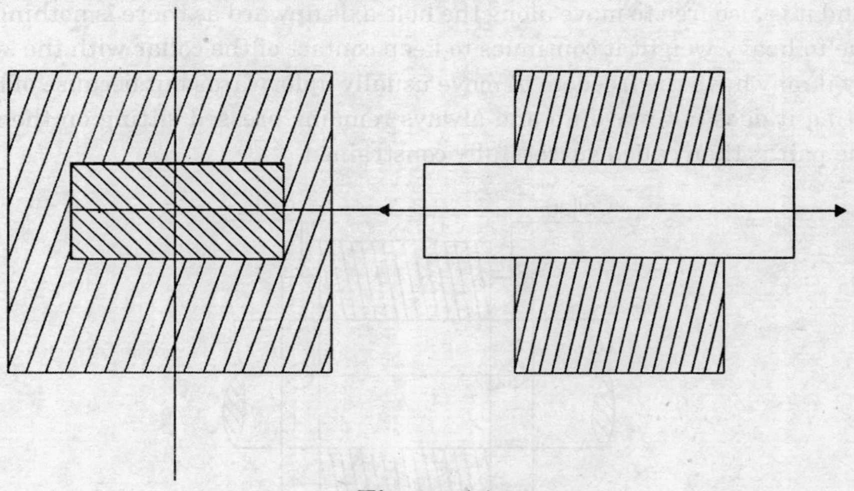

Fig. 9.2 (a)

(*a*) Completely Constrained Motion

When the motion between a Kinematic pair is limited to a definite direction whatever may be the direction of the force applied, then it is said to be completely constrained motion. For example a rectangular bar moving in a rectangular hole. The block can have motion only along its longitudinal axis.

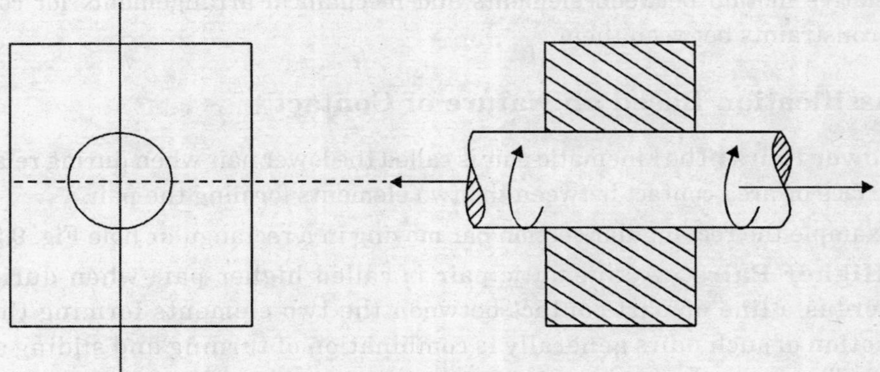

Fig. 9.2 (b)

(*b*) Incompletely Constrained Motion

When the motion can take place in more than one direction independently it is called incompletely constrained. For example the motion of round bar in a round hole.

The round bar can move along its longitudinal axis as well as it can rotate in the round hole. These two motions are independent of each other.

(*c*) Successfully Constrained Motion

Heavily loaded round shaft is supported through a collar on a supporting belock. The shaft can rotate and it is also free to move along the hole axis upward as there is nothing to check its. However due to heavy weight it continues to keep contact of the collar with the support. Thus theoretically it may have the freedom to move usually upforwards but because of heavy weight it is supporting it does not move up and always remains pressed sitting on the support. The motion of the pair is then called successfully constrained.

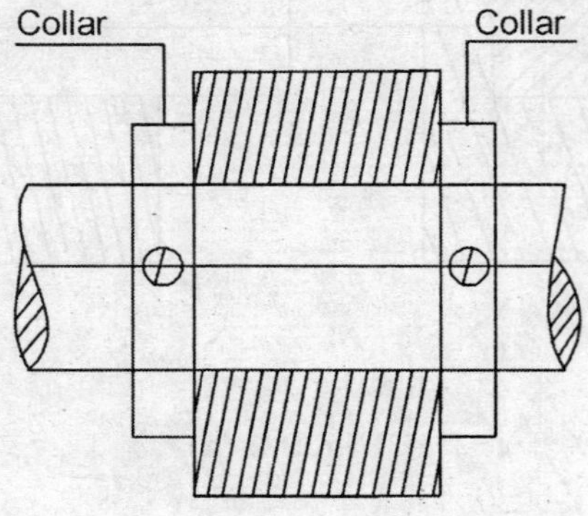

Fig. 9.2 (c)

9.8 CLASSIFICATION OF KINEMATIC PAIRS

Classification of kinematic pairs is done in three different ways depending of the nature of contact, relative motion between elements and mechanical arrangements for complete and successful constraints between them.

9.8.1 Classification Based on Nature of Contact

(*i*) Lower Pairs : The kinematic pair is called the lower pair when during relative motion there is surface or area contact between the two elements forming the pair.

For example the rectangular section bar moving in a rectangular hole Fig. 9.2 (a).

(*ii*) Higher Pairs : A kinematic pair is called higher pair when during relative motion there is a line or part contact between the two elements forming the pair. The relative motion of such pairs generally is combination of turning and sliding as shown in Fig. 9.2 (b). The examples are

(*a*) Belt and Roller learnings.

(*b*) Cams and follower pair Fig. (9.4).

(*c*) Belt and rope drives.

(d) Toothed gears.

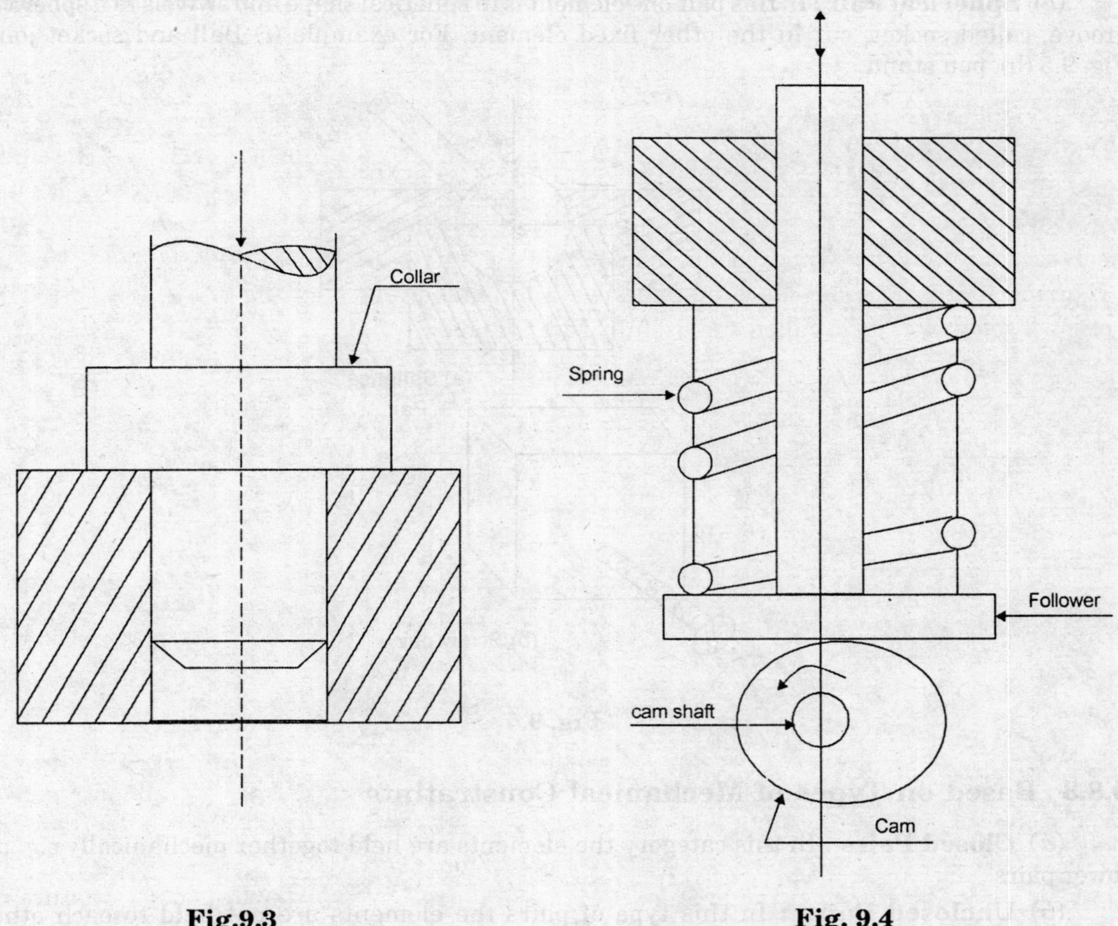

Fig.9.3

Fig. 9.4

9.8.2 Based on Nature of Relative Motion

***(a)* Sliding Pairs :** The links of the sliding pair have completely constrained sliding relative motion e.g. cross head in guides of the steam engine Fig. (9.1) piston and cylinder in steam engine Fig. (9.1). Rectangular section rod in rectangular hole Fig. 9.2 (a).

***(b)* Turning Pair :** In turning pair one element or link is constrained to revolve about a fixed axis of the other link. For example (i) wheel of car about the axle, shaft with collars on both ends, Fig. 9.2 (c), *(ii)* Crank shaft turning in its bearings Fig. (9.1), crank and big end of connecting rod Fig. (9.1).

***(c)* Rolling Pair :** In rolling pair one of the links is fixed and the other rolls in it. For example ball and roller bearings, Railway engine wheel rolling on rail.

***(d)* Screw Pair :** In the screw pair one element of the pair has both sliding and turning motion relative of the other link and the sliding motion is obtained by means of the screw fixed into the first link and threads cut in the other link. For example nut and bolt as shown in Fig. 9.5 (a).

(e) Spherical Pair : In this pair one element is in spherical shape and swivels in a spherical groove, called socket, cut in the other fixed element. For example (*i*) Ball and socket joint Fig. 9.5 (b), pen stand.

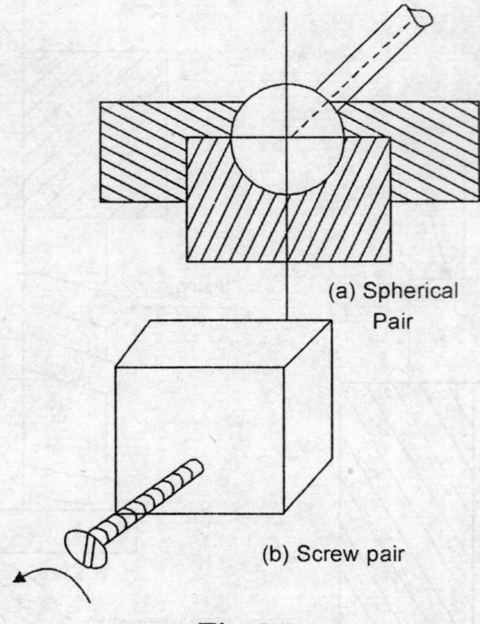

Fig. 9.5

9.8.3 Based on Types of Mechanical Constratints

(a) Closed Pairs : In this category the elements are held together mechanically *e.g.* all lower pairs.

(b) Unclosed Pairs : In this type of pairs the elements are not held to each other mechanically rather they are made to stay in contact with each other by spring force or by gravity. For example in cam and follower mechanism the contact between the cam face and the follower is maintained with the help of spring force Fig. (9.4) and hence is called Force-Closed Pair.

9.9 TYPES OF LINKS AND JOINTS

(a) Binary Links and Joints. When two links are joined at the same connection it is called a Binary Joint e.g. in Fig. 9.6 (*a*), 9.6 (*b*) the joints shown at points A, B, C and D are Binary joints and the links 1, 2 and 3, 4 are Binary Links.

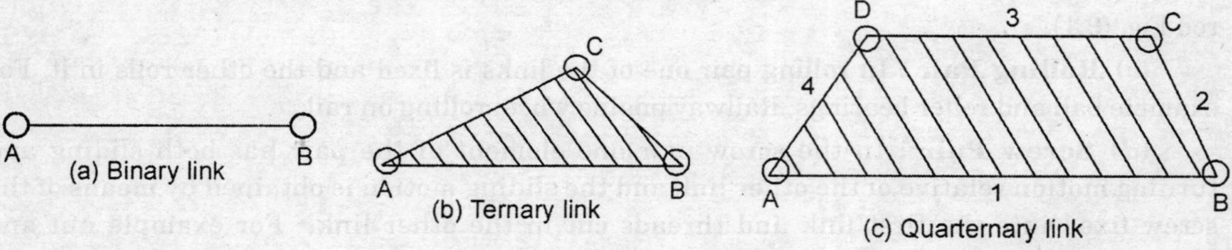

Fig. 9.6

Thus a link is classified into binary, ternary and quarternary depending upon the ends on which turning pairs can be placed. The links shown in Fig. 9.6 (*a*) and (*b*) are rigid links and there is no relative motion between the joints within the link.

(*b*) Ternary Joint and Links. When three links are jointed together at one connection, it is known as Ternary Joint *e.g.* in Fig. 9.7 (*a*) joint *D* is a ternary joint while joints *A*, *B*, *C*, and *E*, are binary joints. Similarly a body with three elements is called ternary link *e.g.* elements 2 and 3 in Fig. 9.7 (*b*) and 2 and 3 in Fig. 9.7 (*c*) are example of ternary links.

(*c*) Quarternary Link or Joint. When four links are jointed at the same connection, it is called a quaternary joint and a body with four elements is defined as quarternary link e.g. of link in Fig. 9.7 (*c*).

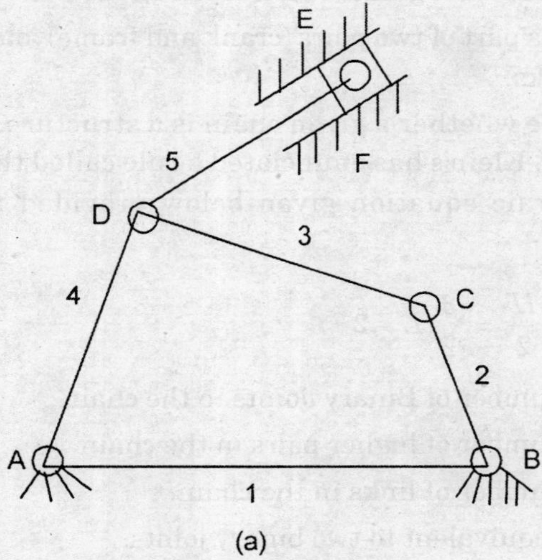

(a)

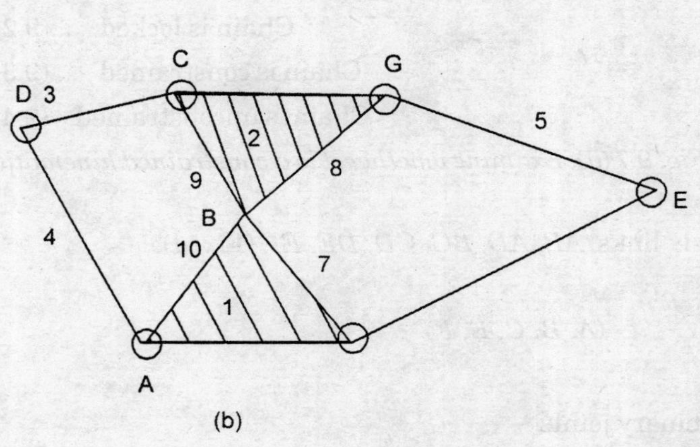

(b)

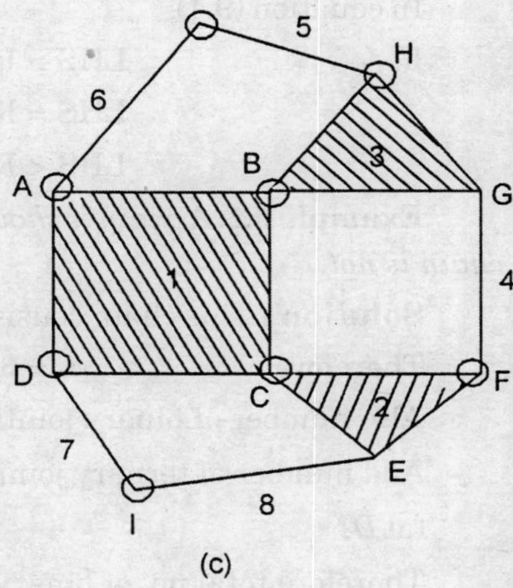

(c)

Fig. 9.7

9.10 KINEMATIC CHAIN

When a number of kinematic pairs are joined together in such a way that each link forms part of two pairs and the relative motion between the links or elements is completely or successfully constrained and the last link joins the first link, then this arrangement or combination of links is called a Kinematic chain.

In the light of the above statement we now proceeds to examine the reciprocating engine mechanism Fig. (9.1) In this the pairs are listed below

(*i*) Crank shaft with bearing fitted in the housing frame.

(*ii*) Crank with connecting rod.

(*iii*) Connecting rod with crosshead and piston.

(*iv*) Crosshead and piston with guides fitted in the frame.

Evidently crank forms part of two pairs (crank and frame) and (crank and connecting rod) and so on for the other links.

In order to determine whether a given chain is a structure, or a kinematic chain or an unconstrained chain, A.W. Klein's has enunciated a rule called the "Criterion of constraints" in the form of an algebraic equation given below provided the chain is having plane motion.

$$J + \frac{H}{2} = \frac{3}{2} L - 2 \quad \text{...(9.1)}$$

where J = Number of Binary Joints in the chain

H = Number of higher pairs in the chain

L = Number of links in the chain.

(*i*) A Ternary joint is equivalent to two binary joints.

(*ii*) A Quarternary joint is equivalent to three binary joints.

In equation (9.1).

If LHS > RHS Chain is locked ...(9.2)

LHS = RHS Chain is constrained ...(9.3)

LHS < RHS Chain is unconstrained (9.4)

Example 9.1. *A chain is shown in Fig. 9.7 (a). Examine whether it is a constrained kinematic chain is not.*

Solution : This chain consists of six links, *AB, AD, BC, CD, DE, EF* (Fixed).

Therefore $L = 6$

Also number of binary joints = 5 (*A, B, C, E, F*)

And number of ternary joints = 1

(at *D*) $\simeq$ 2 binary joints

Therefore total no. of binary joints

$$= 5 + 2 = 7$$

Further $H = 0$

as there is no higher pair. Thus substituting the values in Klein equation (9.1)

$$\text{LHS} = J + \frac{H}{2}$$

$$= 7 + 0 = 7$$

$$\text{RHS} = \frac{3}{2} \times L - 2$$

$$= \frac{3}{2} \times 6 - 2 = 7$$

That is LHS = RHS

Therefore in the light of expression (9.3) the chain is constrained kinematic chain.

Example 9.2. *A chain is shown in Fig. 9.7 (b). Examine as per Klein equation whether this is constrained Kinematic chain or not.*

Solution : Here the number of links = 10

Joints 1 Binary (at $B + R$) = 2

Ternary (A, E, C, G) = $4 \approxeq 4 \times 2$

= 8 binary

Quarternary B = $1 \approx 1 \times 3$

= 3 binary

Therefore Binary joints = 2 + 8 + 13

= 13

There is no higher pair J or $H = 0$

Therefore LHS of Klein's equation (9.1)

$$\text{LHS} = \text{J} + \frac{H}{2}$$

$$= 13 + 0 = 13$$

$$\text{RHS} = \frac{3}{2}\ \text{L} - 2$$

$$= \frac{3}{2} \times 8 - 2$$

$$= 12 - 2 = 10$$

We see that LHS > RHS

Therefore as per equation (9.2) the given chain is locked.

Example 9.3. *In Fig. 9.8 one shown four different chains in the shape of polygons. Examine as per Klein's equation whether they are constrained kinematic chains.*

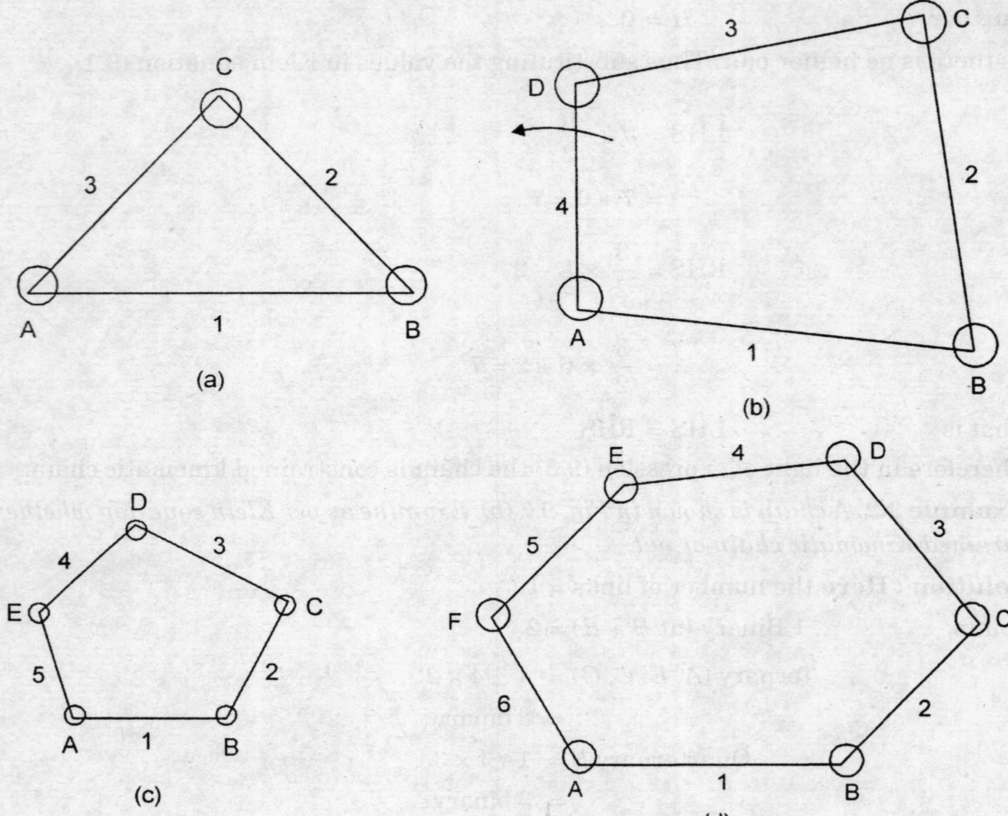

Fig. 9.8

Solution : (*a*) In Fig. 9.8 (a)

$$L = 3$$

$$J = 3$$

$$H = 0$$

as there is no higher pair.

Therefore LHS of equation (9.1)

$$\text{LHS} = 3$$

$$\text{RHS} = \frac{3}{2}\ L - 2$$

$$= \frac{3}{2} \times 3 - 2 = 4.5 - 2$$

$$= 2.5$$

That is $\text{LHS} > \text{RHS}$

Thus the chain is locked

(*b*) In Fig. 9.8 (*b*), $L = 4$

$$J = 0$$

$$H = 4$$

LHS of equation (9.1)

$$\text{LHS} = J + \frac{H}{2}$$

$$= 4 + 0 = 4$$

$$\text{RHS} = \frac{3}{2} \times L - 2$$

$$= \frac{3 \times 4}{2} \times -2$$

$$= 6 - 2 = 4$$

or $\qquad$ LHS = RHS

Thus as per equation (9.3), the chain in the shape quadrilateral having 4 links is a constrained kinematic chain.

(*c*) In the chain $\qquad L = 5$

$J = 5$

$H = 0$

Therefore LHS of equation (9.1)

$$= J + \frac{H}{2}$$

$$= 5 + 0 = 5$$

RHS of equation (9.1) $\qquad = \frac{3}{2} L - 2$

$$= \frac{3}{2} \times 5 - 2$$

$$= 7.5 - 2 = 5.5$$

or $\qquad$ LHS < RHS

That is the chain is unconstrained. Thus as per equation (9.3).

(*d*) Here $\qquad L = 0$

$J = 0$

$H = 0$

i.e. $\qquad \text{LHS} = J + \frac{H}{2}$

$$= 6 + 0 = 6$$

$$\text{RHS} = \frac{3}{2} L - 2$$

$$= \frac{3}{2} \times 6 - 2$$

$$= 9 - 2 = 7$$

Again or $\qquad$ LHS < RHS

That's the chain is unconstrained. Thus as per equation (9.4).

Example 9.4. *In Fig. 9.9 are given a chain having 6 links. In Fig. 9.9 (a) the joint E is made some where on the link BC (within BC or on extended BC) similarly the joint G is made on the link AB.*

Show that this six element chain is constrained kinematic chain.

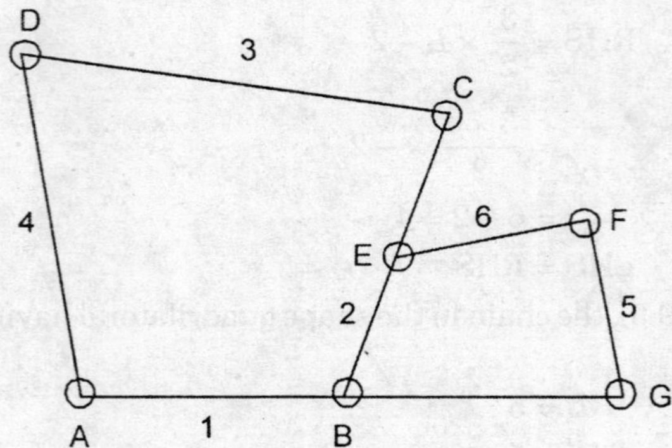

Fig. 9.9

Solution. Here $L = 6$

$J = 7$

$H = 0$

Therefore LHS of equation (9.1)

$$\text{LHS} = J + \frac{H}{2}$$

$$= 7 + 0 = 7$$

$$\text{RHS} = \frac{3}{2}\,\text{L} - 2$$

$$= \frac{3}{2} \times 6 - 2 = 7$$

i.e. LHS = RHS

Thus from equation (9.3), it is a constrained kinematic chain.

9.10.1 Basic Kinematic Chain, Compound Kinematic Chain

From the example 9.1 to 9.4 we conclude that the chain is the shape of a polygon is possible to be kinematic constrained chains only if it has 4 links. Due to this reason 4 link or 4 bar chain is called the basic kinematic chain. However in case we want to have chains with more member of links then in each step 2 links are added and that too in a fashion explained in example (9.4) i.e. one additional link is connected with one existing link and similarly the other additional link is attained/connected with another link of the existing chain. The two additional links on the other end are connected with each other forming a pair. Thus we can have kinematic constrained chains having 6, 8, links attached to the basic chain in the manner explained above. Such chains with number of links 6.8 are called the compound chains.

9.11 MECHANISMS AND MACHINES

In a constrained kinematic chain if one of the links is fixed then this results in what is called a mechanism. If different links of kinematic chain are fixed, one at a time it will result in different mechanisms. The main function of the mechanism is transmit and/or modify motion. Whereas the machine is to obtain mechanical advantage if it modifies not only the motion but forces are also modified, considered and transmitted by the machines.

Thus primary function of the mechanism is to transmit and modify motion whereas machine modifies and transmits mechanical work.

9.12 FOUR BAR CHAIN OR FOUR LINK CHAIN

A four bar chain is shown in Fig. 9.8 (b). It consists of four binary links. The link *AB* is fixed which is called the frame link *AD* is the crank or driver link, while the link CD is the connecting rod (or couples) and *BC* is the lever (or follower link or rocker).

9.12.1 Grashof's Law

Statement : Grashof's law states the for continuous relative motion between the links of a four bar mechanism the sum of the lengths of the longest and shortest links must be less than the sum of the length of the other two links.

i.e. $$(l + s) < (a + h) \qquad ...(9.5)$$

where
l = length of longest link
s = length of shortest link
a, b = lengths of this other two links.

9.12.2 Inversion of Mechanism

As stated above a kinematic chain becomes a mechanism when one of its links to fixed. Therefore, as many number of mechanisms can be obtained as are number of the links in the kinematic chain. Different mechanisms so obtained by fixing different links, one at a time, of the chain are known as Inversions. The relative motion between various links remains the same in all the inversions of a chain. But the shortest motion of various links with respect to the fixed link in the respective inversion many not be the same in all the inversions.

In case there is a 4-bar chain, we can obtain 4 different mechanisms by fixing each of the 4 links in them.

9.13 FOUR BAR CHAIN AND ITS INVERSIONS

In (Fig. 9.10) is shown the simplest basic kinematic chain having 4-pinned joints. The lengths of the four links conform to the Grashof's Law stated above so that the Link *AB* (which is the shortest link) can make a complete rotation) is made the crank.

(*i*) Link (no. 1) *AD* is fixed and is part and parcel of the frame.

(*ii*) Link *AB* (no. 2) is called crank and makes complete rotation.

(*iii*) Link BC (no. 3) connects the crank and the lever and is called the connecting rod or coupler.

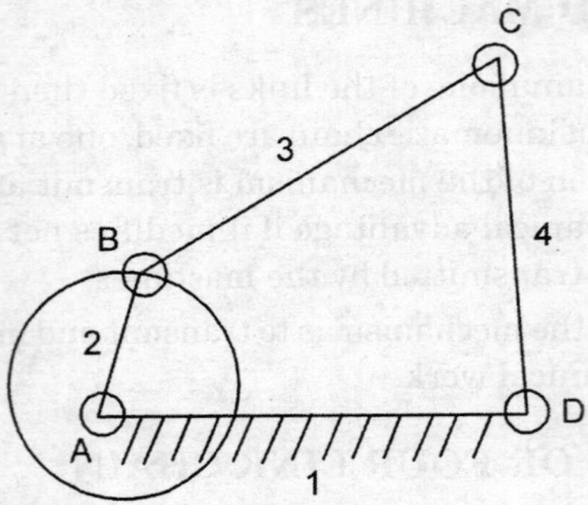

Fig. 9.10

(*iv*) Link *CD* (no. 4), makes a partial to and for rotation or oscillates about the point *D* and is known as rocker or follower.

Thus we see that the rotary motion of the link 2, has been transformed with oscillatory motion of the link 4.

9.13.1 Inversions of 4-Bar Chain

Following are the important inversions of the 4-bar chain.

(*i*) Beam Engine (Also known as Crank and Lever Mechanism) : The mechanism of the beam engine is shown in (Fig. 9.11). It consists of 4 links. The crank *AB* rotates about the fixed point *A,* the lever link *CDE* oscillates about the fixed hinge point *D.* The end *E* of the link *CDE* is connected to the piston rod end *F* (which is constrained to move vertically up and down) through a small link *EF* which is joined to the lever *CE* at one end and the piston rod on the other through pin joints so as to take care of the slight angular movement of the end *E* of the lever link *CDE* because piston along with piston rod are constrained to reciprocate up and down in a straight path. The resultant reciprocting up and down motcon of piston in cylinder is utilised as beam engine.

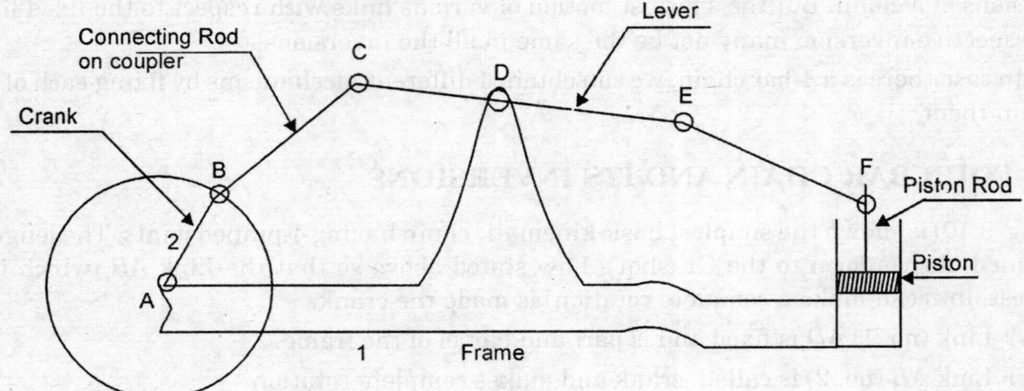

Fig. 9.11.

***(ii)* Coupled Wheels of Locomotive :** In Fig. (9.12) is shown the schematic diagram of coupled wheels of locomotive. The links *AB* and *CD* are equal in length and so are the links *BC* and *AD*. The links 2 and 4 act as cranks and this mechanism is, therefore known as double crank mechanism. Link 1 is fixed and is part of the frame of locomotive and the link 3 acts as a coupler. This mechanism transmits rotary motion from one wheel to other wheel.

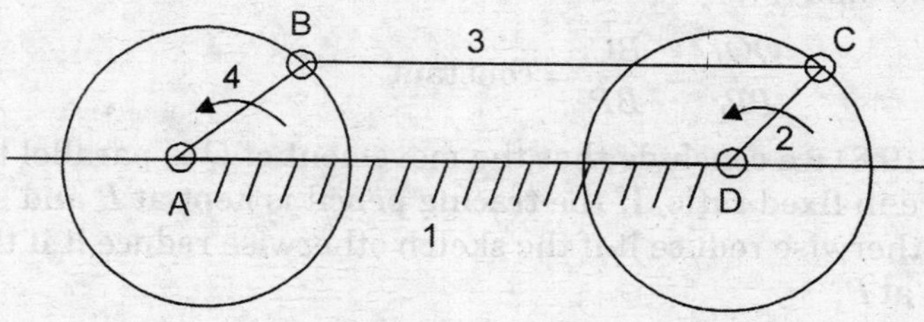

Fig. 9.12

***(iii)* Pantograph (Double Lever Mechanism) :** It is again a quadratic cycle (4 bar chain) in the form of a parallelogram which is used to produce a displacement exactly in enlarged or reduced scale. In Fig. 9.13 is shown schematic sketch of Pantograph *ABCD* is the 4 bar chain with 4 turning pairs. The mechanism is fixed to the frame at the point *A* (say) thought a pin joint to from a turning pairs. This device is used for duplicating the drawings on an enlarged or reduced scale as desired. The link *BC* is extended to *P*. Let *Q* be the point of intersection of the line *AP* and link *CD*. Then it is shown below that the movement of the point *Q* is similar to that of *P* and vice versa.

Proof

Geometrically : Triangles *AQD* and *CQP* are similar because *AD* and *CP* are parallel.

Therefore $$\frac{AQ}{PQ} = \frac{AD}{CP}$$

$$= \frac{BC}{CP} = \text{constant} \qquad \text{...(9.6)}$$

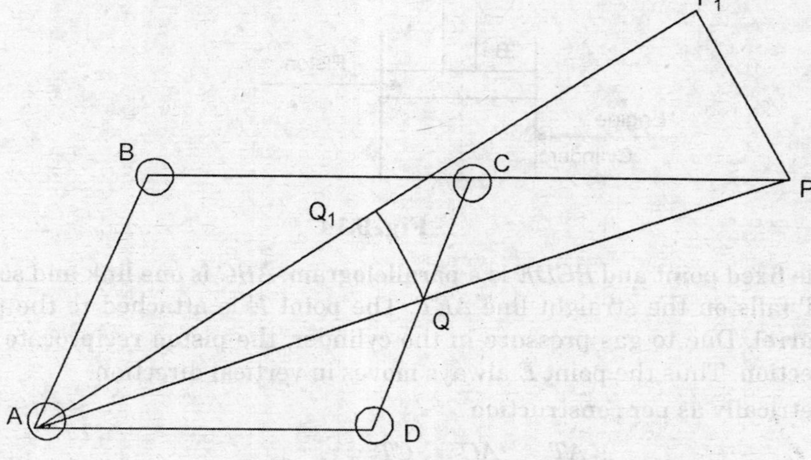

Fig. 9.13

Let the point P move to P_1 join A to P_1 and drawn $Q\,Q_1$ line parallel to PP_1 to meet AP_1 in Q_1, then the triangles APP_1 and AQQ_1 are similar.

$$\therefore \quad \frac{AQ}{PQ} = \frac{QQ_1}{PP_1} \qquad \text{...(9.7)}$$

For equation (9.6) and (9.7)

$$\frac{QQ_1}{PP_1} = \frac{BC}{BP} = \text{constant} \qquad \text{...(9.8)}$$

From equation (9.8) we conclude that the movement of Q is parallel to that of P and also these movements are in fixed ratio. If the tracing pencil is kept at P and stylus at Q at it will enlarge the sketch otherwise reduce it if the sketch otherwise reduce it if the training pencil is kept at Q and stylus at P.

***(iv)* Watt's Engine Indicator Mechanism :** Engine indicator is primarily used to draw the pressure piston displacement or *P-V* diagram corresponding to the pressure in the cylinder, which varies depending upon the location of the piston. In such a device the piston is subjected of varying gas or steam pressure. The piston is loaded with a spring and its deflection is proportional to pressure in cylinder as the area of the piston is constant.

Watt used the principle of Pantograph to record the displacement of the piston (which is proportional to the pressure).

In Fig. 9.14 is shown the schematic diagram of the watts engine indicator.

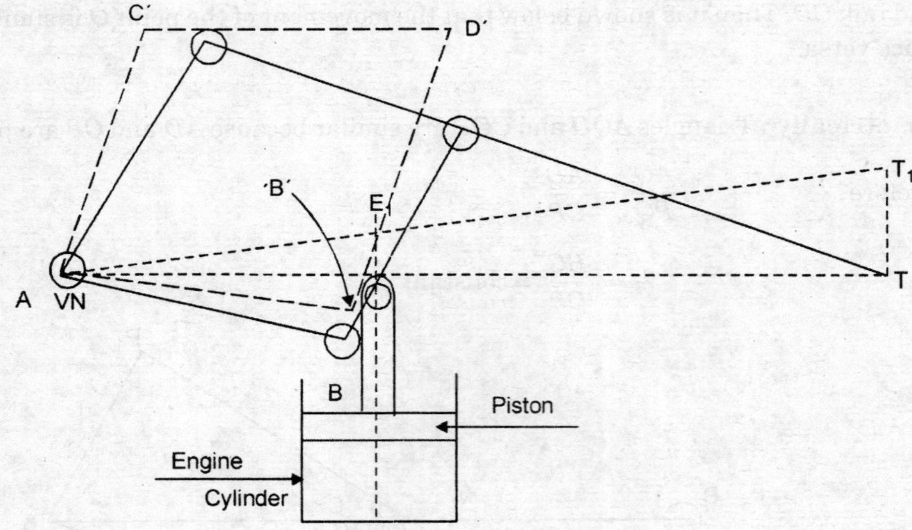

Fig. 9.14

A is the fixed point and *BCDE* is a parallelogram. *ABC* is one link and so is the link *CDT*. The point T falls on the straight line *AET*. The point *E* is attached to the piston rod of the indicator barrel. Due to gas pressure in the cylinder, the piston reciprocate up and down in vertical direction. Thus the point *E* always moves in vertical direction.

Geometrically as per construction

$$\frac{AT}{AE} = \frac{AC}{AB} = \frac{CT}{CD} \qquad \text{...(9.9)}$$

also the ATT_1 and AEE_1'

$$\frac{TT_1}{FA} = \frac{AT_1}{AF_1} = \frac{C_1T_1}{CD} = \frac{CT}{CD} \quad ...(9.10)$$

Thus $$\frac{EE_1}{TT_1} = \frac{CT}{CD} = \text{constant}$$

This ratio gives the magnification scale

9.14 BASIC KINEMATIC CHAIN ITS APPLICATION FOR OBTAINING OSCILLATORY OR ROTARY MOTION

9.14.1 For Oscillatory Output Motion

In article 9.10 it has been shown that four bar kinematic chain is the basic-kinematic chain having four links and foru turning pairs. However depending upon the length of individual links one can get either oscillatory or rotory motion of the other links.

9.14.1.1 Crank and Lever Mechanism

In Figure (9.15) is shown a 4 bar chain in which AD is the crank and rotates about the point A while the link AB is fixed and is part of the frame. The crank AD is smallest in length and the coupler DC is quite long such that the crank AD_1 can rotate through 360°. However in the event when the crank is in position AD_1 the connecting rod DC is also along the line AD_1 as it now occupies the position D_1C_1 and the link BC will be along BC_1 Then for further rotation of the crank AD_1 to the position AD_2 such that AD_2 and C_2D_2 are over-lapping each other then BC_1 and BC_2 are the extreme angular positions of the crank BC because additional rotation of the crank AD from piston AD_2 the and C_2 of the link BC will starts turning backward from the position BC_2. Thus the link BC will be oscillating from position BC_1 to BC_2 and back even though the crank AD will be revolving completly this mechanism is known as Crank and Lever Mechanism.

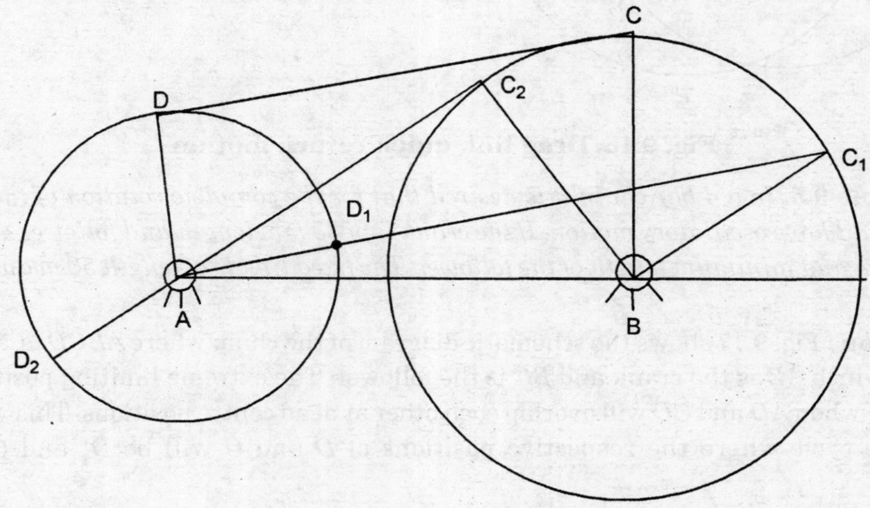

Fig. 9.15

9.14.1.2 Double Crank Mechanism (Drags Link Quick Return Motion)

In Fig. 9.16 is shown the schematic diagram of Drag link quick return motion mechanism in which there is complete rotation of both the crank and the follower. In this the length of the fixed link is much smaller, at the ends of which are pinned the crank and the follower respectively. However the lengths of the links conform to the Grashof's Law.

The shortest link AB is fixed. The lengths of the links are such that for full rotation of crank AD the follower BC also makes complete rotation. The pin point B is connected through a connecting rod BE, to a reprocating block E along the extended line AB to E. When the crank AD is in the position AD_1 the link CE is in portion C_1E_1, where C_1 lies on the line ABE and it is the right end of the diameter of the circular path of movement of point C. The correspouding position of block E is E_1 from this position *i.e* AD_1 the crank AD rotates through an angle θ_f anticlockwise to reach the position D_2 when the corresponding position of the block E is then E_2 such that it has moved to the left most position. The distance E_1E_2 is called the stroke of the block E. Again when the crank AD moves ahead through an angle θ_r to reach the position AD_1, the block E will go to the position E_1' from E_2. Since $\theta_r < \theta_f$ and the crank AD is rotating with uniform angular speed the time taken by E for the return stroke E_2E_1 is shorter than that taken during the forward stroke. That is why it is called drag link quick return motion.

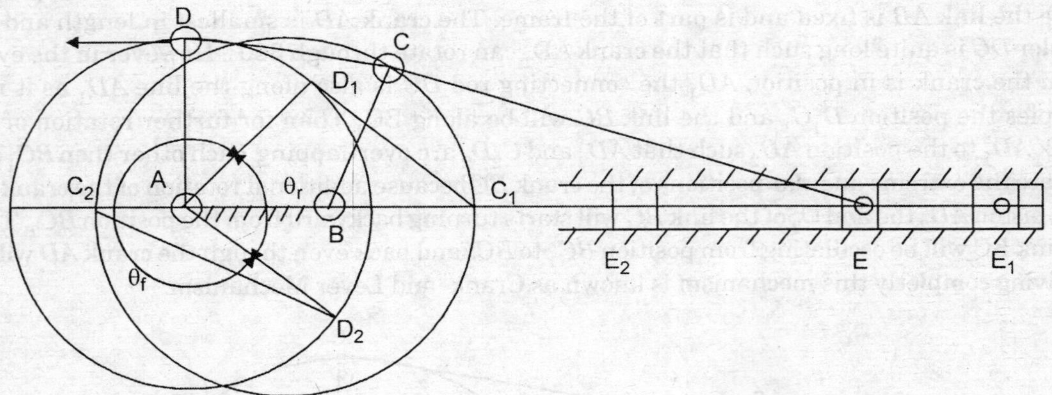

Fig. 9.16. Drag link quick return motion

Example 9.5. *In a 4 bar chain it is desired that for the complete rotation of the crank the follower should have oscillatory motion. If the crank is of 33 mm length and follower 46 mm, find the maximum and minimum length of the follower. The fixed link is of length 50 m and at its end are pin joint.*

Solution : Fig. 9.17 shows the schematic diagram of the chain where $ABCD$ is any general position in which AD is the crank and BC is the follower. The extreme limiting positions of the crank will be when AD and CD will overlap each other at dead centre positions. This will happen twice in the cycle where the respective positions of D and C will be D_1 and C_1, and D_2 and C_2.

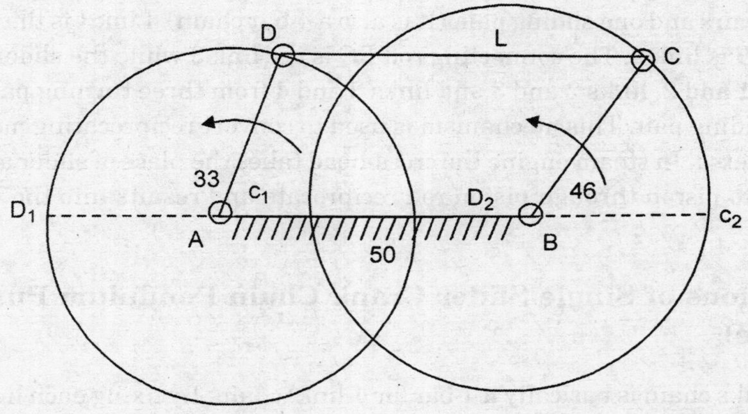

Fig. 9.17

In the first case length of the link *CD* will be

$$l_1 = D_1C_1$$
$$= D_1A + AC_1$$
$$= D_1A + AB - BC1$$
$$= 33 + 50 - 46 = 37 \text{ mm}$$

In this second case $l_2 = D_2C_2$

$$= C_2B + BD_2$$
$$= C_2B + BA - AD_2$$
$$= 46 + 50 - 33$$
$$= 63 \text{ mm}$$

Thus l_1 and l_2 are the minimum and maximum limit of the link couple link.

9.15 SINGLE SLIDER CRANK MECHANISM AND ITS INVERSIONS

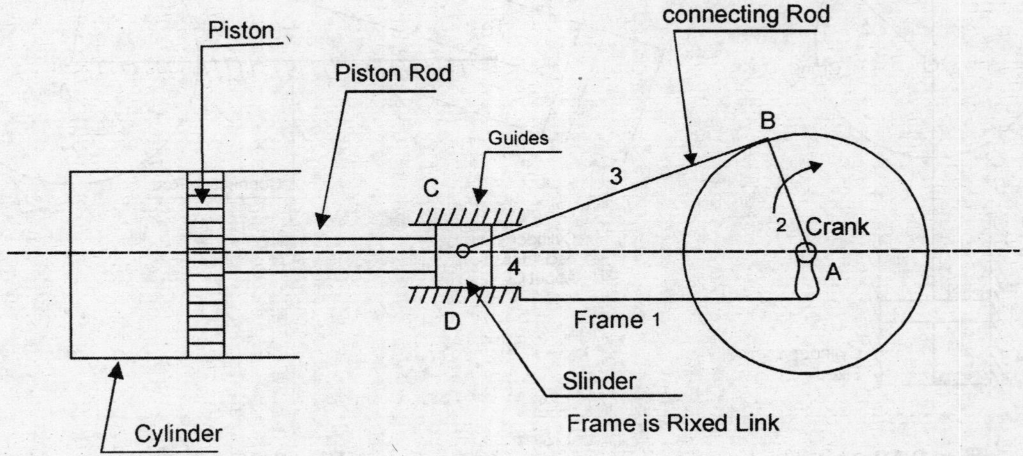

Fig. 9.18.

In Fig. 9.18 is shown the schematic diagram of single slider crank mechanism. It consists of three turning pairs and one sliding pair. (It is also a 4-bar chain). Link 1 is the frame *DA* which is fixed. Crank *AB* is link 2. The connecting rod *BC* is the link 3 while the slider constitutes the link 4. The links 1 and 2, links 2 and 3 and links 3 and 4 from three turning pairs while links 4 and 1 from the sliding pair. This mechanism is used to convert reciprocating motion into rotory motion and vice versa. In steam engine the cross-head takes the place of slider and cross-head is being connected to piston through piston rod reciprocate and results into the rotory motion of the crank shaft.

9.15.1 Inversions of Single Slider Crank Chain Pendulum Pump (Bull Engine)

Single slider cranks chain is basically a 4-bar (or 4-link) chain. By fixing each link one at time in turn four different inversion are obtained. In preceeding paragraph the link 1 which is frame of the engine was fixed and inversion called engine mechanism resulted. Similarly if the link 4 *i.e.* slider is fixed (*i.e.* no sliding action between link 3 and 4) another inversion called the pendulum pump is obtained. In schematic diagram is shown in Fig. 9.19 when the crank (links 2 rotates, the connecting rod (link 3) oscillates about the pin pivoted to the fixed link 4 at point *C* and the piston along with piston rod reciprocates in the fixed cylinder (link 4).

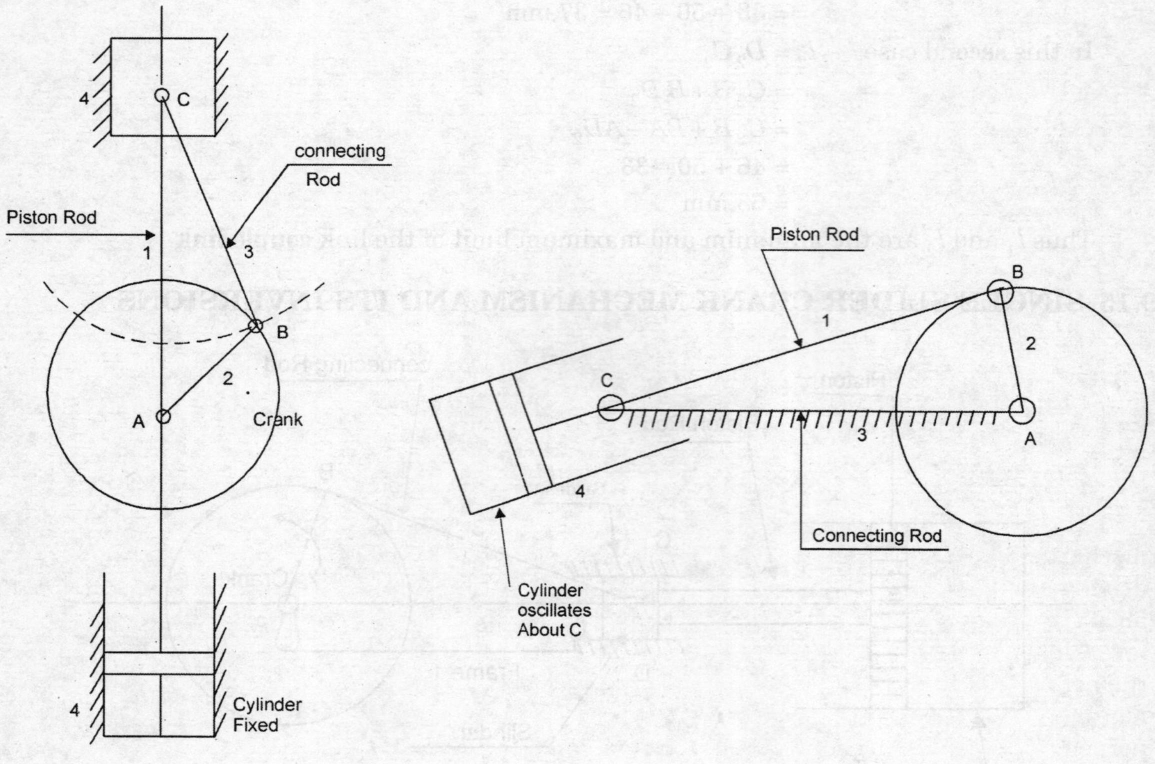

Fig. 9.19 **Fig. 9.20**

9.15.2 Oscillating Cylinder Engine

Yet another inversion of the basic slider-crank chain is obtained by fixing the connecting rod *i.e.*, link 3. This inversion is called the oscillating cylinder engine the schematic diagram of which is shown in Fig. 9.20.

The piston along with piston rod (link 1) reciprocating in the cylinder (link 4). The cylinder itself oscillates about a pin pivoted to the fixed link at *A*.

9.15.3 Crank and Slotted Lever Quick-Return Motion Mechanism

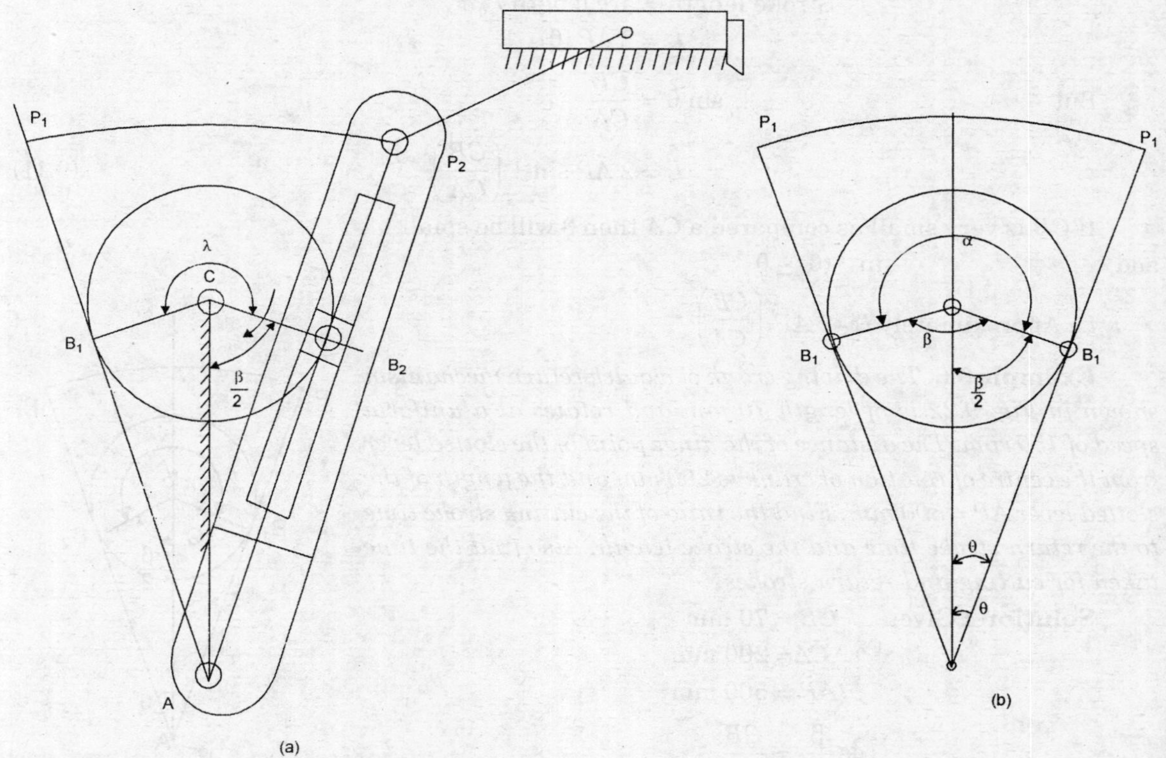

Fig. 9.21

In Fig. 9.21 is shown the schematic diagram of the crank and slotted lever quick return mechanism which corresponds to connecting rod or link 3 of the basic 4-bar chain, fixed. The driving crank *CB* (link no. 2) which has its length smaller than that of fixed link *CA*, revolves about the point *C*, *A* slider (*i.e.* link 1) attached to the crank pin at *B*, slides along the slotted lever (link 4) and makes the slotted lever oscillates about the point *A* which is a pin joint, With the help of a small link *PE* the motion is transmitted from *AP* to the ram which along with the tool mounted on this reciprocates along the line, which is called the line of stroke and it is perpendicular and the link *AC* that is the fixed link.

When the crank *CB* is in position CB_1 *i.e* when CB_1 is perpendicular to AP_1 this slotted link will be in its leftmost extreme position. Similarly the right extreme position of *AP* will be AP_2.

For the crank to move from position CB_1 to CB_2 it will travel through an angle α and for moving from position CB_2 to CB_1 it has to rotate through an angle β. Evidently sine $\alpha < \beta$ during the forward stroke of the ram (during the cutting stroke) it will take longer time than the time during return stroke when the crank, rotating at uniform angular speed, has to move only through an angle β and $\beta < \alpha$.

Therefore the ratio of

$$\frac{\text{Time taken during cutting stroke}}{\text{Time taken during return stroke}} = \frac{\alpha}{\beta}$$

For Fig. 9.21 (*b*)

$$\text{Stroke length} = \text{arc length } P_1 P_2$$

$$L = 2\,AP\,(\theta)$$

But $$\sin\theta = \frac{CB}{CA}$$

∴ $$L = 2\,AP \sin^{-1}\left(\frac{CB}{CA}\right) \quad ...(9.11)$$

If *CB* is very small as compared a *CA* then θ will be small

and $\sin^{-1}(\theta) \simeq \theta$

i.e. Approximately $L \simeq 2\,AP\left(\frac{CB}{CA}\right)$

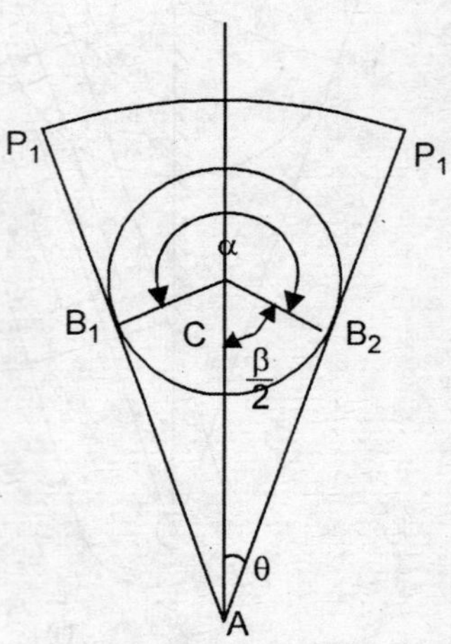

Fig. 9.22

Example 9.6. *The driving crank of a quick return mechanism shown in Fig. 9.22 is of length 70 mm and rotates at a uniform speed of 150 rpm. The distance of the hinge point of the slotted lever from the centre of rotation of crank is 200 mm and the length of the slotted lever AP = 500 mm. Find the ratio of the cutting stroke time to the return stroke time and the stroke length. Also find the time taken for cutting and return strokes.*

Solution : Given $CB = 70$ mm

$CA = 200$ mm

$AP = 500$ mm

$$\cos\frac{\beta}{2} = \frac{CB}{CA}$$

$$= \frac{70}{200}$$

$$\frac{\beta}{2} = \cos^{-1}\left(\frac{70}{200}\right)$$

$$= 69.5°$$

$$\beta = 139°$$

and $$\alpha = 360 - 139$$

$$= 221°$$

$$\frac{\text{cutting stroke time}}{\text{return stroke low}} = \frac{\alpha}{\beta}$$

$$= \frac{221}{139} = 1.59$$

From ΔCAB_2 $\quad \theta = 90 - \frac{\beta}{2}$

$$= 90 - 69.5°$$

$$= 20.5°$$

Stroke $\quad L = 2(OP_2)$

$$= 2\, AP_2 \sin\theta$$

$$= 2 \times 5.2 \times \sin(20.5°)$$

$$= 100 \times .35$$

$$= 350 \text{ mm}$$

Cutting stroke angle of rotation of crank

$$\alpha° = 221°$$

Let time taken for cutting stroke = T sec at 150 rpm

$$\therefore \quad \frac{221}{T} = \frac{150 \times 360}{60}$$

$$T = \frac{150 \times 360}{60 \times 221} \text{ sec}$$

$$= \frac{150 \times 60}{221}$$

$$T = \frac{60 \times 221}{150 \times 360} = 2.46 \text{ secs}$$

Time taken for return stroke

$$\text{Angle } \beta = 360 - \beta = 360 - 221$$

$$= 139°$$

$$\therefore \quad T_r = \text{Time taken for return stroke}$$

$$= \frac{139 \times 60}{15 \times 360} = 1.54 \text{ sec}$$

Alternatively, ratio of these turns has already been calculated

i.e. $\quad \frac{T}{T_r} = 1.59$

$$= T_r = \frac{2.46}{1.59} = 1.540 \text{ sec}$$

9.15.4 Whitworth Quick Return Motion Mechanism

In Fig. 9.22 is shown the schematic diagram of the whitworth quick return motion mechanism alongwith the basic line diagram of the 4-bar chain.

In the Whitworth mechanism link 2 of the basic 4-bar chain is kept fixed and the link 3 is *BC* (original connecting rod in the basic chain) is made the driver rotating link. It may be noted that the length of link $BC > AB$. The link BC is made the driver rotating at uniform angular speed. The slider block attached to the pin C slides to and fro along the slotted link 1 (here CD) which itself is pinned at point A and will revolve about A at varying angular velocity. The slotted lever extends to the point D where one end D of the connecting rod DE (link 5) is pinned. The end E of the link DE is pinned to a reciprocating ram on which is mounted the cutting tool.

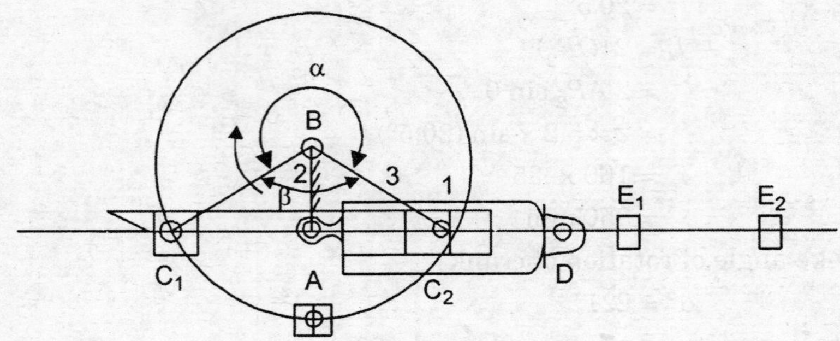

Fig. 9.23

When the rotating link BC occupies the position BC_1, the slotted link is along $D_1 A$ which is normal to the fixed link BA and the cutting tool will be at its extreme left position E_1. When the rotating link BC occupies the position BC_2 the cutting tool will be at its right most position E_2. The distance $E_1 E_2$ will be the cutting stroke length and the time taken for cutting stroke will be proportional to angle α. However during the return stroke the time taken is proportional to angle β.

Since $\alpha > \beta$

$T_e > T_r$

The ratio of cutting stroke time T_c to return stroke time T_r is then

$$\frac{T_c}{T_r} = \frac{\alpha}{\beta} = \frac{\alpha}{(360 - \alpha)} \quad ...(9.12\,a)$$

$$= \frac{(360 - \beta)}{\beta} \quad ...(9.12\,b)$$

Geometrically $\cos\left(\frac{\beta}{2}\right) = \frac{AB}{BC}$

$$\frac{\beta}{2} = \cos^{-1}\left(\frac{AB}{BC}\right) \quad ...(9.13)$$

Length of cutting stroke = Return stroke

$= C_1 C_2$

But $C_1 C_2 = 2AC_2$

And $$L = 2 \times BC \sin\left(\frac{\beta}{2}\right) \quad ...(9.14)$$

Example 9.7. *In the Whitworth quick return mechanism (Fig. 9.23) following data is given*

$$BA = 20 \text{ mm}$$
$$BC = 40 \text{ mm}$$
$$AD = 75 \text{ mm}$$

Returned velocity of $BC = 15$ *rpm*

Calculate the rate $\frac{T_c}{T_r}$, *stroke length cutting stroke and return stroke times.*

Solution. Refer to Fig. (9.23),

Given. data

$$AB = 20 \text{ mm}$$
$$BC = 40 \text{ mm}$$
$$AD = 75 \text{ mm}$$

$$\cos\frac{\beta}{2} = \frac{AB}{BC}$$

$$= \frac{20}{40} > \frac{1}{2}$$

Therefore $$\frac{\beta}{2} = 60°$$

or $$\beta = 120° = \frac{2}{3}\pi \text{ re}$$

$\therefore$ $$\frac{T_c}{T_r} = \left(\frac{360 - \beta}{\beta}\right) = \frac{360 - 120}{120} = 2$$

Now $$\omega = \frac{2\pi \times 15}{60} = \frac{\pi}{2} = 1.57 \text{ rad/sec}$$

$$T_c = \frac{\alpha}{W}$$

$$= \frac{4\pi}{3} \times \frac{2}{\pi} = \frac{8}{3} \text{ sec}$$

$$= 2.66 \text{ sec}$$

i.e. $$T_r = \frac{T_c}{2} = \frac{8}{3} \times \frac{1}{2} = \frac{4}{3} \text{ sec}$$

$$= 1.33 \text{ sec}$$

$$\text{Stroke} = C_1 C_2$$

From equation (9.14)

$$L = 2\,BC \sin\left(\frac{\beta}{2}\right)$$

$$= 2 \times 40 \sin 60 = 80 \times \frac{\sqrt{3}}{2}$$

$$= 40\sqrt{3} = 69.2\% \text{ mm}$$

9.16 DOUBLE SLIDER CRANK CHAIN

It is a Four bar kinematic chain consisting of two turning pairs and two sliding pairs, such that two pairs of the same kind are adjacent.

In Fig. 9.24 is shown the schematic arrangement of the double slider crank chain. Two sliding blocks, links 1 and 3, slide along their the respective slots, cut at right angle, to each other in the frame (link 4) of the mechanism. The turning pairs are formed at points *A* and *B*. The sliding blocks individually form sliding pairs with the frame (link 4) and turning pair with the link. There are three inversions of this kinematic chain which are described below.

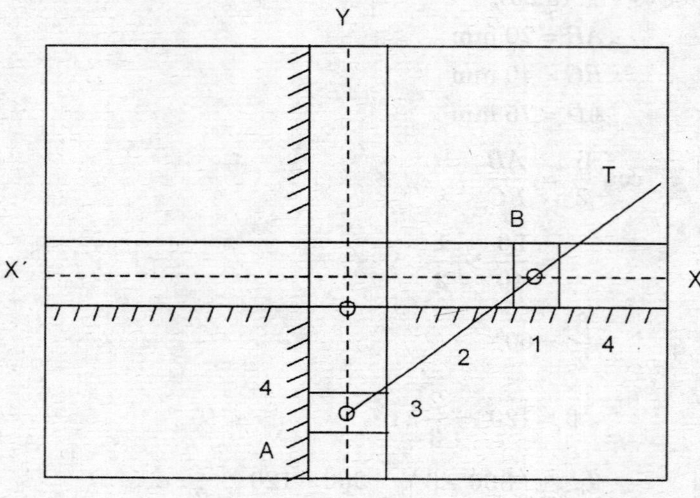

Fig. 9.24

9.16.1 Elliptical Trammel

First Inversion Called Elliptical Trammel is obtained by fixing the Frame *i.e.* link 4. Then the two blocks *i.e.* links 1 and 3 make sliding pairs with the frame and turning pairs with the link 2.

When the links 1 and 3 slide along their respective slots any point such as *T* on the link 4 (other than the mid point between *A* and *B*) will trace an ellipse on the surface of plate (link 4) on which two perpendicular slots have been cut. The proof is given below.

Let the co-ordinates of the tracing point *T* be (x, y) with respect to the reference axes chosen along two centre lines of the slots (Fig. 9.24).

Then
$$x = AT \cos \theta \qquad ...(9.15)$$
$$y = BT \sin \theta \qquad ...(9.16)$$

By squaring and adding

so
$$\left(\frac{x}{AT}\right)^2 + \left(\frac{y}{BT}\right)^2 = \cos^2 \theta + \sin^2 \theta = 1$$

or
$$\frac{x^2}{(AT)^2} + \frac{y^2}{(BT)^2} = 1 \qquad ...(9.17)$$

Equation (9.17) in the equation of an ellipse with AT as the semi-major axis and BT as the semi-minor axis along with and YY axis respectively. This is the reason if this inversion of the mechanism is called Elliptical Trammel.

9.16.2 Scotch Yoke

In this inversion called Scotch Yoke one of the two sliding blocks (either link 3 or link 1) is fixed. This arrangement will result is reciprocating motion of the frame or link 4. The corresponding connection and resulting motion is shown in Fig. 9.25.

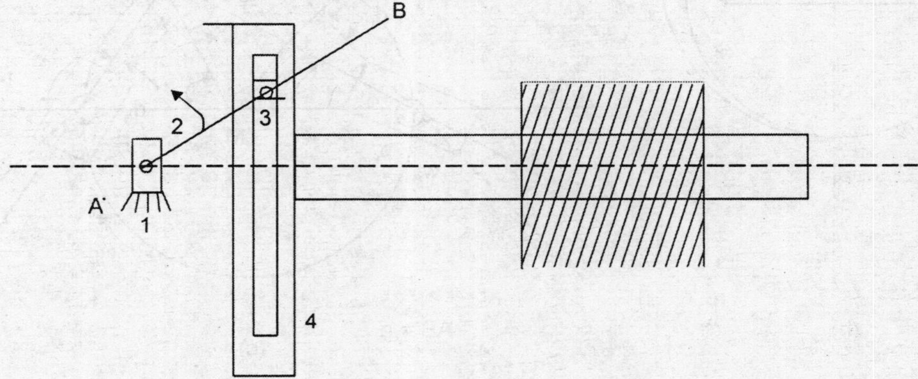

Fig. 9.25

9.16.3 Oldham's Coupling

The third inversion of the basic double slider crank chain is obtained by keeping the link 2 fixed. Since the two blocks are constrained to move along the respective slots and they can also have rotary motion with respect to the pins at A and B, The location of points A and B is fixed in space as link 2 which is link AB is fixed.

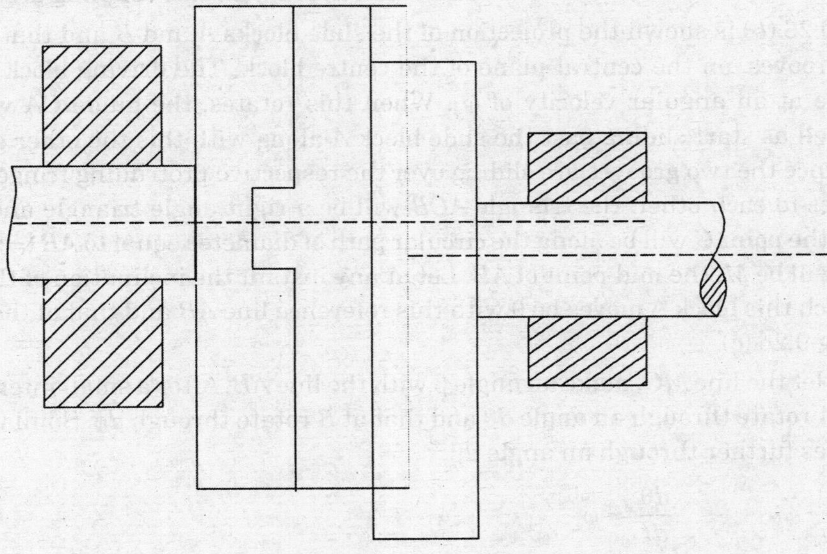

(a)

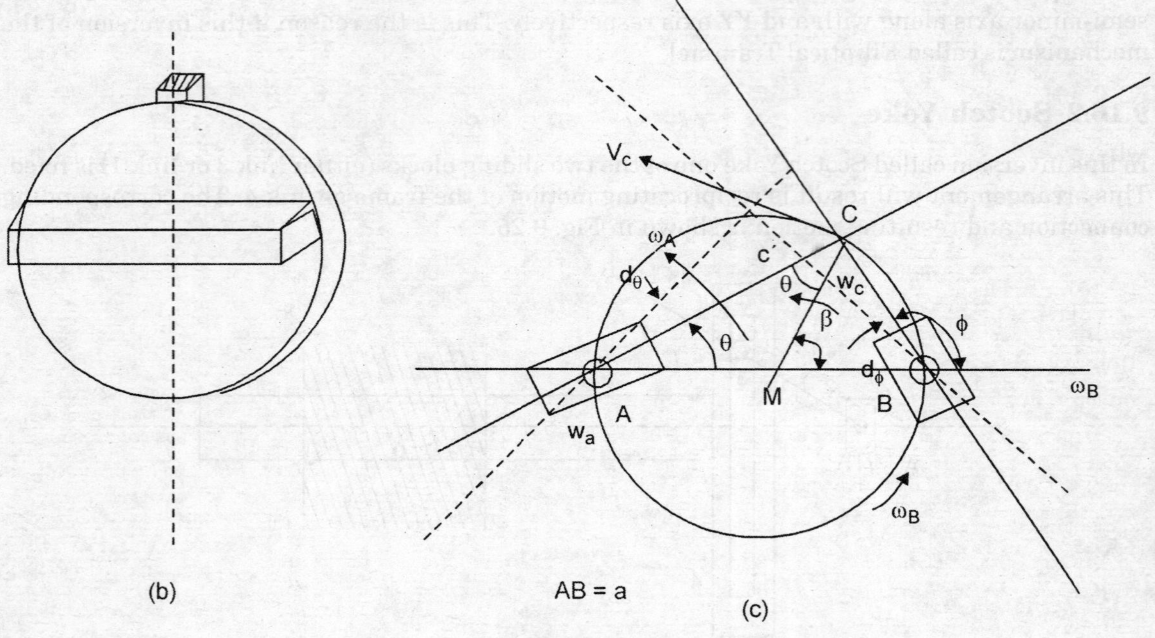

Fig. 9.26

The coupling is used for connecting two parallel shafts. When the distance between them is small. The two shafts to be connected have discs rigidly fixed at the ends of the shafts (generally they are forged along with the shafts). Two grooves are cut in these discs so as to accommodate the two protruding perpendicular fingers attached on the opposite faces of a central block Fig. 9.26 (a) so that they can slide within the grooves cut in the discs attached to the shaft ends.

In the following paragraph it is proved that the two shafts will be rotating at the same angular speeds. Expressions for other variables associated with this coupling are also derived.

In Fig. 9.26 (*b*) is shown the projection of the slide blocks *A* and *B* and that of the centre lines of the grooves, on the central plane of the centre block. The driving block *A* is rotating anticlock wise at an angular velocity of ω_a. When this rotates, the guide *CA* will also start rotating as well as start sliding past the slide block *A* along with this the other guide *CB* will follow suit. Since the two grooves are sliding over the respective protruding fringers, which are at right angles to each other, the triangle *ACB* will be a right angle triangle and as such the movement of the point *C* will be along the circular path of diameter equal to *AB* (= a). The centre of this circle will be *M*, the mid point of *AB*. Let at any instant the inclination of the, axis of the groove in which this block A moves be θ with this reference line *AB* and that of the block *B* be φ as shwn is Fig 9.26 (*c*).

Further let the line *MC*, make an angle β with the line *AB*. After a small interval of time *dt* let the block *A* rotate through an angle *d*θ and that at *B* rotate through *d*φ. Similarly the radial line *MC* rotates further through an angle *d*β.

Then $$\frac{d\theta}{dt} = \omega_a \qquad ...(9.18\,a)$$

$$\frac{d\phi}{dt} = \omega_b \quad \text{...(9.18 } b\text{)}$$

and
$$\frac{d\beta}{dt} = \omega_c \quad \text{...(9.18 } c\text{)}$$

where ω_a = angular speed of shaft A

ω_b = angular speed of rotation of shaft B

ω_c = angular speed of rotation of point C about the centre M of the circular path of motion of C.

Then form ΔABC and ΔACM we have

$$\phi = \frac{\pi}{2} + \theta \quad \text{...(9.19)}$$

and
$$\beta = \theta + \theta = 2\theta \quad \text{...(9.20)}$$

Differentiating equation (9.19) and (9.20) w.r.t. '*t*' we get

$$\omega_b = \frac{d\phi}{dt} = \frac{d\theta}{dt} = \omega_a \quad \text{...(9.21)}$$

Also from
$$\omega_c = \frac{d\beta}{dt} = 2\,\frac{d\theta}{dt} = 2\,\omega_a \quad \text{...(9.22)}$$

From equation (9.21) we conclude that both the shafts are rotating with the same angular speed as if they were coupled to each other. This the reason that this inversion is named Oldhams coupling after the name of its inventer.

From equation (9.22) it is concluded that the point C moves in a circular path about the centre M with twice the angular speed of rotation of block A.

Thus the peripheral velocity of the point C moving along the circular path is

$$v_c = (MC)\,\omega_c$$

$$= \left(\frac{a}{2}\right)(2\,\omega_a) = a\,\omega_c \quad \text{...(9.23)}$$

Sliding Velocities of the Blocks in Respective Grooves

Let it the assumed that in the small interval of time dt, the groove CA has slided past the block A by a small distance $[d\,(CA)]$. Similarly the groove BC has done so past the block B by an amount $[d\,(BC)]$ such that the respective sliding velocities past the blocks A and B are respectively

$$(v_s)_A = \frac{d\,(CA)}{dt} \quad \text{...(9.24)}$$

$$(v_s)_B = \frac{d\,(BC)}{dt} \quad \text{...(9.25)}$$

But Geometrically $CA = AB\cos\theta$

$= a\cos\theta$

or $$\frac{d\,(CA)}{dt} = -a \sin\theta \frac{d\theta}{dt} \qquad ...(9.26)$$

Thus from equation (9.24) and (9.26) we get

$$(v_s)_A = -a\,\omega_a \sin\theta \qquad ...(9.27)$$

Similarly $$BC = AB \sin\theta$$

or $$\frac{d\,(BC)}{dt} = d \cos\theta \frac{d\theta}{dt}$$

or $$(v_s)_B = a\,\omega_a \cos\theta \qquad ...(9.28)$$

The – ve sign the equation (9.27) indicates that the distance CA is decreasing. In other words C is moving towards A. The maximum sliding velocity of the groove past the block A will be obtained from equation (9.27) when $\theta = 90°$ or at the instant when the point C will just be passing over the point A and then from equation (9.27)

$$|(v_s)_A|_{max} = a\omega_a \qquad ...(9.29a)$$

when q = 0. That is when this

point C is just passing over in block B, then from equation (9.27) $|(v_s)_A|_{max} = a$ ...(9.29*b*)

Similarly the maximum sliding velocity of the respective groove past the block B will be obtained from equation (9.28) when $\theta = 0$. That in when the point C is just passing over the point B. Therefor from equation (9.28) at $\theta = 0$.

$$|(v_s)_B|_{max} = a\,\omega_a \qquad ...(9.30a)$$

Again from equation (9.28) we see that the sliding velocity of groove past block B s minimum when $\theta = 90°$. That is when the point C is just passing over the block A.

Therefore from equation (9.28) at $\theta = 90°$.

We have $$|(v_s)_B|_{min} = 0 \qquad ..(9.30b)$$

Example 9.8. *Two parallel shafts are connected by Oldham's coupling. The distance between their axes is 18 mm and the driving shaft is rotating at a uniform angular speed of 300 rpm. Find the maximum sliding speed of the tounges of the intermediate block in the respective slots cut in the flanges connected with the two shafts. During one revolution of drirring block at what angular position of this block does the this velocity become the maximum, also find the peripheral velocity of the point of intersection of the axes, of the two perpendicular fingers attached on opposite faces of the intermediate block while moving on a circular path of diameter equal to the distance between the axes of the two shafts.*

Solution. Given $a = 18$ mm

Na = 300 rpm

$$\omega_a = \frac{2\pi \times 300}{60} = 10\,\pi \text{ rad/sec}$$

Maximum sliding speed of slide block in the respective grooves

$$= a\,\omega_a = 18 \times 10\,\pi = 565.48 \text{ mm/s}$$

$$= 565.48 \text{ m/sec}$$

The sliding speed of the slide blocks in their respective grooves is maximum when the point C is just over the respective slide block.

i.e. when $\theta = 90°$ for $(v_s)_A$ to be maximum and

$\theta = 0°$ for $(v_s)_B$ to be maximum.

Peripheral velocity of point C, from equation (9.23)

$$v_C = a\omega_a$$
$$= 18 \times 10\,\pi \text{ mm/sec}$$
$$= 0.56548 \text{ m/sec}$$

PROBLEMS

1. Define the term 'Rigid Body'.
2. Differentiate between
   (*a*) Mechanism and machine
   (*b*) Mechanism and structure
   (*c*) Kinematic chain and mechanism
3. Explain completely constrained, successfully constrained and incompletely constrained motions. Support your answer with neat sketches.
4. Define kinematic pair.
5. Differentiate between lower pairs and higher pairs.
6. Is it possible to have a kinematic chain with (*a*) Three turning pairs (*b*) Four turning pairs (*c*) Five turning pairs and two sliding pairs.
7. Describe the construction and operation of (*i*) Elliptical Trammel (*ii*) Oldham's coupling.
8. In a crank and slotted lever mechanism used in shaper, the distance between the centre of rotation of crank and the centre of oscillation of the slotted lever is 200 mm and the crank length is 60 mm. The length of the slotted lever measured from the hinge point is 600 mm. Find the ratio of the cutting stroke time to return stroke time. If the speed of crank is 20 rpm find the time taken during the cutting stroke and return stroke. What is the stroke length ? (**Ans.** 1.98; 1.79 sec, 1.219 sec, 360 mm)
9. In the Whitworth quick return mechanism shown in Fig. 9.23 following data is given $BA = 25$ mm, $BC = 50$ mm, $AD = 80$ mm, rotational velocity of crank $BC = 30$ rpm. Calculate the time taken during cutting and return strokes, the stroke length. (**Ans.** 1.33 sec, 0.667 sec, 86.6 mm)

❑❑❑

CHAPTER 10

LIFTING MACHINES

10.1 INTRODUCTION

A machine is a device which receives energy in some available form and uses it to do some assigned work. For example a lathe or a shaper in a workshop receives electrical energy to run the motor attached to it and this energy is used to cut the metal to give proper shape to the raw material. In general the machine performs the following functions .

(*i*) To magnify or multiply force with the help of a crowbar.

(ii) To apply force at a particular point and in a particular direction.

(*iii*) To lift heavy weight or to support heavy weight.

(*iv*) To gain speed as in bicycle.

10.2 CLASSIFICATION OF MACHINES

(*i*) Simple Machines : The machine is classified as simple machine if there is only one point of application of effort and one point for load. For example, lever, screw jack, inclined plane, pulley, wheel and axle.

(*ii*) Compound Machines are machines which have multiple points for application of efforts and one point of load. Their mechanism is complicated. A number of simple machines are used to *make* a compound machine. For example, slotting machine, grinding machine, broaching machine, lathe etc.

(*iii*) Lifting machines **:** The machine which are used to left-heavy loads by applying small effort are called Lifting machines.

10.3 BASIC CONCEPTS AND DEFINITIONS ASSOCIATED WITH LIFTING MACHINES

(*a*) Load : The weight or load to be lifted or supported by the machine is called the load on the machine. It is represented by the symbol W.

(b) Effort : It is the force applied to the machine and lift the load as well as an to overcome the resistance. It is represented by P.

(c) Input to the Machine : It is the work done by the effort, on the machine. If the effort P moves through a distance 'y', the work done will be Py.

Thus Input = $P.y$ ...(10.1)

(d) Output of the Machine : The useful work done by the machine, to lift load W, through or distance x

Then Output = $W.x$ ...(10.2)

(e) Mechanical Advantage (MA) : It is defined as the ratio of the load lifted to the effort applied. Thus if MA is the mechanical advantage

$$MA = \frac{W}{P} \quad ...(10.3)$$

It depends upon the design of the machine and the frictional resistance.

(f) Velocity Ratio (VR) : It is defined is the ratio of the distance moved by the effort to the distance moved by the load. If VR represents velocity ratio, then

$$\text{V.R.} = \frac{y}{x} \quad ...(10.4)$$

(g) Efficiency : Efficiency of the machines is defined as the ratio of output of the machine to the input to the machine. Thus if η represents the efficiency

$$\eta = \frac{\text{Output}}{\text{Input}} = \frac{W.x}{Py}$$

$$= \frac{MA}{VR} \quad \text{[From 10.3 \& 10.4]}$$

$$\% \eta = \frac{MA}{VR} \times 100 \quad ...(10.5\,a)(10.5\,b)$$

Due to the presence of friction, output of a machine will always to less than input from the efficiency of an actual machine will never be 100%.

Thus, in an actual machine

$$\eta = \frac{\text{MA}}{\text{VR}} < 1 \text{ or } VR < MA$$

Since MA is usually > 1 VR would then all the greater than unity VR > 1 or in simple machine a small force applied through large distance overcomes (10.6 large form overcomes distances).

or

(h) Ideal or Perfect machine : An ideal or perfect machine is one in which output equals the input and the efficiency is equal 100% or 1 for ideal machines

$$\eta = \frac{\text{Output}}{\text{Input}} = 1$$

i.e.

$$\text{Input} = \text{Output}$$

$$Py = W.x$$

$$\frac{W}{P} = \frac{y}{x}$$

$$MA = \text{VR}$$

In actual practice no machine is perfect a part of the work done on the machine is lost in overcoming the friction. Thus the work done on the machine is always greater than output. Further mechanical advantage of actual machine will be less than that of an ideal machine but the velocity ratio is same for actual machine as well as ideal machine.

(*i*) Ideal Effort : If W is the load lifted by an ideal machine in which the effort is P_i (say) and the corresponding velocity ratio is VR then

$$(MA)_i = \frac{W}{P_i}$$

or
$$P_i = \frac{W}{(MA)_i} \quad ...(10.7)$$

But for ideal machine

$$(VR)_i = (MA)_i \quad ...(10.8)$$

Also
$$(VR)_i = (VR)_a = VR \quad ...(10.9)$$

∴ From (10.7) through (10.9)

$$P_i = \frac{W}{(VR)_i} = \frac{W}{(VR)_a} \quad ...(10.10)$$

Also for actual machine in which friction is present

$$\eta = \frac{(WA)_a}{(VR)_a} = \frac{W}{P_a} \cdot \frac{1}{(VR)}$$

∴
$$P_a = \frac{W}{(VR)} \cdot \frac{1}{\eta} \quad ...(10.11)$$

Thus comparing (10.10) and (10.11) as η is less than 1

$$P_a > P_i \quad ...(10.12)$$

That is actual effort will be greater than P_i. Alternatively in difference between P_a and P_i is the difference between P_a and P_i is the effort wasted to overcome friction while lifting the load W.

i.e.
$$P_{friction} = P_a - P_i$$

$$= \frac{W}{\eta} \cdot \frac{1}{(VR)} - \frac{W}{(VR)} \quad \text{...(From 10.10 and 10.11)}$$

$$= \frac{W}{(VR)}\left(\frac{1}{\eta} - 1\right) \quad ...(10.13)$$

Equation (10.13) give the values of effort to overcome friction.

(*j*) Ideal Load : Similarly for a given effort P are can estimate the frictional load.

Let W_i be ideal load lifted, the

$$W_i = P\,(VR) \quad ...(10.14)$$

and
$$W_{actual} = W_a = (P\eta)\,(VR) \quad ...(10.15)$$

Therefore decrease in lifted load due to friction

$$W_f = W_i - W_a$$

$$= P\,(VR) - (P\eta)\,(VR)$$

$$= P\,(VR)\,(1 - \eta) \quad ...(10.16)$$

10.3.1 Efficiency of a machine in Terms of Actual Effort and Ideal Effort

We have from equation (10.11)

$$P_a = \frac{W}{(VR)} \cdot \frac{1}{\eta} \qquad ...(10.11)$$

and from equation (10.14)

$$P_i = \frac{W}{VR} \qquad ...(10.10)$$

Dividing equation (10.10) by (10.11)

$$\eta = \frac{P_i}{P_a} \qquad ...(10.17)$$

$$= \frac{\text{Ideal Effort}}{\text{Actual Effort}}$$

Example 10.1. *An effort of 70 N is required by a* machine *to lift a load of 630 N. The distance moved by the effort is 570 mm for the load movement of 60 mm. Calcula‸‸ the MA, VR and efficiency of the* machine.

Solution :

$$MA = \frac{W}{P} = \frac{630}{70} - 9$$

$$(VR) = \frac{570}{60} = 9.5$$

$$\therefore \quad \eta = \frac{MA}{VR} - \frac{9}{9.5}$$

$$= 0.947 \text{ i.e. } 94.7\%.$$

Example 10.2. A machine *has an efficiency of 70% and its velocity ratio is 20. Estimate the load that can be lifted by applying an effort of 70 N.*

Solution : Given $\eta = 0.7$

and $(VR) = 20$

We know $\eta = \frac{MA}{VR}$

or $20 \times .7 = (MA)$

i.e. $MA = 14$

Now $MA = \frac{W}{P}$

or $W = P.(WA)$

$= 70 \times 14$

$= 980$ N

Example 10.3. *In a lifting* machine *20% of the effort is lost to overcome friction. Find the load that can be lifted through a height of 1 m if an effort of 250 N is moved through 12 m. Also calculate the MA and efficiency of this* machine.

Solution : Giving $P_a = 250$

Let load to lifted $= W$

$$VR = \frac{12.0}{1.0} = 12$$

Effort lost in friction $P_f = 0.2 \times 250 = 50$ N

We
$$P_f = P_a - P_i = P_a - \frac{W}{VR}$$

or
$$50 = 250 - \frac{W}{VR}$$

$$= 250 - \frac{W}{12}$$

i.e.
$$\frac{W}{12} = 200$$

or
$$W = 200 \times 12$$
$$= 2400 \text{ N} \quad \textbf{Ans.}$$

$$\therefore \quad MA = \frac{W}{P} = \frac{2400}{250} = 9.6$$

$$\therefore \quad \eta = \frac{MA}{VR} = \frac{9.6}{12} = 0.8 \text{ or } 80\% \quad \textbf{Ans.}$$

Example 10.4. *In a simple* machine *an effort of 180 N is needed to lift a load of 1600 N. If its velocity ratio is 22 find (a) the efficiency (b) frictional force in terms of effort (c) frictional force in terms of load.*

Solution : Given data

$$W = 1600 \text{ N}$$
$$P = 100 \text{ N}$$
$$MA = \frac{W}{P} = \frac{1600}{100} = 16$$
$$VR = 22$$
$$\therefore \quad \eta = \frac{(MA)}{(VR)} = \frac{16}{22} = 0.727$$

(*b*) Frictional force in terms of effort

$$P_f = P_a - P_i$$

where
P_f = Effort lost in frictional
P_c = Actual effort
P_i = Ideal effort

$$\therefore \quad P_f = P_a - P_i$$

But
$$P_i = \frac{W}{VR} \qquad \left[\because \eta_i = \frac{MA}{VR} = 1\right]$$

$$= \frac{1600}{22} = 72.727 \qquad \therefore \begin{bmatrix} or 1 = \frac{W}{P_i}\left(\frac{1}{VR}\right) \\ \therefore p_i = \frac{W}{VR} \end{bmatrix}$$

and $\qquad P_a = 100$

$\therefore \qquad P_f = 100 - 72.723$

$\qquad = 27.272$ N

(*c*) Frictional force is terms load

$$W_f = W_i - W_a$$

But $\qquad \eta_i = \frac{MA}{(VR)} = \frac{W}{P\,(VR)} = 1$

or $\qquad W_i = P.(VR)$

$\qquad = 100 \times 22$

$\qquad W_a = 1600$

$\therefore \qquad W_f = 2200 - 1600$

$\qquad = 600$ N

10.4 REVERSIBLE AND IRREVERSIBLE MACHINE

Assume an effort *P* applied through a distance *y* lifts a load *W* through a distance *x*. After this let the effort be removed then two situations can arise viz. the work is done by the machine and the load falls back or the machine is not capable to do work in reverse direction and the load stays where it had been left after lifting.

In the first case when the machine does work in reverse direction and load falls, then the machine is called reversible machine. In second case when the load stays at the lifted position and does not fall, it is called irreversible machine.

Mathematical Condition for Reversibility

Let the effort *P* applied through a distance *y* lifts the load *W* through a distance *x*.

That $\qquad$ Input - Py

and $\qquad$ Output Wx

Since $\qquad$ Input > Output, the difference will be the frictional work

i.e. $\qquad$ Frictional work $(Py - Wx)$ ...(10.1)

In case on removal of effort after lifting the load *W*, the load falls (*i.e* for a reversible machine,) the output of the machine must be more than the frictional work

i.e. for reversible machine

$$Wx > (Py - Wx)$$

or $\qquad 2Wx > Py$

$$\frac{Wx}{Py} - \frac{1}{2}$$

$$\frac{\frac{W}{P}}{\frac{y}{x}} > \frac{1}{2}$$

or $$\frac{MA}{VR} > \frac{1}{2}$$

i.e. $$\eta > 0.5 \text{ or } 50\%$$

Similarly in case the load does not fall on removal of effort (*i.e.,* the machine is irrecverssibli)

$\therefore$ $$\eta = \frac{MA}{VR} = \frac{8}{20} = 0.4 \text{ i.e. } 40\%$$

As the efficiency is less than 50%, the machine is irreversible or self locking

$$w_f = w_i - w_a$$

where w_a = Actual load

w_a = 1600 N

w_i = Ideal load

$= P\,.\,(VR) = 200 \times 20 = 4000$ N

$\therefore$ $w_f = 4000 - 1600 = 200$ N

10.5 LAW OF MACHINE

The equation governing the relationship between the load lifted and the effort applied, for a certain velocity ratio is called the Law of machine. For constant velocity ratio and assuming the constant coefficient of friction the applied effort P will be proportional to the load being lifted.

In case this is no friction *i.e.* for ideal machine the effort will be directly proportional to the load. For no load condition effort will be zero. Thus the law for ideal machine will be represented by straight line as shown in Fig. 10.1 and will pass through origin.

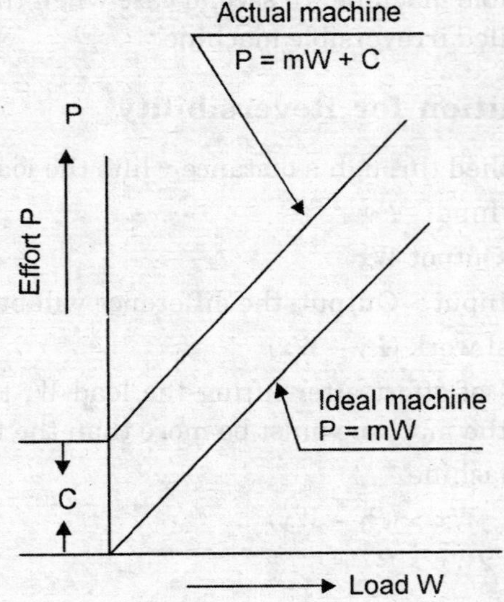

Fig. 10.1.

However for actual machine, the curve for the law will not pass through origin because of friction for which some effort will be lost to overcome it. Thus if it is assumed that coefficient of friction is constant and the velocity ratio does not change, the curve for the actual machine will shift upwards as shown in Fig. 10.1 such that the intercept on y-axis (here P-axis) by the straight line will be equal to C where C is a function of coefficient of friction

$$P = mW + C \quad \text{...(10.19)}$$

The values of m and C is determined by actually measuring the efforts to lift two different loads. Let P_1 and P_2 be the efforts corresponding to loads W_1 and W_2 respectively.

Then $$P_1 = mW_1 + C \quad \text{...(10.20)}$$

$$P_2 = mW_2 + C \quad \text{...(10.21)}$$

Substituting (10.21) for (10.20)

$$P_1 - P_2 = m\,(W_1 - W_2)$$

or $$m = \left(\frac{P_1 - P_2}{W_1 - W_2}\right) \quad \text{...(10.22)}$$

Equation (10.22) gives the slope of the line

Then $$C = P_1 - mW_1$$

$$= P_1 - \left(\frac{P_1 - P_2}{W_1 - W_2}\right) W_1 \quad \text{...(10.23)}$$

Equation (10.23) gives the value of the friction constant C.

Mechanical Advantage, its Maximum Value

Again $$MA = \frac{W}{P} \quad \text{...(1.24)}$$

Substituting value of P from equation (10.19) with equation (10.24)

$$MA = \frac{W}{mW + C}$$

or $$MA = \frac{1}{m + \frac{C}{W}} \quad \text{...(10.25)}$$

From equation (10.25), it is obvious that when W increases $\frac{C}{W}$ will decrease and so does the demoninator of RHS of equation (10.25), and MA will go on increasing. When W is very large then $\frac{C}{W}$ will become very small.

or mathematically when $W \rightarrow \infty$, then $\frac{C}{W} \rightarrow 0$

Thus the maximum value of MA will thin approach.

$$(MA)_{max} = \frac{1}{m}$$

This fact is depicted in Fig. (10.2) where the variation of *MA* with *W* is shown.

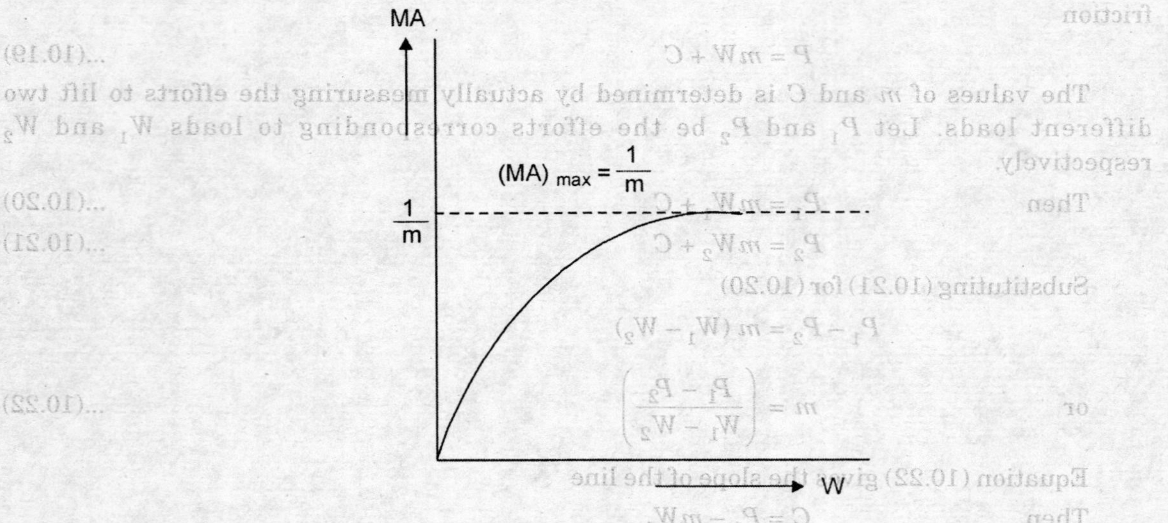

Fig. 10.2

This curve touches horizontal line at distance of '$\frac{1}{m}$' along *y*-axis, asymptotically.

Thus $$(M.A)_{max} = \frac{1}{m} = \left(\frac{W_1 - W_2}{P_1 - P_2}\right) \qquad ...(10.26)$$

Maximum efficiency

We know $$\eta = \frac{MA}{VR}$$

$$= \frac{W}{P.(VR)}$$

$$= \frac{W}{(mW + C)(VR)}$$

or $$\eta = \left[\frac{1}{\left(m + \frac{C}{W}\right)}\right]\left[\frac{1}{(VR)}\right] \qquad ...(10.27)$$

For (10.25) or (10.27) $$\eta = \left(\frac{MA}{VR}\right) \qquad ...(10.28)$$

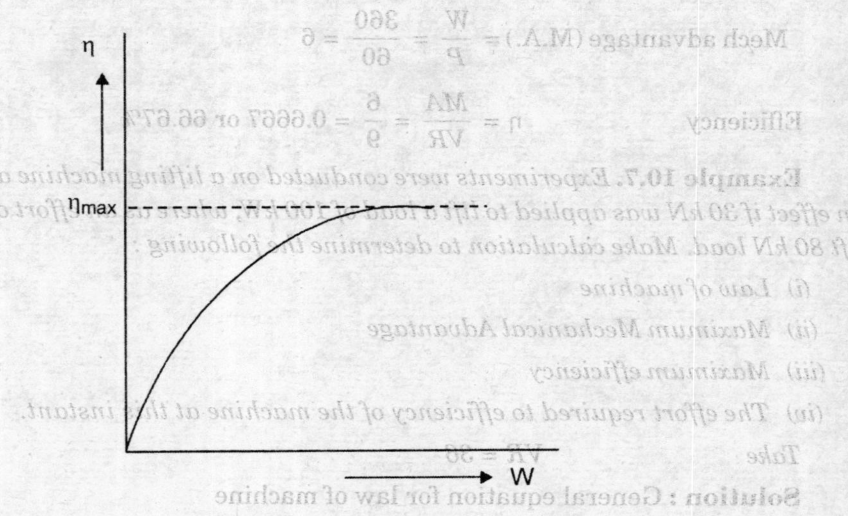

Fig. 10.3

Fig. 10.3 shows the variation of efficiency with load. As the load W increases the demoninator of RHS of equation (10.23) will go on decreasing which in turn will increase the value of efficiency such that

when $W \to \infty$

$$\frac{C}{W} \to 0 \quad \text{Then } \eta_{max} \to \frac{1}{m} \cdot \frac{1}{VR} \qquad \text{...(10.29 a)}$$

That is the curve touches $\left(\frac{1}{m}\right)\left(\frac{1}{VR}\right)$ line asymptotically.

Evidently from equation (10.27) and (10.28)

We get
$$\eta_{max} = \frac{(MA)_{max}}{(VR)} \qquad \text{...(10.29 b)}$$

Example 10.6. *A simple lifting machine raised a load of 30 N through a distance of 200 mm. The effort of 60 N moved 1.8 m during the process. Calculate the velocity ratio, mechanical advantage and efficiency of the machine.* *(PTU Dec. 2005)*

Solution : Given data

$P = 60$ N

$W = 360$ N

$x = 200$ mm

$y = 1.8$ m

$= 1800$ mm

$$VR = \text{Velocity ratio} = \frac{\text{Distance moved by effort}}{\text{Distance moved by load}}$$

$$= \frac{y}{x} = \frac{1800}{200} = 9$$

Mech advantage (M.A.) $= \frac{W}{P} = \frac{360}{60} = 6$

Efficiency $\eta = \frac{MA}{VR} = \frac{6}{9} = 0.6667$ or 66.67%

Example 10.7. *Experiments were conducted on a lifting machine and it was observed that an effect if 30 kN was applied to lift a load of 100 kW, where as an effort of 26 kN was required to lift 80 kN load. Make calculation to determine the following :*

(i) Law of machine

(ii) Maximum Mechanical Advantage

(iii) Maximum efficiency

(iv) The effort required to efficiency of the machine at this instant.

Take $VR = 36$

Solution : General equation for law of machine

$$P = mW + C$$

(*i*) For first set of observations

$$P_1 = 30 \text{ kN}$$
$$W_1 = 100 \text{ kN}$$

For second set

$$P_2 = 26 \text{ kN}$$
$$W_2 = 80 \text{ kN}$$

∴ Subtiacting the corresponding values in the law equation

$$30 = 100\,m + C$$
$$26 = 80\,m + C$$

Subtracting one from the other

$$4 = 20 \text{ m}$$

i.e. $m = 0.2$

Then $30 = 200 + C$

or $C = 10$

or $C = 10$

∴ Equation representing law is

$$P = 0.2\,W + 10$$

(*ii*) From equation (10.26) we have

$$(M.A.)_{max} = \frac{1}{m} = \frac{1}{0.2} = 5$$

(*iii*) And from equation (10.29 *a*)

$$\eta_{max} = \frac{1}{m} \cdot \frac{1}{(VR)} = \frac{1}{0.2 \times 30}$$

$$= 0.1667 \text{ or } 16.67\%$$

(*iv*) Let P_4 be the effort required and lift load of 20 kN

$$P_4 = 0.2\ W_4 + C$$
$$= 0.2 \times 20 + 10$$
$$= 14 \text{ kN}$$

Then
$$\text{M.A.} = \frac{W_4}{P_4} = \frac{20}{14} = 1.43$$

$$\eta_4 = \frac{MA}{VR} = \frac{1.43}{30}$$
$$= 0.0476 \text{ or } 4.76\%$$

Example 10.8. *A lifting machine has efficiency of 62% and velocity ratio 16. What effort will be needed to lift a load of 160 N. (b) Also find the equation of the law of machine. (c) Then estimate the effort to run the machine (i) at no load (ii) to lift a load of 100 N.*

Solution : (*a*)
$$VR = 16$$
$$\eta = 0.62$$
$$W = 160 \text{ N}$$
$$P = 7$$

Now
$$\eta = \frac{MA}{(VR)}$$

$\therefore$
$$M.A. = \eta\ (VR) = (.62)\ 16$$
$$= 9.92$$

Now
$$\text{M.A.} = \frac{160}{P_a}$$

$\therefore$
$$9.92 = \frac{160}{P_a}$$

or
$$P_a = \frac{160}{9.92} = \mathbf{16.129\ N\ \ Ans.}$$

(*b*) Law of machine

$$P = mW + C$$

where m = slope of eqn.

and C = Intercept of law curve on P-axis

= Effort lost to overcome friction

$= P_f$

i.e.
$$C = P_f = P_{\text{actual}} - P_{\text{ideal}}$$

But
$$P_{\text{idal}} = \frac{W}{VR} = \frac{160}{16} = 10$$

and
$$P_a = 16.129 \text{ N}$$

$\therefore$
$$C = P_f = 16.129 - 10$$
$$= 6.129 \text{ N}$$

Further when effort

$$P_a = 16.129$$

then $W = 160$

$\therefore$ From machine law equation

$$P = mW + C$$

$$16.129 = m \times 160 + C$$

or $16.129 = 160\,m + 6.129$

$\therefore$ $160\,m = 10$

$$m = \frac{10}{160} = \frac{1}{16}$$

or $m = 0.0625$

Thus the final equation of machine law is

$$P = 1.0625\,W + 6.129$$

(c) When no load is being lifted

i.e. $W = 0$

Effort $P_0 = m \times W + C$

$$= 0 \times m + 6.129$$

or $P_0 = 6.129$ N

Let P_{100} = Effort required to lift low of 100 N

$$P_{100} = mW + C$$

$$= 0.0625 \times 100 + 6.129$$

$$= 6.25 + 6.129$$

$= 12.279$ N **Ans.**

Example 10.9. *In a lifting machine, having VR = 38 an effort of 125 N lifted 3125 N load. Again on the same machine 170 N effort lifted a load of 4258 N. Determine (i), the law of machine (ii) effort required to lift a load of 4500 N (iii) maximum M.A (iv) Maximum efficiency (v) Also find this efficiencies when the lifted loads were 3125 N, 4258 N and 4500 N respectively.*

Let law of machine be

$$P = mW + C$$

But it is given for

$$W_1 = 3125 \text{ N}$$

$$P_1 = 125 \text{ N}$$

and for $W_2 = 4258$ N

$$P_2 = 170 \text{ N}$$

Therefore $125 = 125\,m + C$

$$170 = 4258\,m + C$$

Subtracting $45 = 1133\,m$

$$m = \frac{45}{1133} = 0.0394$$

and $125 = 3125 \times 0.0394 + C$

$$C = 125 - 124.117$$

$$= 0.883$$

Thus the law of machine is

(*i*) $P = 0.0394\,W + 0.883$

(*ii*) Let Effort be $= P$ for $W = 4500$ N

$$P = 0.0394 \times 4500 + 0.883$$

$$= 177.3 + 0.883$$

$$= 178.183 \text{ N}$$ **Ans.**

(*iii*) $$(MA)_{max} = \frac{1}{m} = \frac{1}{0.0394}$$

$$= 25.38$$

(*iv*) $$\eta_{max} = \frac{1}{m \times VR}$$

$$= \frac{25.38}{VR} = \frac{25.38}{38}$$

$$= 0.6079 \text{ or } 60.79\%$$

(*v*) When $W_1 = 3125$

$$P_1 = 125$$

$\therefore$ $$MA = \frac{3125}{125}$$

$\therefore$ $$\eta_1 = \frac{3125}{125 \times 38} = 0.6579$$ or **65.79%** Ans.

$$VR = 38$$

When $W_2 = 4258$ N

$$P_2 = 170$$

$$MA = \frac{4258}{170}$$

$$\eta_2 = \frac{4258}{170} \times \frac{1}{38} = 0.6591$$ or **65.91%**

$$W_3 = 4500 \text{ N}$$

$$P_3 = 178.183 \text{ N}$$

$$VR = 38$$

$\therefore$ $$MA = \frac{4500}{178.183}$$

$$\eta_3 = \frac{4500}{178.183} \times \frac{1}{38}$$

$$= 0.6646 \text{ or } 66.46\%$$ **Ans.**

Example 10.10. *During an experiment on a lifting machine with VR = 40 following two sets of observations were made. For load 100 N effort was 15 N and for load 200 N effort was 25 N. Establish the equation for machine law and determine the efficiency when the load to be lifted is 300 N.*

Solution : Let equation the machine law equation be $P = mW + C$

Substituting the values from the sets of observation

$$15 = 100\,m + C$$
$$25 = 200\,m + C$$

Sectional one equation from the other

$$10 = 100\text{ m}$$

or $$m = \frac{1}{10}$$

$\therefore$ $$C = 15 - 100\text{ m}$$
$$= 15 - \frac{100}{10} = 5$$

Thus the equation for machine law is

$$P = 0.1\,W + 5$$

At load = 300 N, therefore

$$P = 0.1 \times 300 + 5 = 35\text{ N}$$

Mechanical advantage $$MA = \frac{300}{35}$$

and $$VR = 40$$

$\therefore$ $$\eta = \frac{mA}{VR} = \frac{300}{35 \times 40} = 0.2142$$

i.e. = 21.42%. **Ans.**

10.6 PULLEYS : FIXED AND MOVABLE

In Fig 10.4 a pulley is shown. It has two components, a sheave and a block. The sheave is essentially a metallic or wooden wheel with a groove cut along its periphery over which a rope can pass. In case chain replaces the rope, then instead of groove sprocket teeth are cut on the periphery of the wheel. The sheave is mounted on an axle which is fixed to frame called the block. The shieve can rotate freely about the axle.

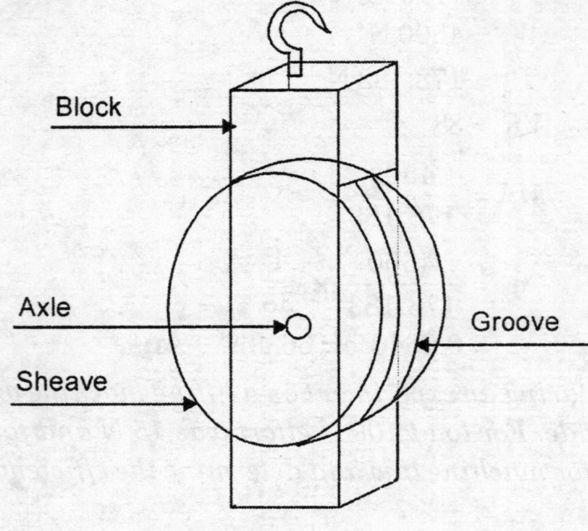

Fig. 10.4

The pulley is known as fixed pulley if the block is fixed. In this case the axis of rotation the heave is fixed in space it can not move up and down. On the other hand the pulley is known or movable pulley if the block is movable. In this case the sheave can rotate about it one's and this axis itself can move up and down.

While analysing various pulley arrangements in the forthcoming paragraphs following assumptions will be made.

(*i*) The weight of the pulley is negligibly small as compared to the weight lifted.

(*ii*) The pulley is smooth. In other words the tension is the rope passing around the pulley is same throughout.

10.6.1 Single Fixed Pulley

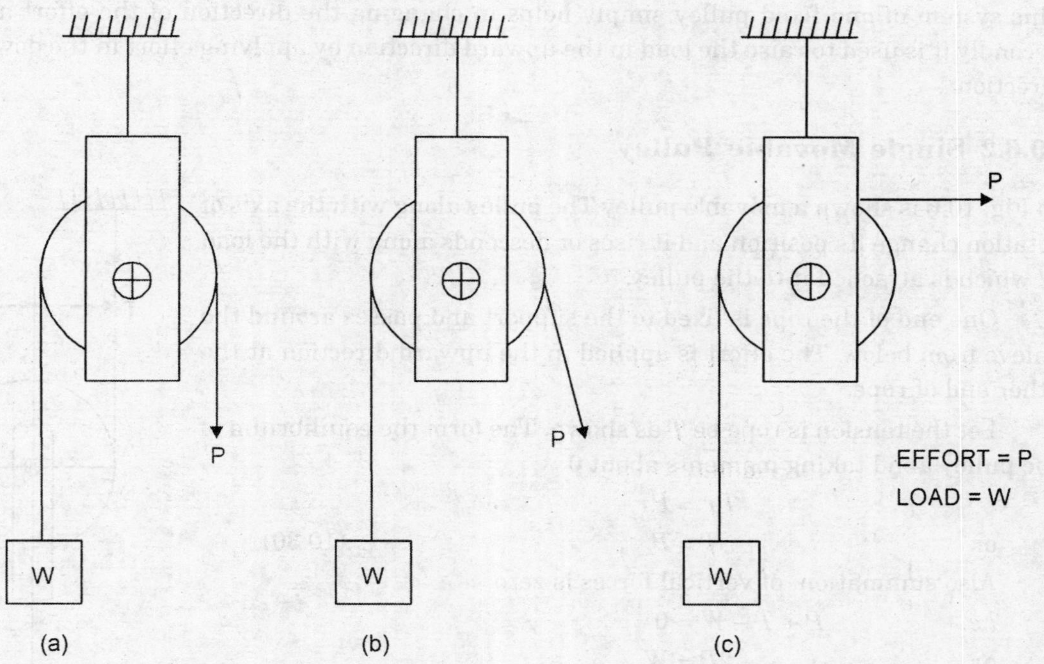

Fig. 10.5

In Fig. 10.5 is shown fixed pulley the position of the axle does not change position. The load which has to be lifted is attached on one end of the rope which passes around the shieve in its groove and the effort is applied to the other end of the rope. The effort can be applied in any direction *e.g.* vertically downward Fig. 10.5 (*a*), inclined to vertical its Fig. 10.5 (*b*), or even in the horizontal direction Fig. 10.5 (*c*). In all of these cases the tension in the rope will be same.

For the equilibrium of the sheave, taking moments of the force W and P about the axis of rotation

$$\Sigma M = 0$$

or $$P.r = W.r$$

where r is the radius of the pulley

or $$P = W$$

Therefore mechanical advantage

$$MA = \frac{W}{P} = 1 \qquad ...(10.28)$$

Also for no friction, the distance moved by W and P will be equal

i.e. $$VR = 1$$

Therefore $$\eta = \frac{MA}{VR} = \frac{1}{1} = 1 \text{ or } 100\% \qquad ...(10.29)$$

From equation (10.28) and (10.29), it is concluded that there is no mechanical advantage. This system of one fixed pulley simply helps in changing the direction of the effort applied. Secondly it is used to raise the load in the upward direction by applying effort in the downward direction.

10.6.2 Single Movable Pulley

In Fig. 10.6 is shown a movable pulley The pulley along with the axis of rotation change its position and it rises or descends along with the load W which is attached it to the pulley.

One end of the rope is fixed to the support and passes around the shieve from below. The effort is applied in the upward direction at the other end of rope.

Let the tension is rope be T as shown. The form the equilibrium of the pulley abnd taking moments about 0

$$T.r = P.r$$

or $$T = P \qquad ...(10.30)$$

Also summation of vertical forces is zero

i.e. $$P + T - W = 0$$

or $$2P = W$$

or $$P = \frac{W}{2} = T \qquad ...(10.31)$$

$\therefore$ $$MA = \frac{W}{P}$$

$$= \frac{W}{W} \times 2 = 2$$

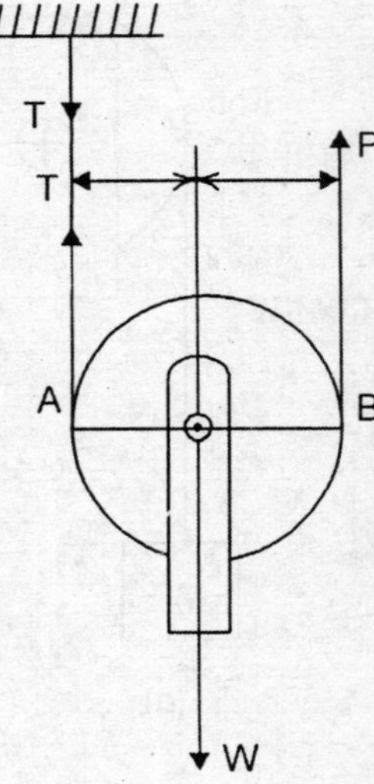

Fig. 10.6

Due to practical difficulties encountered in applying the effort P in the vertically upward direction as in Fig. 10.6, single movable pulley is not used alone. In order to overcome this practical difficulty a combination of fixed and movable pulley shown in Fig. 10.7 is used. The direction of the effort applied is changed with the help of fixed pulley. It is very convenient to apply the effort P in the downward direction.

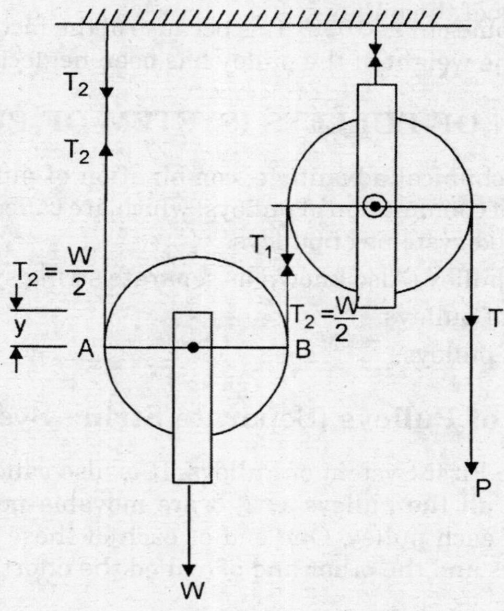

Fig. 10.7

For raising the load W, upward through a distance y, the rope or string is pulled by effort P through a distance

$$(AA' + BB') = 2\,y$$

Evidently the effort moves through a distance which is double the distance moved by the load W

$$\therefore \qquad VR = \frac{\text{(Distance moved by effort)}}{\text{(Distance moved by load)}}$$

$$\text{or} \qquad VR = \frac{2y}{y} = 2$$

For the equilibrium of movable pulleys

$$2\,T_2 = W$$

$$\text{or} \qquad T_2 = \frac{W}{2}$$

Further for the equilibrium of the shieve of fixed pulley

$$T_1 = T_2 = \frac{W}{2}$$

$$\text{and} \qquad P = T_1$$

$$\therefore \qquad P = T_1 = T_2 = \frac{W}{2}$$

$$MA = \frac{W}{P} = \frac{W}{\frac{W}{2}} = 2$$

$$\eta = \frac{MA}{VR} = \frac{2}{2} = 1 \quad i.e.\ 100\%$$

The efficiency has come out as 100%. It is because of the fact that it has been assumed that there is no friction and the weight of the pulley has been neglected.

10.7 COMBINATION OF PULLEYS (SYSTEM OF PULLEYS

In order to get move mechanical advantage, combination of pulleys is resorted to. There are three different manners of combination of pulleys, which are called systems of pulleys. Following are three most widely used systems of pulleys.

1. First system of pulley (Also known as separate string system of pulleys).
2. Second system of pulleys.
3. Third system of pulleys.

10.7.1 First System of Pulleys (Separate String System of Pulleys)

In Fig. 10.8 is shown the First System of Pulleys. It is also called "Separate String System of Pulleys". In this system all the pulleys 1, 2, 3 are movable pulleys. A separate rope passes around the periphery of each pulley. One end of each of these ropes is attached to the fixed support above the pulleys and the other and of (called the effort end) is fixed to the axle of the next upper pulley.

The load is attached to the body of the bottom most pulley whereas the effort is applied to the effort end of the rope which passes round the uppermost pulley. But for the convenience of the application of the effort, the fixed, pulley is used.

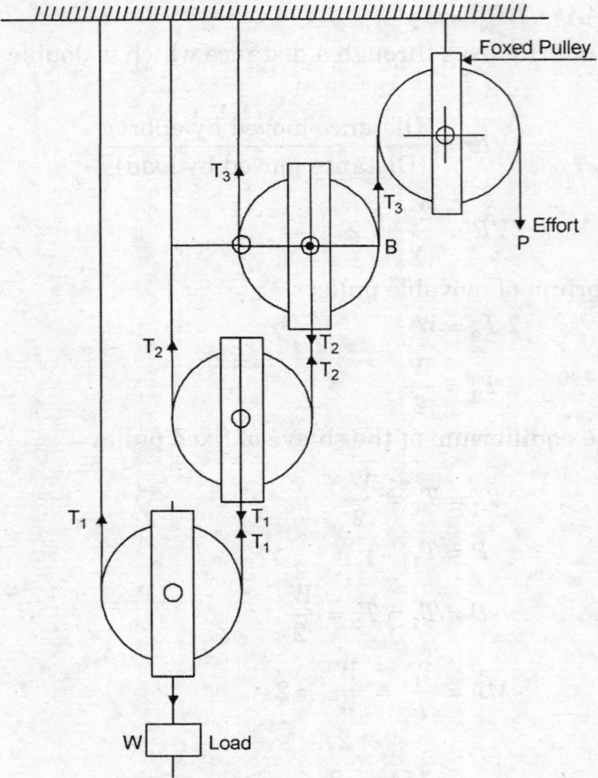

Fig. 10.8. First system of pulley

Analysis

Let T_1, T_2, T_3 etc. be the tensions in the ropes as shown in Fig. 10.8.

Then for the equilibrium of each individual pulley we get (neglecting the weight of original pulleys and assuming no friction).

$$W = 2\,T_1$$

$$T_1 = 2\,T_2 \qquad \text{...(10.32)}$$

$$T_2 = 2\,T_3$$

and $$P = T_3 \qquad \text{...(10.33)}$$

Also $$W = 2\,T_1,\ 2T_2 = T_1,\ 2\,T_3 = T_2$$

$\therefore$ $$W = 2 \times 2\,T_2 \qquad \text{...(10.34)}$$

$$= 4\,(2\,T_3) = 8\,T_3 = (2)^3\,T_3 \qquad \text{...(10.35)}$$

$$MA = \frac{W}{P} = \frac{2^3\,(T_3)}{T_3}$$

Substitute for P from (10.33)

$$MA = 2^3 = 8 \qquad \text{...(10.36)}$$

Since the friction is negl. ted along with the weight of the pulleys, the equations derived above are for ideal first system of pulleys).

$\therefore$ Input = Output

Assuming the effort moves through a distance 'y' corresponding to which the movement of W will be 'x' (say).

Then $$Py = Wx$$

or $$\frac{y}{x} = \frac{W}{P} = MA \qquad \text{...(10.37)}$$

Also $$\frac{y}{x} = VR \qquad \text{...(10.38)}$$

From (10.38) through (10.38)

$$MA = VR = 8 = 2^3$$

or $$= 2^n \qquad \text{...(10.39)}$$

where 'n' is the number of movable pulleys.

Alternatively the velocity ratio can also be found in the following manner.

Let the effort end of the rope be pulled through a distance y. As the last pulley *i.e.* number 3 pulley is supported by 2 segments of rope each segment will get shortened by $\frac{y}{2}$ which in turn will move the axle C by a distance $\frac{1}{2} \times \frac{y}{2} = \frac{y}{2^2}$ and so one for other pulleys. Thus way the lowermost pulley *i.e.* the centre or axle of A will move up by $\frac{y}{2^3}$ or $\frac{y}{2^n}$ if there are 'n' movable pulleys.

Thus the weight W attached to A will move up a distance $x = \frac{y}{2^n}$

$$\therefore \qquad VR = \frac{y}{\frac{y}{2^n}} = 2^n = 2^3$$

Since here $\qquad n = 3$

(b) In case the weight of the pulleys is taken into account the analysis will be as under.

Let each pulley be of weight = W_p then for the equilibrium of each pulley will give in turn (Fig. 10.8).

$$2\,T_1 = W + W_p$$

$$T_1 = \frac{W}{2} + \frac{W_p}{2}$$

and
$$2\,T_2 = T_1 + W_p$$

$$T_2 = \frac{T_1}{2} + \frac{W_p}{2}$$

$$= \frac{1}{2}\left(\frac{W}{2} + \frac{W_p}{2}\right) + \frac{W_p}{2}$$

$$= \frac{W_p}{2^2} + \frac{W_p}{2^2} + \frac{W_p}{2}$$

And
$$2\,T_3 = T_2 + W_p$$

$$T_3 = \frac{1}{2}\,(T_2) + \frac{W_p}{2}$$

$$= \frac{1}{2}\left[\frac{W}{2^2} + \frac{W_p}{2^2} + \frac{W_p}{2}\right] \times \frac{W_p}{2}$$

$$= \frac{W}{2^3} + \frac{W_p}{2^3} + \frac{W_p}{2^2} + \frac{W_p}{2}$$

If there are only 3 pulleys their

$$T_3 = P = \frac{W}{2^3} + \frac{W_p}{2^3} + \frac{W_p}{2^2} + \frac{W_p}{2}$$

Similarly if their are n pulleys

$$T_n = P = \frac{W}{2^n} + \frac{W_p}{2^n} + \frac{W_p}{2^{n-1}} + \frac{W_p}{2^{n-2}} + \frac{W_p}{2^{n-2}} + \ldots\ldots \frac{Wp}{2}$$

Multiplying both sides by 2^n it reduces to

$$2^nP = W + W_p\,(1 + 2 + 2^2 + 2^3 + \ldots + 2^{n-1})$$

or
$$2^nP = W + W_p\,(2^n - 1) \qquad \ldots(10.40)$$

Now mechanical advantage

$$MA = \frac{W}{P}$$

or $$MA = \frac{W}{\frac{1}{2^n}\left[W + W_p(2^n - 1)\right]}$$

Substitute value of P from equation (10.40)

or $$MA = \frac{2^n W}{W + W_p\ (2^n - 1)} \quad ...(10.41)$$

If the pulleys weight is negligible, then putting $W_p = 0$ equation (10.30) reduces to

$$MA = \frac{2^n W}{W} = 2^n \quad ...(10.42)$$

This result is same as in equation (10.39)

$VR = 2^n$ as in equation (10.39)

10.7.2 Second System of Pulleys (Pulley Block System)

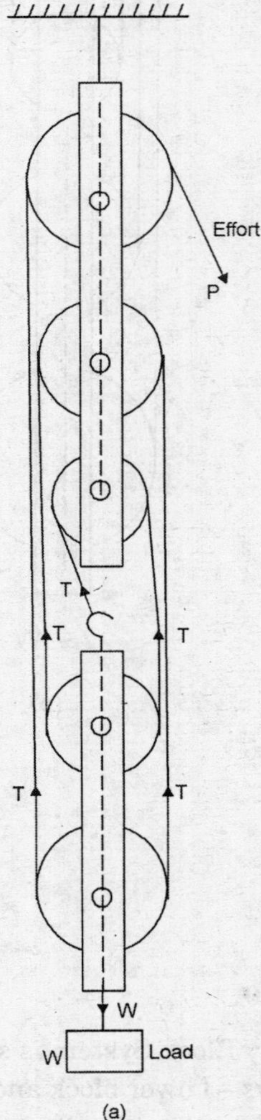

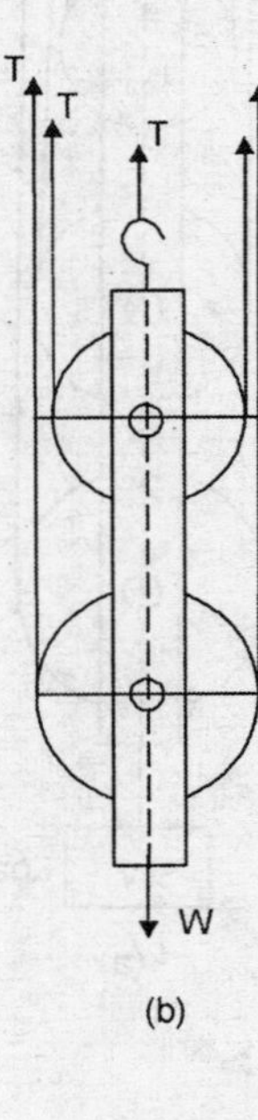

Fig. 10.9

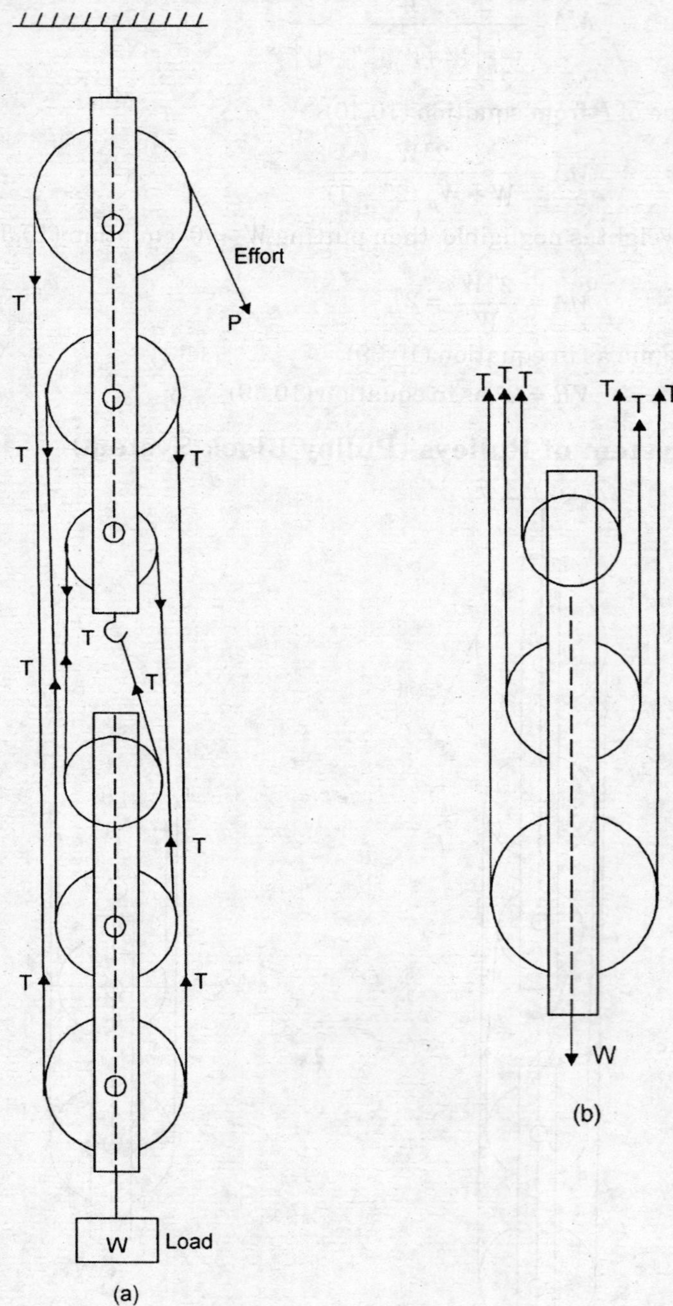

Fig. 10.10

Second system of pulleys also known as Pulley Block System is shown is Figs. (10.9) and (10.10). In this system there are two blocks of pulleys – Lower block and upper block. The upper block is fixed while the lower block is movable. The number of pulleys in the two blocks may be equal or the upper block may contain one pulley more than the number of pulleys in the lower block. In either case the lower block is movable but upper block is fixed and it supports the lower block.

Arrangement of pulleys when both the blocks have the same number of pulley is shown in Fig. 10.10 but when the number of pulleys in upper block is more their the arrangement is shown in Fig. 10.9.

In both the arrangements there is only one string. In Fig. 10.10, one end of the string is fixed to the upper block and after taking the rope round each pulley as shown in Fig. 10.10, the other end of the slope is pulled by the effort P. However in the case when the number of pulleys is more in the upper block (Fig. 10.9), one end of the rope is fixed to lower block and the other end after taking the rope around each pulley is manner shown in Fig. 10.9, is pulled by the effort P.

In the case of arrangement of Fig. 10.9 (*a*) the free body diagram of the lower block is given in Fig. 10.9 (*b*). Evidently there are five segments of rope each carrying tension T to support load W (neglecting the weight of pulleys). Therefore under ideal conditions where there is no friction then efficiency being 100% we have

$$VR = MA$$

$$P = T$$

But $$W = 5\,T$$

$$= 5\,P$$

or $$\frac{W}{P} = 5$$

i.e. $$MA = 5$$

$\therefore$ $$MA = VR = 5 = \text{Total no. of pulleys} = n \qquad ...(10.43a)$$

In the second case *i.e.* for arrangement shown in Fig. 10.10. The free body diagram A lower block gives.

$$W = 6\,T$$

and $$P = T$$

$\therefore$ $$W = 6T = 6\,P$$

$$MA = \frac{W}{P} = 6$$

For ideal system $$MA = VR = 6$$

$$= \text{Total no. of pulleys} = n \qquad ...(10.43b)$$

10.7.3 Third System of Pulleys (Separate String Attached to Weight System)

Third system of pulleys also known as separate string attached to weight system is shown in Fig. 10.11.

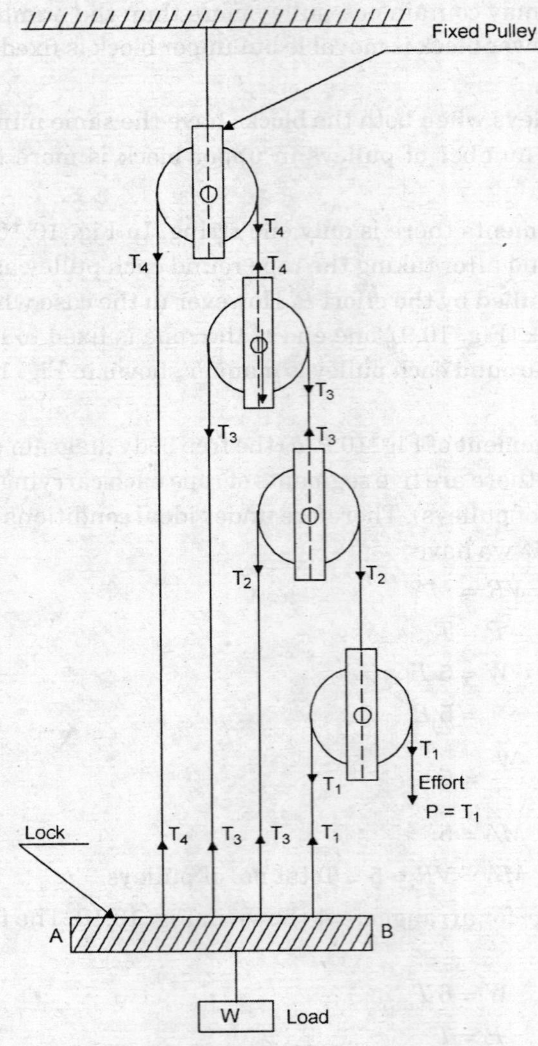

Fig. 10.11

In this system of pulleys, the number of strings is equal to this number of pulleys. One end of each string is attached to the movable block AB to which the load is connected the other and of each rope passing around the periphery of upper pulley is fastened to the next lower pulley as shown in Fig. 10.11.

If T_1, T_2, T_3 are the tensions in the strings passing round the movable pulleys then for equilibrium of the pulleys 1, 2, 3 .... is turn we get

$$T_1 = P$$

$$T_2 = T_1 = 2\,P$$

$$T_3 = 2\,T_2 = 2 \times 2\,T_1 = 2^2\,T_1 = 2^2\,P$$

$$T_4 = 2\,T_3 = 2 \times 2^2\,P = 2^3\,P$$

and $$W = T_1 + T_2 + T_3 + T_4$$

$$= P + 2\,P + 2^2\,P + 2^3\,P$$
$$= P\,(1 + 2 + 2^2 + 2^3\,P)$$
$$= P\,(2^4 - 1)$$

$$\therefore \qquad MA = \frac{W}{P} = 2^4 - 1$$

In case their are n pulley then

$$MA = \frac{W}{P} = 2^n - 1$$

For ideal condition i.e. weightless pulleys and no friction.

$$VR = MA$$

or $$\eta = 100\%$$

$$= 2^n - 1 \qquad \text{...(10.44)}$$

(*b*) Weight of Pulleys is Not Negligible

Let each of the pulleys be of weight W_P. Then for the equilibrium of each pulley in turn we have

Pulley No. 1 (Fig. 10.11)

$$P = T_1$$
$$T_2 = 2\,T_1 + W_P = 2\,P + W_P$$

Pulley No. 2

$$T_3 = 2\,T_1 + W_P = 2\,(2\,P + W_P) + W_P$$
$$= 2^2\,P + 2\,W_P + W_P$$

Pulley No. 3

$$T_4 = 2\,T_3 + W_P = 2\,[2^2P + 2W_P + WP] + WP$$
$$= [2^3\,P + 2^2\,W_P + 2W_P] + W_P$$

In case there are n pulleys ($n - 1$) movable and one fixed), then

$$T_n = [2^{n-1}\,P + 2^{n-2}\,W_P + 2^{n-3}\,W_P + \ldots + 2\,W_P] + W_P$$

But $$W = T_1 + T_2 + T_3 \ldots. + T_n$$

or $$W = P + (2P + W_P) + (2^2\,P + 2\,W_P + W_P) + (2^3\,P + 2^2\,W_P + 2\,W_P + W_P)$$
$$+ \ldots - 1\,(2^{n-1}\,P + 2^{n-2}\,W_P + 2^{n-3}\,W_P + \ldots\,2\,W_P + W_P)$$

or $$W = P\,(1 + 2 + 2^2 + 2^3 + \ldots + 2^{n-1}) + W_P\,(1 + 2 + 2^2 + 2^3 + \ldots + 2^{n-2})$$
$$+ W_P\,(1 + 2 + 2^2 + 2^3 + \ldots + 2^{n-3}) + \ldots + 2\,W_P + W_P$$

or $$W = P\,(2^{n-1}) + W_P\,(2 + 2^2 + 2^3 + \ldots + 2^{n-2} + 2^{n-1}) - W_P\,(n - 1)$$

or $$W = P\,(2^n - 1) + W_P\,(2^n - n - 1) \qquad \text{...(10.45 a)}$$

or $$\frac{W}{P} = (2^n - 1) + \frac{W_p}{P}\,(2^n - n - 1) \qquad \text{...(10.45 b)}$$

If W_P is negligible then equation (10.45) reduces is

$$MA = \frac{W}{P} = 2^n - 1$$

Thus equation is same as equation (10.44).

or $$\frac{W}{P} = MA$$

VR. Let the load W be lifted through a distance 'x'. Then the downward moved by pulley 3

$$x = (2^1 - 1)\,x$$

After the pulley 3 moves down by distance x along with the moving up of the load by distance x, then the slackness $2x$ so produced will lower the next pulley no. 2 by

$$2x + x = 3x$$

is distance moved by pulley no. 2

$$= 3x = (2^2 - 1)\,x$$

Again due to the lowering down of pulley 2 by $3x$ and the moving up of load by x, the slackness $4x$ so produced will lower the next pulley no. 1 by

$$4x + 3x + x = 7x$$

$\therefore$ Distance moved by pulley no. 1 $= 7x$

$$= (2^3 - 1)\,x \qquad \text{...(10.46)}$$

Similarly if their one n pulleys in the system, the last pulleys will move $(2^n - 1)\,x$

$$\therefore \qquad VR = \frac{\text{Distance moved by effort}}{\text{Distance moved by load}}$$

$$= \frac{(2^n - 1)x}{x} = 2^n - 1 \qquad \text{...(10.47)}$$

Example 10.11. *The efficiency of a two sheave pulley block is 70%. A load of 1000 N is to be lifted. What effort will be required to lift this load ? Find the effort lost due to friction.*

Solution : No. of sheave in the pulley block = 2

$$\therefore \qquad VR = 2 \times \text{no. of sheave}$$

$$= 2 \times 2 = 4$$

Now $$MA = \frac{W}{P}$$

$$\eta = \frac{MA}{(VR)} = \frac{W}{P \times 4}$$

$$\therefore \qquad P = \frac{W}{4x} = \frac{1000}{4 \times .7} = 357.14 \text{ N}$$

$$\therefore \qquad P = 357.14 \text{ N}$$

$$P_f = P_{\text{acted}}\, P_{\text{ideal}} = 357.14 - \frac{W}{(VR)}$$

$$= 357.14 - \frac{100}{4} = 107.14 \text{ N}$$

Example 10.12. *In the first system of pulleys there are three pulleys and an effort of 60 N can just support a load of 360 N. Find the efficiency of the system and find the effort lost in friction.*

Solution : No. of pulley = 3

$\therefore \quad VR = 2^3 = 8$

$$\text{MA} = \frac{W}{P} = \frac{360}{60} = 6$$

$$\therefore \quad \eta = \frac{MA}{VR} = \frac{6}{8} = 0.75 = 75\%$$

Effort lost in friction $P_f = P_a - P_i = P_a - \dfrac{MA}{VR}$

$$= 60 - \frac{360}{8} = 60 - 45 = 15 \text{ N}$$

Similarly friction in terms of dual

$$W_f = W_i - W$$
$$= P \times VR - W$$
$$= 60 \times 8 - 360$$
$$= 480 - 360 - 120 \text{ N}$$

Example 10.13. *In the first system of pulleys there are 4 movable pulleys. The weight of each is 15 N. Find (a) the effort required to lift load of 10000 N. (b) Recalculate the effort if the weight of the pulleys is negligible.*

Solution : $n = 4$

$$W_P = 15 \text{ N}$$
$$W = 10000 \text{ N}$$

From equation (10.39)

(*a*) $$MA = \frac{2^n W}{W + W_P (2^n - 1)}$$

$$= \frac{2^4 \times 10000}{10000 + 15 (2^4 - 1)}$$

$$= \frac{160000}{10225} = 15.648$$

But $$MA = \frac{W}{P} = \frac{10000}{P}$$

$$\therefore \quad P = \frac{10000}{15.648} = 639.06 \text{ N}$$

(*b*) When $W_P = 0$

$$VR = 2^n = 2^4 = 16$$

$$MA = \frac{2^n W}{W + 0} = 2^n = \frac{W}{P}$$

$$\therefore \quad P = \frac{W}{2^n} = \frac{10000}{16} = 625 \text{ N} \quad \textbf{Ans.}$$

Example 10.14. *In the second system of pulleys there are 2 pulleys in the lower block and 3 pulleys in the upper block. An effort of 125 N lifted a load of 500 N. Find the efficiency of the system.*

Solution : Total number of pulleys in the second system

$$= 2 + 3 = 5$$

Therefore $VR = 5$

New Effort = 125 N

$$W = 500 \text{ N}$$

$$MA = \frac{W}{P} = \frac{500}{125} = 4$$

$$\therefore \quad \text{Efficiency} = \frac{MA}{VR} = \frac{4}{5} = 0.4 \text{ or } 80\%$$

Example 10.15. *In the second system of pulley, there are three pulleys each in both the blocks. It is observed that when not loaded the effort requred is 30 N. It is also estimated that further wastage in friction increases the pulls at a rate of 2% of load lifted. Find (a) the effort to lift a load of 3600 N when it is assumed that there is no friction (b) The effort required to lift the load when the friction is taken into consideration (c) Efficiency of the system at this load.*

Solution : (*a*) Under frictionless conditions. In the second system of pulley we have from equation (10.4).

$$MA = VR = n$$

(For no friction)

Here $n = 6$

$\therefore$ $MA = 6$

and $W = 3600$

$$\frac{W_i}{P_i} = 6$$

or $P_i = \dfrac{3600}{6} = 600$ N **Ans.**

(*b*) When friction is taken into consideration

Effort required when not loaded

$$= 30 \text{ N}$$

Additional effort lost when loaded

$$= 2\% \text{ of } 3600 \text{ N}$$

$$= \frac{2}{100} \times 3600 = 72 \text{ N}$$

$\therefore$ Total effort lost due to friction

$$= 30 + 72 = 102$$

Thus total effort required to lift 3600 kg

$= P_i + 102 = 600 + 102 = 702$ N **Ans.**

Therefore efficiency at this load

$$= \frac{\text{Effort without friction}}{\text{Effort with friction}}$$

$$= \frac{600}{702} = 0.8547 \text{ or } 85.47\% \textbf{ Ans.}$$

Example 10.16. *A worker weighing 700 N in a factory sits on a platform attached to the lower block of the system of six pulleys arranged in the second system of pulleys. The collective weight of the lower block and the platform is 100 N. The worker supports himself by exerting a downward pull at the free end of the rope. If friction is neglected find the minimum pull the man should apply.*

Solution : Given data : Second system of pulleys

Total number of pulleys = 6 = n

Weight of worker + weight of platform (in second system of pulleys)

$$= 700 \text{ N} + 100 \text{ N}$$

$$= 800 \text{ N}$$

$$MA = VR = \text{N} = 6$$

Let pull exerted by the worker = P

As the worker sitting on the platform pulls the rope with a pull of P the effective load will then be

i.e. $(800 - P) = W_e$ (Say) ...(*A*)

Now $MA = \frac{W_e}{P} = 6$ *i.e* $W_e = 6P$

W_e = effective load

$= (800 - P)$

$6P = W_e$

$= 800 - P$ [From eqn. (*A*)]

or $7P = 800$

$$P = \frac{800}{7} = 114.28 \text{ N} \quad \textbf{Ans.}$$

Example 10.17. *In a third order pulley system having three pulleys, an effort of 250 N is required to lift a load of 1250 N. Calculate the efficiency of the system. Also find the effort lost in friction.*

Solution : Given data $n = 3$

$P = 250$ N

and $W = 1250$

We have for 3rd order system of pulleys (equation 10.42)

$$VR = 2^n - 1 = 2^3 - 1 = 7$$

$$MA = \frac{W}{P} = \frac{1250}{250} = 6$$

$$\eta = \frac{MA}{VR} = \frac{6}{7} = 0.8571 \text{ or } 85.71\%.$$

Let E_f be the effort lost in friction

then $E_f = P_{actual} - P_{ideal}$

Now $P_{ideal} = \frac{W}{VR} = \frac{1250}{7}$

$= 178.57$ N

$\therefore$ $E_f = 250 - 178.57$

$= 71.43$ N **Ans.**

Example 10.18. *In a third order of pulley system there are two pulleys and its efficiency is 70%. Find the value of the effortn to lift a load of 2100 N.*

Solution : Given data $\eta = 0.7$

$n = 2$

$\therefore$ $VR = 2^n - 1 = 22 - 1 = 3$

$\eta = \frac{MA}{VR}$

i.e. $MA = 0.7 \times (VR) = 0.7 \times 3 = 2.1$

$\frac{W}{P} = 2.1$

$\therefore$ $P = \frac{2100}{2.1} = 1000$ N **Ans.**

Example 10.19. *In a third order of pulley system there are four pulleys. Weight of each pulley is 5 N. How much load can be supported by an effort of 150 N. Also find the load that can be supported by the system without applying any effort.*

Solution : (*a*) Given data

$n = 4$

$W_P = 5$ N

P = 150 N

Let W be the load that can be lifted by the effort *P* of 150 N. We have from equation (10.43 *a*).

$W = P(2^n - 1) + W_P(2^n - n - 1)$

$= 150(2^4 - 1) + 5(2^4 - 4 - 1)$

$= 150 \times 15 + 5 \times 11$

$= 2250 + 55 = 2305$ N **Ans.**

(*b*) Here $P = 0$

Again from equation (10.43 *a*)

$W = P(2^n - 1) + W_P(2^n - n - 1)$

$= 0 + 5(2^4 - 4 - 1)$

$= 55$ N **Ans.**

10.8 WESTON'S DIFFERENTIAL PULLEY BLOCK

Weston's differential pulley block consists of two blocks (Refer to Fig. 10.12) *A* and *B*.

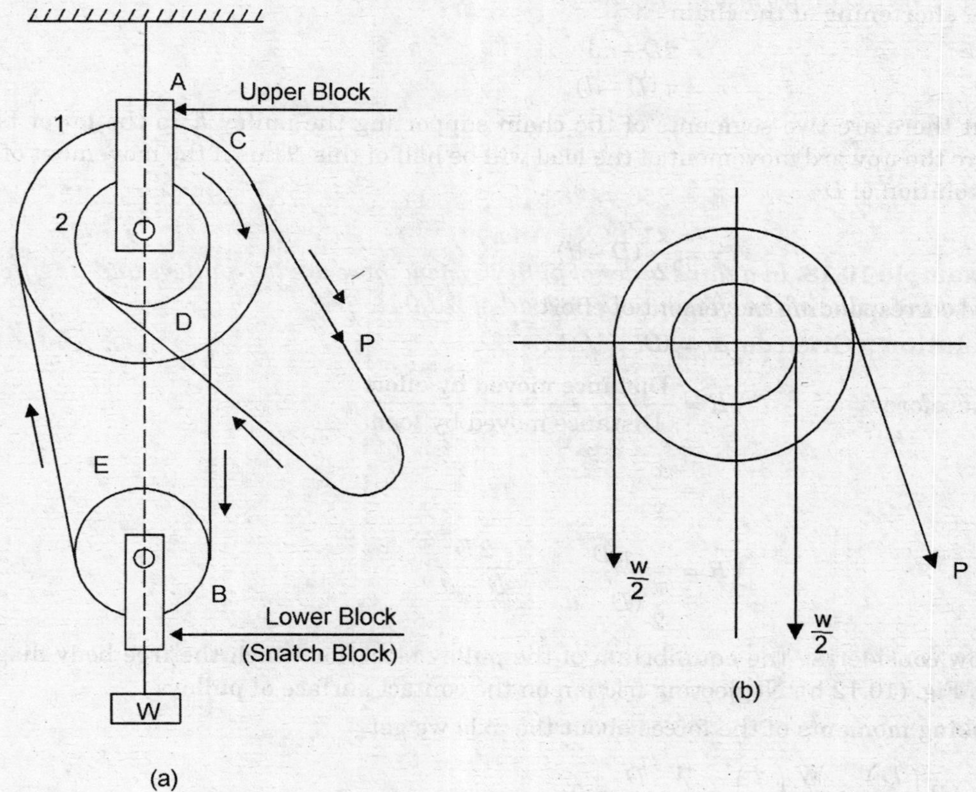

Fig. 10.12.

The upper block A is fixed and consists of two pulleys C and D both fixed to the common axle but are of different diameters. The lower block B called the snatch block, carries a movable pulley to which load is attached.

Endless chain passes over pulleys number 1 of the upper block A, then under the pulley of snatch block and finally over the smaller pulleys number 2 of the upper block. The effort P is applied to that part of chain which passes around the bigger pulley of block A, *i.e.* pulley number 1. In order to prevent chain from slipping, teeth are provided on the pulleys.

Analysis : Let the diameter of the bigger pulley, C be $= D$

and the dia of the smaller pulley, $D = d$

$$D = d$$

Load to be lifted $= W$

and Effort applied $= P$

Then in one revolution of the pulley C the effort will move through a distance $= \pi D$ clockwise (unwinding of chain from C)

Since both the pulleys C and D are fixed to the common axle simultaneously a length $= (\pi d)$ passes over the periphery of pulley D (winding the chain on pulley D). Thus the lengthening of chain on the effort side and its shortening on the load side will result in lifting of the load

Net shortening of the chain

$$= \pi D - \pi d$$

$$= \pi (D - d)$$

But there are two segments of the chain supporting the pulley E in the lower block B. Therefore the upward movement of the load will be half of this. Thus if the movement of load is y for revolution of D

$$y = \frac{\pi}{2} (D - d)$$

and corresponding movement of effort

$$x = \pi D$$

Therefore $$VR = \frac{\text{Distance moved by effort}}{\text{Distance moved by load}}$$

$$= \frac{x}{y}$$

or $$VR = \frac{\pi D}{\frac{\pi}{2} (D - d)} = \frac{2D}{D - d} \qquad ...(10.46)$$

Now considering the equilibrium of the pulley block for which the free body diagram is given in Fig. (10.12 b). Neglecting friction on the contact surface of pulleys.

Taking moments of the forces about the axle we get

$$P\left(\frac{D}{2}\right) + \frac{W}{2}\left(\frac{d}{2}\right) - \frac{W}{2} \cdot \frac{D}{2} = 0$$

$$\frac{PD}{2} = \frac{W}{4} (D - d)$$

or $$\frac{W}{P} = \frac{2D}{(D - d)}$$

i.e. $$MA = \frac{W}{P} = \left(\frac{2D}{D - d}\right) \qquad ...(10.47)$$

From equation (10.46) and (10.47)

$$MA = VR$$

$\therefore$ $$\eta = \frac{MA}{VR} = 1 \text{ or } 10\%$$

The value of efficiency comes out to be 100%. This is so as the friction has been neglected. When friction is taken into account the value of MA will be less than that given by equation (10.47) and the efficiency will be less than 100%.

Example 10.20. *A Weston's differential pulley consists of upper block and lower block. There are two cogged grooves out on the upper block their diameters are 400 mm and 300 mm respectively. Calculate the effort required to lift a load of 2000 N. Take the efficiency of the* machine *as 40%.*

Solution : Given data

$$D = 400 \text{ mm}$$
$$d = 320 \text{ mm}$$
$$\eta = 0.4$$
$$W = 2000 \text{ N}$$
$$P = ?$$

From equation (10.46)

$$VR = \frac{2D}{D-d} = \frac{2 \times 400}{400 - 320}$$
$$= \frac{2 \times 400}{80} = 10$$

Now $$\eta = \frac{MA}{VR}$$

$\therefore$ $$MA = 0.4 \times 10 = 4$$

But $$MA = \frac{W}{P}$$

$\therefore$ $$P = \frac{W}{MA} = \frac{2000}{4} = 500 \text{ N} \quad \textbf{Ans.}$$

Example 10.21. *In a Weston's differential pulley block, the ratio of number of recesses in peripheries of smaller and bigger pulleys is 8 : 9. If the efficiency of the* machine *is 65% of find the effort required to lift, load of 2000 N.*

Solution : Given data

$$\eta = 65\%$$

Let N_1 = No. of recesses in longer pulley

N_2 = No. of recesses in smaller pulley

$$\frac{N_2}{N_1} = \frac{8}{9}$$

or $$9\,N_2 = 8\,N_1$$

or $$N_1 = \frac{9}{8}\,N_2$$

$$VR = \frac{2D}{D-d} = \frac{2\,N_1}{N_1 - N_2}$$
$$= \frac{2\,N_1}{N_1 - \frac{8}{9}N_1} = \frac{2\,N_1}{N_1} \times 9$$
$$= 18$$

Now $$\eta = \frac{MA}{VR}$$

$$0.65 = \frac{MA}{18}$$

or $$MA = 18 \times .65 = 11.7$$

or $$\frac{W}{P} = 11.7$$

or $$P = \frac{W}{11.7} = \frac{2000}{11.7} = 170.94 \text{ N } \textbf{Ans.}$$

Example 10.22. *In a Weston's differential pulley block, the difference in this number of recesses in the two pulleys is 4. An effort of 200 N just lifts a load of 4 kN at an efficiency of 40%. Find (b) number of recesses in both the pulleys.*

Solution : Let N_1 = No. of recesses in bigger pulley

and N_2 = No. of recesses in smaller pulley

Then $$N_1 - N_2 = 4 \text{ (given)}$$

i.e. $$N_1 = N_2 + 4$$

$$\eta = 60\% = 0.6$$

$$W = 4000 \text{ N}$$

$$P = 200 \text{ N}$$

$$MA = \frac{W}{P} = \frac{4000}{200} = 20$$

$\therefore$ $$\eta = \frac{MA}{VR}$$

or $$VR = \frac{M}{\eta} = \frac{20}{0.4} = 50.00$$

But $$VR = \frac{2N_1}{N_1 - N_2}$$

$$= \frac{2N_1}{N_1 - (N_1 - 4)} = \frac{2N_1}{4}$$

$$= 0.5 N_1$$

i.e. $$0.5 N_1 = 50$$

or $$N_1 = \frac{50}{0.5} = 100$$

$\therefore$ $$N_2 = 100 - 4 = 96 \textbf{ Ans.}$$

Example 10.23. *In a differential pulley block diameters of the concentric pulleys are 350 mm and 250 mm. It is observed that an effect of 450 N just supports a load 2250 N. Calculate (a) the efficiency of the* machine *(b) Effort lost in frictions (c) frictional load.*

Solution : (*a*) Given data

$$D = 350 \text{ mm}$$

$$d = 250 \text{ mm}$$

$$W = 2250 \text{ N}$$

$$P = 450 \text{ N}$$

$$MA = \frac{2250}{450} = 5$$

$$VR = \frac{2D}{D-d} = \frac{2 \times 350}{350 - 250}$$

$$= \frac{700}{100} = 7$$

$$\therefore \qquad \eta = \frac{MA}{VR} = \frac{5}{7} = 0.7143 \text{ or } 71.43\%$$

(*b*) Effort lost in friction $\quad P_f = P_a - P_i$

Now $\quad P_i = P_{\text{ideal}} = \frac{W}{VR} = \frac{2250}{7} = 321.428 \text{ N}$

$$P_f = 450 - 321.428 = 188.572 \text{ N}$$

(*c*) $\quad W_f = W_i - W_a = P \times VR - 2250$

$$= 450 \times 7 - 2250 = 900 \text{ N}$$

10.9 SIMPLE WHEEL AND AXLE

As the very name of the machine suggests it consists of a wheel and an axle which are keyed to same shaft (Fig. 10.13). The shaft is supported on ball bearings bearings to reduce friction One end an of the rope which is wound (clockwise as seen in the direction *B* to *A*) on the axle is fixed on the axle and the free and is connected to the load *W*. On the wheel whose diameter is greater than that of axle is wound a rope (anticlockwise) as seen from *B* towards *A* with one end fixed to the wheel and on the other end is applied the effort so that when effort is applied downward the load will be lifted up.

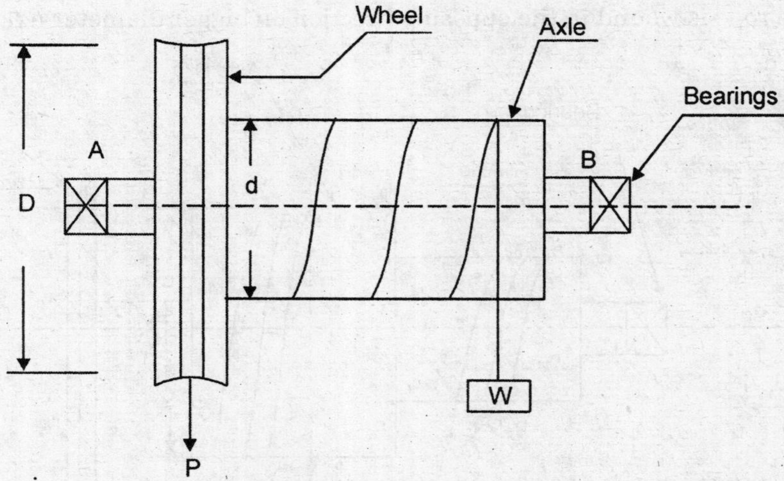

Fig. 10.13. Simple wheel and axle

Analysis. As the wheel and axle are mounted as the same spindle in one revolution (provided as the thickness of ropes neglected.)

Distance moved down by effort

$$= \pi D$$

Distance the load will move up

$$= \pi d$$

where D = dia. of wheel

d = dia. of axle

$$\therefore \quad VR = \frac{\pi D}{\pi d} = \frac{D}{d} \quad \text{...(10.48)}$$

In case the diameter of rope (or its thickness) is to be taken into account, then assuming the thicknesses of ropes as t_1 and t_2 respectively for the two ropes used with wheel and axle,

then
$$VR = \frac{\pi(D+t_1)}{\pi(d+t_2)}$$

$$= \frac{D+t_1}{d+t_2} \quad \text{...(10.49)}$$

$$MA = \frac{W}{P}$$

$$\therefore \quad \eta = \frac{MA}{VR} = \frac{W(D+t_1)}{P(d+t_2)} \quad \text{...(10.50)}$$

10.10 DIFFERENTIAL WHEEL AND AXLE

In this lifting machine, the axle used is a compound axle. The axle is made in two parts of different diameters. The two axles and a wheel are all mounted on the same shaft (Fig. 10.14). The effort is applied to the free and of the rope wound on wheel. Another rope is wound as the two axle and it carries the load through a pulley. The rope on the wheel and on the smaller diameter axles are wound in the same direction (say clockwise as seen from the direction B to A where as the rope is wound in the opposite direction on bigger diameter axle.

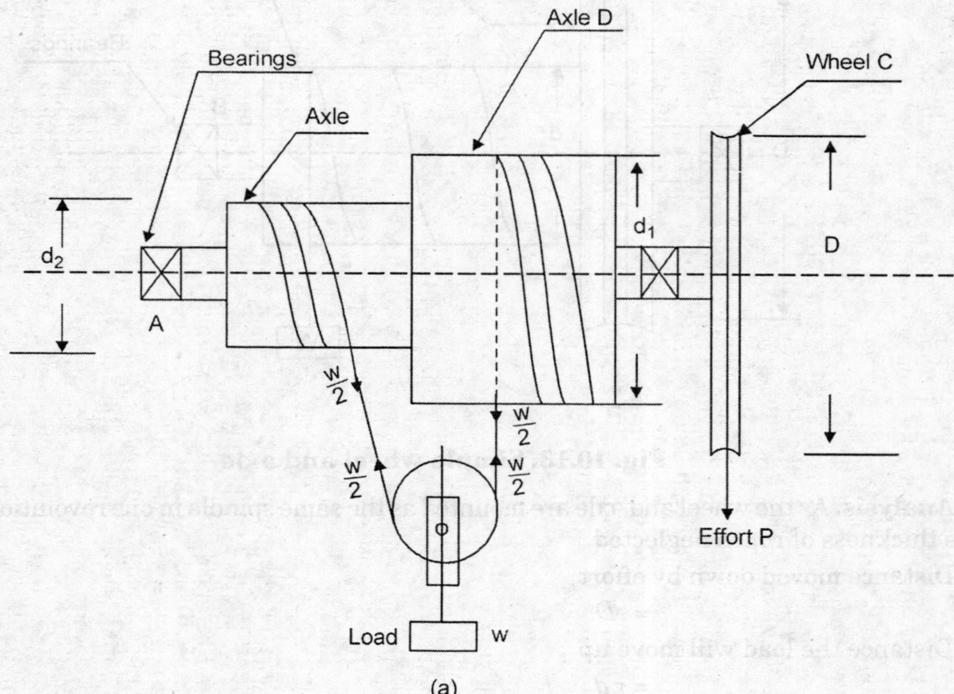

(a)

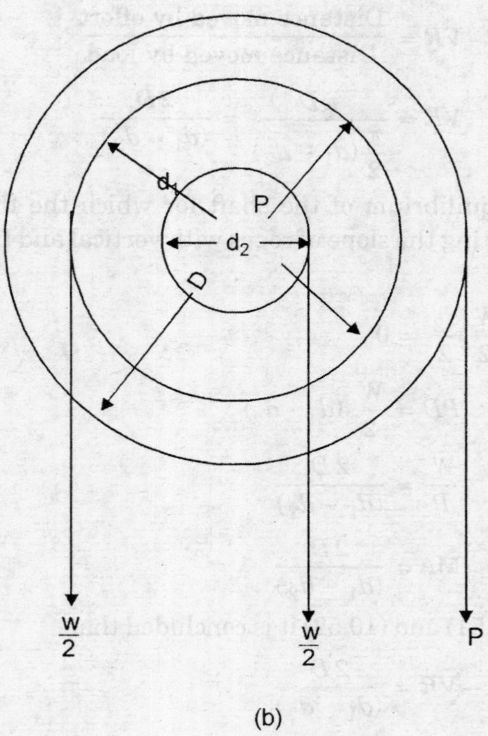

(b)

Fig. 10.14

Analysis : Let D = Diameter of wheel C

d_1 = diameter of bigger axle, D

d_2 = diameter of smaller diameter axle E.

Since the wheel and two axles are keyed to same shaft, their rotation will be same. Thus for one revolution. Downward *Distance* moved by effort

$$P = \pi D$$

Length of the rope that will wind on bigger axle D

$$= \pi d_1$$

Length of rope that will unwind on small direction axle E

$$= \pi d_2$$

$\therefore$ Net length of rope wheel will wind on bigger axle

$$= (\pi d_1 - \pi d_2)$$

Since there are two segments supporting the load pulley, the load will be lifted up by a distance

$$= \frac{1}{2}\,(\pi d_1 - \pi d_2)$$

$$= \frac{\pi}{2}\,(d_1 - d_2)$$

$$\therefore \qquad VR = \frac{\text{Distance moved by effort}}{\text{Distance moved by load}}$$

or
$$VR = \frac{\pi D}{\frac{\pi}{2}(d_1 - d_2)} = \frac{2D}{(d_1 - d_2)} \qquad ...(10.51)$$

Considering the equilibrium of the shaft for which the free body diagram is shown in Fig. 10.14 (b) and neglecting the slope of ropes with vertical and friction. Taking moments about this shaft axis.

$$\frac{PD}{2} - \frac{W}{2}\frac{d_1}{2} + \frac{W}{2}\frac{d_2}{2} = 0$$

$$PD = \frac{W}{2}(d_1 - d_2)$$

or
$$\frac{W}{P} = \frac{2D}{(d_1 - d_2)} \qquad ...(10.52\,a)$$

or
$$\text{MA} = \frac{2D}{(d_1 - d_2)} \qquad ...(10.52\,b)$$

From equation (10.51) and (10.52) it is concluded that

$$MA - VR = \frac{2D}{(d_1 - d_2)}$$

i.e.
$$\eta = \frac{MA}{VR} = 1 \quad \text{or } 10\%$$

The efficiency has come out to be as 100%. It is due to the fact that friction has been neglected. In actual practice the efficiency will be lower than 100%.

In case the thickness of the ropes is to be taken into account the expression for *VR* and *MA* will be modified as given below.

Let t_1 = Thickness order of rope as the wheel

t_2 = Thickness (diameter) of ropes as axles.

Then
$$VR = \frac{2(D + t_1)}{(d_1 + t_2) - (d_2 + t_2)}$$

$$VR = \frac{2(D + t_1)}{(d_1 - d_2)} \qquad ...(10.53)$$

$$MA = \frac{2(D + t_1)}{(d_1 - d_2)} \qquad ...(10.54)$$

Example 10.24. *In simple wheel and axle system the diameter of wheel is 450 mm and that of axle 150 mm. It is used to lift a load 2000 N for which an effort of 1000 N was applied.*

(a) What is the efficiency of the machine.

(b) Had there been no friction, how much more load could be lifted with this effort ?

(c) What will be the efficiency if the thickness of ropes is taken into account and

Take $t_1 = t_2 = 20$ *mm*

Solution : Given data

$$D = 450 \text{ mm}$$
$$d = 150 \text{ mm}$$
$$W = 2000 \text{ N}$$
$$P = 1000 \text{ N}$$

From equation 10.46

(*a*)
$$VR = \frac{D}{d} = \frac{450}{150} = 3$$

$$MA = \frac{W}{P} = \frac{2000}{1000} = 2$$

$\therefore$
$$\eta = \frac{MA}{VR} = \frac{2}{3} = 0.6667$$

i.e. 66.67%

When friction is neglected

(*b*)
$$\eta = 100\% = 1.0$$

i.e.
$$\frac{MA}{VR} = 1 \quad \text{or } MA = VR$$

$$\frac{W_i}{P} = VR$$

$$W_i = 1000 \times 3 = 3000$$

$\therefore$ Additional load that can be lifted

$$= W_i - W_{\text{actual}}$$
$$= 3000 - 2000$$
$$= 1000 \text{ N} \quad \textbf{Ans.}$$

(*c*) From equation (10.47)

$$VR = \frac{D + t_1}{d + t_2} = \frac{450 + 20}{150 + 20} = \frac{470}{170}$$

$$= \frac{47}{17}$$

$$MA = \frac{2000}{1000} = 2$$

Then
$$\eta = \frac{MA}{VR}$$

$$= \frac{2 \times 17}{47} = \frac{34}{47} = 72.34 \quad \text{or } 72.34\%$$

Example 10.25. *In a simple wheel and axle arrangement wheel diameter is 455 mm and axle diameter 110 mm. The thickness of two ropes is 5 mm what is the value of VR. If the efficiency of the machine, which lifting is load of 2200 N at lifting speed of 15 m/min. is 65%, calculate what work must be applied to the machine per minute.*

Solution : Given data

$$D = 455 \text{ mm}$$
$$d = 110 \text{ mm}$$
$$t_1 = t_2 = 5 \text{ mm}$$

From equation (104) we have

$$VR = \frac{D + t_1}{d + t_2} = \frac{455 + 5}{110 + 5}$$
$$= \frac{460}{115} = 4.0 \textbf{ Ans.}$$

Now $$\eta = \frac{\text{Output work}}{\text{Input work}}$$

$$0.65 = \frac{2200 \times 15}{\text{Input work}} \text{ Nm/m}$$

$\therefore$ $$\text{Input work} = 0.65 \times 2200 \times 15$$
$$= 21450 \text{ N-m/mm } \textbf{Ans.}$$

Example 10.26. *In a differential wheel and axle arrangement wheel diameter is 260 mmand the axle diameters are 100 and 80 mm respectively. If an effort of 60 N lifts a load of 900 N find (a) out the VR, MA and efficiency of the machine (b) Also find the number of revolutions, the wheel will make to lift the load through a distance of 100 mm.*

Solution : Given data

$$D = 260 \text{ mm}$$
$$d_1 = 110 \text{ mm}$$
$$d_2 = 80 \text{ mm}$$
$$P = 60 \text{ N}$$
$$W = 900 \text{ N}$$

(*a*) From equation (10.51)

$$VR = \frac{2D}{d_1 - d_2} = \frac{2 \times 260}{110 - 80}$$
$$= \frac{520}{30} = 17.33$$
$$MA = \frac{W}{P} = \frac{900}{60} = 15$$

$\therefore$ $$\eta = \frac{15}{17.33} = 0.8655 \quad \textit{i.e.}\ 86.55\%$$

(*b*) Let Distance moved up by load be by and that by effort be x

$$y = 100 \text{ mm}$$

Now $$VR = \frac{\text{Distance moved by efforts}}{\text{Distance moved by load}}$$

$$\text{or} \quad = \frac{x}{y}$$

$$\text{or} \quad 17.33\,y = x$$

$$\text{i.e.} \quad x = 17.33 \times 100 = 1733 \text{ mm}$$

If N is the number of revolution of wheel, then

$$N(\pi D) = 1733$$

$$N = \frac{1733}{\pi \times 260} = \frac{1733}{876.814}$$

$$= 2.12 \text{ revolutions.}$$

Example 10.27. *A differential wheel and axle lifting machine has following specifications. Wheel diameter = 600 mm Axles diameters are 225 mm and 170 mm respectively. Rope diametr every where is 10 mm. If the efficiency of the machine is 65% find the load that can be lifted by an effort of 80 N.*

Solution : Given data

$$D = 600 \text{ mm}$$

$$d_1 = 225 \text{ mm}$$

$$d_2 = 170 \text{ mm}$$

$$\eta = 65\%$$

$$t_1 = t_2 = 10 \text{ mm}$$

From equation (10.53) we have

$$VR = \frac{2(D + t_1)}{(d_1 - d_2)} = \frac{2(600 + 10)}{(225 - 170)}$$

$$= \frac{2 \times 610}{55} = 22.18$$

$$\text{Now} \quad \eta = \frac{MA}{VR}$$

$$\therefore \quad MA = 0.65 \times 22.18 = 14.417$$

$$= \frac{W}{P}$$

$$W = P \times 14.417$$

$$= 80 \times 14.417$$

$$= 1153.36 \text{ N} \quad \textbf{Ans.}$$

10.11 WORM AND WORM WHEEL (WORM AND WORM GEAR)

Worm and Worm Gear (Also called a worm and worm wheel) is eventually a scrent meshing with a special helical gear. In Fig. 10.15 are shown worm and worm gear. The worm shown has two threads but any number upto six sometimes more may be used depending upon the design. The geometry of the worm is similar to that of power serew, e.g. lead serew of the lathe, the screw of the screw jack. Rotation of the worm simulates a linearly advancing involute rack (as in the case of a pinion and rack gear). Since the axes of worm and worm wheel are perpendicular but skew, the rotation of worm about its axis will generate rotary motion of the worm wheel such that is one revolution of the worm (worm single start) it will push the worm wheel through one tooth. If the number of teeth on the worm gear are T wound revolution of the worm the worm gear will rotate $\frac{1}{T}$ revolution.

In the present case when worm and worm wheel has to be used as a lifting machine, a wheel is attached to worm shaft so that one end of a rope is fixed to the wheel and the other end is used to apply effort. This wheel is called the effort wheel. Similarly on the worm wheel shaft is fixed the load drum on which a rope is wound the other end of rope is fixed with the load so that when the worm wheel rotates the load drum will also rotate thin by the rope will get wound on the drum and lift the load.

Analysis

Let the effort wheel diameter

$$= D_p$$

$$\text{Effort} = P$$

$$T = \text{No. of teeth as worm gear}$$

Thus in one revolution of the wrom x = The distance moved by $P = \pi D_P$

And corresponding to this the distance mvoed by load = y

$$= \pi D_W \times \frac{1}{T}$$

where D_w = worm wheel diameth

$$\therefore \quad VR = \frac{x}{y}$$

$$\text{or} \quad VR = \frac{\pi D_P}{\pi D_W \frac{1}{T}} = \frac{T D_P}{D_w}$$

If friction is neglected then $\eta = 100\%$

$$\text{Then} \quad MA = VR = \frac{T D_P}{D_W}$$

If friction is included

$$MA = \frac{W}{P} \quad \text{...(10.55)}$$

$$\eta = \frac{MA}{VR} = \frac{W}{P\,(VR)}$$

If there are multiple threads on the worm *e.g.* 2 threads then in one revolution of worm the worm wheel will advance by 2 teeth or corresponding rotation of worm wheel will be due to 2 teeth.

That is worm wheel Rotaion = $\frac{2}{T}$ revolution

$\therefore$ Movement of effort = πD_P

$$\text{Rotation of laod diamete} = \frac{2}{T} \text{ revolution}$$

$$\text{Movement of load} = \pi D_W \frac{2}{T}$$

$$VR = \frac{\pi D_P}{\pi D_W \frac{1}{T}} = \frac{T D_P}{D_W} \qquad \text{...(10.56)}$$

$$MA = \frac{W}{P}$$

$$\eta = \frac{W}{P\,(VR)} \qquad \text{...(10.57)}$$

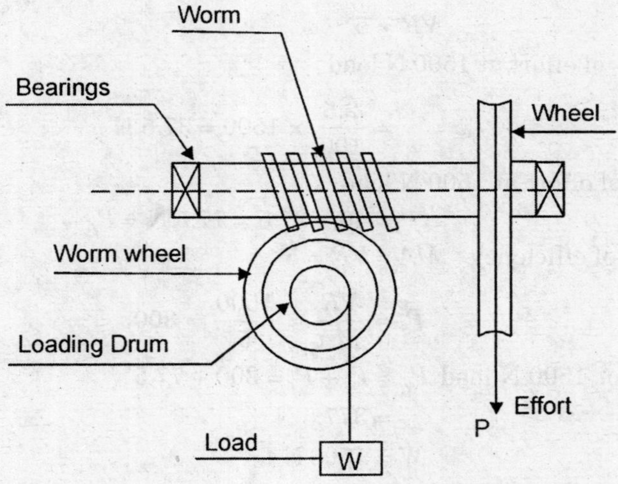

Fig. 10.15.

Exampe 10.28. *A worm and worm wheel arrangement has following specifications* ***worm*** *: single start thread*

No. of teeth on worm wheel = 100

$$D_P = \textit{Effort wheel dia} = 300 \textit{ mm}$$

$$D_W = \textit{Load wheel dia} = 150 \textit{ mm}$$

If the efficiency of the machine is 20% what effort is needed to lift a load of 100 N

Solution : From equation (10.55)

$$VR = \frac{T D_P}{D_W} = \frac{100 \times 300}{150}$$

$$= 200$$

$$\eta = \frac{MA}{VR} = 0.2$$

$$MA = 0.2 \times 200 = 400$$

But $$MA = \frac{W}{P}$$

i.e. $$40 = \frac{1000}{P}$$

or $$P = \frac{1000}{40} = 25 \text{ N} \quad \textbf{Ans.}$$

Exmaple 10.29. *Five pulleys are arranged in the second order of system of pulleys. When the system is not loaded, the effort required to raise the lower movable block is 40 N. Further loss of effort due to friction, increases the pull at the rate of 2.5% of the load lifted. Find the effort required to lift a load of 1500 N and the efficiency of the system at this load.*

Solution : Given $n = 5$

P_f at zero load = 40 N

$\therefore$ $VR = 5$

Additional loss of effort at 1500 N load

$$= \frac{2.5}{100} \times 1500 = 37.5 \text{ N}$$

$\therefore$ Total loss of effort at 1500 N load

$$= 40 + 37.5 = 77.5 \text{ N} = P_f$$

Now for 100% of efficiency $MA = VR = 5$

or $$P_i = \frac{VR}{MA} = \frac{1500}{5} = 300$$

Therefore P_a for 1500 N load $P_a = P_i + P_f = 300 + 77.5$

$$= 377.5 \text{ N}$$

(b) At $W = 1500$ N

$$\eta = \frac{\text{Effort without friction}}{\text{Effort with friction}}$$

$$= \frac{300}{377.5} = 0.7947 \text{ or } 79.47\% \quad \textbf{Ans.}$$

PROBLEMS

1. Explain the terms, (*a*) Lifting machine (*b*) Mechanical advantage (*c*) Velocity ratio (*d*) efficiency.
2. When is a machine an ideal lifting machine ?
3. Explain the difference between the actual effort and ideal effort.
4. State the law of machines.
5. What does the intercept of the machine law curve with *y*-axis represent ? Under what circumstances the intercept is zero.
6. What is the relationship of the slope of machine law curve and the maximum *MA*.
7. For an ideal machine, give the relationship between *MA* and *VR*.

8. What is the limiting values of efficiency for a self-locking machine ?
9. In second order system of pulleys, what is the relationship between *VR* and the number of pulleys in both the blocks.
10. In the third order system of pulleys give the expression for the *VR* in terms of the number of pulleys. What will be the effort required to raise is load of 611 N.
11. What is the relationship between *MA* and *VR* in an ideal machine ?
12. In first order system of pulleys give the expression for VR in terms of the number of pulleys .
13. Describe briefly the following lifting machines. Support your answer with neat diagram.
    (*a*) Simple wheel and axle.
    (*b*) Wheel and differental axle.
    (*c*) Differential pulley block.
    (*d*) Worm and worm wheel.
14. A simple lifting machine raised a load of 360 N through a distance of 200 mm. The effort of 66 N moved 118 m during the process. Calculate the velocity ratio, mechanical advantage and efficiency of the machine. [(*PTU 2005*)o (**Ans.** (9, 6 66.67%))
15. In a lifting machine an effort of 30 N is required to raise a load of 1.0 kN. If the efficiency of the machine is 0.75, what is the velocity ratio ? If on this machine an effort of 65 N raised a loadof 2 kN. What is now the efficiency ? (**Ans.** (*a*) 61.5%)
16. A simple wheel and axle arrangement is used to lift a weight of 460 N by applying an effort of 120 N. The wheel diameter is 400 mm and axle is of 100 mm diameter. Find MA, VR and efficiency of the machine. (**Ans.** 383,4 95.75)
17. The diameter of the effort wheel and axles in a wheel and differential axle arrangement are 400 mm, 300 mm and 200 mm respectively. Find the load that can be lifted by an effort of 100 N assuming efficiency as 0.75. Take thickness of rope as 10 mm. (**Ans.** 615 N)
18. A Weston's differential pulley consists of two pulley blocks. The upper block has two cogged grooves, whose diameters are 260 mm and 240 mm respectively. Find the effort required to lift a load of 9100 N if efficiency is 35%. (**Ans.** 100 N)
19. Six movable pulleys are arranged in first order of system of pulleys. Calculate the effort required to lift a load of 4800 N if the efficiency of the system is 75%. (**Ans.** 1000 N)
20. Five pulleys are arranged in second system of pulleys. At no load, the effort required to just lift the lower block is 25 N. If due to the frictional force the effort increases at the rate of 2% of the load find the effort requied to lift a load of 3000 N and the efficiency of the system at this load. (**Ans.** 685N, 87.6%)
21. A worm and worm wheel arrangement has the following specification
    (*i*) Worm threads – Single start
    (*ii*) No. of teeth or worm wheel = 80
    (*iii*) Diameterof effort wheel = 450 mm
    (*iv*) Diameter of load wheel = 225 mm
    (*v*) Efficiency of the machine = 30%
    Find the effort required to raise a load of 2400 N. (**Ans.** 50 N)

❑❑❑

CHAPTER 11

CONCEPT OF MECHANICAL BEHAVIOUR OF ENGINEERING MATERIALS

11.1 INTRODUCTION

In the present day world, the production and consumption of steel is considered to be index of development of some country. Not only the items that are made of steel and its alloys but for the non-metal items that we see around indirectly lead us to the fact that for their production steel has extensively been used in the form of machinery which is made of steel. Be it paper, a book plastic toys, household goods, furniture as a matter of fact any item we see around, tools and machinery made of steel have been employed for their manufacture.

Steel which is extracted from iron ore is subjected to various manufacturing processes to ultimately acquire the shape of some component or the other so as to form part of an assembly which is called a machine. Although as part of a particular machine these components are subjected to various types of loads but they do not change their configuration and size so that the machine continues, to perform its assigned duty for which it has been designed. This is possible only if the behaviour of the engineering materials used, is throughly investigated and understood by the designer. This objective is fulfilled by the subject called "Mechanics of Solids "for which more popular prevalent name is "Strength of Materials.

Before topics related to 'Mechanics of solid' is taken up it is considered essential to review the basic knowledge already gained by the reader during his pre-engineering course particularly about, free body digarms, equations of static equilibrium and other associated terms in the following parapraphs.

11.2 FREE BODY DIAGRAM

Free body diagram of a member of a structure, or of an element of a member in static equilibrium, is the diagram of only that member or the element, as if made free from the rest of the structure, with all the external forces acting on it duly marked. These external forces will include those due to the presence of adjacent members in contact with the one in question, the gravitational effect and the inertia force if the member is in motion. Fig. 11.1(a) shows the free body diagram of a weight hanging from the end of a crane rope, the gravitational force is balanced by the tension T in the rope. The Fig. 11.1(b) shows the free body diagram, of the two elements of a tension test specimen under the action of two equal and opposite forces P, obtained by taking a section CC. The force, F_1 on the element AC_1 is due to the cohesion of the plane C_2, while F_2 on the element C_2B is due to the plane C_1. Since the test specimen AB is in equilibrium any portion of that will also be in equilibrium. For the equilibrium of individual elements AC_1 and C_2B, F_1 and F_2 should both be equal to P.

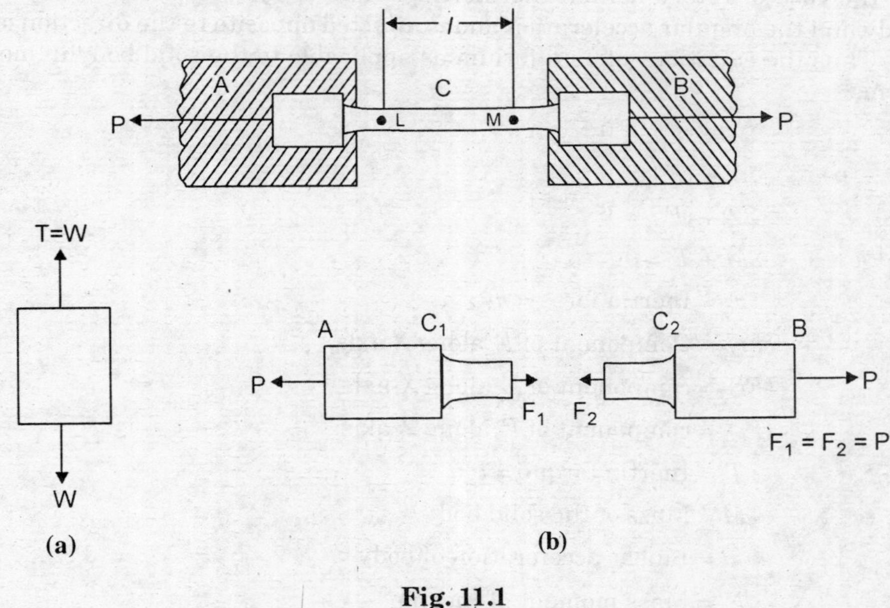

Fig. 11.1

11.2.1 Equations of Static Equilibrium

Despite the fact that the reader would have been familiar with the equations of static equilibrium, still it is felt necessary to re-write these equations here so that the reader is not at loss to understand the basic concepts involved in this subject properly. Whenever a solid body under the action of various forces is in static equilibrium or in the state of rest then the algebraic sum of the components of these forces along any chosen three perpendicular axes of reference are separately zero. In addition the algebric sum of the moments of these forces about any point is also zero, if it is not rotating Mathematically these results are expressed in the form of following four equations

$$\Sigma \mathbf{X} = 0 \quad ..(A)$$

$$\Sigma Y = 0 \quad ...(B)$$

$$\Sigma Z = 0 \quad ...(C)$$

$$\Sigma M = 0 \quad ...(D)$$

where ΣX, ΣY and ΣZ represent the algebraic sum of the components of forces along X, Y and Z direction respectively and ΣM the algebraic sum of moments of these individual forces about any point.

11.2.2 Equations of Static Equilibrium and Body in Motion

In article 11.2.1 the equations of equilibrium have been introduced and it was pointed out that these equations are derived for the state of rest of the solid body. However D'Alembert pointed out that these equations can be applied to a solid body in motion (linear or rotory) as well, provided the inertia force in the case of linear motion or inertia torque in the case of rotory motion is superimposed on the body in addition to the given forces. The inertia force is equal to the product of mass and acceleration and is directed opposite to the direction of acceleration. Similarly in the case of rotory motion the inertia torque is the product of mass moment of inertia of body and the angular acceleration and is directed opposite to the direction of angular acceleration. Thus the equations of equilibrium as applicable to the solid body in motion take following forms

$$\Sigma X + (F_i)_x = 0 \quad ...(E)$$

$$\Sigma Y + (F_i)_y = 0 \quad ...(F)$$

$$\Sigma Z + (F_i)_z = 0 \quad ...(G)$$

$$\Sigma M + T_i = 0 \quad ...(H)$$

where F_i = Inertia force = $m \times a$

$(Fi)_x$ = Component of F_i along X axis

$(Fi)_y$ = component of F_i along X-axis

$(Fi)_z$ = component of F_i along Z-axis

T_i = Inertia torque = $I_m \alpha$

M = Mass of the solid body

a = Linear acceteration of body

I_m = Mass moment of Inertia

α = Angular acceleration of body

These equations are known as *D'* Alemberts equations of equilibrium for solid bodies in motion and this is known as *D'* Alemberts' Principle.

As a special case if the body is having uniform linear motion or uniform angular motion then both linear acceleration 'a' and angular acceleration 'α' will be zero and the equations (E), (F), (G) and (H) reduce to (A), (B), (C) and (D) respectively. Thus we conclude that when a solid body is at rest or in the state of uniform motion the equations (A) through (H) apply However when it is accelerating or decelerating then edquations (E) through (H) will apply.

The utility of the *D'* Alemberts principle as applicable to bodies in motion can better be understood with the help of following example. Let a passanger train consisting of 16 bogies be pulled by an electric locomotive. If it is assumed that the mass of each bogie be M and the air resistance experienced by each bogie be F_a then it is desired to calculate the stress in each coupler which connects two adjacent bogies. The area of cross-section of each coupler is A. For evaluating the stress in the last coupler connecting 16th bogie with 15th bogie we take a section any where along the length of the coupler and draw the free body diagram of the last bogie. Two situations may arise firstly the train be moving with uniform speed and secondly when the train is accelerating.

In first case there will be no inertia force as the acceleration is zero. Thus the free body diagrams of the last two bogies will be as given in figure 11.2. Then according to *D'* Alemberts' principle considering the equalibrium of bogie number 16, we have for the horizontal forces

$$F_a = P_{16} \quad \text{...(I)}$$

Similarly for the equalibrium of 15th bogie we will get

$$P_{16} + F_a = P_{15} \quad \text{or} \quad P_{15} = 2F_a$$

and so on

$$P_{15} + F_a = F_{14} \quad \text{or} \quad P_{14} = 3F_a$$

$$\vdots \qquad P_2 = 15F_a$$

$$P_2 + F_a = P_1 \quad \text{or} \quad P_1\ 16F_a$$

Thus the first coupler which connects the first bogie with the locomotive will be stressed maximum and it will be equal to $\frac{16F_a}{A}$. While the last coupler will be stressed least $\left(= \frac{F_a}{A}\right)$. The max stress $= \frac{16F_a}{A}$.

In the second case when the train is accelerating with an acceleration equal to '*a*' then according to *D'* Alemberts principle each bogie will be accelerating and on each the inertia force which is equal to $(M \times a)$ will be acting and it will be in the opposite direction to that of the acceleration. The free body diagrams of a few bogies is given in Fig (11.3)

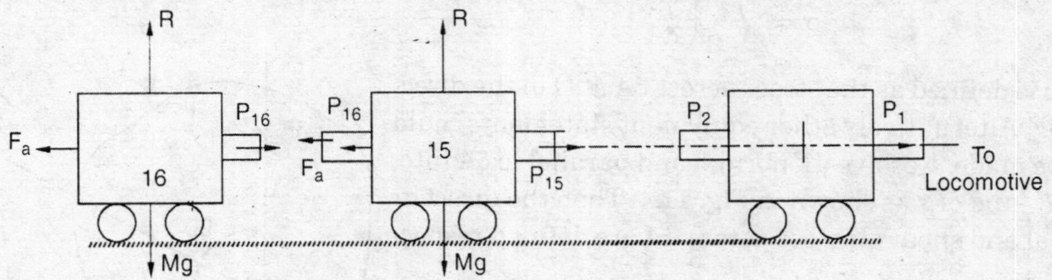

Fig. 11.2

Then for the equilibrium of bogies number 16

$$F_a + F_i = P_{16}$$

or

$$P_{16} = \boldsymbol{F_a} + M.a$$

Similarly for bogie no 15

$$M.a + F_a + P_{16} = P_{15}$$

or

$$(M.a + F_a) + (Fa + M.a) = P_{15}$$

or

$$2M.a + 2F_a = P_{15}$$

and so on $F_i + F_a + P_2 = P_1$

or $16F_a + 16M.a = P_1$

The stress in each coupler can then be evaluated by dividing the force transmitted by each coupler with area of cross section A.

Thus the max stress will occur in the first coupler which will be equal to $(16F_a + 16M.a)/A$ and likewise the least stressed coupler will be the 16th coupler connecting the 15th and 16th bogies and will be equal to $(F_a + M.a)/A$.

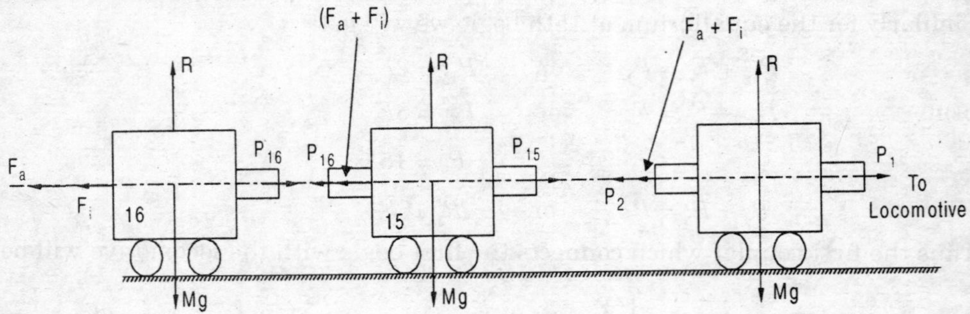

Fig. 11.3

11.3 DIRECT STRESS

Let O be a point in a solid body which is in equilibrium under the action of a number of forces. Let δA be a small plane area with ON normal to it. Across area δA a force δF, say, is being transmitted or sustained by the body to give a quantity

$$\sigma = \underset{\delta A \to 0}{Lt} \left(\frac{\delta F}{\delta A} \right)$$

which is defined as the stress across δA at O in the direction δF. Alternatively other equivalent statements could also be made. Resolve dF normal and parallel to δA into, say δF_n and δFs as shown in Fig. 11.4. Then the direct or normal and shear stresses across dA are defined respectively as

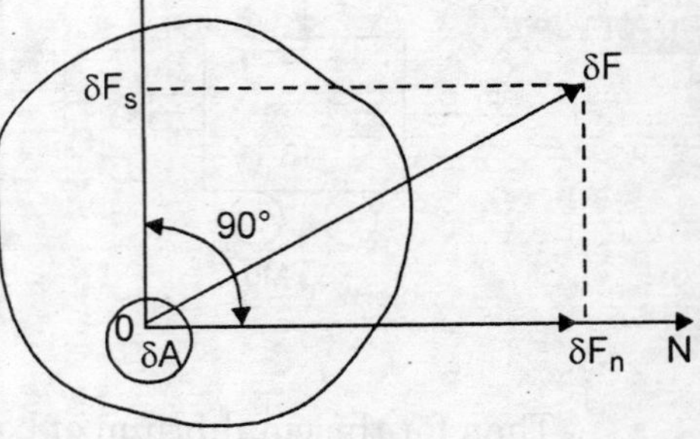

Fig. 11.4

$$\sigma_n = \underset{\delta A \to 0}{Lt} \left(\frac{\delta F_n}{\delta A} \right)$$

$$\tau = \underset{\delta A \to 0}{Lt} \left(\frac{\delta F_s}{\delta A} \right)$$

In case δF is acting along ON, that is normal to δA, the stress σ itself will be the normal or direct stress and the shear stress τ will be zero. Thus the direct or normal stress at a point of a plane section is defined as the force sustained per unit area of cross-section normal to it.

It is once again emphasized that the solid body must be in equilibrium and it should sustain or transmit force. Evidently to generate stress in a solid body there must be two equal and opposite forces acting on the body and these forces must have their respective points of application separated by some distance. The intervening material of the solid body will then be stressed since that portion of the body will be sustaining the force. As an example in the test specimen of Fig. 11.1(*b*) the direct stress over the section *CC* will be F_1/A_0 (= P/A_0) where A_0 is the area of cross-section of the specimen since A_0 is the area which is sustaining the normal force *P*. The stress is usually measured in Newton/Metre Square (N/m^2) or Pascal (Pa). Since Pascal is a very small unit it is not uncommon to use Mega Pascal (MPa) or Giga Pascal (GPa) as units of stress, where

$$1 \text{ MPa} = 10^6 \text{ Pa}$$

and

$$1 \text{ GPa} = 10^9 \text{ Pa}$$

As a further illustration take the case of a rod *AB* Fig. 11.5 of area of cross-section, A_0, lying on a smooth table, is acted upon by a force *P* along its axis. This external force *P* will not generate any stress in the rod. Rather it will set the rod in motion causing acceleration according to Newton's second law of motion. The point to note here is that the force *P* is not being transmitted or sustained by the rod *AB*.

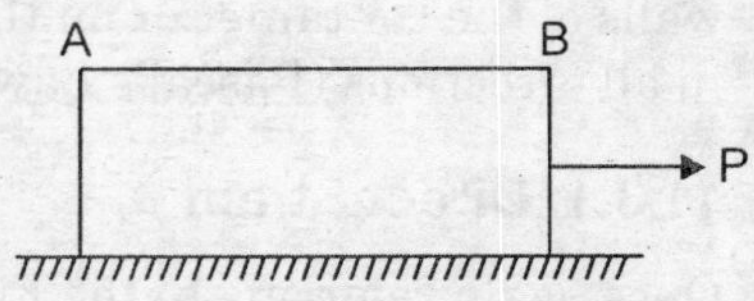

Fig. 11.5

In order to generate stress in it, would require equal force *P* to be applied at section *A* in the opposite direction. In that event the rod would have not been set into motion but would have been at rest or in state of equilibrium. It may further he noted that a solid body need not necessarily be in static equilibrium for it to be in the state of stress. It could be stressed while in motion as well. For example if the rod *AB*, Fig. 11.5, is acted upon by the force $F(<P)$ on the section *A* towards left, then force *F* will be sustained by the rod *AB* to generate normal stress F/A_0 in the rod. However the force $(P - F)$ will set the rod into motion according to Newtons second law of motion.

In order to elaborate this point further let the rod *AB* be acted upon by an other force *P* acting on the section *B* but in leftward direction as shown in fig. 11.6 and we intend to find the stress at any section *CC* lying in between *A* and *B*. Then taking a section *CC* at the point, the free body diagrams of the two elements AC_1 and C_2B are drawn as shown in Fig. 11.5. Evidently if the body *AD* is in static equilibrium so will be the elements AC_1 and C_2B. If we consider the static equilibrium of individual elements AC_1 and C_2B and resolve the force, along the direction of *P* we have

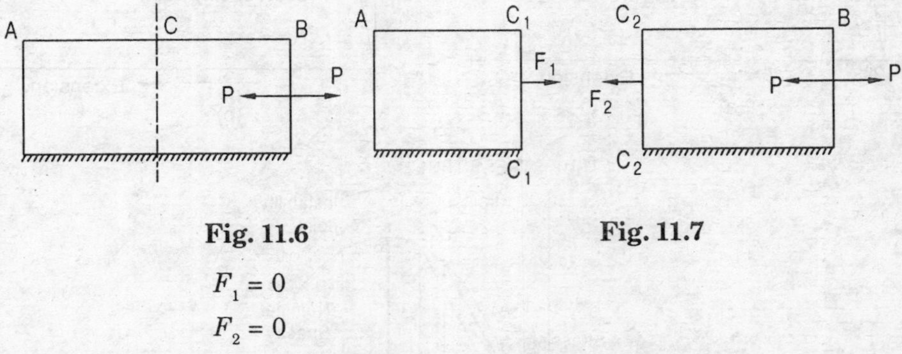

Fig. 11.6 **Fig. 11.7**

$$F_1 = 0$$

and

$$F_2 = 0$$

where F_1 is the reaction of C_2 on C_1 and F_2 the reaction of C_1 on C_2 and $F_1 = F_2 = 0$

Thus the force sustained by any cross-section lying between *A* and *B* will be zero. In other words the rod material will not be stressed. This proves the statement made above in the previous paragraph *i.e.*, to generate stress two equal and opposite forces must be applied at points at finite distance apart.

A beginner in the subject is sometimes at loss to distinguish between stress and pressure as both have the same units. It is to be noted that stress is encountered in solids and is associated with the direction of the sustained force with respect to the plane passing through the point under consideration. The magnitude of stresses will be different on different planes passing through the point under consideration. This aspect will be dealt in more detail in Chapter 2. On the contrary pressure is the term associated with fluids (liquids and gases) and the force exerted per unit area due to the impact of the moving molecules of the fluid on the walls of the container or on the body immersed in the fluid and is same at a point in the fluid, in all directions (Pascal's Law).

11.3.1 Direct Strain

Once again referring to the Fig 11.1(*b*), let there be two points L and M on the specimen which are at a distance l_0 called the gauge length, before the application of the force **P**. After the force is applied the specimen will stretch and the distance between L and M will increase to say l. The change in length per unit gauge length $(l - l_0) / l_0$, is called the axial or direct strain.

11.4 THE TENSION TEST

Consider a tension test specimen of round section as shown in Fig. 11.1(*b*) being stretched in a testing machine. The load recorded by the universal testing machine is the reaction of the test specimen. At room temperature the load extension diagram in general, at strain rate less than 2×10^{-3}/sec, will tend to follow the curve of Fig. 11.8(*a*), which is typical for an annealed metal like aluminium or copper.

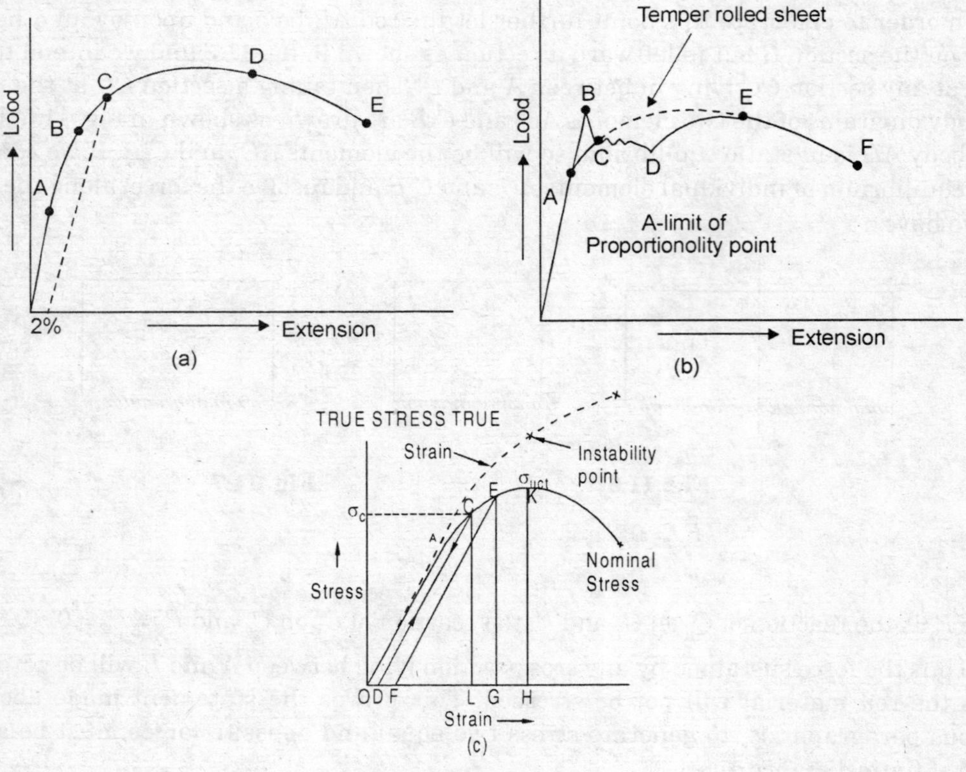

Fig. 11.8

Initially the relation between the load and extension is essentially linear—that is, portion *OA* of the curve, where *A* defines the limit of proportionality. On further straining, the relation between the load and extension no longer remains linear but the material is still elastic. That is, if the load is released the specimen will revert back to its original length. The maximum load which can be applied without causing permanent deformation is known as the 'elastic limit' represented by point *B* in Fig. 11.8(*a*). Usually the points *A* and *B* are very near to each other and the real difference can be achieved only by using highly sensitive measuring devices. At the point *B* ends the elastic straining and the 'inelastic' or plastic deformation ensues. Beyond the point B there is a general permanent extension of the test piece until the load attains a maximum value, point *D*, corresponding to which the stress is called the 'ultimate tensile stress' σ_{ult} (= P_{max}/A_0). At this load the specimen necks down locally rendering the deformation non-uniform and the load falls off with increasing extension, thereby causing the fracture at a load, point *E*, which is less than that corresponding to the ultimate tensile stress point. In addition another term, called the '*proof stress*' is usually quoted in structural design. This refers to the stress, σ_c (= P_c/ A_0) required to produce a small strain of 0.2%. This is represented by point C in the Fig. 11.8 (CA_0)

11.4.1 St. Venants' Principle of Uniform Stress

In figure 11.1 is illustrated the equilibrium of the left portion AC_1 of the test specimen under the action of the external force *P* at the left end and the tensile stress acting on the cross-section plane C_1. For the equilibrium of this portion the stress at this section is calculated as $\sigma = P/A$ and it is assumed to be uniform over the cross-section. But St. Venant observed that this formula is correct only if the cross section under consideration, such as *C*, here, is at least three diameters away from the ends of the straight portion where the fillets start. This is due to the fact that the "lines of force flow" are not evenly spaced over different cross-sections taken along the length Fig. 11.9. It was observed by St. Venant that they become substantially uniformly distributed at points about three diameters away from the end fillets. The statement of St. Venants principle is as follows.

The stress distribution over the cross-section will be uniform, for essentially concentrically applied load, over sections which are at least there diameter away from the ends where the fillet starts.

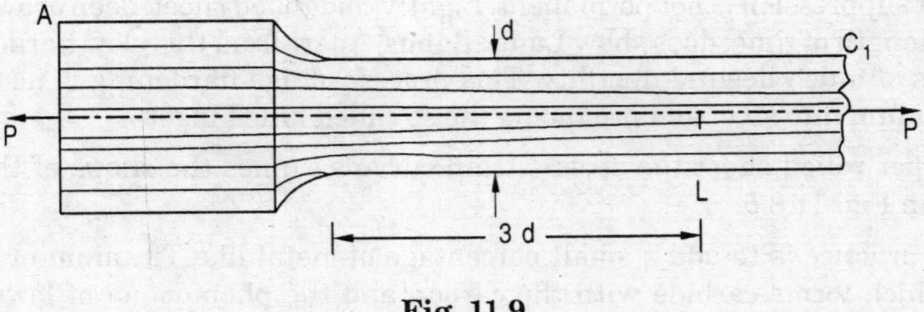

Fig. 11.9

Thus while performing the tensile test the gauge length on the tensile test piece is taken in the central portion and its two ends such as *L* and *M* in figure 11.9 must be at least three diameters from the respective ends where the fillet starts and on both sides the machine grips should be at equal distance from the ends. Then and then only the stress on any cross-section taken any where along the length *LM* will be uniform.

11.4.2 The Yield Point In Low Carbon Steel

Typical load extension curve after Murray (1) for annealed low carbon steel (earlier called mild steel) is shown in Fig. 11.8 (*b*). The marked difference between this and two diagrams of the materials of the type discussed above is the exhibition of pronounced yield point. Upto the point *B*, the load rises steadily and the material remains elastic. The extension per unit gunge length at this point is of the order 10^{-3}. At low carbon or nitrogen levels in mild steels, at the point *B* the load suddenly drops and then fluctuates about some constant value of load. The specimen extends considerably at this almost constant load. This sudden extension in mild steels is localized to certain zones in the metal and surface markings, known as "Luder bands" or "Strecteher strains" are produced. Under this situation we have some material which has yielded and some which has not. As the stretching continues upto the point *D* almost the entire specimen gets uniformly deformed. The extension *CD* is of the order of 10^{-2} per unit gunge length. Beyond the ponit *D* the load rises again as the specimen work hardens.

The stress at the point *B* at which the load suddenly drops is called the 'upper yield point' and the average constant load in the range CD to which the load has suddenly dropped is called the 'Lower yield point'. This gives rise to a peculiar feature that for low carbon steel (mild steels) that the stress required to maintain plastic flow immediately after yielding has started, is lower than that required to start it. Cottrell (2) has put forth his theory of "Formation of cottrell's Atmosphere" around the dislocation to explain the occurrence of this phenomenon in low carbon steels. According to this theory carbon or nitrogen atoms, in *BCC* ion *i.e.,* in Ferrite phase, diffuse the positions of low energy. The dislocations are pinned or locked with these interstitial solutes (carbon or nitrogen) and additional stress, over that normally required for the movement of dislocations, is needed in order to tear some of the dislocations away from the restraining impurity atoms. This results in increase in stress. When the dislocation line is pulled free from the influence of the solute atoms, it can slip to a lower stress called lower yield point stress. The significant point is that a very small concentration of solute is required to lock or pin along the whole length of dislocation. This phenomenon gives rise to rough surface during drawing or stamping processes.

In the field of metal working, such as in deep drawing of steel blanks, for making steel utensils etc. the presence of Luder Bands is not desirable. The modern method to suppress this phenomenon is to "temper roll" it. The soft annealed sheet of mild steel is subject to light cold rolling (½ to 2 %) reduction in thickness) not more than 24 hours before deep drawing because the yield point suppression is not permanent. Lightly cold rolled sheet, deep drawn after standing for sufficient length of time, does show Luder Bands. Apart from this slow hardening occurs and results in loss of much desired ductility. This process of age-hardening is usually checked by adding aluminium into steel and producing the so called killed ingot.

For temper rolled sheet the stress-strain curve assumes the shape of the one shown in broken lines in Fig. 11.8 *b*.

Modern practice is to add a small percentage of metal like Titanium or columbium into mild steel, which forms carbide with the carbon and the phenomena of lower yield point is eliminated completely and no Luder Bonds are produced in the finished product.

---

1. Murry 6, "The yield point in mild steel, engineering vol. 178, Pp 366, 1954
2. Cotrell; Physical Metallurgy, *Mc Graw* hill.
3. Vijendra Singh, Physical Mettillurgy, Standard Publishers and Distributers, New Delhi.

Due to wide use of mild steel in industry where permanent change in shape in brought about, the complete stress-strain diagram for the metal is invariably considered. On the other hand its use in structures and machine elements is limited only upto the limit of proportionality point where the emphasis is only on small strains.

11.4.3 True Stress Strain Curve

If we now divide the load by the original area of cross-section A of the test specimen and the extension by the original length to (which is usually called the gauge length), then the curves shown in Fig. (11.8) will be called the nominal or Engg stress-strain curve. These are called nominal or Engg stress and strain because both the parameters *i.e.* stress and strain are obtained by dividing with constant quantities respectively. Actually as the load is slowly or gradually increased the area of cross-section will continuously go on reducing (poissons ratio effect, see chapter 2). Similarly at any instant we take the extension with respect to the original length ℓ_o. But actually the stretch at any particular instant occurs over the current length. Thus if we consider the current area of cross-sections and divide the current load with it we will get the "TRUE STRESS". Likewise the change in length over the current length should be taken into account for calculating the TRUE STRAIN.

In the light of above, if P is the current load and 'd' he current diameter of the specimens then

$$\text{True stress } \sigma = \frac{4P}{\pi d^2}$$

Similarly if dl is the elemental elongation in current length l, at which point the current load is P, then the incremental strain $d\varepsilon$ is defined as

$$d\varepsilon = \frac{dl}{l}$$

Integrating this equation we get the total current strain ε

$$\int_0^{\varepsilon} d\varepsilon = \int_{l_0}^{l} \frac{dl}{l}$$

i.e., $$\varepsilon = \ln\left(\frac{l}{l_0}\right)$$

Thus at any current load P the true strain will be

$$\varepsilon = \ln\left(\frac{l}{l_0}\right)$$

In view of the logarithm value of (l/l_0) true strain is also called logarithmie strain or natural strain and associated with this are the names True stress or natural stress of the value $\frac{4P}{\pi d^2}$

The true stress-strain curve will be a monotonically rising curve and it will be higher than the nominal or Engg. Stress strain curve as shown in Fig. (11.8C)

Upto the elastic limit point, these curves are very near to each other. However after elastic limit point, the true stress-strain curve is much higher than nominal stress-strain curve.

11.4.4 Yield Envelope and Bauschinger Effect

In the Fig. (1.2*c*) and portion *AK* of the stress-strain curve is known as the plastic portion of the curve. If the stress on specimen corresponding at any point *C* in this portion of curve σ_C is reduced gradually to zero value this reverse stress-strain curve will be a straight line *CD* parallel to the original elastic line *OA*. This phenomenon was observed by Bauschinger and is known as Bauschinger effectAt the point *D* the stress will be zero but residue strain *OD* will be left in the specimen. This strain *OD* is also known as the 'plastic strain' component, ε_p, of the total strain *EC* at *C* which is equal to *OL* the portion *DL* is called the elastic component , ε_e, of total strain

Thus
$$\varepsilon_C = OL = OD + DL$$
$$= \varepsilon_p + \varepsilon_e$$

Once again if we restart increasing the stress on the specimen it will follow the path ***DC*** and then along the curve *EK*. In other words *C* becomes the new yield point and the corresponding stress, σ_c, the new yield stress. Similarly if this process is repeated when the stress level is σ_E then *E* will yet become another new yield point stress, This will continue upto the point *K*, which is the ultimate stress point.

Therefore the portion *AK* is also known as "yield envelope". Another name assigned to the curve *AK* is strain hardening portion of the stress-strain curve.

11.4.5 Ductility and Malleability

Ductility hitherto, had been defined as the property of the material due to which it could be drawn into wire. It was very well known to the people that gold is more ductile than copper, copper is more ductile than Aluminium and so on. that is, this term was known qualitatively and had not been difined quantitatively. It was in late fifties of the last century that during an international meet, the scientists put their stamp on the proper definition of this term. According to this the strain to which a uniaxially stressed specimen can be subjected prior to instability point or ultimate stress point is called the ductility of the material. That is *OH* (Fig. 11.8 *c*) is the measure of ductility.

Malleability

Similarly the malleability which is the property due to which the material can be deformed into plate in quantitatively defined as the strain at instability point for that particular material.

11.5 UNIFORM STRESS

To generate a uniform stress over a section the force must be applied at its centroid. To prove this statement assume a section shown in Fig.. 11.10, with the centroid at C, under a uniform stress σ. Let the total force acting on the section be P.

If *dA* be an elemental area at a distance *r* from *C*.

Then, we have
$$\int \sigma \, dA = P \quad \text{...(11.1)}$$

Taking moments of elemental forces about the centroid we have if *P* is at a distance *a* from *C*

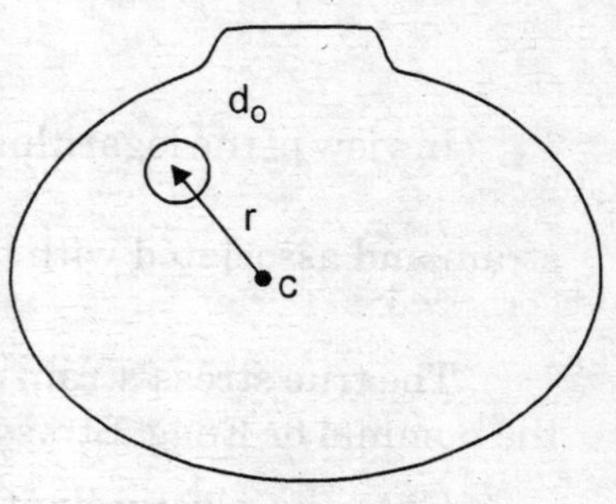

Fig. 11.10

$$P.a = \int r\sigma \, dA = \sigma \int r \, dA \quad \text{(as } \sigma \text{ is constant)}$$
$$= \sigma \times \text{First moment of area about centroid.}$$
$$= 0.$$

Now since P is not zero, therefore

$$a = 0.$$

Thus the line of action of P must pass through the centroid of the section.

11.6 SHEAR STRESS AND STRAIN

Consider a rectangular block $ABCD$ (Fig. 11.11) of dimensions $a \times b$ and thickness t. Let it be subjected to two equal and opposite forces P on the faces AB and CD as shown. Taking a section XX at a height 'y' from the face DC the free body diagrams of

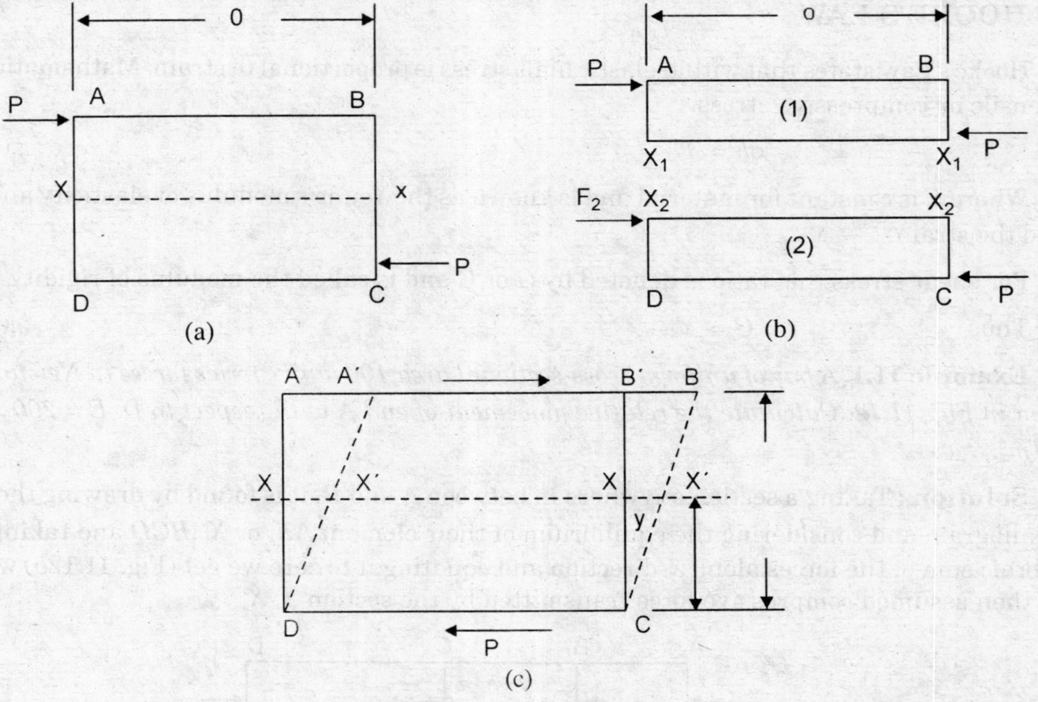

Fig. 11.11

the two elements number (1) and (2) are shown in Fig. 11.11(b). The face X_1X_1 offers a cohesive force F_2 on to the section X_2X_2 and in its turn the face X_2X_2 gives a reaction F_1 onto the face X_1X_1. For the equilibrium of the two elements separately which follows from that of the original block, we conclude that $F_1 = F_2 = P$. The effect of the force F_1 and F_2 on the respective faces is to induce a tendency for the face X_1X_1 to slide over the face X_2X_2 in the direction of the arrowhead F_2. The forces F_1 and F_2 are called the shearing forces and the force per unit area of the cross-section transmitting this force is known as the shear stress. It is represented by τ and in this case is given by

$$\tau = F_1 / \text{at} \; (= P/at) \qquad \text{...(11.2)}$$

It is evident that the shear stress, in this case, is independent of the distance of the section *XX* from *CD*. That is shear stress on any horizontal section will be constant and that section will have a tendency to slide to the right with respect to its adjacent plane immediately below it. The cumulative effect will be to displace the face *AB* to the position *A'B'* prime with respect to *DC*. Likewise then, section *XX* will occupy the position *X'X'* such that at any height *y* from *DC* the displacement at right angles to *y* is *XX'*. This displacement *XX'* over unit height is called the shear strain and is represented by symbol ϕ Fig. 11.11(*c*) such that

$$\phi = \frac{XX'}{Y} = \tan ADA' \simeq \text{angle } ADA' \quad ...(11.3)$$

That is, the shear strain is an angular displacement measured in radian and is constant for any value of *y* between zero and *b*.

11.7 HOOKE'S LAW

Hooke's Law states that within elastic limit stress is proportional to strain. Mathematically, for tensile or compressive stress.

$$\sigma/\varepsilon = E \quad ...(11.4)$$

Where *E* is constant for material and is known as the Young's modulus of elasticity and ε is called the strain.

For shear stress this ratio is denoted by *G* or *C* and is called the modulus of rigidity.

Thus $$G = \tau/\phi \quad ...(11.5)$$

Example 11.1. *A bar of uniform cross-sectional area 100 mm² carries forces in Newtons as shown in Fig. 11.12. Calculate the relative movement of end A with respect to D.* $E = 200 \times 10^3$ MN/m^2.

Solution : Taking a section anywhere in between *A* and *B*, it is found by drawing the free body diagram and considering the equilibrium of their element *AX*, or X'_1BCD and taking the algebraic sum of the forces along *X*-direction and equating it to zero we get (Fig. 11.12*b*) where *F*, is then assumed compressive force transmitted by the section *X, X,*

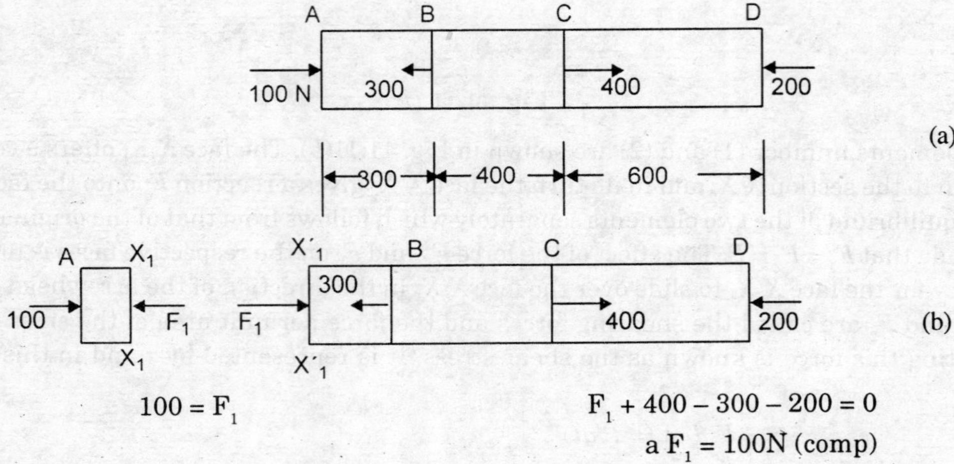

$$100 = F_1 \qquad F_1 + 400 - 300 - 200 = 0$$

$$\therefore F_1 = 100\text{N (comp)}$$

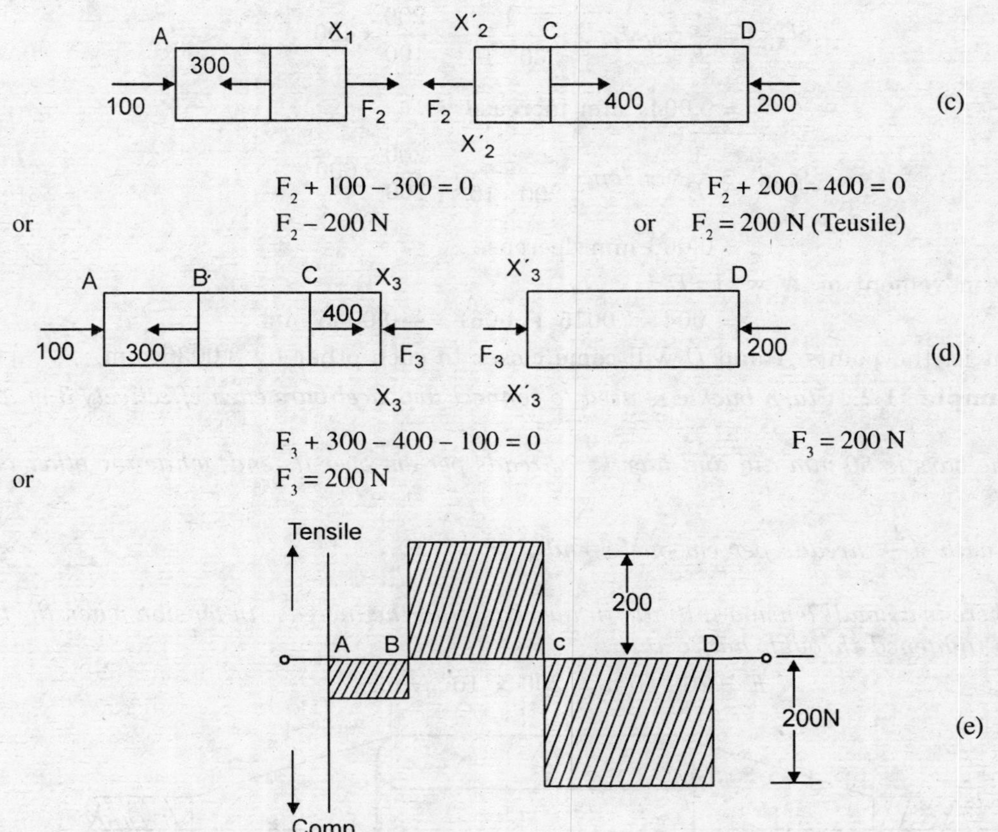

Fig. 11.12

$$F_1 = 100 = 0$$

or $$F_1 = 100 \text{ N comp.}$$

(Tensile)

Similarly if F_2 is therefore transmitted by any section X_2X_2, between B and C, then considering the equilibrium of either of the elements ABX_2 or X'_2CD by drawing their. Free body diagram and equating the algebraic sum of the forces to zero (Fig. 11.12*c*) we have

$$F_2 + 100 - 300 = 0$$

$$F_2 = 200 \text{ N Tensile}$$

Again of F_3 (comp.) is assumed to be the transmitted force over any section X_3X_3 between C and D, then considering the equilibrium of either of the elements ABC X_3 or X'_3D by drawing their free body diagrams and equating their algebraic sum of forces to zero we get (Fig. 11.12*d*)

$$F_3 = 200 \text{ comp.}$$

Accordingly the force distribution diagram is drawn in Fig. 11.12(*e*).

Thus the respective deformations are

$$\delta l_{AB} = \frac{1}{E}\sigma_{AB}.l_{AB} = \frac{1}{200 \times 10^3} \times \frac{100}{100} \times 300$$

$$= 0.0015 \text{ mm decrease}$$

$$\delta l_{AB} = \frac{1}{E}\sigma_{BC}.l_{BC} = \frac{1}{200 \times 10^3} \times \frac{200}{100} \times 400$$

$$= 0.004 \text{ mm increase}$$

$$\delta l_{AB} = \frac{1}{E}\sigma_{CD}.l_{CD} = \frac{1}{200 \times 10^3} \times \frac{200}{100} \times 600$$

$$= 0.006 \text{ mm decrease}$$

The movement of '*A*' w.r.t. '*D*'

$$= .004 - (.0015 + .006) = -0.0035 \text{ mm}$$

That is, the points *A* and *D* will come closer to each other by 0.0035 mm.

Example 11.2. *A turn buckle is used to connect two steel bars each effectively 3 m long. One of the bars is 50 mm dia and has* $1\frac{4}{5}$ *threads per cm over its end, while the other is 75 mm dia with* $1\frac{2}{5}$ *threads per cm on its end.*

If there is a small tension initially in the rods, find the increase in tension when the turn buckle is tightened through half a turn.

Take $E = 200 \; GPa = 200 \times 10^9 \; N/m^2$.

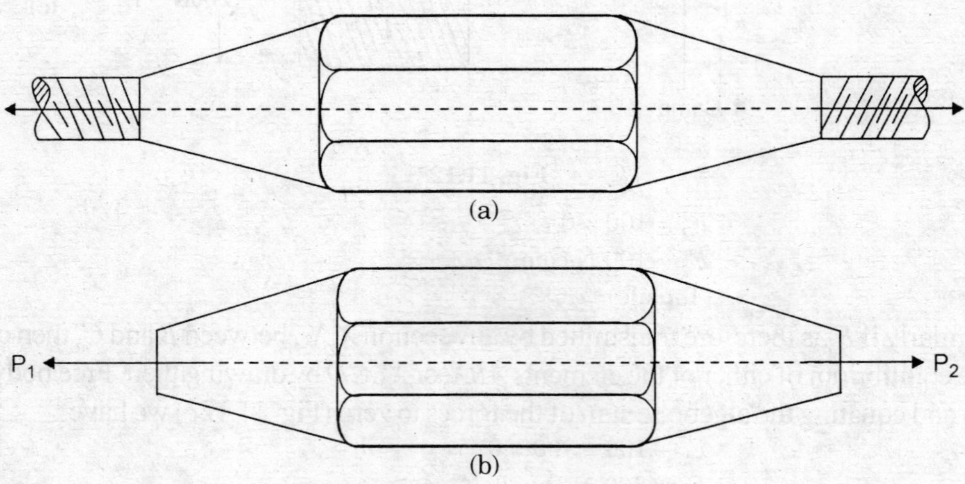

Fig. 11.13

Solution : In Fig. 11.13(*b*) is drawn the free body diagram of the turn buckle with P_1 and P_2 respectively as the reactions of the two bars. Evidently for the equilibrium of the turn buckle

$$P_1 = P_2 = P(\text{say}) \qquad ...(11.6)$$

Where *P* is the increase in tension required to be evaluated.

Total stretch in the bars = sum of the movements of the ends due to rotation of turn buckle

$$= \left(\frac{5\times10}{9}\times\frac{1}{2}\times\frac{5\times10}{7}\times\frac{1}{2}\right)\text{mm}$$

$$= \frac{25\times16}{63}\text{ mm} \qquad ..(11.7)$$

If ε_1 and ε_2 are the respective strains in the two bars, the total stretch is also given by $(\varepsilon_1 l_1 + \varepsilon_2 l_2)$ where l_1 and l_2 are the respective lengths of the bars. Thus with the help of equations (1.6) and (1.7), we have

$$\frac{25\times16}{63} = \varepsilon_1 l_1 + \varepsilon_2 l_2$$

$$= \frac{P_1}{A_1E}l_1 + \frac{P_2}{A_2E}l_2$$

$$= \frac{P}{E}l_1\left(\frac{1}{A_1}+\frac{1}{A_2}\right) \quad \text{as} \quad l_1 = l_2$$

$$= \frac{P\times3000}{200\times10^3}\left[\frac{4}{\pi(50)^2}+\frac{4}{\pi(75)^2}\right]$$

Thus $\qquad P = 0.575$ MN .

11.8 STRESS IN A BAR DUE TO ITS OWN WEIGHT

Consider a bar of length l metre and uniform cross-section area A m^2 hanging from the top, Fig. 11.14. The weight of the portion *BC* of the bar will act on the cross section *CC*.

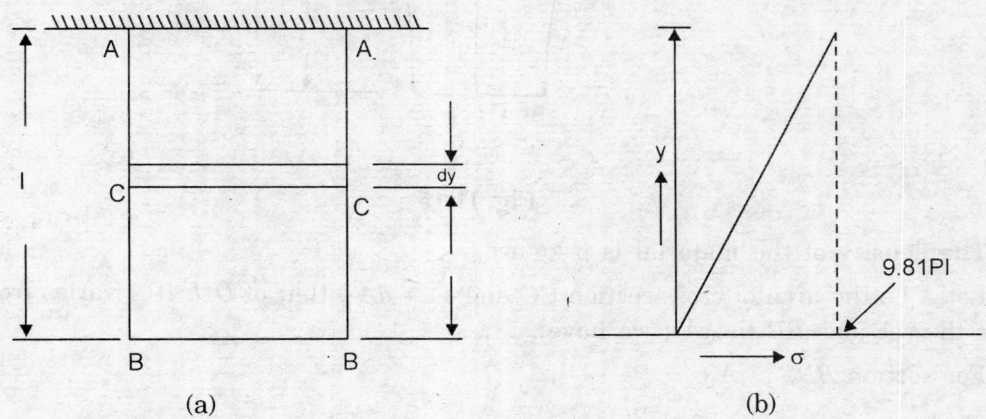

Fig. 11.14

If ρ kg/m^3 is the density of the material, the stress at a height *y* metre from *B* is given by

$$\sigma = \frac{9.81\times Ay\rho}{A}$$

$$= 9.81\ \sigma y\ \text{N/m}^2 \qquad ...(11.8)$$

Evidently the stress rises linearly with y and

$$\sigma_{m\,ax} = \sigma_A = 9.81\ \sigma l \text{ N/m}^2 \text{ or Pa} \qquad ...(11.9)$$

The stretch dl in the length dy will be given by

$$dl = \frac{\sigma}{E} dy \qquad ...(11.10)$$

Where E is the Young's modulus of elasticity.

Total elongation in the bar

$$= \int_0^l (\sigma / E)\, dy = \int_0^l \frac{9.81}{E} \rho y dy$$

$$= \frac{9.81}{2E} \rho l^2 \qquad ...(11.11)$$

11.8.1 Tie Bar of Uniform Strength

A tie bar of uniform strength carrying a load of P Newton is shown hanging from the top in the Fig. 1.8. Let the stress everywhere by σ N/m^2.

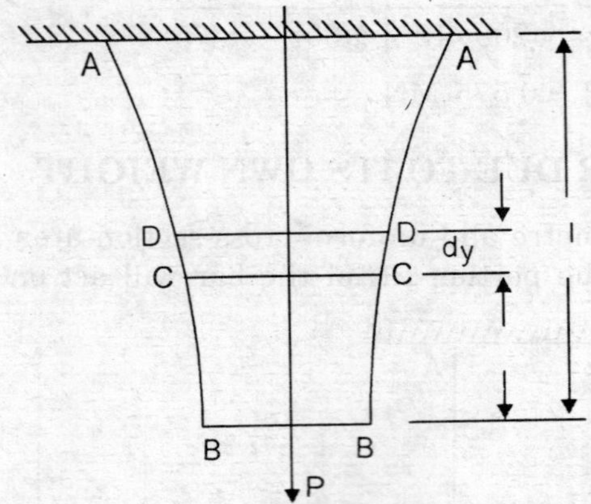

Fig. 11.15

The density of the material is ρ kg/m^3.

Let A be the area of cross-section CC and $(A + dA)$ that of DD. If A varies from A_1 to A_2 for the sections BB to AA ; we have

For section BB, $A_1\sigma = P$

or $\sigma = P/A_1$ ...(11.12)

For section CC, $A\sigma = P + 9.81 \times$ mass of BC in kg ...(11.13)

For section DD,

$(A + dA)\, \sigma = P + 9.81$ mass of portion BC in kg

$+ 9.81$(mass of portion CD in kg)

$= A\sigma + 9.81$ (mass of portion CD in kg)

...from equation (11.13)

That is $A\sigma + dA\sigma = A\sigma + 9.81 \times (Ady)\rho$

or $$\frac{dA}{A} = \frac{9.81\,\rho}{\sigma}dy \quad \text{...(11.14)}$$

Integrating $$[\ln A]_{A_1}^{A} = \left[\left(\frac{9.81\,\rho}{\sigma}\right)y\right]_0^y$$

or $\ln(A/A_1) = 9.81\ \rho y/\sigma$

or $A = A_1\ e^{9.81\ \rho y/\sigma}$

At $y = l\ ,\ A = A_2$

Therefore $$A_2 = A_1 e^{\left[\frac{9.81\,\rho\ell}{\sigma}\right]} \quad \text{...(11.15)}$$

11.9 STRESS IN A BAR DUE TO ROTATION

In the Fig. 11.16 is shown a bar of length l revolving about Y-axis at an angular speed of omega rad/sec.

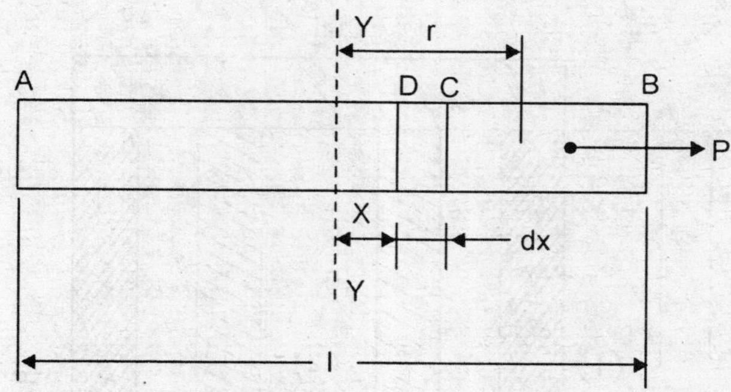

Fig. 11.16

Tensile force on the element DC = Centrifugal force P on the part $CB = \omega^2\ rM$, where M is the mass of part CB.

Thus if A is the area of cross-section in m^2 of the bar and ρ the density of the material in kg/m^3, then

$$P = \rho A\left(\frac{\ell}{2} - x\right)\omega^2\left[x + \frac{1}{2}\left(\frac{l}{2} - x\right)\right]$$

or $$P = \frac{1}{2}\rho A\omega^2\ (l^2/4 - x^2)\ \text{Newton}$$

i.e. $$\sigma_c = \frac{1}{2}\rho\omega^2\ (l^2/4 - x^2)\ \text{N/m}^2 \quad \text{...(11.16)}$$

Maximum stress will occur at

$$x = 0$$

i.e. $$\sigma_{max} = \frac{1}{8}\rho\omega^2 l^2 \quad \text{...(11.17)}$$

Extension of the element DC

$$= \frac{\sigma_c}{E}dx$$

Thus substituting the value of σ_c from equation (1.16) and integrating, the total extension in the bar is given by

$$dl = 2\int_0^{l/2} \frac{1}{2E}\rho\omega^2 (l^2/4 - x^2)\,dx$$

$$= \frac{\rho\omega^2}{E}\left[\frac{l^2}{4}x - \frac{x^3}{3}\right]_0^{l/2}$$

$$= \rho\omega^2\, l^3/(12\,E) \text{ metre.}$$

11.10 COMPOSITE BARS

A load carrying member which is made up of two different materials is called a composite or compound bar.

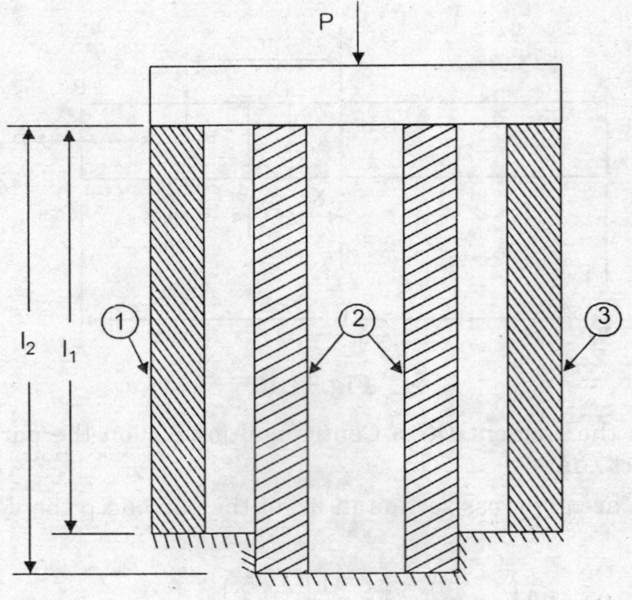

Fig. 11.17

In Fig. 11.17 the two materials having the cross-sectional areas A_1 and A_2, the corresponding modulii of elasticity as E_1 and E_2 constitute a composite bar.

11.10 COMPOSITE BAR UNDER THE EFFECT OF EXTERNAL LOAD

If the upper ends are rigidly connected to each other, the compression in each will be same

i.e. $$\delta l_1 = \delta l_2$$

or $$\frac{\sigma_1}{E_1}l_1 = \frac{\sigma_2}{E_2}l_2 \quad ...(11.19)$$

Also the applied load will be equal to the sum of loads sustained by each part

i.e. $P_1 + P_2 = P$

or $\sigma_1 A_1 + \sigma_2 A_2 = P$ ...(11.19)

Then σ_1 and σ_2 can be evaluated by solving the simultaneous equations (11.18) and (11.19) to give

$$\sigma_1 = \left(\frac{P}{A_1}\right) \Bigg/ \left[1 + \frac{E_2 l_1}{E_1 l_2} \frac{A_2}{A_1}\right] \quad ...(11.20a)$$

$$\sigma_1 = \left[\frac{P}{A_2}\right] \Bigg/ \left[1 + \frac{E_1 l_2}{E_2 l_1} \cdot \frac{A_1}{A_2}\right] \quad ...(11.20b)$$

If $l_1 = l_2$, equations (1.20) reduce to

$$\sigma_1 = \left(\frac{P}{A_1}\right) \Bigg/ \left[1 + \frac{E_2}{E_1} \frac{A_2}{A_1}\right] \quad ...(11.21a)$$

and

$$\sigma_2 = \left(\frac{P}{A_2}\right) \Bigg/ \left[1 + \frac{E_1}{E_2} \frac{A_1}{A_2}\right] \quad ...(11.21b)$$

Both σ_1 and σ_2 will be compressive in nature.

11.11 EFFECT OF TEMPERATURE RISE ON COMPOSITE BAR

To start with, the upper ends of both the bars are rigidly connected to each other so that they are at level *A*. However if the ends were free to expand, the end *A* of the first bar would have moved to the position *BB* and that of second one to *CC*, due to temperature rise of *t*°C, (Fig. 11.18) such that

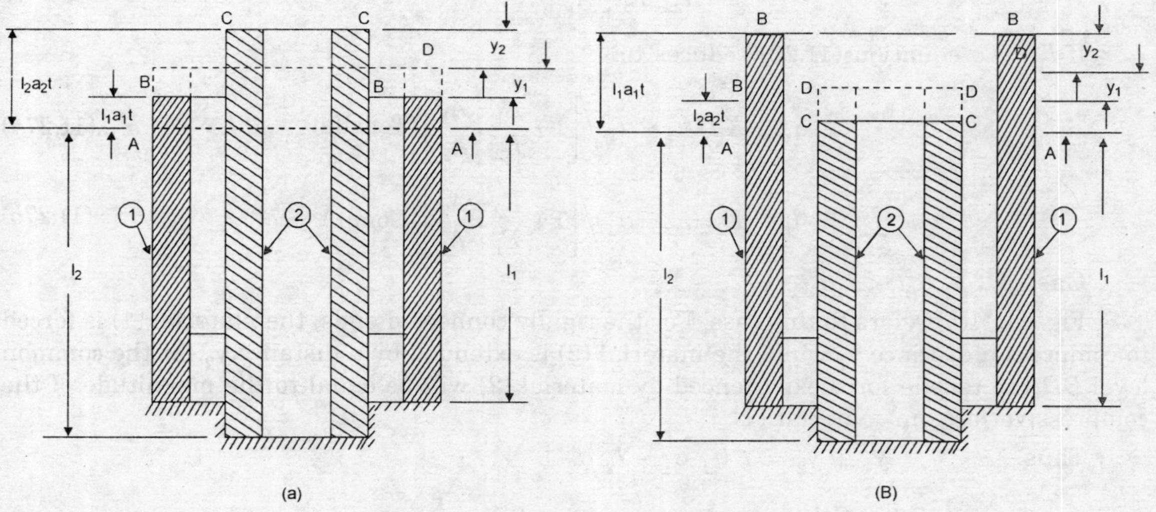

Fig. 11.18

$$AB = l_1 \ \alpha_1 t \quad ...(11.22)$$

$$AC = l_2 \ \alpha_2 t \quad ...(11.23)$$

where α_1 and α_2 are the respective coefficients of linear thermal expansion.

Case I. $l_2 \alpha_2 t > l_1 \alpha_1 t$

Since the two ends are rigidly connected to each other material (1) is forced to extend through a distance y_1 to the level D while the material (2) is forced to compress through a distance y_2 to the level D Fig. 11.11(a)

Geometrically, from the Fig. (11.11a)

$$y_1 + y_2 = (l_2\alpha_2 - l_1\alpha_1)t$$

or $$\frac{\sigma_1}{E_2} l_1 + \frac{\sigma_2}{E_2} l_2 = (l_2\alpha_2 - l_1\alpha_1) \qquad ...(11.24)$$

Where α_1 and α_2 are respectively the stresses in the two materials.

Also as no external force is applied at the level D when equilibrium is attained :

Tensile force in material (1) = Compressive force in material (2)

i.e. $$P_1 = P_2$$

$$\sigma_1 A_1 = \sigma_2 A_2 \qquad ...(11.25)$$

Solving the simultaneous equations (11.24) and (11.25), we get

$$\sigma_1 = \frac{E_1 t(\alpha_2 - \alpha_1 l_1 / l_2)}{\left[\frac{l_1}{l_2} + \frac{E_1}{E_2} \cdot \frac{A_1}{A_2}\right]} \text{ (Tensile)} \qquad ...(11.26a)$$

$$\sigma_2 = \frac{E_2 t \left(\alpha_2 - \alpha_1 \frac{l_1}{l_2}\right)}{\left[1 + \frac{l_1}{l_2} \cdot \frac{E_2}{E_1} \cdot \frac{A_2}{A_1}\right]} \text{ (Compressive)} \qquad ...(11.26b)$$

If $l_1 = l_2$, equation (11.26) reduces to

$$\sigma_1 = E_1 t\,(\alpha_2 - \alpha_1) \Big/ \left[1 + \frac{E_1}{E_2} \cdot \frac{A_1}{A_2}\right] \text{ (Tensile)} \qquad ...(11.27a)$$

$$\sigma_1 = E_2 t\,(\alpha_2 - \alpha_1) \Big/ \left[1 + \frac{E_2}{E_1} \cdot \frac{A_2}{A_1}\right] \text{ (Comp.)} \qquad (11.27b)$$

Case II. $l_2 \alpha_2 t < l_1 \alpha_1 t$

Fig. 11.11(b) refers to this case. For the rigidly connected ends, the material (1) is forced to compress a distance y_1 while the material (2) is extended by a distance y_2. At the common level DD the tensile force experienced by material (2) will be equal to the magnitude of the compressive force in material (1).

Thus $$y_1 + y_2 = t\,(l_1 \sigma_1 - l_2 \alpha_2)$$

i.e. $$\frac{\sigma_1 l_1}{E_1} + \frac{\sigma_2 l_2}{E_2} = t\,(l_1\alpha_1 - l_2 a_2) \qquad ...(11.28)$$

Equation (1.25) holds good here also. Solving the equations (1.25) and (1.28) for σ_1 and σ_2 we get

$$\sigma_1 = \frac{E_1 t\left(\alpha_1 - \frac{l_2}{l_1}\alpha_2\right)}{\left[1 + \frac{E_1 l_2 A_1}{E_2 l_1 A_2}\right]} \text{ (Compressive)} \qquad ...(11.29a)$$

$$\sigma_1 = \frac{E_2 t\left[\alpha_1 - \left(\frac{l_2}{l_1}\alpha_2\right)\right]}{\left[1 + \frac{E_1 A_2}{E_2 A_1}\right]} \text{ (Tensile)} \qquad ..(11.30a)$$

If $l_1 = l_2$ these reduce to

$$\sigma_1 = \frac{E_1 t(\alpha_1 - \alpha_2)}{\left[1 + \frac{E_1}{E_2}\cdot\frac{A_1}{A_2}\right]} \text{ (Compressive)} \qquad ...(11.29b)$$

$$\sigma_1 = \frac{E_2 t(\alpha_1 - \alpha_2)}{\left[1 + \frac{E_1 A_2}{E_2 A_1}\right]} \text{ (Tensile)} \qquad ...(11.30b)$$

11.12 EFFECT OF BOTH THE TEMPERATURE RISE AND EXTERNAL LOAD

The stresses due to individual effects are evaluated as per the method explained above. The resultant stresses are then calculated by adding the stresses algebraically in each material.

Example 11.3. *A steel bar of uniform diameter of 40 mm is heated to 80°C and then clamped at the ends with the help of two fixtures 4 m apart, and left to cool down to room temperature of 20° C at which temperature the distance between the fixtures was found to be 1.0 mm shorter than that at 80° C. Determine the stress in the bar when it has cooled down to room temperature and the reaction at the fixtures.*

Take $E = 200\ GN/m^2$ and

$\alpha = 1.1 \times 10^{-5}$/deg.C.

Solution : $l = 4000$mm

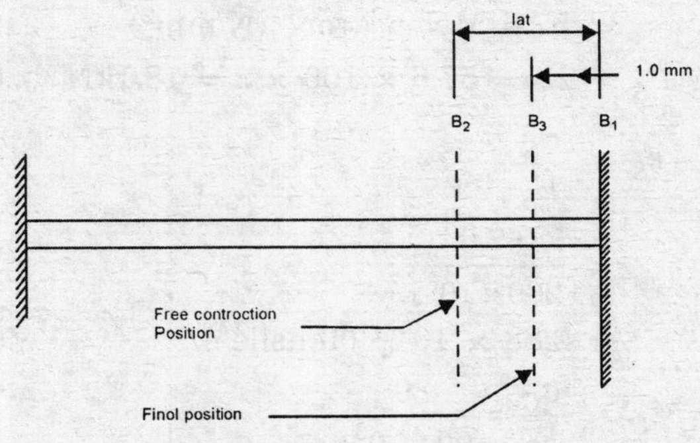

Fig. 11.19

Temperature drop $t = 80 - 20 = 60°C$

B_1B_2 = Free contraction $= l \alpha t$

$= 4000 \times 1.1 \times 10^{-5} \times 60 = 2.64$ mm

Therefore B_2B_3 = Stretch in the bar

$= B_1B_2 = - B_1B_3$

$= 2.64 - 1.0 = 1.64$ mm

Thus the stress in the bar is given by

$$\sigma = \frac{E}{l} \times 1.64$$

$$= \frac{200 \times 10^3}{4000} \times 1.64 = 82 \text{ MN/m}^2 \text{ or N/mm}^2 \text{ (Tensile)}$$

Reaction of Supports $= \sigma \times A = 82 \times \frac{\pi}{4}(40)^2$

$= 10^3 \times 10^3$ N

$= 103$ kN.

Example 11.4. *A copper sleeve, 21 mm internal and 27 mm external diameter, surrounds a 20 mm steel bolt, one end of the sleeve being in contact with the shoulder of the bolt. After putting a rigid washer on the other end of the tube, a nut is screwed on the bolt until the compressive stress in the sleeve is 80 MN/m². Find the range of the external axial load that can be applied to this assembly if the stress in the bolt is never to be compressive and in the sleeve never to be zero. Take $E_S = 200$ GN/m² and $E_C = 90$ GN/m².*

Solution : $A_S = \frac{\pi}{4} \times 20^2 = 10\pi \text{ mm}^2$;

$$A_C = \frac{\pi}{4}(27^2 - 21^2) = 72\,\pi\,\mu\mu^2$$

When the tube is in compression due to the tightening of the bolt, bolt will be under tension and

$$P_C = P_S$$

or $\sigma_C A_C = \sigma_S A_S$

or $\sigma_S = \frac{\sigma_C A_C}{A_S} = \frac{80 \times 72\pi}{100\,\pi}$

$= 57.6$ MN/m² or (N/mm²) ...(11.31)

$P_S = P_C = 57.6 \times 100 \times \pi = 18.1$kN ...(11.32)

$$\varepsilon_S = \frac{\sigma_S}{E_S}$$

$$= \frac{57.6}{200 \times 10^3}$$

$= 2.88 \times 10^{-4}$ (Tensile) ...(11.33)

$$\varepsilon_C = \frac{\sigma_C}{E_C} = \frac{80}{90 \times 10^3}$$

$= 8.89 \times 10^{-4}$ (Compressive) ...(11.34)

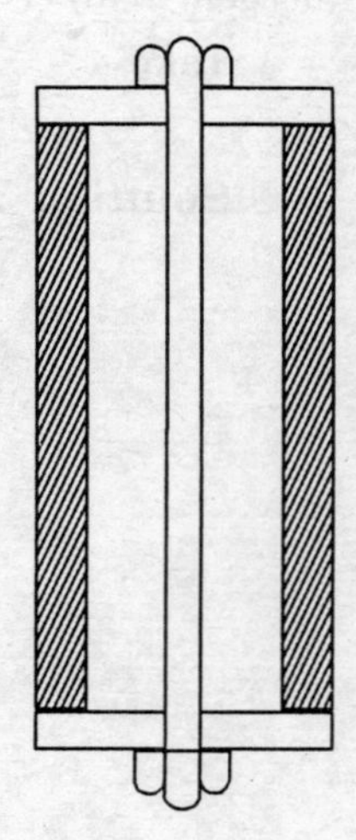

Fig. 11.20

(*b*) Two cases can arise depending on whether the load, compressive or tensile, is applied on the bolt or on the tube.

(*i*) **Tensile load on the bolt :** In this case the limiting condition will be when the compressive stress already existing in the tube is just brought down to zero. This will happen when the bolt is stretched further to the extent that the additional stretch or strain induced in it is equal to the compression or compressive strain in tube as calcualted in equation (11.34) (as these lengths are equal)

$$\varepsilon_{S_2} = \varepsilon_C = 8.89 \times 10^{-4} \quad \text{...(11.35)}$$

$$P_2 = P_{S_2} = A_S E_S$$

$$= 100\ \pi \times (200 \times 10^{-4}) (8.89 \times 10^{-4})\text{N}$$

$$= 5.9\ \text{N (Tensile)}\} \quad \text{...(11.37)}$$

(*ii*) **Compressive load on the bolt :** Here the appleid load will be shared by both bolt and tube and the limiting case will be when the additional induced compressive stress in the bolt becomes equal in magnitude to the existing tensile stress in the bolt as given by equation (11.31).

i.e. $$\sigma_{S_3} = \sigma_S \quad \text{...(11.38)}$$

$$P_3 = P_{S_3} + P_{C_3} \quad \text{...(11.39)}$$

and $$\varepsilon_{S_3} = \varepsilon_{C_3}$$

or $$\frac{\sigma_S}{E_S} = \frac{p_{C_3}}{A_C E_C}$$

i.e. $$P_{C_3} = \sigma_S A_C E_C / E_S \quad \text{...(11.40)}$$

Thus substituting values in equation (11.39)

$$P_3 = \sigma_S A_S + \sigma_S A_C / E_S$$

$$= \sigma_S [A_S + A_C E_C / E_S]$$

$$= 57.6 \left[100\pi + 72\pi \frac{90}{200}\right]$$

i.e $$P_3 = 24\ \text{kN (Compressive)} \quad \text{...(11.41)}$$

(*iii*) **Tensile load on tube :** Here again the load will be shared by both the bolt and the tube and the stress in the tube will be the limiting case and the additional tensile stress induced in the tube will become equal to 80 N/mm^2 or 80 MN/m^2.

i.e. $$\sigma_{C_4} = 80\ \text{N/mm}^2 \quad \text{...(11.42)}$$

$$P_4 = P_{S_4} + P_{C_4} \quad \text{...(11.43)}$$

$$\sigma_{S_4} = \varepsilon_{C_4}$$

or $$\sigma_{S_4} / E_S = \sigma_{C_4} / E_C$$

Thus $$P_4 = \sigma_{S_4} A_S + \sigma_{C_4} A_C$$

$$= \sigma_{C_4}\left[A_C + \frac{E_S}{E_C}A_S\right]$$

$$= 80\left[72\pi + \frac{200}{90}\times 100\pi\right] \text{N}$$

i.e. $P_4 = 73.9$ kN (Tensile) ...(11.44)

(*iv*) **Compressive load applied on tube :** In this case the load with be borne only by tube and the limiting condition will be when the stress in bolt reduces to zero. This will happen when the compressive strain induced in tube due to external load is just equal in magnitude to the strain ε_s in bolt as given in equation (11.33).

Thus $\varepsilon_{C_5} = \varepsilon_S$ ...(11.45)

$$P_5 = A_{C_5} A_C$$

$$= 72 \times (2.88 \times 10^{-4}) \times (90 \times 10^3) \text{ N}$$

i.e. $P_5 = 5.86$ kN (Compressive) ...(11.46)

From equations (11.37), (11.41), (11.44) and (11.46) it is evident that the given conditions to be fulfilled the allowable range of the external load is from 5.86 kN (Compressive) to 55.9 kN (tensile).

Example 11.5. *If instead of applying the external load to the bolt and tube assembly of Example 11.4, the temperature is raised through 50°C, find the resultant stresses in the bolt and the tube.*

Take $\alpha_{steel} = 11 \times 10^{-6}$/°C

$\alpha_{copper} = 18 \times 10^{-6}$/°C .

Solution : In this case since the lengths are equal, we have from equation (11.24)

$$\frac{\sigma_S}{E_S} = \frac{\sigma_S}{E_S} + \frac{\sigma_C}{E_C} = t\,(\alpha_C - \alpha_S) \quad ...(11.47)$$

Where t is the temperature rise.

Also in equilibrium state the compressive load in copper will be equal to tensile load in bolt, due to the temperature rise effect.

i.e $\sigma_S A_S = \sigma_c A_c$

or $\sigma_S = (72/100)\sigma_c$ ...(11.48)

Substituting equation (11.48) in equation (11.47)

$$\sigma_c\left[\frac{0.72}{E_S} + \frac{1}{E_C}\right] = t(\alpha_c - \alpha_S)$$

or
$$\sigma_c\left(\frac{0.72}{200\times 10^3} + \frac{1}{90\times 10^3}\right) = 50\ (7 \times 10^{-6})$$

$$= 23.8 \text{ N/mm}^2 \text{ or MPa (Comp.)} \quad ...(11.49a)$$

and from eqn. (1.48)

$$\sigma_S = .72 \times 23.8$$

$$= 17.2 \text{ MN/m}^2 \text{ or MPa (Tensile)} \quad ...(11.49b)$$

Thus from equations (11.31) and (11.49)

Resultant stress in bolt

$$= 57.6 + 17.2 = 74.8 \text{ MN/m}^2 \text{ (Tensile)}$$

Resultant stress in tube

$$= 80.0 + 23.8 = 103.8\text{MPa (Comp.)}$$

Example 11.6. *Two bars one of aluminum and the other of steel, joined end to end, are placed lengthwise between two rigid supports 600 mm apart. Their axes are horizontal and are along the same straight line. Aluminium bar is 200 mm long and steel bar is 400 mm long. Their cross-sectional areas are 1000 mm² and 200 mm² while Young's modulii are 0.6 × 10⁶kg/cm² (588.6 MN/m²) and 2 × 10⁶ kg/cm² (1962 MN/m²) respectively. The whole system is in un-stressed condition at 10°C. The system is then heated to a temperature of 60°C. Find the stresses developed in the two materials if*

$$a_{steel} = 12 \times 10^{6}/°C$$

and $$\alpha_{al} = 24 \times 10^{-6}/°C$$ *(Punjab University)*

Solution : If the composite bar were free to expand, the corresponding increase in length of the two portions will be added up to give the total increase in length δl as

$$\begin{aligned} \delta l &= \delta l_s + \delta l_a \\ &= l_s\, \alpha_s (T_2 - T_1) + l_\alpha (T_2 - T_1) \\ &= (T_2 - T_1)\, [\alpha_s l_s + \alpha_a l_a] \\ &= (60 - 10)\, (12 \times 10^{-6} \times 400 + 24 \times 10^{-6} \times 200) \\ &= 50 \times 9600 \times 10^{-6} \\ &= 0.48 \text{ mm.} \end{aligned}$$

Since both the metals have the same axis, they will carry the same compressure axial load, P, if the free expansion is checked by rigid supports. Thus the compression in both the metals due to the load P will add up to *dl*

i.e., $$\frac{P}{A_s S_s} l_s + \frac{P}{A_a E_a} l_a = \delta l$$

or $$P\left[\frac{400 \times 10^2}{(200)\,(2 \times 10^6)} + \frac{200 \times 10^2}{(1000)\,(0.6 \times 10^6)}\right] = 0.48$$

or $$P = 11076.9 \text{ kg}$$

Therefore $$\sigma_s = \frac{P}{A_S} = \frac{110076.9}{200} \times 10^2$$

$$= 553.85 \text{ kg/ cm}^2 \text{ (comp.)}$$

and $$\sigma_a = \frac{P}{A_a} = \frac{11076.9}{1000} \times 10^2$$

$$= 1107.70 \text{ kg/cm}^2 \text{ (comp.)}$$

Example 11.7. *A brass rod of 6 mm diameter and 1 m long is joined at one end to a rod of steel 6 mm diameter and 1.3 m long. The compound rod is placed in a vertical position with the steel rod at the top and connected top and bottom to rigid fixings in such a way that it is carrying a tensile load of 3.5 kN.*

An attachment is fixed at the junction of the two rods and to this a vertical axial load of 1.3 kN is applied downwards. Calculate the stresses in the steel and brass. The temperature is then raised through 30°C. What are the final stresses in steel and brass. Take

$$E_s = 200 \ GN/m^2$$
$$E_b = 85 \ GN/m^2$$
$$\alpha_s = 12 \times 10^{-6}/°C,$$
$$\alpha_b = 19 \times 10^{-6}/°C.$$

(Punjab University)

Solution : (*a*) In the first instance, when the composite bar *ABC* is held in fixtures, both steel and brass portions of the rod carry the tensile load of 3500 *N*. Thus the initial tensile stresses in each of the metals will be equal, as their areas of cross sections are equal.

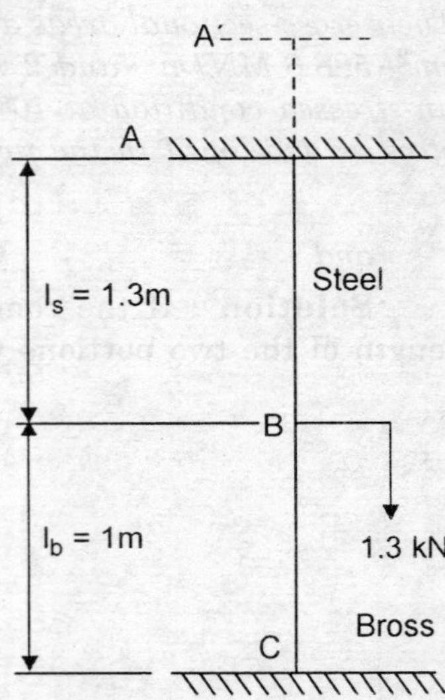

Fig. 11.21

$$A_s = A_b = \frac{\pi}{4}(6)^2$$
$$= 28.274 \text{mm}^2$$

$$\sigma_{s_1} = \sigma_{b_1} = \frac{3500}{28.274}$$
$$= 123.79 \text{N/mm}^2$$

or $\quad$ MN/m² (Tensile)

(*b*) The load of 1300 *N* applied at the junction point *B* (Fig. 11.14), will induce compressive stress in brass and tensile stress in steel. The magnitudes of the loads P_s and P_b due to the action of 1300 *N* load at *B*, will be such that the stretch in steel will be equal in magnitude to the compression in brass, because the total length *AC* remains constant.

That is $\quad P_s + P_b = 1300$

and $\quad \delta l_s = \delta l_b$

or $$\frac{P_s l_s}{A_s E_s} = \frac{P_b l_b}{A_b E_b}$$

or $$P_s = \frac{E_s l_b}{E_b l_s} P_b \qquad (\because A_s = A_b)$$

$$= \frac{200}{85} \times \frac{1}{1.3} P_b = 1.81 P_b$$

Therefore

$$1.81 P_b + P_b = 1300$$

or $\quad P_b = 461 \text{ N}$

Then $\quad P_s = 1300 - 461$

$$= 839 \text{ N}$$

Thus the corresponding additional stresses in steel and brass, due to the application of 1300 N load are

$$\sigma_{s2} = \frac{P_s}{A_s} = \frac{839}{28.274}$$

$$= 29.67 \text{ MN/m}^2 \text{ (Tensile)}$$

$$\sigma_{b2} = \frac{P_b}{A_b} = \frac{461}{28.274}$$

$$= 16.3 \text{ MN/ m}^2 \text{ or MPa (Comp.)}$$

Then the resultant stresses are

$$\sigma_b = \sigma_{b1} - \sigma_{b2}$$

$$= 123.79 - 16.3$$

$$= 107.49 \text{MN/m}^2 \text{ or MPa (Tensile)}$$

$$\sigma_s = \sigma_{s1} + \sigma_{s2} = 123.79 + 29.67$$

$$= 153.46 \text{MN/m}^2 \text{ or MPa (Tensile)}$$

(c) In the absence of fixtures, the rod AC would have expanded freely due to temperature rise to CA'. But the presence of fixtures effectively compresses the rod AC by an amount AA' which is equal to $(l_s\,\alpha_s t + l_b\,\alpha_b t)$. Thus if σ N/m² is the compressive stress in steel and brass (it will be same as the cross section areas are same) due to temperature rise, then we have

$$\frac{\sigma}{E_s} l_s + \frac{\sigma}{E_b} l_b = l_s\,\alpha_s t + l_b \alpha_b t$$

or

$$\left[\frac{1.3}{200\times10^9} + \frac{1}{85\times10^9}\right] = 30\ (1.3 \times 12 \times 10^{-6} + 1 \times 9 \times 10^{-6})$$

or

$$\sigma = 56.9 \times 10^6 \text{ N/mm}^2$$

$$= 56.9 \text{MN/m}^2 \text{ or MPa}$$

Therefore, the final stresses in steel and brass are

$$\sigma_{s3} = \sigma_s - \sigma = 143.46 - 56.9$$

$$= 96.56 \text{ MN/m}^2 \text{ or MPa (Tensile)}$$

$$\sigma_{b3} = \sigma_b - \sigma = 107.49 - 56.9$$

$$= 0.59 \text{MN/ m}^2 \text{ or MPa (Tensile)}$$

Example 11.8. *A bar AD (Fig. 11.22) is pinned at A, and supported by a copper rod 40 mm² in area and 1.5 m long at B, also supported at C by a steel rod of 30 mm² in area and 1 m length. The bar AD is rigid and horizontal before applying a load of 1000 kg (9810 N) at the end D. Determine the stresses and elongations produced in copper and steel rods after the application of above load. Take*

$$E_c = 10 \times 10^3 \ kg/mm^2 \ (98.1 GN/m^2)$$

$$E_s = 20 \times 10^3 \ kg/mm^2 \ (196.2 \ GN/m^2).$$

(Punjabi University)

Solution : Let the tensile loads sustained by copper and steel rods be P_c and P_s respectively. Then for the equilibrium of the rod ***AD***, taking moments of P, P_c and P_s about A we have

$$P_c \times 1 + P_s \times 2 = P \times 2.5$$

or $$P_c + 2P_s = 9810 \times 2.5 = 24525 \quad \text{...(A)}$$

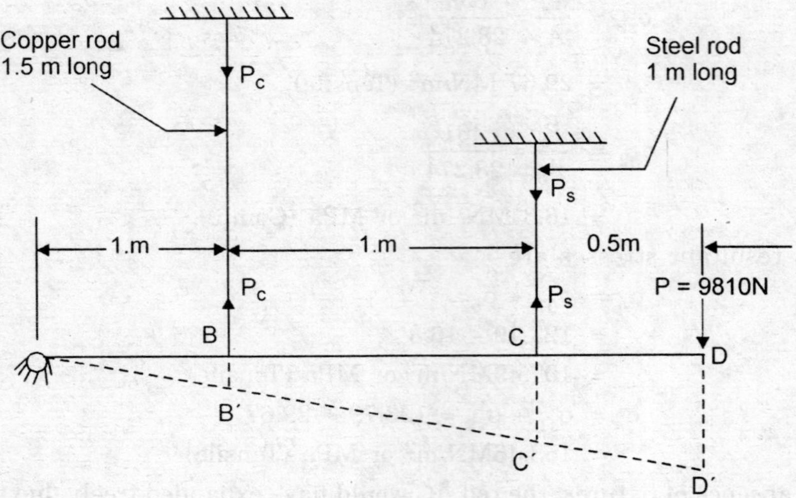

Fig. 11.22

Further due to the loads P_c and P_s the copper and steel rods will get stretched by *BB′* and *CC′* respectively. However, the extensions will be such that *A*, *B′*, *C′* and *D′* will lie in one straight line, as the rod *AD* is rigid. Therefore from similar triangles *ABB′* and *ACC′* we have

$$\frac{CC'}{BB'} = \frac{AC}{AB} = \frac{2}{1}$$

or $$CC' = 2BB'$$

or $$\frac{P_S}{A_S E_S} l_s = 2 \frac{P_C}{A_C E_C} l_c$$

or $$P_s = 2\left(\frac{A_s}{A_C}\right)\left(\frac{E_s}{E_C}\right)\left(\frac{l_c}{l_s}\right) P_c$$

$$= 2\left(\frac{30}{40}\right)\left(\frac{196.2}{98.1}\right)\left(\frac{1.5}{1.0}\right) P_c$$

or $$P_s = 4.5\, P_c \quad \text{...(B)}$$

Substituting (*B*) into (*A*), we get

$$P_c = 2452.5 \text{ N}$$

and $$P_s = 11036.25 \text{ N}$$

Therefore $$\sigma_c = \frac{2452.5}{40}$$

$$= 61.31 \text{ N/mm}^2 \text{ or MPa (Tensile)}$$

and $\sigma_s = \dfrac{11036.25}{30}$

$= 367.88 \text{N/mm}^2$ or MPa (Tensile)

Also $\delta l_c = BB' = \dfrac{\sigma_c}{E_c} \times l_c$

$= \dfrac{61.31 \times 10^6 \times 1.5}{98.1}$ m

$= 0.9375$ mm

Similarly $\delta l_s = CC' = \dfrac{\sigma_s}{E_s} l_s$

$= \dfrac{367.88 \times 10^6 \times 1.0}{196.2 \times 10^9}$ m

$= 1.875$ mm.

Example 11.9. *Fig. 11.23 shows a round steel rod supported in a recess and surrounded by a coaxial brass tube. The upper end of the rod is 0.1 mm below that of the tube and an axial load is applied to a rigid plate resting on the top of the tube.*

(a) Determine the magnitude of the maximum permissible load if the compressive stress in the rod is not to exceed 110 MN/m² and that in the tube is not to exceed 80 MN/m².

(b) Find the amount by which the tube will be shortened by the load if the compressive stress in the tube is the same as that in the rod,

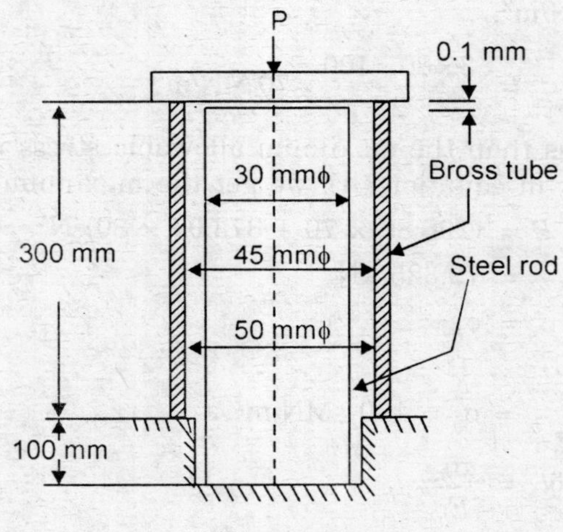

Fig. 11.23

Take $Es = 200 \text{ GN/m}^2 = 200 \times 10^3$ MPa

$E_b = 100 \text{ GN/m}^2 = 100 \times 10^3$ MPa *(Punjab University)*

Solution. (*a*) $A_s = \dfrac{\pi}{4} \times 706.86 \text{ mm}^2$

$A_b = \dfrac{\pi}{4}(50^2 - 45)^2 = 373.07 \text{ mm}^2$

Let σ_s and σ_b MN/m² or N/mm² be the stresses induced in the steel rod and brass tube respectively. Since the sum of the loads carried by the steel and brass is equal to the total load

$$P = \sigma_s \ A_s + \sigma_b \ A_b$$

$$= 706.86 \ \sigma_s + 373.07\sigma_b \text{ N} \quad ...(A)$$

When the top plate makes contact with steel and compresses it, the compression of the brass will exceed that of steel by 0.1 mm.

That is $\delta l_b = \delta l_s + 0.1$

or $$\frac{\sigma_b}{E_b} l_b = \frac{\sigma_s}{E_s} l_s + 0.1$$

or $$\frac{\sigma_b}{100 \times 10^3} \times 300 = \frac{\sigma_s}{200 \times 10^3} \times 400 + 0.1$$

or $$3\sigma_b = 3\sigma_s + 100 \quad ...(B)$$

The maximum compressive stress of 110 MN/m² in rod and 80 MN/m² in tube will not occur simultaneously, rather the magnitudes of induced stresses in the two materials will be related to each other by the equation (*B*). Thus if it is assumed that the stress in steel is the limiting case, then from equation (*B*) for $\sigma_s = 110$ MN/m²,

$$\sigma_b = \frac{2 \times 110 + 100}{3}$$

$$= 106.7 \text{ MN/m}^2$$

which is more than the allowable stress in brass leading to the conclusion that the assumption is wrong. Then again assuming the stress in brass to be the limiting case, then from (*B*) for $\sigma_b = 800$ MN/m²,

$$\sigma_s = \frac{3 \times 80 - 100}{2} = 70 \text{ MN/m}^2$$

which is safe being less than the maximum allowable stress in steel. Thus substituting these values of σ_b and σ_g in equation (*A*), we get the maximum permissible load *P* as

$$P = (706.86 \times 70 + 373.07 \times 80) \text{ N}$$

$$= 79.325 \text{ kN}$$

(*b*) In this case $\sigma_s = \sigma_b$

Then from equation (*B*)

$$\sigma_b = \sigma_s = 100 \ \text{MN/m}^2$$

Therefore $$\delta l_b = \frac{\sigma_b}{E_b} \times l_b$$

$$= \frac{100}{100 \times 10^3} \times 300 \text{ mm}$$

$$= 0.3 \text{ mm.}$$

Example 11.10. *A reinforced concrete column is 300 mm in diameter and has 4 steel bars of 12 mm diameter embedded in it. If the allowable stresses in steel and concrete are 65.0 MN/m² and 4.40 MN/m² respectively, calculate the safe axial load which the column can carry. Take* $E_s/E_c = 15$.

Solution : Total area of column $= \frac{\pi}{4}(300)^2 = 70686 \text{ mm}^2$

$$A_s = \text{Area of steel} = 4 \times \frac{\pi}{4}(12)^2 = 452.4 \text{ mm}^2$$

Therefore area of concrete portion

$$= A_c = 70686 - 452.4$$

$$= 70233.6 \text{ mm}^2$$

Since both steel and concrete portions will compress by the same amount and also their original lengths are same, the compressive strains produced by the corresponding loads sustained by each will be same

i.e., $\varepsilon_c = \varepsilon_s$

or $$\frac{P_c}{A_c E_c} = \frac{P_s}{A_s E_s} \quad \text{...}(A)$$

Also from (*A*) $$\frac{\sigma_c}{E_c} = \frac{\sigma_s}{E_s}$$

or $$\frac{\sigma_s}{\sigma_s} = \frac{E_s}{E_c} = 15 \quad \text{...}(B)$$

Thus if it is assumed that the stress in concrete is the limiting case, then for 4 MN/m² stress to be induced in concrete the stress induced in steel will be from (*B*)

$$\sigma_s = 15 \times \sigma_c = 15 \times 4 = 60 \text{ MN/m}^2$$

(which is less than the allowable stress in steel)

Thus our assumption of limiting stress in concrete is valid. Corresponding loads sustained by concrete and steel are

$$P_c = \sigma_c A_c = 4 \times 70233.6 = 280934 \text{ N}$$

$$= 280.934 \text{ kN}$$

$$P_s = \sigma_s \times A_s = 60 \times 452.4 = 27144 \text{ N}$$

$$= 27.144 \text{ kN}$$

Thus total load which the column can carry

$$= P_c + P_s$$

$$= 280.934 + 27.144$$

$$= 308.078 \text{ kN}.$$

Example 11.11. *Two steel plates, each 25 mm thick, are held together with the help of 20 mm diameter steel bolts. If the pitch of the threads is 2.5 mm, find the increase in stress in the shank of the bolt when the nut is turned through* $\frac{1}{200}$ *of a turn with respect to the head of the bolt, assuming that the plates do not deform. Take* $E_s = 205 \text{ GN/m}^2$.

Solution : Length of the bolt shank = 25 + 25 = 50 mm.

Axial movement of the nut corresponding to rotation of $\frac{1}{200}$ of a turn $= \frac{2.5}{200} = 0.0125$ mm.

Thus that rotation of the nut will cause 0.0125 mm stretch in 50 mm length of the shank giving rise to strain e given by

$$\varepsilon = \frac{0.0125}{50}$$

Therefore $\sigma = \varepsilon E = \frac{0.0125}{50} \times (205 \times 10^3)$

$$= 51.25 \text{ MN/m}^2.$$

Example 11.12. *A copper sleeve, 21 mm internal and 27 mm external diameter, surrounds a 20 mm steel bolt, one end of the sleeve being in contact with the shoulder of the bolt. The sleeve is 60 mm long. After putting a rigid washer on the other end of the sleeve, a nut is screwed on the bolt through 10 degrees. If the pitch of the threads is 2.5 mm, find the stresses induced in the copper sleeve and steel bolt. Take E_s = 200 GN/m² and E_c = 90 GN/m².*

(Punjab University)

Solution : Due to 10° rotation of the nut, it will move along the axis of the bolt through a distance δ (say), such that

$$\delta = \frac{10}{360} \times \text{pitch}$$

$$= \frac{10}{360} \times 2.5 = 0.0694 \text{ mm}$$

Further the axial movement of the nut, δ, will be equal to the compression in the copper sleeve plus stretch in the steel bolt.

i.e. $\delta = \varepsilon_c l_c + \varepsilon_s l_s$

or $0.0694 = 60(\varepsilon_c + \varepsilon_s)$ $[\because l_c = l_s = 60 \text{ mm}]$

or $$\frac{0.0694}{60} = \frac{\sigma_c}{E_c} + \frac{\sigma_s}{E_s} \quad ...(A)$$

Also the tube and the bolt, in equilibrium, will carry the same load.

i.e., $P_c = P_s$

$$\sigma_c A_c = \sigma_s A_s$$

or $$\sigma_c = \frac{\sigma_s A_s}{A_c} = \sigma_s \frac{\frac{\pi}{4}(20)^2}{\frac{\pi}{4}(27^2 - 21^2)} = 1.39\sigma_s \quad ...(B)$$

From (*A*) and (*B*), we get

$$\frac{0.0694}{60} = \left(\frac{1.39}{90 \times 10^3} + \frac{1}{200 \times 10^3}\right)\sigma_s$$

$$= 0.2044 \times 10^{-4} \, \sigma_s$$

or $$\sigma_s = \frac{0.0694}{60} \times \frac{1}{0.2044 \times 10^{-4}} = 56.59 \text{ (Tensile)}$$

From (*B*) $$\sigma_c = 1.39 \times 56.59$$
$$= 78.66 \text{ MN/m}^2 \text{ (compressive)}$$

11.13 MECHANICAL PROPERTIES OF MATERIALS AND THEIR CLASSIFICATION

11.13.1 Homogeneous and Isotropic material

A homogeneous material is one in which the physical properties like the density, yield stress, modulus of elasticity. Poisons ratio, modulus of rigidity etc. are same at each and every point in the body. Isotropic means that these poperties do not change with the change in direction *e.g.* the yield point will have the same value at a point even if it is measured in any direction. In the case of rolled metallic sheet the value of youngs modulus will be more if the test piece is cut out of sheet along the diection of rolling as compared to the test piece cut along the direction perpendicular to its direction of rolloing which will mean that the rolled sheet is not isotropic.

11.13.2 Elastic Plastic Materials

A material whose stress strain curve is linear upto elastic limit point and plastic beyond elastic limit point is called the elastic plastic material, steel falls in this category and its stress-strain curve is as shown in Fig. 11.8. These are also called a elastic-strain hardening materials.

11.13.3 RIGID PEREFECTLY PLASTIC MATERIAL

Rigid material will have stress-strain curve on shown in Fig. 11.24 (*a*) such material will not be strained upto yield point. It will be a vertical line along y-axis and then horizontal at σ_y level which will correspond to perfectly plastic region where there will be constant stress level but strain will increase.

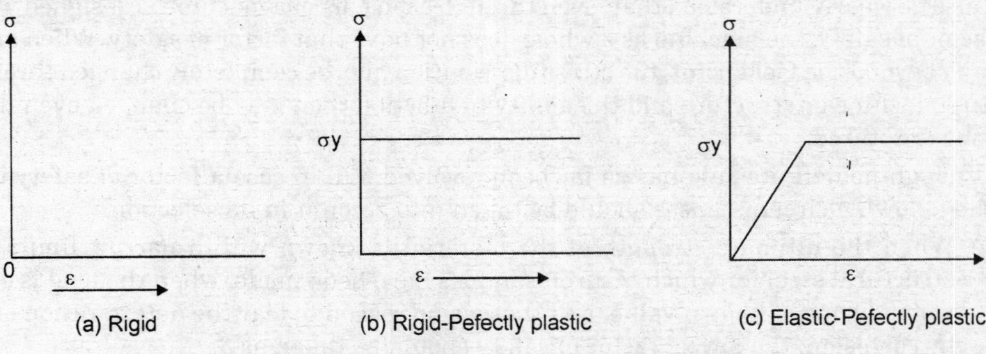

(a) Rigid (b) Rigid-Pefectly plastic (c) Elastic-Pefectly plastic

Fig. 11.24

11.13.4 Ductile and Brittle Materials

Ductile materials under tensile load show large permanent strain upto the instability point. Thus these materials, can be drawn into thin wises. There are two thoughts to quantify ductility. According to one it is defined as the % elongation of the specimen after it breaks. The length *l* of the specimen after the specimen has fractured is measured between the two points, which had been used to originally measure l_0 before yielding, by placing the two fractured pieces together.

Then percentage elongation

$$= \frac{l - l_0}{l_0} \times 100$$

However the latest thinking which (has now come up) is to stretch the specimen upto instability point, when the neck formation has just started, the length l is measured. Then the total stress upto instability point which is defined below, gives the measure of ductability

$$(\varepsilon_t)_{\text{wet}} = \frac{l - l_0}{l_0}$$

11.13.5 Working Stress or Allowable Stress

Is the maximum stress to which, it is considered safe to subject a member in service. The term "allowable stress" is preferable to "working stress", since the latter often is used to indicate the actual stress in a material when in service. Allowable stresses for different materials for different conditions of service are specified by different authorities on the basis of tests and on experience. In general, for ductile materials, allowable stress is considerably less than elastic limit.

11.13.6 Factor of Safety

Factor of safety is ratio of ultimate strength of the material to allowable stress. The terms was originated for determining the allowable stress. The ultimate strength of a given material is divided by an arbitrary factor, called the safety factor, dependent on material and the use to which it has to be put, gives the allowable stress. In present day design practice, it is customary to use allowable stress as specified by recognised authorities, or codes rather than arbitrary factor of safety. One reason for this is that the factor of safety is misleading, in that, it implies a greater degree of safety than actually exists. For example, a factor of safety of 4 does not mean that a member can carry load four times as great as that for which it has been designed. It should also be clearly understood that, even through each part of a machine is designed with the same factor of safety, the machine as a whole does not have that factor of safety. When one part is stressed beyond the yield point, the stress distribution may be completely changed throughout the entire machine or structure and the ability to function then may be changed even through no part has ruptured.

Although no definite rule has so far been evolved still in case a factor of safety is to be used, the following circumstances should be taken into account in its selection.

(*a*) When the ultimate strength of the material is known within narrow limits, as for structural steel for which tests on samples have been made, when the load is entirely a steady one of known value, and there is no reason to fear the deterioration of metal by corrosion, the lowest factor of safety should be taken as 3.

(*b*) When the circumstances of (a) are modified by a portion of the load being variable, as in floors of warehouses, the factor of safety should not be less than 4.

(*c*) When the whole or nearly whole load is likely to be put on and taken off, as the suspension rods of floors of bridges, the factor of safety should be 5 or 6.

(*d*) When the stresses are reversed in direction from tension to compression as in bridges and parts of machines, the factor of safety should not be less than 6.

(e) When the part or component is subjected to repeated shocks, the factor of safety should not be less than 10.

(f) When there is probability of property loss caused by failure of the part he large or even iff loss of life may result as in a derrick hoisting materials over a crowded street, the factor of safety should be very large ever as high as 50.

In mathematical terms

$$\text{Safety factor} = \frac{\text{Ultimate stress}}{\text{Working stress}}$$

11.13.7 Hardness

It is the resistance to indentation or to seratehing, abrasion or cutting. Indentation hardness is usually measued by indentation with hardened steel balls, Diamond cone or Diamond pyramid. Most popular tests are Brinell hardness test or Rockwell hardness test.

11.13.8 Fatigue

Fatigue of metals is a phenomenon leading to the fracture under the repeated cyclic or fluctuating stresses wheel are much below the elastic limit of the material. It is a progressive phenomenon and initiates as a minute crack developing under the fluctuating stresses.

(*b*) **Endurance limit.** It is the maximum stress level below which the material will not fracture or fail even infinite number of cycles the stress may be repeated. Fig. 11.26 shows the plot on log-log graph paper of $\sigma\ v_s$ N, are numbe of cycles to cause failure.

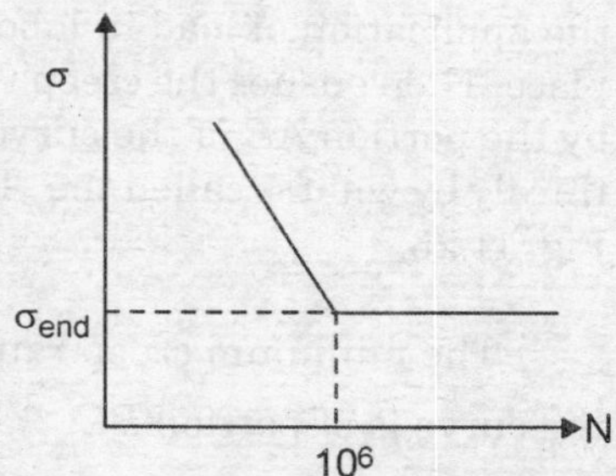

Fig. 11.25 No. of cycles to cause failure

It is observed that as the stress level decrease the number of cycles requied to cause failure increased. However when the stress level is σ_e it is noticed that the curve becomes horizontal. The knee of the curve corresponds to 10^6 cycles and the stress level is σ_e, called Endurance limit. Evidently at σ_e stress, the number of cycles may even be very large still the failure does not occur. That is why the limit if 10^6 cycles is in a way pertains to infinite life and σ_e the endurance limit for the particular type of loading. (That is when the stress is due bending, direct stress, repeated or cycle etc.)

11.13.9 Fatigue life

At any particular stress level which is less than elastic limit the number of cycles of application of stress required to cause failure is called the fatigue life at that particular stress level. If the stress level is equal to the endurance limit, then the life is called the infinite life.

11.13.10 Fatigue or Endurance Ratio

The ratio of endurance limit to the tensile strength or ultimate stress is called the Endurance or Fatigue ratio.

11.13.11 Fatigue Strength

The limiting stress below which a material will withstand a specified number of cycles of stress without fracture, is called the fatigue strength for the given life in number of cycles.

11.13.12 Creep

Creep or flow of metals in a phase of plastic or inelastic action. Some solids such as asphalt or paraffin, flow specially at room temperature index extremely small stress, zinc, lead and tin show signs of creep at room temperature under extremely small stresses; zinc. lead and tin show signs of creep at room temperature under moderate stresses. At sufficiently high temperatures, practically all metals creep under stresses which vary with temperature, the higher the temperatures lower being the stress at which creep takes place. The deformation due to creep continues to increase indefinitely and becomes the extreme importance in members subjected to high temperatures, as parts in turbines, boilers, super heaters etc.

Creep limit is the maximum unit stress under which unit distorsion wil not exceed a specified value during a given period of time at a specified temperature. A value much used in tests and suggested as a standard for comparing materials, is the maximum unit stress at which creep does not exceed 1% in 100,000 hours.

(*b*) **Creep graphs :** In Fig. 11.26 is shown the type of the creep strain *VS* time curve. After the application of load in laboratory test, the instantaneous elastic deformation $OA = E_i$ takes place. Then ensues the creep. At first it is "primary transient creep" in Fig. 11.26 it is represented by the portion *AB* of the curve which is followed by the "steady State Creep" shown as *BC* and thirdly by what is called the "Tertiary Creep" or "Accelerated Creep" which is shown as CD in the Fig. 11.26.

The minimum creep rate $\frac{\Delta\varepsilon}{\Delta t}$ is determined by the slope of the secondary creep portion of the curve is of portion *BC*

i.e. $$\frac{\Delta\varepsilon}{\Delta t} = \tan\phi = \phi \qquad \text{...(11.50)}$$

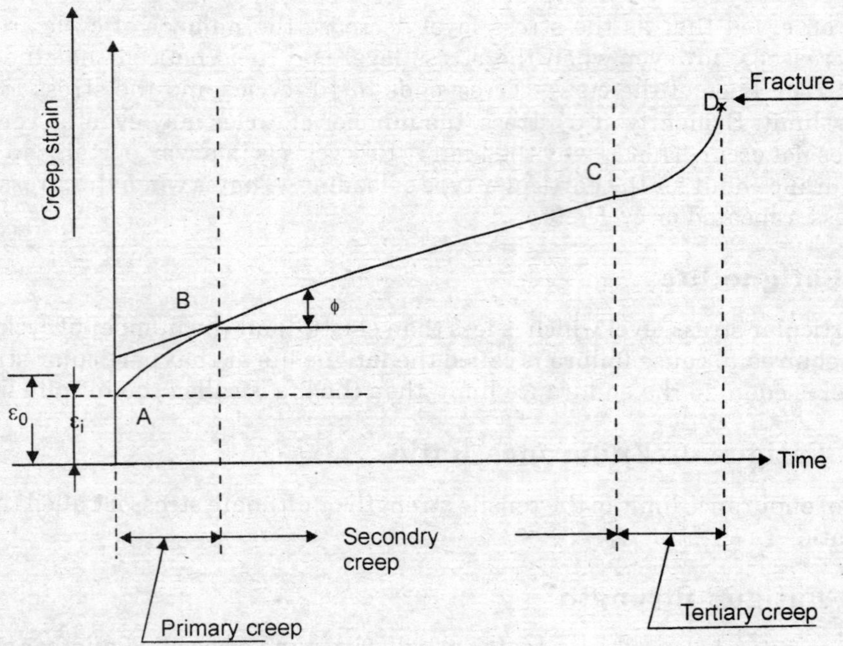

Fig. 11.26

Then at any time t, with in the secondary creep range the total creep strain ε_t.

$$\varepsilon_t = \varepsilon_0 + t \tan \phi$$

$$\simeq \varepsilon_0 + t\,\phi \qquad ...(11.51)$$

where $\varepsilon_0 = OE$

= Intercept on y-axis of the line CB produced

(*c*) Again primary creep or cold creep is a transient phenomenon as the work hardening of the material due to deformation, starts side by side and therefore the primary creep has a decreasing rate *i.e.* the slope of the curve AB goes on decreasing. Because of its nature it is also called the transient creep.

(*d*) Secondary creep curve BC is a straight line and has steady rate or constant slope. Therefore it is also called steady state or hot or viscous creep. The slope remains constant due to the equilibrium attained between the work hardening processes and annealing effect due to high temperature.

(*e*) **Tertiariy Creep** or Accelerated creep. This is the final stage where the creep is at an accelerating rate leading to fracture which represents the process of progressive damage which ends up into a fracture.

(*f*) **Creep Ductility :** Elongation at rupture, of the creep specimen is known as creep ductivity.

(*g*) **Creep Strength :** Is defned as the highest stress which the material can withstand for a specified length of time without excessive strain.

(*h*) **Creep Rupture Strength :** Is defined as the highest stress which a material can withstand for a specified length of time without rupture.

(*i*) **Creep Limit :** It is the maximum stress which a metal can sustain indefinitely without deforming which is generally taken as 0.001% strain after between the 25th and 35th hours and 0.2% strains after 45 hours.

I1.13.13 Brittleness

Brittleness is the property of a material that permits it to be slightly deformed without rupture. No material is perfectly brittle, that is capable of no deformation before rupture. Therefore brittceness is a relative term, glass being one of the most brittle material. Brittle materials have relatively short stress-strain curves the common structural materials, cast iron , brick and stone are brittle in comparison with steel.

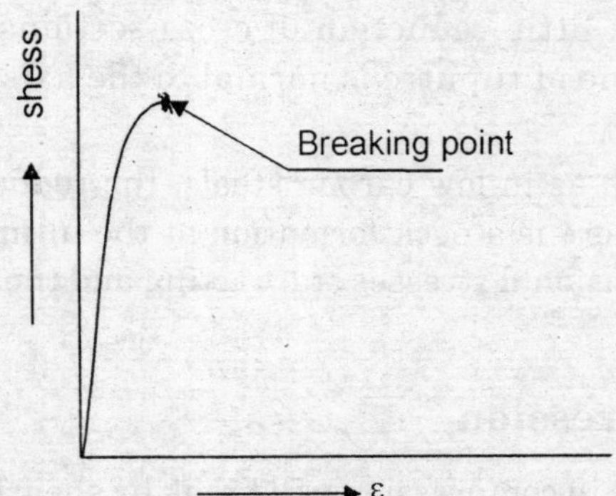

Fig. 11.27. **Stress-strain curve of brittle material**

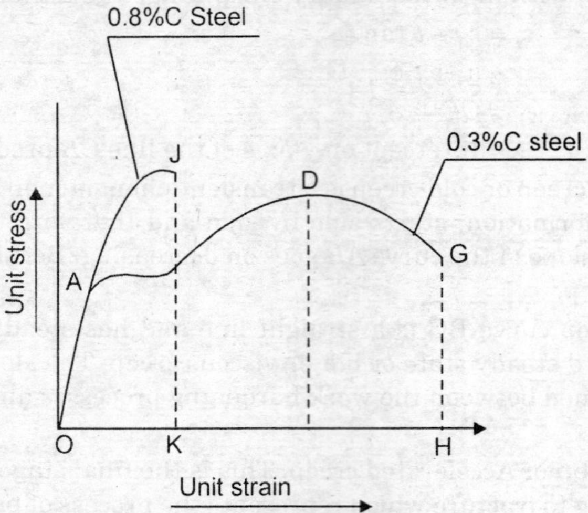

Fig. 11.28. **Thoughness comparison.**

11.13.14 Toughness

Toughness is ability of the material to withstand high unit stress together with great unit strain without complete fracture. The area *OAGH* or *OJK*, under the curve of stress-strain (Fig. 11.28) is the measure of the toughness of the material.

11.13.15 Difference between Thoughness and Ductility

The distinction between toughness and ductility is that ductility deals only with the ability to deform and is represented by the strain at instability point, whereas toughness considers both the ability to deform and the stress developed during deformation that is the energy content.

11.14 TYPES OF FRACTURE

11.14.1 Under Tension

A rod of brittle material, as of cast iron will rupture in a tension test in a clean sharp fracture with very little reduction of cross-sectional area and very little elongation. The plane of ruputre is normal to the axis of the load (Fig. 11.30 a).

In a ductile material, as in low carbon steels, the reduction of area and elongation are greater. There is a neck formation at the ultimate stress point, at which point three dimensional stresses start acting and the fracture is in the form of cup and cone.

(a) (b)

Fig. 11.29

11.14.2 Under Compression

A prism of brittle material, in compression will break by shearing along oblique planes. Greater the brittleness of the material, the more nearly will these planes be parallel to the direction of applied force Fig. 11.30 (*a*) and 11.30 (*b*) and 11.30 (*c*) for brick, concrete and timber respectively in descending order of brittleness In Fig. 11.30 (*d*) is shown the deformation of lead (or aluminium or mild steel) a plastic material which flattens out under load, and ends up into a barrel with crack, all over the circumference, parallel to the line of the loading.

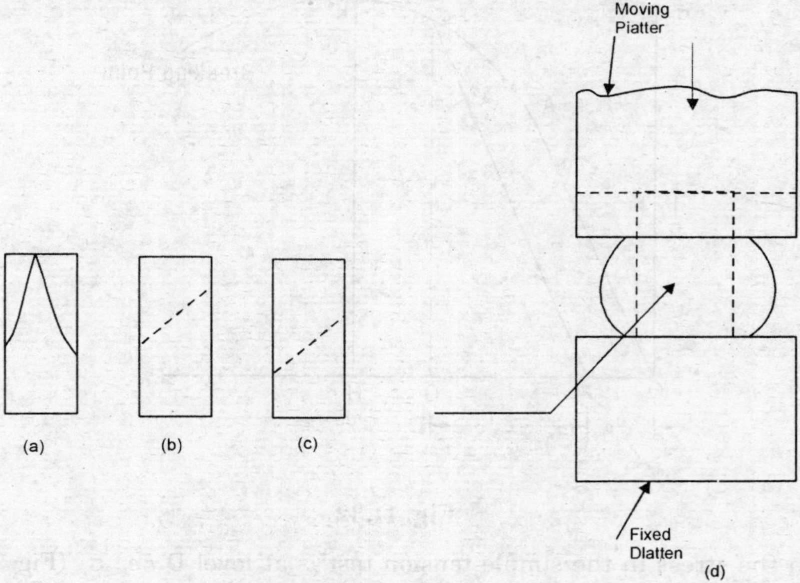

Fig. 11.30

11.14.3 Under Twisting Moment

(a) When a ciruclar shaft of ductile material like low carbon steel is subjected to twisting moment, the shearing along the cross-sectional plane, normal to the axis of the twisting moment, will take place Fig. 11.32 (*a*).

(*b*) In the case of brittle material like *C.I.* put to Twisting Moment the fracture takes place along the helix inclined to the cross-sectional. plane at 45°. The failure is actually due to tensile stress acting normal to plane which is inclined to cross-sectional plane at 45°. Fig. 11.32(*b*).

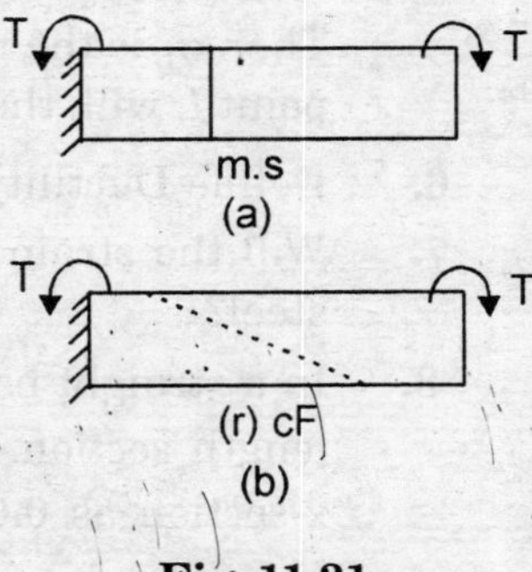

Fig. 11.31

REVIEW QUESTIONS

1. Define the terms, "stress", "pressure".
2. Explain the difference between nominal strain and True Strain in a simple Tension Test.
3. Explain the difference between Nominal stress and True Stress in a uniaxial stress, test price.
4. State whether it is true or false : "If the tensile test piece while in the plastic range of the curve, is unloaded it follows a straight path parallel to the elastic line.

   [Ans. True]

5. (*a*) State whether true or false :

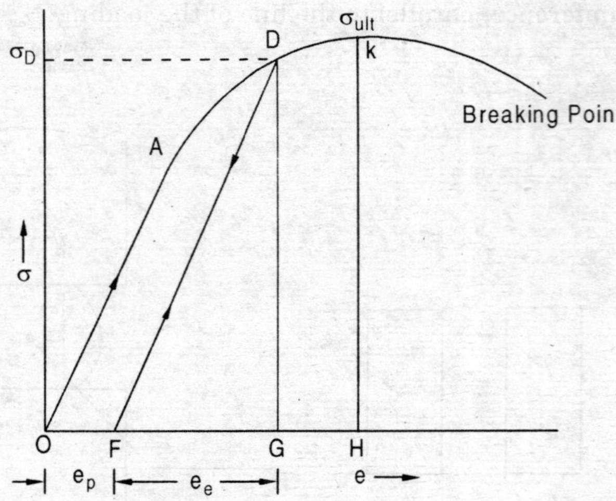

Fig. 11.32

When the stress in the simple tension test is at level D *i.e.,* σ_D (Fig. 11.32) and the specimen is unloaded to no stress state the left over strain OF is equal to the plastic strain component of the total strain of D and FG is the elastic strain component or elastic recovery. [**Ans.** True]

(*b*) If the test piece is reloaded or restressed from the point *F* it follows the path *FD*. Then σ_D is the new yield stress. Similarly if this process is repeated with reference to point *L* will, then the new yield point corrspond to σ_L? [Yes]

6. Define Ductility and Malleability.

7. Will the strain at Instability point in the case of Aluminium be more then that for steel? [Yes]

8. In a straight bar two equal and opposite forces of 100 kN each are acting at the mid length section. What is stress in the cross section at ¾th length. Take the area of x-section as 0.01 m^2. [**Ans.** Zero]

9. (*a*) State *D'* Alemberts Principle.

(*b*) Can a body in linear motion be in a state of stress ? State yes of No.

If yes, give example. [yes]

10. Can the equilibrium equations for a body in static equilibrium be applied to the body in motion? If so explain the phenomenon. [yes]

11. State Hookes law.

12. Define proof stress.

13. Define quantitatively the terms, ductility and malleability.

14. What is meant by "Homogeneous materials" and "Isotropic materials"?

15. (*a*) A copper tube 25 mm internal diameter and 30 mm external diameter surrounds a 20 mm steel bolt Fig. (11.33). After putting rigid washers *A* and *B* on two sides of the tube the nut is just tightened with hand so that the bolt head side washer is in contact with end *C* of the tube and other washer is just in contact with end *D* of the tube. The nut is further tightened so that it rotates ¼th of a rotation with respect to the bolt head. If the pitch of the threads is 1.5 mm find the stresses induced in bolt and copper tube. Take E_S = 200 × 10^3 MPa and E_C = 90 × 10^3 MPa. Take length of tube = 300 mm.

(*b*) The washers *A* and *B* are then pulled apart with the help of wire ropes. Find the value of the pull such that the stress in copper take becomes zero. What is then the stress in steel bolt ?

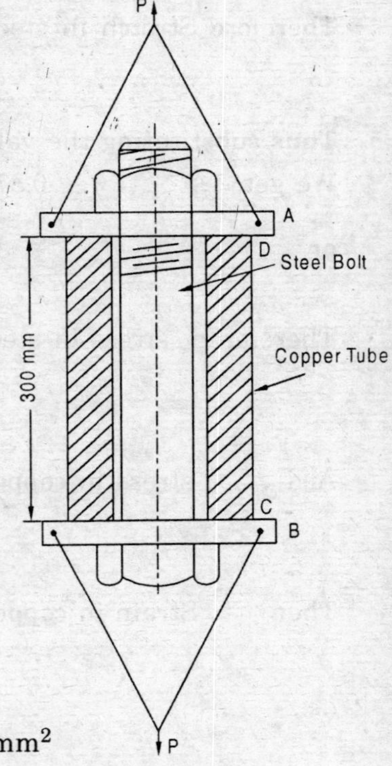

Fig. 11.33

Solution : Area of cross-section of bolt

$$A_S = \frac{\pi}{4} \times (20)^2 = 100\pi \text{ mm}^2 = 314.159 \text{ mm}^2$$

Area of cross-section of the tube

$$A_C = \frac{\pi}{4}(30^2 - 25^2)$$

$$= \frac{\pi}{4} \times 275 = 215.98 \text{ mm}^2$$

Free movement of nut with respect of head

$$= \text{Pitch} \times \text{no of terms}$$

$$= 1.5 \times \frac{1}{4} = 0.375 \text{ mm}$$

$$= \text{compression in copper} + \text{stretch in steel} \qquad ...(A)$$

If *W* is the axial compressive load in copper

Then compression in Cu = $\varepsilon_C l_C$

$$= \left(\frac{W}{A_C}\right)\frac{l_C}{E_C}$$

$$= \frac{W}{215.98} \times \frac{300}{90 \times 10^3} \text{ mm}$$

$$= 0.0154 \times 10^{-3} \text{ W mm}$$

At equilibrium tensile load in steel = compressive load in Cu = W

Therefore Stretch in steel $= \dfrac{W}{A_S}\dfrac{1}{E_S}l_s = \dfrac{W}{314.119} \times \dfrac{1}{200 \times 10^3} \times 300$ mm

$= 0.00477 \times 10^{-3}$ W mm

Thus substituting the values in equation (*A*)

We get $0.375 = 0.154 \times 10^{-3}$ W + $.00477 \times 10^{-3}$ W

or $W = \dfrac{0.375 \times 10^3}{0.02017} = 18.59 \times 10^3\text{N} = 18.59$ kN

Thereforeσ_s stress in steel $= \dfrac{W}{A_S} = \dfrac{18.59 \times 10^3}{314.159} = 0.0592 \times 10^3$ N/mm²

$= 59.2$ MPa (Tensile)

And stress in copper $= \sigma_c = \dfrac{W}{A_c} = \dfrac{18.59 \times 10^3}{215.98} = 0.08614 \times 10^3$ N/mm²

$= 86.14$ MPa (comp.)

Then Strain in copper $= \dfrac{\sigma_C}{E_C} = \dfrac{86.14 \text{ MPa}}{90 \times 10^3 \text{ MPa}}$

i.e., $\varepsilon_c = \dfrac{86.14}{90 \times 10^3} = 0.957 \times 10^{-3}$

$\delta l_c = 300 \times 0.957 \times 10^{-3}$

$= 0.287$ mm

Solution : (*b*) In the part (*b*) the load *P* will act as pull only on the bolt. Its value is to be found such that the resultant stress in copper is zero. In other words if the bolt stretches further by an amount 0.287 mm then resultant compression stress in copper will become zero because when additional stretch in steel equals 0.287 mm at that time the washers will just leave contact with tribe.

Thus additional stress in steel will be due stretch of 0.287 mm

$\delta'_S = \varepsilon'_S \, l_s$

i.e., $\varepsilon'_S = \dfrac{\delta'_S}{l_s} = \dfrac{0.287}{300}$

Therefore $\sigma'_S = \varepsilon'_S E_S = \dfrac{.287}{300} \times 200 \times 10^3$ MPa

$= 191.33$ MPa (tensile)

That is additional load $P'_S = \sigma'_S A_s = \left(\dfrac{0.287}{300} \times 200 \times 10^3\right) 314.159$

$= 60 \times 10^3$ N $= 60$ kN

Thus we conclude that 60 kN pull applied in the bold will bring the stress in copper to zero.

Then the resultant stress in steel bolt

$= 59.2 + 191.33 = 250.53$ MPa (tansile)

16. A brass tube 20 mm internal dia and 20 mm external diameter surrounds a 18 mm shank of a steel bolt. After putting rigid washers on the two sides of the tube the nut is rotated such that it is just tight and no load is exerted on the tube by the washers. The length of the tube is 200 mm. The temperature is then raised to 270°C. If the room temperature is 20°C, find the increase in the stresses in brass and steel.

Take $\alpha_b = 18.7 \times 10^{-6}/°C$

$\alpha_s = 11.6 \times 10^{-6}/°C$

$E_b = 100$ GPa

$E_S = 200$ GPa

Solution : Rise in temperature = t = 220 – 20 = 200°C

Free expansion of brass = $l_b \alpha_b t = \delta_b$

or $\delta_b = 200 \times 18.7 \times 10^{-6} \times 200 = 0.748$ mm

Free expansion of steel = $l_S \alpha_S t = \delta_S$

or $\delta_S = 200 \times 11.6 \times 10^{-6} \times 200 = 0.464$ mm

Because of this difference in the free expressions steel will get pulled up while brass will get compressed due to temperature rise.

Thus if σ_S and σ_b are the stresses induced in two materials then

$$(\delta_b - \delta_s) = \frac{\sigma_b}{E_b} l_b + \frac{\sigma_s}{E_S} l_s$$

But $l_b = l_s$ $$(0.748 - 0.464) = \left(\frac{\sigma_b}{E_b} + \frac{\sigma_S}{E_S}\right) l_S$$

or $$0.284 = \left[\frac{\sigma_b}{E_b} + \frac{\sigma_S}{E_S}\right] 200$$

or $$\frac{\sigma_b}{E_b} + \frac{\sigma_S}{E_S} = \frac{.254}{200} = 0.00142 \qquad ...(A)$$

Also at equilibrium $$P_S = P_b$$

$$\sigma_S A_S = \sigma_b A_b$$

or $$\sigma_S = \left(\frac{A_b}{A_S}\right) \sigma_b \qquad ...(B)$$

But $$A_b = \frac{\pi}{4}(25^2 - 20^2)$$

$$= \frac{\pi}{4} \times 227 = 176.71 \text{ mm}^2$$

$$A_S = \frac{\pi}{4} \times 18^2 = 254.469 \text{ mm}^2$$

$\therefore$ $$\sigma_S = \frac{176.71}{254.469} \times \sigma_b = 0.6938\ \sigma_b$$

Thus from equations (*A*) and (*B*) we get

$$\frac{\sigma_b}{(100 \times 10^3)} + \frac{0.6938\,\sigma_b}{(200 \times 10^3)} = 0.00142$$

or $$\sigma_b\,[2 + 0.6938] = 0.00142 \times 200 \times 10^3$$

$$\sigma_b = \frac{0.00142 \times 200 \times 10^3}{2.6938} = 0.528 \text{ MPa}$$

Then $$\sigma_S = 0.6938\,\sigma_b$$

$$= 0.6938 \times 0.528 = 0.3663 \text{ MPa}$$

PROBLEMS

1. A rigid cross bar is supported horizontally by two vertical bars, *A* and *B* of equal length and hanging from their tops. The bars *A* and *B* are 0.6 m apart. The cross bar stays horizontal even after a vertical force of 6000 *N* is applied to it at a point 0.4 m from *B*. If the stress in *A* is 200 MN/m², find the stress in B and the areas of cross section of the two rods.

Take $E_A = 200$ GN/m²

and $E_B = 130$ GN/m² (**Ans.** : 130 MN/m², 20.0 mm² ; 15.385 mm²)

2. A rigid bar *AD*, 1 m long, is hinged at *A* and is held horizontal by supporting it at *B* and *C* as shown in the Fig. 11.34 The area of cross section of the steel rod *BE* is 16 mm² and that of copper rod *CF* is 12 mm². The bar *AD* is acted upon by a vertical force of 500 N at *D*. Find the stresses induced in the rods *BE* and *CF*. Take

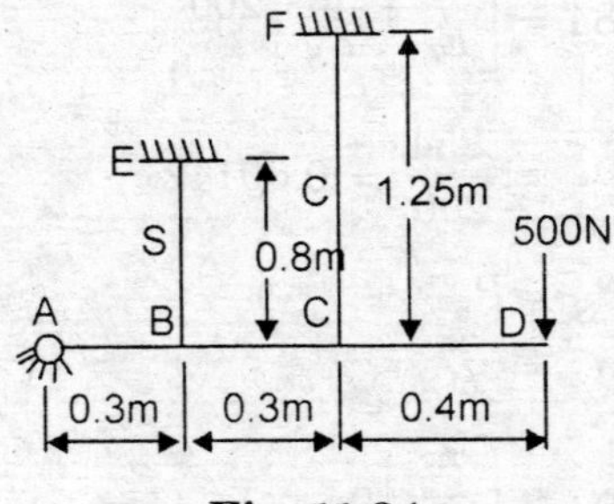

Fig. 11.34

$E_S = 200$ GN/m²

and $E_C = 90$ GN/ m² . (**Ans.** 56.0 MN/m², 32.2 MN/m²)

3. A steel rod, 20 mm diameter, passes centrally through a steel tube 25 mm internal diameter and 40 mm external diameter. The tube, 750 mm long, is closed by two rigid washers of negligible thickness which are fixed onto the tube by tightening the nuts on the threaded rod. If the pitch of the threads is 2.5 mm, find the increase in the stresses in the tube and the rod when one nut is turned through one quarter of a turn with respect to the nut on the other end. Take

$E_{steel} = 205$ GN/m² . (**Ans.** 49.8 MN/m² ; 121.2 MN/m²)

4. A cast iron bracket as shown in Fig. 11.35 is held fast to the foundation with the help of 4 steel bolts. Find the increase in tensile stresses of the four bolts due to the load P of magnitude 1000 N. The bolts are all 12 mm in diameter.

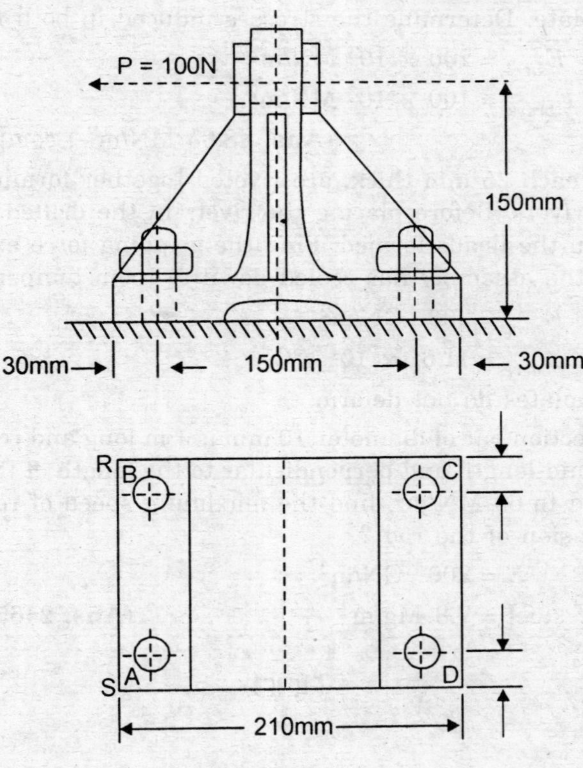

Fig. 11.35

(**Ans.** For A and B 0.596 MN/m^2 ; For C and D 3.576 MN/m^2)

[**Hint** : The bracket will have a tendency of turning about the line RS and the bolts will have a tendency to stretch proportional to their distances from RS].

5. A 25 mm diameter brass bar is enclosed in a steel tube of 25 mm internal diameter and 50 mm external diameter. Both of them are 1m long at room temperature and fastened rigidly to each other at the ends.

If the room temperature is 20°C, find to what temperature the assembly should be heated so as to generate a compressive stress of 48.7 MN/m^2 in brass. What is then the stress in steel ?

If the composite bar is then pulled with a force of 50000 N, find the resulting stresses in both the metals.

Take $E_{steel} = 200$ GN / m^2

$E_{brass} = 100$ GN / m^2

$\alpha_{steel} = 11.6 \times 10^{-6}$ /°C ; $\alpha_{brass} = 18.7 \times 10^{-6}$/°C .

(**Ans.** 80°C, 16.23 MN/m^2 tensile ; 34.17 MN/m^2 comp ; 45.3 MN/m^2 tensile)

6. A 28 mm diameter steel bar, 0.4 long, is placed centrally within a brass tube having an inside diameter of 30 mm and outside diemeter of 40 mm. The bar is shorter in length than the tube by 0.12 mm. While the bar and the tube are held vertically on a rigid horizontal platform, a compressive force of 60000 N is applied at the top of the tube through a rigid plate. Determine the stresses induced in both the bar and the tube.

Take $E_{steel} = 200 \times 10^3$ MN/m^2

$E_{brass} = 100 \times 10^3$ MN/m^2.

(**Ans.** 48.85 MN/m^2 ; comp. 54.42 MN/m^2 comp.)

7. Two steel plates, each 25 mm thick, are riveted together forming a lap joint by using 25 mm dia steel rivets. Before placing the rivets in the drilled holes, they are heated to 600°C and then the heads formed. Find the gripping force exerted by each rivet on the plates when the assembly has cooled down to room temperature of 20°C.

Take $E_{steel} = 200$ GN/m^2

$\alpha_{steel} = 11.6 \times 10^{-6}$/°C .

Assume that the plates do not deform. (**Ans.** 660.52 kN)

8. A uniform cross section bar of diameter 10 mm is 1 m long and revolving at a high speed about an axis at mid-length and perpendicular to the length of the bar. If the maximum stress is restricted to 65 MN/m^2, find the maximum speed of rotation. What will then be the total extension of the rod ?

Take $E = 200$ GN/m^2

Density of steel = 7.8 Mg/m^3. (**Ans.** 2465.7 r.p.m. ; 0.2167 mm)

□□□

CHAPTER 12

RELATION BETWEEN ELASTIC CONSTANTS

12.1 INTRODUCTION

By elastic constants of a material we mean four constants of the material namely, youngs modulus or modulus of elasticity E, modulus of rigidity or shear Modulus G, Bulk Modulus K and Poisson's ratio ν.

While discussing the Hooke's Iaw the Youngs Modulus E has been introduced in Chapter 11. It deals with the direct stress or normal stress and the corresponding direct strain. The stress and the corresponding strain may be tensile or compressive. In case it is tensile stress it will give rise to positive strain or alternatively the dimension will increase, (+ sign) . Similarly in the case of compressive stress the dimension along the direction of the stress will decrease (– ve sign). Accordingly, for tensile and compressive direct stresses, the young modulus of elasticity, E, had been introduced as per Hookes Law. Similarly for the shear stresses, the modulus of rigidity or shear modulus G, had been defined and introduced. In the following paragraphs other two constants namely the Poisson's ratio and Bulk modulus are one introduced and the relationship between these constants has been derived.

12.2 POISSON'S RATIO

When a body is acted upon by a direct stress, tensile or compressive, the dimensions will increase or decrease along the line of action of the stress applied. The change in length per unit length along this direction is known as longitudinal or primary strain. Accompanied with the strain in the longitudinal direction there are strains in the lateral directions (*i.e.* along both y and z directions if the primary strain is along x-axis) known as lateral or secondary strains but these strains are of the opposite sign. That is if their is tensile stress σ_x acting the x-axis then there will be positive strain, in the primary direction and negative strains in the lateral directions and vice versa. For a material stressed within elastic limit, thé ratio of the magnitudes of lateral strains to primary strain is constant. This ratio is known as Poisson's ratio which is represented by ν.

Thus if ε_x is the strain along the direction of the stress σ_x (Fig. 12.1), which is along *x*-axis or the longitudinal axis of the circular bar *AB*, then longitudinal strain ε_x, along *x*-axis is given by Hooke's Law.

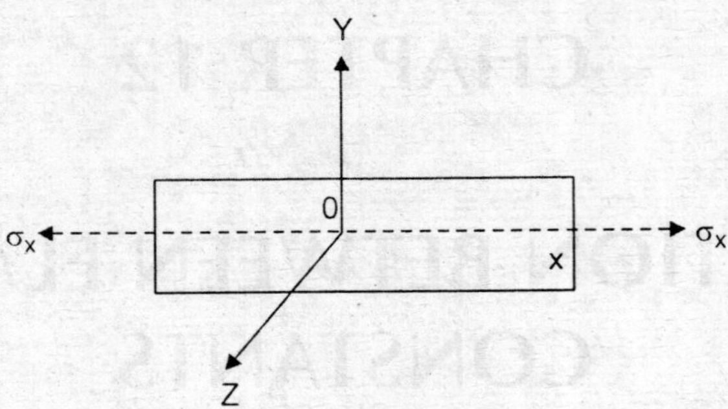

Fig. 12.1

$$\varepsilon_x = \frac{\sigma_x}{E} \qquad \text{...(12.1)}$$

when E = Youngs modulus of the material and if along the lateral directions *YY* and *ZZ* the lateral strains are ε_y and ε_z. These lateral strains, will be negative if ε_x is positive.

The poison's ratio ν is given by

$$\nu = \frac{\text{(Lateral strain)}}{\text{(Longitudinal strain)}}$$

$$\nu = \left(\frac{-\varepsilon_y}{\varepsilon_x}\right) = -\frac{\varepsilon_y}{\varepsilon_x} \qquad \text{...(12.2 a)}$$

Also $$\nu = \left(\frac{-\varepsilon_z}{\varepsilon_x}\right) \qquad \text{...(12.2 b)}$$

or $$\varepsilon_y = -\nu\,\varepsilon_x \qquad \text{...(12.3 a)}$$

and $$\varepsilon_z = -\nu\,\varepsilon_n \qquad \text{...(12.3 b)}$$

From equation (12.2) we observe that ν is – ve but this negative sign is not reported in the data books where only the numerical value is recorded for different materials. But certainly while calculating the lateral strains the negative sign must be taken into account.

12.2.1 Volumetric Strain

Consider a rectangular parallel piped *ABCDEFGH*, (Fig. 12.2) under the action of three direct stresses, σ_x, σ_y and σ_z acting along the three axes, *xx*, *yy* and *zz* normal to the plane, *ABCD*, *ABGF* and *ADEF* respectively.

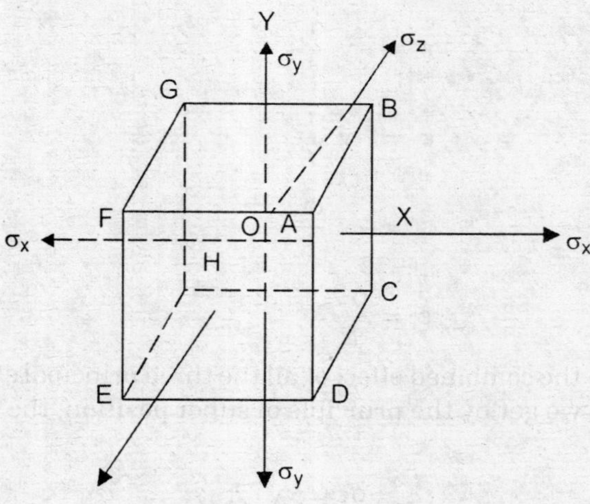

Fig. 12.2

On these planes there is no shear stress acting that is these planes have only normal or direct stresses acting on them. Those planes on which there is no shear stress are called principal planes. The corresponding direct stresses on these planes are called the principal stresses. Let the youngs modulus or modulus of elasticity and Poisson's ratio of the material are E and ν respectively.

Due to the tensile stress σ_x there will be longitudinal strain ε_n along the x-axis which will be equal to $\frac{\sigma_x}{E}$. Simultaneously, along the lateral directions yy and zz there will be contraction and the corresponding lateral strains will be $(-\gamma\varepsilon_x)$ along both YY and ZZ axes.

i.e. $$\varepsilon_y = -\nu\,\varepsilon_x = -\nu\frac{\sigma_x}{E}$$

and $$\varepsilon_z = -\nu\,\varepsilon_x = -\nu\frac{\sigma_x}{E}$$

Thus due to σ_x, σ_y and σ_z the strains along the three directions are as under

Due to σ_x $$\varepsilon_x = \frac{\sigma_x}{E}$$

$$\varepsilon_y = -\nu\frac{\sigma_x}{E}$$

$$\varepsilon_z = -\nu\frac{\sigma_x}{E} \qquad \text{...(12.4 } a\text{)}$$

Due to σ_y $$\varepsilon_x = -\nu\frac{\sigma_y}{E}$$

$$\varepsilon_y = \frac{\sigma_y}{E}$$

$$\varepsilon_z = -\nu \frac{\sigma_y}{E} \qquad \text{...(12.4 a)}$$

Due to σ_z

$$\varepsilon_x = -\nu \frac{\sigma_z}{E}$$

$$\varepsilon_y = -\nu \frac{\sigma_z}{E}$$

$$\varepsilon_z = \frac{\sigma_z}{E} \qquad \text{...(12.4 a)}$$

Therefore due to the combined effect of all the three principal stresses acting simultaneously on the parallel piped, we get by the principle of super position, the strains in the three direction as

$$\varepsilon_x = \frac{\sigma_x}{E} - \frac{\nu}{E}(\sigma_y + \sigma_z) \qquad \text{...(12.5 a)}$$

$$\varepsilon_y = \frac{\sigma_y}{E} - \frac{\nu}{E}(\sigma_z + \sigma_x) \qquad \text{...(12.5 b)}$$

$$\varepsilon_z = \frac{\sigma_z}{E} - \frac{\nu}{E}(\sigma_x + \sigma_y) \qquad \text{...(12.5 c)}$$

Let x, y and z be the original lengths of the three sides of the parallel piped along the xx, yy and zz directions respectively

Then volume V is given by

$$V = xyz \qquad \text{...(12.6)}$$

Again due to the three stresses σ_x, σ_y and σ_z let three sides increase respectively by dx, dy and dz due to which the corresponding increase in volume be dV. Then differentiating equation (12.6) w.r.t. V we get

$$1 = xy\frac{dz}{dV} + yz\frac{dx}{dV} + zx\frac{dy}{dV}$$

or

$$dV = xy\,dz + yz\,dx + zx\,dy \qquad \text{...(12.7)}$$

Dividing equation (12.7) with equation (12.6) we get

$$\frac{dV}{V} = \frac{dx}{x} + \frac{dy}{y} + \frac{dz}{z} \qquad \text{...(12.7)}$$

But

$$\frac{dV}{V} = \text{Volumetric strain} = \varepsilon_V$$

$$\frac{dx}{x} = \text{Strain along x direction} = \varepsilon_x$$

Similarly

$$\frac{dy}{y} = \varepsilon_y$$

or

$$\frac{dz}{z} = \varepsilon_z$$

Therefore equation (12.7) reduces to

$$\varepsilon_V = \varepsilon_x + \varepsilon_y + \varepsilon_z \qquad ...(12.8)$$

Therefore substituting equation (12.5) into equation (12.8) we get

$$\varepsilon_V = \left[\frac{\sigma_x}{E} - \frac{\nu}{E}(\sigma_y + \sigma_z)\right] + \left[\frac{\sigma_y}{E} - \frac{\nu}{E}(\sigma_z + \sigma_x)\right] + \left[\frac{\sigma_z}{E} - \frac{\nu}{E}(\sigma_x + \sigma_y)\right]$$

$$= \frac{1}{E}(\sigma_x + \sigma_y + \sigma_z) - \frac{2\nu}{E}(\sigma_x + \sigma_y + \sigma_z)$$

$$\varepsilon_V = \frac{dV}{V} = \frac{(1-2\nu)}{E}(\sigma_x + \sigma_y + \sigma_z) \qquad ...(12.9)$$

12.2.2 HYDROSTATIC STRESS AND BULK MODULUS AND RELATION BETWEEN *E*, *K* AND υ

As special case when the parallel piped in Fig. (12.2) is acted in all the directions by the same stress σ

i.e. when $\sigma_x = \sigma_y = \sigma_z = \sigma$

then σ is known as hydrostatic stress as it is synonymous with hydrostatic pressure p acting on it when the body is immensed deep in water.

Then in that case

$$\frac{dV}{V} = \varepsilon_V = \frac{(1-2\nu)}{E}(3\sigma) \qquad ...(12.10)$$

The ratio $\dfrac{\sigma}{\left(\dfrac{dV}{V}\right)}$ *i.e.* hydrostate static stress divided by volumetic strain is called the Bulk Modulus of the material represented by K and is constant for any material

i.e. $$K = \frac{\sigma}{\left(\dfrac{dV}{V}\right)} \qquad ...(12.11)$$

Therefore from (12.10) and (12.11) we get

$$E = 3(1-2\nu)\frac{\sigma}{\left(\dfrac{dV}{V}\right)}$$

or $$E = 3(1-2\nu)\,K \qquad ...(12.12)$$

Equation (12.12) gives the relationship between the youngs modulus, Poisons ratio and back modulus.

12.3 COMPLEMENTARY SHEAR STRESSES

Consider a rectangular block of length x, height y and thickness z under the action of shear stresses τ and τ_1 as shown in Fig. 12.3 *i.e* on the plane *BCHG* normal to x axis the shear stress is τ and that on the plane normal to y axis ($ABGF$) is τ_1. For the static equilibrium of the block and equating

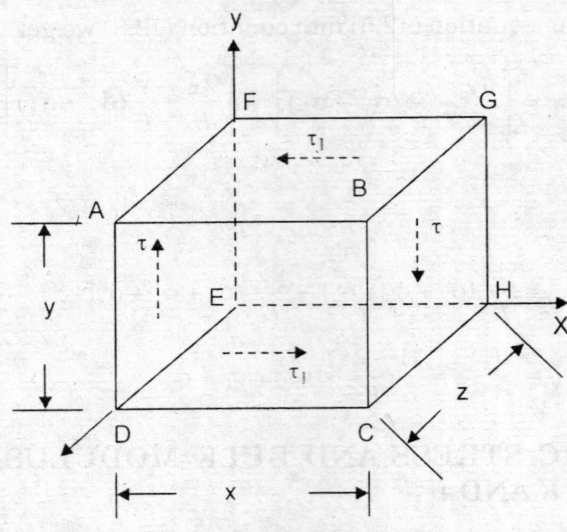

Fig. 12.3

$$JSM = 0$$

i.e. Algebraic of the movements due to the forces must be zero.

Let the corresponding forces due to the shear stress t and t_1 he P and P_1

Then $\quad P = \tau\,(\text{area } BCHG) = \tau \times (yz)$ ...(12.13)

and $\quad P_1 = \tau_1\,(\text{area } DEHC) = \tau_1\,(xz)$ ...(12.14)

Taking moments due to forces P acting on faces $BGHC$ and $DEFA$ at a distance x which are equal and opposite

$$= P \times x \text{ (clockwise).}$$

Similarly the moment of faces P_1 equal and opposite acting of faces $ABGF$ and $DCHE$.

$$= P_1\,y \text{ (anticlockwise)}$$

$\therefore$ For $\quad \Sigma M = 0$

$$P_x - P_1 y = 0$$

or $\quad (\tau\, yz)\, x = (\tau_1\, xz)\, y$

or $\quad \tau = \tau_1$ ...(12.15)

Equation ((12.15) shows that a shear stress τ must be accompanied by another shear stress τ_1 of equal magnitude but of opposite sense, acting on planes at right angles to the planes of original shear stress τ. Then shear stress τ_1 is called the complementary shear stress. Thus system is known as system under pure shear stress.

12.3.1 Cube Under Pure Shear

Consider a parallel piped $ABCOEPG$ of square section under the action of complementary shear stress as shown in the Fig. 12.4 and it is in static equilibrium. If the entire block is in static equilibrium, then a part of it will also be in equilibrium. Therefore taking a section along the diagonal plane $ACHF$ and assuming the direct and shear stress in the plane $ACHF$ of which the side view is shown in Fig. (12.4 b), are σ_{n_1} and τ_1 resolving the faces along the normal to the face AC we get

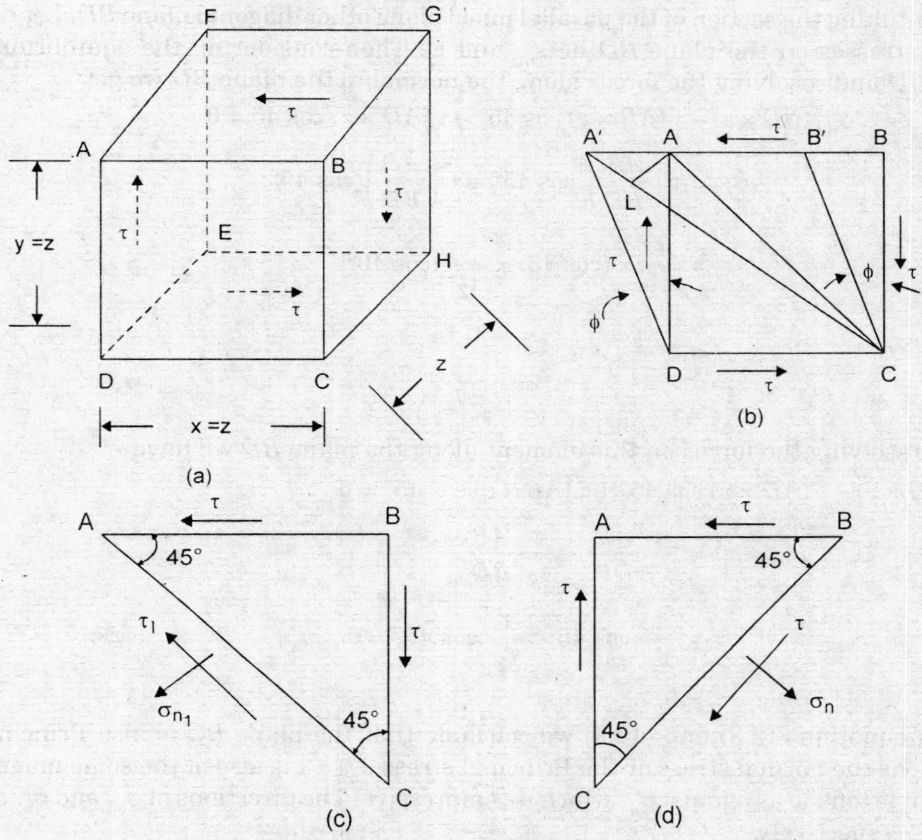

Fig. 12.4

$$\sigma_{n_1}(AC \times z) + \tau(BC \times z)\cos 45° + \tau(AB \times z)\cos 45° = 0$$

or $$\sigma_{n_1} = -\tau\frac{BC}{AC}\cos 45° - \tau\frac{BC}{AC}\cos 48°$$

$$= -\tau(\cos 45)^2 - \tau(\cos 45)^2$$

$$= -2\tau\left(\frac{1}{2}\right) = -\tau$$

That is $$\sigma_{n_1} = -\tau \quad ...(12.16)$$

Similarly resolving the forces along the plane CA

$$\tau_1(CA \times z) + \tau(AB \times z)\cos 45 - \tau(BC \times z)\cos 45° = 0$$

Since $AB = BC$ being side of square it reduces to

$$\tau_1(CA \times z) = 0$$

or $$\tau_1 = 0 \quad ...(12.17)$$

From equation (12.16) and (12.17) we conclude that the plane AC is a Principal plane on which the direct stress or Principal stress is $-\tau$ *i.e.* compressive in nature and of magnitude τ which is the magnitude of the applied shear stresses on the planes AB and BC as shown.

Again taking the section of the parallel piped along other diagonal plane BD. Let the direct and shear stresses on the plane BD he s_{n2} and t_2. Then considering the equilibrium of the element ABD and resolving the forces along the normal to the plane BD we get

$$\sigma_{n2}(BD \times z) - \tau(AB \times z)\cos 45° - \tau(AD \times z)\cos 45 = 0$$

or $$\sigma_{n2} = \tau\left(\frac{AB}{BD}\right)\cos 45° + \tau\left(\frac{AD}{BD}\right)\cos 45°$$

$$= \frac{\tau}{\tau_2} \times (\cos 45) \times \frac{\tau}{\tau_2}(\cos 40°$$

$$= 2\frac{\tau}{\tau_2} = \tau$$

or $$\tau_{n2} = \tau \qquad ...(12.18)$$

Now resolving the forces on this element along the plane BD we have

$$\tau_2(BD \times z) - \tau(AD \times z)\cos 45° + \tau(AB \times z)\cos 45° = 0$$

or $$\tau_2 = \frac{\tau}{\tau_2}\frac{AB}{DB} - \frac{\tau}{\tau_2}\frac{AB}{BD}$$

$$= \frac{\tau}{\tau_2}\cos 45 - \frac{\tau}{\tau_2}\cos 45° = 0$$

i.e. $$\tau_2 = 0 \qquad ...(12.19)$$

From equation (12.8) and (12.9) we conclude that the plane BC is also Principal plane having $+\tau_x$ as the normal stress or the Principal stress $\sigma_{n2} = \tau$ is also of the some magnitude as $\sigma_{n1} = \tau$ but it is tensile as against σ_{n1} which is compressive. The directions of σ_{n1} and σ_{n2} are along DB and AC respectively.

$\therefore$ $$\text{Strain along } CA = \frac{\sigma_{n_2}}{E} - \nu\frac{\sigma_{n_1}}{E}$$

or $$\varepsilon_{CA} = \frac{\tau}{E} - \frac{\nu(-\tau)}{E} = \frac{\tau}{E} + \frac{\nu\tau}{E}$$

or $$\varepsilon_{CA} = \frac{\tau}{E}(1 + \nu) \qquad ...(12.20)$$

And $$\text{Strain along } BD = \frac{\sigma_{n_1}}{E} - \nu\frac{\sigma_{n_2}}{E}$$

$$= \frac{(-\tau)}{E} - \frac{\nu}{E}\tau$$

$$= -\frac{\tau}{E}(1 + \nu) \qquad ...(12.11)$$

Again considering the effect of shearing stresses the parallel piped. The shear stress will distort the square section to shape $A'B'CD$ (if we assume the force AB moves relative to CD) Then angle $A'DA'$ will be the shear strain angle f as per its definition

$$\tan\phi = \frac{AA'}{AD} \simeq \phi \text{ (as } \phi \text{ is small)} \qquad ...(12.21)$$

which G is shear modulus or modulus of rigidity

Also if AC is perpendicular on CA'

Shear $$AA' = \frac{AL}{\cos(AA'L)}$$

Now $$\angle AA'L \simeq 45°$$

Also $$\angle ACD = 45°$$

$\therefore$ $$LA' = AA' \cos 45 = \frac{AA'}{\sqrt{2}}$$

Now $$\varepsilon_{CA} = \frac{AL - AC}{AC} = \frac{A'L}{DA\sqrt{2}} = \frac{AA'}{\sqrt{2}} \times \frac{1}{DA\sqrt{2}}$$

$$= \frac{AA'}{2\,DA}$$

$$= \frac{\phi}{2} \quad ...(12.21)$$

$$\varepsilon_{CA} = \frac{1}{2}\left(\frac{\tau}{G}\right) \quad ...(12.22)$$

Therefore from equations (12.10) and (12.13)

We get $$\varepsilon_{CA} = \frac{\tau}{E}(1 + \nu) = \frac{\tau}{2G}$$

or $$E = 2(1 + \nu)G \quad ...(12.23)$$

Thus from equation (12.12) and (12.23) we have the final relationship between the four static constants E, K, G and ν as

$$E = 3(1 - 2\nu)K = 2(1 + \nu)\text{G} \quad ...(12.24)$$

Again From (12.24 a) $$\frac{E}{3K} = 1 - 2\nu$$

or $$\nu = \frac{1}{2}\left[1 - \frac{E}{3K}\right]$$

and $$\frac{E}{2G} = 1 + \nu$$

or $$\nu = \frac{E}{2G} - 1$$

$\therefore$ $$\nu = \frac{1}{2}\left[\frac{3K - E}{3K}\right] = \left(\frac{E - 2G}{2G}\right)$$

By cross-multiplication we have

or $$3GK - GE = 3KE - 6KG$$

$$3GK + 6KG = E(3K + G)$$

$$E = \frac{9GK}{3K + G} \quad ...(12.35\,b)$$

12.4 LIMITING VALUES OF POISSON'S RATIO

For an isotropic (a material which does not have a preferred orientation) perfectly plastic material the stress-strain curve will be horizontal. Thus the value of the Youngs' modules, E, which is equal to the slope of the stress-strain curve will be zero.

Therefore, from equation 12.4 we get from the first part of this equation.

$$\nu = -1$$

From the second part of this equation

$$\nu = 0.5$$

Thus the limiting values of n are – 1 to 0.5. In other words 0.5 is the maximum value of the Poisson's ratio.

Since ordinary materials contract laterally when stretched and expand laterally when compressed axially, Poisson's ration for such materials is positive. It is 0.45 for lead, 0.33 for aluminium and 0.27 to 0.33 for common steels. For cellular solids such as typical foams its value lies within the range 0.1 to 0.4.

For cork the value of Poisson's ratio is nearly zero. It is due to this property that cork stopper is used while bottling champagne. The cork must be easily inserted and removed, yet it also must withstand the pressure from within the bottle. Rubber, with a poisson's ratio of nearly 0.5 could not be used for this purpose because it would expand when compressed in to the neck of the bottle and would jam. Cork, by contrast with Poisson's ratio of nearly zero, is ideal in this application.

Biological tissues and various rubbery materials have values of Poisson's ratio approaching 0.5. Therefore from equation (12.24), it is seen that

$$G = \frac{3K(1-2\nu)}{[2(1+\nu)]} \qquad \text{...(12.25)}$$

And when $\nu \rightrightarrows 0.5$ the value of G approaches zero.

Thus for rubbery materials the shear modulus G, is much less than the bulk modulus, K. In physical sense these materials readily undergo shear deformation but resist volumetric (or bulk) deformation.

On the other extreme a solid with Poisson's ratio close to – 1 would have the value of K from equation (2.71) approaching to zero because

$$K = \frac{2G(1+\nu)}{[3(1-2\nu)]} \qquad \text{...(12.26)}$$

When $\nu \rightrightarrows -1, K \rightrightarrows 0$

In physical sense their behaviour will be opposite of rubber. That is they are very difficult to shear but easy to deform volumetrically. Mathematically for such solids the value of shear modulus, G will be much greater than the bulk modulus K. Therefore these materials may be called ANTI-RUBBER materials.

12.4.1 Negative Poisson's Ratio Materials

Till 1968 it was believed by many that materials with negative values of Poisson's ratio were unknown (Young 1968), (Love 1944) had reported a single example of "Single Crystal" pyrite in having a Poisson's ratio of –0.14 in his book published in 1944. He further suggested that this effect might have been result from a twinned crystal. Analysis of the tensorial elastic constants (*R*. Hill) of anisotropic single crystal cadmium suggests Poisson's ratio may attain negative values in some directions (Li 1976). Anislotropic macroscopic, two-dimensional flexible models of certain honeycomb structures (not materials) may be used in such applications as sponges, shock absorbing materials, air filters, fasteners, car and air craft seats. For having additional information the reader is advised to refer to the article entitled "Advances in negative Poisson's ratio materials" published in "Advaznced Materials, 5, pp 229–296 (1993).

12.5 SHEARING FORCE AND BENDING MOMENT IN BEAMS

12.5.1 Beams and their Classification

Any member of a machine or structure under the action of lateral forces in the axial plane is called a beam. The effect of these forces will be to cause bending of the member. Depending upon the end conditions there are different types of beams. A beam with both the ends rigidly fixed is known as a fixed or encastre beam. If one of the ends is rigidly fixed and the other is absolutely free it is known as a cantilever beam. If both the supports at the ends are such that the beam is capable of taking any angular position at the ends and not the lateral movement, the beam is called to be a freely or simply supported beam ; in actual practice this requirement is met by providing a well lubricated hinge at one end and a support on rollers at the other.

If such a beam has ends extending beyond the supports, it is called an overhanging beam. A beam which is supported at more than two supports is called a continuous beam.

As beams are invariably horizontal, and the external forces are weights, the beams will be taken horizontal and the external forces as vertical, although the same conclusions will hold good in other cases. Members of structures and machines transmit axial loads in addition to lateral forces acting upon them. However in the present chapter only transverse forces will be considered.

A concentrated or point load is one which is applied to the beam through a knife edge. A uniformly distributed load is that which is spread evenly over a certain length of the beam so that the load carried by a portion of the beam is proportional to the length of that portion. A distributed load will be called a varying load when the rate of loading varies from one point to another.

12.6 MOMENT OF A FORCE, PLANE OF MOMENT AND AXIS OF MOMENT

Moment of a force about a point is the product of the force and the perpendicular distance of the point from the force vector. Plane containing the force vector and the point about which moment is taken is called in PLANE OF MOMENT. And the axis passing through the point and normal the plane of moment is called the axis of the moment. Since force is a vector quantity the moment will also be a vector quantity as the distance is a scalar quantity. The force vector is represented by an arrowhead the length of which to some chosen scale represents the magnitude of the force and the arrow head indicates the direction of the force. Similarly to represent a moment vector with an arrow head right handed screw rule is followed. The length of the arrowhead will represent the magnitude of the moment to some scale. However before we describe the method to decide the placement of the arrow and the direction to which it is to point we have to see whether the moment about the given point is clockwise moment a anticlockwise moment depending upon the location of the point with respect to the direction of the arrow head representing the force vector. If the point lies on the left side of force vector arrow head as seen along the arrow head then the moment will be reckoned as anticlockwise. Similarly if the point in question lies on the right hand side of the force vector then it will be called the clockwise moment because this moment will tend to rotate the minutes arm of the clock, placed at the point where the moment is taken, in the clockwise direction or vice versa Fig. (12.5). After having decided whether the moment is clockwise or anticlockwise, we proceed to apply the right hand screw rule. The right handed screw rule states that with respect to a fixed nut placed in the plane of the moment, the direction of axial movement of the right handed screw when acted upon by the given moment, will be the direction of the arrow head representing the moment vector. Evidently the axial movement of the screw with respect to fixed nut which is placed in the plane of the moment, will be normal to the plane of the moment. Thus if the plane of the moment is vertical and is visible in elevation, the normal will be visible in the plan view and will be normal to plane of the moment. It may be noted that the arrow head will be normal to the plane of the moment and it will be parallel to the axis of the moment.

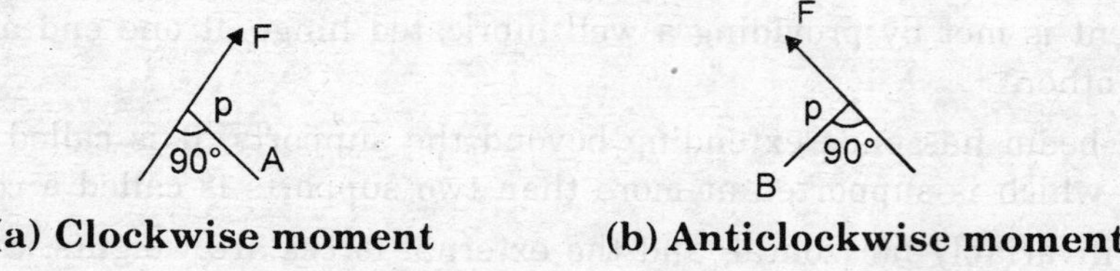

(a) Clockwise moment **(b) Anticlockwise moment**

Fig. 12.5

12.7 CLASSIFICATION OF MOMENT : BENDING MOMENT AND TWISTING MOMENT

The next question arises as to whether the moment acting on any section of a beam or shaft is bending moment or twisting moment. When the plane of the moment is normal to the cross sectional plane of the beam or shaft on which the moment is acting, the moment is called Bending moment as it will tend to bend the beam or shaft. Alternately moment will be a Bending Moment if the axis of the moment is parallel to or lies in the cross sectional plane of the beam.

Similarly if the plane of the moment is parallel to or contains the cross sectional plane of the beam or shaft it will be called twisting moment as it tends to twist the shaft. Alternately the moment will be a twisting moment if the axis of the moment is normal to the cross-sectional plane of the beam or shaft or if the axis of the moment coincides with the longitudinal axis if the shaft.

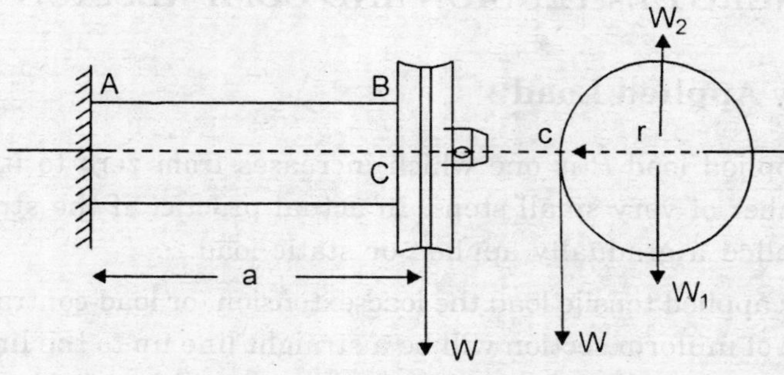

Fig. 12.6

As an illustration, in Fig. 12.6 is shown a circular bar *AB* with end *A* fixed in wall and at the end *B* is attached a pulley over which a rope passes around its circumference. Let a load *W* be hanging from be rope.

That is load *W* is applied vertically along the tangent to the circumference of the pulley at *C*. Let the radius of the pulley be *r*.

From our own side we apply equal and opposite vertical loads W_1 and W_2 at the point *B*. Also both W_1 and W_2 are equal to *W*.

That is $W_1 = W_2 = W$, then they will not alter the given load condition.

(a) **Effect of W and W_2.** These loads coustitute a momemt of magnitude $(W \times r) = T$ (say). The plane of the moment *T* is the plane of the pulley and is visible in the side view. Evidently the axis of the moment, which is normal to the plane of the movement will coincide with the axis of the rod. Thus moment *T* is the Twisting Moment" as per the definition. It will try to twist the axis *B* of the rod with respect to the end *A*.

(b) Effect W_1 will give rise to the moment $W_1 \times a = W_1$ a about the end *A* of this rod which is a cantilever beam. The plane of the moment is the vertical plane passing through the points *A* and *B*. Evidently the axis of the rod *AB* lies in the plane of the moment itself or the axis of the moment which is normal to this plane of moment will be normal to axis of rod. Thus as per definition this moment $W_1 a = M$ (say) will be the Bending moment and it will tend to bend the rod.

12.8 STRAIN ENERGY AND RESILIENCE

When a bar is acted upon by a tensile or compressive load there is a change in the length and the point of application of the load moves against the resistance offered by the material which gets strained under the action of the load. In so doing the work is done on the bar. If the bar had been loaded within the elastic limit of the material the work done on the bar will be recovered when the load is removed. In other words, the work done on the bar, while loading, is stored in the bar as strain energy which can be recovered wholly or partially depending upon whether the bar was loaded up to or beyond its elastic limit. The strain energy stored per unit volume is called the resilience of the body. Further the *'Proof Resilience'* is the strain energy per unit volume stored in the body at its elastic limit point Strain energy is always a positive quantity and, being work units, will be expressed in Nm (or Joules).

If the material is uniformly stressed everywhere the strain energy per unit volume will be constant. However if the stress varies from point to point the strain energy per unit volume will vary from point to point and will depend upon the stress distribution and shape of the body.

12.9 STRAIN ENERGY IN TENSION AND COMPRESSION

12.9.1 Gradually Applied Load

A gradually applied load P is one which increases from zero to its current value P in infinitely large number of very small steps. In actual practice if the strain rate is less than 2×10^{-3}/sec. it is called a gradually applied or static load.

For a gradually applied tensile load the load-extension (or load-contraction for compressive load) curve for a bar of uniform section will be a straight line up to the limit of proportionality point Fig. 12.7. Corresponding to the load P the stretch will be x and the work done will be represented by the shaded area. Thus the strain energy U is given by

$$U = \frac{1}{2} Px \quad \text{...(12.27)}$$

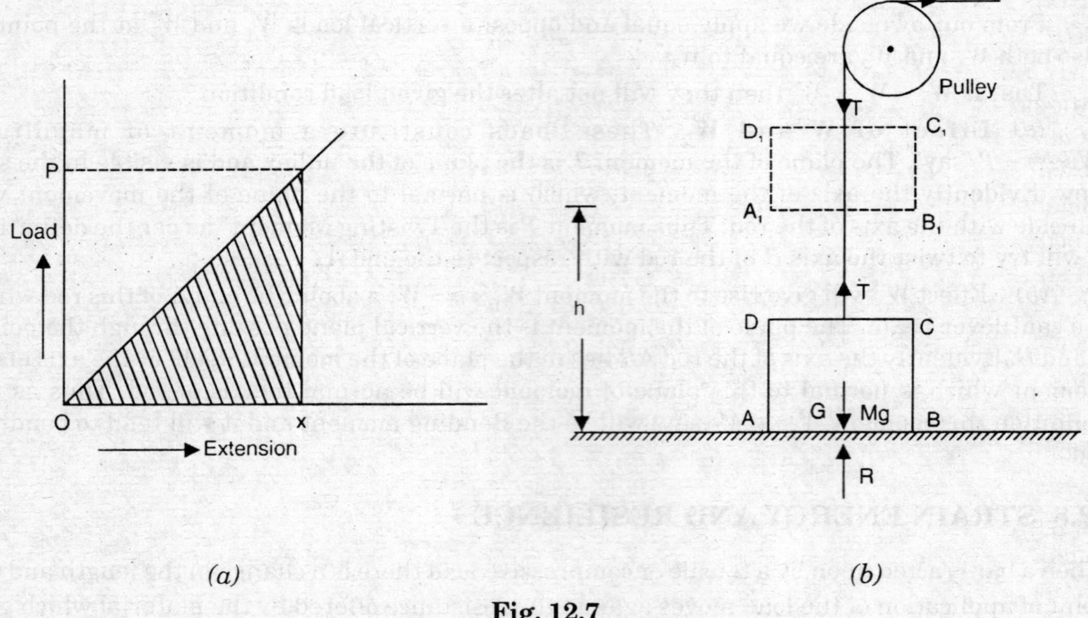

Fig. 12.7

(*i*) Bottom of body AB, just touching table and

$$R = 0$$

$$T = \text{Mg}$$

Rope snapped, then sudden load on table

(*ii*) Rope tight but carries tension T, (< Mg)

$$R + T = \text{Mg}$$

(*iii*) Rope just loose so that R = Mg and Tension = 0 gradually applied load.

(iv) Bottom of body, AB at height h above table rope snapped → shock on impact load.

If the length of the bar is l and its area of cross-section A, then the strain energy can be expressed in terms of stress s and Young's modulus E of the material

$$U = (A\sigma)\frac{1}{2}\left(\frac{\sigma}{E}l\right) = \frac{\sigma^2}{2E}(Al)$$

or $$U = \frac{\sigma^2}{2E}\text{(Volume)} \qquad ...(12.28)$$

For equation (12.28) the resilience or the strain energy per unit volume in simple tension or compression is $\sigma^2/2E$. Further if σ_e is the elastic limit of the material, the proof resilience will be $\sigma_e^2/(2E)$ which is constant for a material.

12.2.2 Suddenly Applied Load

A suddenly applied load P_1 is one which is applied in one step of value P_1, that is, it remains constant during the period the stretch or contraction attains the final value. Consider a body $ABCD$ of mass M hanging from the end of a rope as shown in Fig. 12.7*b* and is being lowered slowly by releasing the rope with the help of a worm and worm wheel arrangement attached to the other end of the rope. When the body is completely in air and not touching the table, the tension T in the rope will be equal to Mg. When the body is lowered slowly its lower face AB will come close to the top face of the table and it will be just above the table surface at a height zero but not touching the table then again $T = Mg$ and the reaction, R of the table on body will be euqal to zero. But when the rope is further released slowly, the reaction R will start increasing and the tension T will simultaneously start decreasing at the same rate and after some time R will become equal Mg and T will reduce to zero and the rope will become loose. Evidently the load on the table will be equal to R which in turn will be equal to Mg.

Thus a load of Mg has been applied which has increased from zero to current value Mg gradually and hence it will be called gradually applied load. Secondly when the bottom of the body was just above the table at zero height but $T = Mg$ and $R = 0$ at that time had we cut the rope, the entire load Mg would have come on table in one go. It would then be called a suddenly applied load.

Thirdly when the body was hanging in the air and at height 'h' above the table, if we cut the rope, the body will fall and come in contact with the table with certain velocity. We call this a shock or impact load. In the light of above, in the case of a suddenly applied load, P_1, the strain energy is given by

$$U = P_1 x \qquad ...(12.29)$$

If the deformation x is same as in the gradually-applied-load case the bar will experience the same strain and will therefore be under the action of same stress which in turn will result in the same strain energies in the two cases. Then comparing equations (12.28) and (12.29)

$$\frac{1}{2}Px = P_1 x$$

or $$P = 2P_1 \qquad ...(12.30)$$

From equation (12.30) it is observed that in order to produce the same strain and same stress a gradually applied load should be twice the value of the suddenly applied load. Alternatively a suddenly applied load will induce stress in the bar which is twice the one induced by same load on the bar but applied gradually. It may be noted that in the case of the suddenly applied load the strain and stress considered are instantaneous maximum ones obtained immediately after the application of the sudden load. As a matter of fact in the case of suddenly applied load the system is set into free damped vibrations and the stress in the bar changes in magnitude and nature from tension to compression to tension and so on. The maximum value of the stress on each side progressively goes on diminishing till it ultimately settles down to a value equal to that due to gradually applied load* when the vibrations die out.

12.9.3 Impact Load

An impact load is one which strikes the body with certain velocity as is the case of a falling mass M through height h (Fig. 12.8) above the collar attached to the lower end of the rod AB of length l.

It is assumed that the mass M falls freely without any loss of energy in impact and produces an instantaneous extension x and due to it instantaneous stress s is induced in this rod. Then the loss of potential energy of the mass will be stored in the form of strain energy in the rod.

$$Mg\,(h + x) = \frac{\sigma^2}{2E}(Al) \qquad \text{...(12.31)}$$

or $$Mg\left(h + \frac{\sigma l}{E}\right) = \frac{\sigma^2}{2E}(Al)$$

or $$\sigma^2 - \sigma\left(\frac{2Mg}{A}\right) - \frac{2\,MghE}{Al} = 0 \qquad \text{...(12.32)}$$

Taking only positive root of equation (12.32)

$$\sigma = \frac{Mg}{A}\left[1 + \sqrt{1 + \frac{2AEh}{Mgl}}\right] \qquad \text{...(12.33)}$$

Fig. 12.8

Case I.

When $h = O$, the impact velocity will be zero and the load will be a suddenly applied load. Then the equation (12.33) gives $\sigma = \frac{Mg}{A} = 2\frac{P}{A}$ which is same as obtained in article 12.9.2.

Case II.

When x is very small as compared to h, then from equation (12.31)

$$Mgh = \frac{\sigma^2}{2E}(Al) \qquad \text{...(12.31)}$$

*Note : The reader is urged to refer to any good book on vibrations.

or $$\sigma = \sqrt{\frac{2\,Mgh\,E}{Al}}$$

Example 12.1. *A round bar of steel 1 m long is 20 mm diameter for half its length and 25 mm diameter over the rest. A gradually applied axial load of 1 kN acts on it. Find the total strain energy and strain energy per unit volume. E = 200 GN/ m²*

Solution : $$A_1 = \frac{\pi}{4}(0.02)^2 = \pi \times 10^{-4}\ \text{m}^2$$

$$\sigma_1 = \frac{1000}{\pi \times 10^{-4}} = 3.18 \times 10^6\ \text{N/m}^2$$

$$A_2 = \frac{\pi}{4}(.025)^2 = 4.91 \times 10^{-4}\ \text{m}^2$$

$$\sigma_2 = \frac{1000}{4.91 \times 10^{-4}} = 2.035 \times 10^6\ \text{N/m}^2$$

$$\text{Total strain energy} = \frac{\sigma_1^2}{2E}(A_1 l_1) + \frac{\sigma_2^2}{2E}(A_2 l_2) = \frac{l_1}{2E}\left[\sigma_1(\sigma_1 A_1) + \sigma_2 A_2)\right]$$

$$= \frac{0.5}{2E}(P\sigma_1 + P\sigma_2)$$

[Because $P = \sigma_1 A_1 = \sigma_2 A_2 = 1000$ N and $l_1 + l_2 = .5$]

$$= \frac{100 \times 0.5}{2 \times 200 \times 10^9}(3.18 \times 10^6 + 2.035 \times 10^6) = 0.006515\ J$$

Total volume $= A_1 l_1 + A_2 l_2 = 0.5\ (\text{p} \times 10^{-4} + 4.91 \times 10^{-4})$
$= 4.025 \times 10^{-4}\ \text{m}^3$

Therefore average strain energy per unit volume

$$U/V = \frac{0.006515}{4.025 \times 10^{-4}} = 16.2\ \text{J/m}^3.$$

Example 12.2. *A wagon, of mass 5 t, is attached to a steel wire rope, the other end of which is wound round a brake drum. The wagon is descending a slope at a uniform speed of 5 km/h and the rope is taut. By the action of the brake the drum is suddenly stopped and the wagon is brought to rest. If the length of the rope between the wagon and the drum at that instant was 250 m and its cross-sectional area is 300 mm², find the maximum stress in the rope. E = 200 GN/m².*

Solution : Strain energy of rope = Kinetic energy of wagon

$$\frac{\sigma^2}{2E}Al = \frac{1}{2}Mv^2$$

or $$\sigma^2 = 5000 \times \left(\frac{5000}{3600}\right)^2 \times \frac{200 \times 10^9}{(300 \times 10^{-6}) \times 250}$$

or $$\sigma^2 = \frac{(10)^{10}}{36\sqrt{3}}$$

i.e., $$\sigma = 160 \times 10^6 \text{ N/m}^2 = 160 \text{ MN/m}^2$$

Example 12.3. *A bar of certain material, 40 mm diameter and 1.2 m long, has a collar securely fitted to one end. It is suspended vertically with the collar at the lower end and a mass of 2000 kg is gradually lowered on to the collar, producing an extension in the bar of 0.25 mm. Find the height from which this load could be dropped on to the collar if the maximum tensile stress in the bar is to be 100 MN/m².*

Solution. In the first case when the load is applied gradually

$$\sigma_1 = \frac{(2000 \times 9.81) \times 4}{\pi \times (0.040)^2} = 15.6 \times 10^6 \text{ N/m}^2$$

and $$\varepsilon = \frac{0.25 \times 10^{-3}}{1.2}$$

Therefore, $$E = \frac{\sigma_1}{\varepsilon} = \frac{15.6 \times 10^6 \times 1.2}{(0.25 \times 10^{-3})} = 75.0 \times 10^9 \text{ N/m}^2$$

For the second case $\sigma_2 = 100 \times 10^6$ N/m²

Corresponding extension

$$x = \frac{(100 \times 10^6)}{75 \times 10^9} \times 1.2 = 1.6 \times 10^{-3} \text{ m}$$

If h is the height through which the mass has to be dropped, then from equation (12.31), we have

$$(2000 \times 9.81)(h + 1.6 \times 10^{-3}) = \frac{(100 \times 10^6)}{75 \times 10^{-3}} \times 1.2$$

from which $h = 3.52$ mm.

12.10 STRAIN ENERGY DUE TO SIMPLE SHEAR

Referring to Fig. 12.9 if a, b and t are the length, breadth and thickness of the rectangular block under the action of gradually applied shearing force P due to which the movement of the face AB is $AA¢$. Then the work done, which is stored in the block as shear strain energy is given by the expression given below, where ϕ is the shear strain angle

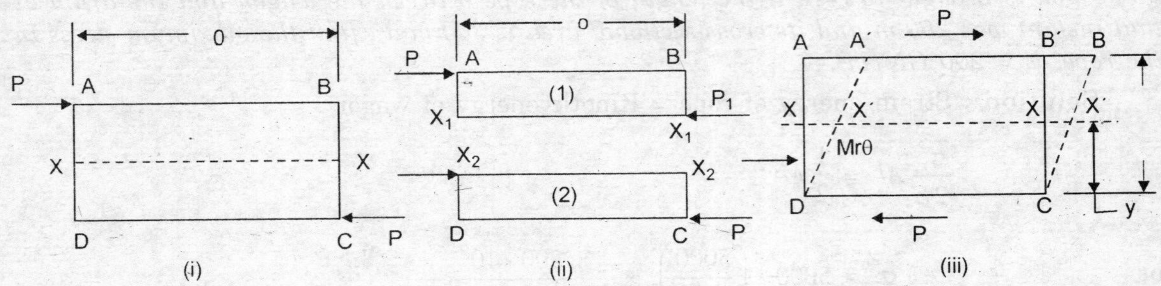

Fig. 12.9

$$U = \frac{1}{2} P(AA') = \frac{1}{2}(\tau . a . t)(b\,\phi)$$

$$= \frac{1}{2}\tau\phi\,(\text{volume}) = \frac{1}{2}\tau\left(\frac{\tau}{G}\right)\text{volume}$$

i.e. $$U = \frac{\tau^2}{2G}(\text{volume}) \quad \text{...(12.34)}$$

Example 12.4. *On A bar of steel 1m long 20 mm diameter a load of 1 kN is applied suddenly. Calculate the instantaneous maximum stress and the elongation ? What will be stress and elongating after some time when the initial vibrations have died out. Take E = 200 GPa.*

Solution : d = 20 mm

l = 1 m

E = 200 GPa

W = Suddenly applied load = 1000 N

Area of cross-section = A = $\frac{\pi}{4} \times (20)^2$

= 100 N mm²

Instantaneous $\sigma_{max} = \frac{2W}{A}$

$= \frac{2 \times 1000}{100\,\pi} - \frac{20}{\pi}$ N/mm²

= 6.366 MPa

Given E = 200 GPa = 200 × 10³ MPa

Corresponding $\varepsilon_{max} = \frac{\sigma_{max}}{E}$

$= \frac{6.366}{200 \times 10^3}$

∴ Instantaneous stretch = δl

$= l\,(\varepsilon_{max})$

$= (1000) \times \frac{6.366}{200 \times 10^3}$

= 0.0318 mm

(*b*) When the sudden load is applied the system is set into vibration mode. These vibrations die out after some time and the system comes to static state.

Then the stress $\sigma = \frac{W}{A}$

$= \frac{1000}{100\,\pi} = \frac{10}{\pi}$ N/mm²

= 3.183 MPa

Corresponding stretch = $\left(\dfrac{\sigma}{E} \times l\right)$

$$= \frac{3.183}{200 \times 10^3} \times 1000 \text{ mm}$$

$$= 0.0159 \text{ mm}$$

Example 12.5. *A circular bar 50 mm diameter, 0.5 m long is subjected to a gradually applied load of 60 kN. Find the strain energy and the stain energy per unit volume.*

Take $E = 200\,GPa$

Solution :

$$l = 500 \text{ mm}$$

$$d = 50 \text{ mm}$$

$$W = 60 \times 1000\,N$$

$$A = \frac{\pi}{4} \times (50)^2 \text{ mm}^2$$

$$\sigma = \frac{60 \times 1000 \times 4}{\pi \times (50)^2} \text{ N/mm}^2$$

$$= 30.5577 \text{ MPa}$$

$$U = \text{Strain energy} = \frac{\sigma^2}{2\,E} \times \text{vol}$$

$$= \frac{(30.5577)^2}{2\,E} \times \left(\frac{\pi}{4}(50)^2 \times 500\right)$$

$$= \frac{(30.5577)^2 \times \pi \times 2500 \times 500}{8 \times 200 \times 10^3}$$

$$= 2291.8 \text{ N-mm}$$

$$= 22918 \text{ N-m or Joules}$$

$$\text{Vol.} = \frac{\pi}{4}(50)^2 \times 500 = 981.74 \text{ mm}^3$$

$$\text{U/unit vol.} = \left[\frac{2.2918}{(981.74 \times 10^{-9})}\right] \text{J/m}^3$$

$$= 2.3344 \times 106 \text{ J/m}^3$$

$$= 2334.4 \text{ k/m}^3 \text{ or } 2.3344 \text{ kJ/m}^3.$$

Example 12.6. *A mild steel test piece was tested in tension and the following readings were obtained :*

Diameter of specimen = 20 mm

Length of specimen = 20 cm

Extension under 30 kN load = 0.08 mm

Load at yield point = 150 kN

Maximum load = 225 kN

Length of specimen after fracture = 25 cm

Calculate the values of the following :

(a) Youngs modulus

(b) Yield point stress

(c) Ultimate strengths

(d) Percentage elongation *[P.T.U. 2001]*

Solution : Diameter of specimen

$$d = 20 \text{ mm}$$

$$l = 200 \text{ mm}$$

$$dl = \text{extension } 0.08 \text{ mm at load}$$

$$P = 30 \text{ kN}$$

$$\therefore \quad \varepsilon = \frac{\delta l}{l} = \frac{0.08}{200}$$

and

$$\sigma = \frac{30 \times 1000 \times 4}{\pi \times (20)^2} = \frac{300}{\pi} \text{ N/mm}^2$$

$$\therefore \quad E = \frac{\sigma}{\varepsilon} \times \frac{300}{\pi} \times \frac{200}{0.08} = 238732.4 \text{ N/mm}^2 \text{ or MPa}$$

$$= 238.73 \text{ MPa or}$$

(*b*)

$$\text{Yield stress} = \sigma_y = \frac{\text{Load at yield}}{\text{Area}}$$

$$\sigma_y = \frac{150 \times 10^3 \times 4}{\pi \times (20)^2} = 477.46 \text{ N/mm}^2$$

$$= 477.46 \text{ MPa}$$

(*c*) Ultimate strength

$$= \frac{\text{Maximum load}}{\text{Area}}$$

$$= \frac{225 \times 10^3 \times 4}{15 \times (20)^2} \text{ N/mm}^2$$

$$= 716.2 \text{ N/mm}^2 \text{ or MPa}$$

(d)

$$\% \text{ elongation} = \frac{25 - 20}{10} \times 100$$

$$= \frac{5}{20} \times 100 = 25\%$$

Example 12.7. *A cylindrical test piece of diamonds 30 mm was put to tension. When the load was 60 kN the measured extension was 0.08 mm over the gauge length of 200 mm and the diameter reduced by 0.0045 mm. Calculate the youngs modulus and Poisson's ratio of the material. Also calculate the value of the Bulk Modulus and Modulus of rigidity.*

Solution :

$$d = 30 \text{ mm}$$

$$A = \frac{\pi}{4} \times (30)^2 = 225\,\pi \text{ mm}^2$$

$$P = 60 \text{ kN}$$

$$\therefore \quad \sigma = \frac{P}{A} = \frac{60 \times 1000}{225\,\pi}\ \text{N/mm}^2$$

$$\varepsilon = \frac{dl}{l} = \frac{0.08}{200} = 0.0004$$

E = Young's Modulus

$$= \frac{\sigma}{\varepsilon} = \frac{60 \times 1000}{225\,\pi} \times \frac{1}{0.0004}$$

$= 0.212 \times 106$ N/mm^2 or MPa

= 212 GPa

$$\text{Lateral stress} = \frac{\delta d}{d} = \frac{0.0045}{30}$$

$$\therefore \quad \text{Poisson's ratio} = \nu = \left(\frac{\frac{\delta d}{d}}{\varepsilon} \right)$$

$$= \frac{0.0045}{30} \times \frac{1}{0.0004}$$

$= 0.375$

(b) From equation (12.24) we have

$$E = 3k(1 - 2\nu)$$

or

$$k = \frac{E}{3(1 - 2\nu)}$$

$$= \frac{212 \times 10^9}{3(1 - 2 \times 0.375)}\ N\text{/mm}^2$$

$= 282.6 \times 10^9\ N$/mm^2

= 282.6 GPa

(c) Again from equation (12.24) we have

$$E = 2G(1 + \nu)$$

or

$$G = \frac{E}{2(1 + \nu)}$$

$$= \frac{212 \times 10^9}{2(1 + 0.375)} = 77.09 \times 10^9\ \text{N/mm}^2$$

= 77.09 GPa

Example 12.8. *A bar 4 m long and 1200 mm² area of cross-section is subjected to a sudden load which produces an instantaneous stretch of 1 mm. Find the magnitude of the sudden load and the maximum stress induced in the bar. Take E = 200 GPa*

Solution : Given data

$$\delta l = 1.0 \text{ mm}$$

$$l = 4000 \text{ mm}$$

Then $\qquad \varepsilon_{max} = \dfrac{1.0}{4000} = 0.25 \times 10^{-3}$

Therefore $\qquad \sigma_{max} = E\,\varepsilon_{max}$

$= (200 \times 10^3)(0.25 \times 10^{-3})$

$= 50$ MPa

(b) Let the suddenly applied load = W N

Then the max stress = $\sigma_{max} = \dfrac{2W}{A}$ N/mm^2

or $\qquad W = \dfrac{A}{2}\sigma_{max}$

$= \dfrac{1200}{2} \times 50$

$= 30000\,N$

$= 30\,kN$

Example 12.9. A steel rod 4m long and 1000 mm^2 area of cross-section is hanging from the top and has a rigid collar at the lower end. A 10kg moving mass falls on the collar at the lower end from a height of 1cm. calculate the instantaneous maximum stress in the rod. Take E = 200 GPa.

Solution : Given data

$M = 10$ kg

$h = 10$ mm

$E = 200 \times 10^3$ MPa

$l = 400$ mm

$A = 1000$ mm^2

$g = 9.827$ m/sec^2 = 9827 mm/sec^2

We have from equation **(12.32)**

$$\sigma = \frac{Mg}{A}\left(1 + \sqrt{1 + \frac{2\,A\,E\,h}{Mgl}}\right)$$

$$= \frac{0 \times 98.27}{1000}\left[1 + \sqrt{1 + \frac{2 \times 1000 \times (200 \times 10^3) \times 10}{10 \times 9827 \times 400}}\right]$$

$$= 98.27\left[1 + \sqrt{102.76}\right]$$

$= 1094.4$ N/mm^2 or MPa

Example 12.10. *A steel bar of square section 10 × 10 mm, 1 m long, when subjected to a gradually applied load of 200 N stretches 0.01 mm and the sides of the square section reduce by 2.7 × 10^{-5} mm each. Determine the Poisons ratio, youngs modulus, bulk modulus and shear modulus of the material.*

Solution : $\qquad A = 10 \times 10 = 100$ mm^2

ε = Longitudinal strain = $\dfrac{\delta l}{l} = \dfrac{0.01}{1000} = 1 \times 10^{-5}$

$$\text{Applied stress } \sigma = \frac{W}{A} = \frac{200}{100} = 2 \text{ N/mm}^2 \text{ or MPa}$$

$$\therefore \quad E = \frac{\sigma}{\varepsilon} = \frac{2}{1 \times 10^{-5}} = 2 \times 10^5$$

$$= 200 \times 10^3 \text{ MPa}$$

$$= 200 \text{ GPa}$$

Lateral strain $= \dfrac{\delta a}{a}$

where a = side of the square section

$= 10$ mm

and $\delta a = 2.7 \times 10^{-5}$

$$\therefore \quad \text{Lateral strain} = \frac{\delta a}{a} = \frac{2.7 \times 10^{-5}}{10}$$

$$= 2.7 \times 10^{-6}$$

$$\therefore \quad \text{Poisson's ratio } \nu = \frac{\left(\dfrac{\delta a}{a}\right)}{\left(\dfrac{\delta l}{l}\right)}$$

$$= \frac{2.7 \times 10^{-6}}{10 \times 10^{-5}}$$

$$= 0.27$$

From equation (12.24) $E = 3K(1 - 2\nu) = 2G(1 + \nu)$

$$K = \frac{E}{3(1 - 2\nu)}$$

$$= \frac{200 \times 10^3}{3(1 - 2 \times .27)} = \frac{200 \times 10^3}{3 \times 0.46}$$

$$= 144.9 \times 10^3 \text{ N/mm}^2$$

$$= 144.9 \times 10^3 \text{ MPa}$$

$$= 144.9 \text{ GPa}$$

Again from equation (12.24)

$$E = 2G(1 + \nu)$$

or

$$G = \frac{E}{2(1 + 0.27)} = \frac{200 \times 10^3}{2 \times 1.27} \text{ N/mm}^2$$

$$= 0.78 \times 10^5$$

$$= 78 \times 10^3 \text{ MPa}$$

$= 78$ GPa **Ans.**

Example 12.11. *A gradually applied tensile load of 1500 N acts on the specimen of shape and give shown in Fig. 12.10.*

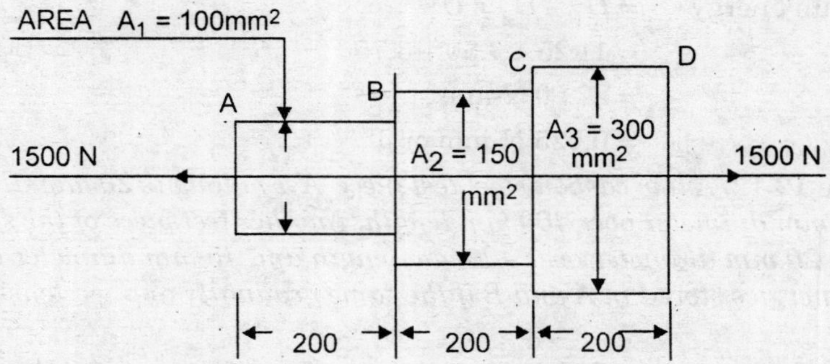

Fig. 12.10

Calculate the strain energy stored in the bar AD

Take $E = 200\,GPa$

Solution : stress in $AB = \sigma_1 = \dfrac{1500}{100} = 15\ \text{N/mm}^2$

Strain in $BC = \sigma_2 = \dfrac{1500}{150} = 10\ \text{N/mm}^2$

Stress in $CD = \sigma_3 = \dfrac{1500}{300} = 5\ \text{N/mm}^2$

$$U_1 = \text{Strain Energy in } AB = \frac{1}{2}\frac{\sigma_1^2 V_1}{E}$$

where V_1 = volume of portion AB

$$\therefore \quad U_1 = \frac{1}{2E}\sigma_1^2 A_1 l_1$$

$$= \frac{1}{2E}(15)^2 \times 100 \times 200$$

$$= \frac{225}{E} \times 10^4 \text{ N-mm}$$

$$= \frac{225 \times 10^4}{200 \times 10^3} = 11.25 \text{ Nmm}$$

Similarly $U_2 = \dfrac{1}{2E}\sigma_2^2 A_2 l_2$

$$U_2 = \frac{(10)^2 \times 150 \times 200}{2 \times 200 \times 10^3} = 7.5 \text{ N-mm}$$

and $U_3 = \dfrac{1}{2E}\sigma_3^2 A_3 l_3$

$$= \frac{1 \times (5)^2 \times 300\ (200)}{2 \times 200 \times 10^3} = 3.75 \text{ Nmm}$$

Total strain energy $= U_1 + U_2 + U_3$

$= 11.25 + 7.50 + 3.75$

$= 22.50$ N-mm

$= 0.025$ N-mm or J

Example 12.12. *A low carbon steel test piece 'A' 1 m long is 20 mm diameter over 600 mm length and 15 mm diameter over 400 mm length. Another test piece of the same material and of same length is 20 mm diameter over 400 mm length and 15 mm diameter over 600 mm length. Compare the energies stored in A and B if the same gradually applied tensile load is applied on them.*

Solution. For A $l_1 = 600$ mm

$d_1 = 20$ mm

$l_2 = 400$ mm

$d_2 = 15$ mm

For B $L_1 = 400$ mm

$D_1 = 20$ mm

$L_2 = 600$ mm

$D_2 = 15$ mm

Let the tensile gradually applied tensile load be P

For test piece *A*

$$\sigma_1 = \frac{4\,P}{\pi\,d_1^{\,2}}\ ;\ l_1 = 600\ d_1 = 20$$

$$V_1 = \frac{\pi}{4}\,d_1^{\,2}\,l_1$$

$$U_1 = \frac{\sigma_1^{\,2}}{2\,E} \times V_1$$

$$= \frac{1}{2E}\left[\frac{4P}{\pi\,(d_1)^2}\right]^2 \left(l_1\,\frac{\pi}{4}\,d_1^{\,2}\right)$$

$$= \frac{1}{2\,E}\,\frac{16\,P^2}{(\pi\,d_1^{\,2})^2} \times l_1 \times \frac{\pi\,d_1^{\,2}}{4}$$

$$= \frac{1}{2\,E}\,\frac{4\,P^2}{\pi\,d_1^{\,2}}\,l_1$$

$$= \frac{2\,P^2\,l_1}{\pi\,d_1^{\,2}\,E} = \frac{2\,P^2\,(600)}{\pi\,(20)^2\,E}$$

Similarly
$$U_2 = \frac{2\,P^2\,l_2}{\pi\,d_2^{\,2}\,E} = \frac{2\,P^2\,(400)}{\pi\,(15)^2\,E}$$

$$\therefore \quad U_A = \frac{2P^2}{\pi E}\left[\frac{l_1}{d_1^2} + \frac{l_2}{d_2^2}\right]$$

$$= \frac{2P^2}{\pi E}\left[\frac{600}{20^2} \times \frac{400}{15^2}\right]$$

$$= \frac{2P^2}{\pi E}[1.5 + 1.7777]$$

$$= \frac{2P^2}{\pi E}(3.2777)$$

For the test piece B

$$D_1 = 20 \text{ mm}$$
$$D_2 = 15 \text{ mm}$$

$$U_B = \frac{2P^2}{\pi E}\left[\frac{L_1}{D_1^2} + \frac{L_2}{D_2^2}\right]$$

$$= \frac{2P^2}{\pi E}\left[\frac{400}{(20)^2} \times \frac{600}{(15)^2}\right]$$

$$= \frac{2P^2}{\pi E}[1 + 2.6666]$$

$$= \frac{2P^2}{\pi E}[3.6666]$$

$$\therefore \quad \frac{U_A}{U_B} = \frac{\frac{2P^2}{\pi E}(3.2777)}{\frac{2P^2}{\pi E}(3.6666)}$$

$$= 0.8939$$

Example 12.13. *Two bars A and B each 0.6 m long are subjected to the same axial blow. 'A' is 40 mm over 0.4 m length and 20 mm over 0.2 m length and B is 40 mm diameter over 0.2 m length and 20 mm diameter over 0.4 m length. If the material of the both the rods is same, find the ratio of the maximum stress in A to maximum stress in B.*

Solution. Since the axial blow is same as both the rods. The energy emitted by the blow for both the rods will be same. Thus the energy contents of the rods after they receive the same blow will be same.

i.e. $U_A = U_B$

Let σ_a be the maximum stress induced in rod A and σ_b that in the rod B. Since the maximum stress induced in any bar will occur at that portion at which the area if cross-section is least.

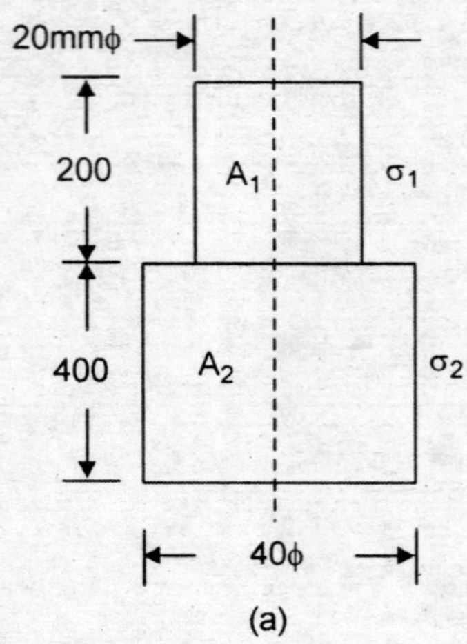

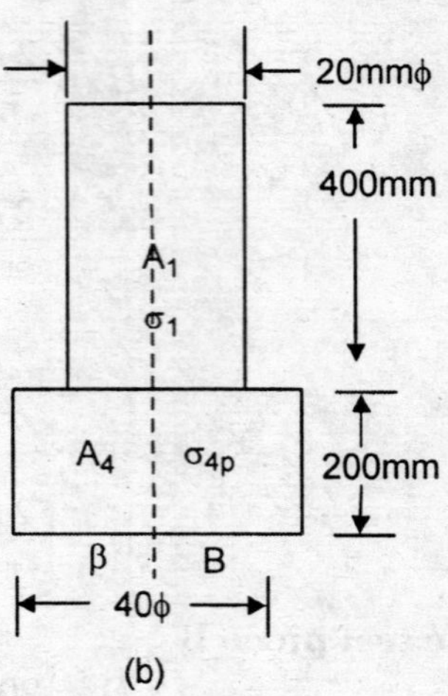

Fig. 12.11

$$U_A = U_1 + U_2$$

$$= \frac{\sigma_1^2}{2E} A_1 l_1 + \frac{\sigma_2^2}{2E} A_2 l_2 \qquad \text{(since } s_a = s_1\text{)}$$

Now $\sigma_1 A_1 = s_2 A_2$

$$\therefore \quad \sigma_2 = \frac{\sigma_1 A_1}{A_2} = \frac{\sigma_a \frac{\pi}{4} d_1^2}{\frac{\pi}{4}(d_2)^2} = \frac{\sigma_a \cdot (20)^2}{40^2}$$

$$= \frac{\sigma_a}{4} = \frac{\sigma_1}{4}$$

R\ $$U_A = \frac{\sigma_1^2}{2E} A_1 l_1 + \frac{\sigma_2^2}{2E} A_2 l_2$$

$$= \frac{1}{2E}\left[\sigma_a^2 \frac{\pi}{4}(20)^2\ 200 + \left(\frac{\sigma_a}{4}\right)^2 \left(\frac{\pi}{4}\right) \times (40)^2 \times 400\right]$$

$$= \frac{\sigma_a^2}{2E}\left(\frac{\pi}{4}\right)(80000 + 40000)$$

$$= \frac{\sigma_a^2}{2E}\left(\frac{\pi}{4}\right)(120000)$$

$$U_B = \frac{\sigma_3^2}{2E} A_3 l_3 + \frac{\sigma_4^2}{2E} A_4 l_4$$

Since $\sigma_3 = \sigma_b$

and $\sigma_3 A_3 = \sigma_4 A_4$

$$\sigma_b = \sigma_4 \left(\frac{40}{20}\right)^2$$

or $\sigma_4 = \frac{\pi}{4}$

$$U_B = \frac{1}{2E}\left[\sigma_b^2 \frac{\pi}{4}(20)^2 \times 400 + \left(\frac{\sigma_b}{4}\right)^2 \left(\frac{\pi}{4}\right)(400)^2 (200)\right]$$

$$= \left(\frac{\sigma_b^2}{2E}\right)\left(\frac{\pi}{4}\right) \times (20 \times 40)\left[20 \times 10 + \left(\frac{1}{4}\right)^2 40 \times 10\right]$$

$$= \frac{\sigma_b^2}{2E}\left(\frac{\pi}{4}\right)(8000)\left[20 + \frac{40}{16}\right]$$

$$U_B = \frac{\sigma_b^2}{2E}\left(\frac{\pi}{4}\right)(8000)(22.5)$$

Since $U_A = U_B$

Therefore substituting the values of UA and UB

$$\frac{\sigma_a^2}{2E}\left(\frac{\pi}{4}\right)(120000) = \frac{\sigma_b^2}{2E}\left(\frac{\pi}{4}\right)(8000)(22.5)$$

or $$\left(\frac{\sigma_a}{\sigma_b}\right)^2 = \frac{8000 \times 22.5}{120000} = \frac{3}{2}$$

$$\frac{\sigma_a}{\sigma_b} = \sqrt{\frac{3}{2}}$$

or $\sigma_a : \sigma_b = \sqrt{3} : \sqrt{2}$

Example 12.14. *A steel bar of area of cross-section 1200 mm² and length 1.0 m is subjected to a gradually applied of 120 kN. Find (a) the strain energy stored in the bar (b) If the elastic unit of steel is 24 MPa find the proof resilience.*

Take $E = 200\ GPa$

Solution : $A = 1200\ \text{mm}^2$

$l = 1000$ mm

$W = 120000$ N

$$\sigma = \frac{120000}{1200} = 100\ \text{N/mm}^2$$

$$\therefore \quad U = \frac{\sigma^2}{2E} \times Al = \frac{(100)^2 \times 1200 \times 1000}{2 \times (200 \times 10^3)}$$

$= 3 \times 10^5$ N-mm

$= 300$ Nm (J)

$\sigma_e = 240$ MPa or N/mm²

$$E = 200 \times 10^3 \text{ N/mm}$$

$$\text{Proof resilience} = \frac{\sigma_e^2}{2E} \times Al$$

$$= \frac{(240)^2 \times 1200 \times 1000}{2\,E\,(200 \times 10^3)}$$

$$= (240)^2 \times 3$$
$$= 172800 \text{ N-mm}$$
$$= 172.8 \text{ N-m or J}$$

$$\text{Modulus of resilience} = \frac{\sigma_e^2}{2E} = \text{Proof resilience per unit vol.}$$

$$= \frac{(240)^2}{2 \times 200 \times 10^3}$$

$$= \frac{240 \times 240}{400 \times 10^3}$$

$$= 144 \times 10^{-3} \text{ N/mm}^2$$
$$= 0.144 \text{ N/mm}^2$$

Example 12.15. *Three bars A, B and C of the same material and of same length have their areas of cross sections in the ratio 1 : 2 : 3. Find the ratio of the strain energies of A, B and C when*

(a) the three bars are subjected to same gratually applied load

(b) when the maximum stress induced in all the three rods is same.

(c) If these bars receive the same below find the ratio of the maximum stresses induced in the three bars.

Solution. (*a*) Gradually applied load is same let the areas of cross-sections are A_a, A_b are A_c

Then $\quad A_a : A_b : A_c :: 1 : 2 : 3$

i.e. $\quad A_b = 2\,A_a$

$$A_e = 3\,A_a$$

and $\quad l_a = l_b = l_c = l \text{ (say)}$

$$\sigma_a = \frac{W}{A_a}$$

$$\therefore \quad U_a = \frac{1}{2}\frac{\sigma_a^2}{E} \times A_a\,l_a = \frac{1}{2}\frac{W^2}{EA_a}\,.\,l$$

$$U_b = \frac{1}{2}\frac{\sigma_b^2}{E}\,A_b\,l_b$$

$$= \frac{1}{2E}\frac{W^2}{A_b}\,l = \frac{lW^2}{2E}\cdot\frac{1}{A_b} = \frac{lW^2}{2E(2A_a)}$$

Similarly $$U_c = \frac{1}{2}\frac{W^2}{A_c}\,l = \frac{lW^2}{2E}\frac{1}{A_c} = \frac{lW^2}{2E(3Aa)}$$

$\therefore$ $$\frac{U_b}{U_a} = \frac{\frac{lW^2}{2E}\left(\frac{1}{2A_a}\right)}{\frac{lW^2}{2E}\frac{1}{A_a}} = \frac{1}{2}$$

and $$\frac{U_c}{U_a} = \frac{\frac{lW^2}{2E}\left(\frac{1}{3A_a}\right)}{\frac{lW^2}{2E}\left(\frac{1}{A_a}\right)} = \frac{1}{3}$$

$\therefore$ $$U_a : U_b : U_c = 1 : \frac{1}{2} : \frac{1}{3}$$

(*b*) Let the loads he such that

$$\sigma_a = \sigma_b = \sigma_c = \sigma$$

But $$A_c : A_b : A_c : : 1 : 2 : 3$$

Then $$U_a = \frac{\sigma_a^{\,2}}{2E}A_a l$$

$$U_b = \frac{\sigma_b^{\,2}}{2E}A_b l$$

$$U_c = \frac{\sigma_c^{\,2}}{2E}A_c l$$

$$\frac{U_b}{U_a} = \frac{A_b}{A_a} = 2 \qquad \text{(given } A_b = 2A_a\text{)}$$

and $$\frac{U_c}{U_a} = \frac{A_c}{A_a} = 3 \qquad \text{(given } A_c = 3A_a\text{)}$$

$\therefore$ $$U_a : U_b : U_c : : 1 : 2 : 3$$

(*c*) When the three bars receive the same below

$$U_a = U_b = U_c$$

Let the three corresponding maximum stresses be σ_a, σ_b, σ_c

Then $$U_a = \frac{\sigma_a^{\,2}}{2E}A_a l$$

$$U_b = \frac{\sigma_b{}^2}{2E} A_b l = \frac{\sigma_a{}^2}{2E} A_a l$$

Since $\quad U_b = U_a$

$$\therefore \quad \frac{U_b}{U_a} = \frac{2\sigma_b{}^2}{2E} A_a l \cdot \frac{2E}{(\sigma_a^2 A_a l)} = 2\frac{\sigma_b{}^2}{\sigma_a{}^2} \text{ [Because } A_a : A_b : A_c :: 1 : 2 : 3]$$

or $\quad \sigma^2 = 2\, s_1{}^2$

or $\quad \sigma_a = \sqrt{2}\ s_1$

and $\quad U_c = \frac{\sigma_c{}^2}{2E} A_c l = \frac{\sigma_c{}^2}{2E} 3 A_a l$

$$\therefore \quad \frac{\dfrac{\sigma_c{}^2}{2E} 3 A_a l}{\dfrac{\sigma_2{}^2}{2E} A_a l} = 3\frac{\sigma_c{}^2}{\sigma_a{}^2}$$

or $\quad \sigma_a{}^2 = 3\sigma_c{}^2$

or $\quad U_c = U_c$

or $\quad \sigma_a = \sqrt{3}\ \sigma_c$

$$\therefore \quad \sigma_a : \sigma_b : \sigma_c :: 1 : \sqrt{2} : \sqrt{3}$$

Example 12.16. *Three test pieces of the same material and of same length have their dimensions as shown in Fig. 12.12. They are subjected to the same gradually applied load find the ratio of the strain energies stress in them.*

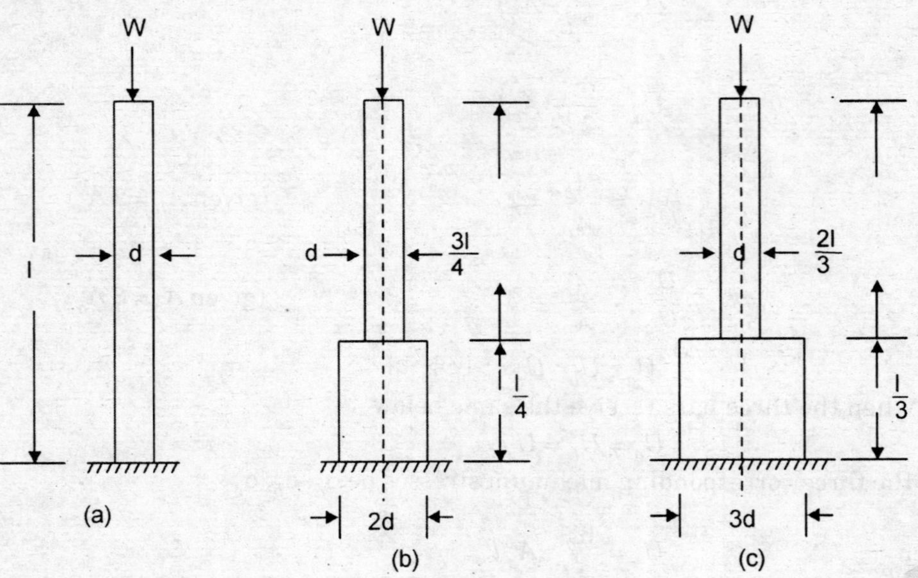

Fig. 12.12.

Solution. $\sigma_a = \dfrac{W}{A}$

For test piece *A*

$$U_a = \frac{\sigma^2}{2E} \times \left(\frac{\pi}{4} d^2 l\right)$$

$$= \left(\frac{4W}{\pi d^2}\right)^2 \frac{1}{2E}\left(\frac{\pi}{4} d^2 l\right)$$

$$= \frac{1}{2E}\left[\frac{W^2 \times 4}{\pi d^2} l\right] = \frac{W^2}{2E(\pi d^2)}(4l)$$

For test piece *B*

$$U_b = U_1 + U_2$$

For the portion of dia 'd'

$$U_1 = \left(\frac{4W}{\pi d^2}\right)^2 \frac{1}{2E} \times \frac{\pi}{4} d^2 \left(\frac{3l}{4}\right) = \frac{4W}{\pi d^2}\,\frac{1}{2E}\,\frac{3l}{4}$$

For portion of diameter = $2d$

$$U_2 = \left[\frac{4W}{\pi (2d)^2}\right]^2 \left(\frac{1}{2E}\right) \times \frac{\pi}{4} (2d)^2 \frac{l}{4}$$

$$= \frac{4W^2}{\pi (2d)^2}\,\frac{1}{2E} \times \frac{l}{4} = \frac{W^2}{\pi (2d)^2}\,\frac{l}{2E}$$

$\therefore$
$$U_b = U_1 + U_2$$

$$= \frac{4w^2}{\pi d^2}\left(\frac{1}{2E}\right)\frac{3l}{4} + \frac{4w^2}{\pi (2d)^2} \cdot \frac{1}{2E} \cdot \frac{l}{4}$$

$$= \left(\frac{1}{2E}\right)\frac{W^2}{\pi d^2}\left[3l + \frac{l}{4}\right]$$

$$= \frac{1}{2E}\left(\frac{W^2}{\pi d^2}\right)\left(\frac{13}{4} l\right)$$

$\therefore$
$$\frac{U_B}{U_A} = \frac{\left(\dfrac{1}{2E}\,\dfrac{W^2}{\pi d^2}\right)\left(\dfrac{13l}{4}\right)}{\left(\dfrac{1}{2E}\,\dfrac{W^2}{\pi d^2}\right)(4l)} = \frac{13}{4 \times 4}$$

$$= \frac{13}{16}$$

For test piece *C*

For portion of diameter

$$U_1 = \frac{\sigma^2}{2E} \times \frac{\pi}{4} d^2 \left(\frac{2l}{3}\right)$$

$$= \frac{\sigma^2}{2E} \frac{1}{2E} \left(\frac{4W}{\pi d^2}\right)^2 \cdot \left(\frac{\pi}{4} d^2\right) \frac{2l}{3}$$

$$= \left(\frac{1}{2E}\right) \frac{4W^2}{\pi d^2} \left(\frac{2l}{3}\right) \cdot \left(\frac{4^2}{\pi d^2}\right) \frac{1}{2E} \left(\frac{2l}{3}\right)$$

For portion of diameter = 3*d*

$$U_2 = \frac{1}{2E} \sigma^2 \times \frac{\pi}{4} (3d)^2 \left(\frac{l}{3}\right)$$

$$= \frac{1}{2E} \left[\frac{4W}{\pi (3d)^2}\right]^2 \left[\frac{\pi}{4} (3d)^2\right] \frac{l}{3}$$

$$= \frac{1}{2E} \frac{4W}{\pi (3d)^2} \frac{l}{3}$$

$$= \frac{1}{2E} \left(\frac{W^2}{\pi d^2}\right) \frac{4}{9} \cdot \frac{l}{3}$$

$$= \frac{W^2}{2E(\pi d^2)} \cdot \frac{4l}{27}$$

$$U_c = \frac{1}{2E} \left(\frac{W^2}{\pi d^2}\right) \left[\frac{2l}{3} + \frac{4l}{27}\right]$$

$$= \frac{1}{2E} \left(\frac{W^2}{\pi d^2}\right) \left(\frac{22}{27}\right)$$

$$\therefore \quad \frac{U_C}{U_A} = \frac{\frac{1}{2E} \left(\frac{W^2}{\pi d^2}\right) \frac{22}{27} l}{\frac{1}{2E} \left(\frac{W^2}{\pi d^2}\right) (4l)}$$

$$= \frac{22}{27 \times 4} = \frac{11}{54}$$

or $$U_A : U_B : U_C :: 1 : \frac{13}{16} : \frac{11}{54}$$

Example 12.17. *A round bar 2 m long and 25 mm diameter hangs vertically from one end. At the lower end a rigid collar is attached. A mass of 2000 kg falls from a height of 100 mm on the rigid collar. Find (a) the instantaneous maximum stress induced in the bar. (b) If the stretch in the bar is neglected, then what will be its value?*

Take $E = 200\ GPa$

Solution :

$$g = 9.827\ \text{m/sec}^2 = 9827\ \text{mm/sec}^2$$

$$M = 2000\ \text{kg}$$

$$h = 100\ \text{mm}$$

$$d = 25\ \text{mm}^2$$

$$\therefore \quad A = \frac{\pi}{4} \times (25)^2 = 490.874\ \text{mm}^2$$

$$l = 2000\ \text{mm}$$

From equation (12.32)

$$\sigma = \frac{Mg}{A}\left[1 + \sqrt{\frac{1 + 2\ A\ E\ h}{Mgl}}\right]$$

$$= \frac{2000 \times 98.27}{490.874}\left[1 + \sqrt{\frac{1 + 2 \times 490.874 \times 200 \times 10^3 \times h}{2000 \times 9827 \times 2000}}\right]$$

$$= 40.038 \times 10^3 \left[1 + \sqrt{1.07067}\right]$$

$$= 40.038 \times 10^3\ (2.034)$$

$$= 81.557 \times 10^3\ \text{N/mm}^2$$

(*b*) When the sketch in the bar is neglected, then we have from equation (12.33)

$$\sigma = \left[\frac{2\ Mg\ h\ E}{A\ l}\right]^{\frac{1}{2}}$$

$$= \left[\frac{2 \times 2000 \times 9827 \times 100 \times 200 \times 10^3}{490.874 \times 2000}\right]^{\frac{1}{2}}$$

$$= 28.297\ \text{N/mm2 or MPa}$$

Example 12.18. *A load W acts at end of a cantilever. AB of length L (Fig. 12.13) State whether the moment of this load W at the section A is bending movement or twisting moment. Explain why ?*

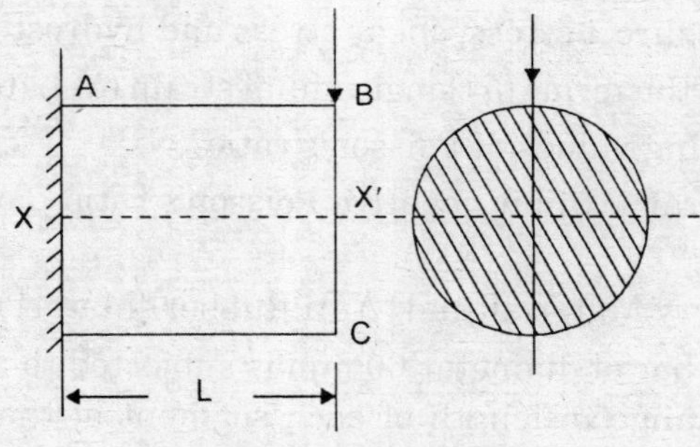

Fig. 12.13

Solution : $W \times L = M$ is the bending movement because the plane of the moment is the plane of the paper. Then the axis will be normal to the plane. Since longitudinal axis of the leam is in the plane of the paper, the axis of the moment is the axis of the beam, therefore it is a bending moment.

Example 12.19. *In Figure 12.14 is shown we circular section bar AB with its end A fixed in the wall so that AB acts like a cantilever. At the end B is fixed a wheel of radius 'r' and at the periphery of the wheel is fixed a rope from which a load W is hanging. State whether at the section A of the cantilever there is a BM or twisting moment and what are their values?*

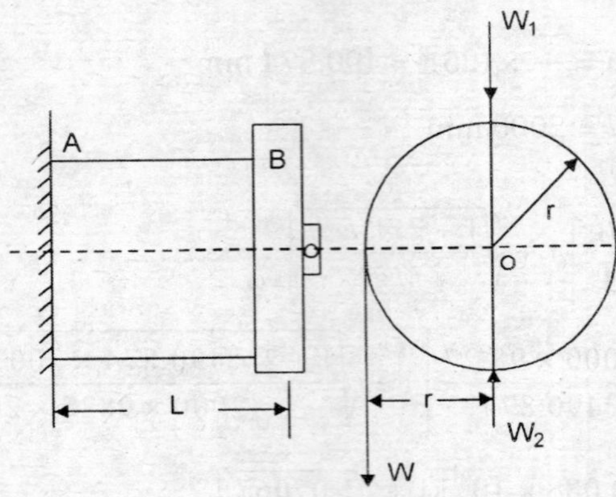

Fig. 12.14

Solution. The system is in static equilibrium condition. At the point *O*, centre of the wheel from our own side we apply two equal and opposite loads. W_1 and W_2

Such that $W_1 = W_2 = W$

Now *W* and W_2 being equal and opposite in direction and are at a distance equal to '*r*' they constitute a couple the moment of which is $(W \times r)$ with the plane of the wheel as its plane. Naturally the axis of this moment will be normal to the plane of the wheel as seen in side view. Thus, it axis will coincide with axis of the bar *AB*. Thus (*Wr*) moment is the twisting moment.

In addition the load *W*, at the end *B* will give rise to moment $W \times L$ which will be bending moment at section A as explained in. Example 12.18.

PROBLEMS

1. Define and explain direct stress, shear stress and hydrostatic stress.
2. Define and explain the terms (*a*) longitudinal strain (*b*) Lateral strain (*c*) Poisson's ratio.
3. What are the limiting values of Poisson's ratio?
4. What type of materals exhibit negative Poisson's ratio ? Where do such materials find their utility ?
5. Explain the 'Rubbery Material' and "Anti Rubber "Materials" Wehre are these used ?
6. A low carbon steel bar of diameter 60 mm is subjected to axial forces, as shown in Fig. 12.15. Find the change in length of each segment of the bar and then calculate the relative movement of the end *D* with respect to end A of the bar.

Take $E = 200$ GPa

Draw the force distribution diagram. (**Ans.** $d_{AB} = 0.53.5$ mm, $d_{BC} = 40.4244$ mm, $d_{CD} = -0.3536$ mm, $d_{AD} = -0.3697$ mm, *i.e.* *A* and *D* conclusion)

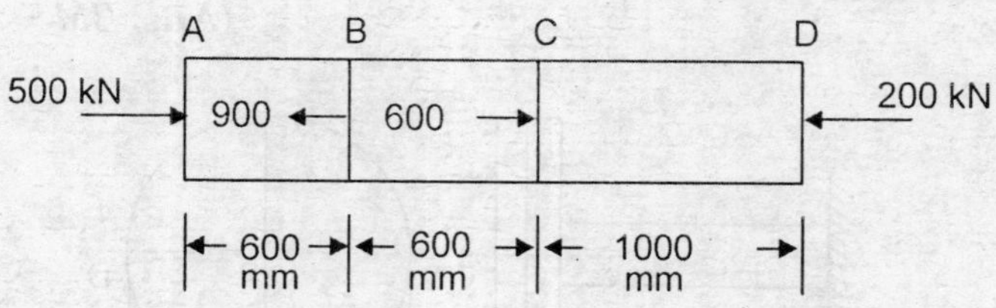

Fig. 12.15

7. A steel piece of the shape, shown in Fig. 12.16 is subjected to an axial pull of 10 kN.

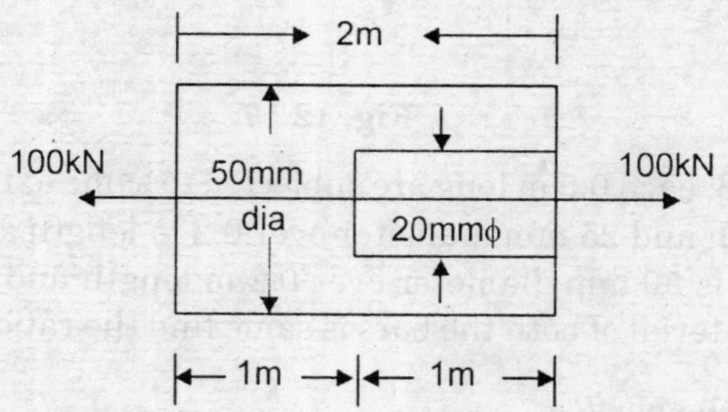

Fig. 12.16

Find the extension of the piece *AC* Take E = 200 GPa

8. A hole of 20 mm diameter is to be punched in a plate of 25 mm thickness (*i*) If the ultimate shear stress of the plate material is 360 MPa find the force required to punch the hole. (**Ans.** 565.5 kN)

9. A steel bar 1 m long and 20 mm diameter is subjected to an axial pull of 25 kN. Find the change is length, change in diameter and volumetric strain.

Take $E = 200$ GPa

and Poisson ratio = 0.25 (**Ans.** $dl = 3.979 \times 10^{-4}$ mm,

$d = 2.4864 \times 10^{-6}$ mm, $\frac{dV}{V} = 5.9655 \times 10^{-7}$)

10. A cylindrical test piece of diameter 25 mm was subjected to a tensile load of 40 kN. The measured extension was 0.08 mm over a gauge length of 200 mm and the reduction in diameter was 0.0025 mm. Find youngs modulus, Poisson's ratio, bulk modulus and shear modulus.

(**Ans.** $E = 203.72 \times 10^3$ MPa $n = 0.25$, $K = 135.8 \times 10^3$ MPa $G = 81.49 \times 10^3$ MPa)

11. A steel ball of 100 mm diameter is inmersed in sea water to a depth where the hydrostalic pressure is 100 MPa. If the value of E = 200 GPa and n = 0.27 for steel, calculate the change in volume (decrease) of steel ball. (**Ans.** 1.021 mm^3)

12. A steel bar AB, 2m long is fixed at the end A and at end B, a rigid circular disc. of diameter 400 mm is keyed. At the periphery of the disc a force of 10 kN is applied in the vertically downward direction as shown in the figure (12.17). Find the values of bending moment and twisting moment at the fixed end *A* of the bar.

(**Ans.** BM = 20 kN-m, T = 2 kN-m)

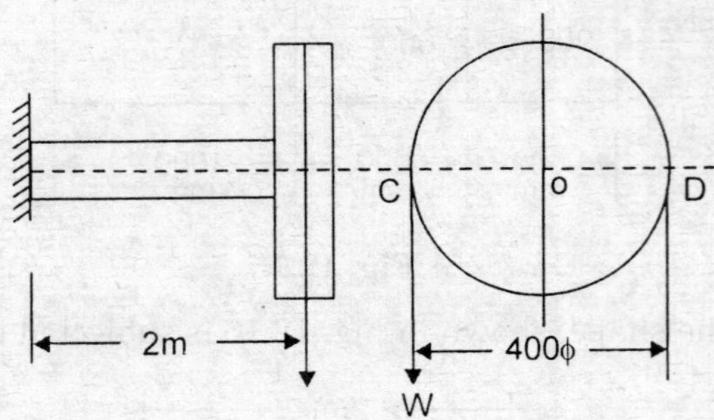

Fig. 12.17

13. Two bars A and B, each 0.5 m long are subjected to same axial blow. A is 50 mm diameter over 0.3 m length and 25 mm diameter over 0.2 m length and 25 mm diameter over 0.2 m length and B is 50 mm diameter over 0.2 m length and 25 mm diameter over 0.3 m length. If the material of both the bars is same find the ratio of maximum induced in A to that induced in the bar *B*.

(**Ans.** $\sqrt{\frac{14}{11}}$)

13. A bar 10 mm diameter and 3 m long is fixed the top and hangs vertically. At the lower and is fixed is rigid disc and on this disc a mass of 1 kg is allowed to be fall freely through a distance of 50 mm. Calculate the maximum stress induced in the rod

Take E = 200 GPa. (**Ans.** 92.5 MPa)

CHAPTER 13

ENGINEERING MATERIALS

13.1 MATERIAL AND CIVILIZATION

Knowledge of the existing materials and the capacity to develop new materials is the basic strength of a nation. Although a larger segment of Indian population is engaged in the field of agriculture still, not to be left behind the developing countries. The Govt. of India after achieving independence started Five Year Plans for all round development in terms of infrastructure, communication and power. In order to achieve success in these fields not only the knowledge of the existing materials but development of new ones, which are stronger cheaper and lighter in weight, is essential. An engineer starts with the idea for a better part or for a whole new concept, such as an airplane that can take off on land or water in a vertical direction. Eventually, a particular design approach is selected and its manufacture executed. This prototype is then put to tests to assess its performance. In order to optimize its performance various design parameters are changed. In so doing the major role is played by material to be used. It is not only the cost of material but the ease with which it can be subjected to various operations so as to get the desired shape of the part.

13.2 CLASSIFACATION OF MATERIALS

In general the engineering materials are classified as under:

(a) Ferrous metals and their alloys

(b) **Non-ferrous metals and their alloys:** These are further sub classified as Aluminium and/or copper based alloys like brass and bronze.

(c) **Polymers:** Natural and Synthetic. Thermoplastic and thermosetting

(d) **Ceramics:** Glass. Optical fiber glass

(e) **Composites:** Fiber reinforced metal matrix.

(f) **Smart materials:** Piezoelectric, Shape memory, Thermo-chromic, photo-chromic and magneto-chromic

13.3 DEFINITION OF BASIC TERMS ASSOCIATED WITH PROPERTIES

(a) **ELASTICITY:** When a body is stressed in tension or compression up to a level less than the elastic limit and the stress is removed the body regains back the original shape. This property is known as elasticity. Refer to fig.11.8 Article 11.4 of this book). The Stress strain curve while increasing load and while decreasing load is same. The theories developed for this property are applied in the design of machine and structure members.

(b) **PLASTICITY:** It is property of the material to undergo permanent change in shape when the stress applied is more than that at elastic limit point and less than the ultimate stress. The strain at any point within these limits is the sum of elastic and plastic components of strain. When the external stress is removed the stress strain curve follows a straight line path on return journey parallel to the original elastic line. (Refer to Fig. 11.8 c of article 11.4) .When the load is removed there will be left some strain equal to the plastic strain component and the elastic strain component will be recovered. This is also called the spring back. The theories developed for the property are used in the processes where the shape change of the part is required, for example when a circular sheet blank is formed into a tumbler with the help of a die.

(c) **STRENGTH:** It is the maximum load which the material can sustain without breaking. This is equal to the value of the stress at ultimate point (Refer to Fig.11.8)

(d) **DUCTILITY:** It has, hither to, been defined as the property of the material due to which could be drawn into wire. It is very well known to the people that gold is more ductile than copper, copper is more ductile than aluminium and so on . In other words this property was known qualitatively and had not been defined quantitatively. It was in late fifties of the last century that during an international meet the scientists put their stamp on the proper definition of this term. According to this the strain to which a uni-axially stressed specimen can be subjected prior to the instability point (Ultimate limit Point) is called the ductility of the material. In fig. 11.8, OH is the measure of ductility.

(e) **MALLEABILITY:** Similar to ductility, malleability is the property of the material due to which it can be deformed into plate. Quantitatively it is defined as the strain at instability point.

(f) **BRITTLENESS:** It is defined as the tendency of a material to fracture or fail upon application of a small amount of force, impact or shock. A material is brittle if, when subjected to stress, it breaks without significant deformation (strain) Brittle materials absorb relatively very little energy. Examples are Glass, cast iron.

(g) **TOUGHNESS:** It is the property of the material body to absorb energy and plastically deform without fracture.

(h) **RESILIENCE:** Resilience is the strain energy which can be recovered from a deformed body when the load causing the stress in the body is removed, provided the stress was less than the limit of proportionality. When the stress is equal to the limit of proportionality then the energy stored per unit volume of the body is called proof resilience or modulus of resilience for that material.

(i) **HARDNESS:** Hardness of a material is the resistance to localized deformation due to indentation, scratching or bending.

(j) **MACHINABILITY:** Machinability is an involved term with many ramifications. However simply stated, it is the ease with which the metal can be removed in operations like turning, milling, drilling, reaming, sawing and threading. Ease of metal removal implies, among other things, that the forces acting against the cutting tool will be relatively low, that the chip will be easily broken up, that a good finish will result, and the tool will last a reasonable good period of time before it has to be replaced or resharpened Another way of expressing this is to give each material a " machinability rating" In USA it has been done for ferrous metals, using the American Iron and Steel Institute (AISI) B1112 as the basis of 100 % machinability. For example the machinability rating of one type of stainless steel is 60 % and for an aluminium alloy it is 240%.

(k) **FORMABILITY:** Formability is the ability of metal to be formed It is based on the ductility of the metal, which in turn is related to its crystal structure. Face-centered cubic crystals have the greatest opportunity for slip resulting in greater ductility. Formability is also dependent upon grain size. Large grain size materials are more ductile because they have relatively less grain boundary area and are used for tough jobs like deep drawing.

(l) **WELDABILITY:** Weldability is defined as the ease with which the quality welds can be made.

13.4 CLASSIFICATION OF CARBON STEELS

Following is the classification of carbon steels based on the Carbon content:

(1) LOW CARBON STEEL: Carbon percentage in low carbon steels varies from 0.05 to Approximately 0.23 %. These can be case- carburized to increase carbon percentage In the surface layers to increase surface hardness. Composition of a few such steels are given in table 13.1.

TABLE 13.1 Part (A) for low carbon steels

C%	Mn %	S%	HOT Yield Stress MPa	ROLLED Ultimate stress MPa	COLD Yield stress MPa	DRAWN Ultimate stress MPa
0.13to 0.15	0.7to 1.0	0.10 to 0.15	532	942	835	988
0.14 to 0.2	1.0 to 1.3	0.08 to 0.13	623	1033	912	1064
0.14 to 0.23	1.3 to 1.6	0.08 to 0.13	654	1064	988	1140

TABLE 13.2 PART (B) for medium and high carbon srteels

0.32 to 0.39	1.31 to 1.65	0.08 to 0.13	836	1444	1216	1368
0.37 to 0.45	1.35 to 1.65	0.04 to 0.15	988	1520	1292	1444
0.45 to 0.55	0-.70 to 1.00	0.05 to 0.14	988	1672	1368	1596

(2) MEDIUM CARBON STEEL: Carbon percentage in medium Steels varies from 0.30 to approximately 0.60

(3) HIGH CARBON STEEL: Carbon percentage in high carbon steel varies from 0.60 to approximately 0.95%.

The machinability of low carbon steels is approximately 55% to 60 % that of B1112. These steels are soft and ductile these steels can easily be hot rolled and are commercially marketed as' Hot Rolled Steels' (HRS) of different sections such as I, C, T, Round, Rectangular and Angle sections of different sizes are available. Similarly these are also available as Cold Rolled Steels (CRS)., which essentially is due to their good qualities of machinability and formability. As they are soft, they have a tendency to build up on the cutting edge of the tool which in turn renders the cutting process inefficient. Evidently free machining steel bars are annealed and then machined as per the design. These components are heat treated by heating and quenching to harden the medium and high carbon steels. Medium-carbon steels are better in this respect, even though the cutting forces are higher. For this reason their use in making machine parts or structure members is much more than low-cartbon steels. Their hardness is more than that of the low-carbon steels.High-carbon steels, when heat treated are very hard.The machinability is very low as compared to medium Carbon steels. Even heat treated hardened high carbon steel tools are used to cut low carbon steels. Carbon content of the upper layers of a low carbon steel part can be increased by the process known as 'Case-carburizing'. In so doing the surface of the part becomes hard to render it wear-resistant. The inner portion or the core remains tough. However in the case of high carbon steel the Hardness is increased throughout the entire depth after Heat treatment.

13.4.1 Free-machining Steels, Alloy Steels

In order to optimise the total fabrication cost by faster cutting-speeds and greater feed rates, free-machining Steels have been developed. They include alloy and stainless steels. The machinability is affected by a large number of factors including the normal constituents of steel, the process of its making and subsequent treatment. However, for the present text following few main additives which distinguish free-machining steel from the others. Most common additives are sulfur, phosphorus, lead and selenium for carbon steels. For each of these individual additives, American Iron and Steel Institute has published data for their machinability and other properties. Similarly chromium, molybdenum, nickel are added individually or in combination for making free -machining alloy steels.

13.5 HIGH-STRENGTH LOW-ALLOY STEELS

High-strength low- alloy steels are defined as steels having higher yield strength than plain carbon steels. The essential contents of these steels are C = 0.05 to 0.2 %, Mo = 0.1 to 0.2%

M*n* = 2.0 %. However in certain steels, C*u*, C*r*, N*i*, Titanium, Z*n* and Vanadium are added individually or in combination in small quantities. Maximum copper content is 0.5 % and that of Vanadium is 0.005 %. The yield strength of such steels can be raised upto a level of 1075 MPa while the yield strength of plain low-carbon steels is of the order of 350 MPa. For this reason they are extensively used in the construction of bridges and other steel structures.

13.5.1 Ultra-High-Strength Steels

The addition of N*i*, C*r*, and Mo as alloying elements individualy or in combination, in medium-carbon steels results in raising the yield strength to a level of 1380 MPa. Depending upon the content of alloying elements these are named as follows:

(a) Ni- steel.

(b) Ni- Cr steel.

(c) Mo- steel

(d) Cr- Mo steel

The carbon content in these steels is 2.0% and they are tempered and quenched to achieve the desired strength.

13.6 TOOL STEELS

Following are the types of tool steels

***(a)* Carbon Tool Steels:** These steels contain no alloying metals and the carbon content is 1.10 to 1.30%. This is used for making tools to be used for cutting operations where a hard cutting edge is required at low cutting speed. It is used for forming tools, particularly where it is essencial to form the tool before hardening.

***(b)* HIGH SPEED STEEL (HSS):** High speed steel is a steel containing at least 12% Tungsten, 7% molybdenum or equivalent combination of these two, together with appreciable percentage of chromium and Vanadium and, at times, cobalt. HSS tools are much more Generally used than those of other tol material. They combine Superior cutting qualities, strength, toughness, and relatively Low first cost for a wide range general machining operations.

Composition of various high speed steels is given in table 13.1

13.7 STAINLESS STEELS

Carbon steels tend to corrode in damp atmospheric conditions. An addition of a minimum of 17% chromium and a minimum of 7% nickel in low carbon steels will not allow it to corrode. Such steels are known as stainless steels. A typical composition of 18% chromium, 8% nickel. is popularly known as 18-8 stainless steel.

13.7.1 Stainless Steels (Magnetic Properties)

All austenitic grade stainless fully annealed steels are non- Magnetic. The situation is, however, different when these steels are subjected to cold working by processes such as Rolling, shot blasting, grinding etc, their permeability starts increasing and start behaving as if they have acquired magnetic properties. Further if the cold worked stainless steel is subjected to stress relieving process (heating upto 537 degrees Celsius followed by slow cooling in furnace), the magnetic properties are lost. And the steel becomes non-magnetic. As a matter of fact with cold working due to deformation is martensite is induced, which is ferromagnetic phase. Further ferrite phase stainless steels are ferromagnetic and are used for soft magnetic components such as solenoid cores.

13.8 CAST IRON: CLASSIFICATION

The distinguishing feature of all cast irons (CI) is that they have a relatively high carbon content. Steels range up to 2% carbon. Cast irons overlap with steels somewhat and range from 1.5% upto 5% carbon. It is principally the form of the carbon which is governed by thermal conditions and alloying elements, that provides various structures. These are classified into the following five main types:

(1) GRAY CAST IRON

(2) WHITE CAST IRON

(3) MALLEABLE CAST IRON

(4) DUCTILE CAST IRON (NODULAR CAST IRON)

(5) COMPACTED GRAPHITE CAST IRON.

GRAY CAST IRON: The terms gray and white CI refer to the appearance of the fractured section. Gray CI has grayish appearance and in the later it is whitish. The grayish appearance is due to the large amount of graphite flakes on the broken surface and the large size flakes are the result of slow cooling rate of the solidifying molten metal. If the cooling rate is faster it will result in leaving the carbon in the combined form as cement for which the common name is' iron carbide 'and its appearance is whitish.

WHITE CAST IRON: AS pointed above the white cast iron is produced by cooling the cast iron rapidly.The carbon is in the combined form known as cementite. Commonly it is also refered to as' chilled cast iron'. This structure is very hard and brittle. The surface is very shining. If the alloying elements such as Chromium, Molybdenum and nickel are added martensite will be formed. Irons of this type are very hard and wear resistant and find their use in the making of crusher jaws, balls, crusher rollers, hammer mills and the like. The compressive strength of gray CI is its outsranding feature. For example the average tensile strength of CI is 250 MPa and its compressive strength is 1040 MPa. Gray cast iron is most widely used of all types of CI. typical applications are: Engine blocks of automobiles, pipe fittings, agricultural equipment, bath tubs, household appliances, machine tools, housings of gear boxes, etc.

MALLEABLE IRON: Malleable iron is made from white CI. The casting of white cast iron is put in furnace, where temperature is maintained at 870 degrees C, for 48 to 60 hours. Still keeping it in furnace it is cooled at a rate 8 to 10 degrees per hour until the temperature falls to a level in the range 677 to 704 degrees C. It is held at this temperature for a period of 24 hours. The furnace is then opened and allowed to cool to room temperature. During this time the combined carbon of white CI separates into graphite and ferrite. In case pearlitic malleable iron is required the rate of cooling is Is made faster and adjusted so as to get the desired structure.As the very name implies, malleable irons are softer than gray iron and are, therefore, used where greater toughness and shock resistance are required such as in automobile parts, Farm implements, small tools, hardware, and pipe fittings.

DUCTILE IRONS (NODULAR IRON), (SPHEROIDAL IRON): As pointed earlier cast iron is brittle. However if a small quantity of magnesium, about 453 gram per 1000 kg of iron is added, it causes the flake graphite to take on a spherical shape graphite in a spherical shape has minimum surface for a given volume. Therefore, there are fewer discontinuities in the surrounding metal, thereby giving it far more strength, ductility., toughness and wear resistance. Ductile iron goes by other names such as "Nodular iron" and "Spheroidal iron".

COMPACTED GRAPHITE IRON (CGI): CGI became commercially available as dependable methods for its production were developed. Its graphite occurs as blunt flakes that are interconnected within each cell. This graphite structure and the resulting properties of the iron are intermediate between gray and ductile irons. The compacted graphite shape has been known for some time and also has been called quasiflake, aggregated flake, seminodular and vermicular graphite.

Its production is similar to that for ductile iron in requiring close metallurgical control and rare earth element additions, but it also requires another alloying element such as titanium to minimize the formation of spheroidal graphite. CGI retains much of the castability of gray iron, but has a higher tensile strength and some ductility. Its matrix structure can be adjusted by alloying or heat treatment.

TABLE 13.3 High speed steels

TYPE	C%	Tungston%	Mo%	Cr%	Va%	Cobalt%
18-4-1	0.7 to 0.75	18.0	--	4.0	1.0	--
18-4-2	0.8 to 0.85	18.0	0.75	4.0	2.0	--
18-4-3	0.85 to 1.1	18.0	--	4.0	3.0	--
14-4-2	0.7 to 0.75	14.0 --	--	4.0	2.0	

TABLE 13.4 Cobalt Steels

18-4-1+4 % cobalt						
	0.7 to 0.75	18.0	0.5	4.0	1.0	4.0
18-4-2+ 7 % cobalt						
	0.75 to 0.85	18.0	0.75	4.0	2.0	7.0

13.9 NON-FERROUS METALS (LIGHT METATS AND ALLOYS)

The age of air travel and mass transportation of men and material has attracted the attention of engineers to develop light but strong materials. Metals commonly classed as light metals are those having density less than that of steel (7800 kg per cubic metre). Common metals in this class are aluminum, magnesium, beryllium and titanium.

13.9.1 Aluminum Alloys

Aluminium alloys are alloys in which aluminium (Al) is the predominant metal. The typical alloying elements are copper, magnesium, manganese, silicon, and zinc. There are two principal classifications, namely casting alloys and wrought alloys, both of which are further subdivided into the categories heat-treatable and non-heat-treatable. About 85% of aluminium is used for wrought products, for example rolled plate, foils and extrusions. Cast aluminium alloys yield cost effective products due to the low melting point, although they generally have lower tensile strengths than wrought alloys. The most important cast aluminium alloy system is Al-Si, where the high levels of silicon (4.0% to 13%) contribute to give good casting characteristics. Aluminium alloys are widely used in engineering structures and components where light weight or corrosion resistance is required.[1]

Alloys composed mostly of the two lightweight metals aluminium and magnesium have been very important in aerospace manufacturing since somewhat before 1940. Aluminium-magnesium alloys are both lighter than other aluminium alloys and much less flammable than alloys that contain a very high percentage of magnesium.

Aluminium alloy surfaces will keep their apparent shine in a dry environment due to the formation of a clear, protective layer of aluminium oxide. In a wet environment, galvanic corrosion can occur when an aluminium alloy is placed in electrical contact with other metals with more negative corrosion potentials than aluminium.

Aluminium alloy compositions are registered with The Aluminum Association. Many organizations publish more specific standards for the manufacture of aluminium alloy, including the Society of Automotive Engineers standards organization, specifically its aerospace standards subgroups,[2] and ASTM International.

Alloy designations

Wrought and cast aluminium alloys use different identification systems. Wrought aluminium is identified with a four digit number which identifies the alloying elements.

Cast aluminium alloys use a four to five digit number with a decimal point. The digit in the hundreds place indicates the alloying elements, while the digit after the decimal point indicates the form (cast shape or ingot).

Temper designation

The temper designation follows the cast or wrought designation number with a dash, a letter, and potentially a one to three digit number, e.g. 6061-T6. The definitions for the tempers are:[4][5]

-F

As fabricated

-H

Strain hardened (cold worked) with or without thermal treatment

-H1

Strain hardened without thermal treatment

-H2

Strain hardened and partially annealed

-H3

Strain hardened and stabilized by low temperature heating

Second digit

A second digit denotes the degree of hardness

-HX2 = 1/4 hard

-HX4 = 1/2 hard

-HX6 = 3/4 hard

-HX8 = full hard

-HX9 = extra hard

-O

Full soft (annealed)

-T

Heat treated to produce stable tempers

-T1

Cooled from hot working and naturally aged (at room temperature)

-T2

Cooled from hot working, cold-worked, and naturally aged

-T3

Solution heat treated and cold worked

-T4

Solution heat treated and naturally aged

-T5

Cooled from hot working and artificially aged (at elevated temperature)

-T51

Stress relieved by stretching

-T510

No further straightening after stretching

-T511

Minor straightening after stretching

-T52

Stress relieved by thermal treatment

-T6

Solution heat treated and artificially aged

-T7

Solution heat treated and stabilized

-T8

Solution heat treated, cold worked, and artificially aged

-T9

Solution heat treated, artificially aged, and cold worked

-T10

Cooled from hot working, cold-worked, and artificially aged

-W

Solution heat treated only.

Note: W is a relatively soft intermediary designation that applies after heat treat and before aging is completed. The -W condition can be extended at extremely low temperatures but not indefinitely and depending on the material will typically last no longer than 15 minutes at ambient temperatures.

Wrought alloys

The International Alloy Designation System is the most widely accepted naming scheme for wrought alloys. Each alloy is given a four-digit number, where the first digit indicates the major alloying elements.

- 1000 series are essentially pure aluminium with a minimum 99% aluminium content by weight and can be work hardened.
- 2000 series are alloyed with copper, can be precipitation hardened to strengths comparable to steel. Formerly referred to as duralumin, they were once the most common aerospace alloys, but were susceptible to stress corrosion cracking and are increasingly replaced by 7000 series in new designs.
- 3000 series are alloyed with manganese, and can be work hardened.
- 4000 series are alloyed with silicon. They are also known as silumin.
- 5000 series are alloyed with magnesium.
- 6000 series are alloyed with magnesium and silicon, are easy to machine, and can be precipitation hardened, but not to the high strengths that 2000 and 7000 can reach.
- 7000 series are alloyed with zinc, and can be precipitation hardened to the highest strengths of any aluminium alloy.
- 8000 series is a category mainly used for lithium alloys.

Wrought aluminium alloy composition limits (% weight)

Alloy	Si	Fe	Cu	Mn	Mg	Cr	Zn	V	Ti	Bi	Ga	Pb	Zr	Limits††		Al
														Each	Total	
1050[6]	0.25	0.40	0.05	0.05	0.05			0.05						0.03		99.5 min
1060	0.25	0.35	0.05	0.03	0.03	0.03	0.05	0.05	0.03	0.03	0.03	0.03	0.03	0.03		99.6 min
1100	0.95 Si+Fe	0.05-0.20	0.05				0.10							0.05	0.15	99.0 min
1199[6]	0.006	0.006	0.006	0.002	0.006		0.006	0.005	0.002		0.005			0.002		99.99 min
2014	0.50-1.2	0.7	3.9-5.0	0.40-1.2	0.20-0.8	0.10	0.25		0.15					0.05	0.15	remainder
2024	0.50	0.50	3.8-4.9	0.30-0.9	1.2-1.8	0.10	0.25		0.15					0.05	0.15	remainder
2219	0.2	0.30	5.8-6.8	0.20-0.40	0.02		0.10	0.05-0.15	0.02-0.10				0.10-0.25	0.05	0.15	remainder
3003	0.6	0.7	0.05-0.20	1.0-1.5			0.10							0.05	0.15	remainder
3004	0.30	0.7	0.25	1.0-1.5	0.8-1.3		0.25							0.05	0.15	remainder
3102	0.40	0.7	0.10	0.05-0.40			0.30		0.10					0.05	0.15	remainder
4043	4.5-6.0	0.80	0.30	0.05	0.05		0.10		0.20					0.05	0.15	remainder
5052	0.25	0.40	0.10	0.10	2.2-2.8	0.15-0.35	0.10							0.05	0.15	remainder
5083	0.40	0.40	0.10	0.40-1.0	4.0-4.9	0.05-0.25	0.25		0.15					0.05	0.15	remainder
5086	0.40	0.50	0.10	0.20-0.7	3.5-4.5	0.05-0.25	0.25		0.15					0.05	0.15	remainder
5154	0.25	0.40	0.10	0.10	3.10-3.90	0.15-0.35	0.20		0.20					0.05	0.15	remainder
5356	0.25	0.40	0.10	0.10	4.50-5.50	0.05-0.20	0.10		0.06-0.20					0.05	0.15	remainder
5454	0.25	0.40	0.10	0.50-1.0	2.4-3.0	0.05-0 20	0.25		0.20					0.05	0.15	remainder
5456	0.25	0.40	0.10	0.50-1.0	4.7-5.5	0.05-0.20	0.25		0.20					0.05	0.15	remainder
5754	0.40	0.40	0.10	0.50	2.6-3.6	0.30	0.20		0.15					0.05	0.15	remainder
6005	0.6-0.9	0.35	0.10	0.10	0.40-0.6	0.10	0.10		0.10					0.05	0.15	remainder
6005A†	0.50-0.9	0.35	0.30	0.50	0.40-0.7	0.30	0.20		0.10					0.05	0.15	remainder
6060	0.30-0.6	0.10-0.30	0.10	0.10	0.35-0.6	0.05	0.15		0.10					0.05	0.15	remainder
6061	0.40-0.8	0.7	0.15-0.40	0.15	0.8-1.2	0.04-0.35	0.25		0.15					0.05	0.15	remainder
6063	0.20-0.6	0.35	0.10	0.10	0.45-0.9	0.10	0.10		0.10					0.05	0.15	remainder
6066	0.9-1.8	0.50	0.7-1.2	0.6-1.1	0.8-1.4	0.40	0.25		0.20					0.05	0.15	remainder
6070	1.0-1.7	0.50	0.15-0.40	0.40-1.0	0.50-1.2	0.10	0.25		0.15					0.05	0.15	remainder
6082	0.7-1.3	0.50	0.10	0.40-1.0	0.60-1.2	0.25	0.20		0.10					0.05	0.15	remainder
6105	0.6-1.0	0.35	0.10	0.10	0.45-0.8	0.10	0.10		0.10					0.05	0.15	remainder

Alloy	Si	Fe	Cu	Mn	Mg	Cr	Zn	V	Ti	Bi	Ga	Pb	Zr	Limits††		Al
														Each	Total	
6162	0.40-0.8	0.50	0.20	0.10	0.7-1.1	0.10	0.25		0.10					0.05	0.15	remainder
6262	0.40-0.8	0.7	0.15-0.40	0.15	0.8-1.2	0.04-0.14	0.25		0.15	0.40-0.7		0.40-0.7		0.05	0.15	remainder
6351	0.7-1.3	0.50	0.10	0.40-0.8	0.40-0.8		0.20		0.20					0.05	0.15	remainder
6463	0.20-0.6	0.15	0.20	0.05	0.45-0.9		0.05							0.05	0.15	remainder
7005	0.35	0.40	0.10	0.20-0.70	1.0-1.8	0.06-0.20	4.0-5.0		0.01-0.06				0.08-0.20	0.05	0.15	remainder
7022	0.50	0.50	0.50-1.00	0.10-0.40	2.60-3.70	0.10-0.30	4.30-5.20		0.20					0.05	0.15	remainder
7072	0.7 Si+Fe	0.10	0.10	0.10		0.8-1.3								0.05	0.15	remainder
7075	0.40	0.50	1.2-2.0	0.30	2.1-2.9	0.18-0.28	5.1-6.1		0.20					0.05	0.15	remainder
7079	0.3	0.40	0.40-0.80	0.10-0.30	2.9-3.7	0.10-0.25	3.8-4.8		0.10					0.05	0.15	remainder
7116	0.15	0.30	0.50-1.1	0.05	0.8-1.4		4.2-5.2	0.05	0.05		0.03			0.05	0.15	remainder
7129	0.15	0.30	0.50-0.9	0.10	1.3-2.0	0.10	4.2-5.2	0.05	0.05		0.03			0.05	0.15	remainder
7178	0.40	0.50	1.6-2.4	0.30	2.4-3.1	0.18-0.28	6.3-7.3		0.20					0.05	0.15	remainder

†Manganese plus chromium must be between 0.12-0.50%.

††This column lists the limits that apply to all elements, whether a table column exists for them or not, for which no other limits are specified.

5000 series

5005

Aluminium alloy 5005 is used in decorative and architectural applications that require an anodized finish.[7]

13.9.2 Fabrication Characteristics of Aluminium

MACHINABILITY: Aluminium alloys are too soft for good machinability. Due to lower value of modulus of elasticity the deflection will be much higher under heavy cuts. Similar effect will be there while clamping the work piece. Both these leed to in accuracies in dimentions of finished product.

FORMABILITY: Most aluminium alloys can readily be formed cold. They consume less energy.

WELDABILITY: Since Aluminium has very high thermal conductivity, more heat will be required to weld two pieces as lot of heat is lost to air around.

CASTABILITY: Aluminium is one of the more easily cast material. Alloying elements which are added in molten aluminium, to facilitate casting are Silicon, magnesium, and copper. Siliucon increases the fluidity of molten aluminium allowing it to flow farther in thin walls of the cavity. It also reduces the coefficient of expansion.Aluminium castings containing 8 % magnesium respond very well to Heat treatment. However casting becomes more difficult.

Copper when added to Al ,will harden it. It increases the strength tremendously in both the heat treated and non-heat treated conditions.

Duralumin is an alloy made up of 90% aluminum, 4% copper, 0.5%-1% magnesium

13.10 COPPER AND COPPER BASED ALLOYS

Commercially pure copper is available in several grades, All of which have the same mechanical properties. "Electrolytic tough pitch copper" is susceptible to embrittlement when heated in reducing atmosphere, but it has high electrical conductivity. "Deoxidised copper" has lower electrical conductivity but very good cold working characteristics without getting brittle. Modified coppers are as follows:

(a) Tellurium copper. It contains tellurium 0.5 % and induces free cutting characteristics. Selenium and lead are also used for this purpose)

(b) Tellurium-nickel copper is an age hardening alloy as high strength, high conductivity and excellent machinability.

13.11 BRASSES

Brasses are principally alloys of copper and zinc. They are referred to by percentage of each, for example 70-30 brass means 70 % copper and 30 % zinc. Small quantities of other elements such as lead, tin, or aluminium are added to obtain the desired colour, strength, ductility, machinability, corrosion resistance or a combination of these properties.

13.12 BRONZE: BRONZES ARE MAINLY COPPER- TIN ALLOYS

13.12.1 Classification of Brasses

***(a)* Alfa brasses:** They contain less than 36 % zinc and are single phase alloys.

***(b)* Alfa-beta brasses:** Contain more than 36 % zinc and are two phase structure.

***(c)* Leaded brasses:** Contain upto 88.5 % copper and upto 3.25 % lead.

***(d)* Tin brasses:** The percentage of tin is low, 0.75 to 1.0 % Known as admiralty and naval brass.

***(e)* Nickel silver:** contains copper 65 %, nickel 18 %, and the remainder is zenc. Its colour is silvery white hence its name is nickel silver.

13.13 COMPOSITES

Composite materials in short known as composites, are made by the combination of two or more materials, usually composed of a reinforcing material and a compatible binder material, to obtain specific properties and characteristics. This idea is not new, villagers in India used to mix straw in clay matrix to plaster the walls of their jhugis. Today composites have moved through high technology and emerged as engineering materials or as 12.212.3 .Most advanced composites today are composed of graphite fibers with an epoxy binder. Boron fibers are also used. There are 24 different matrix materials including epoxies, urethanes, silicones, polyimides and phenolics. Both cut and continuous glas fibers have been used for Composite materials, but due to their greater mass and Lower tensile strength, people do not classify this group with advanced composites.

13.13.1 Thermoset and Thermoplastics

Thermoset and thermoplastics are two main types of composite matrixes. Although there are numerous combinations of different fibers and epoxy binders, there are two major types, viz. thermoset and thermoplastic, which determine the methods of lay-up and , forming and cutting etc.

(a) **THERMOSET:** Thermoset binders are cured once and can never be re-formed. Thus they have a one-time use. These are principally epoxy, phenolic, bismaleimide, and polyimide. Thermoset curing starts immediately the epoxy is formulated. At room temperature the curing process takes two weeks. However at zero degree farnheit it will take more than two years.

(b) **THERMOPLASTIC:** In contrast to thermoset where it is subject to cycle of forming, heating and curing only once, in thermoplastics the sequential cycle can be repeated as many number of times as are required to achieve the desired characteristics Thermoplastic matrixes are many and varied, including nylon (polyamide), polysulfone, polyphenylene sulfide, polyether, and ethelketone. Although most composites are being produced with resin matrix materials, but recently metal matrix composites have been introduced and their extensive use is an index of their popularity which is primarily due to their high strength, the most frequently used metal is aluminum , although alloys of titanium, magnesium, and copper are also being used.

13.13.2 Fiber Science

Depending upon the reqirement, proper combination of the Two factors, namely, (i) percentage of fiber in whole Composite and (ii) orientation of placement of the fiber in the matrix during fabrication of the part during its production. The fibers can be placed along the longitudinal direction of the Part (called the Wrap in the fiber) or the fibers can be placed in the matrix transversely at right angles to the direction of Linear axis of the work piece, which is called weft. With New modern weaving techniques, fibers can be layed in the matrix with some bias to suit the requirement.

13.13.3 Composite Construction

Metal matrix composites (MMC) posses exceptionly high strength-to-weight and stiffness-to-weight ratios. They provide greater heat resistance and higher thermal and electrical conductivity. Further their performance does not get degraded due to moisture absorption. Carbon or graphite fibers are regularly produced which have tensile strength of 4800 MPa. For the purpose of estimating stresses, it is assumed that the fibers carry the load and the matrix merely holds the fiber together. This assumption is far from satisfactory. In addition the orientation of fibers has its own part to play, which is not considered.

13.14 SMART MATERIALS

Smart materials are materials that have one or more properties that can be significantly changed in a controlled fashion by external stimuli, such as stress, temperature, moisture, pH, electric or magnetic fields.

There are a number of types of smart material, some of which are already common. Some examples are as following:

- **Piezoelectric** materials are materials that produce a voltage when stress is applied. Since this effect also applies in the reverse manner, a voltage across the sample will produce stress within the sample. Suitably designed structures made from these materials can therefore be made that bend, expand or contract when a voltage is applied.
- **Shape memory** alloys and shape memory polymers are materials in which large deformation can be induced and recovered through temperature changes or stress changes (pseudoelasticity). The large deformation results due to martensitic phase change.
- **Magnetostrictive** materials exhibit change in shape under the influence of magnetic field and also exhibit change in their magnetization under the influence of mechanical stress.
- **Magnetic shape** memory alloys are materials that change their shape in response to a significant change in the magnetic field.
- **pH-sensitive polymers** are materials that change in volume when the pH of the surrounding medium changes.
- **Temperature-responsive** polymers are materials which undergo changes upon temperature.
- **Halochromic materials** are commonly used materials that change their colour as a result of changing acidity. One suggested application is for paints that can change colour to indicate corrosion in the metal underneath them.
- **Chromogenic systems** change colour in response to electrical, optical or thermal changes. These include electrochromic materials, which change their colour or opacity on the application of a voltage (e.g. liquid crystal displays), thermochromic materials change in colour depending on their temperature, and photochromic materials, which change colour in response to light-for example, light sensitive sunglasses that darken when exposed to bright sunlight.
- **Ferrofluid**
- **Photomechanical** materials change shape under exposure to light.
- **Self-healing** materials have the intrinsic ability to repair damage due to normal usage, thus expanding the material's lifetime
- **Dielectric elastomers (DEs)** are smart material systems which produce large strains (up to 300%) under the influence of an external electric field.

13.15 OPTICAL FIBERS

An optical fiber is a flexible, transparent fiber made of very pure glass (silica) not much wider than a human hair that acts as a waveguide, or "light pipe", to transmit light between the two ends of the fiber.[1] The field of applied science and engineering concerned with the design and application of optical fibers is known as fiber optics. Optical fibers are widely used in fiber-optic communications, which permits transmission over longer distances and at higher bandwidths (data rates) than other forms of communication. Fibers are used instead of metal wires because signals travel along them with less loss and are also immune to electromagnetic interference. Fibers are also used for illumination, and are wrapped in bundles so they can be used to carry images, thus allowing viewing in tight spaces. Specially designed fibers are used for a variety of other applications, including sensors and fiber lasers.

An optical fiber junction box. The yellow cables are single mode fibers; the orange and blue cables are multi-mode fibers: 50/125 µm OM2 and 50/125 µm OM3 fibers respectively.

Optical fiber typically consists of a transparent core surrounded by a transparent cladding material with a lower index of refraction. Light is kept in the core by total internal reflection. This causes the fiber to act as a waveguide. Fibers that support many propagation paths ortransverse modes are called multi-mode fibers (MMF), while those that only support a single mode are called single-mode fibers (SMF). Multi-mode fibers generally have a larger core diameter, and are used for short-distance communication links and for applications where high power must be transmitted. Single-mode fibers are used for most communication links longer than 1,050 meters (3,440 ft).

Joining lengths of optical fiber is more complex than joining electrical wire or cable. The ends of the fibers must be carefully cleaved, and then spliced together either mechanically or by fusi

13.15.1 Plastic Optical Fibers

Plastic optical fiber (POF) (or Polymer optical fibre) is an optical fiber which is made out of plastic. Traditionally PMMA (acrylic) is the core material, and fluorinated polymers are the cladding material. Since the late 1990s however, much higher-performance POF based on perfluorinated polymers (mainly polyperfluorobutenylvinylether) has begun to appear in the marketplace.

In large-diameter fibers, 96% of the cross section is the core that allows the transmission of light. Similar to traditional glass fiber, POF transmits light (or data) through the core of the fiber. The core size of POF is in some cases 100 times larger than glass fiber.

POF has been called the "consumer" optical fiber because the fiber and associated optical links, connectors, and installation are all inexpensive. The traditional PMMA fibers are commonly used for low-speed, short-distance (up to 100 meters) applications in digital home appliances, home networks, industrial networks(PROFIBUS, PROFINET), and car networks (MOST). The perfluorinated polymer fibers are commonly used for much higher-speed applications such as data center wiring and building LAN wiring.

In relation to the future request of high-speed home networking, there has been an increasing interest in POF as a possible option for next-generation Gigabit/s links inside the house. To this end, several European Research projects are active, such as POF-ALL and POF-PLUS.

For telecommunications, the more difficult-to-use glass optical fiber is more common. This fiber has a core made of germania-doped silica. Although the actual cost of glass fibers are lower than plastic fiber, their installed cost is much higher due to the special handling and installation techniques required.

One of the most exciting developments in polymer fibers has been the development of microstructured polymer optical fibers (mPOF), a type of photonic crystal fiber.

POF in short:

1. PMMA & Polystyrene are used as fiber core, with refractive indices of 1.49 & 1.59 respectively.
2. Generally, fiber cladding is made of silicone resin (refractive index ~1.46).
3. High refractive index difference is maintained between core and cladding.
4. POF have high numerical aperture.
5. Have high mechanical flexibility and low cost.

POF fiber also has applications in sensing. It is possible to write Fiber Bragg grating in single and multimode mode POF. There are advantages in doing this over using silica fiber since the POF can be stretched further without breaking, some applications are described in the PHOSFOS project page.

13.16 SEMICONDUCTORS

A semiconductor is a material with electrical conductivity due to electron flow (as opposed to ionic conductivity) intermediate in magnitude between that of a conductor and an insulator. This means a conductivity roughly in the range of 103 to 10?8 siemens per centimeter. Semiconductor materials are the foundation of modern electronics, including radio, computers, telephones, and many other devices. Such devices include transistors, solar cells, many kinds of diodes including the light-emitting diode, the silicon controlled rectifier, and digital and analog integrated circuits. Similarly, semiconductor solar photovoltaic panels directly convert light energy into electrical energy. In a metallic conductor, current is carried by the flow of electrons. In semiconductors, current is often schematized as being carried either by the flow of electrons or by the flow of positively charged "holes" in the electron structure of the material. Actually, however, in both cases only electron movements are involved.

Common semiconducting materials are crystalline solids, but amorphous and liquid semiconductors are known. These include hydrogenated amorphous silicon and mixtures of arsenic, selenium and tellurium in a variety of proportions. Such compounds share with better known semiconductors intermediate conductivity and a rapid variation of conductivity with temperature, as well as occasional negative resistance. Such disordered materials lack the rigid crystalline structure of conventional semiconductors such as silicon and are generally used in thin film structures, which are less demanding for as concerns the electronic quality of the material and thus are relatively insensitive to impurities and radiation damage. Organic semiconductors, that is, organic materials with properties resembling conventional semiconductors, are also known.

Silicon is used to create most semiconductors commercially. Dozens of other materials are used, including germanium, gallium arsenide, and silicon carbide. A pure semiconductor is often called an "intrinsic" semiconductor. The electronic properties and the conductivity of a semiconductor can be changed in a controlled manner by adding very small quantities of other elements, called "dopants", to the intrinsic material. In crystalline silicon typically this is achieved by adding impurities of boron or phosphorus to the melt and then allowing the melt to solidify into the crystal. This process is called "doping".

Energy bands and electrical conduction

In classic crystalline semiconductors, the electrons can have energies only within certain bands (i.e. ranges of levels of energy). Energetically, these bands are located between the energy of the ground state, the state in which electrons are tightly bound to the atomic nuclei of the material, and the free electron energy, the latter describing the energy required for an electron to escape entirely from the material. The energy bands each correspond to a large number of discrete quantum states of the electrons, and most of the states with low energy (closer to the nucleus) are full, up to a particular band called the valence band. Semiconductors and insulators are distinguished from metals because the valence band in any given metal is nearly filled with electrons under usual operating conditions, while very few (semiconductor) or virtually none (insulator) of them are available in the conduction band, the band immediately above the valence band.

The ease with which electrons in a semiconductor can be excited from the valence band to the conduction band depends on the band gap between the bands. The size of this energy bandgap serves as an arbitrary dividing line (roughly 4 eV) between semiconductors and insulators.

With covalent bonds, an electron moves by hopping to a neighboring bond. The Pauli exclusion principle requires the electron to be lifted into the higher anti-bonding state of that bond. For delocalized states, for example in one dimension - that is in a nanowire, for every energy there is a state with electrons flowing in one direction and another state with the electrons flowing in the other. For a net current to flow, more states for one direction than for the other direction must be occupied. For this to occur, energy is required, as in the semiconductor the next higher states lie above the band gap. Often this is stated as: full bands do not contribute to the electrical conductivity. However, as the temperature of a semiconductor rises above absolute zero, there is more energy in the semiconductor to spend on lattice vibration and on exciting electrons into the conduction band. The current-carrying electrons in the conduction band are known as "free electrons", although they are often simply called "electrons" if context allows this usage to be clear.

Electrons excited to the conduction band also leave behind electron holes, i.e. unoccupied states in the valence band. Both the conduction band electrons and the valence band holes contribute to electrical conductivity. The holes themselves don't actually move, but a neighboring electron can move to fill the hole, leaving a hole at the place it has just come from, and in this way the holes appear to move, and the holes behave as if they were actual positively charged particles.

One covalent bond between neighboring atoms in the solid is ten times stronger than the binding of the single electron to the atom, so freeing the electron does not imply destruction of the crystal structure.

Holes: electron absence as a charge carrier

The concept of holes can also be applied to metals, where the Fermi level lies within the conduction band. With most metals the Hall effect indicates electrons are the charge carriers. However, some metals have a mostly filled conduction band. In these, the Hall effect reveals positive charge carriers, which are not the ion-cores, but holes. In the case of a metal, only a small amount of energy is needed for the electrons to find other unoccupied states to move into, and hence for current to flow. Sometimes even in this case it may be said that a hole was left behind, to explain why the electron does not fall back to lower energies: It cannot find a hole. In the end in both materials electron-phonon scattering and defects are the dominant causes for resistance.

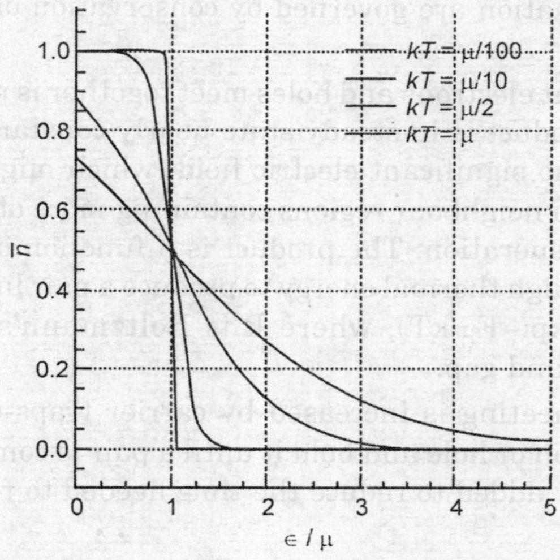

Fig. 13.1

Fermi-Dirac distribution. States with energy ε below the Fermi energy, here, have higher probability n to be occupied, and those above are less likely to be occupied. Smearing of the distribution increases with temperature.

The effective mass is important as it affects many of the electrical properties of the semiconductor, such as the electron or hole mobility, which in turn influences the diffusivity of the charge carriers and the electrical conductivity of the semiconductor.

Typically the effective mass of electrons and holes are different. This affects the relative performance of p-channel and n-channel IGFETs.[2]

The top of the valence band and the bottom of the conduction band might not occur at that same value of k. Materials with this situation, such as silicon and germanium, are known as indirect band gap materials. Materials in which the band extrema are aligned in k, for example gallium arsenide, are called direct band gap semiconductors. Direct gap semiconductors are particularly important in optoelectronics because they are much more efficient as light emitters than indirect gap materials.

Carrier generation and recombination

For more details on this topic, see Carrier generation and recombination.

When ionizing radiation strikes a semiconductor, it may excite an electron out of its energy level and consequently leave a hole. This process is known as electron-hole pair generation. Electron-hole pairs are constantly generated from thermal energy as well, in the absence of any external energy source.

Electron-hole pairs are also apt to recombine. Conservation of energy demands that these recombination events, in which an electron loses an amount of energy larger than the band gap , be accompanied by the emission of thermal energy (in the form of phonons) or radiation (in the form of photons).

In some states, the generation and recombination of electron-hole pairs are in equipoise. The number of electron-hole pairs in the steady state at a given temperature is determined by quantum statistical mechanics. The precise quantum mechanical mechanisms of generation and recombination are governed by conservation of energy and conservation of momentum.

As the probability that electrons and holes meet together is proportional to the product of their amounts, the product is in steady state nearly constant at a given temperature, providing that there is no significant electric field (which might "flush" carriers of both types, or move them from neighbour regions containing more of them to meet together) or externally driven pair generation. The product is a function of the temperature, as the probability of getting enough thermal energy to produce a pair increases with temperature, being approximately $\exp(-E_G/kT)$, where k is Boltzmann's constant, T is absolute temperature and E_G is band gap.

The probability of meeting is increased by carrier traps-impurities or dislocations which can trap an electron or hole and hold it until a pair is completed. Such carrier traps are sometimes purposely added to reduce the time needed to reach the steady state.

Semi-insulators

Some materials are classified as **semi-insulators.** These have electrical conductivity nearer to that of electrical insulators. Semi-insulators find niche applications in micro-electronics, such as substrates for HEMT. An example of a common semi-insulator is gallium arsenide.[3]

Doping

For more details on this topic, see Doping (semiconductor).

The property of semiconductors that makes them most useful for constructing electronic devices is that their conductivity may easily be modified by introducing impurities into their crystal lattice. The process of adding controlled impurities to a semiconductor is known as doping. The amount of impurity, or dopant, added to an intrinsic (pure) semiconductor varies its level of conductivity. Doped semiconductors are often referred to as extrinsic. By adding impurity to pure semiconductors, the electrical conductivity may be varied not only by the number of impurity atoms but also, by the type of impurity atom and the changes may be thousand folds and million folds. For example, a 1 cm^3 specimen of a metal or semiconductor has of the order of 10^{22} atoms. In a metal every atom donates at least one free electron for conduction, thus 1 cm^3 of metal contains on the order of 10^{22} free electrons. Whereas a 1 cm^3 of sample pure germanium at 20 °C, contains about

4.2×10^{22} atoms and 2.5×10^{13} free electrons and 2.5×10^{13} holes (empty spaces in crystal lattice having positive charge). The addition of

13.125.2 A P-type semiconductor (P for Positive) is obtained by carrying out a process of doping: that is, adding a certain type of atoms to the semiconductor in order to increase the number of free charge carriers (in this case positive).

When the doping material is added, it takes away (accepts) weakly bound outer electrons from the semiconductor atoms. This type of doping agent is also known as an acceptor material and the vacancy left behind by the electron is known as a hole.

The purpose of **P-type doping** is to create an abundance of holes. In the case of silicon, a trivalent atom (typically from Group 13 of the periodic table, such as boron or aluminium) is substituted into the crystal lattice. The result is that one electron is missing from one of the four covalent bonds normal for the silicon lattice. Thus the dopant atom can accept an electron from a neighboring atom's covalent bond to complete the fourth bond. This is why such dopants are called acceptors. The dopant atom accepts an electron, causing the loss of half of one bond from the neighboring atom and resulting in the formation of a "hole". Each hole is associated with a nearby negatively charged dopant ion, and the semiconductor remains electrically neutral as a whole. However, once each hole has wandered away into the lattice, one proton in the atom at the hole's location will be "exposed" and no longer cancelled by an electron when you have 3 electrons and 1 hole surrounding a particular nucleus with 4 protons. For this reason a hole behaves as a quantity of positive charge. When a sufficiently large number of acceptor atoms are added, the holes greatly outnumber the thermally excited electrons. Thus, the holes are the majority carriers, while electrons are the minority carriers in P-type materials. Blue diamonds (Type IIb), which contain boron (B) impurities, are an example of a naturally occurring P-type semiconductor.

Therefore, to a first approximation, sufficiently doped P-type semiconductors can be thought of as only conducting holes.

13.16.3 Basic science of n-type semiconductors

Semiconductors are defined by their unique electric conductive behavior. Metals are good conductors because at their Fermi level, there is a large density of energetically available states that each electron can occupy. Electrons can move quite freely between energy levels without a high energy cost. Metal conductivity decreases with temperature increase because thermal vibrations of crystal lattice disrupt the free motion of electrons. Insulators, by contrast, are very poor conductors of electricity because there is a large difference in energies (called a band gap) between electron-occupied energy levels and empty energy levels that allow for electron motion.

Insulator conductivity increases with temperature because heat provides energy to promote electrons across the band gap to the higher electron conduction energy levels (called the conduction band). Semiconductors, on the other hand, have an intermediate level of electric conductivity when compared to metals and insulators. Their band gap is small enough that small increase in temperature promotes sufficient number of electrons (to result in measurable currents) from the lowest energy levels (in the valence band) to the conduction band. This creates electron holes, or unoccupied levels, in the valence band, and very loosely held electrons in the conduction band.[2][3] An intrinsic semiconductor is made up ideally of one pure element, typically silicon. At room temperature, the conductivity of intrinsic semiconductors is relatively low. Conductivity is greatly enhanced by a process called doping, in which other elements are added to the intrinsic crystal in very small amounts to create what is called an extrinsic semiconductor. When the dopant provides extra electrons to the host, the product is called an n-type semiconductor. The process of doping is described as it introduces energy levels into band gap; those levels are filled with electrons and lie close to the conduction band so that even slight thermal agitation can release them into the conduction band.[2]

It should be noted, that the negative charge of the electrons is balanced by an equivalent positive charge in the center of the impurity atoms. Therefore, the net electrical charge of the semiconductor material is not changed.[4]

Dopants

The doping atoms usually have one more valence electron than one type of the host atoms. The most common example is atomic substitution in group IV solids (silicon, germanium, tin which contain four valence electrons) by group V elements (phosphorus, arsenic, antimony) which contain five loosely bound valence electrons. The situation is more uncertain when the host contains more than one type of atoms. For example, in III-V semiconductors like gallium arsenide, silicon can be a donor when it substitutes for gallium and acceptor when it replaces arsenic. Some donors have fewer valence electrons than the host, such as alkali metals, which are donors in most solids

In the preceding description an important fact is ignored for the sake of simplicity: the dispersion of the energy. The reason that the energies of the states are ergy-momentum dispersion broadened into a band is that the energy depends on the value of the wave vector, or k-vector, of the electron. The k-vector, in quantum mechanics, is the representation of the momentum of a particle.

The dispersion relationship determines the effective mass, m^*, of electrons or holes in the semiconductor, according to the formula:

$$m^* = \hbar^2 \cdot \left[\frac{d^2 E(k)}{dk^2}\right]^{-2}$$

REVIEW QUESTIONS

1. What is the percentage carbon content of the High carbon steels ? Which category of these steels *(a)* Low carbon steels, *(b)* Medium carbon steels, and *(c)* can be case- hardened?
2. Explain the effect of thermal conductivity on weldability.
3. Define ductility.
4. How the machinability is rated?
5. What is the most common additive for free-machining alloy steels? stainless steels?
6. What is the percentage content of Cr and Ni in 18-8 Stainless steel? What are the main alloying elements in these ?

7. Compare the main characteristics of Austenitic, Ferritic And Martensitic stainless steels vis-à-vis magnetic, Corrosion resistance, formability and heat treatment ability.

Ans. Ferritic Stainless steels:

(a) They are magnetic.

(b) Good resistance to corrosion.

(c) Good formability but moderately hardened by cold working.

(d) They cannot be hardened by heat treatment.

Austenitic Stainless steels:

(a) They are non magnetic.

(b) All are excellent corrosion resistant.

(c) Have very good formability. They even gain strength by forming process.

(d) They get hardened by cold working

Martensitic steels

(a) They are magnetic.

(b) They are corrosion resistant.

(c) They are fairly ductile. But can be cold- worked to a small extent.

(d) They can be hardened by heat treatment.

8. What for the abbreviations HSLA and HSS stand?
9. Which are the main additive elements in martensitic alloy steel? Give atleast 4 examples where these are used?
10. What is the commercial name of 18-4-1 steel? What do these numerical figures represent?
11. Name the alloying elements and their percentage in 'headfield steel'?
12. Give the %age of alloying elements in duralumin?
13. Give the %age of metal contents in the following:
(a) Alpha brasses, *(b)* Alpha-Beta brasses, *(c)* Leaded brasses, and *(d)* Tin brasses.
14. What are Nickel silvers? What does 65-18 Nickel-Silver Indicate ?
15. Explain the main difference between thermoset and Thermoplastic types of composites.
16. What is meant by fiber-science?

CHAPTER 14

CENTROID, CENTRE OF GRAVITY, SECOND MOMENT OF AREA AND MASS MOMENT OF INERTIA

14.1 CENTRE OF GRAVITY

A solid body is built-up of a large number of very small particles. Each and every particle is attracted by the earth and the force of attraction is towards the centre of earth. Since the size of the body is very small as compared to the radius of earth, these individual forces acting on the particles of which the body is made of, will be parallel to each other. The force on each particle will be proportional to the mass of that particle. If we find the resultant of all these individual small forces it will give the total weight of the body *i.e.*, the force with which it is being attracted by the earth and it is directed towards the centre of earth. Let us imagine a supporting force which is equal and opposite to the resultant force of attraction of the earth, then the body will be in equilibrium. In otherwards it is meant to say that the resultant upward force is passing through the centre of gravity. Since the body can move or rotate again the resultant force of attraction lay earth on the body will pass through the C.G. and its magnitude will be equal to the weight of the body. This force of attraction will again represent the total weight of the body and its line of action will pass through the centre of gravity. The point of intersection of these two resultant forces in two positions will give the location of centre of gravity (C.G. or supply G.). Since the centre of gravity is one and only one of a solid body it will not change is location irrespective of the rotation or angular position of the solid body.

Now we proceed to locate the centre of gravity by using Varignon's theorem for different simple cases.

14.1.1 C.G. of a Straight Wire of Length l

Let AB be the wire of uniform section and w be the weight per unit length of wire. Take A as the origin and consider a small length dx situated at a distance x from A.

Fig. 14.1

The weight of the element CD of length dx will be

$$dw = w\,dx$$

Total weight of wire AB

$$W = \int_0^P w\,dx = wl \qquad \text{...(14.1)}$$

Let x be the distance of C.G. of wire AB through which the line of action of the resultant force of attraction by earth will pass. Thus, taking moments about 'A' of the various small forces acting on individual particles

$$M = \int_0^l (w\,dx)\,x$$

$$= \left(w\frac{x^2}{2}\right)_0^l = \boldsymbol{wl^2}$$

$$= \frac{1}{2}W(l)$$

From Varignon's theorem this moment is equal to the moment of the resultant force of attraction of earth which will be equal to W and its location is assumed to be at distance $\bar{x}$ from A

i.e., $$W\bar{x} = \frac{W}{2}l$$

$$\bar{x} = \frac{l}{2} \qquad \text{...(14.2)}$$

14.1.2 Corollary

In case we consider a line only. Theoretically the line does not have mass but still we take a thin wire of very small diameter. Its centre of gravity, will be at a distance of $\frac{l}{2}$ from one end. However, in the case of line we use the word centroid rather than the centre of gravity which term is used when the force of attraction of earth is considered.

14.1.3 Centre of Gravity of Plane Lamina of Uniform Thickness

As the plane lamina has uniform thickness t and we take ρ as the density of the material.

Let dA be the elemental area around the point (x, y) and its thickness is t along z axis.

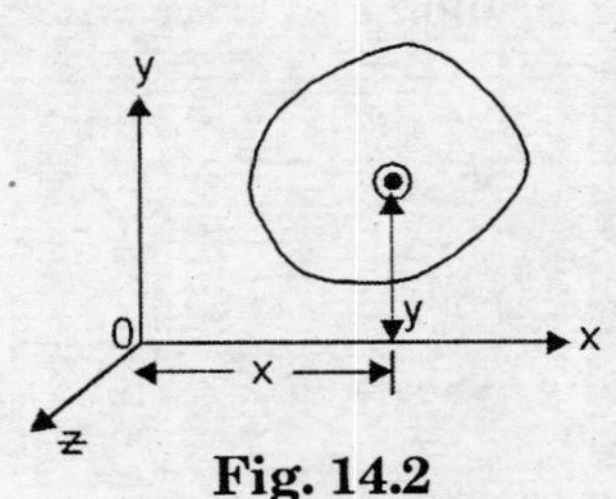

Fig. 14.2

$$dv = \rho\, t(dA)$$

Weight of the element dw, and if 'g' acceleration due to gravity

$$dw = (\rho\, dv)\, g$$

$$w = \int \rho g\, dv = \int \rho g\, dA t$$

$$= \rho g t \int dA$$

$$= \rho\, gt\, A \quad ...(14.3)$$

Where $A = \int dA$

$= \text{Total area}$

Then from Varignon's theorem taking moment about X-X-axis.

$$W\bar{x} = \oint y\, dw$$

$$= \int y \rho g t\, dA$$

$$(\rho g t A)\bar{y} = \rho g t \int y\, dA$$

$$\bar{y} = \frac{\int y\, dA}{\int dA} \quad ...(14.4)$$

Similarly if we take moments about $\bar{y}$-axis and $\bar{x}$ is the distance of C.G. from $\bar{y}$-axis

$$(\rho g t A)\bar{x} = \frac{\int x\, dA}{\int dA} \quad ...(14.5)$$

Along the x-direction the thickness in uniform and equal to 't' the C.G. will be at chief their thickness. Thus the cordinates of the C.G. are fully known as $\left(\bar{x}, \bar{y} \text{ and } \bar{z} = \frac{f}{z}\right)$.

Corollary

If a plane lamina of almost zero thickness is considered we will be get what is known as the cordinates of the point which will be known as centroid of the plane figure.

Thus $$\bar{x} = \frac{\int x\, dA}{\int dA} \quad ...(14.5a)$$

$$\bar{y} = \frac{\int y\, dA}{\int dA} \quad ...(14.5b)$$

14.1.4 Centroid of An ARC of *A* Circle

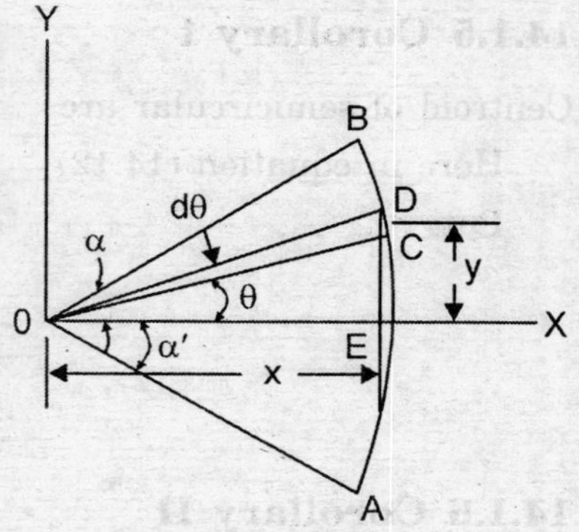

Let *AB* be an arc of a circle of radius *R* and subtending an angle $z\propto$ at the centre.

Let a small element *CD* of the arc which subtends an angle $d\theta$ at the centre be at angular position θ from the central line *ox* which divides the angle $z\propto$.

The if $(\bar{x}, \bar{y})$ are the co-ordinates of the centroid (C.G.) of the arc *AB*, we have using Varignon's theorem.

Moment of the resultant must be equal to the sum of moments of weights of elemental lengths about *Y* and *X* axes respectively.

Fig. 14.3

$$\bar{x} \times m\int_{-\propto}^{+\propto} R\,d\theta = \int_{-\propto}^{+\propto} m\,dl\,R\cos\theta$$

$$= \int_{-\propto}^{+\propto} m(R\,d\theta)R\cos\theta$$

or

$$m\bar{x}\,R[\theta]_{-\propto}^{\propto} = mR^2\int_{-\propto}^{\propto}\cos\theta\,d\theta \qquad ...(14.6a)$$

$$mR\bar{x}\;z\propto = mR^2[\sin\theta]_{-\propto}^{\propto}$$

$$= m\;R^2(\sin^{\propto} + \sin\propto)$$

$$= 2m\;R^2\;\sin\propto$$

$$\bar{x} = \frac{R\sin\propto}{\propto} \qquad ...(14.6b)$$

For distance of C.G. from *x*-axis. Let it be $\bar{y}$. Then taking moments what *X*-axis and using Varignon's theorem

$$\bar{y}\int_{\propto}^{\propto} mR\,d\theta = \int_{-\propto}^{\propto} y(R\,d\theta)m$$

$$\bar{y}_m R(\theta)_{-\propto}^{\propto} = \int_{-\propto}^{\propto}[m(R\,d\theta)]R\sin\theta$$

$$= mR^2\int_{-\propto}^{\propto}\sin\theta\,d\theta$$

$$= mR^2[-\cos\theta]_{-\propto}^{\propto}$$

$$\bar{y}\,mR\;2\propto = mR^2[-\cos\propto + \cos\propto]$$

i.e.,

$$\bar{y} = 0 \qquad ...(14.7)$$

Thus the C.G. lies on the central axis and is at distance

$$\bar{x} = \frac{R\sin\propto}{\propto} \qquad ...(14.8)$$

14.1.5 Corollary I

Centroid of semicircular arc

Here in equation (14.12)

Put

$$z\propto \;=\; \pi$$

$$\propto \;=\; \pi/2 \qquad \text{...(14.8)}$$

$$\therefore \qquad \bar{x} \;=\; R\frac{\sin \pi/2}{\pi/2} = \frac{2R}{\pi} \qquad \text{...(14.9)}$$

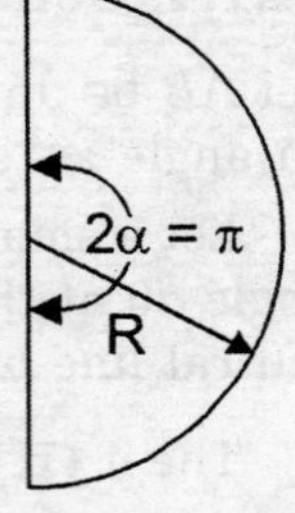

Fig. 14.4

14.1.6 Corollary II

Centroid of a equation of circular

Put $\quad 2\propto \;=\; \pi/2$

or $\quad \propto \;=\; \pi/4$

Therefore

$$\bar{x} \;=\; R\frac{\sin \pi/4}{\pi/4} = \frac{4R}{\pi} \times \frac{1}{\sqrt{2}}$$

$$= \frac{2\sqrt{2}}{\pi}R \qquad \text{...(14.10)}$$

Fig. 14.5

Example 14.1. *Delermine the co-ordinates of the centroid of the lines shown in Fig. 14.6.*

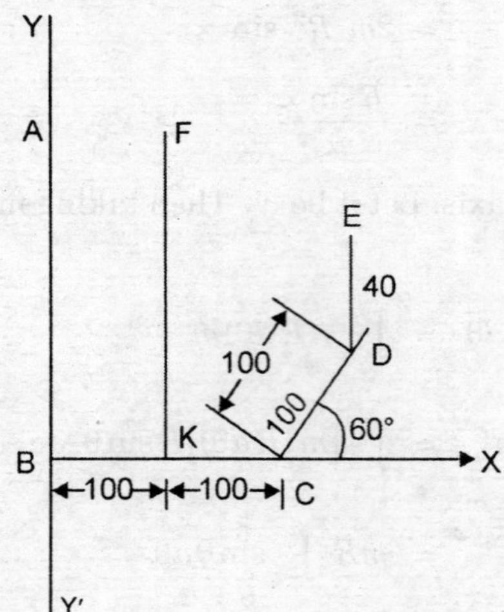

Fig. 14.6

Solution: Taking the centroids of the respective lengths at their mid points and taking moments of these individual length segments about Y axis we get by applying Varignon's theorem.

$$AB \times 0 + (FK \times BK) + \left(BC \times \frac{1}{2}BC\right) + CD\left(BC + \frac{DC}{2}\cos 60\right) + ED(BC + CD\cos 60)$$

$$= \bar{x} \times 740$$

or $$0 + 400\times100+\left(200\times\frac{200}{2}\right)+100\left(200+\frac{100}{2}\times\frac{1}{2}\right)+40\left(200+100\frac{1}{2}\right)$$

$$= \bar{x}\times740$$

or $$4 \times 10^4 + 1 \times 10^4 + 100\,(225) + 40\,(250)$$

$$= 740\bar{x}$$

$$5 \times 10^4 + 2.25 \times 10^4 + 10000 = 740\bar{x}$$

$$\bar{x} = \frac{8.25\times10^4}{740} = 1.115 \times 10^2 = 111.5$$

Taking moments of length segments about X-axis and applying variguous theorem we get

$$740\bar{y} = (400\times200)+(400\times200)+(BC\times0)+$$

$$\left(CD\times\frac{CD}{2}\sin60\right)+DE\left(CD\sin60+\frac{DE}{2}\right)$$

$$= 8\times10^4+8\times10^4+0+\left(100\times50\times\frac{\sqrt{3}}{2}\right)+40\left(100\frac{\sqrt{3}}{2}+\frac{1}{2}\times40\right)$$

$$= 16\times10^4+2500\sqrt{3}+40\left(50\sqrt{3}+200\right)$$

$$= 160000+4500\sqrt{3}+800$$

$$= 24594.3$$

$$\therefore \quad \bar{y} = \frac{24594.3}{740} = 33.235 \text{ mm } \textbf{Ans.}$$

Example 14.2. *A clip E ABCD made of spring steel is shown in Fig. 14.7. The radius R of the circular position is 100 mm and the curved portion subtends an angle equal to 270° degrees at its centre. The straight legs EA and DC are each = 100 mm and they are inclined at 45° OY line and subtend an angle of 90° at O. Find the distance of the centroid of the clip wire from XX line.*

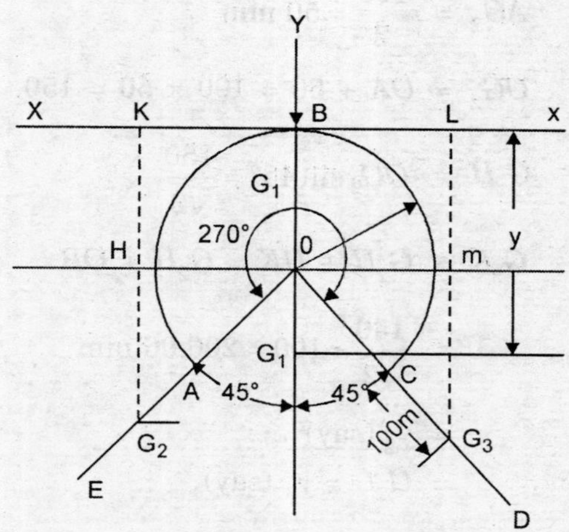

Fig. 14.7

Solution : Let G_1 be the centroid of the curved position *ABC*. We have from equation 14.8

$$OG_1 = \frac{R\sin\alpha}{\alpha}$$

Where

$$R = 100 \text{ mm}$$

$$2\alpha = 270°$$

$$= \frac{3\pi}{2} \text{ radius}$$

or

$$\alpha = \frac{3\pi}{4}$$

$$\therefore \quad OG_1 = \frac{100 \times \sin\left(\frac{3\pi}{4}\right)}{\left(\frac{3\pi}{4}\right)}$$

$$= 100 \times \frac{1}{\sqrt{2}} \times \frac{4}{3\pi} = 30.01 \text{ mm}$$

Also length of arc

$$ABC = l_1 = R\,(2\alpha)$$

$$= R \times \frac{3\pi}{2}$$

$$= 150\,\pi = 471.24 \text{ mm}$$

Geometrically

If G_2 and G_3 an C.G. of *EA* and *DC* respectively

$$AG_2 = \frac{100}{2} = 50 \text{ mm}$$

$$OG_2 = OA + 50 = 100 \times 50 = 150$$

$$\therefore \quad G_2H = OG_2 \sin 45° = \frac{150}{\sqrt{2}}$$

$$G_2K = G_2H + HK = G_2H + OB$$

$$= \frac{150}{\sqrt{2}} + 100 = 206.06 \text{ mm}$$

$$= y_2 \text{ (say)}$$

$$= G_2L = y_3 \text{ (say)}$$

Taking moments about the line $X - X$, if $\bar{y}$ is the distance of C.G. of the clip wire from XX

Total length l = are $ABC + AE + CD$

$$= l_1 + l_2 + l_3$$

$$= R\,(2\alpha) + 100 + 100$$

$$= 100\left(\frac{3\pi}{2}\right) + 200$$

$$l = 671.24 \text{ mm}$$

$$\therefore \quad l\bar{y} = l_1 y_1 + l_2 y_2 + l_3 y_3 \quad \text{...(B)}$$

where y_1 = distance of G_1 from $XX = G_1B$

y_2 = distance of G_2 from $XX = G_2K$

y_3 = distance of G_3 from $XX = G_3L$

Now as her Varignon's theorem

$$l\bar{y} = l\bar{y}_1 + l_2\bar{y}_2 + l_3\bar{y}_3$$

And

$$l_1 = l_1 l\bar{y}_1 + l_2\bar{y}_2 100\frac{3\pi}{2} = 150\pi = 471.24$$

$$l_2 = l_3 = 100$$

$$y_1 = G_1B = OB - OG_1 = 100 - 30.01 = 69.09 \text{ mm}$$

$$y_2 = G_2K = 206.06 \text{ mm}$$

Therefore substituting values in equation (*B*) we get

$$l\bar{y} = 471.24 \times 69.09 + 2(100)\,(206.06)$$

or

$$671.24\bar{y} = 32939.7 + 41212$$

$$= 74151.7$$

$$\bar{y} = \frac{74151.7}{671.24} = 110.47 \text{ mm} \quad \textbf{Ans.}$$

14.2 LOCATION OF CENTRE OF GRAVITY AND CENTROID

We now proceed to develop the method to locate the centre of gravity of a body whose mass is M and density of the material in r (say). The solid body can be considered to be built up of large number of elemental masses of weights W_1, W_2 ... which are located at points $(x_i, y_i, z_i)_{i = 1 \text{ to } n}$ with respect to certain dosen axis of references. Let the co-ordinates of the centre of gravity be $(\bar{x}, \bar{y}, \bar{z})$. It W is the total weight of the body. Then W being the resultant of the elemental weights

$$W = W_1 + W_2 + W_3$$

or

$$W = \Sigma W_i \quad \text{...(14.11)}$$

Taking moments about axes of reference in turn and applying Varignon's theorem we get

$$Mg\bar{x} = \Sigma g\ m_i\ x_i$$

or

$$M\bar{x} = \oint x\,dm$$

It ρ is the density and V the volume the rances to

$$\rho V\bar{x} = \oint \rho x\,dV$$

or

$$V\bar{x} = \oint x\,dV \qquad ...(14.11a)$$

Similarly for y-and z axis

$$V\bar{y} = \oint y\,dV \qquad ...(14.11b)$$

$$V\bar{z} = \oint z\,dv \qquad ...(14.11c)$$

Also

$$V = \int dv$$

Then equation (14.11) reduce to

$$\bar{x} = \frac{\oint x\,dV}{\oint dV} \qquad ...(14.12a)$$

$$\bar{y} = \frac{\oint y\,dV}{\oint dV} \qquad ...(14.12b)$$

$$\bar{z} = \frac{\oint z\,dV}{\oint dV} \qquad ...(14.12c)$$

$(\bar{x}, \bar{y}, \bar{z})$ also the co-ordinates of the centre of gravity of the total mass who density is uniform.

Further if we have a body consisting of plane laminas of uniform thickness of plane areas A_1, A_2 ... then equation (14.12) reduce to the following equation

For figures consisting of plane areas as

$$\bar{x} = \frac{\oint x\,dA}{\oint dA} \qquad ...(14.13a)$$

$$\bar{y} = \frac{\oint y\,dA}{\oint dA} \qquad ...(14.13b)$$

However, in the case of figures consisting of plane area $(\bar{x}, \bar{y})$ will be (co-ordinates of what is called centroid. Thus centre of gravity refers to the mass of the three dimensional body while a centroid pertains to plane areas

14.2.1 Difference Betwen Centre of Gravity and Centroid

From the analysis done in article 14.2 it is concluded that

(*a*) The term centre of gravity applies is the bodies having mass so that the resultant force of attraction due to the gravitational pull of earth wit pass through the centre of gravity irrespective of the mentation of the body.

(*b*) On the other hand centroid is associated with lines, plane areas having no mass and the moment of area about any axis through the centroid is zero.

14.3 AXIS OF SYMMETRY AND LOCATION OF CENTROID

In case the area has an axis of symmetry the centroid will be on it. The proof of this statement is given below.

Let the *y*-axis be the axis of symmetry of the given area. Consider any two elemental areas A_1 and A_2 which are at the same distance from *y*-axis *i.e.*, x_1 and x_2 and $x_1 = x_2 = a$. The entire area on the left of *y*-axis is of the same shape as that on the right of *y*-axis. Therefore for every elemental area on the left side we will have similar elemental area on the right side of *y*-axis. The moment of such similar elemental area about *y*-axis will cancel each other when the summition of $\Sigma x_i A_i$ is taken.

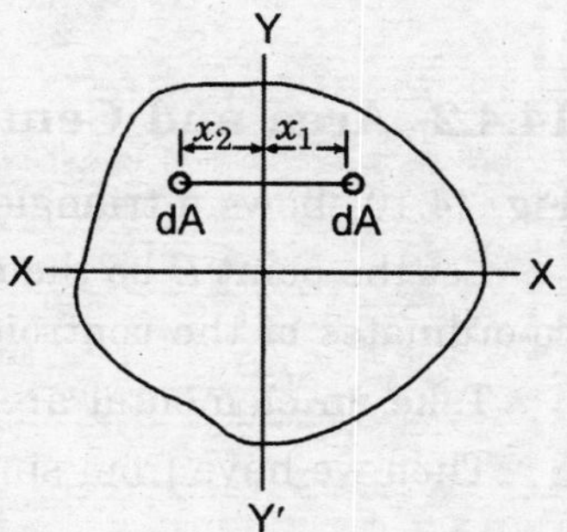

Fig. 14.8

Thus $$\oint x\,dA = 0$$

Therefore from equation (14.13*a*) we conclude that

$$\bar{x} = \frac{\oint x\,dA}{\oint dA} = 0$$

From this it is concluded that the centroid will be on *Y*-axis. Similarly if *X*-axis is also the axis of symmetry the centroid will also be on *x*-axis which leads to the conclusion if both *X-X* and *Y-Y* are the axis of symmetry then the origin, the point of intersection of axis of reference is the centroid of the section.

14.4 AREAS AND CENTROID OF PLANE FIGURES

14.4.1 Rectangle

In Fig. 14.9 is drawn a rectangle *ADCD* of width '*b*' and depth '*d*'. If we take the *X-X* and *Y-Y* axes midway between *AB* and *DC* and between *AD* and *BC* mespectively, then these will be axes of symmetry. Therefore the origin *O* is the centroid of the section

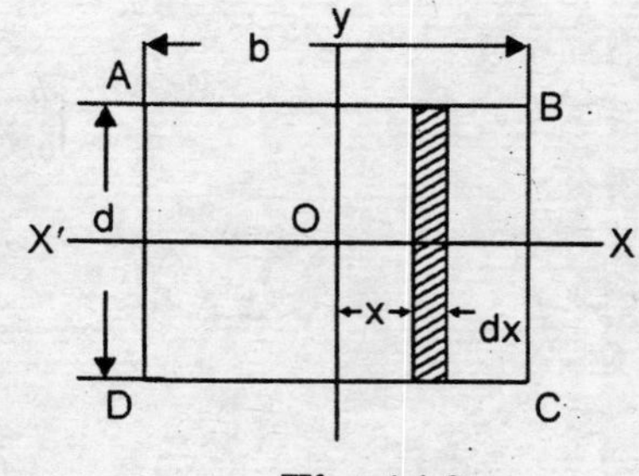

Fig. 14.9

For finding the area of rectangle *ABCD* take an element of width *dx* at a distance *x* from the *Y*-axis. The length of the element is *d*.

$\therefore$ Elemental area $= dA = d(dx)$

Integrating we get the total area

$$A = \int_{-\frac{b}{2}}^{b/2} d(dx)$$

$$= d \int_{-\frac{b}{2}}^{b/2} dx = d[x]_{-b/2}^{b/2}$$

$$= d\left(\frac{b}{2} + \frac{b}{2}\right) = bd \qquad \text{...(14.14)}$$

14.4.2 Area and Centroid of Triangle

Fig. 14.10 shows a triangle whose base *BC* of length *b* and height *h*.

Let the point *B* be the origin and take the *X*-axis along *BC* and *OY* as the *Y*-axis. Let the co-ordinates of the centroid be $(\bar{x}, \bar{y})$.

Take an elemental area at heigh *y* of depth *dy* parallel to *BC* or *X*-axis.

Then we have from similar triangles *ADE* and *ABC*

$$\frac{DE}{BC} = \frac{AF}{AH}$$

$$= \frac{h-y}{h}$$

That

$$DE = (b)\frac{(h-y)}{h}$$

$$= \left(b - \frac{by}{h}\right) \qquad \text{...(14.15)}$$

Fig. 14.10

Therefore from equation (14.13b)

$$\bar{y} = \frac{\oint y\,dA}{\oint dA} = \frac{\int y(DE)\,dy}{\int dA} \qquad \text{...(14.16)}$$

$$\int_0^h y\,dA = \int_0^h y\left(b - \frac{by}{h}\right)dy$$

$$= \left[\frac{y^2}{2}b - \frac{b}{h}\frac{y^3}{3}\right]_0^h$$

$$= \frac{bh^2}{2} - \frac{1}{3}\frac{bh^3}{h}$$

$$= \frac{bh^2}{2} - \frac{bh^2}{3} = \frac{bh^2}{6} \qquad \text{...(14.17)}$$

Also $$A = \int dA = \int_0^h (DE)\,dy$$

$$= \int_0^h \left(b - \frac{by}{h} \right) dy$$

$$= \left[by - \frac{by^2}{2h} \right]_0^h = bh - \frac{bh^2}{2h}$$

$$= \frac{bh}{2} \qquad \text{...(14.18)}$$

∴ **From** equation (14.16) and (14.17) and (14.18) we get

$$\bar{y} = \frac{bh^2}{6} \times \frac{2}{bh} = \frac{b}{3} \qquad \text{...(14.19)}$$

From (14.19) it is concluded that the centroid of a triangle is at $\frac{2}{3}$ *rd* of its height from the base.

Similarly if the side *AB* is taken as base and the perpendicular distance of *C* from *AB* as height, the centroid will lie at $\frac{1}{3}$ of height of *C* from base *AB*. The intersection of these will give the location of centroid *G*.

14.4.3 Centroid of A Sector of a Cricle

Fig. 14.11 shows a sector *AOB* subtending an angle 2∝ at the centre. Let the axis be taken along the bisector of the angle *AOB* and *Y* axis *OY* perpendicular to *OX*.

An elemental area ($r\,d\theta$) dr which lies at an angular position θ from *OX* is shown

$$dA = (r\,d\theta)\,dr$$

Then $$A = \int_{\theta=-\propto}^{\propto} \int_{r=0}^{R} dA$$

$$= \int_{\theta=-\propto}^{\propto} \int_{r=0}^{R} (r\,dr)\,d\theta$$

$$= \int_{\theta=-\propto}^{\propto} \left[\frac{r^2}{2} \right]_0^R d\theta$$

$$= \left[\frac{R^2}{2} \theta \right]_{-\propto}^{\propto} = \frac{R^2}{2}[\propto + \propto] = R^2 \propto \qquad \text{...(14.20)}$$

Fig. 14.11

and if $\bar{x}$ is the distance of centroid from 0, along OX the axis of symmetry

$$\bar{x} = \frac{\int x\,dA}{\int dA}$$

$$= \frac{\int \text{moment about } y\text{-axis}}{\text{Area}}$$

$$\int \bar{x}\,dA = \int_{\theta=-\propto}^{\propto} \int_{r=0}^{R} (r\,d\theta\,dr)\,r\cos\theta$$

$$= \int_{\theta=-\propto}^{\propto} \left[\int_{r=0}^{R} r^2\,dr \right] \cos\theta\,d\theta$$

$$= \frac{R^3}{3} \int_{-\propto}^{\propto} \cos\theta\,d\theta = \frac{R^3}{3} [\sin\theta]_{-\propto}^{\propto}$$

$$= \frac{R^3}{3} 2\sin\propto = \frac{2}{3} R^3 \sin\propto \quad \text{...(14.21)}$$

From equation (14.20) and (14.21)

$$\bar{x} = \left(\frac{2}{3} R^3 \sin\propto \right) \frac{1}{R^2 \propto}$$

$$= \frac{2}{3} \frac{R\sin\propto}{\propto} \quad \text{...(14.22a)}$$

Since OX axis divides the angle $2\propto$ subtended by the arc AB at the centre, this is the axis of symmetry for the area and the centroid is at a distance of $\bar{x}$ given by equation (14.22*a*)

i.e., $$OG = \frac{2}{3} \frac{R\sin\alpha}{a} \quad \text{...(14.22b)}$$

14.4.4 Centroid of a Quarter of a Circle of Radius *R*

In Fig. 14.12 is shown a quarter of circular area CD. The axis of reference are along OX and OY where OX the centre of the circle of which the given area is a part. Let G be the centroid of the area lying on line OA which bisects the angle COD Joining OG we get the axis of symmetry.

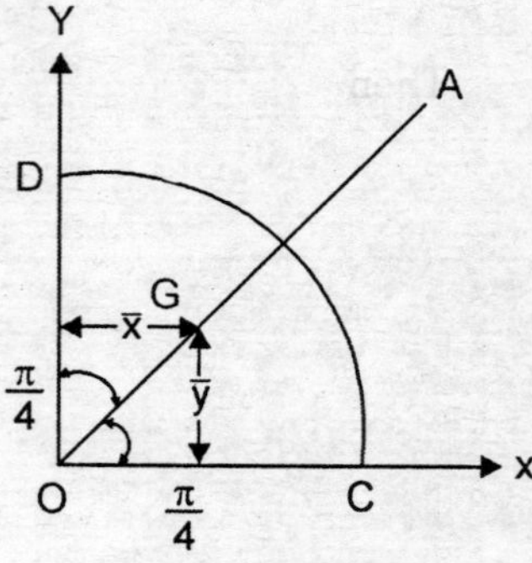

Fig. 14.12

Thus from equation (14.22*b*) by putting $2\propto = \pi/2$ we get in the present case

or $$\propto = 45° = \pi/4$$

$$OG = \frac{2}{3} \frac{R\sin\left(\frac{\pi}{4}\right)}{\pi/4}$$

$$= \frac{8R}{3\pi} \frac{1}{\sqrt{2}} = \frac{8}{3} \frac{R}{\pi\sqrt{2}}$$

Thus if $(\bar{x}, \bar{y})$ we the co-ordinates of G with respect to the axis OX and OY.

$$\bar{x} = OG \cos\frac{\pi}{4} = \frac{8R}{3\pi\sqrt{2}} \cdot \frac{1}{\sqrt{2}} = \frac{4R}{3\pi} \quad \text{...(14.23a)}$$

Also

$$\bar{y} = OG \sin\frac{\pi}{4} = \frac{8R}{3\pi}\frac{1}{(\sqrt{2})}\frac{1}{\sqrt{2}} = \frac{4R}{3\pi} \quad \text{...(14.23b)}$$

And from equation (14.20)

$$A = R^2 \propto = R^2\,\pi/4 = \pi R^2/4$$

14.4.5 Centroid of A Semi Circular *AREA*

In Fig. 14.13 is shown a semi circular area ABC or radius R. How the arc ABC subtends angle of π of the centre 0. Also G lies x Y-axis as it is an axis of symmetry.

Thus $\quad 2\propto = \pi$ or $\propto = \pi/2$

$\therefore$ From equations (14.22)

Fig. 13.13

$$OG = \frac{2}{3}R\frac{\sin\frac{\pi}{2}}{\pi/2} = \frac{4R}{3\pi} \quad \text{...(14.25)}$$

Also from equation (14.20) area of semi circle A dy putting $\propto = A_2$

We get

$$A = \frac{\pi}{2}R^2 \quad \text{...(14.26)}$$

14.4.6 Centroid and Area of Circle

In the case of a circular area both X and Y axes are axes of symmetry. Therefore the centre O of the circle itself will be the centroid being the put of intersection of areas of symmetry. Ref. Fig. 14.14.

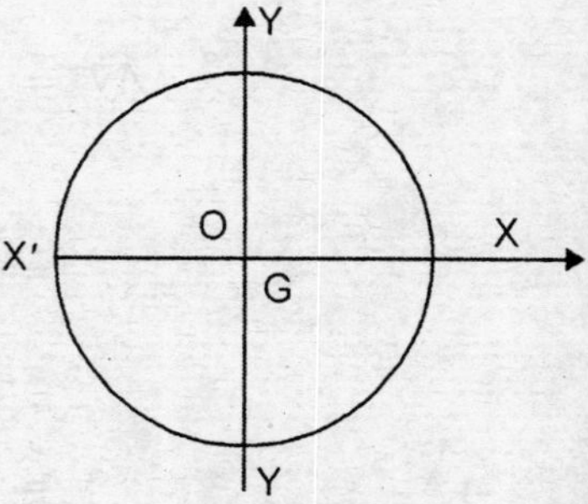

Fig. 14.14

For evaluating area put $2\propto = 2\pi$ in equation (14.22)

or

$$\propto = \pi$$

$$A = R^2\propto = \pi R^2 \quad \text{...(14.27)}$$

14.4.7 Area under Parabolic Curve $y = kx^2$ and its Centroid

In Fig. 14.15 is shown a parabolic curve OA of equation

$$y = kx^2 \quad \text{...(14.28)}$$

Elemental area of width (dx) and length y is taken at a distance x from the origin

i.e.,

$$dA = y\,dx$$

$$= kx^2\,dx \quad \text{...(14.29)}$$

It is desired to find the area under the curve OA where the coordinates of A are $(x_1\ y_1)$. Integrating equation (14.28) we get μx area under the curve OA as

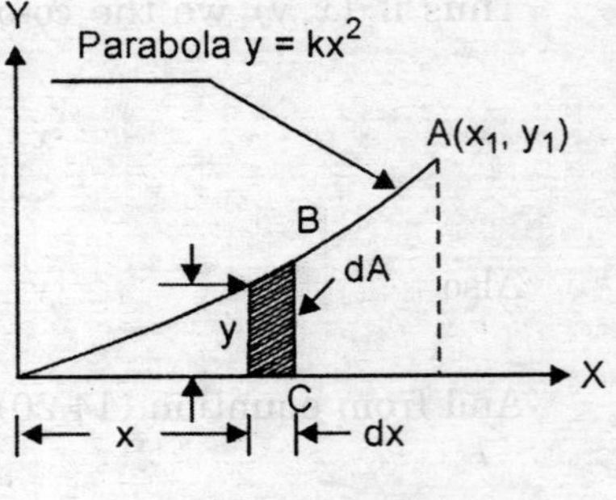

Fig. 14.15

$$A = \int_{x^2 a}^{x_1} kx^2 dx$$

$$= k\left[\frac{x^3}{3}\right]_0^{x_1} = \frac{kx_1^3}{3} \quad \text{...(14.30)}$$

If $(\bar{x}, \bar{y})$ are the co-ordinates of centroid G then taking moments about Y-axis give

$$A\bar{x} = \text{moment of area about } Y\text{-axis}$$

$$= \int_0^{x_1} x\,dA = \int_0^{x_1} (x)(y\,dx) = \int_0^{x_1} (x)(kx^2)\,dx$$

or $$A\bar{x} = k\int_0^{x_1} x^3 dx = k\left[\frac{x^4}{4}\right]_0^{x_1} = \frac{kx_1^4}{4} \quad \text{...(14.31)}$$

Therefore, Fig. (14.29), (14.30)

$$\bar{x} = \frac{kx_1^4}{4} \times \frac{3}{kx_1^3} = \frac{3}{4}x_1 \quad \text{...(14.32)}$$

Again taking moments of elemental area about X-axis we get

$$A\bar{y} = \int_{x=0}^{x_1} dA \cdot \left(\frac{y}{2}\right)$$

$$= \int_{x=0}^{x_1} (y\,dx)\frac{y}{2} = \int_{x=0}^{x_1} kx^2 \cdot \frac{kx^2}{2}\,dx$$

$$= \int_0^{x_1} \frac{k^2 x^4}{2}\,dx = \frac{k^2}{2}\left[\frac{x^5}{5}\right]_0^{x_1}$$

$$= \frac{k^2}{10}x_1^5$$

$$\therefore \quad \bar{y} = \frac{k^2}{10}x_1^5 \times \frac{3}{kx_1^3} = \frac{3}{10}k\,x_1^2 \quad \text{...(14.33)}$$

Since the point $(x_1\ y_1)$ lies on parabola we have Fig. (14.28)

$$y_1 = kx_1^2 \quad i.e. \quad k = \frac{y_1}{x_1^2} \quad \text{...(14.34a)}$$

$$\therefore \quad \bar{y} = \frac{3}{10}\frac{y_1}{x_1^2}x_1^2 = \frac{3}{10}y_1 \qquad ...(14.34b)$$

and from (14.32) $\quad \bar{x} = \frac{3}{4}x_1$

$\therefore$ co-ordinates of G are $\quad \left[\frac{3}{4}x_1, \frac{3}{10}y_1\right]$

And $\quad$ Area A $= \frac{kx_1^3}{3} = \frac{y_1}{x_1^2}\cdot\frac{x_1^3}{3} = \frac{x_1 y_1}{3}$ $\qquad$ ...(14.35)

14.4.8 Area under Parabolic curve $y^2 = kx$ and its Centroid

In Fig. 14.16 is shown the parabolic curve

$$y^2 = kx \qquad ...(14.36)$$

Elemental area $\quad dA = y\,dx$

$$= \sqrt{k}\,x^{\frac{1}{2}}\,dx$$

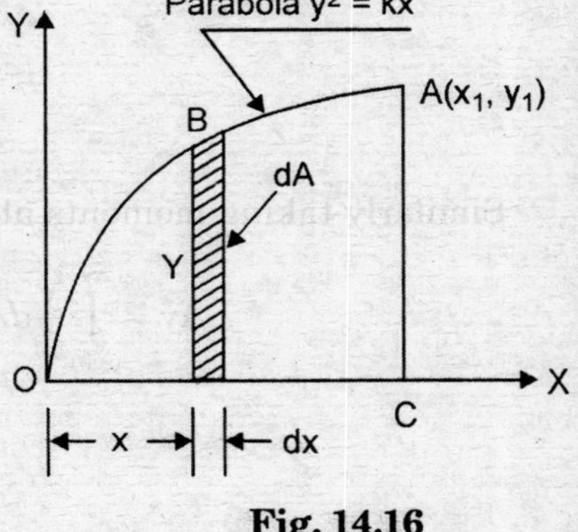

Fig. 14.16

Therefore Area under the curve *OBA* when the co-ordinates of point *A* are (x_1, y_1)

$$A = \int_{x=0}^{x_1} dA = \int_{x=0}^{x_1} y\,dx$$

$$= \int_{x=0}^{x_1} \sqrt{k}\,x^{\frac{1}{2}}\,dx = \sqrt{k}\left[\frac{2x^{3/2}}{3}\right]_0^{x_1}$$

$$A = \frac{2}{3}\sqrt{k}\,x_1^{3/2} \qquad ...(14.37a)$$

Since point $A\ (x_1, y_1)$ lies on curve therefore from equation (14.36)

$$y_1^2 = kx_1$$

or $\quad k = \frac{y_1^2}{x_1}$

Then equation (14.37*a*) reduces to

$$A = \frac{2}{3}\sqrt{\frac{y_1^2}{x_1}}\,x_1^{3/2}$$

$$= \frac{2}{3}\frac{y_1}{x^{1/2}}x_1^{3/2} = \frac{2}{3}x_1 y_1 \qquad ...(14.37b)$$

If G is the centroid of the area under the curve OBA, the co-ordinates of A being (x_1, y_1) then moment about Y-axis gives

$$A\bar{x} = \int_{x=0}^{x_{1x}} x\,dA = \int_{x=0}^{x_1} x\,y\,dx$$

$$= \int_{x=0}^{x_1} x\sqrt{k}\,x^{\frac{1}{2}}\,dx$$

$$= \sqrt{k}\int_{x=0}^{x_1} x^{3/2}\,dx = \sqrt{k}\left[\frac{2x^{5/2}}{5}\right]_0^{x_1}$$

$$= \frac{2}{5}\sqrt{k}\,x_1^{5/2}$$

Substituting value of A from equation (14.37a)

$$\therefore \quad \bar{x} = \frac{\frac{2}{5}\sqrt{k}\,x_1^{5/2}}{\frac{2}{3}\sqrt{k}\,x_1^{3/2}} = \frac{3}{5}x_1 \qquad \text{...(14.38)}$$

Similarly taking moments about X-axis we get

$$A\bar{y} = \int_{x=0}^{x_1} \frac{y}{2}\,dA = \frac{1}{2}\int_{x=0}^{x_1} y\cdot(y\,dx)$$

$$= \frac{1}{2}\int_{x=0}^{x_1} y^2\,dx \text{ from equation (14.36)}$$

$$= \frac{1}{2}\int_{x=0}^{x_1} kx\,dx$$

or

$$A\bar{y} = \frac{1}{2}k\cdot\left(\frac{x^2}{2}\right)_0^{x_1} = \frac{1}{4}kx_1^2$$

$$\therefore \quad \bar{y} = \frac{\frac{1}{4}kx_1^2}{\frac{2}{5}\sqrt{k}\,x_1^{3/2}} = \frac{5\sqrt{k}\sqrt{x_1}}{8} \qquad \text{...(14.38)}$$

Also

$$k = \frac{y_1^2}{x_1}$$

$$\sqrt{k} = \frac{y_1}{\sqrt{x_1}} \qquad \text{...(14.39)}$$

Therefore from (14.38) and (14.39)

$$\bar{y} = \frac{5}{8}\frac{y_1}{\sqrt{x_1}}\sqrt{x_1} = \frac{5}{8}y_1$$

Therefore the centroid G has co-ordinates

$$(\bar{x}, \bar{y}) \text{ as } \left(\frac{3}{5}x_1, \frac{5}{8}y_1\right) \quad ...(14.40)$$

and $$A = \frac{2}{3}x, y_1 \quad ...(14.41)$$

14.4.9 Area and Centroid of an Ellipse

Let the ellipse be represented by the equation

$$\frac{x^2}{a^2} + \frac{y^2}{b^2} = 1 \quad ...(14.42)$$

The curve for this equation is drawn in Fig. (14.17).

Since the area is symmetrical about both X and Y-axes the centroid will be the origin O.

Let an elemental area be the strip AB of width dx at a distance x from origin O.

Then $$dA = y\,dx$$

But from equation (14.42) $$y = \frac{b}{a}(a^2 - x^2)^{\frac{1}{2}}$$

$$\therefore \quad dA = \frac{b}{a}(a^2 - x^2)^{\frac{1}{2}}\,dx$$

and total area $$A = 4\int_{x=0}^{a} dA$$

$$= 4\int_{x=0}^{a} \frac{b}{a}(a^2 - x^2)^{\frac{1}{2}}\,dx$$

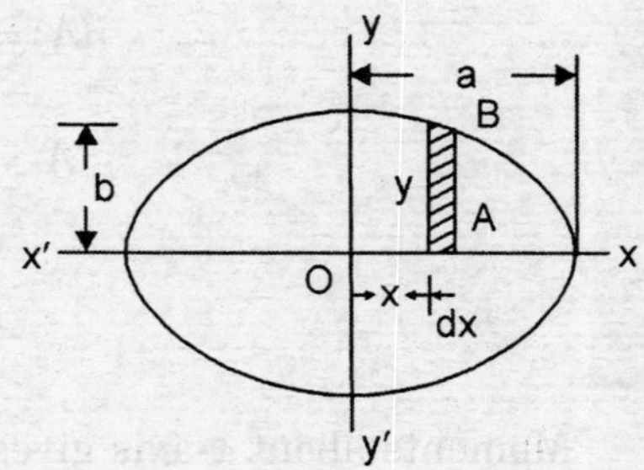

Fig. 14.17

Substitute $$x = a \sin \theta$$

$$dx = a \cos \theta \, d\theta$$

$$\therefore \quad A = \frac{4b}{a}\int_{0}^{\pi/2} (a^2 - x^2 \sin^2 \theta)^{\frac{1}{2}}\, a \cos\theta\, d\theta$$

$$= \frac{4b}{a}\int_{0}^{\pi/2} a^2 \cos^2 \theta\, d\theta$$

$$= \frac{4ba}{2}\int_0^{\pi/2} 2\cos^2\theta\, d\theta$$

$$= 2ba\int_0^{\pi/2} (1+\cos 2\theta)\, d\theta$$

$$= 2ba\left[\theta + \frac{\sin 2\theta}{2}\right]_0^{\pi/2}$$

$$= 2ba\left(\frac{\pi}{2}\right) = \pi ab \qquad ...(14.43)$$

14.4.10 Area and Centroid of Quarter Ellipse

Let the curve be defined by equation

$$\frac{x^2}{a^2}+\frac{y^2}{b^2} = 1$$

Since the curve is neither symmetrical about X-axis nor Y-axis the centroid of this area will not lie at O.

Let the elemental area CD be of width dx lying at a distance of x from OY.

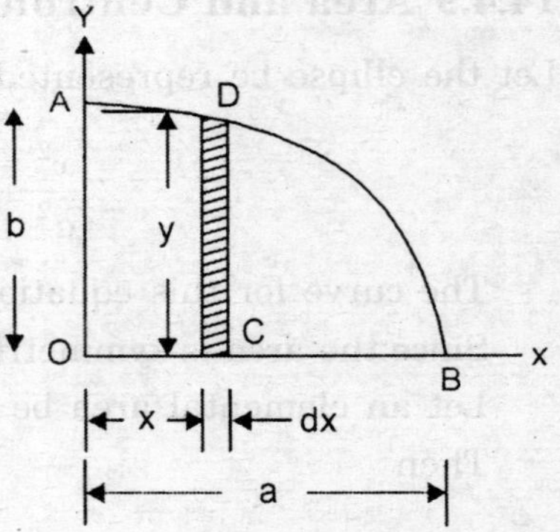

Then the area

Fig. 14.18

$$dA = y\ dx$$

$$A = \int_{x=0}^{a} y\,dx = \frac{1}{4} \text{ (area of entire ellipse as found in equation 14.43)}$$

$$= \frac{\pi ab}{4} \qquad ...(14.44)$$

Moments about x-axis gives $\bar{y}$, so that

$$A\bar{y} = \int_{x=0}^{a} (y\,dx)\frac{y}{2} = \frac{1}{2}\int_{x=0}^{a} y^2\,dx$$

But

$$\frac{y^2}{b^2} = 1-\frac{x^2}{a^2}$$

$$y^2 = \frac{b^2(a^2-x^2)}{a^2}$$

$\therefore$

$$A\,\bar{y} = \frac{1}{2}\int_{x=0}^{a}\frac{b^2}{a^2}(a^2-x^2)\,dx$$

$$= \frac{1}{2}\int_0^a \left(b^2 - \frac{b^2}{a^2}x^2\right)dx$$

$$= \frac{1}{2}\left(b^2 x - \frac{b^2}{a^2}\frac{x^3}{3}\right)_0^a$$

$$= \frac{1}{2}\left(b^2 a - \frac{b^2}{3a^2}a^3\right) = \frac{1}{2}\frac{2}{3}ab^2 = \frac{b^2}{3}a$$

$\therefore$
$$\bar{y} = \frac{ab^2}{3}\cdot\frac{1}{A} = \frac{ab^2}{3}\cdot\frac{4}{\pi ab} = \frac{4b}{3\pi} \quad \textbf{Ans.} \qquad ...(14.45)$$

For finding $\bar{x}$ take moments about y-axis

$$\bar{x}A = \int_{x=0}^{a} (y\,dx)\,x$$

Now
$$y^2 = b^2\left(1 - \frac{x^2}{a^2}\right)$$

$$= \frac{b^2}{a^2}(a^2 - x^2)$$

or
$$y = \frac{b}{a}(a^2 - x^2)^{\frac{1}{2}}$$

$\therefore$
$$\bar{x}A = \int_{x=0}^{a} \frac{b}{a}(a^2 - x^2)^{\frac{1}{2}}\,x\,dx$$

Put
$$x = a\sin\theta$$
$$dx = a\cos\theta\,d\theta$$

$\therefore$
$$\bar{x}A = \int_0^{\pi/2} \frac{b}{a}(a\cos\theta)(a\sin\theta)(a\cos\theta)\,d\theta$$

$$= a^2 b\int_0^{\pi/2} \cos^2\theta\sin\theta\,d\theta$$

$$= -a^2 b\int_0^{\pi/2} \cos^2\theta(-\sin\theta)\,d\theta$$

$$= -a^2 b\left(\frac{\cos^3\theta}{3}\right)_0^{\pi/2}$$

$$= -\frac{a^2 b}{3}[0-1] = \frac{a^2 b}{3}$$

$$\therefore \quad \bar{x} = \frac{a^2 b}{3} \times \frac{1}{A} = \frac{a^2 b}{3} \frac{4}{\pi ab} = \frac{4a}{3\pi} \quad ...(14.46a)$$

Therefore the coordinates of the centroid are

$$(\bar{x}, \bar{y}) \rightarrow \left(\frac{4a}{3\pi}, \frac{4b}{3\pi}\right) \quad ...(14.46b)$$

and

$$A = \frac{\pi ab}{4} \quad ...(14.47)$$

14.4.11 Centroid and Area of Semi-elliptic Area

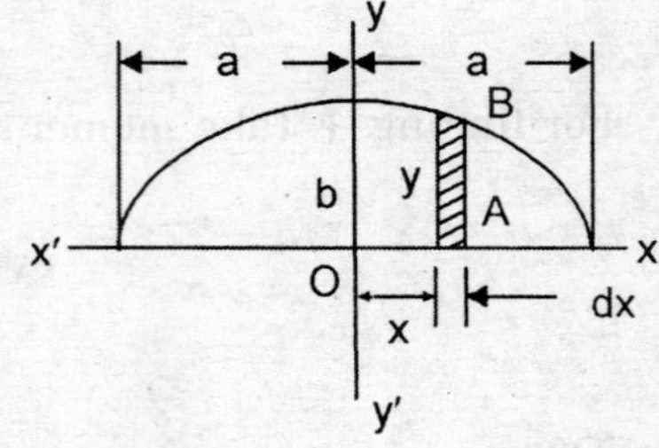

Fig. 14.19

$$\frac{x^2}{a^2} + \frac{y^2}{b^2} = 1$$

or

$$y^2 = \left(1 - \frac{x^2}{a^2}\right) b^2$$

As the Y-axis is the axis of symmetry the C.G. will be an it

For finding the value of $\bar{y}$ take moments about X-axis. For this elemental area ($y\,dx$) as shown in Fig. 14.19.

$$\bar{y} \int_{x=-a}^{a} y\,dx = \int_{x=-a/1}^{a} (y\,dx)\left(\frac{y}{2}\right)$$

$$2\bar{y}\int_0^a \left(1 - \frac{x^2}{a^2}\right)^{\frac{1}{2}} b\,dx = \frac{1}{2}\int_{x=-a}^{a} y^2\,dx$$

$$\bar{y}\,\frac{2b}{a}\int_0^a \sqrt{a^2 - x^2}\,dx = \frac{1}{2} \times 2\int_0^a y^2\,dx$$

or

$$\frac{2b}{a}\bar{y}\int_0^a \sqrt{a^2 - x^2}\,dx = \int_0^a b^2\left(1 - \frac{x^2}{a^2}\right)dx$$

$$= \frac{b^2}{a^2}\int_0^a (a^2 - x^2)\,dx$$

$$= \frac{b^2}{a^2}\left[a^2 x - \frac{x^3}{3}\right]_0^a = \frac{2}{3}\frac{b^2 a^3}{a^2} = \frac{2}{3}ab^2 \quad ...(14.48)$$

For LHS put $x = a \sin \theta$

$$dx = a \cos \theta \, d\theta$$

$$\text{LHS} = \frac{2b}{a}\bar{y}\int_0^{\pi/2} \sqrt{a^2 - a^2 \sin^2 \theta}\ (a\cos\theta\, d\theta)$$

$$= \frac{2b}{a}\bar{y}\int_0^{\pi/2} (a\cos)(a\cos\theta)\, d\theta$$

$$= 2ab\bar{y} \times \frac{1}{2}\int_0^{\pi/2} 2\cos^2\theta\, d\theta$$

$$= \frac{2ab}{2}\bar{y}\int_0^{\pi/2} (1+\cos 2\theta)\, d\theta$$

$$= ab\bar{y}\left(\theta + \frac{\sin 2\theta}{2}\right)_0^{\pi/2}$$

$$= \frac{\pi ab\bar{y}}{2}$$

Above expression, then reduces to

$$\frac{\pi ab\bar{y}}{2} = \frac{2}{3}ab^2$$

$$\therefore \qquad \bar{y} = \frac{2}{3}\frac{ab^2 \times 2}{\pi ab} = \frac{4b}{3\pi} \qquad \text{...(14.49)}$$

Thus the centroid of semi-ellipse lie at a distance $\frac{4b}{3\pi}$ from the central axes (*i.e.*, *OX*) and on the *Y*-axis.

14.5 CENTRE OF GRAVITY OF SOLIDS

Centre of gravity or in short (C.G.) is the point where the Resultant Earth's gravitational force of attraction is passing. The method to be followed for locating the C.G. will be same as that followed while finding the co-ordinates of centroids of plane figures by involving the Varignon's theorem *i.e.*, by taking moments of the individual forces about the three axes of reference in turn.

Table 14.1. Areas and Location of Centroid of Common Plane Figures.

	Figure	Area	Centroid $(\bar{x}, \bar{y})$
Rectangle	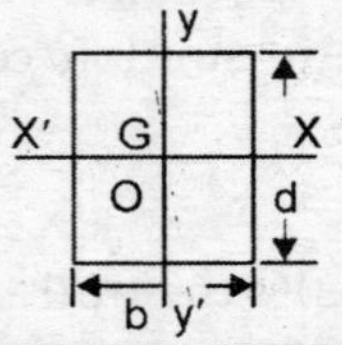	$b \times d$	(0, 0)
Triangle	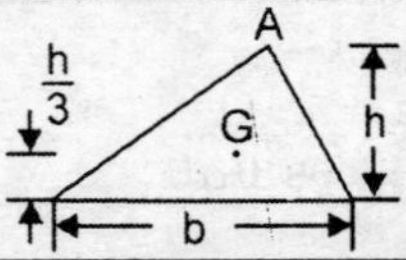	$\frac{1}{2}bh$	$\bar{y}$ of $G = \frac{h}{3}$
Semi Circle	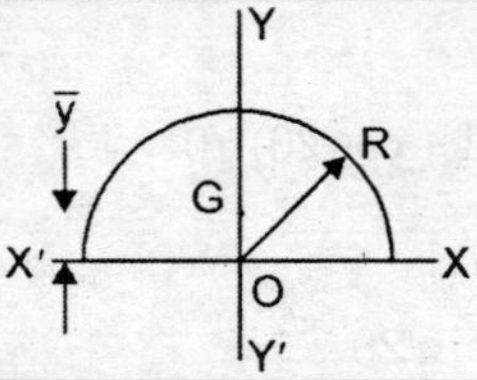	$\frac{\pi}{2}R^2$	$\bar{y} = \frac{4R}{3\pi}$
Quarter Circle	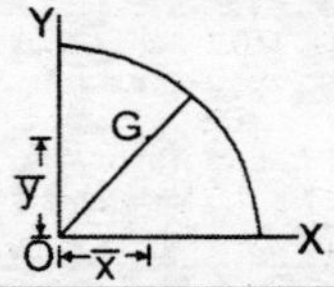	$\frac{\pi R^2}{4}$	$\bar{x} = \frac{4R}{3\pi}$ $\bar{y} = \frac{4R}{3\pi}$
Parabola $y = kx^2$	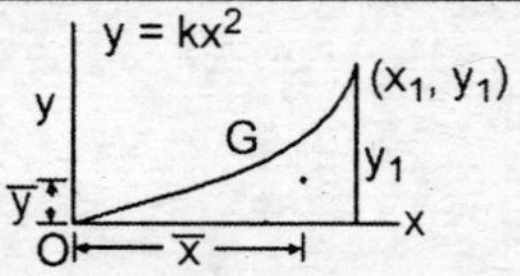	$\frac{x_1 y_1}{3}$	$(\bar{x}, \bar{y})$ $\left(\frac{3}{4}x_1, \frac{3}{10}y_1\right)$
Parabola $x = ky^2$	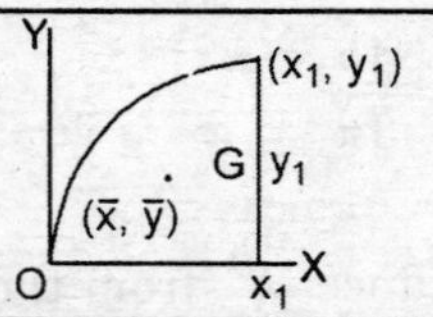	$\frac{2}{3}x_1 y_1$	$\left(\frac{3}{5}x_1, \frac{5}{8}y_1\right)$
Ellipse	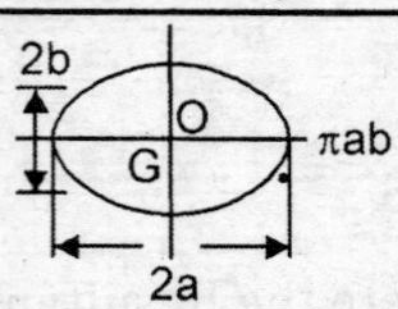	πab	(0, 0)
Quarter of Ellipse	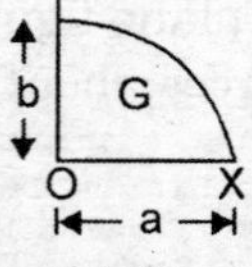	Area $= \frac{\pi ab}{4}$	centroid $\left(\frac{4a}{3\pi}, \frac{4b}{3\pi}\right)$
Semi Ellipse	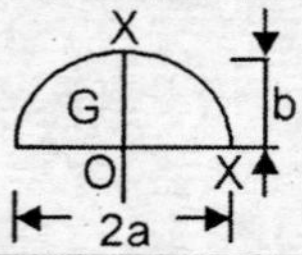	$\frac{\pi ab}{2}$	$\left(\frac{4b}{3\pi}\right)$

14.5.1 C.G. of a Solid Sphere

Since the sphere is symmetrical about the three axes of reference *OX*, *OY* and *OZ*. Thus their common point *i.e.*, *O*, the origin is C.G. of the solid sphere. (Fig. 14.20)

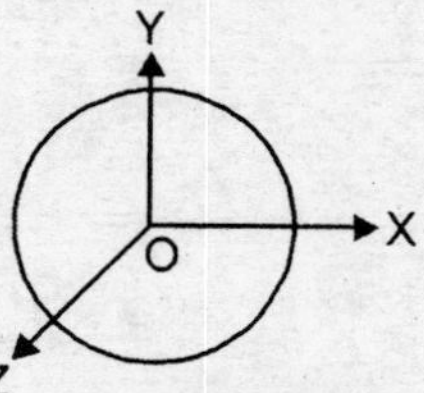

Fig. 14.20

14.5.2 Centre of Gravity of a Right Circular Cone

In Fig. 14.21 is shown a right-circular curve of height h and circular base of radius r.

Evidently *AO* in the axis of symmetry and the C.G. will be x the axis *i.e.*, *OY*. Chose the axis of reference *OX*, *OY* and *OZ* with origin at *O*, the centre of base circle.

Let an elemental volume in the form of a thin circular disc *DFE* of rader *DF* and thickness dy be lying at a height of $FO = y$ from the base circle

Then $$dV = \pi(DF)^2\, dy \qquad ...(14.50)$$

From Fig. (14.32) symmetrically we have

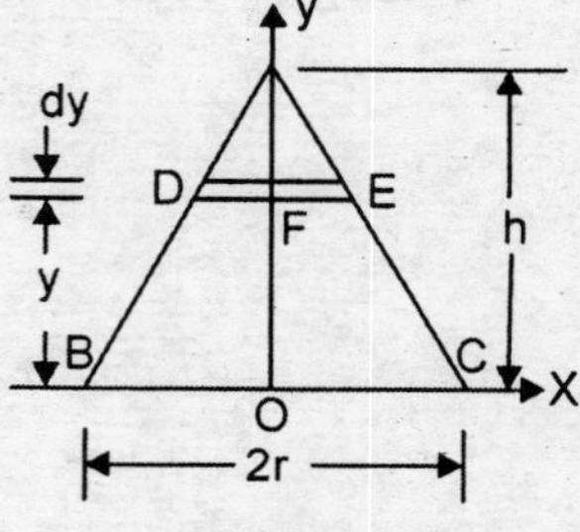

Fig. 14.21

$$\frac{DF}{BO} = \frac{AF}{AO}$$

or $$DF = \frac{AF}{AO}(BO)$$

$$= \frac{(h-y)}{h} r$$

or $$DF = r\left(\frac{h-y}{h}\right)$$

$$= r\left(1-\frac{y}{h}\right) \qquad ...(14.51)$$

Thus from equation (14.50) and (14.51)

$\therefore$ $$dV = \pi\left(\frac{r^2}{h^2}(h-y)^2\right)dy$$

$$= \pi r^2\left[\left(1-\frac{y}{h}\right)^2\right]dy$$

$$V = \int_{y=0}^{h} \pi r^2\left\{1+\frac{y^2}{h^2}-\frac{2y}{h}\right\}dy$$

$$= \pi r^2\left[y+\frac{y^3}{3h^2}-\frac{2y^2}{2h}\right]_0^h$$

$$= \pi r^2 \left[h + \frac{h^3}{3h^2} - \frac{h^2}{h} \right]$$

$$= \pi r^2 \left[h + \frac{h}{3} - h \right]$$

$$= \frac{\pi r^2 h}{3} \qquad \text{...(14.52)}$$

If the density of the cone material is ρ, then

Weight of cone $= \rho \times V$

$$W = \frac{\pi}{3} r^2 h \rho$$

Taking moments about *X*-axis

$$\bar{h} W = \int_{y=0}^{h} (\rho dV) y$$

$$= \int_{y=0}^{h} \rho y \left\{ \pi (DF)^2 dy \right\}$$

$$= \pi\rho \int_{y=0}^{h} y \left\{ r^2 \left(1 - \frac{y}{h} \right)^2 dy \right.$$

$$= \pi r^2 \rho \int_0^h y \left(1 + \frac{y^2}{h^2} - \frac{2y}{h} \right) dy$$

$$= \pi r^2 \rho \int_0^h \left(y + \frac{y^3}{h^2} - \frac{2y^2}{h} \right) dy$$

$$= \pi r^2 \rho \left[\frac{y^2}{2} + \frac{y^4}{4h^2} - \frac{2y^3}{3h} \right]_0^h$$

$$= \pi r^2 \rho \left[\frac{h^2}{2} + \frac{h^2}{4} - \frac{2}{3} h^2 \right]$$

$$= \rho \frac{\pi r^2}{12} h^2 [6 + 3 - 8]$$

or

$$\bar{h}\frac{\pi}{3}r^2h\rho = \frac{\pi r^2h^2}{12}\rho$$

$$\bar{h} = \frac{h}{4} \qquad \text{...(14.53)}$$

That is the C.G. lies on the axis of cone and it is at a distance of $\frac{h}{4}$ from the base.

14.5.3 Volume of a Solid Sphere of Radius *r*

In Fig. 14.22 section of the sphere by $X - Z$ plane is shown.

ABC is disc of radius x and it is at a distance of z from the diametral plane *OXY* and of thickness dz.

$$dV = \pi x^2\,dz$$

$$= \pi(r^2 - z^2)\,dz$$

$$V = 2\int_{z=0}^{r} \pi(r^2 - z^2)dz$$

$$= 2\pi\int_{z=0}^{r} (r^2 - z^2)dz$$

$$= 2\pi\left[r^2z - \frac{z^3}{3}\right]_0^r$$

$$= 2\pi\left[r^3 - \frac{r^3}{3}\right]$$

$$= 2\times\frac{2\pi r^3}{3} = \frac{4}{3}\pi r^3 \qquad \text{...(14.54)}$$

Fig. 14.22. Section of sphere in $x - z$ plane.

Since the sphere is symmetrical about the three axes, the C.G. coincides with 0.

14.5.4 To Find the Location of the C.G. of a Solid Hemisphere

In Fig. 14.23 a section by zx plane of the hemisphere is shown. *ABC* is a disc of thickness dz at a distance of z from the base of the hemisphere

$$BC^2 = x^2 = r^2 - z^2$$

$$dV = \pi x^2\,dz$$

$$= \pi(r^2 - z^2)\,dz$$

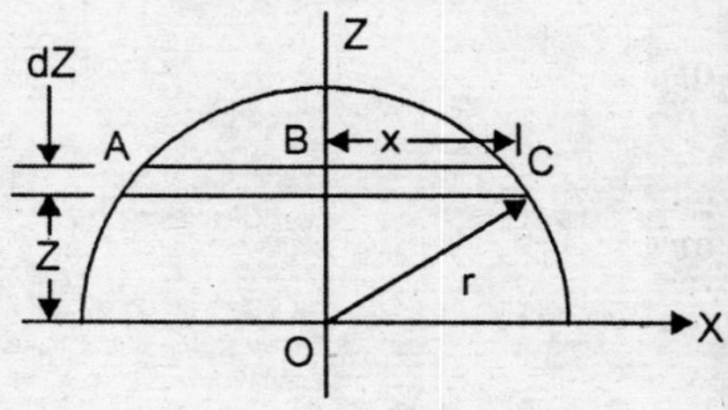

Fig. 14.23

$$V = \int_{z=0}^{r} \pi(r^2 - z^2)\,dz$$

$$= \pi\left[r^2 z - \frac{z^3}{3}\right]_0^r$$

$$= \pi\left[r^3 - \frac{r^3}{3}\right] = \frac{2}{3}\pi r^3 \qquad \text{...(14.55)}$$

For finding the location of C.G. which will be lying on the axis of symmetry OZ. If $\bar{z}$ is the height of the C.G. from the base. Then using Varignon's theorem and taking moments about OX we get

$$(\rho V)\bar{z} = \int_{z=0}^{r} (\rho\, dV)\, z$$

$$= \int_{z=0}^{r} \rho \pi x^2 z\, dz$$

$$= \int_{z=0}^{r} \pi\rho(r^2 - z^2) z\, dz$$

$$= \pi\rho \int_{z=0}^{r} (r^2 z - z^3)\, dz$$

$$= \pi\rho\left[\frac{r^2 z^2}{2} - \frac{z^4}{4}\right]_0^r$$

Substitute value of V from equation 14.55.

or $$\bar{z}\left(\frac{2}{3}\pi r^3 \rho\right) = \frac{\pi\rho}{4}[2r^4 - r^4]$$

or $$= \frac{\pi\rho r^4}{4}$$

or $$\bar{z} = \frac{\pi\rho r^4}{4} \times \frac{3}{2\pi r^3 \rho}$$

$$\bar{z} = \frac{3r}{8} \qquad \text{...(14.56)}$$

CENTROID OF COMPOSITE AREAS AND C.G. OF COMPOSITE BODIES

Example 14.1. *Find the location of centroid of a T-section shown in Fig. 14.24.*

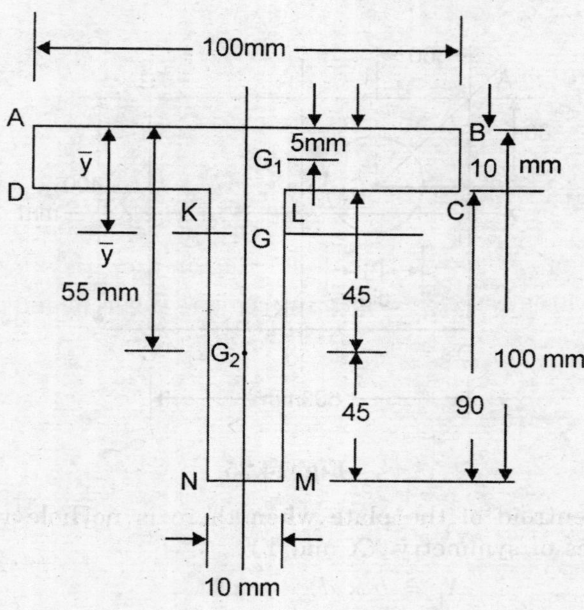

Fig. 14.24

Solution: The compsite section consists of two rectangular areas *ABCD* and *KLMN*.

Let these areas be represented by A_1 and A_2 with their centroids G_1 and G_2 such that the distances of G_1 and G_2 from the reference line *AB* are y_1 and y_2. Let the distance of centroid *G* of the composite section be $\bar{y}$ from *AB*. Then taking moments of the respective areas about *AB* we will get by using Varignon's theorem.

These $\quad A_1 = 100 \times 10 = 1000 \text{ mm}^2$

$$y_1 = \frac{1}{2} \times 10 = 5 \text{ mm}$$

$$A_2 = 90 \times 10 = 900 \text{ mm}^2$$

$$y_2 = 100 - 45 = 55 \text{ mm}$$

Then $\quad (A_1 + A_2)\bar{y} = A_1 y_1 + A_2 y_2$

$$(1000 + 900)\bar{y} = 1000 \times 5 + 900 \times 55$$

$$1900\bar{y} = 5000 + 99500$$

$$= 54500$$

$$\bar{y} = \frac{54500}{1800} = 28.68 \text{ mm}$$

Example 14.2 : *A plate ABCD is 600 mm wide and 400 mm deep A hole of diameter 15 mm is drilled at a point G_2 which is 60 mm from AB and 60 mm from AD. Find the location of the centroid of the plate.*

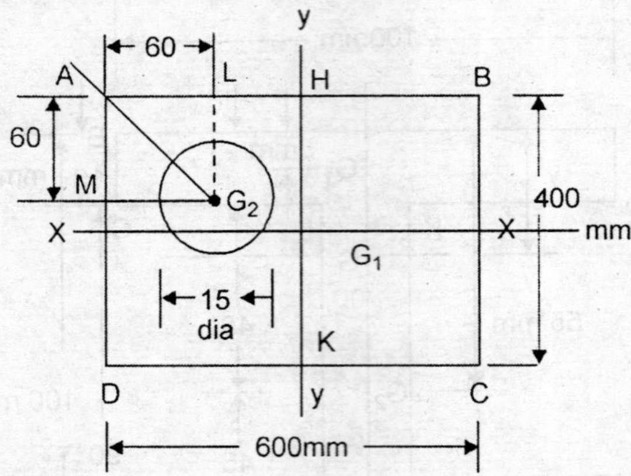

Fig. 14.25

Solution : The centroid of the plate when there is no hole will be G_1 the point of intersection of the axes of symmetry *XX* and *YY*.

Then $A_1 = b \times d$

where $b = 600$

$d = 400$

$= 600 \times 400$

$= 24 \times 10^4 \text{ mm}^2$

The C.G. of the hole will be at G_2 and its area is

$$A_2 = \frac{\pi}{4} \times (15.0)^2 = 176.71 \text{ mm}^2$$

∴ Resultant area

$$A_1 - A_2 = 240000 - 176.71$$

$$= 23.98 \times 10^4 \text{ mm}^2$$

Let the centroid of the final area be at a distance $\bar{y}$ from *AB*.

Taking moments about *AB*

$$(A_1 - A_2)\bar{y} = A_1 \times (HG_1) - A_2(LG_2)$$

$$= 24 \times 10^4 \times 200 - 176.71 \times 60$$

$$= 48 \times 10^6 - 1.06 \times 10^4$$

or $23.98 \times 10^4\ \bar{y} = 47.989 \times 10^6 \text{ mm}^3$

$$\therefore \quad \bar{y} = \frac{47.989 \times 10^6}{23.98 \times 10^4} = 200.1 \text{ mm } \textbf{Ans.}$$

If $\bar{x}$ is the distance of G from AD, then taking moment about AD gives

$$(A_1 - A_2)\bar{x} = A_1 \times (NG_1) - A_2\,(M\,G_2)$$
$$= 24 \times 10^4 \times 300 - 176.71 \times 60$$
$$= 72 \times 10^6 - 1.06 \times 10^4$$
$$= 7198.94 \times 10^4$$

or $$23.98 \times 10^4\, x = 7198.94 \times 10^4$$

$$\bar{x} = \frac{7198.94}{23.989}$$
$$= 300.2 \text{ mm } \textbf{Ans.}$$

Example 14.3. *In Fig. 14.26 is shown a composite area which consists of a rectangle ABCD 600 × 300 mm and a right angled triangle BEC of base BC = 300 nm and height BE = 120 mm.*

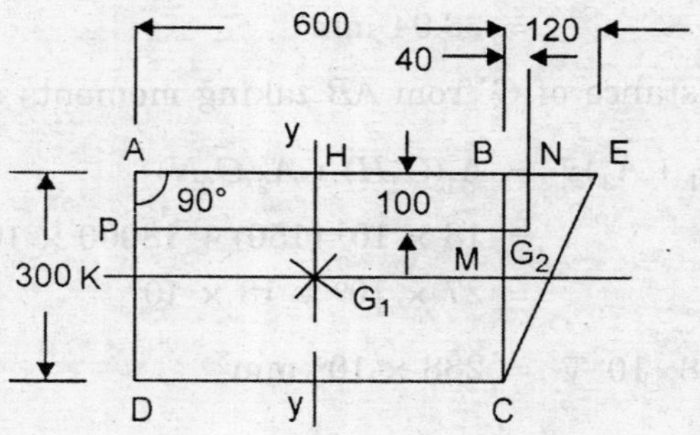

Fig. 14.26

Find the distances of the centroid of the given figure from AD and AB.

Solution :

$$A_1 = \text{Area of rectangle } ABCD$$
$$= 600 \times 300$$
$$= 18 \times 10^4 \text{ mm}^2$$
$$G_1H = 150 \text{ mm}$$
$$KG_1 = 300 \text{ mm}$$
$$A_2 = \text{Area of triangle } BEC$$
$$= \frac{1}{2} \times 300 \times 120 = 18000 \text{ mm}^2$$
$$MG_2 = \frac{1}{3}(BE) = \frac{1}{3} \times 120 = 40 \text{ mm}$$

$\therefore$ $$PG_2 = 600 + 40 = 640 \text{ mm}$$

and $$NG_2 = \frac{1}{3}(BC) = \frac{1}{3} \times 300 = 100 \text{ mm}$$

Therefore total area = $(A_1 + A_2)$

$$= 18 \times 10^4 + 18000$$

$$= 198 \times 10^3 \text{ mm}^2$$

Taking moments of areas about AD and applying Varignon's theorem, if $\bar{x}$ is the distance of G from AD.

$$(A_1 + A_2)\bar{x} = A_1(KG_1) + A_2(PG_2)$$

$$= 18\times10^4 \times \frac{600}{2} + 18000\times640$$

or $$198\times10^3\bar{x} = 54 \times 10^6 + 11.526 \times 11^6$$

i.e., $$198\times10^3\bar{x} = 65.526 \times 10^6$$

$$\bar{x} = \frac{65.526\times10^6}{198\times10^3}$$

$$= 33.94 \text{ mm}$$

Again if $\bar{y}$ in the distance of G from AB taking moments about AB of areas

$$(A_1 + A_2)\bar{y} = A_1(G_1H) + A_2(G_2N)$$

$$= 18 \times 10^4 (150) + 18000 \times 100$$

$$= 27 \times 10^6 + 18 \times 10^5$$

$$198\times10^3\bar{y} = 288 \times 10^5 \text{ mm}^3$$

$$\therefore \quad \bar{y} = \frac{288\times10^5}{198\times10^3}$$

$$= 145.95 \text{ mm}$$ **Ans.**

Example 14.4. *In Fig. 14.27 is shown a trapezium of height 'h' and base of width 'b'. Find the height of the centroid of the trapezium from AB.*

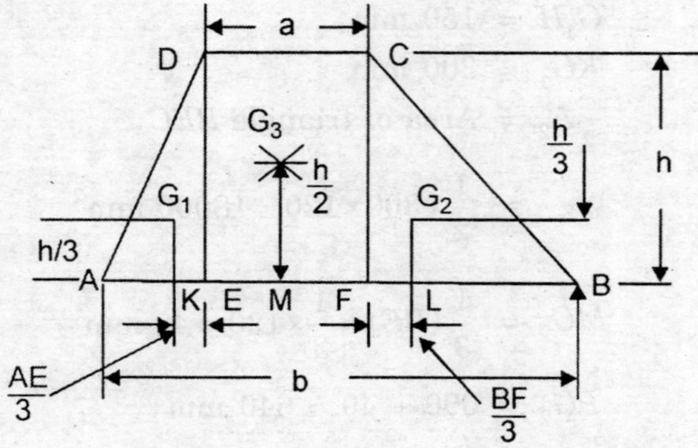

Fig. 14.27

Solution : Given data

$$AB = b$$

$$CD = a$$

Height $$DE = h$$

The given area can be split into 3 areas *viz* triangle *ADE*, Rectangle *DCFE* and triangle *CFB* having their centroids at G_1, G_3 and G_2 respectively such what G_1 and G_2 are at heights $\frac{h}{3}$ from *AB* and G_3 is at height $\frac{h}{2}$ from *AB*.

Let the controid *G* of the trapezium be at height $\bar{y}$ from *AB* taking moments of areas *ADE*, *DEFC* and *CFB* about the base *AB* we get

$$(\text{Area } ABCD)\,\bar{y} = (\text{Area } AED)\,(G_1K) + (\text{Area } DEFC)\,(G_3A) + (\text{Area } CFB)\,(G_2L)$$

$$\left[\frac{1}{2}(a+b)h\right]\bar{y} = \left[\frac{1}{2}h(AE)\right]\frac{h}{3} + (ah)\frac{h}{2} + \left[\frac{1}{2}h(FB)\right]\frac{h}{3}$$

$$= \frac{h^2}{6}(AE + FB) + \frac{ah^2}{2}$$

$$= \frac{h^2}{6}(b-a) + \frac{ah^2}{2}$$

$$= \frac{h^2}{6}[b - a + 3a]$$

$$= \frac{h^2}{6}(b + 2a)$$

Therefore $$\bar{y} = \frac{h^2}{6}(b+2a)\cdot\frac{2}{(a+b)h}$$

or $$\bar{y} = \frac{h}{3}\left(\frac{b+2a}{b+a}\right) \ldots \quad \textbf{Ans.}$$

Example 14.5. *In Fig. 14.28 is shown a plane lamina in which ABCD is rectangle with side AB = 100 mm and the portion AEB is semicircular area with AB as diameter. The height of the rectangular portion i.e., BC is such that the centroid of the composite section coincides with O. Find the value of h.*

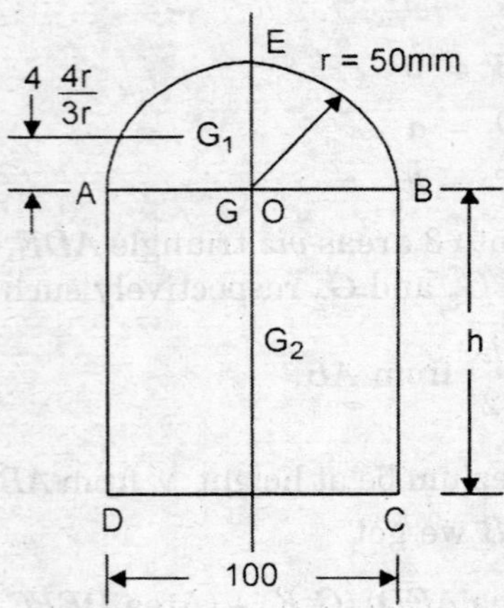

Fig. 14.28

Solution : Given data Radius of semicircular area

$$r = \frac{100}{2} = 50 \text{ mm}$$

∴ Area of semicircular portion

$$A_1 = \frac{1}{2}(\pi r^2) = \frac{1}{2}\pi(50)^2$$

$$= 3928.99 \text{ mm}^2$$

From equation (14.25) we have the distance of centroid of semicircular area from O.

$$OG_1 = \frac{4r}{3\pi} = \frac{4 \times 50}{3 \times \pi} = 21.22 \text{ mm}$$

For this rectangular portion *ABCD*

$$A_2 = AB \times h = 100\,h$$

and $$OG_2 = \frac{1}{2}h$$

If we take *AB* as the axis of reference and since the centroid of the composite section is to be on it the moments of area A_1 and A_2 which are a opposite sides of *AB* must balance

i.e., $$A_1(OG_1) = A_2(OG_2)$$

or $$(3926.99 \times 21.22) = (100\,h)\frac{4}{2}$$

$$= 50\,h^2$$

∴ $$h^2 = \frac{3928.99 \times 21.22}{50} = 1660.614$$

∴ $$h = 40.824 \text{ mm } \textbf{Ans.}$$

Example 14.6. *Using analytical method determine the centre of gravity of the plane uniform lamina shown in Fig. 14.29. (U.P. Tech Un. 2001-02)*

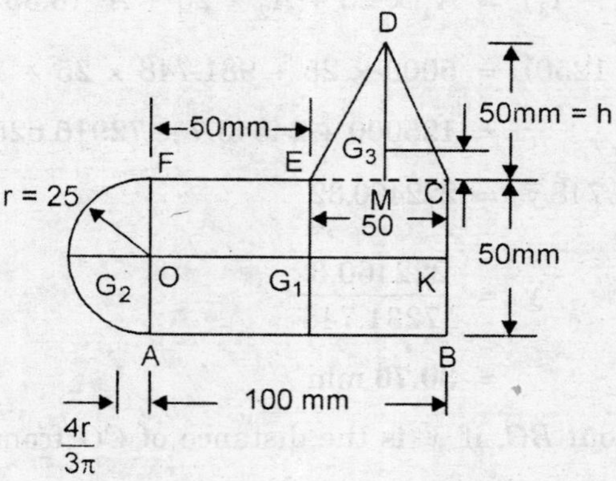

Fig. 14.29

Solution : The lamina contains 3 areas *viz* rectangular portion A_1 semicircular portion A_2 and triangular portion A_3.

Given Data $\quad r = \dfrac{50}{2} = 25$ mm

$$A_1 = 100 \times 50 = 5000 \text{ mm}^2$$

$$AG_1 = 500$$

$$A_2 = \frac{1}{2}(\pi r^2) = \frac{\pi(25)^2}{2} = 981.748 \text{ mm}^2$$

$$OG_2 = \frac{4R}{3\pi} = \frac{4 \times 25}{3\pi} = 10.01 \text{ mm}$$

$$G_2K = 10.01 + 100 = 110.01 \text{ mm}$$

$$A_3 = \frac{1}{2}(CE)\,DM$$

Now $\quad CE = \text{CF} - EF = 100 - 50 = 50$

and $\quad A_3 = \dfrac{1}{2}(CE) \times \text{height of } \Delta CDE$

$$= \frac{1}{2} \times 50 \times 50 = 1250 \text{ mm}^2$$

$$G_3M = \frac{1}{3} \times \text{height of } \Delta DEC = \frac{1}{3} \times 25 = \frac{25}{3}$$

$$= 8.3333 \text{ mm}$$

Taking moments about AB, if $\bar{y}$ is the distance of $C.G$ *from* AB

$$\bar{y}(A_1 + A_2 + A_3) = A_1 \times 25 + A_2 \times 25 + A_3\ (8.3333 + 50)$$

$$\bar{y}(500 + 981.748 + 1250) = 5000 \times 25 + 981.748 \times 25 + 1280 \times 58.3333$$

$$= 125000 + 24543.7 + 72916.625$$

$$7231.748\,\bar{y} = 222460.32$$

$$\therefore \quad \bar{y} = \frac{222460.32}{7231.748}$$

$$= \mathbf{30.76}\ \text{mm}$$

Now take moments about BC, if $\bar{x}$ is the distance of CG from BC

$$\bar{x}(5000 + 981.748 + 1250) = A_2 \times (G_2K) + A_3(MC) + A_1 \times (G_1K)$$

$$= (981.748 \times 110.01) + (1250 \times 25) + (5000 \times 50)$$

$$7231.746\,\bar{x} = 108002.1 + 31250 + 250000$$

$$= 389252.1$$

$$\bar{x} = \frac{389252.1}{7231.746}$$

$$= 53.825\ \text{mm} \quad \textbf{Ans.}$$

Example 14.7. *Determine the location of centroid of the I-section shown in Fig. 14.30.*

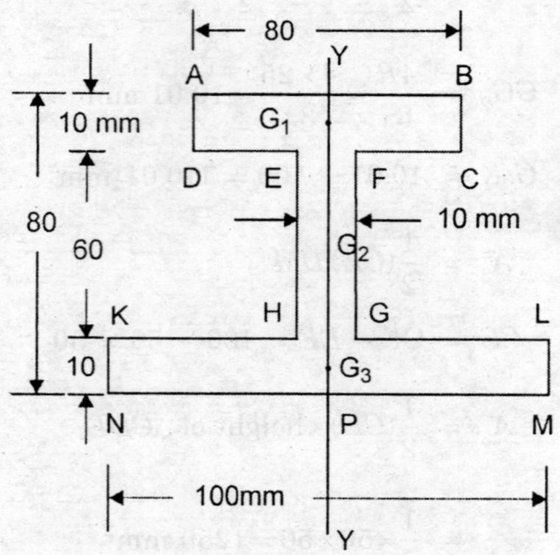

Fig. 14.30

Solution : Since YY is the axis of symmetry the centroid will be on it. Therefore it is required to find the distance of G from the line NM.

Given data $\quad A_1 = \text{Area } ABCD = 81 \times 10 = 8000\ \text{mm}^2$

$$A_2 = \text{Area } EFGH = 60 \times 10 = 600 \text{ mm}^2$$

$$A_3 = \text{Area } KLMN = 100 \times 10 = 1000 \text{ mm}^2$$

Let $\bar{y}$ be the distance of CG from NM. Then moments about NM give

$$(A_1 + A_2 + A_3)\bar{y} = A_1(G_1P) + A_2(G_2P) + A_3(G_3P)$$

$$(800 + 600 + 1000)\bar{y} = 800\ (80 - 5) + 600\ (10 + 30) + 1000 \times 5$$

or $$2400\bar{y} = 60000 + 24000 + 5000$$

$$= 89000$$

$\therefore$ $$\bar{y} = \frac{89000}{2400} = 37.08 \text{ mm}$$

Example 14.8. *Find the co-ordinates of the centroid of the plane area shown in Fig. 14.31 if the axes of reference are taken along FA and AD.*

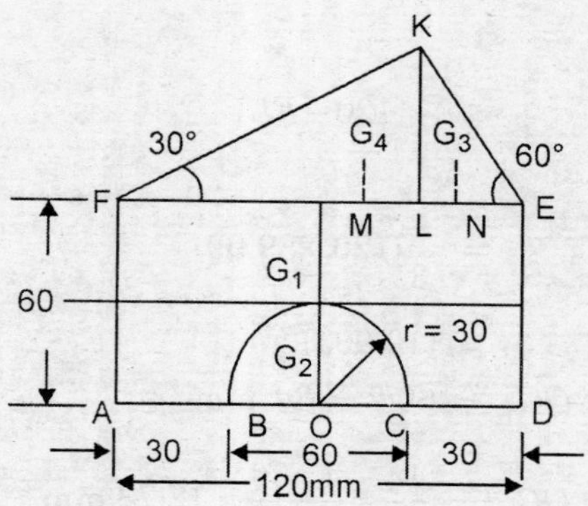

Fig. 14.31

Solution : The Fig. 14.31 consists of rectangle *ABCDEF* out of which a semicircular area is removed and the triangles *FKL* and *KLE*. Let G_1, G_2, G_3 and G_3 be the centroids of these respective areas

G_1 of area *ADEF*

G_2 of semi circle of radius 30 mm

G_3 of ΔKLE

G_4 of ΔKLF

Given data $FE = 120$ mm

$$FK = FE \cos 30° = 120 \times \frac{\sqrt{3}}{2} = 103.92$$

and $$KE = EF \sin 30 = 120 \times \frac{1}{2} = 60$$

$$\therefore \quad FL = FK\cos 30 = 103.92\times\frac{\sqrt{3}}{2} = 89.99 \text{ mm}$$

$$KL = FK\sin 30 = 103.92\times\frac{1}{2} = 51.96$$

$$\therefore \quad G_4M = G_3N = \frac{1}{3}\times KL = \frac{51.96}{3} = 17.32$$

and

$$ML = \frac{1}{3}FL = \frac{1}{3}\times 89.99 = 29.9967$$

$$FM = FL - ML$$

$$= 89.99 - 29.9967 = 59.99$$

$$LN = \frac{1}{3}LE$$

$$= \frac{1}{3}[120 - FL]$$

$$= \frac{1}{3}(120 - 89.99)$$

$$= (10.003) \text{ mm}$$

$$OG_1 = 60/2 = 30 \text{ mm}$$

$$OG_2 = \frac{4r}{3\pi} = \frac{4\times 30}{3\times\pi} = 12.73 \text{ mm}$$

For finding $\bar{x}$ take moments about AF

$$(A_1 - A_2 + A_3 + A_4)\bar{x} = A_1x_1 - A_2x_2 + A_3x_3 + A_4x_4 \quad ...(A)$$

Now

$$A_1 = 120 \times 60 = 7200 \text{ mm}^2$$

$$A_2 = \frac{1}{2}\pi r^2 = \frac{1}{2}\times\pi\times 30^2 = 1413.72$$

$$A_3 = \Delta KLE = \frac{1}{2}\times LE\times KL$$

$$= \frac{1}{2}(120 - FL)\times 51.96$$

$$= \frac{1}{2}(120 - 89.99)\times 51.96$$

$$= 779.66 \text{ mm}^2$$

$$A_4 = \text{Area } FKL$$

$$= \frac{1}{2} FL \times KL$$

$$= \frac{1}{2}(89.99)51.96$$

$$= 2337.94 \text{ mm}^2$$

$$x_1 = \frac{1}{2} \times 120 = 60$$

$$x_2 = 60$$

$$x_3 = FN = FL + LN$$

$$= 89.99 + 100.003$$

$$= 99.993 \text{ m}$$

$$x_4 = FM = 59.993$$

Also

$$y_1 = G_1O = \frac{1}{2} \times 60 = 30$$

$$y_2 = OG_2 = 12.73$$

$$y_3 = 60 + G_3N$$

$$= 60 + 17.32 = 77.32$$

$$y_4 = 60 + GM$$

$$= 60 + 17.32 = 77.32$$

Then from equation (*A*)

$$\bar{x}(7200 - 1413.72 + 779.66 + 2337.94)$$

$$= 7200 \times 60 - 1413.72 \times 60 + 779.66 \times 99.99 + 2337.94 \times 59.993$$

i.e.,

$$8903.8\bar{x} = 432000 - 848232 + 77911.4 + 140112.74$$

$$= 495200.9$$

$$\bar{x} = \frac{495200.9}{8903.8}$$

$$= 55.61 \text{ mm}$$

Taking moments about *ABD*

$$(A_1 - A_2 + A_3 + A_4)\bar{y} = A_1y_1 - A_2y_2 + A_3y_3 + A_4y_4$$

$$8903.8\bar{y} = 7200 \times 30 - 1413.72 \times 12.73 + 779.66 \times 77.32 + 2337.94 \times 77.32$$

$$= 216000 - 17996.6 + 60283.3 + 18076.9$$

$$= 276363.6$$

$$\bar{y} = 31.039 \text{ mm}$$

14.6 SECOND MOMENT OF PLANE *AREA*

14.6.1 (*AREA* Moment of Inertia)

Definitions and theorem of perpendicular axes

Consider on elemental area dA Fig. (14.32 *a*) at a distance r from the line AB. Then the term rdA is called the first moment of area about AB just similar to the term moment of a force. The first moment of the entire area will be $\oint r\,dA$. Similarly the term $\oint r^2\,dA$ is known as the second moment of area. However, it is a popularly called moment of inertia even through it is a misnomer as the term moment of intertia is to be associated with the mass of the body. Therefore, in order to avoid confusion it is called 'second moment of area" or "Area moment of Inertia.

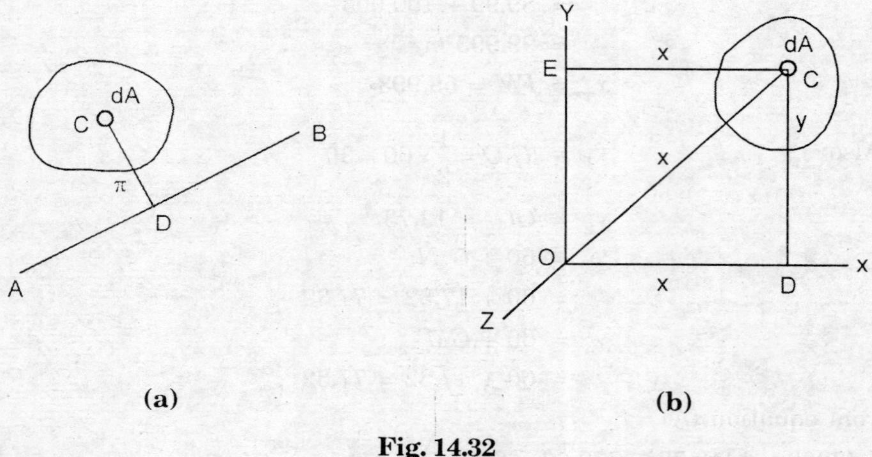

Fig. 14.32

Now consider the area dA with reference to axes OX and OY. That is dA has co-ordinates (x, y).

Then $\oint x^2\,dA$ is called the second moment of area about y-axis and is represented by I_{yy} similarly $\oint y^2\,dA$ is known as second moment of area about x-axis and is represented by I_{xx} when the area lies in XY plane.

However if dA is situated at a distance r from z-axis which is normal to plane in which the area lies, then the term $\oint r^2\,dA$ is called the "Polar second moment area" or "Area Polar Moment of inertia".

Evidently $\quad r^2 = x^2 + y^2$

$$\therefore \quad \oint r^2 dA = \oint (x^2 + y^2)\,dA$$

$$I_{ZZ} = \oint x^2 dA + \oint y^2 dA$$

$$= I_{yy} + I_{xx}$$

i.e.,
$$I_{ZZ} = I_{XX} + I_{YY} \qquad ...(14.57)$$

Equation (14.57) is known as theorem of perpendicular axes

14.6.2 Theorem of Parallel Axes

Statement

Second moment of area about an axis in the plane of the area is equal to the sum of second moment of area about parallel centroidal axis and the product of area and square of the distance between the two parallel axes. Mathematically the statement is

$$I_{XX'} = I_{XX} + Aa^2 \qquad ...(14.58)$$

where GG passes through centroid G and $X'X'$ is any other parallel axis at a distance 'a' from the centroidal axis.

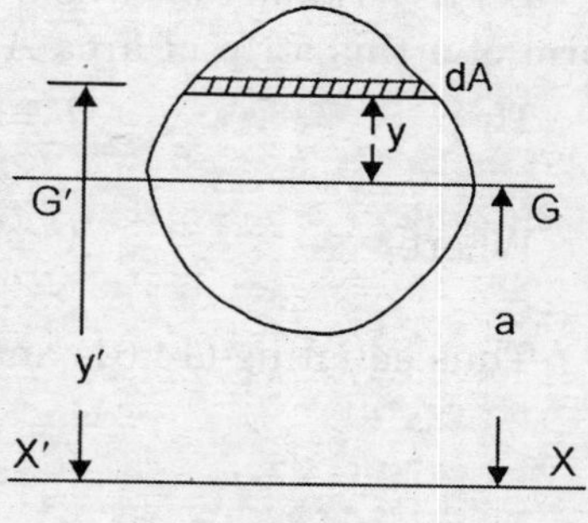

Fig. 14.33

Proof

Let the distance of an elemental area dA parallel to XX *be* y and the distance from $X'X'$ is y'.

Evidently
$$y' = y + a$$

Then
$$I_{XX'} = \oint (y')^2 dA$$
$$= \oint (y+a)^2 dA$$
$$= \oint [y^2 + a^2 + 2aydA]$$

or
$$I_{XX'} = \oint y^2 dA + a^2 \int dA + 2a \int dA\, y \qquad ...(14.59)$$

The term $\oint y\,dA$ is nothing but first moment of area about GG which is positing through centroid G that its distance is zero from CG.

Thus
$$\oint y\,dA = 0$$

Therefore equation (14.59) reduces to

$$I_{XX'} = \oint y^2 \cdot dA + a^2 A$$
$$= I_{XX} + a^2 A \qquad ...(14.60a)$$

or
$$= I_{GG} + a^2 A \text{ Hence the Proof} \qquad ...(14.60b)$$

Note: While using the expression (14.60 a) sometimes the students commit a serious mistake and they use I_{XX} for any axis other than centroidal axis. This will be wrong. Therefore the form (14.60 b) is better as I_{GG} will automatically indicate that it is second moment of area about an axis passing through CG.

14.7 RADIUS OF GYRATION

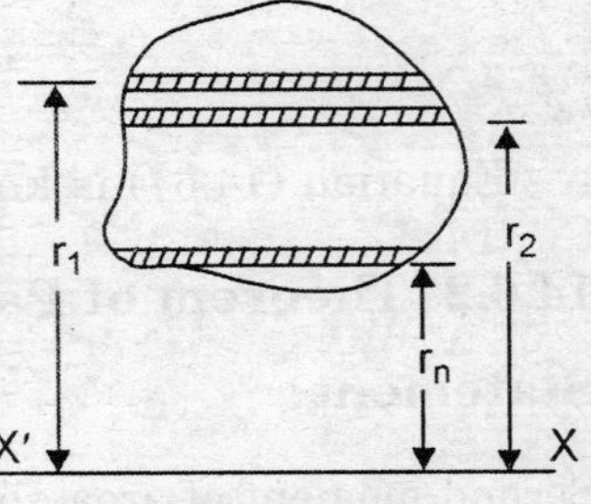

Fig. 14.34

Suppose the given area is split up into small strips parallel to the axis about which second moment of are is to be calculated.

Then $$I = A_1r_1^2 + A_2r_2^2 + A_3r_3^2 \quad ...(14.61)$$

Let us assume that whole (mass) or area is concentrated in the form of a this strip of area A at distance k

Then $$I = Ak^2 \quad ...(14.62)$$

Where $$A = A_1 + A_2 + A_3. = \sum_{i=1}^{n} A_1$$

Thus equating (14.61) and (14.62)

$$k = \sqrt{\frac{I}{A}} \quad ...(14.63)$$

This k is known as radius of gyration. The units of I will be mm^4 and or A = mm^2. Evidently k will be in mm.

14.8 SECOND MOMENT OF AREA OF STANDARD CASES (MOMENT OF INERTIA)

14.8.1 Rectangle

In Fig. 14.35 is shown a rectangle $ABCD$ of length 'l' and depth 'd'. It has two axes of symmetry XX, YY with 0 considering with the centroid a.

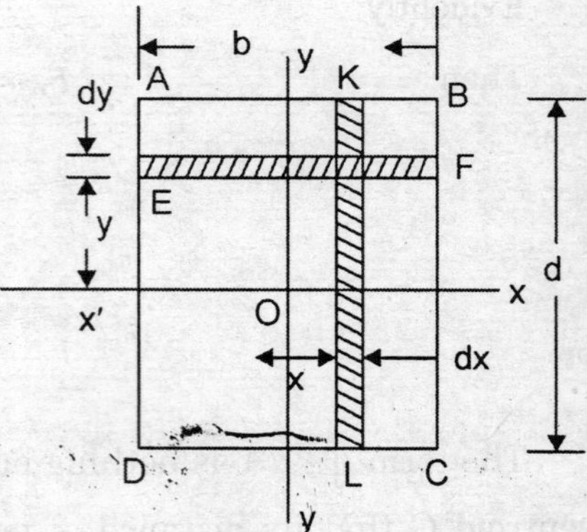

Fig. 14.35

We now proceed to find the second moment of area about XX axis. Take a strip EF of depth dy at a height 'y' from the x-axis.

Then $$d\,I_{xx} = dA\,y^2$$

or $$I_{XX} = \int_{d/2}^{-d/2} dA\,y^2 = \int_{-d/2}^{d/2} y^2(b\,dy)$$

$$= \left[\frac{by^3}{3}\right]_{-\frac{d}{2}}^{d/2} = \frac{b}{3}\left[\frac{d^3}{8} - \left(-\frac{d^3}{8}\right)\right]$$

$$= \frac{bd^3}{3\times 4} = \frac{bd^3}{12} \quad ...(14.64)$$

For calculating I_{yy} take elemental strip of width dx parallel to y-xis at a distance x.

Then $$I_{yy} = \int_{-h/2}^{h/2} x^2\, dA = \int_{-h/2}^{h/2} x^2\, d(dx) = d\left[\frac{2^3}{3}\right]_{-h/2}^{h/2} = \frac{dh^3}{12}$$

14.8.1.1 Second Moment of Area of a Rectangle about its sides

We have from equation (14.61)

$$I_{XX} = \frac{BD^3}{12}$$

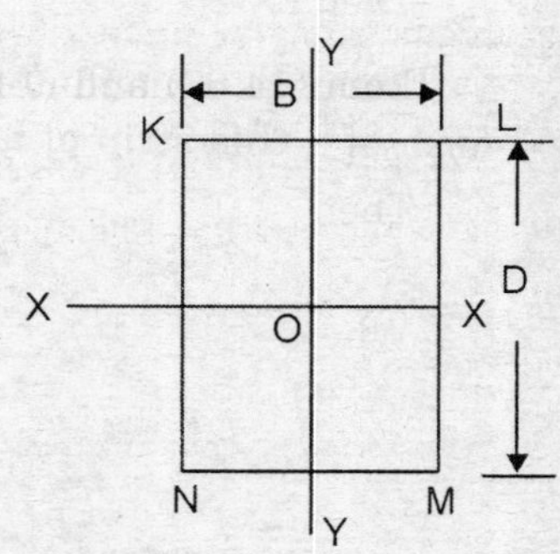

Fig. 14.36 (a)

Using theorem of parallel axes

$$I_{NM} = I_{XX} + A\left(\frac{1}{2}LM\right)^2$$

$$= \frac{BD^3}{12} + (BD)\left(\frac{1}{2}D\right)^2$$

$$= \frac{BD^3}{12} + \frac{BD^3}{4} = \frac{BD^3}{3} \quad ...(14.65)$$

Similarly $$I_{YY} = \frac{DB^3}{12}$$

$$I_{KN} = I_{YY} + A\left(\frac{1}{2}B\right)^2 = I_{yy} + BD\left(\frac{B}{2}\right)^2$$

$$I_{KN} = \frac{DB^3}{12} + \frac{B^3D}{4} = \frac{B^3D}{3} \quad ...(14.66)$$

Note: While using equations (14.64) and (14.65) one has to be very careful. Most of the time the students use the wrong value for I_{XX} and I_{YY} as the expressions are simular. It may be noted that the side parallel to the axis about which area moment of inertia is being calculated has power 3.

14.8.2 Second Moment of Square Area

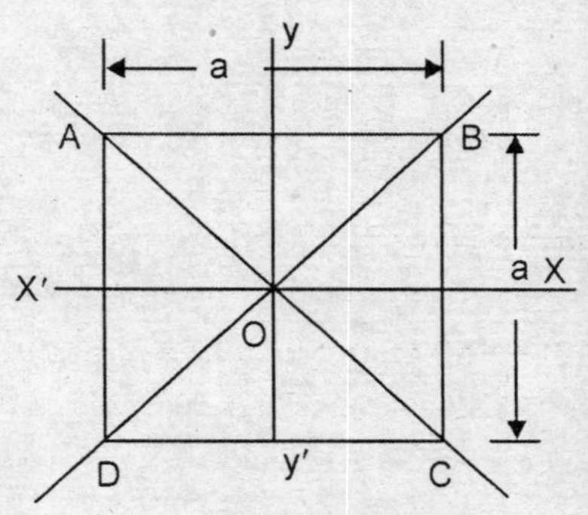

Fig. 14.36 (b)

In equation (14.64) and (14.63)

If we put $b = d = a$ (say)

Then $$I_{XX} = I_{YY} = \frac{a^4}{12}$$

If I_{zz} is moment of inertia about zz axis, then from the theorem of perpendicular axes, we have

$$I_{ZZ} = I_{XX} + I_{YY}$$
$$= 2I_{XX} \qquad [\because I_{XX} = I_{YY}] \qquad ...(14.67)$$

Further for a square section diagonals also meet at 90°

$$\therefore \quad I_{ZZ} = I_{DB} + I_{CA}$$
$$= 2I_{DB} = 2I_{CA} \qquad ...(14.68)$$

$\therefore$ From (14.67) and (14.68)

We get
$$I_{XX} + I_{YY} = I_{ZZ}$$
or
$$2I_{XX} = I_{ZZ} = 2I_{CA}$$
or
$$I_{XX} = I_{YY} = I_{AC} = I_{BD}$$
$$= \frac{a^4}{12} \qquad ...(14.69)$$

14.8.3 Second Moment of Area of a Triangle About its Base

In Fig. 14.37 is shown a triangle *ABC* whose base *BC* is of length '*l*' and height of the vertex *A* is '*h*'.

Geometrically
$$AD = h$$
$$DF = y$$
$$AF = (h - y)$$

EFK is an elemental area *dA* of thickness *dy* at a distance *y* from the base *BC*.

From similar triangles *ABC* and *AEK*

$$\frac{EK}{BC} = \frac{AF}{AD}$$

$$\therefore \quad EK = \left(\frac{h-y}{h}\right) \times BC$$

$$= b\left(1 - \frac{y}{h}\right)$$

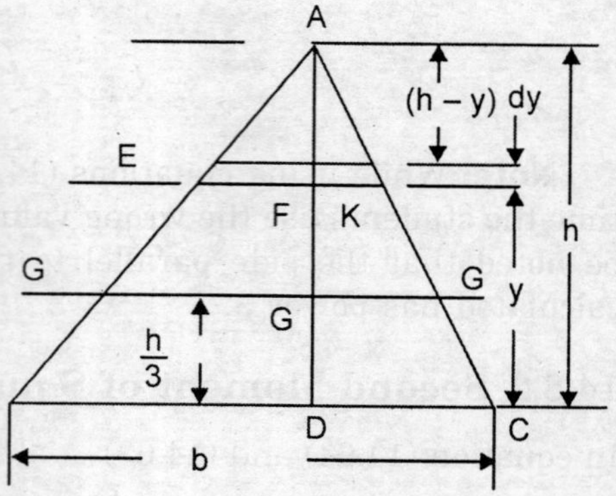

$$\therefore \quad I_{BC} = \int_{y=0}^{h} y^2 dA$$

$$= \int_{y=0}^{h} y^2 \cdot (EK) dy$$

$$= \int_{0}^{h} y^2 \left[b\left(1 - \frac{y}{h}\right)\right] dy$$

Fig. 14..37

$$= b\int_{y=0}^{h}\left(y^2 - \frac{y^3}{h}\right)dy$$

$$= h\left[\frac{y^3}{3} - \frac{y^4}{4h}\right]_0^h$$

i.e.,

$$I_{BC} = b\left[\frac{h^3}{3} - \frac{h^4}{4h}\right]$$

$$= bh^3\left(\frac{1}{3} - \frac{1}{4}\right)$$

$$= \frac{bh^3}{12} \qquad \text{...(14.70)}$$

14.8.3.1 Corollary

It has been proved earlier in equation (14.19) that centroid of a triangle is at a height of $\frac{h}{3}$ from the base and the area of the triangle is $\frac{1}{2}bh.$

Therefore of I_{GG} which is called the second moment of triangular area about centroidal axes parallel to base then we have from principle of parallel axes.

$$I_{BC} = I_{GG} + A(\bar{y})^2$$

or

$$I_{GG} = I_{BC} - A(\bar{y})^2$$

Now

$$I_{BC} = \frac{bh^3}{12} \qquad \text{...(14.70)}$$

$$A = \frac{1}{2}bh$$

$\therefore$

$$I_{GG} = I_{BC} - A(\bar{y})^2$$

or

$$I_{GG} = \frac{bh^3}{12} - \left(\frac{bh}{2}\right)(\bar{y})^2$$

$$= \frac{bh^3}{12} - \frac{bh}{2}.\left(\frac{h}{3}\right)^2$$

$$= \frac{bh^3}{12} - \frac{bh^3}{18} = bh^3\left(\frac{b}{12+18}\right)$$

or

$$I_{GG} = \frac{bh^3}{36} \qquad \text{...(14.71)}$$

14.8.4 Second Moment of Area of Hollow Rectangular Section

In Fig. 14.38 is shown a hollow rectangular section which is symmetrical about both the *XX* and *YY* axes.

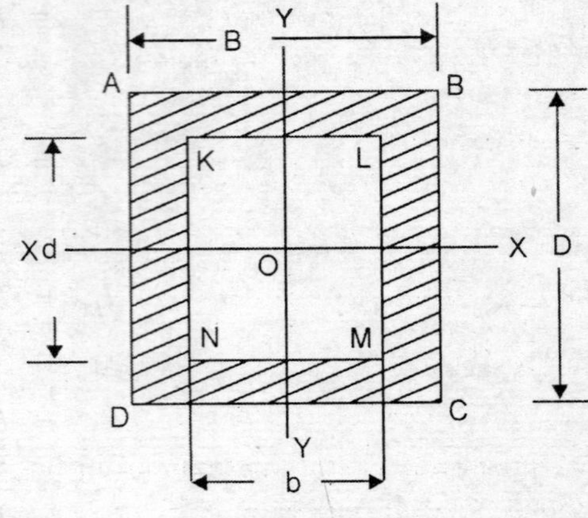

Fig. 14.38

$$I_{XX} \text{ of solid rectangle } ABCD = \frac{BD^3}{12}$$

$$I_{XX} \text{ of solid rectangule } KLMN = \frac{bd^3}{12}$$

$\therefore I_{XX}$ of the given hollow section

$$= \frac{BD^3}{12} - \frac{bd^3}{12} \qquad ...(14.72)$$

Similarly

$$I_{YY} \text{ of the hollow section} = \frac{DB^3}{12} - \frac{db^3}{12} \qquad ...(14.73)$$

14.8.5 Second Moment of Area of Plane Circular Section First Method

In Fig. 14.39 is shown a circular section of radius *R*. Since the area is symmetrical about any diameter whether along *X*-axis or along *Y*-axis. The origin itself will be the *CG* of the section. Take an elemental area *ABC* parallel to *X*-axis of thickness *dy* and at a distance of *y* from *X*-axis. Then

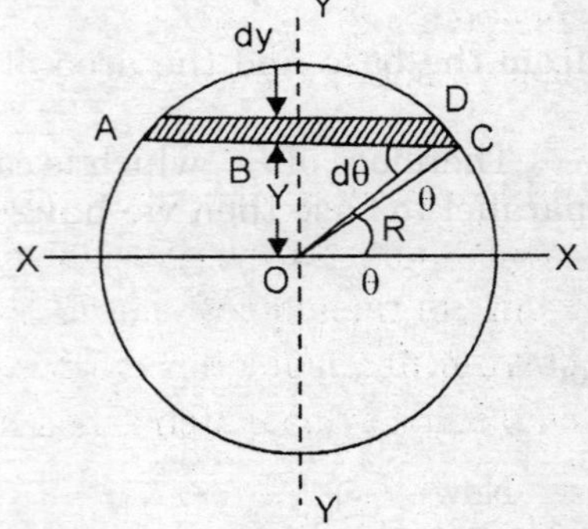

Fig. 14.39

$$I_{xx} = \int_{y=-d/2}^{d/2} y^2 dA$$

$$= \int_{y=-\frac{d}{2}}^{d/2} (AC)\, dy\, y^2$$

$$I_{XX} = \int_{y=-\frac{d}{2}}^{d/2} 2(BC)\, dy\, y^2 \qquad ...(14.74)$$

If we assume that the radial line *OC* is inclined at an angle θ with *X*-axis

$$y = R \sin \theta \qquad ...(14.75a)$$

$$dy = R \cos \theta \, d\theta \qquad ...(14.75b)$$

Also $\quad BC = R \cos \theta$

Then in the light of equation (14.75), equation (14.74) reduces to

$$I_{XX} = 2\int_0^{\pi/2} (2BC)\, y^2 dy \quad (\because \text{ section is symmetrical about } Y\text{-axis})$$

$$= 4\int_0^{\pi/2} (R\cos\theta)(R^2 \sin^2\theta)(R\cos\theta)\, d\theta$$

$$= 4\int_0^{\pi/1} R^4 \sin^2\theta \cos^2\theta\, d\theta$$

$$= R^4 \int_{\theta=0}^{\pi/2} (4\sin^2\theta \cos^2\theta)\, d\theta$$

$$= \frac{R^4}{2}\int_0^{\pi 2} 2(\sin 2\theta)^2 d\theta$$

$$= \frac{R^4}{2}\int_{\theta=0}^{\pi/2} (1-\cos 4\theta)\, d\theta$$

$$= \frac{R^4}{2}\left[\theta + \frac{\sin 4\theta}{4}\right]_0^{\pi/2} = \frac{\pi R^4}{4}$$

$$I_{XX} = \frac{\pi}{4}\left(\frac{D}{2}\right)^4 = \frac{\pi D^4}{64} \qquad \text{...(14.76)}$$

Second Method

In this method we take elemental area in the form of a ring of radial thickness dr situated at a distance of r from ZZ-axis.

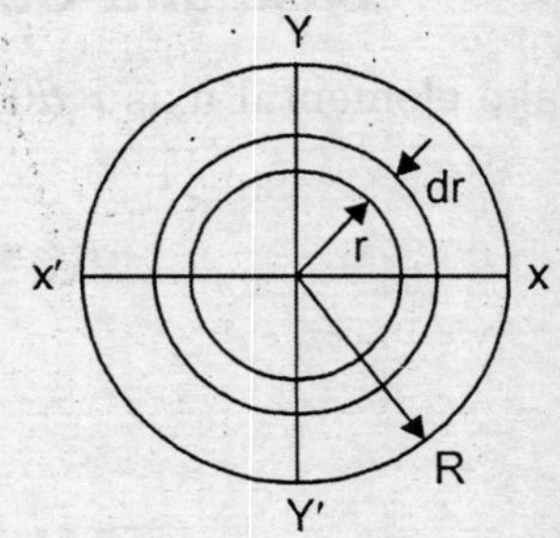

Fig. 14.40

Then $$I_{zz} = \int_0^R r^2 dA = \int_{r=0}^R r^2 (2\pi r\, dr)$$

$$= 2\pi \int_{r=0}^R r^3\, dr$$

$$\therefore \quad I_{ZZ} = 2\pi\left[\frac{r^4}{4}\right]_0^R = \frac{2\pi}{4}R^4 = \frac{\pi R^2}{2}$$

But we know from the theorem of perpendicular axes that

$$I_{ZZ} = I_{XX} + I_{YY}$$

Also $$I_{XX} = I_{YY}$$

$\therefore$ $$I_{ZZ} = 2I_{XX}$$

or $$I_{XX} = \frac{1}{2}(I_{ZZ})$$

$$= \frac{1}{2}\left(\frac{\pi R^4}{2}\right) = \frac{\pi R^4}{4}$$

$$= \frac{\pi D^4}{64} \quad \text{...(14.78)}$$

Note: Evidently the second method is much simpler and less time consuming. Therefore it is suggested that the students follow this method in the examination unless it is stated otherwise in the question.

14.8.5.1 Corollary

In case of a hollow plane circular section as shown in Fig. 14.41.

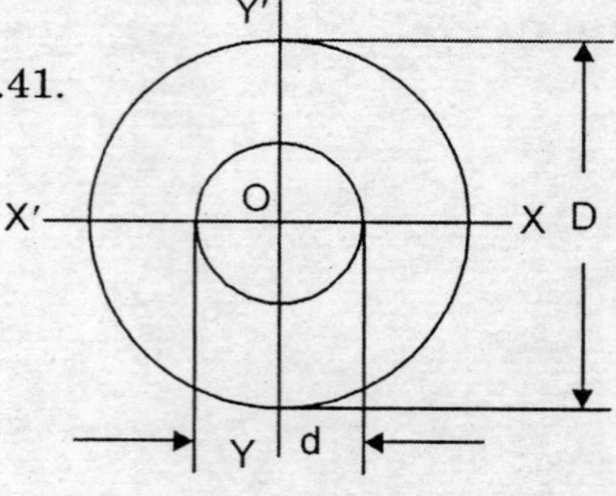

Fig. 14.41

$$I_{XX} = I_{YY} = \frac{\pi D^4}{64} - \frac{\pi d^4}{64}$$

and $$I_{ZZ} = I_{XX} + I_{YY}$$

$$= 2\,I_{XX}$$

$$= 2\left[\frac{\pi D^4}{64} - \frac{\pi d^4}{64}\right]$$

$$I_{zz} = \frac{\pi}{32}(D^4 - d^4) \quad \text{...(14.79)}$$

14.8.5.2 Second Moment of Area of a Semicircular Area about base and Centroidal axis

Take elemental axis $r\,d\theta\,dr$ as shows in Fig. 14.42

$$I_{XX} = \int_{r=0}^{R}\int_{\theta=0}^{\pi}(r\,d\theta\,dr)(r\sin\theta)^2$$

$$= \int_{r=0}^{R}\int_{\theta=0}^{\pi} r^3\sin^2\theta\,d\theta\,dr$$

$$= \int_{r=0}^{R}\left[\frac{r^3}{2}\int_{0}^{\pi}(2\sin^2\theta)\,d\theta\right]dr$$

Fig. 14.42

$$= \int_{r=0}^{R} \frac{r^3}{2} \int_0^{\pi} (1 - \cos 2\theta)\, d\theta\, dr$$

$$= \int_{r=0}^{R} \frac{r^3}{2} \left[\theta - \frac{\sin 2\theta}{2} \right]_0^{\pi} dr$$

$$= \int_0^{R} \frac{r^3}{2} [\pi - 0 + 0]\, dr$$

$$= \frac{\pi}{2} \left[\frac{r^4}{4} \right]_0^{R} = \frac{\pi R^4}{8}$$

$$= \frac{\pi}{8} \left(\frac{D}{2} \right)^4 = \frac{\pi D^4}{128} \qquad \text{...(14.80)}$$

From equation (14.25) we know that the *CG* of semicircle is at a distance $\frac{4R}{3\pi}$ from *XX*.

We have from theorem of parallel axes.

$$I_{XX} = I_{GG} + A\left(\frac{4R}{3\pi} \right)^2$$

Now $$A = \frac{\pi R^2}{2}$$

$\therefore$ $$I_{GG} = I_{XX} - A\left(\frac{4R}{3\pi} \right)^2$$

$$= \frac{\pi R^4}{8} - \frac{\pi R^2}{2} \times \frac{16 R^2}{9\pi^2}$$

$$= \frac{\pi R^4}{8} - \pi R^4 \cdot \frac{8}{9\pi}$$

$$I_{GG} = \frac{\pi R^4}{8} \left[1 - \frac{64}{9\pi^2} \right] \qquad \text{...(14.81a)}$$

$$= \frac{\pi R^4}{8} \left[1 - \frac{64}{9\pi^2} \right]$$

$$= \left[\frac{\pi D^4}{128} - \frac{D^4}{18\pi} \right] \qquad \text{...(14.81b)}$$

14.8.5.3 Area Moment of Inertia of Quarater Circle

In Fig. 14.43 is shown elemental area

$$dA = (r\,d\theta)\,dr$$

Located at an angle θ with OX. Therefore

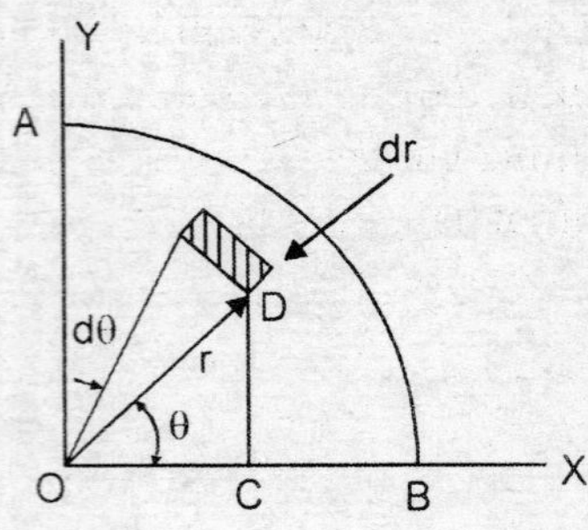

Fig. 14.43

$$I_{XX} = \int_{r=0}^{R}\int_{\theta=0}^{\pi/2} (r\,d\theta)\,dr\,(r\sin\theta)^2$$

$$= \int_{r=0}^{R} r^3 \int_{\theta=0}^{\pi/2} \sin^2\theta\,d\theta\,dr$$

$$= \int_{r=0}^{R} \frac{r^3}{2} \int_{0}^{\pi/2} 2\sin^2\theta\,d\theta\,dr$$

or
$$I_{XX} = \int_{r=0}^{R} \frac{r^3}{2} \int_{0}^{\pi/2} (1-\cos 2\theta)\,d\theta\,dr$$

$$= \int_{r=0}^{R} \frac{r^3}{2}\left[\theta - \frac{\sin 2\theta}{2}\right]_0^{\pi/2} dr$$

$$= \int_{r=0}^{R} \frac{r^3}{2}\left[\frac{\pi}{2} - 0 + 0\right] dr$$

$$= \int_{r=0}^{R} \frac{\pi r^3}{4}\,dr = \frac{\pi}{16}\left[r^4\right]_0^R$$

$$I_{XX} = \frac{\pi R^4}{16} \qquad \text{...(14.82a)}$$

$$= \frac{\pi}{16} \times \left(\frac{D}{2}\right)^4 = \frac{\pi D^4}{256} \qquad \text{...(14.82b)}$$

For finding I_{GG} parallel to X-X

We have $\quad A = \dfrac{\pi R^2}{4}$

Distance of G from X-axis $\quad \bar{y} = \dfrac{4R}{3\pi}$

Now $$I_{XX} = I_{GG} + A(\bar{y})^2$$

$\therefore$ $$I_{GG} = I_{XX} - A(\bar{y})^2$$

$$= \frac{\pi R^4}{16} - \frac{\pi R^2}{4}\left(\frac{4R}{3\pi}\right)^2$$

$$= \frac{\pi R^4}{16} - \pi R^4 \frac{4}{9\pi^2}$$

or $$I_{GG} = \frac{\pi R^4}{16} - \pi R^4 \times \frac{4}{9\pi^2}$$

or $$I_{GG} = \frac{\pi R^4}{16}\left[1 - \frac{64}{9\pi^2}\right] \quad \text{...(14.83}a\text{)}$$

$$= \frac{\pi D^4}{256} - \frac{\pi D^4}{256} \cdot \frac{64}{9\pi^2}$$

$$I_{GG} = \left(\frac{\pi D^4}{256} - \frac{D^4}{36\pi}\right) \quad \text{...(14.83}b\text{)}$$

Table 14.2. Second Moment of Area of some Standard Plane Figures

Figure	I
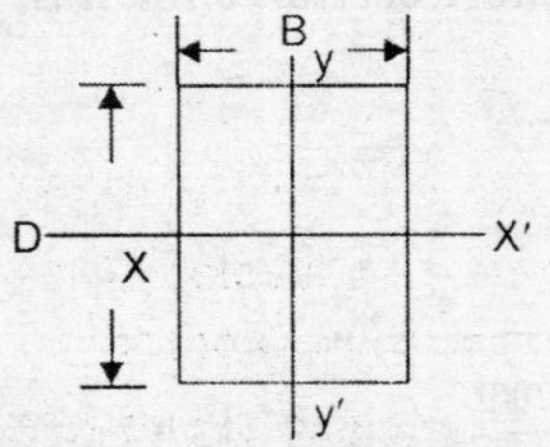 B, y, y′, D, X, X′	$I_{XX} = \frac{BD^3}{12}$ $I_{YY} = \frac{DB^3}{12}$
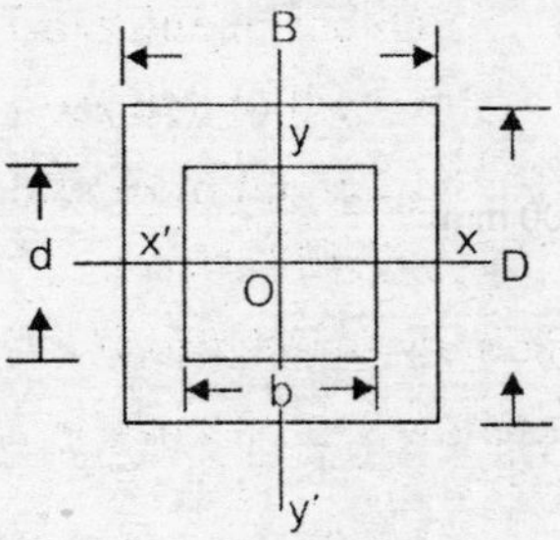 B, b, D, d, x, x′, y, y′, O	$I_{XX} = \frac{BD^3}{12} - \frac{bd^3}{12}$ $I_{XX} = \frac{DB^3}{12} - \frac{db^3}{12}$

contd...

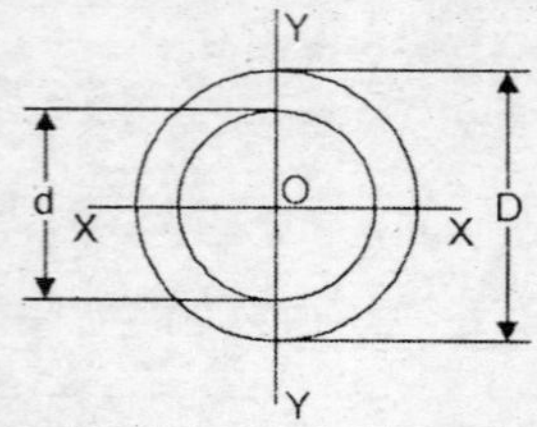 Y, d, X, O, X, D, Y	$I_{XX} - I_{YY} = \frac{\pi D^4}{64} - \frac{\pi D^4}{64}$ $I_{ZZ} = \frac{\pi D^4}{32} - \frac{\pi d^4}{32}$
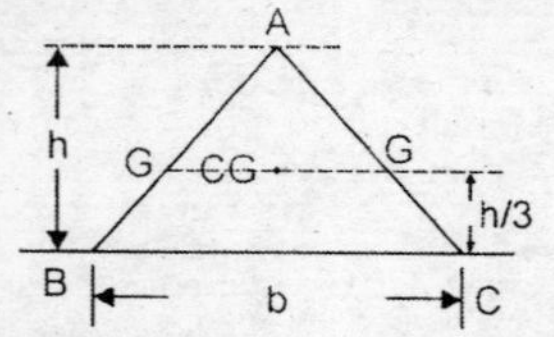 A, h, G, CG, G, h/3, B, b, C	$I_{BC} = \frac{bh^3}{12}$ $I_{GG} = \frac{bh^3}{36}$
Semi circle 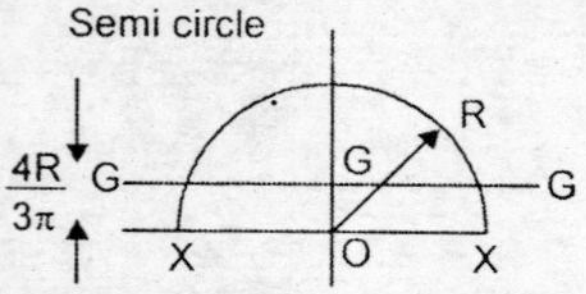 $\frac{4R}{3\pi}$, G, G, R, G, X, O, X	$I_{XX} = \frac{\pi D^4}{128}$ $I_{GG} = \left(\frac{\pi D^4}{128} - \frac{D^4}{18\pi}\right)$
Quarter circle 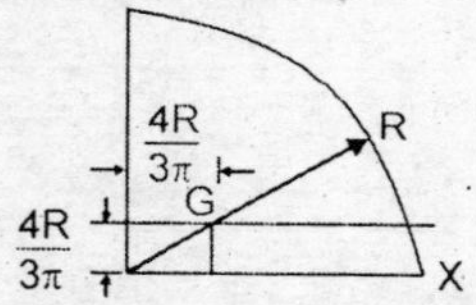 $\frac{4R}{3\pi}$, G, R, $\frac{4R}{3\pi}$, X	$I_{XX} = \frac{\pi D^4}{256}$ $I_{GG} = \left(\frac{\pi D^4}{256} - \frac{D^4}{36\pi}\right)$

14.9 AREA MOMENT OF INERTIA OF COMPOSITE SECTIONS

14.9.1 Area Moment of Inertia of T-section

Example 14.9. *Find the centroidal moment of Inertia of the T-section shown in Fig. 14.44 about horizontal centroidal aixs.*

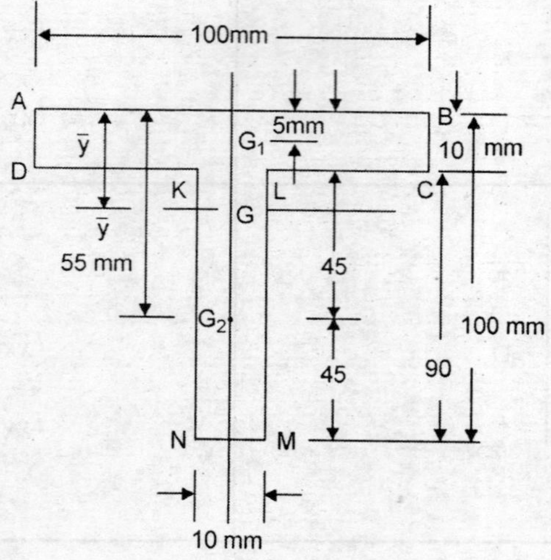

Fig. 14.44

Solution: Given composite section consists of the rectangular areas namely *ABCD* and *KLMN*. Let the areas of these rectangles be A_1 and A_2 with their centroids at G_1 and G_2 respectively. Assume that the centroid of the whole composite area be of point *G* which is at a distance of $\bar{y}$ from *AB*.

Then taking moments about *AB* of the respective areas we have

$$A_1 = 100 \times 10 = 1000 \text{ mm}^2$$

$$y_1 = \frac{1}{2} \times 10 = 5 \text{ mm}$$

$$A_2 = 90 \times 10 = 900 \text{ mm}^2$$

$$y_2 = 100 - 45 = 55 \text{ mm}$$

Total area $A = A_1 + A_2$

$$= 1000 + 900 = 1900 \text{ mm}^2$$

$$A\bar{y} = A_1 y_1 + A_2 y_2$$

or $1900\,\bar{y} = 1000 \times 5 + 900 \times 55$

$$1900\,\bar{y} = 5000 + 49500$$

or $\bar{y} = \dfrac{54500}{1900} = 28.68 \text{ mm}$

Applying the theorem of parallel axis

$$I_{GG} = I_{GG} \text{ of } A_1 + I_{GG} \text{ of } A_2 \qquad ...(A)$$

But I_{GG} of $A_1 = I_{G_1}$ of $A_1 + A_1 (G_1 G)^2$

$$= \frac{AB \cdot (BC)^3}{12} + (100 \times 10)\,(\bar{y} - y_1)^2$$

$$= \frac{100 \times 10^3}{12} + 1000(28.68 - 5)^2$$

$$= 8.33 \times 10^3 + 560.74 \times 10^3$$

$$= 569.07 \times 10^3 \text{ mm}^4$$

And I_{GG} of $A_2 = I_{G_2G_2}$ of $A_2 + A_2 (GG_2)^2$

$$= \frac{10 \times 90^3}{12} + (90 \times 10)(y_2 - \bar{y})^2$$

$$= 607.5 \times 10^3 + 900\,(55 - 28.68)^2$$

$$= 607.5 \times 10^3 + 623.468 \times 10^3$$

$$= 1230.968 \times 10^3 \qquad ...(B)$$

Therefore Adding (*A*) and (*B*) we get

$$I_{GG} = 569.07 \times 10^3 + 1230.968 \times 10^3$$

$$= 1800.0 \times 10^3 \text{ mm}^4$$ **Ans.**

Example 14.10. *Find the Area moment of inertia of the section shown in Fig. 14.45 about centroidal axis parallel to AB.*

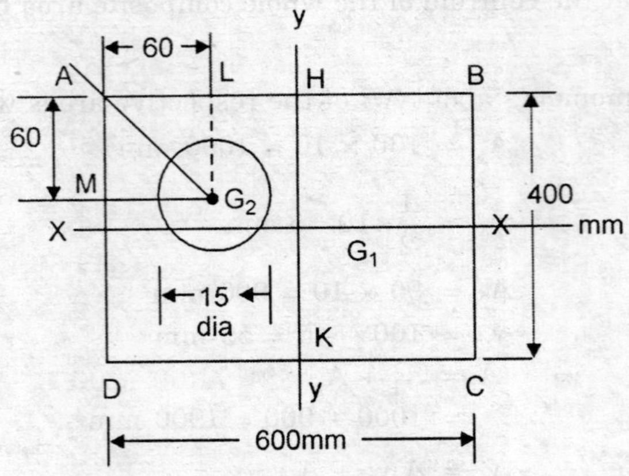

Fig. 14.45

Solution : Given data Delta size 600 × 400 mm, *Ra* of hole = 180 mm

A_1 = Area of whole rectangle when there is no hole

$= b \times d$

Where $b = 600$ mm

$d = 400$ mm

$= 600 \times 400$

$= 24 \times 10^4 \text{ mm}^2$

Let *G* be the location of section without hole it will be of the mid point G_2 the *CG* of circular hole above will be at its centre.

$$A_2 = \frac{\pi}{4}(15)^2 = 176.71 \text{mm}^2$$

Not area $A = A_1 - A_2 = 240000 - 176.71$

$= 23.98 \times 10^4 \text{ mm}^2$

If the centroid *G* is at distance of $\bar{y}$ from the line *AB*, the taking moments about *AB*

$$A\bar{y} = A_1 \times HG_1 - A_2(LG_2)$$

$$= 24 \times 10^4 \times 200 - 176.71 \times 60$$

$$= 48 \times 10^6 - 1.06 \times 10^4$$

or $(23.98 \times 10^4)\,\bar{y} = 47/98 \times 10^6$ ...(A)

$$\therefore \quad \bar{y} = \frac{47.989 \times 10^6}{23.98 \times 10^4}$$

$= 200.1$ mm

$$I_{GG} = I_G \text{ of rectangle } ABCD - R_G \text{ of circular area}$$

Now $\qquad I_{GG}$ of $ABCD = I_{G_1} + A_1(y_{G_1} - \bar{y})^2$

$$= \frac{600 \times (400)^3}{12} + (600 \times 400)[200 - 200.1]^2$$

$$= 32 \times 10^8 + [24 \times 10^4 \times 0.001]$$

$$= 320000 \times 10^4 + 240.0$$

$$= 32.0024 \times 10^8$$

After I_{GG} of circular hole $= I_{G_2} + A_2(GG_2)^2$

$$= \frac{\pi \times (15)^2}{64} + 176.71(60 - 200.1)^2$$

$$= 2485.05 + 3468465$$

$$= 3410950 = 3.411 \times 10^6$$

Thus of given are $\qquad I_G = I_G\ ABCD - I_{GG}$ of hole

$$= 32.0024 \times 10 - 3.411 \times 10^6$$

$$= 3200.24 \times 10^8 \text{ mm}^4 \quad \textbf{Ans.}$$

Example 14.11. *Find the area moment of inertia about centroidal axis parallel to AB inertia of the composite section shown in Fig. 14.46.*

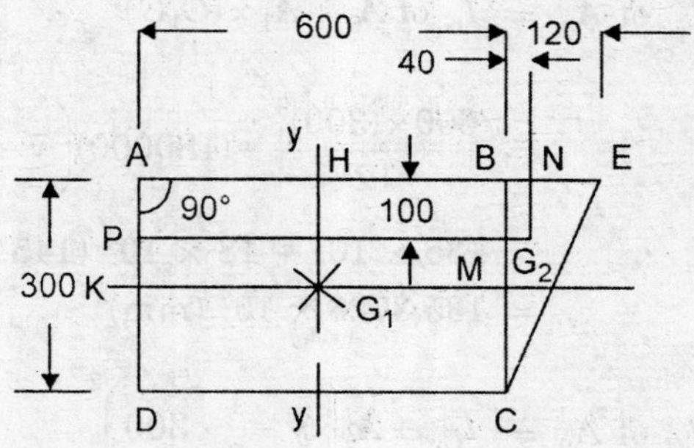

Fig. 14.46

Solution : Given data

$$AB = 600,$$

$$AD = 300,$$

$$G_1K = \frac{1}{2}300 = 150$$

$$A_1 = AB \times AD$$

$$= 600 \times 300 - 180000 \text{ mm}^2$$

$$BE = 129,\ G_2N$$

$$= \frac{1}{3} \times 300 = 100$$

$$A_2 = \text{Area of } DBEC$$

$$= \frac{1}{2} \times 300 \times 120$$

$$= 1800 \text{ mm}^2$$

$$A = A_1 + A_2$$

$$= 18 \times 10^4 + 18000$$

$$= 198 \times 10^3 \text{ mm}^2$$

$$G_2N = \frac{1}{3} \times BC = \frac{1}{3} \times 3N = 100$$

Taking moments about AB

$$198 \times 10^3 \bar{y} = A_1 y_1 + A_2 y_2$$

$$= A_1(G_1K) + A_2(G_2N)$$

$$= (18 \times 10^4)\ 150 + 18000 \times 100$$

$$= 27 \times 10^6 + 18 \times 10^5$$

$$= 288 \times 10^5$$

$$\therefore \quad \bar{y} = \frac{288 \times 10^5}{198 \times 10^3} = 145.45 \text{ mm}$$

Now $$I_G \text{ of } A_1 = I_{G_1} \text{ of } A_1 + A_1 \times (G_1G)^2$$

$$= \frac{600 \times (300)^3}{12} + (180000)\left(\bar{y} - \frac{300}{2}\right)^2$$

$$= 135 \times 10^7 + 18 \times 10^4\ (145.45 - 150)^2$$

$$= 135.3726 \times 10^7\ (\text{mm})^4$$

$$I_G \text{ of } A_2 = I_{G_2} + A_2\left(\bar{y} - \frac{1}{3} \times 300\right)^2$$

$$= \frac{BE \times BC^3}{12} + (18000)(45.45 - 100)^2$$

$$= 27 \times 10^7 + 3.718 \times 10^7$$

$$= 30.718 \times 10^7\ (\text{mm})^4$$

$\therefore$ Total $$I_G = I_G \text{ of } A_1 + I_G \text{ of } A_2$$

$$= 135.3726 \times 10^7 + 30.718 \times 10^7$$

$$= 136.09 \times 10^7 \textbf{ Ans.}$$

AREA MOMENT OF INERTIA OF TRAPEZIUM

Example 14.12. *A trapezium ABCD is shown in Fig. 14.47. Find the area moment of inertia of the trapezium about the base AB also about the centroidal axis parallel to AB.*

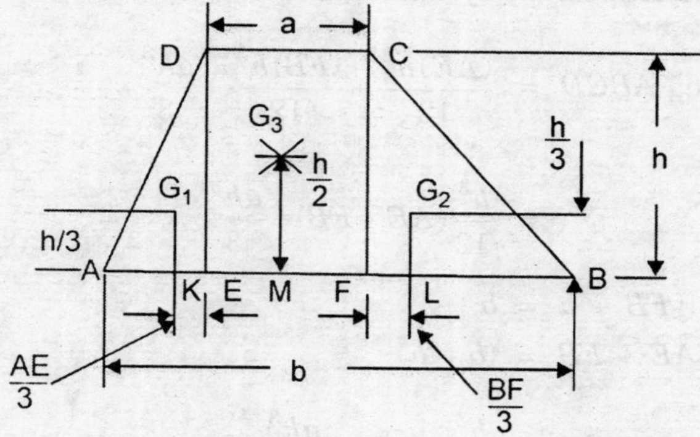

Fig. 14.47

In example 14.4, the centroid of this very trapezium has been found. The results obtained there are reproduced below. Total area of the trapezium has been split into three areas consisting of the triangle *DAE*, rectangle *DCFE* and trianlge *CFB*. The results obtained there are as below

$$A_1 = \text{Area of } \Delta DAE = \frac{1}{2}h(AE)$$

$$\text{Height of centroid } G_1 \text{ of area } \Delta DAE = \frac{h}{3}$$

$$A_2 = \text{Area of } \Delta CFB = \frac{1}{2}h(FB)$$

$$\text{Height of centroid } G_2 \text{of area } \Delta\ CFB = \frac{1}{3}h$$

From *AB* $\quad A_3 = \text{Area of rectangle } DCFE = ah$

$$\text{Height of centroid } G_3 \text{ for } AB = \frac{h}{2}$$

With the help of equation (14.70) we have for areas A_1 and A_2.

$$I_{AB} \text{ of area } A_1 = \frac{(AE)h^3}{12} \qquad ...(A)$$

$$I_{AB} \text{ of area } A_2 = \frac{(FB)\times h^3}{12} \qquad ...(B)$$

And with the help of equation (14.64*b*)

$$I_{AB} \text{ of area } A_3 = \frac{ah^3}{3} \quad ...(C)$$

I_{AB} of the whole trapezium by adding (*A*), (*B*) an (*C*)

$$I_{AB} \text{ of area } ABCD = \frac{(AE)h^3}{12} + \frac{(FB)h^3}{12} + \frac{ah^3}{3}$$

$$= \frac{h^3}{12}(AE + FB) + \frac{ah^3}{3}$$

But $AE + FB + a = b$

$\therefore \quad AE + FB = (b - a)$

$$\therefore \quad I_{AB} = \frac{h^3}{12}(b-a) + \frac{ah^3}{3}$$

$$= \frac{h^3}{12}(4a + b - a)$$

$$= \frac{h^3}{12}(3a + b) \quad ...(D)$$

Again from the theorem of parallel axis

$$I_{AB} = I_G + A(\bar{y})^2$$

Now from example (14.4), the distances of *C.G.* from *AB* is $\bar{y}$

$$\bar{y} = \frac{h}{3}\left(\frac{b+2a}{b+a}\right) \quad ...(E)$$

and Area of trapezium $= \left(\dfrac{a+b}{2}\right)h \quad ...(F)$

$$I_{GG} = I_{AB} - A(\bar{y})^2$$

$$= \frac{h^3}{12}(3a+b) - \frac{1}{2}(a+b)h\left[\frac{h}{3}\frac{(b+2a)}{(a+b)}\right]^2$$

$$= \frac{h^3}{12}(3a+b) - \frac{h^3}{18}\frac{(b+2a)^2}{(b+a)}$$

$$= \frac{h^3}{36}\left[3(3a+b) - \frac{2(b+2a)^2}{(b+a)}\right]$$

$$= \frac{h^3}{36(b+a)}[3(3a+b)(b+a) - 2(b+2a)^2]$$

or
$$I_{GG} = \frac{h^3}{36(b+a)}[9ab + 9a^2 + 3b^2 + 3ab - 2b^2 - 8a^2 - 8ab]$$

$$= \frac{h^3}{36(b+a)}[4ab + b^2 + a^2]$$

$$= \frac{h^3}{36(b+a)}[(b+a)^2 + 2ab]$$

$$I_{GG} = \frac{h^3}{36}\left[(a+b) + \frac{2ab}{(a+b)}\right] \quad ...(G)$$

14.10 PRODUCT OF INERTIA

In Fig. 14.48 is shown an area A of which an elemental area dA is situated at a point P whose co-ordinates with respect to X-Y areas are (x, y). In other words its distance from X-axis is y and that from Y-axis is x. Then the term $xy\, dA$ is called the product of inertia of elemental area dA and $\int xy\, dA$ is known as the product of inertia of the whole area and is represented by I_{xy}.

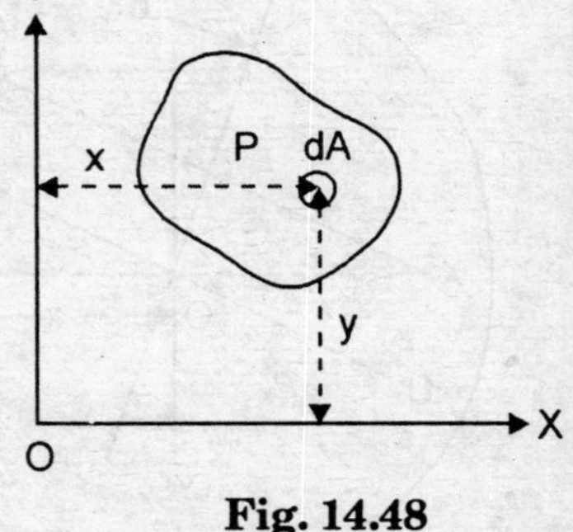

Fig. 14.48

That is
$$I_{xy} = \int xy\, dA \quad ...(14.84)$$

Mathematicaıly it is expressed as

$$Ax_G y_G = \int xy\, dA = \sum_{i=1}^{n} x_i y_i A_i$$

$$= x_1 y_1 A_1 + x_2 y_2 A_2 + ... + x_n y_n A_n \quad ...(14.85)$$

where (x_1, y_1) is CG of area A_1, (x_2, y_2) that of area A_2 and so on.

It may be noted that the product of inertia may be positive, negative or even zero.

In case the area is symmetrical about one or both the areas. Then two equal elemental area dA on be located at a distance x, $-x$ (of symmetry axis X-axis) such that the product inertia will have the times $\int xy\, dA$ and $\int (x) y\, dA$ each integrated for half μ_x area. These times will cancel each other and

$$I_{xy} = 0$$

In addition the product of inertia with respect to centroidal axis will also be zero. Let UU' and VV' be new set of axes which pass through the centroid of the area G. Then if

I_{uv} = 0 the Product of inertia I_{uv} is called the Principal Product moment of inertia. The product inertia with respect to centroidal axes will also be zero.

The value of I_{uv} will depend upon the orientation of new set of axes with respect to the original axes *i.e.*, *OX, OY*.

Let us suppose that the new set of axis *UU'* and *VV'* are inclined at an angle θ with the original axes *OX, OY*. As θ will change, so will the values of I_{uv}. At some stage when $I_{uv} = 0$ then at that angle θ, the new axes will be called the Principal Axes and I_{uv} will be called the principal moment of Inertia.

We now proceed to determine the principal moment of inertia and other corresponding location of the axes *UU* and *VV*.

Let (*x, y*) be the co-ordinates of an elemental area *dA* with respect to axes *XOX'* and *YOY'* (Fig. 14.62) with origin at *CG*. Let *UOU'* and *VOV'* be another set of axes inclined at θ (anticlockwise) with the original axes *OX, OY*. Let the same elemental area *dA* have (*u, v*) co-ordinates with respect to *OU, OV* axes. (Fig 14.49)

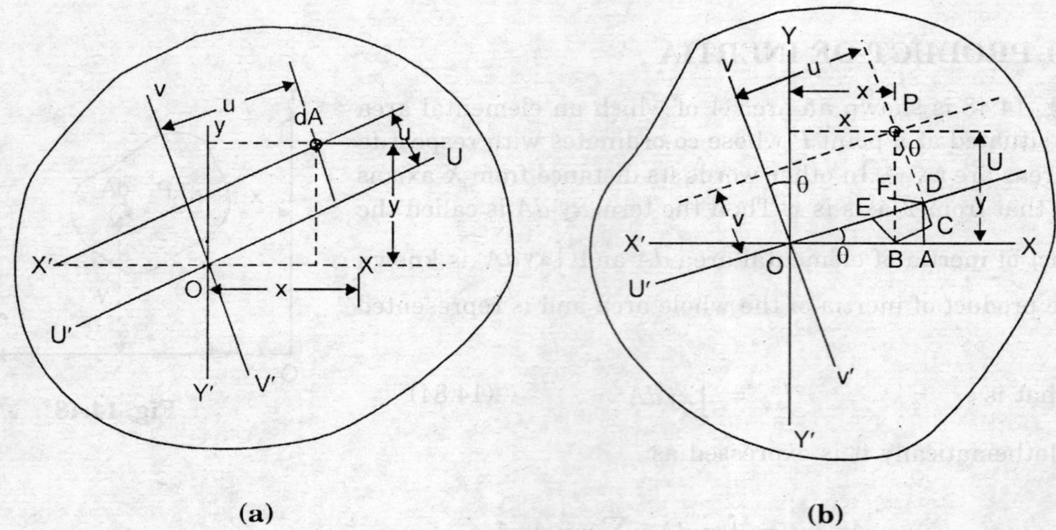

Fig. 14.49

Then Geometrically $\quad OB = x,\ BP = y,\ OD = u,\ PD = v$

The $\quad u = GD = OD = u = GE + ED$

$$= GB \cos\theta + BC$$

$$= GB \cos\theta + PB \sin\theta$$

$$u = x \cos\theta + y \sin\theta \qquad \text{...(14.86)}$$

Also $\quad PD = v = PC - CD$

$$= y \cos\theta - BE$$

$$= y \cos\theta - x \sin\theta \qquad \text{...(14.87)}$$

Now $\quad I_{UU} = \int v^2 dA$

$$= \int (y\cos\theta - x\sin\theta)^2\,dA$$

$$= \int (y^2\cos^2\theta + x^2\sin^2\theta - 2xy\sin\theta\cos\theta)\,dA$$

$$= \cos\theta\int y^2 dA + \sin^2\theta\int x^2 dA - \sin 2\theta\int xy\,dA$$

$$I_{UU} = I_{xx}\cos^2\theta + I_{yy}\sin^2\theta - I_{XY}\sin 2\theta$$

$$= \frac{I_{xx}}{2}(1+\cos 2\theta) + \frac{I_{yy}}{2}(1-\cos 2\theta) - I_{XY}\sin 2\theta$$

$$= \frac{1}{2}(I_{xx}+I_{yy}) + \frac{I_{xx}-I_{yy}}{2}\cos 2\theta - I_{xx}\sin 2\theta \qquad \text{...(14.88}b\text{)}$$

Similarly

$$I_{VV} = \int u^2 dA$$

$$= \int (x\cos\theta + y\sin\theta)^2\,dA\,\text{z}$$

$$= \int [x^2\cos^2\theta + y^2\sin^2\theta + xy\sin\theta\cos\theta]\,dA$$

$$= \int x^2\cos^2\theta\,dA + \int y^2\sin^2\theta\,d\theta + \int xy\sin 2\theta\,dA$$

$$= \cos^2\theta\, I_{yy} + \sin^2\theta\, I_{xx} + (\sin 2\theta)\, I_{xy}$$

or

$$I_{VV} = I_{xx}\sin^2\theta + I_{yy}\cos^2\theta + I_{xy}\sin 2\theta$$

$$= \frac{I_{xx}}{2}(1-\cos 2\theta) + \frac{I_{yy}}{2}(1+\cos^2\theta) + I_{xy}\sin^2\theta \qquad \text{...(14.89}a\text{)}$$

$$= \frac{1}{2}(I_{xx}+I_{yy}) - \left(\frac{I_{xx}-I_{yy}}{2}\right)\cos^2\theta - I_{yy}\sin^2\theta$$

Also

$$I_{uv} = \int uv\,dA \qquad \text{...(14.89}b\text{)}$$

$$= \int (x\cos\theta + y\sin\theta)(y\cos\theta - x\sin\theta)\,dA$$

$$= \int [xy\cos^2\theta - x^2\sin\theta\cos\theta + y^2\sin\theta\cos\theta - xy\sin^2\theta]\,dA$$

$$= \int [xy(\cos^2\theta - \sin^2\theta) + (y^2 - x^2)\sin\theta\cos\theta)\,dA$$

$$= \int [xy\cos 2\theta)\,dA + \int \left(\frac{1}{2}\sin 2\theta\right)(y^2 - x^2)\,dA$$

or

$$I_{uv} = \cos 2\theta\int xy\,dA + \frac{\sin 2\theta}{2}\left\{\int y^2 dA - \int x^2 dA\right\}$$

$$I_{uv} = I_{xy}\cos 2\theta + \left(\frac{I_{xx}-I_{yy}}{2}\right)\sin 2\theta \qquad \text{...(14.90)}$$

For finding principal moment of inertia put

$$I_{uv} = 0$$

∴ from equation (14.90) θ for Principal moment of Inertia we get by equating I_{uv} to zero.

$$0 = I_{xy}\cos 2\theta + \frac{I_{xx} - I_{yy}}{2}\sin 2\theta$$

or

$$\tan 2\theta = \frac{2I_{xy}}{I_{yy} - I_{xx}} \qquad \text{...(14.91)}$$

Therefore if q_1 and q_2 are two values of q from equation (14.91)

$$\theta_1 = \frac{1}{2}\tan^{-1}\left[\frac{2I_{xy}}{I_{yy} - I_{xx}}\right] \qquad \text{...(14.92}a\text{)}$$

$$\theta_2 = \frac{\pi}{2} + \theta_1$$

Corresponding to there values of q as q_1 and q_2 the values of the Principal Moment of Inertia can be obtained from the equations (14.88) and (14.89).

For value of $q = q_1$ we have from equation (14.92*a*) after constructing the basic trignometric triangle as in Fig. 14.50.

$$\sin 2\theta_1 = \frac{I_{xy}}{\sqrt{\left(\frac{I_{yy} - I_{xx}}{2}\right)^2 + I_{xy}^2}} \qquad \text{...(14.93}a\text{)}$$

$$\cos 2\theta_1 = \frac{\frac{1}{2}(I_{yy} - I_{xx})}{\sqrt{\left(\frac{I_{yy} - I_{xx}}{2}\right)^2 + I_{xy}^2}} \qquad \text{...(14.93}b\text{)}$$

Fig. 14.50

Therefore substituting values from equations (14.93*a*) and (14.93*b*) into equation (14.88) and (14.89) we get for $\theta = \theta_1$.

$$(I_{uu})_{\theta_1} = \frac{1}{2}I_{xx}(2\cos^2\theta_1) + \frac{1}{2}I_{yy}(2\sin^2\theta_1) - I_{xy}\sin 2\theta_1$$

$$= \frac{1}{2}I_{xx}(1 + \cos 2\theta_1) + \frac{1}{2}I_{yy}(1 - \cos 2\theta_1) - I_{xy}\sin 2\theta_1$$

$$= \left(\frac{I_{xx} + I_{yy}}{2}\right) + \left(\frac{I_{xx} - I_{yy}}{2}\right)\cos 2\theta_1 - I_{xy}\sin 2\theta_1$$

$$= \left(\frac{I_{xx} + I_{yy}}{2}\right) + \frac{\left(\frac{I_{xx} - I_{yy}}{2}\right)\left(\frac{I_{yy} - I_{xx}}{2}\right)}{\sqrt{\left(\frac{I_{yy} - I_{xx}}{2}\right)^2 + I_{xy}^2}} - I_{xy}\frac{I_{xy}}{\sqrt{\left(\frac{I_{yy} - I_{zz}}{2}\right)^2 + I_{xy}^2}}$$

$$= \frac{1}{2}(I_{xx} + I_{yy}) - \frac{\left(\frac{I_{yy} - I_{xx}}{2}\right)^2}{\sqrt{\left(\frac{I_{yy} - I_{xx}}{2}\right)^2 + I_{xy}^2}} - \frac{I_{xy}^2}{\sqrt{\left(\frac{I_{yy} - I_{xx}}{2}\right)^2 + I_{xy}^2}}$$

$$(I_{uu})_{\theta_1} = \frac{1}{2}(I_{yy} + I_{xx}) - \sqrt{\left(\frac{I_{yy} - I_{xx}}{2}\right)^2 + I_{xy}^2} \qquad ...(14.94)$$

Again substituting values from equation (14.93*a*) and (14.93*b*) into equation (14.89) we get

$$(I_{VV})_{\theta_1} = \frac{1}{2}\left(I_{xx} + I_{yy}\right) - \left(\frac{I_{xx} - I_{yy}}{2}\right)\cos 2\theta_1 + I_{xy} \sin 2\theta_1$$

$$= \frac{1}{2}(I_{xx} + I_{yy}) - \frac{\left(\frac{I_{xx} - I_{yy}}{2}\right)\frac{1}{2}(I_{yy} - I_{xx})}{\sqrt{\left(\frac{I_{yy} - I_{xx}}{2}\right)^2 + I_{xy}^2}} + I_{xy}\frac{I_{xy}}{\sqrt{\left(\frac{I_{yy} - I_{xx}}{2}\right)^2 + I_{xy}^2}}$$

$$= \frac{1}{2}(I_{xx} + I_{yy}) + \frac{\left(\frac{I_{yy} - I_{xx}}{2}\right)^2 + I_{xy}^2}{\sqrt{\left(\frac{I_{yy} - I_{xx}}{2}\right)^2 \times I_{xy}^2}}$$

i.e., $$(I_{vv})_{\theta_1} = \frac{1}{2}(I_{xx} + I_{yy}) + \sqrt{\left(\frac{I_{yy} - I_{xx}}{2}\right)^2 \times I_{xy}^2} \qquad ...(14.95)$$

Note: Since the value of q as q_1 is the solution of the equation (14.91) which was obtained by equating $I_{uv} = 0$, the corresponding values of $(I_{UU})_{\theta_1}$ and $(I_{VV})\theta_1$ refer to principal moment of inertia and angles q_1 and q_2 refer the directions of Principal Axes.

14.10.1 Maximum and Minimum Values of I_{UU} and I_{VV}

For maximum or minimum values of I_{uu} and I_{vv} we must have

$$\frac{dI_{uu}}{d\theta} = \frac{dI_{vv}}{d\theta} = 0$$

From equation (14.88*b*) and we get after differentiation to *r* to 'θ'.

$$-\left(\frac{I_{xx} - I_{yy}}{2}\right)(2\sin 2\theta) - 2I_{xy}\cos 2\theta = 0$$

$$\tan 2\theta = \frac{I_{xy}}{\frac{1}{2}(I_{yy} - I_{xx})} \qquad ...(14.96)$$

This equation is same as equation (14.91). This leads to the conclusion that. Principal moment of inertia are also maximum and minimum. A perusal of equations (14.94) and (14.95) shows that $(I_{vv})_{\theta_1}$ is greater than $(I_{uu})_{\theta_1}$. Thus $(I_{vv})_{\theta_1}$ will be maximum and $(I_{uu})_{\theta_1}$ will be minimum value of the moment of inertia.

Also $$\theta_2 = \frac{\pi}{2} + \theta_1$$

or $$2\theta_2 = \pi + 2\theta_1$$

$\therefore$ $$\sin 2\theta_2 = \sin(\pi + 2\theta_1) = -\sin 2\theta_1 \quad \text{...(14.97a)}$$

and $$\cos 2\theta_2 = \cos(\pi + 2\theta_1) = -\cos 2\theta_1 \quad \text{...(14.97b)}$$

Therefore substituting these values in equation (14.89) we get

$$(I_{vv})_{\theta_2} = I_{xx}\sin^2\theta_2 + I_{yy}\cos^2\theta_2 + I_{xy}\sin 2\theta_2$$

$$= \frac{1}{2}I_{xx}(2\sin^2\theta_2) + \frac{1}{2}I_{yy}(2\cos^2\theta_2) + I_{xy}\sin 2\theta_2$$

$$= \frac{1}{2}I_{xx}(1-\cos 2\theta_2) + \frac{1}{2}I_{yy}(1+\cos 2\theta_2) + I_{xy}\sin 2\theta_2$$

$$= \frac{1}{2}(I_{xx}+I_{yy}) - \frac{1}{2}(I_{xx}-I_{yy})\cos 2\theta_2 + I_{xy}\sin 2\theta_2$$

$$= \frac{1}{2}(I_{xx}+I_{yy}) + \frac{1}{2}(I_{xx}-I_{yy})\cos 2\theta_1 - I_{xy}\sin 2\theta_1 \quad \text{(From 14.95)}$$

$$= \frac{1}{2}(I_{xx}+I_{yy}) + \frac{1}{2}(I_{xx}-I_{yy})\frac{(I_{yy}-I_{xx})/2}{\sqrt{\left(\frac{I_{yy}-I_{xx}}{2}\right)^2 + I_{xy}^2}} - I_{xy}\frac{I_{xy}}{\sqrt{\left(\frac{I_{yy}-I_{xx}}{2}\right)^2 + I_{xy}^2}}$$

or $$(I_{VV})_{\theta_2} = \frac{1}{2}(I_{xx} \times I_{yy}) - \sqrt{\left(\frac{I_{yy}-I_{xx}}{2}\right)^2 + I_{xy}^2} \quad \text{...(14.98)}$$

Similarly from equations (14.88*a*) and (14.97)

$$(I_{uu})_{\theta_2} = \frac{1}{2}(I_{xx}+I_{yy}) + \left(\frac{I_{xx}-I_{yy}}{2}\right)\cos 2\theta_2 - I_{xy}\sin 2\theta_2$$

or $$(I_{uu})_{\theta_2} = \frac{1}{2}(I_{xx}+I_{yy}) + \left(\frac{I_{xx}-I_{yy}}{2}\right)\cos 2\theta_1 - I_{xy}(-\sin 2\theta_1)$$

$$= \frac{1}{2}(I_{xx}+I_{yy}) - \left(\frac{I_{xx}-I_{yy}}{2}\right)\frac{\left(\frac{I_{yy}-I_{xx}}{2}\right)}{\sqrt{\left(\frac{I_{yy}-I_{xx}}{2}\right)^2 + I_{xy}^2}} + \frac{I_{xy}\times I_{xy}}{\sqrt{\left(\frac{I_{yy}-I_{xx}}{2}\right)^2 + I_{xy}^2}}$$

i.e.,
$$(I_{uu})_{\theta_2} = \frac{1}{2}(I_{xx}+I_{yy}) + \sqrt{\left(\frac{I_{yy}-I_{xx}}{2}\right)^2 + I_{xy}^2} \qquad ...(14.98)$$

From equations (14.94), (14.95), (14.97) and (14.98) we see that

$$(I_{uu})_{\theta_2} = (I_{vv})_{\theta_1} \qquad ...(14.99a)$$

and

$$(I_{vv})_{\theta_2} = (I_{uu})_{\theta_1} \qquad ...(14.99b)$$

Also
$$(I_{uu})_{\theta_1} + (I_{vv})_{\theta_1} = (I_{uu})_{\theta_2} + (I_{vv})_{\theta_2} = (I_{xx}+I_{yy}) \qquad ...(14.100)$$

14.11 THEOREM OF PARALLEL AXES FOR PRODUCT OF INERTIA

In the Fig. 14.51 is shown an area whose product of inertia about *XX* and *yy* axes which are passing through the centre of gravity *G* are known.

An elemental Area *dA* has co-ordinates (*x*, *y*) with respect to the original areas *OX*, *OY*. passing through *CG*.

$$\therefore \quad I_{XX} = \oint dA\, y^2$$

$$I_{YY} = \oint dA\, x^2$$

$$I_{XY} = \oint dA\, xy$$

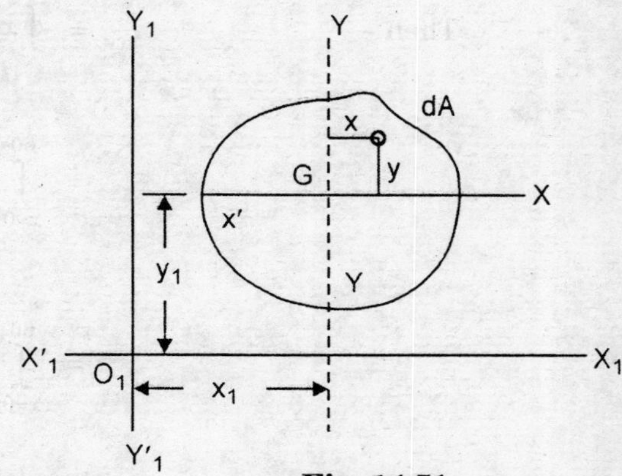

Fig. 14.51

Let X_1X_1 and Y_1Y_1 be another set of axes parallel to the given set *i.e.,* $X_1'O_1X'_1$ is parallel to *X'OX* and $Y_1O_1Y'_1$ is parallel to *YOY'* such that the co-ordinates of *O* will respect to new origin O_1 are say (x_1, y_1). Therefore the co-ordinates of *dA* with respect to $X_1^1O_1X_1$ and $Y_1^1O_1Y_1$ axes are $(x_1 + x)$ and $(y_1 + y)$.

$$\therefore \quad I_{X_1Y_1} = \oint (x_1+x)(y+y_1)\,dA$$

$$= \int (x_1y + x_1y_1 + xy + xy_1)\,dA$$

$$= x_1\int y\,dA + x_1y_1\int dA + \int xy\,dA + y_1\int x\,dA \qquad ...(14.101)$$

Since $\int xdA$ and $\int y(dA)$ are first moments of area about centroidal axes OY and OX respectively. They must be zero. Then equation (14.101) reduces to

$$I_{X_1Y_1} = \oint xy\,dA + x_1y_1A$$

$$= I_{XY} + x_1y_1A \qquad ...(14.102)$$

Example 14.13. *Find the product moment of inertia of the right angled triangle OAB about XOY axes as shown in Fig. 14.52.*

Solution : Let the elemental area $dA = dxdy$ be taken at any point P whose co-ordinates are (x, y)

Then $$I_{XY} = \oint xy\,dA$$

Now the stop of the OB is

$$\frac{40}{60} = \frac{2}{3}$$

Therefore equation of OB is

$$y = \frac{2}{3}x$$

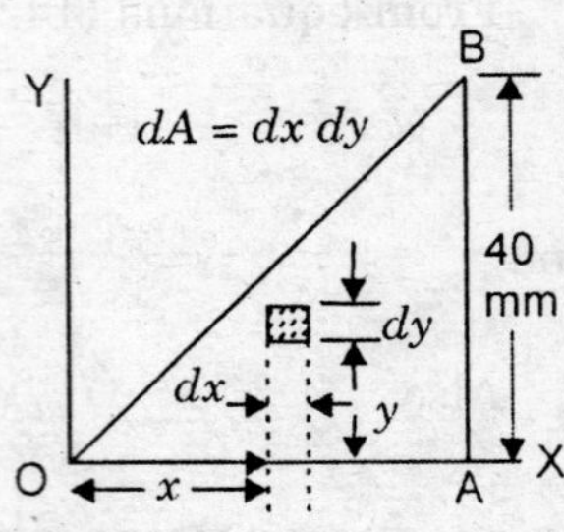

Fig. 14.52

Then $$I_{xy} = \oint xy\,dx\,dy$$

$$= \int_{x=0}^{60}\int_{y=0}^{\frac{2}{3}x} xy\,dy\,dx$$

$$= \int_{x=0}^{60} x\left[\frac{y^2}{2}\right]_{y=0}^{\frac{2}{3}x} dx$$

or $$I_{xy} = \int_{x=0}^{60} \frac{x}{2}\left[\frac{4x^2}{9} - 0\right]dx$$

$$= \int_{x=0}^{60}\left[\frac{2}{9}x^3\right]dx = \frac{2}{9}\left[\frac{x^4}{4}\right]_0^{60}$$

$$= \frac{1}{18}[60]^4$$

$$= 72 \times 10^4 \text{ mm}^4$$

Example 14.14. *Find the product moment of inertia of a quadrant OAB of a circle about the perpendicular axes OX and OY shown in Fig. 14.53. Take the radius of quadrant as r.*

Solution : Let the elemental are $dx\ dy$ be located at p having co-ordinates (x, y)

Equation of curve AB is

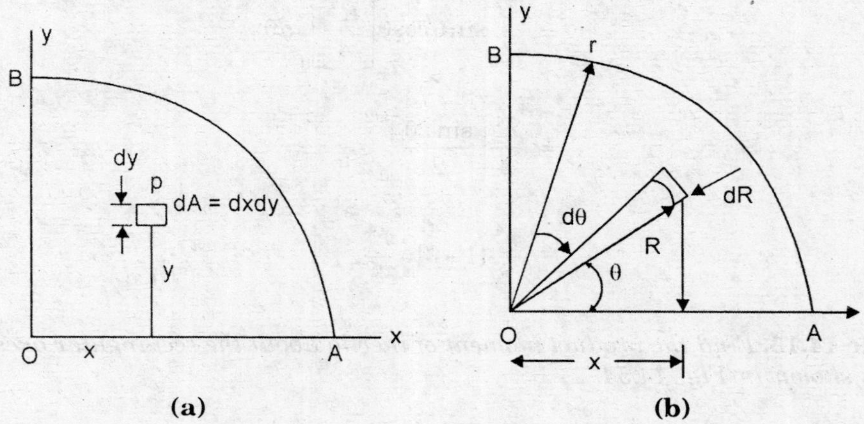

Fig. 14.53

$$x^2 + y^2 = r^2$$

or
$$y = \sqrt{r^2 - x^2}$$

$\therefore$
$$I_{xy} = \int_{x=0}^{r}\left\{\int_{y=0}^{\sqrt{r^2-x^2}} xy\, dy\right\}dx = \int_{x=0}^{r}\left\{\left[\frac{xy^2}{2}\right]_0^{\sqrt{r^2-x^2}}\right\}dx$$

or
$$I_{xy} = \int_{x=0}^{r}\frac{x}{2}(r^2 - x^2)\,dx$$

$$= \frac{1}{2}\left[r^2\frac{x^2}{2} - \frac{x^4}{4}\right]_0^r$$

$$= \frac{1}{2}\left(\frac{r^4}{2} - \frac{r^4}{4}\right) = \frac{r^4}{8}$$

Alternatively in polar co-ordinates

$$I_{xy} = \iint xy\, dA$$

From Fig. (14.61b) $\quad dA = R\ d\theta\ dR$

$$x = R\cos\theta,\ y = R\sin\theta$$

Then
$$I_{xy} = \int_{\theta=0}^{\pi/2}\int_{R=0}^{r}(R\cos\theta)(R\sin\theta)\,R\,dR\,d\theta$$

$$= \int_{\theta=0}^{\pi/2}\left[\int_{R=0}^{r}(\sin\theta\cos\theta)r^3\, dR\right]d\theta$$

$$= \int_{\theta=0}^{\pi/2}\sin\theta\cos\theta\left[\frac{R^4}{4}\right]_0^r d\theta$$

$$= \frac{r^4}{4}\left[\frac{\sin 2\theta}{2}\right]_0^{\pi/2}$$

$$= \frac{r^4}{8}[1-0] = \frac{r^4}{8}$$

Examle 14.15. *Find the product moment of inertia about the rectangular axes OX and OY of the area shown in Fig. 14.54.*

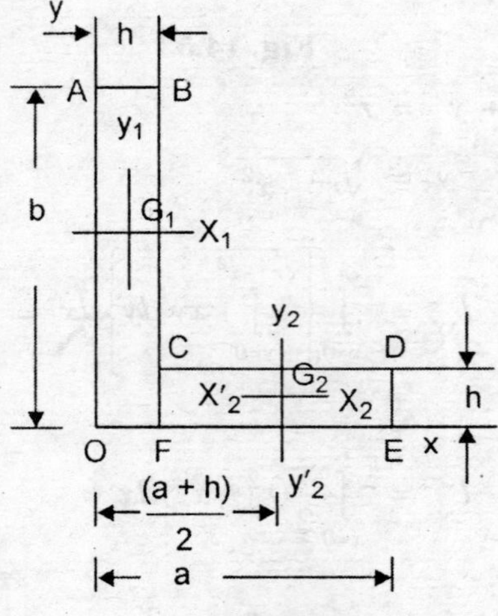

Fig. 14.54

Solution : Let G_1 and G_2 be the centroids of the areas *ABFO* and *CDEF* respectively having the corresponding areas A_1 and A_2. Let the co-ordinates of G_1 and G_2 with respect to *XOY* axes be (x_1, y_1) and (x_2, y_2)

Then for the area *ABFO* from equation (14.102) *i.e.*, theorem of parallel axes for product moment of inertia

$$I_{xy} = I_{x_1y_1} + A_1x_1y_1$$

Now $I_{x_1y_1}$ of area *ABFO* = 0 since G_1X_1 and G_1V_1 are the principal axes of the area *ABRO* being axes of symmetry.

And $$A_1 = b\cdot h; x_1 = \frac{h}{2} \text{ and } y_1 = \frac{b}{2}$$

Therefore I_{xy} of area

$$ABFO = 0 + (bh)\left(\frac{h}{2}\right)\left(\frac{b}{2}\right) = \frac{b^2h^2}{4}.$$

Again for the Area *CDEF*

$$I_{xy} = I_{x_2y_2} + A_2x_2y_2$$

Here also $I_{x_2y_2} = 0$ as G_2X_2 and G_2Y_2 are axes of symmetry.

and

$$A_2 = (a-h)h$$

$$x_2 = OF + \frac{1}{2}(FE) = h + \frac{a-h}{2} = \frac{(a+h)}{2}$$

$$y_2 = \frac{h}{2}$$

Thereore I_{xy} of area

$$CDEF = 0 + (a-h)h\left[\frac{a+h}{2}\right]\left(\frac{h}{2}\right)$$

$$= (a^2 - h^2)\frac{h^2}{4}$$

∴ Total product of inertia of entire area

$$I_{xy} = \frac{b^2h^2}{4} + \frac{h^2}{4}(a^2h^2)$$

$$= \frac{h^2}{4}(b^2 + a^2 - h^2) \textbf{ Ans.}$$

Example 14.16. *Find the centroidal principal moment of inertia of an equal angle section shown in Fig. 14.55.*

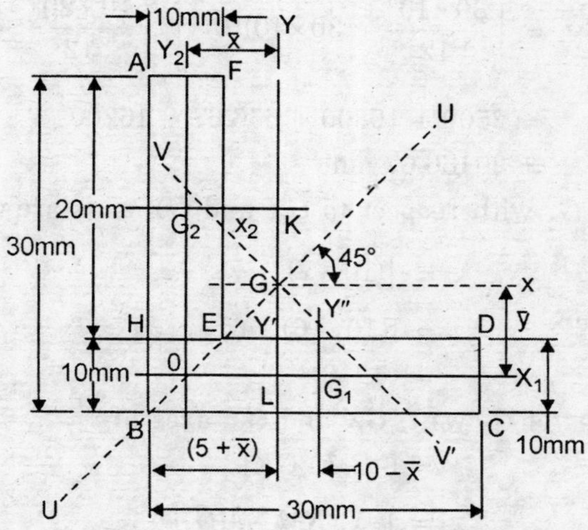

Fig. 14.55

Solution : Let $(\bar{x}, \bar{y})$ be the co-ordinates of the centroid G, with respect to the rectangular axes OX_1 and OY_2. Since this is an equal angle

$$\bar{x} = \bar{y}$$

For finding the value of $\bar{y}$, take moments of areas about the live OX_1, the centroidal axes of area $BCDH$.

$\therefore$ $(30 \times 10 + 20 \times 10)\bar{y}$ = moment of area $HDCB$ about OX_1 + moment of area $AFEH$

That is

$$500\bar{y} = (30 \times 10 \times 0) + (20 \times 10 \times 15)$$

$$= 3000$$

$$\therefore \quad \bar{y} = \frac{3000}{500} = 6 \text{ mm}$$

Also because of equal angle specifications

$$I_{xx} = I_{yy}$$

$$= \left[I_{G_1X_1} \text{ of } HDCB + A_{HDCB} \times (\bar{y})^2\right] + \left[I_{G_2X_2} \text{ of } AFEH + A_{AFEH} \times (KG)^2\right]$$

Now

$$KG = KM - MG$$

$$= \frac{1}{2}(AH + HB) - \bar{y}$$

$$= \frac{1}{2}(20 + 10) - 6 = 9$$

$$I_{xx} = I_{yy} = \left[\frac{30 \times 10^3}{12} + (30 \times 10)(6)^2\right] + \left[\frac{10 \times 20^3}{12} + (10 \times 20)(9)^2\right]$$

$$= 2500 + 10800 + 666.67 + 16200$$

$$= 30166.67 \text{ mm}^4$$

Now co-ordinates of G_1 with respect to GX and GY axes are $(MG_1, -GM)$

$$= [(10 - \bar{x}), -\bar{y}]$$

$$= [(10 - 6), -6]$$

$$= (4, -6)$$

Similarly co-ordinates of G_2 w.r.t. GX and GY axes are

$$= [-G_2K, GK]$$

$$= [-\bar{x}, (KM - MG)]$$

$= [-\bar{x}, 9]$

$= [-6, 9]$

$\therefore \quad I_{xy}$ of area $HDCB$ $= I_{X_1Y_1}$ of Area $HDCB + A_{HDCB}(4)(-6)$

$= 0 + (30 \times 10)(-24)$

(Because OX_1, is the axis of symmetry $\therefore$ ($I_{x_1y_1} = 0$)

$= -7200 \text{ mm}^4$

Now I_{XY} of area

$AFEH = I_{x_2y_2}$ of $A_{AFEH} + A_{AFEH}(-6)(9)$

$= 0 + 10 \times 20\,(-54)$

$= -10800 \text{ mm}^4$

Total $\quad I_{XY} = -7200 - 10800$

$= -18000 \text{ mm}^4$

If q is the inclination of principal axes GU an GV with GX passing through G then

$$\tan 2\theta = \frac{2I_{XY}}{I_{XX} - I_{YY}} = \infty \; (\because I_{XX} = I_{YY})$$

i.e., $\quad 2\theta_1 = 90°$ or $\theta_1 = 45°$

and $\quad \theta_2 = 90 + 45 = 135°$

Thus the Principal axes GU and GV are inclined with GX at 45° and 135°.

For $\theta = 45°$ $\quad I_{UU} = \frac{1}{2}(I_{XX} \times I_{YY}) + \frac{1}{2}(I_{XX} - I_{YY})\cos 90° - I_{XY}\sin 90°$

$= I_{XX} + 0 - I_{XY}$

$= 30166.67 + 0 - (-18000)$

$= 84166.67 \text{ mm}^4$

Also

$I_{ZZ} = I_{UU} + I_{VV}$

$= I_{XX} \times I_{YY}$

or $\quad I_{VV} = I_{XX} + I_{YY} - I_{UU}$

$= 2I_{XX} - I_{UU} \qquad (\because I_{XX} = I_{YY})$

$= 2(30166.6) - 48166.67$

$= 12166.67 \text{ mm}^4$

Example 14.17. *Find the centroidal principal moment of inertia of the section shown in Fig. 14.56.*

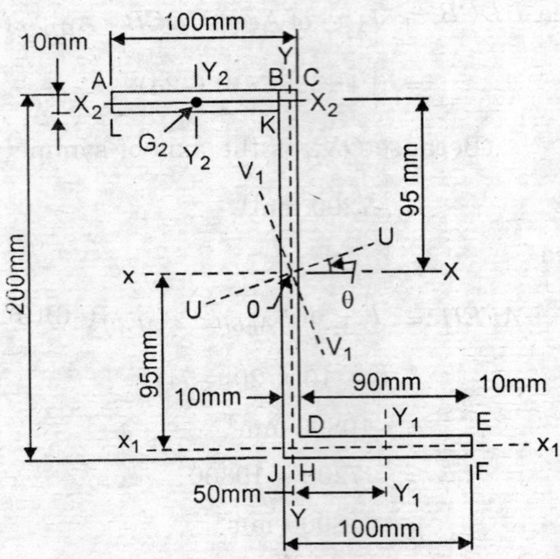

Fig. 14.56

Solution : Evidently the centroid G of the given Z-section is in the middle of web. Let the centroidal principal axes be GU and GV.

Further let G_1 and G_2 be the centroids of the rectangular areas $BCHJ$ and $ABKL$ respectively.

$$I_{yy} = I_{yy} \text{ of area } BCHJ + I_{yy} \text{ of area } DEFH + I_{yy} \text{ of area } ABKL$$

$$= \left(\frac{200 \times 10^3}{12}\right) + \left[\frac{10 \times 90^3}{12} + (90 \times 10)(50)^2\right] + \left[\frac{10 \times 90^3}{12}(90 \times 10)50^2\right]$$

$$= 573.167 \times 10^4 \text{ mm}^4$$

$$I_{XX} = I_{XX} \text{ of area } BCHJ + I_{XX} \text{ of area } DEFH + I_{XX} \text{ of area } ABKL$$

$$= \frac{10 \times 200^3}{12} + \left[\frac{90 \times 10^3}{12} + (90 \times 10)(95)^2\right] + \left[\frac{90 \times 10^3}{12} + (90 + 10)(95)^2\right]$$

$$= 2292.6 \times 10^4 \text{ mm}^4$$

I_{xy} of area $BCHJ$ = 0 since GX and GV are the axes of symmetry.

Now co-ordinates of G_1 wrt GX, GY axes are

$$I_{xy} \text{ of area } DEFH = I_{x_1y_1} \text{ of } DEFH + A_{DEFH}\,(50)\,(-95)$$

$$= .0 + (90 + 10)\,(50)\,(-95)$$

$$= -425000 \text{ mm}^4$$

Again the co-ordinates of G_2 w.r.t. GX, GY axes are (–50, 95)

$$I_{xy} \text{ of area } ABKL = I_{x_2 y_2} \text{ of } ABKL + A_{ABKL}\,(-50)\,(95)$$

$$= 0 + (90 \times 10)\,(-50)\,(95)$$

$$= -4275000 \text{ mm}^4$$

$$\therefore \text{Total } I_{XY} \text{ of the } z\text{-section} = 0 - 4275000 - 4275000$$

$$= -\,855 \times 10^4 \text{ mm}^4$$

$$\therefore \qquad \tan 2\theta = \frac{2I_{xy}}{I_{yy} - I_{xx}}$$

$$= \frac{2(-855 \times 10^4)}{573.167 \times 10^4 - 2292.6 \times 10^3}$$

$$= 0.9945$$

$$\therefore \qquad 2\theta = \tan^{-1}(0.9945)$$

$$= 44.84°$$

or
$$\theta = \frac{44.84}{2} = 22.42°$$

Then
$$I_{UU} = I_{XX}\cos^2\theta + I_{XY}\sin^2\theta - I_{XY}\sin 2\theta$$

$$= 2292.6 \times 10^4 \cos^2(22.42) + 573.167 \times 10^4 \cos^2(22.42) - (855 \times 10^4)\sin(44.84°)$$

$$= 2645.37 \times 10^4 \text{ mm}^4 \textbf{ Ans.}$$

Now
$$I_{UU} + I_{VV} = I_{XX} + I_{YY}$$

$$\therefore \qquad I_{VV} = (I_{XX} + I_{YY}) - I_{UU}$$

$$= (2292.6 \times 10^4 + 573.16) \times 10^4) - (2645.37 \times 10^4\_$$

$$= 220.397 \times 10^4 \text{ mm}^4 \textbf{ Ans.}$$

14.12 MASS MOMENT OF INERTIA AND RADIUS OF GYRATION

In Fig. 14.57 is shown a body of mass M with centre of gravity at $G(\bar{x}, \bar{y})$.

Let dm be an elemental mass having co-ordinates (x, y) with respect to XOY axes. Then the second moment of mass dm about axis OX

$$dI_{XX} = (dm)\,y^2$$

$$(I_m)_{xx} = \sum_{i=1}^{n} dm_i y_i^2$$

Integrating
$$(I_m)_{xx} = dm_1 y_1^2 + dm_2 y_1^2 \ldots dm_n y_n^2$$

$$(I_m)_{XX} = M(\bar{y})^2 = \sum_{i=1}^{n} (dm_1)\,y_i^2$$

Again the second moment of dm_i about Y-axis

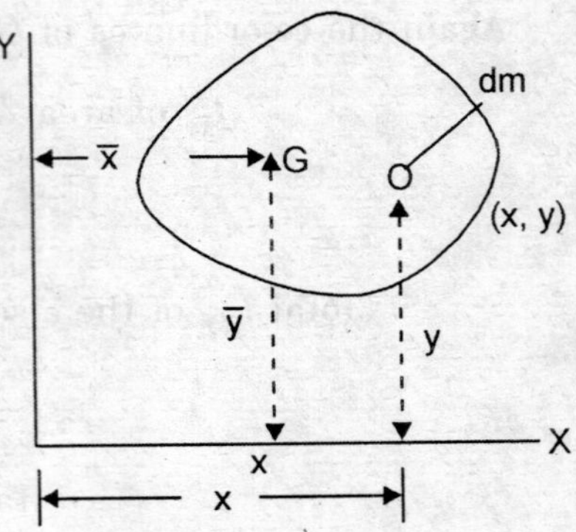

Fig. 14.57

$$d\,(I_m)_{yy} = dm_i x_i^2$$

or
$$(I_m)_{yy} = \sum_{i=1}^{n} dm_i\, n_i^2$$
$$= dm_1 x_1^2 + dm_2 x_2^2 + \ldots + dm_n x_n^2$$

$$(I_m)_{yy} = M(\bar{x})^2 = \sum_{0=1}^{n} dm_i x_i^2 \qquad \ldots(14.103)$$

Radious of Gyration

It k is the radius of gyration of a body of mass M, it is defined by the expression.

Then
$$Mk^2 = I_m$$
$$k = \sqrt{\frac{I_m}{M}}$$

14.12.1 Mass Moment of Inertia of Rectangular Plate

In Fig. 14.58, is shown a rectangular plate $ABCD$ of width 'b' and depth 'd'. The thickness of the plate is 't'. The axes of reference OX and OY are taken through the CG which is coinciding with 0.

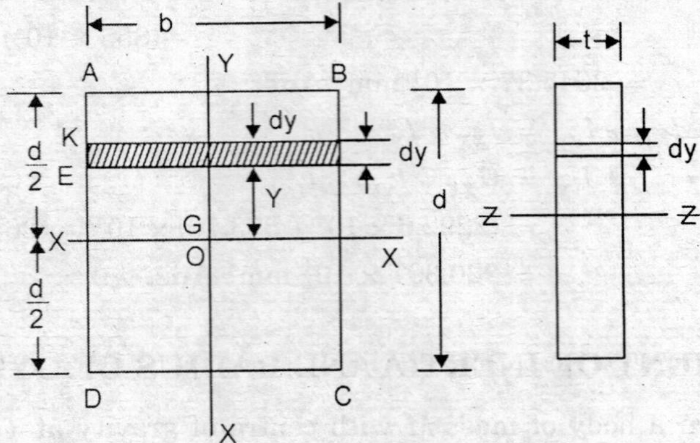

Fig. 14.58

$LKEF$ is an elemental strip of width 'b', depth dy and thickness t. It is situated at a height 'y' from the axis OX.

$$\therefore \quad d(I_m)_{XX} = (dm)\, y^2$$

Where dm = mass of the elemental strip

$$= \text{density} \times \text{vol of strip}$$
$$= \rho\,(dV) = \rho\,(bt\;dy)$$

$$\therefore \quad d(I_m)_{XX} = \rho\; bt\; dy\; y^2$$

Integrating we get $$(I_m)_{XX} = \rho bt \int_{-d/2}^{d/2} y^2 dy = \rho bt \left[\frac{y^3}{3}\right]_{-\frac{d}{2}}^{d/2}$$

$$= \frac{\rho bt}{3}\left[\frac{d^3}{8} + \frac{d^3}{8}\right] = \frac{\rho bt d^3}{12}$$

$$= \rho t\left(\frac{bd^3}{12}\right) = \rho t\, I_{XX} \qquad ...(14.104)$$

But I_{XX}, the second moment of area of rectangle about XX is given by

$$I_{XX} = \frac{bd^3}{12}$$

$$\therefore \quad (I_m)_{XX} = \rho t\ I_{XX}$$

$$= \rho td \frac{bd^2}{12}$$

$$= \rho(btd) \cdot \frac{d^2}{12} = \rho V \cdot \frac{d^2}{12}$$

$$= M\frac{d^2}{12}$$

V = Vol. of plate

Where M = Mass of plate = ρV

$$\therefore \quad (I_m)_{XX} = \frac{Md^2}{12} \qquad ...(14.105a)$$

Similarly by taking the elemental strip parallel to y-axis we get

$$(I_{mm})_{yy} = \frac{Mb^2}{12} \qquad ...(14.105b)$$

14.12.1.1 Mass Moment of Inertia of Rectangular Plate about base

From equation (14.105a)

$$(I_m)_{XX} = \frac{Md^2}{12}$$

Now from theorem of parallel axis

$$(I_m)_{DC} = (I_m)_{XX} + M\left(\frac{d}{2}\right)^2$$

$$= \frac{Md^2}{12} + \frac{Md^2}{4} = \frac{Md^2}{3} \quad \text{...(14.106)}$$

14.12.2 Mass Moment of Inertia of Circular Disc

In Fig. 14.59 is shown a disc of Radius R and thickness t. Let ρ be density of the material.

Let there be an elemental area of ring shape whose radial thickness is dx and is in the shape of a circular ring of radius x.

Then dv = vol. of elemental ring of thickness t

$$= (2\pi\, x\, dx)\, t$$

$$(I_m)_{zz} = x^2 dm$$

$$= x^2(\rho\, dv)$$

$$= \rho x^2(2\pi\, x\, dx\, t)$$

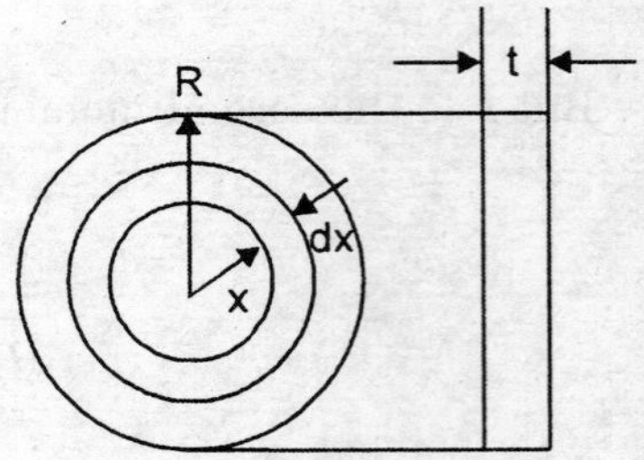

$$\therefore \quad (I_m)_{zz} = 2\pi\rho t \int_0^R x^3\, dx$$

$$= 2\pi\rho t \left[\frac{x^4}{4}\right]_0^R$$

$$= \frac{2\pi\rho R^4 t}{2} = (\pi R^2 \rho t) \times \left(\frac{R^2}{2}\right)$$

$$= \frac{MR^2}{2} \quad \text{...(14.107)}$$

Fig. 14.59

Where ρ = density of material

M = mass of disc

Now $(I_m)_{YY} = (I_m)_{XX}$

and by the theorem of perpendicular axes

$$(I_m)_{ZZ} = (I_m)_{XX} + (I_m)_{YY}$$

$$= 2(I_m)_{XX}$$

$$\therefore \quad (I_m)_{XX} = (I_m)_{YY} = \frac{1}{2}(I_m)_{ZZ}$$

$$= \frac{1}{2}\frac{MR^2}{2} = \frac{MR^2}{4} \quad \text{...(14.108)}$$

14.12.3 Mass Moment or Inertia of a Cylinder

Let the hollow cylinder be of outer radius R and inner radius r and length L (Fig. 14.60).

Let an elemental annular cylinder of radius x and radial thickness dx be taken. Then mass moment of inertia of an annular cylincler.

$$d(I_m)_{ZZ} = dm\ x^2$$
$$= (\rho\ dv)\ x^2$$
$$= \rho\ (2\pi\ x\ dx\ L)\ x^2$$
$$(I_m)_{ZZ} = 2\rho\pi L \int_r^R x^3\, dx$$
$$= 2\pi\rho L \left(\frac{x^4}{4}\right)_r^R$$
$$= \frac{\pi\rho L}{2}(R^4 - r^4)$$
$$= \frac{\pi\rho L}{2}(R^2 - r^2)(R^2 + r^2)$$
$$= \frac{\rho}{2}[\pi(R^2 - r^2)L](R^2 + r^2)$$
$$= \frac{1}{2}\rho(\text{vol of cylinder})\ (R^2 + r^2)$$

Fig. 14.60

$$\therefore \quad = \frac{1}{2}\rho\, dv(R^2 + r^2)$$
$$= \frac{M}{2}(R^2 + r^2) \qquad ...(14.109)$$

Now $(I_m)_{ZZ} = (I_m)_{XX} + (I_m)_{YY}$

Also $(I_m)_{XX} = (I_m)_{YY}$

$$\therefore \quad (I_m)_{XX} = (I_m)_{XX} = \frac{1}{2}(I_m)_{ZZ}$$
$$= \frac{M}{4}(R^2 + r^2) \qquad ...(14.110)$$

14.12.4 Mass Moment of Inertia of a Ring of Uniform Cross-section

Let the mass of unit length of wire of given section be m *i.e.*, $m\ 2\pi R = M$ when M is total mass of ring and R the radius.

$$d_m = (R\ d\theta)\ m$$
$$d(I_m)_{ZZ} = (m\ R\ d\theta)\ R^2$$

Integrating $(I_m)_{ZZ} = \int_{\theta=0}^{2\pi} m R^3\, d\theta = mR^3 [\theta]_0^{2\pi}$

$$= mR^3 [\theta]_0^{2\pi}$$

$$= 2\pi\, mR^3 = (2\pi R\, m)\, R^2$$

$$= MR^2$$

Now from theorem of perpendicular are axes

$$(I_m)_{ZZ} = (I_m)_{XX} + (I_m)_{YY}$$

$$= 2(I_m)_{XY}$$

$$(I_m)_{XX} = (I_m)_{YY}$$

$$= \frac{1}{2}(I_m)_{ZZ}$$

$$= \frac{1}{2}MR^2 \qquad ...(14.111a)$$

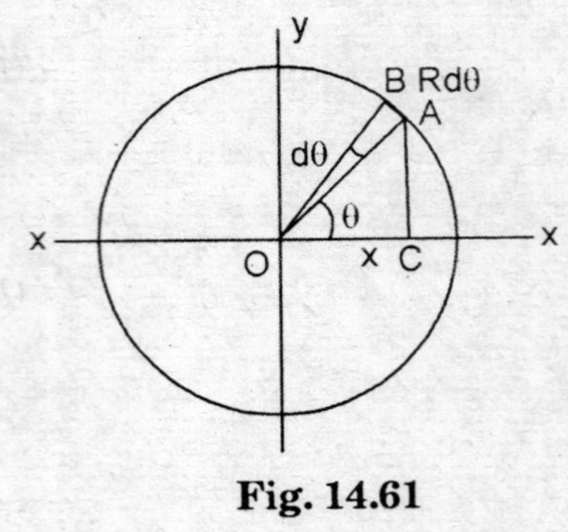

Fig. 14.61

Alternatively

From Fig. (14.61)

$$d(I_m)_{XX} = dmy^2$$

$$= (m\; Rd\theta)\; y^2$$

$$= (m\; R\; d\theta)\; (R\; \sin\, \theta)^2$$

$$= mR^3 \int_0^{2\pi} \sin^2\theta\, d\theta$$

$$= \frac{mR^3}{2}\int_0^{2\pi}(2\sin^2\theta)\, d\theta$$

$$= \frac{mR^3}{2}\int_0^{2\pi}(1-\cos 2\theta)\, d\theta$$

$$= \frac{mR^3}{2}\left[\theta - \frac{\sin 2\theta}{2}\right]_0^{2\pi}$$

$$= \frac{mR^3}{2}[(2\pi - 0) - (0)]$$

$$= \frac{2\pi mR^3}{2} = (2\pi Rm)\frac{R^2}{2}$$

or

$$(I_m)_{xx} = \frac{MR^2}{2} \qquad ...(14.111b)$$

14.12.5 Mass Moment of Inertia of a Right-circular cone about its axis of Symmetry

In Fig. 14.62 is shown a section of right circular cone of height h and base of radius r.

Let the axis of reference be taken as AX, AZ such that origin coincides with the vertex A of the cone. Then Z-Z is the axis of symmetry.

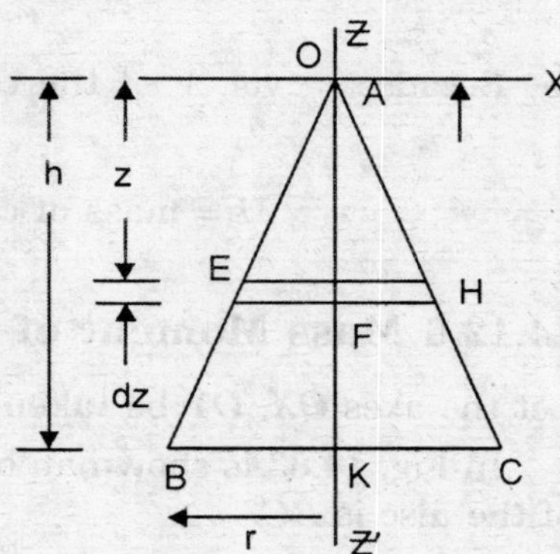

Fig. 14.62

Let an element in the form of a disc EFH be taken whose radius is EF and thickness dz at distance of z from OX. Then from similar triangles ABK and AEF

$$\frac{EF}{AF} = \frac{BK}{AK}$$

or

$$EF = \frac{r}{h}\cdot z \quad ...(14.112)$$

Then mass of the elemental disc of radius EF

$$dM = \rho \times dV$$

$$= \rho(\pi\,(EF)^2\,dz)$$

$$= \rho\pi\left(\frac{r}{h}z\right)^2 dz$$

Now from equation (14.10) I_m of a disc about, normal to the section, centroidal axis is

$$d(I_m)_{ZZ} = dm\cdot\frac{(EF)^2}{2}$$

$$= \rho\pi\left(\frac{r}{h}z\right)^2 dz\frac{(EF)^2}{2}$$

$$= \frac{\rho\pi}{2}\frac{r^2z^2}{h^2}dz\frac{r^2z^2}{h^2}$$

$$= \frac{\rho\pi}{2h^4}r^4\,z^4\,dz$$

$$= \frac{\pi\rho}{2}\frac{r^4}{h^4}\int_0^h z^4\,dz$$

$$= \frac{\pi}{4}\rho\frac{r^4}{h^4}\left[\frac{z^5}{5}\right]_0^h$$

$$= \frac{\pi}{2}\rho\frac{r^4}{h^4}\frac{h^5}{5}$$

$$= \frac{\pi}{2}\rho\frac{r^4h}{5}$$

$$= \left(\frac{\pi r^2h}{3}\rho\right)\frac{3r^2}{10} = \frac{3Mr^2}{10}$$

Because vol. V of the cone = $V = \dfrac{\pi r^2 h}{3}$ ...(14.113)

$\therefore$ M = mass of cone = $\rho V = \left(\rho \dfrac{\pi}{3} r^2 h\right)$

14.12.6 Mass Moment of Inertia of a sphere of Radius *R*

Let the axes *OX, OY* be taken though *O* the *CG* of sphere.

In Fig. 14.63 is shown an elemental disc of thickness dy at a right y from *OX*. The radius of the disc is *BC*.

$$\therefore \quad dM = Mm \text{ of elemental disc}$$

$$= \rho\pi(BC)^2 dy$$

$$\text{Now} \quad BC^2 = (R^2 - z^2)$$

$$\therefore \quad d(I_m)_{ZZ} = \frac{dM\,(BC)^2}{2}$$

$$= \frac{1}{2}\rho\pi(BC)^2 dz\,(BC)^2$$

$$(I_m)_{ZZ} = 2\int_0^R \frac{\rho\pi}{2}(x^2 - z^2)^2\, dz$$

$$= \int_{z=0}^{R} \rho\pi[R^4 + z^4 - 2R^2z^2]dz$$

$$= \rho\pi\left[R^4 z + \frac{z^5}{5} - \frac{2R^2z^3}{3}\right]_0^R$$

$$= \rho\pi\left[R^5 + \frac{R^5}{5} - \frac{2}{3}R^5\right]$$

$$= \frac{\rho\pi R^5}{15}[15 + 3 - 10] = \frac{8}{15}\rho\pi R^5$$

$$= \left(\frac{4}{3}\pi R^3\right)\rho\cdot\frac{2}{5}R^2$$

$$= \frac{2}{5}MR^2$$

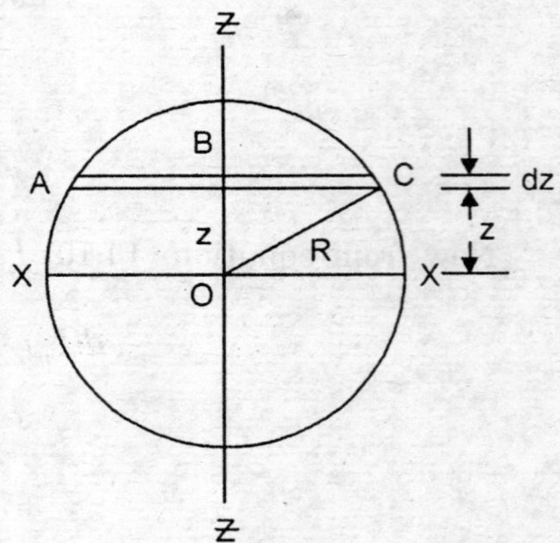

Fig. 14.63

PROBLEM

1. Explain the difference between centroid and centre of gravity.
2. Define centre of gravity and centroid.
3. Define the terms "Area moment of Inertia" and "Mass Moment of Inertia".
4. State and prove the theorems of parallel axes and perpendicular axes in the case of "Area Moment of Inertia".
5. Derive the expressions for second moment of area of a rectangle of width B and depth D about centroidal axis parallel to the depth.
6. Define the term "Radius of Gyration".
7. Show that the second moment of area of a circular section of diameter D about a diametric is $\frac{\pi D^4}{64}$.
8. Derive the expression for second moment of area of a triangle of height 'h' and base 'b' about the centroidal axis parallel to base.
9. A hollow rectangular section has wall thickness of 100 mm all around. The outside dimensions are width 1300 mm and depth 1200 mm. Calculate the second moment of area about both the centroidal axes XX and YY.

   (**Ans.** $I_{XX} = 955.3 \times 10^8$ mm^4, $I_{YY} = 1087.8 \times 10^8$ mm^4)
10. A cast iron channel section is shown in Fig. 14.64. Find the distance of centroid of the section from the base DF. Then calculate the value of second moment of area about centroidal axis parallel to DF.

    (**Ans.** $\bar{y} = 97.3$ mm, $I_{GX} = 2299.6971 \times 10^5$ mm^4)

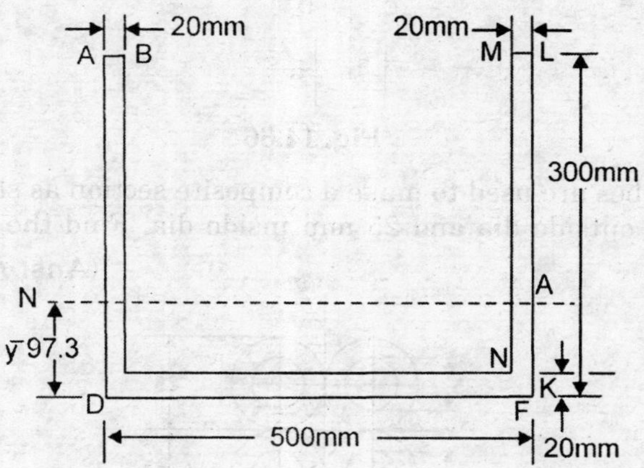

Fig. 14.64

11. Find the value of second moment of area of an I-section shown in Fig. 14.65 about XX. (**Ans.** 0.297×10^{10} mm^4)

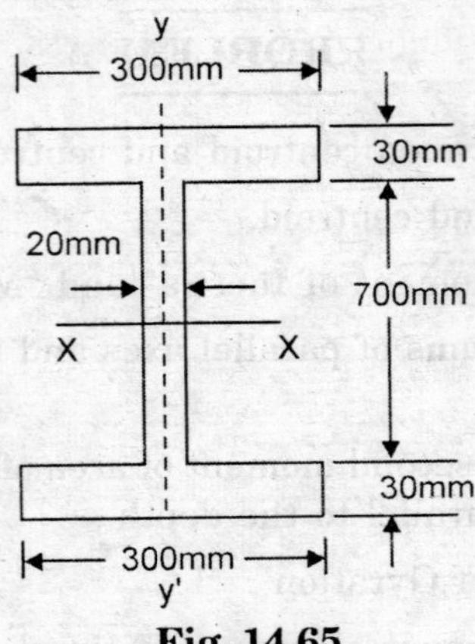

Fig. 14.65

12. Find the distance of centroid of the area of T-section shown in Fig. 14.66 from AB and then calculate Area Moment of Inertia about centroidal axis parallel to AB.

(**Ans.** $\bar{y}$ = 60 mm I_{XY} = 31.487 × 10^6 mm^4)

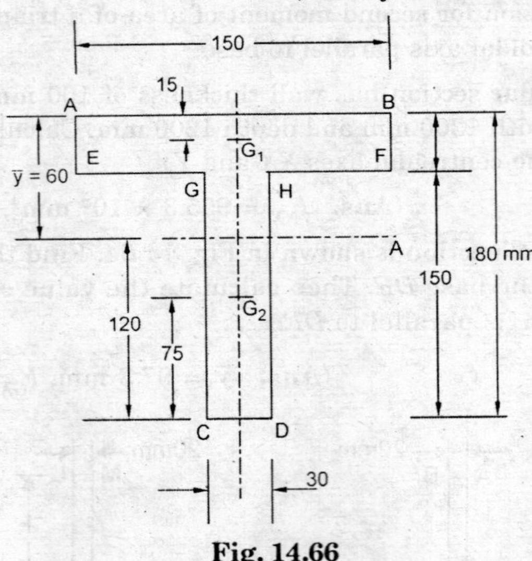

Fig. 14.66

13. Four similar tubes are used to make a composite section as shown in Fig. 14.67. Each tube is 35 mm outside dia and 25 mm inside dia. Find the value of I_{XX}.

(**Ans.** I_{XX} = 78.575 × 10^4 mm^4)

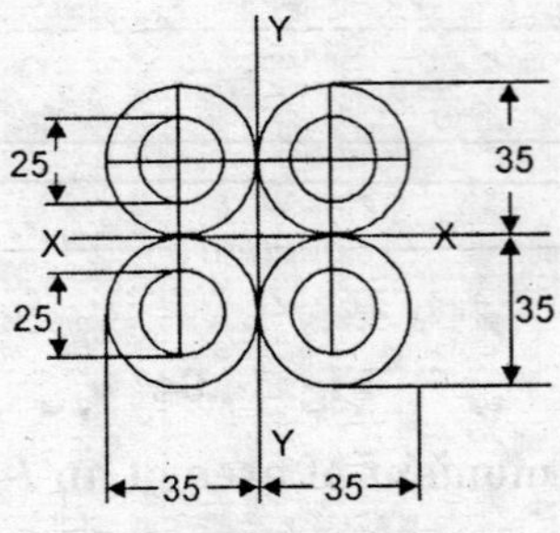

Fig. 14.67

14. In Fig. 14.68 is shown a hollow rectangular section which symmetrical about *XX*. Find the distance of centroid from the edge *BC* and then calculate I_{YY} about an axis passing through centroid and parallel to *BC*.

(Ans. $\bar{y}$ = 708 mm , $(I_G)_{YY}$ = 25.586 × 10⁶ mm⁴)

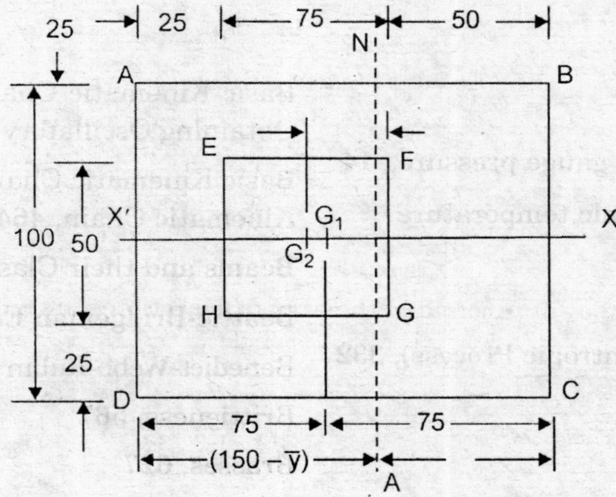

Fig. 14.68

15. A wooden beam rectangular section has a hole of 40 mm diameter as shown in Fig. 14.82. Find the distance of centroid from the base *DC* and then calculate $(I_{GG})_{XX}$.

(Ans. $(I_G)_{XX}$ = 24.95 × 10⁶ mm⁴, $\bar{y}$ = 75.64 mm)

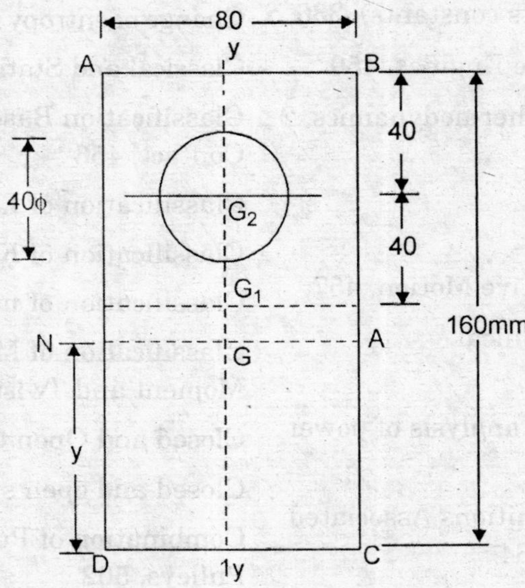

Fig. 14.69

□□□

INDEX

D

E

F

G

N

O

P

Q

R

S

Y

■■■

APPENDIX

(1) TRICONOMETRIC FORMULAE

$$\sin(A+B) = \sin A \cos B + \cos A \sin B$$

$$\sin(A-B) = \sin A \cos B - \cos A \sin B$$

$$\cos(A+B) = \cos A \cos B - \sin A \sin B$$

$$\cos(A-B) = \cos A \cos B + \sin A \sin B$$

$$\tan(A+B) = \frac{\tan A \quad \tan B}{1 \quad \tan A \tan B}$$

$$\tan(A-B) = \frac{\tan A - \tan B}{1 + \tan A \tan B}$$

$$\tan 2A = \frac{2\tan A}{1 - \tan^2 A}$$

$$\sin 2A = 2\sin A \cos A = \frac{2\tan A}{1 \quad \tan^2 A}$$

$$\cos 2A = 2\cos^2 A - \sin^2 A = 1 - 2\sin^2 A$$

$$= 2\cos^2 A - 1 = \frac{1 - \tan^2 A}{1 + \tan^2 A}$$

$$2 \sin A \cos B = \sin(A+B) + \sin(A-B)$$

$$2 \cos A \sin B = \sin(A+B) - \sin(A-B)$$

$$2 \cos A \cos B = \cos(A+B) + \cos(A-B)$$

$$2 \sin A \sin B = \cos(A-B) - \cos(A+B)$$

$$\sin C + \sin D = 2\sin\frac{C+D}{2}\cos\frac{C-D}{2}$$

$$\sin C' - \sin D = 2\cos\frac{C+D}{2}\sin\frac{C-D}{2}$$

$$\cos C + \cos D = 2\cos\frac{C+D}{2}\cos\frac{C-D}{2}$$

$$\cos C - \sin D = -2\sin\frac{C+D}{2}\sin\frac{C-D}{2}$$

(2) DIFFERENTIATION

$$\frac{du}{dx} = \frac{du}{dy}\cdot\frac{dy}{dx}$$

$$\frac{d}{dx}(uv) = u\frac{dv}{dx} + v\frac{du}{dx}$$

$$\frac{d}{dx}e^x = e^x$$

$$\frac{d}{dx}(\ln x) = \frac{1}{x}$$

$$\frac{d}{dx}\sin x = \cos x$$

$$\frac{d}{dx}\cos x = -\sin x$$

$$\frac{d}{dx}\tan x = \sec^2 x$$

$$\frac{d}{dx}(\sec x) = \sec x \tan x$$

$$\frac{d}{dx}(\sin^{-1} x) = \frac{1}{\sqrt{1-x^2}}$$

$$\frac{d}{dx}(\tan^{-1} x) = \frac{1}{1+x^2}$$

$$\frac{d}{dx}(\sec^{-1} x) = \frac{1}{x\sqrt{x^2-1}}$$

$$\frac{d}{dx}(\sinh x) = \cosh x$$

$$\frac{d}{dx}(\tanh x) = \text{sech}^2 x$$

(3) INTEGRATION

$$\int x^n dx = \frac{x^{n+1}}{n+1} \quad (\text{where } n \neq -1)$$

$$\frac{1}{x}dx = \ln(x)$$

$$\int e^x dx = e^x$$

$$\sin x dx = -\cos x$$

$$\int \cos x dx = \sin x$$

$$\tan x dx = -\ln(\cos x)$$

$$\sec x dx = \ln(\sec x + \tan x)$$

$$\int \sec^2 x dx = \tan x$$

$$\int \frac{dx}{a^2 + x^2} = \frac{1}{a}\tan^{-1}\left(\frac{x}{a}\right)$$

$$\int \frac{dx}{a^2 - x^2} = \frac{1}{2a}\ln\left(\frac{a+x}{a-x}\right)$$

$$\int \frac{dx}{x^2 = a^2} = \frac{1}{2a}\ln\left(\frac{x-a}{x+a}\right)$$

$$\int \sqrt{a^2 - x^2}\,dx = \frac{1}{2}x\sqrt{a^2 - x^2} + \frac{a^2}{2}\sin^{-1}\left(\frac{x}{a}\right)$$

$$\int \sqrt{a^2 + x^2}\, dx = \frac{x}{2}\sqrt{a^2 + x^2} + \frac{a^2}{2}\sinh^{-1}\left(\frac{x}{a}\right)$$

$$\int \sqrt{x^2 - a^2}\, dx = \frac{x}{2}\sqrt{x^2 - a^2} - \frac{a^2}{2}\cosh^{-1}\left(\frac{x}{a}\right)$$

$$\int e^{ax} \sin bx dx = \frac{e^{ax}}{a^2 + b^2}(a \sin bx - b \cos bx)$$

$$\int e^{ax} \cos bx dx = \frac{e^{ax}}{a^2 + b^2}(a \cos bx + b \sin bx)$$

$$\int \sinh x dx = \cosh x$$

$$\int \cosh x dx = \sinh x$$

$$\int \tanh x dx = \ln(\cosh x)$$

$$\int \coth x dx = \ln(\sinh x)$$

$$\int \sec h^2 x dx = \tanh x$$

$$\int F\left(\frac{dF}{ax}\right) dx = \frac{F^{n+1}}{n+1} \quad \text{(where } n \neq -1\text{)}$$

and F is a function of x

Integration By Parts

$$\int F_1 F_2 dx = F_1 \int F_2 dx - \int \left[\frac{dF_1}{dx}\left\{\int F_2 dx\right\}\right] dx$$

$$\int_{-\pi/2}^{\pi/2} \frac{d\theta}{a + b\sin\theta} = \left[\frac{2}{\sqrt{a^2 - b^2}} \tan^{-1}\left\{\frac{a\tan\frac{\theta}{2} + b}{\sqrt{a^2 - b^2}}\right\}\right]_{-\pi/2}^{\pi/2} \quad \text{(where } a > b\text{)}$$

Definite integral in turms of gama function

$$\int_0^{\pi/2} \sin x^n \cos_x^m dx = \frac{\left|\frac{n+1}{2}\right.\left|\frac{m+1}{2}\right.}{2\left|\frac{m+n+2}{2}\right.} \quad \text{(where } n \text{ and } m \text{ are +ve integers)}$$

where,

$$\left|n+1\right. = n\left|n\right.$$

$$\left|1\right. = 1$$

$$\left|\frac{1}{2}\right. = \sqrt{\pi}$$